6/06

ENCYCLOPEDIA OF

MEDICAL DEVICES AND INSTRUMENTATION

Second Edition
VOLUME 3

Echocardiography and Doppler Echocardiography – Human Spine, Biomechanics of

ENCYCLOPEDIA OF MEDICAL DEVICES AND INSTRUMENTATION, SECOND EDITION

Editor-in-Chief
John G. Webster
University of Wisconsin–Madison

Editorial Board
David Beebe
University of Wisconsin–Madison

Jerry M. Calkins
University of Arizona College of Medicine

Michael R. Neuman
Michigan Technological University

Joon B. Park
University of Iowa

Edward S. Sternick
Tufts–New England Medical Center

Editorial Staff
Vice President, STM Books: **Janet Bailey**
Associate Publisher: **George J. Telecki**
Editorial Director: **Sean Pidgeon**
Director, Book Production and Manufacturing:
Camille P. Carter
Production Manager: **Shirley Thomas**
Illustration Manager: **Dean Gonzalez**
Senior Production Editor: **Kellsee Chu**
Editorial Program Coordinator: **Surlan Murrell**

ENCYCLOPEDIA OF

MEDICAL DEVICES AND INSTRUMENTATION

Second Edition
Volume 3

Echocardiography and Doppler Echocardiography – Human Spine, Biomechanics of

Edited by

John G. Webster

University of Wisconsin–Madison

The *Encyclopedia of Medical Devices and Instrumentation* is available online at
http://www.mrw.interscience.wiley.com/emdi

WILEY-INTERSCIENCE

A John Wiley & Sons, Inc., Publication

Library of Congress Cataloging-in-Publication Data:

Encylopedia of medical devices & instrumentation/by John G. Webster,

    editor in chief. – 2nd ed.
      p. ; cm.
    Rev. ed. of: Encyclopedia of medical devices and instrumentation. 1988.
    Includes bibliographical references and index.
    ISBN-13 978-0-471-26358-6 (set : cloth)
    ISBN-10 0-471-26358-3 (set : cloth)
    ISBN-13 978-0-470-04068-3 (v. 3 : cloth)
    ISBN-10 0-470-04068-8 (v. 3 : cloth)

    1. Medical instruments and apparatus–Encyclopedias. 2. Biomedical engineering–Encyclopedias. 3. Medical physics–Encyclopedias. 4. Medicine–Data processing–Encyclopedias. I. Webster, John G., 1932- . II. Title: Encyclopedia of medical devices and instrumentation.

    [DNLM: 1. Equipment and Supplies–Encyclopedias–English. W 13

E555 2006]
R856.A3E53 2006
610.2803–dc22
2005028946

ENCYCLOPEDIA OF

MEDICAL DEVICES AND INSTRUMENTATION

Second Edition
VOLUME 3

Echocardiography and Doppler Echocardiography – Human Spine, Biomechanics of

E

ECG. See ELECTROCARDIOGRAPHY, COMPUTERS IN.

ECHOCARDIOGRAPHY AND DOPPLER ECHOCARDIOGRAPHY

PETER S. RAHKO
University of Wisconsin Medical
School
Madison, Wisconsin

INTRODUCTION

Echocardiography is a diagnostic technique that utilizes ultrasound (high frequency sound waves above the audible limit of 20 kHz) to produce an image of the beating heart in real time. A piezoelectric transducer element is used to emit short bursts of high frequency, low intensity sound through the chest wall to the heart and then detect the reflections of this sound as it returns from the heart. Since movement patterns and shape changes of several regions of the heart correlate with cardiac function and since changes in these patterns consistently appear in several types of cardiac disease, echocardiography has become a frequently used method for evaluation of the heart. Echocardiography has several advantages over other diagnostic tests of cardiac function:

1. It is flexible and can be used with transducers placed on the chest wall, inside oral cavities such as the esophagus or stomach, or inside the heart and great vessels.
2. It is painless.
3. It is a safe procedure that has no known harmful biologic effects.
4. It is easily transported almost anywhere including the bedside, operating room, cath lab, or emergency department.
5. It may be repeated as frequently as necessary allowing serial evaluation of a given disease process.
6. It produces an image instantaneously, which allows rapid diagnosis in emergent situations.

The first echocardiogram was performed by Edler and Hertz in 1953 (1) using a device that displayed reflected ultrasound on a cathode ray tube. Since that time multiple interrogation and display formats have been devised to display reflected ultrasound. The common display formats are

M-mode: A narrow beam of reflected sound is displayed on a scrolling strip chart plotting depth versus time. Only a small portion of the heart along one interrogation line ("ice pick" view) is shown at any one time.

Two-Dimensional Sector Scan (2D): A sector scan is generated by sequential firing of a phased array transducer along different lines of sight that are swept through a 2D plane. The image (a narrow plane in cross-section) is typically updated at a rate of 15–200 Hz and shown on a video monitor, which allows real time display of cardiac motion.

Three-Dimensional Imaging (3D): Image data from multiple 2D sector scans are acquired sequentially or in real time and displayed in a spatial format in three dimensions. If continuous data is displayed, time becomes the fourth dimension. The display can be shown as a loop that continuously repeats or on some systems in real time. Software allows rotation and "slicing" of the display.

Doppler: The Doppler effect is used to detect the rate and direction of blood flowing in the chambers of the heart and great vessels. Blood generally moves at a higher velocity than the walls of cardiac structures allowing motion of these structures to be filtered out. Blood flow is displayed in four formats:

Continuous Wave Doppler (CW): A signal of continuous frequency is directed into the heart while a receiver (or array of receivers) continuously processes the reflected signal. The difference between the two signals is processed and displayed showing direction and velocity of blood flow. All blood velocities in the line of sight of the Doppler beam are displayed.

Pulsed Wave Doppler (PW): Many bursts of sound are transmitted into the heart and reflected signals from a user defined depth are acquired and stored for each burst. Using these reflected signals an estimate is made of the velocities of blood or tissue encountered by the burst of sound at the selected depth. The velocity estimates are displayed similar to CW Doppler. The user defined position in the heart that the signal is obtained from stipulate the time of the acquisition of the reflected signal relative to transmission and length of the burst, respectively. The difference between transmitted and reflected signal frequencies is calculated, converted to velocity, and displayed. By varying the time between transmission and reception of the signal, selected velocities in small parts of the heart are sampled for blood flow.

Duplex Scanning: The 2D echo is used to orient the interrogator to the location of either the CW or PW signal allowing rapid correlation of blood flow data with cardiac anatomy. This is done by simultaneous display of the Doppler positional range gate superimposed on the 2D image.

Color Doppler: Using a complex array of bursts of frequency (pulsed packets) and multiple ultrasound acquisitions down the same beam line, motion of blood flow is estimated from a cross-correlation among the various acquired reflected waves. The data is combined as an overlay onto the 2D sector scan for anatomical orientation. Blood flow direction, and an estimate of flow velocity are displayed simultaneously with 2D echocardiographic data for each point in the sector.

CLINICAL FORMATS OF ULTRASOUND

Current generation ultrasound systems allow display of all of the imaging formats discussed except 3D/4D imaging that is still limited in availability to some high end systems.

Specialized transducers that emit and receive the ultrasound have been designed for various clinical indications. Four common types of transducers are used (Fig. 1):

Transthoracic: By far the most common, this transducer is placed on the surface of the chest and moved to different locations to image different parts of the heart or great vessels. All display formats are possible (Fig. 2).

Transesophageal: The transducer is designed to be inserted through the patient's mouth into the esophagus and stomach. The ultrasound signal is directed at the heart from that location for specialized exams. All display formats are possible.

Intracardiac: A small transducer is mounted on a catheter, inserted into a large vein and moved into the heart. Imaging from within the heart is performed to monitor specialized interventional therapy. Most display formats are available.

Intravascular: Miniature sized transducers are mounted on small catheters and moved through arteries to examine arterial pathology and the results of selected interventions. Limited 2D display formats are available some being radial rather than sector based. The transducers run at very high frequencies (20–30 MHz).

Ultrasound systems vary considerably in size and sophistication (Fig. 3). Full size systems, typically found in hospitals display all imaging formats, accept all types of

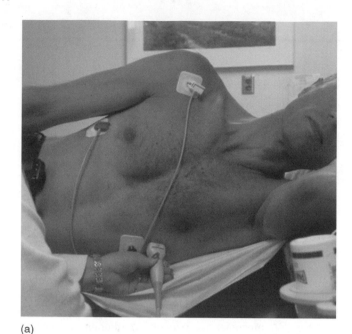

(a)

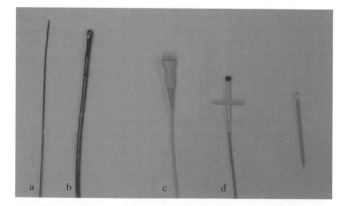

Figure 1. Comparative view of commonly used ultrasound transducers: (a) intracardiac, (b) transesophageal, (c) transthoracic, and (d) Pedof (special Doppler) transducer. A pencil is shown for size reference.

Figure 2. (a) Phased array transducer used for transthoracic imaging. The patient is shown, on left side, on exam table, typical for a transthoracic study. The transducer is in the apical position. (b) Diagram of the chest showing how the heart and great vessels are positioned in the chest and the locations where transthoracic transducers are placed on the chest wall. The four common positions for transducer placement are shown. The long axis (LAXX) and short axis (SAXX) orientations of the heart are shown for reference. These orientations form a reference for all of the views obtained in a study. Abbreviations are as follows: aorta (Ao), left atrium (LA), left ventricle (LV), pulmonary artery (PA), right atrium (RA), and right ventricle (RV).

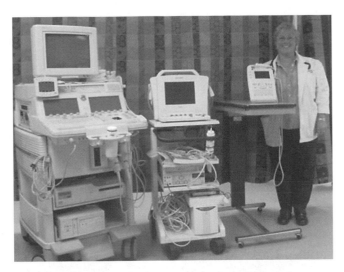

Figure 3. Picture showing various types of ultrasound systems. At left a large "full-size" system is shown that can perform all types of imaging. At center is a miniaturized system that has most of the features of the full service system, but limited display, analysis and recording formats are available. The smallest "hand-held" system shown at right has basic features and is battery operated. It is designed for rapid screening exams integrated into a clinical assessment at the bedside, clinic, or emergency room.

transthoracic and transesophageal transducers, allow considerable on line image processing and support considerable analytic capacity to quantify the image.

Small imaging systems accept many but not all transducers and produce virtually all display formats, but lack the sophisticated array of image processing and analysis capacity found on the full systems. These devices may be used in ambulatory offices or other specialized circumstances requiring more basic image data. A full clinical study is possible.

Portable hand-held battery operated devices are used in limited circumstances, sometimes for screening exams or limited studies. Typically transducers are limited, image processing is rudimentary and analysis capacity very limited.

PRINCIPLES OF ULTRASOUND

Prior to discussing the mechanism of image production some common terms that govern the behavior of ultrasound in soft tissue should be defined. Since ultrasound is propagated in waves, its behavior in a medium is defined by

$$\lambda = \frac{c}{f}$$

where f is the wave *frequency*, λ is the *wavelength*, and c is the *acoustic velocity* of ultrasound in the medium. The acoustic velocity for most soft tissues is similar and remains constant for a given tissue no matter what the frequency or wavelength (Table 1). Thus in any tissue frequency and wavelength are inversely related. As frequency increases, wavelength decreases. As wavelength decreases, the minimum distance between two structures, that allows them to be characterized as two separate structures, also decreases. This is called the *spatial resolution* of the instrument. One might conclude that very high frequency should always be used to maximize resolution. Unfortunately, as frequency increases, penetration of the ultrasound signal into soft tissue decreases. This serves to limit the frequency and, thus, the resolving power of an ultrasonic system for any given application.

Sound waves are emitted in short bursts from the transducer. As frequency rises, it takes less time to emit the same number of waves per burst. Thus, more bursts of sound can be emitted per unit of time, increasing the spatial resolution of the instrument (Fig. 4). The optimal image is generated by using a frequency that gives the highest possible resolution and an adequate amount of penetration. For transthoracic transducers expected to penetrate up to 24 cm into the chest, typical frequencies used are from 1.6 to 7.0 MHz. Most transducers are broadband in that they generate sound within an adjustable range of frequency rather than at a single frequency. Certain specialized transducers such as the intravascular transducer may only need to penetrate 4 cm. They may have a frequency of 30 MHz to maximize resolution of small structures.

The term *ultrasonic attenuation* formally defines the more qualitative concept of tissue penetration. It is a complex parameter that is different for every tissue type and is defined as the rate of decrease in wave amplitude per distance penetrated at a given frequency. The two important properties that define ultrasonic attenuation are reflection and absorption of sound waves (2). Note in Table 1 that ultrasound easily passes through blood and soft tissue, but poorly penetrates bone or air-filled lungs.

Acoustic impedance (z) is the product of acoustic velocity (c) and tissue density (ρ); thus this property is tissue specific but frequency independent. This property is important because it determines how much ultrasound is

Table 1. Ultrasonic Properties of Some Selected Tissues[a]

Tissue	Velocity of Propagation, 10^3 m·s^{-1}	Density, g·mL^{-1}	Acoustic Impedance, 10^6 rayl[b]	Attenuation at 2.25 MHz, dB·cm^{-1}
Blood	1.56	1.06	1.62	0.57
Myocardium	1.54	1.07	1.67	3
Fat	1.48	0.92	1.35	1.7
Bone	~3–4	1.4–1.8	4–6	37
Lung (inflated)	0.7	0.4	0.26–0.46	62

[a]Adapted from Wilson D. A., Basic principles of ultrasound. In: Kraus R., editor. The Practice of Echocardiography, New York: John Wiley & Sons; 1985, p 15. This material is used by permission of John Wiley & Sons, Inc.
[b]1 rayl = 1 kg·m^{-2}·s^{-1}.

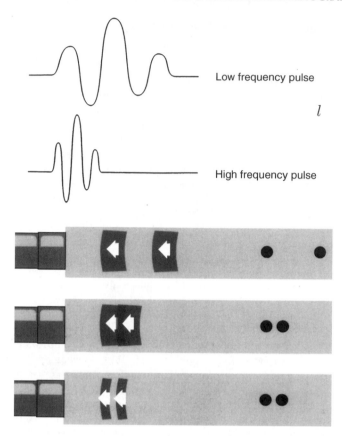

Figure 4. Upper panel Two depictions of an emitted pulse of ultrasound. Time moves horizontally and amplitude moves vertically. Note the high frequency pulse of three waves takes less time. Lower panel Effect of pulse duration on resolution. One echo pulse is delivered toward two reflectors and reflections are shown. In the top panel, the reflectors are well separated from each other and distinguished as two separate structures. In the middle panel, pulse frequency and duration are unchanged, but the reflectors are close together. The two returning reflections overlap, the instrument will register the two reflectors as one structure. In the lower panel the pulse frequency is increased, thus pulse duration is shortened. The two objects are again separately resolved. (Reprinted from Zagzebski JA Essentials of Ultrasound Physics. St. Louis: Mosby-Year Book; copyright © 1996, p 28, with permission from Elsevier.)

reflected at an interface between two different types of tissue (Table 1). When a short burst of ultrasound is directed at the heart, portions of this energy are reflected back to the receiver. It is these reflected waves that produce the image of the heart. A very dense structure such as calcified tissue has high impedance and is a strong reflector.

There are two types of reflected waves: *specular reflections* and *diffuse reflections* (Fig. 5). Specular reflections occur at the interface between two types of tissue. The greater the difference in acoustic impedance between two tissues, the greater the amount of specular reflection and the lower the amount of energy that penetrates beyond the interface. The interface between heart muscle and blood produces a specular echo, as does the interface between a heart valve and blood. Specular echoes are the primary

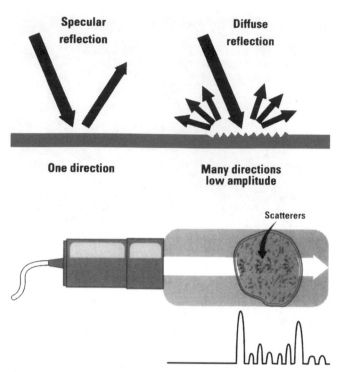

Figure 5. Upper panel Comparison between specular and diffuse reflectors. Note the diffuse reflector is less angle dependent than the specular reflector. Lower panel Example of combined reflections (shown at bottom of figure) returning from a structure, typical of reflections coming back from heart muscle. The large amplitude specular echo corresponds to the border of the structure. The interior of the structure produces low amplitude scattered reflections. (Reprinted from Zagzebski JA, Essentials of Ultrasound Physics. St. Louis, Mosby-Year Book; copyright © 1996 p 12 with permission from Elsevier.)

echoes that are imaged by M-mode, 2D echo, and 3D echo and thus primarily form an outline of the heart. Diffuse reflected echoes are much weaker in energy. They are produced by small irregular more weakly reflective objects such as the myocardium itself. Scattered echoes "fill in the details" between the specular echoes. With modern equipment scattered echoes are processed and analyzed providing much more detail to tissue being examined.

Doppler echocardiography uses *scattered echoes* from red blood cells for detecting blood flow. Blood cells are Rayleigh scatterers since the diameter of the blood cells are much smaller than the typical wavelength of sound used to interrogate tissue. Since these reflected signals are even fainter than myocardial echoes, Doppler must operate at a higher energy level than M-mode or 2D imaging.

Harmonic imaging is a recent addition to image display made possible by advances in transducer design and signal processing. It was first developed to improve the display of contrast agents injected intravenously as they passed through the heart. These agents, gas filled microbubbles 2–5 μm in diameter, are highly reflective Rayleigh scatterers. At certain frequencies within the broadband transducer range the contrast bubbles resonate, producing a relatively strong signal at multiples of the fundamental interrogation frequency called harmonics. By using a high pass filter (a system that blanks out all frequency below a certain level), to eliminate the fundamental frequency reflectors, selective reflections from the second harmonic are displayed. In second harmonic mode, reflections from the resonating bubbles are a much stronger than reflections from soft tissue and thus bubbles are preferentially displayed. This allows selective analysis of the contrast agent as it passes through the heart muscle or in the LV cavity (3).

Harmonic imaging has recently been applied to conventional 2D images without contrast. As the ultrasound wave passes through tissue, the waveform is modified by nonlinear propagation through tissue causing a shape change in the ultrasound beam. This progressively increases as the beam travels deeper into the heart. Electronic canceling of much of the image by filtering out the fundamental frequency allows selective display of the harmonic image, improving overall image quality by elimination of some artifacts. The spatial resolution of the signal is also improved since the reflected signal analyzed and displayed is double that of the frequency produced by the transducer (4).

ECHOCARDIOGRAPHIC INSTRUMENTATION

The transducer is a piezoelectric (pressure electric) device. When subjected to an alternating electrical current, the ceramic crystal (usually barium titanate, lead zirconate titanate, or a composite ceramic) expands and contracts producing compressions and rarefactions in its environment, which become waves. Various transducers produce ultrasonic waves within a frequency range of 1.0–40 MHz. The waves are emitted as brief pulses lasting ~ 1 μs out of every 100–200 μs. During the remaining 99–199 μs of each interval the transducer functions as a receiver that detects specular and diffuse reflections as they return from the heart. The same crystal, when excited by a reflected sound wave, produces an electrical signal and sends it back to the echocardiograph for analysis and display. Since one heartbeat lasts somewhere from 0.3 to 1.2 s, the echocardiographic device sends out a minimum of several hundred impulses per beat allowing precise tracking of cardiac motion throughout the beat.

After the specular and diffuse echoes are received they must be displayed in a usable format. The original ultrasound devices used an A-mode format (Fig. 6) that displayed depth on the y axis and amplitude of the signal on the x-axis. The specular echoes from boundaries between cardiac chambers register as the strongest echoes. No more than 1D spatial information is obtained from this format.

In a second format, B-mode, the amplitudes of the returning echoes are displayed as dots of varying intensity on a video monitor in what has come to be called a gray scale (Fig. 6). If the background is black (zero intensity), then progressively stronger echoes are displayed as progressively brighter shades of gray with white representing the highest intensity. Most echocardiographic equipment today uses between 64 and 512 shades of gray in its output display. The B-mode format, by itself, is not adequate for cardiac imaging and must be modified to image a continuously moving structure.

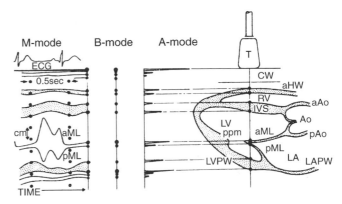

Figure 6. Composite drawing showing the three different modes of display for a one-dimensional (1D) ultrasound signal. In the right half of the figure is a schematic drawing of a cross-section through the heart. The transducer (T) sits on the chest wall (CW) and directs a thin beam of ultrasonic pulses into the heart. This beam traverses the anterior wall (aHW) of the right ventricle (RV), the interventricular septum (IVS), the anterior (aML), and posterior (pML) leaflets of the mitral valve, and the posterior wall of the left ventricle (LVPW). Each dot along the path of the beam represents production of a specular echo. These are displayed in the corresponding A-mode format, where vertical direction is depth and horizontal direction is amplitude of the reflected echo and B-mode format where again vertical direction is depth but amplitude is intensity of the dot. If time is added to the B-mode format, an M-mode echo is produced, which is shown in the left panel. This allows simultaneous presentation of motion of the cardiac structures in the path of the echo beam throughout the entire cardiac cycle; measurement of vertical depth, thickness of various structures, and timing of events within the cardiac cycle. If the transducer is angled in a different direction, a distinctly different configuration of echoes will be obtained. In the figure, the M-mode displayed is at the same beam location as noted in the right-side panel. Typical movement of the AML and PML is shown. ECG = electrocardiogram signal. (From Pierand L., Meltzer RS., Roelandt J, Examination techniques in M-mode and two-dimensional echocardiography. In: Kraus R editor, The Practice of Echocardiography, New York: John Wiley & Sons; copyright © 1985, p 69. This material is used by permission of John Wiley & Sons, Inc.)

To image the heart, the M-mode format (M for motion) was devised (Fig. 6). With this technique, the transducer is pointed into the chest at the heart and returning echoes are displayed in B-mode. A strip chart recorder (or scrolling video display) constantly records the B-mode signal with depth of penetration on the y-axis and time the parameter displayed on the x-axis. By adding an electrocardiographic signal to monitor cardiac electrical activity and to mark the beginning of each cardiac cycle, the size, thickness, and movement of various cardiac structures throughout a cardiac cycle are displayed with high resolution. By variation of transducer position, the ultrasound beam is directed toward several cardiac structures (Fig. 7).

The M-mode echo was the first practical ultrasound device for cardiac imaging and has produced a considerable amount of important data. Its major limitation is its limited field of view. Few spatial relationships between cardiac structures can be displayed that severely limits diagnostic capability. The angle of interrogation of the heart is also

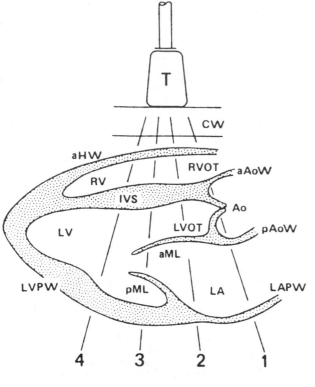

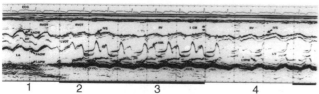

Figure 7. Upper panel Schematic diagram of the heart as in Fig. 6. The principal M-mode views are labeled 1–4. The corresponding M-mode image from these four views is shown in lower panel Abbreviations as in Fig. **6a** Additional abbreviations: AV = Aortic valve, AAOW = anterior aortic wall, LA = left atrial posterior wall, LVOT = left ventricular outflow tract, RVOT = right ventricular outflow tract. (From Pierand L, Meltzer RS, Roelandt J, Examination techniques in M-mode and 2D echocardiography. In: Kraus R editor, The Practice of Echocardiography, New York, John Wiley & Sons; copyright © 1985, p 71. This material is used with permission of John Wiley & Sons, Inc.)

difficult to control. This can distort the image and render size and dimension measurements unreliable.

Since the speed of sound is rapid enough to allow up to 5000 short pulses of ultrasound to be emitted and received each second at depths typical for cardiac imaging, it was recognized that multiple B-mode scans in several directions could be processed rapidly enough to display a "real-time" image. Sector scanning in two dimensions was originally performed by mechanically moving a single element piezo-electric crystal through a plane. Typically, 128 B-mode scan lines were swept through a 60–90° arc 30 times · s⁻¹ to form a video composite B-mode sector (Fig. 8). These mechanical devices have been replaced by transducer arrays that place a group of closely spaced piezoelectric elements, each with its own electrical connection to the ultrasound system, into a transducer. The type of array used depends on the structure

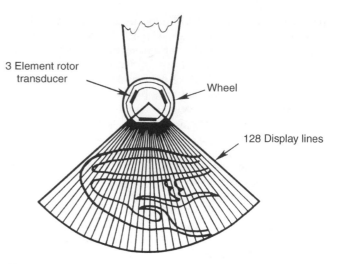

Figure 8. Diagram of a 2D echo mechanical transducer with three crystals. As each segment sweeps through the 90° arc, an element fires a total of 128 times. The composite of the 128 B-mode lines form a 2D echo frame. Typically there are 30 frames/s of video information, a rate rapid enough to show contractile motion of the heart smoothly. (From Graham PL, Instrumentation. In: Krause R. editor. The Practice of Echocardiography, New York: John Wiley & Sons; copyright © 1985, p 41. This material is used by permission of John Wiley & Sons, Inc.)

being imaged. For cardiac ultrasound, a phased array configuration is used, typically consisting of 128–1024 individual elements. In a phased array transducer, a portion of the elements are fired to produce each sector scan line. The sound beams are electronically steered through the sector by changing the time delay sequence of the array elements (Fig. 9). In a similar fashion, all elements are electronically sequenced to receive reflected sound from selected parts of the sector being scanned (5).

Use of a phased array device allows many other modifications of the sound wave in addition to steering. Further sophisticated electronic manipulations of the time of sound transmission and delays in reception allow selective focusing of the beam to concentrate transmit energy that enhance image quality of selected structures displaced in the sector. Recent design advances have markedly increased the sector frame rate (number of displayed sectors/second) to levels beyond 220 Hz, markedly increasing the time resolution of the imaging system. While the human visual processing system cannot resolve time at such a rapid rate, high frame rates allow for sophisticated quantitation based on the 2D image, such as high-resolution graphs of time based indexes.

DOPPLER ECHOCARDIOGRAPHY

Application of the Doppler effect allows analysis of blood flow within the heart and great vessels. The Doppler effect, named for its discoverer Christian Doppler, describes the change in frequency and wavelength that occurs with relative motion between the source of the waves and the receiver. If a source of sound remains stationary with respect to its listener, then the frequency and wavelength of the sound will also remain constant. However, if the

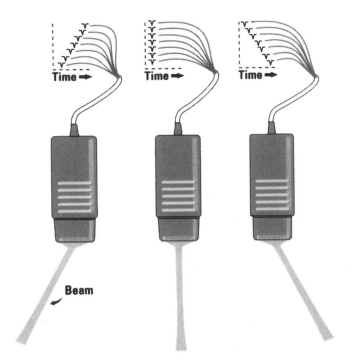

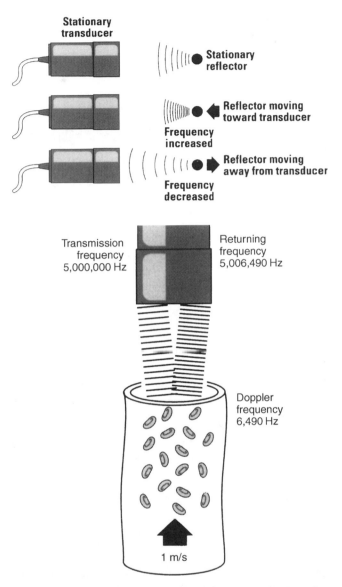

Figure 9. Diagram of a phased array transducer. Beam direction is varied by changing the delay sequence among the transmitted pulses produced by each individual element. (From Zagzebski JA. Essentials of Ultrasound Physics. St. Louis: Mosby-Year Book Inc.; copyright © 1996, with permission from Elsevier.)

sound source is moving away from the listener wavelength increases and frequency decreases. The opposite will occur if the sound source is moving toward the listener (Fig. 10).

The Doppler principle is applied to cardiac ultrasound in the following way: A beam of continuous wave ultrasound is transmitted into the heart and reflected off red blood cells as they travel through the heart. The reflected impulses are then detected by a receiver. If the red blood cells are moving toward the receiver, the frequency of the reflected echoes will be higher than the frequency of the transmitted echoes and vice versa (Fig. 10).

The difference between the frequency of the transmitted and received echoes (usually called the Doppler shift) can be related to the velocity of blood flow by the following equation:

$$V = \frac{c(f_r - f_t)}{2(f_t)(\cos\theta)30}$$

where V is the blood flow velocity, c is the speed of sound in soft tissue (1540 m·s^{-1}), f_r is the frequency of the reflected echoes, f_t is the frequency of the transmitted echoes, and θ is the intercept angle between the direction of blood flow and the ultrasound beam. Thus, flow toward the transducer will produce a positive Doppler shift ($f_r > f_t$), while flow away from the transducer will produce a negative Doppler shift ($f_r < f_t$). The only variable that cannot be directly measured is θ. Since $\cos 0° = 1$, it follows that maximal flow will be detected when the Doppler beam is parallel to blood flow. Since blood flow cannot be seen with 2D echo, at best, θ can only be estimated. Fortunately, if θ is within 20° of the direction of blood flow, the error introduced by angulation is small. Therefore,

Figure 10. Upper panel The Doppler effect as applied to ultrasound. The frequency increases slightly when the reflector is moving toward the transducer and decreases when the reflector is moving away from the transducer. Lower panel Relative magnitude of the Doppler shift caused by red blood cells moving between cardiac chambers. In this example the frequency shift corresponds to a movement rate of 1 m·s^{-1}. (Reprinted from Zagzebski JA. Essentials of Ultrasound Physics. St. Louis: Mosby-Year Book Inc.; copyright © 1996, with permission from Elsevier.)

most investigators do not formally correct for θ. Instead, the Doppler beam is aligned as closely as possible in the presumed direction of maximal flow and then adjusted until maximal flow is detected (6).

Doppler echo operates in two basic formats. Figure 10 depicts the CW method. An ultrasound signal is continuously transmitted into the heart while a second crystal (or array of crystals) in the transducer continually receives reflected signals. All red blood cells in the overlap region between the beam patterns of the transmit and receive crystals contribute to the calculated signal. The frequency

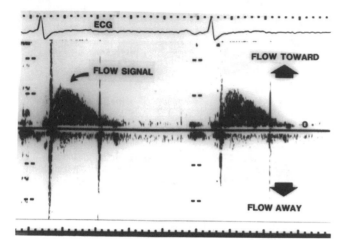

Figure 11. Example of a continuous wave Doppler signal taken from a patient. Flow toward the transducer is a positive (upward) deflection from the baseline and flow away from the transducer is downward. ECG = electrocardiogram signal.

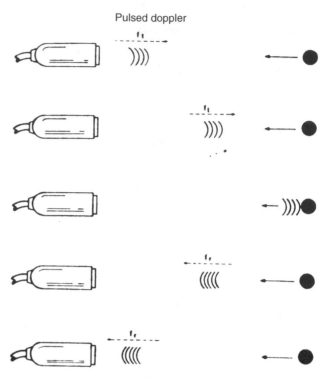

Figure 12. Pulsed Doppler echocardiography. In place of a continuous stream of ultrasound, brief pulses are emitted similar to M-mode or 2D imaging. By acquiring reflected signal data over a limited time window following each pulse, reflections emanating only from a certain depth may be received. (From Feigenbaum H, Echocardiography (5th ed), Philadelphia, PA: Lea & Febiger; 1994, p 29, with permission from Lippincott Williams & Wilkins ©.)

content of this signal, combined with an electrocardiographic monitor lead, is then displayed on a strip chart similar to an M-mode echo (Fig. 11).

The advantage of the CW method is that it can detect a wide range of flow velocities encompassing every possible physiologic or pathologic flow state. The main disadvantage of the CW format is that the site of flow cannot be localized. To overcome the lack of localization of CW Doppler, a second format was developed called pulsed Doppler (PW). In this format, similar to B-mode echocardiographic imaging, brief bursts of ultrasound are transmitted at a given frequency followed by a silent interval (Fig. 12). Since the time it takes for a reflected burst of sound waves to return to the receiving crystal is directly related to the distance the reflecting structure is from the receiver, the position in the heart from which blood flow is sampled can be precisely controlled by limiting the time interval during which reflected ultrasound is received. This is known as range gating and allows the investigator to limit the area sampled to small portions of the heart or great vessel. There is a price to be paid for sample selectivity, however. The maximal detectable velocity PW Doppler is able to display is equal to one-half the pulse repetition frequency (frequently called the Nyquist limit). This reduces the number of situations in which flow velocity samples unambiguously display the flow phenomenon. A typical PW Doppler display shows flow both toward and away from the transducer (Fig. 13).

As one interrogates deeper structures progressively further from the transducer, the pulse repetition frequency must, of necessity, be decreased. As a result, the highest detectable velocity of the PW Doppler mode becomes progressively smaller. Due to attenuation, the sensitivity for detecting flow becomes progressively lower at greater distances from the transducer. Despite these limitations, selective sampling of blood flow allows interrogation of a wide array of cardiac structures in the heart.

When a measured flow has a velocity in a particular direction greater than the Nyquist limit, not all of the

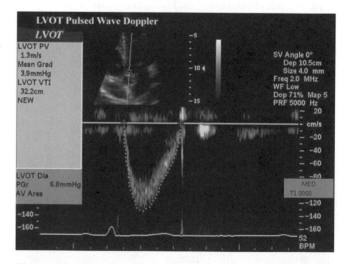

Figure 13. Example of a pulsed wave Doppler tracing. The study was recorded in duplex mode from the LVOT. The small upper insert shows the 2D image which guides positioning of the sample volume (arrow). The Doppler signal is shown below and has been traced by the sonographer using electronic analysis system. Data automatically detected from tracing the signal are shown on the left and include the peak velocity (PV) of flow and the integral of flow velocity (VTI) that can be used for calculations of cardiac output.

spectral envelope of the signal is visible. Indeed, the velocity estimates "wrap-around" to the other side of the velocity map and the flow appears to be going in the opposite direction. This phenomenon where large positives velocities are displayed as negative velocities is called aliasing. There are two strategies to mitigate or eliminate aliasing. The zero shift line (no velocity) may be moved upward or downward, effectively doubling the display range in the desired direction. This may be sufficient to "un-wrap" the aliased velocity and display it entirely in the appropriate direction. Some velocities may still be too high for this strategy to work. To display these higher velocities an alternative method called high pulse repetition frequency (high PRF) mode is employed. In this mode, sample volumes at multiples of the main interrogation sample volume are also interrogated. This is accomplished by sending out bursts of pulse packets at multiples of the burst rate necessary to sample at the desired depth. The system displays multiple sites from which the displayed signal might originate. While this creates some ambiguity in the exam, the anatomy displayed by the 2D exam usually allows a correct delineation as to which range gate is creating the signal (Fig. 14).

By itself, Doppler echo is a nonimaging technique that only produces flow patterns and audible tone patterns (since all Doppler shifts fall within the audible range). Phased array transducers, however, allow simultaneous display of both 2D images and Doppler in a mode called duplex Doppler echocardiography. By using this combination, the PW Doppler sample volume is displayed as an overlay on the 2D image and is moved to a precise area in the heart where the flow velocity is measured (Fig. 13). This combination provides both anatomic and physiologic information about the interrogated cardiac structure. Most commonly, Duplex mode is used with the PW wave format of Doppler echo. However, it is also possible to position the continuous wave beam anywhere in the 2D sector by superimposing the Doppler line of interrogation on top of the 2D image.

Just as changing from an M-mode echo to a 2D sector scan markedly increases the amount of spacial data simultaneously available to the clinician, Doppler information can be expanded from a single a PW wave sample volume or CW line to a full sector array. Color Doppler echocardiography displays blood flow within the heart or blood vessel as a sector plane of velocity information. By itself, a color flow scan imparts little information so the display is always combined with the 2D image as an overlay so blood flow may be instantly correlated with anatomic structures within the heart or vessel.

Color Doppler uses the same transmission principles as B-mode 2D imaging and PW Doppler. Brief transmit pulses of sound are steered along interrogation lines in a sector simultaneously with usual B-mode imaging pulses (Fig. 15). In place of just one pulse of sound, multiple pulses are transmitted. The multiple bursts of sound, typically 4–8 in number, are referred to as packets or ensembles of pulses. The first signal stores all reflected echoes along each scan line. Reflectors from subsequent pulses in the packet are received, stored, and rapidly compared to the previous packets. Reflected waves that are identical during each burst in the packet are canceled

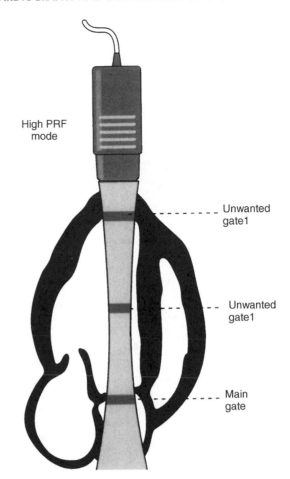

High PRF mode

Unwanted gate1

Unwanted gate1

Main gate

Figure 14. Example of high PFR mode of pulsed Doppler. The signal is used for velocity detected at the main gate because pulse packets are sent out more frequently at multiples of the frequency needed to sample at the main gate. Information can also be acquired at other gates that are multiples of the main gate. While the Nyquist limit is higher due to a higher sampling rate some signal ambiguity may occur due to information acquired from the unwanted gates. (Reprinted from Zagzebski JA. Essentials of Ultrasound Physics. St. Louis: Mosby-Year Book Inc. copyright © 1996 with permission from Elsevier.)

out and designated as stationary. Reflected waves that progressively change from burst to burst are acquired and processed rapidly for calculation of the phase shift in the ultrasound carrier. Both direction and velocity of movement are proportional to this phase change. Estimates for the average velocity are assigned to a pixel location on the video display. The velocity is estimated by an auto correlator system. On the output display, velocity is typically displayed as brightness of a given color similar to B-mode gray scale with black meaning no motion and maximum brightness indicating the highest velocity detected. Two contrasting colors are used to display direction of flow, typically a red-orange group for flow toward the transducer and a blue group away from transducer. Since the amount of data processed is markedly greater than a typical B-Mode, maximum frame rates of sector scan displays tend to be much lower. This limitation is due both to the speed of sound and the multiple packets of ultrasound evaluated in each interrogation line. To maximize

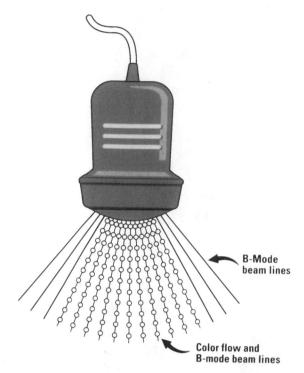

B-Mode
beam lines

Color flow and
B-mode beam lines

Figure 15. Diagram showing method of transmission of color flow Doppler ultrasound signals. Packets or ensembles of pulses represented by the open circles in the figure are sent out along some of the scan lines of the image. The reflected waves are analyzed in an autocorrelation circuit to allow display of the color flow image. (Reprinted from Zagzebski JA. Essentials of Ultrasound Physics. St. Louis: Mosby-Year Book Inc. copyright © 1996, with permission from Elsevier.)

time resolution, the color flow sector used to display data during a 2D echo may be considerably reduced in size compared to a usual 90° 2D sector scan. Some systems can only display color flow data at relatively slow frame rates of 6–10 Hz. Recent innovations, in which there are multiple receive lines for each transmit line allow much higher frame rates giving excellent time resolution on some systems (7).

Clinically, precise measurement of flow velocity is usually obtained with PW or CW Doppler. Color is used to rapidly interrogate a sector for the presence or absence of blood flow during a given part of the cardiac cycle. Another important part of the color exam is a display of the type of flow present. Normal blood flow is laminar; abnormal blood flow caused by valve or blood vessel pathology is turbulent. The difference between laminar and turbulent flow is easily displayed by color Doppler. With laminar flow, the direction of flow is uniform and variation in velocity of adjacent pixels of interrogation is small. With turbulent flow, both parameters are highly variable. The auto correlation system analyzing color Doppler compares the variance in blood flow between different pixels. The display can be set to register a third color for variance such as green, or the clinician may look for a "mosaic" pattern of flow in which non uniform color velocities and directions are scattered through the sector areas of color interrogation (Figs. 16 and 17).

As with pulsed Doppler there is a Nyquist limit restriction on maximal velocity than can be displayed. The zero flow position line may be adjusted as with PW Doppler to maximize the velocity limit in a given direction. High PRF is not possible with color Doppler. The nature of the color display is such that aliasing results in a shift from one color sequence to the next. Thus, in some situations a high velocity shift can be detected due to a clear shift in color (e.g., from a red-orange sequence to a blue sequence). This phenomenon has been put to use clinically by purposely manipulating the velocity range of color display to force aliasing to occur. By doing this, isovelocity lines are displayed outlining a velocity border of flow in a particular direction and a particular velocity (8).

Thus far, all discussion of Doppler has been confined to interrogation and display of flow velocity. Alternate modes of interrogation are possible. One mode, called power Doppler (or energy display Doppler) assesses the amplitude of the Doppler signal rather than velocity. By evaluating amplitude in place of velocity, this display becomes proportional to the number of moving blood cells present in the interrogation field rather than the velocity. This application is particularly valuable for perfusion imaging when the amount of blood present in a given area is of primary interest (9).

ULTRASOUND SIGNAL PROCESSING, DISPLAY, AND MANAGEMENT

Once each set of the reflected ultrasound data returns to the transducer, it is processed and then transmitted to video display. The information is first processed by a scan converter, which assigns video data to a matrix array of picture elements, "pixels." Several manipulations of the image are possible to reduce artifacts, enhance information in the display, and analyze the display quantitatively.

The concept of attenuation has been introduced earlier. In order to achieve a usable signal, the returning reflections must be amplified. The amplification can be done in multiple ways. Similar to a sound system, overall gain may be adjusted to increase or decrease the sensitivity of the received signal. More important, however, is the progressive loss of signal strength that occurs with reflections from deeper structures due to attenuation. To overcome this issue, ultrasound systems employ a variable gain circuit that selectively allows gain control at different depths. The applied gain is changed as a function of time (range) in the gain circuit, hence the term time gain compensation (TGC) is used to describe the process. This powerful tool can "normalize" the overall appearance of the image helping make much weaker returning echoes from great depth appear equal to near-field information (Fig. 18). The user also has slide pot control of gain as a function of depth. Some of this user-defined adjustment is applied as part of the TGC function or later as part of digital signal processing.

Manipulation of data prior to writing into the scan converter is called preprocessing. An important part of preprocessing is data compression. The raw data received by the transducer encompasses such a broad energy range

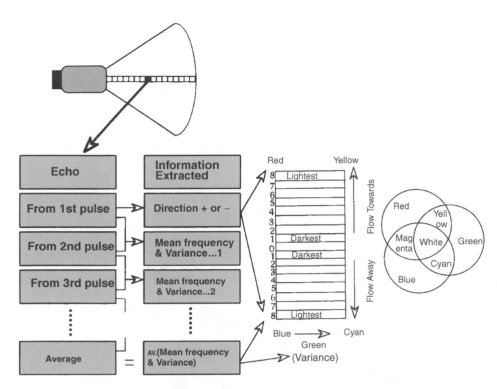

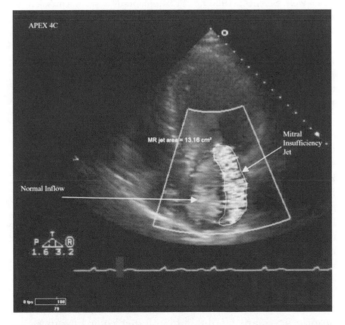

Figure 16. Diagram of color Doppler signal processing. At top, each box in the insonification line indicates a pulse of sound. A packet is made up of 4–8 pulses. The steps in the analysis cycle of a packet are shown in the "Echo" column and represent the comparison function of the autocorrelation system. From the first pulse the analysis determines the appropriate color representing direction. Comparisons between subsequent packets detect the velocity of blood flow. The brightness of the color selected corresponds to velocity. Comparisons of the variability of velocity are also done. Green is added proportional to the amount of variance. The final pixel color represents an average of the packet data for direction, velocity and variance. The right-hand column is the "color bar" that summarizes the type of map used, displays the range of color and brightness selected, and depicts how variance is shown. (From Sehgal CM, Principles of Ultrasound and Imaging and Doppler Ultrasound. In: Sutton, MG et al. editors: Textbook of Echocardiography and Doppler Ultrasound in Adults and Children (2nd ed). Oxford: Blackwell Science Publishing; 1996. p 29. Reprinted with permission of Blackwell Publishing, LTD. p 29)

that it cannot be adequately shown on a video display. Therefore, the dynamic range of the signal is reduced to better fit visual display characteristics. Selective enhancement of certain parts of the data is possible to better display borders between structures. Finally, persistence may be added to the image. With this enhancement, a fraction of data from the previous video frames at each pixel location may be added and averaged together. This helps define weaker diffuse echo scatterers and may work well in a static organ. However, with the heart in constant motion, only modest amounts of persistence add value to the image. Too much persistence reduces motion resolution of the image.

Once the digital signal is registered in the scan converter, further image manipulation is possible. This is called postprocessing of the image. Information in digital memory has a 1:1 ratio between ultrasound signal amplitude and video brightness. Depending on the structure imaged, considerable information may not be discernible in the image. With postprocessing, the relationship of video brightness to signal strength can be altered, frequently to enhance weaker echos and suppress high amplitude echoes. This may result in a better appreciation of less echogenic structures such as myocardium and the edges of the myocardium. The gray scale image may be transformed to a pseudo-color display that adds color to video amplitude data. In certain circumstances this may allow differentiation of pathologic changes from normal. Selective magnification of the image is also possible (Fig. 19).

Figure 17. Color Doppler (here shown in gray scale only) depicting mitral valve regurgitation. The large mosaic mitral insufficiency jet is caused by turbulent flow coming through the mitral valve. The turbulent flow area has been traced to measure its area. The more uniform flow in the left atrium is caused by normal flow in the chamber. It is of uniform color and of much lower velocity.

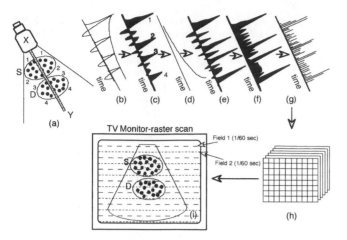

Figure 18. Processing steps during transmission of sound data to the scan converter. Raw data emerges from the transducer (a) that is imaging two objects that produce specular border echoes at points 1, 2, 3, and 4 and scattered echoes in between. (b) Specular echo raw data. (c) Low noise amplification is applied. Both specular and scattered echoes are now shown. Relative signal strength declines with depth due to attenuation of the signal, making signal 4 weaker than signal 1 even though the material border interface at borders 1 and 4 are identical. (d) Time gain compensation is applied proportionally by depth to electronically amplify signals received progressively later after the pulse leaves the transducer. This helps equalize perceived signal strength (e). The signal is then rectified (f) and smoothed (g) before entering the scan converter. (h) The process is repeated several times per second, in this case all new data appears every 1/30 of a second. The end result is a "real-time" video display of the two structures. (From Sehgal SH, Principles of Ultrasound Imaging and Doppler Ultrasound. In: Sutton MG et al. editors. Textbook of Echocardiography and Doppler Ultrasound in Adults and Children. Oxford: Blackwell Science; 1996. p 11. Reprinted with permission of Blackwell Publishing, LTD.)

Signal processing of PW and CW Doppler data includes filtering and averaging, but the most important component of the analysis is the computation of the velocity estimates. The most commonly used method of analysis is the fast Fourier transform analyzer, which estimates the relative amplitude of various frequency components of the input signal. This system (Fig. 20) divides data up into discreet time segments of very short duration (1–5 ms). For each segment, amplitude estimates are made for each frequency that corresponds to different velocity components in the flow and the relative amplitude of each frequency is recorded on gray scale display. Laminar flow typically has a narrow, discrete velocity range on PW Doppler while turbulent flow may be composed of the entire velocity range. Differentiation of laminar from turbulent flow may help define normal from abnormal flow states. Similar analysis is used to display the CW Doppler signal. Color Doppler displays can be adjusted by using multiple types of display maps. Each manufacturer has basic and special proprietary maps available to enhance color flow data.

All Doppler data can be subjected to selective band pass filters. For conventional Doppler imaging, signals coming from immobile structures or very slow moving structures such as chamber walls and the pericardium, are effectively blanked out. The range of velocity filtered can be changed

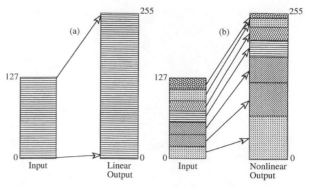

Figure 19. B-mode post processing occurs in which signal input intensity varies from 0 to 127 units. On the left side of the figure a linear output is equally amplified, the signal intensity range is now 0–255. On the right, nonlinear amplification is applied. The output is manipulated to enhance or compress the relative video intensity of data. In this example high energy specular reflection video data is relatively compressed (high numbers) while low energy data (from scattered echoes and weak specular echoes) is enhanced. Thus relatively more of the video range is used for relatively weaker signals in the final product. This postprocessing is in addition to time gain compensation done during preprocessing. (From Sehgal SH. Principles of Ultrasound Imaging and Doppler Ultrasound, In: Sutton MG et al. editors: Textbook of Echocardiography and Doppler Ultrasound in Adults and Children. Oxford: Blackwell Science; 1996. p 12. Reprinted with permission of Blackwell Publishing, LTD.)

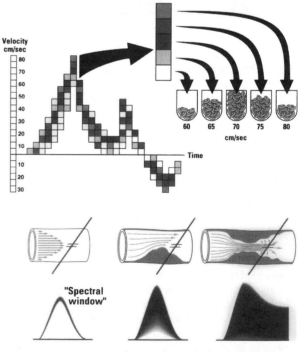

Figure 20. Upper panel Build-up of a spectral Doppler signal by Fast-Fourier analysis. The relative amplitude of the signal at each pixel location is assigned a level of gray. With laminar flow, the range of velocities is narrow resulting in a narrow window of displayed velocity. Lower panel As flow becomes more turbulent the range of velocities detected increases to the point that very turbulent signals may display all velocities. (Reprinted from Zagzebski JA. Essentials of Ultrasound Physics. St. Louis: Mosby-Year Book Inc. copyright © 1996, p 100,101, with permission from Elsevier.)

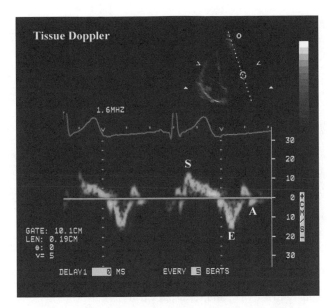

Figure 21. Example of tissue Doppler imaging. Duplex mode is used; the 2D image is shown in the upper insert. The sample volume has been placed outside the left ventricular cavity over the mitral valve annulus and is being used to detect movement of that structure. The three most commonly detected waveforms are shown: (S) systolic contraction wave, (E) early diastolic relaxation wave, and (A) atrial contraction wave.

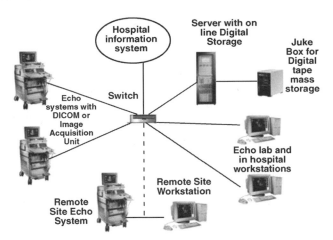

Figure 22. Organization chart showing a digital echo image storage system. Images generated by the ultrasound systems in the hospital and remote sites are input via high speed links through a switch to the server. The server has on line digital storage and is linked to a jukebox mass storage device. From the server, data may be viewed at any workstation either in the hospital or remotely. The hospital information system is linked to the system. This can allow electronic ordering of studies, downloading of demographic data on patients and interface with the workstations. Typically, images are reviewed at a workstation; a report is composed and is electronically signed. The report is stored in the server but may also be sent to the Hospital Information System electronic record or put out on paper as a final report.

for different clinical circumstances. In some situations, not all of the slower moving structures can be fully eliminated without losing valuable Doppler data, resulting in various types of artifact.

New techniques of Doppler analysis focus on analyzing wall motion with Doppler and displaying the data either in color or with Fourier transform analysis. In this setting, the band pass filters are set to eliminate all high velocity data from blood flow and only analyze very low velocity movement coming from the wall of the ventricles or other structures such as valve tissue. Analysis of tissue motion measures the velocity of wall movement at selected locations in the heart using the typical PW sample volume. Conventional strip chart display is used and velocity can be displayed to represent both the velocity of contraction and the velocity of relaxation (Fig. 21). Color Doppler uses the auto correlator system to calculate velocity of large segments of myocardium at once. Systems that record data at high color Doppler frame rates store sufficient information to allow selective time—velocity plots to be made at various locations of the cardiac chamber walls. Color Doppler may also be utilized to calculate strain and strain rate, another alternate display mode being investigated as a parameter of ventricular function (10,11).

Once the video signal processing is complete, it is displayed on a monitor. Until recently, most monitors were analogue and used standard NTSC video display modes. The composite (or RGB) signal was sent to a videocassette recorder for long-term storage and playback. The study could then be played on a standard video playback system for review and interpretation. Many laboratories continue to use this method of recording and storage of images.

Recently, many echo laboratories have changed to digital image display and storage. In some labs, the ultrasound system directly records the study in digital format, storing data on a large hard disk in the ultrasound system. The data is then output, typically using a DICOM standard signal to a digital storage and display system. Other laboratories may attach an acquisition box to the ultrasound system that digitally records and transmits the RGB signal to a central image server.

A digital laboratory set up is shown in Fig. 22. It has several advantages over videotape. Image data is sent to a server and on to a digital mass storage device. Depending on the amount of digital storage and volume of studies in the lab, days to months of image data may be instantly available for review. More remote data is stored in a mass storage system (PACS system) on more inexpensive media such as digital tape. Data in this remote storage may be held online in a second "juke box" server or be fully off line in storage media that must be put back on line manually.

Digital systems link the entire lab together and are typically integrated with the hospital information system. Common studies may be organized by patient and displayed on multiple workstations within the laboratory, hospital, and at remote clinics. This is a marked improvement compared to having each study isolated to one videotape. The quality of the image is superior. High-speed fiber optic links allow virtually simultaneous image retrieval at remote sites. Lower speed links may need several minutes to transmit a full study. To reduce the amount of storage, studies are typically compressed. Compression ratios of 30:1 can be used without evidence of any significant image degradation.

THE ECHOCARDIOGRAPHIC EXAMINATION

The Transthoracic Exam

A full-featured cardiac ultrasound system is designed to allow the operator to perform an M-mode echo, 2D echo, CW Doppler, PW Doppler, color Doppler, and tissue Doppler examination. Except for the specialized CW Doppler transducer, the entire exam is usually performed with a broadband multipurpose transducer. Occasionally, a specialized transducer of a different frequency or beam focus must be utilized to interrogate certain types of suspected pathology. The examination is performed in a quiet, darkened room with the patient supine or lying in a left lateral position on a specialized exam bed. The transducer is covered with a coupling medium (gel-like substance) that allows a direct interface between the transducer head and the patient's skin. If this coupling is broken the signal will be lost since ultrasound reflects strongly from tissue–air interfaces (Table 1). The transducer is then placed between the ribs, angled in the direction of the heart (ultrasound penetrates bone poorly and bone causes image artifacts), and adjusted until a satisfactory image is obtained (Fig. 2a). Electrodes for a single channel electrocardiogram are applied to the patient. The electrocardiogram signal is displayed continuously during the exam.

A standard examination begins in the left parasternal position (i.e., in the fourth and fifth rib interspaces just left of the sternum) (Fig. 2b). By orienting the 2D plane parallel to the long axis of the heart, an image of the proximal aorta, aortic valve, left atrium, mitral valve, and left ventricle can be obtained (Fig. 23). The standard M-mode echocardiographic views for dimension measurement are also obtained from this view (Fig. 6). Color Doppler is activated to examine flow near the mitral and aortic valves. By angulation from this position the tricuspid valve and portions of the right atrium and right ventricle are brought into view (Fig. 24). By rotation of the transducer ~90° from the long axis, the parasternal short-axis view is obtained.

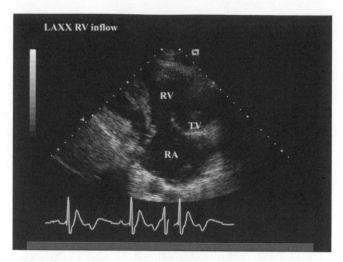

Figure 24. Parasternal long axis view angled toward the right ventricle. Right-sided structures are shown: right atrium (RA), right ventricle (RV) and tricuspid valve (TV).

By progressively changing the angle of the transducer it is possible to obtain a series of cross-sectional views through the left ventricle from near the apex to the base of the heart at the level of the origin of the great vessels (Figs. 25 and 26). In an optimal study, several cross-sections through the left ventricle, mitral valve, aortic valve, and to a lesser degree the tricuspid valve, pulmonic valve, and right ventricular outflow tract can be obtained. Since conventional 1D arrays yield only 2D images, only small slices of the heart are examined at any one time.

The transducer is then moved to the mid-left chest slightly below the left breast where the apex of the heart touches the chest wall (Fig. 2). The 2D transducer is then angled along the long axis toward the base of the heart. The result is a simultaneous display of all four chambers of the heart and the mitral and tricuspid valves (Fig. 27). No M-mode views are taken from this position. Using duplex

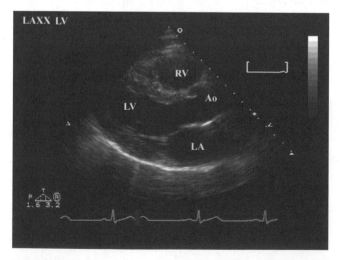

Figure 23. Parasternal long axis view of the heart (LAXX) similar to orientation shown in Fig. 6. Abbreviations are as follows: aorta (Ao), left ventricle (LV), left atrium (LA), and right ventricle (RV). This is a 2D image.

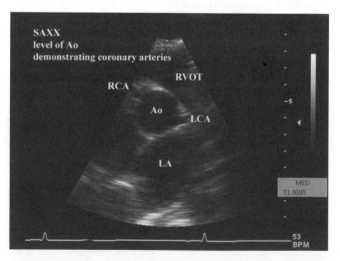

Figure 25. Parasternal short axis view at the level of the great vessels. Shown is the aorta (Ao), a portion of the left atrium (LA), the origin of the left (LCA), and right (RCA) coronary arteries, and a part of the right ventricular outflow tract.

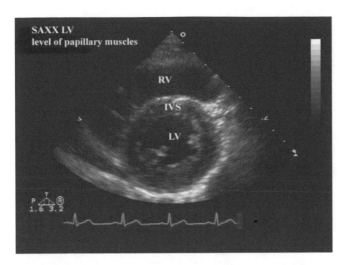

Figure 26. Short axis orientation at the transducer (SAXX) showing a cross-sectional view of the left ventricle (LV). The muscle appears as a ring. The septum (IVS) separates the LV from the right ventricle (RV).

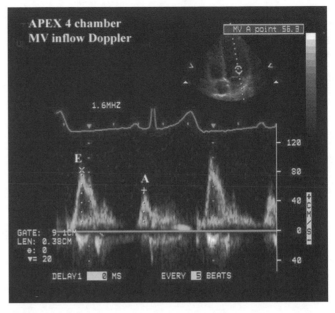

Figure 28. By using the same transducer orientation as Fig. 27, several other types of data are obtained. In this case, the system has been switched to Duplex mode. The 2D echo is used as a guide to position the mitral sample volume (small upper image). Flow is obtained across the mitral valve (MV) with waveforms for early (E), and atrial (A) diastolic flow shown.

mode, PW Doppler samples of blood flow are taken to show the forward flow signal across the mitral and tricuspid valves (Fig. 28). By further anterior angulation, the left ventricular outflow tract and aortic valve are imaged (Fig. 29) and forward flow velocities are sampled at each site. The 2D image allows precise positioning of the Doppler sample volume. By rotation of the transducer, further selective portions of the walls of the left ventricle are obtained (Fig. 30). Color Doppler is then used to sample blood flow across each heart valve, screening for abnormal turbulent flow, particularly related to valve insufficiency (leakage during valve closure). The transducer is then moved to a location just below the sternum (Fig. 2b) and aimed up toward the heart where the right atrium and atrial septum can be further interrogated, along with partial views of the other chambers (Fig. 31). In some

patients, cross-sectional views equivalent to the short-axis view may be obtained.

However, this view in most patients is less useful because the heart is further from the transducer causing reduced resolution and increased attenuation of the echo signals. Finally, the heart may be imaged from the suprasternal approach (Fig. 2b), which generally will allow a view of the ascending aorta and aortic arch (Fig. 32).

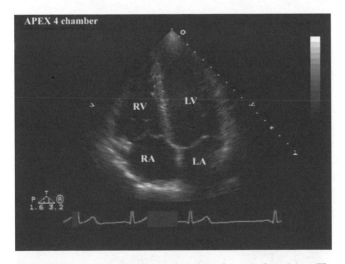

Figure 27. The transducer is placed in the apical position. The view shown is the apical "four chamber" view since all four chambers of the heart can be imaged simultaneously. Abbreviations are as follows: left ventricle (LV), left atrium (LA), right atrium (RA), and right ventricle (RV).

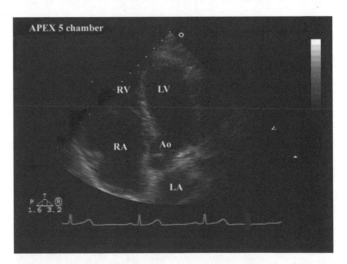

Figure 29. From the view shown in Fig. 27, the transducer has been tilted slightly to show the outflow region of the left ventricle (LV) where blood is ejected toward the aorta (Ao). Abbreviations are as follows: left atrium (LA), right atrium (RA), and right ventricle (RV).

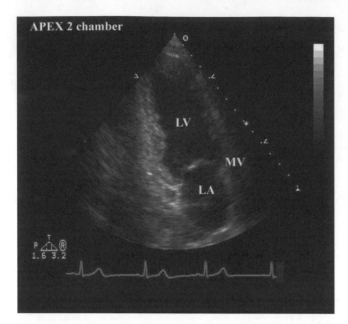

Figure 30. From the position shown in Fig. 27, the transducer is rotated to show a different perspective from the apical view. The "two-chamber" view shows only the left-sided chambers and the mitral valve (MV) that controls flow between these two chambers. Abbreviations are as follows: left ventricle (LV) and left atrium (LA).

The Transesophageal Exam

About 3–10% of hospital-based echocardiograms are performed using a specialized transesophageal (TEE) device. The ultrasound transducer, smaller but otherwise of virtually equal capability to the transthoracic device, is attached to the end of an endoscope. The patient is prepared using a topical anesthetic agent in the mouth and pharynx to eliminate the gag reflex and given conscious sedation to increase patient comfort. Patient status is monitored by a nurse, the ultrasound system operated by a sonographer

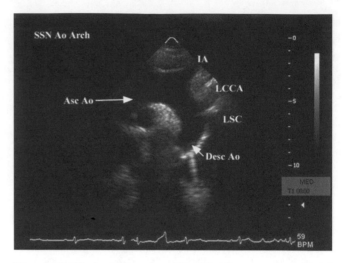

Figure 32. The transducer is positioned at the suprasternal notch (see Fig. 2b). A portion of the aorta (Asc Ao and Dsc Ao) is visible along with the origins of three branches coming off of this structure: innominate artery (IA), left common carotid artery (LCCA) and left subclavian (LSC).

and the transducer inserted by a physician who personally performs the exam. As the transducer is passed down in the esophagus, multiple imaging planes are obtained. These planes are obtained by a combination of movement of the transducer to different levels of the esophagus, changing the angle of the transducer and controlling the transducer head. Originally, the transducer was fixed in one location on the endoscope. Virtually all devices now made allow the operator to rotate the transducer through a 180° arc markedly increasing the number of imaging planes in the exam (Figs. 33–35). Using this method multiple views of structures in both a long and short axis configuration are possible. The transducer may also be passed into the stomach where additional views are possible.

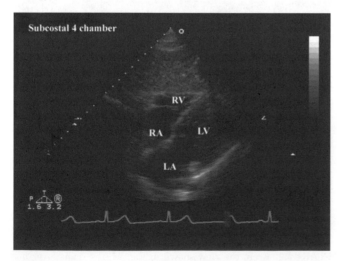

Figure 31. The transducer is moved to the subcostal position (see Fig. 2b). Portions of all four chambers are visible. Abbreviations are as follows: left atrium (LA), left ventricle (LV), right atrium (RA), and right ventricle (RV).

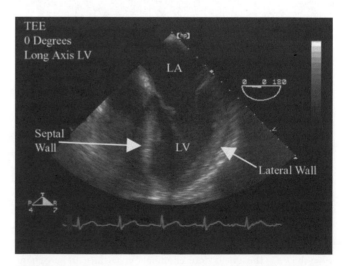

Figure 33. Image generated by a transducer placed in the esophagus. Abbreviations are as follows: left ventricle (LV) and left atrium (LA).

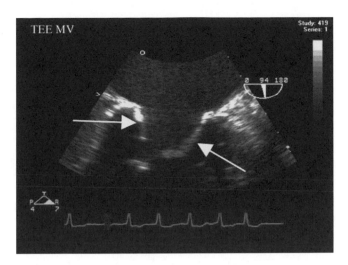

Figure 34. Image from a transesophageal transducer. This is a magnified view of the mitral valve leaflets (arrows).

The exam is performed in a similar fashion to the transthoracic exam in that multiple views of cardiac chambers and valves are taken in an organized series of views to achieve a complete examination of the heart. Many planes are similar to the transthoracic exam except for location of the transducer. Other views are unique to the TEE exam and may be the clinical reason the exam was performed. Once the exam is complete, the transducer is removed, and the patient is monitored until recovered from the sedation and topical anesthesia.

The Intracardiac Exam

During specialized procedures it may be valuable to examine the heart from inside. This is usually done during specialized sterile cardiac procedures when a transthoracic transducer would be cumbersome or impossible to use and a TEE also would be impractical.

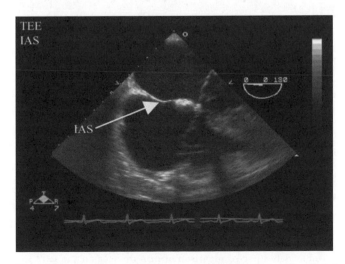

Figure 35. Image from a transesophageal transducer. The septal structure (IAS) that separates the two atrial chambers is shown (arrow).

A catheter with a miniaturized transducer is passed under fluoroscopic guidance up large central veins from the groin to the right side of the heart. For most applications, the device may be placed in the right atrium, or infrequently, in the right ventricle. The transducer is a miniature phased array device that can provide 2D and Doppler information. In most cases, the clinical diagnosis is already known and the patient is undergoing a specialized cardiac procedure in the catheterization laboratory to close a congenital defect or monitor radio frequency ablation during a complex electrophysiologic procedure, attempting to eliminate a cardiac rhythm disorder. The device contains controls that allow change of angle in four directions. This, combined with positional placement of the catheter in different parts of the cardiac chambers, allows several different anatomic views of cardiac structures (Fig. 36).

The Stress Echo Exam

Combining a transthoracic echo with a stress test was first attempted in the 1980s, became widely available in the mid-1990s, and is now a widely utilized test for diagnosis of coronary artery disease. The echo exam itself is purposely brief, usually confined to 4–6 2D views.

A patient having a stress echo is brought to a specialized stress lab that contains an ultrasound system set up to acquire stress images, an exam table, and a stress system that most commonly consists of a computerized electrocardiography system that runs a motorized treadmill (Fig. 37). The patient is connected to the 12-lead electrocardiogram stress system and a baseline electrocardiogram is obtained. The baseline echo is next performed, recording 2D views (Fig. 38). Then the stress test is begun. It can be performed in two different formats:

1. Exercise: If the patient is able to walk on a treadmill (or in some labs, pedal a bike), a standard maximal exercise test is performed until maximum effort has been achieved. Immediately upon completion of exercise, the patient is moved back to the echo exam table and repeat images are obtained within 1–2 min of completion of exercise.
2. Pharmacologic stimulation: If the patient is unable to exercise, an alternative is stimulation of the heart with an intravenous drug that simulates exercise. Most commonly, this is dobutamine, which is given, in progressively higher doses in 3–5 min stages. The patient remains on the exam table the entire time, connected to the electrocardiographic stress system. 2D echo images are obtained; at baseline, low dose, intermediate dose, and peak dose of drug infusion during the exam.

In both types of tests, clinical images are recorded and then displayed so that pretest and posttest data can be examined side by side. Comparisons of cardiac function are carefully made between the prestress and poststress images. The test relies on perceived changes in mechanical motion of different anatomic segments of the heart. Both inward movement of heart muscle and thickening of the walls of the ventricles are carefully evaluated. The normal

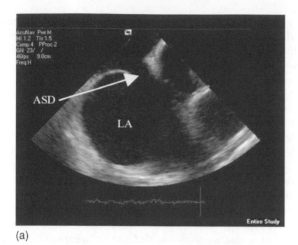

(a)

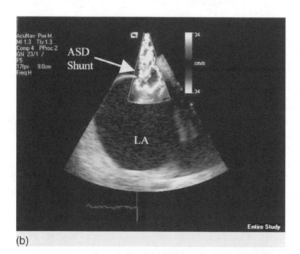

(b)

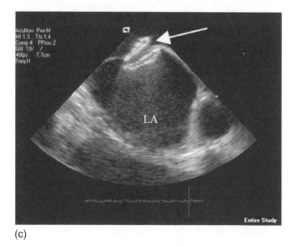

(c)

Figure 36. (a) Image of the atrial septum in a patient with an atrial septal defect (arrow). This image was taken with an intracardiac transducer placed in the right atrium (RA). The image is of similar quality to that seen in Fig. 35. (b) Image of same structure as (a) demonstrating color flow capabilities at the intracardiac transducer. Color Doppler (seen in gray scale) confirms the structure is a hole with blood passing through the hole (ASD shunt). (c) Image taken from same position as (a) and (b). The atrial septal defect has been closed with an occluder device (arrow). Doppler was used (not shown) to confirm that no blood could pass through the previously documented hole.

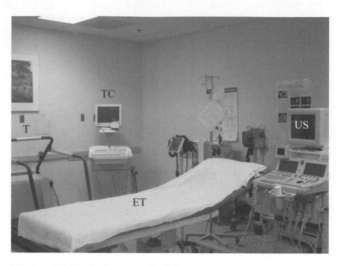

Figure 37. Typical set-up of a laboratory to perform stress echocardiograms. Shown are the echocardiographic examination table (ET), the treadmill (T), the treadmill computer control system (TC), and the ultrasound system (US).

heart responds by augmenting inward motion and wall thickening in all regions. If blood supply to a given segment is not adequate, mechanical motion and wall thickening either fails to improve or deteriorates in that segment, but improves in other segments. This change defines an abnormal response on a stress echo (Fig. 38). The location and extent of abnormal changes in wall motion and thickening are reported in a semiquantitative manner. In addition, the electrocardiogram response to stress is also compared with the baseline exam and reported along with patient symptoms and exercise capacity.

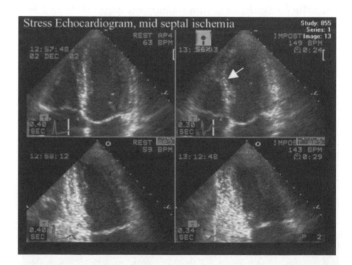

Figure 38. Quad screen format of view obtained from a stress echocardiography examination. Four views are displayed simultaneously, and the particular views shown can be changed to accommodate direct comparison of pre- (images on the left) and posttest images on the right views of the same part of the heart as shown in the example. The arrow indicates an area of the heart that failed to respond normally.

CLINICAL USES OF ECHOCARDIOGRAPHY

M-Mode Echocardiography

The M-mode echo was the original cardiac exam and for years was the dominant cardiac ultrasound study performed. The 2D echo has superseded M-mode echo as the primary examination technique and in most circumstances is superior. However, many laboratories continue to perform at least a limited M-mode exam because dimension measurements are well standardized. In addition, due to its high sampling rate, M-mode echo is superior to 2D echo for timing of events within the cardiac cycle and for recording simultaneously, with other physiologic measurements such as phonocardiograms and pulse tracings. Certain movement patterns of heart valves and chamber walls are only detected by M-mode, thus providing unique diagnostic information (Fig. 39).

Most M-mode echo data is obtained from multiple different "ice pick" views, all obtained from the parasternal position (Fig. 7). The 2D image is used to direct the cursor position of these views. The first view angles through a small portion of the right ventricle, the ventricular septum, the left ventricular chamber, the posterior wall of the left ventricle, and the pericardium. From this view, left ventricular wall thickness and the short-axis dimensions of this chamber can be measured. By calculating the change in dimension between end diastole (at the beginning of ejection) and end systole (at the end of ejection), the fractional shortening can be measured using the equation:

$$\frac{LVEDD - LVESD}{LVEDD} = \text{fractional shortening}$$

where LVEDD is the left ventricular end diastolic dimension and LVESD is the left ventricular end systolic dimension. This measurement estimates left ventricular function (Fig. 40). Dimension measurements give a relatively precise estimate of left ventricular chamber size to calculate whether the ventricle is inappropriately enlarged. In a similar manner wall thickness measurements can be utilized to determine if the chamber walls are inappropriately thick (left ventricular hypertrophy), asymmetrically thickened (hypertrophic cardiomyopathy), or inappropriately thin (following a myocardial infarction).

The second ice pick view passes through the right ventricle, septum, mitral valve leaflets, and posterior wall. This view is used primarily to evaluate motion of the mitral valve. Certain types of mitral valve disease alter the pattern of motion of this valve (Fig. 39). Other abnormal patterns of leaflet motion may indicate dysfunction elsewhere in the left ventricle.

The third ice pick view passes the echo beam through the right ventricle, aortic valve, and left atrium. From this view, analogous to the mitral valve, the pattern of aortic valve motion will change in characteristic ways allowing the diagnosis of primary aortic valve disease or diseases that cause secondary aortic valve motion changes. Also, from this view the diameter of the left atrium is measured and whether this structure is of normal size or enlarged can be of considerable importance in several circumstances. Other views are possible, but rarely used.

Measurements taken from M-mode may be performed during the exam by the sonographer. Usually, the scrolling M-mode video display is saved in digital memory on the system. The saved data is then reviewed until the best depiction of the various views discussed above is displayed. The sonographer then uses electronic calipers, automatically

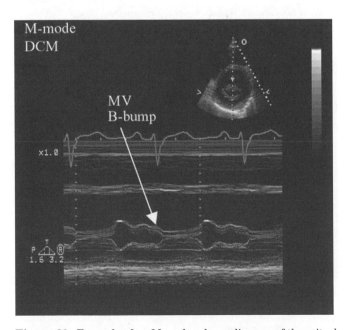

Figure 39. Example of an M-mode echocardiogram of the mitral vale showing abnormal motion, in this case a "B-bump" indicating delayed closure of the valve. DCM = dilated cardiomyopathy.

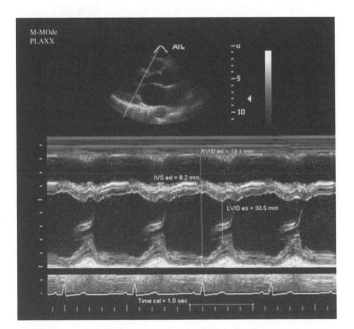

Figure 40. Example of an M-mode echocardiogram of the left ventricle. The typical measurements of wall thickness (IVSed and LVPWed) and chamber dimensions (RVIDed, LVIDed and LVIDes) are shown. Several calculated parameters are possible from these values.

calibrated by the system, to record dimension measurements (Fig. 40). Basic measures are made and formulas for derived data such as the % FS are automatically calculated. Measurements can also be made "off line" using special workstations that display the video information at the time the study is interpreted by the physician.

Two-Dimensional Echocardiography

The 2D exam gives information about all four cardiac chambers and all four cardiac valves. It also serves as the reference point for positioning all Doppler sample volumes and the color Doppler exam. The heart is imaged from multiple positions, which not only improves the chance of useful information being obtained, but also allows better characterization of a given structure because the structure is seen in several perspectives. The primary role of a 2D echo is to characterize the size of the left and right ventricles as normal or enlarged, and if enlarged, estimate the severity of the problem. Second, the 2D exam evaluates pump function of the two ventricles. Function is characterized globally (i.e., total ventricular performance) or regionally (i.e., performance of individual parts of each ventricle). Some types of disease affect muscle function relatively equally throughout the chambers. Other forms of disease, most notably coronary artery atherosclerosis, which selectively changes blood supply to various parts of the heart, cause regional changes in function. In this disease, some portions of the heart may function normally while other areas change to fibrous scar and decrease or stop moving entirely. Global function of the left ventricle can be characterized quantitatively. The most common measurements of function use calculations of volume during the cardiac cycle. This is quantified when the heart is filled maximally just before a beat begins and minimally just after ejection of blood has been completed. The volume calculations are used to determine of ejection fraction. The equation is

$$\text{Ejection fraction} = \frac{\text{LVEDV} - \text{LVESV}}{\text{LVEDV}} \times 100$$

where LVEDV = left ventricular end diastolic volume and LVESV = left ventricular end systolic volume.

The 2D echo is sensitive for detecting abnormalities within the chambers, such as blood clots or vegetations (infectious material attached to valves). Abnormalities surrounding the heart, such as pericardial effusions (fluid surrounding the heart), metastatic spread of tumors to the heart and pericardium, and abnormalities contiguous to the heart in the mediastinum or great vessels can be readily imaged. Most of this information is descriptive in nature (Figs. 41 and 42).

The ability to directly and precisely make measurements from 2D echo views for quantitative measurements of dimensions, areas, and volumes is built into the ultrasound system and is typically done by the sonographer during the exam in a similar fashion to M-mode. Further measurements may be performed off line on dedicated analysis computers. Many parts of the interpretation of the exam, however, remain primarily descriptive and are usually estimated by expert readers.

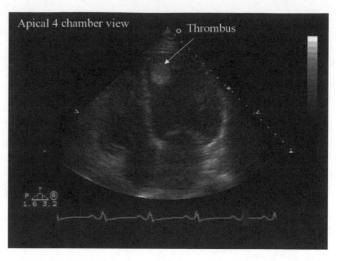

Figure 41. Example of detection of a blood clot (thrombus) in the left ventricular chamber.

Doppler Echocardiography

While the 2D echo has considerably expanded the ability to characterize abnormalities of the four heart valves, it has not been possible to obtain direct hemodynamic information about valve abnormalities by imaging alone. Doppler imaging provides direct measurement of hemodynamic information.

By using Doppler, six basic types of information can be obtained about blood flow across a particular region:

1. The direction of the blood flow.
2. The time during the cardiac cycle during which blood flow occurs.
3. The velocity of the blood flow.
4. The time the peak velocity occurs.
5. The rate at which velocity changes.

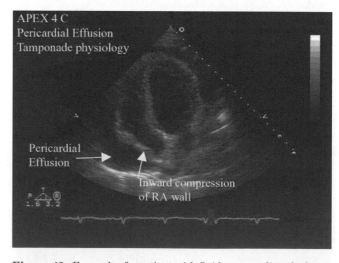

Figure 42. Example of a patient with fluid surrounding the heart. The dark area surrounding the heart (pericardial effusion) is shown. In this case, increased pressure caused by the effusion compresses part of the right atrium (RA).

6. The pressure drop or gradient across a particular valve or anatomic structure.

Data about pressure gradients is derived from the velocity measurement using the Bernoulli equation:

$$P_1 - P_2 = \underbrace{1/2\rho(V_2^2 - V_1^2)}_{\substack{\text{Convective} \\ \text{acceleration}}} + \underbrace{\rho \int_1^2 \frac{dv}{dt} ds}_{\substack{\text{Flow} \\ \text{acceleration}}} + \underbrace{R\,(v)}_{\substack{\text{Viscous} \\ \text{friction}}}$$

where $P_l - P_2$ is the pressure drop across the structure V_2 and V_1 being blood flow velocity on either side of the structure, and ρ is the mass density of blood (1.06×10^3 kg/m^3). For applications in the heart, the contributions by the flow acceleration and viscous friction terms can be ignored. In addition, V_1 is generally much less than V_2 (thus, V_1 can usually be ignored), and ρ is a constant for the mass density of blood (6). Combining all these changes together results in the final "simplified" form of the Bernoulli equation:

$$P_1 - P_2 = 4\,V^2$$

Cardiac Output. When the heart rate, blood flow velocity integral, and cross-sectional area of the region across which the blood flow is measured are known, cardiac output can be estimated using the following equation:

$$\text{CO} = A \times V \times \text{HR}$$

where CO is the cardiac output, A is the cross-sectional area, V is the integrated blood flow velocity, and HR is the heart rate.

The Character of the Blood Flow. The differentiation between laminar and turbulent blood flow can be made by observation of the spectral pattern. In general, laminar flow (all recorded velocities similar) occurs across normal cardiac structures of the heart, while disturbed or turbulent flow (multiple velocities detected) occurs across diseased or congenitally abnormal cardiac structures (Fig. 20).

Doppler is most valuable in patients with valvular heart disease and congenital heart disease. In the case of valve stenosis (abnormal obstruction to flow), use of Doppler echocardiography allows quantification of the pressure gradient across the valve (Fig. 43). Using the continuity principle, which states that the product of cross-sectional area and flow velocity must be constant at multiple locations in the heart, it is possible to solve for the severity of valve stenosis. The equation may be written as noted and then manipulated to solve for the area at the stenotic valve (A_2).

$$A_1 V_1 = A_2 V_2$$
$$\frac{A_1 V_1}{V_2} = A_2$$

For valvular insufficiency, Doppler echocardiography is most useful when the color Doppler format is used in

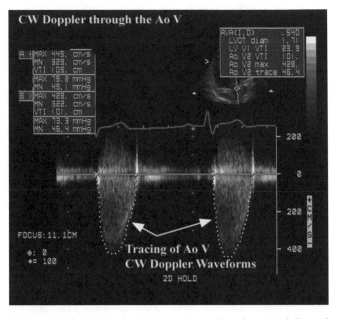

Figure 43. Example of a continuous wave Doppler signal through the aortic valve (AoV). Two tracings are shown along with measurement technique also demonstrated.

conjunction with 2D echo. Since an insufficient valve (i.e., a valve that allows backward leakage of blood when closed) produces turbulent flow in the chamber behind the valve, color Doppler immediately detects its presence. The extent to which turbulent flow can be detected is then graded on a semiquantitative basis to characterize the amount of valve insufficiency (Fig. 17). Since only 2D are interrogated at any given time, the best results are obtained when Doppler sampling is done from more than one view.

In patients with congenital heart disease, Doppler echocardiography allows the tracing of flow direction and velocity across anatomic abnormalities, such as holes between various cardiac chambers (i.e., atrial or ventricular septal defects). It can also display gradients across congenitally malformed valves and great vessels and also determine the direction and rate of flow through anatomically mal positioned chambers and great vessels.

In addition to direct interrogation of heart valves for detection of primary valve disease, Doppler flow sampling is used to evaluate changes in flow across normal valves that may indicate additional pathology. For example, the systolic blood pressure in the lungs (pulmonary artery pressure) may be estimated by quantifying the velocity of flow of an insufficiency jet across the tricuspid valve (another application of the Bernoulli equation). This calculated value, when added to the estimated central venous pressure (obtained in a different part of the exam) gives an excellent estimate of pulmonary artery pressure.

A second important measurement involves characterizing the way the left ventricle fills itself after ejecting blood into the aorta. There is a well-described set of changes in the pattern of flow across the mitral valve, changes in flow into the left atrium from the pulmonary veins and changes in the outward movement in the muscle itself characterized by tissue Doppler that can help categorize the severity of changes in filling of the left ventricle.

SPECIALIZED CLINICAL DATA

Transesophageal Echocardiography

The TEE exam is used when a transthoracic exam either cannot be performed or gives inadequate information. One limitation of echo is the great degree of variability in image quality from patient to patient. In some circumstances, particularly in intensive care units, the TEE exam may provide superior image quality since its image quality is not dependent on patient position or interfered with by the presence of bandages, rib interfaces, air or other patient dependent changes. Similarly, during open heart surgery a TEE is routinely used to assess cardiac function pre- and postintervention and pre- and postheart valve replacement or repair. Since the TEE probe is in the esophagus, outside of the surgeon's sterile field images are obtained even when the patient's chest is open. This capability allows the surgeon to evaluate the consequences of, for example, a surgical repair of a heart valve, when the heart has been restarted, but before the chest is sutured closed. The TEE exam also visualizes parts of the heart not seen by any transthoracic view. A particular example of this is the left atrial appendage, a part of the left atrium. This structure sometimes develops blood clots that can only be visualized by TEE.

Three-Dimensional Reconstruction

While the 2D exam displays considerable data about spatial relationships between structures and quantification of volume, there is still considerable ambiguity in many circumstances. One way to further enhance the exam is to use 3D reconstruction.

Its use has been a significant challenge. All early methods developed computerized routines that characterized the movement of the transthoracic transducer in space. Images were acquired sequentially and then reconstructed first using geometric formulae and later using more flexible algorithms without geometric assumptions. The data, while shown to be useful for both adding new insight into several cardiac diseases and improving quantitation, did not achieve practical acceptance due to the considerable operator time and effort required to obtain just one image (12).

Recently innovations in image processing and transducer design have produced 3D renditions of relatively small sections of the heart in real time. Use remains limited at present but further development is expected to make 3D imaging a practical reality on standard ultrasound systems (Fig. 44).

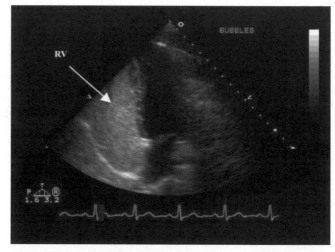

(a)

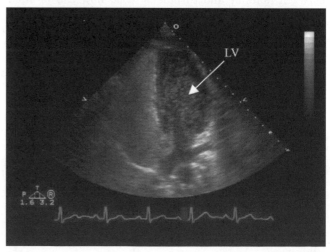

(b)

Figure 45. (a) Example of injection of agitated saline into systemic veins. The contrast moves into the right atrium (RA) and right ventricle (RV), causing transient full opacification. The left-sided chambers are free of contrast. (b) Similar to (a). However, some of the contrast bubbles have crossed to the left ventricle (LV), proving a communication exists between the right and left sides of the heart.

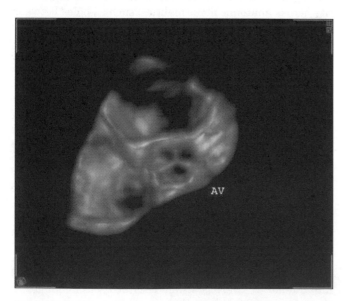

Figure 44. Example of on-line 3D reconstruction. The heart is imaged at the level of the aortic valve where all three leaflets are shown.

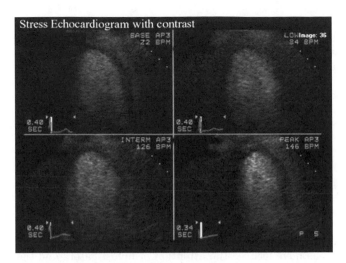

Figure 46. Example of contrast use to improve visualization of the walls of the left ventricle. The contrast agent used in this case is a gas-filled microbubble that fills the left ventricle during a stress test. The entire chamber is white with full contrast enhancement.

Contrast Imaging

Contrast agents are combined with certain 2D echo exams to enhance the amount of diagnostic information available on the exam, or improve the quality of the exam.

There are two types of contrast agents. One type is made from normal saline solution. An assistant generates the contrast for injection during regular 2D imaging. Typically 0.5 mL of air is added to a 10 mL syringe of saline and vigorously hand agitated between 2 syringes for about 15 s. This agitation causes production of several million small bubbles. The bubbles are too large to pass through capillaries, thus when injected into a vein, they are all filtered out by the passage of venous blood through the lungs. Thus when injected into a normal heart, saline contrast passes into the right atrium and right ventricle, produces a transient intense response and disappears, not making it to the left ventricle. This property makes saline injection ideal for detecting an abnormal hole, or shunt passage across the atrial septum or ventricular septum. When a communication of this type is present bubbles cross directly from the right side to the left side of the heart. This is an extremely sensitive method for making this diagnosis (Fig. 45).

A second type of contrast agent, developed and now commercially marketed, is a gas filled microbubble smaller than a red blood cell. This type of agent, made from perfluorocarbons covered by an albumen or lipid shell, when intravenously injected passes through the lungs and opacifies the left side of the heart as well as the right side. The bubbles are gradually destroyed by ultrasound and blood pressure, lasting for several minutes in ideal circumstances. When combined with harmonic imaging these contrast agents markedly improve the quality of the 2D image, particularly for the evaluation of left ventricular wall motion (Fig. 46). The contrast agent markedly enhances the border between blood and the chamber wall. Use of these agents is variable among laboratories, but its use substantially decreases the number of nondiagnostic studies. Still under investigation is whether these same contrast agents can be used for evaluation of blood flow within the heart muscle. This information could be of particular value for patients with coronary artery disease either during acute episodes of ischemia when a new coronary stenosis is suspected or as an enhancement to the stress echocardiogram (13).

BIBLIOGRAPHY

Cited References

1. Edler I, Hertz CH. The use of ultrasonic reflectoscope for the continuous recording of the movements of heart walls. Kungl Fysiografiska sällskapets I lund forhandlingar 1954;24(5):1–19.
2. Zagzebski JA. Physics of Diagnostic Ultrasound. In: Rowland J, editor. Essentials of Ultrasound Physics. St. Louis: Mosby; 1996. p 1–18.
3. Hancock J, Dittrich H, Jewitt DE, Monaghan MJ. Evaluation of myocardial, hepatic and renal perfusion in a variety of clinical conditions using an intravenous ultrasound contrast agent (Optison) and second harmonic imaging. Heart 1999;81:636–641.
4. Spencer KT, Bednarz J, Rafter PG, Korcarz C, Lang RM. Use of harmonic imaging without echocardiographic contrast to improve two-dimensional image quality. Am J Cardiol 1998; 82:794–799.
5. Kisslo J, VonRamm OT, Thurstone FL. Cardiac imaging using a phased array ultrasound system II: clinical technique and application. Circulation 1976;53:262–267.
6. Nishimura RA, Miller FA, Callahan MI, Benassl RC, Seward JB, Tajik AJ. Doppler echocardiography: Theory, instrumentation, technique and application. Mayo Clinic Proc 1985;60:321–343.
7. Omoto R, Kasai C. Physics and instrumentation of Doppler color mapping. Echocardiography 1987;4:467–483.
8. Vandervoort PM, Rivera JM, Mele D, Palacios IF, Dinsmore RE, Weyman AE, Levine RA, Thomas JD. Application of color Doppler flow mapping to calculate effective regurgitant orifice area. Circulation 1993;88:1150–1156.
9. Burns PN. Instrumentation for contrast echocardiography. Echocardiography 2002;19:241–258.
10. Sade LE, Severyn DA, Kanzaki H, Dohi K, Gorcsan J. Second generation tissue Doppler with angle corrected color codes wall displacement for quantitative assessment of regional left ventricular function. Am J Cardiol 2003;92:554–560.
11. Urhelm S, Edvardson T, Torpi H, Angelsen B, Smiseth O. Myocardial strain by Doppler echocardiography: Validation of a new method to quantify regional myocardial function. Circulation 2000;102:1158–1164.
12. Mele D, Levine RA. Quantitation of ventricular size and function: Principles and accuracy of transthoracic rotational scanning. Echocardiography 2000;17:749–755.
13. Porter TR, Cwajg J. Myocardial contrast imaging: A new gold standard for perfusion imaging. Echocardiography 2001; 18:79–87.

Reading List

Anderson B. Echocardiography: The Normal Examination and Echocardiographic Measurements. Brisbane: Fergies; 2000.
Goldberg BB. Ultrasound Contrast Agents. St. Louis: Mosby Year Book; 1997.
Feigenbaum H. Echocardiography. 5th ed. Philadelphia: Lea & Febiger; 1994.

McB Hodgson J, Sheehan HM. Atlas of Intravascular Ultrasound. New York: Raven Press; 1994.

Oh JK, Seward JB, Tajik AJ. The Echo Manual. Philadelphia: Lippincott-Raven; 1999.

Otto CM. Textbook of Clinical Echocardiography. 2nd ed. Philadelphia: W.B. Saunders; 2000.

Weyman AE. Principles and Practice of Echocardiography. 2nd ed. Philadelphia: Lea & Febiger; 1994.

Zagzebski JA. Essentials of Ultrasound Physics. St. Louis, Mosby; 1994.

Hatle L, Angelsen B. Doppler Ultrasound in Cardiology. Philadelphia: Lea and Febiger; 1985.

Marwick TH. Stress Echocardiography: Its role in the diagnosis and evaluation of coronary artery disease. Dordrecht: Kluwer Academic Publishers; 1994.

Freeman WK, Seward JB, Khandheria BK, Tajik AJ. Transesophageal Echocardiography. Boston: Little Brown; 1994.

See also BIOIMPEDANCE IN CARDIOVASCULAR MEDICINE; ULTRASONIC IMAGING.

ECT. See ELECTROCONVULSIVE THERAPY.

EDUCATION, BIOMEDICAL ENGINEERING. See BIOMEDICAL ENGINEERING EDUCATION.

EDUCATION, COMPUTERS IN. See MEDICAL EDUCATION, COMPUTERS IN.

EEG. See ELECTROENCEPHALOGRAPHY.

EGG. See ELECTROGASTROGRAM.

ELECTRICAL TREATMENT OF BONE NONUNION. See BONE UNUNITED FRACTURE, ELECTRICAL TREATMENT OF.

ELECTROANALGESIA, SYSTEMIC

AIME LIMOGE
The René Descartes
University of Paris
Paris, France

TED STANLEY
Salt Lake City, Utah

INTRODUCTION

Electroanalgesia, electroanesthesia, neurostimulation, neuromodulation, and other physical methods of producing analgesia, anesthesia, and/or decreased sensitivity to painful stimuli are old concepts that are beginning to be revitalized in the recent past. For > 40 years, there has been a revival of electrotherapy in the treatment of pain. Analgesia by electrical current is now based on transcutaneous or percutaneous nerve stimulation, deep stimulation, posterior spinal cords stimulation, and transcutaneous cranial electrical stimulation (1–8). One reason for this has been the increased awareness of spinal and supraspinal opioid analgesic mechanisms, including the precise pathways, receptors, and neurotransmitters involved in pain perception, recognition, modulation, and blockade. Another reason is the renewed belief that nonpharmacological manipulation of these receptors and transmitters should be possible with electricity since numerous progress have been made in the development of electric current waveforms that result in significant potentiation of the analgesic and hypnotics action of many intravenous and inhaled anesthetics without producing significant side effects (9–20). Finally, recent successes of transcutaneous electrical nerve stimulation (TENS) in the treatment of pain and transcutaneous cranial electrical stimulation (TCES) as a supplement during anesthesia to obtain postoperative analgesia by potentiating the anesthetic agents used during the intra- and postoperative phases. The popularity of electroacupuncture in a variety of pain and pain related areas have focused the attention of investigators and the public on electricity as a beneficial medical therapy. In this article, some of the most recent developments in nerve and brain stimulatory techniques using electrical stimulation to produce analgesia are addressed.

HISTORY

Alteration of pain perception utilizing forms of electrical stimulation dates back to the Greco-Roman period. Electrostimulation to decrease the pain started with the "electric fish" (torpedo marmorata), as 46 years after Jesus Christ, Scribonius Largus, physician to emperor Claudius, recommended the analgesic shock of the Torpille in the treatment of the pain (21,22). Unfortunately, in those days attempts were crude and success was limited for many reasons none-the-least of which was a poor understanding of the fundamentals of electricity. Interest in electroanalgesia was renewed in the seventeenth century when Von Guericke built the first electrostatic generator to apply locally to relieve pain; however, results were still marginal. At the beginning of the twentieth century, Leduc reawakened interest in the idea of producing sleep and local and general anesthesia with low frequency impulsional electrical current. He used unidirectional rectangular intermittent current of 100 Hz with an ON-time of 1 ms and OFF-time of 9 ms with a moderate amperage (0.5–10 mA) on a variety of animals and on himself to evaluate the effects of electricity on the central nervous system (23,24). Electrodes were placed on the forehead and kidney areas and electrostimulation resulted in apnea, cardiac arrhythmias, cardiac arrest, and convulsions in dogs and a "nightmarelike state" in which the subject was aware of pain. Despite these inauspicious beginnings, studies continued. In 1903, Zimmern and Dimier produced postepilectic coma with transcerebral currents, and in 1907, Jardy reported the first cases of surgery in animals with electroanesthesia. Between 1907 and 1910, Leduc and other performed a number of surgical operations on patients with electricity as an anesthetic supplement (1–5).

In the early decades of the twentieth century, electroanalgesia was always associated with intense side effects including muscle contractures, prolonged coma (cerebral shock), cerebral hemorrhage, hyperthermia, cardiac arrhythmias, and convulsions. Because of these difficulties, interest waned. In 1944, Frostig and Van Harreveld began experimenting with an alternating current from 50 to 60 mA and a variable voltage bitemporally. The advantage of this more complex method of stimulation was less muscular spasm and contraction (4,5). Unfortunately, these approaches still resulted in transient periods of apnea, cardiac arrhythmias, and standstill as well as fecal and urinary soillage. These problems could be reduced, but not eliminated, by decreasing amperage. Numerous other investigators began using many diverse currents without much success. The most interesting results were obtained in 1951 by Denier (25,26) and in 1952 by Du Cailar (4). Denier began experimenting with high frequency (90 kHz) rectified sinusoidal current with a pulse duration of 3 ms (on time) and a resting time of 13 ms (OFF time), knowing that the effects of modulation at a high frequency current are those of the envelope of its waves. Du Cailar introduced the idea of electropharmaceutical anesthesia by utilizing a premedication of morphin–lobelin in association with a barbituric induction, along with the electrical current. This idea of electropharmaceutical anesthesia that was taken up again in the Soviet Union in 1957 by Ananev et al. using the current of Leduc combined with a direct current (1–3), and in the United States by Hardy et al. using an alternating sinusoidal current of 700 Hz current of Knutson (27–29), and during the same period by Smith using the current of Ananev (1). But with these currents the experimenters were always bothered by side effects (muscle contractions of the face with trismus, of the body with apnea, etc.) that required the use of curare and for all practical purposes made this approach to anesthesia more complicated that conventional anesthesia.

Other investigators began studying mixtures of pharmaceutical agents, including opioids, barbiturates, and later benzodiazepines and butyrophenones in combination with electric currents to reduce and hopefully eliminate these problems, which were often attributed to "the initial shock of the electric current". Others began studying the shape of the current waveform and its frequency. Sances Jr., in United States, used the current of Ananev associated with white noise (5,30) while Shimoji et al. in Japan, used a medium frequency (10 kHz) monophasic or biphasic current with sinusoidal or rectangular waves (31,32). Many were able to produce impressive analgesia and anesthesia in animals, but significant problems (apnea, hypersialorrhea, muscular contractures, convulsions, cardiac arrhythmias) continued to occur in humans. As a result, from the 1950s until the present time, many investigators focused on appropriate electrode placement. It was Djourno who thought that the principal problem to resolve was to find the ideal position for the electrodes to determine the trajectory of the electric current so as to touch precise zones of the brain. This is why he advocated electrovector anesthesia applied with three electrode pairs (vertex-palate, temporal-temporal, fronto-occipital) (4,33). During this time the Soviets Satchov et al. preferred interferential currents of middle frequencies (4000–4200 Hz) associated with barbiturates transmitted by two pairs of crossed electrodes (left temporal–right retromastoid and right temporal–left retromastoid). Others suggested that the problems can be minimized by using mixtures of sedative, hypnotic, and analgesic drugs, plus low amperage electrical currents to produce the ideal effect.

The result of all this activity is that there is still no general agreement on the importance of electrode placement (although frontal and occipital are probably most popular), waveform, wave frequency, current strength, interference currents, or the role of supplemental pharmacotherapy (4,34–38). What was agreed was that it appeared impossible to reliably produce problem-free "complete anesthesia" in humans using any available electrical generators and associated apparatus. Instead, the most successful approaches to electroanesthesia have used waveforms, frequencies, and currents that produce few, if any, side effects (and result in significant analgesia), but must be supplemented with pharmacological therapies to be a "complete anesthetic". While some may scoff at these modest gains, others remain optimistic because using a variety of neurostimulatory approaches, reproducible and quantifiable analgesia was now possible without pharmaceutical supplementation.

Analgesia and Electroneurostimulation

The advancement of the spinal gate control theory of pain by Melzach and Wall (39,40), the discovery of central nervous system opiate receptors, and the popularity and apparent effectiveness of acupuncture in some forms of pain management have given support to the basis that neurostimulatory techniques can produce analgesia via readily understandable neurophysiological changes rather than mysterious semimetaphysical flows of mysterious energy forces (41,42). It is now clear that electrical stimulation of the brain and peripheral nerves can markedly increase the concentration of some endogenous opiates (β-endorphin, δ-sleep producing factor, etc.) in certain areas of the brain and produce various degrees of analgesia. It is proposed that pain relief from electrical stimulation also results from a variety of other mechanisms including alteration in central nervous system concentrations of other neurotransmitters (serotonin, substance P), direct depolarization of peripheral nerves, peripheral nerve fatigue, and more complex nervous interactions (43–46).

Whatever the mechanisms producing analgesia with electrical stimulation, many clinicians are beginning to realize the advantages of these techniques. Neurostimulatory techniques are relatively simple, devices are often portable, their parameters (controls) are easy to understand and manipulate, and application usually requires minimal skills. Moreover, there are few, if any, side effects, addiction is unheard of, if a trial proves unsuccessful little harm is done, the techniques reduce requirements for other analgesics, and usually the stimulation itself is pleasant.

Transcutaneous Electrical Nerve Stimulators

Transcutaneous electrical nerve stimulation, currently called TENS, is the most frequently used device for treatment of acute postoperative and chronic pain of most etiologies. The first portable transcutaneous electrical stimulators were produced in the 1970s with controllable wave forms and modulable patterns of stimulation. The goal was to produce a compact, lightweight, portable miniaturized current generator to provide stimulation by means of skin contacting electrodes, and able to be used as the patient went about normal daily activities. To that end, as well as safety reasons, the devices were battery powered. A plethora of electrical nerve stimulators can be found on the market. Dimensions are approximately the size of a pack of cigarettes and can be worn by the patient by use of straps or belts. These stimulators, that have one or more adjustable electric parameters that provide no ease of operation, deliver biphasic waves of low frequency of 1–250 Hz with current intensity from 50 to 100 mA. These electrical stimulations result in a tingling or vibrating sensation. Patients are able to adjust the dial settings with respect to frequency and intensity of the stimulus.

The stimulation electrodes must permit uniform current density and have a stimulation surface > 4 cm^2 in order to avoid cutaneous irritation caused by elevated current densities. The material must be hypoallergenic, soft, and flexible to allow maximal reduction of any discomfort while providing for lengthy stimulation in diverse situations. The impedance at the biologic electrode–skin interface can be minimized by the choice of material as well as the use of a conducting gel. Materials used to make the electrodes can be carbon-based elastomeres as well as malleable metals. Most recent developments use adhesive-type ribbons impregnated with silver and are activated by a solvent and provide improved conductibility. For a clinician who is inexperienced in electronics or electroneurophysiology, it is difficult to choose wisely as parameters available for use are created by inventors or producers with absolutely no scientific basis. Analysis of results obtained with the majority of these devices is based on subjectivity of the physician or the patient. The domain is merely empiric. It is a pity that the parameters chosen in the production and use of these devices is by researchers that have not taken advantage of the available scientific works in electrophysiology, notably those of Willer (47,48) on the nociceptive reflex of exercise in humans. A neurostimulator must be selected that will provide proper nerve excitation that is reproducible and durable and that does not cause lesions from burns or electrolysis. Consequently, all those stimulators that deliver direct or polarized current should be used carefully as well as those that deliver a radio frequency (RF) in excess of 800 kHz. One must chose stimulators that deliver a constant biphasic asymmetric current, that is, one that delivers a positive charge that is equal to the negative charge providing an average intensity of zero. To guide the clinician, it must be recalled that current always takes the path of least resistance, and therefore a current of low frequency can only be peripheral the more one increases the frequency. Otherwise, undesirable effects will be produced under electrodes. It is know that a sensation of numbness appears from 70 to 100 Hz and that a motor action appears from 1 to 5 Hz.

Implantable Electrical Nerve Stimulators

Other forms of stimulation consist of implanted neurostimulators, spinal cord stimulation (SCS) (dorsal column stimulators), and deep brain stimulation (DBS). Peripheral nerve neurostimulation implants are also often used for chronic pain but may be employed for acute ulnar, brachial plexus, or sciatic pain in critically ill patients (8).

There are two types of implantable electrical stimulators: Passive-type stimulator with RF made up of as totally implantable element (receptor) and an external element (transmitter) that supplies the subcutaneous receiver through the skin using an RF modulated wave (500 kHz–2 MHz). Active-type totally implantable stimulator, supplied by two mercury batteries (which lasts for 2–4 years) or a lithium battery, which lasts for 5 or 10 years. These devices enable several parameters to be controlled (amplitude peak, wave width, frequency gradient). The variation of these parameters obviously depends on the patient, the region stimulated and the symptom which it is desired to modify.

ACTUAL CLINICAL NEUROSTIMULATORY TECHNIQUES

Certain precautions must be taken and the patient must be well advised as to the technique, the principles of stimulation, and all desired effects. These techniques should not be used on patients wearing a cardiac pacemaker, pregnant women, or in the vicinity of the carotid sinus. The methods demand the utmost in patience, attention to detail, and perseverance. It must be regularly practiced by medical or paramedical personnel.

The most important application of neurostimulatory techniques in clinical use today is in management of acute postoperative and chronic pain, however, since 1980 numerous terms are used in the articles to describe the diverse techniques for electrical stimulation of nervous system. Certain words do not harmonize with reality, such as TransCranial Electrostimulation Treatment (TCET) or Transcranial Electrostimulation (TE). In reality, the microamperage and low frequency used do not enable penetration of the current into the brain, they correspond to a peripheral electrostimulation, which is a bad variant of Transcutaneous Electrical Nerve Stimulation, now being used for certain painful conditions.

Transcutaneous Electrical Methods

Transcutaneous Electrical Nerve Stimulation (TENS). The purpose of this method is to achieve sensitive stimulation, by a transcutaneous pathway, of the tactile proprioceptive fibers of rapid conduction with minimal response of nociceptive fibers of slow conduction and of efferent motor fibers. Numerous studies have documented that TENS in the early postoperative period reduces pain, and thus the need for narcotic analgesics, and improves pulmonary function as measured by functional residual capacity. TENS is also frequently applied in chronic unremitting pain when other approaches are less effective or ineffective. This method is the simplest technique, and appears to be effective by alleviating the appreciation of pain (6,49).

The points of stimulation and the stimulation adjustments must be multiple and carefully determined before concluding that the effect is negative. Different stimulation points are used by the various authors: One can stimulate either locally by placing the electrodes in the patient at the level of the painful cutaneous area and more particularly on the trigger point that may be at times some distance from the painful zone (50), or along a nerve pathway "upstream" away from the painful zone to cause parasthesia in the painful area, or an acupuncture point corresponding to the points depicted an acupuncture charts (41,42). The stimulation time is usually 20–30 min and repeated at fixed hourly intervals, and discontinued when the pain is relieved. Whatever method is used, one must avoid the production of harmful stimulations or muscle contractions and the stimulation must be conducted with the patient at rest.

In acute injury states where pain is localized, TENS can produce analgesia in up to 80% of patients (51), but this percentage decreases to ~ 20% effectiveness at the end of a year. In order to obtain this result, this stimulation has the sensation of "pins and needles" in the area of the cutaneous stimulation. This phenomenon appears to be in part similar to a placebo effect estimated at 33% regardless of the type of current employed or the location of applied current. As the affected area increases in size, TENS is less likely to be sufficient and is also less effective in chronic pain, especially if the cause of the pain itself is diffuse.

The mechanism by which TENS suppresses pain is probably related to spinal and/or brain modulation of neurotransmitter and/or opiate or other γ-aminobutyric acid (GABA) receptor function. This method works best with peripheral nerve injuries and phantom and stump pains. Transcutaneous nerve stimulators are usually less effective in low back pain or in patients who have had multiple operations. It is often totally unsatisfactory for pain (particularly chronic pain) that does not have a peripheral nerve cause such as pain with a central nervous system etiology or an important psychological component (depression and anxiety) (52–55).

Transcutaneous Acupoint Electrical Stimulation (TAES). Acupuncture, in its traditional form, depends on the insertion of needles into specific acupuncture points in the body as determined by historical charts. Electrical Acupuncture (EA) or TAES employs Low Frequency (LF) stimuli of 5–200 Hz in the needles inserted at the classical acupuncture points. Occasionally, nontraditional acupuncturists use the needles at or near the painful area. Usually these types of treatments produce mild degrees of analgesia. Electrical acupuncture is essentially as benign as TENS and produces its effects by similar mechanisms (42,53,56). Unfortunaly, EA is more expensive toperform than TENS because it necessitates the presence of an acupuncturist clinician. Thus, it is likely that EA will not become as popular as TENS for treatment of most pain problems.

Transcutaneous Cranial Electrical Stimulation (TCES). This method is a special form of electrical stimulation that employs a stimulator that gives a complex current (specific waveforms and high frequency). It was developed by a French group headed by Limoge (35–38,57–59). The TCES method has been used for analgesia during labor pain and before, during, and after surgery, and has recently been shown to be effective to potentiate the analgesic drugs for

major surgery, and cancer pain (60). With TCES two electrodes are placed in back of the ear lobe and behind the mastoid bone and one electrode at intersection of the line of the eyebrowns and the sagittal plane. The resulting analgesia is systemic rather than regional (see the section Electrical Anesthesia for a more complete description of the current).

Neurosurgical Methods

Percutaneous Electrical Nerve Stimulation (PENS). This method consists of an electric stimulation by means of a surgically implanted electrode (subcutaneous) coupled by RF induction to an external stimulator nerve. This surgical technique produces long-term positive results of ~ 70% (61,62). It is possible to carryout this procedure quite simply by temporarily implanting needle electrodes at the acupuncture points or auriculotherapy points. This technique produces results similar to those of classic TENS (63,64).

Spinal Cord Stimulation (SCS). This is a neurosurgical method utilized in cases of failure of simple pharmacological or physical treatment where the percutaneous test was positive. As with PENS an RF stimulator is implanted, which this time is connected to electrodes at a level with the posterior spinal cord. The electrodes are actually placed in the epidural space, to provide a percutaneous pathway, under local anesthesia and radiological control. It is often difficult to obtain good electrode position and electrodes can easily become displaced. This technique is reserved for desperate cases as the results are of long term. Approximately 30% are discouraging results (8).

Deep Brain Stimulation (DBS). This method is a complicated and awkward procedure bringing to mind stereotaxis (8). It consists of implanting electrodes at the level of the Ventral Postero-Lateral (VPL) nucleus of the thalamus, which is in relation to afferent posterior cords at the level of PeriAcqueductal Grey Matter (PAGM) or at the level of the PeriVentricular Grey Matter (PVGM), where endorphin and serotonin neurons are found at the motor cortex, which is the start of the pyramidal fascia (10–13). Results obtained are encouraging in cases of consecutive pains at the deafferentation (72%), but of no value in case of pains of nociception. Deep brain stimulation is employed in patients when pain is severe, when other approaches have failed, and when there is a desire to avoid a "drugged existence" and life expectancy is at best a few months. It is often an approach to patients with metastatic cancer. This method is less successful when pain originates from the central nervous system (secondary to stroke, trauma, quadriplegia). The DBS-stimulating probes are usually targeted for the periaqueductal gray matter when pain is deep seated, or for the sensory thalamus or medial lemniscus when is superficial.

Electrical Anesthesia

As mentioned previously, it has never been nor is not now possible to produce, "complete anesthesia" with electricity alone in humans without producing serious side effects. On the other hand, work by numerous investigators has demonstrated that one or more methods of electropharmaceutical anesthesia (anesthesia consisting of a combination of an electric current with anesthetic agents) is not only possible, but also desirable because of the lack of side effects and reduced requirements for neurodepressants. During past years, progress in chemical anesthesia has been so successful that the objective was not to replace classical anesthesia, but to more precisely confirm studies performed on animals, potentiation of anesthetic drugs by Transcutaneous Cranial Electrical Stimulation (TCES) to obtain postoperative electromedicinal analgesia, to the end that toxicity induced by chemical drugs could be dramatically reduced. The use of TCES is not without considerable supporting data, as from 1972 to 2000, many clinical trials involving TCES had been carried out on patients under electromedicinal anesthesia and provide > 20 specific references (7). During those clinical trials, anesthetists noticed that complaints of patients operated under TCES were less numerous in the recovery room than complaints of patients operated with chemical anesthesia. It seems that a state of indifference and reduction of painful sensation persisted in TCES-treated patients. These observations were scientifically confirmed in a study (59) in which 100 patients operated under electroanesthesia (EA) was compared to another 100 patients submitted to narco-neurolept-analgesia, a classical anesthesia (CA): the head nurses were ordered to administered 15 mg (i.m.) of pentazocine in case of patient complaints. It is worth noting that the first 16 postoperative hours, the average intake of pentazocine for the patients of the EA group was 8.1 mg/ patient, whereas it was 29.7 mg/patient (3.67 time higher) for the patients of the CA group. This difference between groups is highly statistically significant ($p < 0.001$) (Fig. 1).

This residual and prolonged analgesia is surely one of the most important advantages of TCES, but few clinicians benefit from its advantages at the present time. The most likely reason that few clinicians benefit from TCES is that it is not yet approved for use in the United States, Canada, and many countries in Europe by the respective regulatory agencies. Recent research carried on in numerous laboratories has increased our knowledge of the neurobiological

effects of these currents, and allowed the establishment of serious protocols dedicated to new clinical applications (7).

Nature of the Limoge's Current. Limoge et al. demonstrated that complex currents of their design are capable of producing profound analgesia without provoking initial shock, pain, or unpleasant sensations, burns, other cutaneous damage, muscular contractures, cerebral damage or convulsions, and respiratory or circulatory depression (58). The Limoge current consists of high frequency (HF) biphasic asymmetrical wave trains composed of modulated high frequency (166 kHz) pulse trains, regularly interrupted with a repetition cycle of 100 Hz (7,57). These wave trains are composed of successive impulsional waves of a particular shape: one positive impulse of high intensity and short duration (2 μs), followed by a negative impulse of weak intensity and long duration (4 μs) adjusted in such a way that the positive surface is equal to the negative surface. The average intensity of this current equals 0 mA. The use of such a negative phase makes it possible to eliminate all risk of burns. The "on-time" of the low frequency (LF) wave trains is 4 ms, followed by a 6 ms "OFF-time" (Fig. 2).

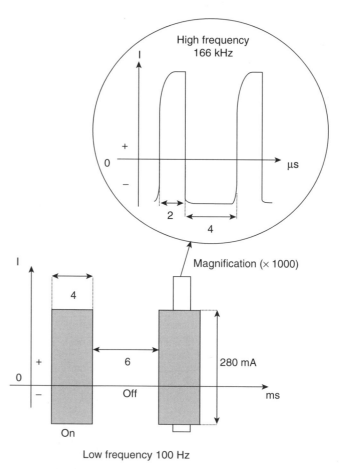

Figure 2. The Limoge waveform pattern: a modulated HF (166 kHz) pulse trains (top) regularly interrupted with a repetition cycle of 100 Hz. Concerning the high frequency, note the exponential ascent and acute fall of the waveform, and also note the area of the positive deflection is equal to that of negative deflection.

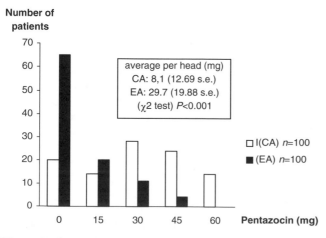

Figure 1. Comparison of two groups receiving pentazocin during the first 16 h after surgery.

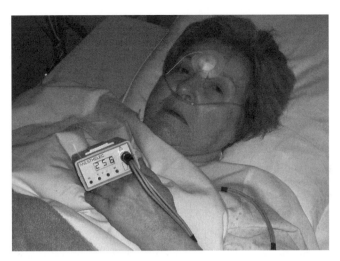

Figure 3. Application of the device on a patient during post operative period. See the placement of the frontal electrode.

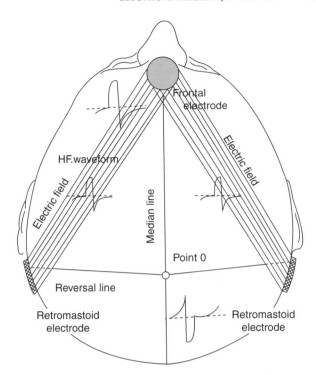

Figure 4. Location of the electrodes and shape of wave on the scalp. The center of the frontal electrode is situated at the intersection of the line of the eyebrowns and the sagittal plane. The center of the two retromastoid electrodes is localized in the retromastoid fossa. On the scalp the amplitude of HF waves diminish in measurement as the point O (occipital line) is approached, and behind that point there is an inversion of the wave form. The lines joining the frontal electrode to retromastoid electrodes represent the projected distribution of Limoge currents through the brain with action at the level of the periacqueductal gray matter and the limbic system.

This type of current was gradually developed over ∼20 years through numerous human clinical studies. The shape and cyclic ration of the HF waves are felt to be of utmost importance in the production of analgesia. Various shapes of waves have been tested (triangular, rectangular, exponential). Clinical impressions suggest that the most profound analgesia occurs with HF waveforms having an exponential ascent and acute fall. The most effective cyclic ratios are 2:5 with LF waves and 1:3 with HF waves with peak-to-peak intensity between 250 and 300 mA and peak-to-peak voltage between 30 and 40 V.

Electrodes. Three electrodes are used. One frontal electrode is placed between the eyebrows and two posterior electrodes are placed behind the mastoid process on each side of the occiput (Fig. 3). It is hoped that the intracerebral electric field thus obtained spreads on each side of the median line and it thus successful in stimulating opioid receptors surrounding the third and fourth ventricles and the paraventricular areas of the brain. In addition, some of the electric current spreads over the scalp, thus provoking peripheral electrostimulation (Fig. 4). The use of HF biphasic current permits employment of self-sticking electrodes made of silver (active diameter 30 mm), without risk of burns and without unpleasant sensations under electrodes.

Transcutaneous Cranial Electrical Stimulators Using Limoge Currents. Until now three types of devices only give Limoge currents: two American devices called Foster Biotechnology Neurostimulation Device (FBND) and Electro-Analgesia Stimulation Equipment (EASE) and one French device called Anesthelec (Fig. 5). The electrical stimulator must abide by general safety rules. The use of electrosurgical units during TCES requires excellent electrical isolation of the generator to avoid any risk of return of the current or skin burns under electrodes, a fault in the electrosurgical unit. These portable devices of type LF with isolated output are composed of one HF oscillator, one oscillator with LF relaxation with internal power supply generating HF (166 kHz), and LF (100 Hz) currents for

therapeutic use and one delay circuit, to stop output when its level is too high. The battery pack must be protected against short circuit as well as polarity inversion and detachable as it is rechargeable. The patient cable must

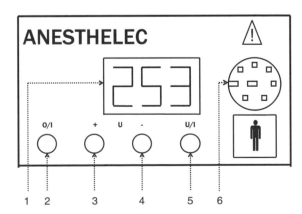

Figure 5. Front view of the Anesthelec generator box 1 → 3 digits displayer: Current intensity (from 000 to 300 mA); 2 → Pushing button On/Off; 3 → Pushing button for increment of the output current intensity; 4 → Pushing button for decrement of the output current intensity; 5 → Pushing button to select display (intensity of voltage); 6 → Connector for the three electrodes.

be an interlocking type preventing accidental disconnection and the functioning of the device must be simplified with detectors to measure the intensity of the current and the voltage applied to the patient to confirm proper contact between skin and electrodes. Concerning electromagnetic compatibility, the device must be autonomous with no possible direct or indirect link mains supplies. Mini box and manipulation components must be made in isolated material and the patient cables must be shrouded and the applied elements must be protected against overvoltage.

Clinical Usage of TCES

The TCES method being used with increasing frequency in France, and many other european countries, in Russia, in Mexico, and in Venezuela. It is not yet approved for use in the United States, but is being evaluated both in patients and in volunteers. Numerous studies have demonstrated that TCES is particularly effective in urologic, thoracic, and gastrointestinal surgery, but is not limited to these types of operative procedures. Patients receiving TCES require less nitrous oxide (N_2O) (30–40% less) to prevent movement in response to a pain stimulus. This method potentiates both the amnestic and analgesic effects of N_2O and prolongs residual postanesthetic analgesia at sites of trauma. The mechanism of analgesia resulting from TCES during administration of N_2O is unknown. Volunteers getting TCES without N_2O for 1 h are not sleepy or amnesic, but do report a warm and tingling sensation all over their body, and are objectively analgesic to many forms of painful stimulation (14–16). Similar results have been obtained with some TENS units in patients with chronic pain and after operation in patients with acute postoperative pain (54,55). As mentioned previously, some have suggested that receptor sites situated in the central gray area of the brain, the spinal cord, and other areas in the central nervous system regulate the effects of painful stimulation, analgesia, and the perception of somatic pain. Electrical stimulation of these receptor sites has been shown to result in relief from pain and can be antagonized by narcotic antagonists. Furthermore, the analgesic actions of TENS can be reversed with antagonists like naloxone (9,10). This suggests that TENS and TCES may be producing analgesia by stimulating increased production and/or release of the body's endogenous analgesics, the endorphins, enkephalins, serotonin and/or other neurotransmitters and neuromodulators (5,11–13,43–46,63–65).

To separate facts from empiricism and anecdotal information for several years, teams of researchers and clinicians attempted to show in animals and in humans what are the neurobiological mechanisms brought into pay by the TCES with currents of Limoge. For that reason, a study was conducted in France on rats on TCES potentiation of halothane-induced anesthesia and the role of endogenous opioid peptides was addressed (19). Carried out in double blind for 10 h prior to tracheotomy and the inhalation of halothane, the TCES provoked in the stimulated rats (TCES group, $n = 10$), a significant decrease ($p < 0.001$) in the Minimum Alveolar Concentration of Halothane (MACH) in comparison with the nonstimulated rats (con-

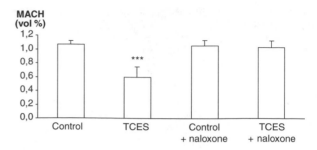

Figure 6. Effects of TCES on halothane requirements in rats. *** indicate significant difference between TCES and CONTROL groups (ANOVA, $p < 0.001$).

trol group, $n = 10$). This effect was completely inhibited by a subcutaneous injection of 2 mg/kg of naloxone (antagonist of morphine), which restored the MACH to its initial value in the TCES group without affecting the control group (Fig. 6). Moreover, TCES potentiation of halothane-induced anesthesia was dramatically increased by inhibition of enkephalin degradation. Thus the decrease of the MACH is associated with the potentiation the analgesic action of enkephalins released in the cellular space by TCES. These results demonstrate the direct involvement of endogenous opioid peptides on therapeutic effects of TCES.

In addition, a double-blind study carried out during labor and delivery on parturients to provide evidence of a mode of action of TCES on maternal plasma secretion of β-endorphins (66). To evaluate the rate of β-endorphins, blood samples were drawn from two groups of voluntary women in parturition (a TCES group, $n = 23$, and a control group, $n = 17$) at four precise stages: at the moment the electric generator was attached, after 1 h of the current application, at the time of complete dilatation, and finally after the delivery. The dosages were achieved by the radio-immuno enzymatic method. The plasmatic rate of β-endorphins was identical in the beginning for the two groups as those described in the literature, but this rate was progressively augmented in a significant fashion during the course of the labor from the first hour ($p < 0.05$) for the TCES group (Table 1).

It is more interesting to know the rate of endorphins produced in the cerebral structures known for their abundance of opiate receptors, more so than in the plasma. The exploration of the effects of TCES on brain opioid peptides was conducted at the Vishnevski Institute in Moscow by dosing endorphins in the cerebral spinal fluid (CSF) before cardiac surgery and after 30 min of TCES. The dosage showed that TCES augmented significantly ($p < 0.01$) the rate of β-endorphins in the CSF when compared to the control group and the effects of TCES reversed by naloxone (49,67,68).

These studies can partially explain the mode of action of TCES with currents of Limoge in the brain and permit not only rectification of protocols for clinical trials already carried out, but also provide better indication for utilization of TCES.

For all clinical applications, it must be keep in mind that the currents of Limoge provoke endogenous neurosecretions (7), which are not immediate, they require a certain

Table 1. Evaluation of β-Endorphin During Labor and Delivery on Parturients[a]

	Medication	Labor Time	β-Endorphin Plasmatic Rate, $pg \cdot mL^{-1}$			
			Installation	After 1 h	Dilatation	Delivery
Control ($n = 17$)	Peridural: 1 patient Morphine: 11 patients None: 5 patients	<1 h 30 min: 4 patients >1 h 30 min: 13 patients	123 (±12)	127 (±10)	124	160
TCES ($n = 23$)	Peridural: 2 patient Morphine: 3 patients None: 18 patients	<1 h 30 min: 11 patients >1 h 30 min: 12 patients	133 (±11) N.S.[b]	167 (±12)[c]	186	182

[a]Results of β-endorphin rates are expressed as mean ± s.e.m. (when available).
[b]N.S. indicates no difference between β-endorphin rates of control and TCES groups when measured at the installation of labor.
[c]Indicates significant difference between β-endorphin rates of control and TCES groups (t-test, $p < 0.05$) when measured 1 h after the installation of labor.

amount of time for their induction, and then are maintained all along the stimulation application. In consequence, the utilization of this technique in classical anesthesia is not the best indication except for major interventions of long duration. One must also remember that during > 10,000 major surgical interventions carried out under classical anesthesia combined with TCES it has been proven that the Limoge currents has a potentiation effect on opioid and non-opioid analgesics, morphinomimetics, psychotropes, and psycholeptics, (14–20) and this potentiation allows a decrease in drug doses, and therefore a decrease in toxicity. But one must admit objectively that, during past years, progress in chemical anesthesia has been so successful that the TCES will not replace classical anesthesia. The potentiation of drugs nevertheless by TCES can open new perspectives in the treatment of pain whether postoperative or chronic. To be precise the potentiation of opioid analgesia by TCES under specific conditions, was demonstrated by Stinus et al. (17). The authors showed that potentiation was a function of (a) the intensity of the stimulation, (b) the opioid dose administered, (c) the duration of TCES applied preceding opioid administration, and (d) the position and the polarity of the electrodes. This experimental approach was of prime importance as it allowed determination of the most efficient parameters, studied the therapeutic effects of TCES in humans, and increased our knowledge of the effects of TCES on neurobiological substrates.

Taking account of animal experimentation and clinical trials, one must know that to be successful in clinical applications, a correct basal protocol for TCES use should be followed. The main parameters are, the correct placement of the electrodes, starting electrostimulation no < 2 h prior to the introduction of drugs and continuation of TCES delivery during the pharmacokinetic action of drugs.

Abolition of Postoperative Pain (Fig. 3). Patients operated under TCES associated with pharmaceutical anesthesia complain strikingly less often about pain than those operated with a classical anesthesia. The TCES method induces a postoperative analgesia for an average of 16 h. A double-blind study has been made during per and postoperative period on 39 patients (TCES group $n = 20$ and control group $n = 19$) undergoing an abdominal surgery (20). Upon arrival in the recovery room, patients were given a computerized, patient-controlled analgesia (PCA) device to deliver IV buprenorphine (50 µg boluses, 30 min lock-out) during the first four postoperative hours. The recorded variables included postoperative requirements, pain scores with pain visual analogue scale (VAS) (from 0 = no pain to 10 = worst), sedation, (from 0 = not arousable to 4 = awake) awake) and were collected hourly from the first to the sixth postoperative hour by a blinded investigator. There was a highly significant reduction of cumulative buprenorphine requirements in the TCES group compared with the control group (2.36 ± 0.19 vs. 3.43 ± 0.29 $µg \cdot kg^{-1} h^{-1}$; $p < 0.01$) (Table 2). At each postoperative hour, patients required less buprenorphine in the TCES group. These results indicate that TCES reduces narcotic requirements for postoperative analgesia. TCES may have potential to facilitate early postoperative analgesia in patients undergoing major surgery. Therefore this technique allows a maximal restriction of pharmaceutical contribution.

Obstetric Electroanalgesia (66,69). In order to test the analgesic efficacy of TCES with Limoge currents during

Table 2. Buprenorphine Consumption[a]

Postoperative hours (H)	TCES	Control
H1	1.35 ± 0.15	1.57 ± 0.13
H2	0.90 ± 0.16	1.21 ± 0.18
H3	0.60 ± 0.15	1.10 ± 0.16[b]
H4	0.60 ± 0.18	1.00 ± 0.15[b]
Total dose ($µg \cdot kg^{-1} \cdot h^{-1}$)	2.36 ± 0.19	3.43 ± 0.29[c]

[a]Data are expressed as mean ± SEM.
[b]$p < 0.05$.
[c]$p < 0.01$.

labor and delivery, a double-blind study was performed with "anesthelec" on 20 cases for whom analgesia was necessary (TCES group I, current "on", $n = 10$, and control group II, current "off ", $n = 10$). Labor and delivery were carried out by a medical team different from those using the anesthelec. The results showed that TCES, with or without nitrous oxide inhalation, decreases by 80% the number of epidural analgesia or general anesthesia that would otherwise have been unavoidable. To define the effects of TCES, maternal and fetal parameters of 50 deliveries carried out under TCES were compared with 50 deliveries carried out under epidural analgesia (70).

TCES was used only if analgesia was required. These clinical trials were a retrospective comparison between two similar nonpaired series. Despite the fact that analgesia obtained with TCES was less powerful than with epidural analgesia, this method showed many advantages: total safety for the child and the mother, easy utilization, shorter labor time, decreased number of instrumental extractions and potentially reduced costs. Good acceptance and satisfaction for the mother should stimulate a rapid evolution and acceptance of this new method.

The TCES method should be applied following the first contractions. Analgesia is established after 40 min of stimulation. A diminution of pain is achieved that is comparable to that obtained after an injection (IV) of morphine (but it is less profound than with epidural analgesia), a decrease in vigilance with euphoria is obtained without inducing sleep, but allowing compensatory rest between contractions. The pupils are enlarged. Stimulation is applied throughout the birthing procedure and residual analgesia persists for several hours following delivery. Results are best if the expectant mother participates in a preparatory course for the birthing experience or if she uses musicotherapy in conjunction with TCES. If analgesia is insufficient it is possible to have patients breath nitrous oxide and oxygen (50:50) or to administer an epidural analgesia for the remainder of procedure. Thus obstetrical analgesia utilizing the currents of Limoge allows a reduction of labor time in all primapares ($p < 0.001$) and is without risk to the mother or child. Mothers in labor appreciate this simple, nonmedicinal, nonpainful technique that allows them to actively participate in the delivery.

Electropharmaceutical Anesthesia in Long Duration Microsurgery. For major operations and those of long duration the results are most encouraging as TCES permits a reduction of anxiolytics and neuroleptics by 45% and reduction of morphinomimetics by 90% and demonstrates the possibilities of drug potentiation to prolong analgesia while at the same time providing a less depressive general anesthetic (7,58,59,68). Early results have improved thanks to animal research and revision of protocols more particularly (17–19). In 1972, it was not know to begin electrostimulation three hours prior to medicinal induction (4).

Potentiation of Morphine Analgesia for Patients with Chronic Pain and Associate Problems (71). For all neurophysiological applications, a basic protocol must be followed. This protocol is as follows: If the patient is being treated pharmacologically, for the first time, never stop the che-

mical medication but diminish the dosage each day until a threshold dose is obtained according to the particular pathology and the patient. Begin TCES at least 1 h before medication whether it be on awakening in the morning or 2 h prior to going to bed. (There is no contraindication in maintenance of stimulation all-night long.)

If the patient is not being treated chemically, the effect of the current is best if there is a "starter dosage" of medicine. It is therefore recommended that a weak medicinal dose be prescribed according to the pathology and begin the TCES 1 h before the patient takes the dose, and continue stimulation during the time of pharmacocinetic action of the medicine.

This protocol will permit treatment of cancer patients at home whenever possible under medical supervision: This is a less traumatizing course of action than having the patients come into hospital every day. In the beginning, one must maintain the standard pain medication therapy and the patient should be connected to the Limoge Current generator for 12 h (during the night, if possible); the potentiometer is turned clockwise to a reading of 35 V and 250–300 mA, peak to peak. After this first treatment phase, the patient can use the machine for 3 h whenever they feels the need. The analgesic effect of TCES may not appear until the third day of treatment. Then TCES is initiated upon awakening. After 1 h, standard pain medication is given and TCES therapy is continued for another hour. Three hours before bedtime, TCES is again administered for 2 h, then standard pain medication is given and TCES therapy continued for another hour. The patient should enjoy restful sleep. After 8 days, the standard pain medication therapeutic dose should be decreased gradually, but not totally terminated. After this status has been achieved, patients may use the machine whenever they feel the need, for 3 h preferably with the reduced dose of the standard pain medication. The minimal therapeutic dose of the pain medication, however, may have to be adjusted upward somewhat due to individual differences in some patients.

CONCLUSION

All numerous and previous clinical trials have demonstrated that TCES reduces narcotic (fentanyl) requirements in patients undergoing urologic operations with pure neuroleptanesthesia (droperidol, diazepam, fentanyl, and air-oxygen) (20,36–38). Use of TCES in a randomized double-blind trial of these patients resulted in a 40% decrease in fentanyl requirements for the entire operation. Unfortunenately, while available TCES units (using currents of 250–300 mA peak to peak, with an average intensity of zero) provide analgesia and amnesia, they do not produce complete anesthesia. Whether any form of TCES or the use of very high frequency (VHF) will provide more analgesia and amnesia, that is, amounts sufficient to result in complete anesthesia without need for pharmaceutical supplementation, without problems has yet to be carefully evaluated but obviously needs to be studied. Considerable research must continue in this area.

Theoretically, lower doses of narcotics or lower concentrations of inhalation anesthetics should result in fewer alterations in major organ system function during anesthesia. This could mean that anesthesia with TCES produces less physiological insult than more standard anesthetic techniques and results in a shorter postoperative recovery period. It has been observed that TCES plus N_2O results in analgesia that persists after stimulation is terminated and N_2O is exhaled (7,14,15,20,58–60). This suggests that intraoperative use of TCES might reduce postanesthetic analgesic requirements, and that future clinical trials must be initiated to confirm this suggestion.

The 30.000 plus major interventions realized under TCES in France and in Russia since 1972 and the > 5000 drug withdrawals undertaken in opioid addicted patients at the Medical Center of the University of Bordeaux since 1979 without even the most minor incident permits us to conclude that the currents of LIMOGE are absolutely innocuous and cause no side effects. This simple technique reduced the use of sedative medicaments such as psychotropes or psycholeptics that often lead to "legal" addiction. The TCES is atoxic, reproductible, causes no personality change and is without habituation. Briefly, this technique fits perfectly into the domaine of all aspects of classical and alternative medicine as well as human ecology.

BIBLIOGRAPHY

Cited References

1. Smith RH. Electrical Anesthesia. Springfield (IL): CC Thomas Publ.; 1963.
2. Smith RH, Tatsuno J, Zouhar RL. Electroanesthesia: a review-1966. Anesth Analg 1967;40:109–125.
3. Smith RH. Electroanesthesia Review article. Anesthesiology 1971;34:61–72.
4. Limoge A. An Introduction to Electroanesthesia. Baltimore: University Park Press; 1975. p 1–121.
5. Sances Jr A, Larson SJ. Electroanesthesia. New York: Academic Press; 1975. p 1–367.
6. Shealy CN, Maurer D. Transcutaneous nerve stimulation for control pain. Surg Neurol 1974;2:45–47.
7. Limoge A, Robert C, Stanley TH. Transcutaneous cranial electrical stimulation (TCES): a review 1998. Neurosci Biobehav Rev 1999;23:529–538.
8. White PF, Li S, Chiu JW. Electroanalgesia: Its role in acute and chronic pain management. Anesth Analg 2001;92:505–513.
9. Adams JE. Naloxone reversal of analgesia produced by brain stimulation in the human. Pain 1976;2:161–166.
10. Hosofrichi Y, Adams JE, Linchitz R. Pain relief by electrical stimulation of the central gray matter and its reversal by naloxone. Science 1977;197:183–186.
11. Snyder SH, Goodman RR. Multiple neurotransmitter receptors. Neurochemistry 1980;35:5–15.
12. Snyder SH. Brain peptides as neurotransmitters. Science 1980;209:976–983.
13. Pasternack GW. Opiate enkephalin and endorphin analgesia: Relations to a single subpopulation of opiate receptors. Neurology 1981;31:1311–1315.
14. Stanley TH, et al. Transcutaneous cranial electrical stimulation increases the potency of nitrous oxide in humans. Anesthesiology 1982;57:293–297.
15. Stanley TH, et al. Transcutaneous cranial electrical stimulation decreases narcotic requirements during neuroleptanesthesia and operation in man. Anesthol Analg 1982;61:863–866.
16. Bourke DL, et al. TENS reduces halothane requirements during hand surgery. Anesthesiology 1982;61:769–772.
17. Stinus L, et al. Transcranial electrical stimulation with high frequency intermittent current (Limoge's) potentiates opiate-induced analgesia: blind studies. Pain 1990;42:351–363.
18. Auriacombe M, et al. Transcutaneous electrical stimulation with Limoge current potentiates morphine analgesia and attenuates opiate abstinence syndrome. Biol Psychiat 1990;28:650–656.
19. Mantz J, et al. Transcutaneous cranial electrical stimulation with Limoge's currents decreases halothane requirements in rats: evidence for involvement of endogenous opioids. Anesthesiology 1992;76:253–260.
20. Mignon A, et al. Transcutaneous cranial electrical stimulation (Limoge's currents) decreases bupremorphine analgesic requirements after abdominal surgery. Anesth Analg 1996;83:771–775.
21. Scribonius L. Compositiones medicae. Padua: Frambottus; 1655. Chapts. 11 and 162.
22. Kane K, Taub A. A history of local electrical anesthesia. Pain 1975;1:125–138.
23. Leduc S. Production du sommeil et de l'anesthésie générale et locale par les courants électriques. C R Acad Sci Paris 1902;135:199–200.
24. Leduc S. L'électrisation cérébrale. Arch Electr Med 1903; 11:403–410.
25. Denier A. Electro-anesthésie. Anesth Analg Réan 1951;8(1): 47–48.
26. Denier A. Anesthésie électrique. EMC 36550 A 10 1958;4:1–8.
27. Knutson RC. Experiments in electronarcosis. A preliminary study, Anesthesiology 1954;15:551–558.
28. Knutson RC, et al. The use of electric current as an anesthetic agent. Anesthesiology 1956;17:815–825.
29. Knutson RC, Tichy FY, Reitman J. The use of electrical current as an anesthetic agent. Anesthesiology 1966;17: 815–825.
30. Cara M, Cara-Beurton M, Debras C, Limoge A, Sances Jr A, Reigel DH. Essai d'anesthésie électrique chez l'Homme. Ann Anesth Franç 1972;13:521–528.
31. Shimoji K, et al. Clinical electroanesthesia with several methods of current application. Anesth Analg 1971;50:409–416.
32. Shimoji K, Higashi H, Kano T. Clinical application of electroanesthesia. In: Limoge A, Cara M, Debras Ch, editors. Electrotherapeutic Sleep and Electroanesthesia. Volume IV, Paris: Masson; 1978. p 96–102.
33. Djourno A, Kayser AD. Anesthésie et sommeil électriques. Paris: P.U.F.; 1968.
34. Sachkov VI, Liventsev NM, Kuzin MI, Zhukovsky VD. Experiences with interference currents in clinical surgery. In: Wageneder FM, Schuy S, editors. Electrotherapeutic Sleep and Electroanesthesia. Excerpta Medica Foundation; 1967. p 321–326.
35. Limoge A. The use of rectified high frequency current in electrical anaesthesia. In: Wageneder FM, Schuy St, editors. Electrotherapeutic sleep and electroanaesthesia. Volume I, Amsterdam: Excerpta Medica Foundation; 1967. p 231–236.
36. Cara M, Debras Ch, Dufour B, Limoge A. Essais d'anesthésie électromédicamenteuse en chirurgie urologique majeure. Bull Acad Méd 1972;156:352–359.
37. Debras C, Coeytaux R, Limoge A, Cara M. Electromedicamentous anesthesia in Man. Preliminary results Rev I E S A 1974; 18–19, 57–68.
38. Limoge A, Cara M, Debras C. Electrotherapeutic Sleep and Electroanesthesia. Paris: Masson; 1978.

39. Melzack R, Wall PD. Pain mechanism: a new theory. Science 1965;150:971–79.
40. Wall PD. The gate control theory of pain mechanisms. A re-examination and re-statement. Brain 1978;101:1–18.
41. Sjölund B, Ericson R. Electropuncture and endogenous morphines. Lancet 1975;2:1085.
42. Fox EJ, Melzach. Transcutaneous nerve stimulation and acupuncture. Comparison of treatment for low back pain. Pain 1976;2:141–149.
43. Akil H, et al. Encephaline-like material elevated in ventricular cerebrospinal fluid of pain patient after analgesic focal stimulation. Science 1978;201:463.
44. Henry JL. Substance P and pain: An updating. Trends Neurosc 1980;3:95–97.
45. Le Bars D. Serotonin and pain. In: Osbone NN, Hamon M, editors. Neuronal Serotonin. New York: John Wiley & Sons Ltd.; 1988. Chapt. 7. p 171–229.
46. Bailey PL, et al. Transcutaneous cranial electrical stimulation, experimental pain and plasma β-endorphin in man. Pain 1984;2:S66.
47. Willer JC, Boureau F, Albe-Fessard D. Role of large diameter cutaneous afferents in transmission of nociceptive messages: electrical study in man. Brain Res 1978;132:358–364.
48. Willer JC, Boureau F, Albe-Fessard D. Human nociceptive reactors; effects of spacial sommation of afferent input from relatively large diameter fibers. Brain Res 1980;201:465–70.
49. Pederson M, McDonald S, Long DM. An investigation determining the efficacy of TENS and the use of analgesia during labor in groups of women. Pain 1984;2:S69.
50. Melzack R, Stillwell D, Fox E. Trigger points and acupuncture points for pain: Correlations and implications. Pain 1977;3:23–28.
51. Hanai F. Effect of electrical stimulation of peripheral nerves on neuropathic pain. Spine 2000;25:1886–1892.
52. Campbell JN, Long DM. Peripheral nerve stimulation in the treatment of intractable pain. J Neurosurg 1976;45:692–699.
53. Woolf CJ. Transcutaneous electrical nerve stimulation and the reaction to experimental pain in human subjects. Pain 1979;7:115–127.
54. Kim WS. Clinical study of the management of postoperative pain with transcutaneous electrical nerve stimulation. Pain 1984;2:S73.
55. Park SP, et al. Transcutaneous electrical nerve stimulation. (TENS) for postoperative pain control. Pain 1984;2:S68.
56. Wilson OB, et al. The influence of electrical variables on analgesia produced by low current transcranial electrostimulation of rats. Anesth Analg 1989;68:673–681.
57. Limoge A, Boisgontier MT. Characteristics of electric currents used in human anesthesiology. NATO-ASI Series. In: Rybak B, editor. Advanced Technobiology. Germantown (MD): Sijthoff & Noordhoff; 1979. p 437–446.
58. Debras C, et al. Use of Limoge's current in human anesthesiology. NATO-ASI Series. In: Rybak B, editor. Advanced technobiology. Germantown (MD): Sijthoff & Noordhoff; 1979. p 447–465.
59. Limoge A, et al. Electrical anesthesia. In: Spiedijk J, Feldman SA, Mattie H, Stanley TH, editors. Developments in Drugs Used in Anesthesia. The Boerhave Series. Leiden: University Press; 1981. p 121–134.
60. Limoge A, Dixmerias-Iskandar F. A personal experience using Limoge's current during a major surgery. Anesth Analg 2004;99:309.
61. Shealy CN, Mortimer JT, Reswick JB. Electrical inhibition of pain by stimulation of the dorsal columns. Anesth Analg 1967;46:489–91.
62. Burton CV. Safety and clinical efficacy of implanted neuroaugmentive spinal devices for the relief of pain. Appl Neurophysiol 1977;40:175–183.
63. Prieur G, et al. Approche mathématique de l'action biologique des courants de Limoge. J Biophys Biomec 1985;9(2):67–74.
64. Malin DH, et al. Auricular microelectrostimulation: naloxone reversible attenuation of opiate abstinence syndrome. Biol Psychiat 1988;24:886–890.
65. Malin DH, et al. Augmented analgesic effects of enkephalinase inhibitors combined with transcranial electrostimulation. Life Sci 1989;44(19):1371–1376.
66. Stimmesse B, et al. β-endorphines plasmatiques et analgésie électrique durant l'accouchement. Cahiers d'Anesth 1986;34:641–642.
67. Schloznikov BM, Kuzin MI, Avroustsky M. Influence de l'E.S.C.T. sur le contenu de β-endorphines dans le liquide céphalo-rachidien et dans le plasma sanguin. Nouvell Biol Expér 1984;5:515–516.
68. Kuzin MI, Limoge A. Electrischer strom und schmerzausschaltung. Berlin: Wissenschaft und Menschheit; 1985. p 50–57.
69. Kucera H, et al. The effects of electroanalgesia in obstetrics. In: Limoge A, Cara M, Debras Ch, editors. Electrotherapeutic Sleep and Electroanesthesia. Volume IV, Paris: Masson; 1978. p 73–77.
70. Champagne C, et al. Electrostimulation cérébrale transcutanée par les courants de Limoge au cours de l'accouchement. Ann Fr Anesth Réanim 1984;3:405–413.
71. Lakdja F. (personal communication).

See also ELECTROPHYSIOLOGY; SPINAL CORD STIMULATION; TRANSCUTANEOUS ELECTRICAL NERVE STIMULATION (TENS).

ELECTROCARDIOGRAPHY, COMPUTERS IN

HOMER NAZERAN
The University of Texas
El Paso, Texas

INTRODUCTION

Digital computers have the ability to store tremendous amount of data and retrieve them with amazing speed for further processing and display. These attributes of computers make them extremely useful in a modern clinic or hospital environment. Computers play a central role in medical diagnosis and treatment as well as management of all processes and information in the hospital including patient data. Computers greatly facilitate the recording and retrieval of patient data in a simple way. All areas of hospital including patient admittance and discharge, the wards, all specialty areas, clinical and research laboratories are now interconnected through computer intranet and even nationally or globally connected through computer internet networks. This arrangement provides for better coordination of patient management throughout various hospital departments, and reduces patient waiting times. Obviously, computers also play a very important role in all aspects of administration and management of the hospital like other sophisticated institutions. Therefore, they greatly facilitate the overall planning and operation of the hospital resulting in improved healthcare services.

As the electrocardiographic (ECG) signal is one of the most, if not the most, measured and monitored vital signs, computers have had a tremendous impact in electrocardiography. One of the most well known areas of application of computers in medical diagnosis is their use in recording, monitoring, analysis and interpretation of ECG signals. Computers reduce interpretation time and ensure improved reliability and consistency of interpretation. Computers assist cardiologists by providing cost-effective and efficient means in ECG interpretation and relieve them from the tedious task of reviewing large numbers of ECG recordings. Computers also increase the diagnostic potential of ECG signals. Of course, cardiologists consider patient history, the details of the morphology of the ECG signals and other pertinent patient data backed by their clinical knowledge and experience to make a complete diagnosis. It should be clearly stated and emphasized that computers do in no way relieve the physicians of forming a complete clinical decision. However, computers provide them with information on the ECG examination in a clearer fashion and save them from the burden of routine, repetitive, and subjective calculations. An additional advantage of using computers in electrocardiography is the ease of storage and retrieval of ECG data for further study and analysis.

Computer processing of ECG signals, which are analogue in nature with amplitudes in low millivolt range and low frequency content (0.05–150 Hz), involves digitization with a typical sampling frequency of 500 Hz and 12 bit resolution in analog-to-digital conversion process. Further processing involves digital filtering, removal of powerline and biological artifacts (like EMG interference), averaging and automatic measurement of amplitudes and durations of different parts of the ECG signal. Computer analysis programs developed based upon the interpretative experience of thousands of experts performed on millions of ECG records, provide important ECG waveform components, such as amplitude, duration, slope, intervals, transform domain features, and interrelationships between individual waves of the ECG signal relative to one another. These parameters are then compared with those derived from normal ECGs to decide whether there is an abnormality in the recordings. There are a wealth of algorithms and methods developed over several decades to assist with computer-aided analysis and interpretation of ECG signals.

Computer-assisted ECG analysis is widely performed in ECG monitoring and interpretation. This method is effectively deployed in routine intensive care monitoring of cardiac patients, where the ECGs of several patients are continuously monitored to detect life-threatening abnormalities. The purpose of monitoring is not only to detect and treat ventricular fibrillation or cardiac arrests, but also to detect the occurrence of less threatening abnormalities like heart blocks and arrhythmias. The occurrence of such episodes and their timely detection helps clinicians to make early diagnostic decisions and take appropriate therapeutic measures. The accurate detection of trends resulting in dangerous cardiac abnormalities by visual inspection of ECG displays or chart recorders is a difficult task. Computer-assisted analysis of ECG signals to extract baselines, amplitudes, slopes, and other important parameters to establish minimum and maximum values for these parameters provides an efficient and accurate means to track the nature of the ECG rhythms.

The ECG acquisition and analysis systems like many other biomedical devices are designed to measure physiological signals of clinical relevance and significance. To design and effectively use appropriate instrumentation to measure and process such signals, an understanding of the origin and properties of these signals are of prime importance. This article starts off with an attempt to present a distilled overview of the origin of bioelectric signals in general, and the electorcardiographic signal in particular. This overview sets the basis to briefly describe the cardiac vector and its projection along specific directions (leads) providing the surface ECG recordings and review the basic instrumentation necessary to record these signals. A detailed description of a high end computer-based 12 lead clinical ECG acquisition and analysis system provides an example to appreciate the role of computers in diagnostic electrocardiography. An overview of the role of computers in high resolution electrocardiography (body surface potential mapping) and other ECG-based diagnostic devices then follows. A brief introduction to computer-based ECG monitoring systems and a block diagram description of a QRS detection algorithm illustrate how computer programming serves as a basis to detect cardiac arrhythmias. The article ends with a web-based ECG telemonitoring system as an example.

REVIEW OF BASIC CONCEPTS

From cellular physiology, we recall that biopotentials are produced as a consequence of chemical activity of excitable or irritable cells. Excitable cells are components of the neural, muscular, glandular as well as many plant tissues. More specifically, biopotentials are generated as a consequence of ionic concentration difference of electrolytes (mainly Na^+, K^+, Cl^- ions) across the cellular membrane of excitable cells. Ionic differences are maintained by membrane permeability properties and active transport mechanisms across the cellular membrane (ionic pumps.)

The cellular membrane is a semipermeable lipid bilayer that separates the extracellular and intracellular fluids having different ionic concentrations. As a consequence of semipermeability and differences in concentration of ions, electrochemical gradients are set up across the membrane. Ionic transfer across the membrane by diffusion and active transport mechanisms results in the generation of a voltage difference (membrane potential), which is negative inside. The resting membrane potential is mainly established by the efflux of K^+ ions due to diffusion and is balanced by the consequent inward electric field due to charge displacement. This equilibrium voltage can be estimated by the Nernst equation (1), which results from application of electric field theory and diffusion theory. If we consider the effects of the three main ions, potassium

(K^+), sodium (Na^+), and chloride (Cl^-), the Goldman–Hodgkin and Katz equation can be used to calculate the resting membrane potential (1).

The smallest sources of bioelectric signals (biosources) are single excitable cells. These cells exhibit a quiescent or resting membrane potential across the cellular membrane of several millivolts (mV) (−90 mV with several hundred milliseconds in duration for ventricular myocytes). When adequately stimulated, the transmembrane potential in excitable cells becomes positive inside with respect to outside (depolarization) and action potentials are generated (at the peak of the action potential in ventricular

myocytes, the membrane potential reaches about +20 mV). Action potentials are produced by sudden permeability changes of cellular membrane to ions: primarily sodium and potassium ions. Action potentials are all-or-none monophasic waves of depolarization that travel unattenuated with a constant amplitude and speed along the cellular membrane (Fig. 1).

The excitable cells function in large groups as a single unit and the net effect of all stimulated (active) cells produces a time-varying electric field in the tissue surrounding the biosource. The surrounding tissue is called a volume conductor. The electric field spreads in the volume

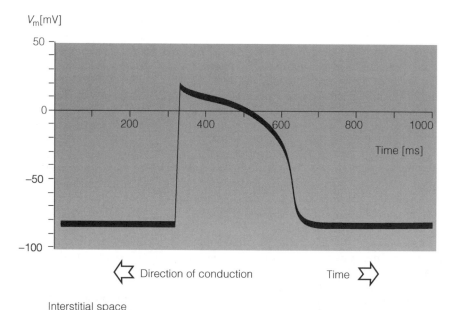

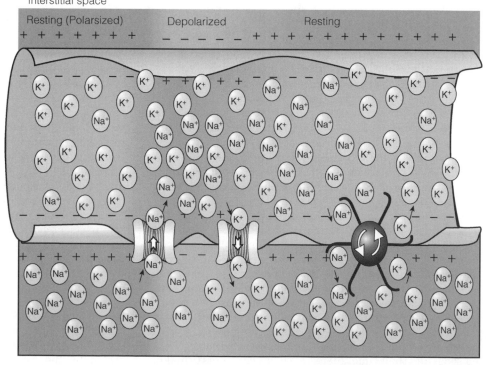

Figure 1. The monophasic action potential, direction of conduction of the action potential, and movement of ions across the cellular membrane. (Courtesy of Ref. 2.)

conductor and can be detected as small voltages by means of bioelectrodes or simply electrodes placed in the tissue or on the skin. Electrodes are sensors, which convert ionic current flow in the living tissue to electronic current flow in the electromedical instrument.

To understand the origin (electrogenesis) of biopotential signals like ECG, we should consider the following:

1. Electrical activity (bioelectric phenomena) at the cardiac cellular level and the extracellular potentials generated as the result of the electrical activity of single cardiac cells placed in a large homogeneous bathing (conducting) medium with the same composition as body fluids (volume conductor fields of simple bioelectric sources).

2. Extracellular potentials generated as the result of the electrical activity of a large number of myocardial cells (tissues) placed in a large conducting medium with the ionic composition of body fluids (volume conductor fields of complex bioelectric sources).

3. The relationship between these extracellular potentials and the gross electrical activity recorded on the body surface as ECG signals.

A simplified version of the volume conductor problem at the cellular level can be considered as follows. If a single excitable cell is placed in a bathing conductive medium, it acts like a constant current source. When the biosource becomes adequately depolarized, an action potential is generated across its membrane and it injects a current to the surrounding medium. The conductive medium presents as a load with a long range of loading conditions depending on its geometry, temperature, and so on. The lines of current flowing out of the excitable cell into the volume conductor with a specific resistance r, gives rise to an extracellular field potential proportional to the transmembrane current (i_m) and the medium resistance (r) according to Ohm's law. Obviously, the extracellular field potential increases with higher values of membrane current or tissue resistance.

There has been considerable debate about the exact relationship between the action potential across the cellular membrane and the shape of the extracellular field potential. However, the work of many researchers with different types of excitable cells has confirmed that the extracellular field potential resembles the second derivative of the transmembrane action potential. This means that a monophasic action potential creates a *triphasic* extracellular field potential (1). It has also been shown that the extracellular field potential is shorter in duration and much smaller in magnitude (μV compared to mV). Of course, this relationship has been established for cases when the geometry of the biosource and its surrounding environment is simple and the volume conductor is isotropic (uniform in all directions).

More realistically, when a piece of excitable tissue in the living organism becomes electrically active it becomes depolarized, acts like a point current source and injects a current into the anisotropic volume conductor comprised of tissues and body fluids surrounding it with different conductances (resistances). Consequently, the spatial distribution of current will not resemble that of a simple dipole placed in an isotropic volume conductor. However, it has been shown that an active nerve trunk (comprised of thousands of sensory and motor nerve fibers simultaneously stimulated) placed in a large homogeneous volume conductor generates an extracellular field potential which is quite similar in shape to that of a single nerve fiber (1). It is concluded that, the extracellular field potential is formed from the contributions of superimposed electric fields of the component biosources in the nerve trunk. The general form of the extracellular field potential of a nerve trunk in response to electrical stimulation is triphasic, it has amplitude in the microvolt range and it loses both amplitude and high frequency content at large radial distances from the nerve trunk. It is observed that the major contribution to the triphasic extracellular field potential is from the motor nerves in the trunk. It has also been shown that with a change in the volume conductor load (e.g., an increase in the specific resistance of the volume conductor or a decrease in the radial distance from the complex biosource) the amplitude of the recorded extracellular field potential increases (1).

The concepts discussed above are directly applicable to explain the relationship between the extracellular field potentials generated by complex and distributed biosources (current generators) like the cardiac tissue, the muscles, and the brain and their electrical activities recorded on the body surface as ECG, electromyogram (EMG) and electroencephalogram (EEG) signals.

In summary, excitable cells and tissues (biosources), when adequately stimulated, generate monophasic action potentials. These action potentials cause the injection of constant currents into a large bathing medium surrounding the biosource (considered as a point current source). As a result of the current flow in the volume conductor with specific resistance, extracellular field potentials are generated in the medium. These field potentials are triphasic in shape, of shorter duration and smaller amplitude compared to the transmembrane action potential. As the resistivity of the medium increases and the radial distance from the biosource decreases, the field potential increases. These field potentials are recorded as clinically useful signals on the body surface.

The biopotentials most frequently measured and monitored in modern clinics and hospitals are electrocardiogram (ECG: a recording of the electrical activity of the heart), electromyogram (EMG: a recording of the electrical activity of the muscle), electroencephalogram (EEG: a recording of the electrical activity of the brain), and others. Based on the physiological concepts reviewed above, now we present a brief overview of the electrogenesis of the ECG signals that carry a wealth of information about the state of health and disease of the heart. Having done this, we look at basics of electrocardiography.

BASICS OF ELECTROCARDIOGRAPHY

The conduction system of the heart consists of the sinoatrial (SA) node, the internodal tracts, the atrioventricular

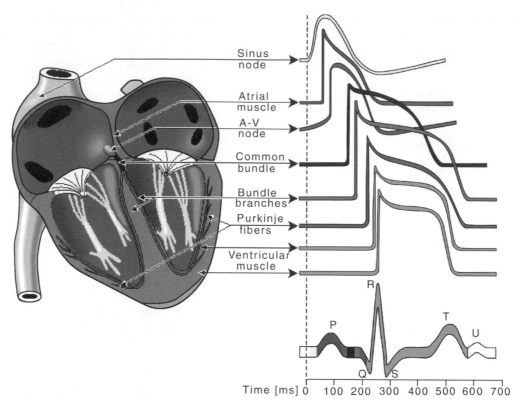

Figure 2. Wavesforms of action potentials in different specialized cells in the conductive pathway of a normal heart and their contribution with color coding to the surface ECG. (Courtesy of Ref. 2.)

(AV) node, the bundle of histidene (His), the right bundle branch (RBB), the left bundle branch (LBB), and the Purkinjie network. The rhythmic electrical activity of the heart (cardiac impulse) originates in the SA node. This node is known as the natural pacemaker of the heart, approximately the size of the tip of a pencil, located at the junction of the superior vena cava and the right atrium. The impulse then propagates through internodal and interatrial (Buchmans's bundle) tracts. As a consequence, the pacemaker activity reaches the AV node by cell-to-cell atrial conduction and activates the right and left atrium in an organised manner. The pacemaker action potential has a fast activation phase, a very short steady recovery phase, followed by a fairly rapid recovery phase and a characteristic slow depolarization phase leading to self-excitation (Fig. 2). The pacemaker cells of the SA node act as a biological oscillator.

As atria and ventricles are separated by fibrous tissue, direct conduction of cardiac impulse from the atria to the ventricles can not occur and activation must follow a path that starts in the atrium at the AV node. The cardiac impulse is delayed in the AV node for ~ 100 ms. It then proceeds through the bundle of His, the RBB, the LBB, and finally to the terminal Purkinjie fibers that arborize and invaginate the endocardial ventricular tissue. The delay in the AV node is beneficial since electrical activation of cardiac muscle initiates its successive mechanical contraction. This delay allows enough time for completion of atrial contraction and pumping of blood into the ventricles. Once the cardiac impulse reaches the bundle of His, conduction

is very rapid, resulting in the initiation of ventricular activation over a wide range. The subsequent cell-to-cell propagation of electrical activity is highly sequenced and coordinated resulting in a highly synchronous and efficient pumping action by the ventricles.

Essentially, an overall understanding of the genesis of the ECG waveform (cardiac field potentials recorded on the body surface) can be based on a cardiac current dipole model placed in an infinite (extensive) volume conductor. In this model, an active (depolarizing) region of the tissue is considered electronegative with respect to an inactive (repolarizing) region. Therefore, a boundary or separation exists between negative and positive charges. This is regarded as a current dipole: a current source and sink separated by a distance. According to the dipole concept, a traveling excitation region can be considered as a dipole moving with its positive pole facing the direction of propagation. Thus a nearby recording electrode placed in the surrounding volume conductor (referenced to an indifferent electrode placed in a region of zero potential) will detect a positive-going field potential as excitation approaches and a negative-going field potential as it passes away. Repolarization (recovery) is considered as a dipole with its negative pole facing the direction of propagation. Therefore, the propagation of excitation can be considered as the advance of an array of positive charges with negative charges trailing and the recovery could be considered as the approach of negative charges with positive ones trailing. Consequently, an upward deflection in the biopotential recording indicates the approaching of excitation

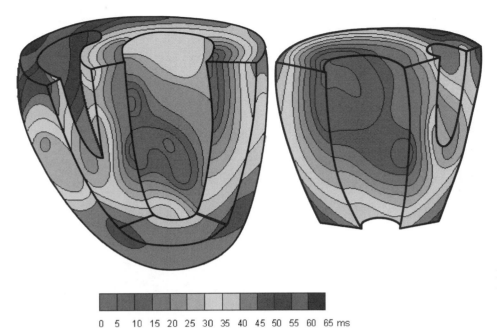

Figure 3. Isochrone surfaces in ventricular activation color coded to show spatiotemporal propagation. (Courtesy of Ref. 2.)

(depolarization) toward the positive (recording) electrode and a downward deflection indicates a recovery (depolarization) in the recorded signal.

As the wave of excitation (depolarization) spreads throughout the conductive pathways and tissues, specific excitation regions called isochrones are synchronously excited. In the ventricles, these synchronous activation regions propagate in a temporally and spatially orderly fashion from the endocardial to the epicardial direction (Fig. 3.)

In a localized region of the heart many cells are simultaneously activated because of the high electrical and mechanical coupling (functional syncytium) between the myocardial cells. Each activation region can be viewed as an elementary dipole, and all elementary dipoles could be vectorially added to all others to form a single net dipole. (For more details on simple and multiple dipole models see the section Electrocardiography.) Therefore, at each instant of time, the total cardiac activity can be represented by a net equivalent dipole current source. The electric field produced by this dipole source represents the total electrical activity of the heart and is recorded at the body surface as the ECG signal (Fig. 4). (For quantitative details see chapter 6 in Ref. 2.)

In the early 1900, Einthoven postulated that the cardiac excitation could be viewed as a vector. He drew an equilateral triangle with two vertices at two shoulders and one at the navel (representing the left leg). With the cardiac vector representing the spread of cardiac excitation inside the triangle, the potential difference measured between two vertices of the triangle (known as the limb leads) with respect to right leg, is proportional to the projection of the vector on each side of the triangle (Fig. 5).

In summary, based on the aforementioned concepts, electrocardiographers have developed an oversimplified model to explain the electrical activity of the heart. In this model, the heart is considered as an electric dipole (points of equal positive and negative charges separated from one another by a distance), denoted by a spatiotemporally changing dipole moment vector **M**. This dipole moment (amount of charge times distance between positive and negative charges) is called the cardiac vector. As the wave of depolarization spreads throughout the cardiac cycle, the magnitude and orientation of the cardiac vector changes and the resulting bioelectric potentials appear throughout the body and on its surface. The potential differences (ECG signals) are measured by placing electrodes on the body surface and connected to biopotential amplifier. (For details, see the section Bioelectrodes, and Electrocardiographic Monitors.) In making these potential measurements, the amplifier has a very high input impedance to minimally disturb the cardiac electric field that produces the ECG signal. As it was discussed before, when the depolarization wavefront points toward the recording positive electrode (connected to the + input terminal of the bioamplifier), the output ECG signal will be positive going, and when it points toward the negative electrode, the ECG signal will be negative going. The time varying cardiac vector produces the surface ECG signal with its characteristics P wave, QRS complex, and T wave during the cardiac cycle. These field potentials are measured by using bioelectrodes and biopotential amplifiers to record the ECG tracings.

BASIC INSTRUMENTATION TO RECORD ECG SIGNALS

As the electrical events in the normal heart precede its mechanical function, ECG signals are of great clinical value in diagnosis and monitoring of a wide variety of cardiac abnormalities including myocardial infarction and chamber enlargements. Therefore, ECG signal acquisition systems are widely used in cardiology, cardiac catheterization laboratories, intensive and cardiac care units, and at patient's bedside, among other areas.

As it was described earlier, the electrical activity of the heart can be best modeled and characterized by vector quantities. However, it is easier to measure scalar

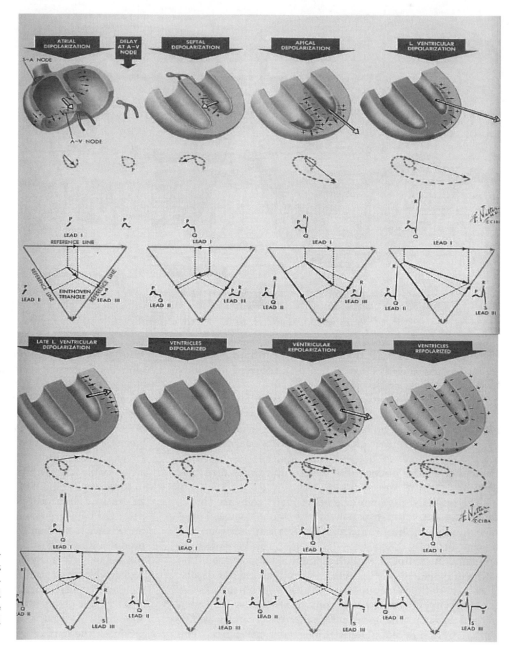

Figure 4. The total cardiac electrical activity represented by the net dipole (cardiac vector) during different phases of the cardiac cycle and its projection along the frontal plane electrocardiographic leads (I, II, and III). (Courtesy of Ref. 3.)

quantities, such as potential differences between specified points on the torso known as surface ECGs. These ECG signals have a diagnostically significant frequency content between 0.05 and 150 Hz. To ensure stability of the baseline, a good low frequency response is required. The instabilities in the baseline recordings originate from changes in the electrode–electrolyte interface at the point of contact of the bioelectrode with the skin. (For details see the section Bioelectrodes.) To faithfully record fast changes in the ECG signals and distinguish between other interfering signals of biological origin, adequate high frequency response is necessary. This upper frequency value is a compromise between several factors including limitation of mechanical recording parts of ECG machines using direct writing chart recorders.

To amplify the ECG signals and reject nonbiological (e.g., powerline noise) as well as biological interferences (e.g., EMG), differential amplifiers (DAs) with high gains (typically 1000 or 60 dB) and excellent common mode rejection capabilities must be used. Typically, common mode rejection ratios (CMMRs) in the range of 80–120 dB with 5 KΩ imbalance between differential amplifier input leads provide a desirable level of environmental and biological noise and artifact rejection in ECG acquisition systems. In addition to this, in very noisy environments it becomes necessary to engage a notch (band-reject) filter centered at 60 Hz or 50 Hz (in some countries) to reduce powerline noise further. A good review of adaptive filtering method applied to ECG powerline noise removal is given in Adaptive Filter Theory by Simon Haykin (4).

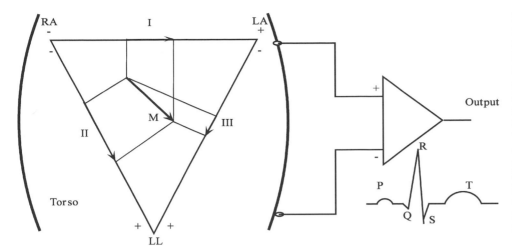

Figure 5. Einthoven equilateral triangle. The vertices are LA (left arm), RA (right arm), and LL (left leg). The *RL* (right leg) is used as a reference for potential difference measurements and is not shown. I, II and III represent electrocardiographic frontal limb leads. The + and − represent connection to the terminals of an ECG (biopotential) amplifier. Lead I is the potential difference between LA and RA. Lead II is the potential difference between LL and RA. Lead III is the potential difference between LL and LA. (The input polarity of the amplifier shown is for recording: III limb.)

Many modern biomedical instruments use instrumentation amplifiers (IAs). These amplifiers are advanced versions of differential amplifiers enhanced with many additional desirable characteristics such as very high input impedance >100 MΩ) to prevent loading the small ECG signals to be picked up from the skin. The final stages of the ECG amplifier module limit the system's response (band pass filtering) to the desirable range of frequencies for diagnostic (i.e., 0.05–150 Hz) or monitoring (i.e., 0.5–40 Hz) purposes. A more limited bandwidth in the monitoring mode provides improved signal to noise ratio and removes ECG baseline drift due to half-cell potentials generated at the electrode/electrolyte interface and motion artifacts. Driven right-leg amplifiers improve the CMRR. The amplified and adequately filtered ECG signals are then applied to display, recording or digitization modules of a computer-based ECG acquisition system. Detailed specifications for diagnostic ECGs have been developed by the American National Standards Institute (5). For more detailed specification and design as well as other aspects of ECG instrumentation and measurements, see the Appendix, Bioelectrodes, Electrocardiography, and Electrocardiographic Monitors.

COMPUTER SYSTEMS IN ELECTROCARDIOGRAPHY

Diagnostic Computer-Based ECG Systems

Heart diseases cause a significant mortality rate in the world. Accurate diagnosis of heart abnormalities at an early stage could be the best way to save patients from disability or death. The ECG signal has considerable diagnostic value and ECG monitoring is a very well established and commonly used clinical method. Diagnostic ECG testing is performed in a doctor's office, in a clinic or hospital as a routine check up. In this test, a full 12-lead ECG (to be described later) is acquired from a resting subject and displayed on a chart recorder or a computer screen to diagnose cardiac diseases. In cardiac care units (CCUs), a patient's single lead ECG may be continuously acquired and displayed on a cathode ray tube (CRT) or a computer screen for signs of cardiac beat abnormalities. The ECG

monitoring capabilities are now an integral part of a number of other medical devises, such as cardiotachometers, Holter monitors, cardiac pacemakers, and automatic defibrillators.

Most of the traditional clinical ECG machines used a single channel amplifier and recording system. The recoding was achieved using a direct writing chart recorder. In modern systems, however, computer memory, display screen and printing capabilities are deployed to record, display and report the ECG data. Traditional single channel systems used a multiposition switch to select the desired lead connection (I, II, III, aV$_R$, aV$_L$, aV$_F$, V$_1$, V$_2$, V$_3$, V$_4$, V$_5$, and V$_6$) and apply it to the biopotential amplifier and chart recorder. Only one ECG lead at a time could be selected and recorded with these machines. The block diagram of a modern single channel computer-based ECG acquisition system is shown in Fig. 6.

Most of the modern ECG machines are multichannel systems. They include several amplifier channels and record several ECG leads simultaneously. This feature enables them to considerably reduce the time required to complete a set of standard clinical ECG recordings. As the ECG leads are recorded simultaneously, they can be shown in their proper temporal relationship with respect to each other. These systems use microprocessors to acquire the cardiac signals from the standard 12-lead configuration by sequencing the lead selector to capture four groups of three lead signals, and switch groups every few seconds. The high end computer-based systems capture all 12-leads simultaneously and are capable of real-time acquisition and display of the standard 12-lead clinical ECG signals (see Example below.)

Modern ECG acquisition systems achieve certain desirable features, such as removal of artifacts, baseline wander, and centering of the ECG tracings by using specialized computer algorithms. These systems perform automatic self-testing on power up and check for lead continuity and polarity and indicate lead fall-off or reversal. They deploy digital filters implemented in software to considerably improve the ECG signal quality and automatically remove baseline drift and reduce excessive powerline and biological noise. Powerful software programs not only minimize

Figure 6. Block diagram of a three-lead electrocardiograph. The normal locations for surface electrodes are right arm (RA), right leg (RL = ground or reference), left arm (LA), and left leg (LL). In 12-lead electrocardiographs, six electrodes are attached on the chest of the patient as well. (Courtesy of Ref. 6.)

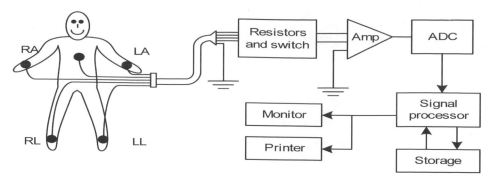

baseline drift without signal distortion during rest, they also produce high quality ECG tracings during patient monitoring, exercise and ambulation.

12-Lead Clinical Electrocardiography. The most commonly used or standard clinical ECG system is the 12-lead system. This system is comprised of three bipolar limb leads (I, II, III) connected to the arms and legs; three augmented leads (aV_R, aV_L, aV_F); and six unipolar chest or precordial leads (V_1, V_2, V_3, V_4, V_5, V_6).

The connections to measurement points in the 12-lead system are shown in Fig. 7. Six of these 12 leads are frontal leads (the bipolar and the augmented leads) and six of them are transverse leads (the precordial leads). The frontal leads are derived from three measurement points RA, LA, LL with reference to RL. Therefore, any two of these six leads contain exactly the same information as the other four. The dipole source model can be used to explain the total electrical behaviour of the heart (See the section Electrocardiography and Ref. 2.)

Basically, any two of the three I, II, and III leads could represent the cardiac activity in the frontal plane and only one chest lead could be used to represent the transverse activity. Chest lead V_2 is a good representative of electrical activity in the transverse plane as it is approximately orthogonal (perpendicular) to the frontal plane. Overall, as the cardiac electrical activity could be modeled as a dipole, the 12-lead system could be considered to have three independent leads and nine redundant leads. However, since the chest leads also detect unipolar elements of the cardiac activity directed toward the anterior region of the heart, they carry significant diagnostic value in the transverse plane. As such, the 12-lead ECG system can be considered to have eight independent and four redundant leads. Now the question arises why all the 12-leads are recorded then. The main reason is that the 12-lead system enhances pattern recognition and allows cardiologists to compare the projections of the resultant vectors in the two orthogonal planes and at different angles. This facilitates and validates the diagnostic process. Figure 8 shows the surface anatomical positions for the placement of the bioelectrodes in 12-lead electrocardiography.

A Detailed Example: The Philips PageWriter Touch 12-Lead ECG System. Currently, there are a number of highly advanced 12-Lead ECG data acquisition, analysis and interpretative systems on the market. A detailed description and comparison of all these systems are beyond the scope of this chapter (for information on these systems see manufacturers web sites). Due to space limitation, only one representative system will be described in detail as an example. The Philips PageWriter Touch (Philips Medical, MA) is an advanced 12-lead ECG acquisition and analysis system with many user-friendly features. It has a compact design and is best suited for busy hospitals and fast-paced clinical environments. It has an intuitive touch screen which is fully configurable and it has a powerful build-in interpretative 12-lead ECG signal analysis algorithm developed for rapid and accurate interpretation of ECG signals (Fig. 9).

The PageWriter supports a variety of ECG data acquisition modes, display settings and reporting schemes. It provides real-time color-coded ECG signals enabling the user to perform quality control checks on the acquired ECG data. Quality control features include surface anatomical diagrams that alert the user to the location of loose or inoperable electrodes (Fig. 8). It is equipped with a full-screen preview display of the ECG report before print out. It has an ergonomic design that facilitates ECG data collection from the patient bedside (Fig. 10).

The PageWriter Touch has an alphanumeric keyboard, an optional barcode scanner, and an optional magnetic card reader to increase the speed and accuracy of patient data entry and management. Data handling features include indexed ECG storage and quick transmission of stored ECG data to an ECG data management system. By using indexed thumbnail images of each ECG report the user can instantly view and print the stored reports. Multiple reporting formats may be applied to any saved ECG records.

In brief, the PageWriter Touch features accurate and fast 12-lead ECG signal acquisition and analysis capabilities for both adults and children patients. It has real-time color-coded display facilities and provides instantaneous snapshots of stored ECG data on the touch screen. It has a variety of ECG data review and printing capabilities. The touch screen provides easy and flexible configurability of features (Fig. 11). (For more details on Technical Specifications of PageWriter Touch see the Appendix.)

Overview of System Description. The PageWriter Touch electrocardiograph performs acquisition, analysis, presentation, printing, storage, and transfer of ECG signals as well as other patient data. A generalized block diagram of the system is shown in Fig. 12.

The system is comprised of three major subsystems equipped with an LCD display and touch screen module. A brief and high level description of these subsystems is as follows.

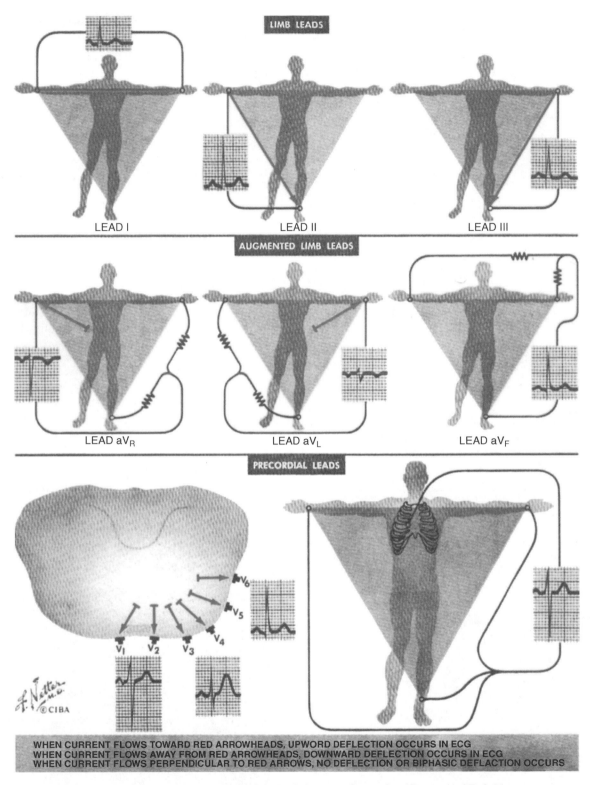

Figure 7. Lead connections in standard 12-lead electrocardiography. (Courtesy of Ref. 3.)

1. Main Controller Board. This is a single board computer (SBC) with extensive Input–Output facilities, running Windows CE 3.0. The PageWriter application software controlling the overall operation of the electrocardiograph runs on the Main Controller Board, which includes the display and user-input subsystems. The application software interacts with numerous hardware and software subsystems. The Main Controller Board SBC contains loader software and Windows CE kernel image in its

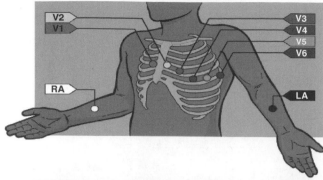

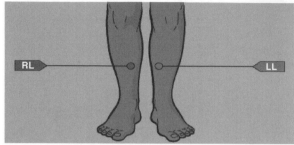

Figure 8. Surface anatomical positions for application of bioelectrodes in the 12-lead ECG system. (Courtesy of Philips Medical Systems, USA.) (Illustration courtesy of Mark Wing, Illustrator.)

internal flash memory (32 MB). At system boot, a system RAM test is performed by the loader (64 MB RAM onboard). The Windows CE kernel loads next. After loading of the CE, the application launcher runs, verifying system and executable images before loading the SierraGUI (Graphical User Interface) application. All interactions with the system operator (user) are implemented through the SierraGUI application. The application software and all ECG archives are stored on a separate 128 MB Compact-Flash (CF) card installed on the Main Controller Board (Fig. 12.)

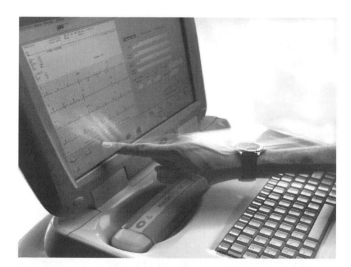

Figure 9. PageWriter Touch, a high-end 12-lead electrocardiograph with built-in 12-lead ECG signal interpretation capabilities. (Courtesy of Philips Medical Systems, USA.)

Figure 10. Easy data collection set up at bedside. (Courtesy of Philips Medical Systems, USA.)

2. Printer Controller Board. This is a controller board that provides all the real-time management of the printer. The Printer Controller Board communicates with the Main Controller Board through a USB port. The Printer Controller Board is a microprocessor-based control board for the electrocardiograph thermal printer mechanism. This board is connected by a USB port to the Main Controller Board and is powered by the power circuit of the Main Controller Board. It provides ECG waveform rendering and basic bitmap imaging operations, and uses a PCL-like control language API for page description and feed control. It controls the print head, motor, and detects drawer-open and top-of-form.

3. Patient Input Module (PIM). This is a controller running Windows CE 3.0, coupled with a signal acquisition board employing mixed-signal Application Specific Integrated Circuit (ASIC) technology developed for ECG data acquisition. The PIM communicates with the Main Controller Board through a USB port (Fig. 12.)

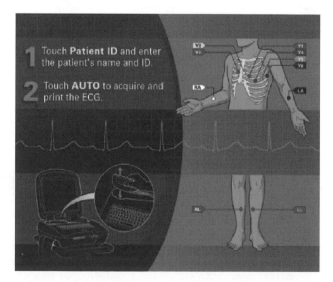

Figure 11. PageWriter Touch electrocardiograph has a user-friendly interface. (Courtesy of Philips Medical Systems, USA.) (Illustration courtesy of Mark Wing, Illustrator.)

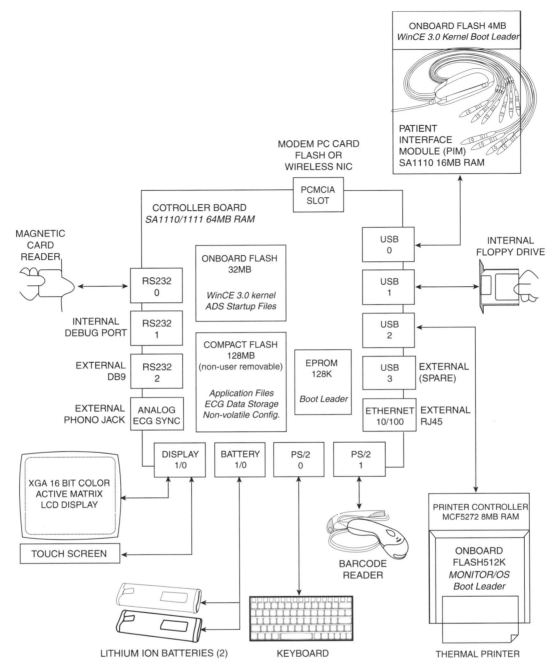

Figure 12. A generalized block diagram for PageWriter Touch. (Courtesy Philips Medical Systems, USA.)

The LCD display module for the PageWriter Touch electrocardiograph is an extended graphic array or XGA-compatible, full-color with backlight and overlaid touch screen (Fig. 9.) It is driven by the Main Controller Board using a graphics accelerator chip and dedicated touch screen support hardware. The touch screen provides finger-tap input substituting for the normal Win32 mouse-click input.

ECG Data Flow and Storage in PageWriter Touch. The ECG data flow and storage in the PageWriter Touch electrocardiograph is shown in Fig. 13.

The ECG signals are picked up at the body surface using electrodes attached to properly prepared skin at specified anatomical sites. The ECG data stream is then routed in real-time to the Main Controller Board, where it is written into the application buffers in RAM. These buffers are used to present the ECG data stream on the LCD screen in real-time. When the user initiates an AUTO report print, presses the ACTION button on the Patient Input Module or the Snapshot button on the display, or uses the Timed ECG acquisition, the corresponding 10 s segments of the ECG data are then copied to the temporary ECG storage in RAM. These 10 s segments are named ECG reports that

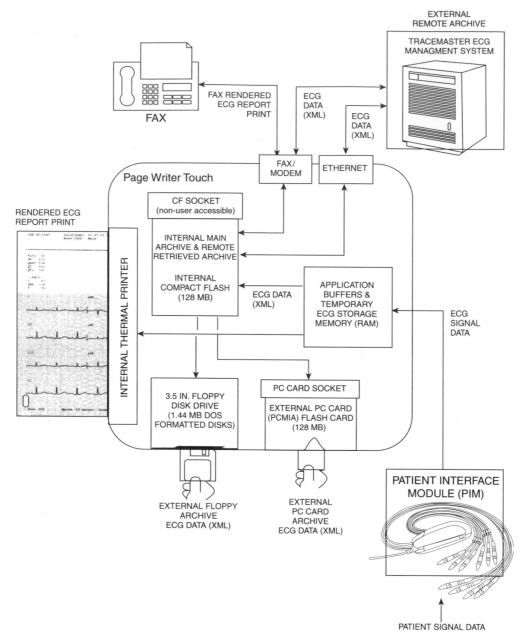

Figure 13. The ECG data flow and storage in PageWriter Touch electrocardiograph. (Courtesy of Philips Medical systems, USA.)

can be previewed and printed. In AUTO mode, the ECG report may be automatically printed after copying to storage. An ECG report contains waveforms, analysis information, patient demographics, and acquisition information, along with operator and device information. Figure 14 shows a rendered ECG report print sample.

The PageWriter Touch 12-Lead ECG Data Analysis Program. This program analyzes up to 12 simultaneously acquired ECG waveforms over a 10 s period using interpretive criteria by patient-specific information. For examples of ECG diagnostic criteria see Ref. (7). The analysis algorithm analyzes ECG data and produces an interpretive report to assist the clinician make more informed patient

assessment in a broad range of clinical settings. The analysis program is developed to allow clinicians read and interpret ECG findings more quickly and efficiently, provide accurate, validated ECG measurements to facilitate physician overreading and improve treatment decision making, generate detailed findings that highlight areas of interest for physicians review, provide high levels of reproducibility for more consistent ECG interpretation.

The algorithm monitors the quality of the ECG waveforms acquired from 10 electrodes (12 leads), recognizes patterns, and performs basic rhythm analysis. Using advanced digital signal processing techniques, the algorithm removes biological and non-biological noise and artifacts while minimizing distortion of the ECG waveforms

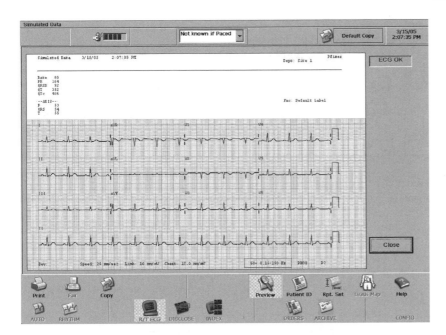

Figure 14. A 12-lead ECG report sample for printing or copying. (Courtesy of Philips Medical systems, USA.)

and preserving their diagnostic quality. By using a set of diagnostic criteria, the 12-lead ECG analysis program generates interpretative statements, summarizes the findings for the ECG signals, and highlights key areas of concern for physician review. The analysis program has a built-in pacemaker pulse detector and paced rhythm classification routine. The program distinguishes a variety of atrial, ventricular and atrioventricular (AV) sequential pacing modes to recognize asynchronous pacing typically seen with a magnet in place. In addition to automated detection capabilities, the algorithm provides user selected configuration of "pacemaker patient" or "non-pacemaker patient" for more accurate analysis.

The 12-lead analysis program also incorporates factors, such as age and gender that impact a patient's risk for developing specific forms of cardiac disease. It uses gender-specific interpretation criteria to take into account key physiological differences between males and females. For example, it applies gender-specific evaluation of Q waves for improved detection of Acute Myocardial Infarction (AMI), which is often missed in female patients. It also uses more gender-specific criteria for detection of axis deviation, Left Ventricular Hypertrophy (LVH), and prolonged QT segment.

The 12-lead algorithm also includes an advanced Pediatric Criteria Program, which uses age to select clinically relevant interpretative statements related to cardiac rhythm and morphology. If a patient's age is < 16 years, the algorithm automatically uses pediatric ECG interpretation criteria, which accounts for higher rates and narrower QRS complexes in this patient population. The Pediatric Criteria Program recognizes 12 distinct age groups to ensure most relevant-age interpretation criteria are applied for analyzing the ECG data. In fact, patient age is used to define normal limits in heart rate, axis deviation, ECG segment time intervals, voltage values for interpretation accuracy in tachycardia, bradycardia, prolongation or shortening of PR and QT intervals, hypertrophy, early

repolarization, myocardial ischemia and infarct, as well as other cardiac conditions. The pacemaker detection and classification algorithm built into the Pediatric Analysis Program has the ability to reliably distinguish pacemaker pulses from the very narrow QRS complexes often produced by neonatal and pediatric patients. It also reduces the likelihood of false diagnosis in non-paced patients, while enabling more accurate paced rhythm analysis for the pediatric age group.

To improve on the classification accuracy of a diagnostic dilemma in pediatrics to distinguish between mild Right Ventricular Hypertrophy (RVH) and Incomplete Right Bundle Branch Block (IRBBB), the algorithm combines 12-lead synthesized vectorcardiogram transverse plane measurements with scalar ECG measurements. In addition, the pediatric analysis program provides improved QT measurements in pediatric patients.

The ST segment elevation in ECG tracings is an effective indicator of acute myocardial infraction. The Page-Writer Touch 12-Lead ECG Analysis Program provides advanced software for detection of ST Elevation Acute Myocardial Infarction (STEMI). This software feature provides clinicians with a valuable decision support tool when working with patients presenting symptoms that suggest accurate coronary syndromes.

Other cardiac conditions, such as benign early repolarization and acute pericarditis, tend to mimic the ECG diagnosis of STEMI, and degrade algorithm detection accuracy. To address this difficulty, the ECG Analysis Program separates the confounders by examining the patterns of ST elevation. Improved measurements in ST deviation enables the algorithm to achieve both high sensitivity and specificity in more accurate detection of STEMI condition.

Exercise (Stress) Electrocardiography. Exercise (Stress or Treadmill) electrocardiography is a valuable diagnostic and screening procedure primarily used to diagnose coronary artery disease (CAD). Exercise on treadmill may induce ischemia that is not present at rest. Exercise electrocardiography is usually performed to screen for the presence of undiagnosed CAD, to evaluate an individual with chest pain, to clarify abnormalities found on a resting ECG test, and to assess the severity of known CAD.

Accurate interpretation of stress ECG is dependent on a number of factors including a detailed knowledge of any prior medical condition, presence of chest pain and other coronary risk factors. A modern exercise ECG system includes a 12-lead ECG diagnostic system with accurate and advanced ECG signal processing capabilities. It also includes appropriate interfaces for treadmills, ergometers as well as motion-tolerant noninvasive blood pressure (NIBP) monitors.

Vectorcardiography. In 12-lead clinical electrocardiography described above, the scalar ECG signals are recorded from 10 electrodes placed on specific sites on the body surface. These tracings show detailed projections of the cardiac vector as a function of time. However, they do not show the underlying cardiac vector. Vectorcardiography is a method in which the cardiac electrical activity along three orthogonal axes (x, y, z) in the principal planes (frontal, transverse, sagittal) are recorded and the activity of any two of the three is displayed on a CRT. This display is a closed loop showing the locus of the tip of the cardiac vector during the evolution of atrioventricular depolarization–repolarization phases in one cardiac cycle (Fig. 4.) These closed loops are known as vectorcardiogram

(VCG). Figure 15 shows the 3D representation of the cardiac electrical activity and its projections onto the principal planes recorded as vectorcardiogram. Figure 16 shows the basic principles of vectorcardiography and the VCG.

In contrast to ECG tracings that show the detailed morphology of the electrical activity of the heart in any one single lead direction, the VCG is the simultaneous plot of the same electrical events in two perpendicular (orthogonal) lead directions. This gives a detailed representation of the cardiac vector and produces loop-type patterns on the CRT. The magnitude and orientation of the P, QRS, and T loops are then determined from the VCG. The VCG plots provide a clear indication of the cardiac axis and its deviation from normal. Each VCG exhibits three loops, showing the vector orientation of the P waves, the QRS complex and the T wave during the cardiac cycle. The high amplitude of the QRS complex loop predominates the VCG recordings and sensitivity adjustments need to be made to adequately display the loops resulting from the P and T waves.

The VCG is superior to the clinical 12-lead scalar ECG in recognition of undetected atrial and ventricular hypertrophy and some cases of myocardial infarction and the capability to diagnose multiple infarctions in the presence of bundle branch blocks. For example, in most infarcts the cardiac vector orientation moves away from the area of infarction. The advantage of VCG is that it only requires three orthogonal leads to provide full information on cardiac electrical activity. This translates into simpler algorithms and less computation. However, the full 12-lead

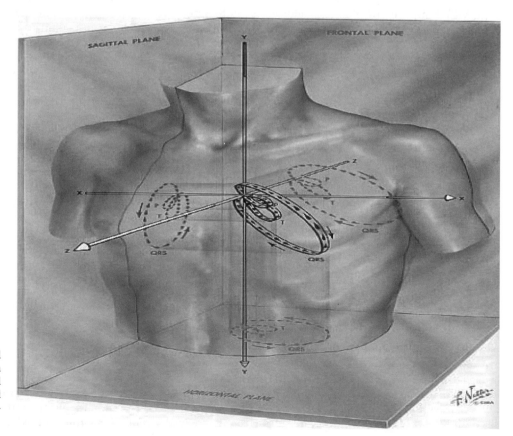

Figure 15. The three-dimensional (3D) representation of the evolution of the cardiac vector for P, QRS, and T loops during one cardiac cycle and their projections onto the three principal planes. (Courtesy of Ref. 3.)

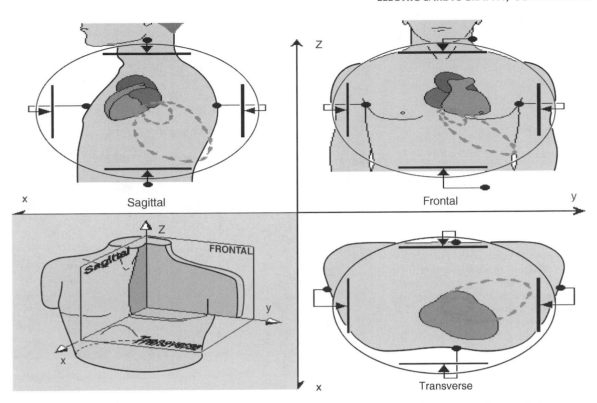

Figure 16. Vectorcardiography based on mutually orthogonal uniform lead fields set up by parallel electrode arrangements on opposite sides of the torso in three principal planes. (Courtesy of Ref. 2.)

scalar ECG provides detailed time plots of the ECG signals and highlights the changes in waveform morphology. Decades of clinical experience has made the 12-lead electrocardiography the standard method of practice in diagnosis of heart disease and favors that method over vecorcardiography. For a comprehensive and detailed description of Vectorcardiography see Refs. 2 and 8.

High Resolution Electrocardiography (Body Surface Potential Mapping). The electrical activity of the heart triggering its pumping action is a spatiotemporally distributed process. As discussed above, electrocardiography provides temporal information on timing of cardiac cycle. It can provide information on chamber enlargement and is an established tool for diagnosis of cardiac arrhythmias. Exercise testing using 12-lead clinical electrocardiography is the most common means of clinical investigation in evaluation of patients with chest pain. Even though clinical 12-lead ECG provides excellent temporal resolution in assisting clinical diagnosis of heart disease, it is limited in providing spatial information on the evolution of the electrical activity during the cardiac cycle.

The 12-lead ECG is now a standard diagnostic tool in clinical practice, but it is limited in detecting some cardiac abnormalities due to it sparse spatial sampling of the cardiac vector. Cardiac electrical imaging or enhanced electrocardiography alleviates this limitation by acquiring more information through deployment of a larger array (hundreds) of electrodes installed in a jacket to record more spatial samples of cardiac electrical activity.

As discussed in the sections above, the heart is a spatially and temporally distributed biosource embedded in an irregularly shaped volume conductor with conductivity inhomogenities. In standard electrocardiography one can measure heart rate and detect arrhythmias, determine the location of cardiac ischemia and infarction, localize some sites of conduction disorders between atria and ventricles or areas of transient ischemia or even detect chamber hypertrophy or congenital heart defect. However, standard ECG is specifically limited in the following: accurate detection of events in the non-anterior regions of the heart, underestimation of the deflections of the ST segment during ischemia if they are weak or nonexistent on the regions of the torso sampled by the ECG leads, inaccuracy of ECG amplitudes in reflecting the extend of physiological alterations due to the spatial integration effects of many simultaneous events, some of which partially cancel so that small defects can result in large changes in the ECG signal or vice versa.

Simply stated, standard ECG is limited in spatial resolution as it is used to describe the complex cardiac fields generated by the electrical activity of a spatiotemporally distributed biosource like the heart. Therefore, the need for enhanced resolution electrocardiography exists. Recent developments in cardiac imaging techniques have been largely driven by this need. A number of active research centers around the world are dedicated to the development of this technology (9–12). As with other imaging methods used in medicine, recent developments in cardiac electrical imaging have greatly benefited from better data

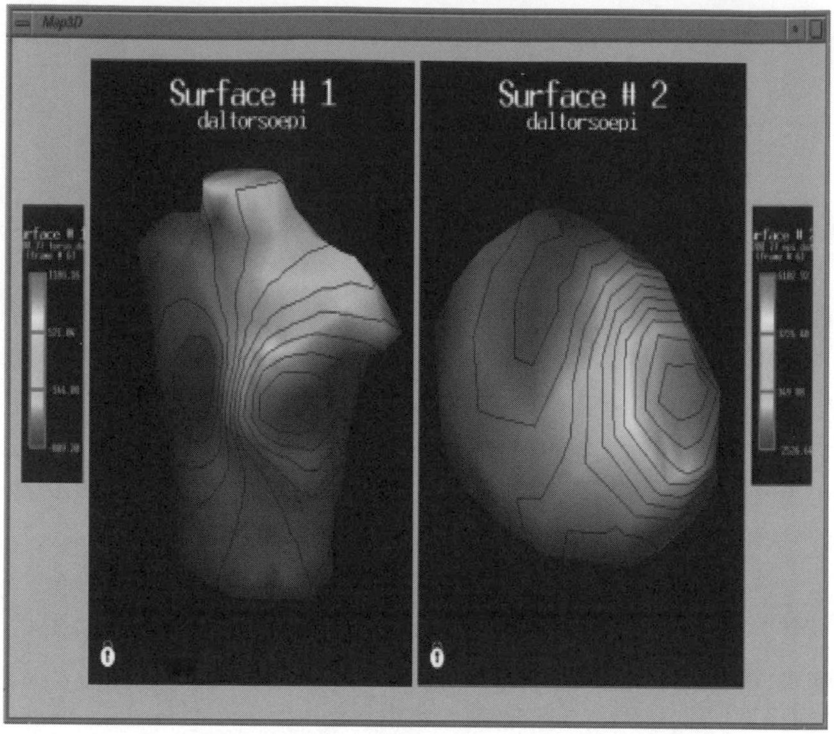

Figure 17. Color-coded BSPMs and computed potential distributions at the surface of the heart during the peak of the QRS complex using 120 ECG electrodes. The surface on the right shows the computed potential distribution on the surface of the heart at the same instant. The relative size of the two surfaces is distorted but their orientations are aligned as in the geometric model used for the inverse calculation. The scaling and isopotential contours are local to teach surface. Scale bars on both sides indicate the voltage (in microvolt) to color mapping. (Courtesy of Dr. Robert MacLeod, Scientific and Imaging Institute, University of Utah, Ref. 9.)

acquisition capabilities and advancements in electronics and computer hardware and software capabilities. Figure 17 shows color-coded BSPMs and computed potential distributions at the surface of the heart during the peak of the QRS complex using 120 ECG electrodes as an example of how advanced computing methods can improve the spatial resolution of electrocardiography.

In summary, a popular modern approach to extend the capabilities of the standard 12-lead ECG system in mapping the electrical activity of the heart is high resolution electrocardiography or body surface potential mapping (BSPM). This approach attempts to provide spatial information by including a larger number of recording electrodes covering the entire torso. The BSPM facilitates the spatial visualization of the cardiac electrical activity. It has been shown that BSPM is superior to 12-lead electrocardiography in detection and localization of acute and remote myocardial infarction (MI). However, the full capability of BSPM to localize acute MI is as yet, not established.

Computers play a central role in fast acquisition and real-time processing of a large number of ECG signal channels, as well as in modeling and visualization of the electrical images of the heart and eventual construction of 3D animated representations of body surface potential maps. These topics are active areas of research and deserve separate chapters in their own right.

Computer-Based Monitoring ECG Systems

All modern diagnostic and monitoring ECG acquisition and analysis systems are equipped with state-of-the-art computer technology. ECG monitoring devices constitute an essential part of the patient monitoring systems. Patient monitoring systems are used to continuously or intermittently measure the vital signs (generally ECG, heart rate, pulse rate, blood pressure, and temperature) and perform automatic analysis on them. Monitoring is carried out at the bedside, or a central station. Bedside monitors are now widely used in intensive care, cardiac care and operating rooms. Computer-based bedside monitors perform ECG analysis and generate alarms if life-threatening arrhythmias occur. The ECG monitoring systems can be networked to share common computing and analysis resources. Telecommunication technologies are now used to transmit ECG signals back and forth between computing facilities. For a detailed discussion of ECG-based monitoring systems see Ambulatory (Holter) Monitoring and Electrocardiographic Monitors.

Computer-Based Heart Beat Detection. Computer-based arrhythmia detectors that provide accurate detection of prevalent heart diseases could greatly help cardiologists in early and efficient diagnosis of cardiac diseases. Such devices now constitute an integral part of computer-based cardiac monitoring systems. The morphologic and rhythmic character of the ECG signal can be interpreted from its patterns, which normally have periodic waveforms such as PQRSTU-waves. Rhythmical patterns deviating from normal ECG (cardiac arrhythmias) have correlation with heart injuries or its malfunction in pumping the blood, such as premature beats, flutter and fibrillation patterns. The most serious pattern is ventricular fibrillation, which may be life threatening due to deficient supply of oxygen by the blood to the vital organs, including the heart itself. Considering other factors influencing the ECG patterns such as medications taken by patients, their physiological and psychological condition as well as their health records, an accurate diagnosis can be made based on cardiac arrhythmia classification.

Accurate determination of the QRS complex and more specifically, reliable detection of the R-wave peak plays a central role in computer-based ECG signal analysis, Holter data analysis and robust calculation of beat-to-beat heart rate variability (13). Fluctuations in the ECG signal baseline drift, motion artifact due to electrode movement and electromyographic (EMG) interference due to muscular activity frequently contaminate the ECG signal. In addition, morphological variations in the ECG waveform and the high degree of heterogeneity in the QRS complexes make it difficult to distinguish them from tall peaked P and T waves.

Many techniques have therefore been developed to improve the performance of QRS detection algorithms. These techniques are mainly based on band pass filtering, differentiation techniques, template matching and others. Recently, it has been shown that the first differential of the ECG signal and its Hilbert Transform can be used to effectively distinguish the R-waves from large, peaked T and P waves with a high degree of accuracy (14). Even though this method provides excellent R peak detection performance, it lacks automatic missing-R-peak correction capability and has not been tested on pediatric ECG data. Manual correction of missing R peaks in the ECG signal is time consuming and tedious and could play a critical role in error-free derivation of HRV signal. Figure 18 shows the block diagram for an Enhanced Hilbert Transform-based (EHT) method with automatic missing-R-peak correction capability for error-free detection of QRS complexes in the ECG signals.

In this algorithm, filtered ECG signals are differentiated first to reduce baseline drift and motion artifacts. This step sets the ECG peaks to zero. Hilbert Transform of the differentiated signal is then calculated and the conjugate of the Hilbert Transform is obtained. With this operation, the zero crossings in the differentiated signal becomes prominent peaks in the Hilbert transformed conjugate of the differentiated signal. A threshold is then selected based upon the normalized RMS value (as it gives a measure of the noise content in the signal) of the Hilbert transformed ECG signal. The time instants for which the signal amplitude is greater than the threshold value are stored in an array. Peak detection is performed on the original signal. The peak values and the time at which they occur are stored in separate arrays. The R–R intervals are then calculated. If an R–R interval is > 130% of the previous R–R interval, then the algorithm is considered to have missed an R-peak from the ECG signal. Hence a correction for the missing beat is made based upon an updated moving average of the previous R–R intervals. The amplitude of the missing peak is estimated as the average of the two previous R-peaks adjacent to it (13).

ECG Signal Telemonitoring

With advancements in bioinstrumentation, computer, and telecommunications technologies now it is feasible to design vital sign telemonitoring systems to acquire, record, display, and transmit physiological signals (e.g., ECG) from the human body to any location. At the same time, it has become more practical and convenient for medical and paramedical personnel to monitor vital signs from any computer connected to the Internet.

Telemonitoring can improve the quality, increase the efficiency, and expand access of the healthcare delivery system to the under-staffed, remote, hard-to-access, or under-privileged areas where there is a paucity of medical practitioners and facilities. It seems reasonable to envision that a telemonitoring facility could significantly impact areas where there are needs for uniform healthcare access such as under-served populations of rural areas, developing countries, space flights, remote military bases, combat zones, and security healthcare facilities. Mobile patient telemonitoring (i.e., emergency medicine), posthospital patient monitoring, home care monitoring, patient education and continuing medical education all will benefit from telemonitoring. Figure 19 shows the graphical user interface of a Web-based vital sign telemonitor accessed from a client site (15).

APPENDIX

The PageWriter Touch Technical Specifications. (Courtesy of Philips Medical Systems, USA.)

ECG Acquisition

R/T (real-time) ECG (12 leads)
AUTO (12 leads)
RHYTHM (up to 12 leads)
DISCLOSE (1–3 leads)

Keyboard

Full alphanumeric

Touch Screen Display

1024 × 768 pixel resolution
30.4 × 22.8 cm (15 in. diagonal) color liquid crystal touch screen display with backlight

Patient Module

Action button allows user to take ECG snapshots from the bedside

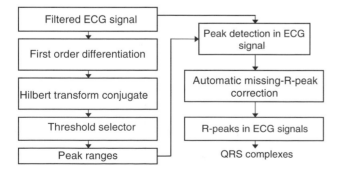

Figure 18. Block diagram for the Enhanced Hilbert Transform-based (EHT) QRS detection algorithm with automatic built-in missing R-peak correction capability. (Courtesy of Ref. 13.)

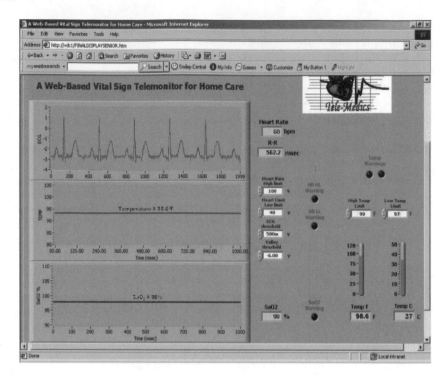

Figure 19. The graphical user interface of a web-based telemonitor. (Courtesy of Ref. 15.)

Signal Processing/Acquisition
Sampling Rate

1000 samples per second per electrode/lead
12 bit A/D conversion provides 5 μV resolution

Auto Frequency Response

0.05–150 Hz, 0.15–150 Hz, 0.5–150 Hz,
0.05–100 Hz, 0.15–100 Hz, 0.5–100 Hz,
0.05–40 Hz, 0.15–40 Hz, 0.5–40 Hz

Rhythm Frequency Response

0.05–150 Hz, 0.15–150 Hz, 0.05–100 Hz,
0.15–100 Hz, 0.05–40 Hz, 0.15–40 Hz

Filters

AC noise
Baseline wander
Artifact

Printer

Printer Resolution

High resolution, digital-array printer using thermal-sensitive paper
200 dpi (voltage axis) by 500 dpi (time axis)

Report Formats

3 × 4 (1R, 3R)
6 × 2
Panoramic 12 (Cabrera)
12 × 1

Rhythm (up to 12 selected leads)
Extended Measurements
One Minute Disclose (1 lead)
Full disclosure (5 min, 1 to s selected leads)

Battery Operation

Capacity

Typically 50 ECGs and copies on a single charge or 40 min of continuous rhythm recording
Recharge
Seven hours in standby mode to >90% capacity (typical)

Network Connections

10/100 Base-T IEEE 802.3 Ethernet Via RJ45 connector (standard)
Optional software required for Proxim Range LAN 27410 CE PC card wireless LAN connection

FAX Capability (optional)

Group 3, Class 1 or 2 fax

Modem (optional for USA & Canada)

V.90, K56 flex, enhanced *V*.34, *V*32 bits, *V*.32, *V*.22 bits and below

Barcode Reader (optional)

Reads Code 39 (standard and full ASCII)

Magnetic Card Stripe Reader (optional)

Reads cards adhering to ISO 7810, 7811-1, -2, -3, -4, and JIS X6301 and X6302

ECG Storage

XML File Format
150 ECGs to internal flash memory
2–3 ECGs typical per 1.4 MB floppy disk
150 ECGs per 128 MB PCMCIA card (optional)

ECG File Formats

XML and XML SVG

Power and Environment

Line Power 100–240 Vac, 50/60 Hz, 150 VA max

Environmental Operating Conditions

15–35 °C (50–104 °F)
15–70% relative humidity (noncondensing)
Up to 4550 m (15,000 ft.) altitude

Environmental Storage Conditions

0–40°C (32–122 °F)
15–80% relative humidity (noncondensing)
Up to 4550 m (15,000 ft) altitude

Cardiograph Dimensions

~45 × 45.8 × 16 cm (17.7 × 18.0 × 6.34 in.)

Cardiograph Weight

~13 kg (28 lb.) including accessories

Patient Interface Module

Remote, microprocessor-controlled module

Safety and Performance

Meets the following requirements for safety and performance:

IEC 60601-1:1988 + A1: 1991 + A2: 1995 General Requirements for Safety including all National Deviations
IEC 60601-1-2: 1993 General Requirements for Safety Electromagnetic Compatibility
IEC 60601-2-25: 1993 + A1: 1999 Safety of Electrocardiographs
IEC 55011: 1998 Radio Frequency disturbance, Limits and Methods of Test
AAMI EC11: 1991 Diagnostic Electrocardiographic Devices
JIST 1202: 1998 Japanese Industrial Standard for Electrocardiographs

ACKNOWLEDGMENT

The excellent support of Ms Mary-Lou Lufkin with the Diagnostic ECG Division at Philips Medical Systems, USA is greatly appreciated.

BIBLIOGRAPHY

Cited References

1. Clark JW. The origin of biopotentials. In: Webster JG, editor. Medical Instrumentation. 3rd ed. New York: John Wiley & Sons; 1998, p 121–182.
2. Malmivuo J, Plonsey R. Bioelectromagnetism: Principles and Applications of Bioelectric and Biomagnetic Fields. New York: Oxford University Press; 1995.
3. Netter FH. The Heart, Vol. 5. The Ciba Collection of Medical Illustrations. Ciba Pharmaceutical Company; 1971.
4. Haykin S. Adaptive Filter Theory. 4th ed. New York: Prentice Hall; 2001.
5. Anonymous. American National Standard for Diagnostic Electrocardiographic Devices. ANSI/AAMI EC 13, New York: American National Standards Institute; 1983.
6. Webster JG, editor. Bioinstrumentation. New York: Wiley; 2004.
7. Available at http://medstat.med.utah.edu/kw/ecg/ACC_AHA. html.
8. Macfarlane PW, Lawrie TDV, editors. Comprehensive Electrocardiology: Theory and Practice in Health and Disease. 1st ed. Vols. 1–3. New York: Pergamon Press; 1989. 1785 p.
9. Available at http://www.sci.utah.edu.
10. Available at http://butler.cc.tut.fi/~malmivuo/bem/index.htm.
11. Available at http://rudylab.wustl.edu.
12. Available at http://www.bioeng.auckland.ac.nz/projects/ei/eimaging.php.
13. Chatlapalli S, et al. Accurate Derivation of Heart Rate Variability Signal for Detection of Sleep Disordered Breathing in Children. Proc 26th Annu Int Conf IEEE EMBS San Francisco, (CA) Sept., 2004.
14. Benitez D, Gaydecki PA, Zaidi A, Fitzpatric AP. The use of Hilbert Transform in ECG signal analysis. Comput Biol Med Sept.2001;31(5):399–406.
15. Mendoza P, et al. A Web-based Vital Sign Telemonitor and Recorder for Telemedicine Applications. Proc IEEE/EMBS, 26th Annu Int Conf, San Francisco (CA). Sept. 1–4, 2004.

See also ARRHYTHMIA ANALYSIS, AUTOMATED; GRAPHIC RECORDERS; PHONOCARDIOGRAPHY.

ELECTROCONVULSIVE THERAPHY

MILTON J. FOUST, JR
MARK S. GEORGE
Medical University of South Carolina
Charleston, South Carolina

INTRODUCTION

Electroconvulsive therapy (ECT) is a technique for the treatment of severe psychiatric disorders, which consists of the deliberate induction of a generalized tonic-clonic seizure by electrical means. Contemporary ECT devices typically deliver bidirectional (alternating current) brief-pulse square-wave stimulation through a pair of electrodes that are applied externally to the patient's scalp. The procedure is now almost always performed under general anesthesia, although, in some unusual situations, such as in developing countries with limited medical resources, it may be occasionally done without anesthesia (1). As with other convulsive therapies that historically preceded ECT, the goal is to produce a seizure. The presence of seizure activity appears to be essential; stimuli that are below the seizure threshold appear to be clinically ineffective. However, although the production of a seizure appears to be necessary, a seizure alone is not sufficient. Some forms of seizure induction are, in fact, clinically ineffective. A variety of psychiatric and neurological conditions exist that respond favorably to ECT, although the majority of patients treated with ECT have mood disorders, such as unipolar or bipolar depression, particularly when severe or accompanied by psychotic symptoms. Certain other conditions, such as mania, schizoaffective disorder, catatonia, neuroleptic malignant syndrome, Parkinson's disease, and intractable seizures, may respond to ECT as well. Schizophrenia has also been treated with ECT, although the results tend to be less favorable than those obtained in patients with mood disorders. Those patients with schizophrenia who also have a prominent disturbance of mood probably respond best to ECT (2,3). Typically, a series or a course of treatments is prescribed. By convention, ECT treatments are usually given two to three times per week. A course usually consists of around six to eight treatments, which may then be followed by maintenance treatment in the form of either medication, additional ECT given at less frequent intervals, or both. A number of questions still remain regarding the most effective methods for performing ECT, the mechanism of action of ECT, and what role there may be in the future for ECT and other forms of brain stimulation, such as repeated transcranial magnetic stimulation (rTMS), magnetic seizure therapy (MST), and vagus nerve stimulation (VNS).

HISTORY

ECT was first used by the Italian psychiatrists Ugo Cerletti (Fig. 1) and Lucio Bini to treat a disorganized, psychotic man found wandering the streets of Rome in 1938. The results were dramatic, with complete recovery reported (4). The treatment was developed as an alternative to other, higher risk forms of artificial seizure induction, specifically Ladislas von Meduna's convulsive therapy involving the use of stimulants such as camphor, strychnine, and Metrazol (pentylenetetrazol) (4–6). ECT was welcomed due to its effectiveness with otherwise treatment-resistant and profoundly disabled patients. However, the procedure was at risk of being abandoned due to the incidence of fractures (up to 40%) caused by uncontrolled seizure activity. This problem was resolved by the introduction of muscle relaxation (originally in the form of curare, and later with depolarizing muscle relaxants such as succinylcholine) and general anesthesia (7). The use of anesthesia and muscle relaxation was one of the most important innovations in ECT treatment, another being the use of brief-pulse square-wave stimulation in place of sine-wave alternating current. Brief-pulse ECT was found to cause less cognitive impairment compared with sine-wave ECT, as well as less disruption of the EEG (8,9). The routine use of oxygen and monitoring of vital signs, cardiac rhythm, pulse oximetry,

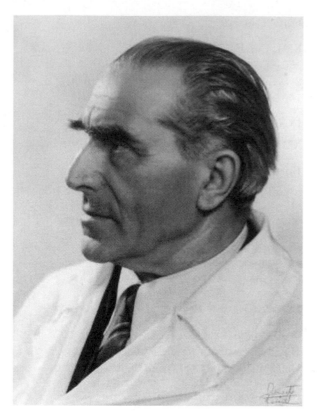

Figure 1. Ugo Cerletti 1877–1963. Reprinted with permission from the American Journal of Psychiatry, (Copyright 1999). American Psychiatric Association.

electromyography (EMG), and electroencephalography (EEG) have also helped to reduce the risks associated with the procedure.

PRE-ECT EVALUATION

In order to receive ECT, a patient must first be evaluated by a physician (typically a psychiatrist) who is trained and credentialed in the procedure and who agrees that the patient is a suitable candidate, based on psychiatric history and examination, physical condition, and capacity to consent. If a patient is unable to consent, a suitable substitute decision-maker must be identified, usually a family member (in some states, a court-order may be required). The process of evaluation typically consists, at a minimum, of a psychiatric interview and mental status examination, a medical history (including a past medical history, family and social history, and review of systems) and physical examination (including a screening neurological examination and fundoscopic examination to exclude papilledema) (1,2,10). It is necessary to review the patient's current medications, including those that are prescribed for concurrent medical conditions as well as psychiatric medications, and to obtain some information about previous medication trials for the psychiatric condition under consideration. Usually, it is desirable to obtain some basic laboratory studies (such as complete blood count, serum electrolytes, BUN, glucose and creatinine, urinalysis, liver function tests, and thyroid function tests) both to screen for medical conditions that may cause depressive or psychotic symptoms and to identify conditions of increased ECT or anesthesia risk. Most patients, and especially older patients or patients with a history of cardiovascular disease, should have an electrocardiogram (ECG). The use of brain imaging is controversial; many practitioners prefer some form of pre-ECT brain imaging to identify or exclude the possibility of an intracranial mass, one of the few conditions that may be associated with a high risk of mortality with ECT (11). However, it has been argued that a neurologic examination should be sufficient to screen for this particular risk factor (12).

CONSENT

Prior to undergoing ECT or any other procedure, the patient (or patient's surrogate decision-maker) must demonstrate a satisfactory understanding of the nature of the procedure, its risks, benefits, and alternatives. In some states, the physician may need to petition the court for permission to perform ECT on patients who lack the capacity to consent. The issue of consent is complex, as those patients who are most in need of ECT are generally those who are the most ill, and are often the most vulnerable and impaired. It is possible to have a severe psychiatric illness and yet still retain the capacity to rationally evaluate the necessary issues involved in making a decision as to whether to have ECT, but one can be seriously incapacitated as well. Evaluating the patient's capacity to give informed consent is one of the most important parts of the pre-ECT consultation process, along with establishing the presence or absence of an appropriate indication for treatment and identifying concurrent medical conditions and medications that may increase the risk of treatment. It is part of the consultant's responsibility to educate the patient about what is, for many, an unfamiliar or frightening treatment. Often, much of what a patient or family may understand of ECT consists of disturbing images, such as those presented in films like "One Flew over the Cuckoo's Nest" (13). In most cases, the patient will be able to understand the information presented and engage in a rational decision-making process despite their illness. However, some patients may be able to express a superficial understanding of the facts at hand and yet be impaired in the ability to make rational decisions (14), which may be demonstrated through self-destructive behavior, lack of self-care, or irrational refusal of necessary treatment. In these cases, it becomes necessary to seek a substitute, usually a family member, who can make medical decisions on the patient's behalf. State laws differ regarding the details and circumstances under which another person can make these types of treatment decisions, and it is necessary to become familiar with the particular local laws governing consent.

INDICATIONS

The most common indication for ECT is severe, treatment-resistant depression, either of the unipolar or bipolar type

Table 1. Indications for ECT

Unipolar or bipolar depression
Mania
Schizoaffective disorder
Schizophreniform disorder
Schizophrenia
Catatonia
Neuroleptic malignant syndrome

(Table 1). The syndrome of depression is characterized by a sad or depressed mood, as well as disturbances in energy level, sleep, and the capacity to experience pleasure (anhedonia) (15). It may include psychotic symptoms such as delusions (fixed, false beliefs that are held despite evidence to the contrary) or hallucinations. Patients with so-called "unipolar" depression exhibit one or more episodes of depression without episodes of mania. Such patients will usually be formally diagnosed as having a major depressive disorder (MDD). Patients with so-called "bipolar" depression have also suffered from one or more episodes of mania, frequently in a cyclical pattern of alternating mania followed by depression. Those patients who are most severely ill, particularly those who are delusional or catatonic, will typically respond best to ECT. Although usually reserved for those patients who have not had a successful response to one or more medication trials, ECT is an appropriate first treatment when the patient's life is threatened by severe illness in the form of aggressively self-destructive behavior, refusal or inability to eat or drink, or extreme agitation. Mania, a condition characterized by an abnormally elevated or irritable mood, hyperactivity, agitation, impulsivity, and grandiosity, also responds well to ECT. Schizophrenia is often treated with ECT, particularly in some European and Asian countries, but may respond less well, unless mood disturbance is a prominent component of the patient's illness. Neuroleptic malignant syndrome (an antipsychotic drug-induced syndrome that shares many of the characteristics of catatonia) is a less common indication for ECT and may be used when the syndrome persists despite the usual interventions such as discontinuing neuroleptics and treatment with dopamine agonists. Catatonia (which may be an expression of either a mood disorder or schizophrenia) is characterized by mutism and immobility, sometimes with alternating periods of agitation. These patients may respond to treatment with benzodiazepines, but if they do not, ECT is indicated and frequently effective. Recurrent, treatment-refractory seizures may respond to ECT as well, suggesting an anticonvulsant mechanism of action for ECT (16). Patients with Parkinson's disease may improve with ECT, possibly due to the dopaminergic effect of ECT.

Certain conditions exist, such as personality disorders, that do not respond well to ECT or may even reduce the likelihood of successful treatment when they coexist with a more suitable indication, such as a mood disorder (17–19). In some cases, the burden of disability and suffering may be so great (and the risk of serious complications so low) that ECT may reasonably be offered even if the patient's diagnosis is not one of those generally considered a standard indication for the treatment (20).

Table 2. Conditions of Increased Risk with ECT

Increased intra-cranial pressure
Recent myocardial infraction or stroke
Unstable angina
Severe cardiac valvular disease
Severe congestive heart failure
Unstable aneurysms
Severe pulmonary disease
Pheochromocytoma
Retinal detachment
Glaucoma

Table 3. Side Effects and Complications with ECT

Common:
Cognitive side effects
    Transient postictal confusion or delirium
    Retrograde and anterograde amnesia
Headaches
Muscle soreness
Nausea
Less common or rare:
Death (1/10,000 patients)
Aspiration
Brochospasm or laryngospasm
Cardiovascular
    Arrhythmias
    Severe hypertension or hypotension
    Cardiac ischemia
    Myocardial infarction
    Cardiac arrest
Neurological
    Prolonged or tardive seizures
    Nonconvulsive status epilepticus
    Stroke
Prolonged apnea due to pseudocholinesterase deficiency
Malignant hyperthermia of anesthesia

CONDITIONS OF INCREASED RISK

As will be discussed later in this article, ECT as currently practiced is a relatively low risk procedure. It can be safely used to treat all patient groups including children and adolescents (21–25), pregnant women (26,27), and the elderly (28). In particular, age alone should not be considered a barrier to treatment; elderly patients are among those who often have the most dramatic and favorable responses to ECT (29). However, certain medical conditions exist that may, to a greater or lesser degree, contribute to an increase in risk of morbidity or mortality with the procedure (Table 2). Most significant among these would be severe or unstable cardiac disease or the presence of a space-occupying lesion (such as a tumor) within the cranial cavity, resulting in increased intracranial pressure. Very small tumors without a visible mass effect on computerized tomography (CT) or magnetic resonance imaging (MRI) do not appear to pose a high risk with ECT (30). Detecting such masses and making the distinction between low and high risk lesions may help to support a rationale for pre-ECT brain imaging as a screening tool (11).

COMPLICATIONS

Serious complications with ECT are rare (Table 3). The risk of death has been estimated at 4 per 100,000 treatments (31). The risk of other potentially life-threatening complications, such as myocardial infraction and stroke, is also very low, although the risk of cardiac arrhythmias appears to be higher in persons with pre-existing cardiac disease (32,33). The introduction of general anesthesia and muscle relaxation has almost eliminated the risk of fractures with ECT. Both retrograde amnesia and anterograde amnesia are common, but it is unusual for cognitive impairment to be severe or prolonged. Minor side effects such as headaches, muscle aches, and nausea frequently occur; these side effects are usually transient and easily managed with symptomatic treatment.

The amnesia or memory loss that occurs with ECT typically takes two forms: loss of memory for past or previously learned information (retrograde amnesia) as well as difficulty in learning new information (anterograde amnesia). The retrograde amnesia associated with ECT tends to be greater for "public" or "impersonal" knowledge about the world than for autobiographical or personal memories. Memories for remote events also tend to be better preserved than that for more recent events. Bilateral ECT appears to produce greater and more persistent memory deficits than right-unilateral ECT (34). ECT-induced anterograde amnesia is typically greatest immediately following treatment and tends to rapidly resolve in the weeks following the last treatment of a series. Recovery also typically occurs over time with retrograde amnesia, although patients may notice some persistent gaps in memory for past events. Although anecdotal, it may be reassuring to patients to be made aware of the stories of psychologists and physicians who have had ECT, benefited, and resumed their professional activities (35). Uncommon exceptions to these rules exist, however. A few patients complain of severe and persistent problems with memory that cause them much distress (36). No satisfactory explanation exists for this phenomenon of severe ECT-induced memory loss (37). No reliable evidence exists that ECT causes damage or injury to the nervous system.

MEDICATIONS AND ECT

Experts in the past have recommended stopping antidepressants prior to ECT (38), although a recent study now suggests that tricyclic antidepressants (TCAs) and selective serotonin reuptake inhibitors (SSRIs) may be safe in combination with ECT (39). Evidence exists that certain antipsychotic medications (such as haloperidol, risperidone, and clozapine) may have beneficial effects in combination with ECT (40–44). Many drugs that were originally developed and used as anticonvulsants (such as carbamazepine, valproic acid, and lamotrigine) are frequently used as either mood stabilizers or as adjunctive agents in antidepressant regimens. As these drugs, by definition, can be expected to inhibit seizure activity, they are generally tapered and discontinued prior to beginning

ECT. An exception would be those patients for whom these drugs are prescribed for a concurrent seizure disorder. In these cases, anticonvulsant drugs should usually be continued in order to minimize the risk of uncontrolled seizure activity between treatments. Lithium, a commonly used mood stabilizer, has been associated with prolonged delirium following ECT. In general, it should be avoided, but, in special circumstances and with careful monitoring, its use in combination with ECT may be justified (45,46). Both chlorpromazine and reserpine, an antipsychotic and antihypertensive agent that acts through the depletion of neuronal dopamine, have been associated with severe hypotension and death when combined with ECT (47).

Certain drugs that are prescribed for concurrent medical conditions (and not primarily for psychiatric conditions) help to reduce the risks associated with ECT and anesthesia. Patients who have gastro-esophageal reflux disease (GERD) should be treated with a suitable medication (such as a histamine-2 receptor blocker or proton pump inhibitor) in the morning prior to ECT to reduce the risk of reflux and aspiration. Patients who may be especially prone to aspiration can be treated with intravenous metoclopramide to accelerate gastric emptying. Pregnant women may be given sodium citrate/citric acid solution by mouth. Patients with known hypertension should receive their usual antihypertensive regimen prior to ECT; an exception would be made for diuretics, which should be delayed until after ECT to avoid bladder filling before or during the procedure. Some patients may have an exaggerated hypertensive response to the outpouring of catecholamines that occurs with seizure response to ECT, which can usually be controlled with intravenous beta-adrenergic antagonists such as esmolol or labetalol. A patient with a pheochromocytoma (catecholamine-secreting tumor) is at especially high risk of severe hypertension (48). Such a patient may require more aggressive measures (such as arterial blood-pressure monitoring and intravenous sodium nitroprusside) in order to maintain satisfactory blood pressure control (49). Patients with pulmonary disease should have their pulmonary status assessed prior to treatment and should receive their usual medications, including inhaled beta-agonists or steroids. As ECT and the resulting seizure produces a transient increase in intraocular pressure (50), patients with retinal detachment may be at risk of further eye injury and should have this condition treated prior to ECT. Similarly, patients with glaucoma should have their intraocular pressures controlled with suitable medications prior to ECT (51).

ANESTHESIA

Methohexital was previously a very popular anesthetic drug for ECT, but manufacturing problems made it essentially unavailable for several years, forcing changes in ECT anesthesia practice (52). Anesthesia for ECT may be induced with an intravenous injection of other short-acting anesthetic agents such as propofol and etomidate. Etomidate may be substituted for methohexital or propofol in an effort to produce seizures of longer duration (53) or to

stimulate seizures in those uncommon situations of no seizure activity despite maximum stimulus and bilateral electrode placement. Once anesthesia is induced (typically within seconds following the injection), a muscle relaxant is injected, typically succinylcholine, but a nondepolarizing muscle relaxant such as mivacurium may be used when there are coexisting medical conditions that increase the risk of exaggerated hyperkalemic response with succinylcholine, such as burns, renal failure, or neurologic disease, or if a history of malignant hyperthermia exists (54,55). The ECT stimulus may elicit a vagal (parasympathetic) response, which can lead to bradycadia, transient heart block, or even asystole, which has been explained as the result of forced expiration against a closed glottis during the stimulus, or it may be a direct effect of the stimulus on the central nervous system. Bradycardia and asystole have been observed in the postictal period as well (56). An anticholinergic compound such as glycopyrrolate or atropine may be injected proximate to the anesthetic to reduce the bradyarrhythmias that may occur with ECT (57). If the stimulus is successful in producing a seizure, it results in an outpouring of catecholamines and a sympathetic response with resulting tachycardia and hypertension as noted above, which is usually transient and without clinical significance, but in some patients, especially those with pre-existing hypertension or cardiovascular disease, it may be desirable to limit this response using an antihypertensive agent (54,55,58).

Patients are unable to breathe without assistance when anesthetized and fully relaxed; ventilation and oxygenation are provided by means of positive-pressure ventilation with 100% oxygen through a bag and mask. Endotracheal intubation is rarely required. In some patients (such as pregnant women), the risk of reflux may be higher and intubation may be the preferred option for airway management (27). Some patients may have abnormalities of the face and upper airway that interfere with mask ventilation. Other options for airway management including intubation, laryngeal mask airway (59,60), or even tracheostomy may be considered.

MONITORING

The type of medical monitoring that is used during ECT includes ECG blood pressure, pulse oximetry, EMG, and EEG. Modern ECT devices typically include the capacity to monitor and record ECG, EMG, and EEG; commonly, a paired right and left fronto-mastoid EEG placement is used. Two-channel EEG monitoring is helpful both to ensure that the seizure generalizes to both hemispheres and to evaluate the degree of inter-hemispheric EEG symmetry or coherence, as well as postictal EEG suppression, all factors that may predict the therapeutic efficacy of the seizure (61). Two EMG electrodes are placed on a foot, usually the right, which is isolated from the rest of the patient's circulation with a blood pressure cuff acting as a tourniquet, which minimizes the effect of the muscle relaxant and permits the observation and recording of motor seizure activity in the foot, even with complete paralysis otherwise.

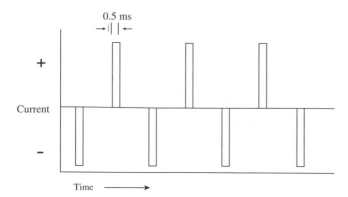

Figure 2. Brief-pulse square-wave ECT stimulus.

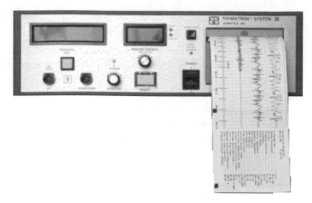

Figure 3. Somatics Thymatron system IV ECT device (Image provided courtesy of Somatics, LLC).

ECT STIMULUS

As noted previously, the ECT stimulus itself typically consists of a series of brief electrical pulses. Early ECT devices provided stimulation with sine-wave alternating current; essentially the same type of current that is distributed through public utilities for use in household appliances and lighting. Most modern ECT devices now provide a stimulus that consists of brief pulses with a bidirectional square-wave pattern (Fig. 2). (One exception is the Ectron series 5A, manufactured by Ectron, Ltd, Letchworth, Hertfordshire, England. This device delivers unidirectional pulses.) The American Psychiatric Association has recommended the use of "constant-current" devices; so-called because a relatively constant (rather than continuously varying) current is maintained for the duration of the pulse. The advantage of using a constant-current rather than constant-voltage or constant-energy device is that the clinician is able to deliver a predetermined quantity of charge by varying the time interval of exposure to the current (62).

At 100% of available output, a typical ECT device such as the Thymatron System IV (Somatics, LLC, Lake Bluff, IL) can provide approximately 504 mC of charge (Table 4) (64), which will be delivered as brief pulses of 0.5 ms each at a frequency of 70 Hz for a total stimulus duration of 8 s. Members of the Thymatron (Fig. 3) series (DX, DGX, and System IV) and similar devices such as those of the Spectrum series (4000M, 4000Q, 5000M, and 5000Q manufactured by MECTA, Lake Oswego, OR) are calibrated in such a way

that the user can select incremental quantities of charge. The device labels and manuals will usually refer to these as increments of "percent energy," although the energy required to deliver the charge is actually dependent on the particular impedance of the patient being treated. The user (clinician) may have the option of individually adjusting such parameters as "percent energy" (charge), pulse frequency, pulse-width, and total duration of pulse train. Both voltage and energy will be altered by the device to adjust for variations in impedance. The average impedance for a human receiving ECT is approximately 220 Ω, and the energy required to deliver 504 mC is 99.4 J (63).

Over an 8 s total stimulus duration, the average power for that interval is 99.4 J/8 s = 12.4 W, although the power during each discrete electrical pulse is obviously much greater. Assuming a pulse frequency of 70 Hz (140 pulses/s with bidirectional current), delivered over a total stimulus duration (stimulus train) of 8 s, with a pulse-width of 0.5 ms (or 0.0005 s), then the power during each discrete electrical pulse of this stimulus would be:

$$99.4\,J/(140\,\text{pulses/s}*8\,s*0.0005\,\text{s/pulse})$$

$$= 99.4\,J/0.56\,s = 177.5\,W$$

The stimulus is transmitted to the head of the patient (and, indirectly, to the brain) through externally applied scalp electrodes. These electrodes may either be stainless-steel discs held in place with an elastic band or self-adhesive flexible conductive pads. The stimulus electrodes can be applied in several configurations, including bilateral (also called bifronto-temporal), right unilateral, and bifrontal. In bilateral ECT, the electrodes are placed approximately 3 cm above the midpoint of a line between the canthus of the eye and tragus of the ear. With right unilateral ECT, the right electrode is placed in the conventional position for bilateral ECT, with the second electrode just to the right of the vertex of the skull. Bifrontal electrode placement is the newest of these various methods of performing ECT. Bifrontal placement is 5 cm above the angle of the orbit of the eye, which is usually found near the outer edge of the eyebrow. Comparative studies suggest that bilateral ECT may be more effective at lower energy levels, although right unilateral and bifrontal ECT may result in less cognitive impairment (64–68).

Table 4. ECT Device Specifications (Thymatron System IV)

Current:	0.9 A (fixed)
Frequency:	10 to 140 Hz
Pulsewidth:	0.25 to 1.5 ms
Duration:	0.14 to 7.99 s
Output:	approx. 25 to 504 mC
	(5 to 99.4 J at 220 Ω)
	1008 mC (188.8 J at 220 Ω)
	with double-dose option
Input:	120 V AC, 60 Hz
Recording:	4 channel (2 EEG, EMG, ECG)

We consider the patient and device (including electrodes and cables) to be a circuit, and, for purposes of simplification, impedance in this circuit is usually treated as more or less equivalent to resistance. However, it should be recognized that impedance in the circuits being discussed has both resistive and reactive components (62). These reactive components are capacitance and inductance. Both inductance effects and capacitance in tissue are assumed to be low, but the treatment cables and electrode–skin interface may make a more significant contribution, altering both the amplitude and frequency of the stimulus. Complicating matters further, it should also be noted that the circuit impedance varies with the intensity of the stimulus. ECT devices typically provide a "static impedance" reading prior to treatment, which can vary widely, but is usually around 2000 Ω. The "dynamic impedance" can be measured during the stimulus and is substantially less, an average of about 200 Ω. The static impedance is measured with a very small test stimulus (insufficient to cause a seizure or even be detected by the patient). This static impedance is used primarily to test the integrity of the circuit. If the impedance is greater than 3000 Ω, it is assumed that either a break in the circuit occured or inadequate contact exists at the skin-electrode interface. Similarly, an excessively low impedance suggests shunting through a very low impedance channel (short-circuit) such as saline, conductive gel, or skin between closely applied electrodes. Interestingly, the seizure threshold (quantity of charge required to stimulate a seizure) is typically lower for unilateral than for bilateral ECT, despite the lower interelectrode distance with potential for lower impedance and greater shunting though scalp tissues with unilateral placement, which is thought to be a result of differing patterns of charge density, with maximum charge density in the frontal region for bilateral electrode placement and maximum charge density for unilateral placement in the region of the motor strip, an area of lower intrinsic seizure threshold (62).

SEIZURE RESPONSE

A seizure is characterized neurophysiologically by the paroxysmal synchronous discharge of populations of neurons. It is recognized clinically as abnormal behavior associated with abnormal neuronal activity and may present in a variety of different forms such as simple partial seizures, complex partial seizures, or generalized tonic-clonic seizures (69). During ECT, the patient is anesthetized and relaxed, so many of the clinical characteristics are altered. However, the observable characteristics of the ECT seizure are consistent with the generalized tonic-clonic type of seizure. The ECT stimulus predictably results in a pattern of spike and slow-wave activity that is distinctive and recognizable on EEG (70). The seizure activity rapidly spreads throughout both hemispheres and usually lasts for less than 60 s. The immediate goal of ECT is to produce such a seizure; stimuli that do not result in seizure activity appear to lack therapeutic benefit. However, it is possible to produce seizures, specifically with right unilateral ECT at or only slightly above the seizure threshold, that have a relatively weak therapeutic effect (64). Most experts recommend that right unilateral ECT be given with a stimulus that is at least five times the seizure threshold. Threshold may either be determined by titration (i.e., by giving a series of increasing stimuli), or the stimulus may be given at maximum energy. If bilateral or bifrontal electrode placement is used, energies just above the seizure threshold may be sufficient, and seizure threshold can be estimated based on the patient's age (71).

Certain characteristics of the seizure activity may be associated with therapeutic efficacy, including coherence or symmetry between hemispheres of high amplitude slow waves on EEG and marked postictal suppression following the end of the seizure (61). Adjustments can be made in technique to try to improve response based on these characteristics and other seizure characteristics, including overall seizure duration.

MECHANISM OF ACTION OF ECT

The precise mechanisms of action of ECT are not well understood. A number of biochemical and physiological changes exist that have been detected following both ECT in humans and electroconvulsive shock (ECS) in animals. Some of these parallel those of antidepressant drugs, but others do not. In vivo studies of long-term ECS (and repeated tricyclic antidepressant drug administration) show increases in postsynaptic 5-hydroxytryptamine type 1a (5-HT-1a) receptor sensitivity. Long-term ECS increases 5-HT-2a receptors, but antidepressant administration in animals decreases 5-HT-2a receptors. Both antidepressant treatment and ECS reduce beta-adrenergic receptors (72). Evidence for increased dopaminergic activity with ECS and ECT exists, probably as a result of increased dopamine synthesis (73). Proton magnetic resonance spectroscopy measurements following ECT have demonstrated increases in concentrations of cortical gamma-amino butyric acid (GABA), an inhibitory neurotransmitter (74). This observation, as well as the finding that increasing seizure threshold during a course of ECT is associated with clinical response, has led to the hypothesis that a linked anticonvulsant and antidepressant response to ECT exists (61).

Magnetic resonance imaging (MRI) scans obtained before and after ECT have failed to demonstrate any structural changes in the brain. Positron emission tomography (PET) and single-photon emission computerized tomography (SPECT) scans of the brain during the ECT seizure show marked increases in global and regional metabolic activity and blood flow (75). However, PET scans taken several hours or days following ECT have shown a marked reduction in absolute and prefrontal activity (76). The majority of functional imaging studies with depressed patients have shown pretreatment deficits in similar regions (77). Paradoxically, those subjects with the greatest post-ECT reduction in the left prefrontal cortex show the greatest clinical response (78,79). One possible explanation for this may be that specific areas of the brain exist that are hypermetabolic in the depressed state and are suppressed by the effects of ECT, although, as mentioned above, most scans show resting prefrontal hypoactivity in depression.

At the level of gene expression, ECS appears to induce brain-derived neurotrophic factor (BDNF) and its receptor, protein tyrosine kinase B (TrkB). Using our current understanding of the effects of ECS and neurotransmitter signal transduction, a model can be constructed that traces the effects of ECS/ECT-induced neuronal depolarization through pathways of norepinephrine (NE), 5-HT and glutamate release, monoamine receptors and ionotropic glutamate receptor binding by these neurotransmitters, cyclic adenosine monophosphate (cAMP) and other second-messengers coupled to these receptors, and protein kinases stimulated by these second-messengers. Protein kinase A (PKA), protein kinase C (PKC), and calcium/calmodulin-dependent protein kinase (CaMK) phosphorylate and activate cAMP response element binding protein (CREB). CREB is a transcription factor for BDNF. BDNF then induces neuronal sprouting and may have other beneficial effects that reverse stress-induced neuronal atrophy (80) and, presumably, depression.

FUTURE DIRECTIONS FOR ELECTROMAGNETIC BRAIN STIMULATION

A number of unanswered questions regarding ECT still exist, including such questions as:

- What are the most effective methods of maintenance therapy (medications, ECT, or both) following an acute course of treatment?
- Which medications (including antidepressants, antipsychotics, and mood stabilizers) can be successfully and safely combined with ECT?
- What is the best electrode placement from a treatment efficacy and side-effect point of view?
- How can the ECT stimulus be modified to make it more therapeutically effective and diminish side effects (especially cognitive effects)?
- What is the actual mechanism (or mechanisms) by which ECT exerts its antidepressant, antipsychotic, and other beneficial effects?

ECT itself is no longer the only form of electromagnetic brain stimulation. Both rTMS (often referred to simply as TMS) and VNS have shown promise for the treatment of mood disorders (79,81). As it does not involve the production of seizures and therefore requires no anesthesia, TMS is a particularly attractive alternative to ECT, especially for mood disorders resistant to treatment with medication. MST involves magnetic stimulation at frequencies designed to provoke seizures. This technique obviously requires general anesthesia and muscle relaxation just as ECT does. The hope is that MST may be as effective as ECT, but with less cognitive impairment due to the more focal nature of the stimulus.

ECT will remain an important option for severe and treatment-refractory psychiatric illness for the foreseeable future. An improved understanding of the pathophysiology of psychiatric illness as well as the mechanism of action of

electromagnetic brain stimulation will lead to further refinements in technique that will make ECT and related therapies safer, more effective, and more acceptable to patients and their families. The availability of alternative methods of brain stimulation will provide a wider range of choices and will create opportunities for combining therapies in ways that will be more compatible with individual patients' needs. Although changing social and political conditions may affect its image and public acceptance, it is unlikely that ECT will disappear or be replaced for some time to come.

ACKNOWLEDGMENTS

The authors would like to express their appreciation to Carol Burns, RN, ECT Program Coordinator of the Medical University of South Carolina Department of Psychiatry and Behavioral Sciences and to Drs. Melinda Bailey and Gary Haynes of the Medical University of South Carolina Department of Anesthesiology for their helpful advice during the preparation of this article.

BIBLIOGRAPHY

Cited References

1. Mudur G. Indian group seeks ban on use of electroconvulsive therapy without anesthesia. Br Med J 2002;324(6):806.
2. American Psychiatric Association. The Practice of Electroconvulsive Therapy: Recommendations for Treatment, Training and Privileging, A Task Force Report of the American Psychiatric Association. 2nd ed. Washington (DC): American Psychiatric Press, Inc.; 2000.
3. Kellner CH, Pritchett JT, Beale MD, Coffey CE. Handbook of ECT. Washington (DC): American Psychiatric Press, Inc.; 2001.
4. Fink M. Meduna and the origins of convulsive therapy. Am J Psychiatry 1984;141(9):1034–1041.
5. Fink M. Convulsive therapy: A review of the first 55 years. J Affect Disord 2001;63:1–15.
6. Fink M. Induced seizures as psychiatric therapy: Ladislas Meduna's contributions in modern neuroscience. J ECT 2004; 20:133–136.
7. Bennett AE. Curare: A preventive of traumatic complications in convulsive shock therapy. Convulsive Ther 1997;13:93–107. Originally published in the Am J Psychiatry 1941; 97:1040–1060.
8. Weiner RD, Rogers HJ, Davidson JR, Kahn EM. Effects of electroconvulsive therapy upon brain electrical activity. Ann NY Acad Sci 1986;462:270–281.
9. Weiner RD, Rogers HJ, Davidson JR, Squire LR. Effects of stimulus parameters on cognitive side effects. Ann NY Acad Sci 1986;462:315–325.
10. Klapheke MM. Electroconvulsive therapy consultation: An update. Convulsive Ther 1997;13:227–241.
11. Coffey CE. The role of structural brain imaging in ECT. Psychopharmacol Bull 1994;30:477–483.
12. Kellner CH. The CT scan (or MRI) before ECT: A wonderful test has been overused. Convulsive Ther 1996;12:79–80.
13. McDonald A, Walter G. The portrayal of ECT in American movies. J ECT 2001;17(4):264–274.
14. Appelbaum PS, Grisso T. Assessing patients' capacities to consent to treatment. N Engl J Med 1988;319(25):1635–1638.

15. Diagnostic and Statistical Manual of Mental Disorders, 4th ed. Text Revision (DSM-IV-TR®), Washington (DC): American Psychiatric Press, Inc.; 2000.

16. Griesemer DA, Kellner CA, Beale MD, Smith GM. Electroconvulsive therapy for treatment of intractable seizures: Initial findings in two children. Neurology 1997;49.

17. DeBattista C, Mueller K. Is electroconvulsive therapy effective for the depressed patient with comorbid borderline personality disorder? J ECT 2001;17(2):91–98.

18. Feske U, Mulsant BH, Pilkonis PA, Soloff P, Dolata D, Sackeim HA, Haskett RF. Clinical outcome of ECT in patients with major depression and comorbid borderline personality disorder. Am J Psychiatry 2004;161(11):2073–2080.

19. Sareen J, Enns MW, Guertin J. The impact of clinically diagnoses personality disorders on acute and one-year outcomes of electroconvulsive therapy. J ECT 2000;16(1):43–51.

20. Fink M. The broad clinical activity of ECT should not be ignored. J ECT 2001;17(4):233–235.

21. Ghaziuddin N, Kutcher SP, Knapp P, Bernet W, Arnold V, Beitchman J, Benson RS, Bukstein O, Kinlan J, McClellan J, Rue D, Shaw JA, Stock S, Kroeger K. Practice parameter for use of electroconvulsive therapy with adolescents. J Am Acad Child Adolesc Psychiatry 2004;43(12):1521–1539.

22. Cohen D, Paillere-Martinot M-L, Basquin M. Use of electroconvulsive therapy in adolescents. Convulsive Ther 1997;13:25–31.

23. Moise FN, Petrides G. Case study: Electroconvulsive therapy in adolescents. J Am Acad Child Adolesc Psychiatry 1996;35:312–318.

24. Schneekloth TD, Rummans TA, Logan K. Electroconvulsive therapy in adolescents. Convulsive Ther 1993;9:158–166.

25. Bertagnoli MW, Borchardt CM. A review of ECT for children and adolescents. J Am Acad Child Adolesc Psychiatry 1990;29:302–307.

26. Yonkers KA, Wisner KL, Stowe Z, Leibenluft E, Cohen L, Miller L, Manber R, Viguera A, Suppes T, Altshuler L. Management of bipolar disorder during pregnancy and the postpartum period. Am J Psychiatry 2004;161:608–620.

27. Walker R, Swartz CM. Electroconvulsive therapy during high-risk pregnancy. Gen Hosp Psychiatry 1994;16:348–353.

28. Kelly KG, Zisselman M. Update on electroconvulsive therapy (ECT) in older adults. J Am Geriatr Soc 2000;48(5):560–566.

29. O'Connor MK, Knapp R, Husain M, Rummans TA, Petrides G, Snyder GK, Bernstein H, Rush J, Fink M, Kellner C. The influence of age on the response of major depression to electroconvulsive therapy: A C.O.R.E. report. Am J Geriatr Psychiatry 2001;9:382–390.

30. McKinney PA, Beale MD, Kellner CH. Electroconvulsive therapy in a patient with cerebellar meningioma. J ECT 1998; 14(1):49–52.

31. Abrams R. The mortality rate with ECT. Convulsive Ther 1997;13:125–127.

32. Nuttall GA, Bowersox MR, Douglass SB, McDonald J, Rasmussen LJ, Decker PA, Oliver WC, Rasmussen KG. Morbidity and mortality in the use of electroconvulsive therapy. J ECT 2004;20(4):237–241.

33. Zielinski RJ, Roose SP, Devanand DP, Woodring S, Sackeim HA. Cardiovascular complications in depressed patients with cardiac disease. Am J Psychiatry 1993;150:904–909.

34. Lisanby SH, Maddox JH, Prudic J, Davanand DP, Sackeim HA. The effects of electroconvulsive therapy on memory of autobiographical and public events. Arch Gen Psychiatry 2000;57:581–590.

35. Fink M. A new appreciation of ECT. Psychiatric Times 2004; 21(4):

36. Donahue AB. Electroconvulsive therapy and memory loss. J ECT 2000;16(2):133–143.

37. Abrams R. Does brief-pulse ECT cause persistent or permanent memory impairment? J ECT 2002;18:71–73.

38. American Psychiatric Association. The Practice of Electroconvulsive Therapy: Recommendations for Treatment, Training and Privileging, A Task Force Report of the American Psychiatric Association. Washington, DC: American Psychiatric Press, Inc.; 1990.

39. Lauritzen L, Odgaard K, Clemmesen L, Lunde M, Öhrström J, Black C, Bech P. Relapse prevention by means of paroxetine in ECT-treated patients with major depression: A comparison with imipramine and placebo in medium-term continuation therapy. Acta Psychiatrica Scandanavica 1996;94:241–251.

40. Hirose S, Ashby CR, Mills MJ. Effectiveness of ECT combined with risperidone against aggression in schizophrenia. J ECT 2001;17:22–26.

41. Sjatovic M, Meltzer HY. The effect of short-term electroconvulsive treatment plus neuroleptics in treatment-resistant schizophrenia and schizoaffective disorder. J ECT 1993;9:167–175.

42. Chanpattana W, Chakrabhand MLS, Kongsakon R, Techakasem P, Buppanharun W. Short-term effect of combined ECT and neuroleptic therapy in treatment-resistant schizophrenia. J ECT 1999;15:129–139.

43. Tang WK, Ungvari GS. Efficacy of electroconvulsive therapy combined with antipsychotic medication in treatment-resistant schizophrenia: A prospective, open trial. J ECT 2002; 18:90–94.

44. Frankenburg FR, Suppes T, McLean P. Combined clozapine and electroconvulsive therapy. Convulsive Ther 1993;9:176–180.

45. Kellner CH, Nixon DW, Bernstein HJ. ECT-drug interactions: A review. Psychopharmacol Bull 1991;27(4):

46. Mukherjee S. Combined ECT and lithium therapy. Convulsive Ther 1993;9(4):274–284.

47. Klapheke MM. Combining ECT and antipsychotic agents: Benefits and risks. Convulsive Ther 1993;9:241–255.

48. Carr ME, Woods JW. Electroconvulsive therapy in a patient with unsuspected pheochromocytoma. Southern Med J 1985; 78(5):613–615.

49. Weiner R. Electroconvulsive therapy in the medical and neurologic patient. In: Stoudemire A, Fogel BS, editors. Psychiatric Care of the Medical Patient. New York: Oxford University Press; 1993.

50. Edwards RM, Stoudemire A, Vela MA, Morris R. Intraocular changes in nonglaucomatous patients undergoing electroconvulsive therapy. Convulsive Ther 1990;6(3):209–213.

51. Good MS, Dolenc TJ, Rasmussen KG. Electroconvulsive therapy in a patient with glaucoma. J ECT 2004;20(1):48–49.

52. Kellner C. Lessons from the methohexital shortage. J ECT 2003;19(3):127–128.

53. Stadtland C, Erhurth A, Ruta U, Michael N. A switch from propofol to etomidate during an ECT course increases EEG and motor seizure duration. J ECT 2002;18(1):22–25.

54. Folk JW, Kellner CH, Beale MD, Conroy JM, Duc TA. Anesthesia for electroconvulsive therapy: A review. J ECT 2000;16:157–170.

55. Ding Z, White PF. Anesthesia for electroconvulsive therapy. Anesthesia Analgesia 2002;94:1351–1364.

56. Bhat SK, Acosta D, Swartz C. Postictal systole during ECT. J ECT 2002;18(2):103–106.

57. Rasmussen KG, Jarvis MR, Zorumski CF, Ruwitch J, Best AM. Low-dose atropine in electroconvulsive therapy. J ECT 1999;15(3):213–221.

58. McCall WV. Antihypertensive medicines and ECT. Convulsive Ther 1993;9:317–325.

59. Nishihara F, Ohkawa M, Hiraoka H, Yuki N, Saito S. Benefits of the laryngeal mask for airway management during electroconvulsive therapy. J ECT 2003;19:211–216.

60. Brown NI, Mack PF, Mitera DM, Dhar P. Use of the Pro-Seal™ laryngeal mask airway in a pregnant patient with a difficult airway during electroconvulsive therapy. Br J Anaesthesia 2003;91:752–754.

61. Sackeim HA. The anticonvulsant hypothesis of the mechanism of action of ECT: Current status. J ECT 1999;15(1):5–26.

62. Sackeim HA, Long J, Luber B, Moeller J, Prohovnik I, Davanand DP, Nobler MS. Physical properties and quantification of the ECT stimulus: I. Basic principles. Convulsive Ther 1994;10(2):93–123.

63. Abrams R, Swartz CM. Thymatron® System IV Instruction Manual. 8th ed. Somatics, LLC; 2003.

64. Sackeim HA, Prudic J, Devanand DP, Kiersky JE, Fitzsimons L, Moody BJ, McElhiney MC, Coleman EA, Settembrino JM. Effects of stimulus intensity and electrode placement on the efficacy and cognitive effects of electroconvulsive therapy. N Engl J Med 1993;328:839–846.

65. Delva NJ, Brunet D, Hawken ER, Kesteven RM, Lawson JS, Lywood DW, Rodenburg M, Waldron JJ. Electrical dose and seizure threshold: Relations to clinical outcome and cognitive effects in bifrontal, bitemporal, and right unilateral ECT. J ECT 2000;16:361–369.

66. Bailine SH, Rifkin A, Kayne E, Selzer JA, Vital-Herne J, Blieka M, Pollack S. Comparison of bifrontal and bitemporal ECT for major depression. Am J Psychiatry 2000;157:121–123.

67. Letemendia FJJ, Delva NJ, Rodeburg M, Lawson JS, Inglis J, Waldron JJ, Lywood DW. Therapeutic advantage of bifrontal electrode placement in ECT. Psycholog Med 1993;23:349–360.

68. Lawson JS, Inglis J, Delva NJ, Rodenburg M, Waldron JJ, Letemendia FJJ. Electrode placement in ECT: Cognitive effects. Psycholog Med 1990;20:335–344.

69. Benbadis S. Epileptic seizures and epileptic syndromes. Neurolog Clin 2001;19(2):251–270.

70. Beyer JL, Weiner RD, Glenn MD. Electroconvulsive Therapy: A Programmed Text. 2nd ed. Washington (DC): American Psychiatric Press; 1985.

71. Petrides G, Fink M. The "half-age" stimulation strategy for ECT dosing. Convulsive Ther 1996;12:138–146.

72. Newman ME, Gur E, Shapira B, Lerer B. Neurochemical mechanisms of action of ECT: Evidence from in vivo studies. J ECT 1998;14(3):153–171.

73. Mann JJ. Neurobiological correlates of the antidepressant action of electroconvulsive therapy. J ECT 1998;14(3):172–180.

74. Sanacora G, Mason GF, Rothman DL, Hyder F, Ciarcia JJ, Ostroff RB, Berman RM, Krystal JH. Increased cortical GABA concentrations in depressed patients receiving ECT. Am J Psychiatry 2003;160(3):577–579.

75. Nobler MS, Teneback CC, Nahas Z, Bohning DE, Shastri A, Kozel FA, Goerge MS. Structural and functional neuroimaging of electroconvulsive therapy and transcranial magnetic stimulation. Depression Anxiety 2000;12(3):144–156.

76. Nobler MS, Oquendo MA, Kegeles LS, Malone KM, Campbell CC, Sackeim HA, Mann JJ. Decreased regional brain metabolism after ECT. Am J Psychiatry 2001;158(2):305–308.

77. Drevets WC. Functional neuroimaging studies of depression: The anatomy of melancholia. Annu Rev Med 1998;49:341–361.

78. Nobler MS, Sackeim HA, Prohovnik I, Moeller JR, Mukherjee S, Scnur DB, Prudic J, Devanand DP. Regional cerebral blood flow in mood disorders, III. Treatment and clinical response. Arch Gen Psychiatry 1994;51:884–897.

79. George MS, Nahas Z, Li X, Kozel FA, Anderson B, Yamanaka K, Chae J-H, Foust MJ. Novel treatments of mood disorders based on brain circuitry (ECT, MST, TMS, VNS, DBS). Sem Clin Neuropsychiatry 2002;7:293–304.

80. Duman RS, Vaidya VA. Molecular and cellular actions of chronic electroconvulsive seizures. J ECT 1998;14(3):181–193.

81. George MS, Nahas Z, Kozel FA, Li X, Denslow S, Yamanaka K, Mishory A, Foust MJ, Bohning DE. Mechanisms and state of the art of transcranial magnetic stimulation. J ECT 2002;18: 170–181.

Reading List

Abrams R. Electroconvulsive Therapy. 4th ed. New York: Oxford University Press; 2002.

Sackeim HA, Devanand DP, Nobler MS. Electroconvulsive therapy. In: Bloom FE, Kupfer DJ, editors. Psychopharmacology: The Fourth Generation of Progress. 4th ed. New York: Raven Press; 1995.

Rasmussen KG, Sampson SM, Rummans TA. Electroconvulsive therapy and newer modalities for the treatment of medication-refractory mental illness. Mayo Clin Proc 2002;77:552–556.

Dolenc TJ, Barnes RD, Hayes DL, Ramussen KG. Electroconvulsive therapy in patients with cardiac pacemakers and implantable cardioverter defibrillators. PACE 2004;27:1257–1263.

Electroconvulsive Therapy. NIH Consensus Statement, June 10–12, 1985, National Institutes of Health, Bethesda, MD Online. Available at http://odp.od.nih.gov/consensus/cons/051/051_statement.htm.

Information about ECT. New York State Office of Mental Health. Online. Available at http://www.omh.state.ny.us/omhweb/ect/index.htm. and http://www.omh.state.ny.us/omhweb/spansite/ect_sp.htm (Spanish version).

See also ELECTROENCEPHALOGRAPHY; REHABILITATION, COMPUTERS IN COGNITIVE.

ELECTRODES.   See BIOELECTRODES; CO_2 ELECTRODES.

ELECTROENCEPHALOGRAPHY

DAVID SHERMAN
DIRK WALTERSPACHER
The Johns Hopkins University
Baltimore, Maryland

INTRODUCTION

The following section will give an overview of the electroencephalogram (EEG), its origin, and its validity for diagnosis in clinical use. Since Berger (1) demonstrated in 1929 that the activity of the brain can be measured using external electrodes placed directly on the intact skull, the EEG has been used to study functional states of the brain. Although the EEG signal is the most common indicator for brain injuries and functional brain disturbances, the complicated underlying process, creating the signal, is still not well understood.

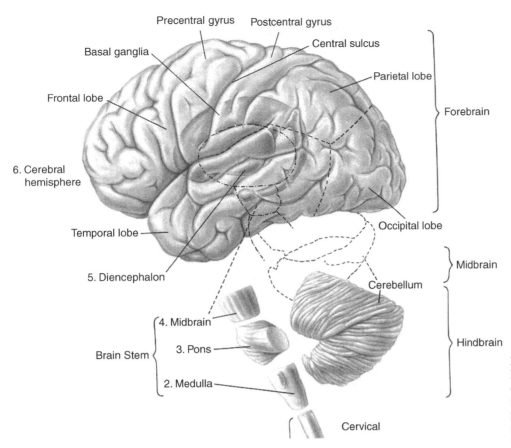

Figure 1. The major portions of the human cerebrum called lobes. Areas external to the cerebrum include the midbrain areas such as the diencephalon and the hindbrain areas such as the cerebellum, medulla, and pons. (Adapted from Ref. 5.)

The section is organized as follows. The biophysical basis of the origin of the EEG signal is described first, followed by EEG recordings and classification. Afterward, the validity and scientific basis for using the EEG signal as a tool for studying brain function and dysfunction is presented. Finally, logistical and technical considerations as they have to be made in measuring and analyzing biomedical signals are mentioned.

ORIGIN OF EEG

Valid clinical interpretation of the electroencephalogram ultimately rests on an understanding of the basic electrochemical and electrophysical processes through which the patterns are generated and the intimate nature of the brain's functional organization, at rest and in action. Therefore, the succeeding parts of the following discussion deal with the gross organization of the cortex, which is generally assumed to be the origin of brain electrical activity that is recorded from the surface of the head (2–4), the different kinds of electrical activity and resulting potential fields developed by cortical cells. Figure 1 shows the general organization of the human brain.

Organization of the Cerebral Cortex

Even though different regions of the cortex have different cytoarchitectures and each region has its own morphological patterns, aspects of intrinsic organization of the cortex are general (6,7). Most of the cortical cells are arranged in the form of columns, in which the neurons are distributed with the main axes of the dendritic trees parallel to each other and perpendicular to the cortical surface. This radial orientation is an important condition for the appearance of powerful dipoles. Figure 2 shows the schematic architecture of a cortical column. It can be observed that the cortex, and within any given column, consist of different layers. These layers are places of specialized cell structures and within places of different functions and different behaviors in electrical response. An area of very high activity is, for example, layer IV, which neurons function to distribute information locally to neurons located in the more superficial (or deeper) layers. Neurons in the superficial layers receive information from other regions of the cortex. Neurons in layers II, III, V, and VI serve to output the information from the cortex to deeper structures of the brain.

Activity of a Single Pyramidal Neuron

Pyramidal neurons constitute the largest and the most prevalent cells in the cerebral cortex. Large populations of these pyramidal neurons can be found in layers IV and V of the cortex. EEG potentials recorded from electrodes placed on the scalp represent the collective summation of changes in the extracellular potentials of pyramidal cells (2–4).

The pyramidal cell membrane is never completely at rest because it is continually influenced by activity arising in other neurons with which it has synaptic connections (8).

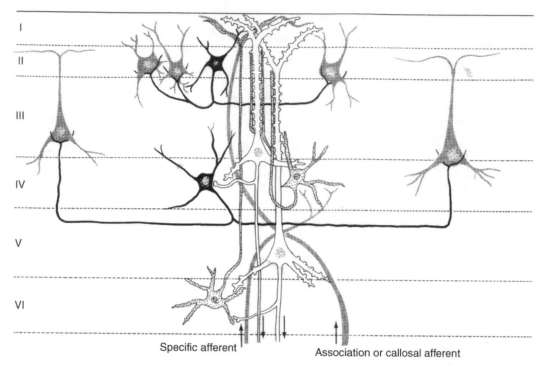

Specific afferent   Association or callosal afferent

Figure 2. Schematic of the different layers of the cerebral cortex. Pyramidal cells that are mainly in layers III and V are mainly responsible for the generation of the EEG. (Adapted from Ref. 5.)

Such synaptic connections may be excitatory or inhibitory and the respective transmitter changes the permeability of the membrane for K^+ (and/or Cl^-), which results in a flow of current (for details, please see Ref. 8).

The flow of current in response to an excitatory postsynaptic potential at the site on the apical dendrite of a cortical pyramidal neuron is shown in Fig. 3. The excitatory postsynaptic potential (EPSP) is associated with an inward current at the postsynaptic membrane carried by positive ions and an outward current along the large expanse of the extra-synaptic membrane. For simplicity, only one path of outward current is illustrated through the soma membrane. This current flow in the extracellular space causes the generation of a small potential due to extracellular resistance (shown by R in Fig. 3).

As an approximation it is possible to estimate the extracellular field potential as a function of the transmembrane potential (9)

$$\phi_e = \frac{a^2 \sigma_i}{4\sigma_e} \int \frac{\partial^2 v_m / \partial x^2}{r} dx \qquad (1)$$

where ϕ_e is the extracellular potential, a is the radius of axon or dendrites, v_m is the transmembrane potential, σ_i, is the intracellular conductance, and σ_e is the extracellular conductance. For a derivation of the above mentioned equation, please see Ref. 9.

Although these extracellular potentials individually are small, their sum becomes significant when added over many of cells. This is because the pyramidal neurons are more or less simultaneously activated by this way of synaptic connections and the longitudinal components of their extracellular currents will add, whereas their transversal components will tend to cancel out.

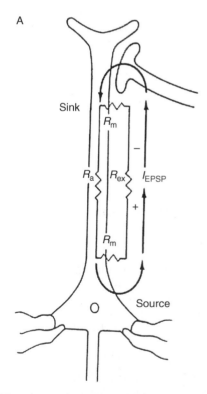

Figure 3. The current flow within a large pyramidal cell. Ionic flow is established to enable charge balance. (Adapted from Ref. 2.)

Generation of EEG Potentials

The bulk of the gross potentials recorded from the surface of the scalp results from the extracellular current flow associated with summated postsynaptic potentials in synchronously activated vertically oriented cortical pyramidal neurons. The exact configuration of the gross potential is related in a complex fashion to the site and the polarity of the postsynaptic potentials. Considering two cortical pyramidal neurons (shown in Fig. 3), the potential measured by a microelectrode at the location P is given by (4)

$$\Phi_p(t) = \frac{1}{4\pi\sigma}\left[\frac{I_a}{R_1}\cos(2\pi f_a t + \alpha_a) - \frac{I_a}{R_2}\cos(2\pi f_a t + \alpha_a) \right.$$
$$\left. + \frac{I_b}{R_3}\cos(2\pi f_b t + \alpha_b) - \frac{I_b}{R_4}\cos(2\pi f_b t + \alpha_b) \right]$$
$$+ \text{ Similar contributions from other dipoles} \qquad (2)$$

where I_a and I_b are the peak magnitudes of current for each dipole with phases α_a and α_b, respectively, and is the conductivity of the medium. "Similar contribution from other dipoles" refers to the contribution of dipoles other than the two shown, most significantly from the dipoles physically close to the electrode. Dipoles that are farther from the recording electrode but in synchrony ($\alpha_a = \alpha_b = \ldots = \alpha$), and aligned in parallel, contribute an average potential proportional to the number of synchronous dipoles $|\overline{\Phi}_p| \approx m$. Dipoles that are either randomly oriented or with a random phase distribution contribute an average potential proportional to the square root of their number $\overline{\Phi}_p| \cong \sqrt{m}$. Thus, the potential measured by a microelectrode can be expressed by the following approximate relation (4):

$$\overline{\Phi}_p| \sim \frac{1}{8\pi\sigma}\left[\sum_{i=1}^{l} \frac{I_i}{R_i} + m\frac{I_s d}{R_s^2} + \sqrt{n}\frac{I_a d}{R_a^2} \right] \qquad (3)$$

Here the subscripts i s, and a refer to local, remote synchronous, and remote asynchronous dipoles. I_s And I_a are the effective currents for remote synchronous and remote asynchronous sources, which may be less than the total current, depending on the orientation of the sources with respect to the electrode. Also l, m, and n are the numbers of local, remote synchronous, and remote asynchronous sources, which are located at average distances R_i, R_s, and R_a respectively. Note that a microelectrode like the scalp electrode used for EEG recordings measures a field averaged over a volume large enough to contain perhaps 10^7–10^9 neurons.

Contribution of Other Sources

A decrease in the membrane potential to a critical level of approximately 10 mV less than its resting state (depolarization) initiates a process that is manifested by the action potential (10). Although it might seem that action potentials traveling in the cortical neurons are a source of EEG, they contribute little to surface cortical records, because they usually occur asynchronously in time in large numbers of axons, which run in many directions relative to the surface. Other reasons are that the piece of membrane that is depolarized by an action potential at any instant of time

is small in comparison with the portion of membrane activated by an EPSP and that action potentials are of short durations (1–2 ms) in comparison with the duration of EPSPs or IPSPs (10–250 ms). Thus, their net influence on potential at the surface is negligible. An exception occurs in the case of a response evoked by the simultaneous stimulation of a cortical input (11), which is generally called a compound action potential.

Other cells in the cortex like the glial cells are unlikely to contribute substantially to surface records because of their irregular geometric organization, such that produced fields of current flow sum to a small value when viewed from a relatively great distance on the surface. There is also activity in deep subcortical areas, but the resulting potentials are too much attenuated at the surface to be recordable.

The discussed principles show that the surface-recorded EEG can be observed as the result of many active elements, where the postsynaptic potentials from cortical pyramidal cells are the dominant source.

Volume Conduction of EEG

Recording from the scalp has the disadvantage that there are various layers with different conductivities between the electrode and the area of cortical potential under the electrode. Therefore, potentials recorded at the scalp are not only influenced by the above mentioned patterns, but also by regions with different conductivities. Layers lying around the brain are such regions. These include the cerebrospinal fluid (CSF), the skull, and the scalp. These layers account, at least in part, for the attenuation of EEG signals measured at the surface of the scalp, as compared with those recorded with a microelectrode at the underlying cortical surface or with a grid of electrodes directly attached to the cortex (ECoG). These shells surrounding the brain account for an attenuation factor of 10 to 20 (12). This attenuation mainly affects the high-frequency–low-voltage component (the frequency range above 40 Hz), which has been shown to carry important information about the functional state of the brain, but it is almost totally suppressed at the surface.

EEG SIGNAL

The EEG signal consists of spontaneous potential fluctuations that also appear without a sensory input. It seems to be a stochastic signal, but it is also composed of quasi-sinusoidal rhythms. The synchrony of cerebral rhythms may occur from pacemaker centers in deeper cortical layers like the thalamus or in subcortical regions, acting through diffuse synaptic linkages, reverberatory circuits incorporating axonal pathways with extensive ramifications, or electrical coupling of neuronal elements (13). The range of amplitudes is normally from 10 to 150 μV, when recorded from electrodes attached to the scalp. The EEG signal consists of a clinical relevant frequency range of 0.5–50 Hz (10).

EEG Categories

Categorizing EEG signals into waves of a certain frequency range has been used since the discovery of the electrical

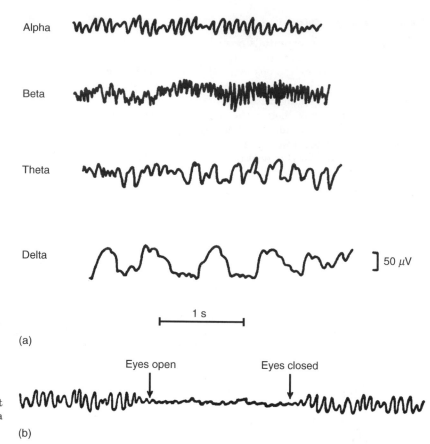

Figure 4. (a) Examples of EEG from different frequency bands. (b) The phenomenon of alpha desynchronization. (From Ref. 14).

activity of the brain. Therefore, these different frequency bands have been the most common feature in EEG analysis. Although this feature contains a lot of useful information as presented below, its use is not with criticism. It can be observed that there is some physiological and statistical evidence for the independence of these bands, but the exact boundaries vary between people and change with the behavioral state of each person (age, mental state, etc.). In particular, between human EEG and EEG signals recorded from different species of animals one can find different EEG patterns and frequency ranges. Nevertheless, the different frequency bands for human EEG are described below, because of their importance as an EEG feature. Most of the patterns observed in human EEG could be classified into one of the following bands or ranges:

Delta	below 3.5 Hz (usually 0.1–3.5 Hz)
Theta	4–7.5 Hz
Alpha	8–13 Hz
Beta	usually 14–22 Hz

A typical plot of these frequency bands is shown in Fig. 4, and the different bands are described below.

Delta Waves. The appearance of delta waves is normal in neonatal and infants' EEGs and during sleep stages in adult EEGs. When slow activity such as the delta band appears by itself, it indicates cerebral injury in the waking adult EEG. Dominance of delta waves in animals that have had subcortical transections producing a functional separation of cerebral cortex from deeper brain regions suggests that these waves originate solely within the cortex, independent of any activity in the deeper brain regions.

Theta Waves. This frequency band was included in the delta range until Walter and Dovey (15) felt that an intermediate band should be established. The term "theta" was chosen to allude to its presumed thalamic origin.

Theta frequencies play a dominant role in infancy and childhood. The normal EEG of a waking adult contains only a small amount of theta frequencies, mostly ovserved in states of drowsiness and sleep. Larger contingents of theta activity in the waking adult are abnormal and are caused by various forms of pathology.

Alpha Waves. These rhythmic waves are clearly a manifestation of the posterior half of the head and are usually found over occipital and parietal regions. These waves are best observed under conditions of awakeness but during physical relaxation and relative mental inactivity. The posterior alpha rhythm can be temporarily blocked by mental activities, or afferent stimuli such as influx of light while eye opening (Fig. 4b). This alpha blocking response was discovered by Berger in 1929 (1). Mainly thalamocortical feedback loops are believed to play a significant role in the generation of the alpha rhythm (16).

Beta Waves. Beta activity is found in almost every healthy adult and is encountered chiefly over the frontal and central regions of the cortex. The voltage is much lower than in alpha activity (seldom exceeds 30 μV). Beta

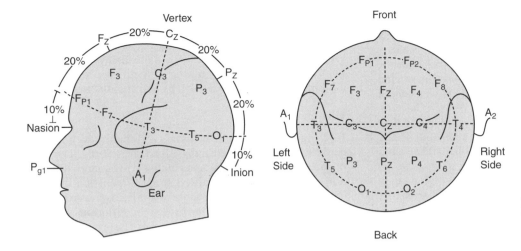

Figure 5. (a)and (b) The side and top view of the layout of a standardized 10–20 electrode system. Adapted from Ref. 14.

activity shows considerable increase in quantity and voltage after the administration of barbiturates, some non-barbituric sedatives, and minor tranquilizers. It also appears during intense mental activity and tension.

Clinical EEG

To obtain EEG recordings, there are several standardized systems for electrode placement on the skull. The most common are those of the standard 10–20 system of the International EEG Federation, which uses 30 electrodes placed on four landmarks of the skull as observed in Fig. 5.

It is now possible to obtain unipolar (or monopolar) and bipolar derivations from these electrodes. Using a bipolar derivation, one channel is connected between a pair of electrodes and the resultant difference in the potential between these two electrodes is recorded. Therefore, bipolar derivations give an indication of the potential gradient between two cerebral areas. Unipolar (monopolar) derivations can either be obtained by recording the potential-difference between the "active" electrodes and one "indifferent" electrode, placed elsewhere on the head (ear, nose), or with respect to an average reference, by connecting all other leads through equal-valued resistances (e.g., 1 MΩ) to a common point (17). The advantages of unipolar derivations are that the amplitude of each deflection is proportional to the magnitude of the potential change that causes it and the demonstration of small time differences between the occurrence of a widespread discharge at several electrodes. Small, nonpolarizable, disk Ag-AgCl electrodes are used together with an electrode paste. Recorded potentials are amplified using a high gain, differential, capacitivly coupled amplifier. The output signals are displayed on a chart recorder or a monitor screen. For more details about unipolar or bipolar derivations and EEG-amplifiers, please see (Ref. 11).

SCIENTIFIC BASIS FOR EEG MONITORING

The scientific basis for using EEG as a tool for studying brain function and dysfunction rests on the following four neurobiologic qualities of EEG.

Link With Cerebral Metabolism

The above presented discussion on the origin of potential-differences, recorded from the brain, shows that the EEG can be observed as a result of the synaptic and cellular activity of cortical pyramidal neurons. These neuronal and synaptic activities are directly linked to cerebral metabolism. Cerebral metabolic activity in turn depends on multiple factors including enzyme synthesis, substrate phosphorylation, axonal transport, and adenosine triphosphate (ATP) production from mitochondrial and glytolytic pathways (18). Thus, the EEG is a composite phenomenon reflecting complicated intracellular, intraneuronal, and neuro-glial influences. Although this multifaceted system makes it obvious that a selection of any single mechanism underlying the electrocortical manifestations may not be possible, the EEG is still a highly sensitive indicator of cerebral function (19).

Sensitivity to Most Common Causes of Cerebral Injury

The most common causes of cerebral injury are hypoxia and ischemia. It can be observed that hypoxia-ischemia causes a severe neuronal dropout in the cortical layers 3 and 5, leading to well-known hisyophatologic patterns of laminar necrosis. As pyramidal neurons that occupy the cortical layers are the main source for EEG, this loss of neuronal activity changes the cortical potentials and therefore makes EEG very sensitive to these common insults.

Correlation With Cerebral Topography

The standardized systems for electrode placement (Jung, international 10–20 system, etc.) establish a consistent relationship between electrode placement and underlying cerebral topography (20). Therefore, changes in EEG recorded from these electrodes of different areas of the skull reflect a consistent topographical relationship with underlying cerebral structures and allows useful inferences about disease localization from abnormalities in EEG detected at the scalp.

Ability to Detect Dysfunctions at a Reversible Stage

Heuser and Guggenberger (21) showed in 1985 that EEG deteriorates before the disruption of neuronal membrane and before significant reduction of cellular ATP levels. Siesjo and Wieloch (22) demonstrated in 1985 that during cerebral ischemia, changes in EEG correlate with elevated tissue lactate levels while ATP levels remain normal. Astrup (23) showed that a reduction in cerebral blood flow (CBF) affects EEG much before it causes neuronal death. These and several other reports make it clear that EEG offers the ability to detect injury at a reversible stage.

EEG also allows a prediction of recovery after brain dysfunctions like cerebral ischemia after cardiac arrest (24). Various studies in this report show that EEG recordings at several stages during recovery allow the prediction of whether the patient has a favorable outcome. Goel (25) showed that parameters obtained from EEG recordings may serve as an indicator for the outcome after hypoxic-asphyxic encephalophaty (HAE). These attributes make the EEG a very attractive neurologic observational tool.

LOGISTICAL AND TECHNICAL CONSIDERATIONS

Although the last sections have shown that EEG signal detection may serve as an important indicator for detection of neurological status and disorders, its clinical use and acceptance under a neurological care regime is limited. This is due to the complicated nature of the EEG signal and because of the difficulties regarding the interpretation of the signals. Some challenges of EEG analysis are as follows.

Artifacts

Recordings of physiological signals, especially from the surface of the body, have the problem that they are superimposed or distorted by artifacts. EEG signals are especially prone to artifact distortions due to their weak character. Therefore, a knowledge about the possible sources of distortion is necessary to estimate the signal-to-noise ratio (SNR). These artifacts are mostly generated from various kinds of sources. The sources of artifacts can be devided into two mayor groups: the subject-generated artifacts and the artifacts generated by the equipment. Subject-generated artifacts include EMG artifacts like body movement, muscle contraction of the neck, chewing, swallowing, coughing, involuntary movements (like myoclonic jerks, palotal myoclonus, nystagmus, asymmetric oculomotor paralysis, and decerebrate or decorticate posturing), and eye movements. Scalp edema can produce artifactual reductions in amplitude regionally or over an entire hemisphere. Pulse and EKG could also contribute as artifacts in EEG.

Artifacts generated by the equipment include ventilator artifacts that typically appear as slow wave like activity, vibrations of electrical circuitry around the subject, and power line interference. By taking the appropriate methods, a lot of these artifacts can be prevented [for more details, see Mayer-Waarden (10)].

Another method, to eliminate both artifacts generated by the equipment and the subject, is to use a differential amplifier for the recording between two electrodes. The assumption here is that the transmission time of artifacts between two electrodes can be neglected, and therefore, the artifacts at both electrodes are in-phase. The signal to be recorded is assumed to have a time delay from one to another electrode, and taking the difference therefore eliminates the artifact but keeps the signals' nature.

Inter-User Variability

Interpreting physiological signals is difficult, even for specialists, because of their subject-specific nature. The complicated nature of EEG signals makes it even more difficult, and data generated by different EEG analysis methods (especially techniques like feature analysis, power spectral analysis, etc.) may be interpreted in different ways by different analysts. An analysis of inter-user-variability of clinical EEG interpretation was presented by Williams et al. (26), which showed that even EEG data interpretation by EEG analysts could be different. Therefore, more standardized and objective methods of EEG analysis are extremely desirable.

Inter-Individual Variability

As mentioned, the consistency of the human EEG is influenced by many parameters and makes EEG unique for a certain person and for a specific point-in-time. Intrinsic parameters are the age and the mental state of the subject (degree of wakefulness, level of vigilance), the region of the brain, hereditary factors, and influences on the brain (injuries, functional disturbances, diseases, stimuli, chemical influences, drugs, etc.). To detect deviations from "normal" EEG, it would be necessary to compare this "abnormal" EEG with the "normal" EEG as a reference. Therefore, attempts have been made to obtain normative EEG for each of the classes discussed above (i.e, normative EEG for various age groups, normative EEG under the influence of varying amounts of different drugs, etc.), but these databases of normative data are still not sufficient to cover the variety of situations possible in real-life recordings. On the other hand, these normative data vary too much for considering them as one person's "normal" EEG.

Labor-Intensive and Storage Problems

For patient monitoring in the operating room or for chronic monitoring tasks that are necessary for cases of gradual insults and injuries, the EEG recordings can be extremely labor intensive. This makes it necessary to have either efficient means of collecting, storing, and displaying the long-term recordings or to come up with new techniques of compressing the EEG data, while preserving its characteristic features. Better methods to overcome these problems have been developed in both directions, although primarily toward efficient storage and display techniques. Methods of compressed spectral array representation overcome the limitation of compressed display to a great extent.

DISCUSSION

The review presented above emphasizes that the EEG is sensitive to different states of the brain and therefore may serve as a useful tool for neurological monitoring of brain function and dysfunction. In various clinical cases, the EEG has been used to observe patients and to make critical decisions. Nevertheless, its complicated nature and difficulty of interpretation has limited its clinical use. The following section should give an overview of the common techniques in analyzing EEG signals.

TECHNIQUES OF EEG ANALYSIS

Introduction

The cases presented above show that EEG has significant clinical relevance in detecting several diseases as well as in different stages of recovery. Still the complex nature of EEG has so far restricted its use in many clinical situations. The following discussion should give an overview of the state-of-the-art EEG monitoring and analysis techniques. The presented EEG analysis methods are divided into two basic categories, parametric and nonparametric, respectively, assuming that such a division is conceptually more correct than the more common differentiation between frequency and time-domain methods because they represent two different ways of describing the same phenomenon.

NonParametric Methods

In most of these analysis methods, the statistical properties of EEG signals are considered realizations of a Gaussian random process. Thus, the statistics of an EEG signal can be described by the first-and second-order moments.

These nonparametric time-domain and frequency-domain methods have been the most common way in analyzing EEG signals. In the following description of the different methods, it is also mentioned whether the technique is still being used in clinical settings.

Clinical Inspection. The most prevalent method of clinically analyzing the EEG is the visual inspection of chart records obtained from EEG machines. It uses the features observed in real-time EEG (like low frequency-high amplitude activity, burst suppression activity, etc.) for diagnostic and prognostic purposes. Several typical deviations from the normal EEG are related to different stages of the brain. This method suffers from the limitations like inter-user variability, labor intensiveness, and storage problems. A detailed description of logistical and technical considerations faced with EEG analysis is given later in this section.

Amplitude Distribution. Amplitude distribution is based on the fact that a random signal can be characterized by the distribution of its amplitude and accompanying mean, variance, and higher order moments. It can be observed that the amplitude distribution of an EEG signal most of the time can be considered as Gaussian (27) and the deviations from Gaussianity and its time-varying properties have been clinically used to detect and analyze different sleep stages (28,29). This method is now less popular because of more powerful and sophisticated EEG analysis techniques.

Interval Distribution. This is one of the earliest methods of quantitating the EEG (30). The method is based on measuring the distribution of intervals between either zero or other level crossings, or between maxima and minima. Often, the level crossings of the EEG's first and second derivatives are also computed to obtain more information about the spectral properties of the signal. Due to its ease of computation, the method has been shown to be useful in monitoring long-term EEG changes during anesthesia or sleep stages. Although simple, some theoretical problems are associated with this technique: It is extremely sensitive to high-frequency noise in the estimation of zero crossings and to minor changes in EEG. Also the zero crossing frequency (ZXF), the number of times the EEG signal crosses the zero voltage line, is not unique to a given waveform. Very different waveforms could give rise to the same ZXF. Despite the limitations, modified versions of period analysis are still used for clinical applications (30).

Interval-Amplitude Analysis. Interval-amplitude analysis is the method by which the EEG decomposed in waves or half-waves, both defined in time, by the interval between zero crossings, and in amplitude by the peak-to-through amplitudes. The amplitude and the interval duration of a half-wave are defined by the peak through differences in amplitude and time; the amplitude and the interval duration of a wave are defined by the mean amplitude and the sum of the interval durations of two consecutive half-waves (31,32). This method has been used clinically for sleep monitoring and depth of anesthesia studies (33).

Correlation Analysis. The computation of correlation functions constituted the forerunner of contemporary spectral analysis of EEG signals (34,35). The correlation function for random data describes the general dependence of the values of the data at one time on the values of the same data in the case of autocorrelation analysis (or of different data in the case of cross-correlation analysis) at another time. The cross-correlation between two signals x and y is defined as

$$\Phi_{xy}(\tau) := E\{x(t)y(t+\tau)\} \qquad (4)$$

where τ is the lag time (note that $\Phi_{xy}(\tau)$ becomes the autocorrelation function for $x = y$ and it can be estimated for discrete data by

$$\hat{\Phi}_{xy}(m) = \frac{1}{N-|m|} \sum_{n=0}^{N-|m|-1} x(n)y(n+m),$$
$$m \, \varepsilon \{0, 1, 2, \ldots, M << N\} \qquad (5)$$

and m is the lag number, M is the maximum lag number and $\hat{\Phi}_{xy}(m)$ is the estimate of the correlation function at lag number m. This estimation is unbiased but not consistent (36). The following modifications of this method have been used clinically:

1. Polarity coincidence correlation function: In this method, the signals are replaced by their signum equivalents, where $\text{sign}[x(t)] = +1$ for $x(t) > 0$ and $\text{sign}[x(t)] = -1$ for $x(t) < 0$. This modification achieves

computational simplification and has been shown (37) to be useful for EEG analysis.

2. Auto- or cross-averaging: This method consists of making pulses at a certain phase of the EEG (e.g., zero-crossing, peak, or through) that are then used to trigger a device that averages the same signal (auto-averaging) or another signal (cross-averaging). In this way, rhythmic EEG phenomena can be detected (38,39).

3. Complex demodulation: This method is related to correlation functions and allows one to detect a particular frequency component and to follow it over time. This is done by multiplying EEG signals with a sine wave of a desired frequency to give a product at 0 Hz. The 0 Hz component is then retained using a low-pass filter, obtaining the frequency component of interest. This method has been used to analyze visual potentials (40) and sleep spindles (41). However, correlation analysis has lost much of its attractiveness for EEG analysis since the advent of the Fourier transformation (FT) computation of power spectra.

Power Spectra Analysis. The principal application for a power spectral density function measurement of physical data is to establish the frequency composition of the data, which in turn bears an important relationship to the basic characteristics of the physical or biological system involved. The power spectrum provides a statement of the average distribution of power of a signal with respect to frequency. The FT serves as a bridge between the time domain and the frequency domain by identifying the frequency components that make up a continous waveform. An equivalent of the FT for discrete time signals is the discrete Fourier transform (DFT), which is given by

$$X(\omega) = \sum_{n=-\infty}^{+\infty} x(n)\exp(-j\omega n) \qquad (6)$$

An approximation of this DFT can be easily computed using an algorithm, developed in 1965 by Cooley and Tukey (42) and known as the fast Fourier transform (FFT).

An estimation of the power spectrum can now be obtained by Fourier-transforming (using FFT/DFT) either the estimation of the autocorrelation function, as developed in the previous section (Correlogram), or the signal and calculating the square of the magnitude of the result (Periodogram). Many modifications of these methods have been developed to obtain unbiased and consistent estimates (for details, please see Ref. 43). One estimator for the power spectrum, developed by Welch (44), will be used in the last section of this work.

Based on the frequency contents, human EEG has been classified into different frequency bands, as described. Correlations of normal function as well as dysfunctions of the brain have been made with the properties (frequency content, powers) of these bands. Time-varying power spectra have also been used to analyze time variations in EEG frequency properties (45). One of the main advantages of this kind of analysis is that it retains almost all the information content of EEG, while separating out the low-frequency artifacts into a small band of frequencies.

On the other hand, it suffers from some of the limitations of feature analysis, namely, inter-user variability, labor intensiveness, and storage problems. There have been attempts to reduce the labor intensiveness by creating displays like linear display of spectral analysis and grays-cale display of spectral analysis (30), which compromises the amount of information presented.

Cross-Spectral Analysis. This kind of analysis allows quantification of the relationship between different EEG signals. The cross-power spectrum $\{P_{xy}(f)\}$ is the product of the smoothed DFT of one signal and the complex conjugate of the other [see for details, Jenkins and Watts (46)]. As $P_{xy}(f)$ is a complex quantity, it has a magnitude and phase and can be written as

$$P_{xy}(f) = |P_{xy}(f)|\exp[j\phi_{xy}(f)] \qquad (7)$$

where $j = \text{sqrt}(-1)$, and $\phi_{xy}(f)$ is the phase spectrum. With the cross-power spectrum, a normalized quantity, the coherence function, can be defined as follows:

$$\text{coh}_{xy}(f) = \frac{|P_{xy}(f)|^2}{P_{xy}(f)P_{yy}(f)} \qquad (8)$$

where $P_{xx}(f)$ and $P_{yy}(f)$ are the autospectral densities of $x(t)$ and $y(t)$. The spectral coherence can be observed as a measurement of the degree of the "phase synchrony" or "shared activity" between spatially separated generators. Therefore, unity in this quantity indicates a complete linear relationship between two electrode sites, whereas a low value for the coherence function may indicate that the two EEG locations are connected via a nonlinear pathway and that they are statistically mostly independent.

Coherence functions have been used in several investigations of the EEG signal generation and their relation to brain functions, including studies of hippocampal theta rhythms (47), on limbic structures in humans (48), on thalamic and cortical alpha rhythms (49), on sleep stages in humans (29), and in EEG development in babies (50).

A more generalized form of coherence is the so called "spectral regression-amount of information analysis" [introduced and first applied to EEG analysis by Gersch and Goddard (51)] which expresses the linear relationship that remains between two time series after the influence of a third time series has been removed by a partial regression analysis. If the initial coherence decreases significantly, one can conclude that the coherence between the two initially chosen signals is due to the effect of the third one. The partial coherence between the signals x and z, when the influence of y is eliminated, can be derived from

$$P_{zz,y}(f) = P_{zz}(f)(1 - \text{coh}_{zy}(f)) \qquad (8)$$

and

$$P_{xx,y}(f) = P_{xx}(f)(1 - \text{coh}_{xy}(f)) \qquad (9)$$

$P_{xz,y}(f)$ is the conditioned cross-spectral density and can be calculated as

$$P_{xz,y}(f) = P_{xz}(f)(1 - \frac{P_{xy}(f)P_{yz}(f)}{P_{yy}(f)P_{xz}(f)}) \qquad (10)$$

This method has been mainly used to identify the source of EEG seizure activity (51,52).

Bispectrum Analysis. The power spectrum essentially contains the same information as autocorrelation and hence provides a complete statistical description of a process only if it is Gaussian. In cases where the process is non-Gaussian or is generated by nonlinear mechanisms, higher order spectra defined in terms of higher order moments or cumulants provide additional information that cannot be obtained from the power spectrum (e.g., phase relations between frequency components). There are situations, due to quadratic nonlinearity, in which phase coupling between two frequency components of a process results in a contribution to the power at a frequency equal to their sum. Such coupling affects the third moment sequence, and hence, the bispectrum is used in detecting such nonlinear effects. Although used in experimental settings (53,54), bispectral analysis techniques have not yet been used in clinical settings, probably due to both the complexity of the analysis and the difficulty in interpreting results.

The bispectrum of a third-order stationary process can be estimated by smothing the triple product

$$B(f_1, f_2) = E\{F_{xx}(f_1)F_{xx}(f_2)^*F_{xx}(f_1 + f_2)\} \quad (11)$$

where $F_{xx}(f)$ represents the complex FT of the signal and $F_{xx}(f)^*$ is the complex conjugate of $F_{xx}(f)$ [for details, please see Huber et al. (55) and Dumermuth et al. (56)].

Hjorth Slope Descriptors. Hjorth (57) developed the following parameters, also called descriptors, to quantify the statistical properties of a time series:

$$\text{activity,} \quad A = a_0$$

$$\text{mobility,} \quad M = \left[\left(\frac{a_2}{a_0}\right)\right]^{\frac{1}{2}}$$

$$\text{complexity,} \quad C = \left[\left(\frac{a_4}{a_2}\right) - \left(\frac{a_2}{a_0}\right)\right]^{\frac{1}{2}}$$

where

$$a_n = \int_{-\infty}^{+\infty} (2\pi f)^n S_{xx}\,(df)$$

Note here that a_0 is the variance of the signal ($a_0 = \sigma^2$), a_2 is the variance of the first derivative of the signal, and a_4 is the variance of the signal's second derivative. Hjorth also developed a special hardware for real-time computation of these three spectral moments, which allows the spectral moments to vary as a function of time. Therefore, this form of analysis can be applied to nonstationary signals, and it has been used in sleep monitoring (58) and in quantifying multichannel EEG recordings (59). It should be noted that Hjorth's descriptors give a valid description of an EEG signal only if the signals have a symmetric probability density function with only one maximum. As this assumption cannot be made in general practice, the use of the descriptors is limited.

Parametric Methods

The motivation for parametric models of random processes is the ability to achieve better power spectrum density (PSD) estimators based on the model, than produced by classical spectral estimators. In the last section, the PSD was defined as the FT of an infinite autocorrelation sequence (ACS). This relationship may be considered as a nonparametric description of the second-order statistics of a random process. A parametric description of the second-order statistic may also be devised by assuming a time-series model of the random process. The PSD of the time-series model will then be a function of the model parameters (and not of the ACS). A special class of models, driven by white noise processes and processing rational system functions, is the autoregressive (AR), the moving average (MA), and the autoregressive moving average (ARMA) model.

One advantage of using parametric estimators is, for example, better spectral resolution. Periodogram and correlogram methods construct an estimate from a windowed set of data or ACS estimates. The unavailable data or unestimated ACS values outside the window are implicitly zero, which is an unrealistic assumption, that leads to distortions in the spectral estimate. Some knowledge about the process from which the data samples are taken is often available. This information may be used to construct a model that approximates the process that generated the observed time sequence. Such models will make more realistic assumptions about the data outside the window instead of the null data assumption. Thus, the need for window function can be eliminated. Therefore, a parametric PSD estimation method is useful in real-time estimation because a short data sequence is sufficient to determine the model. The following parametric approaches have been used to analyze EEG signals.

ARMA Model. The ARMA model is the generalized form of the AR and MA model, which represents the time series $x(n)$ in the following form:

$$x(n) + a(1)x(n-1) + a(2)x(n-2)\ldots + a(p)x(n-p)$$
$$= w(n) + b(1)w(n-1) + b(2)w(n-2)\ldots$$
$$+ b(q)w(n-q) \quad (12)$$

where $a(n)$ are the AR parameters, $b(n)$ are the MA parameters, $w(n)$ is the error in prediction, and p,q are the model orders for the AR and MA model, respectively.

The power spectrum $P_{xx}(z)$ of this time series $x(n)$ can be obtained by using the ARMA parameters in the following fashion:

$$P_{xx}(z) = |\frac{\sum_{i=0}^{q}1 + b(1)z^{-1} + b(2)z^{-2} + \ldots b(q)z^{-q}}{\sum_{i=0}^{p}1 + a(1)z^{-1} + a(2)z^{-2} + \ldots a(p)z^{-p}}|^2 W(z)$$
$$(13)$$

where $W(z)$ is the z-transform of $w(n)$. Note here that if we set all $b(q)$ equal to zero, we obtain an AR model, represented by poles close to the unit circle only and therefore an all-pole-system, and if we set all $a(p)$ equal to zero, we obtain an MA model. The ARMA spectrum can model both sharp peaks as they are obtained from an AR spectrum and deep nulls as they are typical for an MA spectrum (60). Although ARMA is a more generalized form of the AR model, in most EEG applications, it is sufficient to compute the AR model becuase EEG signals have been found to be represented effectively by such a model (45). The AR model will be described in more detail in the following section.

Inverse AR Filtering. Assuming that an EEG signal results from a stationary process, it is possible to approximate it as a filtered noise with a normal distribution. Consequently, passing such an EEG signal through the inverse of its estimated autoregressive filter could be performed to obtain the generator noise (also called the residues) of the signals, which is normally distributed with mean zero and variance σ^2. The deviation from a noise with a normal distribution can be used as an important tool to detect nonstationarity and nonlinearities in the original signal. This method has been used to detect transient nonstationarities present in epileptiform EEG (45).

Kalman Filtering. A method of analyzing time-varying signals consists of applying the so-called Kalman estimation method of tracking the parameters describing the signal (61,62). The Kalman filter recursively obtains estimates of the parametric model coefficients (such as those of an AR model) using earlier as well as current data. These data are weighted by the Kalman filter, depending on the signal-to-noise ratio (SNR) of the respective data. For the estimation of the parametric model, coefficients data with a high SNR are weighted higher than data with a lower SNR (37).

This method is not easy to implement due to its sensitivity to model order and initial conditions; it also tends to be computationally extensive. Despite these limitations, recursive Kalman filtering has been used in EEG analysis for deriving a measure of how stationary the signal is and for EEG segmentation signal (61,62). This segmentation of the EEG signal into quasi-stationary segments of variable length is necessary and useful in reducing data for the analysis of long EEG recordings under variable behavioral conditions. Adaptive segmentation based on Kalman filtering has been used to analyze a series of clinical EEGs to show a variety of normal and abnormal patterns (63).

BURST AND ANALYZING METHODS

Introduction

This article has shown that EEG signals are sensitive to various kinds of diseases and reflect different stages of the brain. Specific EEG patterns can be observed after ischemic brain damage and during deep levels of anesthesia with volatile anesthetics like enflurane, isoflurane, or babiturate anesthesia (64). The patterns are recognized as periods of electrical silence disrupted by bursts of high-voltage activity. This phenomenon has been known since Derbyshire et al. (65) showed that wave bursts separated by periods of electrical silence may appear under different anesthetics. The term "burst suppression" was introduced to describe the occurrence of alternating wave bursts and blackout sequences in narcotized animals (66), in the iso-

lated cerebral cortex (67), during coma with dissolution of cerebral functions (68), after drama associated with cerebral anoxia (69), and in the presence of a cortical tumor (70). Other bursting-like patterns in the EEG are seizures as they occur during epilepsy. Although also episodes of high voltage, the background EEG is not suppressed in the presence of seizures.

The knowledge about occurrence of these bursts and perods of electrical silence in the EEG is of important clinical value. Although burst suppression during anesthesia with modern anesthetics is reversible and harmless, it often is an ominous sign after brain damage (71). Frequently occurring seizures may indicate a severe injury state. Thus, it is of great interest to detect these burst and burst suppression sequences during surgery or in other clinical settings. We have already presented several methods to analyze EEG signals, their advantages and disadvantages. In the case of short episodes of burst suppression or spikes, however, methods that maintain the time-varying character of the raw EEG signal are necessary.

In this section, we want to present the mechanisms of the underlying processes, which cause burst-suppression or spiking. Methods that show the loss of EEG signal power during the occurrence of burst suppression and methods that can follow the time-varying character of the raw input signal are presented. Finally, we will present some methods that have been used to detect bursts and seizures based on detection of changes in the power of the signal.

Mechanisms of Bursts and Seizures

Bursts can be observed as abrupt changes in the activity of the entire cortex. These abrupt changes led to the assumption that a nonlinear (ON-OFF or bang-bang control system) inhibiting mechanism exists in the central nervous system (CNS) that inhibits the burst activity in the EEG. Recent studies confirm this theory and have shown that during burst-suppression, the heart rate also is decreased (72,73). At the end of the suppression, this inhibition is released abruptly, permitting burst activity in EEG and increase in heart rate. The task of such a control system in the CNS may be to decrease the chaotic activity in a possibly injured or intoxicated brain. As cortical energy consumption is correlated with the EEG, decreased cortical activity also avoids excessive, purposeless energy consumption (74). Studies on humans under isoflurane anesthesia have shown that increased burst-suppression after increased anesthesia concentration does correlate with cerebral oxygen consumption (75).

Another interesting observation is the quasi-sinusoidal character of the EEG signal during bursting. This has been shown by Gurvitch et al. (76) for the case of hypoxic and posthypoxic EEG signals in dogs. In contrast to anesthesia evoked bursts, which also contain higher frequency components up to 30 Hz, these hypoxic and posthypoxic bursts are high-voltage slow-wave signals, with frequency components in the delta range (77). Figure 6 shows two typical cortical EEG recordings from an isoflurane-anesthetized dog and a piglet after hypoxic insult. The power spectrum of the first burst in each recording is shown, respectively. Bispectral analysis as described has shown that there is

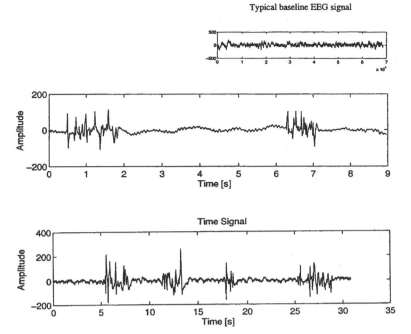

Typical baseline EEG signal

Figure 6. Burst suppresion under the inluence of isoflurane and after hypoxic insult. (a) Cortical EEG from a
dog anesthetized with isoflurane/DEX. (b) Cortical EEG
from a piglet during recovery from a hypoxic insult. Note
the similarities in electrically silent EEG interrupted by
high voltage EEG activity.

significant phase coupling during bursting (78). Due to
these observations, we assume the EEG signal to be qua-
siperiodic during bursting and seizure sequences. Note
that this observation is an important characteristic and
will be used in the next section as a basic assumption for
the use of an energy estimation algorithm.

The first cellular data on EEG burst suppression patterns were presented by Steriade et al. in 1994 (79). This
study examined the electrical activity in cells in the thalamus, the brain stem, and the cortex during burst suppression in anesthetized cats. They showed that although the
activity of intracellularly recorded cortical neurons matches
the cortical EEG recording, the recording from thalamic
neurons displays signs of activity during the periods of
electrical silence in the cortex and the brain stem. But it
has also been observed that the cortical neurons are not
unresponsive during periods of electrical silence. Thalamic
volleys delivered during the epochs of electrical silence were
able to elicit neuronal firing or subthresholding depolarizing
potentials as well as the revival of EEG activity. This
observation led to the assumption that full-blown burst
suppression is achieved through complete disconnection
within the prethalamic, thalamocortical, and corticothala-
mic brain circuits and indicates that, in some instances, a
few repetitive stimuli or even a single volley may be enough
to produce recovery from the blackout during burst suppression. Sites of disconnection throughout thalamocortical systems are mainly inhibited synaptic transmissions due to an
increase in GABAergic inhibitory processes at both thalamic
and cortical synapses. Note that we showed that postsynaptic extracellular potentials at cortical neurons are the origin
of the EEG signal. Therefore, this failure of synaptic transmission explains the flatness in the EEG during burst
suppression. The spontaneous recurrence of cyclic EEG
wave bursts may be observed as triggered by remnant
activities in different parts of the affected circuitry, mainly
in the dorsothalamic-RE thalamic network in which a sig-

nificant proportion of neurons remains active during burst
suppression. However, it is still unclear why this recovery is
transient and whether there is a real periodicity in the
reappearance of electrical activity. According to the state
of the whole system, the wave bursts may fade and be
replaced by electrical silence or may recover toward a
normal pettern.

Seizures are sudden disturbances of cerebral function.
The underlying causes of these disorders are heterogenous
and include head trauma, lesions, infections, and genetic
predisposition. The most common injury that causes seizures is epilepsy. Epileptic seizures are short, discrete
episodes of abnormal neuronal activity involving either a
localized area or the entire cerebrum. The abnormal time
series may demonstrate abrupt decreases in amplitude,
simple and complex periodic discharges, and transient
patterns such as spikes (80) and large amplitude bursts.

Generalized seizures can be experimentally induced by
either skull shocks to the animal or through numerous
chemical compounds like pentylenetetrazol (PTZ). Several
studies have shown that there are specific pathways
through which the seizure activity is mediated from deeper
cortical areas to the superficial cortex (81). Figure 7 shows a
cortical EEG recording from a PTZ-treated rat. Nonconvul-
sive seizures are not severe or dangerous . In contrast,
convulsive seizures like the seizures caused by epilepsy
might be life threatening, and a detection of these abnormalities in the EEG at an early stage of the insult is desirable.

Reasons for Burst and Seizure Detection

We have seen in the previous section that there are various
possible sources that can cause EEG abnormalities, like
seizures or burst suppression interrupted by spontaneous
activity outbreaks. In this section, now we want to describe
why it is of importance to detect these events. Reasons for
detecting bursts or seizures are as follows.

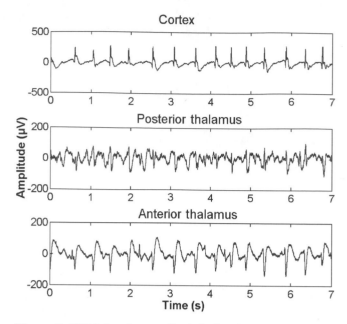

Figure 7. PTZ-induced generalized clonic seizure activity in the cortex and the thalamus of a rat. The figures from the top to the bottom show seizure activity recorded from the trans-cortex, the hippocampus, the posterior thalamus, and the anterior thalamus. Note the occurrence of spikes in the cortical recording before the onset of the seizure. At time point 40, one can see the onset of the seizure in the cortical recording, whereas the hippocampus shows increased activity already at time point 30. Such recordings can be used to study the origin and the pathways of seizures.

Confirmation of the Occurrence. In the case of seizures, it is obvious that it is desirable to detect these seizures as an indicator of possible brain injuries like epilepsy. Epileptic or convulsive seizures might be life-threatening, and detection at an early stage of the injury is necessary for medication. The frequency with which seizures occur in a patient is the basis on which the diagnosis of epilepsy is made. No diagnosis for epilepsy will be made based only on the occurrence of occasional seizures. Burst suppression under anesthesia is an indicator for the depth of anesthesia (82), and the relationship between the duration of burst suppression parts and bursting episodes is therefore desirable. In the case of hypoxia, however, burst suppression indicates a severe stage of oxygen deficiency and a possible risk of permanent brain damage or brain death. In the recovery period, in post-hypoxic analysis, the occurance of bursts might be of predictive value whether or not the patient has a good outcome (83–85).

To Further Analyze Seizure or Burst-Episodes. Not only the presence of bursts or seizures can serve as an physiological indicator, furthermore special features or characteristics of these EEG abnormalities are of importance. Intracranial EEG patterns at seizure onset have been found to correlate with specific pathology (86), and it has been suggested that the different morphologies of intracranial EEG seizure onset have different degrees of localizing value (87).

In the case of anesthesia or hypoxia-induced bursts, the duration of the bursts and the burst suppression parts may indicate the depth of anesthesia or the level of injury, respectively. Frequency or power analysis of these bursts may help discriminating these two kinds of bursts from one another (77). This is important, for example, in open heart surgery to detect reduced blood flow to the brain at a reversible stage.

Localization. Localization of the source of an injury or an unknown phenomenon is always desirable. This is valid especially in the case of epilepsy, where the injured, seizure-causing part of the brain can be operatively removed. Detecting the onsets of bursts or seizures in different channels from different regions of the brain may help us to localize the source of these events. In particular, recordings from different regions of the thalamus and the cortex have been used to study pathways of epileptic seizures.

Ability to Present Signal-Power Changes During Bursting

We have already mentioned that bursts can be observed as a sequence in the EEG signal with increased electrical activity and within sequences of increased power or energy. Therefore, looking at the power in the EEG signal can give us an idea about the presence of bursts and burst suppression episodes in the signal. Looking at the power in different frequencies of a signal is classically done by estimating the PSD. We will present three methods here that have already been used in EEG signal analysis. First is a method to estimate the PSD by averaging over a certain number of periodograms, which is known as the Welch method. After obtaining the power spectrum over the entire frequency range, the total power in some certain frequency bands can then be obtained by summing together the powers in the discrete frequencies that fall in this frequency band. For this method, the desired frequency bands have to be known in advance. One method to obtain the knowledge where the dominant frequencies may be found in the power spectrum is to model the EEG signal with an AR model. Beside the fact that this method calculates the dominant frequencies, we also obtain the power in these dominant frequencies and can use this method directly to follow the power in the dominant frequencies. The third method will be a method to perform time-frequency analysis as a method to obtain the energy of a signal as a function of time as well as a function of frequency. We will present the short-time Fourier transform (STFT) as such a time-frequency distribution. As mentioned, these methods have been already used in EEG signal processing.

Feature Extraction. One major problem with these methods is the large amount of data that become available. Therefore, attempts have been made to extract spectral parameters out of the power spectrum that for themselves contain enough necessary information about the nature of the original EEG signal. The classic division of the frequency domain in four major subbands (called alpha, beta, theta, and delta waves) as described in the first section, has been one possibility of feature extraction and data reduction. However, we have also observed that these subbands may vary among the population and the major frequency components of a human EEG might be different from the predominant frequency components of an EEG recorded from animals. Furthermore, some specific EEG changes typically involve an alteration or loss of power in specific frequency components of the EEG (88) or a shift in power over the frequency domain from one frequency range to another. This observation led to the assumption that the pre-devision of the frequency domain into four fixed subbands may not give features that are sensitive to such kinds of signal changes. We therefore propose the use of "dominant frequencies" as parameters; these are frequencies at which one can find a peak in the power spectrum, and therefore, these dominant frequencies can be observed as the frequencies with an increased activity. Recent studies (25) have shown that following the power in these dominant frequencies over time has a predictive value after certain brain injuries, whether or not the patient has a good outcome. In fact, detecting changes in power in dominant frequencies may be used as a method to visualize changes in the activity in certain frequency ranges. Another method to reduce some of the information of the power spectrum to one single value is to calculate the mean frequency of the spectrum at an instant point of time. This value can be used as a general indicator for changes in the power spectrum from one instant time point to another. Other spectral parameters that will not be described here are, for example, peak frequency or various different defined edge frequencies like the medium frequency. However, the effectiveness of these EEG parameters in detecting changes in the EEG, especially in detecting injury and the level at which they become sensitive to injury, has not been well defined. After the description of each method, we will present how we can obtain the power in the desired frequency bands and the mean frequency.

Power Spectrum Estimation Using the Welch-Method.
We have already observed the use of the FT and its discrete performance in the DFT in the first section. In this section, we now want to show how we can use the DFT to obtain an estimator for the PSD.

To estimate the PSD there are two classic possibilities. The first and most direct method is the periodogram built by using the discrete-time data sequence and transforming it with DFT/FFT. We describe the algorithm in detail:

$$I(N) = \frac{1}{N}|\sum_{n=1}^{n=N} x(n)\exp(-j\omega n)| \tag{14}$$

where $x(n)$ is the discrete time signal and N is the number of FFT points. It can be observed that this basic estimator is not statistically stable, which means that the estimation has a bias and is not consistent because the variance does not tend to be zero for large values of N. The second method to achieve the PSD estimation is more indirect, in which the autocorrelation function of the signal is estimated and transformed via DFT/FFT. This estimation is called a correlogram:

$$I_N(\omega) = \sum_{m=-(N-1)}^{N-1} \hat{\Phi}_{xx}\exp(-j\omega m) \tag{15a}$$

where $\hat{\Phi}_{xx}(m)$ is the estimated autocorrelation function of a time signal $x(n)$:

$$\hat{\Phi}_{xx}(m) = \frac{1}{N}\sum_{n=0}^{N-|m|-1} x(n)x(n+|m|) \tag{15b}$$

To avoid these disadvantages of the periodogram as an estimator for PSD, many variations of this estimator were developed, reducing the bias and variance of the estimation. The most popular method among these estimators is the method of Welch (44). The given time sequence is devided into k overlapping segments of L points each, and the segments are windowed and transformed via DFT/FFT. The estimator of the PSD is then obtained by the mean of these spectra. It can be observed that as more spectral samples are used to build this estimator, the more the variance is reduced. Assuming a given sequence length of N points, the variance of the estimate will decrease if the number of points in each segment decreases. Note that a decrease in number of points results in a loss of good spectral resolution. Therefore, a compromise has to be found to achieve a small variance and a sufficient spectral resolution. To increase the number of segments, which are used to build the mean, an overlap of the segments of 50% is used.

Use of a finite segment length, $n = 0,\ldots,N-1$, of the signal $x(n)$ for computation of the DFT is equivalent to multiplying the signal $x(n)$ by a rectangular window $w(n)$. Therefore, due to the filtering effects of the window function, sidelobe energy is generated where the spectrum is actually zero. The window function also causes some smoothing of the spectrum when N is sufficiently large. To reduce the amount of sidelobe leakage caused by windowing, a nonrectangular window that has smaller sidelobes may be used. Examples of such windows include the Blackman, Hamming, and Hanning windows. However, use of these windows for reduction of sidelobe leakage also causes an increase in smoothing of the spectrum. Figure 8 shows the difference of sidelobe leakage effects between a rectangular and a Blackman window. In our case, a Tukey window is used in respect to a sufficient suppression of sidelobes and to obtain sharp mainlobes at the containing frequencies (90):

$$u(x) = \begin{cases} 0.5(1-\cos(\pi x/d)) & 0 \le x \le d \\ 1 & d \le x \le 1-d \\ 0.5(1-\cos(\pi(1-x)/d)) & 1-d \le x \le 1 \end{cases}$$

The resultant estimator of PSD is obtained by using the equation:

$$\hat{S}_{xx}(\exp(j\Omega)) = \frac{1}{kA}\sum_{i=1}^{K}\frac{1}{L}\left|\sum_{k=0}^{L-1}x_i(k)\,f_l(k)\exp(-j\Omega k)\right|^2 \tag{16}$$

where k is the number of segments, L is the number of points in each segment, $f_L(k)$ is the data window function, and

$$A = \frac{1}{L}\sum_{k=0}^{L-1} f_L^2(k) \tag{17}$$

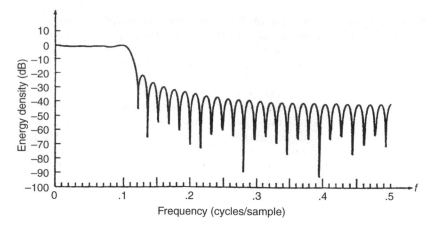

Figure 8. Comparison of sidelobe leakage effects in the spectrum using (a) rectangular window versus (b) Blackman window. Spectra are computed for a signal with (voltage spectrum $X(f) = 1$, abs(f) < 1; $X(f) = 0$ otherwise. The Blackman window reduces sidelobe effects and increases smoothing of the spectrum. (Adapted from Ref. 89.)

which is a factor to obtain an asymptotically unbiased estimation. Even if the PSD estimator (using the Welch method) is a consistent estimator, we have to note that it is only an approximation of the real PSD. Beside the above-mentioned limitations, unwanted effects also result from using DFT/FFT. These include aliasing, leakage, and the picket fence effect. Most of these effects may be avoided by using a window of appropriate characteristic and by fulfilling the Nyquist criterion, which is that the highest signal frequency component has to be less than one half the sampling frequency.

The FT and autocorrelation method can compute the power spectrum for a given segment of the EEG signal. The spectrum over this segment must therefore be assumed to be stationary. Loss of information will occur if the spectrum is changing over this segment, because temporal localization of spectral variations within the segment is not possible. Because burst suppression violates the assumptions (91) underlying power spectrum analysis and may cause misleading interpretations (92,93), it is necessary to increase time resolution. To track changes in the EEG spectrum over time, spectral anlysis can be performed on succesive short segments (or epochs) of data. Note that we used the spectral analysis of such epochs for our method above. We therefore may expect that the spectral analyses for the short segments are less consistent and that the effects of signal windowing will play a more important role.

Parameter Extraction. For selected sequences of EEG at each stage during a recording, spectral analysis might be

performed using the Welch method. To obtain the power in the dominant frequencies, the powers in the average power spectrum are summed together over the frequency range of interest. This summation is made because the dominant frequency may vary in a small frequency range:

$$P(f_d) = \sum_{k=n}^{n+1} S(k)n + \ell < N/2 \qquad (18)$$

where $S(k)$ is the average power spectrum, N is the FFT length, f_d is the dominant frequency, and $tN/2f_s$ is the bandwidth of the frequency band. Following the power in these specific frequency bands over time, we obtain a trend-plot of the power in different dominant frequency bands. This is shown in Fig. 9, where three sequences of 30 s are presented, which are recorded from a dog during different stages of isoflurane anesthesia. The dominant frequencies have been found using an AR model and are in the range of 0.5–5 Hz, 10–14.5 Hz, and 18–22.5 Hz. The recorded data are sampled with $f_s = 250$, and the sampled sequence is divided into segments of 128 points each with an overlap of 50%. The PSD estimator is obtained as described in Eq. 16.

The mean frequency (MF) of the power spectrum is computed from the following formula:

$$\text{MF} = \frac{\sum_{K=1}^{N/2} S(K)(KF_s/N)}{\sum_{K=1}^{N/2} S(K)} \qquad (19)$$

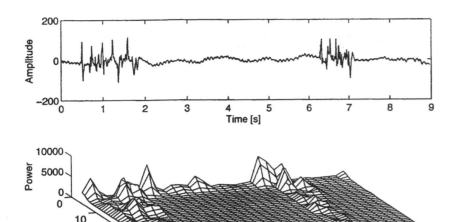

Figure 9. Trend in power in dominant frequency bands. The top row shows three input signals as they are obtained at different stages of anesthesia in a dog. The recordings show from left to right a baseline EEG followed by an EEG epoch obtained after administration of a sedative drug and finally the EEG after reversing the process. The three epochs were recorded in a distance of 1000 s to one another. The second row shows the averaged power spectra obtained with the Welch method, respectively. The data were sampled with $f = 250$ π and the FFT length is 128 points. The bottom row shows the trend in the three dominant frequency bands 0.5–5, 10–14.5, and 18–22.5 Hz. Note that this method does not provide the possibility to visualize the bursts in the EEG recording, but it can give a trend in power changes in dominant frequency bands.

where $S(K)$ is the average power spectrum, N is the FFT length, and F_s is the sampling frequency. The mean frequency can be observed as the frequency instant at which one can find the "center of mass" in the power spectrum.

Short-Time Spectral Analysis

The Algorithm. The STFT is one of the most used time-frequency methods. Time-frequency analysis is performed by computing a time-frequency distribution (TFD), also called a time-frequency representation (TFR), for a given signal. The main idea of a TFD is to allow determination of signal energy at a particular time as well as frequency. Therefore, these TFDs are functions of two dimensions, time and frequency, which have an inherent tradeoff between time and frequency resolution that can be obtained. This tradeoff between time and frequency resolution arises due to the required windowing of the signal to compute the time-frequency distribution. For good time resolution, a short time window is necessary; meanwhile a good frequency resolution requires a narrowband filter, which corresponds to a long time window. But these two conditions, a window with arbitrarily small duration and arbitrarily small bandwidth cannot be fulfilled at the same time. Thus, a compromise has to be found to achieve sufficiently good time and frequency resolution.

The STFT as one possible realization of a TFD is performed by sliding an analysis window across the signal time series and computing the FT for the current time point. The STFT for a continous-time signal $x(t)$ is defined as follows:

$$\text{STFT}_x^{(y)}(t, f) = \int_{t'} [x(t')\gamma^*(t' - t)]' e^{-j2\pi f t'} dt' \quad (20)$$

where $\gamma(t')$ is the analysis window and * denotes the complex conjugate. As discussed the analysis window chosen for the STFT greatly influences the result. Looking at the two extremes shows this influence best. Consider the case of the delta function as analysis window: $\gamma(t) = \delta(t)$. In this case, the $\text{STFT} = \sum_{t=0}^{N-1} x(t)\exp(-j2\pi ft)$, which is essen-

tially $x(t)$ and yields perfect time resolution, but no frequency resolution. On the other hand, if the analysis window is chosen to be a constant value $\gamma(t) = 1$ for all time, then the STFT becomes the Fourier transform $X(f)$, with perfect frequency resolution but no time resolution. Therefore, an appropriate window to provide both time and frequency resolution lies somewhere between these two extremes. In our case, a Hanning window is chosen as the analysis window. Figure 10 shows a plot of the STFT as it is obtained by transforming segments of the data, sampled with 250 points. The segments of 128 points each and an overlap of 50% are transformed via a 256 point FFT.

Feature Extraction. Calculating the power of the three dominant frequency bands in each segment, as described in equation 18, we obtain a contour plot of the power in these bands as shown in Fig 11. Also the mean frequency can be calculated in each segment, as described in Eq. 19.

Power Spectrum Estimation using AR-Model

The Algorithm. In the last section, the PSD was defined as the FT of an infinite ACS. This relationship may be considered a nonparametric description of the second-order statistics of a random process. A parametric description of the second-order statistic may also be devised by assuming a time-series model of the random process. The PSD of the time-series model will then be a function of the model parameters (and not of the ACS). A special class of models, driven by white noise processes and processing rational system functions, is AR, ARMA, and MA. One advantage of using parametric estimators is, for example, better spectral resolution. Periodogram and correlogram methods construct an estimate from a windowed set of data or ACS estimates. The unavailable data or unestimated ACS values outside the window are implicitly zero, which is an unrealistic assumption that leads to distortion in the spectral estimate. Some knowledge about the process from which the data samples are taken is often available. This information may be used to construct a model that approximates the process that generated the observed time sequence. Such models will make more realistic

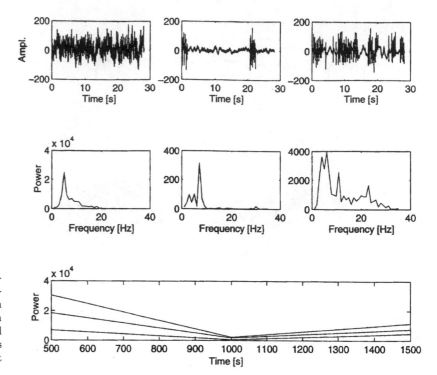

Figure 10. STFT. (a) Epoch of an EEG signal recorded from an isoflurane/DEX-treated dog. The time-varying character of this signal can be presented in the power spectrum using the STFT as shown in (b).STFT was performed with a 256 point FFT and a window length (Hamming window) of 128 points with an overlap of 50%. Data were sampled at $f = 250$ samples.

assumptions about the data outside the window instead of the null data assumption. In our case, the AR-modeling is used instead of AM or ARMA because of the advantage that the model parameter can be obtained by solving linear equations. The assumption is that if the model order p is chosen correctly, and the model parameters are calculated correctly, we obtain a PSD estimation with $p/2$ or $(p+1)/2$ sharp peaks in the spectrum at the so-called dominant frequencies.

The AR parameters have been shown to follow a recursive relation (94):

$$a_p(n) = a_{p-1}(n) + K_p a_{p-1}^*(p-n) \quad \text{for } n = 1 \ldots (p-1) \quad (21)$$

where $a_p(n)$ are the parameters for model order p, and $a_{p-1}(n)$ for model order $(p-1)$. K_p is the reflection coefficient for order p, and in Burg's maximum entropy method, K_p is determined by minimizing the arithmetic mean of the forward and backward linear prediction error power, i.e.,

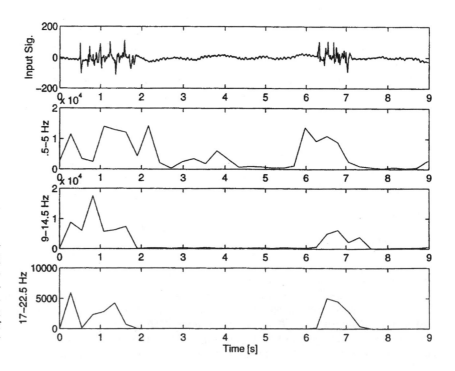

Figure 11. Energy profile in dominat frequency bands using the STFT. (a) EEG epoch recorded from isoflurane/DEX-treated dog. Summing together the energy in the frequency bands 0.5–5, 9–14.5, and 17–22.5 Hz, we obtain the energy profiles in the dominant frequency bands as shown in (b). Note that burst sequences and sequences of electrical silence in the EEG signal can be clearly distinguished with this method.

minimizing

$$\rho_p^{\text{fb}} = \frac{1}{2N}\left(\sum_{n=p+1}^{N}|e_p^f(n)|^2 + \sum_{n=p+1}^{N}|e_p^b(n)|^2\right) \quad (22)$$

where e^f and e^b are the forward and backward linear prediction errors. Minimizing this equation gives us the reflection coefficients K_p for model order p:

$$K_p = \frac{-2\sum_{n=p+1}^{N}e_{p-1}^f(n)e_{p-1}^{b*}(n-1)}{\sum_{n=p+1}^{N}|e_{p-1}^f(n)|^2 + \sum_{n=p+1}^{N}|e_{p-1}^b(n)|^2} \quad (23)$$

To choose the right model order, we use the Akaike criterion (95), which is based on the maximum likelihood approach and is termed the Akaike Information Criterion (AIC) (95):

$$\text{AIC}(p) = N^*\ln(\rho_p) + 2p \quad (24)$$

where ρ_p is the error variance for model order p. The error variance follows the relation:

$$\rho_p = \rho_{p-1}(1 - |K_p^2|) \quad (25)$$

The optimum model order p has to minimize AIC. With the obtained model parameter, it is now possible to present the data sequence $x(n)$ in the following way (44):

$$x(n) = w(n) - a(1)x(n-1) - a(2)x(n-2)$$
$$- \ldots - a(p)x(n-p) \quad (26)$$

where the $a(i)$ are the model parameters, p is the model order, and $w(n)$ is the error in prediction. If we now choose the model order and the model parameters correctly for our estimation, $w(n)$ turns out to be zero. Taking the z-transform of this equation, we obtain

$$X(z) = \frac{W(z)}{1 + a(1)z^{-1} + a(2)z^{-2} + \ldots + a(p)z^{-p}} \quad (27)$$

where $W(z)$ is the z-transform of $w(n)$. Squaring the absolute value of $X(z)$, we obtain the estimated power spectrum:

$$P(z) = |\frac{W(z)}{1 + a(1)z^{-1} + a(2)z^{-2} + \ldots + a(p)z^{-p}}|^2 \quad (28)$$

Parameter Extraction. From equation 27, we can now obtain the dominant frequencies in the estimated power spectrum. The poles of $X(z)$ are obtained from the equation:

$$z^p + a(1)z^{p-1} + \ldots + a(p) \quad (29)$$

Evaluating this expression at the unit circle, we get frequencies ω at which there is a peak in the frequency spectrum of the data sequence and the analog frequencies of the spectral peaks are

$$F_{\text{do min ant}} = \frac{F_{\text{sampling}}}{2\pi}\omega_{\text{do min ant}} \quad (30)$$

It is now possible to evaluate the power in these frequencies either by integrating the power spectrum between desired frequencies or by the method of Johnsen and Anderson (96), which uses the residues to find the power in the peaks.

Thus, AR modeling provides the possibility to estimate the power spectrum of a signal, to calculate the frequencies at which we find a peak in the spectrum, and to obtain the power in these dominant frequencies. Note that an important assumption for a correct use of AR-modeling is a stationary signal. As this assumption cannot be made for EEG signals in long data sequences, the sample data have to be divided into small overlapping segments, in which the signal can be observed as quasi-stationary. The right segment length can be found using AIC criterion and that segment length is taken that minimizes the variance for the calculated model order. In our case for anesthetized dogs, a model order of eight was found to be appropriate. This leads to four dominant frequencies in the following frequency ranges: 0.5–5, 10–14.5, 18–22.5 Hz, and 27–31 Hz. However, it has been observed that the power in the highest dominant frequency is very small in comparison with the power in the lowest three dominant frequencies and this band is therefore ignored in our study.

Another problem with using AR models for single-channel EEG analysis during nonstationary events like burst suppression or seizures is the possibly change in model order (97). Therefore, AR models can be observed to be more appropriate for multichannel analysis.

Feature Extraction. The dominant frequencies and the power in the dominant frequencies are calculated in each segment, respectively. Therefore, a summation of power in a certain frequency band is not necessary. Figure 12 shows the power in the dominant frequencies over time. The input signal is sampled with 125 Hz and subdivided into segments of 50 points each. The segments are allowed to have an overlap of 50%.

Burst Detection Methods

The computerized detection of seizures and bursts requires differentiation of bursts from episodes of electrical silence and ictal (signal during seizure) from interictal (normal signal activity) parts. We already mentioned the characteristics of bursts and seizures, like high amplitude, high energy, and an increase in phase coupling. The change in energy at the onset and offset of bursts and seizures has been used to detect bursts in many applications. Babb et al. (98) constructed a circuit that signaled seizure detection when high amplitude–high frequency activity was observed for at least 5 s. Ives et al. (99) employed amplitude discrimination after summing and band pass filtering the EEG from 16 channels. Gotman (100) employs a detection paradigm based on measures of amplitude and time period obtained after a "half-wave decomposition" of the signals. Gotman has tested this method on numerous cases; many false positive are generated by this method. Murro et al. (101) used a discriminant function based on signal spectra. The advantage of this method is that it does not rely on visual determination of normal and abnormal signal characteristics. Recent studies have used wavelet analysis to detect the onset of bursts or seizures (102). The detection scheme is based on monitoring the variance structure of the wavelet coefficients over a selected scale range and power fluctuations in these scales individually. Webber et al. (103)

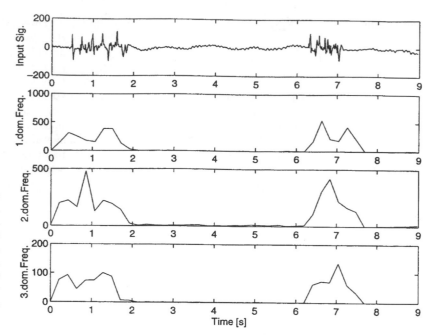

Figure 12. Energy profiles in dominant frequencies using an AR model. (a) EEG epoch recorded from an isoflurane/DEX-treated dog. An AR model (model order 8) was fitted to the EEG sample using segments of 50 points each and an overlap of 50%. The power in the three lowest dominant frequencies (ranges from 0.5–5, 10–14.5, and 18–22.5) was calculated with the residues. Note that this method provides the possibility to distinguish bursts from periods of electrical silence in the EEG signal.

presented in 1994 a detector for epileptiform discharges using an artificial neural network. The detector can detect seizure onsets and offsets in real time and was tested for raw EEG signals as well as for parametrized signals. Similar detectors have been developed by Oezdamer et al. (104) and Yaylali et al. (105). Bullmore et al. (106) presented in 1994 a detection method of ictal events based on fractal analysis of the EEG signal. The basic idea of this detector is that the EEG signal during bursts or seizures tends to have relatively low values for fractal dimensions. However, displaying the onsets and offsets of bursts and seizures very well, this method requires a visual inspection of the obtained image. Lehnerts and Elger (107) presented a detection algorithm in 1995 based on the neuronal complexity loss during ictal signal events. Alarkon et al. (108) presented a detector of seizure onsets in partial epilepsy based on feature changes extracted from the power spectrum. Features used were the total amplitude within specific frequency bands and the parameters activity, mobility, and complexity as developed by Hjorth (58) and presented earlier. The time course of these parameters was then displayed, and the onsets of ictal events were assessed when the parameter changes were above a preset threshold for more than four consecutive epochs. Franaszczuk et al. (109) presented a method in 1994, that allows not only the visualization of seizure onsets but also the flow of activity through different regions of the brain caused by epileptic discharges. The DFT method, a multichannel parametric method of analysis based on an AR model, was used for the analysis. Note that for localization of seizure onsets, this method requires recordings from different regions of the brain, especially recordings made with subdural grid and depth electrode arrays. A study that also used energy measurements in different frequency bands is that of Darcey and Williamson (110). The energy in different frequency bands is obtained from sequentially calculated power spectra, using short analysis windows to provide good time resolution. This method has been described in the previous section as STFT.

The energy ratio of the energy in the ictal EEG with respect to the energy of the preictal baseline EEG was calculated, and detection was performed by comparing this quantity with a threshold. However, it has also been observed in their study that the time resolution obtained with the STFT cannot be increased upon a certain limit, due to simultaneous decrease in frequency resolution.

Agarwal and Gotman present a method of segmentation and clustering based on the nonlinear Teager Energy algorithm or TEA (111). The Teager Energy algorithm is a critical advance in energy measurement methods; based on the nonlinear operators or the second order Volterra kernel (112), it measures the frequency-dependent energy in a signal. The method uses multipliers beyond the simple square law detector ($x^2(t)$) to capture the energy in a filtered portion of the signal. It is based on the energy in a spring concept. The TEA is a two-term time-domain Agarwal, and Gotman uses a TEA variant that does not depend on a square law term. This enables studying the energy function without the evident emphasis of the equivalent zero-lagged autocorrelation term. This would cause undo emphasis of apparent white noise contaminant. For sleep staging and burst detection studies, the TEA turns out to be an essential preliminary analysis component. Sherman et al. (113) use it for burst detection, burst counting, and burst duration calculation in a cardiac arrest (CA) and recovery model in fetal pigs. This study paralleled earlier results highlighting the fact that early bursting in the recovery EEG after CA is indicative of potentially favorable outcomes. High burst counts and time spent in bursting was shown to afford group separability based on a neuro-deficit score (NDS).

CONCLUSIONS

We have presented that bursts during burst-suppression and seizures are signals of short duration with high amplitude and therefore episodes of high energy. This led to the assumption that a discrimination of bursting episodes from burst-suppression parts could be made based on detecting a change in the energy in dominant frequency bands. However, representing the energy of a time-varying signal has its difficulties. The abrupt change from bursting parts to burst-suppression parts and the possibly short duration of these sequences make it impossible to visualize the change in energy at a certain point of time in the average power spectrum. Other methods like the STFT or parametric methods like the AR model provide the possibility to obtain both a time and a frequency resolution. Nevertheless, both methods still require a certain window length to estimate the power spectrum, and the estimated energy in each window represents the averaged energy over a certain number of sampling points.

BIBLIOGRAPHY

Cited References

1. Berger H. Ueber das Elektroenzephalogramm des Menschen. Arch Psychiat Nervenkrankheiten 1929;87:527–570.
2. Speckmann EJ, Elger CE. Introduction in the neurophysiological basis of the EEG and DC potentials. In: Niedermeyer E, daSilva FL, editors. Electroencephalography, Basic Principles Clinical Applications and Related Fields. Urban & Schwarzenberg.
3. Martin JH. Cortical neurons, the EEG, and the mechanisms of epilepsy. In: Kandel ER, Schwartz JH, editors. Principles of Neural Science. New York: Elsevier.
4. Nunez PL. Electric Fields of the Brain: The Neurophysics of EEG. New York: University Press; 1981.
5. Kandel ER, Schwartz JH, editors. Principles of Neural Science. New York: Elsevier; 1985.
6. Mountcastle VB. An organizing principle for cerebral function: The unit module and the distributed system. In: Edelman GM, Mountcastle VB, editors. The Mindful Brain. Cambridge (MA): MIT Press.
7. Hubel DH, Weisel TN. Functional architecture of macaque monkey visual cortex. Proc R Soc London Biol Sci 1978;198:1–59.
8. Kiloh, McComas, Osselton,. Clinical Electroencephalography. New York: Appleton-Century-Crofts; 1972.
9. Plonsey R, Barr RC. Bioelectricity. A Quantitative Approach. New York: Plenum Press.
10. Meyer-Waarden K. Biomedizinische Messtechnik I. Karlsruhe: Institut fuer Biomedizinische Technik; 1994.
11. Webster JG. ed. Encyclopedia of Medical Devices and Instrumentation. Vol. 1. New York: John Wiley & Sons; 1988.
12. daSilva FL, Rotterdam AV. Biophysical aspects of EEG and magnetoencephalogram generation. In: Niedermeyer E, daSilva FL, editors. Basic Principles, Clinical Applications and Related Fields. Urban & Schwarzenberg.
13. Kooi KA, Tucker RP, Marshall RE. Fundamentals of Electroencephalography. 2nd ed. New York; 1978.
14. Clark JW. Origin of Biopotentials. In: Webster JW, editor. Medical Instrumentation: Application and Design. 3rd ed., New York: John Wiley & Sons; 1998.
15. Walter WG, Dovey VJ. Electroencephalography in cases of sub-cortical tumour. J Neurol Neurosurg Psychiat 1944;7:57–65.
16. Speckmann E.-J, Caspers H. Origin of Cerebral Field Potentials. Stuttgart: Thieme; 1979.
17. Goldman D. The Clinical Use of the "Average" Reference Electrode in Monopolar Recording. Electroenceph Clin Neurophysiol 1950;2:209.
18. Siegel G, et al. Basic Neurochemistry. 4th ed. New York: Raven Press; 1989.
19. Jordan KG. Continuous EEG and evoked potential monitoring in the neuroscience intensive care unit. J Clin Neurophysiol 1993;10(4):445–475.
20. Homan RW, Herman J, Purdy P. Cerebral location of international 10-20 system electrode placement. Electroencephalogr Clin Neurophysiol 1987;66:376–382.
21. Heuser D, Guggenberger H. Ionic changes in brain ischemia and alterations produced by drugs. A symposium on brain ischemia. Br J Anesthesia 1985;57:23–33.
22. Siesjo BK, Wieloch T. Cerebral metabolism in ischemia: Neurochemical basis for therapy. A symposium on brain ischemia. Br J Anesthesia 1985;57:47–62.
23. Astrup J, Simon L, Siesjo BK. Tresholds in cerebral ischemias—the ischemia penumbra. Stroke 1981;12:723–725.
24. Eleff SM, Hanley DH, Ward K. Post-resuscitation prognostication, declaration of brain death and mechanisms of organ salvage. In: Paradis NA, Halperin HR, Nowak RM, editors. Cardiac Arrest: The Pathophysiology & Therapy of Sudden Death Philadelphia (PA): Williams and Wilkens; 1995.
25. Goel V. A Novel Technique for EEG Analysis:Application to Neonatal Hypoxia-Asphyxia. In: Biomedical Engineering. Baltimore: Johns Hopkins University; 1995.
26. Williams GW, et al. Inter-observer variability in EEG interpretation. Neurology 1985;35:1714–1719.
27. Elul R. Gaussian behaviour of the electroencephalogram: changes during performance of mental task. Science 1969;164:328–331.
28. Dumermuth G, Gasser T, Lange B. Aspects of EEG analysis in the frequency domain. In: Dolce G, Kunkle H, editors. Computerized EEG Analysis Stuttgart Fischer: 1975.
29. Dumermuth G, et al. Spectral analysis of EEG activity during sleep stages in normal adults. Eur Neurol 1972;7:265–296.
30. Levy WJ, et al. Automated EEG processing for intraoperative monitoring. Anesthesiology 1980;53:223–236.
31. Leader HS, et al. Pattern reading of the electroencephalogram with a digital computer. Electroenceph Clin Neurophysiol 1967;23:566–570.
32. Legewie H, Probst W. On line analysis of EEG with a small computer. Electroenceph Clin Neurophysiol 1969;27:533–535.
33. Harner RN, Ostergren KA. Sequential analysis of quasistable and paroxysmal activity. In: Kellaway P, Petersen I, editors. Quantitative Analytic Studies in Epilepsy New York: Raven Press; 1975.
34. Brazier MAB, Barlow JS. Some applications of correlation analysis to clinical problems in electroencephalography. Electroenceph Clin Neurophysiol 1956;8:325–331.
35. Barlow JS, Brown RM. An analog correlator system for brain potentials. Cambridge (MA): Res Lab Electronics at MIT; 1955.
36. Kiencke U, Dostert K. Praktikum: Digitale Signalverarbeitung in der Messtechnik. Karlsruhe: Institut fuer Industrielle Informationstechnik; 1994.
37. Kaiser JF, Angell RK. New techniques and equipment for correlation computation. Cambridge (MA): Servomechanisms Lab; 1957.
38. Kamp A, Storm VLM, Tielen AM. A method for auto- and cross- relation analysis of the EEG. Electroenceph Clin Neurophysiol 1965;19:91–95.

39. Remond A, et al. The alpha average. I. Methodology and description. Electroenceph Clin Neurophysiol 1969;26:245–265.

40. Regan D. Evoked potentials in basic and clinical research. In: Remond A, editor. EEG Informatics. Didactic Review of Methods and Applications of EEG Data Processing. Amsterdam: Elsevier; 1977.

41. Campbell K, Kumar A, Hofman W. Human and automatic validation of a phase locked loop spindle detection system. Electroenceph Clin Neurophysiol 1980;48:602–605.

42. Cooley JW, Tukey JW. An algorithm for the Mchine Calculation of Complex Fourier Series. Math Computation 1965;19:297–301.

43. Kammeyer KD, Kroschel K. Digitale Signalverarbeitung. 2nd ed. Stuttgart: B.G. Teubner; 1992.

44. Welch PD. The use of Fast Fourier Transform for the estimation of power spectra. IEEE Trans Audio Electroacoust; 1970. AU-15:70–73.

45. daSilva FL. EEG analysis: Theory and practice. In: Niedermeyer E, daSilva FL, editors. Electroencephalography, Basic Principles Clinical Applications and Related Fields. Urban & Schwarzenberg.

46. Jenkins GM, Watts DG. Spectral analysis and its applications. San Fransisco: HoldenDay; 1968.

47. Walter DO, Adey WR. Analysis of brain wave generators as multiple statistical time series. IEEE Trans Biomed Eng 1965;12:309–318.

48. Brazier MAB. Studies of the EEG activity of the limbic structures in man. Electroenceph Clin Neurophysiol 1968; 25:309–318.

49. daSilva FL, et al. Organization of thalamic and cortical alpha rhythms: Spectra and coherence. Electroenceph Clin Neurophysiol 1973;35:627–639.

50. Vos JE. Representation in the frequency domain of nonstationary EEGs. In: Dolce G, Kunkel H, editors. CEAN-Computerized EEG Analysis. Stuttgart: Fischer; 1975.

51. Gersch W, Goddard G. Locating the site of epileptic focus by spectral analysis methods. Science 1970;169:701–702.

52. Tharp BR, Gersch W. Spectral analysis of seizures in humans. Comput Biomed Res 1975;8:503–521.

53. Sherman DL. Novel techniques for the detection and estimation of three wave coupling with applications to brain waves. Purdue University; 1993.

54. Braun JC. Neurological monitoring using time frequency and bispectral analysis of evoked potential and electroencephalogram signals. The Johns Hopkins University; 1995.

55. Huber PJ, et al. Statistical methods for investigating the phase relations in stationary stochastic processes. IEEE Trans Audio-Electroacoustics AU 1971;19:78–86.

56. Dumermuth G, et al. Analysis of interrelations between frequency bands of the EEG by means of the bispectrum. Electroenceph Clin Neurophysiol 1971;31:137–148.

57. Hjorth B. EEG analysis based on time domain properties. Electroenceph Clin Neurophysiol 1970;29:306–310.

58. Caille EJ, Bassano JL. Value and limits of sleep statistical analysis. Objective parameters and subjective evaluations. In: Dolce G, Kuekel H, editors. CEAN—Computerized EEG Analysis. Stuttgart: Fischer; 1975.

59. Luetcke A, Mertins L, Masuch A. Die Darstellung von Grundaktivitaet, Herd, und Verlaufsbefund sowie von paroxysmalen Ereignissen mit Hilfe der von Hjorth ausgegebenen normierten Steiheitsparameter. In: Matejcek M, Schenk GK, editors. Quantitative Analysis of the EEG. Konstanz: AEG-Telefunken; 1973.

60. Kay SM. Modern Spectral Estimation. Englewood Cliffs (NJ): Prentice-Hall, Inc.; 1987.

61. Duquesnoy AJ. Segmentation of EEGs by Means of Kalman Filtering. Utrecht: Inst. of Medical Physics TNO; 1976.

62. Isaksson A. On time variable properties of EEG signals examined by means of a Kalman filter method. Stockholm: Telecommunication Theory; 1975.

63. Creutzfeld OD, Bodenstein G, Barlow JS. Computerized EEG pattern classification by adaptive segmentation and probability density function classification. Clinical evaluation. Electroenceph Clin Neurophysiol 1985;60:373–393.

64. Clark DL, Rosner BS. Neurophysiologic effects of general anesthetics. Anesthesiology 1973;38:564.

65. Derbyshire AJ, et al. The effect of anesthetics on action potentials in the cerebral cortex of the cat. Am J Physiol 1936;116:577–596.

66. Swank RL. Synchronization of spontaneous electrical activity of cerebrum by barbiturate narcosis. J Neurophysiol 1949;12:137–160.

67. Swank RL, Watson CW. Effects of barbiturates and ether on spontaneous electrical activity of dog brain. J Neurophysiol 1949;12:137–160.

68. Bauer G, Niedermeyer E. Acute Convulsions. Clin Neurophysiol 1979;10:127–144.

69. Stockard JJ, Bickford RG, Aung MH. The electroencephalogram in traumatic brain injury. In: Bruyn PJVaGW, editor. Handbook of Clinical Neurology. Amsterdam: North Holland; 1975. 217–367.

70. Fischer-Williams M. Burst suppression activity as indication of undercut cortex. Electroenceph clin Neurophysiol 1963;15:723–724.

71. Yli-Hankala A, et al. Vibration stimulus induced EEG bursts in isoflurane anaesthesia. Electroencephal Clin Neurophysiol 1993;87:215–220.

72. Jaentti V, Yli-Hankala A. Correlation of instantaneous heart rate and EEG suppression during enflurane anaesthesia: Synchronous inhibition of hart rate and cortical electrical activity? Electroencephal Clin Neurophysiol 1990;76:476–479.

73. Jaentti V, et al. Slow potentials of EEG burst suppression pattern during anaesthesia. Acta Anaesthesiol Scand 1993; 37:121–123.

74. Ingvar DH, Rosen I, Johannesson G. EEG related to cerebral metabolism and blood flow. Pharmakopsychiatrie 1979;12: 200–209.

75. Schwartz AE, Tuttle RH, Poppers P. Electroencephalografic Burst Suppression in Elderly and Young Patients Anesthetized with Isoflurane. Anesth Analg 1989;68:9–12.

76. Gurvitch AM, Ginsburg DA. Types of Hypoxic and Posthypoxic Delta Activity in Animals and Man. Electroencephal Clin Neurophysiol 1977;42:297–308.

77. Ginsburg DA, Pasternak EB, Gurvitch AM. Correlation analysis of delta activity generated in cerebral hypoxia. Electroencephal Clin Neurophysiol 1977;42:445–455.

78. Muthuswamy J, Sherman D, Thakor NV. Higher order spectral analysis of burst EEG. IEEE Trans Biomed Eng 1999;46:92–99.

79. Steriade M, Amzica F, Contraras D. Cortical and thalamic cellular correlates of electroencephalographic burst-suppression. Electroenceph Clin Neurophysiol 1994;90:1–16.

80. Spehlmann R. EEG Primer. Amsterdam: Elsevier; 1985.

81. Mirski M. Functional Anatomy of Pentylenetetrazol Seizures. Washington University; 1986.

82. Schwilden H, Stoeckel H. Quantitative EEG analysis during anesthesia with isoflurane in nitrous oxide at 1.3 and 1.5 MAC. Br J Anaesth 1987;59:738.

83. Holmes GL, Rowe J, Hafford J. Significance of reactive burst suppression following asphyxia in full term infants. Clin Electroencephal 1983;14(3).

84. Watanabe K, et al. Behavioral state cycles, background EEGs and prognosis of newborns with perinatal hypoxia. Electroenceph Clin Neurophysiol 1980;49:618–625.

85. Holmes G, et al. Prognostic value of the electroencephalogram in neonatal asphyxia. Electroenceph Clin Neurophysiol 1982;53:60–72.

86. Lieb JP, et al. Neurophatological findings following temporal lobectomy related to surface and deep EEG patterns. Epilepsia 1981;22:539–549.

87. Ojemann GA, Engel JJ. Acute and chronic intracranial recording and stimulation. In: Engel JJ, editor. Surgical Treatment of the Epilepsy. New York: Raven Press; 1987. 263–288.

88. Vaz CA, Thakor NV. Monitoring brain electrical magnetic activity. IEEE Eng Med Biol Mag 1986;Sept.:9–15.

89. Proakis JG, Manolakis DG. Introduction to Digital Signal Processing. New York: Macmillan; 1988.

90. Dumermuth G, Molinari L. Spectral analysis of EEG background activity. In: Remond A, editor. Methods of Analysis of Brain Electrical and Magnetic Signals. Amsterdam: Elsevier; 1987.

91. Schwilden H, Stoeckel H. The derivation of EEG parameters for modelling and control of anesthetic drug effect. In: Stoeckel H, editor. Quantitation, Modelling and Control in Anaesthesia. Stuttgart: Thieme; 1985.

92. Schwilden H, Stoeckel H. Untersuchungen ueber verschiedene EEG parameter als Indikatoren des Narkosezustandes. Anaesth Intesivther Notfallmed 1980;15:279.

93. Levy WJ. Intraoperative EEG Patterns: implications for EEG monitoring. Anesthesiology 1984;60:430.

94. Isaakson A, Wennberg A, Zetterberg LH. Computer analysis of EEG signals with parametric models. Proc IEEE 1981;69:451–461.

95. Akaike H. Recent Development of Statistical Methods for Spectrum Estimation. In: Yamaguchi NF, editor. Recent Advances in EEG and EMG Data Processing. Amsterdam: North-Holland Biomedical Press; 1981.

96. Johnsen SJ, Andersen N. On power estimation in maximum entropy spectral analysis. Geophysics 1978;43:681–690.

97. Hilfiker P, Egli M. Detection and evolution of rhythmic components in ictal EEG using short segment spectra and discriminant analysis. Electroenceph Clin Neurophysiol 1992;82:255–265.

98. Babb TL, Mariani E, Crandall PH. An electronic circuit for detection of EEG seizures recorded with implanted electrodes. Electroenceph Clin Neurophysiol 1974;37:305–308.

99. Ives JR, et al. The on-line computer detection and recording of spontaneous temporal lobe epileptic seizures from patients with implanted depth electrodes via a radio telemetry link. Electroenceph Clin Neurophysiol 1974;73:205.

100. Gotman J. Automatic seizure detection. Electroenceph Clin Neurophysiol 1990;76:317–324.

101. Murro AM, et al. Computerized seizure detection of complex partial seizures. Electroenceph Clin Neurophysiol 1991;79:330–333.

102. Mehtu S, Onaral B, Koser R. Detection of seizure onset using wavelet analysis. In: Proc 16th Annual Int. Conf. IEEE-EMBS. Maryland, 1994.

103. Webber WRS, et al. Practical detection of epileptiform discharges (EDs) in the EEG using an artificial neural network: A comparison of raw and parameterized EEG data. Electroenceph Clin Neurophysiol 1994;91:194–204.

104. Oezdamer O, et al. Multilevel neural network system for EEG spike detection. In: Tsitlik JE, editor. Computer-Based Medical Systems. Washington: IEEE Computer Society Press; 1991. p 272–279.

105. Yaylali I, Jayakar P, Oezdamer O. Detection of epileptic spikes using artificial multilevel neural networks. Electroenceph clin Neurophysiol 1992;82.

106. Bullmore ET, et al. Fractal analysis of electroencephalographic signals intracerebrally recorded during 35 epileptic seizures: Evaluation of a new method for synoptic visualisation of ictal events. Electroenceph Clin Neurophysiol 1994;91:337–345.

107. Lehnerts K, Elger CE. Spatio-temporal dynamics of the primary epileptogenic area in temporal lobe epilepsy characterized by neuronal complexity loss. Electroenceph Clin Neurophysiol 1995;95:108–117.

108. Alarkon G, et al. Power spectrum and intracranial EEG patterns at seizure onset in partial epilepsy. Electroenceph Clin Neurophysiol 1995;94:326–337.

109. Franaszczuk PJ, Bergey GK, Kaminski MJ. Analysis of mesial temporal seizure onset and propagation using the directed transfer function method. Electroenceph Clin Neurophysiol 1994;91:413–427.

110. Darcey TM, Williamson PD. Spatio-Temporal EEG Measures and their Application to Human Intracranially Recporded Epileptic Seizures. Electroenceph Clin Neurophysiol 1985; 61:573–587.

111. Agarwal R, et al. Automatic EEG Analysis during long-term monitoring in the ICU. Electroencephal Clin Neurol 1998;107:44–58.

112. Bovik AC, Maragos P, Quatieri TF. AM-FM Energy Detection and Separation in Noise Using Multiband Energy Operators. IEEE Trans Signal Processing 1993;41(12):3245–3265..

113. Sherman DL, et al. Diagnostic Instrumentation for Neural Injury. IEEE Instrum Measure 2002;28–35.

See also Evoked potentials; monitoring in anesthesia; rehabilitation, computers in cognitive; sleep studies, computer analysis of.

ELECTROGASTROGRAM

DZ Chen
University of Kansas Medical Center
Kansas City, Kansas

Zhiyue Lin
University of Texas Medical Branch
Galveston, Texas

INTRODUCTION

Electrogastrography, a term similar to electrocardiography (ECG), is usually referred to as the noninvasive technique of recording electrical activity of the stomach using surface electrodes positioned on the abdominal skin (1–3). The cutaneous recording obtained using the electrogastrographic technique is called electrogastrogram (EGG). In this article, both electrogastrography and electrogastrogram are abbreviated to EGG. Due to the noninvasive nature and recent advances in techniques of EGG recording and computerized analysis, EGG has become an attractive tool to study the electrophysiology of the stomach and pathophysiology of gastric motility disorders and is currently utilized in both research and clinical settings (4–7).

Although there are now several commercially available hardware–software packages making recording and analysis of EGG relatively easy to perform, many centers still use home-built equipment because the interpretations of specific frequency and EGG amplitude parameters are still debated and the clinical utility of EGG is still under investigation (6–10). Therefore, there are definite needs for better definition of the normal frequency range of EGG and dysrhythmias as well as standardization of EGG recording and advanced analysis methods for extraction and interpretation of quantitative EGG parameters. More outcome studies of EGG are also needed to determine the usefulness of EGG in the clinical settings. This article covers the following topics: a brief historic review of EGG, basics of gastric myoelectrical activity, measurement and analysis of EGG including multichannel EGG, interpretation of EGG parameters, clinical applications of EGG and future development of EGG.

HISTORIC REVIEW OF EGG

Electrogastrography was first performed and reported by Walter Alvarez back in the early 1920s (1,11). On October 14, 1921, Walter Alvarez, a gastroenterologist recorded the first human EGG by placing two electrodes on the abdominal surface of "a little old woman" and connected them to a sensitive string galvanometer. A sinusoid-like EGG with a frequency of 3 cycles/min (cpm) was then recorded. As Alvarez described in his paper, "the abdominal wall was so thin that her gastric peristalsis was easily visible" (1). Alvarez did not publish any other paper on EGG probably because of a lack of appropriate recording equipment.

The second investigator to discover the EGG is I. Harrison Tumpeer, a pediatrician who probably performed the first EGG in children (12). In a note in 1926 (12) and in a subsequent publication (13), Tumpeer reported the use of limb leads to record the EGG from a 5 week old child who was suffering from pyloric stenosis and observed the EGG as looking like an ECG (electrocardiogram) with a slowly changing baseline (11).

However, it took ~ 30 years for EGG to be recovered by R.C. Davis, a psychophysiologist in the mid-1950s (14). Davis published two papers on the validation of the EGG using simultaneous recordings from needle electrodes and a swallowed balloon (14,15). Although Davis made only slow progress in EGG research, his two papers had stimulated several other investigators to begin doing EGG research, such as Dr. Stern who started working in Davis' lab in 1960 (11).

Stevens and Worrall (1974) were probably the first ones who applied the spectral analysis technique to EGG (16). They obtained simultaneous recordings from a strain gauge on the wall of the stomach and EGG in cats to validate the EGG. They not only compared frequencies recorded from the two sites visually in the old fashion way, but also used a fast paper speed in their polygraph and digitized their records by hand once per second, and then analyzed EGG data using Fourier transform (11).

Beginning in 1975, investigators in England published a number of studies on frequency analysis of the EGG signal and made numerous advances in techniques for analysis of the EGG, including fast Fourier transform (FFT) (17),

phase-lock filtering (18), and autoregressive modeling (19). In some of their studies, they compared the EGG with intragastric pressure recordings and reported their findings similar to those of Nelson and Kohatsu (20). They found that there was no 1:1 correlation between the EGG and the contractions. The EGG could be used to determine the frequency of the contractions, but could not be used to determine when contractions were occurring (21).

During this same time, Smout and co-workers at Erasmus University in Rotterdam, The Netherlands, conducted several validation studies of the EGG and made major contributions in the area of signal analysis. In their landmark 1980 paper (22), they were the first ones who showed that the amplitude of the EGG increases when contractions occur. In 1985, Dr. Koch and Dr. Stern reported their study on simultaneous recordings of the EGG and fluoroscopy (23). They repeatedly observed the correspondence between EGG waves and antral contractions during simultaneous EGG-fluoroscopy recordings.

To extract information about both the frequency of EGG and time variations of the frequency, a running spectral analysis method using FFT was introduced by van der Schee and Grashus in 1987 (24), later used by some others (2,25,26) and now still used in most laboratories (5). To avoid the averaging effect introduced by the block processing of the FT, Chen et al. (27,28) developed a modern spectral analysis technique based on an adaptive autoregressive moving average model. This method yields higher frequency resolution and more precise information about the frequency variations of the gastric electrical activity. It is especially useful in detecting dysrhythmic events of the gastric electrical activity with short durations (29).

In 1962, Sobakin et al. (30) performed the EGG in 164 patients and 61 healthy controls and reported that ulcers caused no change in the EGG, but that pyloric stenosis produced a doubling of amplitude, and stomach cancer caused a breakup of the normal 3 cpm rhythm. This was probably the first large-scale clinical use of the EGG. In the past two decades, numerous studies have been reported on the clinical use of the EGG including understanding the relationship between the EGG and gastric motility (22,23,31–37), gastric myoelectrical activity in pregnant women (38–40), gastric myoelectrical activity in diabetics or gastroparetic patients (41–44), gastric myoelectrical activity in patients with dyspepsia (45–50), and prediction of delayed gastric emptying using the EGG (42,46,47,49,51).

As Dr. Stern wrote in 2000, "the history of EGG can be described as three beginnings, a length period of incubation, and a recent explosion" (11). It is beyond the scope of this article to cover every aspect of EGG studies. For more information about the EGG, readers are referred to some excellent articles, reviews, dissertations, and chapters (2,5,6,21,27,52–56).

ELECTROPHYSIOLOGY OF THE STOMACH

Normal Gastric Myoelectrical Activity

Along the gastrointestinal tract, there is myoelectrical activity. *In vitro* studies using smooth muscle strips of

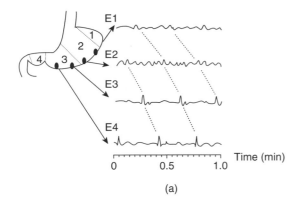

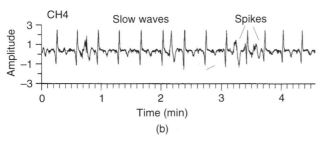

Figure 1. (a) Anatomy of the stomach (1–fundus, 2–body (corpus), 3–antrum, 4–pylorus) and the origin and propagation of the electrogastric signal from proximal (E1) to distal (E4) electrodes measured from the serosa of the human stomach. The dotted lines show the detection of a gastric slow wave traveling distally along the stomach wall. (b) Serosal recording obtained from the distal stomach in a patient. The trace shows slow waves of ∼ 3 cpm with and without superimposed spike potentials.

the stomach have revealed independent gastric myoelectrical activity (GMA) from different regions of the stomach. The highest frequency of the gastric myoelectrical activity was recorded in the corpus and the lowest frequency in distal antrum. However, *in vivo* studies demonstrated a uniform frequency in the entire stomach under healthy conditions, because the myoelectrical activity in the corpus with the highest frequency drives or paces the rest of stomach into the same higher frequency (see Fig. 1).

Gastric myoelectrical activity is composed of slow waves and spike potentials. The slow wave is also called the pacesetter potential, or electrical control activity (57–59). The spike potentials are also called action potentials or electrical response activity (57–59). While slow waves are believed originated from the smooth muscles, recent *in vitro* electrophysiological studies suggest that interstitial cells of Cajal (ICC) are responsible for the generation and propagation of the slow wave (60). The frequency of normal slow waves is species-dependent, being ∼ 3 cpm in humans and 5 cpm in dogs, with little day-to-day variations. The slow wave is known to determine the maximum frequency and propagation of gastric contractions. Figure 1 presents an example of normal gastric slow waves measured from a patient. Normal 3 cpm distally propagated slow waves are clearly noted.

Spike potentials are known to be directly associated with gastric contractions, that is, gastric contractions occur when the slow wave is superimposed with spike potentials.

Note, however, that *in vivo* gastric studies have failed to reveal a 1:1 correlation between spike potentials measured from the electrodes and gastric contractions measured from strain gauges although such a relationship does exist in the small intestine. In the stomach, it is not uncommon to record gastric contractions with an absence of spike potentials in the electrical recording. Some other forms of superimposed activity are also seen in the electrical recording in the presence of gastric contractions.

Abnormal GMA: Gastric Dysrhythmia and Uncoupling

Gastric myoelectrical activity may become abnormal in diseased states or upon provocative stimulations or even spontaneously. Abnormal gastric myoelectrical activity includes gastric dysrhythmia and electromechanical uncoupling. Gastric dysrhythmia includes bradygastria, tachygastria, and arrhythmia. Numerous studies have shown that gastric dysrhythmia is associated with gastric motor disorders and/or gastrointestinal symptoms (4–7,20,61,62).

A recent study has revealed that tachygastria is ectopic and of an antral origin (63). In > 80% of cases, tachygastria is located in the antrum and propagates retrogradely toward the pacemaker area of the proximal stomach. It may partially or completely override the normal distally propagating slow waves. However, most commonly it does not completely override the normal gastric slow waves. In this case, there are two different slow wave activities: normal slow waves in the proximal stomach and tachygastrial slow waves in the distal stomach. A typical example is presented in Fig. 2.

Unlike tachygastria, bradygastria is not ectopic and reflects purely a reduction in frequency of the normal pacemaking activity. That is, the entire stomach has one single frequency when bradygastria occur (63). Bradygastria is originated in the corpus and propagates distally toward the pylorus. The statistical results showing the origins of tachygastria and bradygastria are presented Fig. 3. The data was obtained in dogs and gastric dysrhythmias were recorded postsurgically or induced with various drugs including vasopressin, atropine and glucagon (63).

MEASUREMENT OF THE EGG

Gastric myoelectrical activity can be measured serosally, intraluminally, or coutaneously. The serosal recording can be obtained by placing electrodes on the serosal surface of the stomach surgically. The intraluminal recording can be acquired by intubating a catheter with recording electrodes into the stomach. Suction is usually applied to assure a good contact between the electrodes and the stomach mucosal wall. The serosal and intraluminal electrodes can record both slow waves and spikes, since these recordings represent myoelectrical activity of a small number of smooth muscle cells. These methods are invasive and their applications are limited in animals and laboratory settings.

The EGG, a cutaneous measurement of GMA using surface electrodes, is widely used in humans and clinical settings since it is noninvasive and does not disturb ongoing activity of the stomach. A number of validation studies have documented the accuracy of the EGG by

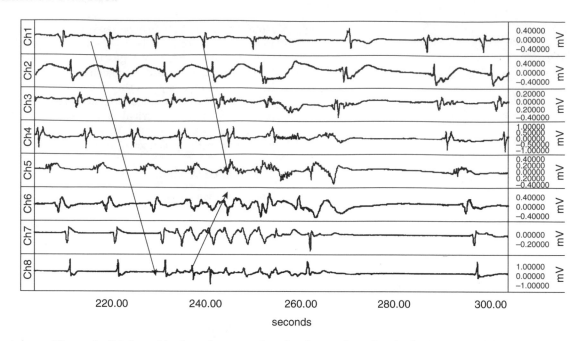

Figure 2. Origin and backward propagation of tachygastria and arrhythmia measured from the sorosa of the canine stomach. An ectopic pacemaker was located at the antrum (channel 8), which propagated both forward and backward. It changed the normal rhythm of original pacemaker that propagated down from the proximal part of the stomach (channel 1) to the distal part of the stomach (channel 8) and almost affected the entire rhythm of the stomach for a minute. A transient bradygastria was observed in all channels before normal rhythm changed back.

comparing it with the recording obtained from mucosal and serosal electrodes (19,22,31,61,64–66). Reproducibility of the EGG recording has been demonstrated, with no significant day-to-day variations (67). In adults, age and gender do not seem to have any influences on the EGG (68–71).

EGG Recording Equipment

The equipment required to record the EGG includes amplifiers, an analog-to-digital (A/D) converter and a personal computer (PC) (Figs. 4 and 5). The EGG signal must be amplified because it is of relatively low amplitude (50–500 μV). An ideal EGG amplifier should be able to

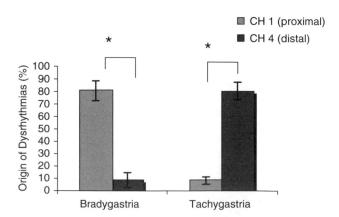

Figure 3. Statistical results showed that bradygastria originated primarily from the proximal body of the stomach, while tachygastria originated primarily from the distal stomach.

enhance the gastric signal and effectively reduce interferences and noise. Abnormal frequencies of gastric slow waves may be as low as 0.5 and as high as 9 cpm. To effectively record the gastric slow wave, an appropriate recording frequency range is 0.5–18 cpm (5,6,72). It is recommended that a good choice of the sampling frequency should be three to four times of the highest signal frequency of interest (73,74). Therefore, a sampling rate for digitization of the EGG signal ≥ 1 Hz (60 cpm) is a proper choice.

A typical EGG recording system is shown in Fig. 4. It is composed of two parts: data acquisition and data analysis. Venders who currently offer or have offered EGG equipment in the past included 3CPM Company, Medtronic/Synectics, Sandhill Scientific, Inc., RedTech, and MMS (The Netherlands), and so on (2,6). To date, there are two U.S. Food and Drug Administration (FDA)-approval EGG systems: one from Medtronic Inc (Minneapolis, MN) and the other from 3CPM Company (Crystal Bay, NV). The 3CPM Company's EGG device is a work station that consists of an amplifier with custom filters, strip chart recorder, and computer with proprietary software—the EGG Software Analysis System (EGGSAS). However, this device is only to record and analyze single-channel EGG.

The newly FDA-approved Medtronic's ElectroGastro-Graphy system provides multichannel EGG recordings and analysis (75–78). It can be either running on Medtronic's Gastro Diagnostic Workstation or consisting of the Medtronic's Polygraf ID with a laptop to make a portable system (see Fig. 5). This system provides an Automatic Impedance Check function and optional Motion Sensor. With the Automatic Impedance Check, all EGG electrodes

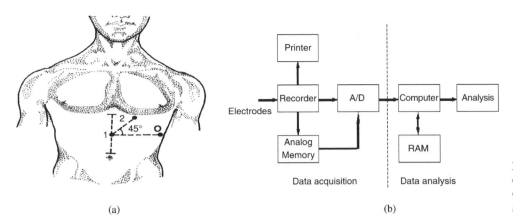

(a)   (b)

Figure 4. (a) Position of abdominal electrodes for the measurement of one-channel EGG. (b) Block diagram of an EGG recording system.

are verified to be in good electrical contact with the skin within 10 s. The optional Motion Sensor can record respiration and patient movements during data capture. This assists physicians to more easily identify motion artifacts, which can then be excluded from subsequent analysis. Currently, this system has been configured to make four-channel EGG recordings with placement of six surface electrodes on the subject's abdomen (75–78) (see Fig. 5a).

An ambulatory recording device is also available and has been used frequently in various research centers (42,46,48,71,76). For example, the ambulatory EGG recorder (Digitrapper EGG) developed by Synectics Medical Inc. (Shoreview, MN) is of the size and shape of a "walkman" (79). It contains one channel amplifier, an A/D conversion

unit, and memories. It can be used to record up to 24-h one-channel EGG with a sampling frequency of 1 Hz. Information colleted during recording can be downloaded into a desktop computer for data storage and analysis (42,79).

Procedures for Recording EGG

Due to the nature of cutaneous measurement, the EGG is vulnerable to motion artifacts. Accordingly, a careful and proper preparation before the recording is crucial in obtaining reliable data.

Skin Preparation. Since the EGG signals are very weak, it is very important to minimize the impedance between

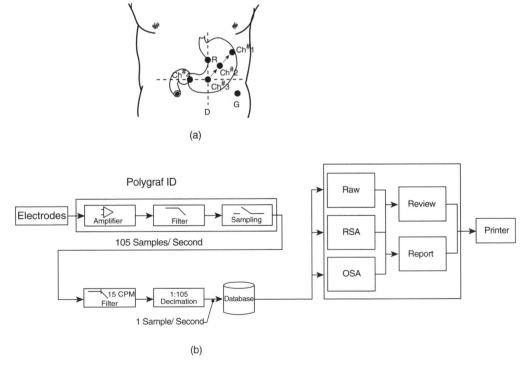

Figure 5. (a) Electrogastrogram electrodes placement for making four-channel EGG recordings consists of placing four active electrodes along the antral axis of the stomach, a reference electrode over the xyphoid process, and a ground electrode on the patient's left side. (b) Block diagram of Medtronic POLYGRAM NETTM ElectroGastroGraphy System (RSA: Running Spectral Analysis; OSA: Overall Spectral Analysis). (Reprinted with permission of Medtronic, Inc.).

Table 1. List of Systems and Procedures for Recording EGG Used by Different Groups

References	Hardware System	Analysis Method	Procedure (Duration and Meal)
Dutch research groups (8,83)	Custom-built four-channels (band-pass filter: 0.01–0.5 Hz, sampling rate: 1 Hz)	Running spectral analysis by short-time Fourier transform (STFT)	2 h before and 3 h after meal (a pancake, 276 kcal)
McCallum and Chen (42,79), Parkman et al. (46,71) Chen et al. (75,76)	MicroDigitrapper (Synectics Medical, Inc.): single channel (cut-off frequency: 1–16 cpm, sampling rate: 1 or 4 Hz). Commercial four-channel EGG recording device (cut-off frequencies: 1.8–16 cpm) (Medtronic-Synectics, Shoreview, MN)	Running spectral analysis by STFT (Gastrosoft Inc., Synetics Medical) or adaptive analysis method)	(1) 30 min before and 2 h after meal (turkey sandwich, 500 kcal) (2) 60 min before and 60 min after meal (two scrambled egg with two pieces of toasted bread 200 mL water, 282 kcal)
Penn State groups (2,84)	An amplifier with custom filters, a strip chart recorder (cut-off frequency: 1–18 cpm, sampling frequency: 4.27 Hz)	Running spectral analysis by STFT and a data sheet with percentage distribution of EGG power in four frequency ranges	Water load test (45 min)

the skin and electrodes. The abdominal surface where electrodes are to be positioned should be shaved if necessary, cleaned and abraded with some sandy skin-preparation jelly (e.g., Ominiprep, Weaver, Aurora, CO) in order to reduce the impedance between the bipolar electrodes to > 10 kΩ. The EGG may contain severe motion artifacts if the skin is not well prepared.

Position of the Electrodes. Standard electrocardiographic-type electrodes are commonly used for EGG recordings. Although there is no established standard, it is generally accepted that the active recording electrodes should be placed as close to the antrum as possible to yield a high signal-to-noise ratio (80). The EGG signals can be recorded with either unipolar or bipolar electrodes, but bipolar recording yields signals with a higher signal-to-noise ratio. One commonly used configuration for recording one-channel EGG is to place one of two active electrodes on the midline halfway between the xiphoid and umbilicus and the other active electrode 5 cm to the left of the first active electrode, 30 cephalad, at least 2 cm below the rib cage, in the midclavicular line. The reference electrode is placed on the left costal margin horizontal to the first active electrode (42,81) (Fig. 4a). One commonly used configuration of electrodes for making four-channel EGG recordings is shown in Fig. 4a, including four active electrodes along the antral axis of the stomach, a reference EGG electrode over the xyphoid process, and a ground EGG electrode on patient's left side (75–78).

Positioning the Patient. The subject needs to be in a comfortable supine position or sit in a reclining chair in a quiet room throughout the study. Whenever possible, the supine position is recommended, because the subject is more relaxed in this position, and thus introduces fewer motion artifacts. The subject should not be engaged in any conversations and should remain as still as possible to prevent motion artifacts (7,8,79).

Appropriate Length of Recording and Test Meal

The EGG recording is usually performed after a fast of 6 h or more. Medications that might modify GMA (prokinetic and antiemetic agents, narcotic analgesics, anticholinergic drugs, non-steroidal anti-inflammatory agents) should be stopped at least 48 h prior to the test (6,7). The EGG should be recorded for 30 min or more (no < 15 min in any case) in the fasting state and for 30 min or more in the fed state. A recording < 30 min may not provide reliable data and may not be reproducible attributed to different phases of migrating motor complex (82) in the fasting state.

The test meal should contain at least 250 kcal with no > 35% of fat (82). Solid meals are usually recommended although a few investigators have used water as the test "meal" (see Table 1).

EGG DATA ANALYSIS

In general, there are two ways to analyze EGG signals. One is time-domain analysis or waveform analysis, and the other is frequency-domain analysis. Numerous EGG data analysis methods have been proposed (18,19,24,27,53,55,56,84–97).

Time-Domain Data Analysis

Like other surface electrophysiological recordings, the EGG recording contains gastric signal and noise.

Table 2. Composition of the EGG

	Components	Frequency (cpm)
Signal	Normal slow wave	2.0–4.0
	Bradygastria	0.5–2.0
	Tachygastria	4.0–9.0
	Arrhythmia	NA[a]
Noise	Respiratory	12–24
	Small bowel	9–12
	ECG	60–80
	Motion artifacts	Whole range

[a]Not available

Compared with other surface recordings, such as ECG, the quality of EGG is usually poor. The gastric signal in the EGG is disturbed or may even be completely obscured by noise (see Table 2). The frequency of gastric signals is from 0.5 to 9.0 cpm, including normal (regular frequency of 2–4 cpm) and abnormal frequencies. The gastric signals with abnormal frequencies may be divided further into bradygastria (regular frequency of 0.5–2.0 cpm), tachygastria (regular frequency of 4–9 cpm) and arrhythmia (irregular rhythmic activities) (62).

The noise consists of respiratory artifacts, interferences from the small bowel, ECG, and motion artifacts (see Table 2). The respiratory artifact has a frequency from 12 to 24 cpm. It is a common and thorny problem. It is superimposed upon almost every EGG recording if not appropriate processed. Occasionally, the slow wave of the small intestine may be recorded in the EGG. The frequency of intestinal slow waves is 12 cpm in duodenum and 9 cpm in the ileum. The intestinal slow wave is usually weaker than the gastric slow wave. One can avoid recording intestinal slow waves by placing electrodes in the epigastric area. The frequency of ECG is between 60 and 80 cpm. It can be eliminated using conventional low pass filtering because its frequency is much higher than that of the gastric signal component. The frequency of motion artifacts is in the whole recording frequency range. To minimize motion artifacts, the subject must not talk and should remain still during recording.

The time-domain analysis methods with the aid of computers that were introduced to facilitate the EGG data analysis include (1) adaptive filtering. It is used to reduce noise such as respiratory artifacts with minimal distortion of the gastric signal component of interest (27,90,91), (2) coherent averaging. It is applied to filter out random noise by averaging a large number of EGG waves (85), (3) use of feature analysis and artificial neural networks to automatically detect and delete motion artifacts (93,94), and (4) use of independent component analysis to separate gastric signals from multichannel EGGs (97).

When noise level is low, it is possible to visually analyze the raw EGG tracing (3,6,84,98) to identify periods of artifact and provide a qualitative determination of recording segments with normal frequencies of ∼ 3 cpm and those of abnormally high (tachygastria) or low (bradygastria) and the presence or absence of a signal power increase after eating a test meal. Artifacts usually are readily recognized visually as sudden, high amplitude off-scale deflections of the EGG signal (see Fig. 6). Artifactual periods must be excluded before analysis. This is because (a) they are usually strong in amplitude and may completely obscure the gastric signal; (b) they have a broad-band spectrum and their frequencies overlap with that of the gastric signal; therefore they are not separable using even spectral analysis method, and jeopardize any kind of quantitative analyses of the EGG data (79).

EGG Parameters (99)

Although a noise-free EGG signal is attainable by means of advanced signal processing techniques (27,86), the waveform analysis of the EGG has rarely been used, because the waveform of the EGG is related to many factors, including the thickness of the abdominal wall of the subject, skin preparation, position of the electrodes, and characteristics of the recording equipment (100). Furthermore, the number of specific characteristics of the EGG is limited. With single-channel EGG recording, only frequency and

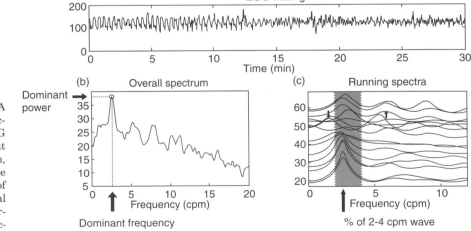

Figure 6. The EGG data analysis. (a) A 30 min EGG recording. (b) The power spectra of the 30 min EGG data. The EGG dominant frequency (DF) and power at DF can be determined from the spectrum, (c) adaptive running spectra. Each curve (from bottom to top) is the spectrum of 2 min EGG data. The percentage of normal slow waves (or dysrhythmias) can be determined from the spectra by counting spectral peaks in each frequency band.

amplitude can be measured. Recent computer simulations and experiments have shown that the propagation of the gastric slow wave can be identified from the multichannel EGG recordings (27,100), it is, however, difficult to get this information from waveform analysis (27). Accordingly, quantitative data analyses of the EGG are mostly based on spectral analysis methods. Some important EGG parameters obtained by the spectral analysis methods are described as in the following sections.

EGG Dominant Frequency and Dominant Power. The frequency believed to be of gastric origin and at which the power in the EGG power spectrum has a peak value in the range of 0.5–9.0 cpm is called the EGG dominant frequency. The dominant power is the power at the dominant frequency. The EGG power can be presented in a linear or decibel (dB) unit. The dominant frequency and power of the EGG are often simplified as EGG frequency and EGG power. Figure 6 shows the definition of the dominant frequency and power of the EGG. Simultaneous cutaneous and serosal (13,17–19) or mucosal (11,16) recordings of GMA have shown that the dominant frequency of the EGG accurately represents the frequency of the gastric slow wave. The dominant power of the EGG reflects the amplitude and regularity of gastric slow waves. The gastric slow wave is regarded as abnormal if the EGG dominant frequency is not within a certain frequency range (e.g., 2–4 cpm). Although there is no established definition for the normal range of the gastric slow wave, it is generally accepted that the dominant frequency of the EGG in asymptomatic normal subjects is between 2.0 and 4.0 cpm (5,6,41,72). The EGG, or a segment of the EGG, is defined as tachygastria if its frequency is > 4.0 cpm, but < 9.0 cpm, bradygastria if its frequency is < 2.0 cpm and arrhythmia if there is a lack of a dominant frequency (see Table 2).

Power Ratio or Relative Power Change. As the absolute value of the EGG dominant power is related to many factors, such as the position of the electrodes, the preparation of skin and the thickness of the abdominal wall, it may not provide much useful information. One of the commonly used EGG parameters associated with the EGG dominant power is the power ratio (PR) or the relative power change after an intervention such as meal, water, or medication. Note that the power of the EGG dominant frequency is related to both the amplitude and regularity of the EGG. The power of the EGG dominant frequency increase when EGG amplitude increases. It also increases when the EGG becomes more regular. Previous studies have shown that relative EGG power (or amplitude) change reflects the contractile strength of the gastric contractions (22,33,37).

Percentage of Normal Gastric Slow Waves. The percentage of normal slow waves is a quantitative assessment of the regularity of the gastric slow wave measured from the EGG. It is defined as the percentage of time during which normal gastric slow waves are observed in the EGG.

Percentage of Gastric Dysrhythmias Including Bradygastria, Tachygastria, and Arrhythmia. The percentage of gastric dysrhythmia is defined as the percentage of time during which gastric dysrhythmia is observed in the EGG. In contrast to the percentage of normal gastric slow waves in an EGG recording, this parameter represents the abnormality of the EGG or gastric slow waves.

Instability Coefficients. The instability coefficients are introduced to specify the stability of the dominant frequency and power of the EGG (99). The instability coefficient (IC) is defined as the ratio between the standard deviation (SD) and the mean:

$$IC = SD/mean \times 100\%$$

The clinical significance of the instability coefficient has been demonstrated in a number of previous studies (37,40,99). The instability coefficients defined by Geldof et al. is slightly different from the one defined above. More information can be found in Refs. 83,101.

Percentage of EGG Power Distribution. The percentage of EGG power distribution was introduced by Koch and Stern (102) and is defined as the percentage of total power in a specific frequency range in comparison with the power in the total frequency range from 1 to 15 cpm. For example;

% of (2.4–3.6 cpm) = the power within 2.4–3.6 cpm/(the total power from 1 to 15 cpm) × 100%

Using this parameter, Koch et al. (102) found that the percentage of power in the 3-cpm range was significantly lower in patients with idiopathic gastroparesis compared to patients with obstructive gastroparesis. They also found that the percentage of power in the tachygastria range (3.6–9.9 cpm) correlated significantly with the intensity of nausea reported during vector-induced motion sickness (103). The advantage of this method is that it is easy for computation. We should be aware, however, that only relative values of this parameters in comparison with the control data should be used. Even in normal subjects, the percentage of normal EGG activity computed in this way will never be 100%. Note that this parameter is sensitive to noise, since any noise component in the frequency band of 1–15 cpm affects the computation of this parameter. Harmonics of the fundamental 3 cpm slow wave may be computed as tachygastria (8,74).

Percentage of Slow Wave Coupling. Slow wave coupling is a measure of the coupling between two EGG channels. The percentage of slow wave coupling is defined as the percentage of time during which the slow wave is determined to be coupled. The slow waves between two EGG channels are defined as coupled if the difference in their dominant frequencies is < 0.2 (77,78,95) or 0.5 cpm (76).

Methods to Obtain EGG Parameters

Spectral analysis methods are commonly used for calculation of the EGG parameters, including power spectral analysis and running spectral analysis (RSA) or time-frequency analysis. The frequency and power of the EGG can be derived from the power spectral density. The periodogram is one of the commonly used methods for the calculation of the power spectrum density (73). In this method, EGG data samples are divided into consequent segments with certain overlap. A FT is performed on each data segment, and the resultant functions of all segments are averaged. The periodogram method is more appropriate for the computation of the power spectrum of a prolonged EGG recording. Whenever there are enough data samples, the periodogram method instead of the sample spectrum should be used for the calculation of the dominant frequency and power of the EGG (21). Another method to estimate the frequency and power of EGG is to use a parametric method such as autoregressive modeling (AR) parameters (19). These AR parameters are initially set at zeros and are iteratively adjusted using the EGG samples. After a certain number of iterations, the EGG signal can be represented by these AR parameters. That is, the power spectrum of the EGG signals can be calculated from these parameters (19,27). The problem is the appropriate selection of the model order or the number of parameters. Too few parameters reduce the accuracy, and too many increasing the computing time (98). Although this method is somewhat time consuming, the advantage is that compared to FFT-based methods, the period over which the signal is analyzed can be much shorter (19).

To extract not only information about the frequency of the EGG, but also information about time variations of the frequency, a running spectral analysis method using FFT was first introduced by a Dutch group (24) and later used by others (25,26,39). This method consists of a series of sequential sample spectra. It is calculated as follows: For a given data set of EGG, a time window (e.g., Hanning window) with a length of D samples is applied to the first D samples, a FFT with the same length is calculated, and a sample spectrum is obtained for the first block of data. The sample spectrum of the next time step is obtained in the same way by shifting the windows of some samples forward. The advantage of this method is easy for implantation. Its drawback is that it may not be able to provide accurate time frequency estimations when the characteristics of the EGG signal change rapidly (72,104).

To avoid the averaging effect introduced by the FFT, Chen et al.(28) developed an adaptive spectral analysis based on the adaptive autoregressive moving average model (27). The main advantage of this method is that it is able to provide the instantaneous frequency of an EGG signal with short duration (29). Thus it is very useful for the detection of gastric dysrhythmias with brief duration, but may not be a good choice for the estimation of the EGG power (104). Recently, an exponential distribution (ED) method was also introduced for representation of EGG (92). The performance of the ED method is in between the RSA and the adaptive method. The cross-terms may deteriorate the performance of the ED method if the EGG signal contains several different frequency components (104). Time–frequency analysis methods other than those mentioned above have also been used, such as wavelet transform and fast Hartley transform (89). The description of these methods is mathematically complex and beyond the scope of this article. The detailed information can be found in (5) and (89).

An example of the calculation of the EGG parameters is shown in Fig. 6. The upper panel presents an EGG recording obtained in a human subject. The power spectrum of this 30-min EGG is shown in the lower left panel. The lower right panel shows the power spectra of the 30-min EGG calculated by the adaptive spectral analysis method (28). Each line in Fig. 6c represents the power spectrum of 2-min nonoverlap data (from bottom to top). The percentage of normal slow wave and dysrhythmias can be calculated from these spectra. Of 15 spectra, 12 have peaks in the 2–4 cpm range, that is, 80% of the EGG recording has normal slow waves. Three spectra (two in the bradygastria range and one in the tachygastria range) have peaks outside the normal slow wave range. The percentage of dysrhythmias is then 20%.

EGG IN ADULTS

EGG in Healthy Subjects

Definitions of what constitutes a normal EGG have been provided by careful analysis of EGG recordings from normal volunteers (6).

Normal EGG Frequency Range. Several studies (4,41,49,68,70,71) in healthy adults have shown that EGG in the fasting state is characterized by a stable slow wave dominant frequency (DF) (median: 3.0 cpm; range: 2–4 cpm) with a relatively small amplitude. Immediately after the test meal, the EGG frequency decreases from the baseline for a short period [~5 min. (4)] and then gradually increases to above the baseline level (media: 3.2 cpm; range: 2–4 cpm). It has been shown that the postprandial EGG DF is also dependent on the type and specific qualities of the ingested test meal (99). Solid meals slightly, but significantly, increase EGG DF, whereas liquid meals temporarily reduce the EGG DF.

Based on the normal EGG frequency range of 2–4 cpm, the overall results of four studies in 189 normal subjects suggest that 70% is an appropriate lower limit of normal for the percentage of the recording time for the EGG rhythm to be in the 2.0–4.0-cpm range (4,41,49,68,70,71).

Note that the definition of normal EGG frequency range reported in the literature varies considerably (see Table 1). The Penn State group defined the percentage of normal EGG activity as the percentage of the power in the frequency range of 2.4–3.6 cpm compared to the total power from 1 to 15 cpm (84). Accordingly, dysrhythms are considered present when too much power is in the low frequency range (bradygastria) or in the high frequency range (tachygastria). This approach is debatable due to the following reasons: (1) The EGG is not sinusoid and thus its power spectrum contains harmonics that are related to the waveform, but not at all associated with gastric rhythmicity. In this method, however, the harmonics are considered as tachgastria (8). (2) The method is very sensitive to motion artifacts that can result in abnormal frequency spectra with significant power in the low frequency range (8,9). Apparently, the differences in the definitions of the normal frequency range of EGG or dysrhthmias are at least related the following two factors: (1) Relative small numbers of subjects were included in the above EGG studies; (2) different analysis methods were used to analyze the EGG data. To establish better definitions of normal frequency rang and dysrhythmias, an international multicenter EGG study with a relative large sample size is needed and the currently used analysis methods should be applied to compare the results.

EGG Power in the Fasting and EGG Power Ratio. Absolute values of EGG power during fasting and the postprandial period are affected by a number of variables including body habitus, electrodes placement, and body position (6,105). However, these factors do not influence the relative value in EGG power, that is, the power ratio between the pre- and postprandial powers. Depending on meals consumed, 90–95% of healthy volunteers exhibit increased postprandial power at DF (6,41,71). Note that different meals may have different effects on the EGG. The main results for the effects of different meals on the EGG power are summarized as follows (99):

Water: Water induces an increase in EGG dominant power and a decrease in EGG dominant frequency. In a study with 10 normal subjects drinking 140 mL water (106), it was found that the EGG dominant frequency was slightly, but significantly, lower than the baseline in the fasting state during the first 10 min after the drink (2.95 vs. 2.73 cpm, $p < 0.05$). The power of the EGG at the dominant frequency was significantly higher after the drink than the baseline. A 3 dB increase in EGG dominant power (equivalent to 41% increase in 3 cpm amplitude) was observed. Similar observations were reported by several other investigators (66,107). In a recently performed study, simultaneous EGG and serosal recordings of gastric myoelectrical activity were made in patients before and after a drink of water (66). Statistical analysis demonstrated that the EGG dominant power change after a drink of water was correlated with that observed in the serosal recording (Spearman's correlation coefficient: $r = 0.757$, $p = 0.007$) and the change of EGG dominant frequency was the same as that from serosal recordings.

Milk: To investigate whether there is a different effect between non-nutritive (water) and a nutritive liquid meal, Chen et al.repeated the study procedure mentioned above in Ref. 106 by asking volunteers to drink 140 mL of 2% milk. The results showed that milk decreases the amplitude of EGG. The average decrease in EGG power in the 10 subjects within the first 10 min was 3.8 dB (equivalent to ~50% decrease in amplitude) (108).

Solid meal: The effects of solid meal on the EGG have been studied by numerous investigators (5,16,33,101,106). More significant increase in EGG dominant power was observed after the solid meal than after a drink of water. For example, the average EGG dominant power after a solid meal over 10 subjects was 6 dB higher than the preprandial value, equivalent to a 100% increase in amplitude (106) (see Fig. 7a and b). The actual amount of increase in EGG dominant power is believed to be associated with the volume and content of the meal (see next section). The dominant frequency of the EGG seems to increase as well after a test meal. Similar to the change in EGG dominant frequency after a drink of water, this increase is often small, but significant (106).

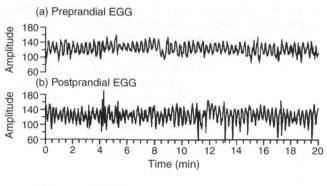

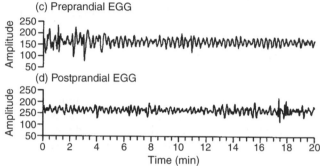

Figure 7. Pre- and postprandial EGG recordings in humans. (a) and (b): normal EGG patterns that show an increase in EGG amplitude postprandialy; (c) and (d): dysryhthmic EGG pattern and a substantial decrease in EGG amplitude after the meal.

EGG IN PATIENTS

Abnormal EGG

A variety of abnormalities have been described on EGG recordings from patients with motility disorders. For example, abnormal EGGs are noted with nausea, vomiting, early satiety, anorexia, and dyspepsia including gastroparesis (41–44,102), nonulcer or functional dyspepsia (46–50,76,109), motion sickness (5,25,110), pregnancy (38–40), and eating disorders (35). Typical EGG abnormalities in patients with motility disorders or symptoms include (*1*) absence of normal slow waves, which is shown in the EGG power spectra as a lack of peaks in the 2–4 cpm frequency range; (*2*) gastric dysrhythmias, including bradygastria, tachygastria, and arrhythmia (see Fig. 8); (*3*) deterioration of the EGG after a test meal, which is shown as a decrease in EGG power in the 2–4 cpm frequency range (see Fig. 7c and d), (*4*) slow wave uncoupling between different gastric segments detected from a multichannel EGG recording (76–78,95,96).

Definition of an abnormal EGG is mainly determined by comparison of EGG findings in healthy volunteers and symptomatic patients (41). At present, it is widely accepted that an EGG is considered as abnormal if the DF is in the tachy- and/or bradygastric frequency ranges for > 30% of the time. This number takes into account the observation that healthy volunteers exhibit periods of time representing up to 30% of recording time in which recognizable EGG rhythms are not distinguishable from background electrical noise either on visual inspection or computer analysis.

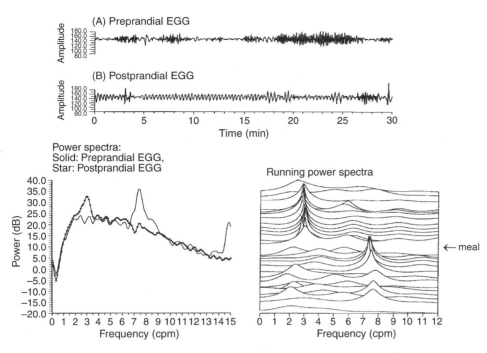

Figure 8. The EGG recordings in a patient with gastroparesis: (a) 30-min preprandial EGG recording, (b) 30-min postprandial EGG. Lower left panel: Power spectra of preprandial EGG (solid line) and postprandial EGG (line with star) shows abnormal response to a meal (decrease in postprandial EGG power) and tachgastria after meal (EGG dominant frequency: 7.4 cpm). Lower right panel shows the running spectra of 30-min preprandial EGG and 30-min postprandial EGG demonstrating the presence of 7–8 cpm tachygastrial peaks before meal and normal 3-cpm peaks after meal.

In addition, a decreased power ratio after a solid meal is also an indication of an abnormal EGG (6,7,41,68–71).

Some institutions have advocated the use of percentage distribution of EGG power in the three major frequency bands to summarize the absolute signal amplitude in the bradygastric, normal rhythm, and tachygastria ranges (2,84,111). For this parameter, an EGG is considered abnormal if the percentage distribution of total EGG power in the tachygastric range is > 20% (6,112). Power distributions in the bradygastric frequency range are highly variable and may be affected by minor variations in the signal baseline or subtle motion artifacts. Thus the calculation of the percentage of the total EGG power in the bradygastric frequency range may not be reliable for the determination of bradygastria (6).

Clinical Role of EGG. The FDA approved EGG as a test for patient evaluation in 2000. The FDA statement on EGG concluded that EGG is a noninvasive test for detecting gastric slow waves and is able to differentiate adult patients with normal myoelectrical activity from those with bradygastrias and techygastrias. The EGG can be considered as part of a comprehensive evaluation of adult patients with symptoms consistent with gastrointestinal motility disorders (6).

The members of the American Motility Society Clinical GI Motility Testing Task Force proposed the following indications for EGG as a diagnostic study to noninvasively record gastric myoelectrical activity in patients with unexplained persistent or episodic symptoms that may be related to a gastric motility and/or myoelectrical disorder (6). The EGG can be obtained: (1) to define gastric myoelectric disturbances in patients with nausea and vomiting unexplained by other diagnostic testing or associated with functional dyspepsia and (2) to characterize gastric myoelectric disturbances associated with documented gastroparesis.

The future clinical applications of EGG are in three main areas: (1) To assist in the clinical evaluation and diagnosis of patients with gastric motility disorders. (2) To determine the gastric response to either caloric stimuli or exogenous stimuli, such as pharmacologic and prokinetic agents or gastrointestinal hormones or gastric electrical stimulation or for patients before and after kidney–pancreas (KP) transplant (113–115). (3) To further evaluate the role of EGG in research and clinical work in infants and children (6,7,116).

EGG IN INFANTS AND CHILDREN

Although the majority of EGG studies are being performed in adults, there is an increased interest for the clinical application of EGG to pediatric patients. In infants, current diagnostic methods for the assessment of gastric motility, such as intraluminal manometry and radionuclide isotope study, are very much limited. Consequently, little is known on gastric myoelectrical in infants since mucosal–serosal recordings are not feasible, and much less information is available in infants than adults on gastric motility. The EGG is therefore an attractive noninvasive alternative to study gastric myoelectrical and motor activities in infants and children. In recent years, a small number of pediatric gastroenterologists and researchers, including Peter Milla, Alberto Ravelli, Salvatore Cucchiara, Giuseppe Riezzo, and Jiande Chen, et al. have began to use the EGG to study the pathophysiology of gastric motility in infants and children (11).

Patterns of GMA in Healthy Pediatric Subjects with Different Ages

To investigate whether EGG patterns are associated with ages, Chen et al. (117) performed EGG studies in five groups of healthy subjects including 10 preterm newborns,

8 full-term newborns, 8 full-term infants (ages 2–6 months), 9 children (ages 4–11 years), and 9 adults. The Digitrpper EGG recorder was used to record EGG signals for 30 min before and 30 min after a test meal in each subject. Spectral analysis methods were applied to computer EGG parameters. The results showed that the percentage of 2–4 cpm slow waves was 26.6±3.9% in the preterm newborns, 30.0±4.0% in full-term newborns, 70±6.1% in 2–6-months old infants ($P < 0.001$ compared with newborns), 84.6±3.2% in 4–11-year old children ($P < 0.03$ compared with infants), and 88.9±2.2% in the adults ($P > 0.05$ compared with children). This study has shown that regular gastric slow waves (2–4 cpm) are absent at birth, present at age of 2–4 months, and well developed at the age of 4–11 years. The EGG in healthy children is similar to that in healthy adults.

Using the percentage of total EGG power in the frequency range 2.5–3.6 cpm as a measure of normal gastric slow waves, Koch et al. reported similar findings in preterm and full-term infants with ages from 3 to 50 days: a low percentage of normal gastric slow waves, no difference between preterm and fulterm infants, and no difference between fasting EGG and fed EGG (111). These studies suggest that gastric slow waves are largely absent at birth, and there is a maturing process after birth.

To study the development or maturation of gastric slow waves in preterm infants, Liang et al. (118) performed a follow-up EGG study in 19 healthy preterm infants at postnatal ages of 1 and 2 weeks and 1, 2, 4, and 6 months (gestational age at birth: 33.5 ± 2.6 week). The results showed that the percentage of normal slow waves was low at birth and there was a progressive increase with age during the first 6 months of life (see Fig. 9). These results suggest that normative EGG data should be established for different age groups and age-matched controls

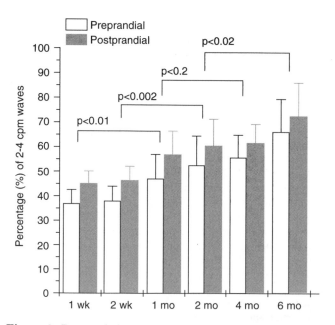

Figure 9. Preprandial and postprandial percentages of the normal 2–4-cpm slow waves in preterm infants at the ages of 1 and 2 weeks and 1, 2, 4, and 6 months. The P values resulted from comparison between paired fasting data or paired fed data.

are necessary for the interpretation of EGG from diseased neonates and infants.

EGG Norms in Healthy Children and Effects of Age, Gender, and BMI

As with any novel technique, the establishment of normal values is a prerequisite for reliable application across populations. Studies on EGG performed in healthy adults have found no major differences in EGG characteristics among age group (68,69,71). Using the same EGG recording and analysis system as utilized in healthy adults, Riezzo et al. performed EGG studies before and after a meal in 114 healthy children (age range: 6–12 years) (69) and Levy et al.conducted EGG studies in 55 healthy children (age range: 6–18 years) for a 1 h baseline preprandial period and a 1 h postprandial period after consummation of a standard 448 kcal meal (119). These studies have shown that the EGG patterns in the healthy children ages from 4 to 18 years are very similar to those in the healthy adults and the key normative values are not influenced by age and gender.

Applications in Pediatric Patients

Electrogastrogram has also been applied to evaluate pediatric patients with various diseases that are associated with gastric motility disorders. Functional dyspepsia presents as a challenge to clinicians with its debilitating features and no organic findings. Invasive diagnostic tests are limited in pediatric practice. Whereas, noninvisive EGG has been used and will be increasingly used to identify possible malfunctioning of the stomach, Cucchiara et al. (120) detected abnormal patterns of the EGG, encompassing all range of dysrhythmia, in 12 out of 14 patients with functional dyspepsia. Abnormalities in gastric myoelectrical activity were also observed from the EGG in pediatric patients with central nervous system disorders, chronic renal failure, and intestinal pseudoobstruction (98,121,122).

FUTURE PROSPECTS

The recording and analysis of the EGG are well established although not yet completely standardized. It is clear that the EGG is a reliable measurement of gastric slow waves and reflects relative contractile activity of the stomach. Clinical applications of the EGG are indicated in numerous studies. However, cautions should be made during recording to minimize motion artifact and in the interpretation of the EGG parameters. Future development in EGG methodology should be focused to reveal more information regarding spatial abnormalities of gastric slow waves and relevant information directly related to gastric contractility. In this regard, multichannel EGG may play a more important role in both electrophysiological studies of the stomach and clinical applications. Recently, several studies have been performed using a four-channel EGG system to derive spatial information from the multichannel EGG. These include spatial distribution of slow-wave frequency and amplitude, slow-wave coupling and propagation (75–78). The gastric slow wave propagation was measured by analyzing the phase shifts or time lags of the EGG signal among different recording channels using cross-covariance analysis (75). Slow-wave coupling was defined as similar frequencies of the EGG signals in different channels and a cross-spectral analysis method has been established to compute the percentage of slow-wave coupling (93). Two single-center studies have suggested that patients with functional dyspepsia have impaired slow-wave propagation and coupling (76,78). Further multicenter studies are needed to determine how the expanded utility of the multichannel EGG.

The EGG is not only of great clinical significance in diagnosing gastric motility disorders, but may also play an important role in predicting and validating gastric electrical stimulation (GES) (116,123). Previous studies have established the efficacy of GES in reducing symptoms of gastroparesis. However, approximately one-third of the patients were shown to have poor response to GES therapy and the mechanism by which GES acts to improve gastroparetic symptoms is currently unclear (124,125). Physiological studies have demonstrated that gastric slow wave activity depends on the function of interstitial cells of Cajal (ICC) in the stomach (126–128). The ICC networks in the myenteric region of the gastric corpus and antrum (IC–MY) have been identified as the source of electrical slow waves that control the maximum frequency and propagation direction of gastric contractions. Ordog et al. (127) demonstrated that the loss of gastric IC–MY resulted in the loss of electrical slow waves and eliminated the ability the musculature to generate slow waves in response to depolarizing stimuli. Thus functional ICC is critical for the production and maintenance of normal slow wave activity and gastric motility. Degeneration of ICC is often found in some GI tract conditions characterized by ineffective motility of the affected segment. Recently, preliminary experiments have demonstrated (a) a normal baseline EGG is significantly correlated with an improved symptomatic outcome after chronic high frequency GES for refractory gastroparesis compared to an abnormal baseline EGG with which $\sim 50\%$ of patients were nonresponders to the high frequency GES therapy; (b) the absence of ICC was correlated with abnormal EGG and patients with an absence of ICC showed less symptom improvement (123,129). These data suggest that baseline EGG status could be used to predict long-term symptom improvement in gastroparetic patients treated with GES. Specifically, marked gastric dysrhythmias as determined by the baseline EGG study is a predictor to a poor response to high frequency GES. Further studies are required to determine the role of EGG in the management of GI motility disorders.

BIBLIOGRAPHY

Cited References

1. Alvarez WC. The electrogastrogram and what it shows. JAMA 1922;78:1116–1118.
2. Stern RM, Koch KL, Stewart WR, Vasey MW. Electrogastrography: Current issues in validation and methodology. Psychophysiology 1987;24:55–64.
3. Abell TL, Malagelada J-R. Electrogastrography: Current assessment and future perspectives. Dig Dis Sci 1988;33: 982–992.
4. Geldof H, Van Der Schee EJ. Electrogastrography: Clinical applications. Scand J Gastroenterol 1989;24(Suppl. 171): 75–82.
5. Chen JZ, McCallum RW. Electrogastrography. Principles and Applications. New York: Raven Press; 1994.
6. Parkman HP, Hasler WL, Barnett JL, Eaker EY. Electrogastrography: a document prepared by the gastric section of the American Motility Society Clinical GI Motility Testing Task Force. Neurogastroenterol Motil 2003;15:89–102.
7. Camilleri M, Hasler W, Parkman HP, Quigley EMM, Soffer E. Measurement of gastroduodenal motility in the GI laboratory. Gastroenterology 1998;115:747–762.
8. Verhagen MAM, VanSchelven L J, Samsom M, Smout AJPM. Pitfalls in the analysis of electrogastrographic recordings. Gastroenterology 1999;117:453–460.
9. Mintchev MP, Kingma YJ, Bowes KL. Accuracy of coutaneous recordings of gastricelectrical activity. Gastroenterology 1993;104:1273–1280.
10. Lorenzo CD, Reddy SN, Flores AF, Hyman PE. Is electrogastrography a substitute for manometric studies in chidren with functional gastrointestinal disirders? Dig Dis Sci 1997;42(1):2310–2316.
11. Stern RM. The history of EGG. Neurogastroenterologia 2000;1:20–26.
12. Tumpeer IH, Blitzsten PW. Registration of peristalsis by the Einthoven galvanometer. Am J Dis Child 1926;11:454–455.
13. Tumpeer IH, Phillips B. Hyperperistaltic electrographic effects. Am J Med Sci 1932;184:831–836.
14. Davis RC, Galafolo L, Gault FP. An exploration of abdominal potentials. J Com Psysiol Psychol 1957;50:519–523.
15. Davis RC, Galafolo L, Kveim K. Conditions associated with gastrointestinal activity. J Com Physiol Psychol 1959;52: 466–475.
16. Stevens LK, Worrall N. External recording of gastric activity: the electrogastrogram. Physiol Psychol 1974;2:175–180.
17. Brown BH, Smallwood RH, Duthie HL, Stoddard CJ. Intestinal smooth muscle electrical potentials recorded from surface electrodes. Med Biol Eng Comput 1975;13:97–102.
18. Smallwood RH. Analysis of gastric electrical signals from surface electrodes using phase-lock techniques. Med Biol Eng Comput 1978;16:507–518.
19. Linkens DA, Datardina SP. Estimations of frequencies of gastrointestinal electrical rhythms using autoregressive modeling. Med Biol Eng Comput 1978;16:262–268.
20. Nelsen TS, Kohatsu S. Clinical electrogastrography and its relationship to gastric surgery. Am J Surg 1968;116:215–222.
21. Chen J, McCallum RW. Electrogastrography: measurement, analysis and prospective applications. Med Biol Eng Comput 1991;29:339–350.
22. Smout JPM, van der Schee EJ, Grashuis JL. What is measured in electrogastrography? Dig Dis Sci 1980;25:179–187.
23. Koch KL, Stern RM. The relationship between the cutaneously recorded electrogastrogram and antral contractions in men. In: Stern RM, Koch KL, editors. Electrogastrography. New York: Praeger; 1985.
24. van der Schee EJ, Grashuis JT. Running spectral analysis as an aid in the representation and interpretation of electrogastrographical signal. Med Biol Eng Comput 1987;25:57–62.
25. Stern RM, Koch KL, Stewart WR, Lindblad IM. Spectral analysis of tachygastria recorded during motion sickness. Gastroenterology 1987;92:92–97.
26. Pfister CJ, Hamilton JW, Nagel N, Bass P, Webster JG, Thompkins WJ. Use of spectral analysis in the detection of frequency differences in the electrogastrograms of normal and diabetic subjects. IEEE Trans Biomed Eng 1988;BME-35:935–941.

27. Chen J. Adaptive filtering and its application in adaptive echo cancellation and in biomedical signal processing, Ph.D. dissertation, Department of Electrical Engineering. Katholieke Universiteit Leuven, Belgium; 1989.

28. Chen J, Vandewalle J, Sansen W, Vatrappen G, Jannssens J. Adaptive spectral analysis of cutaneous electrogastric signals using autoregressive moving average modelling. Med Biol Eng Comput 1990;28:531–536.

29. Chen JDZ, Stewart WR, McCallum RW. Spectral analysis of episodic rhythmic variations in the cutaneous electrogastrogram. IEEE Trans Biomed Eng 1993;BME-40:128–135.

30. Sobakin MA, Smirnov IP, Mishin LN. Electrogastrography. IRE Trans Biomed Electron 1962;BME-9:129–132.

31. Hamilton JW, Bellahsene B, Reichelderfer M, Webster JG, Bass P. Human electrogastrograms: Comparison of surface and mucosal recordings. Dig Dis Sci 1986;31:33–39.

32. Bortolotti M, Sarti P, Barbara L, Brunelli F. Gastric myoelectrical activity in patients with chronic idiopathic gastroparesis. J Gastrointest Motility 1990;2:104–108.

33. Koch KL, Stewart WR, Stern RM. Effect of barium meals on gastric electromechanical activity in man. Dig Dis Sci 1987;32:1217–1222.

34. Chen J, Richards R, McCallum RW. Frequency components of the electrogastrogram and their correlations with gastrointestinal motility. Med Biol Eng Comput 1993;31:60–67.

35. Abell TL, Malagelada J-R, Lucas AR, Brown ML, Camilleri M, Go VL, Azpiroz F, Callaway CW, Kao PC, Zinsmeister AR, et al. Gastric electromechanical and nerohorminal function in anorexia nervosa. Gastroenterology 1987;93:958–965.

36. Geldof H, van der Schee EJ, Grashuis JL. Electrogastrographic characteristics of interdigestive migrating complex in humans. Am J Physiol 1986;250:G165–171.

37. Chen JDZ, Richards RD, McCallum RW. Identification of gastric contractions from the cutaneous electrogastrogram. Am J Gastroenterol 1994;89:79–85.

38. Abell TL. Nausea and vomiting of pregnancy and the electrogastrogram: Old disease, new technology. Am J Gastroenterol 1992;87:689–681.

39. Koch KL, Stern RM, Vasey M, Botti JJ, Creasy GW, Dwyer A. Gastric dysrhythmias and nausea of pregnancy. Dig Dis Sci 1990;35:961–968.

40. Riezzo G, Pezzolla F, Darconza G, Giorgio I. Gastric myoelectrical activity in the first trimester of pregnancy: A cutaneous electrogastrographic study. Am J Gastroenterol 1992;87:702–707.

41. Chen J, McCallum RW. Gastric slow wave abnormalities in patients with gastroparesis. Am J Gastroenterol 1992;97:477–482.

42. Chen JZ, Lin Z, Pan J, McCallum RW. Abnormal gastric myoelectrical activity and delayed gastric emptying in patients with symptoms suggestive of gastroparesis. Dig Dis Sci 1996;41(8):1538–1545.

43. Abell TL, Camilleri M, Hench VS, et al.Gastric electromechanical function and gastric emptying in diabetic gastroparesis. Eur J Gastroenterol Hepatol 1991;3:163–167.

44. Koch KL, Stern RM, Stewart WR, Dwyer AE. Gastric emptying and gastric myoelectrical activity in patients with diabetic gastroparesis: Effect of long-term domperidone treatment. Am J Gastroenterol 1989;84:1069–1075.

45. Cucchiara S, Riezzo G, Minella R, et al. Electrogastrography in non-ulcer dyspepsia. Arch Disease Childhood 1992;67(5):613–617.

46. Parkman HP, Miller MA, Trate D, Knight LC, Urbain J-L, Maurer AH, Fisher RS. Electrogastrography and gastric emptying scintigraphy are complementary for assessment of dyspepsia. J Clin Gastroenterol 1997;24:214–219.

47. Pfaffenbach B, Adamek RJ, Bartholomaus C, Wegener M. Gastric dysrhythmias and delayed gastric emptying in patients with functional dyspepsia. Dig Dis Sci 1997;42 (10):2094–2099.

48. Lin X, Levanon D, Chen JDZ. Impaired postprandial gastric slow waves in patients with functional dyspepsia. Dig Dis Sci 1998;43(8):1678–1684.

49. Lin Z, Eaker EY, Sarosiek I, McCallum RW. Gastric myoelectrical activity and gastric emptying in patients with functional dyspepsia. Am J Gastroenterol 1999;94(9): 2384–2389.

50. Leahy A, Besherdas K, Clayman C, Mason I, Epstein O. Abnormalities of the electrogastrogram in functional dyspepsia. Am J Gastroenterol 1999;94(4):1023–1028.

51. Hongo M, Okuno Y, Nishimura N, Toyota T, Okuyama S. Electrogastrography for prediction of gastric emptying state. In: Chen JZ, McCallum RW, editors. Electrogastrography: Principles and Applications. New York: Raven Press; 1994.

52. Smallwood RH. Gastrointestinal electrical activity from surface electrodes. Ph.D. dissertation Sheffield (UK).

53. Smout AJPM. Myoelectric activity of the stomach: gastroelectromyography and electrogastrography. Ph.D. Dissertation, Erasmus Universiteit Rotterdam, Delft University Press, 1980.

54. Kingma YJ. The electrogastrogram and its analysis. Crit Rev Biomed Eng 1989;17:105–132.

55. van der Schee EJ. Electrogastrography: signal analysis aspects and interpretation. Ph.D. Dissertation, Erasmus Universiteit Rotterdam, Delft University Press, 1984.

56. Stern RM, Koch KL, editors. Electrogastrography: Methodology, Validation and Application. New York: Praeger; 1985.

57. Sarna SK, Daniel EE. Gastrointestinal electrical activity: Terminology. Gastroenterology 1975;68:1631–1635.

58. Hinder RA, Kelly KA. Human gastric pacemaker potential: Site of orgin, spread, and response to gastric transsaction and proximal gastric vagotomy. Am J Surg 1977;133:29–33.

59. Kelly KA. Motiltiy of the stomach and gastroduodenal junction. In: Johnson IA, editor. Physiology of the Gastrointestinal Tract. New York: Raven; 1981.

60. Sanders KM. A case for interstitial cells of Cajal as pacemakers and mediators of neurotransmission in the gastrointestinal tract. Gastroenterology 1996;112(2):492–445.

61. Abell TL, Malagelada J-R. Glucagon-evoked gastric dysrhythmias in humans shown by an improved electrogastrographic technique. Gastroenterology 1985;88:1932–1940.

62. Chen JDZ, Pan J, McCallum RW. Clinical significance of gastric myoelectrical dysrhythmas. Dig Dis 1995;13:275–90.

63. Qian LW, Pasricha PJ, Chen JDZ. Origin and patterns of spontaneous and drug-induced Canine Gastric Myoelectrical Dysrhythmias. Dig Dis Sci 2003;48:508–515.

64. Familoni BO, Bowes KL, Kingma YJ, Cote KR. Can transcutaneous recordings detect gastric electrical abnormalities? Gut 1991;32:141–146.

65. Chen J, Schirmer BD, McCallum RW. Serosal and cutaneous recordings of gastric myoelectrical activity in patients with gastroparesis. Am J Physiol 1994;266:G90–G98.

66. Lin Z, Chen JDZ, Schirmer BD, McCallum RW. Postprandial response of gastric slow waves: correlation of serosal recordings with the electrogastrogram. Dig Dis Sci 2000; 45(4):645–651.

67. Riezzo G, Pezzolla F, Thouvenot J, et al. Reproducibility of cutaneous recordings of electrogasography in the fasting state in man. Pathol Biol 1992;40:889–894.

68. Pfaffenbach B, Adamek RJ, Kuhn K, Wegneer M. Electrogastrography in health subjects: Evaluation of normal values, influence of age and gender. Dig Dis Sci 1995;40:1445–1450.

69. Riezzo G, Chiloiro M, Guerra V. Electrogastrography in health children: Evaluation of normal values, influence of age, gender and obesity. Dig Dis Sci 1998;43:1646–1651.

70. Riezzo G, Pezzolla F, Giorgio I. Effects of age and obesity on fasting gastric electrical activity in man: a cutaneous electrogastrographic study. Digestion 1991;50:176–181.

71. Parkman HP, Harris AD, Miller MA, Fisher RS. Influence of age, gender, and menstrual cycle on the normal electrogastrogram. Am J Gastroenterol 1996;91:127–133.

72. Chen JDZ, McCallum RW. Clinical application of electrogastrography. Am J Gastroenterol 1993;88:1324–1336.

73. Oppenheim AV, Schafer RW. Digital Signal Processing. New Jersey: Prentice Hall; 1975.

74. Mintchev MP, Rashev PZ, Bowes KL. Misinterpretation of human electrogastrograms related to inappropriate data conditioning and acquisition using digital computers. Dig Dis Sci 2000;45(11):2137–2144.

75. Chen JDZ, Zhou X, Lin XM, Ouyang S, Liang J. Detection of gastric slow wave propagation from the cutaneous electrogastrogram. Am J Physiol 1999;227:G424–G430.

76. Lin XM, Chen JDZ. Abnormal gastric slow waves in patients with functional dyspepsia assessed by multichannel electrogastrography. Am J Physiol 2001;280:G1370–G1375.

77. Simonian HP, Panganamamula K, Parkman HP, Xu X, Chen JZ, Lindberg G, Xu H, Shao C, Ke M-Y, Lykke M, Hansen P, Barner B, Buhl H. Multichannel electrogastrography (EGG) in normal subjects: A multicenter study. Dig Dis Sci 2004;49:594–601.

78. Simonian HP, Panganamamula K, Chen JDZ, Fisher RS, Parkman HP. Multichannel electrogastrography (EGG) in symptomatic patients: A single center study. Am J Gastroenterol 2004;99:478–485.

79. Chen JZ, Lin Z, McCallum RW. Toward ambulatory recording of electrogastrogram. In: Chen JZ, McCallum RW, editors. Electrogastrography: Principles and Applications. New York: Raven Press; 1994.

80. Mirizzi N, Scafoglieri U. Optimal direction of the electrogastrographic signal in man. Med Biol Eng Comput 1983; 21:385–389.

81. Patterson M, Rintala R, Lloyd D, Abernethy L, Houghton D, Williams J. Validation of electrode placement in neonatal electrogastrography. Dig Dis Sci 2001;40(10):2245–2249.

82. Levanon D, Zhang M, Chen JDZ. Efficiency and efficacy of the electrogastrogram. Dig Dis Sci 1998;43(5):1023–1030.

83. Smout AJPM, Jebbink HJA, Samsom M. Acquisition and analysis of electrogastrographic data. In: Chen JDZ, McCallum RW, editors. Electrogastrography: principles and applications. New York: Raven Press; 1994. p 3–30.

84. Koch KL, Stern RM. Electrogastrographic data and analysis. In: Chen JDZ, McCallum RW, editors. Electrogastrography: Principles and Applications. New York: Raven Press; 1994. p 31–44.

85. Volkers ACW, van der Schee EJ, Grashuis JL. Electrogastrography in the dog: Waveform analysis by a coherent averaging technique. Med Biol Eng Comput 1983;21:51–64.

86. Chen J. A computerized data analysis system for electrogastrogram. Comput Biol Med 1992;22:45–57.

87. Familoni BO, Kingma YJ, Bowes KL. Study of transcutaneous and intraluminal measurement of gastric electrical activity in human. Med Biol Eng Comput 1987;25:397–402.

88. Familoni BO, Kingma YJ, Bowes KL. Noninvasive assessment of huma gastric motor function. IEEE Trans Biomed Eng 1987;BME-34:30–36.

89. Mintchev MP, Bowes KL. Extracting quantitative information from digital electrogastrograms. Med Biol Eng Comput 1996;34:244–248.

90. Chen J, Vandewalle J, Sansen W, et al. Adaptive method for cancellation of respiratory artifact in electrogastric measurements. Med Biol Eng Comput 1989;27:57–63.

91. Chen JZ, Lin Z. Comparison of adaptive filtering in time-, transforem- and frequency-domain: A electrogastrographic study. Ann Biomed Eng 1994;22:423–431.

92. Lin Z, Chen JDZ. Time-frequency representation of the electrogastrogram–application of the exponential distribution. IEEE Trans Biomed Eng 1994;41(3):267–275.

93. Lin Z, Chen JDZ. Applications of feed-forward neural networks in the electrogastrograms. In: Akay M, editor. Nonlinear Biomedical Signal Processing. Piscataway (NJ): IEEE Press; 2000. p 233–255.

94. Wang ZS, He Z, Chen JDZ. Filter banks and neural network-based future extraction and automatic classification of electrogastrogram. Ann Biomed Eng 1999;27:88–95.

95. Wang ZS, Elsenbruch S, Orr WC, Chen JDZ. Detection of gastric slow wave uncoupling from multi-channel electrogastrogram: validations and applications. Neurogastroenterol Motil 2003;15:457–465.

96. Liang J, Cheung JY, Chen JDZ. Detection and deletion of motion artifacts in electrogastrogram using feature analysis and neural networks. Ann Biomed Eng 1997;25:850–857.

97. Wang ZS, Cheung JY, Chen JDZ. Blind separation of multichannel electrogastrograms using independent component analysis. Med Biol Eng Comput 1999;37:80–86.

98. Ravelli AM, Milla PJ. Electrogastrography in vomiting chidren with disorders of the central nervous system. In: Chen JDZ, McCallum RW, editors. Electrogastrography: Principles and Applications. New York: Raven Press; 1994. p 403–410.

99. Chen JDZ, McCallum RW. Electrogastrographic parameters and their clinical significance. In: Chen JDZ, McCallum RW, editors. Electrogastrography: Principles and Applications. New York: Raven Press; 1994. p 45–73.

100. Liang J, Chen JDZ. What can be measured from surface electrogastrography? Dig Dis Sci 1997;42(7):1331–1343.

101. Geldof H, van der Schee EJ, Smout AJPM. Myoelectrical activity of the stomach in gastric ulcer patients: An electrogastrographic study. J Gastrointest Motil 1989;1:122–130.

102. Koch KL, Bingaman S, Sperry N, Stern RM. Electrogastrography differentiates mechanical vs. idiopathic gastroparesis in patients with nausea and vomiting. Gastroenterology 1991;100:A99, (Abstract).

103. Xu L, Koch KL, Summy-Long J, Stern RM, Demers L, Bingaman S. Hypothalamic and gastric myoelectrical responses to vection in healthy Chinese subjects. Am J Physiol 1993;265:E578–E584.

104. Lin Z, Chen JDZ. Time-frequency analysis of the electrogastrogram. In: Akay M, editor. Time-Frequency and Wavelets in Biomedical Engineering. Piscataway, NJ: IEEE Press; 1996. 147–181.

105. Sanaka MR, Xing JH, Soffer EE. The effect of body posture on electrogastrogram. Am J Gastroenterol 2001;96:S73. (Abstract).

106. Chen J, McCallum RW. The response of electrical activity in normal human stomach to water and solid meal. Med Biol Eng Comput 1991;29:351–357.

107. Watanabe M, Shimada Y, Sakai S, Shibahara N, Matsumi H, Umeno K, Asanoi H, Terasawa K. Effects of water ingestion on gastric electrical activity and heart-rate variability in healthy human subjects. J Autonomic Nervous System 1996;58:44–50.

108. Chen J, McCallum RW. Effect of milk on myoelectical activity in normal human stomach: An electrogastrographic study. Med Biol Eng Comput 1992;30:564–567.

109. Lin ZY, Chen JDZ, McCallum RW, Parolisi S, Shifflett J, Peura D. The prevalence of electrogastrogram abnormalities in patients with non-ulcer and H. pylori infection: results of H. pylori eradication. Dig Dis Sci 2001;46(4):739–745.

110. Stern RM, Koch KL, Leibowitz HW, Lindblad I, Shupert C, Stewart WR. Tachygastria and motion sickness. Aviat Space Environ Med 1985;56:1074–1077.

111. Koch KL, Tran TN, Stern RM, Bringaman S, et al. Gastric myoelectrical activity in premature and term infants. J Gastrointest Motil 1993;5:41–47.

112. Koch KL, Bringaman S, Tran TN, Stern RM. Visceral perceptions and gastric myoelectrical activity in healthy women and in patients with bulimia nervosa. Neurogastronetrol Motil 1998;10:3–10.

113. Hathaway DK, Abell T, Cardoso S, Heartwig MS, Gebely S, Gaber AO. Improvement in autonomic and gastric function following pancreas-kidney versus kidney-alone transplantation and the correlation with quality of life. Transplantation 1994;57:816.

114. Gaber AO, Hathaway DK, Abell T, Cardoso S, Heartwig MS, Gebely S. Improved autonomic and gastric function in pancreas-kidney vs. kidney-alone transplantation contributes to quality of life. Transplant Proc 1994;26:515.

115. Cashion AK, Holmes SL, Hathaway DK, Gaber AO. Gastroparesis following kidney/pancreas transplant. Clin Transplant 2004;18:306–311.

116. Levanon D, Chen JDZ. Electrogastrography: its role in managing gastric disorders. (Invited Review). J Pediatr Gastroenterol Nutr 1998;27:431–443.

117. Chen JDZ, Co E, Liang J, Pan J, Sutphen J, Torres-Pinedo RB, Orr WC. Patterns of gastric myoelectrical activity in human subjects of different ages. Am J Physiol 1997;272: G1022–G1027.

118. Liang J, Co E, Zhang M, Pineda J, Chen JDZ. Development of gastric slow waves in preterm infants measured by electrogastrography. Am J Physiol (Gastrointest Liver Physiol) 1998;37:G503–G508.

119. Levy J, Harris J, Chen J, Sapoznikov D, Riley B, De La Nuez W, Khaskeberg A. Electrogastrographic norms in children: toward the development of standard methods, reproducible results, and reliable normative data. J Pediatr Gastroenterol Nutr 2001;33(4):455–461.

120. Cucchiara S, Riezzo G, Minella R, et al. Electrogastrography in non-ulcer dyspepsia. Arch Disease Childhood 1992;67(5): 613–617.

121. Ravelli AM, Ledermann SE, Bisset WM, Trompeter RS, Barratt TM, Milla PJ. Gastric antral myoelectrical activity in children wit chronic renal failure. In: Chen JDZ, McCallum RW, editors. Electrogastrography: Principles and Applications. New York: Raven Press; 1994. p 411–418.

122. Devane SP, Ravelli AM, Bisset WM, Smith VV, Lake BD, Milla PJ. Gastric antral dysrhythmias in children wit chronic idiopathic intestinal pseudoobstruction. Gut 1992; 33:1477–1481.

123. Forster J, Damjanov I, Lin ZY, Sarosiek I, Wetzel P, McCallum RW. Absence of the interstitial cells of Cajal in patients with gastroparesis and correlation with clinical findings. J Gastrointest Sur 2005;9:102–108.

124. Lin ZY, Chen JDZ. Advances in electrical stimulation of the gastrointestinal tract. Crit Rev Biomed Eng 2002;30(4–6): 419–457.

125. Abell T, McCallum RW, Hocking M, Koch K, Abrahamssion H, LeBlang I, Lindberg G, Konturek J, Nowak T, Quigley EMM, Tougas G, Starkebaum W. Gastric electrical stimulation for medically refractory gastroparesis. Gastroenterology 2003;125:421–428.

126. Dickens EJ, Hirst GD, Tomita T. Identification of rhythmically active cells in guinespig stomach. J Physiol (London) 1999;514:515–531.

127. Ordog T, Ward SM, Sanders KM. Interstitial cells of Cajal generate slow waves in the murine stomach. J Physiol 1999; 518:257–269.

128. Hanani M, Freund HR. Interstitial cells of Cajal—their role in pacing and signal transmission in digestive system. Acta Physiol Scand 170;177–190.

129. Lin ZY, Sarosiek I, Forster J, McCallum RW. Association between baseline parameters of the electrogastrogram and long-term symptom improvement in gastroparetic patients treated with gastric electrical stimulation. Neurogastroenterol Motil 2003;15:345–346.

Reading List

Stern RM, Koch KL. Using electrogastrography to study motion sickness. In: Chen JDZ, McCallum RW, editors. Electrogastrography: Principles and Applications. New York: Raven Press; 1994. p 199–218.

See also GASTROINTESTINAL HEMORRHAGE; GRAPHIC RECORDERS.

ELECTROMAGNETIC FLOWMETER.   See FLOWMETERS, ELECTROMAGNETIC.

ELECTROMYOGRAPHY

CARLO DE LUCA
Boston University
Boston, Massachusetts

INTRODUCTION

Electromyography is the discipline that deals with the detection, analysis, and use of the electrical signal that emanates from contracting muscles.

This signal is referred to as the electromyographic (EMG) signal, a term that was more appropriate in the past than in the present. In days past, the only way to capture the signal for subsequent study was to obtain a "graphic" representation. Today, of course, it is possible to store the signal on magnetic tape, disks, and electronics components. Even more means will become available in the near future. This evolution has made the graphics aspect of the nomenclature a limited descriptor. Although a growing number of practitioners choose to use the term "myoelectric (ME) signal", the term "EMG" still commands dominant usage, especially in clinical environments.

An example of the EMG signal can be seen in Fig. 1. Here the signal begins with a low amplitude, which when expanded reveals the individual action potentials associated with the contractile activity of individual (or a small group) of muscle fibers. As the force output of the muscle contraction increases, more muscle fivers are activated and the firing rate of the fibers increases. Correspondingly, the amplitude of the signal increases taking on the appearance and characteristics of a Gaussian distributed variable.

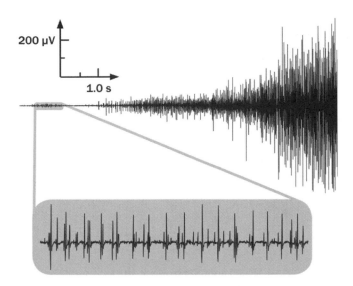

Figure 1. The EMG signal recorded with surface electrodes located on the skin above the first dorsal interosseous muscle in the hand. The signal increases in amplitude as the force produced by the muscle increases.

The novice in this field may well ask, why study electromyography? Why bother understanding the EMG signal? There are many and varied reasons for doing so. Even a superficial acquaintance with the scientific literature will uncover various current applications in fields such as neurophysiology, kinesiology, motor control, psychology, rehabilitation medicine, and biomedical engineering. Although the state of the art provides a sound and rich complement of applications, it is the potential of future applications that generates genuine enthusiasm.

HISTORICAL PERSPECTIVE

Electromyography had its earliest roots in the custom practiced by the Greeks of using electric eels to "shock" ailments out of the body. The origin of the shock that accompanied this earliest detection and application of the EMG signal was not appreciated until 1666 when an Italian, Francesco Redi, realized that it originated from muscle tissue (1). This relationship was later proved by Luigi Galvani (2) in 1791 who staunchly defended the notion. During the ensuing six decades, a few investigators dabbled with this newly discovered phenomenon, but it remained for DuBois Reymond (3) in 1849 to prove that the EMG signal could be detected from human muscle during a voluntary contraction. This pivotal discovery remained untapped for eight decades awaiting the development of technological implements to exploit its prospects. This interval brought forth new instruments such as the cathode ray tube, vacuum tube amplifiers, metal electrodes, and the revolutionary needle electrode which provided means for conveniently detecting the EMG signal. This simple implement introduced by Adrian and Bronk (4) in 1929 fired the imagination of many clinical researchers who embraced electromyography as an essential resource for diagnostic procedures. Noteworthy among these was the contribution of Buchthal and his associates.

Guided by the work of Inman et al. (5), in the mid-1940s to the mid-1950s several investigations revealed a monotonic relationship between the amplitude of the EMG signal and the force and velocity of a muscle contraction. This significant finding had a considerable impact: It dramatically popularized the use of electromyographic studies concerned with muscle function, motor control, and kinesiology. Kinesiological investigations received yet another impetus in the early 1960s with the introduction of wire electrodes. The properties of the wire electrode were diligently exploited by Basmajian and his associates during the next two decades.

In the early 1960s, another dramatic evolution occurred in the field: myoelectric control of externally powered prostheses. During this period, engineers from several countries developed externally powered upper limb prostheses that were made possible by the miniaturization of electronics components and the development of lighter, more compact batteries that could be carried by amputees. Noteworthy among the developments of externally powered prostheses was the work of the Yugoslavian engineer Tomovic and the Russian engineer Kobrinski, who in the late 1950s and early 1960s provided the first examples of such devices.

In the following decade, a formal theoretical basis for electromyography began to evolve. Up to this time, all knowledge in the field had evolved from empirical and often anecdotal observations. De Luca (6,7) described a mathematical model that explained many properties of the time domain parameters of the EMG signal, and Lindstrom (8) described a mathematical model that explained many properties of the frequency domain parameters of the EMG signal. With the introduction of analytical and simulation techniques, new approaches to the processing of the EMG signal surfaced. Of particular importance was the work of Graupe and Cline (9), who employed the autoregressive moving average technique for extracting information from the signal.

The late 1970s and early 1980s saw the use of sophisticated computer algorithms and communication theory to decompose the EMG signal into the individual electrical activities of the muscle fibers (10–12). Today, the decomposition approach promises to revolutionize clinical electromyography and to provide a powerful tool for investigating the detailed control schemes used by the nervous system to produce muscle contractions. In the same vein, the use of a thin tungsten wire electrode for detecting the action potential from single fibers was popularized for clinical applications (13,14). Other techniques using the surface EMG signal, such as the use of median and mean frequencies of the EMG signal to describe the functional state of a muscle and the use of the conduction velocity of the EMG signal to provide information on the morphology of the muscle fibers began to take hold. For a review, see De Luca (15).

The 1990s saw the effective application of modern signal processing techniques for the analysis and use of the EMG signal. Some examples are the use of time and frequency analysis of the surface EMG signal for measuring the relative contribution of low back muscles during the presence and absence of low back pain (16); the use of

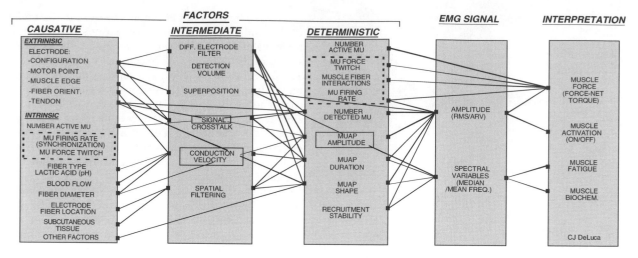

Figure 2. Relationship among the various factors that affect the EMG signal. [Reproduced with Permission from C. J. De Luca, "The Use of Surface Electromyography in Biomechanics." In the Journal of Applied Biomechanics, Vol. 13(No 2): p 139, Fig. 1.]

systematic measurements of the muscle fiber conduction velocity for measuring the severity of the Duchenne Dystrophy (17); the analysis of motor unit action potential delay for locating the origin, the ending and the innervation zone of muscle fibers (18); and the application of time–frequency analysis of the EMG signal to the field of laryngology (19).

New and exciting developments are on the horizon. For example, the use of large-scale multichannel detection of EMG signals for locating sources of muscle fiber abnormality (20); application of neural networks to provide greater degrees of freedom for the control of myoelectric prostheses (21), and for the analysis of EMG sensors data for assessing the motor activities and performance of sound subjects (22) and Stroke patients (23). Yet another interesting development is the emerging use of sophisticated Artificial Intelligence techniques for the decomposing the EMG signal (24). The reader who is interested in more historical and factual details is referred to the book *Muscles Alive* (25).

DESCRIPTION OF THE EMG SIGNAL

The EMG signal is the electrical manifestation of the neuromuscular activation associated with a contracting muscle. The signal represents the current generated by the ionic flow across the membrane of the muscle fibers that propagates through the intervening tissues to reach the detection surface of an electrode located in the environment. It is a complicated signal that is affected by the anatomical and physiological properties of muscles and the control scheme of the nervous system, as well as the characteristics of the instrumentation used to detect and observe it. Some of the complexity is presented in Fig. 2 that depicts a schematic diagram of the main physiological, anatomical and biochemical factors that affect the EMG signal. The connecting lines in the diagram show the interaction among three classes of factors that influence the EMG signal. The causative factors have a basic or elemental effect on the signal. The intermediate factors represent physical and physiological phenomena that are

influenced by one or more of the causative factors and in turn influence the deterministic factors that represent physical characteristics of the action potentials. For further details see De Luca (26).

In order to understand the EMG signal, it is necessary to appreciate some fundamental aspects of physiology. Muscle fibers are innervated in groups called motor units, which when activated generate a motor unit action potential. The activation from the central nervous system is repeated continuously for as long as the muscle is required to generate force. This continued activation generates motor unit action potential trains. These trains from the concurrently active motor units superimpose to form the EMG signal. As the excitation from the Central Nervous System increases to generate greater force in the muscle, a greater number of motor units are activated (or recruited) and the firing rates of all the active motor units increases.

Motor Unit Action Potential

The most fundamental functional unit of a muscle is called the motor unit. It consists of an α-motoneuron and all the muscle fibers that are innervated by the motoneuron's axonal branches. The electrical signal that emanates from the activation of the muscle fibers of a motor unit that are in the detectable vicinity of an electrode is called the motor unit action potential (MUAP). This constitutes the fundamental unit of the EMG signal. A schematic representation of the genesis of a MUAP is presented in Fig. 3. Note the many factors that influence the shape of the MUAP. Some of these are (1) the relative geometrical relationship of the detection surfaces of the electrode and the muscles fibers of the motor unit in its vicinity; (2) the relative position of the detection surfaces to the innervation zone, that is, the region where the nerve branches contact the muscle fibers; (3) the size of the muscle fibers (because the amplitude of the individual action potential is proportional to the diameter of the fiber); and (4) the number of muscle fibers of an individual motor unit in the detectable vicinity of the electrode.

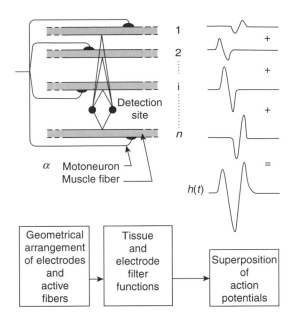

Figure 3. Schematic representation of the generation of the motor unit action potential.

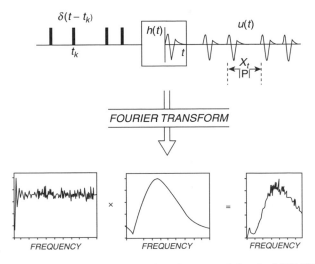

Figure 4. Model for a motor unit action potential train (MUAPT) and the corresponding Fourier transform of the interpulse intervals (IPIs), the motor unit actions potentials (MUAP), and the MUAPT.

The last two factors have particular importance in clinical applications. Considerable work has been performed to identify morphological modifications of the MUAP shape resulting from modifications in the morphology of the muscle fibers (e.g., hypertrophy and atrophy) or the motor unit (e.g., loss of muscle fibers and regeneration of axons). Although usage of MUAP shape analysis is common practice among neurologists, interpretation of the results is not always straightforward and relies heavily on the experience and disposition of the observer.

Motor Unit Action Potential Train

The electrical manifestation of a MUAP is accompanied by a contractile twitch of the muscle fibers. To sustain a muscle contraction, the motor units must be activated repeatedly. The resulting sequence of MUAPs is called a motor unit action potential train (MUAPT). The waveform of the MUAPs within a MUAPT will remain constant if the geometric relationship between the electrode and the active muscle fibers remains constant, if the properties of the recording electrode do not change, and if there are no significant biochemical changes in the muscle tissue. Biochemical changes within the muscle can affect the conduction velocity of the muscle fiber and the filtering properties of the muscle tissue.

The MUAPT may be completely described by its interpulse intervals (the time between adjacent MUAPs) and the waveform of the MUAP. Mathematically, the interpulse intervals may be expressed as a sequence of Dirac delta impulses $\delta i(t)$ convoluted with a filter $h(t)$ that represents the shape of the MUAP. Figure 4 presents a graphic representation of a model for the MUAPT. It follows that the MUAPT, $u_i(t)$ can be expressed as

$$u_i(t) = \sum_{k=1}^{n} h_i(t - t_k)$$

where

$$t_k = \sum_{l=1}^{k} x_l \quad \text{for} \quad k, l = 1, 2, 3, \ldots, n$$

In the above expression, t_k represents the time locations of the MUAPs, x represents the interpulse intervals, n is the total number of interpulse intervals in a MUAPT, and i, k, and l are integers that denote specific events.

By representing the interpulse intervals as a renewal process and restricting the MUAP shape so that it is invariant throughout the train, it is possible to derive the approximations

Mean rectified value
$$= E\{|u_i(t, F)|\} \cong \lambda_i(t, F) \int_0^{\infty} |h_i(t)| dt$$

Mean squared value
$$= MS\{|u_i(t, F)|\} \cong \lambda_i(t, F) \int_0^{\infty} h_i^2(t) dt$$

where F is the force generated by the muscle and is the firing rate of the motor unit.

The power density spectrum of a MUAPT was derived from the above formulation by LeFever and De Luca [(27) and independently by Lago and Jones (28)]. It can be expressed as

$$S_{u_i}(\omega, t, F) = S_{\delta_i}(\omega, t, F)|H_i(j\omega)|^2$$
$$= \frac{\lambda_i(t,F) \cdot \{1 - |M(j\omega,t,F)|^2\}}{1 - 2 \cdot \text{Real}\{M(j\omega,t,F)\} + |M(j\omega,t,F)|^2} \{|H_i(j\omega)|^2\}$$

for $\neq 0$

where is the frequency in radians per second, $H_i(j\omega)$ is the Fourier transform of $h_i(t)$, and $M(j\omega, t, F)$ is the Fourier transform of the probability distribution function, $px(x, t, F)$ of the interpulse intervals.

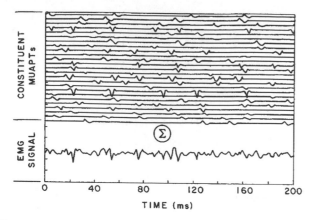

Figure 5. An EMG signal formed by adding (superimposing) 25 mathematically generated MUAPTs.

The EMG Signal

The EMG signal may be synthesized by linearly summing the MUAPTs. This approach is expressed in the equation

$$m(t,F) = \sum_{i=1}^{p} u_i(t,F)$$

and is displayed in Fig. 5, where 25 mathematically generated MUAPTs were added to yield the signal at the bottom. This composite signal bears striking similarity to the real EMG signal.

From this concept, it is possible to derive expressions for commonly used parameters: mean rectified value, root-mean-squared (rms) value, and variance of the rectified EMG signal. The interested reader is referred to *Muscles Alive* (25).

Continuing with the evolution of the model, it is possible to derive an expression for the power density spectrum of the EMG signal:

$$S_m(\omega,t,F) = R(\omega,d)\left[\sum_{i=1}^{p(F)} S_{u_i}(\omega,t) + \sum_{\substack{i,j=1 \\ i \neq j}}^{q(F)} S_{u_i u_j}(\omega,t)\right]$$

where $R(\omega,d) = K \sin^2(\omega d/2v)$ is the bipolar electrode filter function; d is the distance between detection surfaces of the electrode; is the angular frequency; v is the conduction velocity along the muscle fibers; $S_{u_i}(\omega)$ is the power density of the MUAPT, $u_i(t)$; $S_{u_i u_j}(\omega)$ is the cross-power density spectrum MUAPTs $u_i(t)$ and $u_j(t)$; p is the total number of MUAPTs that constitute the signal; and q is the number of MUAPTs with correlated discharges.

Lindstrom (8), using a dipole model, arrived at another expression for the power density spectrum:

$$S_m(\omega,t,F) = R(\omega,d)\left[1v^2(t,F)G(\omega d2v(t,F))\right]$$

This representation explicitly denotes the interconnection between the spectrum of the EMG signal and the conduction velocity of the muscle fibers. Such a relationship is implicit in the previously presented modeling approach because any change in the conduction velocity would directly manifest itself in a change in the time duration

of $h(t)$ as seen by the two detection surfaces of a stationary bipolar electrode.

ELECTRODES

Two main types of electrodes are used to detect the EMG signal: one is the surface (or skin) electrode and the other is the inserted (wire or needle) electrode. Electrodes are typically used singularly or in pairs. These configurations are referred to as monopolar and bipolar, respectively.

Surface Electrodes

There are two categories of surface electrode: passive and active. Passive electrode consists of conductive (usually metal) detection surface that senses the current on the skin through its skin electrode interface. Active electrodes contain a high input impedance electronics amplifier in the same housing as the detection surfaces. This arrangement renders it less sensitive to the impedance (and therefore quality) of the electrode–skin interface. The current trend is towards active electrodes.

The simplest form of passive electrode consists of silver disks that adhere to the skin. Electrical contact is greatly improved by introducing a conductive gel or paste between the electrode and skin. The impedance can be further reduced by removing the dead surface layer of the skin along with its protective oils; this is best done by light abrasion of the skin.

The lack of chemical equilibrium at the metal electrolyte junction sets up a polarization potential that may vary with temperature fluctuations, sweat accumulation, changes in electrolyte concentration of the paste or gel, relative movement of the metal and skin, as well as the amount of current flowing into the electrode. It is important to note that the polarization potential has both a direct current (dc) and an alternating current (ac) component. The ac component is greatly reduced by providing a reversible chloride exchange interface with the metal of the electrode. Such an arrangement is found in the silver–silver chloride electrodes. This type of electrode has become highly popular in electromyography because of its light mass (0.25 g), small size (< 10 mm diameter), and high reliability and durability. The dc component of the polarization potential is nullified by ac amplification when the electrodes are used in pairs. This point is elaborated upon in later sections of this article.

The active surface electrodes have been developed to eliminate the need for skin preparation and conducting medium. They are often referred to as "dry" or "pasteless" electrodes. These electrodes may be either resistively or capacitively coupled to the skin. Although the capacitively coupled electrodes have the advantage of not requiring a conductive medium, they have a higher inherent noise level. Also, these electrodes do not have long term reliability because their dielectric properties are susceptible to change with the presence of perspiration and the erosion of the dielectric substance. For these reasons, they have not yet found a place in electromyography.

An adequately large input impedance is achieved when resistance is on the order of 10 TΩ and capacitance is small

Figure 6. Examples of active surface electrode in bipolar configurations from Delsys Inc. The spacing between the bars is 10 mm, the length of the bars is 10 mm and the thickness is 1 mm. These electrodes do not require any skin preparation or conductive paste or gels.

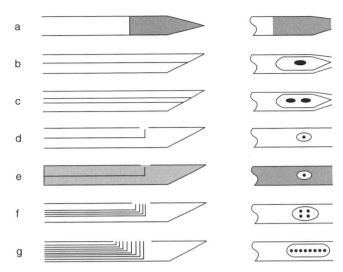

Figure 7. Examples of various needle electrodes: (a) A solid tip single-fiber electrode. If it is sufficiently thin, it can be inserted into a nerve bundle and detect neuroelectrical signals. (b) Concentric needle with one monopolar detection surface formed by the beveled cross-section of centrally located wire typically 200 μm in diameter. Commonly used in clinical practice. (c) Bipolar needle electrode with two wires exposed in cross-section, typically 100 μm in diameter. Used in clinical practice. (d) Single-fiber electrode with 25 μm diameter wire. Used to detect the activity of individual muscle fibers. (e) Macroelectrode with 25 μm diameter wire and with the cannula of the needle used as a detection surface. Used to detect the motor unit action potential from a large portion of the motor unit territory. (f) Quadrifilar planar electrode with four 50 μm wires located on the corners of a square 150 μm apart (center to center). Used for multiple channel recordings and in EMG signal decomposition technique. (g) Multifilar electrode consisting of a row of wires, generally used to study the motor unit territory.

(typically, 3 or 4 pF). The advent of modern microelectronics has made possible the construction of amplifiers housed in integrated circuitry which have the required input impedance and associated necessary characteristics. An example of such an electrode is presented in Fig. 6. This genre of electrodes was conceptualized and first constructed at the NeuroMuscular Research Laboratory at Children's Hospital Medical Center, Boston, MA in the late 1970s. They each have two detection surfaces and associated electronic circuitry within their housing.

The chief disadvantages of surface electrodes are that they can be used effectively only with superficial muscles and that they cannot be used to detect signals selectively from small muscles. In the latter case, the detection of "cross-talk" signals from other adjacent muscles becomes a concern. These limitations are often outweighed by their advantages in the following circumstances:

1. When representation of the EMG signal corresponding to a substantial part of the muscle is required.
2. In motor behavior studies, when the time of activation and the magnitude of the signal contain the required information.
3. In psychophysiological studies of general gross relaxation of tenseness, such as in biofeedback research and therapy.
4. In the detection of EMG signals for the purpose of controlling external devices such as myoelectrically controlled prostheses and other like aids for the physically disabled population.
5. In clinical environments, where a relatively simple assessment of the muscle involvement is required, for example, in physical therapy evaluations and sports medicine evaluations.
6. Where the simultaneous activity or interplay of activity is being studied in a fairly large group of muscles under conditions where palpation is impractical, for example, in the muscles of the lower limb during walking.
7. In studies of children or other individuals who object to needle insertions.

Needle Electrodes

By far, the most common indwelling electrode is the needle electrode. A wide variety is commercially available. (see Fig. 7). The most common needle electrode is the "concentric" electrode used by clinicians. This monopolar configuration contains one insulated wire in the cannula. The tip of the wire is bare and acts as a detection surface. The bipolar configuration contains a second wire in the cannula and provides a second detection surface. The needle electrode has two main advantages. One is that its relatively small pickup area enables the electrode to detect individual MUAPs during relatively low force contractions. The other is that the electrodes may be conveniently repositioned within the muscle (after insertion) so that new tissue territories may be explored or the signal quality may be improved. These amenities have naturally led to the development of various specialized versions such as the multifilar electrode developed by Buchthal et al. (29), the planar quadrifilar electrode of De Luca and Forrest (30), the single fiber electrode of Ekstedt and Stålberg (13), and the macroelectrode of Stålberg (14). The single-fiber electrode consists of a thin, stiff, sharpened metal filament, usually made of tungsten. When inserted into a muscle it detects the action potentials of individual fibers. This electrode has proven to be useful for neurological examinations of dein-nervated muscles. Examples of these electrodes may be seen in Fig. 7.

Wire Electrodes

Since the early 1960s, this type of electrode has been popularized by Basmajian and Stecko (31). Similar electrodes that differ only in minor details of construction were developed independently at about the same time by other researchers. Wire electrodes have proved a boon to kinesiological studies because they are extremely fine, they are easily implanted and withdrawn from skeletal muscles, and they are generally less painful than needle electrodes whose cannula remains inserted in the muscle throughout the duration of the test.

Wire electrodes may be made from any small diameter, highly nonoxidizing, stiff wire with insulation. Alloys of platinum, silver, nickel, and chromium are typically used. Insulations, such as nylon, polyurethane, and Teflon, are conveniently available. The preferable alloy is 90% platinum, 10% iridium; it offers the appropriate combination of chemical inertness, mechanical strength, stiffness and economy. The Teflon and nylon insulations are preferred because they add some mechanical rigidity to the wires, making them easier to handle. The electrode is constructed by inserting two insulated fine (25–100 μm in diameter) wires through the cannula of a hypodermic needle. Approximately 1–2 mm of the distal tips of the wire is deinsulated and bent to form two staggered hooks (see Fig. 8 for completed version). The electrode is introduced into the muscle by inserting the hypodermic needle and then withdrawing it. The wires remain lodged in the muscle tissues. They may be removed by gently pulling them out: They are so pliable that the hooks straighten out on retraction.

In kinesiological studies, where the main purpose of using wire electrodes is to record a signal that is proportional to the contraction level of muscle, repositioning of the electrode is not important. But for other applications, such as recording distinguishable MUAPTs, this limitation is counterproductive. Some have used the phrase "poke and hope" to describe the standard wire electrode technique for this particular application. Another limitation of the wire electrode is its tendency to migrate after it has been inserted, especially during the first few contractions of the muscle. The migration usually stops after a few contractions. Consequently, it is recommended to perform a half dozen or so short duration contraction before the actual recording session begins.

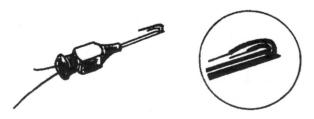

Figure 8. A bipolar wire electrode with its carrier needle used for insertion.

Electrode Maintenance

Proper usage of wire and needle electrodes requires constant surveillance of the physical and electrical characteristics of the electrode detection surfaces. Particular attention should be given to keeping the tips free of debris and oxidation. The reader is referred to the book *Muscles Alive* (25) for details on these procedures as well as suggestions for sterilization.

How to Choose the Proper Electrode

The specific type of electrode chosen to detect the EMG signal depends on the particular application and the convenience of use. The application refers to the information that is expected to be obtained from the signal; for example, obtaining individual MUAPs or the gross EMG signal reflecting the activity of many muscle fibers. The convenience aspect refers to the time and effort the investigator wishes to devote to the disposition of the subject or patient. Children, for example, are generally resistant to having needles inserted in their muscles.

The following electrode usage is recommended. The reader, however, should keep in the mind that crossover applications are always possible for specific circumstances.

Surface Electrodes

Time force relationship of EMG signals.
Kinesiological studies of surface muscles.
Neurophysiological studies of surface muscles.
Psychophysiological studies.
Interfacing an individual with external electromechanical devices.

Needle Electrode

MUAP characteristics.
Control properties of motor units (firing rate, recruitment, etc.).
Exploratory clinical electromyography.

Wire Electrodes

Kinesiological studies of deep muscles.
Neurophysiological studies of deep muscles.
Limited studies of motor unit properties.
Comfortable recording procedure from deep muscles.

Where to Locate the Electrode

The location of the electrode should be determined by three important considerations: (*1*) signal/noise ratio, (*2*) signal stability (reliability), and (*3*) cross-talk from adjacent muscles. The stability consideration addresses the issue of the modulation of the signal amplitude due to relative movement of the active fibers with respect to the detection surfaces of the electrode. The issue of cross-talk concerns the detection by the electrode of signals emanating from adjacent muscles.

For most configurations of needle electrodes, the question of cross-talk is of minor concern because the electrode is so selective that it detects only signals from nearby muscle fibers. Because the muscle fibers of different motor units are scattered in a semirandom fashion throughout the muscle, the location of the electrode becomes irrelevant from the point of view of signal quality and information content. The stability of the signal will not necessarily be improved in any one location. Nonetheless, it is wise to steer clear of the innervation zone so as to reduce the probability of irritating a nerve ending.

All the considerations that have been discussed for needle electrodes also apply to wire electrodes. In this case, any complication will be unforgiving in that the electrode may not be relocated. Since the wire electrodes have a larger pickup area, a concern arises with respect to how the location of the insertion affects the stability of the signal. This question is even more dramatic in the case of surface electrodes.

For surface electrodes, the issue of cross-talk must be considered. Obviously, it is not wise to optimize the signal detected, only to have the detected signal unacceptably contaminated by an unwanted source. A second consideration concerns the susceptibility of the signal to the architecture of the muscle. Both the innervation zone and the tendon muscle tissue interface have been found to alter the characteristics of the signal. *It is suggested that the preferred location of an electrode is in the region halfway between the center of the innervation zone and the further tendon. See the review article by De Luca (12) for additional details.*

SIGNAL DETECTION: PRACTICAL CONSIDERATIONS

When attempting to collect an EMG signal, both the novice and the expert should remember that the characteristics of the observed EMG signal are a function of the apparatus used to acquire the signal as well as the electrical current that is generated by the membrane of the muscle fibers. The "distortion" of the signal as it progresses from the source to the electrode may be viewed as a filtering sequence. An overview of the major filtering effects is presented in Fig. 9. A brief summary of the pertinent facts follows. The reader interested in additional details is referred to *Muscles Alive* (25).

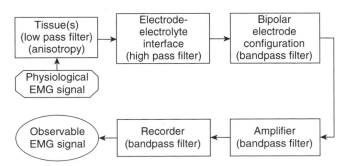

Figure 9. Block diagram of all the major aspects of the signal acquisition procedure. Note the variety of physical properties that act as filters to the EMG signal before it can be observed. The term "physiological EMG signal" refers to the collection of signals that emanate from the surface of the muscle fibers. These are not observable.

Electrode Configuration

The electrical activity inside a muscle or on the surface of the skin outside a muscle may be easily acquired by placing an electrode with only one detection surface in either environment and detecting the electrical potential at this point with respect to a "reference" electrode located in an environment that either is electrically quiet or contains electrical signals unrelated to those being detected. ("Unrelated" means that the two signals have minimal physiological and anatomical associations.) A surface electrode is commonly used as the reference electrode. Such an arrangement is called monopolar and is at times used in clinical environments because of its relative technical simplicity. A schematic arrangement of the monopolar detection configuration may be seen in Fig. 10. The monopolar configuration has the drawback that it will detect all the electrical signals in the vicinity of the detection surface; this includes unwanted signals from sources other than the muscle of interest.

The bipolar detection configuration overcomes this limitation (see Fig. 10). In this case, two surfaces are used to detect two potentials in the muscle tissue of interest each with respect to the reference electrode. The two signals are then fed to a differential amplifier which amplifies the

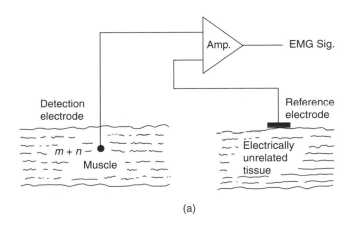

(a)

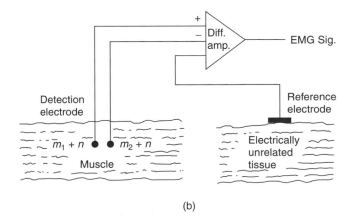

(b)

Figure 10. (a) Monopolar detection arrangement. (b) Bipolar detection arrangement. Note that in the bipolar detection arrangement, the EMG signals are considered to be different, whereas the noise is similar.

difference of the two signals, thus eliminating any "common mode" components in the two signals. Signals emanating from the muscle tissue of interest near the detection surface will be dissimilar at each detection surface because of the localized electrochemical events occurring in the contracting muscle fibers, whereas "ac noise" signals originating from a more distant source (e.g., 50 or 60 Hz electromagnetic signals radiating from power cords, outlets, and electrical devices) and "dc noise" signals (e.g., polarization potentials in the metal electrolyte junction) will be detected with an essentially similar amplitude at both detection surfaces. Therefore, they will be subtracted, but not necessarily nullified prior to being amplified. The measure bf the ability of the differential amplifier to eliminate the common mode signal is called the common mode rejection ratio.

Spatial Filtering

1. As the signal propagates through the tissues, the amplitude decreases as a function of distance. The amplitude of the EMG signal decreases to approximately 25% within 100 μm. Thus, an indwelling electrode will detect only signals from nearby muscle fibers.
2. The filtering characteristic of the muscle tissues is a function of the distance between the active muscle fibers and the detection surface(s) of the electrode. In the case of surface electrodes, the thickness of the fatty and skin tissues must also be considered. The tissues behaves as a low pass filter whose bandwidth and gain decrease as the distance increases.
3. The muscle tissue is anisotropic. Therefore, the orientation of the detection surfaces of the electrode with respect to the length of the muscle fibers is critical.

Electrode Electrolyte Interface

1. The contact layer between the metallic detection surface of the electrode and the conductive tissue forms an electrochemical junction that behaves as a high pass filter.
2. The gain and bandwidth will be a function of the area of the detection surfaces and any chemical electrical alteration of the junction.

Bipolar Electrode Configuration

1. This configuration ideally behaves as a bandpass filter; however, this is true only if the inputs to the amplifier are balanced and the filtering aspects of the electrode electrolyte junctions are equivalent.
2. A larger interdetection surface spacing will render a lower bandwidth. This aspect is particularly significant for surface electrodes.
3. The greater the interdetection surface spacing, the greater the susceptibility of the electrode to detecting measurable amplitudes of EMG signals from adjacent and deep muscles. Again, this aspect is particularly significant for surface electrodes.

4. An interdetection surface spacing of 1.0 cm is recommended for surface electrodes.

Amplifier Characteristics

1. These should be designed and/or set for values that will minimally distort the EMG signal detected by the electrodes.
2. The leads to the input of the amplifier (actually, the first stage of the amplification) should be as short as possible and should not be susceptible to movement. This may be accomplished by building the first stage of the amplifier (the preamplifier) in a small configuration which should be located near (within 10 cm) the electrode. For surface EMG amplifiers the first stage is often located in the housing of the electrodes.
3. The following are typical specifications that can be attained by modern day electronics. It is worth noting that the values below will improve as more advanced electronics components become available in the future.

(a) Common-mode input impedance: As large as possible (typically $> 10^{15}$ Ω in parallel with < 7 pF).
(b) Common mode rejection ratio: > 85 dB.
(c) Input bias current: as low as possible (typically < 5 fA).
(d) Noise (shorted inputs) < 1.5 μV rms for 20–500 Hz bandwidth.
(e) Bandwidth in hertz (3 dB points for 12 dB/octave or more rolloff):

Surface electrodes	20–500
Wire electrodes	20–2,000
Monopolar and bipolar needle electrodes for general use	20–5,000
Needle electrodes for signal decomposition	1,000–10,000
Single fiber electrode	1,000–10,000
Macroelectrode	20–5,000

An example of an eight-channel modern surface EMG amplifier is presented in Fig. 11. Such systems are

Figure 11. An eight-channel surface EMG system from Delsys Inc. The dimensions of this device (205 $\times$ 108 $\times$ 57 mm) are typical for current day units. Note that the active electrodes connect to an input unit that is separate from the body of the amplifier and can be conveniently attached to the body of the subject.

available in configurations of various channels up to 32, but 8 and 16 channel versions are most common.

Recording Characteristics

The effective or actual bandwidth of the device or algorithm that is used to record or store the signal must be greater than that of the amplifiers.

Other Considerations

1. It is preferable to have the subject, the electrode, and the recording equipment in an electromagnetically quiet environment. If all the procedures and cautions discussed in this article are followed and heeded, high quality recordings will be obtained in the electromagnetic environments found in most institutions, including hospitals.

2. In the use of indwelling electrodes, great caution should be taken to minimize (eliminate, if possible) any relative movement between the detection surfaces of the electrodes and the muscle fibers. Relative movements of 0.1 mm may dramatically alter the characteristics of the detected EMG signal and may possibly cause the electrode to detect a different motor unit population.

SIGNAL ANALYSIS TECHNIQUES

The EMG signal is a time and force (and possibly other parameters) dependent signal whose amplitude varies in a random nature above and below the zero value. Thus, simple average aging of the signal will not provide any useful information.

Rectification

A simple method that is commonly used to overcome the above restriction is to rectify the signal before performing mode pertinent analysis. The process of rectification involves the concept of rendering only positive deflections of the signal. This may be accomplished either by eliminating the negative values (half-wave rectification) or by inverting the negative values (full-wave rectification). The latter is the preferred procedure because it retains all the energy of the signal.

Averages or Means of Rectified Signals

The equivalent operation to smoothing in a digital sense is averaging. By taking the average of randomly varying values of a signal, the larger fluctuations are removed, thus achieving the same results as the analog smoothing operation. The mathematical expression for the average or mean of the rectified EMG signal is

$$\overline{|m(t)|}\Big|_{t_j - t_i} = 1 t_j - t_i \int_{t_i}^{t_j} |m(t)| dt$$

where t_i and t_j are the points in time over which the integration and, hence, the averaging is performed. The shorter the time interval, the less smooth the averaged value will be.

The preceding expression will provide only one value over the time window $T = t_j - t_i$. To obtain the time varying average of a complete record of a signal, it is necessary to move the time window T duration along the record. This operation is referred to as moving average.

$$\overline{|m(t)|} = 1 T \int_{t}^{t+T} |m(t)| dt$$

Like the equivalent operation in the analogue sense, this operation introduces a lag; that is, T time must pass before the value of the average of the T time interval can be obtained. In most cases, this outcome does not present a serious restriction, especially if the value of T is chosen wisely. For typical applications, values ranging from 100 to 200 ms are suggested. It should be noted that shorter time windows, T, yield less smooth time dependent average (mean) of the rectified signal.

Integration

The most commonly used and abused data reduction procedure in electromyography is integration. The literature of the past three decades is swamped with improper usage of this term, although happily within the past decade it is possible to find increasing numbers of proper usage. When applied to a procedure for processing a signal, the temp integration has a well-defined meaning that is expressed in a mathematical sense. It applies to a calculation that obtains the area under a signal or a curve. The units of this parameter are volt seconds (V·s). It is apparent that an observed EMG signal with an average value of zero will also have a total area (integrated value) of zero. Therefore, the concept of integration may be applied only to the rectified value of the EMG signal.

$$I\{|m(t)|\} = \int_{t}^{t+T} |m(t)| dt$$

Note that the operation is a subset of the procedure of obtaining the average rectified value. Since the rectified value is always positive, the integrated rectified value will increase continuously as a function of time. The only difference between the integrated rectified value and the average rectified value is that in the latter case the value is divided by T, the time over which the average is calculated. If a sufficiently long integration time T is chosen, the integrated rectified value will provide a smoothly varying measure of the signal as a function of time. There is no additional information in the integrated rectified value.

Root-Mean-Square (rms) Value

Mathematical derivations of the time and force dependent parameters indicate that the rms value provides more a more rigorous measure of the information content of the signal because it measures the energy of the signal. Its use in electromyography, however, has been sparse in the past. The recent increase is due possibly to the availability of analog chips that perform the rms operation and to the increased technical competence in electromyography. The time-varying rms value is obtained by performing the operations described by the term in reverse order; that is,

$$\text{rms}\{m(t)\} = \left(1T \int_t^{t+T} m^2(t)dt\right)^{1/2}$$

This parameter is recommended above the others.

Zero Crossings and Turns Counting

This method consists of counting the number of times per unit time that the amplitude of the signal contains either a peak or crosses a zero value of the signal. It was popularized in electromyography by Williston (32). The relative ease with which these measurements could be obtained quickly made this technique popular among clinicians. Extensive clinical applications have been reported, some indicating that discrimination may be made between myopathic and normal muscle; however, such distinctions are usually drawn on a statistical basis.

This technique is not recommended for measuring the behavior of the signal as a function of force (when recruitment or derecruitment of motor units occurs) or as a function of time during a sustained contraction. Lindström et al. (33) showed that the relationship between the turns or zeros and the number of MUAPTs is linear for low level contractions. But as the contraction level increases, the additionally recruited motor units contribute MUAPTs to the EMG signal. When the signal amplitude attains the character of Gaussian random noise, the linear proportionality no longer holds.

Frequency Domain Analysis

Analysis of the EMG signal in the frequency domain involves measurements and parameters that describe specific aspects of the frequency spectrum of the signal. Fast Fourier transform techniques are commonly available and are convenient for obtaining the power density spectrum of the signal.

Three parameters of the power density spectrum may be conveniently used to provide useful measures of the spectrum. They are the median frequency, the mean frequency, and the bandwidth of the spectrum. Other parameters, such as the mode frequency and ratios of segments of the power density spectrum, have been used by some investigators, but are not considered reliable measures given the inevitably noisy nature of the spectrum. The median frequency and the mean frequency are defined by the equations:

$$\int_0^{f_{\text{med}}} S_m(f)df = \int_{f_{\text{med}}}^{\infty} S_m(f)df$$

$$f_{\text{mean}} = \int_0^f fS_m(f)df \Big/ \int_0^f S_m(f)df$$

where $S_m(f)$ is the power density spectrum of the EMG signal. Stulen and De Luca (34) performed a mathematical analysis to investigate the restrictions in estimating various parameters of the power density spectrum. The median and mean frequency parameters were found to be the most reliable, and of these two the median frequency was found to be less sensitive to noise. This quality is particu-

larly useful when a signal is obtained during low level contractions where the signal to-noise ratio may be < 6.

The above discussion on frequency spectrum parameters removes temporal information from the calculated parameters. This approach is appropriate for analyzing signals that are stationary or nearly stationary, such as those emanating from isometric, constant-force contractions. Measurement of frequency parameters during dynamic contractions requires techniques that retain the temporal information. During the past decade time–frequency analyses techniques have evolved in the field of Electromyography, as they have in the realm of other biosignals such as ECG and EEG. Early among the researchers to apply these techniques to the EMG signal were Contable et al. (35) who investigated the change in the frequency content of EMG signals during high jumps, and Roark et al. (19) who investigated the movement of the thyroarytenoid muscles during vocalization. In both these applications, the time–frequency techniques were essential because they investigated muscles that contracted dynamically and briefly.

Much of the work presented here is adapted, with permission, from Refs. 25, pp. 38, 58, 68, 74, and 81. The author thanks Williams & Wilkens for permission to extract this material.

BIBLIOGRAPHY

Cited References

1. Biederman W. Electrophysiology. 1898.
2. Galvani L. De Viribus Electricitatis. (R. Green, Transl.) London and New York: Cambridge University Press; 1953.
3. Du Bois RE. Untersuchungen uber theirische electricität. 2, 2nd P. Berlin: Verlag von G. Reimer; 1849.
4. Adrian ED, Bronk DW. J Physiol (London) 1929;67:19.
5. Inman VT, Sauders JBCM, Abbott LC J. Bone Jt Surg 1944;26:1.
6. De Luca CJ. MS [dissertation]. University of New Brunswick; 1968.
7. De Luca CJ. Biol Cybernet 1975;19:159.
8. Lindstrom LR. On the Frequency Spectrum of EMG Signals. Technical Report, Research Laboratory of Medical Electronics. Gothenburg, Sweden: Chalmers University of Technology; 1970.
9. Graupe D, Cline WK. IEEE Trans Syst Man Cybernet SMC 1975;5:252.
10. LeFever RS, De Luca CJ. Proceedings of the 8th Annual Meeting of Social Neuroscience; 1985. p 299.
11. LeFever RS, De Luca CJ. IEEE Trans Biomed Eng BME 1982;29:149.
12. McGill KC, Cummins KL, Dorfman LJ. IEEE Trans Biomed Eng 1985;32:470–477.
13. Ekstedt J, Stålberg E. In: Desmedt JE, editor. New Development EMG Clinical Neurophysiology 1. S. Karger; 1973. p 84.
14. Stålberg EJ. Neurol Neurosurg Psychiat 1980;43:475.
15. De Luca CJ. CRC Crit Rev Biomed Eng 1984;11:251–279.
16. Roy SH, De Luca CJ, Emley MC. J Rehab Res Dev 1997;34(4): 405–414.
17. Knaflitz M, Balestra G, Angelini C, Cadaldini M. Basic App Myol 1996;6(2):70,115.
18. Masuda T, Miyano H, Sadoyama T. EEG Clin Neurophysiol 1983;55(5):594–600.
19. Roark RM, Dowling EM, DeGroat RD, Watson BC, Schaefer SD. J Speech Hear Res 1995;38(2):289–303.

20. Zwarts MJ, Stegeman DF. Muscle Nerve 2003;28(1):1–17.
21. Light CM, Chappell PH, Hudgins B, Engelhart K. Med Eng Technol 2002;26(4):139–146.
22. Nawab SH, Roy SH, De Luca CJ. The 26th Int Conf IEEE Eng Med Biol Soc; San Francisco; 2004. p 979–982.
23. Roy SH, Cheng MS, De Luca CJ. Boston: ISEK Congress; 2004.
24. Nawab SH, Wotiz R, Hochstein L, De Luca CJ. Proceedings of the Second Joint Meeting of the IEEE Eng Med and Biol Soc and the Biomed Eng Soc; Houston: 2002. p 36–36.
25. Basmajian JV, De Luca CJ. Muscles Alive. 5th ed. Baltimore: Williams & Wilkins; 1985.
26. De Luca CJ. Muscle Nerve 1993;16:210–216.
27. LeFever RS, De Luca CJ. Proc Annu Conf Eng Med Biol 1976;18:56.
28. Lago P, Jones NB. Med Biol Eng Comput 1977;15:648.
29. Buchthal F, Guld C, Rosenfalck P. Acta Physiol Scand 1957;39:83.
30. De Luca CJ, Forrest WJ. IEEE Trans Biomed Eng BME 1972;19:367.
31. Basmajian JV, Stecko GA. J Appl Physiol 1962;17:849.
32. Willison RG. J Physiol (London) 1963;168:35.
33. Lindstrom LR, Broman H, Magnusson R, Petersen I. Neurophysiology 1973;7:801.
34. Stulen FB, De Luca CJ. IEEE Trans Biomed Eng BME 1981;28:515.
35. Constable R, Thornhill RJ, Carpenter RR. Biomed Sci Instrum 1994;30:69.

See also ELECTROPHYSIOLOGY; REHABILITATION AND MUSCLE TESTING.

ELECTRON MICROSCOPY. See MICROSCOPY, ELECTRON.

ELECTRONEUROGRAPHY

THOMAS SINKJÆR
KEN YOSHIDA
WINNIE JENSEN
VEIT SCHNABEL
Aalborg University
Aalborg, Denmark

INTRODUCTION

Recording techniques developed over the past three decades have made it possible to study the peripheral and central nervous system (CNS) in detail and often during unconstrained conditions. There have been many studies using the ElectroNeuroGram (ENG) to investigate the physiology of the neuromuscular system, in particular, chronic studies in freely moving animals (1,2). Other studies relate to monitoring the state of the nerve (e.g., in relation to axotomized nerves and regeneration of nerve fibers). Clinically, the ENG is used to measure the conduction velocities and latencies in peripheral nerves by stimulating a nerve at different points along the nerve. Extracellular potentials can be recorded by either con-

centric needle electrodes or surface electrodes. The potentials can be derived from purely sensory nerve, from sensory components of mixed nerve, or from motor nerves (3). The study of extracellular potentials from sensory nerves in general has been shown to be of considerable value in diagnosing peripheral nerve disorders. For an indebt description of the ENG in clinical neurophysiology see Ref. 4. Several studies pertain to the use of sensory signals as feedback information to control neuroprosthetic devices. Studies (5) have shown that the application of *closed-loop* control techniques can improve the regulation of the muscle activation. Techniques using an electrical interface to nerves innervating natural sensors (6–12), such as those found in the skin, muscles, tendons, and joints are an attractive alternative to artificial sensors. Since these natural sensors are present throughout the body, remain functional after injury, and are optimally placed through evolution to provide information for natural feedback control, a rich source of information that could be used to control future FES devices exists as long as a method can be found to access them.

Interestingly, much of the peripheral sensory apparatus in spinal and brain-injured human individuals is viable. This means that the natural sensors are transmitting relevant nerve signals through the peripheral nervous system. Therefore, if the body's natural sensors are to provide a suitable feedback signal for the control of FES systems in paralyzed subjects, the challenge is to be able to extract reliable and relevant information from the nerve innervating the sensors over extended periods.

The nerve cuff electrode still has an unrivaled position as a tool for recording ENG signals from peripheral nerves in chronic experiments (13,14) and as means to provide information to be used in neural prosthesis systems (9,15,16). Other kinds of electrodes are to challenge the cuff electrode, such as intra-fascicular electrodes (6,10,18) or multisite electrodes with hundreds of contacts within a few cubic millimeters (2,10,11,17–19). These types of nerve-interface provide advantages with respect to selectivity and number of sensors.

This article describes the characteristics of the peripheral nerve signals, principals of neural recordings, and the signals obtained with different electrodes in long-term implants. An essential part in the success of recording peripheral neural signals or activating peripheral neural tissue is the neural interface.

THE PERIPHERAL NERVOUS SYSTEM

A peripheral nerve contains thousands of nerve fibers, each of them transmitting information, either from the periphery to the CNS or from the CNS to the periphery. The efferent fibers transmit information to actuators; mainly muscles, whereas afferent fibers transmit sensory information about the state of organs and events (e.g., muscle length, touch, skin temperature, joint angles, nociception, and several other modalities of sensory information). Most of the peripheral nerves contain both afferent and efferent fibers, and the peripheral nerve can thus be seen as a bidirectional information channel.

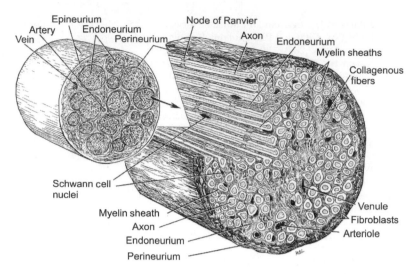

Figure 1. Figure showing the structure and components of the peripheral nerve trunk. (Adapted from Crouche, 1969, with permission.)

Anatomical Definitions and Structures

If the CNS is defined as the tissues and neural circuitry that comprise the brain and spinal cord, the peripheral nervous system can be defined as the part of the nervous system that lies outside of the CNS. Given some generalization of this definition, the peripheral nervous system consists of the spinal roots, dorsal root ganglions, axons/dendrites, support cells, and sensory end organs. The structure of the peripheral nerve trunk is illustrated below showing the following: nerve trunk, nerve fascicle, axon, schwann cells, epineurium, perineurium, and endoneurium. Though not generally considered part of the peripheral nerve, the nerve can also contain resident macrophages, leucocytes, and other cell types involved in the inflammatory response (Fig. 1).

THE NEURO-ELECTRONIC INTERFACE

The ENG can be described as the extracellular potential of an active peripheral nerve recorded at some distance from the nerve or from within the nerve trunk. The extracellular potential is formed from the contributions of the superimposed electrical fields of the active sources within the nerve. The general form of the extracellular response of a whole nerve to electrical stimulation is triphasic, it is in the lower end of the microvolt scale in amplitude, and looses both amplitude and high frequency content at larger radial distances from the nerve trunk (20).

Principles of Nerve Recordings

All cells of the body have a membrane potential. The membrane potential is a consequence of the cell membrane being selectively permeable to different ion species, resulting in different ion concentrations at the intracellular and extracellular space. The difference in ion concentrations causes an electrochemical voltage difference across the membrane, called the membrane potential. Under normal physiological conditions, the interior of the cell is negative with respect to the extracellular space without a net electric current flowing through the membrane. Therefore, if a macroscopic view of the cell membrane is justified (i.e., at distances that are large with respect to the thickness of the membrane), the existence of the resting membrane potential cannot be detected at the extracellular space.

Action Potentials. The change in membrane permeability is achieved by opening and closing of ion channels. The kinetics of the opening and closing of the channels determines the shape of the membrane current, which is characteristic for different types of nerve and muscle cells. The membrane potential changes as current flows through the membrane. The course of the membrane potential, from its resting state over an entire cycle of opening and closing of ion channels back to the initial state, is called the action potential. The associated current flowing through the membrane is the action current (Fig. 2).

A widely used model of peripheral myelinated nerve fibers is based on the model of single action potentials of fibers from rabbit sciatic nerve by Chiu et al. (22), comprising only sodium current and a leakage current. This model was adapted to 37 °C and to human peripheral nerve conduction velocity by Sweeney et al. (23).

An extensive review of different membrane models can be found in Varghese (24). More recent modeling work of human fibers includes Wesselink et al. (25) and McIntyre et al. (25).

Extracellular Currents and Potential Distribution. During the action potential, current is entering and leaving the cell through the membrane. Since there are no sources or sinks of electricity in the nerve fibers, the current flows in closed loops, and the current entering the cell at one site has to leave at a remote location. Likewise, the current leaving the cell has to reenter at a distant location, resulting in the redistribution and conduction of the current through the extracellular space. The extracellular conduction current (or, more precisely, its associated electric or magnetic field) can be detected and measured. The electroneurogram measures the electric potential field generated by the ensemble of extracellular conduction currents originating from active sites at many different fibers.

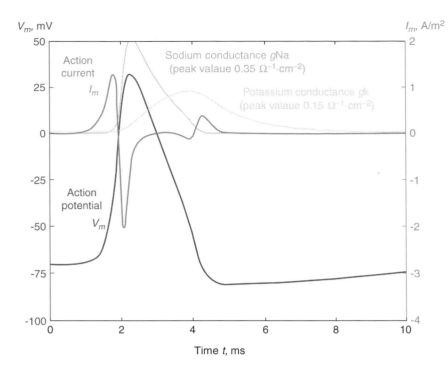

V_m, mV

I_m, A/m²

Action
current
I_m

Sodium conductance gNa
(peak valaue 0.35 $\Omega^{-1} \cdot cm^{-2}$)

Potassium conductance gk
(peak valaue 0.15 $\Omega^{-1} \cdot cm^{-2}$)

Action
potential
V_m

Time t, ms

Figure 2. Action potential and action current of the squid giant axon (unmyelinated nerve fiber). Time course of the propagating action potential Vm and action current Im according to the Hodgkin and Huxley model (21). After initial depolarization, the sodium channels open, resulting in a 100-fold increase of sodium conductance, and thus inflow of sodium ions, which depolarizes the membrane potential even further. The delayed opening of the potassium channels (potassium outflow) and closing of sodium channels repolarizes the membrane potential, with a phase of hyperpolarization (below resting potential).

For the case of a single active myelinated fiber in an unbound homogeneous medium, the fiber can simply be modeled as a series of point current sources, representing the nodes of Ranvier, where current is entering and leaving the fiber (if myelin is assumed to be a perfect insulator). The extracellular potential φ of an active fiber can then easily be calculated by superposition of the electric potential fields of point current sources as follows (26):

$$\varphi(x,y,t) = \frac{1}{4\pi\sigma} \sum_n \frac{I_m(t - x_n/v)}{\sqrt{y^2 + (x - x_n)^2}}$$

where σ is the conductivity, I_m the action current (Fig. 2), v the propagation velocity, and $(x_n, 0)$ the position of the node of Ranvier n.

Figure 2 shows the extracellular potential $\varphi(0,y,t)$ of a single active myelinated fiber (10 µm) in unbound homogeneous medium for different distances y from the fiber. For very short distances from the fiber ($y = 1$ mm, corresponding to one internodal distance), the extracellular potential follows mostly the shape of the action current from the closest node of Ranvier. With increasing distance from the fiber, not only does the amplitude of the extracellular potential drop rapidly down, but also its shape changes as the potential is averaged over increasingly numbers of active nodes. The influence of fiber diameter and distance from the fiber on the amplitude and duration of the extracellular potential has been discussed in detail in Struijk (26).

The amplitude and shape of the extracellular potential are also strongly influenced by inhomogeneities in the surrounding medium, called the volume conductor (e.g., the highly resistive perineurium, surrounding tissue, or cuff electrodes). For a more realistic (complex) volume conductor than the above infinitely large, homogeneous case, numerical methods are required to calculate the distribution of the extracellular potential, such as harmonic series expansion, finite difference, or finite element methods.

Electrodes measure the extracellular potential by physically sampling the electric potential in the extracellular space. Depending on the electrode type, this yields either point-wise information (intrafascicular electrodes, needle electrodes, electrode arrays) or measurements averaged over a larger volume (cuff electrodes, surface electrodes). The configuration (monopolar, bipolar, tripolar) and spacing of the electrodes further affect the recording characteristics (e.g., noise rejection, spatial selectivity, diameter selectivity). Computer models are helpful in understanding these recording characteristics and for the development of electrodes with improved selectivity (27–29).

Neural Signals

Neural Signal Characteristics. The previous section, gives the theoretical basis of how the action potential is generated, and how this, in turn, results in a change in the electric field in the extracellular space. These potentials can be detected using electrodes in various recording configurations to maximize the neural signal detected and minimize the noise pickup. These configurations will be discussed in the next section. Here, the ENG signal is characterized. A starting point is to take recordings made by electrodes placed within the nerve fascicle, such as a longitudinal intra–fascicular electrode (LIFE).

Intrafascicular electrodes place their recording sites within the nerve fascicle, but outside of the nerve fiber. The site can be potentially adjacent to an active nerve fiber. An ENG record from an intrafascicular electrode placed in a branch of the tibial nerve in the rabbit model is shown Fig. 3. The record shows the activity from many active nerve fibers firing asynchronously from one another and superimposed upon one another.

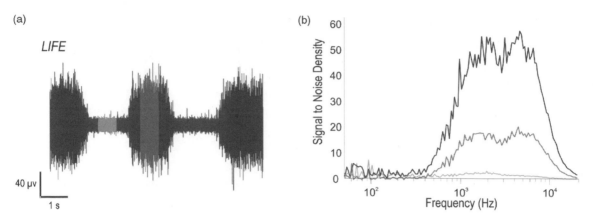

Figure 3. (a) Shows the raw record from a LIFE implanted in a medial gastrocnemius nerve branch of the tibial nerve in response to ankle flexion and extension. (b) Shows the spectral composition of the modulated LIFE activity for three different levels of activity starting from baseline, the lowest, to maximum, the highest. It illustrates that the amplitude of the spectral composition varies with the amount of activity, but the distribution remains constant.

If a closer look is taken of this multiunit activity recording, it is possible to see the spiking activity of single nerve fibers, or single unit activity. Analysis on the largest of these single spikes reveals that the amplitude of the activity is on the order of $\sim 10\text{--}100\ \mu V_{pp}$. The spectral components of intrafascicular ENG records show that, similar to extrafascicular ENG records, the activity has energy starting at ca. $\sim 100\ Hz$. However, unlike the unimodal extrafascicular ENG spectrum, the intrafascicular ENG spectrum is bimodal, with higher frequency components in the $5\text{--}\sim 10\ kHz$ range (30) (Fig. 4).

Whether the ENG signals are recorded from extrafascicular or intrafascicular electrodes, it can be seen that ENGs are low amplitude signals and require amplification of between $1000\times$ and $100000\times$ to bring the amplitude into the $\sim 1\ V$ range where they can be adequately digitally sampled and stored, or recorded on analog tape. Approximate orders of magnitudes of various signals and their approximate spectral frequencies that might appear in an ENG record are shown in the Table 1.

Given the amount of amplification required and the relative amplitudes of the ambient noise that could be picked-up by the recording electrodes, recording configurations and filtering must be considered to minimize the noise pick-up and maximize the neural signals.

Recording Configurations

Components of a Recording Setup. The previous section gave the theoretical basis of how the action potential is generated and how this, in turn, results in a change in the electric field in the extracellular space. These potentials can, in turn, be detected using a combination of electrodes placed in or around the nerve, and recording instrumentation. A typical recording setup consists of the following components starting (from the input biological input end and working toward the data storage or output):

Electrodes: The active electrode(s), the ground electrode.
Differential preamplifier.
Main-amplifier and signal conditioning.
Digitization, Interpretation, and storage.
Visualization and audio monitoring.

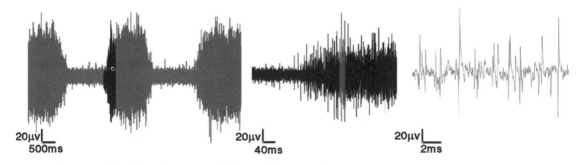

Figure 4. The figure shows an ENG recording in response to repeated mechanical activation of muscle afferents by stretching the muscle at various time scales. The amplitudes in all three traces are the same. (Left panel) Trace is the lowest time resolution and the right trace has the highest time resolution. The darker portion of the left trace is shown in the middle trace, which is zoomed in time by a factor of 12.5. The highlighted portion of the middle trace is represented in the right trace, which is zoomed in time by a factor of 20. (Right panel) Trace single spikes are now resolved, while in the other two traces only the gross mass activity can be resolved.

Table 1. Relative Amplitudes and Frequency Ranges of Components of Signals and Noise that Could Be Present in ENG Records

Source	Amplitude Order	Frequency Range
Electrochemical potential	1 V	dc[a]
Motion artefact	100 mV	Broad
Line noise	10 mV	50–60, 100–120
Thermal (electronics)	10 μV	Broad
Electrocardiogram (ECG)	1 mV	0.1–100 Hz
Electromyogram (EMG)	1 mV	1–1 kHz
Electroneurogram (ENG)	10 μV	100–10 kHz
Electroencephologram (EEG)	1 μV	dc–1 Hz

[a]Direct current = dc.

This instrumentation chain is represented schematically below in Fig. 5.

The components of the instrumentation chain can be blocked into four blocks: Preamplifier, Universal Amplifier/Filter, Computer, and Monitor. The preamplifier accepts signals from the electrode. Its function and configuration will be discussed in detail below. The Universal Amplifier/Filter block performs the analog filtering and main amplification, to reduce out of band noise, antialias and further amplify the signal and prepare the signal for the next block. The Computer block performs the digitization interpretation and depending on the application, storage of the neural data. It extracts and stores the information in the neural signal. Parallel to this is a monitoring block, which provides real-time feedback to the experimenter on the quality of the data being collected. In some cases, the monitoring block could be part of the Computer block. Neural signals are conveniently in the audio range, and the quality of the recording can be very effectively evaluated by simply listening to the signal through an audio monitor.

Differential Preamplifier. The differential preamplifier is a key component used in all measurement schemes. This component can also be called the headstage amplifier, preamplifier, or bioamplifier, and so on. In the context of biopotential recordings, the preamplifier is a low noise amplification stage used to amplify biopotentials to voltage levels in which signal conditioning can be performed using standard filters, amplifiers, discriminators, and so on. It

also serves the purpose of impedance matching between the relatively high impedance electrodes at its input and the relatively low impedance input of standard instrumentation.

Differential preamplifiers can be described as an active (powered), three input (± reference), one output device. Its main function is to measure a potential at each of its two active input relative to a reference potential measured at the ground input terminal, take the difference of the measured potentials and multiply by a fixed gain. In equation form, it performs the following function:

$$v_{out} = A_2 \times [A_1(v_+ - v_{ref}) - A_1(v_- - v_{ref})]$$

A typical differential preamplifier fitted with a standard connector is shown in Fig. 6 (b), though in practice packaging, connector and form factor for different preamplifiers vary widely. Common to differential preamplifiers are the three input terminals, and one output terminal. Although they can be realized using discrete components, most differential preamplifiers are realized using low noise instrumentation amplifier chips. Since the electrodes attached to the v_+, v_-, and v_{ref} terminals of the instrumentation amplifier are physically located in different places; this equation implies that a spatial derivative is performed. It can also be seen that the voltage at the reference terminal is common to both active input terminals and should ideally be rejected along with the common mode rejection capabilities of the amplifier, measured by the common mode rejection ratio, CMRR.

One factor, which has a strong influence on the recording performance, is the impedance match of the electrodes at the active input terminals. In voltage controlled voltage source amplifiers the input impedance is much larger than the impedance of components (the electrode + tissue) preceding it so that most of the voltage drops across the input impedance of the preamplifier, and the voltage is adequately measured without attenuation. At the same time, there must be a finite dc current path from the input of the amplifier to ground to supply the required bias current to the input of the amplifier. Since there are no dc components in the extracellular nerve action potential, a commonly used scheme is to capacitively decouple the input to the amplifier. However, mismatches in the impedance of the electrode seen at each terminal of the amplifier can result in degradation of the CMR capacity of the amplifier.

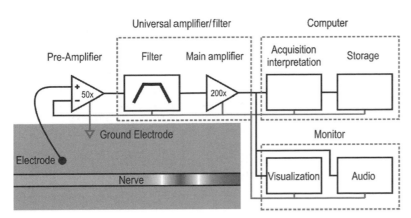

Figure 5. A generic instrumentation chain consisting of a preamplifier that impedance matches the recording electrode to the relatively low impedance filter–amplifier chain, a bandpass filter, main amplifier, data storage and monitoring. Shown also is the grounding configuration of the instrumentation attached to a ground electrode in the tissue.

(a)

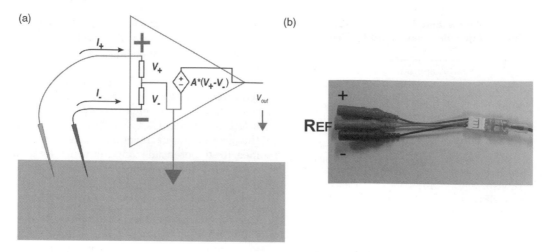

(b)

Figure 6. (a) Shows a schematic representation of the differential preamplifier and its electrode connections when used as a bioamplifier on tissues. (b) Shows a low noise differential preamplifier that could be used to record neural activity with a low impedance electrode.

Though the voltage, the controlled voltage source preamplifier scheme is the most commonly used type in current systems, current mode amplifiers, such as the application of microphone transformers have also been successfully used (31). The main limitations to the transformer amplifier scheme are the physical size of the transformer, which precludes their use in multichannel recording systems, and the necessity to carefully match the electrode impedance and transformer bandwidth.

Monopolar. Figure 7 shows the simplest recording configuration, the monopolar recording configuration where the activity recorded from the + terminal of the preamplifier is amplified relative to a common reference electrode. Though the recording, configuration is commonly used for recording EEG, *in vitro* electrophysiological preparations and implanted cortical signals, the monopolar recording configuration would be sufficient to record activity from the peripheral nerve only in the ideal noise-free case.

In practice, peripheral nerve electrodes, unlike the surface EEG electrodes on the scalp, or intracortical electrodes, are implanted and used on or in nerves that are

generally surrounded by large biological noise sources, such as nearby skeletal muscles or the heart. As seen in the previous section, these bioelectrical sources have amplitudes that are orders of magnitude larger than the nerve signal, whose spectral domain can overlap that of the nerve activity making it difficult to filter the unwanted EMG activity out of the recording. Altering the recording configuration can reduce the pick-up of these unwanted activities. The monopolar recording configuration could be modified so that the reference electrode is place very close to the recording electrode, but far enough away from the nerve so that it only samples the unwanted EMG activity in the vicinity of the recording electrode.

The pick-up of unwanted radio-frequency (RF) noise, such as line noise, through the lead wires between the electrode and the amplifier, is a second problem. To adequately ground the recording system, the reference must have a low impedance dc path between the tissue and the electronic ground points. By necessity, the reference electrode should have a large area. The recording electrode, on the other hand, must sample the electric potential from a small area of nerve in order not to short out the nerve activity. The relative impedance of the recording and reference electrodes is not balanced, making the RF pick-up in each of these two electrodes leads different, and not cancelled by the CMR of the differential amplifier.

Another problem with the monopolar recording configuration arises when one wants to record from more than one channel. To minimize the pick-up of local EMG activity, one might use multiple recording and ground electrodes. However, in general, it is not good practice to have multiple ground electrodes in a single recording set-up (see Fig. 12). Each ground electrode represents a low impedance electrical pathway, through which large stray ground loop currents might pass. Such currents could exist during electrical stimulation or accidental electrocution.

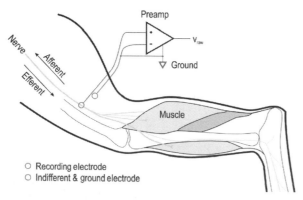

Figure 7. The monopolar recording configuration consisting of a recording electrode on or in the peripheral nerve and a common indifferent/ground electrode.

Bipolar. The bipolar recording configuration overcomes many of the problems of the monopolar recording configuration and is the most commonly used recording configuration for most neural electrodes excluding cuff

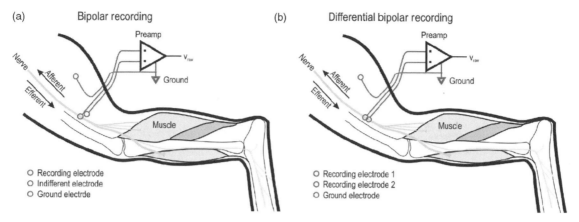

Figure 8. Bipolar recording configurations. (a) Shows a standard bipolar configuration where the signals from one recording electrode are amplified with respect to a second electrode that samples the ambient noise, the indifferent electrode. A separate ground electrode is placed somewhere in the tissue. The differential bipolar recording configuration (b) places a second recording electrode in or on the nerve and the difference of the signals from two active electrodes is amplified.

electrodes. The bipolar recording configuration requires two recording electrodes and a low impedance reference electrode. The second recording electrode is matched to the first recording electrode and is introduced to sample the local noise in the vicinity of the first recording electrode. The two recording electrodes are attached to a differential preamplifier, which amplifies the relative difference of the potentials seen by the first electrode and those seen by that of the second electrode. Thus, the local EMG activity is seen by the differential preamplifier as common mode and is rejected. Similarly, since the two recording electrodes are matched, the RF noise induced on their lead wires is similar and is rejected by the preamplifier. The second electrode effectively samples the noise, but not the neural signal, and is referred to as the indifferent electrode (Fig. 8).

A variation on the bipolar recording configuration is the differential bipolar recording configuration. In this configuration, a second active recording electrode is introduced in or on the nerve instead of an indifferent electrode reducing the total number of electrodes that needs to be implanted. Additional channels can be introduced with the addition of matched recording electrode pairs and their associated preamplifier. Each additional preamplifier shares a common reference electrode to maintain only one low impedance path to ground, and minimizing the hazards of ground loops (see below).

The neural component of the potentials recorded by the electrodes is generated by a traveling waveform, which appears at the different recording sites along the nerve at different times. Thus, there are some consequences in using the differential bipolar recording configuration to the shape of the recorded action potential. Because the action potential is a traveling waveform, measuring the action potential at two different locations along the nerve and taking the difference, $v(t_1, s_1) - v(t_1, s_2)$, can be related to measuring the action potential measured at one location but at two different times, $v(\boldsymbol{t_1}, s_1) - v(\boldsymbol{t_2}, s_1)$. This relationship assumes that the two electrodes record the action potential equally, and the distance between the recording

sites are appropriate. $v(t_1, \boldsymbol{s_1}) - v(t_1, \boldsymbol{s_2})$ is $\Delta v / \Delta s$, while $v(\boldsymbol{t_1}, s_1) - v(\boldsymbol{t_2}, s_2)$ is $\Delta v / \Delta t$. The relationship between the space interval, Δs, and the time interval, Δt, is the conduction velocity of the action potential, $\Delta s / \Delta t$. Therefore, the shapes of action potentials recorded using the differential bipolar configuration are roughly the derivative of the action potential recorded using the monopolar configuration. This alters the generally monophasic action potentials recorded by monopolar recordings to biphasic action potentials.

There is a dependence on the amplitude of the recorded action potential with the spacing between electrodes. The amplitude increases with electrode spacing until the spacing approaches the wavelength of the action potential. The wavelength of the action potential is dependent on the conduction velocity, which is related to the caliber of the nerve fiber. In general, distances should be kept $> 1\,\mathrm{cm}$ for cuff electrode recordings (32), and $> 2\,\mathrm{mm}$ in the case of differential bipolar recordings with LIFE (33).

Tripolar. The tripolar configuration is the most commonly used recording configuration with the nerve cuff electrode. The nerve cuff electrode is an extrafascicular recording configuration which places the active recording sites outside of the nerve fascicle. Since the perineurium has a relatively high impedance, most of the current generated by the nerve fiber during an action potential is restricted to within the fascicle. To maximize the signals, the nerve cuff design surrounds the nerve bundle with an insulating silicone tube that restricts whatever current leakage from the nerve bundle to the space between the nerve bundle and the insulating cuff. Let us first consider a cross sectional schematic of the electrode structure with a bipolar recording scheme as shown in Fig. 9.

One particular advantage of the cuff electrode is that the electrode is generally immune to noise sources that are lateral to the nerve and cuff electrode since they produce electrical fields that are shielded by the nerve cuff and shorted by the circumferential recording site of the electrode. However, noise sources producing electrical fields

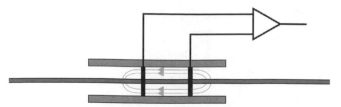

Figure 9. Bipolar differential recording scheme used with the cuff electrode. The insulating cuff restricts the current from the nerve bundle to within the cuff to maximize the voltage generated from the nerve.

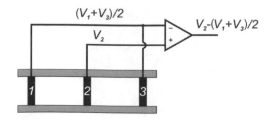

Figure 11. Tripolar recording configuration in a cuff electrode.

that are perpendicular produce a gradient through the nerve cuff, which is amplified by the bipolar recording configuration.

The tripolar recording configuration is a means to reduce or eliminate the longitudinal potential gradients of noise sources outside of the cuff. Two variants of the tripolar configuration are shown below in Fig. 10.

Longitudinal potentials from sources outside of the nerve cuff are influenced by the insulating cuff and linearlized within the cuff. The tripolar recording configuration takes advantage of this property and averages the potentials seen by the outer two electrodes to predict the extracuff noise potential seen at the central electrode, either by directly measuring and averaging the signals, as in the True Tripolar configuration, or by shorting the contacts of the outer two electrodes, as in the Pseudo-Tripolar configuration (shown below in Fig. 11).

The tripolar recording configuration introduces another differentiation operation on the shape of the action potentials recorded, and action potentials recorded with the tripolar recording configuration are triphasic.

Cabling. The cabling between the electrode active site and preamplifier is a major source of noise that can be minimized by taking care during cabling. Line noise and RF noise is induced in the leads between the electrode and the preamplifier since they act as antennae. The effect is similar to that with high impedance wired microphones.

Similar strategies could be used to reduce noise pickup. Lead lengths should be kept as short as possible, though, in the case of implanted electrodes, this may not always be practical or possible since cables from multiple locations are often bundled together and routed to a common percutaneous connector to minimize the number of holes through the skin. Another strategy is to twist the differential lead pairs together so that the noise induced in each pair is nearly the same and can be reduced using the CMR of the differential pre-amplifier. When appropriate, shielded cabling could be further be used, though care should be taken to minimize ground loops when multiple shielded cables are used.

Grounding. A critical factor influencing the quality of the ENG recording is how the system is grounded. A low impedance path to ground must be established between the preamplifier and tissue to keep the input stage of the preamplifier stable and out of saturation. In acute animal preparations, the experiment is commonly performed on a conductive metal surface, which is tied to ground to a common internal ground electrode to reduce line noise. In chronic animal experiments as well as in human implanted systems, this means to reduce noise is not an option since the subject is autonomous and freely moving. In this case, a ground electrode with large surface area, such as braided stainless steel shields, can be used to establish a single ground point to the tissue. Because the ground electrode is intended to provide a low impedance pathway to a ground potential, the ground electrode

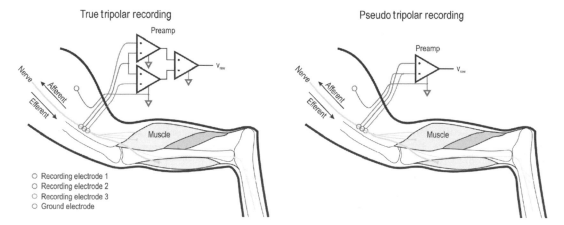

Figure 10. Tripolar recording configurations. (Left panel) Shows the true tripolar recording configuration consisting of a double differential amplification circuit. (Right panel) Shows a pseudo–tripolar recording technique, which approximates the true tripolar recording configuration by shorting the outer two electrodes 1 and 3.

provides a reference 0 potential in the tissue, and the potential of the tissues immediately surrounding the ground electrode forms a 0 isopotential. Multiple ground points to the tissue are typically not a good idea since each ground electrode would have a different chemoelectric potential even if they are made of the same material, and currents would flow from ground electrode to ground electrode to maintain a 0 isopotential at each ground point.

The issue of ground loop currents is particularly a problem when stimulation is performed at the same time as recording and if the stimulator shares the same ground with the recording circuit. Since each ground electrode is at the same potential, the current from a stimulator can flow into one ground electrode and out of another ground electrode, as shown in Fig. 11, making it nearly impossible to predict the current flow and location of stimulation. If the magnitude of stimulation is large, such as with surface stimulation, a significant portion of the current could flow through the ground electrodes since the impedance of the tissue and stimulating electrodes can be considerably higher than the impedance of the ground electrode and ground line. Current flow through the ground line can result in large artifacts and potential damage to high gain amplifiers.

One strategy to reduce ground loop currents especially during stimulation is to electrically isolate different circuits from one another. Circuits, such as the analog amplifier chain that demand low noise, are kept physically in different subcircuits from circuits that are noise-producing, such as stimulator circuits or digital circuits. If these different types of circuits must reside on a common PCB or ASIC, Parallel or Mecca grounding schemes with ground planes surrounding each circuit type can be established to prevent noise from spilling over and interfering with noise sensitive parts of the circuit (34). The strategy applied to Functional Electrical Stimulation (FES) systems where stimulation and recording sites are in close proximity to one another is to electrically isolate the stimulating circuit from the recording circuit. This makes each circuit independent from the other so that currents originating from the stimulation circuit see a high impedance pathway with no return current path in the recording system (Fig. 12).

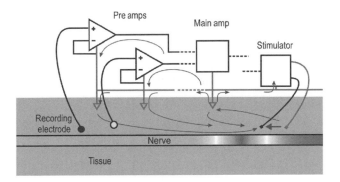

Figure 12. Illustrates potential ground loop currents that can be induced during stimulation if multiple ground electrodes (shown as triangles) are used. Since each ground electrode represents the same potential, the nearest ground electrode can draw current, which passes through the ground plane and out of the other ground electrodes, increasing the uncertainty of the point of stimulation.

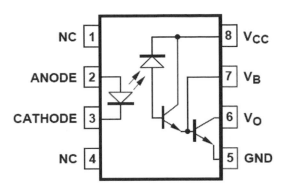

Figure 13. Schematic of a typical optoisolator using diode–diode coupling.

Optoisolators. An effective means to isolate circuits is with optoisolators. Optoisolators consist of two electrically independent halves: a transmitter, and a receiver. The basic analog optoisolator transmitter translates electrical currents into light intensity, typically through an LED. The transmitted light is detected with a phototransistor or photodiode receiver, which gates a current in the receiver that is mapped to the transmitter light intensity (Fig. 13).

If the transmitter and receiver are independently grounded and powered, this combination of optical transmitter/receiver electrically isolates the circuits at its input and output. This, of course, implies that when using optocouplers, each isolated circuit must have its own power supply and ground. The unisolated receivers can share a single power supply. A point should be made here about the number of ground electrodes in a recording set-up using multiple isolated recording/stimulation circuits. Since each isolated circuit must have its own ground, if several optoisolated recording/stimulation circuits are used, then each independent circuit must have its own single ground electrode. Ground loops between the circuits cannot form because the grounds are independent from one another, and there is no low impedance return path for ground loop currents to follow.

The optocoupler shown in the above Fig. 13 is a nonlinear device that is dependent on the coupling of the diode characteristics of the transmitter and receiver. Linear analog optocouplers also exist that linearizes the diode characteristics using feedback with a diode-transistor coupling instead of the diode–diode coupling. The diode–diode coupling, though nonlinear, has a relatively fast response time, with rise and fall times on the order of 2 μs. The diode-transistor coupling scheme, however, is bandwidth limited to considerably < 10 kHz. Newer linear optocouplers attempt to overcome the bandwidth limitation by sampling and flash converting the input voltage to a digital value, transmitting the digital value and reconstructing the analogue value with a DAC (Fig. 14).

Electrode Impedance

Frequency Dependence of the Impedance. The impedance of the electrode is a function of the electrochemical properties of the electrochemical interface. Further discussion on the electrochemical interface and impedance can be

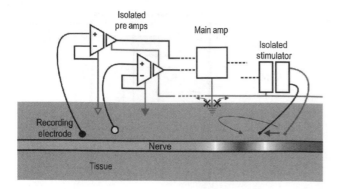

Figure 14. Use of optoisolated amplifiers and an optoisolated stimulator eliminates the ground loops shown in Fig. 8 by eliminating low impedance return paths to the tissue or stimulator circuit. Note that each isolated subcircuit has its own independent ground. Moreover, since there is no current flowing through the main ground from the tissue (shown as a dashed triangle), the main ground does not need to be grounded to the tissue.

found in other sections of this volume and in Grimnes and Martinsen (35).

A typical impedance spectrum for a platinum electrode is shown in Fig. 13. Though the absolute magnitudes of the impedance will vary depending on the total area of the electrode active site and the material used, this figure shows the general shape and characteristics of the impedance spectrum. Depending on the type of electrode used, the upper corner frequency and the lower corner frequency can be shifted. Similarly, the low frequency and high

frequency segments (the segment $< 1\,\mathrm{Hz}$, and $> 1\,\mathrm{kHz}$, respectively, in Fig. 15) can be higher or lower than shown.

As seen earlier, most recording configurations rely on the common mode rejection ratio, CMRR, of the differential preamplifier to reduce noise pick-up and maintain stability of their neural recording. One consequence of the frequency dependence on the electrode impedance is that it influences the common mode rejection ratio. The differential preamplifier and its input can be represented by the following, where V_{cm} represents the common mode noise, which is to be rejected, and V_{dm} represents the differential nerve signal, which is to be amplified (Fig. 16).

Assuming an ideal preamplifier where the two input resistances $R_{\mathrm{in}+}$ and $R_{\mathrm{in}-}$ are equal, and applying the differential preamplifier equation given earlier, the common mode rejection ratio can be represented by

$$\mathrm{CMRR} = \left|\frac{A_{\mathrm{dm}}}{A_{\mathrm{cm}}}\right| = \left|\frac{Z_+ + Z_- + 2R}{Z_+ - Z_-}\right|$$

In the ideal case, where the electrode and tissue impedances are matched, then the CMRR for an ideal differential preamplifier becomes infinite. However, real electrodes are relatively difficult to match, especially when dealing with small, high selectivity active site electrodes. A 10% error in electrode impedance at 1 kHz is common even with matched electrodes. Based on the CMRR equation and typical impedance shown earlier, the CMRR dependence versus, frequency can be calculated (Fig. 17).

It shows that because of the relatively high R_0 impedance at low frequency, there can be degradation in the CMRR, which results in inclusion of some of the relatively

(a)

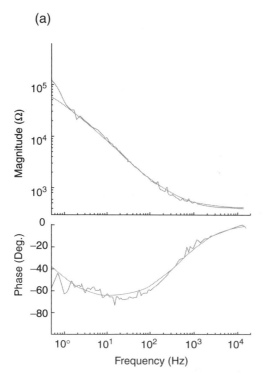

(b)

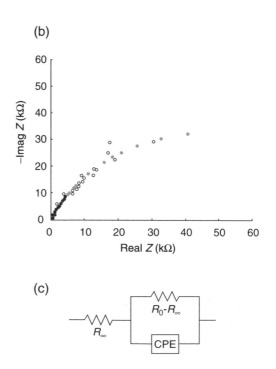

(c)

Figure 15. (a) Shows a typical metal electrode impedance spectrum and its fit to the Cole model shown in (c). (b) is the Warburg representation of the impedance spectrum where the real Z is plotted with respect to the negative of the imaginary Z.

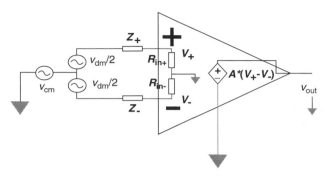

Figure 16. Shows a model of the differential preamplifier and its connections to input impedance loads, common mode, and differential mode inputs. The common mode input can represent noise that we hope to exclude from the ENG while the differential mode input represents the neural signal. The input Z represents the tissue, electrode, and whatever input network is used between the amplifier input and the signal sources.

high amplitude noise that exists at these frequencies. Similarly, not all of the ECG and EMG will be rejected. Therefore, inclusion of noise components from these sources should be expected in the raw ENG record and must be filtered before analysis and use of the ENG can be considered. High pass filtering the output of the differential preamplifier can remove much of the out of band noise not rejected by the CMR of the differential amplifier, but care should be taken to ensure that the differential preamplifier is operating within its linear dynamic range.

EXAMPLES OF LONG-TERM PERIPHERAL NERVE INTERFACES

Peripheral neural electrodes can be grouped according to the way that they interface with the peripheral neural tissue. They can be grouped into three main types: *regenerating*, *intraneural*, and *extraneural electrodes*. Common for all electrode interfaces presented here are that they

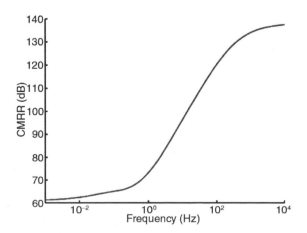

Figure 17. The frequency dependence of the CMRR is shown for an ideal preamplifier using electrodes that are mismatched by 10%. The frequency dependence of the electrodes has a large influence in the CMRR, which resembles the inverse of the electrode impedance.

record the extracellular potential from one or more peripheral nerve axons simultaneously (also referred to as extracellular electrodes). A fourth electrode group aims instead to record the intracellular potential of individual cells (also referred to as intracellular electrodes). In this case, the active recording site is placed inside a cell, so the electrode must therefore be carefully positioned and held in place during the entire duration of the recording. With their use, there is the risk of permanently breaking the membrane and injuring or killing the cell. As such, intracellular electrodes can be used acutely *in vitro*, *in situ*, or *in vivo*, however, they are not suitable for chronic animal, human or clinical use because of mechanical stability issues and the unavoidable damage to the cells that they record from. These electrodes are therefore not considered further here.

The choice of an optimal peripheral neural electrode largely depends on the application or the experimental design. One electrode might prove useful for one purpose, but not for suitable for another. A vast number of parameters have been used in the literature to describe, characterize, and compare neural interfaces. Most electrode interfaces can not only be used for recording from peripheral nerves, but also for applying electrical stimulation to the nervous tissue. However, the focus here will be on peripheral neural interfaces for recording. To provide a base for comparison between electrode types, the discussion will focus on the following issues:

Design parameters, including: The rationale for design, the ease of use and the cost of manufacturing and ease of implantation.

Application of the electrode, including: Recording selectivity and recording stability, expected lifespan, biocompatibility and robustness, and application examples.

Regenerating Electrodes

Introduction. Peripheral nerve axons spontaneously regrow after dissection or traumatic injury. The working principle of a *regenerating electrode* is to fit an electrode in the path between a proximal and distal nerve stump after a deliberate cut or traumatic nerve injury, and to guide the neural regrowth through perforations in the electrode structure. Ideally, the axons will grow through the perforations and allow very high resolution (single axon) recordings.

Sieve Electrodes

Design. Regenerating electrodes are often referred to as *Sieve electrodes*, inspired from their appearance. The structures are typically based on a sieve-shaped or perforated diaphragm, which contains an array of holes. The active sites are placed inside the sidewalls of the hole or in their immediate surroundings (similar to a printed circuit board via) to make physical contact with an axon growing through the hole. They are often designed incorporating a used small flexible tube made of a biocompatibile material (e.g., polyimide or silicone) that is used to hold the two ends of the transected nerve stumps in place with the sieve

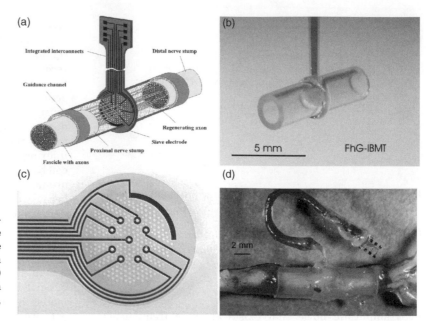

Figure 18. (a) Sketch of a Sieve electrode. The proximal and distal nerve stumps are fed into a guide tube. Neural axons must grow through holes in the Sieve electrode to reconnect with the neural axons on the other side. (b,c) Examples of Sieve electrodes. (d) shows the Sieve electrode *in situ*. (Reproduced with permission from T. Stieglitz, Fraunhofer Inst., Germany.)

electrode structure fixed in place between them. The tubes work to help guide the growing axons and to hold the electrode structure in place (36) (see Fig. 18). The first generations of Sieve electrodes were developed more than two decades ago (37). With the development of micromachining techniques, the devices have become considerably smaller and further enabled more suitable connections between the electrode and the surrounding world (38). These newer generations of sieve devices are made of silicon (39–41), polyimide or silicone substrates (38,42). Sieve electrodes have been designed with various shapes of the holes [round (40), squares (43), or longitudinal slots (39)]. The shape of the hole does not seem to have a major effect on the regrowth of the axons (36), but other design parameters, (such as the hole size, the thickness of the sieve diaphragm, and the relative amount of open versus closed space (also referred to as the *transparency factor*) have been found to be important. The thickness of the sieve diaphragm is typically in the order of 4–10 µm (36,40,41). Several findings suggest that 40–65 µm diameter Sieve holes work well (41,43,44), however, Bradley et al. (41) found that the holes could be as small as 2 µm. This may be explained by the difference in transparency factors of the electrodes. A high transparency factor is needed if the holes are small to secure regrowth (45). Major effort has also been put in to researching the use of neural growth factors (NGF) and how these may assist in promoting the spontaneous regeneration, and this may become an important factor in the future success of the *Sieve* electrodes.

Application. Bradley et al. (46) was among the first that obtained long-term recordings using sieve electrodes in mammals. They recorded from the rat glossophraryngeal nerve, which is mainly sensory, mediating somatosensory and gustatory information from the posterior oral cavity. Spontaneous activity was observed up to 7 weeks after implant. Evoked compound sensory responses were recorded up to 17 weeks after implantation. Other long-term evaluation studies have been reported by, for exam-

ple, Mesninger et al. (42,47,48). Despite the growing number of reports on long-term usage of Sieve electrodes, they have not yet been tested in humans. An ideal axon-to-axon reconnection through a *Sieve electrode* could provide the ultimate interface to monitoring peripheral neural activity, but to date, this concept has not been proven in humans. Deliberate transection of a peripheral nerve to allow implant of a sieve electrode is a highly invasive procedure, and the success of the implant can only be evaluated several weeks after implantation (36). The current designs are not optimal since some axons may stop grow when the meet a wall and not a hole (36). Furthermore, it is currently not possible to control what fiber types regenerate. It has been shown that myelinated fibers are more likely to grow through than unmyelinated fibers, and large fibers are more likely to grow through than small fibers (49). It will be necessary to obtain a detailed map of what kind of sensory or motor axons have reconnected in order to interpret the peripheral neural signals properly (Fig. 19).

Intraneural Electrodes

Introduction. Intraneural electrodes are defined as electrodes that penetrate the endoneurium and perineurium of a peripheral nerve or the dorsal root ganglion (DRG) neuropil. The active sites are placed in the extracelluar space in close proximity with the nerve fibers, and therefore the electrodes aim to record from one single axon or from a small population of axons. The intraneural, penetrating electrodes are dependent on the insulating properties of the epineurium/perineurium to record.

Microneurography. Microneurography is a procedure in which a small needle electrode is placed directly inside a nerve fascicle. The aim is to record from one axon or a small a population of axons to study basic neural coding in the animal or human subjects, or as a diagnostic tool in clinical practice. Sketches of two types of microneurography electrodes are depicted in Fig. 19.

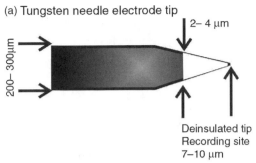

(a) Tungsten needle electrode tip

2– 4 μm

200– 300μm

Deinsulated tip
Recording site
7–10 μm

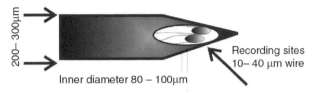

(b) Concentric needle electrode tip

200– 300μm

Recording sites
10– 40 μm wire

Inner diameter 80 – 100μm

Figure 19. Microneurography electrodes. (a) A drawing of the geometry of a typical, commercially available tungsten needle electrode. A separate ground or indifferent electrode must be place in the close surroundings (Drawings appear courtesy of Dr. K. Yoshida.) (b) A drawing of a concentric needle electrode. One or two fine wires are threaded through a hypodermic needle and attached

Design. A classic needle electrode is manufactured from a stiff core material (e.g., tungsten, stainless steel, or carbon fiber) surrounded by a layer of insulating material. The Tungsten needle electrode typically has a 200–300 μm base diameter and a 2–4 μm tip diameter to ensure that the material is stiff enough to be pushed through the skin and muscle tissue and enter the perineurium without damaging the tip. The needle tip is uninsulated to create the active site. A separate ground wire must be placed subcutaneously in the close surroundings of the recording electrode (50,51).

A variation of the classic needle electrode was developed to allow more recording sites to be placed inside the nerve. The concentric needle electrode consists of one or more fine recordings wires threaded through a hypodermic needle. The fine wires are glued in place with an epoxy adhesive. The wires are typically made from tungsten, platinum, or platinum–iridium, and have a 20–25 μm diameter that provides a smaller, and thereby more selective recording, site than with the Tungsten needle electrode. An active site is made at the end of the wire by simply cutting the wire at a straight or slightly oblique angle. The hypodermic needle has a typical outer diameter of 200–250 μm and an inner diameter of 100 μm. The hypodermic needle tip provides a cutting edge for easy insertion of the electrode. Further, the needle shaft works as the ground during recording (52).

The microneurography electrodes are implanted perpendicular to the nerve trunk, and they are typically inserted by hand or using a micromanipulator. These electrodes are inexpensive, and they are typically and both types are easily manufactured by hand.

Application. Since Vallbo and colleagues developed the microneurography technique for more than three decades ago, the technique has proved to be a powerful tool in

clinical experiments on conscious humans in hundreds of studies (51). However, the technique has a number of clear limitations. Placement of one single recording electrode may be extremely time consuming, and the delicate placement may easily be jeopardized if the animal or human subject moves. The microneurography needle electrodes are considered safe for short-term experiments, but they are not applicable for long-term implant and recording in animals or humans.

Longitudinal Intrafascicular Electrodes. The longitudinal intrafascicular electrodes (LIFE) are implanted parallel to the main axis of the peripheral nerve, and the active site of the electrode is placed along recording wires and not at the tip. The active sites typically record from a small population of axons.

Design. One of the first LIFE designs consisted of a simple 300 μm coiled stainless steel wire that was implanted into the nerve using a 30G hypodermic needle (53). The size of the hypodermic needle and implanted electrode may cause unnecessary injury to the nerve, and it also made it unsuitable for implantation in small nerves. Pioneering work was later done at University of Utah, as an attempt to solve these problems. A metal-wire LIFE was developed, see Fig. 21b (54). Here the 1–2 mm recording sites are created on 25 μm insulated, platinum–iridium wire. The recording wires are soldered to larger stainless steel lead-out wires, and the connections are protected by sliding a silicone tube over the solder joint. A polyaramid strand is threaded through this loop and glued to a 50–100 μm electrosharpened Tungsten needle (much similar to the tungsten needle electrode shown in Fig. 19). The needle is used to penetrate the perineurium/epineurium and pull the recording wires into place. The large lead-out wires remain outside the nerve. The electrode is sutured to the epineurium to hold it in place. The lead-out wires are then tunneled subcutaneously through the body to a chosen exit point. These metal-based LIFEs are relatively low cost, and they can be hand built. One of the disadvantages is the limited number of fine wires that can be fitted into one nerve. Further, the hand-building procedure does induce some variability in the size of the recording site, which influences the impedance and signal/noise ratio.

Later generations include polymer-wire LIFEs. Instead of using platinum–iridium wire, the recording wires are here based on 12 μm kevlar fibers. A metal is first deposited onto the kevlar to make the wire conductive, and a layer of silicone insulation is then added only leaving the recording sites exposed (55,56). The purpose of decreasing the intrafascicular wire size was to make the wires more flexible.

Application. The development of the LIFE has mainly been driven by the need for alternative and safe neural interfaces for clinical neuroprosthetic systems. Goodall et al. implanted the 25 μm platinum–iridium wire LIFEs in cat radial nerves and demonstrated that it was possible to obtain selective recordings of afferent activity from up to 6 months (59). Yoshida et al. used the same types of LIFES for recording neural activity in chronically

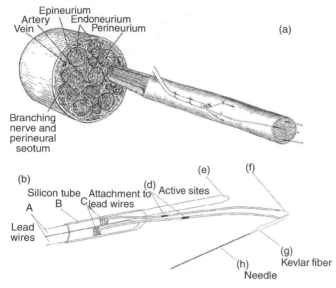

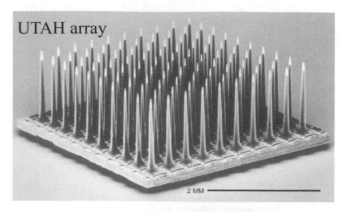

Figure 21. The 100-channel UTAH array. The electrode was originally developed at University of Utah, by Dr. R. A. Normann and colleagues. The electrode is commercially available through Cyberkinetics Inc. (Courtesy of Dr. R. A. Normann, University of Utah.)

Figure 20. (a) Placement of a metal-wire LIFE electrode inside a whole nerve fascicle. Only the fine wires with the active sites are threaded through the nerve whereas the remaining part of the LIFE is located outside the epineurium. (b) is a line drawing of a two-channel metal-wire LIFE electrode. (a) Lead-out wires are located outside the body and tunneled through the body to a chosen exit point. (b,c) A silicone-tube is places over the connection point for protection. (d) Active sites on the recording wires. (f–h) A polyaramid fiber links the recording wire to a tungsten needle. The tungsten needle is used to thread the fine wires through the nerve and is removed after implantation. (Courtesy of Dr. K. Yoshida, Aalborg University, Denmark.)

implanted cats. These studies characterized the effect of electrode spacing for rejecting EMG and stimulus artifacts and explored the possibility of using multichannel recordings for noise reduction. Yoshida and Stein (33) also explored the possibility of extracting joint angle information from muscle afferent signals recorded with LIFEs (6). The information was used as feedback in closed-loop control of the ankle joint in the cats. The chronic recording stability and biocompatibility of the polymer-wire LIFES have been demonstrated by Malstrom et al. (58) in dorsal rootlets and in peripheral nerve (59) (Fig. 20).

Silicon-Based electrodes. The design and manufacturing of silicon-based electrodes take advantage of often sophisticated microfabrication techniques. These electrodes may be inserted transversely into the peripheral nerve as the microneurography needle electrodes, or they may be inserted longitudinally as the LIFE electrodes. Many groups around the world have designed, developed, and tested different silicon-based electrodes, however, the vast majority of those are developed for interfacing with the cerebral cortex tissue. Many of these designs are currently considered unsuitable as chronic peripheral nerve implants because of the risk of mechanical failure of the rigid base structure or small, thin electrode shafts. A notable exception to this view is the Utah array. One electrode that has been used for both peripheral nerve and cerebral cortex implantation is presented here.

Design. The UTAH array was originally developed by Normann and colleagues at University of Utah, and this electrode is now commercially available through Cyberkinetics Inc. The UTAH array is a three-dimensional (3D) silicon structure usually consisting of 100 penetrating electrodes arranged in a 10×10 array (see Fig. 21). Each needle is a long, tapered structure that is 1.5 mm long and has an 80 μm diameter at the base. The distance between the needles is 400 μm. The base substrate for this electrode is silicon, and glass is used as insulation material. Only the very tip of each needle is deposited with gold, platinum, or iridium to form the active site (59). Fig. 22 shows an array with equal height needles, which was originally designed for recording from the cerebral cortex where the neurons of interest are often located at the same depth. A slanted UTAH array was designed with different lengths of the needles to better target different layers in the cerebral cortex or to accommodate for the fact that fascicles in a peripheral nerve are located at different depths. Implantation of the UTAH array requires a special insertion tool that shoots the electrode into the neural tissue in a controlled manner. This is necessary because the high density of the needles increases the amount of force that is necessary to penetrate the dura or the epineurium. Further, the insertion tool avoids the elastic compression of the neural tissue and possibly damage of the neural tissue that was found during manual insertion (60).

Application. The recording properties of the Slanted UTAH array was first tested by Branner and colleagues in acute cat model (61,62) and later in similar chronic cat implants (2). The 100-electrode array was inserted in the cat sciatic nerve using the pneumatic insertion tool (described above). The acute work demonstrated that it was possible to achieve highly selective recordings with the array. In the chronic experiments, the array was held in place using a silicone cuff (self-spiraling or simply oval shaped) to protect the implant and surrounding tissue, to hold the implant in place, and to electrically shield the electrode. This containment was found to be important for the long-term survival of the electrode. In the chronic experiments, electrodes with lead wires were implanted up to 7 months, however, it was only possible to obtain stable recordings for a few days after implant.

The slanted UTAH array has also been implanted in dorsal root ganglions (through the dura) in acute cat model to evaluate the neural coding of the lower leg position (10). In this work, it was possible to obtain selective recordings from the UTAH arrays over the duration of the experiment. The UTAH array has recently been implanted in a human volunteer (in a peripheral nerve in the lower arm), however, the results are pending (63). The UTAH array is also currently under clinical evaluation as a cortical interface in humans by Cyberkinetic, Inc.

Extraneural Electrodes

Introduction. The extraneural electrodes are the least invasive interfacing technique presented here. They are defined as electrodes that encircle the whole nerve or are placed adjacent to the nerve without penetrating the epineurium or dorsal root ganglion neuropil. The electrodes record from a large number of nerve fibers at the same time.

Circumferential Cuff Electrodes. The cuff electrode is probably the peripheral nerve interface that appears with the largest number of variations in designs and configurations. Only the main designs will be presented here. It is the most mature and widely used chronically implanted electrode for use in neuroprosthetics and has been used in a large number of animal studies and implants in human subjects (see section below).

Design. A circumferential cuff electrode consists of a tube or cuff that is placed around the whole nerve. The active sites are attached to the inside of the cuff wall to make contact with the epineurium (see Fig. 22). The tube wall works as an insulating barrier keeping extracellular currents emerging from the nerve within the cuff and blocking electric currents from other surrounding tissue (predominantly muscle activity, but also activity from other nerves) outside of the cuff. The tube wall has found to be essential to obtain measurable voltages from the

nerve trunk and to achieve a good signal/noise ratio. The cuff tube or wall is made of a flexible, biocompatible material, such as silicone, (1,64,65) or polyimide (66,67). The cuffs may be handmade or microfabricated. The microfabrication technique has the advantage that thinner walls can be obtained.

The shape and number of active sites placed inside the cuff wall vary. The active sites are typically made of inert metals, such as platinum or platium–iridium. In the classic configuration, a cuff electrode consists of a silicone tube lined with two or more thin conductive rings that are displaced from one another along their common axis (69). In this design, the number of active sites is usually three, but more may be fitted into the cuff depending on the application. The large conductive rings record from a large number of fibers within the nerve they encircle. Later designs placed several, smaller active sites around the inner cuff wall in an attempt to interface individual fascicles within the nerve trunk, and thereby increase the selectivity of the cuff recording (68,69). The number and size of the active sites have a large impact on the recording and stimulation selectivity, which will be discussed in the next section.

The cuff electrode is fitted onto a nerve by sliding the nerve into a longitudinal slit in the cuff wall while traction is placed on the cuff wall to keep the slit open. The procedure may be made easier for the experimenter/surgeon by gluing a number of sutures to the outside of the cuff wall (see Fig. 22a). The experimenter/surgeon can grab onto these sutures, pull the cuff open, and slide the cuff in place around the nerve. Knots are tied to close the cuff and secure that extracellular current is contained inside the cuff, which is important to achieve good recordings. Other closure mechanisms have been attempted.

1. A hinge or zipper-like closure has been suggested by Dr. M. Haugland, Aalborg University, Denmark [see Fig. 22c and patented by (70)]. Here the cuff is cut open and integrated with a piano hinge structure, that can be zipped up and threaded with a suture for closure. The disadvantage of this

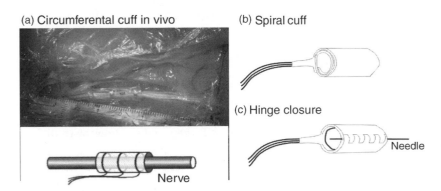

(a) Circumferential cuff in vivo

(b) Spiral cuff

(c) Hinge closure

Nerve

Needle

Figure 22. The sketch representation of the cuff (around the nerve) electrodes (left) and the newly designed 12-contact circular cuff electrode (with a longitudinal slit) for selective stimulation with a four-channel stimulator. In order to minimize contamination of the neural signal from the surrounding biological signals, the electrode must fit tightly around the nerve and close well. An electrode can be closed with the flap covering the longitudinal slit, and kept in position by means of the surgical threads around the cuff (left side). It can be closed by using a "zipper" method where the ends of the cuff have little holes, through which a plastic baton can be pulled (right side). A self-wrapping polyimide multipolar electrode for selective recording/stimulation of the whole nerve is shown in (c). (Part a is courtesy of Dr. W. Jensen, Aalborg University, Denmark.) (Parts b and c are courtesy of Dr. T. Sinkjær and Dr. M. Haugland, Aalborg University, Denmark.)

Figure 23. Sketches of two cuff-electrode types that reshape the nerve. (a) The flat-interface electrode (FINE) has a rectangular shape and flattens the peripheral nerve in order to make selective contact with individual fascicles in the nerve. The active sites are placed on the cuff-wall. (b) The slowly penetrating interfascicular electrode (SPINE) is designed with a number of beams holding the active sites. A closure tube is moved over the beams to force the beams into position. The nerve will reshape around the beams and thereby make selective contact with the fascicles in the nerve. (Redrawn with permission from Dr. Durand, Case Western Reserve University.)

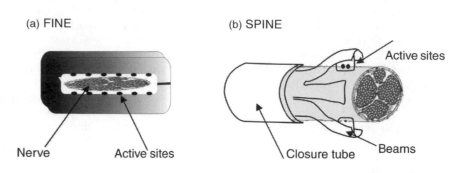

closure mechanism is that it adds several steps to the fabrication process. Furthermore, greater skill is required from the experimenter/surgeon for correct implantation and closure.

2. A second cuff is placed over the first cuff. This is an easy way to securely cover the longitudinal slit, and it is easy to implant. The disadvantage of this method is that overall implant diameter and stiffness increases.

The diameter of a fixed-sized cuff is an important design parameter. The inner diameter cannot be too large since the cuff wall works an electrical shield, and the active sites must make contact with the epineurium. It has been suggested that the fixed-sized cuff should have an inner diameter of $\sim$20–50% larger than the nerve to avoid post-implant nerve damage caused by edema or swelling of the nerve (1,32).

A spiral cuff or a self-wrapping cuff design was originally suggested by Naples et al. (71) to circumvent some of the problems with the fixed-sized cuff. The spiral cuff electrode consists of a planar silicone or polyimide sheeting containing the active sites. The planar sheeting is flexible and is designed to automatically coil around a peripheral nerve (see Fig. 22b). These electrodes are generally handmade, but also have been manufactured by microfabrication techniques. The electrode will make a tight connection with the epineurium, leaving no space between the nerve and the contact sites. The self-sizing property of this design provides several advantages; (1) It is not necessary to determine the cuff diameter in advance as it is with the fixed-sized cuff electrodes. (2) The coiling will automatically and easily close the cuff and hold the cuff into place, and it is therefore very easy to implant. (3) The coiling will provide the necessary constriction of the extracellular current emerging from the nerve fibers without any possible leakages. (4) The coiling will allow the electrode to change size to accommodate for possible edema in the days after implantation.

An example of a cuff-electrode design that incorporates a number of the already discussed principles is a design referred to as the polyimide-hybrid cuff electrode (69) (see Fig. 22d). The electrode is based on a polyimide spiral cuff design containing multiple, platinum contacts on the inner wall. The polyimide-hybrid cuff is manufactured using microfabrication techniques that makes it possible to place a relatively high number of contacts inside the cuff and at

the same time allow space for lead-out wires, which was a problem with the first multicontact cuffs. The microfabrication technique further secures high repeatability in the manufacturing process. Several patents on this type of hybrid fabrication has been issued (72,73).

Application. The first circumferential cuff electrode used for chronic recording was described by Hoffer et al. (74) and later by Stein et al. (75). It is to date the most successful and most widely used chronic peripheral interface in both animals and humans. An overview of current neuroprosthetic applications based on long-term recording from peripheral nerves is given elsewhere in this chapter. Many of these systems use cuff electrodes as their neural interface, and at the time of this writing, is the only implanted nerve interface being used in humans with spinal cord injury or stroke to provide FNS systems with natural sensory feedback information.

Reshaping Cuff Electrodes. The fixed-diameter, circumferential cuffs or the spiral cuffs have their active recording sites placed around the epineurium, and these electrodes therefore mainly record from the most superficial axons or fascicles. The reshaping cuff electrode attempt to accommodate for that by remodeling the shape of the nerve to access individual fascicles.

Design. Two types of reshaping cuff electrodes were developed by Durand and colleagues at Case Western Reserve University. In the slowly penetrating interfascicular electrode (SPINE) the active sites are placed inside the nerve, however, without penetrating the epineurium (76) (see Fig. 23). This is accomplished by a set of blunt penetrating beams that is slowly pushed into the nerve trunk by a closure tube. The active sites are placed on the side of the beams and may therefore access different fascicles of the nerve trunk. A refinement on this concept, the flat interface nerve electrode (FINE) was later developed (77). This electrode has a fixed shape like the fixed diameter, circumferential cuffs, however, the shape is rectangular instead of round (see Fig. 23). Ideally, the individual fascicles will then make contact with different contact sites along the cuff wall.

Application. The reshaping cuff electrodes were mainly developed as a stimulation platform; however,

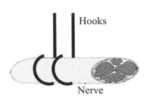

Sketch of hook electrode in place
around a nerve trunk

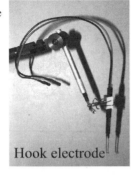

Hook electrode

Figure 24. A hook electrode consists of two simple wires/hooks that are held in close contact with the nerve surface. The picture shows an example of a custom made hook electrode. (Courtesy of Dr. K. Yoshida, Aalborg University, Denmark.)

both types could potentially be used for recording. Acute studies with the SPINE electrode show that the slowly penetrating beams can safely place active sites within the nerve trunk, and it is possible to selectively activate different fascicles of the nerve (77). One of the main design issues with the FINE electrode has been to choose the right size of the electrode without causing nerve damage. Animal studies have shown that the FINE electrode is suitable for both recording and stimulation in chronic animal implants (78,79).

The Hook Electrode. The hook electrode was among the first extraneural electrodes used in experimental neurophysiology and neuroscience. The hook electrode is not useful for long-term nerve implants, however, it is still widely used in acute animal experiments or as an intraoperative tool for nerve identification, and therefore a brief introduction is given here.

Design. The hook electrode is constructed of usually two or more hook-shaped wires that form the recording/stimulation sites (typical materials used are platinum, stainless steel, or tungsten) (see Fig. 24). The nerve trunk or fascicle is placed so the epineurium makes close contact with the hooks. The hook-wires are soldered to insulated lead-out wires that are used to connect with the recording/stimulation equipment. The lead-out wires are usually threaded inside a tube or glued onto a stiff rod to form a base for holding the hook electrode in place. The hook wires must be stiff enough to support the whole nerve or nerve fascicle without yielding. One of the main advantages of the hook electrode is that it is easily handmade at a very low cost.

Application. The distance between the hooks may be varied to change the electrical recording/stimulation properties, however, the relatively large distance between the active sites means that the electrode has a poor selectivity. To use the hook electrode, the whole nerve must first be made accessible in the body, and surrounding fascia must be dissected away to place the hooks underneath the nerve. To avoid electrical shunting between the hooks caused by the extracellular fluid, a paraffin oil or mineral oil pool is often created around the hook electrode and nerve in acute animal studies. For human experiment, the hook electrode and the

nerve may alternatively be suspended into the air to obtain a similar, insulating effect between the hook wires.

USE OF PERIPHERAL NERVE SIGNALS IN NEUROPROSTHETIC APPLICATIONS

Traumatic or nontraumatic injury to nerves at peripheral, spinal, or cortical level may cause permanent loss of sensory and motor functions. The goal of neuroprosthesis applications is to replace, restore, or augment the lost neural function in order to regain the lost mobility or sensation. Popovic and Sinkjær (80) discussed clinical areas where FES is already important or where it holds great potential if the adequate technology will be developed. In the present section, the state-of-the-art of using information provided by peripheral nerve signals in preclinical evaluated and experimental neuroprosthetic applications will be discussed.

Neurorprostheses systems traditionally work by using functional electrical stimulation (FES) to elicit controlled muscle contraction. The functional electrical stimulation can be applied using an open- (feed forward) or a closed-loop (feedback) controller scheme. The open-loop systems have proven to work, however, since no regulation is provided, excessive stimulation and muscle fatigue are commonly observed. Once the stimulation has been determined in the open-loop system, it is not changed, and it is not possible to detect or account for unexpected events or external disturbances. Several studies have shown that the application of closed-loop control techniques can improve the accuracy and stability of FES activated muscles. However, the closed-loop systems are dependent on feedback on the current state of the part of the body under control. The availability of sensors to provide reliable feedback information is, therefore, essential. An illustration of the basic elements in a closed-loop FES control system is given in Fig. 25.

Artificial sensors placed outside the body have been used widely within closed-loop FES applications to provide the necessary feedback signals [e.g., including goniometers

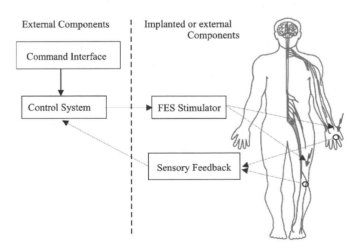

Figure 25. Illustration of the basic elements in a closed-loop neuroprosthesis application. The instrumentation typically includes an external (i.e., outside of the body) command interface and control system and external/implanted interfaces for sensing or activating the biological tissue.

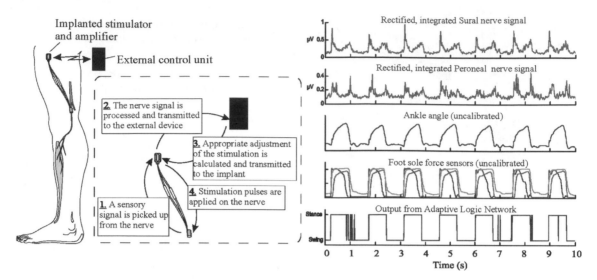

Figure 26. Footdrop correction system, using natural sensory information from the Sural or Peroneal nerve. The nerve signals were processed to determine heel contact with the floor and to decide the timing of the ankle dorsi-flexion stimulation. The stimulation may be applied through a peroneal nerve cuff electrode innervating the Tibials Anterior muscle or by surface electrodes. (Reproduced with permission from Dr. T. Sinkjær, Dr. M. Haugland, and Dr. M. Hansen.)

and accelerometers (80)]. However, the excess time consumed donning and doffing the artificial sensors, and the poor cosmetic visibility of bulky sensors has proven to be a major disadvantage for end user acceptance (1). Artificial sensors have also been placed inside the body; however, they require power to work and can be a problem to replace in case of failure.

The natural sensors, such as those found in, the skin (cutaneous mechanoreceptors, nociceptors), muscles (proprioceptors), or internal organs (visceral neurons) normally providing the CNS with feedback information on the state of the body. They are presently an attractive alternative to the artificial sensors in the cases where the sensors are still viable below the level of brain or spinal cord injury. Unlike artificial sensors, natural sensors are distributed throughout most of the body, and they do not require power or maintenance. Also, efferent motor signals are of interest to monitor the command signals from the CNS to the motor end organs. In many cases, they can be accessed through the same nerve based electrodes used for FES.

Therefore, the peripheral nerve signals of interest here are any neural tissue that is accessible outside the vertebral canal, including afferent and efferent somatic and autonomic pathways (the peripheral nervous system was defined in section The peripheral nervous system and specific peripheral nerve interfaces are described above).

Preclinical Neuroprosthetic Systems

Cutaneous Afferent Feedback for the Correction of Dropfoot. Foot-drop following upper or lower motor neuron deficits is defined as the inability to dorsiflex the foot during the swing phase of the gait (81). Electrical stimulation of the peroneal nerve has proven to be a potentially useful mean for enhancing foot dorsi-flexion in the swing phase of walking and thereby make the patient walk faster and more securely. The stimulator is often located just

distal to the knee and can be either externally mounted or partly implantable. A key element in the system is the external heel-switch placed under the foot, which provides sensory information on heel-to-ground events necessary to switch on the stimulation at the right time. This system has proved clinically functional in large groups of patients. If the external heel switch is replaced with an implantable sensor, it can provide not only the advantages of foot drop corrections systems without the need for footwear, but also eliminates daily problems of mounting the heel switch or swapping them in different pairs of shoes.

The sural nerve is purely sensory and innervates the skin underneath the foot. It has been shown that whole nerve cuff electrode recordings from this nerve can provide information about foot contact events during walking, including heel-to-ground and toe-off-ground events (82). An example of recorded sural nerve activity during walking is shown in Fig. 26. The sural nerve activity has been processed (i.e., filtered, rectified, and bin-integrated). During walking, the nerve signal modulates, and two large spikes are repeatedly observed that correlate with the transitions between swing-to-stance phase and stance-to-swing phase. One of the most significant problem in correlating the nerve signal activity to the actual heel contact during walking is the noise originating from nearby muscle activity. The inclusion of a decision-assistive method like adaptive logic networks as a part of the postprocessing of the neural signal has been used improved the consistency of detecting reliable events (12,83).

The different elements of the drop foot correction system based on natural sensory feedback are shown in Fig. 26, including the external control unit, the implanted sural nerve cuff electrode and amplifier, and the peroneal nerve cuff electrode and stimulator. The sural nerve cuff electrode is connected to an external amplifier, and the signal output is fed to a microprocessor-controlled stimulator that activates the ankle dorsiflexor muscles.

The use of cutaneous afferent feedback in correction of drop-foot has to date been evaluated in several patients. The first evaluation included a 35 year-old subject with a hemiplegic dropfoot. In this case, the cuff electrode was connected to the control system through percutaneous wires, and an external stimulator was placed over the Tibialis Anterior muscle (84). A later version of the system was tested in a 32-year old subject, and the external stimulator was here replaced with an implanted cuff electrode around the peroneal nerve, which innervates the tibialis anterior muscle. An implantable system is currently under development and clinical testing by the Danish company Neurodan A/S.

Cutaneous Afferent Feedback for the Restoration of Hand Grasp. The fingers of the human hand contain an estimated 200–300 touch sensing receptors per square centimeters within the skin. It is among the highest densities of sensory receptors in the body (85). The cutaneous receptors and their responses have been studied extensively using microneurography techniques in normal subjects (85–87). The sensors mainly signal information about indentation and stretch of the skin, and the sensors play an important role in the control of precise finger movements and hand grasp while manipulating objects. It was found that slips across the skin were shown to elicit automatic motor reflex responses that increased the grasp force.

A FES system was developed to evaluate the use of cutaneous sensory information in restoration of hand grasp functions. The system has been evaluated in three spinal cord injured subjects (88–92). Results from a 27 years-old tetraplegic male with a complete C5 spinal cord injury (2 years postinjury) are presented in detail here. Before system was implanted, the patient had no voluntary elbow extension, no wrist function, and no finger function. Furthermore, he had only partial sensation in the thumb, but no sensation in 2nd to 5th fingers. He was implanted with a tripolar nerve cuff electrode on the palmar interdigital nerve (branching of the median nerve). Eight intramuscular stimulation wire electrodes were simultaneously placed in the following muscles: Extensor Pollicis Brevis, Flexor Pollicis Brevis, Adductor Pollicis, and Flexor Digitorum Longus to provide the grip control.

The cuff electrode was implanted around the palmar interdigital nerve, which is a pure cutaneous nerve innervating the radial aspect of the index finger (branching off the median nerve). The cuff electrode recordings were used to detect the occurrence of fast slips (increased neural activity was recorded when objects were slipping through the subject's fingers, and the neural activity showed similar characteristics as the cutaneous sural nerve activity shown in Fig. 26).

The stimulator was controlled by a computer, which also sampled the nerve signals and performed the signal analysis. When stimulation was turned on, the object could be held in a lateral grasp (key grip) by the stimulated thumb. If the object slipped, either because of decreasing muscle force or increasing load, the computer detected this from the processed nerve signal and increased the stimulation intensity with a fixed amount so that the slip was arrested, and the object again held in a firm grip. When extra weight was added, the slipped distance was also comparable to the performance of healthy subjects.

Today this system is developed to an extent where the subject can use it during functional tasks. During an eating session where the subject has a fork in his instrumented hand (grasped between the thumb and index finger), the control system is designed to decrease the stimulation of the finger muscles until the feedback signal from the skin sensors detects a slip between the index finger and the fork. When a slip is detected, the stimulation to the thumb increases automatically proportional to the strength of the sensory feedback, and if no further slips are detected, the controller starts to decrease the stimulation. This continuous tracking of the minimally necessarily needed stimulation means that the hand muscles are only loosely stimulated when the subject decides to rest his arm (which typically happens when they have a conversation with one or more of the persons at a dinner table). The stimulation will automatically increase when they start to eat again by placing the fork in the food. A typically eating session will last 20–30 min. A large fraction of this time is dedicated to "noneating" activities. During such times, the stimulation is at a minimum (keeping the fork in the hand with a loose grasp) and thereby preventing the hand muscles to be fatigued. When the feedback is taken away, the subject will typically leave the stimulation on at high stimulation intensity for the full eating session. This will fatigue the stimulated muscles, and the subject will try to eat their dinner faster, or they will rest their muscles at intervals by manually decreasing the stimulation. It is an effort that requires more attention from the subject than the automatic adjustment of the stimulation intensity.

The natural sensory feedback system was developed at Aalborg University, Denmark, and based on a telemetric system and on the Freehand system (NeuroControl Corp.). The Freehand system consists of a number of epimysial stimulation electrodes to provide muscle activation. More than 200 patients have to date received this system (Fig. 27).

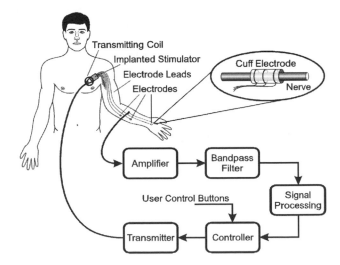

Figure 27. A neural prosthetic system for lateral-key grip restoration after spinal cord injury including sensory feedback. The FreeHand stimulation system (developed at Case Western Reserve University, Ohio) was modified to take advantage of natural sensory feedback signals obtained from a cutaneous, digital nerve in the hand. (Adapted from Ref. 92 with permission.)

Experimental Neural Prosthetic Systems

Somatic Afferent Feedback in Control of Joint Movement.

The natural sensors that carry information about movement and orientation of the body in space (proprioception) are primarily located in the muscles, tendons, joints, and ligaments. The sensory signals are essential for the central nervous system in voluntary control of movement. In contrast to the more on–off type of information provided by the natural sensors located in the skin (e.g., heel contract and toe off events as used in correction of drop foot), the muscle spindle afferents are believed to carry information on muscle displacement (length, velocity, and acceleration). The proprioceptive sensors may therefore be useful in FES systems for control of joint movement. A majority of the research within this area has so far focused on restoration of standing and walking after spinal cord injury, however, the same principles explained here may be applied in control of upper extremity joints. Muscle afferent signals have been studied in animal models using intrafascicular (6,93) and whole nerve cuff electrodes (94,95), and these studies have revealed some characteristic features in muscle afferent responses that must be taken into consideration in a future FES system design.

Muscle afferent signals were obtained from passively stretched agonist-antagonist muscle groups around the ankle joint innervated by the tibial and the peroneal nerve branches in an animal model of spinal cord injury (6,94). The muscle afferent signals were recorded during simple flexion-extension rotations of the ankle joint, and it was found that the muscle afferent activity from the two nerves responded in a complementary manner, see Fig. 28. Thus, the muscle afferent activity increased during muscle stretch, but the activity stopped as soon as the movement was reversed, and the muscle was shortened. In the animal models used here, the muscles are not under normal efferent control from the central nervous system, and the intrafusal muscle fibers (where the muscle spindles are located) therefore become slack during muscle shortening, and the muscle spindles stop responding. The muscle afferents can therefore only provide information about muscle length during muscle stretch. In order to obtain a continuous and reliable muscle afferent feedback signal, it is necessary to use information from more muscles or muscle groups acting around a joint.

In spite of the nonlinear nature of the muscle afferent signals, it has shown to be possible to use the signals as feedback in a closed-loop control system. Yoshida and Horch (6) obtained control of ankle extension (ramp-and-hold movement) against a load and ankle flexion-extension (sinosoidal movement) in a cat model. A look-up table was first established by creating a map between the neural activity and the ankle joint angle. The tibial and peroneal nerve activity was simultaneously recorded using intrafascicular electrodes and decoded according to the look-up table.

To better incorporate the nonlinear behavior, more sophisticated information extraction algorithms have been employed to decode the muscle afferent signals, including a Neuro-Fuzzy network (96), and neural networks (97,98). In all cases, continuous information from both an agonist and antagonist muscle group was used as input to the networks.

The next step is to explore if muscle afferent signals on selected peripheral nerves can be recorded in humans and applied to improve FES standing.

Visceral Neural Signals in Control of Bladder Function.

Normal bladder function (storage and emptying of urine) is dependent on mechanoreceptive sensors detecting bladder volume and efferent control that trigger a bladder contraction. Normal bladder function can be affected by a number of pathological diseases or neural injuries, for example when spinal cord injury occurs at the thoracic or cervial level, the neural pathways normally controlling the bladder are affected. Bladder dysfunction can result in an either overactive bladder (small volume capacity and incomplete emptying, referred to as neurogenic detrusor overactivity) or an underactive bladder (high volume capacity), however, the result of both states is incontinence. Neurogenic detrusor overactivity is the most common form of detrusor dysfunction following spinal cord injury. An overfilling of the bladder can also lead to a possible life-threatening condition of autonomic dysreflexia (overactivity of the autonomic nervous system).

Conventional treatments include suppression of reflex contraction by drugs, manual catheterization, or inhibition of the reflex contraction by surgical intervention, which is also referred to as dorsal rhizotomy. Dorsal rhizotomy

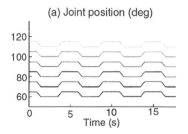
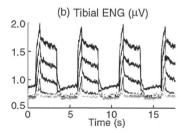
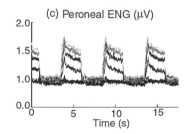

Figure 28. Muscle afferent signals recorded from the Tibial and Peroneal nerve branches during passive joint rotation in a rabbit model. The Tibial and Peroneal nerve activity modulated in a push–pull manner (i.e. the nerves responded to muscle stretch of the Gastrocnemius/Soleus and the Tibialis Anterior muscles, respectively). The magnitude of the response modulated with the initial position of the joint (the amount of prestretch in the muscles). (Courtesy of Dr. Jensen, Aalborg University.)

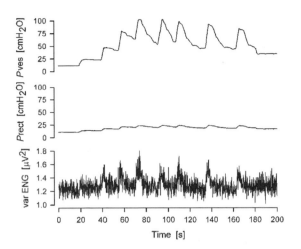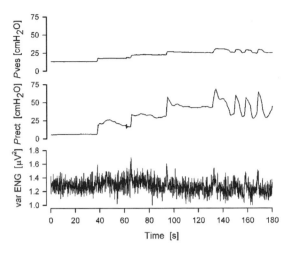

Figure 29. Two examples of recorded neural activity from the sacral roots in an anaesthetized human subject in preparation for surgery for implantation of a FineTech-Brindley Bladder system. The traces show the bladder pressure (top), rectal pressure (middle) and extradural sacral root activity (bottom) over time. (Courtesy of M. Kurstjens.)

increases the bladder capacity, however, reflex erection in male patients is lost, and this is in most cases not an attractive solution. To reestablish continence a closed-loop FES system has been suggested including sensory feedback on bladder volume and/or detrusor activation from bladder nerves.

Work in anaesthetized cats showed correlation between recorded activity from the S2 sacral root and the intravesical pressure associated with fullness of the bladder. Recently, similar information on bladder volume information has also been demonstrated in nerve cuff recordings from the S3 sacral root and the pelvic nerve in anaesthetized pigs (99) and extradural cuff recordings from sacral roots in humans (100) (see Fig. 29). The findings reveal, however, some problems in finding a reliable sensory feedback signal. It proved difficult to detect slow increases in bladder volume (99). Further, recordings from the pelvic nerve or sacral roots, however, are frequently contaminated with activity from other afferents, such as cutaneous and muscle afferents from the pelvic floor, rectum, and anus (100).

A closed-loop FES system for the management of neurogenic detrusor activity would work by using the sensory information from the bladder nerves to inhibit detrusor contractions by stimulating pudendal or penile nerves, or block efferent or afferent pelvic nerve transmission to prevent reflex detrusor contractions. The principle was tested in an anesthetized cat prepration where nerve cuff recordings from the S1 in cats could detect bladder hyper-reflexive like contractions. However, the contractions were detected with a time delay in the order of seconds (using a CUMSUM alogrithm), which decreases the feasibility of the system (101).

Visceral Neural Signals in Respiratory Control. Obstructive sleep apnea is characterized by occlusions of upper airways during sleep. Electrical stimulation of the genioglossus muscle directly or via the hypoglossal nerve can improve the obstructions, however, a signal for detecting the occurrence of the obstructions will be useful in a FES closed-loop application (102).

This principle has been tested in a closed-loop FES system where the response of hypoglossal nerve was recorded during external loading of the pharynx during sleep to simulate upper airway obstructions in a dog model. It was shown that the hypoglossal nerve activity modulated with the pressure both during REM and NREM sleep. Any change in the cyclical rhythm of the recorded activity was used as an indication of the onset of apnea to the controller and triggered the onset of the stimulation of the same hypoglossal nerve to prevent the airway obstruction (103). Stimulation and recording from the same nerve is feasible in this case because the only information needed from the hypoglossal nerve is an indication of the onset of airway obstruction (i.e., no continuous information is needed).

CONCLUSIONS

The goal of the article is to provide the reader with an overview of currently available neural electrodes for long-term interfacing peripheral neural tissue. This is done to evaluate their suitability to monitor the neural traffic in peripheral nerves and their suitability to perform as sensors in neural prosthetic devices.

The role of neural prosthetic systems for increasing the quality of life of disabled individuals is becoming more evident each day. As the demand to develop systems capable of providing expanded functionality increases, so too does the need to develop adequate sensors for control. The use of natural sensors represents an innovation. Future research will show whether nerve-cuff electrodes and other types of peripheral nerve electrodes can be used to reliably extract signals from the large number of other receptors in the body to improve and expand on the use of natural sensors in neural prosthetic systems.

ACKNOWLEDGMENTS

The authors gratefully acknowledge The Danish National Research Foundation and The Danish Research Councils for their financial support.

BIBLIOGRAPHY

Cited References

1. Hoffer JA. Techniques to study spinal-cord, peripheral nerve and muscle activity in freely moving animals. In Boulton AA, Baker GB, Vanderwolf CH, editors. Neural Prostheses. Replacing Motor Function After Disease or Disability. New York Oxford: Oxford University Press; 2005.
2. Branner A, et al. Long-term stimulation and recording with a penetrating microelectrode array in cat sciatic nerve. IEEE Trans Biomed Eng 2004;51(1):146–157.
3. Lee DH, Claussen GC, Oh S. Clinical nerve conduction and needle electromyography studies. J Am Acad Orthop Surg 2004 July–Aug; 12(4):276–287. Review.
4. Fuller G. How to get the most out of nerve conduction studies and electromyography. J Neurol Neurosurg Psychiatry. 2005; 76(Suppl. 2):ii41–46. Review.
5. Crago PE, Nakai RJ, Chizeck HJ. Feedback regulation of hand grasp opening and contact force during stimulation of paralyzed muscle. IEEE Trans Biomed Eng 1991;38(1):17–28.
6. Yoshida K, Horch K. Closed-loop control of ankle position using muscle afferent feedback with functional neuromuscular stimulation. IEEE Trans Biomed Eng 1996;43(2):167–176.
7. Haugland M, Hoffer JA. Slip Information Provided by Nerve Cuff Signals: Application in Closed-Loop Control of Functional Electrical Stimulation. IEEE Trans Rehab Eng 1994;2:29–36.
8. Sinkjaer T, Haugland M, Haase J. Natural neural sensing and artificial muscle control in man. Exp Brain Res 1994;98(3):542–545.
9. Sinkjær T, Haugland M, Struijk J, Riso R. Long-term cuff electrode recordings from peripheral nerves in animals and humans. In: Windhorst U, Johansson H, editors. Modern Techniques in Neuroscience. New York: Springer; 1999. p 787–802.
10. Stein RB, et al. Encoding mechanisms for sensory neurons studied with a multielectrode array in the dorsal root ganglion. Can J Physiol Phramacol 2004;82:757–768.
11. Stein RB, et al. Coding of position by simultaneously recorded sensory neurones in the cat dorsal root ganglion. J Physiol 2004;560(3):883–896.
12. Hansen M, Haugland M, Sinkjær T. Evaluating Robustness of gait event detection based on machine learning and natural sensors. IEEE Trans Neural Syst Rehab Eng 2004;12: 1:81–87.
13. Hoffer JA, Kallesoe K. Nerve cuffs for nerve repair and regeneration. Prog Brain Res 2000;128:121–134.
14. Struijk JJ, Thomsen M, Larsen JO, Sinkjaer T. Cuff electrodes for long-term recording of natural sensory information. IEEE Eng Med Biol Mag 1999;18(3):91–98.
15. Haugland M, Sinkjær T. 2000.
16. Riso RR. Perspectives on the role of natural sensors for cognitive feedback in neuromotor prostheses. Automedica 1998;16:329–353.
17. McDonnall D, Clark GA, Normann RA. Selective motor unit recruitment via intrafascicular multielectrode stimulation. Can J Physiol Pharmacol 2004;82(8–9):599–609.
18. McDonnall D, Clark GA, Normann . Interleaved, multisite electrical stimulation of cat sciatic nerve produces fatigue-resistant, ripple-free motor responses. IEEE Trans Neural Syst Rehabil Eng 2004;12(2):208–215.
19. Aoyagi Y, et al. Capabilities of a penetrating microelectrode array for recording single units in dorsal root ganglia of the cat. J Neurosci Methods 2003;30: 128(1–2):9–20
20. Webster. 1998.
21. Hodgkin AL, Huxley AF. A quantitative description of membrane current and its application to conduction and excitation in nerve. J Physiol (London) 1952;117:500–544.
22. Chiu SY, Ritchie JM, Rogart RB, Stagg D. A quantitative description of membrane currents in rabbit myelinated nerve. J Physiol 1979;292:149–166.
23. Sweeney JD, Mortimer JT, Durand D. Modeling of mammalian myelinated nerve for functional neuromuscular stimulation. Proceedings of the 9th Annual International Conference of the IEEE EMBS 1987; 1577–1578.
24. Varghese A. Membrane Models. In: Bronzino JD, editor. The Biomedical Engineering Handbook: 2nd ed. Boca Raton (FL): CRC Press LLC; 2000.
25. McIntyre CC, Richardson AG, Grill WM. Modeling the excitability of mammalian nerve fibers: influence of afterpotentials on the recovery cycle. J Neurophysiol 2002;87(2):995–1006.
26. Struijk JJ. The extracellular potential of a myelinated nerve fiber in an unbounded medium and in nerve cuff models. Biophys J 1997;72:2457–2469.
27. Perez-Orive , Durand . Modeling study of peripheral nerve recording selectivity. IEEE Trans Rehabil Eng 2000;8(3):320–329.
28. Taylor J, Donaldson N, Winter J. Multiple-electrode nerve cuffs for low-velocity and velocity-selective neural recording. Med Biol Eng Comput 2004;42(5):634–643.
29. Chemineau ET, Schnabel V, Yoshida K. A modeling study of the recording selectivity of longitudinal intrafascicular electrodes. In: Wood D, Taylor J, editors. Getting FES into clinical practice. Proceedings of IFESS-FESnet 2004, 9th Annual Conference of the International Functional Electrical Stimulation Society and the 2nd Conference of FESnet 6–9 September, 2004Bournemouth, UK. 2004 p 378–380.
30. Yoshida K, Struijk JJ. The theory of peripheral nerve recording. In: Horch KW, Dhillon GS, editors. Neuroprosthetics: Theory and Practice World Scientific Publishing Co.; 2004.
31. Nikolic ZM, Popovic DB, Stein RB, Kenwell Z. Instrumentation for ENG and EMG recordings in FES systems. IEEE Trans Biomed Eng 1994;41(7):703–706.
32. Stein RB, et al. Stable long-term recordings from cat peripheral nerves. Brain Res 1977;128(1):21–38.
33. Yoshida K, Stein RB. Characterization of signals and noise rejection with bipolar longitudinal intrafascicular electrodes. IEEE Trans Biomed Eng 1999;46(2):226–234.
34. Ott HW. Noise Reduction Techniques in Electronic Systems. 2nd ed, New York: John Wiley & Sons, Inc.; 1988.
35. Grimnes S, Martinsen ø G. Bioimpedance and Bioelectricity Basics. London: Academic Press; 2000 p 221.
36. Heiduschka P, Thanos S. Implantable Bioelectronic Interfaces for Lost Nerve Functions. Prog Neurobiol 1998;55:433–461.
37. Mannard A, Stein RB, Charles D. Regeneration electrode units: implants for recording from single peripheral nerve fibers in freely moving animals. Science 1974;183(124):547–549.
38. Stieglitz T, et al. A biohybrid system to interface peripheral nerves after traumatic lesions: design of a high channel sieve electrode. Biosensors Bioelectronics 2002;17:685–696.

39. Edell DJ. A peripheral nerve information transducer for amputees: long-term multichannel recordings from rabbit peripheral nerves. IEEE Trans Biomed Eng 1986;33(2):203–214.

40. Akin T, Najafi K, Smoke RH, Bradley RM. A micromachined silicon sieve electrode for nerve regeneration studies. IEEE Trans Biomed Eng 1994;41:305–313.

41. Bradley RM, Smoke RH, Akin T, Najafi K. Functional regeneration of glossopharyngeal nerve through michromachined sieve electrode arrays. Brain Res 1992;594:84–90.

42. Lago N, Ceballos D, RodrÚguez FJ, Stieglitz T, Navarro X. Long-term assessment of axonal regeneration through polyimide regnerative electrodes to interface the peripheral nerve. Biomaterials 2005;26:2021–2031.

43. Kovacs GTA, et al. Silicon-substrate microelectrode arrays for parallel recording of neural activity in peripheral and cranial nerves. IEEE Trans Biomed Eng 1994;41(6):567–577.

44. Zhao Q, Dahlin LB, Kanje M, Lundborg G. Specificity of muscle reinnervation following repair of the transected sciatic nerve. A comparative study of different repair techniques in the rat. J Hand Surg 1992;17:257–261.

45. Yoshida K, Riso R. Peripheral nerve recording electrodes and techniques. In Horch K, Dhillon GS, editors. Neuroprosthetics. Theory and Practice. 1st ed. New York: World Scientific; 2004. p 683–744.

46. Bradley RM, Cao X, Akin T, Najafi K. Long term chronic recordings from peripheral sensory fibers using a sieve electrode array. J Neurosc Methods 1997;73:177–186.

47. Mesinger AF, et al. Chronic recording of regenerating VIIIth nerve axons with a sieve electrode. J Neurophysiol 2000;83:611–615.

48. Shimantani Y, Nikles SA, Najafi K, Bradley RM. Long-term recordings from afferent taste fibers. Physiol Beh 2003;80:309–315.

49. Berthold CH, Lugnegard H, Rydmark M. Ultrastructural morphometric studies on regeneration of the lateral sural cutaneous nerve in the white rat after transection of the sciatic nerve. Scan J Plast Reconstr Surg Supp l 1985;30:1–126.

50. Vallbo Å, Hagbart KE. Activity from skin mechanoreceptors ecorded percutaneously in awake human subjects. Exper Neurol 1968;21:270–289.

51. Vallbo Å, Hagbart KE, Wallin BG. Microneurography: how the technique developed and its role in the investigation of the sympathetic nervous system. J Appl Physiol 2004;96:1262–1269.

52. Hallin RG, Wiesenfeld-Hallin Z, Duranti R. Percutaneous microneurography in man does not cause pressure block of almost all axons in the impaled nerve fascicle. Neurosc Lett 1986;68:356–361.

53. Bowmann BR, Erickson RC. Acute and chronic implanation of coiled wire intraneural electrodes during cyclical electrical stimulation. Ann Biomed Eng 1985;13:75–93.

54. Malagodi MS, Horch K, Schoenberg A. An intrafascicular electrode for recording action potentials in peripheral nerves. Ann Biomed Eng 1989;17:397–410.

55. McNaughton TG, Horch K. Metalized polymer fibers as leadwires and intrafacicular microelectrodes. J Neurosc Methods 1996;8:391–397.

56. Gonzalez C, Rodriguez M. A flexible perforated microelectrode arry probe for action potential recording in nerve and muscle tissue. J Neurosc Methods 1997;72:189–195.

57. Lefurge T, et al. Chronically implanted intrafascicular recording electrodes. Ann Biomed Eng 1991;19:197–207.

58. Malmstrom M, McNaughton TG, Horch K. Recording properties and biocompatibility of chronically implanted polymer-based intrafascicular electrodes. Ann Biomed Eng 1998;26:1055–1064.

59. (a) Jones KE, Campebll PK, Normann RA. A Glass/Silicon Composite Intracortical Electrode Array. Ann Biomed Eng 1992;20:423–437. (b) Lawrence SM, et al. Long-term biocompatibility of implanted polymer-based intrafascicular electrodes. J Biomed Mater Res 2002; 63(5):501–506.

60. Rousche PJ, Normann RA. A method for pneumatically inserting an array of penetrating electrodes into cortical tissue. Ann Biomed Eng 1992;20:413–422.

61. Branner A, Normann RA. A multielectrode array for intrafasciclar recording and stimulation in sciatic nerve of cats. Brain Res Bull 1999;51(4):293–306.

62. Branner A, Stein RB, Normann RA. Selective Stimulation of Cat Sciatic Nerve Using an Array of Varying-Length Microelectrodes. J Neurophysiol 2001;85:1585–1594.

63. Gasson MN, et al. Bi-directional human machine interface via direct neural connection. Proceedings of the IEEE Conference on Robot and Human Interactive Communication, Berlin, Germany; 2002, p 26–270.

64. Stein RB, et al. Principles underlying new methods for chronic neural recording. Can J Neurol Sci 1975;2(3):235–244.

65. Haugland M. A flexible method for fabrication of nerve cuff electrodes. In: Proceedings of the 18th Annual Conference IEEE Engineering in Medicine and Biology. Amsterdam The Netherlands: 1996p 964–965.

66. Stieglitz T, Meyer JU. Implantable microsystems: polyimide-based neuroprostheses for interfacing nerves. Med Device Technol 1999;10(6):28–30.

67. Rodriguez FJ, et al. Polyimide cuff electrodes for peripheral nerve stimulation. J Neurosci Methods 2000;98(2):105–118.

68. Veraat C, Grill WM, Mortimer JT. Selective control of muscle activation with a multipolar nerve cuff electrode. IEEE Trans Biomed Eng 1993;40:640–653.

69. Schuettler M, Stieglitz T. 18 polar hybrid cuff electrodes for stimulation of peripheral nerves. Proceedings of the IFESS. Aalborg, Denmark: 2000. p 265–268.

70. Kallesoe K, Hoffer JA, Strange K, Valenzuela I. Simon Fraser University, Implantable cuff having improved closure. US patent 5,487,756, 1994.

71. Naples GG, Mortimer JT, Schiner A, Sweeney JD. A spiral nerve cuff electrode for peripheral nerve stimulation. IEEE Trans Biomed Eng 1988;35(11):905–916.

72. Grill WM, et al. Thin film implantable electrode and method of manufacture. Case Western Reserve University, US patent 5,324,322. 1994 June 28.

73. Grill WM, Tarler MD, Mortimer JT. Case western Reserve University. Implantable helical spiral cuff electrode. US patent 5,505,201. 1996 April 9.

74. Hoffer JA, Marks WB, Rymer Z. Nerve fiber activity during normal movements. Soc Neurosci Abs 1974: 258.

75. Stein RB, et al. Impedance properties of metal electrodes for chronic recording from mammalian nerves. IEEE Trans Biomed Eng 1978;25:532–537.

76. Tyler DJ, Durand DD. A slowly penetrating interfascicular nerve electrode for selective activation of peripheral nerves. IEEE Trans Rehab Eng 1997;5(1):51–61.

77. Tyler DJ, Durand DD. Functionally Selective Peripheral Nerve Stimulation With a Flat Interface Nerve Electrode. IEEE Trans Neural Syst Rehabil Eng 2002;10(4):294–303.

78. Yoo PB, Sahin M, Durand DD. Selective stimulation of the canine hypoglossal nerve using a multi-contact cuff electrode. Ann Biomed Eng 2004;32(4):511–519.

79. Leventhal DK, Durand DD. Chronic measurement of the stimulation selectivity of the flat interface nerve electrode. IEEE Trans Biomed Eng 2004;51(9):1649–1658.

80. Popovic D, Sinkjær T. Control of movement for the physically disabled: control for rehabilitation technology. 2nd ed. Aalborg: Center for Sensory Motor Interaction, Aalborg University; 2003.

81. Lyons GM, Sinkjær T, Burridge JH, Wilcox DJ. A review of portable FES-based neural orthoses for the correction of drop foot. IEEE Trans Neural Syst Rehabil Eng 2002;10(2): 260–279.

82. Haugland M, Sinkjaer T. Interfacing the body's own sensing receptors into neural prosthesis devices. Technol Health Care 1999;7(6):393–399.

83. Kostov A, Hansen M, Haugland M, Sinkjær T. Adaptive restrictive rules provide functional and safe stimulation patterns for foot drop correction. Art Organs 1999;23(5): 443–447.

84. Haugland M, Sinkjær T. Cutaneous Whole Nerve Recordings Used for Correction of Footdrop in Hemiplegic Man. IEEE Trans Rehab Eng 1995;3:307–317.

85. Johansson RS, Westling G. Tactile sensibility in the human hand: Relative and absolute densities of four types of mechanireceptive units in glabrous skin. J Physiol 1979; 21:270–289.

86. Vallbo Å, Johannson RS. Properties of cutaneous mechanoreceptors in the human hand related to touch sensation. Human Neurobiol 1984;3:3–14.

87. Westling G, Johansson RS. Responses in glabrous skin mechanoreceptors during precision grip in humans. Exper Brain Res 1987;66:128–140.

88. Haugland M, Lickel A, Haase J, Sinkjær T. Control of FES Thumb Force Using Slip Information Obtained from the Cutaneous Electroneurogram in Quadriplegic Man. IEEE Trans Rehab Eng 1999;7:215–227.

89. Haugland M, et al. Restoration of lateral hand grasp using natural sensors. Art Organs 1997;21(3):250–253.

90. Inmann A, Haugland M. Implementation of natural sensory feedback in a portable control system for a hand grasp neuroprosthesis. Med Eng Phy 2004;26:449–458.

91. Inmann A, Haugland M. Functional evaluation of natural sensory feedback incorporated in hand grasp neurprosthesis. Med Eng Phys 2004; 26(6):439–447.

92. Inmann A, et al. Signals from skin mechanoreceptors used in control of hand grasp neurosthesis. Neuroreport 2001;12(13): 2817–2820.

93. Jensen W, Yoshida K. Long-term recording properties of longitudinal intra-fascicular electrodes. 7th Annual Conference of the International Functional Electrical Stimulation Society, IFESS 2002, June 25–29, Ljubljana: Slovenia; 2002 p 138–140.

94. Riso R, Mossallaie FK, Jensen W, Sinkjær T. Nerve Cuff Recordings of Muscle Afferent Activity from Tibial and Peroneal Nerves in Rabbit During Passive Ankle Motion. IEEE Trans Rehab Eng 2000;8(2):244–258.

95. Jensen W, Riso R, Sinkjær T. Position Information in Whole Nerve Cuff Recordings of Muscle Afferents in a Rabbit Model of Normal and Paraplegic Standing. Proceedings of the 20th Annual International Conferences of the IEEE/ EMBS Society. 1998.

96. Micera S, et al. Neuro-fuzzy extraction of angular information from muscle afferents for ankle control during standing in paraplegic subjects: an animal model. IEEE Trans Biomed Eng 2001;48(7):787–789.

97. Jensen W, Sinkjær T, Sepulveda F. Improving signal reliability for on-line joint angle estimation from nerve cuff recordings of muscle afferents. IEEE Trans Neural Syst Rehabil Eng 2002; 10(3):133–139.

98. Sepulveda F, Jensen W, Sinkjær T. Using Nerve Signals from Muscle Afferent Electrodes to Control FES-based Ankle Motion in a Rabbit. Proceedings 23rd Annual International Conference of the IEEE-EMBS. 2001.

99. Jezernik S, et al. Analysis of Bladder Related Nerve Cuff Electrode Recordings from Preganglion Pelvic Nerve and Sacrral Roots in Pigs. J Urol 2000;163:1309–1314.

100. Kurstjens M, et al. Interoperative recording of sacral root nerve signals in humans. Art Organs 2005;29(3):242–245.

101. Jezernik S, Grill WM, Sinkjær T. Detection and inhibition of hyperreflexia-like bladder contractions in the cat by sacral nerve root recording and stimulation. Neurouol Urodyn 2001;20:215–230.

102. Sahin M, Durand DD, Haxhiu MA. Closed-loop stimulation of hypoglossan nerve in a dog model of upper airway obstruction. IEEE Trans Biomed Eng 2000;47(7):919–925.

103. Sahin M, Durand DD, Haxhiu MA. Chronic recordings of hypoglossal nerve activity in a dog model of upper airway obstruction. J App Physi 1999;87(6):2197–2206.

See also ELECTROMYOGRAPHY; EVOKED POTENTIALS; NEUROLOGICAL MONITORS.

ELECTROPHORESIS

JOHN M. BREWER
University of Georgia
Athens, Georgia

INTRODUCTION

The "electrostatic effect" causes particles with the same sign of charge to repel each other, while two particles of opposite charge attract: Charged particles produce electric fields and a charged particle in an electric field experiences a force. Electrophoretic techniques are based on the movement or flow of ions in a solvent produced by an electric field, causing separation of different ions.

THEORY

An external electric field V produces a force on an ion with a charge Q equal to $VQ(1)$. An ion in a solvent will be accelerated, but movement in the solvent will be slowed by an opposite force that comes from the viscosity of the solvent. For small velocities, the opposite (viscous) force is proportional to the velocity of electrophoresis. We call the proportionality constant the "frictional coefficient". The forces from the electric field and viscosity are opposed, so the net acceleration becomes zero and the velocity of electrophoresis (v) constant:

$$v = VQ/\text{frictional coefficient}$$

Mobilities of Ions

The velocity an ion reaches for a given applied field is a specific property of that ion. It is called the "mobility" of

(a)

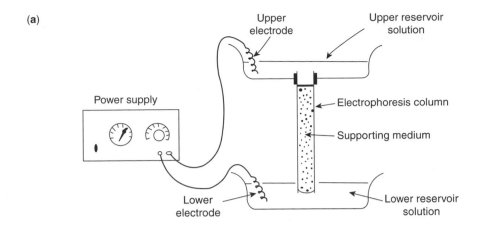

(b)

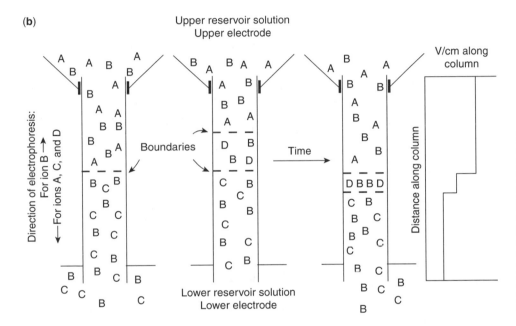

Figure 1. (a) Representation of a vertical electrophoresis experiment. (b) The lower drawing represents the stacking phase of disk electrophoresis or isotachophoresis of a single component, ion D. Charge signs are omitted as cationic and anionic isotachophoretic or disk systems exist. (Reprinted from Clinical Chemistry: Theory, Analysis, Correlation, 4th ed. Kaplan LA, Pesce AJ, Kazmierczak SC, editors, Electrophoresis, p 204, copyright © 2003, with permission from Elsevier.)

that ion:

$$v/V = \text{mobility of ion} = Q/\text{frictional coefficient}$$

For ions such as proteins that are very large compared with solvent molecules, the frictional coefficient is 6π times the product of the solvent viscosity and the effective radius (size) of the ion (1).

The direction the ion moves is determined by the sign of its charge Q and the direction of the electric field. The electric field is produced by two oppositely charged substances (electrodes) placed at opposite ends of the solution the ion is in, and their polarity (charge) determines the field's direction (Fig. 1) (2).

How fast ions in solution move is also affected by some other factors, but these are still hard to calculate (1,3). In aqueous solvents, ions are "hydrated", bound to a number of water molecules (4,5). A protein has many charged groups, both positive and negative, and those in contact with the solvent also will be hydrated (6) and will in addition have hydrated ions of opposite charge, called

"counterions", nearby. In an electrophoresis apparatus (Fig. 1), the protein will move toward the electrode of opposite charge in an irregular fashion because random diffusional motion adds to the uniform electrophoretic movement. After each change of direction, the smaller ions of opposite charge are out of their equilibrium positions and need time(s) to readjust or move back into position. During the readjustment period(s), the effective voltage that the protein is exposed to is lower than the applied voltage, because the smaller counterions, while out of position, produce a field in the opposite direction.

Positively charged ions (cations) move in the opposite direction to negative ions (anions). They all are hydrated and the movement of the protein or other large ion of interest is slowed by a flow of hydrated ions of opposite charge (counterions) in the opposite direction. This is called an "electrophoretic effect" and, together with the "relaxation" or "asymmetry" effect described before, makes it difficult to use the measured velocities of large ions to calculate their real net charge or effective radius (1,3). Electrophoresis is consequently used largely as an empirical technique.

The mobilities of many small ions are known (7). These values are for fully charged ions. The actual (average) charge of ions like acetate depends on the pH, so the observed or actual (effective) mobility of some ions will also be pH dependent.

$$\text{Effective mobility} = (\text{mobility})(\text{average charge})$$

Proteins or polynucleotides have charged groups that determine their ionic properties. These charged groups in proteins are anionic and cationic, with pK values from 3 to 13, so the effective mobilities of proteins are very pH dependent. For proteins, overall (net) charges vary from positive at low pH to zero (isoelectric) to negative at more alkaline pH values. Substances whose charge can vary from positive to negative are called "amphoteric". Routine electrophoretic separations are done at constant pH values, in a "buffered" solution, to minimize effects of pH on velocity of movement.

Electrical Effects

These techniques are electrical in character. An applied voltage V produces movement of oppositely charged ions in opposite directions. These flows of ions are the current i. Solvents resist movement of ions through them and this resistance is R, so that $V = iR$, just as in a wire or any other electrical device. People using electrophoresis are interested in ion movement, so the reciprocal of the resistance, called the "conductivity" (σ), is used instead: $\sigma V = i$.

Several factors determine the conductivity of a solution. The conductivity is proportional to the concentrations of the ions present (actually to their thermodynamic activities). It is also proportional to their mobilities or, if the pH is a factor, to their effective mobilities. Since all ions contribute,

$$\sigma \propto \sum \sigma_i \propto \sum (\text{concentration})_i (\text{mobility})_i (\text{average charge})_i$$

The conductivity is also proportional to the cross-sectional area through which the ions move, but the area is usually constant.

The current must be constant in any tube or channel through which electrophoresis is occurring. There can be no gaps or accumulations of current. Since the conductivity in different parts of the tube may be different (see below), the voltage along a tube or channel may also differ in different parts of the tube or channel.

The purpose of electrophoresis is to separate ions and this may be measured from the movements of a large number of molecules using band broadening of scattered laser light owing to Doppler effects (8), by following the movement of the interface or "boundary" between a solution of a particular ion and the solvent, or by following the movement of a region or "zone" containing an ion of interest through a solvent (Fig. 1). Currently, most workers follow the position(s) of relatively thin zone(s) of material, since complete separation of a mixture takes less time the thinner the original zone of the mixture is. However, even the thinnest zone contains a leading boundary at the front and a trailing boundary at the back, so the following discussion applies to zone and boundary methods. We assume the ion of interest is a large one, say a protein.

The large ion is usually restricted at first to one part of the tube or channel through which electrophoresis occurs. The solvent in other parts must also contain ions, usually small ones, to carry the current where the large ion is not present. In other words, the large ion is part of the electrical system and so carries part of the current and contributes to the conductivity. It will move in the same direction as small ions with the same charge, called "co-ions". So the large ion is competing with the co-ions to carry the current.

The large ions must replace equivalent amounts (in terms of charge) of co-ions so that an excess of ions of one charge never accumulates anywhere. In other words, total positive charges must always equal total negative charges (electroneutrality principle). Large ions often have lower mobilities than co-ions. The co-ions are diluted by the large ions with a lower mobility. The moving zone with the large ions has then a lower average conductivity.

Since gaps in the current cannot occur, the zone with the large ions must keep up with the co-ions it is partially replacing. To do this, the voltage in the large ion-containing zone must increase to keep the current constant. If a single large ion diffuses across the leading large ion–solvent boundary, it has moved into a region of higher conductivity and lower voltage. Consequently, the single large ion slows down until the large ion zone catches up (2,7).

At the trailing boundary, co-ions with higher mobilities are replacing lower mobility large ions. This increases the conductivity in the region the large ions just left. The voltage right behind the trailing boundary is consequently lower. If a single large ion diffuses back into that region, it will be in a lower voltage and will move more slowly than the large ions in the large ion zone.

The effect of this competition is to cause zones of large ions with mobilities lower than those of the co-ions to be sharp at the leading boundary and diffuse at the trailing boundary. In other words, zone broadening due to diffusion of the large ions is either reduced or increased at the front (leading) and rear (trailing) edges, respectively, of the zone, depending on the relative mobilities of large ions and co-ions.

There may be effects of pH as well. The small co-ions and counterions are often also used to maintain the pH. The competition between large ions and co-ions, by reducing co-ion movement, leads to changes in the ratio of co-ion to counterion where the large ions have moved in or out of a region. The pH is affected by the ratio of small conjugate acid and base ions, either of which may be the co-ion, so movement of the large ion is accompanied by changes in pH. Changes in pH can change the net charge on the large ion, especially if it is a protein, and so change its effective mobility. Changes in the pH where electrophoresis is occurring can add to or reduce the competition effects described above.

Enhancing Resolution

Separations of substances produced by any procedure will be counteracted by the resulting diffusion of the separated substances, so diffusion should be reduced as much as possible. The competition between ions with the same charge during electrophoresis can add to or counteract diffusion, and this is the basis of "mobility-based" enhanced resolution techniques.

Mobility-Based Methods. Instead of a zone of solution containing large and small ions set between zones of solvent containing the same small ions, we put different small ions on either side of the large ion zone. This is a "discontinuous" solvent system (2,3,7).

Figure 1b shows an upright cylinder whose lower half is filled with co-ion C and counterion B. The upper half of the cylinder has co-ion A and counterion B. An electric field is applied so that the A ions follow the C ions toward one electrode. If the mobility of the A ion and conductivity of the A solution are less than those of the C solution, the boundary between the solutions will move but diffusion across the boundary by either ion will be restricted. Since the current must be the same all along the tube, the lower conductivity A solution must have a higher voltage making the A ions keep up with the C ions. If A ions diffused ahead into the C solution, the A ions would be in a lower voltage region and would slow until the rest of the A solution caught up. If C ions diffused back into the A solution, the higher voltage there would drive them back to the C solution.

Suppose a solution of ions B and D was inserted between the AB and CB solutions. If ion D has a mobility intermediate between those of A and C, it will remain "sandwiched" between the AB and CB solutions and diffusion will be restricted across the AD and DC boundaries. Suppose further that a mixture of ions DEFG ... of the same charge as A and C were originally present in the intermediate solution and these had mobilities intermediate to those of A and C. Even if DEFG ... were originally mixed together, they would separate and travel, each in separate but adjacent subzones, in order of decreasing mobility going from the C zone to the A zone, all at the same velocity and all with sharply maintained boundaries.

Another effect that occurs in discontinuous systems involves concentrations. If the DEFG ... ions were originally very dilute in their original zone, their zone would compress and each ion's subzone would become relatively thin upon starting the electrophoresis. This is because each ion subzone must carry as much current as the A and C zones, and the higher the conductivities of the AB and CB solutions, the thinner the DEFG ... subzones will get. If DEFG ... are different proteins, which generally have very high molecular weights and relatively low net charges, the subzones will become very thin indeed. The "running" concentrations of DEFG ... will become very high.

If the pattern produced by DEFG ... as they move down the column is determined or if they are made to elute from the column as they emerge from it, the process is called "isotachophoresis", since DEFG ... all move at the same velocity (7). Having DEFG ... unseparated from each other makes analysis or preparation difficult, and isotachophoresis is currently relatively little used.

The most frequently employed procedure to separate DEFG ... is to increase the effective mobility of the A ion so that it runs through the DEFG ... subzones (2,7). These are then in a uniform electric field, that of the AB solution, and so electrophorese independently. While the separating

subzones also begin to diffuse, they were "stacked" in very concentrated, and hence thin subzones. The thinner the subzone is before independent electrophoresis, the thinner it will be at the end of electrophoresis and the better the resolution. This approach is employed in "disk electrophoresis".

The A ion is of lower mobility than the C ion and the pH used lowers its effective mobility further. To increase the effective mobility of the A ion, the original CB solution contains a high concentration of the uncharged conjugate acid or base of the B ion. The ions (and conjugate acid or base) are chosen so that when A replaces C its average net charge increases because of the new or "running" pH, which is controlled by the ratio of B ion to its conjugate acid or base.

It must be emphasized that the running concentrations of the ions and uncharged acids or bases are not arbitrary but are controlled by the physical (electrical) constraints mentioned before and alter if they do not conform initially. The constraints are summarized mathematically in Ref. 7.

The importance of the B ion is emphasized by examination of another electrophoretic technique. Suppose the concentration of the B ion is reduced to zero. Since transport of ions must occur in both directions, reduction of the B ion concentration to zero will reduce migration of A, DEFG ..., and C to zero (or nearly zero: in aqueous solvents, there will always be some movement due to ionization of water) (7,9). A, DEFG ..., and C will be stacked. Since B must be replaced, because of electrical neutrality, with protons or hydroxyl ions, amphoteric ions such as proteins will be "isoionic". The pH of each zone or subzone is that at which each amphoteric ion has zero net charge. This technique is termed "isoelectric focusing".

To separate DEFG ... a mixture of relatively low molecular weight amphoteric substances such as "carrier ampholyte" is added before voltage is applied. These substances have a range of isoelectric points, but the pK values of their acidic and basic groups are close to each other, so they stack to produce a buffered pH gradient (7,10). Indeed, they are prepared and sold on the basis of the pH range they cover.

The proteins "band" or collect at their characteristic isoionic pH values. This state does not last indefinitely, however, since the stacked ampholytes and proteins behave like an isotachophoretic system unless the pH gradient is physically immobilized (7) (see below). Isoelectric focusing is frequently used preparatively, since large amounts of protein may be processed.

Supporting Media-Based Methods. A mechanical problem results from any zone separation procedure. Higher resolution results in thinner separated zones. These are more concentrated, and so are significantly denser than the surrounding solvent, which leads to "convection": the zones fall through the solvent and mix with it. This must be prevented. A preparative isoelectric focusing apparatus uses a density gradient of sucrose or other nonionic substance that contains ions DEFG ... and the carrier ampholytes. The separated zones of D, E, and other ions are buoyed by the increasingly dense solvent below them.

Sucrose density gradients have no mechanical strength, so the electrophoresis pathway must be vertical, and in any case the gradient only reduces convection. Generally, a solid or semisolid supporting medium is used. This can be fibrous (paper, cellulose acetate), particulate (cellulose powder, glass beads), or a gel (starch, agar, polyacrylamide). A gel is a network of interacting or tangled fibers or polymers that traps large quantities of solvent in the network. To increase mechanical strength, some polymers can be covalently crosslinked as is routinely done with polyacrylamide supports. Gels usually have a uniform percentage of gel material, but can be prepared with a gradient of gel material or other substances such as urea (10).

Support media should allow as free a passage as possible to electrophoresing ions while restricting convection. Convection (bulk flow) in a capillary is proportional to the fourth power of the capillary radius (1), but the area available for electrophoretic movement is proportional to the square of the radius. Reducing the radius reduces convection much more than the carrying capacity (area) of the capillary.

Capillary tubes can be used without supporting material inside (capillary electrophoresis) (11), but the other materials operate by offering a restricted effective pore size for electrophoretic transport. The effective pore size varies with the medium: 1–2% agar gels have larger average pore sizes than polyacrylamide gels made from 5 to 10% acrylamide solutions.

The importance of the supporting medium is illustrated by an immunoelectrophoretic technique. Electrophoresis is used to separate antigens before antibody diffuses into the separated antigens (immunodiffusion) (12,13). Immunodiffusion requires supports with average pore sizes large enough to allow antibodies, which are large and asymmetric immunoglobins, to diffuse through them, while still stabilizing the zones of precipitate that form. Agar gels, at 1–2% concentrations, are used.

A support medium is in very extensive contact with the solution. Even a capillary tube has a great deal of surface relative to the volume of solution within. Interaction with the support or capillary surface is often not wanted: chemical interactions with the ions being separated or with the other constituents of the solvent interfere with the ion movement. Adsorption of some substances by the fused silica surfaces of capillaries used in electrophoresis can occur (11). The surface must then be coated with inert material. Support media are often chosen for having as little effect on electrophoresis as possible. Such effects are called "electroosmosis": the solvent flows in an electric field (4). If the medium or capillary inner surface has charged groups, these must have counterions. The charged groups fixed in the matrix of the medium or capillary tube and their counterions are hydrated, but only the counterions can move in an electric field. The counterion movement causes solvent flow next to the capillary wall or support medium.

This case is extreme, but some electroosmosis occurs with any medium. Interaction of any two different substances results in a chemical potential gradient across the interface (4). This is equivalent to an electrical potential difference. Electroosmosis can distort zones of separated ions, reducing separation, but as capillaries become thinner or average pore sizes smaller, sideways (to electrophoretic movement) diffusion tends to overcome effects of electroosmosis (7).

Electrophoretic procedures are to produce separations, and if interactions of any kind improve separations, they are desirable. Separation of serum proteins by capillary electrophoresis in a commercial apparatus (Beckman-Coulter Paragon CZE 2000) depends partly on electroosmotic flow stemming from charged groups on the silica surface of the capillary.

Electrophoresis in solutions containing ionic detergents can separate uncharged substances. Detergents form clusters in water solutions called "micelles" with hydrophilic surfaces and hydrophobic interiors. Micelles of ionic detergents move in an electric field, and uncharged substances, if they partition into the micelles, will be moved along in proportion to the extent of their partitioning. This is "micellar electrokinetic chromatography" (11).

Electrophoresis in support media that reversibly adsorb substances being separated is called "electrochromatography". Unlike conventional chromatography, which uses hydrostatic pressure to move solutions of substances to be separated through the medium, electrochromatography uses an electric field.

The interactions with support media used to increase resolution of electrophoretic separations range from indirect to actual adsorption.

Acrylamide derivatives that will act as buffers over a desired pH range are incorporated into polyacrylamide gels and used in isoelectric focusing (7,9,10). These are called Immobilines. They provide stable pH gradients (7) and control movement of amphoteric ions by controlling the pH. They also stabilize zones of protein that form.

Electrophoretic or diffusional movement of large ions is reduced by supports, since they increase the tortuousness of the path the ions must follow. If the average pore size is made comparable to the effective sizes of the ions undergoing electrophoresis, the ions are literally filtered though the medium so that smaller ions would be least retarded and larger ions retarded more. This "molecular sieving" effect can improve the resolution achievable by electrophoretic techniques: the leading edges of zones of large ions are slowed by the gel matrix allowing the rest of the zones to catch up. This makes zones of larger ions thinner and restricts subsequent diffusion.

This effect can also provide information about the physical characteristics of the ions. The effective mobilities of large ions are determined by their net charges and frictional coefficients. The latter is affected by the shape of the large ion: A very asymmetric molecule has a larger effective size than a more compact one (1,3). Two molecules of identical charge and molecular weight can be separated even in the absence of a support if they have sufficiently different shapes. Effective sizes can be determined from electrophoresis on several gels with different average pore sizes. Effective sizes can be used to calculate diffusion constants (1,3), though they are normally used to obtain molecular weights. Large ions of known molecular weights may be electrophoresed on the same gels to calibrate the pore sizes of the gels (12,13) (see below).

If a mixture of large ions of differing shapes and sizes is converted to a mixture with different sizes but the same shape, then they will migrate according to their net charges and molecular weights. If the charges are made a constant function of the molecular weight then the ions will separate according to their molecular weights alone (3,10,12,13). The degree of separation will also be affected by the average pore size of the medium. Proteins that consist entirely or nearly so of amino acids and that have been treated to reduce any disulfide bonds are usually converted to rod-like structures with a nearly constant charge-to-mass ratio by adding the detergent, sodium dodecyl sulfate (SDS). The detergent binds to such proteins in large and highly uniform amounts (3,12,13). The protein–SDS complexes then migrate in order of decreasing polypeptide molecular weight (see below). It is important to recognize that some proteins may not behave "normally" in these respects, for example, if they contain large amounts of carbohydrate. The SDS–PAGE method, as it is called, is currently probably the single most widely used electrophoresis technique in research laboratories.

Polynucleotide fragments have a relatively constant charge-to-mass ratio at physiological pH values, since each nucleotide is of similar molecular weight and has only a single charged group: the phosphate. If they have the same shape, separation of such fragments can be effected on the basis of molecular weight (i.e., number of nucleotides) by electrophoresis (10,13). This is the basis of the technique for determining polynucleotide sequences (see below).

It is possible to separate and measure molecular weights of native (double-stranded) DNA molecules that are larger than the average pore size of the support medium (14). The electric field pulls on all segments of the DNA equally, but eventually, because of random thermal movement, one end is farther along the field direction and the entire molecule is pulled after that end along a path through the gel. If the field is turned off, the stretched molecule contracts or relaxes into a more compact, unstretched condition. If the electric field is turned on again, the process starts over. The rates of these processes are slower with longer DNAs, so, if the rate of change of field direction is appropriate, the DNAs will separate according to size. If not, they will tend to run together. The change in direction of the applied field and the length of time the field is applied in any direction can be adjusted in commercial instruments. The gel used is agarose (agar without the agaropectin). This is "pulsed-field" gel electrophoresis, and can separate DNA molecules as large as small chromosomes.

Sometimes adsorption strong enough to terminate electrophoretic movement is desired. Substances are separated on a sheet or thin slab of ordinary support medium, then an electric field is applied perpendicular to the face of the sheet and the separated substances are electrophoresed onto a facing sheet of adsorbent material such as nitrocellulose paper. This is "electroblotting" (13,15,16). It concentrates the electroeluted substances on the adsorbent sheet. This makes immunological detection (immunoblotting) or detection by any other method more sensitive (15,16). If the separated substances blotted are DNA, this is "Southern blotting" (17); RNA, "Northern blotting"; protein, "Western blotting" (10,13,15,16). These procedures are to identify particular proteins or polynucleotides with particular sequences.

Cells in living organisms produce tens of thousands of proteins and polynucleotides, and in very different amounts: for example, the concentrations of serum proteins differ by up to 10^7-fold, complicating analysis. Increased resolution has been also achieved through use of two-dimensional (2D) electrophoresis. Some 30–60 proteins can be resolved on a column or sheet of polyacrylamide. If electrophoresis is done in two dimensions, with separation on the basis of different properties of the proteins, 900–3600 proteins could be resolved (18,19). The most popular form of 2D electrophoresis involves isoelectric focusing on a gel strip for one dimension, attaching the gel strip to a gel sheet, and performing SDS–gel electrophoresis for the second dimension. Densitometers to scan the gels and software for computerized mapping and indexing of isolated proteins to help analyze the data are available commercially, for example, from BioRad Corp. and Amersham Biosciences Corp. The procedure is laborious and exacting; many factors affect the patterns obtained (19). Use of Immobiline gel strips for the first dimension separation is very important in improving reproducibility (19). The realized or potential information from 2D electrophoresis is so great that it is a very popular technique in research laboratories.

Voltage-Based Methods. Since diffusion is the enemy of separation and increases with the square root of the time (1), it is obviously better to carry out the electrophoresis faster. Since the velocity of electrophoresis increases with the voltage applied, using a higher voltage would improve resolution. However, increasing the voltage also increases the current and this increases electrophoretic heating, called "Joule heating". Joule heating is proportional to the wattage (volts times current) and can cause convective disturbances in solutions, even if a support medium is present. Joule heating is part of the reason electrophoresis is done in solutions with moderate conductivities, equivalent to 0.1 M NaCl or thereabouts (13): Ion concentrations must be high enough to minimize electrostatic interactions between large molecules (1), but high ion concentrations mean high conductivities, which means high current flow at a given voltage and therefore high heat generation. Joule heating is in fact the ultimate limiting factor in any electrophoretic procedure.

Use of thinner supports or capillaries minimizes the effects of heating. Thin (< 1 mm thick) sheets of polyacrylamide allow more efficient heat dissipation (from both sides) and are also useful when comparing samples. They are necessary for 2D electrophoresis (18,19). Thinner supports use less material but require more sensitivity of analysis. Capillaries are even more easily cooled (from all sides) and very high voltages, of the order of several thousand volts, can be used (11). The Beckman Coulter Paragon CZE 2000 has seven capillaries so can accommodate seven samples at once, and serum protein electrophoresis turnaround times are 10 min. Very small sample volumes, often a few nanoliters, are used. On the other hand, electroosmosis can be a major problem and the detection sensitivity is very low: the separated substances pass a detector, and the capillaries, usually < 0.1 mm in diameter, provide a very short optical pathlength. Detection at very short wavelengths such as 214 nm, where extinction coefficients are higher or use of other measuring methods such as fluorescence or mass spectrometry are alternatives that are employed. Still, the ability to use higher voltages for separation, thus improving resolution, has directed much attention to capillary electrophoresis.

EQUIPMENT AND PROCEDURES

The equipment for electrophoresis includes a power supply, the apparatus on which electrophoresis is actually performed, and the reagents through which electromigration occurs.

Power Supply

This is sometimes combined with the apparatus. The power supply provides a dc voltage that produces the electrophoretic movement. Maximum power outputs should be matched with requirements.

Some power supplies may provide for applying a constant voltage, constant power or constant current. The conductivity of a setup changes with time during electrophoresis because of ion movement, so constant voltage is used for ordinary electrophoresis, constant current for isotachophoresis or disc electrophoresis (to make the zone migration velocity constant) and constant wattage for isoelectric focusing (10). Many power supplies also feature timers to prevent running samples too long. Most have an automatic shutdown of power if the apparatus is accidentally opened. Because even low voltages can be dangerous, safety issues must be addressed.

Apparatus

Many types are available, from simple partitioned plastic boxes (e.g., from Beckman-Coulter) to large and elaborate ones (e.g., from Helena). All basically hold the material through which electromigration occurs, horizontally or vertically, either as a plug or cylindrical rod or, most commonly for routine clinical work, as a sheet. Sizes and shapes of electrophoresis chambers are usually determined by the cooling efficiency of the apparatus. Some have special chambers for circulating cold water and some instruments have incorporated cooling units. These may be able to electrophorese a few samples at a time or many.

Horizontal electrophoresis is preferred when the supporting medium cannot stand the mechanical stress of standing upright. Evaporation from the top surface is often a problem; DNA fragments are separated in "submarine" systems, where the support (an agarose gel) is covered in buffer. The supports used when electrophoresis is done vertically are sometimes encased in glass sheets.

The apparatus also provides for the connection between the electrophoresis chamber and the electrodes. Typically, the connection is made through two relatively large volumes of solution, or "reservoirs". These reservoirs minimize the effects of electrolysis. To reduce contamination from extraneous metal ions, the electrodes are normally made of platinum.

Accessories include drying ovens and containers for staining and washing supports, though these may be integrated with the rest of the apparatus.

Reagents

The reagents constitute the supporting medium and provide the ions for electrical conduction throughout much of the apparatus. The ionic substances should be of good purity, but the major requirement is that the ions should not interact strongly with the substances to be separated (like borate with glycoproteins) unless a specific interaction (e.g., SDS) is desired. This means low heavy metal content and avoidance of large polyanions. Generally, buffers and salts with low ionic charges (no greater than phosphate at pH 7 or so) are preferred.

The reagents for supporting media prepared in the laboratory are usually "electrophoresis grade" for the reasons just given; that is, ionic contaminants such as acrylic acid in acrylamide are preferred to be at low levels. Some workers claim that cheaper grades of many reagents are satisfactory. In our experience, the quality of SDS seems to be important, however.

For research work, gel supports are prepared by heating slurries of agar or agarose in buffer or polymerizing acrylamide and derivatives in buffer and pouring them into a support to set. A "comb" with wide teeth is placed in the cooling or polymerizing liquid to form a series of identical rectangular holes or depressions called "wells" in the support. The samples are mixed with sucrose or glycerol (to make them dense enough to fall into the wells) and also with a low molecular weight dye called "tracking dye" (to visibly mark the progress of electrophoresis), then pipetted into the wells before electrophoresis. For routine clinical use, samples of a few microliters of samples not mixed with sucrose or tracking dye are applied to the surface of the support through a template or mask and allowed to soak into the support a few minutes before electrophoresis.

Commercially prepared supports are convenient and tend to be more uniform, so are employed for routine clinical analysis. These include agarose, cellulose acetate and polyacrylamide. They have plastic sheets as backing for strength. Washing or preelectrophoresing these before use is usually unnecessary.

Measurements

This means determining velocity(ies) or location(s) of the substance(s) electrophoresed. In the cases of isotachophoresis or capillary electrophoresis, measurements of absorbence, fluorescence, heat generation, conductivity, and so on are made while the separation is occurring. If the substances are allowed to elute and are collected in separate fractions, analysis may be done at leisure using any appropriate technique.

Usually, the (hopefully) separated substances remain on the supporting medium; however, diffusion continues after electrophoresis and measurements must be made quickly or diffusion slowed or stopped after electrophoresis ends. Fluorescent labeled single-stranded polynucleotides are measured on DNA sequencing gels (polyacrylamide) using a scanning fluorometer. Double-stranded (native)

DNA is detected using dyes that become fluorescent when the dye molecules move between (intercalate) stacked nucleotide bases, which provides great sensitivity and selectivity. Photographs are made of the patterns. Direct measurement of proteins on the support is difficult to do as the support itself often absorbs in the ultraviolet (UV) or scatters. Transparency of cellulose acetate supports can be increased (clearing) by soaking the support in methanol-acetic acid before measuring absorbence. A scanning spectrophotometer must be used. Otherwise, the support is dried or the separated substances, usually proteins, precipitated in the support.

With autoradiography, radioactivity is measured, most often using photographic film (13) after the support is dried. Autoradiography is potentially the most sensitive method of measurement of distributions, since isotopes of very high specific activity can be used.

Precipitation of proteins in the support matrix is usually done by soaking the matrix in 7–10% acetic acid, though trichloroacetic acid is sometimes used. Any SDS must be removed or precipitation does not occur. Adding methanol (30–40%) to the acetic acid and soaking is usually done to remove SDS. The next step is staining using a dye: This is to enhance visibility, for purposes of inspection, photography, or, if a quantitative record is needed (generally the case in routine clinical work), measurement of the absorbence using a scanning spectrophotometer (densitometry). Some commercial clinical electrophoresis apparatuses (e.g., Helena) have automated staining; otherwise, it is done manually.

The support is soaked in a solution of a dye that strongly and specifically adsorbs to the denatured protein. Excess dye is removed by electrophoresis or by diffusion: subsequent soaking in the same solution but without the dye. This leaves a support with colored bands, dots, or stripes where protein can be found.

The important parameters for detection are specificity and sensitivity. One may be interested in staining specifically for phosphoproteins or lipoproteins, for example. Otherwise, the benchmark parameter is sensitivity. Tables showing frequently used stains are given in Refs. 2 and 13. The sensitivity of soluble protein stains increases with the molar extinction coefficient of the dye (20). Substances with very high effective extinction coefficients are more sensitive, such as metal stains (e.g., silver stain) or colloidal stains such as India ink or colloidal gold. Using silver staining, as little protein as 0.1 ng/band can be detected (13).

Sometimes some specific property must be measured, to identify a particular band or spot. With enzymes, the activity is the most specific property available, and two means of identification can be used.

If a substrate or product of the enzyme has a distinctive color, the support is soaked in an assay medium containing the reactant until a color change due to product (or substrate) appears. Naturally, the support and electrophoresis conditions must be such that the activity was not lost beforehand. If substrate or product is of low molecular weight, some way of restricting its diffusion must be used. Enzyme activity-specific stains are called "zymograms" (21).

Alternatively, the support can be cut into successive sections, incubated in assay medium, and absorbence or other changes resulting from enzyme activity measured. Generally, a duplicate support or half of a single support is stained for protein for comparison.

Recovery of material from supports is sometimes desired, for example, for mass spectrometry. The ease of recovery of large molecules such as proteins by diffusion is proportional to the pore size of the supporting medium. Recovery (especially from acrylamide gels) is sometimes done using electrophoresis.

The times required for electrophoresis, including staining and densitometry, are of the order of hours. This has limited clinical applications of electrophoretic techniques, though capillary electrophoresis techniques may change this.

EVALUATION

The result is a pattern or distribution profile given as a function of distance migrated. Often this is expressed as R_f values, relative to the distance migrated by a known ion, usually tracking dye. Electrophoresis of proteins without denaturants (native gel electrophoresis) on several gels of different percentages (pore sizes) is followed by plotting logarithms of their R_f values versus gel concentration, a "Ferguson plot" (12,13). This plot enables separation of the contributions of size and charge: isozymes, for example, have the same size (slopes of the lines) but different charges (ordinate intercepts). In the presence of SDS, protein complexes are dissociated (3,10,12,13). A protein of unknown subunit molecular weight is electrophoresed alongside of a mixture of proteins of known subunit molecular weight in the presence of SDS. The logarithms of the known molecular weights are plotted versus their R_f values, yielding an approximate straight line, and the unknown molecular weight obtained from its R_f value.

However, the pattern is often presented without further analysis. Electrophoresis experiments are often used to assess the purity or "homogeneity" of a preparation of a macromolecule. Electrophoresis is still the single most widely used criterion of purity. It is best if purity is tested under widely different electrophoresis conditions: different pH values, or the presence of absence of denaturants, such as urea or SDS, in the case of proteins.

Isoelectric points can be obtained most readily by isoelectric focusing, which can also be used as a criterion of purity.

A major use of electrophoresis currently is for DNA sequencing, owing to ongoing determinations of entire genomes of organisms. Capillary electrophoresis using linear (uncross-linked) polyacrylamide has replaced use of polyacrylamide sheets as the method of choice for this purpose, since throughput is increased several-fold because of the higher voltages that can be applied. Applied Biosystems has an instrument with 96 capillaries and Amersham Biosciences has models with up to 384 capillaries. Analysis of the profiles is done by the instruments: each fluorescent peak observed is assigned to adenine, cytosine, guanine, or thymine on the basis of the emission

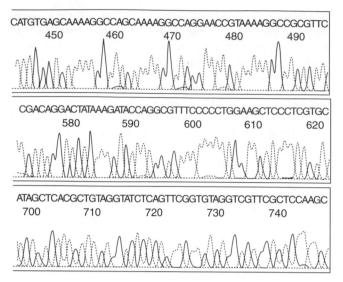

Figure 2. Section of data printout from the Applied Biosystems Prism capillary electrophoresis apparatus. The profile shows signals (fluorescence) given by polynucleotides with a fluorescent nucleotide analogue at the end, obtained as these migrate past the detector. Each peak is produced by a polynucleotide that is one residue (nucleotide) longer than that producing the signal to its left and one residue shorter than the polynucleotide producing the peak to its right. The separation by molecular length is produced by electrophoresis through linear (uncrosslinked) polyacrylamide. Different nucleotide analogues give different emission maxima, so the output identifies these as A, C, G, or T. Note the increasing peak widths due to diffusion. The numbers are the lengths in residues of the polynucleotides. (Courtesy of Dr. John Wunderlich of the Molecular Genetics Instrumentation Facility of the University of Georgia.)

spectra observed as polynucleotides terminated by dideoxynucleotide derivatives (22) with different emission spectra electrophorese past the detector. Data from each capillary are printed separately (Fig. 2).

Routine clinical electrophoresis separations are mostly serum proteins, hemoglobins, or isozymes: creatine kinase, lactic dehydrogenase, and alkaline phosphatase. Then quantitation of the separated proteins is done.

Usually, the proteins are separated on agarose or cellulose acetate sheets, stained, and the resulting pattern scanned using a densitometer. The Beckman-Coulter CZE 2000 instrument with seven capillaries is approved by the FDA for serum protein analysis, and directly measures absorbencies as different proteins move past the detector. In either situation, the electrophoretic patterns from samples (blood serum, urine, or cerebrospinal fluid) obtained from patients are compared with those from samples taken from healthy people (Fig. 3). Higher immunoglobin levels are seen with a number of conditions such as liver disease and chronic infections for example. The profiles may be analyzed visually, but some densitometers have analytical capabilities. Sometimes fluorescence [usually of reduced nicotinamide adenine denucleotide (NADH) in creatine kinase or lactic dehydrogenase isozyme assays] is scanned instead.

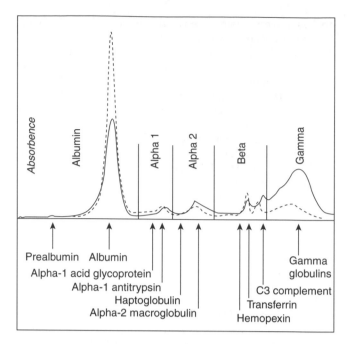

Figure 3. Recordings of absorbence at 214 nm versus time of serum protein electrophoresis on the Beckman-Coulter Paragon CZE 2000. The solid line is the electrophoretogram of a serum protein sample from a person with chronic hepatitis C, relative to a sample from a healthy person (broken line). Characteristic increase in the polyclonal gamma and decreases in albumin and transferrin zones occur. (Courtesy of Beckman-Coulter Inc.)

Reliability

Resolution of proteins whose isoelectric points differ by as little as 0.001 pH unit is claimed for isoelectric focusing using Immobiline gels (13).

The resolving power of electrophoresis on polyacrylamide is sufficiently high that hundreds of polynucleotides, each differing in length from one another by one nucleotide, are separated in DNA sequencing (Fig. 2). Estimates of molecular weights of proteins in SDS–gel electrophoresis is also very widely done. Small differences, of the order of 1000 Da, in molecular weights can be detected (13); absolute molecular weight estimates are less reliable, since the SDS binding depends on the composition of the protein (13). In the author's experience, most protein molecular weights estimated by this technique are not reliable beyond average limits of ±10%. It is best to remember that "molecular sieving" techniques in general measure effective sizes, and estimates of molecular weight involves comparisons with proteins of known mass whose shapes and hydrations are assumed to be similar.

While electrophoresis results are generally fairly reproducible from experiment to experiment, it is best to use internal standards whenever appropriate. This is for location and identification of separated substances and for quantitation.

Electrophoresis is widely used, especially in the biological sciences for investigations of macromolecules. It is an excellent analytical method characterized by high resolution and sensitivity and moderately good reproducibility, a method capable of yielding considerable information about size, shape, composition, and interactions of specific molecules and about distribution of large numbers of molecules at one time.

For clinical purposes, the resolution of isozymes such as lactic dehydrogenase is reliable enough that the technique has been used in critical care diagnoses, that is, life or death decisions.

BIBLIOGRAPHY

Cited References

Note: Developments in this field are covered particularly in the journal Electrophoresis. The catalogues of companies that sell electrophoresis apparatus and supplies, such as Amersham Biosciences, BioRad, Invitrogen, and National Diagnostics are another sources of information.

1. Tanford C. Physical Chemistry of Macromolecules. New York: John Wiley & Sons; 1961.
2. Brewer JM. Electrophoresis. In: Kaplan LA, Pesce AJ, Kazmierczak SC, editors. Clinical Chemistry: Theory, Analysis, Correlation. 4th ed. St. Louis (MO): Mosby; 2003. Chapt. 10, p 201–215.
3. Cantor CR, Schimmel PR. Biophysical Chemistry. San Francisco: W.H. Freeman and Company; 1980.
4. Moore WJ. Physical Chemistry. 4th ed. Englewood Cliffs (NJ): Prentice-Hall; 1972.
5. Kiriukhin MY, Collins KD. Dynamic hydration numbers for biologically important ions. Biophys Chem 2002; 99:155–168.
6. Rupley JA, Gratton E, Careri G. Water and globular proteins. Trends Biochem Sci 1983;8:18–22.
7. Mosher RA, Saville DA, Thormann W. The Dynamics of Electrophoresis. Weinheim, Germany: VCH; 1992.
8. Marshall AG. Biophysical Chemistry: Principles, Techniques and Applications. New York: John Wiley & Sons; 1978.
9. Righetti PG. Immobilized pH gradients, theory and methodology. Amsterdam, The Netherlands: Elsevier; 1990.
10. Hawcroft DM. Electrophoresis: the basics. Oxford (UK): Oxford University Press; 1997.
11. Whatley H. Basic Principles and Modes of Capillary Electrophoresis. In: Petersen JR, Mohammad AA, editors. Clinical and Forensic Applications of Capillary Electrophoresis. Totowa, NJ: Humana Press; 2001. Chapt. 2, p 21–58.
12. Van Holde KE. Physical Biochemistry. 2nd ed. Engelwood Cliffs (NJ): Prentice-Hall; 1985.
13. Dunn MJ. Gel Electrophoresis: Proteins. Oxford (UK): Bios Scientific; 1993.
14. Noolandi J. Theory of DNA gel electrophoresis. In: Chrambach A, Dunn MJ, Radola BJ, editors. Volume 5, Advances in Electrophoresis. New York: VCH; 1992. p 1–57.
15. Gallagher SR, Winston SE, Fuller SA, Harrell JGR. Immunoblotting and immunodetection. In: Ausubel FM, Brent R, Kingston RE, Moore DD, Seidman JG, Smith JA, Struhl K, editors. Current protocols in molecular biology. Unit 10.8, New York: Greene Publishing and Wiley Interscience; 2000.
16. Baldo BA. Protein blotting: Research, Applications and its place in protein separation methodology. In: Chrambach A, Dunn MJ, Radola BJ, editors. Advances in Electrophoresis. Volume 7, New York: VCH; 1994. p 409–478.
17. Highsmith WE, Jr., Constantine NT, Friedman KJ. Molecular Diagnostics. In: Kaplan LA, Pesce AJ, Kazmierczak SC, editors. Clinical Chemistry: Theory, Analysis, Correlation. 4th ed. St. Louis (MO): Mosby; 2003. Chapt. 48, p 937–959.
18. Hochstrasser DF, Tissot JD. Clinical application of high-resolution two-dimensional gel electrophoresis. In: Chrambach A, Dunn MJ, Radola BJ, editors. Advances in Electrophoresis. Volume 6, New York: VCH; 1993. p 270–375.
19. Gorg A, Obermaier C, Boguth G, Harder A, Scheibe B, Wildgruber R, Weiss W. The current state of two- dimensional electrophoresis with immobilized pH gradients. Electrophoresis 2000;21:1037–1053.
20. Merril CR. Gel-staining techniques. Volume 182, Methods Enzymology. New York: Academic Press; 1990. p 477–488.
21. Heeb MJ, Gabriel O. Enzyme localization in gels. Volume 104, Methods Enzymology. New York: Academic Press; 1984. p 416–439.
22. Sanger F, Nicklen S, Coulson AR. DNA sequencing with chain-terminating inhibitors. Proc Natl Acad Sci USA 1977;74:5463–5467.

See also CHROMATOGRAPHY; DNA SEQUENCE.

ELECTROPHYSIOLOGY

SEMAHAT S. DEMIR
The University of Memphis and
The University of Tennessee
Memphis, Tennessee

INTRODUCTION

The Resting Membrane Potential

The cell membrane or sarcolemma, composed of lipid bilayer, is hydrophobic and is highly impermeable to most water-soluble molecules and ions. A potential is developed across the lipid bilayer of the cell membrane due to unequal distribution of charges on the two sides of the membrane, thus the membrane acts as a capacitor. Membrane proteins that span across cell membranes form ion channels allowing transport of small water-soluble ions across the membrane. These channels are highly selective and their selectivity depends on diameter, shape of the ion channel, on the distribution of charged amino acids in its lining (1). The movements of the ions through these channels across the membrane govern its potential.

The transport of the ions across the membrane is either passive or active. The passive mechanism of transport of any ion is governed by its electrochemical gradient, a combination of chemical force exerted by diffusion of ions due to concentration gradient and an electrical force exerted by the electric field developed due the charges accumulated on either side of the membrane (capacitor) (2). Physiologically, cells have a high intracellular potassium concentration, $[K^+]_i$, and a low sodium concentration, $[Na^+]_i$. Conversely, the extracellular medium is high in Na^+ and low in K^+. In cardiac cells at rest, the membrane is mostly permeable to K^+ ions through K^+ leak channels. As K^+ flows out down its concentration gradient, a negative potential is built up inside the cell. This increases as long as it counterbalances the chemical driving force generated by concentration gradient. This potential at which the net ion flux is zero is called Nernst equilibrium potential of that ion. The equilibrium potential for K^+ is given by its

Nernst equation:

$$E_K = \frac{RT}{ZF} \ln \frac{[K^+]_o}{[K^+]_i}$$

where R is the universal gas constant, T is temperature in kelvin, F is Faraday constant, z is the valency of the ion.

Typical resting membrane potentials of excitable cells vary from -90 to -50 mV, depending on the type of the cell. Epithelial cell and erythrocytes have smaller, but still negative membrane potentials. It may tend toward the excitatory threshold for an action potential as in a cardiac pacemaker cell or remain stable with approximately no net ion flux observed in nonpaced cardiac ventricular cells. As the ventricular cell at rest is more permeable to K^+ ions than to any other ion, the resting membrane potential (ca. -84 mV) is close to E_K at $37\,°C$. Due to its permeability to other ions and also due to other transport mechanisms the resting membrane potential does not reach exactly E_K.

The active ionic transport mechanisms maintain the homeostasis of ionic concentrations in both the intra- and extracellular media. These membrane proteins are called carrier (pump) proteins and they utilize energy from hydrolysis of adenosine triphosphate (ATP) to transport ions against their concentration gradient.

EXPERIMENTAL TECHNIQUES TO QUANTIFY IONIC MECHANISMS IN CELLS

The advent of patch clamp technique (3–7) has made it possible to record the current from a single ion channel. The technique involves clamping a patch of the cell membrane and recording either voltage (current–clamp) or current (voltage–clamp or patch–clamp) across the membrane. Using this technique current of order as low as 10^{-12} A can be measured. This could be done using different configurations of patch clamping (7).

1. A freshly made glass pipette with a tip diameter of only a few micrometers is pressed gently on the cell membrane to form a gigohm seal. This is called as cell-attached patch configuration. The pipette solutions form the extracellular solution and the currents across the channel within the patch can be recorded.

2. When gentle suction is applied to the pipette in cell-attached configuration, the membrane ruptures while maintaining the tight seal and the cytoplasm and pipette solution start to mix. After a short time, this mixing is complete and the ionic environment in the cell is similar to the filling solution used in the pipette. This configuration is called whole-cell patch configuration. A recording obtained using this configuration is from whole cell and not from a patch. The advantage of this technique is that the intracellular environment is accessible through the pipette. Current–clamp technique is used in this configuration to measure the action potentials (APs) of excitable cells.

3. Sudden pulling out of the pipette from cell-attached configuration holds the patch that formed the gig-ohm seal giving raise to the inside–out configuration (inside of the cell membrane is exposed to external bath).

4. Slow pulling out of the pipette from whole cell configuration holds the patch that formed the gig-ohm seal giving rise to outside–out configuration. Both inside–out and outside–out configurations allow single channel recordings. Both the intracellular and extracellular baths are accessible in these cases.

5. The fifth configuration is obtained by creating artificial channels (permealizing membrane) on the cell-attached patch by administering antibiotics, like amphotericin. The voltage and current–clamping recordings obtained in this configuration recordings are similar to whole-cell recordings, the advantage being the intracellular medium is not dialyzed.

VOLTAGE-CLAMP TECHNIQUE

The method of voltage clamping has been the primary experimental tool used to reconstruct models of cellular electrical activity. As the behavior of the ion channels is highly nonlinear under changing action potential, this method enables us to quantify their properties by holding the transmembrane potential (membrane potential) at a particular voltage. The basic principle relies on providing current to balance those currents through the ionic channels that are open and thus the transmembrane voltage is clamped at a chosen constant level (clamp voltage) For example, if the Na^+ channel is studied, the membrane potential is initially held at rest. When this potential is changed instantaneously to a depolarized (more positive) potential, sodium channels open and Na^+ ions tends to move in. The voltage amplifier senses these small changes in voltage and a feedback current of equivalent amount is applied in opposite direction of the ion flow. This measurable current changes for different clamp potentials as the driving force ($V_{Clamp}-E_{Na}$), and the gating parameters at that V_{Clamp} changes enabling us to quantify the channel properties. Ion channels conduct ions at a rate sufficiently high that the flux through a single channel can be detected electrically using patch clamp technique. The basic circuit of voltage clamp setup is shown in Fig. 1.

CURRENT–CLAMP TECHNIQUE

The current–clamp technique is used to record action potentials. This technique involves clamping the cell in whole cell configuration and applying a suprathreshold current pulse for a short duration until the Na^+ channels start to activate. The transmembrane change in voltage gives the action potential recording.

A key concept to modeling of excitable cells is the idea of ion channel selectivity of the cell membrane. As the molecular behavior of channels is not known, modeling of

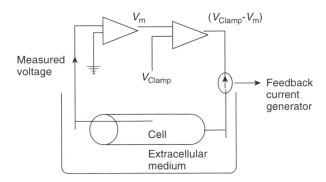

Figure 1. Simplified circuit representation of a voltage–clamp setup (8).

Examples of Simulated Cardiac Ventricular Action Potentials

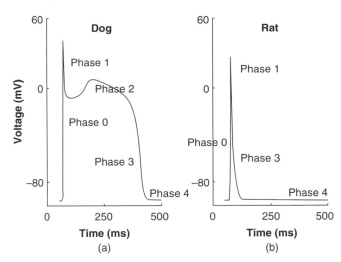

Figure 2. Examples of typical cardiac action potential waveforms of the dog (11) and rat (12) ventricular cell models. The action potential of the human ventricular cell would be similar to that of the dog (panel a). Please consult Tables 1 and 2 for the underlying ionic currents.

nonlinear empirical models of membrane processes help us to understand the role of various currents in the depolarization and repolarization phases and the phenomena that involve the interaction between these processes.

EXAMPLE OF ACTION POTENTIAL AND UNDERLYING IONIC BASIS: THE CARDIAC ACTION POTENTIAL

Membrane excitability is the fundamental property of the nerve and muscle cells, that is, in response to certain environmental stimuli, generates an all-or-none electrical signal or AP. Many different types of cardiac ion channels and ion pumps altogether form a complex process that results in cardiac AP (9). For example, the normal cardiac action potentials can be classified into two broad categories; those that are self-oscillatory in nature, such as pacemaker cells (sinoatrial and atrioventrciular cells) and those that need an external stimulus above a threshold, also called supra threshold, in order to be evoked, such as atrial, Purkinjee fiber, or ventricular cells. An extraordinary diversity in the action potential configurations can be seen in different regions of the heart. The ventricular tissue in particular displays a wide variety of action potential waveforms. These APs include pacemaker potentials in purkinje cells, and disparate action potential durations (APD) and morphologies in cells from the epicardial, mid-myocardial, and the endocardial layers of the ventricle. The ventricular action potential has been studied more frequently than other representative cardiac membrane potentials because ventricular arrhythmias are believed to constitute the majority of reportedly fatal incidences of cardiac arrhythmias (10).

Phases of Cardiac Action Potential

A typical ventricular action potential in higher mammals such as canine and human consists of four distinct phases (Fig. 1a). Phase 0 corresponds to a rapid depolarization or upstroke of the membrane action potential. Phase 1 is the initial rapid repolarization, and is followed by phase 2, which constitutes the action potential plateau. Phase 3 represents the final repolarization, which allows the ventricular cell to return to its resting state in phase 4. In addition to its morphological features, ventricular APs are commonly measured experimentally to determine its char-

acteristics. These include the resting membrane potential (V_{rest}), the peak overshoot (PO) value that is the maximum positive value achieved during the initial phase 0 depolarization, the maximum upstroke velocity (dV/dt_{max}), which occurs during phase 0, and the APDs measured when the APs have repolarized to 50 and 90% of their final repolarization value, also called APD_{50} and APD_{90}, respectively.

One or more of these characteristics is usually altered in the setting of a pathophysiological condition, and helps to quantify the differences between the normal and the abnormal action potentials.

Contributions of Ion Channels to Cardiac Action Potential

The temporal changes in a typical ventricular action potential configuration, that is, depolarization followed by a repolarization (Fig. 2) are governed by the movement of different ions, such as Na^+, K^+, and Ca^{2+} ions across the sarcolemma. These ions are usually transported between the intracellular and the extracellular spaces by means of carrier proteins and channel proteins embedded in the cardiac membrane. These proteins form the passive (ion channel-mediated and carrier-mediated) and active (carrier proteins such as pumps and exchanger) transporters of the cell membrane. The ion channels, the pumps and exchanger are the major ionic currents that form an action potential in mathematical representations. A summary of these currents that are present in a typical cardiac ventricular cell and the role of the currents in action potential generation are summarized in Table 1. The major ionic currents contributing to different phases of the typical action potential are presented in Table 2. The ventricular action potential is the result of a delicate balance of the inward and outward ionic currents and the active transporters (pumps and exchangers).

Table 1. Major Ionic Membrane Mechanisms Underlying a Typical Cardiac Ventricular Action Potential

Membrane Mechanism	Description	Gene (α-subunit)	Role in Action Potential
	Inward Ionic Currents		
I_{Na}	Na$^+$current	SCN5A	Initial depolarization of action potential
I_{CaL}	L-type Ca^{2+} current	α_{1C}, α_{1D}	Maintains plateau phase of action potential
I_{CaT}	T-type Ca^{2+} current	α_{1G}, α_{1H}	Present in the late plateau phase
	Outward Ionic Currents		
I_t	Ca^{2+}-independent transient outward K$^+$ current	Kv4.2 Kv4.3 Kv1.4	Responsible for early repolarization
I_{Kr},I_{Ks}	Rapid and slow delayed K$^+$ rectifier currents	HERG KvLQT1	Aids repolarization during plateau
I_{ss},I_{Kslow}	Slow inactivating K$^+$ currents	Kv2.1 Kv1.5	Aids late repolarization
I_{K1}	Inward rectifier K$^+$ current	Kir2.1 Kir2.2	Late repolarization, helps establish V_{rest}
	Other Ionic Currents		
I_{NaCa}	Na$^+$–Ca^{2+} exchanger current	NCX1 NCX2	Late depolarization
I_{NaK}	Na$^+$–K$^+$ pump current	Na$^+$-K$^+$-ATPase (α)	Late repolarization

To restore the intracellular and extracellular ionic concentrations so that homeostasis is maintained, the ions that cross the cell membrane during an action potential are brought back by active mechanisms like Na$^+$–K$^+$ pump, Ca^{2+} pump and coupled transporters like Na$^+$–Ca^{2+} exchanger, Na$^+$–H$^+$ exchanger. All the active mechanisms utilize hydrolysis of adenosine triphosphate (ATP), the cellular source of energy to achieve this. Of these, Na$^+$–K$^+$ pump, which brings in 2 K$^+$ ions for 3 Na$^+$ ions out per ATP consumed, results in a net positive current (I_{NaK}) in outward direction and Na$^+$–Ca^{2+} exchanger, which exchanges 3 Na$^+$ ions for one Ca^{2+} ion, results in a net positive current (I_{NaCa}) contributing a little to the action potential. In most species, the exchanger current is in its Ca^{2+} influx mode (reverse mode) during depolarization resulting in outward current and is inward during Ca^{2+} efflux mode during repolarization. This is because the equilibrium potential of the current given by ($3E_{Na}$–$2E_{Ca}$) is around −40 mV (13). The current contributed by Ca^{2+} pump is negligible as it pumps few ions across membrane. The Na$^+$–H$^+$ exchanger transports one Na$^+$ for H$^+$ thereby causing no net flux across the membrane.

By convention, any current inward is considered negative and contributes to depolarization and any current outward is considered positive and contributes to repolarization. An inward current of a cation adds positive charge

to the intracellular content, thereby making the transmembrane potential more positive; that is, depolarizes the membrane away from the resting potential.

MATHEMATICAL MODELING OF CARDIAC CELLS

The Hodgkin–Huxley (HH) paradigm (14) type formalism is approached for numerical reconstruction of ventricular AP. At any particular moment, the movement of any particular ion across the membrane depends on the relative density of the channels, the probability of the channel that is selective to the ion being open, the conductance of the ion, and the net driving force of the ion given by the difference (V_m− E_{ion}), where V_m is the transmembrane voltage and E_{ion} is the Nernst potential of the ion (2). Also, it is assumed that ion fluxes are independent of each other, that is, the probability of an ion crossing the membrane does not depend on the probability of a different ion crossing the membrane. Based on this, the cell membrane is modeled as a capacitance in parallel to the resistances that represent the flow of ions through their respective channels along with their driving force. The resistive components are characterized in the original HH model as conductance's (g), the reciprocals of the resistances. The resistive currents can therefore be written

$$I_{ion} = g_{ion} * (V_m - E_{ion})$$

The experiments suggested that these conductances could be voltage and time dependent resulting in a gating mechanism. In HH type models, this gating mechanism is explained by considering the conductance g_{ion} as the product of maximum conductance $g_{ion-max}$ the channel can achieve and gating variables whose value lie between 0 and 1. The behavior of a gating variable x is given by first order differential equation:

$$dx/dt = (x_\infty - x)\tau_x$$

Table 2. The Action Potential Phases and the Major Ionic Current Contributors

Phases of Action Potentials	Description	Major Contributing Ionic Currents
Phase 0	Initial depolarization	I_{Na}
Phase 1	Early repolarization	I_t
Phase 2	Plateau phase	I_{CaL}, I_{Kr}, I_{Ks}, I_{CaT}
Phase 3	Late repolarization	I_{K1}, I_{NaK}, I_{NaCa},I_{ss}, I_{Kslow}, I_{Kr}, I_{Ks},
Phase 4	Resting potential	I_{K1}

where x_∞ is the steady-state value the variable reaches at a particular voltage and τ_x is the time constant at that voltage that determines the rate at which steady state is reached. These variables are voltage dependent. All of these parameters are constrained by experimental data. These models represent the lumped behavior of the channels.

Increased understanding of the behavior of ion channels at a single channel level due to improved patch–clamp techniques lead to the development of state specific Markov models. Based on single channel recordings, it is observed that the channel opening or closing is random. Hence, the conductance $g_{\text{ion-max}}$ is multiplied by the total open channel probability of the ion (P_{ion}). These models represent the channel behavior based on their conformational changes and are capable of reproducing single channel behavior (15).

The rate of change of membrane potential to a stimulus current (I_{st}) is given by

$$(dV/dt) = (-1/C_{\text{m}}) * \left(\sum I_{\text{i}} + I_{\text{St}}\right)$$

where C_{m} is the membrane capacitance, and I_{i} values are different ionic currents.

THEORETICAL RESEARCH AND EXPERIMENTAL RESEARCH IN MURINE CARDIAC VENTRICULAR CELLS

After the first models of the mammalian ventricular cells by Beeler and Reuter (16) and Drouhard and Roberge (17), sophisticated mathematical models that simulate the cardiac action potentials in ventricular cells from different species such as canine (18,19), guinea pig (20–24), human (25,26), frog (27), and rabbit (28) have been published during the past decade. The model equations have usually based on the Hodgkin–Huxley (14) paradigm, wherein an ionic current is described by a set of nonlinear differential equations, and the parameters within these equations are constrained by experimental data obtained via voltage–clamp experiments in ventricular myocytes. There is a growing recognition that it is important to understand the complex, nonlinear interactions between the ionic milieu of the cardiac cell, that ultimately influence the action potential (29).

The mathematical models have demonstrated to be useful didactic tools in research, and have also quantified the important functional differences in the action potential properties between different species. Additionally, the computational models have also provided valuable, semiquantitative insights into the diverse ionic mechanisms underlying the normal/abnormal action potential behavior in different animal models. It is not always possible to make precise experimental measurements regarding the contribution of a particular ionic mechanism to an aberrant action potential. The simulation results from these cardiac models have helped in planning for future experimental studies, and also in making predictions in cases where suitable technology is unavailable (or not developed) to make direct experimental measurements (e.g., visualizing the transmural activity within the ventricular wall). These models will play increasingly important roles in addition to experimental studies in the design and development of future drugs and devices (30). An additional and important feature of these ventricular models has been their ability to simulate intracellular Ca^{2+} transient ($[Ca^{2+}]_i$). Thus these models incorporate the feedback mechanism between the APD and the intracellular calcium $[Ca^{2+}]_i$. The APD is known to influence the amplitude of the $[Ca^{2+}]_i$ in ventricular cells (31), and $[Ca^{2+}]_i$ in turn influences the action potential waveform by Ca^{2+}-induced Ca^{2+} inactivation of I_{CaL}, and by determining the peak magnitude of I_{NaCa} (13).

The previously developed mathematical models of human, dog, guinea pig, rabbit and frog provide a good basis for the understanding of the ionic mechanisms responsible for the generation of the cardiac action potential. However, there are significant differences in the action potential waveforms and their corresponding properties between different species. The unique nature of the rat cardiac action potential, coupled with the recent available experimental data for the ionic mechanisms involved in the genesis of the action potential in isolated rat myocytes, provided the motivation for us to develop the first detailed mathematical model of the rat ventricular action potential. An adult male rat ventricular myocyte model was constructed (32) and utilized this model to study the ionic basis underlying the action potential heterogeneity in the adult rat left ventricle. Important insights into the role of long lasting Ca^{2+} current (I_{CaL}), the Ca^{2+}-independent transient outward K^+ current (I_t), and the steady-state outward K^+ current (I_{ss}) in determining the electrophysiological differences between epicardial and endocardial cells were obtained. This ventricular cell model has been used to investigate the ionic mechanisms that underlie altered electrophysiological characteristics associated with the short-term model of streptozotocin induced, type-I diabetic rats (33) and spontaneously hypertensive rats (34).Our rat ventrcicular myocyte model was further utilized to develop models for the mouse apex and septal left ventricular cells (35–37). Thus these model simulations reproduce a variety of experimental results, and provide quantitative insights into the functioning of ionic mechanisms underlying the regional heterogeneity in the adult rat and mouse ventricle.

The ventricular cell models of dog, guinea pig, human, and rabbit described in the previous section have been mainly used to simulate the so-called spike and dome configurations for action potentials (Fig. 2A) commonly observed in ventricular cells from larger mammalian species (38). However, no mathematical model has been published to represent the murine (rat or mouse) cardiac action potential (Fig. 2B) until our rat ventricular cell (12). The murine ventricular action potentials have a much shorter APD (typically the APD at 90% repolarization (APD$_{90}$) is < 100 ms), and lack a well-defined plateau phase (triangular in shape) (39–41). A comparison of the experimentally recorded ionic currents underlying action potentials in rat–mouse and other mammalian ventricular cells shows that they display markedly different amplitudes and time-dependent behavior. In fact, despite the similarity of action potential waveforms in rat and mouse, the underlying nature of the repolarizing K^+ currents are different (41–43). Thus the unique action potential characteristics, and the lack of models to quantify these membrane properties provided the motivation to develop the rat and mouse ventricular cell models. The other motivation in this case was the widespread use of the murine cardiovascular system for the investigation of the cellular and molecular physiology of the compromised cardiovascular function (44).

Experimental studies indicate that the patterns of action potential waveforms are somewhat similar in rodents (rat or mouse), although the APD is shorter in mouse, and the complement of the K$^+$ currents underlying the cardiac repolarization in mouse are also different than those in rat (45,46). The cardiac repolarization in rat is controlled by two distinct depolarization activated K$^+$ currents, the Ca^{2+}-independent transient outward K$^+$ current (I_t) and the steady-state outward K$^+$ current (I_{ss}), (40,47). In mouse ventricular myocytes, an additional current, the 4-AP sensitive (at concentrations less than 100 μM), slowly inactivating, delayed rectifier K$^+$ current (I_{Kslow}) has been deemed to play an important role (41,48). The properties of the depolarization-activated K$^+$ currents have now been well characterized in rats (40,49) and mouse (43,48), and appear to be significantly different. It is therefore interesting to investigate in computational modeling whether the reported differences in the properties of the depolarization-activated K$^+$ currents can account for the dissimilar nature of the action potential configurations observed in rats and mice.

COMPUTATIONAL MODELING OF THE MURINE VENTRICULAR ACTION POTENTIALS

The goal of my computational modeling laboratory has been to unify different experimental data and to develop biophysically detailed models for the rat and mouse ventricular cells and to determine the underlying ionic channels responsible for differences in cardiac action potential variations in rats and mice under normal and diseased conditions. A computational model has been developed for the rat cardiac ventricular cell based on electrophysiology data. Our control model (12) represents the bioelectric activity in the left ventricular cells in adult male rats. The differences in the membrane properties within the left ventricle to simulate the action potential variations of the endocardial and epicardial cells have been formulated. Also, a right ventricular cell model from our control model was built (the left ventricular cell model) to investigate ionic mechanisms in diabetic rats (33). Our right ventricular cell model was also the template for us to develop a mouse ventricular cell model by utilizing experimental data (8,32).

The left (LV) and right (RV) ventricular cell models for the rat consist of a Hodgkin–Huxley type membrane model that is described by the membrane capacitance, various ionic channels; the fast Na$^+$ current (I_{Na}), long-lasting Ca^{2+} current (I_{CaL}), the 4AP sensitive, Ca^{2+} independent transient outward K$^+$ current (I_t), steady-state outward K$^+$ current (I_{ss}), inward rectifier K$^+$ current (I_{K1}), hyperpolarization activated current (I_f), linear background current

(I_B); the Na$^+$/Ca^{2+} ion exchanger (I_{NaCa}), and the Na$^+$/K$^+$ (I_{NaK}) and Ca^{2+} membrane (I_{CaP}) pumps, that are experimentally observed in rat ventricular cells.

The mouse ventricular cell model was constructed by using the rat right ventricular cell model as the template. The mouse LV apex cell was developed by adding the 4AP sensitive slowly inactivating, delayed rectifier K$^+$ current (I_{Kslow}) based on the data of Fiset et al. (41) and Zhou et al. (48), and by reformulating I_t and I_{ss} based on experiments performed by Agus et al. (35–37) and Xu et al. (43) in mice. Further, a mouse LV septum cell model was developed by formulating a new current I_{tos} based on the data of (Xu et al. (43)), and by reducing the densities of I_{tof}, I_{Kslow}, and I_{ss} by 70, 23, and 30%, respectively, based on data of Gussak et al. (45).

The important results of our simulation studies are

1. The action potential heterogeneity (Fig. 3) in the adult rat LV is mainly due to the changes in the density and recovery kinetics of I_t and due to the altered density of I_{Na} (12).

2. The RV cell model can be developed from the LV cell model by changing the densities of I_t, I_{ss}, I_{CaL}, and I_{NaK} based on experimental data.

3. The changes in the density and the reactivation kinetics of I_t can account for the action potential prolongation differences in RV myocytes of diabetic (type-I, short term) rats (33) and LV myocytes of spontaneously hypertensive rats (35) (Fig. 4).

4. The presence of I_{Kslow} in mouse is one of the main factors contributing to the faster rate of repolarization seen in mouse compared to rats (Fig. 5) (8).

5. The LV septum cell model had more prolonged action potentials than the apex cells and these simulation results (Fig. 6a) are qualitatively similar to the experimental data of (52) (Fig. 6b).

Simulated Action Potentials of the Epicardial and Endocardial Cardiac Ventricular Cells

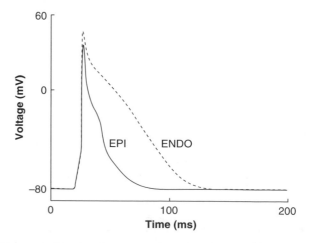

Figure 3. Simulated action potentials of the rat left ventricular (LV) epicardial (EPI) (solid line) and endocardial (ENDO) (dashed line) cells (35–37).

Simulated Action Potentials of the Epicardial Cardiac Ventricular
Cells for Normal and Spontaneously Hypertensive Rats

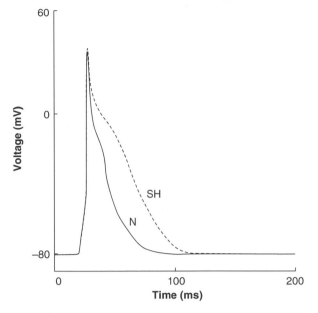

Figure 4. Model generated ventricular action potentials of the epicardial cells for the normal rat (N) (solid line) and spontaneously hypertensive (SH) rat (dashed line) (35–37).

6. The rat epicardial and endocardial ventricular cell models were more rate-sensitive than the mouse ventricular cell model and these simulation data match the experimental data well.

In conclusion, the mathematical modeling study of murine ventricular myocytes complements our knowledge of the biophysical data with simulation data and provide us with quantitative descriptions to understand the ionic currents underlying the cardiac action potential variations in different species. This kind of computational work will

Simulated Cardiac Action Potentials of Rat and Mouse Ventricular Cells

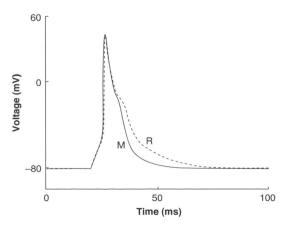

Figure 5. Simulated action potentials of the mouse left ventricular apex cell (M) (solid line) and the rat right ventricular cell (R) (dashed line) (35–37).

Simulated Ventricular Action Potentials for
Mouse LV Apex and Septum Cells

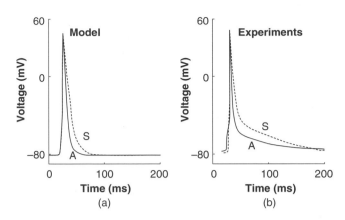

(a)   (b)

Figure 6. Simulated and experimentally recorded action potentials of the mouse left ventricular apex (A) (solid line) and the septum (S) (dashed line) cells (35–37).

enhance our understanding of the ionic mechanisms that contribute to the cardiac action potential variation in normal and diseased animals, and will provide us with better treatments for diseases in humans.

DISSEMINATION OF COMPUTATIONAL MODELS

My computational modeling laboratory has developed an interactive cell modeling web site, iCell (http://ssd1.bme. memphis.edu/icell/) since 1998. iCell, that integrates research and education, was specifically developed as a simulation-based teaching and learning resource for electrophysiology (35–37,51,52). The main objectives for the development of iCell were (1) to use technology in electrophysiology education, (2) to provide a learning and teaching resource via internet for cellular electrophysiology, (3) to allow the user to understand the cellular physiology mechanisms, membrane transport and variations of action potentials and ion channels by running simulations, and (4) to provide a computer platform independent resource for cellular models to be used for teaching, learning and collaboration. The site consists of JAVA applets representing models of various cardiac cells and neurons, and provides simulation data of their bioelectric transport activities at cellular level. Each JAVA-based model allows the user to go through menu options to change model parameters, run and view simulation results. The site also has a glossary section for the scientific terms. iCell has been used as a teaching and learning tool for seven graduate courses at the Joint Biomedical Engineering Program of University of Memphis and University of Tennessee. This modeling tool was also used as a collaboration site among our colleagues interested in simulations of cell membrane activities. Scientists from the fields of biosciences, engineering, life sciences and medical sciences in 17 countries, Argentina, Belgium, Brazil, Canada, China, England, Germany, Greece, Ireland, Japan, Korea, the Netherlands, New Zealand, Spain, Taiwan, Turkey and the United States, have tested and utilized iCell as a simulation-based teaching, learning and collaboration environment. The platform-independent software, iCell, provides us with an interactive and user-friendly teaching and learning resource, and also a collaboration environment for electrophysiology to be shared over the Internet. The usage of simulations for teaching and learning will continue advancing simulation-based engineering and sciences for research and development.

The simulations provided by iCell and other resources, such as CellML (http://www.cellml.org/public/news/index.html), Virtual Cell (http://www.nrcam.uchc.edu/), JSIM (http://nsr.bioeng.washington.edu/PLN/Software/), simBio Cell/Biodynamics Simulation Project of Kyoto University (http://www.biosim.med.kyoto-u.ac.jp/e/index.html), and E-Cell (http://ecell.sourceforge.net/), will continue signifying the important verification and prediction capabilities of the computer models to represent, analyze and complement the physiological data and knowledge. The model development demonstrates that computational models have to be constructed from experimental electrophysiology data, not only to explain and to verify the data that they are based on, but also to predict results for experiments that have not been performed and to guide future experiments.

Overall, computational modeling and simulation results continue to advance our understanding of living systems at the cellular level in cardiac electrophysiology while promoting collaborations and training in interdisciplinary field of bioengineering between scientists in life scientists and engineers. The presentation of computational models in user friendly, interactive and menu driven software are important in bringing collaborators of different disciplines and training scientists and students in cross-disciplinary projects.

IMPACT OF COMPUTATIONAL MODELING IN VENTRICULAR CELLS

The following summarizes impacts of the computational model development of ventricular bioelectric activity and the model-generated data in different disciplines of life sciences. I. *Biophysics and Physiology*: The results of the computational studies expand our knowledge of the living systems at the cellular level in electrophysiology. II. *Clinical Physiology and Medicine*: The insights gained and conclusions derived from the computational studies enhance our understanding of the biocomplexity of the heart, and provide us with better knowledge to be used in the future in treatments for diseases in humans. The cardiac cells' responses to various pathophysiological states with simulation data will also be better understood. III. *Pharmacology*: The differences in ventricular membrane ionic currents, especially outward K^+ currents in different species have very important practical implications. Different drugs are known to affect different ionic currents and to change action potential waveforms in different mammalian heart preparations under various conditions of development, aging and gender. A better understanding of the role of the ionic currents that control repolarization in the ventricular myocytes obtained from various species including rat and mouse, as presented

in this paper, will provide motivation and explanations for species differences in treatment and drug actions, and also promote pharmacological research that may lead to the development of more specific drugs to be used in children and adults.

ACKNOWLEDGMENTS

These computational research projects presented here were funded by the Whitaker Foundation (PI: Dr. S. S. Demir). The author acknowledges the contributions of her former students S. Pandit, S. Padmala, and E. Damaraju to these research projects.

The author thanks her former students, Joe E. McManis, Yiming Liu, Dong Zhang, Srikanth Padmala and Eswar Damaraju for coding JAVA applets in the iCell project, her students Chris Oehmen and Jing Zheng for posting the html pages and Dr. Emre Velipasaoglu, Siddika Demir and Asim Demir for valuable collaborations and discussions. This research was also funded by the Whitaker Foundation (PI, Dr. S. S. Demir).

BIBLIOGRAPHY

Cited References

1. Alberts B, et al. Membrane Transport. Essential Cell Biology: An Introduction to the Molecular Biology of the Cell. New York: Garland Publishing; 1997 p 371–407. Chapt. 12
2. Plonsey R, Barr R. Bioelectricity: A Quantitaive Approach. New York: Kluwer Academic Publications; 2000.
3. Hamill O, et al. Improved patch-clamp techniques for high-resolution current recording from cells and cell-free membrane patches. Pflugers Archiv 1981;391:85–100.
4. Neher E, Sakmann B. Noise analysis of drug induced voltage clamp currents in denervated frog muscle fibres. J Physiol 1976;258(3):705–729.
5. Neher E, Sakmann B. Single-channel currents recorded from membrane of denervated frog muscle fibres. Nature (London) 1976;260(5554):799–802.
6. Neher E, Sakmann B, Steinbach JH. The extracellular patch clamp: a method for resolving currents through individual open channels in biological membranes. Pflugers Arch 1978;375(2):219–228..
7. Neher E, Sakmann B. The patch clamp technique. Sci Am 1992;266(3):44–51.
8. Damaraju E. A Computational Model of Action Potential Heterogeneity in Adult Mouse Left Ventricular Myocytes. M.S. dissertation, University of Memphis, 2003.
9. Fozzard H. Cardiac Electrogenesis and the Sodium Channel. In: Spooner P, Brown A, Catterall W, Kaczorowski G, Strauss H, editors. Ion Channels in the Cardiovascular system:Function and dysfunction. Futura Publishing Company, Inc.; 1994. Chapt. 5.
10. Spooner PM, Rosen MR, editors. Foundations of Cardiac Arrhythmias. 1st ed. New York: Marcel Dekker; 2000.
11. Demir SS, et al. Action Potential Variation in Canine Ventricle: A Modeling Study. Comput Cardiol 1996; 221–224.
12. Pandit SV, Clark RB, Giles WR, Demir SS. A Mathematical Model of Action Potential Heterogeneity in Adult Rat Left Ventricular Myocytes. Biophy J 2001;81:3029–3051.
13. Bers DM. Excitation-Contraction Coupling and Cardiac Contractile Force 2nd ed. The Netherlands: Kluwer Academic Publications; 2001.

14. Hodgkin L, Huxley AF. A quantitative description of membrane current and its application to conduction and excitation in nerve. J Physiol 1952;117:500–544.
15. Irvine L, Jafri M, Winslow R. Cardiac sodium channel Markov model with temperature dependence and recovery from inactivation. Biophysic J 1999;76(4):1868–1885.
16. Beeler GW, Reuter H. Reconstruction of the action potential of ventricular myocardial fibres. J Physiol 1997;268:177–210.
17. Drouhard J, Roberge FA, Revised formulation of the Hodgkin-Huxley representation of the sodium current in cardiac cells. Comput Biomed Res 1987;20:333–350.
18. Winslow RL, et al. Mechanisms of altered excitation-contraction coupling in canine tachycardia-induced heart failure, II: model studies. Circ Res 1999;84:571–586.
19. Fox JJ, McHarg JL, Gilmour RF. Ionic mechanism of electrical alternans. Am J Physiol Heart Circ Physiol 2002;282:H516–H530.
20. Nordin C, Computer model of membrane current and intracellular Ca^{2+} flux in the isolated guinea pig ventricular myocyte. Amer J Physiolo 1993;265:H2117–H2136.
21. Luo C-H, Rudy Y. A model of the ventricular cardiac action potential. Circulation Res 1991;68:1501–1526.
22. Luo C-H, Rudy Y. A dynamic model of the cardiac ventricular action potential. I Simluation of ionic currents and concentration changes. Circulation Res 1994;74:1071–1096.
23. Zeng J, Laurita K, Rosenbaum DS, Rudy Y. Two components of the delayed rectifier K^+ current in ventricular myocytes of the guinea pig type: theoretical formulation and their role in repolarization. Circulation Res 1995;77:140–152.
24. Noble D, Varghese A, Kohl P, Noble P. Improved guinea-pig ventricular cell model incorporating a diadic space, I_{Kr} and I_{Ks}, and length- and tension-dependent processess. Can J Cardiol 1998;14:123–134.
25. Priebe L, Beuckelmann D. Simulation study of cellular electric properties in heart failure. Circ Res 1998;82:1206–1223.
26. ten Tusscher KHWJ, Noble D, Noble PJ, Panfilov AV. A model for human ventricular tissue.Am J Physiol Heart Circ Physiol 2004;286:H1573–H1589.
27. Riemer TL, Sobie A, Tung L. Stretch-induced changes in arrhythmogenesis and excitability in experimentally based heart cell models. Am J Physiol 1998;275:H431–H442.
28. Puglisi JL, Bers DM. LabHEART: an interactive computer model of rabbit ventricular myocyte ion channels and Ca transport. Am J Physiol Cell Physiol 2001;281:C2049–C2060.
29. Winslow RL, et al. Electrophysiological modeling of cardiac ventricular function: from cell to organ. Annu Rev Biomed Eng 2000;2:119–155.
30. Members of the Sicilian Gambit. New approaches to antiarrhythmic therapy, Part I: Emerging therapeutic applications of the cell biology of cardiac arrhythmias. Circulation 2001;104:2865–2873.
31. Bouchard RA, Clark RB, Giles WR. Effects of action potential duration on excitation-contraction coupling in rat ventricular myocytes. Action potential voltage-clamp measurements. Circ Res 1995;76:790–801.
32. Pandit SV. Electrical Activity in Murine Ventricular Myocytes: Simulation Studies. Ph.D. dissertation, University of Memphis; 2002.
33. Pandit SV, Giles WR, Demir SS. A Mathematical Model of the Electrophysiological Alterations in Rat Ventricular Myocytes in Type-I Diabetes. Biophys J 2003;84(2):832–841.
34. Padmala S, Demir SS. A computational model of the ventricular action potential in adult spontaneously hypertensive rats. J Cardiovasc Electrophysiol 2003;14:990–995.
35. Demir SS. Computational Modeling of Cardiac Ventricular Action Potentials in Rat and Mouse. Rev Jne J Physiol 2004;54(6):523–530.
36. Demir SS. An Interactive Electrophysiology Training Resource for Simulation-Based Teaching and Learning. Institute of Electrical and Electronics Engineers (IEEE) Engineering in Medicine & Biology Society Proceedings; 2004. p 5169–5171.
37. Demir SS. The Significance of Computational Modelling in Murine Cardiac Ventricular Cells. J Appl Bionics Biomech 2004;1(2):107–114.
38. Antzelevitch C, Yan G-X, Shimuzu W, Burashnikov A. Electrical Heterogeneity, the ECG, and Cardiac Arrhythmias. In: Zipes DP, Jalife J, editors. Cardiac Electrophysiology: From Cell to Bedside. 3rd ed. Philadelphia: WB Saunders; 1999. p 222–238.
39. Watanabe T, Delbridge LM, Bustamante JO, McDonald TF. Heterogeneity of the action potential in isolated rat ventricular myocytes and tissue. Circ Res 1983;52:280–290.
40. Clark RB, et al. Heterogeneity of action potential waveforms and potassium currents in rat ventricle. Cardiovas Res 1993;27:1795–1799.
41. Fiset C, Clark RB, Larsen TS, Giles WR. A rapidly activating sustained K^+ current modulates repolarization and excitation-contraction coupling in adult mouse ventricle. J Physiol 1997;504:557–563.
42. Nerbonne JM, Nichols CG, Schwarz TL, Escande D. Genetic manipulation of cardiac K^+ channel function in mice: what have we learned, and where do we go from here? Circ Res 2001;89:944–956.
43. Xu H, Guo W, Nerbonne JM. Four kinetically distinct depolarization-activated K^+ currents in adult mouse ventricular myocytes. J Gen Physiol 1999;113:661–678.
44. Chien KR. To Cre or not to Cre: the next generation of mouse models of human cardiac diseases. Circ Res 2001;88:546–549.
45. Gussak I, Chaitman BR, Kopecky SL, Nerbonne JM. Rapid ventricular repolarization in rodents: electrocardiographic manifestations, molecular mechanisms, and clinical insights. J Electrocardiol 2000;33:159–170.
46. Nerbonne JM. Molecular analysis of voltage-gated K^+ channel diversity and functioning in the mammalian heart. In: Page E, Fozzard HA, Solaro RJ, editors. Handbook of Physiology: The Cardiovascular System. New York: Oxford University Press; 2001. 568–594.
47. Shimoni Y, Severson D, Giles WR. Thyroid status and diabetes modulate regional differences in potassium currents in rat ventricle. J Physiol 1995;488:673–688.
48. Zhou J, et al. Characterization of a slowly inactivating outward current in adult mouse ventricular myocytes. Circ Res 1998;83:806–814.
49. Shimoni Y, Light PE, French RJ. Altered ATP sensitivity of ATP-dependent K^+ channels in diabetic rat hearts. Am J Physiol 1998;275:E568–E576.
50. Demir SS. iCell: an Interactive Web Resource for Simulation-Based Teaching and Learning in Electrophysiology Training. Institute of Electrical and Electronics Engineers (IEEE), Engineering in Medicine & Biology Society Proceedings; 2003; p 3501–3504.
51. Demir SS. Simulation-based Training in Electrophysiology by iCell Institute of Electrical and Electronics Engineers, Engineering in Medicine and Biology Engineering in Medicine & Biology Society Proceedings, 4 pages; 2005.

See also BLADDER DYSFUNCTION, NEUROSTIMULATION OF; ELECTROCONVULSIVE THERAPY; PACEMAKERS; SPINAL CORD STIMULATION; TRANSCUTANEOUS ELECTRICAL NERVE STIMULATION (TENS).

ELECTRORETINOGRAPHY

GRAHAM E. HOLDER
Moorfields Eye Hospital
London, United Kingdom

INTRODUCTION

The retina, situated at the back of the eye, is highly complex, consisting of different layers and containing many different cell types. It serves to encode images of the outside world into a suitable form for transmission to the brain for interpretation and the process of "seeing". It is possible to view the retina *in situ* using ophthalmoscopic techniques, and although this may reveal anatomical abnormalities, it may not reveal either the extent or nature of retinal dysfunction. The principal challenge for electroretinography is to provide information regarding retinal function to facilitate patient care.

In essence, a controlled light stimulus is used to stimulate the retina, which responds by generating very small electrical signals that can be recorded, with suitable amplification, using electrodes situated in relation to the eye, usually contacting the cornea. These electrical signals, the electroretinogram (ERG), have defined parameters (timing, shape, size) in normal individuals, and are altered in a predictable manner in disease. In general, the brighter the stimulus, the higher is the amplitude and the shorter the peak time of the ERG. Modification of the adaptive state of the eye (dark adapted or scotopic; light adapted or photopic) facilitate the separation of different cell types and layers within the retina. The objective information provided by electrophysiological examination has a significant effect both on diagnosis and patient management (1).

TECHNIQUES

The main tests of retinal function are the ERG, the massed retinal responses to full-field luminance stimulation, which reflects the function of the photoreceptor and inner nuclear layers of the retina, and the pattern electroretinogram (PERG), which, in addition to being "driven" by the macular photoreceptors, largely arises in relation to retinal ganglion cell function. Knowledge of this latter response can also be particularly useful in improved interpretation of an abnormal cortical visual evoked potential (VEP), but that topic is beyond the remit of this contribution, and the reader is referred elsewhere for a full discussion of the interrelationships between PERG and ERG, and PERG and VEP (2). Brief reference will also be made to the electrooculogram (EOG), which examines the function of the retinal pigment epithelium (RPE) and the interaction between the RPE and the (rod) photoreceptors, and is often used in conjunction with the ERG.

Electrophysiological recordings are affected not only by stimulus and recording parameters, but also by the adaptive state of the eye, and standardization is mandatory for meaningful scientific and clinical communication between laboratories. The International Society for Clinical Electrophysiology of Vision (ISCEV) has published Standards for EOG (3), ERG (4), PERG (5), and the VEP (6). Readers are strongly encouraged not only to adhere to the recommendations of those documents, but also to consider that the Standards are intended as minimum data sets, and that recording protocols in excess of the Standards may be necessary to accurately establish the diagnosis in some disorders. Typical normal traces appear in Fig. 1.

A brief description of each test follows, with emphasis on response generation. Referencing has been restricted; the reader is referred to standard texts for further details (7,8). The multifocal ERG (mfERG) is a relatively recent addition to the diagnostic armamentarium, and although currently more of a research application than a mainstream clinical tool, this is likely to change in the future as more knowledge is gained of the clinical applications and underlying mechanisms. The ISCEV has published guidelines for mfERG, to which the reader is referred (9).

THE ELECTROOCULOGRAM

The EOG enables assessment of the function of the RPE, and the interaction between the RPE and the retinal photoreceptors. The patient makes fixed 30° lateral eye movements during a period of 20 min progressive dark adaptation, followed by a 12–15 min period of progressive light adaptation. The eye movements are made every 1–2 s for ~10 s each minute. The amplitude of the signal recorded between electrodes positioned at medial and lateral canthi reaches a minimum during dark adaptation, known as the dark trough, and a maximum during light adaptation, the light peak. The development of a normal light peak requires normally functioning photoreceptors in contact with a normally functioning RPE, and reflects progressive depolarization of the basal membrane of the RPE. The EOG is quantified by calculating the size of the light peak in relation to the dark trough as a percentage, the Arden index. A normal EOG light rise is >175% for most laboratories.

THE ELECTRORETINOGRAM

The functional properties of the retinal photoreceptors underpin the principles of ERG recording. The retinal rod system, with ~120,000,000 rod photoreceptors, is sensitive under dim lighting conditions, has coarse spatial and poor temporal resolution. The rods adapt slowly to changes in lighting conditions. They do not enable color vision. They have a peak spectral sensitivity in the region of 500 nm. There are three types of retinal cone. (1) Short wavelength (S-cone), (2) medium (M-cone), and (3) long wavelength (L-cone). In the past, they have been referred to as blue, green, and red, respectively. There are perhaps 7,000,000 M- and L-cones and 800,000 S-cones. The relative proportion of L- versus M-cones varies from individual to individual, but approximates to 50% over a population. They are sensitive under bright lighting conditions; their high spatial resolution enables fine visual acuity; they adapt rapidly to changes in lighting conditions and can follow a fast flicker (L- and M-cones). The overall maximum spectral sensitivity is ~550 nm, in the green-yellow

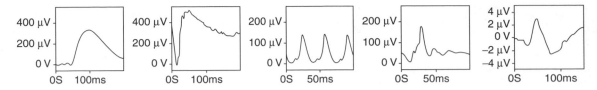

Figure 1. Typical normal ERG recordings. The rod specific ERG consists of the inner-nuclear layer generated b-wave. With a bright flash the waveform now contains an additional a-wave, the first 10–12 ms of which arise in relation to photoreceptor hyperpolarization. The rod specific and bright flash responses are recorded with full scotopic adaptation. After restoration to photopic adaptation the cone flicker and single flash ERGs are recorded. The former consists of a sinusoidal type waveform, the latter containing clear a- and b-waves. The pattern ERG (PERG) is the response of the macula to a reversing black and white checkerboard. See text for further details.

region of the color spectrum. Normal color vision requires all three cone types, but providing at least two cone types are present (there are some disorders in which that is not the case), at least some color vision is enabled. There are no S-cones in the foveola, the very central part of the macula responsible for very fine acuity.

The ERG is recorded using corneal electrodes and is the mass electrical response of the retina using a brief flash of light as a stimulus. The stimuli are delivered using a Ganzfeld bowl, an integrating sphere that enables uniform whole field illumination (Fig. 2a, b). In addition to flash stimulation, the Ganzfeld also allows a diffuse background for photopic adaptation. Some corneal electrodes are bipolar contact lenses with a built-in reference electrode. If such an electrode is not used, the reference electrodes should be sited at the ipsilateral outer canthi. A standard flash is defined by ISCEV as 1.5–3.0 $cd \cdot s \cdot m^{-2}$. The response to this flash under scotopic conditions, with a fully dilated pupil, is the Standard or mixed response (Fig. 1). It is probably this response that may be regarded as the "typical" ERG, but although there is a cone contribution, the standard response is dominated by rod driven activity. The "maximal" ERGs that appear in this article were recorded to ~11.0 $cd \cdot s \cdot m^{-2}$ flash better to view the a-wave. The use of a brighter flash of such intensity is "suggested" in the most recent ISCEV ERG Standard (4). The initial ~10 ms of the a-wave arises in relation to hyperpolarisation of the (rod) photoreceptors and the slope of the a-wave can be

related to the kinetics of phototransduction (10). The larger positive b-wave is generated postreceptorally in the inner-nuclear layer of the retina in relation to depolarization of the ON-bipolar cells (11). The oscillatory potentials, the small wavelets on the ascending limb of the b-wave, are probably generated in relation to amacrine cell activity. When the standard flash is attenuated by 2.5 log units, the stimulus intensity falls below the cone threshold, and a rod-specific b-wave is obtained. At this relatively low luminance there is insufficient photoactivation to record an a-wave (Fig. 1, column A, top).

The ERGs that reflect cone system activity are obtained using a rod-saturating photopic background (17–34 $cd \cdot m^{-2}$) using superimposed single flash and 30 Hz flicker stimulation. The rod system has low temporal resolution and use of a 30 Hz stimulus, combined with a rod-suppressing background, allows a cone-system specific waveform to be recorded. This response is probably the more sensitive measure of cone dysfunction, but is generated at an inner-retinal level (12) and thus does not allow the distinction between cone photoreceptor and cone inner-nuclear layer dysfunction. Although there is a demonstrated contribution from hyperpolarizing (OFF-) bipolar cells to shaping the photopic a-wave (13), this component nonetheless has some contribution from cone photoreceptor function, and some localization within the retina may be obtained with the single flash cone response. The cone b-wave reflects postphototransduction activity, and to a

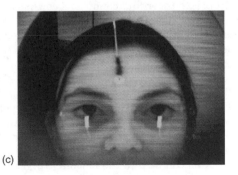

(a)   (b)   (c)

Figure 2. (a) A conventional Ganzfeld used for ERG recording (front view). (b) The subject in position at the Ganzfeld. (c) Photograph taken using an infrared (IR) camera at the back of the Ganzfeld, which is used to monitor eye position and eye opening during both dark adapted and light adapted conditions. The two gold-foil corneal recording electrodes are well seen. The central forehead ground electrode is easily seen; the outer canthus reference electrodes are just visible. (Courtesy of Chris Hogg.)

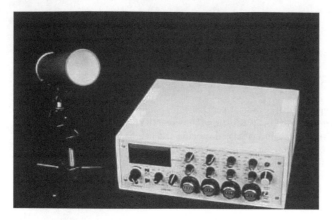

Figure 3. A "mini-Ganzfeld" based on light emitting technology. The device shown has four independent color channels, blue, green, orange and red, each of which can be used as stimulus or background alone or in combination. (Courtesy of Chris Hogg, CH electronics, Bromley, Kent, UK; www.ch-electronics.net.)

short flash stimulus ON and OFF activity within the photopic system is effectively synchronized.

Separation of the cone ON (depolarizing bipolar cells, DBCs) and OFF (hyperpolarizing bipolar cells, HBCs) responses can be achieved using a long duration stimulus with a photopic background (14,15). The stimulus can be generated either via a shutter system or by using light emitting diodes (Fig. 3). Stimulators based on light emitting diodes (LEDs) offer several advantages over standard stimulators. They are of low cost, have a stable output intensity over time (reducing the need for calibration), enable variable and highly accurate stimulus duration, and a have a well-defined narrow band spectral output. Further, being driven by relatively low voltage and current, they are intrinsically safe, and generate low electrical noise. Their use in ERG systems can be expected to increase.

It is also possible to elicit the activity of the S-cone population. In the author's laboratories this is achieved using blue stimuli superimposed upon a bright orange photopic background, again delivered using a LED based device. The background thus serves to suppress activity from rod and L-/M-cone systems. The response under appropriate recording conditions consists of an early component at ~30 ms arising in relation to L-/M-cone systems (there is overlap of the spectral sensitivities of the different cone systems and a small response arises from L-/M-cones with a bright blue stimulus), followed by a component specific for S-cone function at 45–50 ms (16).

The retinal ganglion cells do not significantly contribute to the clinical (flash) ERG. Also, as a mass response, the ERG is normal when dysfunction is confined to small retinal areas, and, despite the high photoreceptor density, this also applies to macular dysfunction; the full-field ERG is normal if dysfunction is confined to the macula (e.g., Fig. 4, column B).

THE PATTERN ELECTRORETINOGRAM

The response of central retina to a structured isoluminant stimulus can be measured, and is known as the pattern

ERG. The stimulus is usually a reversing black and white checkerboard. The PERG has largely inner retinal origins, but is "driven" by the macular photoreceptors, and PERG measurement thus provides both a measure of central retinal function and, in relation to its origins, of retinal ganglion cell function. It is thus of clinical importance not only in the objective assessment of macular function, but also in the electrophysiological differentiation between optic nerve and macular dysfunction by providing a measure of the retinal response to a similar stimulus to that used to evoke the VEP (see Ref. 2 for a comprehensive review). It is a much smaller signal than the (full-field) ERG and computerized signal averaging is used to extract the PERG signal.

The PERG is recorded using noncontact lens electrodes in contact with the cornea or bulbar conjunctiva to preserve the optics of the eye. Suitable electrodes are the gold foil (17), the DTL (18), and the H–K loop (19). Ipsilateral outercanthus reference electrodes are essential to avoid contamination from the cortically generated VEP, such as occurs if forehead or ear "reference" electrodes are used (20). Pupillary dilation is not used.

There are two main components of PERG to a reversing checkerboard with a relatively slow reversal rate (<6 reversals s^{-1}). There is a prominent positive component, P50, at ~50 ms followed by a larger negative component, N95, at ~95 ms (21). Clinical measurement of the PERG usually comprises the amplitude of P50, measured from the trough of the early negative N35 component; the peak latency of P50; and the amplitude of N95, measured to trough from the peak of P50 (Fig. 1). Approximately 70% of P50 is likely to be related to retinal ganglion cell function, but the remainder is not related to spiking cell function and may be generated more distally in the retina (22). The exact origins have yet to be ascertained at the time of writing. The N95 is a contrast-related component generated in the retinal ganglion cells.

An analysis time of 150 ms or greater is usually used for recording the PERG, with ~150 averages per trial needed to obtain a reasonable signal-to-noise ratio. As it is a small response, stringent technical controls are important during recording and are fully discussed elsewhere (8). Binocular stimulation and recording is preferred so the better eye can maintain fixation and accommodation, but it is necessary to use monocular recording if there is a history of squint. P50 is sensitive to optical blur, and accurate refraction is needed. At low stimulus frequencies the amplitude of the PERG is related almost linearly to stimulus contrast. A high contrast black and white reversing checkerboard with 0.8° checks in a 10–16° field is recommended by ISCEV.

MULTIFOCAL ERG

The mfERG attempts to provide spatial information regarding cone system function in central retina. The stimulus usually consists of multiple hexagons displayed on a screen (Fig. 5a) each of which flashes on with its own pseudo-random binary sequence (an M-sequence). A cross-correlation of the local flash sequence with the mass

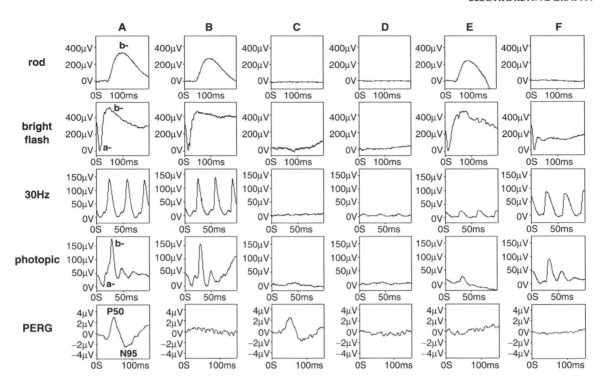

Figure 4. Typical electroretinographic abnormalities in selected diseases compared to those in a normal subject. Column A: Normal subject. Column B: A patient with macular dysfunction; the PERG is undetectable, but full-field ERGs are normal. Column C: "Classical" retinitis pigmentosa; all full-field ERGs are virtually extinguished, but the PERG is normal reflecting sparing of central retinal function. Column D: Rod-cone dystrophy (retinitis pigmentosa); the rod and cone ERGs are markedly abnormal (with the rod ERG being more affected). Note the delayed and reduced cone ERGs, typical abnormalities present when there is generalized cone involvement in the context of photoreceptor degeneration. The abnormal PERG reflects involvement of the macula. Column E: Cone dystrophy; the rod and bright flash ERGs are normal, but the cone single flash and flicker ERGs are delayed and reduced in keeping with generalized cone system dysfunction. The abnormal PERG reflects involvement of the macula. Column F: X-linked congenital stationary night blindness (complete type). The rod specific ERG is undetectable, but the normal a-wave of the bright flash dark adapted ERG confirms the dysfunction to be postphototransduction. There are subtle but significant changes in cone-system derived ERGs (note particularly the broadened trough and reduced amplitude sharply rising peak of the b-wave), and reduction in the PERG.

response derives the responses relating to each individual hexagon thus giving multiple cone system ERG waveforms from a single recording electrode. The mfERG can be of use in disturbances of macular function and to assess the degree of central retinal involvement in generalized retinal disease, but is highly susceptible to poor fixation, and the ability of a patient accurately to maintain good fixation throughout the recording session is a pre-requisite to obtaining clinically meaningful data. Increasing use and development of systems that can control stimulus delivery in relation to eye position can be anticipated. Possibilities include the use of "eye-tracking" devices and direct fundus visualization during stimulation.

CLINICAL APPLICATIONS

EOG

Disorders of rod photoreceptor function can affect the EOG, and the light rise is typically reduced in generalized photo-receptor degenerations such as retinitis pigmentosa (RP,

rod–cone dystrophy), a genetically determined group of disorders. Usually, the reduction in EOG light rise parallels the degree of rod photoreceptor dysfunction, but generalized RPE dysfunction can also manifest a reduced EOG light rise. Indeed, it is the latter property that leads to the main clinical use of the EOG, the diagnosis of Best disease. Best disease, or vitelliform macular dystrophy, is a dominantly inherited macular degeneration related to mutation in the gene VMD2. At presentation there are often distinctive vitelliform lesions at the maculae on funduscopy, but other appearances may occur. The diagnostic findings are of a severely reduced or absent EOG light rise accompanied by normal ERGs. Best disease may present in childhood, but a child may find the repetitive eye movements needed for EOG recording difficult or impossible to maintain for the required 30–40 min. Under such circumstances it is appropriate to test both parents; due to the dominant inheritance pattern one of the disorder, one of the parents will have carry the mutant gene and will manifest a reduced EOG. Adult vitelliform macular dystrophy (pattern dystrophy) may sometimes clinically

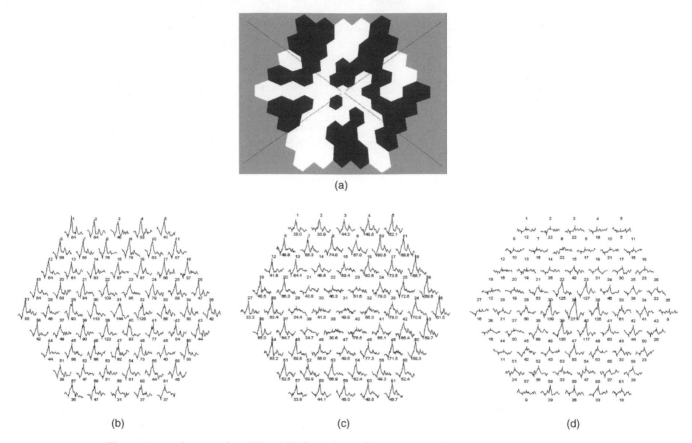

Figure 5. (a) the typical multifocal ERG stimulus; (b) a normal subject; (c) a macular dystrophy. There is loss of the responses to central hexagons but preservation of more peripheral responses. (d) A retinal dystrophy with sparing of central macular function but loss of the responses in the periphery.

be mistaken for Best disease, but although the EOG light rise may be mildly subnormal, it is not reduced to the same extent as in Best disease. The electrophysiological recordings will usually resolve any clinical dilemma in differential diagnosis.

ERG

Although the rod specific ERG b-wave is a sensitive indicator of retinal rod system dysfunction, the fact that it is generated in the inner-nuclear layer of the retina means that reduction in this response does not allow localization of the defect either to those structures or the upstream rod photoreceptors. It is the a-wave of the responses to brighter flashes that directly reflects activity of the photoreceptors and enables the distinction between photoreceptor dysfunction and a primary disorder of inner-retinal function. Genetically determined photoreceptor degenerations, such as the rod–cone (retinitis pigmentosa, RP) or cone–rod dystrophies, thus give overall ERG reduction (Fig. 4, columns C, D). The cone-derived ERGs in generalized photoreceptor degeneration characteristically show abnormalities of both amplitude and timing, particularly evident in the flicker ERG, but RP may occasionally only affect the rod-derived ERGs in the early stages of disease. Truly restricted disease, such as sector RP, is associated

with amplitude reduction, but no implicit time change, whereas diffuse or generalized disease is usually also associated with an abnormally delayed implicit time. Retinitis pigmentosa is associated with pigmentary migration from RPE into retina consequent upon photoreceptor cell death, but the clinical appearance of the ocular fundus may not reflect the severity or nature of the disorder. Electroretinography not only enables accurate diagnosis, when interpreted in clinical context, but may also provide useful prognostic information. There is no rod system involvement in a pure cone dystrophy; such disorders have normal rod responses, but abnormal cone responses, with the 30 Hz flicker response usually showing both amplitude reduction and delay (Fig. 4, column E).

A waveform in which the bright flash a-wave is spared, but there is selective b-wave reduction, is known as a "negative" or electronegative ERG (e.g., Fig. 4, column F, row 2), and is associated with dysfunction postphototransduction, often postreceptoral. For example, in central retinal artery occlusion (CRAO) the finding of a "negative" ERG reflects the duality of the retinal blood supply, with the photoreceptors supplied via choroidal circulation, and the inner-nuclear layer supplied via the central retinal artery. Other causes of negative ERG include X-linked congenital stationary night blindness (CSNB, Fig. 4, column F), X-linked retinoschisis, quinine toxicity, melanoma associated retinopathy (MAR, an autoimmune mediated disorder that can occur in patients with a history of cutaneous malignant melanoma), Batten disease (one of the ceroid neuronal lipofuscinoses), and occasionally cone–rod dystrophy. Carcinoma associated retinopathy (CAR), unlike MAR, usually give profound global ERG reduction in keeping with dysfunction at the level of the photoreceptor rather than a "negative" ERG. A negative ERG is also a relatively common occurrence in Birdshot chorioretinopathy (BCR), an inflammatory disease, but in that disorder such an appearance may normalize following successful treatment, usually with steroids and/or immunosuppressive agents.

The most common ERG abnormality in BCR, or other forms of inflammatory retinal disease, such as uveitis, is a delayed 30 Hz flicker ERG, but there may be much less marked amplitude change than occurs in photoreceptor degeneration. The ERG abnormalities may occur prior to the development of symptoms, and can normalize following treatment. Electrophysiology can thus play an important role not only in the characterization of the disease, but also in the initiation and monitoring of treatment (23). This relatively recent role of the ERG in the management of inflammatory disease can be expected to receive increasing clinical interest in the future.

PERG

Primary Evaluation of Macular Function. Disorders of macular function result in an abnormality of the P50 component of the PERG, often with preservation of the N95/P50 ratio. It is usually P50 amplitude that is affected; latency changes are only occasionally present, particularly in association with macular oedema or serous detachment at the macula. In clinical practice, the PERG and the (full-field) ERG provide complementary information regarding retinal function; the ERG assesses peripheral retinal function, and the PERG the degree of central retinal involvement. For example, dysfunction confined to the macula will have a normal ERG and an abnormal PERG (Fig. 4, column B), a common combination in macular dystrophies, such as Stargardt-fundus flavimaculatus (S-FFM), whereas generalized retinal dysfunction with macular involvement will have both an abnormal ERG and an abnormal PERG. This facilitates the distinction between macular dystrophy, cone dystrophy, and cone–rod dystrophy in a patient with an abnormal macular appearance and a history suggestive of a genetically determined disorder, important to the prognosis and accurate counseling of the patient. In relation to S-FFM, note that some patients have additional full-field abnormalities that may be of prognostic value (24).

The PERG may be normal even when the ERG is almost extinguished in patients with rod–cone dystrophy but normal central retinal function. Further, the objective assessment of macular function provided by the PERG can sometimes demonstrate early central retinal abnormalities prior to the appearance of symptoms or signs of macular involvement.

Ganglion Cell Dysfunction. The PERG will often be normal in disturbance of optic nerve function. However,

there may be retrograde degeneration to the retinal ganglion cells in optic nerve disease and this may selectively affect the ganglion cell derived N95 component. It is N95 loss that is the common abnormality if the PERG is abnormal in optic nerve disease. That is unlike macular dysfunction, where it is the P50 component that is primarily affected. Shortening of P50 latency may also occur in more severe disease, but, again, is not a feature of macular dysfunction. Primary disorders of retinal ganglion cell, such as Leber hereditary optic neuropathy (LHON) and dominantly inherited optic atrophy (DOA), are associated with N95 component loss, marked at presentation in LHON, but often occurring later in the disease process in DOA. There may be additional P50 amplitude reduction in advanced retinal ganglion cell dysfunction, and the associated shortening of P50 latency then becomes an important diagnostic factor. Further, providing there is sufficient vision remaining in at least one eye to maintain fixation for binocular PERG recording, total extinction of the PERG probably does not occur in optic nerve disease. Even in an eye blind from optic nerve disease (no perception of light), a PERG may still readily be detectable (2).

THE PERG IN RELATION TO VEP INTERPRETATION

Although detailed discussion of the VEP is beyond the scope of this article, a short discussion of the use of the PERG in the improved interpretation of VEP abnormality is warranted. The cortically generated VEP to pattern reversal stimulation is a powerful clinical tool in the detection and assessment of optic nerve dysfunction, and pattern VEP latency delay or loss is frequently associated with optic nerve disease. However, the VEP is generated in the occipital cortex, and a delayed PVEP must never be assumed necessarily to indicate optic nerve dysfunction in a visually symptomatic patient. Similar abnormalities can occur either in macular disease or optic nerve disease. The appearance of the macula may be a poor indicator of function, and remember that a normal macular appearance does not necessarily equate to normal macular function. The different types of abnormality present in the PERG in optic nerve and macular diseases usually allow the differentiation between delayed VEP due to retinal macular disease and that due to optic nerve disease. An abnormal VEP with a normal PERG (or a normal P50 component with an abnormality confined to N95) is consistent with optic nerve/ganglion cell dysfunction, whereas pronounced P50 reduction suggests a disturbance of macular function (e.g., Fig. 4, columns B, D, E, F).

MULTIFOCAL ERG

The multifocal ERG can be used to assess the spatial extent of central retinal cone involvement in disease. Normal traces appear in Fig. 5b. Two clinical examples are shown; Fig. 5c shows a patient with a retinal dystrophy in whom there is sparing of central macular function; Fig. 5d shows a patient with a macular dystrophy with loss of the responses to central hexagons, but preservation of more peripheral responses. As a restricted test of central retinal cone function, in clinical circumstances the mfERG should always be taken in conjunction with conventional full-field ERG.

CONCLUSIONS AND FUTURE DIRECTIONS

Diagnosis and management of the patient with visual pathway disease is greatly assisted by the objective functional information provided by electrophysiological examination. Separation of the function of different retinal cell types and layers enables characterization both of acquired and inherited retinal disorders, of great importance when counseling families affected by or at risk of a genetically determined disease. The PERG is complementary to the ERG by providing a measure of central retinal function; the mfERG may play a similar role.

BIBLIOGRAPHY

Cited References

1. Corbett MC, Shilling JS, Holder GE. The assessment of clinical investigations: the Greenwich grading system and its application to electrodiagnostic testing in ophthalmology. Eye 1995;9 (Suppl.): 59–64.
2. Holder GE. The pattern electroretinogram and an integrated approach to visual pathway diagnosis. Prog Retin Eye Res 2001;20:531–561.
3. Marmor MF. Standardization notice: EOG standard reapproved. Doc Ophthalmol 1998;95:91–92.
4. Marmor MF, Holder GE, Seeliger MW, Yamamoto S. Standard for clinical electroretinography (2004 update). Doc Ophthalmol 2004;108:107–114.
5. Bach M et al. Standard for Pattern Electroretinography. Doc Ophthalmol 2000;101:11–18.
6. Odom JV et al. Visual Evoked Potentials Standard (2004). Doc Ophthalmol 2004;108:115–123.
7. Heckenlively JR, Arden GB, editors. Principles and Practice of Clinical Electrophysiology of Vision. St. Louis: Mosby Year Book; 1991.
8. Fishman GA, Birch DG, Holder GE, Brigell MG. Electrophysiologic Testing in Disorders of the Retina, Optic Nerve, and Visual Pathway. 2nd ed. Ophthalmology Monograph 2. San Francisco: The Foundation of the American Academy of Ophthalmology; 2001.
9. Marmor MF, et al. Guidelines for basic multifocal electroretinography (mfERG). Doc Ophthalmol 2003;106:105–115.
10. Hood DC, Birch DG. Rod phototransduction in retinitis pigmentosa: Estimation of parameters from the rod a-wave. Invest Ophthalmol Vis Sci 1994;35:2948–2961.
11. Shiells RA, Falk G. Contribution of rod, on-bipolar, and horizontal cell light responses to the ERG of dogfish retina. Vis Neurosci 1999;16:503–511.
12. Bush RA, Sieving PA. Inner retinal contributions to the primate photopic fast flicker electroretinogram. J Opt Soc Am A 1996;13:557–565.
13. Bush RA, Sieving PA. A proximal retinal component in the primate photopic ERG a-wave. Invest Ophthalmol Vision Sci 1994;35:635–645.
14. Sieving PA. Photopic ON- and OFF-pathway abnormalities in retinal dystrophies. Trans Am Ophthalmol Soc 1993; 91:701–773.
15. Koh AHC, Hogg CR, Holder GE. The Incidence of Negative ERG in Clinical Practice. Doc Ophthalmol 2001;102: 19–30.
16. Arden GB et al. S-cone ERGs elicited by a simple technique in normals and in tritanopes. Vision Res 1999;39:641–650.
17. Arden GB et al. A gold foil electrode: extending the horizons for clinical electroretinography. Invest Ophthalmol Vision Sci 1979;18:421–426.
18. Dawson WW, Trick GL, Litzkow CA. Improved electrode for electroretinography. Invest Ophthalmol Vision Sci 1979;18: 988–991.
19. Hawlina M, Konec B. New noncorneal HK-loop electrode for clinical electroretinography. Doc Ophthalmol 1992;81:253–259.
20. Berninger TA. The pattern electroretinogram and its contamination. Clin Vision Sci 1986;1:185–190.
21. Holder GE. Significance of abnormal pattern electroretinography in anterior visual pathway dysfunction. Br J Ophthalmol 1987;71:166–171.
22. Viswanathan S, Frishman LJ, Robson JG. The uniform field and pattern ERG in macaques with experimental glaucoma: removal of spiking activity. Invest Ophthalmol Vis Sci 2000;41:2797–2810.
23. Holder GE, Robson AG, Pavesio CP, Graham EM. Electrophysiological characterisation and monitoring in the management of birdshot chorioretinopathy. Br J Ophthalmol 2005; in press.
24. Lois N, Holder GE, Bunce C, Fitzke FW, Bird AC. Stargardt macular dystrophy—Fundus flavimaculatus: Phenotypic subtypes. Arch Ophthalmol 2001;119:359–369.

See also BLIND AND VISUALLY IMPAIRED, ASSISTIVE TECHNOLOGY FOR; CONTACT LENSES; VISUAL PROSTHESES.

ELECTROSHOCK THERAPY.   See ELECTROCONVULSIVE THERAPY.

ELECTROSTIMULATION OF SPINAL CORD.   See SPINAL CORD STIMULATION.

ELECTROSURGICAL UNIT (ESU)

JOHN PEARCE
The University of Texas
Austin, Texas

INTRODUCTION

Electrosurgery means the application of radio frequency (RF) current at frequencies between ~ 300 kHz and 5 MHz to achieve a desired surgical result; typically the fusion of tissues or surgical cutting in which the tissue structure is disrupted. In either case, the effect is achieved by heat dissipated in the tissues from the RF current by resistive, or joule, heating. This method has the ability to cut and coagulate tissues simultaneously; and, as a consequence, has made substantial contributions to several branches of clinical medicine since its introduction in the late 1920s.

The tissue effects applied in electrosurgery are typically described as (a) white coagulation, named for its appearance, in which the tissue proteins are degraded at lower temperatures, typically 50–90 °C; (b) black coagulation or carbonization in which tissues are completely dried out (desiccated) and reduced to charred carbonaceous remnants at higher temperatures; and (c) cutting in which tissue structures are separated by the rapid boiling of small volumes of tissue water. These three results usually occur in some combination depending on the applied current and voltage at the so-called active or surgical electrode.

Electrosurgery accomplishes many surgical jobs better than any other device or technique while drastically reducing the morbidity and mortality associated with surgery. It does this by reducing the time under anesthesia and complications due to operative and postoperative hemorrhage. Many of the delicate techniques in neurosurgery would be impossible without electrosurgery—and it is likely that open heart surgery and much of urologic surgery would likewise not be done.

Historical Background

The application of heat for the treatment of wounds dates back to antiquity. According to Major (1), Neolithic skulls unearthed in France show clear evidence of thermal cauterization. The Edwin Smith papyrus (~ 3000 BC) (2) describes the use of thermal cautery for ulcers and tumors of the breast. Licht (3) reports that according to Jee (4) the ancient Hindu god Susruta, the highest authority in surgery, said that "caustic is better than the knife, and the cautery is better than either." Cautery in ancient Hindu medicine included heated metallic bars, boiling liquids, and burning substances.

In cauterization the essential physical mechanism behind the treatment is conduction heat transfer from a hot object placed on the surface to raise the temperature high enough to denature the tissue proteins. Cutting and coagulation by means of electrosurgery is also accomplished by heating tissue to high temperatures, but the essential difference is that the primary mechanism is electrical power dissipation directly in the affected tissues themselves, rather than heat transfer from an external hot object on the tissue surface. It is rather like the difference between heating food in a conventional oven and a microwave oven, in a loose sense. Electrosurgery is sometimes erroneously referred to as electrocautery. Since the physical mechanisms are different it is important to keep these two techniques separate by precise terminology. Electrosurgery is also referred to as surgical diathermy, particularly in Europe. While this term is generally understood, it is a bit of a misnomer since diathermy literally means through-heating, such as might be applied in physical medicine for the relief of pain or in hyperthermia therapy for tumor treatment. Electrosurgical devices in operating rooms are designed and built for surgical use only, and the terminology in this section is standardized on that basis.

Early Experiments with High Frequency Current. The origin of the application of rf current to achieve surgical results is difficult to establish owing to the rapid pace of development in the electrical arts during the late nineenth and early twentieth centuries. Lee De Forrest, the inventor of the vacuum tube (the audion, 1907 and 1908), filed a patent for a spark gap rf generator to be used for electrosurgery in February of 1907 (it was granted in December of 1907) (5). Also, during the same year, Doyen noted that the

effect of a surgical arc on the tissue was not a function of the length of the arc, and that the temperatures of carbonized tissue were as high as 500–600 °C (6). He also found that final temperatures in the range of 65–70 °C resulted in white coagulation while there was no damage to tissues for temperatures < 60 °C (7).

By far, the most effective promoters of electrosurgery were Cushing and Bovie. W. T. Bovie was a physicist attached to the Harvard Cancer Commission. He had developed two electrosurgical units, one for coagulating and one for cutting. Harvey Cushing, the father of neurosurgery, had been concerned for some time with the problem of uncontrolled hemorrhage and *diabetes insipidus*, the often fatal complications of hypophysectomy, among other concerns (8). In 1926, working with Bovie, he applied high frequency current in cerebral surgery with excellent results. They published their work in 1928, emphasizing the three distinct effects of electrosurgery: desiccation, cutting and coagulation (9).

Early Electrosurgical Generators. Cameron-Miller offered the Cauterodyne in 1926, similar to the later model of 1930 that featured both vacuum tube cutting and spark gap coagulation. It came in a burled walnut bakelite case for $200. It is not known if the original 1926 device included tube cutting. The device designed and manufactured by W. T. Bovie was of higher fundamental frequency than the other early devices. The Bovie device used a 2.3 MHz vacuum tube oscillator for pure cutting (i.e., Cut 1) and a 500 kHz fundamental frequency spark gap oscillator for fulguration and desiccation (Cut 2–4 and Coag). The overall size, circuit and configuration of the Bovie device remained essentially constant through the 1970s. Bovie's name is so closely associated with electrosurgery that it is frequently referred to as the Bovie knife in spite of the extensive work which preceded his device and the numerous devices of different manufacture available then and now. In essence, the available electrosurgical generators between ~ 1930 and the 1960s were of the similar design to that used by Bovie, and consisted of a spark gap generator for coagulating and a vacuum tube generator for cutting.

The introduction of solid-state electrosurgical generators in the early 1970s by Valleylab and EMS heralded the modern era of isolated outputs, complex waveforms, more extensive safety features and hand-activated controls. Interestingly, hand-activated controls are actually a recent rediscovery and improvement on those used by Kelly and Ward (10) and those available on the Cauterodyne device. In some ways, higher technology devices of all designs are made inevitable by the recent proliferation of delicate measurement instrumentation that is also attached to a patient in a typical surgical procedure.

Clinical Applications

The topics chosen in this introductory survey are by no means comprehensive. Those readers interested in specific surgical techniques should consult the texts by Kelly and Ward (10) and Mitchell et al. (11) for general, gynecologic, urologic and neurosurgical procedures; Otto (12) for minor electrosurgical procedures; Harris (13), Malone (14), and Oringer (15,16) for dental electrosurgery; and Epstein (17) and Burdick (18), for dermatologic procedures.

When high frequency currents are used for cutting and coagulating, the tissue at the surgical site experiences controlled damage due either to disruptive mechanical forces or distributed thermal damage. Cutting is accomplished by disrupting or ablating the tissue in immediate apposition to the scalpel electrode. Continuous sine waveforms (e.g., those obtained from vacuum tube or transistor oscillators) have proven most effective for cutting. Coagulating is accomplished by denaturation of tissue proteins due to thermal damage. Interrupted waveforms, such as exponentially damped sinusoids (obtained from spark gap or other relaxation-type oscillators) are effective for coagulation techniques requiring fulguration, or intense sparking (*fulgur* is Latin for lightning). However, when no sparks are generated, coagulation is created by tissue heating alone, and the specific waveform is immaterial: only its effective heating power, or root-mean-square (rms) voltage or current determine the extent of the effect. Suffice it to say that the difference between cutting and coagulation is due to combined differences in heat-transfer mechanisms and spatial distribution of mechanical forces. In general, the tissue damage from cutting current is confined to a very small region under the scalpel electrode and is quite shallow in depth. Cells adjacent to the scalpel are vaporized and cells only a few cellular layers deep are essentially undamaged. Though dependent on surgical technique, generally only the arterioles and smaller vessels are sealed when a cut is made using pure sine wave currents. Coagulation currents are used to close larger vessels opened by the incision, to fuse tissue volumes by denaturing the proteins (chiefly the collagen) and to destroy regions of tissue. The tissue damage when coagulating is deeper than when cutting. In the majority of applications for coagulating current, the damage in the tissue is cumulative thermal damage rather than tissue disruption or ablation.

Cutting is a monopolar procedure, although some experiments have been performed with bipolar cutting. That is, the scalpel electrode represents essentially a point current source and the surgical current is collected at a remote site by a large area dispersive, or return electrode. Coagulation may be accomplished using either monopolar or bipolar electrodes. In bipolar applications, both the current source and current sink electrodes are located at the surgical site. A typical bipolar electrode might consist of forceps with the opposing tongs connected the two active terminals of the generator. In both cutting and coagulating processes, whether monopolar or bipolar electrodes are used, a layer of charred tissue often condenses on the cool electrode that must periodically be removed (19).

The histologic effects of cutting current are varied and apparently depend on technique. Knecht et al. (20) found that the healing process in an electrosurgical incision was a bit slower than for incisions made by a cold scalpel. In their study, the wound strength of an electrosurgical cut was less than that of a cold scalpel cut until ~ 21 days after surgery. After 21 days, no difference in wound strength was measurable. Ward found that an electrosurgical cut

generally formed slightly more scar tissue on healing than a cold scalpel cut if the closure of the two wounds was identical (21). The cellular layers within ~ 0.1 mm of the scalpel electrode showed electrodesiccation effects when sine wave cutting was used (21). In a later series of studies on tissues of the oral cavity, Oringer observed that when the cutting current was carefully controlled, the damage was confined to the cut cellular layer, and the layer of cells adjacent to the cut was undamaged (14,15). The cell destruction was apparently self-limiting to the extent that no damage to the cytoplasm or cell nucleus of the cut layer was visible in light or electron micrographs (14). Oringer (16) describes the margin of an excised squamous cell carcinoma that had been removed with electrosurgery. Under the electron microscope at a magnification of 47,400 the margin was seen to contain several clear examples of cells sheared in half with no damage to the remainder. Oringer, and others, observed faster healing with less scar tissue in the electrosurgical incision. The variety of results obtained is likely due to differences in waveform, surgical technique, tissue characteristics and scalpel electrodes used in the studies.

When combined sine wave and interrupted (coagulating) waveforms are used, or when spark gap sources are used for cutting, a coagulum layer extends deeper into the tissues under the desiccated layer (21). Coagulation techniques include (1) fulguration (also called spray coagulation or black coagulation) in which the tissue is carbonized by arc strikes, (2) desiccation, in which the cells are dehydrated resulting in considerable shrinking, and (3) white coagulation, in which the tissue is more slowly cooked to a coagulum. In fulguration techniques, the active electrode is held a few millimeters from the surface of the tissue and arcs randomly strike from the electrode to the tissue. The cell structure is destroyed at a very high temperature resulting in charring of the tissue. In desiccation, the water is evaporated from the cell relatively slowly leaving a white dry flake of powder, the cells appear shrunken and drawn out with elongated nuclei. The overall cellular anatomical characteristics are preserved (21). Desiccation techniques normally take longer to accomplish than fulguration for the same volume of tissue. In white coagulation, the electrode is in intimate contact with the tissue. No arcs strike so the electrode voltage is low by comparison. The total electrode current may be high, but the tissue current density (current per unit area) at all points on the electrode is moderate and the duration of the activation is therefore relatively long. The cellular effects are varied, ranging from a tough coagulum when connective tissue predominates to granular debris easily removed by a curet when the majority of the tissue is epithelial (21). Often the goal of coagulation is to shrink and fuse or thermally damage tissue collagen.

Minor Surgery. Minor surgery may be described as surgery applied to tissues on exterior surfaces of the body under local anesthetic, which includes dermatology. Other external surfaces, which include oral, vaginal and cervical tissues, are also routinely treated as minor surgery cases on an outpatient basis or in the office. Typical minor surgery procedures include the removal of warts, moles

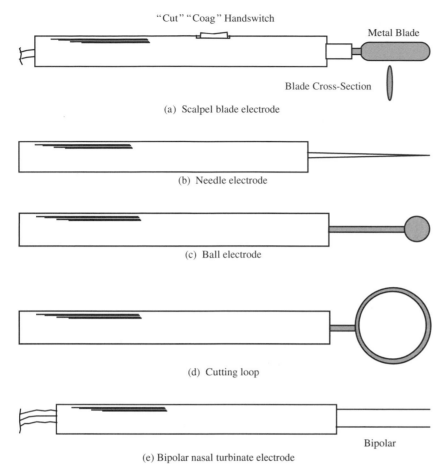

(a) Scalpel blade electrode

(b) Needle electrode

(c) Ball electrode

(d) Cutting loop

(e) Bipolar nasal turbinate electrode

Figure 1. Electrodes typically used for general electrosurgery.

and vascular nevi, the surgical reduction of sebaceous cysts, epiliation, cervical conization, relief of chronic nasal obstruction, and the removal of polyps and tumors. The techniques used in minor surgery are very similar to those of general surgery, although the cutting electrodes come in a wider variety of shapes and sizes.

Figure 1 illustrates some of the electrodes used (19). The standard scalpel blade electrode used for incision is elliptical in cross-section with a cutting edge which is of small radius (not sharp) in order to yield the very high current densities required for tissue ablation. The scalpel electrode is sometimes angled for special techniques. Scalpel electrodes are usually used for incisions, but may also be used for coagulation. The flat side of the blade can be applied to a tissue mass to obtain coagulation. The tip of the electrode may be suspended above the tissue for fulguration. Other electrode shapes accumulate less carbonized tissue residue than the scalpel, and are often preferred for coagulation. The needle, standard coagulation and ball electrodes are used either for desiccation or fulguration. The ball electrode may be used with coagulating current to treat persistent nose bleed that does not respond to other methods.

Bipolar forceps and the turbinate electrode are used in bipolar procedures. The forceps electrode has one electrical connection on each side of the forceps, and is used, for example, to grasp a seeping vessel; current between the forceps electrodes then seals off the vessel end by fusing the vessel walls together. The turbinate electrodes are used to obtain submucous coagulation (desiccation) in the nasal turbinates for relief of chronic vasomotor rhinitis: swelling of the soft tissue in the nasal cavity caused by, for example, allergies or irritants.

Neurosurgery and General Surgery. Blood is toxic to neural tissue. Consequently, blood loss during neurosurgery is to be avoided at all costs. At its inception, electrosurgery presented the first technique that accomplished cutting with negligible blood loss. This was this feature of electrosurgery that so strongly attracted Cushing. Many of the now commonplace neurosurgical procedures would be impossible without some method for obtaining hemostasis while cutting.

White coagulation is generally used in neurosurgery since it does not cause charring. White coagulation takes a relatively long time to obtain compared to fulguration or cutting. Holding the scalpel electrode to a location for long activation times allows deep penetration of the high temperature zone. This effect is used to advantage in rf lesion generation for selective disabling of neural tissue. Bipolar applicators restrict the current to a smaller region and are used extensively for microneurosurgical techniques, since they are more precise and safer. Many of the techniques are properly classed as microsurgery. Often the tissue being cut or coagulated is stabilized with a suction probe while current is applied. The suction probe is usually nonconductive glass or plastic; however, on occasion a metal suction probe is used with monopolar coagulation to localize the current to the immediate area (11). Incisions of the cerebral cortex are usually made with monopolar needle electrodes. Surface tumors are removed by applying a suction probe to the tumor and excising it at the base with a cutting loop.

Dental Electrosurgery. The tissues of the oral cavity are particularly highly vascularized. Also, the mouth and alimentary canal contain high concentrations of bacteria. Since the ability to ingest food is critical to survival these tissues are among the fastest healing of the body. Electrosurgery plays an important role in oral surgery in that it drastically reduces bleeding that would obscure the operative field and the undesirable postoperative effects of pain, edema and swelling in the submaxillary triangle (15). It can be quite difficult to accurately resect redundant tissue masses by cold scalpel techniques in the oral cavity. Electrosurgery allows accurate resection with minimal elapsed time and complications, an important feature when fitting prosthodontic devices. Electrosurgery reduces the hazard of transient bacteremia and secondary infection (14), and the danger of surgical or mechanical metastasis of malignant tumor emboli during biopsy (15). In short, all of the advantages obtained in other types of surgery are experienced in dental surgery as well as some additional beneficial aspects.

The active electrodes used in dental electrosurgery are for the most part similar to those shown in Fig. 2 (19). Several shapes specific to dental procedures are: the open hook electrodes in Fig. 2a are used along with other needle

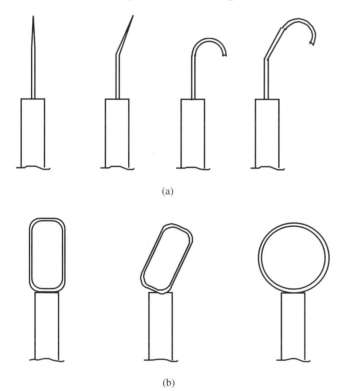

(a)

(b)

Figure 2. Electrodes typically used for dental electrosurgery. (a) Open hooks and needles, straight and angulated to reach difficult locations. (b) Straight and angulated cutting loops.

electrodes to create subgingival troughs, those in Fig. 2b are used with ball electrodes, and loop electrodes for seating orthodontic appliances and denture prostheses and exposing roots. Other interesting applications include the removal of premalignant neoplastic lesions from the surface of the mucosa using a planing loop and the removal of tissue masses from the tongue. Carefully applied electrosurgery can be used to advantage on teeth as well as the soft tissues of the oral cavity.

Urologic Surgery. Electrosurgery is used extensively in urologic procedures. Urologic procedures, specifically transurethral resections of the prostate, utilize by far the highest currents at high voltage for the longest durations and largest number of activations per procedure of any electrosurgical technique. Other urologic applications of electrosurgery include: the resection of bladder tumors, polyp removal by means of a snare electrode or desiccating needle, kidney resection to remove a stone-bearing pocket, and enlarging the urethral orifice in order to pass stones. These other procedures utilize power levels similar to those of general surgery.

A transurethral resection is intended to increase the caliber of a urethra that has been partially closed by an enlarged prostate gland. This procedure is one of the earliest applications of electrosurgery in urology, having been described in considerable detail by Kelly and Ward in 1932 (21). During transurethral resection, the urethra is irrigated with a nonconductive sterile solution, either dextrose or glycerine, while a cutting loop is advanced to remove the encroaching tissue. Surgical cutting accomplished in a liquid medium requires higher currents since the liquid carries heat away and disperses the current (even though it is nonconductive) more than a gaseous environment would. A typical resectoscope used in this procedure is shown diagrammatically in Fig. 3. A long outer metal sheath, which is plastic coated, contains a cutting loop that can be extended from the sheath, a fiber optic cable bundle for viewing the cutting action, an illumination source (either a bulb at the end or an illumination fiber optic bundle), and spaces for influx and efflux of irrigating fluid. The cutting loop (of thin diameter to yield the high current densities required) is moved in and out of the sheath by the surgeon as required to accomplish the resection.

Gynecologic Surgery. One of the common gynecologic applications of electrosurgery is cervical conization. Other common gynecologic applications of electrosurgery include the removal of tumors, cysts, and polyps. The use of electrosurgery in laparoscopic tubal ligations will be treated in some detail since it is also a very common procedure. The laparoscopic tubal ligation is a minimally invasive sterilization procedure typically accomplished by advancing an electrode through a small incision in order to coagulate the Fallopian tube. The position of the uterus is fixed by a cannula inserted through the cervix under the surgeon's control. The abdominal cavity is insufflated with CO_2 gas in order to separate the tissue structures. The coagulating electrode is then advanced through a small incision in the abdominal wall to the Fallopian tube by the surgeon,

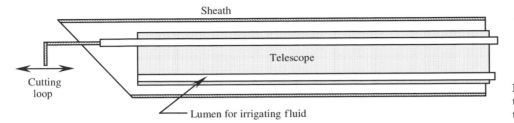

Figure 3. Resectoscope typical for transurethral resections of the prostate (TURP).

observing through an endoscope (laparoscope), which has been inserted through a separate small incision. The isolated Fallopian tube is then coagulated at high current but relatively low voltage.

Both monopolar and bipolar electrosurgical electrodes have been used for this procedure; however, monopolar tubal ligation methods are to be avoided as there have been many instances of bowel wall perforations and other complications following monopolar tubal ligation procedures owing to surgical current flow in the bowel wall. Bipolar techniques confine the current to a small region of tissue and the risk of bowel perforation is minimal in comparison. A bowel wall perforation can still result due to heat transfer from the coagulated tissue, so the coagulating forceps and Fallopian tube must be held away from the bowel wall and allowed to cool before being released. Note that a surrounding CO_2 gas environment will increase the time required for tissue cooling.

FUNDAMENTAL ENGINEERING PRINCIPLES OF ELECTROSURGERY

RF Generators for Electrosurgery

In general, the RF frequencies used for electrosurgery fall between 350 kHz and 4 MHz, depending on manufacturer and intended use. The available output power ranges from ~ 30–300 W. Peak open circuit output voltages vary from < 200 V to 10 kV. Higher open circuit voltages are used to strike longer arcs for fulguration, while the lower voltages are used for bipolar coagulation. Most devices are capable of generating several different waveforms, said to be appropriate for differing surgical procedures.

Vacuum Tube and Spark Gap Oscillators. The original rf generators used in electrosurgery, diathermy, radiotelegraphy, and radar circuits were spark gap oscillators (Fig. 4). The exponentially damped sine wavefrom (Fig. 4b) results from a breakdown of the spark gap (SG in Fig. 4a) that initializes the oscillations. The waveform is often called a Oudin waveform, though it is typical of all spark gap units, Oudin output circuit or not. The RFC is a RF choke to prevent the rf signal from coupling to the power line. Later generator designs utilized vacuum tube oscillator circuits that were typically Tuned-Plate, Tuned-Grid Oscillators (22), as shown in Fig. 5a (Birtcher Electrosectilis), or Hartley oscillators, Fig. 5b (Bovie AG). The output of vacuum tube electrosurgical units is available as either partially rectified [meaning that the RF oscillator is active only on one of the half-cycles of the mains power (i.e., one vacuum tube)] or fully rectified [meaning that the RF

oscillator is active on both half-cycles of the mains power (Fig. 5c)]. In both circuits of Fig. 5 each vacuum tube, V1 and V2, oscillates on opposite half cycles of the mains power, period T. Electrosurgery generator designs varied little from the standard units built by Bovie and Cameron-Miller in the 1920s until $\sim$1970 when solid-state generators became available. Solid-state generators made possible much more sophisticated waveforms and safety devices in a smaller overall package. Though not specific to solid-state technology, isolated outputs became common when solid-state electrosurgery units were introduced. Until $\sim$1995 all electrosurgery generators acted essentially as voltage sources with a typical output resistance in the

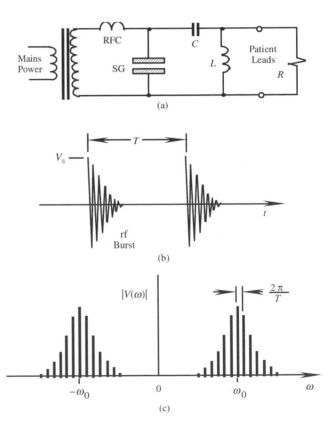

Figure 4. Spark gap oscillator rf generator. (a) Spark gap oscillator circuit. SG = spark gap, which acts as a switch, RFC = radio frequency choke, L and C determine the fundamental rf frequency, f_0 and R the damping (R = the patient). (b) Oudin waveform with amplitude determineed by the supply voltage peak, V_s. (c) Frequency spectrum of the output Oudin waveform has energy centered at $\pm_0$, the fundamental RF oscillation frequency with energy concentrated at harmonics of the repeat frequency to both high and low frequencies.

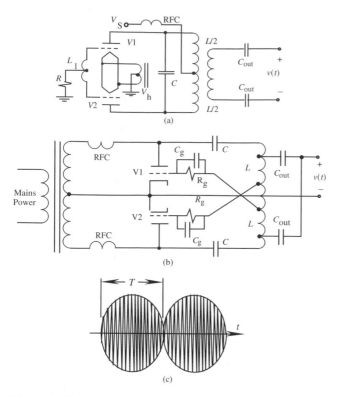

Figure 5. Vacuum tube electrosurgical circuits. (a) Tuned plate, tuned grid oscillator, as used in the Birtcher Electrosectilis. (b) Modified Hartley oscillator, as used in the Bovie AG. (c) Fully rectified output waveform.

neighborhood of 300–500 Ω. Recent generator designs incorporate embedded microprocessors and have constant power delivery modes. Interestingly, both Bovie and Cameron-Miller electrosurgical devices are available in the present day, though the designs are considerably different.

Solid-State Generators.   Solid-state electrosurgical generators generally operate on a different design principle than the vacuum tube devices. Rather than combine the oscillator and power amplifier into one stage, solid-state generators utilize wave synthesis networks that drive a power amplifier output stage. This approach has the advantage that quite complex waveforms may be employed

for cutting and coagulating; although to date the waveforms used vary little, if at all, from those of vacuum tube and spark gap oscillators. Many solid-state generators (chiefly those with bipolar transistors in the output amplifier) do not have as high open circuit output voltages as vacuum tube and spark gap devices. It sometimes appears to the user of those devices that there is less power available for cutting and coagulating since the lower open circuit voltages will not strike arcs as far away from the tissue. This turns out to be a limitation of concern in the high voltage procedures, namely in TURPs, but not in high current procedures, such as laparoscopic tubal ligations. In general, solid-state generators that use bipolar transistors in the high voltage output amplifier stage are vulnerable to transistor failure. The more recently introduced solid-state generators (after ∼ 1985) employ high voltage VMOS or HEXFET field effect transistors (23) in the output stage to give higher open circuit voltages and/or to reduce the stress on the bipolar output transistors.

A general block diagram of a typical solid-state electrosurgical generator is shown in Fig. 6 (24). The fundamental frequency, most often ∼ 500 kHz, is generated by a master oscillator circuit, typically an astable multivibrator. The primary oscillator acts as the clock or timing reference for the rest of the generator. The unmodified master oscillator signal is amplified and used for cutting. An interrupted waveform is formed by gating the continuous oscillator output through an external timing circuit, as shown in the figure. The repeat frequency of the timer is typically on the order of 20 kHz (24), much higher than that of spark gap oscillators. The duty cycle of a waveform is the ratio of duration of the output burst to the time between initiation of bursts. Duty cycles for solid state coagulating waveforms vary, but a duty cycle of 10–20% would be considered typical. This is in sharp contrast to the spark gap devices that have duty cycles often < 1%. The higher duty cycle of solid-state units compensates in part for their lower peak output voltages so the actual maximum available power is similar in both families of devices.

Constant power output is obtained by measuring the output voltage and current and adjusting the drive signal to compensate for changes in the equivalent load impedance (25), as in Fig. 7a. The sampling rate for this adjustment is on the order of hundreds of hertz (∼ 200 Hz for the device depicted). In Fig. 7b, the performance of the example system is compared to a standard voltage source generator

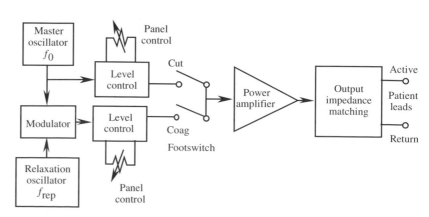

Figure 6. Block diagram for typical solid state ESU. Footswitch (and hand switch) controls simplified by omitting the interlock circuitry that prevents simultaneous activation. Master oscillator sets fundamental RF frequency, f_0. Interrupted waveform repeat frequency, f_{rep}. Provisions for blending cut and coag modes often provided. Power amplifier either bipolar junction transistors or HEXFET or VMOS transistors.

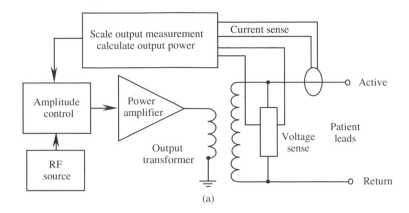

(a)

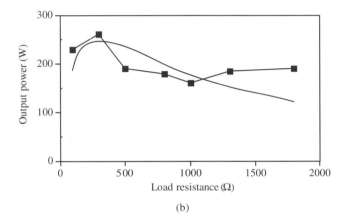

(b)

Figure 7. Constant power generator system. (a) Descriptive block diagram. (b) Typical performance (squares) compared to standard voltage source with fixed source voltage and equivalent output impedance (solid line), Nominal 250 W into 300 Ω (24).

with fixed output impedance 300 Ω and maximum output power of 250 W at 300 Ω load impedance. The differences at high load impedances typical of electrosurgical cutting (up to ~2 kΩ) are easily seen in the figure.

Safety Appliances. Since the electrosurgical unit (ESU) represents a high power source in the operating room, the potential for adverse tissue effects and accidental results must be managed. One of the more common types of accident has historically been one or more skin burns at the dispersive, or return, electrode site (during monopolar procedures). Often these have resulted from the return electrode losing its attachment to the skin. Earlier ESU designs incorporated a so-called circuit sentry, or its equivalent, which ensured that the return electrode cable made contact with the return electrode by means of monitoring current continuity through the cable and electrode connection at some low frequency. This ensured that the electrode was connected to the ESU, but did not ensure that it was connected to the patient.

Modern ESU devices monitor the patient return electrode to ensure its continued connection to the skin. In these devices, the patient return electrode is divided approximately into halves and a small current (typically on the order of 5 mA or less) at high frequency but below the surgical frequency (typically in the neighborhood of 100 kHz) is applied between the electrode halves, the connection being through the skin, only, as in Fig. 8 (25). The impedance between them is monitored, and if it exceeds maximum or minimum limits, or a set fraction of the baseline value (typically 40% or so over the baseline) an alarm sounds and the generator is deactivated. The patient return monitor system resets the baseline impedance if it falls below the initial value.

The high frequency of operation of electrosurgery units also contributes to the hazards associated with its use. At very high frequencies, current may flow in two different ways. First, in good conductors, such as skin and other wet tissues, the current flows by conduction and Ohm's law applies. Second, in insulating dielectric substances, such as air, surgical rubber or plastic, the current flows as so-called displacement current. The value of the displacement current density is linearly related to the frequency, so for the same materials, voltages and geometric arrangement of conductors, higher frequencies conduct more displacement current in dielectric substances than do lower frequencies. A consequence of this relationship is that at electrosurgical frequencies and voltages the capacitance between the wire conductor in a scalpel electrode and tissue over which the wire passes may have low impedance if the insulation is thin. Scalpel electrode wires are covered with thick insulation of high dielectric constant in order to prevent tissue damage at high scalpel electrode voltages. If the wire is inadequately designed or the insulation is damaged, the displacement current in the wire insulation may be dense enough to cause damage to the underlying tissue.

Electrosurgical unit outputs may be isolated, referred to ground, or grounded terminals (Fig. 9). Grounded patient return leads (Fig. 9a) have a hard wired connection to the ESU chassis ground wire. Referred to ground means that

Figure 8. Patient return monitor circuit ensures that the return electrode remains attached to the patient. Split return electrode conforms to patient contours; both halves of return electrode carry surgical rf current. Insulator prevents lower frequency interrogation current from finding a shorter pathway. Current pathway is through the patient's skin, ensuring contact.

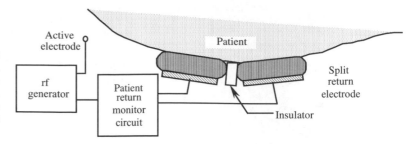

there is a fixed impedance, typically a capacitor, between the patient return lead and ground (Fig. 9b). The capacitance is chosen to represent a low impedance at the ESU rf frequency and a high impedance at power mains frequency and frequencies associated with stimulation of electrically excitable tissues. Isolated outputs (Fig. 9c) have high impedance to earth ground at both output terminals, usually by means of an output transformer; these are almost uniformly used for bipolar modes, and are common in monopolar modes as well. No isolation system is ideal,

however, and all rf currents should be thought of as ground seeking in some sense. In an isolated system, either patient lead can become a source of RF current.

A less obvious consequence of the high frequency is that every object in the room (the surgeon, the patient, the operating table and associated fixtures, the electrosurgical generator, other instrumentation) all have a small but finite parasitic or distributed capacitance to earth. This makes all of the RF currents in the operating room ground seeking, to some extent. Even the best isolated generator will have some ground leakage current at RF, and this value may be much more than the mains frequency leakage current (depending on the design of the output circuitry). Certainly, all of the grounded output generators can have significant differences between the active cable current and the return or dispersive electrode cable current. The difference current flows in all of the parallel pathways, conductive and/or capacitive. If any of these pathways carry too much current in too small an area, a thermal burn could develop. To the extent reasonable, direct conductive pathways through clamps, foot stirrups, and other metallic fixtures should be eliminated. This can be accomplished by using devices that have insulating covers over the pieces likely to contact the tissue, or insulated couplings, connectors or other barriers at some location between the tissue and the clamp which connects the device to the surgical table. Additional safety can be obtained by using monitors and other instruments that have RF isolation built into their electrode leads. These precautions and others will greatly reduce the hazards associated with alternative current pathways. There are International Electrotechnical Commission standards that cover the requirements for electrosurgical applicators, generator output connections, and isolation schemes (26,27). It is important to note that safe operation of electrosurgical devices can be obtained by more than one design strategy. The ESU, patient, and surroundings should be thought of as a system, in order to ensure a safe environment.

Representative Surgical Procedures

The output voltages and currents that are required of an electrosurgical generator depend on the particular procedure for which it is to be used. Fulguration requires high voltages to initiate the arcs, but not large currents, so a generator of high output impedance works quite well. Spray coagulation uses higher currents at the high voltages, the difference between spray coagulation and fulguration being one of degree rather than principle. White coagulation requires relatively high current at low voltage

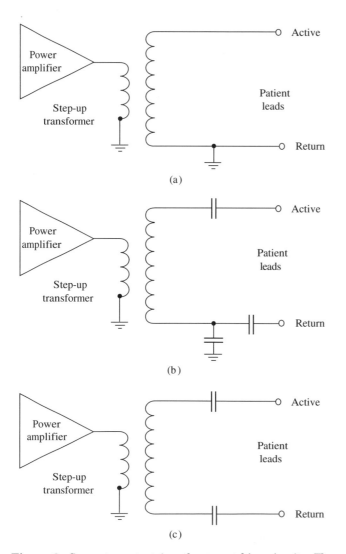

(a)

(b)

(c)

Figure 9. Generator output impedance matching circuits. The step up transformer is used to achieve higher open circuit output voltages. (a) grounded, (b) referred-to-ground, (c) isolated.

since no arc is formed at the scalpel electrode. The highest power, that is voltages and currents together, is required for transurethral resections of the prostate (TURP) procedures. The cutting characteristics of the various generator designs make them more or less optimal for certain procedures. Given the variety of designs available and variations in surgical technique among the users, it is no surprise to find that generator selection often boils down to personal preference.

The two modes used in electrosurgical procedures, bipolar, and monopolar, have quite different current field distributions. The bipolar mode is very effective in coagulating small vessels and tissue masses. Bipolar techniques are especially useful in microsurgical procedures. In bipolar electrosurgery, a two-wire coagulating electrode is clamped around a tissue lump or vessel and electrosurgical current is applied between the electrodes until the desired surgical effect is obtained. In monopolar electrosurgery, there are two separate electrodes, an active or cutting electrode and a large area dispersive or ground or return electrode located at some convenient remote site. Current is concentrated at the active electrode to accomplish the surgery while the dispersive or return electrode is designed to distribute the current in order to prevent tissue damage. The majority of applications of electrosurgery utilize monopolar methods, since that is the traditional technique and is most effective for cutting and excision. There are, however, many situations in which bipolar methods are preferred.

The power needed for a particular cutting or coagulating action depends on whether or not an arc is established at the scalpel electrode, on the volume of tissue, and on the type of electrosurgical action desired. Bipolar actions, which are typified by forceps electrodes grasping a volume of tissue to be coagulated, engage only a very small volume of tissue, since the current field is confined to be near the forceps, and usually white coagulation is desired. Consequently, the current is moderate to high and the voltage is low: tens to hundreds of milliamps (rms) at 40–100 V (rms), typically. In bipolar activations the current is determined primarily by the size of forceps electrodes used, and to a lesser extent by the volume of tissue. The volume of tissue more directly affects the time required to obtain adequate coagulation.

Monopolar actions, which may be for cutting or coagulation, are more variable and difficult to classify. In a study reported in 1973 in *Health Devices* (28), the voltages, currents, resistances, powers, and durations of activation during various monopolar electrosurgical procedures were measured. The resistance measured in this study was the total resistance between the active electrode cable and the dispersive electrode cable. Two types of procedure have significantly different electrical statistics compared to all other surgical cases: laparoscopic tubal ligations and transurethral resections of the prostate. The data in Table 1 have been grouped into general surgery (hernia repair, laparotomies, cholesystectomies, craniotomies etc.), laparoscopic tubal ligations, and TURPs. The table has been assembled from data collected at several different hospitals during procedures performed by different surgeons. For each separate surgical case, the minimum, average and maximum of each variable was recorded. The data in the table represent the means and standard deviations of the recorded minimum, average and maximum value calculated over the cases as originally presented in Ref. 19. While the total

Table 1. Typical Statistics for Representative Surgical Procedures in Which Electrosurgery Is Used[a]

	General Surgery (8 cases)	Laparoscopic Tubal Ligation (19 cases)	Transurethral Resection (8 cases)
Number of activations	22 (s.d = 24)	9 (7.5)	168 (151)
Voltage, V, rms			
Min	118 (58)	82 (20)	212 (43)
Avg	179 (57)	140 (65)	340 (12)
Max	267 (143)	207 (108)	399 (17)
Current, mA, rms			
Min	128 (29)	311 (108)	304 (109)
Avg	243 (116)	423 (73)	600 (30)
Max	423 (240)	615 (202)	786 (47)
Power, W			
Min	18 (7.5)	28 (9.3)	86 (41)
Avg	43 (25)	57 (19)	208 (15)
Max	103 (94)	99 (44)	290 (17)
Resistance, Ω			
Min	620 (720)	200 (50)	400 (40)
Avg	1070 (760)	410 (430)	580 (14)
Max	1960 (890)	660 (670)	1110 (360)
Duration of Activation, s			
Min	1	1	1
Avg	1.6 (1.0)	4.8 (3.3)	1.1 (0.12)
Max	3.6 (3.0)	13.6 (13.1)	2.9 (1.9)

[a]Data collected by ECRI and reported in *Health Devices*(28). Data given include the means of each variable and its associated standard deviation in parentheses. The raw data were given as the minimum, average and maximum value during a particular procedure. The mean is the mean of the recorded minimum, average, or maximum value over all procedures in the study. This table originally appeared in Ref. 19.

number of cases studied under each category is not large, the data do give an overall indication of the range expected in surgical cases.

On the average, laparoscopic tubal ligations required the fewest activations (the range was 5–29), and TURPs by far the most activations (the range was 70–469). The resistances presented by the series combination of the scalpel electrode and tissue were similar for all procedures at 410 Ω for laparoscopic tubal ligations (range 130–1080 Ω), 580 Ω for TURPs (range 340–1800 Ω) and 1070 Ω for general surgery (range 180–2650 Ω). Higher equivalent resistances correlate with arc formation during the cutting process, so the values associated with general surgery might reasonably be expected to be higher.

Ablation, Coagulation and Tissue Fusion

The electrosurgical unit is designed to create irreversible thermal alteration of tissues; that is, controlled thermal damage. The objective is to heat target tissues to temperatures for times sufficient to yield the desired result. All of the physical effects of rf current are the result of elevated temperatures. The key observation is that the degree of alteration depends on both the temperature and the time of exposure. This section describes tissue effects resulting from the rf current from lower to higher temperature ranges.

Kinetic models of thermal damage processes based on an Arrhenius formulation have been used for many years to describe and quantify thermal damage (29):

$$\Omega(\tau) = \int_0^\tau A e^{-[E/RT]} dt \qquad (1)$$

The dimensionless parameter, Ω is an indicator of the relative severity of the thermal damage. Thermal damage is a unimolecular reaction in which tissue proteins change irreversibly from their native ordered state to an altered damage state. In the kinetic model, A is a measure of the molecular collision frequency (s^{-1}), E is an energy barrier the molecules surmount in order to transform from native state to denatured state (J·mol^{-1}), R is the universal gas constant (J·mol^{-1}-K), T is the absolute temperature (K), and t is the time (s). The damage process coefficients, A and

E, must be experimentally determined. This model assumes that only a single first-order process is active: The model can be used on multiple-process damage accumulation if each process is thermodynamically independent with its own set of process coefficients. A damage process may be described by its critical temperature, T_{crit}, defined as the temperature at which $d\Omega/dt = 1.0$.

The physical significance of Ω is that it is the logarithm of the ratio of the initial concentration of undamaged material, $C(0)$, to the remaining undamaged material at the conclusion, $C(\tau)$:

$$\Omega(\tau) = \ln\left\{\frac{C(0)}{C(\tau)}\right\} \qquad (2)$$

However, typical damage end points have been qualitative tissue indicators such as edema formation or hyalinization of collagen (i.e., amorphous collagen as opposed to the normal regular fibrous array). One exception that has proved useful is the birefringent properties of some tissues, primarily muscle and collagenous structures. Birefringent tissue acts similarly to a quarter wave transformer in that polarized light has its polarization rotated as it passes through a regular array. Consequently, when observed through an analyzer filter rotated 90° with respect to the polarizer, birefringent tissue appears bright and nonbirefringent tissue dark (Fig. 10). In muscle, the birefringent properties are due to the regularity and spacing of the actin-myosin array. In collagenous connective tissues, it is the array of collagen fibers that determines birefringence. In both cases, elevated temperatures destroy the regularity of the array and birefringence is lost. In numerical models the transient temperature history of each point may be used along with equations 1 and 2 to predict the extent of thermal damage for comparison to histologic sections.

Ablation. Electrosurgical (RF) current, lasers, ultrasound, and microwaves have been used to obtain coagulative necrosis in myocardium *In vivo* for the elimination of ectopic contractile foci. The first meaning for ablation is to remove, as by surgery. The second meaning, to wear away, melt or vaporize is more familiar in the engineering sense (as in ablative heat transfer). In cardiac ablation, the goal

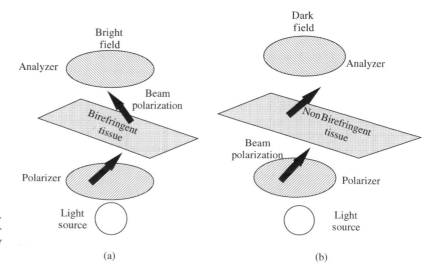

Figure 10. Principle of tissue birefringence. (a) Birefringent tissue is able to rotate the polarization (large arrow) of polarized light. (b) Thermally damaged tissue loses this property.

is to deactivate badly behaved cardiac muscle (reentrant pathways). Though no actual mass defects are created in the tissue structure, the process is termed "ablation" in the sense that the affected tissue is removed from the inventory of electrophysiologically active cardiac muscle. Results are evaluated in clinical use by monitoring electrophysiologic activity during treatment. Additional feedback may be applied to improve the repeatability of results.

Figure 11 illustrates the loss of birefringence at elevated temperatures in cardiac muscle. Figure 11a shows a circular disk active electrode applied to the epicardial surface

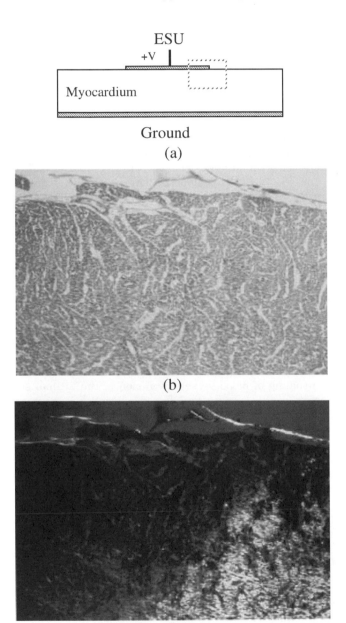

(a)

(b)

(c)

Figure 11. Disk electrode applied to myocardium. (a) Geometry. (b) Light microscopic view of region in dashed rectangle at the edge of the electrode. Original magnification 40×. (c) Transmission Polarizing Microscopy (TPM) view of the same section showing the clear delineation of the zone of birefringence loss. Original magnification 40×.

of excised myocardium. The ground plane is applied to the endocardial surface. Figure 11b is a light microscopic (LM) view of the histologic section at an original magnification of 10× stained with hematoxylin and eosin. Figure 11c is the corresponding Transmission Polarizing Microscopy (TPM) view of the same section. The views are taken from the region shown in dashes at the outer edge of the disk electrode. While the boundary of thermal damage can just be identified by a skilled observer in the LM image (Fig. 11b), a clear line of demarcation is visible in the TPM image (Fig. 11c). For heating times in the range of 1–2 min useful estimates of the kinetic coefficients are $E = 1.45 \times 10^5$ (J·mol^{-1}) and $A = 12.8 \times 10^{21}$ (s^{-1}). These coefficients give erroneous predictions for heating times outside of this range, however.

Birfringence loss identifies irreversible major structural damage in the cardiac myocyte. It is certainly arguable that electrophysiologic function is probably lost at lower temperatures than the 50+ °C required to achieve the result shown in the figure. However, clinically, the 50 °C isotherm seems to correspond with the desired result (30), and also with the birefringence loss boundary for heating times in this range (31).

Tissue Fusion and Vessel Sealing. Irreversible thermal alteration of collagen is apparently the dominant process in coagulation and successful tissue fusion (32–34). Electron microscopic (EM) studies suggest that the end-to-end fusion process in blood vessels is dominated by random reentwinement of thermally dissociated adventitial collagen fibrils (Type I) during the end stage heating and early cooling phases (34). Successful vessel seals depend on the fusion of intimal surface tissues and the support provided by thermal alteration of medial and adventitial tissues.

Collagen is a ubiquitous tissue structural protein consisting of three left-hand α-helices wound in a rope-like molecular form (Fig. 12) with a periodicity of ∼68 nm (35). In this figure a small segment of a typical 300 nm long molecule is shown. The 300 nm molecules spontaneously organize into quarter-staggered microfibrils 20–40 nm in diameter and these coalesce into larger diameter collagen fibers In situ. There are at least 13 different types of collagen fibers that form the structural scaffolding for all tissues, the most common are Types I, II, and III collagen. Collagen In situ is birefringent. One useful measure of irreversible thermal alteration in collagen is that when heated for sufficient time to temperatures in excess of ∼60 °C the regularity of the fiber array is disrupted and collagen loses its birefringent property (see Fig. 13). When viewed through the analyzer, native state collagen shows up as a bright field due to its birefringent properties. Thermally damaged collagen loses this property and is dark in the field. The kinetic coefficients for collagen birefringence loss in rat skin are (36): $A = 1.606 \times 10^{45}$ (s^{-1}) and $E = 3.06 \times 10^5$ (J·mol^{-1}). The birefringence-loss damage process in collagen has a critical temperature of 80 °C. These coefficients have proven useful over a wide range of heating times from milliseconds to hours.

Collagen shrinks in length as well as losing its organized regular rope-like arrangement of fibers. A model for

Figure 12. Sketch of periodic structure of the basic collagen molecule.

collagen shrinkage obtained by Chen et al. (37,38) is a bit different in style from the first-order kinetic model. They measured shrinkage in rat *chordae tendonae* over several orders of magnitude in time for temperatures between 65 and 90 °C and under different applied stresses. They were able to collapse all of their measured data into a single curve, sketched approximately in Fig. 14. In their experiments an initial "slow" shrinkage process (for equivalent exposure time $< \tau_1$) is followed by a rapid shrinkage rate ($\tau_1 < t < \tau_2$) and a final slow shrinkage process. The practical maximum for shrinkage in length, ξ (%), is 60%. Longer equivalent exposures result in gellification of the collagen and complete loss of structural properties. After initial shrinkage, the collagen partially relaxes during cooling, indicated by the shrinkage decay region in Fig. 14. The curve fit functions utilize a nondimensional time axis, t/τ_2, where the fit parameters are expressed in the form of the logarithm of the time ratio:

$$\nu = \ln\left\{\frac{t}{\tau_2}\right\} \tag{3}$$

The shrinkage is obtained by interpolation between the two slow region curves (through the fast region):

$$\xi = (1 - f(\nu))[a_0 + a_1\nu] + f(\nu)[b_0 + b_1\nu] \tag{4}$$

where $a_0 = 1.80 \pm 2.25$; $a_1 = 0.983 \pm 0.937$; $b_0 = 42.4 \pm 2.94$; and $b_1 = 3.17 \pm 0.47$ (all in %). The best-fit interpolation function, $f(\nu)$, is given by

$$f(\nu) = \frac{e^{a(\nu - \nu_m)}}{1 + e^{a(\nu - \nu_m)}} \tag{5}$$

where $a = 2.48 \pm 0.438$, and $\nu_m = \ln\{\tau_1/\tau_2\} = -0.77 \pm 0.26$. Finally, at any temperature τ_2 is given by

$$\tau_2 = e^{[\alpha + \beta P + M/T]} \tag{6}$$

where $\alpha = -152.35$; $\beta = 0.0109$ (kPa^{-1}); $P =$ applied stress (kPa); and $M = 53,256$ (K).

The functional form of τ_2 contains the kinetic nature of the process, but is in the form of an exposure time rather than a rate of formation, as was used in Eq. 2, and so the coefficient, M, is positive. To use the collagen shrinkage model, the shrinkage is referred to an equivalent τ_2. That is, at each point in space and time an equivalent value for the increment in t/τ_2 is calculated and accumulated until shrinkage is calculated.

A representative clinical application of collagen shrinkage is the correction of hyperopia using rf current, termed Conductive Keratoplasty. In this procedure a small needle electrode is inserted into the cornea to a depth of just over one-half of the thickness (see Fig. 15a). A teflon shoulder controls the depth of insertion. The speculum used to retract the eyelids comprises the return electrode in this procedure (Fig. 15b). The RF lesions are placed at 45° increments on circumferences with diameters of 6, 7, or 8 mm—8, 16, or 24 lesions depending on the degree of curvature correction (i.e., diopter change) required (Fig. 15c). Pulsed RF current heats and shrinks the corneal collagen to decrease the circumference and thus the radius of curvature of the cornea. Figure 16 is a histologic cross-section of a typical lesion seen a few minutes after its creation. Figure 16a is a light microscopic section (hematoxylin and eosin stain) and Fig. 16b is the corresponding transmission polarizing microscopy section showing the loss of collagenous birefringence near the electrode path. The effect of shrinkage on the collagen fibers is clearly visible as the normal fibrous wave is stretched in the vicinity of the electrode site (Fig. 16b).

A representative example of tissue fusion processes is the sealing of blood vessels by fusion of the intimal surfaces. In this application, forceps electrodes grasp the vessel and a bipolar rf current is used to heat and fuse the tissue (Fig. 17). An example experimental result is shown in Fig. 18, where a successful seal was obtained *In vivo* in a femoral artery. In that experiment a thermocouple was advanced through the vessel to the site of the electrodes, accounting for the hole in the cross-section.

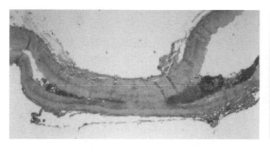

(a)

(b)

Figure 13. Vessel collagen birefringence loss. (a) Thermally fused canine carotid artery, H&E stain (Original magnification 40×). (b) TPM view of same section showing loss of birefringence in adventitial collagen under bipolar plate electrodes.

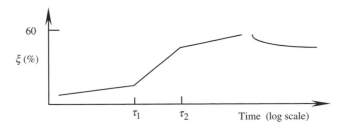

Figure 14. Collagen shrinkage model curve (38). [Reproduced with permission from "Corneal reshaping by radio frequency current: numerical model studies" Proc. SPIE v4247 (2001) pp 109–118.]

Sealing and fusion of vessels by electrosurgical current is strongly influenced by the inhomogeneous architecture of the tissue constituents, particularly in the large arteries. Inhomogeneities in electrical properties of the constituents, specifically smooth muscle, collagen and elastin, lead to sharp spatial gradients in volumetric power deposition that result in uneven heating. The mechanical properties of the various tissue constituents are also of considerable importance. Vascular collagen and elastin distribution varies from vessel to vessel, species to species in the same vessel, and point to point in the same vessel of the same species.

Cutting Processes

The essential mechanism of cutting is the vaporization of water. Water is by far the most thermodynamically active tissue constituent. Its phase transition temperature near 100 °C (depending on local pressure) is low enough that vaporization in tissue exerts tremendous forces on the structure and underlying scaffolding. The ratio of liquid to vapor density at atmospheric pressure is such that the volume is multiplied by a factor of ~1300 when liquid water vaporizes. Expansion of captured water is capable of disrupting tissue structure and creating an incision. The goal in electrosurgical cutting is to vaporize the water in a very small tissue volume so quickly that the tissue structure bursts before there is sufficient time for heat transfer to thermally damage (coagulate) surrounding tissues (39). The same strategy is used in laser cutting.

Cutting electrodes have high rates of curvature to obtain the high electric fields necessary to accomplish cutting. The electrodes are not generally sharp, in a tactile sense, but the edge electrical boundary conditions are such that extremely strong electric fields result. With the possible exception of needle electrodes, the scalpel electrodes of Figs. 1 and 2 are not capable of mechanically damaging tissues. The intense electric fields vaporize tissue water

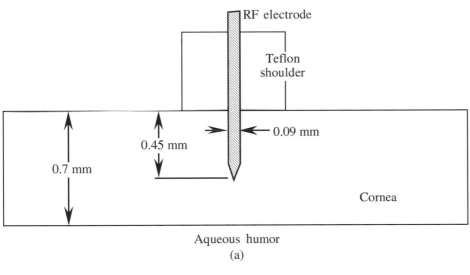

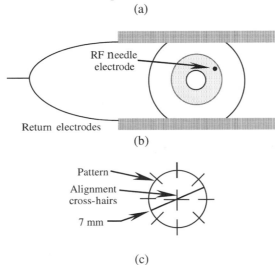

Figure 15. Needle electrode used to shrink corneal collagen and change curvature for correction of hyperopia. (a) Cross-section of electrode in place, (b) View of speculum return electrodes, (c) spot pattern for shrinkage lesion location. [Reproduced with permission from "Corneal reshaping by radio frequency current: numerical model studies" Proc. SPIE v4247 (2001) pp109–118.]

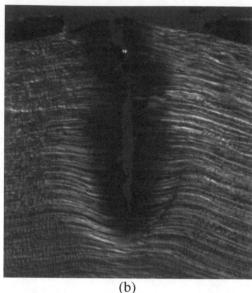

(a)   (b)

Figure 16. Histologic view of collagen shrinkage for Conductive Keratoplasty. (a) Light microscopic view, hematoxylin, and eosin stain. (a) The TPM view, same section. Original magnification 40×.

ahead of the electrode, and the tissue structure parts due to internal tension to allow the electrode to pass essentially without tissue contact. Certainly, if the surgeon advances the electrode too quickly there is a dragging sensation due to perceptible friction, and much more radiating thermal damage around the incision site.

Good cutting technique requires matching the generator output characteristics (primarily the source impedance, but also to some extent the fundamental frequency), open circuit voltage setting, electrode shape and cutting speed to optimize the surgical result. Continuous sine waves are more effective for cutting than interrupted waveforms since the time between vaporization episodes is so much shorter that radiating heat transfer is virtually eliminated. Well-hydrated compartmentalized tissues, like skeletal muscle or gingiva, cut more cleanly than drier tissues, such as skin. Higher fundamental frequencies seem to give cleaner cuts than lower frequencies, but the reason for this is not known.

ADVANCED PRINCIPLES OF ELECTROSURGERY

Hazards of Use and Remedies

Any energy source, including electrosurgery, has hazards associated with its use. The goal of the user is to achieve safe use of the device by minimizing the risk. While it is not possible to completely eliminate hazards, it is possible by

careful technique to reduce them to acceptable levels. Several of the common hazards associated with electrosurgery are alternate site burns, explosions, stimulation of excitable tissues, and interference with monitoring devices and pacemakers. The RF currents are designed to thermally damage tissues, so the possibility of alternate site burns (at sites other than the surgical site) is always present. Explosions of combustible gases were, and still are, a hazard of electrosurgery. Combustible gases include at least two relatively little-used anesthetic agents (ether and cyclopropane) and bowel gas, which has both hydrogen and methane, as a result of bacterial metabolism. Arcs and/or sparks are routinely generated when cutting tissue so there is always an ignition source for a combustible mixture. While the fundamental frequency of electrosurgery is above the threshold for stimulation, arcs or sparks generate low frequency components that can stimulate electrically excitable tissues, that is, muscle and nerve. Radiated rf in the operating room and rf currents in the tissues posed few problems in early days, unless a faulty ground or other machine fault led to a burn at some alternate site. In the present day, electrosurgery raises considerable havoc with instrumentation amplifiers, pacemakers, and other electronic measurement

Figure 18. Histologic cross-section of a vessel sealing experiment in the canine femoral artery. Successful seal obtained with a center temperature of 89 °C for 2 s. Hole in center of section is location of thermocouple used to monitor temperature. Original magnification 40×.

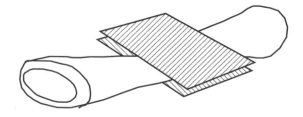

Figure 17. Bipolar forceps electrodes for vessel sealing.

devices. Because so many potentially current-carrying objects and devices are now routinely attached to the patient, the attendant hazards have increased.

The remarkable feature of electrosurgical accidents is that, although they are rare (one estimate puts the probability at $\sim 0.0013\%$, or 200 out of 15 million procedures using electrosurgery, on an annual basis) they are usually a profound and traumatic event for the patient and the surgical team, and often cause for expensive litigation. The non-surgeon might reasonably wonder, in the light of the problems, why electrosurgery is used. The answer is that the tremendous advantages achieved by the use of electrosurgery (the remarkable reduction in morbidity and mortality and the low cost compared to competing technologies) make it a very effective technology for use in clinical medicine. It is important to note that hazards attend any energy source (and electrosurgery is one among several in the operating room) and they can be reduced and managed, but not eliminated. Clever application of appropriate technology can make considerable contributions in this area by reducing device interactions.

Alternate Site Burns

Electrosurgical units in general have very high open circuit output voltages that may appear at the scalpel electrode when it is not in contact with the tissue. When current is flowing the scalpel voltage is much reduced. Standard vacuum tube electrosurgery units may have open circuit voltages approaching 10,000 V peak-to-peak. This high open circuit output voltage is very effective in the initiation of fulguration techniques and in the spray coagulation of large tissue segments, which makes these units popular for certain procedures, especially in urologic surgery. However, the high voltages at the scalpel electrode also can initiate arcs to other objects, and must be handled with caution. This can be an especially difficult problem in minimally invasive surgery through an endoscope since the surgeon's field of view may be limited. The solid-state electrosurgery units usually have lower maximum output voltages (somewhere in the range of 1000–5000 V peak-to-peak depending on design) the exception being recent designs based on HEXFET or VMOS technology, which approach the open circuit voltage of vacuum tube devices.

All electrosurgery units have output voltages that are potentially hazardous.

It is prudent to inspect all surgical cables periodically for damage to the insulation: especially those that are routinely steam sterilized. While in use, it is not a good idea to wrap the active cable around a towel clamp to stabilize it while accomplishing cutting or coagulation: The leakage current to the towel clamp will be concentrated at the tips of the towel clamp and may cause a burn. One should not energize the electrosurgical unit when the scalpel is not being used for cutting or coagulating. This is because when not using electrosurgery, the full open circuit voltage is applied between the active and return electrode cables, and one runs the risk of inadvertent tissue damage. Care should be taken to ensure that an unused scalpel electrode is not placed in a wet environment during surgery. The leakage current of any scalpel electrode may be increased by dampness. Additionally, some of the hand control designs can be activated by moisture at the handle. In one recent case, a hand control scalpel was placed on the drape in between activations. The scalpel was in a valley in the drape sheet in which irrigating fluid collected. The pooled fluid activated the electrosurgery machine and a fairly severe steam or hot water burn resulted on the patient's skin.

The high voltages and frequencies typical of electrosurgery make it essential to use apparatus with insulation of good integrity. It should also be kept in mind that the electric fields extend for some distance around cables in use at high voltage. An effort should be made not to have active and return electrode cables running parallel to each other for any distance in order to reduce the capacitive coupling between them, unless the apparatus is specifically deigned to be used this way, as in bipolar device cables.

There is an additional problem in monopolar endoscopic surgery that must be addressed. It arises when a metal laparoscope with an operating channel is isolated from a voltage reference plane, as depicted in Fig. 19. In this case, the insulating skin anchor isolates the metal laparoscope (or metallic trochar sleeve) so that its potential is determined by parasitic capacitances between the metallic elements. The resulting circuit creates a capacitive voltage divider, diagrammed in Fig. 20. The two parasitic capacitances are unavoidable. The voltage divide ratio for this

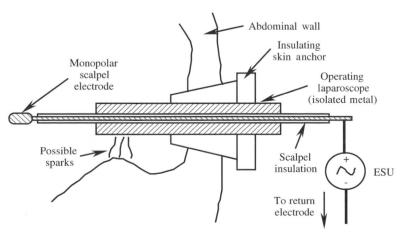

Figure 19. Diagram of isolated laparoscope with active operating channel.

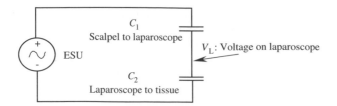

Figure 20. Equivalent capacitive voltage divider circuit for isolated laparoscope with active operating channel.

equivalent circuit is

$$V_L = \frac{C_1}{C_1 + C_2} \qquad (7)$$

The parameter C_1 is always quite a bit larger than C_2, and the surface potential on the laparoscope, V_L, may be as much as 80–90% of the surgical voltage. This induced voltage is capable of generating sparks to the surrounding tissues out of the field of view of the surgeon. Several cases of distributed severe burns have resulted from situations similar to this. There are two remedies. First, the operating laparoscope may be grounded. This does not eliminate the C_1 capacitance, so there will be capacitively coupled currents between the scalpel electrode and the laparoscope. However, the laparoscope will be at zero potential. The second approach involves monitoring the scalpel electrode and return electrode currents. If they are substantially different then a hazard situation exists. There is at least one commercial device designed to do this.

Explosions. The original explosion hazard in the operating room arose from the use of explosive anesthetic agents, such as ether and cyclopropane. Combustible mixtures of these gases can be ignited by any arc discharge of sufficient energy—by static discharges between personnel and grounded objects, by arcs due to power switch openings in mains-powered devices, and by electrosurgical arcs at the cutting site. A few early instances of explosions and ensuing fires in operating rooms stimulated the establishment of requirements for operating room construction and use to lessen this hazard. In particular, operating rooms in the United States were required to have semiconducting flooring that was to be checked periodically for conductivity; operating room personnel were required to wear conductive shoe covers that would drain off accumulated static charge to the conductive flooring; and explosion-proof switches and plugs were required below the five foot level where the relatively heavy combustible gases might collect. Also, surgical table coverings and fixtures are connected to ground through moderate impedance in order to prevent static accumulation. These requirements greatly reduced the explosion risk in the operating room, but lead to some rather amusing *non sequiturs*. For example, it makes little sense to attach an explosion-proof foot switch to an electrosurgical generator (particularly to a spark gap oscillator) since arcing is inherent to the use of electrosurgery. Of course, since the wall plugs for mains power were required to be explosion proof, the power line plug on the generator was also often explosion proof. Because of the inordinate expense of explosion proof construction and the

rare use of combustible gases, the majority of new operating room construction is designed to satisfy the requirements for oxygen use only and specifically designated not for use of explosive gases.

There are, however, other explosion–combustion sources. Two constituents of bowel gas are combustible, hydrogen and methane, and there is almost always sufficient swallowed air to comprise an explosive mixture. Approximately 33% of the adult population are significant methane producers due to the indigenous gut flora and fauna responsible for certain aspects of digestion. One preventive measure that has been used is insufflation of the bowel with carbon dioxide, or some other inert gas, during electrosurgery. A large volume of CO_2 will dilute and displace the combustible mixture, reducing the chance of explosion.

Additionally, care must be taken in pure oxygen environments not to ignite organic materials. Body hair has been ignited in a pure oxygen environment by an electrosurgical arc. In surgery on the larynx, both electrosurgery and CO_2 lasers have ignited plastic endotracheal tubes and surgical gloves in the oxygen-rich environment.

Stimulation of Electrically Excitable Tissues. The stimulation of motor nerves is a well-known phenomenon accompanying electrosurgery. Stimulation of the abdominal muscles is often observed, and the obturator nerve bundle (connecting the spinal column to the leg) is also quite often stimulated. Obturator nerve stimulation is inherent to transurethral resections, and the resulting movement is always undesirable. Part of the stimulation problem arises from the low frequency components of the surgical arc, and part from the confinement of current. For example, the obturator nerve is clustered with the major blood vessels supplying the leg (iliac/femoral artery and vein) and emerges from the pelvis under the inguinal ligament into the upper thigh area. The pelvic bone structures are essentially nonconductive and impede current flow. The nerve–artery–vein cluster represents a relatively high conductivity pathway of small cross-section between the abdomen and the upper thigh, a typical location for the return electrode. As such, the surgical current in TURP procedures is somewhat concentrated in that tissue cluster, and stimulation is often observed. In the case of general abdominal surgery, the majority of surgical current flows in the surface tissue structures, reducing the likelihood of obturator stimulation. In the case of transurethral resections, the current source is deep in the floor of the abdomen, and the surface tissues carry a smaller fraction of the total surgical current. The very high currents in TURPs and the frequent use of spark gap oscillators means that an abundant supply of intense low and high frequency signals is contained in the surgical current. It is a personal observation that motor unit stimulation is closely correlated with relatively intense arc strikes while accomplishing cutting or coagulating. This is to be expected from the frequency spectrum of the surgical signals.

There are, in addition, several reported instances of ventricular fibrillation induced by electrosurgery (40–43). The low frequency arc components and high intensity high frequency signals are capable of stimulating excitable tissues, with ventricular fibrillation as a possible outcome.

Interference with Instrumentation. The cutting current generated by standard oscillators, such as the vacuum tube and solid-state devices, has a well defined, narrow bandwidth. When an arc is struck at the surgical site, extensive signal energy is added at the high frequency end and at the low frequency end of the spectrum. Signals from spark gap oscillators have these components whether or not an arc is established at the tissue. Over the range of physiological frequencies the arc components are of very low amplitude compared to the generator fundamental frequency; but since the surgical currents and voltages are very high, the arc components may be many times larger than physiological signals such as the ECG. This creates a considerable problem for all measuring instruments including demand, or noncompetitive, pacemakers. There are many reported instances of inhibition of demand pacemakers by electrosurgery (44–47).

The population most at risk for pacemaker inhibition is patients undergoing either open heart surgery or transurethral resections of the prostate. Open heart operations put the electrosurgical scalpel electrode in close proximity to the pacemaker electrodes. The TURPs use the highest currents and voltages, and the interference signals are potentially large at the pacemaker electrodes. There is some indication that a higher incidence of prostate problems is associated with long-term pacemaker implantation, so it is to be expected that this problem will continue. The present use of shielded pacemakers virtually eliminates microwave sensitivity, but not electrosurgical effects.

All pacemakers may be damaged to failure by electrosurgery signals, and some caution should be observed when performing electrosurgery in the presence of pacemakers. Bipolar electrosurgery is much less likely than monopolar to cause interference with pacemakers because the surgical current field is confined and the voltages are low. There are many techniques for which bipolar electrosurgery is unacceptable, however, so monopolar methods must be used. In those cases, care should be exercised to encourage the surgical current field to avoid the right ventricle. To the extent that is practical, elevate the pacing electrode site (either the right ventricular wall or the external surface) away from the conductive fluids and tissues when cutting near the heart. Place the dispersive electrode on the left side of the body (this will only help a little bit). Anticipate some form of pacemaker malfunction.

Representative Electric Field Distributions

The effect of rf current on tissues is to heat them according to the first law of thermodynamics:

$$\rho_t c_t \frac{\partial T}{\partial t} = \nabla \bullet (k \nabla T) + q_{gen} + q_m + w_t c_b (T_a - T) - h_{fg} \frac{\partial m}{\partial t} \tag{8}$$

where ρ = density (kg·m^{-3}), c = specific heat (J·kg^{-1}·K^{-1}), T = temperature (K), t = time (s), k = thermal conductivity (W·m^{-1}·K^{-1}), q_{gen} = externally applied rf power (W·m^{-3}), q_m = metabolic heat (W·m^{-3}), w = perfusion

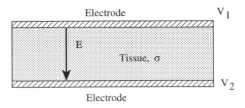

Figure 21. Geometry for infinite flat plate electrode calculation.

(kg$_b$·kg$_t^{-1}$·s^{-1}), h_{fg} = vaporization enthalpy (J·kg^{-1}) and dm/dt is the mass rate of vaporization (kg·s^{-1}). Each term in the relation has units of (W·m^{-3}) and the t subscript refers to tissue, while the b subscript refers to blood properties. This form does not include vaporization processes typical in cutting. Typically, both metabolic heat and perfusion heat transfer are negligible in electrosurgery. It is important to note that all of the tissue properties are dependent on the water content. As tissue dries out the density, electrical, and thermal conductivity decrease. As temperature rises the electrical conductivity increases $\sim$ 1–2%/$^\circ$C, until drying at higher temperatures causes a decrease.

At electrosurgical frequencies the quasistatic assumption applies (simplifying calculations) and heat dissipation is dominated by joule, or resistive, heating (W·m^{-3}):

$$q_{gen} = \mathbf{E} \cdot \mathbf{J} = \sigma |E|^2 = \frac{|J|^2}{\sigma} \tag{9}$$

where $\mathbf{E}$ = electric field vector (V·m^{-1}), $\mathbf{J}$ = current density vector (A·m^{-2}), and σ = electrical conductivity (S·m^{-1}). This section reviews several representative electric field distributions.

Simple Field Geometries: Analytical Solutions. The simplest case is that of the electric field between parallel plates (Fig. 21). For a uniform medium between the plates the electric field is uniform (i.e., V increases linearly from V_2 to V_1) and is equal to the voltage difference divided by the distance between the plates: $(V_1 - V_2)/d = \Delta V/d$. It points from the higher potential (V_1) to the lower potential (V_2). In this case the heating term, q_{gen}, is also simple: $q_{gen} = \sigma \Delta V^2/d^2$. For an operating voltage difference of 50 V(rms) and plate separation distance of 2 cm, the electric field is 2500 (V·m^{-1}), and if the electrical conductivity is 0.3 (S·m^{-1}) the power density would be 1.88 W·cm^{-3}.

Another simple analytical solution is for the case of coaxial cylinders (Fig. 22). In this geometry the potential increases as the natural log of the radius from V_2 to V_1, and the electric field (pointing in the radial direction) is given by

$$E = \frac{I_L}{2\pi\sigma r} \quad a_r \tag{10}$$

where I_L = the current per unit length of coaxial structure (i.e., length into the page, A·m^{-1}). For this geometry the rf power generation decreases as $1/r^2$:

$$q_{gen} = \mathbf{E} \bullet \mathbf{J} = \sigma \left[\frac{I_L}{2\pi\sigma} \right]^2 \frac{1}{r^2} \tag{11}$$

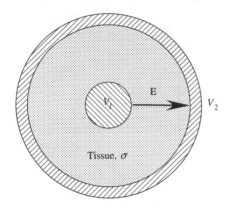

Figure 22. Geometry for coaxial cylinder electrode example. The outer cylinder typically comprises the return electrode, $V_2 = 0$.

For a center cylinder electrode 2 mm in diameter and outer cylinder 2 cm in diameter and tissue electrical conductivity 0.3 (S·m^{-1}), the overall conductance per unit length would be $G = 0.816$ (S·m^{-1})— that is, $G = 2\sigma/(\ln\{b/a\})$. At a voltage difference of 50 V(rms), the current per unit length would be $I_L = 40.9$ (A·m^{-1}) and maximum electric field 21.7 (kV·m^{-1}) at $r = 1$ mm. The maximum volume power density would be 142 W·cm^{-3} at $r = 1$ mm in this example.

These examples are useful in understanding simple surgical fields. The uniform electric field case is not very different from that obtained near the center of a bipolar forceps electrode used to coagulate a tissue volume. The coaxial case is close to the electric field around a needle electrode as long as one is not too close to the tip. In both cases the analytical field expressions are simple because the boundaries are parallel to Cartesian and cylindrical coordinates, respectively. Also, the boundaries are chiefly isopotential, or Dirichlet, boundaries.

Disk Electrode Field. The electric field around a disk electrode applied to a large volume of tissue is a much more complex geometry (Fig. 23). This geometry was solved analytically by Wiley and Webster (48) for the $V = 0$ reference electrode located at $r = +\infty$ and $z = -\infty$. The electrode is an isopotential surface ($\partial V/\partial r = 0$) and the air is insulating (zero flux, $\partial V/\partial z = 0$); and at the edge of the disk ($r = a$) the field must satisfy two mutually exclusive boundary conditions. The consequence is that the electric field and current density both increase without bound at

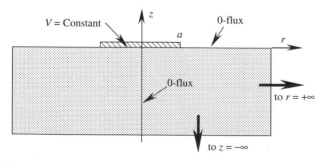

Figure 23. Solution geometry for disk electrode applied to infinite medium. Return electrode, $V = 0$, located at $r = \infty$ and $z = -\infty$.

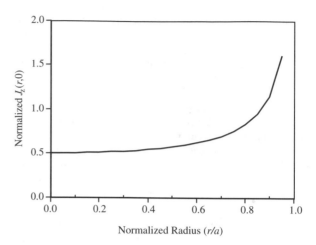

Figure 24. Normalized surface current density under the electrode from $r = 0$–$0.95a$.

the electrode edge, $r = a$ (48):

$$J_z(r, 0) = \frac{J_0}{2\left[1 - (r/a)^2\right]^{1/2}} \tag{12}$$

where $J_0 = I/\pi r^2$, the average current density. Figure 24 plots the normalized surface current ($J_0 = 1$) density versus normalized radius under the electrode. For radii less than $\sim 0.89a$ the current density is less than the average value. This is at first glance an alarming result: It is not obvious why burns do not result at the edge of every electrode. There are three mitigating factors. First, the singularity at $r = a$ is an integrable singularity—meaning that one may integrate $J_z(r, 0)$ from $r = 0$ to $r = a$ and get a finite result (i.e., integral $\{J_z \, dA\} = I$, the total current). Second, actual electrodes do not have mathematically sharp edges, but rather are filleted. Third, heat transfer dissipates the temperature rise at the edge and even though q_{gen} increases without bound (in a microscopic sense) the temperature rise at the edge reaches a finite maximum—still much higher than that at the center of the electrode, however.

The electric fields near a disk electrode, such as that used to ablate myocardium (Fig. 11) is very similar to the analytical solution. Figure 25 is the result of a quasi static Finite Element Method (FEM) calculation for the geometry used to create Fig. 11. The assumed electrode potential is 50 V(rms), and the isopotential lines (25a) are not significantly different in shape from those of the analytical solution (48). The electric field (current density lines) lines (Fig. 25b) are very crowded at the edge of the electrode, as expected from the above discussion. The result is that the volume power density, q_{gen}, (Fig. 25c) is higher by a factor of 1000 or more (see the 12 W·cm^{-3} contour) than about one electrode radius away (cf. to the 0.062 W·cm^{-3} contour).

Needle Electrode Field. Needle electrodes are often used to desiccate or coagulate small lesions. They are also used to shrink corneal collagen to change its curvature, as was discussed above (in the section Ablation, Coagulation and Tissue Fusion). The electric field around the tip of such an electrode is extremely high compared to the average value.

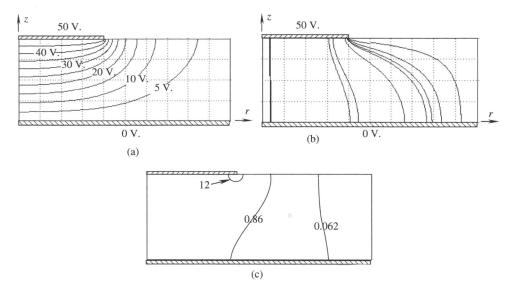

Figure 25. Circular disk electrode FEM model results. Muscle conductivity 90.3 S·m^{-1}. (a) Potential field, $V(r,z)$. (b) Electric field streamlines. (c) Volume power density with contours at 0.062, 0.86, and 12 W·cm^{-3}.

For comparison, the electric field and corresponding volume power density field were calculated around a needle electrode in cornea: Cornea electrical conductivity, at 1.125 S·m^{-1}, is quite a bit higher that the 0.3 S·m^{-1} assumed for muscle in previous calculations. An electrical conductivity of 1.5 S·m^{-1} was assumed for the aqueous humor, accounting for the discontinuity in the constant power density contours at the interface. The results are shown in Fig. 26 for an assumed electrode potential of 50 V(rms).

Note that the volume power densities are extremely high compared to the disk electrode calculation due to the small dimensions of the needle electrode. Power densities of 100 kW·cm^{-3} have adiabatic (no heat transfer) heating rates near 27,000 °C·s^{-1} and will vaporize significant amounts of water in microseconds. In practice, the applied voltage is in the neighborhood of 50–80 V (rms), however, pulsed rf is used with pulse times in the neighborhood of 15 μs. Vaporization begins during the first rf pulses in a very small volume around the electrode tip. Subsequent drying near the tip causes the dominant heat-ing field to advance from the tip toward the surface while the overall impedance decreases due to an average increase in temperature around the electrode. At the higher voltage settings the heating may be sufficient to cause the impedance to increase near the end of rf activation due to drying in the cornea (49).

Bipolar Forceps Electrode Field. The bipolar forceps electrodes typically used in vessel sealing are essentially flat plate electrodes. The electric field between them is nearly uniform, with some small electrode edge effect. Figure 27 is a representative FEM calculation assuming an applied potential of 50 V(rms) and electrode separation of 2 mm. Here the power densities are in the neighborhood of 150–200 W·cm^{-3}. The calculations have again assumed a uniform electrical conductivity of 0.3 S·m^{-1} while the actual value is a complex combination of collagen ($\sigma = 0.26$–0.42 S·m^{-1} depending on orientation), elastin ($\sigma = 0.67$–1.0 S·m^{-1} depending on orientation) and smooth muscle ($\sigma = 0.3$ S·m^{-1}).

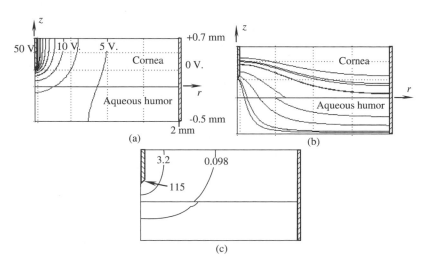

Figure 26. Needle electrode in cornea, FEM model results. Cornea conductivity 1.125 S·m^{-1}, aqueous humor 1.5 S·m^{-1}. (a) Potential field $V(r,z)$ for 50 V (rms) electrode potential. (b) Electric field stream-lines. (c) Volume power density with contours at 115, 3.2, and 0.098 kW·cm^{-3}.

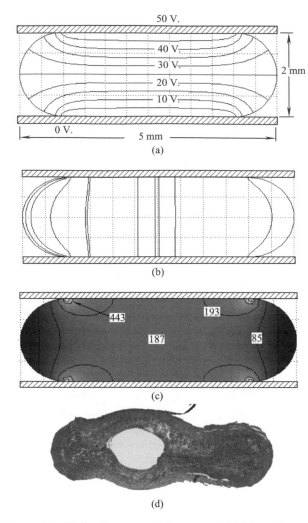

Figure 27. Bipolar forceps applied to a vessel, FEM model results. (a) Potential field, $V(x,y)$ for 50 V(rms) electrode potential. (b) Electric field streamlines. (c) Volume power density, q_{gen}, with contours at 85, 193, and 443 (W·cm^{-3}). (d) Typical result: canine femoral artery temperature 89 °C for 2 s (thermocouple in center of artery; original magnification 40×, Mallory's trichrome stain).

SUMMARY

Radio frequency current has been used for many years to reduce time under anesthesia and improve the clinical result. The ability to simultaneously cut tissue and coagulate blood vessels greatly reduces morbidity and mortality, and enables a large number of procedures that are otherwise not possible, especially in neurosurgery. Electrosurgery is one of several energy sources making substantial contributions in surgery, sources that include: lasers, microwaves, ultrasound, micromechanical disruption, and cryogenics. The major advantage of electrosurgery over the other modes originates from its ability to shape the treated tissue volume by design of the applicator electrode geometry alone. Consequently, a single electrosurgical generator is capable of a wide range of applications from various forms of coagulation and tissue fusion to cutting, either through an endoscope or in open procedures, either microsurgery or, as it were, macrosurgery. Since the

electrodes are in contact with tissue, apposition pressure may be applied to improve the fusion probability. Electrode contact also gives the surgeon tactile feedback not characteristic of many of the other energy sources.

Useful predictions of clinical outcomes may be made with quasistatic electric field models coupled to transient thermal models including kinetic formulations of damage accumulation. Numerical model work can be used (*1*) to study trends due to uncertainty in tissue properties or changes in electrode voltage, current, power, or time of application; (*2*) to reveal discontinuities in the electric field due to boundary, conditions; and (*3*) to study the relative importance of each of the governing physical phenomena individually. When trends in the model match experimental results, one knows that all of the relevant physical phenomena have been included and the surgical process is well understood.

As with any such source of energy there are hazards in the use of electrosurgery. The hazards may be minimized, but not eliminated. Nevertheless, the advantages of use far outweigh the disadvantages, particularly in several very important procedures. The best safeguard is an informed and vigilant user population.

BIBLIOGRAPHY

Cited References

1. Major RA. History of Medicine Vols. I and II. Springfield (IL): Charles C. Thomas; 1954.
2. Breasted JH. The Edwin Smith Surgical Papyrus. Chicago: 1930.
3. Licht S. The history of therapeutic heat. Therapeutic Heat and Cold. 2nd ed. Elizabeth Licht; 1965. Chapt. VI.
4. Jee BS. A Short History of Aryan Medical Sciences. London: 1896.
5. Geddes LA, Roeder RA. DeForest and the first electrosurgical unit. IEEE Eng Med Biol Mag Jan/Feb 2003; 84–87.
6. Doyen E. Sur la destruction des tumeurs cancereuses accessibles; par la methode de la voltaisation bipolaire et de l'electro-coagulation thermique. Archg Elec Med Physiother Cancer 1909;17:791–795.
7. Doyen E. Surgical Therapeutics and Operative Techniques. Vol. 1. New York: Wm. Wood; ~ 1917; p 439–452.
8. Moyer CA, Rhoads JE, Allen JG, Harkins HN. Surgery Principles and Practices. 3rd ed. Philadelphia: Lippincott; 1965.
9. Cushing H, Bovie WT. Electrosurgery as an aid to the removal of intracranial tumors. Surg Gyn Obstet 1928;47: 751–784.
10. Kelly HA, Ward GE. Electrosurgery. Philadelphia: Saunders; 1932.
11. Mitchell JP, Lumb GN, Dobbie AK. A Handbook of Surgical Diathermy. Bristol: John Wright & Sons; 1978.
12. Otto JF, editor. Principles of Minor Electrosurgery. Liebel-Flarsheim Co.; 1957.
13. Harris HS. Electrosurgery in Dental Practice. Philadelphia: Lippincott; 1976.
14. Malone WF. Electrosurgery in Dentistry; Theory and Application in Clinical Practice. Springfield (IL): Chas C Thomas; 1974.
15. Oringer MJ. Electrosurgery in Dentistry. 2nd ed. Philadelphia: Saunders; 1975.
16. Oringer MJ. Color Atlas of Oral Electrosurgery. Chicago: Quintessence; 1984.

17. Epstein E, Epstein E, Jr, editors. Techniques in Skin Surgery. Philadelphia: Lea & Febiger; 1979.
18. Burdick KH. Electrosurgical Apparatus and Their Application in Dermatology. Springfield (IL): Chas C Thomas; 1966.
19. Pearce JA. Electrosurgery. Chichester: Chapman & Hall; 1986.
20. Knecht C, Clark RL, Fletcher OJ. Healing of sharp incisions and electroincisions in dogs. JAVMA 1971;159:1447–1452.
21. Kelly HA, Ward GE. Electrosurgery. Philadelphia: Saunders; 1932.
22. Operating and service instructions for model 770 Electrosectilis, The Birtcher Corporation, El Monte (CA).
23. Operator and service manual, MF-380, Aspen Laboratories Inc., Littleton (CO).
24. Valleylab SSE2-K service manual. Boulder (CO): The Valleylab Corporation;
25. User's guide, Force 2 electrosurgical generator. The Valleylab Corporation, Boulder (CO).
26. Medical electrical equipment part 1: General requirements for safety. IEC/EN 60601-1, International Electrotechnical Commission; 1988.
27. Medical electrical equipment part 2-2: Particular requirements for safety—High frequency surgical equipment. IEC/EN 60601-2-2, International Electrotechnical Commission; 1999.
28. Electrosurgical units. Health Devices. Jun–Jul 1973;2:n8–9.
29. Henriques FC. Studies of thermal injury V: The predictability and the significance of thermally induced rate processes leading to irreversible epidermal injury. Arch Pathol 1947;43:489–502.
30. Panescu D, Fleischman SD, Whayne JG, Swanson DK. Contiguous Lesions by Radiofrequency Multielectrode Ablation. Proc IEEE-Eng Med Biol Soc 17th Annu Meet. 1995; 17, n1.
31. Pearce JA, Thomsen S. Numerical Models of RF Ablation in Myocardium. Proc IEEE-Eng Med Biol Soc 17th Annu Meet; vol. 17, n1, 1995; p 269–270.
32. Lemole GM, Anderson RR, DeCoste S. Preliminary evaluation of collagen as a component in the thermally-induced 'weld'. Proc SPIE 1991;1422:116–122.
33. Kopchock GE, et al. CO_2 and argon laser vascular welding: acute histologic an thermodynamic comparison. Lasers Surg. Med. 1988;8(6):584–8.
34. Schober R, et al. Laser-induced alteration of collagen substructure allows microsurgical tissue welding. Science 1986;232:1421–11.
35. Collagen Volume I Biochemistry. Nimni ME, editors. Boca Raton (FL): CRC Press; 1988.
36. Pearce JA, Thomsen S, Vijverberg H, McMurray T. Kinetic rate coefficients of birefringence changes in rat skin heated in vitro. Proc SPIE 1993;1876:180–186.
37. Chen SS, Wright NT, Humphrey JD. Heat-induced changes in the mechanics of a collagenous tissue: isothermal, isotonic shrinkage. Trans ASME J Biomech Eng 1998;120:382–388.
38. Chen SS, Wright NT, Humphrey JD. Phenomenological evolution equations for heat-induced shrinkage of a collagenous tissue. IEEE Trans Biomed Eng 1998;BME-45:1234–1240.
39. Honig WM. The mechanism of cutting in electrosurgery. IEEE Trans Bio-Med Eng 1975;BME-22:58–62.
40. Geddes LA, Tacker WA, Cabler PA. A new electrical hazard associated with the electrocautery. Biophys Bioengr Med Instrum 1975;9n2:112–113.
41. Hungerbuhler RF, et al. Ventricular fibrillation associated with the use of electrocautery: a case report. JAMA 21 Oct 1974;230n3:432–435.
42. Orland HJ. Cardiac pacemaker induced ventricular fibrillation during surgical diathermy. Anesth Analg Nov 1975;3n4; 321–326.
43. Titel JH, et al. Fibrillation resulting from pacemaker electrodes and electrocautery during surgery. Anesthesiol Jul–Aug 1968;29:845–846.
44. Batra YK, et al. Effect of coagulating and cutting current on a demand pacemaker during transurethral resection of the prostate: a case report. Can Anesthes Soc J Jan 1978;25n1: 65–66.
45. Fein RL. Transurethral electrocautery procedures in patients with cardiac pacemakers. JAMA 2 Oct 1967;202: 101–103.
46. Greene LF. Transurethral operations employing high frequency electrical currents in patients with demand cardiac pacemakers. J Urol Sept 1972;108:446–448.
47. Krull EA, et al. Effects of electrosurgery on cardiac pacemakers. J Derm Surg Oct 1975;1n3:43–45.
48. Wiley JD, Webster JG. Analysis and control of the current distribution under circular dispersive electrodes. IEEE Trans BioMed Eng May 1982;BME-29n5:381–384.
49. Choi BJ, Kim J, Welch AJ, Pearce JA. Dynamic impedance measurements during radio-frequency heating of cornea. IEEE Trans Biomed Eng 2002;49n12:1610–1616.

See also CRYOSURGERY; ION-SENSITIVE FIELD EFFECT TRANSISTORS.

EMERGENCY MEDICAL CARE.   See CARDIOPULMONARY RESUSCITATION.

EMG.   See ELECTROMYOGRAPHY.

ENDOSCOPES

BRETT A. HOOPER
Areté Associates
Arlington, Virginia

INTRODUCTION

The word endoscope is derived from two Greek words, endo meaning "inside" and scope meaning "to view". The term endoscopy is defined as, "using an instrument (endoscope) to visually examine the interior of a hollow organ or cavity of the body." In this second edition of the Encyclopedia of Medical Devices and Instrumentation, we will revisit the excellent background provided in the first edition, and then move on from the conventional "view" of endoscopes and endoscopy to a more global "view" of how light can be delivered inside the body and the myriad light-tissue interactions that can be used for both diagnostic and therapeutic procedures using endoscopes. We will update the medical uses of endoscopy presented in the first edition, and then look at the medical specialties that have new capabilities in endoscopy since the first edition; cardiology and neurosurgery. This will be by no means an exhaustive list, but instead a sampling of the many capabilities now available using endoscopes. We will also present new delivery devices (fibers, waveguides, etc.) that have been introduced since the optical fiber to allow a broader range of wavelengths and more applications that deliver light inside the body.

HISTORY

Pre-1800s

Endoscopy had its beginnings in antiquity with the inspection of bodily openings using the speculum, a spoon-shaped instrument for spreading open the mouth, anus, and vagina. Another instrument of fundamental importance to endoscopy is the catheter, as it has been used to evacuate overfilled bladders for more than 3000 years. Catheters and a rectoscope were used by Hippocrates II (460–377 BC), where inspection of the oral cavity and of the pharynx was routine, including operations on tonsils, uvula, and nasal polyps. This trend continued for the better part of two millennia.

1800–1900

In 1804, Phillip Bozzini came upon the scene with the *Lichtleiter*, and attempts were made to view into the living body through the endoscope's narrow opening (1). It took almost 100 years to achieve this goal. These prototype endoscopes consisted of hollow tubes through which light of a candle, and even sunlight, was projected in order to visualize the inside of a bodily opening. The *Lichtleiter*, the light conductor, consists of two parts: (*1*) the light container with the optical part; and (*2*) the mechanical part, which consists of the viewing tubes fitted to accommodate the anatomical accesses of the organs to be examined. The apparatus is shaped like a vase, is ∼ 35 cm tall, made of hollow lead, and covered with paper and leather. On its front is a large, round opening divided into two parts by a vertical partition. In one half, a wax candle is placed and held by springs such that the flame is always in the same position. Concave mirrors are placed behind the candle and reflect the candle light through the one-half of the tube onto the object to be examined. The image of the object is then reflected back through the other half of the tube to the eye of the observer. Depending on the width of the cavity to be examined, different specula were used. These specula consisted of leaves that could be spread open by use of a screw device in order to expand the channels. An example is shown in Fig. 1.

Figure 1. Phillip Bozzini's *Lichtleiter* endoscope.

In 1828, the physicist C. H. Pfaff mentioned that platinum wires could be made incandescent through electric current. The glowing platinum wire has a desirable side effect; its white-hot heat illuminates the cavity of a surgical field so brightly that it provided the first internal light source. In 1845, the Viennese dentist Moritz Heider was the first to combine the illumination and tissue heating capabilities of the platinum wires when he cauterized a dental pulp with this method. However, the simultaneous heat that is produced by the wire is so intense that it was difficult to attach to the tip of small endoscopes. The term endoscopy was first used in the medical literature in 1865 by two French physicians, Segals and Desmoreaux. In 1873, Kussmaul, successfully passed a hollow tube into the stomach of a sword-swallower. Observation was limited because these long stiff hollow tubes were poorly lit by sun, candlelight, and mirrors. It took the advent of X rays to prove the swallowing of a sword was not a trick (2). Later that year, Gustave Trouvé, an instrument maker from Paris, introduced the polyscope at the World Exhibition in Vienna. His polyscope took many forms; rectoscope, gastroscope, laryngoscope, and cystoscope for looking into the rectum, stomach, larynx, and urinary tract, respectively. Trouvé was responsible for the idea of using electric current successfully for endoscopic illumination by placing the light source at the tip of the instrument, and the first to accomplish internal illumination. He was also the first to utilize a double prism for two observers by splitting the field with a Lutz prism by 90°, and incorporated a Galilean lens system, which magnified 2.5-fold, but did not enlarge the visual field. In 1876, Maximillian Nitze started his work on the urethroscope, and by the fall of 1877 he had instruments for illumination of the urethra, bladder, and larynx that used the platinum wire illumination method. Nitze also reported the first cystoscopy of the urinary bladder in 1877. As early as 1879, Nitze had his instrument patented for bladder, urethra, stomach, and esophagus in both Europe and the United States. The use of fiber optics was foreshadowed by Dr. Roth and Professor Reuss of Vienna in 1888 when they used bent glass rods to illuminate body cavities. With the invention of the light bulb by Thomas Edison, small light sources could be attached to the tip of the endoscopes without the need for cooling. The mignon lamp with a relatively small battery to operate it, led the next resurgence in endoscopy.

1900–Present

It was not until the early 1900s, however, that endoscopy became more than a curiosity. With the advancement of more sophisticated equipment and the discoveries of electrically related technologies, illumination capabilities were greatly improved. New and better quality lenses, prisms, and mirrors dramatically enhanced the potential applications of endoscopy and opened up new avenues of application. During this period, Elner is credited in 1910 with the first report of a technically advanced gastroscope (3) to view the stomach, and only 2 years later, Sussmann reported the first partially flexible gastroscope, which utilized screw systems and levers to contort the scope's shaft (4). The diagnostic merit of endoscopy was beginning to be realized, as an atlas of endoscopic pathologies was being developed (5). Methods for saving images of the scene under investigation ranged from the early metal halide photographic process to 35 mm film cameras. Cameras were attached on view ports to allow simultaneous documentation of the view that the endoscopist had. Only within the last few decades has the use of digital photography been applied to the field of endoscopy, with an enhancement in diagnostic potential because of digital signal processing capabilities. In spite of the technical advances, problems with heat produced at the tip, blind spots in the field of view, and organ perforation limited the widespread use of endoscopes.

The most significant technological development began in the 1950s with the introduction of fiber optics. These thin glass or plastic optical fibers (on the order of 1 mm) allowed for "cold light" illumination at the tip, controlled flexibility to remove blind spots, and with quality image transmission opened up photographic capabilities, thus improving previous limitations. The use of fiber optics also allowed for the incorporation of ancillary channels for the passage of air, water, suction, and implementation of biopsy forceps or instruments for therapeutic procedures. Obviously, the most significant capabilities fiber optics has brought to endoscopes are their small size, flexibility, and ability to deliver laser light. Lasers are now used in both diagnostic and therapeutic procedures. In fact, the overwhelming advantage of using laser light is that it can act as both a diagnostic and therapeutic light source, and often in the same endoscope. Let us now look briefly at the theory behind efficient propagation of light in an optical fiber.

LIGHT DELIVERY WITH OPTICAL FIBERS

Efficient light delivery is made possible in solid-core optical fibers by use of total internal reflection. Total internal reflection (TIR) is achieved at the interface between two media when the incident medium has a refractive index larger than the transmitting medium, $n_i > n_t$, and the angle of incidence is greater than the critical angle. The critical angle is defined as the angle of incidence $_i$ where the transmitted angle $_t$ goes to $90°$, and with no transmittance there is total internal reflection. For a given interface where n_i and n_t are known, the critical angle can be calculated from Snell's law as (6)

$$\theta_{\text{critical}} = \sin^{-1}(n_t \sin \theta_t / n_i) \qquad (1)$$

For the optical fiber, TIR is achieved by designing the core refractive index n_i to be greater than the cladding refractive index n_t, and by focusing the input light so that it propagates along the axis of the fiber thereby maintaining a large angle with respect to the interface between core and cladding (see Fig. 2). Apart from choosing a fiber material that has close to lossless transmission at the wavelength of choice, one needs to focus the light beam so that the diameter of the fiber core is at least three times the radius of the input light beam and the cone of convergent focused light of angle 2 is less than the acceptance cone of the fiber core. The acceptance half-angle of the fiber is related to the

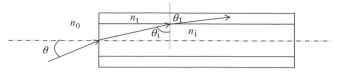

Figure 2. Light coupling into an optical fiber with core refractive index n_i and cladding refractive index n_t, and the associated incident angles (θ and θ_i), refracted angles, and their ray trajectories. The numerical aperture (NA) is a measure of the acceptance angle θ of a fiber.

NA and the refractive indexes by

$$\text{NA} = n_0 \sin \theta = (n_i^2 - n_t^2)^{1/2} \qquad (2)$$

Numerical apertures for solid-core fibers range from 0.2 to 1 for fibers in air, where $n_0 = 1.0$, and are a measure of the light gathering capability of a fiber. An NA of 1 indicates that essentially all light incident at the fiber, even light near a $90°$ angle of incidence, can be launched into the fiber. In practicality, the NA approaches 1 and focusing and transverse mode distribution then play an important role in efficient light coupling. This is the case for unclad single-crystal sapphire fibers where the NA approaches one. For hollow-core waveguides (HWGs) the same rules apply for efficient coupling, that is, core diameter at least three times the radius of the input light beam and the convergence angle of the focused light θ less than the acceptance angle of the waveguide. Waveguiding is accomplished by coating the inside of the HWG with a metallic or dielectric coating depending on the desired wavelength. Fibers are available in diameters from several micrometers, for single-mode transmission, to ~ 1 mm, for multimode fibers. A propagating mode in a fiber is a defined path in which the light travels. Light propagates on a single path in a single-mode fiber or on many paths in a multimode fiber. The mode depends on geometry, the refractive index profile of the fiber, and the wavelength of the light. In the next section, we will look at the many different kinds of "optical fibers" that have become available since the previous edition and the myriad wavelengths that are now available for diagnostic and therapeutic procedures through endoscopes.

NEW "OPTICAL FIBER" DEVICES

There are now available to the endoscopist an array of different light delivery devices that can be incorporated into an endoscope. Conventional silica glass fibers transmit wavelengths from the visible (400 nm) to the mid-infrared (IR) (2.5 μm). New solid-core fibers are available for the transmission of IR wavelengths; these include germanium-oxide glass, fluoride glass, sapphire (out to 5 μm), and silver halide fibers (out to 30 μm) (7–9). In addition to the conventional solid-core optical fiber, there are now fibers available that have a hollow core and fibers that are intentionally "holey", that is to say they have a solid core with an array of periodic holes that run the length of the fiber. The hollow waveguides come in two flavors, but each efficiently transmits IR wavelengths.

The first hollow waveguide is a small silica tube that is metal coated on the inside and efficiently transmits IR light from ~ 2 to $10 \, \mu m$. These HWGs can be optimized for transmission of $2.94 \, \mu m$ Er:YAG laser light or $10.6 \, \mu m$ CO_2 laser light, and can therefore handle high laser power for therapeutic procedures (Polymicro Technologies, LLC). The second HWG is known as a photonic bandgap fiber (PBG) and is made from a multilayer thin-film process, rolled in a tube, and then heat drawn into the shape of a fiber. These PBG fibers are designed to transmit IR wavelengths, but only over a narrow band of $1–2 \, \mu m$, for example, $9–11 \, \mu m$ for delivery of CO_2 laser light (Omniguide Communications, Inc.). The "holey" fibers are silica fibers that have been very carefully etched to create holes in a periodic pattern about the center of the core that span the length of the fiber. The pattern of these holes is chosen to propagate light very efficiently at a particular band of wavelengths, hence these fibers can be designed for a particular application; to-date largely in the visible and near-IR.

With each of these new fibers there are trade-offs in flexibility, transmission bandwidth, and ability to handle high power laser light. There are also differing fiber needs for diagnostic and therapeutic application. Diagnostic imaging fibers, for example, require a large NA for high light gathering capability. Comparatively, a therapeutic fiber endoscope may require a small NA to confine the delivery of the high power laser light to a specific region. This can also be achieved by use of evanescent waves launched at the interface between a high refractive-index optic and the tissue. Evanescent waves are different from the freely propagating light that typically exits fibers in that it is a surface wave. This surface wave only penetrates into the tissue on the order of the wavelength of the light. For typical wavelengths in the visible and IR, this amounts to light penetration depths on the order of microns, appropriate for very precise delivery of light into tissue. Both diagnostic and therapeutic applications have been demonstrated by coupling a high refractive-index conic tip (sapphire and zinc sulfide) to a HWG (see Fig. 3) (10). The

diagnostic capability allows Fourier transform infrared (FTIR) spectroscopy to be performed on living (*In vivo*) tissue, where, for example, fatty tissues can be distinguished from normal intimal aorta tissue. Through the same HWG catheter, tissue ablation of atherosclerotic plaque has proven the therapeutic capabilities. These diagnostic and therapeutic applications can potentially take advantage of evanescent waves, HWGs, and mid-infrared FT spectroscopy in the $2–10 \, \mu m$ wavelength range (10).

A hybrid optical fiber consisting of a germanium trunk fiber and a low OH silica tip has shown the ability to transmit up to 180 mJ of Er:YAG power for applications requiring contact tissue ablation through a flexible endoscope (11). This pulse energy is more than sufficient for ablation of a variety of hard and soft tissues.

Next, we will look at the potential influence these new light delivery systems may have in endoscopy.

MEDICAL APPLICATIONS USING ENDOSCOPY

Endoscopy has had a major impact on the fields of medicine and surgery. It is largely responsible for the field of minimally invasive surgery. The ability to send diagnostic and therapeutic light into the body via minimally invasive procedures has reduced patient discomfort, pain, and length of hospital stay; and in some cases has dramatically changed the length of stay after a procedure from weeks to days. Optical fibers have had a profound effect on endoscopy, and in doing so, dramatically changed medicine and surgery. These small, flexible light pipes allowed physicians to direct light into the body where it was not thought possible, and even to direct laser light to perform microsurgeries in regions previously too delicate or intricate to access.

In this section, we will examine the state of endoscopy in arthroscopy, bronchoscopy, cardiology, cystoscopy, fetoscopy,

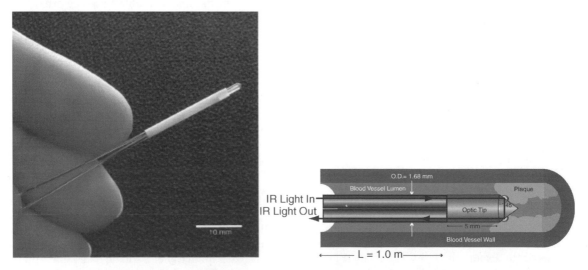

Figure 3. An example of a HWG endoscope. Both diagnostic and therapeutic applications have been demonstrated by coupling a high refractive-index conic tip (sapphire and zinc sulfide) to a HWG. In this geometry with two waveguides, diagnostic spectrometer light can be coupled into the tip in contact with tissue via one waveguide and sent back to a detector via the other waveguide.

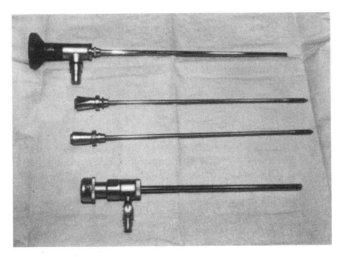

Figure 4. Arthroscope and internal components used to view the interior of knee joint space.

gastrointestinal endoscopy, laparoscopy, neurosurgery, and otolaryngology.

Arthroscopy

Arthroscopy had its birth in 1918 when Takagi modified a pediatric cystoscope and viewed the interior of the knee of a cadaver (12). Today its use is widespread in orthopedic surgery with major emphasis on the knee joint. Arthroscopy has had a major impact on the way knee surgery is performed and with positive outcomes. The surgery is minimally invasive often with only two small incisions; one for an endoscope to visualize the interior of the knee, and another to pass microsurgical instruments for treatment. Presently, endoscopes have the ability to view off-axis at a range of angles from 0 to 180°, with varying field-of-view (see Fig. 4). The larger the field-of-view the more distorted the image, as in a fish-eye lens. A 45° off-axis endoscope, for example, with a 90° field-of-view can be rotated to visualize the entire forward-looking hemisphere.

The most common procedure is arthrotomy, which is simply the surgical exploration of a joint. Possible indications include inspection of the interior of the knee or to perform a synovial biopsy. Synovia are clear viscous fluids that lubricate the linings of joints and the sheaths of tendons. Other indications include drainage of a hematoma or abscess, removal of a loose body or repair of a damaged structure, such as a meniscus or a torn anterior cruciate ligament, and the excision of an inflamed synovium (13,14).

Future directions in orthopedic surgery will see endoscopes put to use in smaller joints as endoscopes miniaturize and become more flexible. The ultimate limit on the diameter of these devices will likely be 10s of micrometers, as single-mode fibers are typically 5–10 μm in diameter. These dimensions will allow for endoscopy in virtually every joint in the body, including very small, delicate joints. Work in the shoulder and metacarpal-phalanges (hand–finger) joints is already increasing because of these new small flexible fiber optic endoscopes.

Bronchoscopy

Bronchoscopy is used to visualize the bronchial tree and lungs. Since its inception in the early 1900s bronchoscopy has been performed with rigid bronchoscopes, with much wider application and acceptance following the introduction of flexible fiber optics in 1968. An advantage of the equipment available is its portability, allowing procedures to be done at a patient's bedside, if necessary. With the introduction of fiber optics, in 1966, the first fiber optic bronchoscope was constructed, based on specifications and characteristics that were proposed by Ikeda. He demonstrated the instrument's use and application and named it the bronchofiberscope. Development over the last several decades has seen the use of fiberoptic endoscopes in the application of fiberoptic airway endoscopy in anesthesia and critical care. These endoscopes have improved the safe management of the airway and conduct of tracheal and bronchial intubation. Fiber optic endoscopy has been particularly helpful in the conduct of tracheal and bronchial intubation in the pediatric population.

Bronchoscopy is an integral part in diagnosis and treatment of pulmonary disease (15,16). Bronchoscopic biopsy of lung masses has a diagnostic yield of 70–95%, saving the patient the higher risk associated with a thoracotomy. Pulmonary infection is a major cause of morbidity and mortality, especially in immuno-compromised patients, and bronchoscopy allows quick access to secretions and tissue for diagnosis. Those patients with airway and/or mucous plugs can quickly be relieved of them using the bronchoscope. Another diagnostic use of the bronchoscope is pulmonary alveolar lavage, where sterile saline is instilled into the lung then aspirated out and the cells in the lavage fluid inspected for evidence of sarcoidosis, allergic alveolitis, for example. Lavage is also of therapeutic value in pulmonary alveolar proteinosis.

Bronchoscopy is usually well tolerated by the patient with complications much less than 1% for even minor complications. Laser use has also allowed for significant relief of symptoms in cancer patients.

Cardiology

Early developments in minimally invasive cardiac surgery included the cardiac catheter (1929), the intra-aortic balloon pump (1961), and balloon angioplasty (1968). Endoscopy in cardiology has largely focused on using intravascular catheters to inspect the inside of blood vessels and more recently the inside of the heart itself. Catheters are thin flexible tubes inserted into a part of the body to inject or drain away fluid, or to keep a passage open. Catheters are similar to endoscopes, and they also have many diagnostic and surgical applications.

For diagnostic purposes, angiography uses X rays in concert with radio-opaque dyes (fluoroscopy) to look for blockages in vessels, usually the coronary arteries that supply oxygenated blood to the heart. A catheter is introduced into the femoral artery and sent up the aorta and into the coronary arteries to assess blood flow to the heart. The catheter releases the dye and real-time X-ray fluoroscopy tracks the dye as it is pumped through the coronary artery. Angiography in concert with intravascular

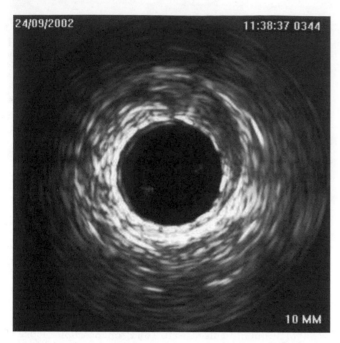

Figure 5. Image of a blood vessel with a stent using IVUS. (Courtesy of LightLab Imaging.)

ultrasound (IVUS) is the currently accepted diagnostic in cardiology (17–21). IVUS emits acoustic energy out the tip of a catheter and listens for echoes to image the inside of coronary arteries (see Fig. 5). An IVUS image is a cross-sectional view of the blood vessel, and complements the X-ray fluoroscopy image. The IVUS has been used to assess atherosclerotic plaques in coronary arteries and has been very successful at guiding the placement of stents. Stents are expandable metal mesh cages in the shape of cylinders that act as scaffolding to open obstructions in vessels caused by atherosclerosis.

Another technique showing promise in endovascular imaging uses light. The light-driven technology, optical coherence tomography (OCT), is successful at detecting fatty plaques, including those that are vulnerable to rupture. Figure 6 compares OCT images of coronary artery to

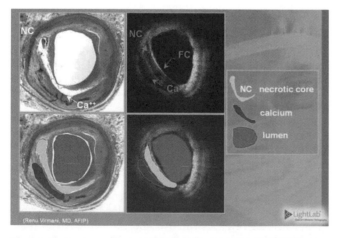

Figure 6. Images of coronary artery comparing OCT with conventional histology. (Courtesy of LightLab Imaging.)

conventional histology (LightLab Imaging, Inc.). Visible are the necrotic core of an atherosclerotic plaque and the thin cap at the intimal surface of the vessel. Near-IR light has been used diagnostically for differentiating between oxygenated and deoxygenated blood in myocardium (22–24). The NIR light in the 2–2.5 μm wavelength range has also been used to characterize myocardial tissue (25). In this wavelength range, the absorption due to water is declining through a local minimum, while there are absorption peaks in fat at 2.3 and 2.4 μm that show distinct spectral signatures. It has also been suggested that IR light might be used to identify atherosclerotic plaques (26–29), including plaques that are at high risk of rupture or thrombosis (30). Arai et al. showed that fibrofatty plaques have characteristic absorbances at mid-IR wavelengths of 3.4 and 5.75 μm that are significantly greater than normal aorta tissue (29). Peaks at these wavelengths, and potentially other subtleties in the absorption spectra [derived from multivariate statistical analysis (MSA)], can be targeted for developing a diagnostic profile similar to that described by Lewis and co-workers for IR and Raman spectroscopic imaging (31,32). These mid-IR frequencies may also be optimized for selective therapy via precise laser surgery (10,33,34).

Surgical laser techniques have become routine over the past two decades in a number of medical specialties such as ophthalmology and dermatology. However, in cardiology initial enthusiasm for fiber optic catheter ablation of atherosclerotic plaque (laser angioplasty) waned in the face of unpredictable vascular perforations, restenosis, and thrombosis (35,36). Therapeutically, IR light has found application in transmyocardial revascularization (TMR) through several partial myocardial perforations (37). Ideally, lasers can ablate tissue with exquisite precision and nearly no damage to the surrounding tissue. This precision requires extremely shallow optical penetration, such that only a microscopic zone near the tissue surface is affected. This has been accomplished by using short laser pulses at wavelengths that are strongly absorbed by proteins (193 nm in the UV) (38) or water (3 μm in the IR) (39).

Therapeutic techniques using catheters recently have targeted the atherosclerotic plaque that is deposited in the vessel wall as we age. The deposition of atherosclerotic plaque in coronary arteries is important for two reasons; the arteries are small (1–3 mm), and they supply blood to the heart muscle itself, so any reduction or blockage of oxygen-supplying blood to the heart shows up symptomatically as angina or worse, a heart attack. Catheters equipped with lasers and inflatable balloons have been used to open these blockages by ablating the plaque or compressing the plaque back into the vessel wall, respectively. Problems with balloon angioplasty have been vessel wall perforation, thrombosis (blood clot formation), and restenosis (reobstruction of the vessel due to an immune response). Recently, balloon angioplasty was augmented by the use of stents, as mentioned above. These expandable cages are opened via a balloon catheter in the vessel lumen and act as a mechanical scaffold to hold the plaque obstruction in place on the vessel wall. Stenting, however, still suffered from restenosis, because of the mechanical injury induced by the stent. Stents coated with small amounts of

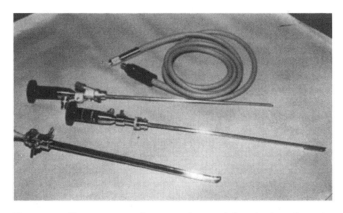

Figure 7. Cystoscope and accessories used for viewing the urinary tract in the medical specialty urology.

drugs (drug-eluding stents) have met with some success, in that they are able to retard the regrowth of intimal hyperplasia responsible for the restenosis.

Cystoscopy

Endoscopy is an integral part of the practice of urology. There are many procedures that look inside the urinary tract using a cystoscope, with typical access through the ureter (see Fig. 7). Common cystoscopy procedures include ureteral catheterization, fulguration and or resection of bladder tumor(s), direct vision internal urethrotomy, insertion of stent, removal of bladder stone, and kidney stone fragmentation (40,41).

A wide variety of therapeutic accessories are available for endoscopic treatment, including snares and baskets for stone removal, and electrohydraulic lithotripsy and laser delivery for fragmenting large stones (42,43). These procedures are minimally invasive and, therefore, can be accomplished with a much lower rate of morbidity and mortality than would be achieved in an open surgical procedure.

Laser incision of urethral, bladder neck, and urethral strictures and fragmentation of kidney stones is being investigated using flexible endoscopes made from germanium fibers to transport Er:YAG laser light. Bladder neck strictures are defined as a narrowing or stenosis of the bladder neck that may result in recalcitrant scarring and urinary incontinence. A significant number of patients undergoing surgery for benign or malignant prostate cancer suffer from bladder neck strictures, and there is no simple and effective minimally invasive treatment. The Er:YAG laser can ablate soft tissue ~ 20–30 times better than a Ho:YAG laser ($2.12\,\mu m$), which is the laser of choice in urology. The absorption coefficient is many orders of magnitude different for these two lasers, $\sim 10,000$ cm^{-1} for Er:YAG versus 400 cm^{-1} for the Ho:YAG. This translates to a $1/e$ depth of optical penetration of 1 versus $25\,\mu m$ for these two lasers. Water is the dominant absorptive chromophore in the tissue in this mid-IR region of the spectrum. Hence, the Er:YAG is better suited to procedures where precision is required in laser ablation of soft tissue.

Fetoscopy

Fetoscopy allows for the direct visualization of a fetus in the womb. It also allows for the collection of fetal blood samples and fetal skin sampling, for diagnosis of certain hemoglobinopathies and congenital skin diseases, respectively (44–46).

The instrument is a small-diameter (1–2 mm) needlescope with typical entry through the abdominal wall under local anesthesia and guided by ultrasound. Optimal viewing is from 18 to 22-weeks gestation when the amniotic fluid clarity is greatest. Once the instrument is introduced, a small-gauge needle is typically used to obtain blood samples or small biopsy forceps can be used for skin biopsy.

Complete fetal visualization is not usually achieved. The procedure has some risk with fetal loss $\sim 10\%$ and prematurity another 10%. The future of this procedure is not clear, as it is not currently in general use, despite the safe and relatively simple prenatal diagnosis it offers. It has been replaced in many circumstances by ultrasound for the fetal visualization aspect, but is still valuable for blood and skin sampling.

Gastrointestinal Endoscopy

The techniques of fiberoptic gastrointestinal (GI) endoscopy were developed in the 1960s, but surgeons were slow to embrace the techniques. These procedures were developed by gastroenterologists who became skilled practitioners and teachers of the art (see Fig. 8). Gradually, GI surgeons adopted these procedures, and in 1980 the American Board of Surgery mandated that endoscopic training be a part of the curriculum in general surgical training. Endoscopy has since become an integral part of surgical education and practice (47). The GI endoscopy is used in the esophagus, stomach, small bowel and liver, biliary, colon, and in pediatric endoscopy. There are numerous accessories available for the GI endoscope. They include biopsy forceps, graspers, cautery tools, and wire snares (see Figs. 9 and 10).

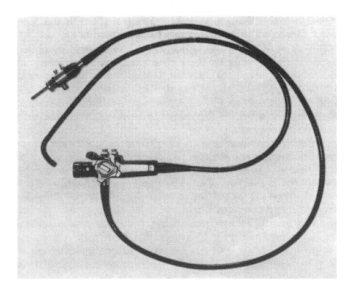

Figure 8. Upper intestinal panendoscope for the adult patient.

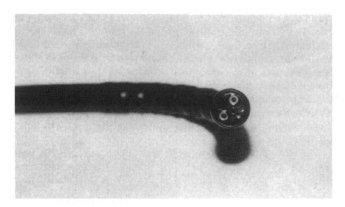

Figure 9. An end view of the distal tip of a panendoscope illustrating the accessory channels and illumination ports.

In the esophagus, endoscopy is used in the dilation of benign esophageal strictures, balloon dilatation of Achalasia, management of foreign bodies and bezoars of the upper GI tract, and endoscopic laser therapy of GI neoplasms (48–50). Endoscopy of benign esophageal strictures is a common procedure performed by gastroenterologists to treat esophageal narrowing and relieve dysphagia. Achalasia is a motility disorder of the esophagus characterized by symptoms of progressive dysphagia for both solids and liquids with (1) aperistalsis in the body of the esophagus, (2) high lower esophageal sphincter (LES) pressure, and (3) failure of the LES to relax. Examples of foreign bodies are coins, meat impaction, frequently in the elderly population, sharp and pointed objects, such as, a toothpick, a chicken or fish bone, needles, and hatpins. Bezoars are a hard mass of material often found in the stomach and are divided into three main types: phytobezoars, trichobezoars, and miscellaneous. These bezoars can be managed by endoscopy dependent on size and location, typically by capture and removal rather than endoscopic fragmentation. Since the 1970s, the incidence of esophageal adenocarcinoma has increased more rapidly than any other form of cancer and now represents the majority of esophageal neoplasms in the West (51). Esophagectomy is considered the gold standard for the treatment of high grade dysplasia in Barrett's esophagus (BE) and for noninvasive adenocarcinoma (ACA) of the distal esophagus (52). Barrett's esopha-

gus is the replacement of native squamous mucosa by specialized intestinal metaplasia and is known to be the major risk factor for the development of adenocarcinoma. A recent study of 45 patients supports the use of endoscopic surveillance in patients who have undergone "curative" esophagectomy for Barrett's dysplasia or localized cancer (53–55). OCT imaging has recently shown the ability to image Barrett's esophagus through a small fiber endoscope. The GI neoplasm treatment is also particularly amenable using neodymium-YAG laser palliative treatment of malignancies of the esophagus and gastroesophageal junction as first described by Fleischer in 1982.

For the stomach, endoscopic therapy for benign strictures of the gastric outlet is but one procedure. These strictures of the gastric outlet are similar to the esophageal stricture mentioned above. They are most frequently caused by peptic ulcer disease in the region of the pylorus, although chronic ingestion of nonsteroidal anti-inflammatory drugs is a frequent cause as well. Other endoscopic procedures in the stomach include sclerotherapy of esophageal varices (abnormally swollen or knotted vessels, especially veins), percutaneous endoscopic gastrostomy and jejunostomy, injection therapy for upper GI hemorrhage, and thermal coagulation therapy for upper GI bleeding.

In the small bowel and liver, enteroscopy is an endoscopic procedure to directly inspect the small bowel mucosa and take biopsies by enteroscopy from selected sites in the jejunum and proximal ileum. Until the recent development of the enteroscope, these parts of the small bowel were not possible to endoscopically evaluate. Enteroscopy is in its technological infancy. It presently allows for access to the most proximal aspects of the small bowel to obtain tissue at bleeding lesions. However, it cannot be utilized to adequately visualize the entire mucosa.

Endoscopy in the biliary system is used to perform sphincterotomies, papillotomies and stent insertions, to manage large bile duct stones and malignant biliary strictures, biliary and pancreatic manometry, and endoscopic retrograde cholangiopancreatography. The use of endoscopes all but changed the surgical technique for gallbladder removal and moved its outcome from a major inpatient surgery to a minimally invasive surgery with only small incisions for the endoscope.

In the colon, nonoperative and interventional management of hemorrhoids is performed with endoscopes. Other procedures include dilatation of colonic strictures, approaches to the difficult polyp and difficult colonic intubation, and clinical approaches of anorectal manometry. Additionally, colonoscopy is used for investigating irritable bowel syndrome, Crohn's disease, and ulcerative colitis (47,56).

Pediatric endoscopy has been used to perform gastroscopy, colonoscopy, and endoscopic retrograde cholangiopancreatography. Advances have largely involved the application of techniques used in adults to the pediatric patient; this was made possible with the introduction of smaller fiberoptic endoscopes.

And finally general endoscopy has been used as a-surveillance program for premalignant lesions, to assess outpatient endoscopy, and endosonography and echo probes.

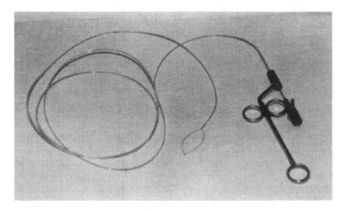

Figure 10. A polyp wire snare device with retraction handle.

Laparoscopy

Laparoscopy or peritoneoscopy is an important diagnostic procedure that allows direct visualization of the surface of many intra-abdominal organs, as well as allowing the performance of guided biopsies and minimally invasive therapy. Landmarks in laparoscopic surgery include the first laparoscopic appendectomy (1983) and the first laparoscopic cholescystectomy (1987). The first laparoscope was an ordinary proctoscope with illumination coming from an electric headlight worn by the endoscopist. Since then, with the advent of newer technologies, a fiber optic cable was added to the rigid telescope making flexible laparoscopes available. Today, despite the advent of various noninvasive scanning technologies, laparoscopy is still clinically useful for visualizing and biopsying intra-abdominal tumors, particularly those that involve the liver, peritoneum, and pelvic organs (57,58).

The laparoscopic procedure begins with the introduction of a trocar into the abdomen at the inferior umbilical crease for insufflation of carbon dioxide gas. A trocar is a sharply pointed steel rod sheathed with a tight-fitting cylindrical tube (cannula), used together to drain or extract fluid from a body cavity. The whole instrument is inserted then the trocar is removed, leaving the cannula in place. The gas acts as a cushion to permit the safe entry of sharp instruments into the peritoneal cavity and enable a better view. Common procedures include laparoscopic cholecystectomy and laparoscopic appendectomy, laparoscopic removal of the gallbladder and appendix, respectively. Laparoscopic cholecystectomy has all but replaced the conventional surgery for removal of the gallbladder. What used to involve opening the abdominal cavity for gallbladder removal and a 5–7 day stay at the hospital, has been transformed to a minimally invasive procedure with only a few days in the hospital. Lastly, laparoscopic sterilization and abortion are also performed with endoscopes.

Neuroendoscopy

Fiber optics is particularly well suited for the field of neuroendoscopy for both diagnostic and therapeutic procedures in the inner brain, because of size, flexibility, visualization, and laser delivery. We briefly review a case study that highlights some of the advantages of fiber optic endoscopes for minimally invasive surgery in the inner brain.

A recent case report on laser-assisted endoscopic third ventriculostomy (ETV) for obstructive hydrocephalus shows the use of diagnostic and therapeutic endoscopy in neurosurgery (59). Under stereotactic and endoscopic guidance, multiple perforations in the ventricular floor using a 1.32 μm neodymium–yttrium–aluminum–garnet (Nd:YAG) or an aluminum–gallium–arsenide (AlGaAs) 0.805 μm diode laser and removal of intervening coagulated tissue ensued with a 4–6 mm opening between third ventricle and basilar cisterns. These perforations allow for the cerebrospinal fluid (CSF) to be diverted so that a permanent communication can be made between the third cerebral ventricle and arachnoid cisterns of the cranial base. In a series of 40 consecutive cases, 79% of the patients had a favorable outcome. This compares well with a recent series summarizing > 100 patients and long-term follow-up with success rates ranging from 50 to 84%.

When the 1.32 μm Nd:YAG laser is used, the high absorption in water requires that the fiber be placed in contact with the ventricular floor. Conversely, the high power diode laser's dispersion-dominant properties can lead to damage to neural structures around the ventricular cavity. Therefore, the 0.805 μm diode laser was used in a contact mode, but only after carbonization of the fiber tip so that thermal increase of the fiber tip allowing ventricular floor perforation was due to absorption of the laser energy by the carbon layer only and not by direct laser–tissue interaction. The 1.32 μm Nd:YAG laser was found to have higher efficiency for coagulation and perforation than the 0.805 μm diode laser, and would appear to be the better choice for neuroendoscopic use in this procedure. The endoscope allows for visualization and treatment of a very difficult part of the brain to access, and the use of lasers in endoscopes is advantageous in cases of distorted anatomy and microstructures and may reduce technical failures.

In another case of endoscopic-delivered laser light for therapeutic purpose, an IR free-electron laser (FEL) was used to ablate (cut) a suspected meningioma brain tumor at Vanderbilt's Keck FEL Center (60). A HWG catheter was used to deliver 6.45 μm IR FEL light to a benign neural tumor in the first human surgery to use a FEL. The 6.45 μm FEL light is a candidate for soft tissue surgery because of its ability to ablate (cut) soft tissue with a minimum of thermal damage to the surrounding tissue; on the order of micrometers of damage. This is obviously very important for surgery in the brain where viable, eloquent tissue may be in contact with the tumorous tissue that is being removed.

The FEL is a research center device that generates laser light over a vast portion of the electromagnetic spectrum; to date FELs have generated laser light from the UV (190 nm) to millimeter (61). The FEL is beneficial in identifying wavelengths, particularly in the IR where there are no other laser sources, for selective laser-tissue interaction.

Otolaryngology

Early use of endoscopes for ear, nose, and throat often focused on the interior of the ear. The benefit of endoscopes for diagnosis and therapy had been recognized early on with the advent of the laser and fiber optics (62–65). Recent investigations at the University of Ulm on the use of an Er:YAG laser with a germanium-oxide fiber delivery system has focused on tympanoplasty and stapedotomy (middle ear surgery) (66). The Er:YAG laser was found to be optimum for operating on the eardrum along the ossicles as far as the footplate without carbonization, and with sharp-edged, 0.2-mm-diameter canals "drilled" through the bone. Using this technique, children with mucotympanon could have their eardrums reopened in the doctor's office without the need for drain tubes.

An endoscope suitable for quantitatively examining the larynx (vocal chords) uses a green laser and a double reflecting mirror (67). The device can be clipped onto the shaft of a commercial rigid laryngoscope. The double reflecting mirror sends out two beams parallel to one another that allows for quantitative morphometry of laryngeal structures such as, vocal cords, glottis, lesions, and polyps.

The miniaturization and flexibility of fiber optics has allowed endoscopes to be applied in the small and delicate organ of the ear with much success. A case in point for the unique capabilities that the fiber optic endoscope has that can be applied to many fields of medicine in a very productive manner.

Future Directions

One pressing issue for "reusable" endoscopes is the ability to guarantee a clean, sterile device for more than one procedure. With the advent of the World Wide Web (WWW), many web sites are available to gain information on endoscopes, as well as the procedures they are used in. The U.S. Food and Drug Administration (FDA) has a site at their Center for Devices and Radiological Health (CDRH) that monitors medical devices and their performance in approved procedures. The FDA has also created guidelines for cleaning these devices.

The results of an FDA-sponsored survey, Future Trends in Medical Device Technology: Results of an Expert Survey in 1998, expressed a strong view that endoscopy and minimally invasive procedures would experience significant new developments during the next 5 and 10 year periods leading to new clinical products (68). The 15 participants included physicians, engineers, healthcare policy-makers and payers, manufacturers, futurists and technology analysts. In interviews and group discussions, survey participants expressed an expectation of continuing advancements in endoscopic procedures including fiber optic laser surgery and optical diagnosis, and a range of miniaturized devices. Clinically, most participants expected an emphasis on minimally invasive cardiovascular surgery and minimally invasive neurosurgery; two new fields we introduced in this edition. Also predicted were continuing advances in noninvasive medical imaging, including a trend to image-guided procedures. Most profound expectations were for developments in functional and multimodality imaging. Finally, participants observed that longer term trends might ultimately lead to noninvasive technologies. These technologies would direct electromagnetic or ultrasonic energy, not material devices, transdermally to internal body structures and organs for therapeutic interventions.

Perhaps a stepping-stone on this path is the PillCam (Given Imaging), a swallowable 11×26-mm capsule with cameras on both ends and a flashing light source, used to image the entire GI tract from the esophagus to the colon. The PillCam capsule has a field of view of $140°$, and enables detection of objects as small as 0.1 mm in the esophagus and 0.5 mm in the small bowel. Figure 11 shows the PillCams used for small bowel (SB) and esophagus (ESO) procedures and samples of the images obtained. Shown are examples of active bleeding, Crohn's disease, and tumor in the small bowel; and normal Z-line, esophagitis, and suspected Barrett's in the esophagus. Patient exam time is 20 min for an esophageal procedure and 8 h for a small bowel procedure. As the PillCam passes through the GI tract images are acquired at 2 Hz and the information is transmitted via an array of sensors secured to the patient's chest and abdomen and passed to a data recorder worn around the patient's waist. The PillCam generates $\sim 57,000$ images in a normal 8 h procedure, while the patient is allowed to carry on their normal activity. An obvious enhancement of patient comfort.

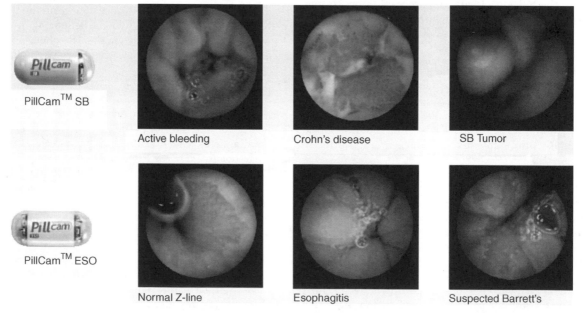

Figure 11. The PillCam from Given Imaging and sample images from the small bowel and esophagus. Shown are examples of active bleeding, Crohn's disease, and tumor in the small bowel; and normal Z-line, esophagitis, and suspected Barrett's in the esophagus. The PillCam is a swallowable 11×26 mm capsule with cameras on both ends and a flashing light source, used to image the entire GI tract from the esophagus to the colon.

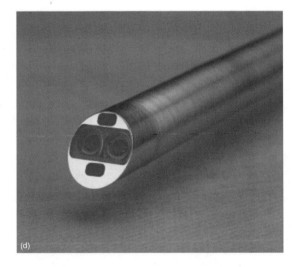

Figure 12. The da Vinci Surgical System consists of (a) a surgeon console, patient-side cart, instruments, and image processing equipment. The surgeon operates while seated at a console viewing a 3D image of the surgical field. The surgeon's fingers (b) grasp master controls below the display that translate the surgeon's hand, wrist, and finger movements into precise, real-time movements of the surgical instruments (c) inside the patient. The patient-side cart provides the robotic arms, with two or three instrument arms and one endoscope arm, that execute the surgeon's commands. The laparoscopic arms pivot at the 1 cm operating ports eliminating the use of the patient's body wall for leverage and minimizing tissue damage. The instruments are designed with seven degrees of motion that mimic the dexterity of the human hand and wrist. Operating images are enhanced, refined, and optimized for 3D viewing through the (d) 3D endoscopic camera system. (Courtesy of Intuitive Surgical, Inc.)

AESOP (formerly Computer Motion and now Intuitive Surgical) was the first robot, FDA approved in 1994, to maneuver a tiny endoscopic video camera inside a patient according to voice commands provided by the surgeon. This advance marked a major development in closed chest and port-access bypass techniques allowing surgeons direct and precise control of their operative field of view. In 1999, one-third of all minimally invasive procedures used AESOP to control endoscopes.

The da Vinci Surgical System (Intuitive Surgical) provides surgeons the flexibility of traditional open surgery while operating through tiny ports by integrating robotic technology with surgeon skill. The da Vinci consists of a surgeon console, a patient-side cart, instruments and image processing equipment (see Fig. 12). The surgeon operates while seated at a console viewing a 3D image of the surgical field. The surgeon's fingers grasp the master controls below the display with hands and wrists naturally positioned relative to their eyes, and the surgeon's hand, wrist, and finger movements are translated into precise, real-time movements of endoscopic surgical instruments inside the patient. The patient-side cart provides up to four robotic arms, three instrument arms, and one endoscope arm that execute the surgeon's commands. The laparoscopic arms pivot at the 1 cm operating ports and are designed with seven degrees of motion that mimic the dexterity of the human hand and wrist. Operating images are enhanced, refined, and optimized using image synchronizers, high intensity illuminators, and camera control units to provide enhanced 3D images of the operative field via a dual-lens three-chip digital camera endoscope. The FDA has cleared da Vinci for use in general laparoscopic surgery, thoracoscopic (chest) surgery, laparoscopic radical prostatectomies, and thoracoscopically assisted cardiotomy procedures. Additionally, the da Vinci System is also presently involved in a cardiac clinical trial in the United States for totally endoscopic coronary artery bypass graft surgery. This technology will likely find application in vascular, orthopedic, spinal, neurologic, and other surgical disciplines that will certainly enhance minimally invasive surgery.

Minimally invasive technologies that enhance the present state of endoscopy will continue. The expectation that microelectromechanical systems (MEMS) technology will add a plethora of miniaturized devices to the armament of the endoscopist is well founded, as it is an extension of the impact that fiber optics had on the field of endoscopy. The MEMS will likely add the ability to have light source and detector at the tip of the endoscope, instead of piping the light into and out of the endoscope. Many other functions including "lab on a chip" MEMS technology may allow for tissue biopsies to be performed *in situ*. This miniaturization will likely lead to more capabilities for endoscopes, as well as, the ability to access previous inaccessible venues in the body. Endoscopy is poised to continue its substantial contribution to minimally invasive procedures in medicine and surgery. This will pave the way for the likely future of noninvasive procedures in surgery and medicine.

BIBLIOGRAPHY

Cited References

1. Matthias Reuter, Rainer Engel, Hans Reuter. History of Endoscopy. Stuttgart: Max Nitze Museum Publications; 1999.

2. Wolf RFE, Krikke AP. The X-files of sword swallowing. Available at www.ecr.org/Conferences/ECR1999/sciprg/abs/p010189.htm.

3. Elner HD. Ein gastroskop. Klin Wochenschr 1910;3:593.

4. Sussmann M. Zur Diptrik des gastroskop. Ther Gegenw 1912;53:115.

5. Schindler R, Lehrbuch U. Atlas D Gastroskop, Munich: Lehmann 1923.

6. Hecht E. Optics. 4th ed. New York: Addison-Wesley; 2002.

7. Harrington JA. A review of IR transmitting, hollow waveguides. Fiber Integr Opt 2000;19:211–227.

8. Rave E, Ephrat P, Goldberg M, Kedmi E, Katzir A. Silver halide photonic crystal fibers for the middle infrared. Appl Opt 2004;43(11):2236–2241.

9. Mackanos MA, Jansen ED, Shaw BL, Sanghera JS, Aggarwal I, Katzir A. Delivery of midinfrared (6 to 7-μm) laser radiation in a liquid environment using infrared-transmitting optical fibers. J Biomed Opt 2003;8(4):583–593.

10. Hooper BA, Maheshwari A, Curry AC, Alter TM. A Catheter for Diagnosis and Therapy using IR Evanescent Waves. Appl Opt 2003;42:3205–3214.

11. Chaney CA, Yang Y, Fried NM. Hybrid germanium/silica optical fibers for endoscopic delivery of erbium:YAG laser radiation. Lasers Surg Med 2004;34:5–11.

12. Altman RD, Kates J. Arthroscopy of the knee. Semin Arthritis Rheum 1983;13:188–199.

13. Drez D, Jr. Arthroscopic evaluation of the injured athelete's knee. Clin Sports Med 1985;4:275–278.

14. Altman RD, Gray R. Diagnostic and Therapeutic Uses of the Arthroscope in Rheumatoid Arthritis and Osteoarthritis. New York: American Journal of Medicine; 1983. p 50–55.

15. Phillon DP Collins, JV. Current status of fiberoptic bronchoscopy. Postgrad Med J 1984;60:213–217

16. Mitchell DM, Emerson CJ, Collyer J, Collins JV. Fiberoptic bronchoscopy: Ten years on. Br Med J 1980;2:360–363.

17. Nissen S. Coronary angiography and intravascular ultrasound. Am J Cardiol 2001;87(4A):15A–20A.

18. Nissen SE. Who is at risk for atherosclerotic disease? Lessons from intravascular ultrasound. Am J Med 2002;(Suppl 8A):27S–33S.

19. Tuzcu EM, De Franco AC, Goormastic M, Hobbs RE, Rincon G, Bott-Silverman C, McCarthy P, Stewart R, Mayer E, Nissen SE. Dichotomous pattern of coronary atherosclerosis 1 to 9 years after transplantation: Insights from systematic intravascular ultrasound imaging. JACC 1996;27(4):839–846.

20. Nadkarni SK, Boughner D, Fenster A. Image-based cardiac gating for three-dimensional intravascular ultrasound imaging. Ultrasound Med Biol 2005;1:53–63.

21. Bourantas CV, Plissiti ME, Fotiadis DI, Protopappas VC, Mpozios GV, Katsouras CS, Kourtis IC, Rees MR, Michalis LK. In vivo validation of a novel semi-automated method for border detection in intravascular ultrasound images. Br J Radiol 2005;78(926):122–129.

22. Jobsis FF. Noninvasive, infrared monitoring of cerebral and myocardial oxygen sufficiency and circulatory parameters. Science 1977;198:1264–1267.

23. Parsons WJ, Rembert JC, Bauman RP, Greenfield Jr JC, Piantadosi CA. Dynamic mechanisms of cardiac oxygenation during brief ischemia and reperfusion. Am J Physiol 1990;259 (Heart Circ. Physiol. 28):H1477–H1485.

24. Parsons WJ, Rembert JC, Bauman RP, Greenfield Jr JC, Duhaylongsod FG, Piantadosi CA. Myocardial oxygenation in dogs during partial and complete coronary artery occlusion. Circ Res 1993;73(3):458–464.

25. Nilsson M, Heinrich D, Olajos J, AnderssonEngels S. Near infrared diffuse reflection and laser-induced fluorescence spectroscopy for myocardial tissue characterisation. Spectrochim Acta Part A-Mol Biomol Spectrosc 1997;51(11):1901–1912.

26. Manoharan R, Baraga JJ, Rava RP, Dasari RR, Fitzmaurice M, Feld MS. Biochemical-analysis and mapping of atherosclerotic human artery using FT-IR microspectroscopy. Atherosclerosis 1993;103(2):181–193.

27. Baraga JJ, Feld MS, Rava RP. Detection of atherosclerosis in human artery by midinfrared attenuated total reflectance. Appl Spectrosc 1991;45(4):709–710.

28. Rava RP, Baraga JJ, Feld MS. Near-infrared Fourier-Transform Raman spectroscopy of human artery. Spectrochim Acta Part A-Mol Biomol Spectrosc 1991;47(3–4):509–512.

29. Arai T, Mizuno K, Fujikawa A, Nakagawa M, Kikuchi M. Infrared absorption spectra ranging from 2.5 to 10 μm at various layers of human normal abdominal aorta and fibrofatty atheroma in vitro. Laser Surg Med 1990;10:357–362.

30. Casscells W, Hathorn B, David M, Krabach T, Vaughn WK, McAllister HA, Bearman G, Willerson JT. Thermal detection of cellular infiltrates in living atherosclerotic plaques: possible implications for plaque rupture and thrombosis. Lancet 1996;347(9013):1447–1449.

31. Colarusso P, Kidder L, Levin I, Fraser J, Arens J, Lewis EN. Infrared spectroscopic imaging: from planetary to cellular systems. Appl Spectrosc 1998;52(3):106A–119A.

32. Kodali DR, Small DM, Powell J, Krishna K. Infrared microimaging of atherosclerotic arteries. Appl Spectosc 1991;45:1310–1317.

33. Edwards G, Logan R, Copeland M, Reinisch L, Davidson J, Johnson J, Maciunas R, Mendenhall M, Ossoff R, Tribble J, Werkhaven J, O'Day D. Tissue ablation by a Free-Electron Laser tuned to the Amide II band. Nature (London) 1994;371:416–419.

34. Awazu K, Nagai A, Aizawa K. Selective removal of Cholesterol Esters in an Arteriosclerotic region of blood vessels with a free-Electron Laser. Laser Surg Med 1998;23:233–237.

35. Holmes DR, Bresnahan JF. Interventional cardiology. Cardiol Clin 1991;9:115–134.

36. Linsker R, Srinivasan R, Wynne JJ, Alonso DR. Far-ultraviolet laser ablation of atherosclerotic lesions. Laser Surg Med 1984;4:201–206.

37. Hughes GC, Kypson AP, Yin B, St Louis JD, Biswas SS, Coleman RE, DeGrado TR, Annex BH, Donovan CL, Lanolfo KP, Lowe JE. Induction of angiogenesis following transmyocardial laser revascularization in a model of hibernating myocardium: a comparison of holmium:YAG, carbon dioxide, and excimer lasers. Surgical Forum L 1999;115–117.

38. Puliafito CA, Steinert RF, Deutsch TF, Hillenkamp F, Dehm EJ, Alder CM. Excimer laser ablation of cornea and lens: experimental studies. Ophthalmology 1985;92:741–748.

39. Cummings JP, Walsh Jr. JT Erbium laser ablation—the effect of dynamic optical properties. Appl Phys Lett 1993;62:1988–1990.

40. Walsh PC, Gittes RF, Perlmutter AD, Stamey TS, editors. Campbell's Urology. Philadelphia: W. B. Saunders; 1986. p 510–540.

41. Segura JW. Endourology. J Urol 1984;132:1079–1084.

42. Powell PH, Manohar V, Ramsden PD, Hall RR. A flexible cytoscope. Br J Urol 1984;56:622–624.

43. Hoffman JL, Clayman RV. Endoscopic visualization of the supravesical urinary tract: Transurethral ureteropuleloscopy and percutaneous nephroscopy. Semin Urol 1985;3:60–75.

44. Benzie RJ. Amniocentesis, amnioscopy, and fetoscopy. Clin Obstet Gynecol 1980;7:439–460.

45. Rodeck CH, Nicolaides KH. Fetoscopy and fetal tissue sampling. Br Med Bull 1983;39:332–337.

46. Rauskolt R. Fetoscopy. J Perinat Med 1983;11:223–231.

47. Sleisenger MH, Fordtran JS, editors. Gastrointestinal Disease. Philadelphia: Saunders; 1983. p 1599–1616.

48. Sleisenger MH, Fordtran JS, editors. Gastrointestinal Disease. Philadelphia: Saunders; 1983. p 1617–1626.

49. Silvis SE, editor. Therapeutic Gastrointestinal Endoscopy. New York: Igaku-Shoin; 1985. p 241–268.

50. Huizinga E. On esophagoscopy and sword swallowing. Ann Otol Rhinol Laryngol 1969;78:32–34.

51. Shaheen N, Ransohoff DF. Gastroesophageal reflux, Barrett esophagus, and esophageal cancer: scientific review. JAMA 2002;287:1972–1981.

52. Spechler SJ. Clinical practice. Barrett's Esophagus. N Engl J Med 2002;346:836–842.

53. Sampliner RE. Updated guidelines for the diagnosis, surveillance, and therapy of Barrett's esophagus. Am J Gastroenterol 2002;97:1888–1895.

54. Wolfsen HC, Hemminger LL, DeVault KR. Recurrent Barrett's esophagus and adenocarcinoma after esophagectomy. BMC Gastroenterology 2004;4:18.

55. Jean M, Dua K. Barrett's Esophagus: Best of Digestive Disease Week 2003. Curr Gastroenterol Rep 2004;6:202–205.

56. Hunt RH, Waye JD, editors. Colonoscopy. London: Chapman & Hall; 1981. p 11–18.

57. Ohligisser M, Sorokin Y, Hiefetz M. Gynecologic laparoscopy, a review article. Obstet Gynecol Surv 1985;40:385–396.

58. Robinson HB, Smith GW. Application for laparoscopy in general surgery. Surg Gynecol Obstet 1976;143:829–834.

59. Devaux BC, Joly L, Page P, Nataf F, Turak B, Beuvon F, Trystram D, Roux F. Laser-assisted endoscopic third ventriculostomy for obstructive hydrocephalus: Technique and results in a series of 40 consecutive cases. Lasers Surg Med 2004;34:368–378.

60. Cram GP, Copeland ML. Nucl Instrum Methods Phys Rev B 1998;144:256.

61. Edwards G et al. Free-electron-laser-based biophysical and biomedical instrumentation. Rev Sci Instrum 2003;74:3207–3245.

62. Paparella MM, Shumrick DA, editors. Otolaryngology Philadelphia: W.B. Saunders; 1980. p 2410–2430.

63. Ballenger JJ, editor. Diseases of the Nose, Throat, Ear, Head, and Neck. Philadelphia: Lea & Febiger; 1985. p 1293–1330.

64. Steiner W. Techniques of diagnostic and operative endoscopy of the head and neck. Endoscopy 1979;1:51–59.

65. Vaughan CW. Use of the carbon dioxide laser in the endoscopic management of organic laryngeal disease. Otolaryngol Clin North Am 1983;16:849–864.

66. Pfalz R, Hibst R, Bald N. Suitability of different lasers for operations ranging from the tympanic membrane to the base of the stapes. Adv in Oto-Rhino-Laryngol 1995;49:87–94.

67. Schade G, Leuwer R, Kraas M, Rassow B, Hess M. Laryngeal morphometry with a new laser 'clip on' device. Lasers Surg Med 2004;34:363–367.

68. Future Trends in Medical Device Technology: Results of an Expert Survey. Available at http://www.fda.gov/cdrh/ost/trends/TOC.html.

Further Reading

Ponsky JL. Atlas of Surgical Endoscopy. St. Louis (MO): Mosby-Year Book; 1992.

Barkin J, O'Phelan C, editors. Advanced Therapeutic Endoscopy. New York: Raven Press; 1990.

Ovassapian A. Fiberoptic Airway Endoscopy in Anesthesia and Critical Care. New York: Raven Press; 1990.

Niemz M. Laser-Tissue Interactions: Fundamentals and Applications. 2nd ed. Berlin Heidelberg: Springer-Verlag; 2002.

Infrared Fiber Systems. Available at http://www.infraredfibersystems.com.

Polymicro Technologies, LLC. Available at http://www.polymicro.com.

Omniguide Communications, Inc. Available at http://www.omniguide.com.

LightLab Imaging, Inc. Available at http://www.lightlabimaging.com.

Given Imaging. Available at www.givenimaging.com.

Intuitive Surgical, Inc. Available at www.intuitivesurgical.com.

See also ESOPHAGEAL MANOMETRY; FIBER OPTICS IN MEDICINE; MINIMALLY INVASIVE SURGERY.

ENGINEERED TISSUE

GREGORY E. RUTKOWSKI
University of Minnesota-Duluth
Duluth, Minnesota

INTRODUCTION

History of Tissue Engineering

In 1988, researchers gathered at the Granlibakken Resort in Lake Tahoe, CA under the sponsor ship of the National Science Foundation (NSF) to develop the fundamental principles of tissue engineering as an emerging technology. Based on an earlier proposal by Dr. Y.C. Fung to develop an Engineering Research Center focused on the engineering of living tissues, NSF held several meetings that led to the decision to designate tissue engineering as a new emerging field. A formal definition was finally agreed upon at the Granlibakken workshop. Based on this meeting, tissue engineering is defined as "the application of the principles and methods of engineering and the life sciences toward the fundamental understanding of structure–function relationships in normal and pathological mammalian tissues and the development of biological substitutes to restore, maintain, and improve function" (1). This was further refined in 1992 by Eugene Bell who developed a list of more specific goals:

1. Providing cellular prostheses or replacement parts for the human body.
2. Providing formed acellular replacement parts capable of inducing regeneration.
3. Providing tissue or organ-like model systems populated with cells for basic research and for many applied uses such as the study of disease states using aberrant cells.
4. Providing vehicles for delivering engineered cells to the organism.
5. Surfacing nonbiological devices (2).

These discussions eventually culminated in the pioneering review article by Langer and Vacanti in 1993 (3). The general strategies to create engineered tissue would include the isolation of cells or cell substitutes, the use of tissue-inducing substances, and development of three-dimensional (3D) matrices on which to grow tissue.

Much of tissue engineering owes its beginnings to reconstructive surgery and internal medicine. In the sixteenth century, Gaspare Tagliacozzi developed a method for nasal reconstruction using flaps of skin taken from the arm and grafted onto the injury site (4). With the scientific development of the germ theory of disease and the sterile techniques that were introduced, modern surgery became established as a means to treat patients with internal injuries. In the late nineteenth century, surgeons used veins and arteries as conduits to enhance the nerve regeneration (5). World War I saw improvements in reconstructive surgery as doctors were able to hone their skills due to the number of soldiers injured in battle. Reconstructive surgery had it limitations in terms of the availability of biological material. By the 1940s, much progress had been made in understanding the function of the immune system is accepting tissue from a donor. This eventually led to the first successful organ transplant (kidney) in 1954. The next 50 years, would see tremendous advances in organ transplants as well as in immunosuppressive drug therapy.

An alternative to organ transplant has been the development of artificial devices to mimic biological function. The mid-nineteenth century also saw the rise in the use of prosthetic limbs that would initiate the use of artificial devices to replace biological functions. Artificial limbs were used as far back as the Dark Ages to assist knights heading off to battle. The intervening centuries saw improvements over such devices through the use of stronger, lighter materials, and a better understanding of biomechanics (6). Besides limbs, artificial devices have also been invented to replace the function of certain internal organs. The dialysis machine was created in the 1940s to assist patients with acute renal failure. In 2001, an implantable artificial heart was first used in a human. While many advances have been made in artificial devices, some of the drawbacks include the breakdown of the artificial materials, a lack of interaction with the human body, and the inability to self-renew.

The modern era of tissue engineers seeks to overcome the limitations of reconstructive surgery, organ transplantations, and prosthetic devices by creating functional, completely biocompatible tissues and organs. Since the late 1980s, the field of tissue engineering has grown exponentially and continues to draw scientists from diverse fields, from the biological and medical sciences to engineering and materials science.

THEORY

Tissue engineering adds to the modern health care system by providing the tools to assist in the repair of tissue and organs damaged by injury and disease. An exact replica of the tissue could potentially be grown in the lab and later inserted into the patient. Alternatively, precursors may be placed in the body with the expectation that it will develop into fully formed functional tissue. Also, devices may be implanted into the body to encourage the regeneration of already existing tissue in the body.

Engineered tissue is created by combining relevant cells and chemical factors within a 3D matrix that serves as a

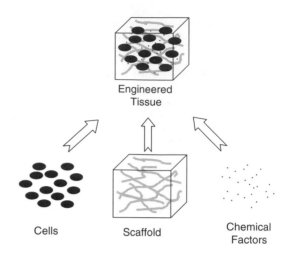

Figure 1. Basic components of engineered tissue.

scaffold (Fig. 1). Sterile cell culture techniques allows for the expansion of cells *in vitro* to obtain sufficient quantities for use in engineered tissue. Cells can originate from the patient, another donor, or even animal tissue. As our understanding of biochemical cues during development expands, tissue formation can be better controlled through the delivery of pertinent growth factors or through the interaction of complex multiple cell cultures. Scaffold materials may be biological (collagen, extra cellular matrix) or synthetic. Synthetic materials can be designed to mimic the extra cellular matrix and may be infused with chemical factors to support tissue regeneration. Since the discovery and availability of synthetic polymers in the 1940s, scientific advances have made these materials biodegradable and biocompatible.

Cellular Processes

During the formation of tissue, either during development or in the lab, cells undertake many different processes. In order to maintain the integrity of the tissue, the cells tissue must adhere to others as well as to the surrounding matrix material. Certain cells must migrate through the 3D space in order to properly position themselves within the tissue. Once in position, they must also continue to multiply to provide adequate tissue growth. Vascularization and innervation of the tissue is required to complete integration of the engineered tissue with its surroundings. Tissue derived from donor material also has immune concerns.

Development. Engineered tissue is best created by mimicking the processes that occur during development (7). Following fertilization, the zygote divides until the blastocyst is formed. The early embryonic tissue is comprised of two cell phenotypes: epithelial cells and migratory mesenchymal cells. The transformation between these cells is regulated by growth factors, the extra cellular matrix, and intracellular signaling. As the embryo develops, these cells will become diversified as they commit to various tissue forms. These cells will eventually differentiate into the various cell types found in the body.

Morphogenesis describes the cascade of events that leads to the spatial pattern formation of the developing embryo. Regeneration of tissue mimics this process. Specific chemical factors, called morphogens, provide information on the pattern via diffusion. The combination of various morphogens due to diffusion leads to complex pattern formation. One well-studied example is the development of bone. Several bone morphogenic factors have been isolated and observed to affect the bone and cartilage morphogenesis and formation. Also, BMPs have been implicated in the development of tissue as diverse as skin, eye, heart, and kidney. These morphogens have the potential to assist in the engineering of bone tissue.

Adhesion. Tissue is held together by the adhesion processes between different cells and the cells and extracellular matrix. The cell–cell interactions consist of tight, anchoring, and communication junctions. Tight junctions are composed of transmembrane proteins that form connections between cells. More contact points will decrease the permeability between the cells so that the tissue can become essentially impermeable. This is typically found with epithelial cells, as in the intestinal, reproductive, and respiratory tract, where the barrier that is formed is essential to function.

Anchoring junctions are loosely held connections that take advantage of the cytoskeleton. With adherens junctions, actin filaments of the cytoskeleton are connected and linked to integrins on the cell exterior. These integrins form focal contacts that interact with cadherins found on the surface of other cells. The focal contacts can also interact with extracellular domains. With the involvement of the cytoskeleton, these junctions can also affect cell function by providing a mechanism to signal changes in cell growth, survival, morphology, migration, and differentiation.

Desmosomes are similar to adherens junctions, but they involve intermediate filament protein, such as vimentin, desmin, and keratin. These junctions connect with other cells via cadherins. A similar junction called the hemidesmosome behaves in the same manner, but connects with basal lamina proteins via integrin.

Communication junctions are those that provide direct communication between cells. These are large proteins that form pore structure that connects the two cells. These large pores (connexons) allow the transport of molecules between cells. These junctions are commonly found between neurons for rapid signal conduction.

In order to connect with other cells or the extracellular matrix, the junction proteins must interact with some receptor molecule. Integrins will bind with the amino acid sequence arginine-glycine-aspartic acid (RGD). This sequence is found in several proteins, such as collagen and fibronectin and these peptides can be incorporated onto other surfaces in order to improve cell adhesion. Cadherin–cadherin binding is mediated via Ca^{2+}. Without the presence of Ca^{2+}, the connection is subject to proteolysis. The Ig-like receptors contain motifs found in immunoglobulins. These receptors can interact with neural cell adhesion molecule (N-CAM) and are present during development. Finally, selectins are specific receptor types that are expressed during the inflammatory response. The lectin domain found in oligosaccharides on neutrophils allows these cells to interact with endothelial cells along the surface of blood vessels.

Migration. During embryogenesis, diffusible factors and ECM composition are important factors in the pattern formation of tissue development. The migration of cells is necessary for the formation of tissue not just for development, but also during regeneration. Cell migration is also observed in other body functions. Angiogenesis, or blood vessel formation, involves the migration of endothelial cells into new tissues. For immune responses, B and T cells patrol the body ready to attack invaders. Tumor invasion and, more importantly, metastasis relies on the migration of cancer cells into other parts of the body. Controlling migration can help to control the spread of cancer.

Signals from various factors can lead to the release of cells from their contact with other cells or with the extracellular matrix. Once released, different mechanisms can affect the migration pattern of the cell depending on the cell type. Cells may move in random directions. The movement can be modeled by Brownian motion. In this case, motion is due to collisions with other particles. This motion is characterized by the time between collisions, the mean free path (average distance before hitting another particle) and the average speed. The characteristics are dependent on the density of the particles.

Of more relevance is the issue of directed migration. Direct migration depends on some sort of gradient. In response to some kind of stimulus, cells may move toward or away from the source of the stimulus. The stimulus may affect speed, direction, or both. Chemotaxis is the general term describing the response of cells to a chemical gradient. The strength of the effect is dependent on the absolute concentration as well as the steepness of the gradient.

Growth. Mitosis is a tightly controlled process to regulate cell growth and depends on signals from the cell's environment. During mitosis, the deoxyribonucleic acid (DNA) is replicated and copies are separated as the cells divide into two exact copies. This process repeats itself depending on the extracellular signaling and the properties of the cell itself. Certain cells, such as pancreatic beta cells, are locked in the cell cycle and mitosis is arrested. In order to overcome this barrier, stem cells can be used to generate new differentiated cells. Also, cells may be converted into a precursor cell form that can undergo proliferation before switching back to the differentiated form.

Cell growth relies on the availability of essential nutrients. As the need for nutrients increases during cells proliferation, the availability becomes reduced barring some mechanism for distributing the nutrients to the growing tissue. Distribution of nutrients occurs naturally in the human body in the form of the network of blood vessels. The lack of such a vasculature for engineered tissue limits the effective size that tissue can grow.

Vascularization. For most engineered tissue to become integrated into the human body, it must become vascularized by the body's own blood vessel network. The body provides a natural mechanism for neovascularization by the process of wound healing. When tissue has been damaged, a cascade of events is initiated to block the bleeding wound and encourage healing. Platelets, fibrin, and fibronectin first form a mesh plug. Mast cells release chemotactic agents to recruit other cells as part of the inflammatory response. Keratinocytes migrate to the site of the injury and begin to proliferate. Vascular endothelial growth factor (VEGF) and fibroblast growth factor (FGF) are released to encourage blood vessel formation. Once the new tissue has been vascularized, remodeling of the tissue occurs to complete the healing process.

For engineered tissue, two options are used to ensure vascularization and promote tissue growth and integration. The engineered tissue can be designed to contain a vascular network *in vitro* that would then be connected to the body's own network when it is implanted. This presents a significant engineering challenge in trying to create several different types of engineered tissue in one construct simultaneously. An alternative method is to engineer the tissue to recruit blood vessels from the body's existing framework. This has been accomplished though the controlled release of VEGF from the biodegradable support matrix of the engineered tissue (8).

Innervation. To become fully integrated into the body, certain tissues and organs must also reconnect with the body's own nervous system. Several organs of the body form connections with the sympathetic nervous system. These connections are important to regulation of the organ. Skeletal muscle tissue makes sensory and motor connections via the peripheral nervous system. In any case, the cells of these engineered tissues need to make synaptic connection with the axons of the relevant neurons. Because the new tissue is being engineered to replace that which has been lost to injury or disease, the neural framework may not be available for integration. If it is present, the neurons may be encouraged to regenerate toward the tissue and form new connections. Another complex, but theoretically possible, option may be to engineer the tissue with neural connections that can be later integrated into the existing nervous system. Also, artificial devices may be integrated into the system to provide the appropriate control of the tissue function.

Immune Concerns. As with organ transplants, engineered tissue also has concerns of rejection by the immune system. The major histocompatibility complex I (MHC I) present on the cells of the engineered tissue are recognized by the T cells of the immune system as a foreign body. This would eventually lead to cell lysis. The primary means of preventing rejection, though, is the use of immune suppressant drug therapy. While this may be acceptable, for a patient receiving a life saving organ transplant, it is not for those receiving engineered tissue. Certain tissues, such as cartilage, may not interact with the immune system. Metabolic tissues that only interact chemically can be physically separated from the environment. Some engineered tissues may contain cells that are only temporary until the body's own cells can take over. In these cases, immune suppressant drug may be a viable option. For most other tissues, methods are being developed for side stepping the immune system.

Because the cells of the immune system are formed within the bone marrow, one method is to transplant the bone marrow from the cell and tissue donor along with the organ. The donated bone marrow can form a chimera with the patient's existing bone marrow to allow the adaptation of the immune system to the new tissue (9).

At the molecular level, the antigen of the foreign cell can be blocked or completely eliminated. In order to block the antigen, a fragment of antibody to the antigen can be added to the tissue to mask the foreign cells from the patient's own immune system (10). This effect is temporary as the fragments will eventually separate from the antigen. Another drawback is that is may not counter all the mechanisms of immune response. The antigen can also be removed by placing an inactive form of the antigen gene into the cells (11). A gene can also be added to the cells to produce a protein that will inhibit rejection (12). Finally, oligonucleotides can be added to hybridize with either ribonucleic acid (RNA) or DNA in order to inhibit the transcription or translation of the antigen molecule (13).

Microencapsulation is a means of physically separating the cells from the environment. For this method, cells are surrounded by a porous synthetic material. As long as the membrane coating remains intact, the cells are isolated from the immune system. The pore size can be adjusted to allow chemical interaction with the environment while preventing cellular interaction. The encapsulation should allow nutrients to permeate through the membrane and reach the cells. This was first used in clinical studies for the encapsulation of xenogenic pancreatic islets (14).

Cell Types. While the response of the immune system is of great importance to the success of the engineered implant, the source of cells and their application will determine the best method for immunomodulation. Cells may come from the patients themselves (autologous), from a human donor (allogeneic), or from another species (xenogeneic). Each type has its own advantages and disadvantages.

Autologous cells are derived from the patient, expanded in culture, and then placed back into the patient. These cells are completely biocompatible with the patient and this eliminates the need for any immune modulation. Genzyme, for example, has developed a successful protocol for the repairing articular cartilage. One drawback to using autologous cells is that it requires a period of time to expand the cells. This would not be acceptable for patients needing an immediate tissue replacement. As a result, the engineered tissue does not have off-the-shelf availability. Depending on the tissue and the amount of damage, the amount of cells that may be harvested from the patient may be insufficient to form tissue.

Allogeneic cells can help to overcome some of the drawbacks to autologous cells because the cells come from donor sources that may be pooled together. This can provide off-the-shelf availability, but at the expense of immune rejection. The cells can be engineered to become immune acceptable or immunosuppressive drug therapy may be used. These cells are also well suited for tissues that do not interact with the patient's vasculature system or may only be used until the native tissue regenerate.

Advanced Tissue Science developed a skin replacement tissue (Dermagraft) with cells derived from circumcision surgeries.

Xenogeneic cells are derived from nonhuman species. An animal source can provide an unlimited amount of cells, but they provoke an acute immune response within minutes of implantation into the body. One useful application of such cells is the encapsulation of pancreatic islets (14,15). The membrane can be adjusted to allow the islets to regulate blood sugar and insulin levels while protecting the cells from the immune system. Another drawback to xenogeneic cells is the threat of transmission of animal viruses and well as endogenous retroviruses that may interact with the patient.

Cells that have been isolated may be modified to alter their characteristics before being incorporated into engineered tissue. Primary cells that are subcultured eventually become cell lines. These cell lines may be finite or continuous. The cells may also be normal or transformed compared to the primary cell from which it is derived. Transformation is associated with genetic instabilities that may lead to a change in the phenotype. These changes may impart different growth characteristics for the cell, such as immortalization, anchorage independence, and overall growth rate. The genetic material may be altered to effect gene expression. Changes in protein expression may effect the secretion of chemical factors or even the formation of extracellular matrix. In the worst case, cells may become cancerous and prove to be detrimental to the patient. Examples of stable, well-characterized cell lines used in tissue engineering include HEK-293 for nerve growth factor secretion (16), and 3T3-L1 for adipose tissue formation (17).

Primary cells as well as cell lines can be artificially transformed by introducing new genes by the process of transfection. In this process, a new gene is carried by some vector, such as a liposome or virus, to the host cells. The gene is transferred into the cells and delivered into the nucleus where it can be transcribed. This method allows the creation of cells with desirable characteristics for engineered tissue. Such cells may secrete growth factor that will enhance tissue formation. Examples of engineered tissue utilizing transfected cells include insertion of hBMP-4 gene in bone marrow stromal cells for bone tissue engineering (18) and aquaporin gene transfection in the LLC-PK1 cell line for a bioartificial renal device (19), secretion of neuronal growth factors from fibroblasts to enhance axonal regeneration (20).

Regardless of the source, primary cells have limitations depending on their age and genetic makeup. Adult cells may only have a limited number of cell divisions before they reach senescence or even become cancerous. Transformed cell lines may allow for continuous growth, but without proper control can become a problem for the patient. These cells are typically used to enhance regeneration of existing tissue, but do not become integrated into the body. Because of these issues, a universal cell that can develop into multiple tissue types has an infinite capacity to proliferate without loss of function, and if immune acceptable would be ideal for engineered tissue. Stem cells come closest to meeting these criteria.

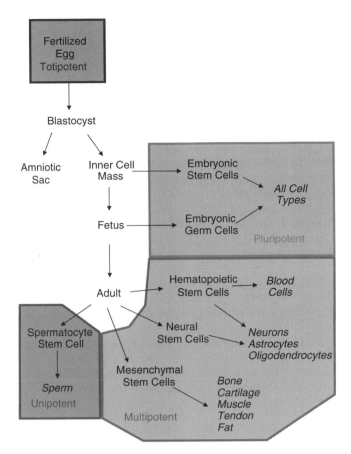

Figure 2. Stem cell potency. Relationship of stem cells to various stages of development. Cells and tissue derived from stem cells are shown in italics.

Stem Cells. The potency of stem cells is defined by their ability to differentiate into one or more cell genotypes (see Fig. 2). Unipotent stem cells give rise to only one cell type (*ex. Spermatogonia*). Multipotent stem cells are more functional and can differentiate into multiple cell types. Embryonic stem cells are considered pluripotent in that they can form all the cell types found in an embryo and adult. The fertilized egg develops into an embryo as well as the amniotic sac and is considered totipotent. Embryonic stem cells hold the most promise for engineering tissue, but more work must be done on the basic principles of stem and progenitor cell biology as well as on the control of differentiation.

Multipotent stem cells obtained from embryonic or adult sources are considered the optimal choice for tissue engineering applications because of their ability to form multiple tissue type. Hematopoietic stem cells (HSCs) are the most well characterized of stem cells. While isolated from bone marrow, HSCs can differentiate to form skeletal muscle, cardiac muscle, hepatocytes, endothelial cells, and epithelial cells. Multipotent stem cells can be found in several other tissues, such as brain, heart, pancreas, retina, liver, and lung, and skin.

To take full advantage of the functionality of stem cells, more work needs to be done to elucidate the mechanisms for differentiation. Once established, stem cells can be encouraged to differentiate into the required cell type.

The mechanism may rely on an environmental cue as to whether they are physical contact, chemical, or chemotactic in nature. The ability of stem cell to proliferate, integrate, and differentiate also depends on the methods of identifying, isolating, and expanding. Protocols for stem cell culture need to be developed and optimized to ensure the cells achieve their full potential.

Since some stem cells are harder to obtain than others, getting stem cells to transdifferentiate from one form to another would allow for further flexibility. Evidence suggests that bone marrow stromal cells may convert to neural stem cells (21) and neural stem cells to hematopoietic stem cells (22). Recent evidence, though, points to the fusion of stem cells with other cells instead of true transdifferentiation (23).

Another source for stem cells is from embryos discarded from *in vitro* fertilization clinics. Human embryonic stem cells have an even greater capacity to form tissues than the multipotent stem cells. These cells have been isolated from humans and protocols for long-term cultures have successfully been developed (24). The use of embryonic stem cells has raised many ethical concerns because of the destruction of the fetus from which they are derived. This has led to legislation tightly regulating their use. To avoid these concerns, several proposals have been suggested to obtain embryonic stem cells without compromising the potential for life (25).

Stem cells have been used successfully for tissue engineering applications. Stem cells have been seeded on scaffolds to form cartilage, small intestine, and bone (26). Neural stem cells have also been used for repair of spinal cord injuries (27). Embryonic stem cells have also been seeded onto scaffolds where the native vasculature integrated into the engineered tissue (28).

Instead of preseeding cells on scaffolds, stems cells may also be injected directly to the site of injury to promote tissue regeneration. Clinical studies in nonhuman primates have shown that neural stem cells can enhance repair of spinal cord injuries (29). Stem cells have also been used to regenerate damaged cardiac muscle tissue (30). Embryonic stem cells may also be used to regenerate pancreatic beta cells in order to alleviate diabetes (31). Mesanchymal stem cells also show some promise for the repair of articular cartilage for those suffering from osteoarthritis (32).

Scaffolds

Primary cells will form monolayer cultures when dissociated from tissue. In order to encourage the cells to form engineered tissue, they must be placed in a 3D matrix that acts as a support scaffold. Ideally, these scaffolds should be compatible with the cells as well as being biodegrade. The scaffold should have a high porosity to ensure the diffusion of nutrients and other chemical factors as well as provide room for cell migration and proliferation. The material should have a higher surface area to ensure adequate room for cell attachment. The matrix should maintain its structural integrity until tissue integration has been completed. For certain applications, the final construct may need to be formed into a specific 3D shape (see Fig. 3).

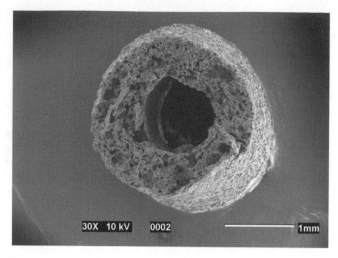

Figure 3. Example of a scaffold for engineered tissue. This scaffold is designed for nerve regeneration. The cylindrical conduit was manufactured by using the solvent casting method (see below) with particle leaching. Poly(vinyl alcohol) rods were coated with a suspension of salt crystals in a poly(lactic acid) (PLA)–chloroform solution. Dissolution of the rods permitted the formation of the large central pore through which nerve tissue could regenerate. A porosity of 85% was obtained by controlling the volume fraction of salt crystals.

For tissue engineering, scaffolds may come in different forms. An acellular matrix can be used to recruit cells from the host tissue (33). With this form, the immune response is not a concern though inflammation may occur. For most applications, cells must be seeded onto the scaffold to ensure successful tissue formation. Cells can be seeded onto collagen gels that mimic the naturally occurring extracellular matrix (34,35). Cells can also be encouraged to self-assemble in culture. These systems are ideal for tissue that forms simple shapes, such as sheets and cylinders (36,37). Biodegradable polymers provide a flexible means of creating scaffolds with properties desirable for tissue formation.

Natural Polymers. Macromolecules found with the extracellular matrix provide chemical and structural support for living tissue. Because of their function in tissue formation, these natural polymers provide an excellent scaffold for engineering new tissue. These natural polymers have some limitations in their application. The mechanical properties and degradation rates cannot be controlled as well as synthetic polymers (38). Also, material derived from donors may elicit an immune response (39). Many of these limitations can be overcome by chemical modification of the materials as well as creating composite materials of natural polymers or natural–synthetic polymers.

Collagen is a large family of proteins that make up much of the extracellular matrix. At least 19 different types have been isolated. Fibrillar collagen is comprised of several collagen types that combine to form extended structures that assist is maintaining tissue integrity. The relative amounts of collagen are characteristic of the tissue type. Collagen has been used for cartilage (40), nerve regeneration (41), gallbladder engineering (42), corneal tissue (43), and skin (44). It can be cross-linked to enhance mechanical properties (45). Magnetically aligned collagen fibers can also be used to enhance nerve regeneration (46).

Matrigel is comprised of proteins found in the basal lamina. The basal lamina, or basement membrane, is a specialized area of the extracellular matrix found at epithelial–stromal boundaries. It is composed of collagen, laminin, nidogen, and perlecan, (a heparin sulfate proteoglycan). It has been used for spinal cord repair (47) vascular network formation (48), and cardiac tissue (49).

Alginates are synthesized naturally by brown seaweeds as well as some bacteria. They are comprised of units of mannuronic and guluronic acid. The proportion and repeat unit sequence is dependent on the organism of origin. Alginates will form a gel in the presence of calcium ions. Alginates have been used for cartilage (50), cardiac tissue (51), and liver engineering (52).

Fibrin is formed from the polymerization of fibrinogen that occurs during the wound healing process. The process is controlled by the presence of thrombin. Fibrin has been used for cardiovascular repair (53), bone (54), cartilage (55), and skin (56). Like collagen, the fibrin chains can be cross-linked to alter the characteristics of the matrix (57).

Chitosan is derived from chitin, a polysaccharide that comprises the exoskeleton of crustaceans. The amide group on chitin is replaced by an amine group. Since chitosan is not protein based, it does not elicit an immune response. Chitosan has been used for the engineering of cartilage (58) and bone (59). It has also been combined with other materials to create novel composite matrices (60,61).

Besides isolating single components as scaffolding material, intact decellularized tissue can also be used. In this method, tissue is obtained from a donor and the cells are destroyed chemically using compounds, such as sodium dodecyl sulfate. This methods is currently used to engineer blood vessels (62) and heart valves (63). Decellularized tissue may also be applied to the engineering of other types of tissues, such as utilizing porcine skin for urinary tract reconstruction (64).

Synthetic Polymers. While natural polymers are an excellent choice for scaffolds because of their biocompatibility, synthetic polymers offer more flexibility. Composites with natural polymers can also be made to take advantage of their properties as well. Biodegradable polymers also have the additional benefit of decomposing into nontoxic metabolites that can be removed from the body as waste. The molecules can be modified to obtain mechanical properties and appropriate degradation rates that are well suited to the specific application.

The most common polymers in use are PLA and poly(glycolic acid) (PGA). These are the first two polymers to obtain approval by the Food and Drug Administration (FDA) for use in medical implants. During degradation, the polymer is hydrolyzed to form lactic and glycolic acids. Since PLA is more hydrophobic than PGA, it has a slower degradation rate. As such, the degradation rate of the polymers can be easily adjusted by changing the ratio of lactic acid to glycolic acid within copolymers of PLA and PGA, poly(lactic-*co*-glycolic acid) (PLGA). The degradation

rate is also affected by the overall molecular weight and extent of crystallinity of the polymer. During degradation, the polymer swells with water. The nonspecific interactions between the water and polymer lead to hydrolysis at random locations along the polymer chain. This leads to bulk degradation of the matrix that can be detrimental to the release profile of drugs. The hydrophobicity of PLA also makes the matrix less amenable for cell adhesion. Adsorption of proteins, such as laminin (65), to the matrix surface encourages cell adhesion. Amines can also be incorporated into the polymer chain to control protein and cell attachment (66), because many types of tissue have been engineered using PLA, PGA, as well as combinations of their stereoisomers and copolymers (see Ref. 67 for a review).

The limitations due to bulk degradation can be overcome using matrices based on polyanhydrides. These polymers degrade only at the surface. When used for drug delivery, the release profiles can be affected just by altering the shape of the matrix. For tissue engineering, these polymers provide a mechanism for renewing the surface to allow new areas for cells to grow. As the surface degrades, cells will slough off providing a fresh surface for cell growth. One example of such a polymer, recently approved by the FDA for use in the treatment of glioblastoma multiformae, is derived from bis(*p*-carboxyphenoxy propane) and sebacic acid.

While PGA, PLA, and their copolymers provide much flexibility for developing scaffolds that have particular degradation properties, they are fairly brittle. This makes them unsuitable for use in certain engineered tissues, such as tendon and ligament, where mechanical stresses may be present. Polycaprolactone (PCL) is also a polyester, but its longer monomer unit provides greater elasticity. The polymer and degradation product is nontoxic, which has led to its FDA approval for use as a long-term contraceptive implant (68). Other elastomeric polymers are poly-4-hydroxybutyrate and polyhydroxyalkanoate, which have been used in various engineered tissues (69).

Certain polyesters have been developed to degrade into biocompatible products that make them suitable for drug delivery and tissue engineering applications. Poly(propylene fumarate) (PPF), for example, will degrade into fumaric acid, a natural component of the Kreb's cycle, and 1,2-propanediol, a common drug diluent. Mechanical properties can be improved through the use of chemical cross-linkers or certain ceramic composites (70). A tyrosine-based polycarbonate is another polyester that has been used in bone engineering (71). This osteoconductive polymer has excellent structural properties and has been shown to promote bone growth (72).

Like polyanhydrides, poly(ortho esters) can be designed for surface degradation. Their use in drug delivery (73) also make them suitable candidates for tissue engineering. Some poly(ortho esters) may degrade into acidic byproducts that can autocatalyze the degradation process, which may be useful for systems where fast degradation of the scaffold is desired.

In contrast to most hydrocarbon-based polymers, poly(phosphazenes) consist of a phosphorous and nitrogen chain. These polymers undergo hydrolysis to form phophate and ammonium salts. Ethyl glycinate substituted poly(phosphazenes) have shown promise for osteoblast attachment (74).

Polyurethanes have long been used for medical implants because of their good biocompatibility and mechanical properties. They have been used in long-term implants, such as cardiac pacemakers and vascular grafts. Though not degradable, these polymers can be cross-linked with degradable compounds to create scaffolds for engineered tissue (75). The main disadvantage is the toxicity of the degradation byproducts especially for cross-linkers based on diisocyanates.

Poly(amino acids) have good potential as a scaffold material because they are biocompatible and release amino acids as their degradation product. Poly-L-lysine is a common example that is adsorbed to surfaces in order to improve cell adhesion. Enzymatic degradation makes the breakdown of the polymer difficult to control. These polymers also have high reactivity and moisture sensitivity. Along with being expensive to produce, these materials have limited use for engineered tissue. An alternative is to create a pseudo-poly(amino acid), where the amino group in the polymer backbone is replaced with another nonamide linkage. This can improve the stability and mechanical properties of the polymer (76).

Hydrogels. Hydrogels are a subcategory of biomaterials defined by their ability to retain water within their polymeric matrix. Because of the high water content, hydrogels have mechanical properties similar to that of soft tissue. This may limit the application of hydrogels to certain tissues, but the environment closely simulates the environment of native tissue. Hydrogels are created by the cross-linking of water soluble polymers while in an aqueous environment. The cross-linking can be initiated chemically or via exposure to light of a particular wavelength. Natural polymers, such as fibrin and collagen, can be used to create hydrogels. For synthetic hydrogels, polyethylene glycol is a popular choice because of its biocompatibility, hydrophilicity and customizable transport properties (77).

Cells can also be entrapped within the matrix during the gelling process that permits a more uniform distribution. Cell entrapped in hydrogels include chondrocytes (78), fibroblasts (79), and smooth muscle (80). Photoinitiated cross-linking can be used to create cell–matrix composites *in situ* (81). In such systems, cells still maintain their viability (82). This process can be used to create a cell–polymer matrix that fits exactly into the shape of the tissue defect.

Scaffold Fabrication. Once a biomaterial has been chosen that has the properties crucial to the particular tissue to be engineered, it must be fabricated into an appropriate matrix in which the cells can survive and form the tissue. Several factors, such as porosity and pore structure, surface area, structural strength, and shape, are relevant to the design of a suitable scaffold. In general, a high porosity is important to the formation of tissue because it provides space for the cells to grow, as well as allows nutrients to diffuse into the matrix and promote cell survival (3). For certain situations, though, the porosity

should be optimized to ensure that growth factors important to tissue regeneration are retained within the matrix (83).

The strength of the scaffold is important to ensure that the scaffold will protect the regenerating tissue until it becomes integrated into the body. Also, some engineered tissue, such as bone and cartilage, may require a strong scaffold in order to retain integrity while functioning under physiological loading conditions. The structural strength is dependent on the mechanical properties of the polymer as well as the processing of the scaffold. A high surface area will ensure adequate contact for the cell adhesion. Depending on the polymer, the surface may have to be modified to improve adhesion. Depending on the application, the scaffold may need to be processed to form a particular 3D shape. For example, cylinders can be used as vascular grafts and for nerve regeneration.

Several processing techniques are available for the fabrication of scaffolds (Table 1). These techniques allow for the control of porosity and pore structure as well as the contact area for cell adhesion. In fiber bonding, a polymer is dissolved in a suitable solvent and fibers from another polymer are suspended within the solution (84). The solvent is then removed by vacuum drying. Heat is slowly applied to the composite material to cause the fibers to bond to one another. The other polymer is again dissolved leaving behind the bonded fiber matrix. To ensure success in this process, solvents must be chosen that do not dissolve the fibers and the fibers must have a melt temperature lower than that of the matrix polymer.

A similar method, called particle leaching, involves the incorporation of particles suspended within the polymer solution (85). As with fiber bonding, the particles are locked within the polymer matrix after vacuum drying is done to remove the solvent. In this method, though, the particles

Table 1. Summary of Various Scaffold Fabrication Methods

Fabrication Method	Characteristics
Solvent casting	No thermal degradation
	Residual solvent may harm cells
Fiber bonding	High porosity
	Thermal degradation
	Limited solvent–polymer combination
Particle leaching	Porosity easy to control
	Entrapment of particle with matrix
	Brittle foams
Gas foaming	No organic solvents
	Noncontinuous pores
Freeze drying	Small pore sizes
	Low temperature
Phase separation	For entrapment of small bioactive molecules
	Low temperature
Extrusion	Long fibers
	Thermal degradation
Membrane lamination	Three dimensional shapes
3D printing	Slow processing
	Control of shapes
In situ polymerization	Limited polymer–cell combinations
	Injectable, shape to fit defect

are removed via dissolution leaving behind a porous matrix. Typically, salt or sugar crystals, which are soluble in water, are suspended within a relatively hydrophobic polymer. The porosity can be controlled by altering the amount of particles suspended in the matrix. If the amount of particles is too low, they may become trapped within the polymer and remain undissolved. The foams that are formed tend to be brittle and also may require prewetting with ethanol to promote fluid infiltration. Instead of suspending the particles in a solution, they can be placed into a polymer melt that is subsequently cooled and the particles later dissolved away. A major drawback to this process is the thermal degradation of the polymer that can greatly affect its mechanical properties.

With gas foaming, carbon dioxide is added to the solid polymer at a very high pressure so that it infiltrates the matrix (86). When the pressure is rapidly dropped, the gas will expand within the polymer creating a porous foam. This method eliminates the need for an organic solvent whose presence may be detrimental to cell survival and tissue formation. One disadvantage is that the pores may not connect. This can be overcome by using this method in combination with particulate leaching.

Freeze drying can also be used to create porous polymer matrices (87). A polymer dissolved in an organic solvent can be mixed with water to form an emulsion. The emulsion is then quenched in liquid nitrogen and the water is removed by freeze drying. This method can create a matrix with 90% porosity and very high surface area, but the pore size may be too small.

Bioactive molecules may be added to the polymer matrix so that their release can enhance tissue formation. In many cases, though, the bioactive molecule may be incompatible with the organic solvent or thermal conditions used in the processing of the polymer matrix. Phase separation may be used to overcome this problem (88). In this method, the bioactive molecule is dispersed within the polymer solution and the solution is cooled until two liquid phases are formed. The two immiscible liquids are quickly frozen and the solvent is removed by sublimation. Removal of the solvent-rich phase leads to the highly porous structure while the bioactive molecule remains trapped within the solid polymer phase. The cold temperatures used in this process ensure that the bioactive molecule is not adversely affected. This method works well with small molecules, but can be difficult with large proteins.

For certain structural applications, such as bone, the mechanical strength of the polymer matrix is important to the success of the engineered tissue. In these cases, the matrix should have a high compressive strength to help the tissue maintain its integrity until the formation of the engineered tissue. To improve the mechanical properties of the polymer, hydroxyapatite fibers may be included to reinforce the polymer matrix (89).

Specific 3D shapes may be required for certain engineered tissue. Simple shapes like rods and cylinders can be formed via thermal extrusion (90). Particulates can be added to create porous structures as with the particulate leaching methods described earlier (91). The extrusion process may also lead to thermal degradation of the polymer matrix. Membranes formed by other techniques may be cut into particular patterns and stacked to form specific 3D shapes (92). The membranes can be bonded through the use of a suitable solvent. The 3D printing is a method akin to rapid prototyping that can also be used to form specific shapes (93). In this method, a polymer powder is spread into an even layer, and solvent is sprayed in a specific pattern to bind the powder together. Another powder layer is added and the solvent is sprayed on to bind the powder to the previous layer. The process is repeated until the entire structure has been formed. Using this method, features as small as 300 μm can be formed.

When specific shapes are not required, *in situ* polymerization can be used to deliver cells and matrix molecules to the site of the of tissue defect. This process can be used to create a cell–polymer matrix to precisely fill the site of the defect. Poly(propylene fumarate) has been used as bone cement (94). Hydrogels containing chondrocytes have been used for cartilage engineering (95).

To ensure proper cell adhesion and growth, the surface of the polymer must be suitable for cell attachment. The wettability and surface free energy greatly influence the extent of cell adhesion. A more hydrophilic surface is necessary to ensure cells will attach to the surface. This must be balanced with the cells needed to interact with surrounding cells in order to form tissue. Surface eroding polymers, such as polyanhydrides, can also renew their surface providing additional area for cell adhesion, which helps to promote cell growth. The surface of the polymer may need to be modified. Amphipathic molecules may be adsorbed to the surface. For example, the polar side chains of poly-L-lysine provide a suitable surface for cell attachment. Extracellular matrix molecules, such as laminin, fibronectin, and collagen, can also be adsorbed to the surface. These molecules contain the amino acid sequence (RGD), which binds to integrins present on the cell surface. The RDG along with IKVAV and YIGSR (found on laminin) can also be covalently bonded to the polymer to create sites for cell adhesion.

The surface of the scaffold can also be modified using microfabrication techniques, these methods can be used to alter the surface chemistry for improved cell adhesion or to create micron scale structural features to affect cell function (96). Besides altering the surface chemistry, microfabrication techniques can also be used to create microscale morphological features on the surface. Microscale features can effect the attachment, motility, and proliferation of fibroblasts in culture (97). The texture of the surface can also influence cardiac myocytes at the molecular level, such as protein localization and gene expression (98). Grooved surfaces can also be used to physically align cells and direct nerve regeneration (99).

Photolithography is a technique commonly used in the production of computer chips. With this method, a special type of photopolymer called a photoresist is coated onto the surface of the scaffold material. Separately, metal is coated onto a glass substrate and laser etched to form a mask in the pattern of interest. The mask is placed over the photoresist and ultraviolet (UV) light is shown through to polymerize the photoresist into the pattern of interest. A solvent is used to remove the unpolymerized photoresist and expose the scaffold. The surface can now be modified by several methods. Compounds, such as collagen and laminin, can be adsorbed to the surface. Small molecules, such as the RGD tripeptide, can be covalently bonded to the surface. The surface can also be exposed to oxygen plasma to make the area more hydrophilic. Once modified, the remaining photoresist can be removed using another solvent. The success of this method depends on the choice of appropriate solvents and photoresist to use in conjunction with the scaffold and surface modifiers. The major drawback of this technique is that it can only be applied to planar surfaces and relies on the use of expensive specialized equipment. Lithography can also be used to create elastomeric stamps. These stamps can be used to place molecules onto the substrate in a defined pattern. The stamp can be used repeatedly, which helps to reduce the cost compared to traditional photolithography.

Reactive ion etching builds on the premise of photolithography to create microstructural features on the surface (see Fig. 4). Using glass or quartz for the substrate, metal can be deposited onto the patterned photoresist. When the photoresist is removed, the surface is exposed to ion plasma that etches into the substrate leaving deep grooves in the surface. The substrate can now be used as a mold. The polymer matrix can be dissolved and cast onto the mold and dried. When the matrix is lifted from the mold, it will contain the micropatterned structure. As with photolithography the matrix can only be planar. Despite this disadvantage, this method has been used to improve nerve regeneration in bioartificial nerve grafts (100).

Surface of the biodegradable matrix can be etched directly using the process of laser ablation. In this method, a laser beam is used to etch grooves into the substrate. The laser is pulsed rapidly to allow the dispersion of thermal energy and thus reduce the degradation of the polymer matrix.

Another method for depositing molecules in specific patterns is through the use of microfluidics. In this method, a microscale channel system is placed on top of the substrate. Various solutions are passed through the channel and molecules are allowed to interact with the substrate surface. The channel system is removed leaving behind the patterned substrate.

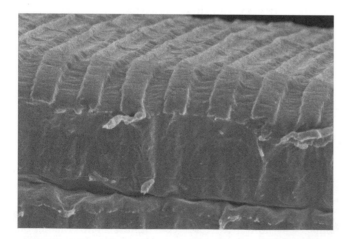

Figure 4. Microfabrication of polymer surface using a combination of reactive ion etching and compression molding. Grooves are 10 μm wide and 3 μm deep.

Signals

Once cells have been seeded onto a biodegradable matrix, signals must be provided to ensure the cells will continue to grow and form fully functional tissue (see Fig. 5). These signals must mimic the environment in which tissue naturally regenerates or develops. Signals may come from contact with the extracellular matrix or other cells. Diffusible factors may be delivered to the cells spatially and transiently. Mechanical and electrical stimuli may also promote tissue formation for certain cell types. Cues may also be provided based on the spatial arrangement of the cells.

Extracellular Matrix. When forming tissue, cells will interact with various proteins that make up the ECM. The ECM is comprised of collagens, proteoglycans, hyaluronic acid, fibronectin, vitronectin, and laminin. Integrins that are found on the surface of cells can interact with short amino acid sequences located on various ECM proteins. For example, cells will adhere to the arginine-glycine-aspartic acid-serine sequence (RGDS) found in fibronectin (101). Integrins are transmembrane proteins that also interact with cytoskeletal proteins like actin. In response to changes in the ECM, they modulate signals to the cells that result in a change in cell function. Certain amino acid sequences have also been implicated in the interactions between integrins and the ECM for specific cell types. These include REDV for endothelial cell adhesion (102), IKVAV for neurite outgrowth (103), and LRE for synaptic development (104).

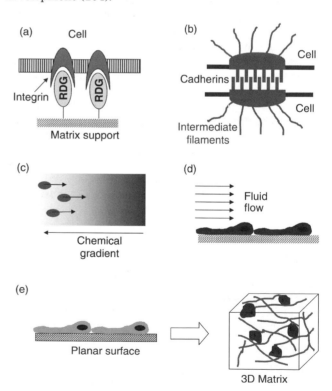

Figure 5. Examples of signals used to control tissue formation. (a) Cell–matrix interaction (RGD–integrin binding). (b) Cell–cell interaction (desmosome). (c) Diffusible factors (chemotaxis). (d) Mechanical forces (shear). (e) Spatial organization.

The primary purpose of the ECM is to anchor the cells so that tissue can form. The connection can ensure that polarized cells are oriented in the correct direction. When cells die, the ECM may retain its integrity until new cells can grow and form tissue. The composition of the ECM varies among tissues. The differences help to define the boundaries between tissue types as well as promote a microenvironment that is best suited for that particular tissue. Soluble regulatory factors may also be stored within the ECM. Disruption of ECM, and thus the tissue, lead to the release of these regulatory factors in order to affect cell function. The ECM can affect functions, such as cell growth, differentiation, and apoptosis. During development, the ECM plays a role in the repatterning of the epithelial-mesanchymal transformation. The physical connect between the cells and the ECM also helps with the modulation of signals due to mechanical stresses.

Diffusible Factors. During development or tissue regeneration, soluble diffusible factor help to control various cell functions, such as proliferation, adhesion, migration, and differentiation. Examples include the various families of growth factors and morphogenic proteins as well as nutrient to ensure cell survival. These factors may be expressed by the cell itself (autocrine) or it may come from a nearby cell (paracrine) or a remote site (endocrine). These factors can elicit changes in the cell cycle, promote differentiation, encourage cell motility and regulate the synthesis of DNA and protein.

During normal development or regeneration, these factors are secreted from various cells and delivered to the target tissue. In order to mimic this effect in engineered tissue, various methods may be used. Prior to implantation, the engineered tissue may be subject to a medium that is supplemented with factor to promote tissue growth. The scaffold for the engineered tissue may have soluble factors incorporated into the matrix. As the substrate degrades, the factors are released in a controlled manner. Molecules may also be immobilized onto the surface of the scaffold in order to alter cell function. Immobilized growth factors can still impart their effect on the cell without being internalized by endocytosis. This provides a mechanism to potentiate the effect of the growth factor.

For controlled release, the scaffold has a dual purpose to provide structural support for the engineered tissue as well as deliver signaling factors at the appropriate time and dose. Other particles that provide controlled release, such as microspheres, may also be incorporated into the engineered tissue. Using a combination of chemical factors with different release profiles can mimic the developmental processes.

Spatial Organization. The spatial arrangement of the cells in culture can have a direct impact on their function. Three-dimensional scaffolds provide support, but also encourage cells to display the appropriate phenotype for tissue formation. Hepatocytes are prone to loss of phenotype when cultured as a monolayer (105). Chondrocytes exhibit fibroblast behavior in monolayers compared to 3D systems. The 3D cultures secrete a great amount of type II collagen that is essential for cartilage formation (106).

The shape of the scaffold is an important consideration for the regeneration of certain tissue types. Nerve tissue engineering uses tubular conduits to help guide the regenerating axons. A luminal diameter that is 2.5 times the diameter of the nerve bundle is optimal for axon extension (107).

Mechanical Modulation. When grown in static cultures, cells tend to settle and form monolayers. This can lead to a loss of phenotype that would be detrimental to tissue formation. To prevent this disaggregation, cells can be cultured in microgravity (108). Microgravity is achieved through the used of specialized bioreactor systems that keeps the cell-scaffold matrix suspended within the media. The net mechanical forces on the tissue are essentially zero.

Mechanical stress may need to be imparted on the engineered tissue in order to ensure the proper development. Shear stresses can encourage the alignment of fibrillar extracellular matrix proteins, such as collagen. This effect will encourage the cell to align with the matrix (109). Alignment of collagen can also be achieved through the use of magnetic forces (46). Pulsatile shear stresses can also effect the orientation of endothelial cells in engineered blood vessels (110). This cyclic stress lead to the orientation of the cells along the circumference of the blood vessel as compared to cells aligned in the axial direction for constant shear loading.

Physical loading of tissue can also impact the mechanical properties of tissues. Signals are modulated via the interaction between the cells and the ECM. Cyclic loading can improve the tensile strength of tissues, such as arteries, heart valves, ligament, muscle, and bone (111). Hydrostatic pressure can also effect the formation of tissue. Cyclic pressure can alter the metabolic function of chondrocytes to produce greater amounts of type II collagen for cartilage formation (112).

Electrical Signals. Besides chemical and mechanical signals, cells will also respond to electrical stimuli, such as electric fields and direct current. Electric fields describe the amount of force exhibited by a charged particle. These forces can have an impact on materials that can be induced to carry a charge, such as cells, or individual ionic molecules. Electric fields can be used to move polarizable materials (dielectrophoresis) or encourage the motion of ions (iontophoresis). Application of direct current (dc) can also be used to mimic the conduction of electrical signals. The effect depends on the tissue type and its conductivity as well as the type of current (alternating or direct), the frequency and magnitude of voltage, and the uniformity of the electric field.

Because the neuromuscular system relies on electrical signaling for its functions, such stimulation can affect tissue formation. Muscle will atrophy when not exposed to an electrical stimulus. Providing an electrical stimulus to muscles until new neural connection can be made will greatly improve the likelihood of success for the nerve regenerative process. Neurons themselves may also respond to electrical stimuli by altering the effect of biochemical cure on growth cone dynamics (113). Myocytes will develop into functional heart tissue under electrical stimulation (114). Muscle progenitor cells also show improved contractility when exposed to electrical stimulation (115).

Other tissues that do not typically exhibit electrical behavior may still respond to such stimuli. Corneal epithelium has been found to emit dc electrical fields with injured tissues. These cells will migrate toward the cathode in the presence of artificial electrical fields (116). Direct and alternating current (ac) electrical fields can also be used to reduce wound area and wound volume comparatively (117,118).

Cell-to-Cell Interactions.   Cells can communicate directly with each other through physical contact via tight junctions, gap junctions, cadherins, and desmosomes or through chemical interactions using soluble factors. Such contact is important for architectural support and for inducing a favorable phenotype. In complex tissue containing multiple cell types, one cell type can act as a physical support for other cells. For example, smooth muscle cells help to support various tubular tissues, such as bladder ducts and blood vessels. Other cells may provide molecular signals to control the metabolic processes in other cells. Schwann cells secrete growth factors that stimulate neurons to regenerate. Without this chemical support, some cells may cease to proliferate and eventually die.

Bioreactor Technology

Once the appropriate cell source, scaffold, and signaling molecules have been selected, the engineered tissue construct may require a specialized bioreactor in which to grow. The bioreactor can provide the vehicle for supplying chemical factors, as well as place mechanical stresses on the tissue if necessary. Liquid media optimized for tissue growth supplies nutrients to the growing tissue and waste products are removed.

A perfusion bioreactor describes the general type of system where cells are retained. The cell-scaffold construct is held in place and media is continuously fed into the bioreactor. The tissue may be physically pinned in place or fixed due to a balance of gravitational and shear forces. Spinner flasks provide mixing through the use of stirrers (119). The engineered tissue is pinned in place while the stirrer keeps the fluid well mixed. The mixing ensured that adequate nutrients are delivered to the cells and wastes are quickly removed to prevent toxic buildup.

Several tissues may require mechanical stimuli in order to achieve full functionality. Novel bioreactor systems have been developed to provide these stresses. Pulsatile shear forces have been used to create engineered arteries with burst strengths comparable to native vessels (110). Placing chondrocytes under cyclic hydrodynamic pressure can enhance the formation of articular cartilage (120). Mechanical cyclic stretching can also improve the strength of engineered ligaments (121).

Tissue Properties

Biomechanics.   Engineered tissue should have the same mechanical properties of native tissue in order to achieve full functionality. This issue is especially important with tissue, such as bone and cartilage. Mechanical properties can also influence the tissue architecture by altering the structure of the extracellular matrix. The cytoskeleton interactions with the extracellular matrix can also be affected by the mechanical properties of the tissue. This can alter cell growth, movement, and metabolic function.

Certain tissues also interact with the nervous system based on mechanical signals. The alveoli and passageways in lungs respond to inhalation (122). Stretch receptors in the urinary bladder can detect when the organ must be emptied. Sensory receptors in the skin can detect touch. Engineered tissue must account for physical interaction with other tissue as well as the outside environment.

The influence of biomechanics can be seen at multiple spatial dimension. At the lowest level, biomechanics is influenced by the forces of individual molecules. The individual proteins of the extracellular matrix determine its mechanical properties. The formation and dissolution of chemical bonds can also alter the strength of the tissue. At the cellular level, the plasma membrane and cytoskeleton influence the physical properties of the cell. Cells can also interact with their environment via mechanotransduction pathways. The adhesion, aggregation, and migration of cells will affect the overall tissue and its mechanical properties. At the tissue level, cells interact with the extracellular matrix to create a material with specific physical properties that are essential to its function. An example of this hierarchy can be seen in tendon tissue (123). Individual collagen molecules are bundled together to form microfibers and fibrils. These fibrils combine with fibroblasts to create fascicles that in turn comprise the tendon tissue.

The mechanical properties of structural tissue can be defined by the relationship between stress and strain. As a certain level of stress is placed on the tissue, it will deform. This deformation relative to the initial length defines the strain. In many cases, stress is proportional to strain. The constant of proportionality is called the Young's modulus. For some tissue-like bone, the Young's modulus may vary from location to location (124).

Tissue exhibits characteristics of viscoelastic behavior. As with most biological materials, tissue is slow to deform in response to stress. This slow deformation is known as creep. When stress is quickly placed on tissue and the tissue deforms, the amount of force needed to maintain the deformation decreases over time. This phenomenon is known as stress relaxation. This viscoelastic behavior may be due to the presence of interstitial fluid or the movement of collagen fiber as with cartilage. Subtissue components, like lamellae in bone, may slip relative to one another. Also, some tissue may undergo internal friction due to repeated motions as is seen in tendons and ligaments.

Besides normal forces due to stress, shear forces can also influence the physical properties of tissue. Shear forces are present in tissue where fluid flow is involved. For example, blood flow will influence the properties of endothelial cells in veins and arteries. Compressive forces cause the flow of interstitial fluid in cartilage.

Structural tissue must be engineered to respond to the shear and normal forces it will encounter when placed into the body. In some cases, such as bone, the material placed into the site of injury must meet the mechanical requirements of the surrounding tissue. If not, the engineered tissue will become damaged thus reducing the chances of recovery.

Biocompatibility. Ideally, the cells and scaffolds that are used in engineered tissue would not provoke an immune or inflammatory response. Unless cells are obtained from the patients themselves, the possibility of rejection remains. The possibility of rejection is dependent on the access that the immune system has with the implanted engineered tissue. For tissue that lacks an established vasculature, for example, cartilage and epidermis, the immune response is limited. Also, certain engineered tissues may utilize cells for a short period of time until the patient's own cells recover. For example, nerve regeneration may be promoted through the use of donor Schwann cells that may become unnecessary once the nerve has recovered. Barring these special cases, methods must be developed to protect the tissue from the immune system.

As with organ transplants, the primary method of controlling the immune response to engineered tissue is through the use of immunosuppressant therapy. These drugs are required for the life of the patient. To avoid side effects and the long-term implications of having a suppressed immune system, methods must be developed to mask the cells. Bone marrow from the donor can be transplanted with the engineered tissue to create a chimera. The combined bone marrow can reduce the likelihood of rejection.

At the molecular level, surface antigens can be altered to prevent rejection. Fragments of antibodies that do not elicit cytotoxic T cell attack may be used to make the antigen from the patients own immune system. This is only temporary and may not prevent all immune mechanisms. Gene ablation can be used to create cells where the antigen coding gene has been inactivated. Alternatively, a gene may be added that is used to synthesize a protein that inhibits rejection. The DNA or RNA that codes for the antigen can be blocked with oligonucleotides that hybridize to prevent transcription or translation.

For cells whose function is primarily metabolic, the cells can be encapsulated to isolate them from the immune system. Encapsulation is accomplished by surrounding the cells with an artificial membrane. The pores in the membrane must be small enough to prevent interaction with the immune system, but large enough to allow nutrients and therapeutic proteins to pass through. The membrane must also be able to withstand any mechanical stress that may disrupt the membrane and allow interaction with the immune system.

Besides the immune system, the implantation of engineered tissue will, to some extent, elicit an inflammatory response. The response initiates a cascade of chemical reactions that alters the gene expression of circulating blood cells (granulocytes, platelets, monocytes, and lymphocytes), resident inflammatory cells (e.g., macrophages, mast cells) and endothelial cells, In conjunction with the cellular activation, many compounds (e.g., growth factors, cytokines, chemokines) are released. The overall effect is characterized by increased blood flow to the tissue that increases temperature and causes redness, swelling, and pain. This creates an environment that isolates the inflamed tissue from the body and promotes wound healing.

During the wound healing, damaged tissue is cleared by the circulating blood cells, A fibrin and fibronectin meshwork is created to act as a substrate for migrating and proliferating cells. Granular tissue containing fibroblasts, myofibroblasts, monocytes, lymphocytes, and endothelial cells forms at the site of the injury. The endothelial cells lead to the formation of new blood vessels (angiogenesis). The tissue undergoes remodeling as tissue regenerates to replace what had been lost. At the end of the process, fibroblast, myofiborblast, and endothelial cells will undergo programmed cells to reduce scarring and return the tissue to its original form.

Engineered tissue and biomaterials implanted in the body can be affected by the inflammatory and wound healing processes. During the granulation process, fibroblasts may migrate to the implant and encase it. This may isolate the engineered tissue from the surrounding tissue and prevent the integration of the engineered tissue with the native tissue. The additional layer of cells will reduce the diffusion of nutrients to the implant. Also, vascularization may be reduced further, hampering regeneration. Though, the layer of fibroblasts may be problematic, the complete lack of fibroblasts may also indicate some level of toxicity from the implant. The ideal implant will enhance the wound healing process while ensuring the engineered tissue does not become isolated.

Cryopreservation

To engineer tissue that can be delivered on demand, methods must be developed for long-term storage. Cryopreservation can be used not only for the storage of the final tissue product, but also for the cellular components. When new cells are isolated for use in engineered tissue, they may be preserved for latter expansion to prevent senescence or DNA mutation. The original tissue from which cells are derived may also be stored for later use. The isolated cells can be preserved while a sample is screened for infectious agents as well as the ability to form functional tissue. Cells must be banked in accordance with FDA regulations to ensure their genetic stability. During the production stage, the engineered tissue may be preserved at various steps for later quality control testing. The engineered tissue can be stored at hospitals to ensure availability

The protocol for cryopreservation depends on the cooling rate as well as the addition of compounds to reduce cell damage. If the cooling rate is too low, osmosis will drive water from the cell leading to severe dehydration. A high cooling rate will promote intracellular ice formation. The expansion of the ice forming inside the cell may lead to damage of the plasma membrane. Additives may be used to prevent cell damage caused by the increased ionic concentration in the unfrozen part of the tissue. Additives may be permeating [dimethyl sulfoxide (dmss) and ethylene glycol] or nonpermeating [poly(vinyl pyrrolidone) and starch]. The use of permeating additives would necessitate additional steps to remove the additive in preparation of the engineered tissue for implantation. One drawback to nonpermeating additives is that the osmotic stress due to high extracellular concentrations may lead to dehydration of the cells. The viability of the cells that have been preserved will depend not just on the method, but also on the cell concentrations and the type of cell.

Regulations

During the 1990s, the FDA recognized the need to develop regulations for products derived from tissue engineering principles. Engineered tissue is renamed by the FDA as tissue engineered medical products (TEMPs), which would include products used to replace damage tissue. Organs for transplantation and blood were excluded from this designation. Examples include artificial skin, bone graft, vascular grafts, nerve grafts and metabolic assist devices.

The primary concerns of TEMPs were disease transmission, controls of product manufacturing to ensure product integrity, clinical safety and efficacy, promotional claims, and monitoring the industry. The characteristics that determine the level risk include the cells source, the viability of cells and tissue, homologous function, manipulation of the implant chemically or genetically, systemic versus local effects, the long-term storage of the device, and the combination of the cells with other cells or drugs.

The first rule (63 FR 26744 Establishment Registration and listing of manufacturers of Human Cellular and Tissue-based Products) proposed the registration of the various established companies involved in the manufacturing of TEMPs in order to provide better communication between the FDA and industry. The second rule (Suitability Determination for Donors of Human Cellular and Tissue-Based Products, Final Rule 5/24/04 changes to 21 CFR 210, 211, 820) is designed to make changes to the regulations for Good Manufacturing Practices to require industry to test cells and tissue in order to prevent the transmission of disease and the unwitting use of contaminated tissue. The thirds rule (Current Good Tissue Practice for Manufacturers of Human Cellular and Tissue-based Products; Inspection and Enforcement, final rule 5/25/05 changes to 21 CFR 820) defines the methods, facilities, and controls used in the manufacturing of TEMPs.

EXAMPLES OF ENGINEERED TISSUE

Engineered tissue implants are designed from a combination of cells, a biodegradable scaffold, and chemical signal. Once a need has been established for an implant, the system must be designed to be compatible with the patient. The implant relies on an adequate source of cells that can form functional tissue. Alternatively, the implant may be designed to recruit cells from the patient. Since the human body is not a static system, the engineered tissue must be able to interact with the patients existing tissue. Cells must continue to survive. The structure of the scaffold must maintain its integrity until the tissue is fully integrated. The tissue structure should be achieved very quickly and be fully accepted by the host. The implant should foster cell migration into and out of the new tissue.

The implant should seamlessly integrate with surrounding tissue.

Tissue can be categorized into three main types: structural, metabolic, or combination. Structural tissue provides physical strength and a stable structure. The strength is determined by the extracellular matrix supported by the cells. Examples of structural tissue include bone, ligament, and vascular grafts. Metabolic-type tissues are defined as having functions based on the secretion or absorption of chemicals. These can be subdivided into tissues that respond to some stimulus (pancreas) and those that function independent of stimuli (liver). Combined tissue exhibit characteristics of both structural and metabolic tissue (skin).

Metabolic

Pancreas. The pancreas controls the blood sugar level through the regulation of insulin. The loss of this function leads to diabetes mellitus and requires daily injection of insulin. Over the long term, diabetes can damage other tissue such as eyes, kidneys, and nerve. Islets of Langerhans comprise only 1–2% of pancreatic tissue, but are the cells that regulate insulin levels. The islets are comprised of different cell types that lack the capacity to expand. As a result, islets must come from donor material.

The number of organ donors cannot supply sufficient islets for fill the demand of diabetes patients. Xenografts offer a viable alternative. Pig pancreatic tissue is currently used for insulin production and so is considered a good candidate for islet transplantation. To prevent an acute immune response, the islets must be encapsulated. Glucose and insulin along with other small molecules would diffuse across the membrane to allow chemical interaction while preventing antibodies from initiating an immune response.

Pancreatic islets have been encapsulated in various polymers. A biodegradable hydrogel-based coating has been developed for timed degradation of the capsule (125). The polymer can be modified to degrade at the same time that the islets reach the end of their functional life. The cells would then be removed by the immune system. Additional islets can be injected into the patient as necessary. A bioartificial device incorporating microencapsulated pancreatic islets can be connected to the vasculature to allow chemical interaction between the islets and the blood stream (126).

An alternative to the bioartificial pancreas is to inject islets directly into the existing pancreatic tissue. Islets have successfully been transplanted into diabetic patients (127). For this method to be successful, the islets must be completely biocompatible with the patient. Progenitor cells have been isolated from pancreatic tissue that exhibits the capacity to differentiate into cells similar to beta cells (128). Evidence indicates that adult stem cells may also develop into pancreatic cells without showing any indication of transdifferentiation (129). More work, though, needs to be done to create a universally acceptable stem cell derived pancreatic islet.

Liver. The primary cause of liver damage is cirrhosis due to alcohol consumption and hepatitis. This damage will prevent the proper metabolism of nutrients as well as toxic substance that must be removed from the body. When damage is severe enough, the only treatment available is an organ transplant, but due to the shortage of liver donors, many patients die waiting. Partial liver transplants use a single donor liver that is divided and transplanted into multiple patients. This technique demonstrates the ability of implanted tissue to assist the native liver (130). Individual cultured hepatocytes may also be transplanted to assist in liver function (131). Cells are placed in organs with a high level of vascularization (liver, pancreas, spleen). This enhances the chemical interaction between the hepatocytes and the blood stream. Like islets, the hepatocytes may also be microencapsulated to eliminate the need for immune suppressant therapy.

Until an organ becomes available, a patient can be sustained through the use of a liver assist device. Several different liver assist devices have been developed (132). In these systems, hepatocytes are placed within a bioartificial device and allowed to contact blood for plasma. The cells are retained behind some form of membrane and small molecules are allowed to diffuse through and become metabolized. The technology behind liver assist devices may eventually lead to the development of a whole engineered organ.

Future research is geared toward creating an engineered organ. Techniques will need to be developed to ensure a high level of vascularization through the engineered organ. Another method is to create a totally implantable liver assist device to act as a permanent liver replacement. The system will need to be engineered to be self-contained and small enough to fit into the abdomen of the patient.

Structural Tissue

Bone. The main purpose of bone is to provide structural support for the human body. Bone also helps to protect the internal organs of the body. It provides attachment sites for muscles to allow locomotion.

The primary cells found in bone are osteoclasts, osteoblasts, and osteocytes. Osteoblasts are responsible for the deposition of bone matrix while osteoclasts erode it. The cell dynamics permits the continuous turnover of bone matrix material that helps the tissue quickly respond to damage. Osteocytes can respond to mechanical stimuli and also promote blood–calcium homeostasis. The bone matrix is comprised of 65–70% hydroxyapatite. The remainder is composed of organic molecules, such as collagen 1, fibronectin, and various glycoproteins, sialoproteins and, proteoglycans (133).

Autologous bone grafts are the primary method for bone repair. This method has been very successful, but is restricted by the amount of tissue that can be obtained from the patient. Allogenic tissue may be used, but its rate of graft incorporation is much lower. Also, rejection of the tissue may be problematic and may introduce pathogens. Metals and ceramics offer an alternative to the grafts, but cannot become fully integrated into the existing tissue. This may lead to fatigue failure at the metal–bone interface or breakup of the ceramic material.

For engineered bone tissue, the graft combines scaffold with bone cells and compounds that promote osteoinduction. The most common scaffold is based on natural or synthetic hydroxyapatite (134). These materials have major drawbacks in terms of their brittleness and rapid dissolution rate. Many synthetic and natural polymers have been considered for use as a scaffold (133). Osteoblasts are the primary cell component of engineered tissue because of their ability to synthesize bone matrix. These cells may be isolated from the patient or donor source. Another option is to isolate mesanchymal stem cells from the bone marrow. The cells have been show to differentiate into the various bone cells when exposed to dexamethasone (135). The primary growth factors that have been found to affect bone tissue formation are bone morphogenetic proteins (BMPs), transforming growth factor beta (TGF-), fibroblast growth factors (FGFs), insulin growth factor I and II (IGF I/II), and platelet derived growth factor (PDGF). The VEGF may also be incorporated into the matrix to promote vascularization of the bone tissue.

The bioreactors used for bone tissue engineering have mostly been confined to spinner flask and rotating wall vessels. The rotating wall vessels help to promote cell interactions, but the microgravity imposed by the bioreactor may actually lead to bone loss (136).

Currently, most engineered bone tissue utilizes a combination of mesanchymal stem cells with a highly porous biodegradable matrix (133). Alternatively, multipotent cells isolated from the periosteum seeded on PLGA scaffolds created bone tissue that was well integrated with the native tissue (137). Transfecting bone progenitor cells with BMP-2 can also enhance their ability to create new tissue (138).

Because of the large number of growth factor that can influence bone formation, more work must be done to understand the underlying mechanisms of how these growth factors influence the various bone cell types. New materials with surface modifications are also being evaluated for their ability to control cell differentiation and migration. Rapid prototyping is also emerging as a processing technique to create scaffolds with control spatial arrangement, porosity, and bulk shape (139).

Cartilage. Cartilage is divided into two types: fibrous and articular. Fibrocartilage is used to provide shape, but flexibility to body parts, such as the ear and nose, while articular cartilage is found in joints where bone meets bone. Damage to articular cartilage can lead to painful arthritis and loss of motion. Traditional surgical repair techniques involve removal of damaged cartilage and reshaping the tissue to prevent further damage. Grafts from autologous tissue have not been successful in repair tissue. One reason for the difficulty in repair is the lack of a vasculature. The wound healing process available to other tissue types will not effect cartilage regeneration.

Genzyme has developed a procedure that involves the use of autologous cells that are expanded in culture and reintroduced into the site of injury. This method has had a fairly high success rate and can be used for defects up to 15 cm^2. Larger defects require the use of a scaffold to hold the engineered tissue in place. Photomonomers can be combined with chondrocytes and other growth factors to engineer tissue *in vivo* (140). Ultraviolet light is then used to initiate polymerization. This method is ideal for irregularly shaped defects and is less invasive. The hydrogel structure of the cross-linked polymer is also ideal for cartilage growth because the chondrocytes have a spherical morphology like that of native cells.

One of the main characteristics of cartilage that gives it strength is the trizonal arrangement of collagen fiber. This characteristic is difficult to mimic *in vitro*. Polyethylene oxide based photopolymers have successfully been used to encapsulate chondrocytes into discrete layers (141). Chondrocytes also respond to mechanical stresses that will affect its production of extracellular matrix. Cyclic hydrostatic pressure that mimics the forces present in knee joints can increase the synthesis of type II collagen (112). Type II collagen is an important protein in the extracellular matrix of articular cartilage. Fibrocartilage ECM consists primarily of type I cartilage.

Cell culture technology has allowed for the rapid expansion of autologous chondrocytes. Problems associated with immune rejection can be avoided, but two surgeries are required to create the engineered cartilage. Inroads have been to engineer cartilage that is structurally equivalent to native tissue. With proper control of differentiation, stems cells may be injected directly into the injured site to form new cartilage. Advances in immunomodulation technology may render stem cells resistant to the immune system that will eliminate the need for acquiring autologous cells. Also, injectable biomaterials may eventually be created that can be spatially arranged *in situ* to mimic the trizonal arrangement of collagen.

Blood Vessels. Blood vessels play a major role in delivering nutrient and removing wastes from all parts of the body. Disruptions to the flow of blood can lead to tissue loss downstream from the site of injury. In response to transection of the blood vessel, a cascade of events leads to the clotting of the blood and wound repair. Beside physical damage to the blood vessel, arthrosclerosis can lead to partial or complete blockage of the blood vessel. Within the heart, this can lead to myocardial infarction and even death unless bypass surgery is performed. Artificial grafts are used to bypass the flow of blood around the site of the blockage. This method has only been successful for grafts >6 mm (142). Small diameter grafts are being engineered to fill this gap (143). The major problem with grafting is restenosis, which can lead to failure.

The artery consists of three layers: intimal, medial, and adventitia. The intimal layer is comprised of the endothelial cells that line the inside of the blood vessel. For engineered arteries, these cells are typically obtained from donor material. Advances still need to be made in stem cell research to create new endothelial cells. The smooth muscle cells of the medial layer control the flow of blood via constriction or dilation of the artery. The adventitia layer is made up of fibroblasts and extra cellular matrix. While this layer is commonly left out of engineered arteries, its presence provides additional mechanical strength (37).

For the scaffold, PLA, PGA and their copolymers as well as poly-4-hydroxybutyrate are most commonly used. Engineered arteries have also been created without the use of synthetic materials (37). In these cases, donor arteries have their cells removed. The remaining matrix structure is reseeded with endothelial cells on the inner-lumen and smooth muscle cells on the exterior. Once the cell–scaffold implant has been created, a mechanical stimulus can be used to confer increased strength. Specialized bioreactors that provide cyclic shear flow that mimic the pulsatile flow of blood have been developed for engineering arteries (144).

The ideal engineered artery would have the mechanical strength to withstand the pulsatile shear stresses and blood pressure, would be biocompatible, and would integrate seamlessly with the existing blood vessel. Since autologous vessels may not be practical for patients suffering from arthrosclerosis, an acellular implant would be better suited. Acellular implants have displayed excellent results in animal models (145). These systems have a greater capacity to remodel their structure to better mesh with the native tissue. Cells are recruited from the existing tissue. Better understanding of the remodeling mechanism in humans would help to improve these implants.

Combined

Skin. Skin is considered to be the largest organ of the body. Its primary function is to protect the body from the environment. It helps to regulate fluid content and body temperature. Skin acts as the first line of defense for immune surveillance. Sensory receptor found in the skin help the body examine the environment. When injured, skin has the capacity to self-repair. The common form of injury to skin is a burn. Ulcerations may also form due to venal stasis, diabetes, and pressure sores. Skin continuity may be disrupted due to accidental physical trauma or the removal of skin cancers. Though the skin has the capacity to regenerate, engineered skin may be required to provide physical protection and metabolic regulation until the native tissue covers the wound.

Skin is comprised of two layers: the dermis and epidermis. The epidermis is the outer layer comprised of keratinocytes. This layer protects the dermis layer and helps to regulate heat and water loss. The dermis contains the vasculature, nerve bundles, and lymphatic systems that connect the skin to the rest of the body.

The traditional method to large area skin repair was to obtain skin from intact portions of the body and spread it around to increase the rate of regeneration. This was problematic for patients with severe injuries. Also, the area for harvesting may become damaged as well. An alternative was to use donor cadaver tissue that usually provoked an immune response. To overcome these obstacles, skin may be engineered to be biocompatible with the patient and help to promote the body's ability to self-repair.

Engineered skin should contain a belayed structure that allows for rapid vascularization and innervation from the body. The dermis layer should promote rapid wound repair. The epidermal layer should have the capacity to protect the body from the environment. The system should become fully integrated into the wound (146).

Initially, engineered skin was designed to act as a wound dressing and not become integrated with the native tissue. Alloderm, from Life Cell Technologies and approved in 1992, was an acellular dermal matrix derived from cadaver tissue. Integra was approved in 1996 and consisted of a silicone sheet coated with collagen and glycoaminoglycans. The silicone sheet helped to prevent fluid loss, but needed to be removed when the tissue eventually healed. Autologous cultured skin substitutes have been used in combination with Integra for engraftment into burn wounds of pediatric patients. The elastic quality of this engineer skin allowed the new tissue to grow with the patient (147). In 1998, Organogenesis received FDA approval for the first tissue engineer product (Apligraf). The product utilized a collagen gel sheet seeded with fibroblast. Dermagraft (Advanced Tissue Sciences) was the first skin substitute to utilize biodegradable polyglactin. The fiber mesh was coated with fibroblasts that secrete growth promoting factors. More advanced cultured skin substitutes, next generation engineered skin, should have a full thickness bilayered structure that can become integrated into the wound for faster recovery.

For full thickness engineered skin to become integrated into the patient's own tissue, the dermis layer must promote immediate vascularization. Without a supply of nutrients to the tissue, the epidermal layer would begin to die and slough off (148). The inclusion of fibroblasts and endothelial cells in the dermis layer promoted the formation of a capillary bed that improved neovascularization (149).

Engineered skin can also be used for other functions besides wound repair. Stem cells that are normally present in the epidermis may be transfected and included in engineered skin to produce therapeutic proteins for patient suffering from chronic disorders (150). Engineered skin may contain hair follicle cells to create tissue for hair implants (151). Future engineered skin should include melanocytes to match the skin to the patient's native tones as well as sweat glands to regulate sweating and sensory receptors that integrate with the existing nervous system.

Nerves. Nerve tissue engineering presents a unique problem not encountered with other tissue. Nerve tissue is comprised of neurons and satellite cells. In the peripheral nervous system, the satellite cells are Schwann cells. These cells secrete growth factors to stimulate nerve regeneration when they lose contact with neurons. In contact with neurons, Schwann cells ensheath or myelinate the axon in order to enhance the conduction of the electrical signal. The central nervous system (brain and spinal column) contains oligodendrocytes that ensheath axons and astrocytes forming the blood–brain barrier. Instead of promoting regeneration, oligodendrocytes are actually inhibitory. Damage to these neurons can lead to the loss of sensory and motor function, paralysis, and even death.

When a nerve becomes severed, the proximal segment of the axon closest to the cell body will extend and enter the degenerated distal portion and continue growing until new synaptic connections are made. If connections are not made in a timely manner, the neuron may lose function and die making the loss of function permanent. The traditional method of nerve repair is to surgically reconnect the two severed ends. When the damage is too extensive to reconnect the tissue, donor tissue from a less critical nerve is used to bridge the gap. As an alternative, an engineered bioartificial nerve graft can be used to bridge the gap to eliminate the need for the donor material. Several nerve grafts have been developed that utilize different scaffolds materials, various cells, and combinations of growth factor in order to enhance nerve regeneration (152). Clinical trials for peripheral nerve repair have shown success using poly(tetrafluoroethylene) to regenerate nerve tissue with gaps up to 4 cm long (153). The FDA has also approved a PGA conduit (Neurotube, Neuroregen LLC, Bel Air, MD) and a collagen-based nerve tube (NeuraGen, Integra Neurosciences, Plainsboro, NJ) for peripheral nerve repair. For the material that is chosen, the physical parameters of the conduit, such as porosity, can be optimized to ensure adequate nutrient reach the regenerating tissue while retaining growth factor with the implant (83).

Various chemical matrices, such as ECM components, can be added to the conduit to further enhance nerve regeneration (152). While the inclusion of these factors may provide incremental improvements in the overall design, they have not surpassed traditional surgical techniques to regenerate nerves (154). In the 1940s, Weiss, a pioneer in artificial nerve graft research, claimed that the ideal substrate for nerve regeneration is degenerated nerve (155). Degenerated nerve contains the matrix molecules and cellular components that have been naturally optimized to promote nerve regeneration.

Non-neuronal cells may be added to the artificial nerve graft in order to create an environment that mimics the natural nerve regeneration process. Schwann cells can be used to enhance the regeneration of both peripheral and central nerve tissue (156). Though they are not found in the central system, Schwann cells can overcome the inhibitory effect of the oligodendrocytes. Alternatively, fibroblasts transfected to produce nerve growth factor and seeded into an artificial nerve graft have been used to regenerate nerve tissue (157). Neural stem cells used in conjunction with a structured polymer scaffold have led to functional recovery from spinal cord injuries in rat models (27).

The next generation of nerve grafts should regenerate nerve tissue over much longer distances than currently achieved. Grafts containing a matrix comparable to degenerated nerve should make this possible. Microfabrication techniques also show promise for control of the growth of axons at the cellular level (100). Eventually, techniques must be developed to transplant new neural tissue into the existing system. This will become necessary for the innervation of engineered tissue and organs that relay on communication with the nervous system. Additional research is needed to understand the impact of additional neural contact on the overall system. These techniques may make possible a cure for paralysis where nerve tissue has degenerated beyond repair.

FUTURE PROSPECTS

The advances made in tissue engineering since the 1980s are set to transform the standard methods for medical care. The previous section is just a small sampling of the tissues currently under investigation. Just about every tissue in the body is currently being studied for repair using engineered tissue or cell therapy. Though much has been accomplished, more work still needs to be done.

Stem cells have shown great promise as a universal cell for developing engineered tissue implants. Research adult stem cells, such as mesanchymal and hematopoietic cells, hints at their capacity to act as a pluripotent cell type. Additional work may reveal more cell types that these can differentiate into. Strict regulations may hinder investigation of embryonic stem cells, but additional research in cultivation methods can alleviate ethical concerns. More work still needs to be done to control the differentiation of these cells to prevent the formation of teratomas.

The variety of tissue types being engineered makes development of a single universal scaffold difficult. The mechanical characteristics of individual tissues require scaffolds with specialized properties. Although a few polymers have been accepted by the FDA, more will need to be evaluated to ensure enough options are available for the engineered tissue. New biomaterials should also be investigated to ensure a wide variety of options. For a given class of biomaterials, though, the polymer should be customizable to ensure the appropriate mechanical and degradation properties. Also, the material should be capable of manipulation to enhance adhesion and control tissue formation.

As stem cells increase in importance, more signaling factors to control differentiation will be needed. Such control should not be limited to diffusible factors, but should also include surface bound molecules that can mimic the cell–cell contact interactions that occur during development. Such molecules may be bound to the surface using microfabrication techniques to encourage differentiation of the stem cells into spatially arranged multiple cell types. With advances in gene transfer and other immunomodulation techniques, stem cells may be rendered fully biocompatible with the patient regardless of the donor source.

Several specific examples of engineered tissues were presented earlier. Each tissue type has its own issues that must be overcome for the engineered tissue to be successful. In general, though, many tissues must interact with the existing vascular and nervous systems. As engineered tissue turns to engineered organs, protocols must be developed to ensure adequate nutrient will reach the cells. Vascular endothelial growth factor provides a means of recruiting blood vessels to the new tissue, but would not be useful in growing whole organs *in vitro*. Bioreactors will need to be designed to accommodate multiple tissue types, including endothelial cells for neovasculature formation. As the mechanisms for stem cell differentiation become better understood, the possibility of growing entire organs may become a reality.

Another general concern for engineered tissues is the connection with the existing nervous system. Muscle and skin interact with the peripheral nervous system. Several internal organs interact with the brain via the autonomic nervous system. For such organs and tissue to be fully integrated and functional, methods must be developed to attract existing neurons. In some cases, nerve tissue may not be present so the engineered tissue may need to include a neural component that can be integrated with the nervous system.

Engineered tissue will ultimately be used to repair practically any tissue in the body. With the almost infinite combinations of biomaterials, growth factors and cells, the only limit to creating new tissue is one's imagination.

BIBLIOGRAPHY

Cited References

1. Skalak R, Fox CF. Tissue Engineering: Proceedings of a Workshop, Held at Granlibakken, Lake Tahoe, California; 1988 Feb. 26–29. New York: Liss; 1988.
2. Bell E. Tissue Engineering: Current Perspectives. Boston: Birkhäuser; 1993.
3. Langer R, Vacanti JP. Tissue engineering. Science 1993;260 (5110):920–926.
4. Zimbler MS. Gaspare tagliacozzi (1545–1599):Renaissance surgeon. Arch Facial Plast Surg 2001;3(4):283–284.
5. Huber GC. A study of the operative treatment for loss of nerve substance inperipheral nerve. J Morph 1895;11:629–740.
6. NUPOC. Prosthetics History. 9/19 2004;2004(12/29).
7. Birchmeier C, Birchmeier W. Molecular aspects of mesenchymal-epithelial interactions. Annu Rev Cell Biol 1993;9:511–540.
8. Ajioka I, Akaike T, Watanabe Y. Expression of vascular endothelial growth factor promotes colonization, vascularization, and growth of transplanted hepatic tissues in the mouse. Hepatology 1999;29(2):396–402.
9. Charlton B, Auchincloss Jr H, Fathman CG. Mechanisms of transplantation tolerance. Annu Rev Immunol 1994;12:707–734.
10. Faustman D, Coe C. Prevention of xenograft rejection by masking donor HLA class I antigens. Science 1991;252(5013):1700–1702.
11. Markmann JF, et al. Indefinite survival of MHC class I-deficient murine pancreatic islet allografts. Transplantation 1992;54(6):1085–1089.
12. Platt JL. A perspective on xenograft rejection and accommodation. Immunol Rev 1994;141:127–149.
13. Ramanathan M, et al. Characterization of the oligodeoxynucleotide-mediated inhibition of interferon-gamma-induced major histocompatibility complex class I and intercellular adhesion molecule-1. J Biol Chem 1994;269(40):24564–24574.
14. Soon-Shiong P, et al. Insulin independence in a type 1 diabetic patient after encapsulated islet transplantation. Lancet 1994;343(8903):950–951.
15. Soon-Shiong P, et al. Long-term reversal of diabetes by the injection of immunoprotected islets. Proc Natl Acad Sci USA 1993;90(12):5843–5847.
16. McConnell MP, et al. In vivo induction and delivery of nerve growth factor, using HEK-293 cells. Tissue Eng 2004;10(9–10):1492–1501.
17. Fischbach C, et al. Generation of mature fat pads in vitro and in vivo utilizing 3-D long-term culture of 3T3-L1 preadipocytes. Exp Cell Res 2004;300(1):54–64.
18. Jiang XQ, et al. The ectopic study of tissue-engineered bone with hBMP-4 gene modified bone marrow stromal cells in rabbits. Chin Med J (Engl) 2005;118(4):281–288.
19. Fujita Y, et al. Transcellular water transport and stability of expression in aquaporin 1-transfected LLC-PK1 cells in the development of a portable bioartificial renal tubule device. Tissue Eng 2004;10(5–6):711–722.
20. Jin Y, Fischer I, Tessler A, Houle JD. Transplants of fibroblasts genetically modified to express BDNF promote axonal regeneration from supraspinal neurons following chronic spinal cord injury. Exp Neurol 2002;177(1):265–275.

21. Mezey E, et al. Turning blood into brain: Cells bearing neuronal antigens generated in vivo from bone marrow. Science 2000;290(5497):1779–1782.

22. Bjornson CR, et al. Turning brain into blood: A hematopoietic fate adopted by adult neural stem cells in vivo. Science 1999; 283(5401):534–537.

23. Ying QL, Nichols J, Evans EP, Smith AG. Changing potency by spontaneous fusion. Nature London 2002;416(6880): 545–548.

24. Thomson JA, et al. Embryonic stem cell lines derived from human blastocysts. Science 1998;282(5391):1145–1147.

25. Holden C, Vogel G. Cell biology. A technical fix for an ethical bind? Science 2004;306(5705):2174–2176.

26. Levenberg S, Langer R. Advances in tissue engineering. Curr Top Dev Biol 2004;61:113–134.

27. Teng YD, et al. Functional recovery following traumatic spinal cord injury mediated by a unique polymer scaffold seeded with neural stem cells. Proc Natl Acad Sci USA 2002;99(5):3024–3029.

28. Levenberg S, et al. Endothelial cells derived from human embryonic stem cells. Proc Natl Acad Sci USA 2002;99(7): 4391–4396.

29. Iwanami A, et al. Transplantation of human neural stem cells for spinal cord injury in primates. J Neurosci Res 2005;80(2):182–190.

30. Smits AM, et al. The role of stem cells in cardiac regeneration. J Cell Mol Med 2005;9(1):25–36.

31. Trucco M. Regeneration of the pancreatic beta cell. J Clin Invest 2005;115(1):5–12.

32. Luyten FP. Mesenchymal stem cells in osteoarthritis. Curr Opin Rheumatol 2004;16(5):599–603.

33. Ellis DL, Yannas IV. Recent advances in tissue synthesis in vivo by use of collagen-glycosaminoglycan copolymers. Biomaterials 1996;17(3):291–299.

34. Bell E, Ivarsson B, Merrill C. Production of a tissue-like structure by contraction of collagen lattices by human fibroblasts of different proliferative potential in vitro. Proc Natl Acad Sci USA 1979;76(3):1274–1278.

35. Weinberg CB, Bell E. A blood vessel model constructed from collagen and cultured vascular cells. Science 1986;231(4736): 397–400.

36. Auger FA, et al. Skin equivalent produced with human collagen. In Vitro Cell Dev Biol Anim 1995;31(6):432–439.

37. L'Heureux N, et al. A completely biological tissue-engineered human blood vessel. FASEB J 1998;12(1):47–56.

38. Lee CH, Singla A, Lee Y. Biomedical applications of collagen. Int J Pharm 2001;221(1–2):1–22.

39. Schmidt CE, Baier JM. Acellular vascular tissues: Natural biomaterials for tissue repair and tissue engineering. Biomaterials 2000;21(22):2215–2231.

40. Wakitani S, et al. Repair of large full-thickness articular cartilage defects with allograft articular chondrocytes embedded in a collagen gel. Tissue Eng 1998;4(4):429–444.

41. Liu S, et al. Axonal regrowth through collagen tubes bridging the spinal cord to nerve roots. J Neurosci Res 1997;49(4):425–432.

42. Atala A. Tissue engineering for bladder substitution. World J Urol 2000;18(5):364–370.

43. Orwin EJ, Hubel A. In vitro culture characteristics of corneal epithelial, endothelial, and keratocyte cells in a native collagen matrix. Tissue Eng 2000;6(4):307–319.

44. Pomahac B, et al. Tissue engineering of skin. Crit Rev Oral Biol Med 1998;9(3):333–344.

45. Elbjeirami WM, Yonter EO, Starcher BC, West JL. Enhancing mechanical properties of tissue-engineered constructs via lysyl oxidase crosslinking activity. J Biomed Mater Res A 2003;66(3):513–521.

46. Ceballos D, et al. Magnetically aligned collagen gel filling a collagen nerve guide improves peripheral nerve regeneration. Exp Neurol 1999;158(2):290–300.

47. Xu XM, Zhang SX, Li H, Aebischer P, Bunge MB. Regrowth of axons into the distal spinal cord through a schwann-cell-seeded mini-channel implanted into hemisected adult rat spinal cord. Eur J Neurosci 1999;11(5):1723–1740.

48. Sieminski AL, Padera RF, Blunk T, Gooch KJ. Systemic delivery of human growth hormone using genetically modified tissue-engineered microvascular networks: Prolonged delivery and endothelial survival with inclusion of nonendothelial cells. Tissue Eng 2002;8(6):1057–1069.

49. Zimmermann WH, Melnychenko I, Eschenhagen T. Engineered heart tissue for regeneration of diseased hearts. Biomaterials 2004;25(9):1639–1647.

50. Stevens MM, Qanadilo HF, Langer R, Prasad Shastri V. A rapid-curing alginate gel system: Utility in periosteum-derived cartilage tissue engineering. Biomaterials 2004;25 (5):887–894.

51. Dar A, Shachar M, Leor J, Cohen S. Optimization of cardiac cell seeding and distribution in 3D porous alginate scaffolds. Biotechnol Bioeng 2002;80(3):305–312.

52. Dvir-Ginzberg M, Gamlieli-Bonshtein I, Agbaria R, Cohen S. Liver tissue engineering within alginate scaffolds: Effects of cell-seeding density on hepatocyte viability, morphology, and function. Tissue Eng 2003;9(4):757–766.

53. Mol A, et al. Fibrin as a cell carrier in cardiovascular tissue engineering applications. Biomaterials 2005;26(16):3113–3121.

54. Karp JM, Sarraf F, Shoichet MS, Davies JE. Fibrin-filled scaffolds for bone-tissue engineering: An in vivo study. J Biomed Mater Res 2004;71A(1):162–171.

55. Fussenegger M, et al. Stabilized autologous fibrin-chondrocyte constructs for cartilage repair in vivo. Ann Plast Surg 2003;51(5):493–498.

56. Bannasch H, et al. Skin tissue engineering. Clin Plast Surg 2003;30(4):573–579.

57. Pittier R, Sauthier F, Hubbell JA, Hall H. Neurite extension and in vitro myelination within three-dimensional modified fibrin matrices. J Neurobiol 2004.

58. Suh JK, Matthew HW. Application of chitosan-based polysaccharide biomaterials in cartilage tissue engineering: A review. Biomaterials 2000;21(24):2589–2598.

59. Seol YJ, et al. Chitosan sponges as tissue engineering scaffolds for bone formation. Biotechnol Lett 2004;26(13):1037–1041.

60. Li K, et al. Chitosan/gelatin composite microcarrier for hepatocyte culture. Biotechnol Lett 2004;26(11):879–883.

61. Li Z, et al. Chitosan-alginate hybrid scaffolds for bone tissue engineering. Biomaterials 2005;26(18):3919–3928.

62. Schaner PJ, et al. Decellularized vein as a potential scaffold for vascular tissue engineering. J Vasc Surg 2004;40(1):146–153.

63. Grabow N, et al. Mechanical and structural properties of a novel hybrid heart valve scaffold for tissue engineering. Artif Organs 2004;28(11):971–979.

64. Kimuli M, Eardley I, Southgate J. In vitro assessment of decellularized porcine dermis as a matrix for urinary tract reconstruction. BJU Int 2004;94(6):859–866.

65. Miller C, et al. Oriented schwann cell growth on micropatterned biodegradable polymer substrates. Biomaterials 2001; 22(11):1263–1269.

66. Cook AD, Hrkach JS, Gao NN, Johnson IM, Pajvani UB, Cannizzaro SM, Langer R. Characterization and development of RGD-peptide-modified poly(lactic acid-co-lysine) as an interactive, resorbable biomaterial. J Biomed Mater Res 1997;35(4):513–523.

67. Webb AR, Yang J, Ameer GA. Biodegradable polyester elastomers in tissue engineering. Expert Opin Biol Ther 2004;4(6):801–812.

68. Pitt CG. Biodegradable polymers as drug delivery systems. In: Chasin M, Langer RS, editors. Poly-e-Caprolactone and its Copolym, Ers. New York: Marcel Dekker; 1990.

69. Martin DP, Williams SF. Medical applications of poly-4-hydroxybutyrate: A strong felxible absorbable biomaterial. Biochem Eng J 2003;16:97–105.

70. Temenoff JS, Mikos AG. Injectable biodegradable materials for orthopedic tissue engineering. Biomaterials 2000;21(23): 2405–2412.

71. Tangpasuthadol V, Pendharkar SM, Kohn J. Hydrolytic degradation of tyrosine-derived polycarbonates, a class of new biomaterials. part I: Study of model compounds. Biomaterials 2000;21(23):2371–2378.

72. Muggli DS, Burkoth AK, Anseth KS. Crosslinked polyanhydrides for use in orthopedic applications: Degradation behavior and mechanics. J Biomed Mater Res 1999;46(2):271–278.

73. Daniels AU, et al. Evaluation of absorbable poly(ortho esters) for use in surgical implants. J Appl Biomater 1994;5(1): 51–64.

74. Laurencin CT, et al. Use of polyphosphazenes for skeletal tissue regeneration. J Biomed Mater Res 1993;27(7): 963–973.

75. Zdrahala RJ, Zdrahala IJ. Biomedical applications of polyurethanes: A review of past promises, present realities, and a vibrant future. J Biomater Appl 1999;14(1):67–90.

76. James K, Kohn J. In: Park K, editor. Pseudo-Poly (Amino Acids):Examples for Synthetic Materials Derived from Natural Metabolites. Washington (DC): American Chemical Society; 1997.

77. Peppas NA, Bures P, Leobandung W, Ichikawa H. Hydrogels in pharmaceutical formulations. Eur J Pharm Biopharm 2000;50(1):27–46.

78. Elisseeff J, et al. Photoencapsulation of chondrocytes in poly(ethylene oxide)-based semi-interpenetrating networks. J Biomed Mater Res 2000;51(2):164–171.

79. Gobin AS, West JL. Cell migration through defined, synthetic ECM analogs. FASEB J 2002;16(7):751–753.

80. Mann BK, et al. Smooth muscle cell growth in photopolymerized hydrogels with cell adhesive and proteolytically degradable domains: Synthetic ECM analogs for tissue engineering. Biomaterials 2001;22(22):3045–3051.

81. Anseth KS, et al. In situ forming degradable networks and their application in tissue engineering and drug delivery. J Control Release 2002;78(1–3):199–209.

82. Elisseeff J, et al. Transdermal photopolymerization for minimally invasive implantation. Proc Natl Acad Sci USA 1999;96(6):3104–3107.

83. Rutkowski GE, Heath CA. Development of a bioartificial nerve graft. II. nerve regeneration in vitro. Biotechnol Prog 2002; 18(2):373–379.

84. Mikos AG, et al. Preparation of poly(glycolic acid) bonded fiber structures for cell attachment and transplantation. J Biomed Mater Res 1993;27(2):183–189.

85. Mikos AG, Lyman MD, Freed LE, Langer R. Wetting of poly-(L-lactic acid) and poly(DL-lactic-co-glycolic acid) foams for tissue culture. Biomaterials 1994;15(1):55–58.

86. Mooney DJ, et al. Novel approach to fabricate porous sponges of poly(D,L-lactic-co-glycolic acid) without the use of organic solvents. Biomaterials 1996;17(14):1417–1422.

87. Whang K, Thomas H, Healy KE. A novel method to fabricate bioabsorbable scaffolds. Polymer 1995;36:837–841.

88. Lo H, Ponticiello MS, Leong KW. Fabrication of controlled release biodegradable foams by phase separation. Tissue Eng 1995;1:15–27.

89. Thomson RC, Yaszemski MJ, Powers JM, Mikos AG. Hydroxyapatite fiber reinforced poly(alpha-hydroxy ester) foams for bone regeneration. Biomaterials 1998;19(21):1935–1943.

90. Widmer MS, et al. Manufacture of porous biodegradable polymer conduits by an extrusion process for guided tissue regeneration. Biomaterials 1998;19(21):1945–1955.

91. Thomson RC, Yaszemski MJ, Powers JM, Mikos AG. Fabrication of biodegradable polymer scaffolds to engineer trabecular bone. J Biomater Sci Polym Ed 1995;7(1):23–38.

92. Mikos AG, Sarakinos G, Leite SM, Vacanti JP, Langer R. Laminated three-dimensional biodegradable foams for use in tissue engineering. Biomaterials 1993;14(5):323–330.

93. Park A, Wu B, Griffith LG. Integration of surface modification and 3D fabrication techniques to prepare patterned poly(L-lactide) substrates allowing regionally selective cell adhesion. J Biomater Sci Polym Ed 1998;9(2):89–110.

94. Peter SJ, Kim P, Yasko AW, Yaszemski MJ, Mikos AG. Crosslinking characteristics of an injectable poly(propylene fumarate)/beta-tricalcium phosphate paste and mechanical properties of the crosslinked composite for use as a biodegradable bone cement. J Biomed Mater Res 1999;44(3):314–321.

95. Bryant SJ, Durand KL, Anseth KS. Manipulations in hydrogel chemistry control photoencapsulated chondrocyte behavior and their extracellular matrix production. J Biomed Mater Res A 2003;67(4):1430–1436.

96. Andersson H, van den Berg A. Microfabrication and microfluidics for tissue engineering: State of the art and future opportunities. Lab Chip 2004;4(2):98–103.

97. Berry CC, Campbell G, Spadiccino A, Robertson M, Curtis AS. The influence of microscale topography on fibroblast attachment and motility. Biomaterials 2004;25(26): 5781–5788.

98. Motlagh D, Senyo SE, Desai TA, Russell B. Microtextured substrata alter gene expression, protein localization and the shape of cardiac myocytes. Biomaterials 2003;24(14): 2463–2476.

99. Miller C, Jeftinija S, Mallapragada S. Micropatterned schwann cell-seeded biodegradable polymer substrates significantly enhance neurite alignment and outgrowth. Tissue Eng 2001;7(6):705–715.

100. Rutkowski GE, Miller CA, Jeftinija S, Mallapragada SK. Synergistic effects of micropatterned biodegradable conduits and schwann cells on sciatic nerve regeneration. J Neural Eng 2004;1:151–157.

101. Pierschbacher MD, Ruoslahti E. Variants of the cell recognition site of fibronectin that retain attachment-promoting activity. Proc Natl Acad Sci USA 1984;81(19):5985–5988.

102. Hubbell JA, Massia SP, Desai NP, Drumheller PD. Endothelial cell-selective materials for tissue engineering in the vascular graft via a new receptor. Biotechnology (NY) 1991;9(6):568–572.

103. Tashiro K, et al. A synthetic peptide containing the IKVAV sequence from the A chain of laminin mediates cell attachment, migration, and neurite outgrowth. J Biol Chem 1989;264(27):16174–16182.

104. Hunter DD, et al. An LRE (leucine-arginine-glutamate)-dependent mechanism for adhesion of neurons to S-laminin. J Neurosci 1991;11(12):3960–3971.

105. Hsiao CC, et al. Receding cytochrome P450 activity in disassembling hepatocyte spheroids. Tissue Eng 1999;5(3):207–221.

106. Ronziere MC, et al. Ascorbate modulation of bovine chondrocyte growth, matrix protein gene expression and synthesis in three-dimensional collagen sponges. Biomaterials 2003;24(5):851–861.

107. Buti M, et al. Influence of physical parameters of nerve chambers on peripheral nerve regeneration and reinnervation. Exp Neurol 1996;137(1):26–33.

108. Yoffe B, et al. Cultures of human liver cells in simulated microgravity environment. Adv Space Res 1999;24(6):829–836.

109. Brown RA, et al. Tensional homeostasis in dermal fibroblasts: Mechanical responses to mechanical loading in three-dimensional substrates. J Cell Physiol 1998;175(3):323–332.

110. Niklason LE, et al. Functional arteries grown in vitro. Science 1999;284(5413):489–493.

111. Fink C, et al. Chronic stretch of engineered heart tissue induces hypertrophy and functional improvement. FASEB J 2000;14(5):669–679.

112. Carver SE, Heath CA. Increasing extracellular matrix production in regenerating cartilage with intermittent physiological pressure. Biotechnol Bioeng 1999;62(2):166–174.

113. Ming G, et al. Electrical activity modulates growth cone guidance by diffusible factors. Neuron 2001;29(2):441–452.

114. Radisic M, et al. Functional assembly of engineered myocardium by electrical stimulation of cardiac myocytes cultured on scaffolds. Proc Natl Acad Sci USA 2004;101(52):18129–18134.

115. Cannon TW, et al. Improved sphincter contractility after allogenic muscle-derived progenitor cell injection into the denervated rat urethra. Urology 2003;62(5):958–963.

116. Zhao M, Dick A, Forrester JV, McCaig CD. Electric field-directed cell motility involves up-regulated expression and asymmetric redistribution of the epidermal growth factor receptors and is enhanced by fibronectin and laminin. Mol Biol Cell 1999;10(4):1259–1276.

117. Bogie KM, Reger SI, Levine SP, Sahgal V. Electrical stimulation for pressure sore prevention and wound healing. Assist Technol 2000;12(1):50–66.

118. Spadaro JA. Mechanical and electrical interactions in bone remodeling. Bioelectromagnetics 1997;18(3):193–202.

119. Sikavitsas VI, Bancroft GN, Mikos AG. Formation of three-dimensional cell/polymer constructs for bone tissue engineering in a spinner flask and a rotating wall vessel bioreactor. J Biomed Mater Res 2002;62(1):136–148.

120. Carver SE, Heath CA. Influence of intermittent pressure, fluid flow, and mixing on the regenerative properties of articular chondrocytes. Biotechnol Bioeng 1999;65(3):274–281.

121. Matsuda N, Yokoyama K, Takeshita S, Watanabe M. Role of epidermal growth factor and its receptor in mechanical stress-induced differentiation of human periodontal ligament cells in vitro. Arch Oral Biol 1998;43(12):987–997.

122. Schumacker PT. Straining to understand mechanotransduction in the lung. Am J Physiol Lung Cell Mol Physiol 2002;282(5):L881–2.

123. Kastelic J, Galeski A, Baer E. The multicomposite structure of tendon. Connect Tissue Res 1978;6(1):11–23.

124. Augat P, et al. Anisotropy of the elastic modulus of trabecular bone specimens from different anatomical locations. Med Eng Phys 1998;20(2):124–131.

125. Lanza RP, et al. Xenotransplantation of cells using biodegradable microcapsules. Transplantation 1999;67(8):1105–1111.

126. Iwata H, et al. Bioartificial pancreas research in japan. Artif Organs 2004;28(1):45–52.

127. Shapiro AM, et al. Islet transplantation in seven patients with type 1 diabetes mellitus using a glucocorticoid-free immunosuppressive regimen. N Engl J Med 2000;343(4):230–238.

128. Seaberg RM, et al. Clonal identification of multipotent precursors from adult mouse pancreas that generate neural and pancreatic lineages. Nat Biotechnol 2004;22(9):1115–1124.

129. Ianus A, Holz GG, Theise ND, Hussain MA. In vivo derivation of glucose-competent pancreatic endocrine cells from bone marrow without evidence of cell fusion. J Clin Invest 2003;111(6):843–850.

130. Chan C, et al. Hepatic tissue engineering for adjunct and temporary liver support: Critical technologies. Liver Transpl 2004;10(11):1331–1342.

131. Selden C, Hodgson H. Cellular therapies for liver replacement. Transpl Immunol 2004;12(3–4):273–288.

132. Kulig KM, Vacanti JP. Hepatic tissue engineering. Transpl Immunol 2004;12(3–4):303–310.

133. Salgado AJ, Coutinho OP, Reis RL. Bone tissue engineering: State of the art and future trends. Macromol Biosci 2004;4(8):743–765.

134. LeGeros RZ. Properties of osteoconductive biomaterials: Calcium phosphates. Clin Orthop 2002;(395)(395):81–98.

135. Pittenger MF, et al. Multilineage potential of adult human mesenchymal stem cells. Science 1999;284(5411):143–147.

136. Droppert PM. The effects of microgravity on the skeletal system–a review. J Br Interplanet Soc 1990;43(1):19–24.

137. Perka C, et al. Segmental bone repair by tissue-engineered periosteal cell transplants with bioresorbable fleece and fibrin scaffolds in rabbits. Biomaterials 2000;21(11):1145–1153.

138. Lee JY, et al. Effect of bone morphogenetic protein-2-expressing muscle-derived cells on healing of critical-sized bone defects in mice. J Bone Joint Surg Am 2001;83-A(7):1032–1039.

139. He C, Xia L, Luo Y, Wang Y. The application and advancement of rapid prototyping technology in bone tissue engineering. Sheng Wu Yi Xue Gong Cheng Xue Za Zhi 2004;21(5):871–875.

140. Elisseeff J. Injectable cartilage tissue engineering. Expert Opin Biol Ther 2004;4(12):1849–1859.

141. Kim TK, et al. Experimental model for cartilage tissue engineering to regenerate the zonal organization of articular cartilage. Osteoarth Cart 2003;11(9):653–664.

142. Quinones-Baldrich WJ, et al. Is the preferential use of polytetrafluoroethylene grafts for femoropopliteal bypass justified? J Vasc Surg 1988;8(3):219–228.

143. Schmedlen RH, Elbjeirami WM, Gobin AS, West JL. Tissue engineered small-diameter vascular grafts. Clin Plast Surg 2003;30(4):507–517.

144. Barron V, et al. Bioreactors for cardiovascular cell and tissue growth: A review. Ann Biomed Eng 2003;31(9):1017–1030.

145. Daly CD, Campbell GR, Walker PJ, Campbell JH. In vivo engineering of blood vessels. Front Biosci 2004;9:1915–1924.

146. Auger FA, et al. Tissue-engineered skin substitutes: From in vitro constructs to in vivo applications. Biotechnol Appl Biochem 2004;39(Pt 3):263–275.

147. Boyce ST, et al. The 1999 clinical research award. cultured skin substitutes combined with integra artificial skin to replace native skin autograft and allograft for the closure of excised full-thickness burns. J Burn Care Rehabil 1999;20(6):453–461.

148. Supp DM, Wilson-Landy K, Boyce ST. Human dermal microvascular endothelial cells form vascular analogs in cultured skin substitutes after grafting to athymic mice. FASEB J 2002;16(8):797–804.

149. Black AF, et al. In vitro reconstruction of a human capillary-like network in a tissue-engineered skin equivalent. FASEB J 1998;12(13):1331–1340.

150. Andreadis ST. Gene transfer to epidermal stem cells: Implications for tissue engineering. Expert Opin Biol Ther 2004;4(6):783–800.

151. Cooley J. Follicular cell implantation: An update on hair follicle cloning. Facial Plast Surg Clin North Am 2004; 12(2):219–224.

152. Belkas JS, Shoichet MS, Midha R. Peripheral nerve regeneration through guidance tubes. Neurol Res 2004;26(2):151–160.

153. Stanec S, Stanec Z. Reconstruction of upper-extremity peripheral-nerve injuries with ePTFE conduits. J Reconstr Microsurg 1998;14(4):227–232.

154. Hentz VR, et al. The nerve gap dilemma: A comparison of nerves repaired end to end under tension with nerve grafts in a primate model. J Hand Surg (Am) 1993;18(3):417–425.

155. Fields RD, Le Beau JM, Longo FM, Ellisman MH. Nerve regeneration through artificial tubular implants. Prog Neurobiol 1989;33(2):87–134.

156. Heath CA, Rutkowski GE. The development of bioartificial nerve grafts for peripheral-nerve regeneration. Trends Biotechnol 1998;16(4):163–168.

157. Patrick Jr CW. et al. Muristerone A-induced nerve growth factor release from genetically engineered human dermal fibroblasts for peripheral nerve tissue engineering. Tissue Eng 2001;7(3):303–311.

Reading List

Atala A, Lanza RP. Methods of tissue engineering. San Diego: Academic Press; 2001. p 1285.

Lanza RP, Langer RS, Vacanti J. Principles of tissue engineering. San Diego: Academic Press; 2000. p 995.

Saltzman WM. Tissue engineering: Engineering principles for the design of replacement organs and tissues. Oxford, New York: Oxford University Press; 2004. p 523.

See also BIOMATERIALS: TISSUE ENGINEERING AND SCAFFOLDS; SKIN TISSUE ENGINEERING FOR REGENERATION.

ENVIRONMENTAL CONTROL

DENIS ANSON
College Misericordia
Dallas, Pennsylvania

INTRODUCTION

Electronic aids to daily living (EADLs) are devices that can be used to control electrical devices in the client's environment (1). Before 1998 (2), these devices were generally known by the shorter term, "Environmental Control Unit" (ECU). Technically, this term should be reserved for furnace thermostats and similar controls. The more generic EADL applies to control of lighting and temperature, but also applies to control of radios, televisions, telephones, and other electrical and electronic devices in the environment of the client (3,4). See Fig. 1.

These systems all contain some method for the user to provide input to the EADL, some means of determining the current state of the device to be controlled (although this is often visual inspection of the device itself, since EADLs are

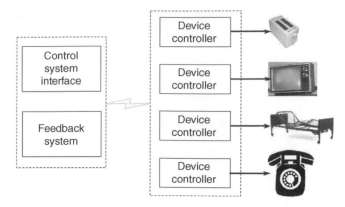

Figure 1. Components of an EADL system.

generally thought of as being applied to the immediate environment), and a means of exerting control over the targeted device. The degree and generalization of control differs among various EADL systems. These systems may provide a means of switching power to the target device, of controlling the features of an external device, or may subsume an external device to provide enhanced control internally.

POWER SWITCHING

The simplest EADLs only provide switching of the electrical supply for devices in a room. Although not typically considered as EADLs, the switch-adapted toys provided to severely disabled children would, formally, be included in this category of EADLs. To adapt a conventional battery powered toy, the therapist inserts a "battery interrupter" to allow an external switch to control the flow of power from the batteries to the workings of the toy. Power switching EADLs operate in precisely the same manner. A switch is placed in series with the device to be controlled, so that the inaccessible power switch of the device can be left in the "ON" position, and the device can be activated by a more accessible external switch. To provide control over appliances and lights in the immediate environment, primitive EADL systems consisted of little more than a set of electrical switches and outlets in a box that connected to devices within a room via extension cords. Control 1(5), for example, allowed the connection of eight devices to the receptacles of the control unit. Such devices are limited in their utility and safety, since extension cords pose safety hazards to people in the environment through risks of falls (tripping over the extension cords) and fires (overheated or worn cords). Because of the limitations posed by extension cords, EADL technology was driven to use remote switching technologies (Fig. 2).

Second generation EADL systems used various remote control technologies to activate power to electrical devices in the environment. These strategies include the use of ultrasonic pulses [e.g., TASH Ultra 4 (6)] (Fig. 4) infrared (IR) light [e.g., Infrared Remote Control (7)], and electrical signals propagated through the electrical circuitry of the home [e.g., X-10 (8) Fig. 3]. All of these switching

Figure 2. Electralink power switching module.

Figure 3. X-10 switching modules.

technologies remain in use, and some are used for much more elaborate control systems. Here we are only considering power switching, however (Fig. 4).

The most prevalent power-switching EADL control system is that produced by the X-10 Corporation. The X-10 system uses electrical signals sent over the wiring of a home to control power modules that are plugged into wall sockets in series with the device to be controlled. (In a series connection, the power module is plugged into the wall, and the remotely controlled device is plugged into the power module.) The X-10 supports up to 16 channels of control, with up to 16 modules on each, for a total of up to 256 devices controlled by a single system. The signals used to control X-10 modules will not travel through the home's power transformer so, in single family dwellings, there is no risk of interfering with devices in a neighbor's home. This is not necessarily true, however, in an apartment setting, where it is possible for two X-10 users to inadvertently control each other's devices. The general set-up of early X-10 devices was to control up to 16 devices on an available channel so that such interference would not occur. In some apartments, the power within a single unit may be on different "phases" of the power supplied to the building. (These phases are required to provide 220-V power for some appliances.) If this is the case, the X-10

signals from a controller plugged into one phase will not cross to the second phase of the electrical wiring. A special "phase cross-over" is available from X-10 to overcome this problem. The X-10 modules, in addition to switching power on and off, can be used, via special lighting modules, to dim and brighten room lighting. These modules work only with incandescent lighting, but add a degree of control beyond simple switching. For permanent installations, the wall switches and receptacles of the home may be replaced with X-10 controlled units. Because X-10 modules do not prevent local control, these receptacles and switches will work like standard units, with the added advantage of remote control.

When they were introduced in the late 1970s, X-10 modules revolutionized the field of EADLs. Prior to X-10, remote switching was a difficult and expensive endeavor, restricted largely to applications for people with disabilities and to industrial applications. The X-10 system, however, was intended as a convenience for able-bodied people who did not want to walk across a room to turn on a light. Because the target audience was able to perform the task without remote switching, the technology had to be inexpensive enough that it was easier to pay the cost than get out of a chair. The X-10 made it possible for an able-bodied

Figure 4. Tash Ultra-4 power switching modules.

person to remotely control electrical devices for under $100, where most disability-related devices cost several thousand dollars.

Interestingly, the almost universal adoption of X-10 protocols by disability related EADLs did not result in sudden price drops in the disability field. In cases where simple power switching provided adequate control, many clinicians continue to adapt mass-market devices for individuals with disabilities. While this may not be a good use of clinician time, it does allow those with limited funding to gain a degree of control over their environments.

FEATURE CONTROL

As electronic systems became more pervasive in the home, simply switching of lights and coffee pots failed to meet the needs of individuals with disabilities who wanted to control the immediate environment. With wall current control, a person with a disability might be able to turn a light or television on and off, but would have no control beyond that. A person with a disability might want to be able to surf cable channels as much as an able-bodied person with a television remote control (9). When advertisements are blaring from the speakers, a person with a disability might want to be able to turn down the sound, or tune to another radio station. Because the ability to control a radio from across the room became a sales advantage when marketing to sedentary, able-bodied adults, nearly all home electronic devices are now delivered with a remote control, generally using IR signals. Most of these remote controls are not usable by a person with a disability, however, due to the small buttons and labels that require fine motor control and good sensory discrimination (Fig. 5).

The EADL systems designed to provide access to the home environment of a person with a disability must provide more than on/off control of home electronics. They must also provide control of the features of home electronic devices. Because of this need, EADL systems frequently have hybrid capabilities. They will incorporate a means of directly switching power to remote devices, often using

Figure 5. Imperium 200H provides infrared remote control of entertainment systems, power switching, and hospital bed control.

X-10 technology. This allows control of devices such as lights, fans, and coffee pots, as well as electrical door openers and other specialty devices (10). They will also typically incorporate some form of IR remote control, which will allow them to mimic the signals of standard remote control devices. This control will be provided either by programming in the standard sequences for all commercially available VCRs, televisions, and satellite decoders, or through teaching systems, where the EADL learns the codes beamed at it by the conventional remote. Preprogrammed control codes allow a simple set-up process to enable the EADL to control various devices, but only those devices whose codes existed prior to the manufacture of the EADL. The advantage of the learning approach is that it can learn any codes, even those that have not yet been invented. The disadvantage is that the controls must be taught, requiring more set-up and configuration time for the user and caregivers. In addition, there are cases where the internal IR switching speed of the EADL differs enough from that of the device to be controlled that some signals cannot be reproduced reliably.

Infrared remote control, as adopted by most entertainment systems controllers, is limited to approximate line of sight control. Unless the controller is aimed in the general direction of the device to be controlled (most have wide dispersion patterns), the signals will not be received. This means that an EADL cannot directly control, via IR, any device not located in the same room. However, IR repeaters, such as the X-10 Powermid (11) can overcome this limitation by using radio signals to send the control signals received in one room to a transmitter in the room of the device to be controlled. With a collection of repeaters, a person would be able to control any infrared device in the home from anywhere else in the home.

One problem that is shared by EADL users and able-bodied consumers is the proliferation of remote control devices. Many homes now are plagued with a remote control for the television, the cable–satellite receiver, the DVD player/VHS recorder (either one or two devices), the home stereo/multimedia center, and other devices, all in the same room. Universal remote controls allow switching from controlling one device to another, but are often cumbersome to control. Some hope is on the horizon for improved control of home audiovisual devices with less difficulty. In November of 1999, a consortium of eight home electronics manufacturers released a set of guidelines for home electronics called HAVi (12). The HAVi specification will allow compliant home electronics to communicate so that any HAVi remote control can operate the features of all of the HAVi devices sharing the standard. A single remote control can control all of the audiovisual devices in the home, through a single interface. Such standards are effective to the extent that they are actually implemented. As of the summer of 2004, the HAVi site lists six products, from two manufacturers, that actually use the HAVi standard.

The Infrared Data Association (13) (IrDA) is performing similar specifications work focusing purely on IR controls. The IrDA standard will allow an IR remote control to operate features of computers, home audiovisual equipment, and appliances equipped with IR controls through a single standard protocol. In addition to allowing a single remote control to control a wide range of devices, IrDA standards will allow other IrDA devices, such as PDAs, personal computers, and augmentative communications systems to control home electronics. Having a single standard for home electronics will simplify the design of EADL systems for people with disabilities.

A more recent standard, V2 (14), offers a much greater level of control. If fully implemented, V2 would allow a single EADL device to control the all of the features of all electronic devices in its vicinity, from the volume of the radio through the setting of the thermostat in the hall, to the "Push to Walk" button on the cross-walk. Using a V2 EADL, a person with a disability could move from environment to environment, and be able to control the V2 enabled devices in any location.

One interesting aspect of feature control by EADLs is the relationship between EADLs and computers. Some EADL systems, such as the Quartet Simplicity (15), include features to allow the user to control a personal computer through the EADL. In general, this is little more than a passthrough of the control system of the EADL to a computer access system. Other EADLs, such as the PROXi (16), are designed to accept control inputs from a personal computer. The goal in both cases is to use the same input method to control a personal computer as to control the EADL. In general, the control demands of an EADL system are much less stringent than those for a computer. An input method that is adequate for EADL control may be very tedious for general computer controls. On the other hand, system that allows fluid control of a computer will not be strained by the need to also control an EADL. The

"proper" source of control will probably have to be decided on a case-by-case basis.

One of the most important features of the environment to be controlled by an EADL is not generally thought of as an electronic device. In a study of EADL users conducted in Finland (17), users reported that the feature that provided the most gain in independence was the ability to open doors independently. While doors are not electronic devices, powered door openers may be, and can be controlled through the EADL.

SUBSUMED DEVICES

Finally, modern EADLs frequently incorporate some common devices that are more easily replicated than controlled remotely. Some devices, such as the telephone, are so pervasive that an EADL system can assume that a telephone will be required. Incorporating telephone electronics into the EADL is actually less expensive, due to telephone standardization, than inventing special systems to control a conventional telephone. Other systems designed for individuals with disabilities are so difficult to control remotely that the EADL must include the entire control system. Hospital bed controls, for example, have no provisions for remote control, but should be usable by a person with a disability. Hence, some EADLs include hospital bed controls internally, so that a person who is in the bed can control its features (Fig. 6).

A telephone conversation may be considered as having several components. The user must "pick up" to connect to the telephone system (able-bodied individuals do this by

Figure 6. Relax II scanning EADL with IR remote control and X-10 switching.

picking up the handset). If the user is responding to an incoming call, the act of picking up initiates the connection. If the user is "originating" a call, the act of picking up will be followed by "dialing", which describes the person to whom the user wishes to speak. When a connection is made (both parties have picked up), a conversation may ensue. At the end of the conversation, the call is "terminated" by breaking the connection.

Many EADL systems include a built-in speakerphone, which will allow the user to originate and answer telephone calls using the electronics of the EADL as the telephone. Because of the existing standards, these systems are generally analogue, single-line telephones, electronically similar to those found in the typical home. Many business settings now use multiline sets, which are not compatible with home telephones. Other settings use digital interchanges, which are also not compatible with conventional telephones. Finally, there is a move currently to use "Voice Over Internet Protocol" (VOIP) to bypass telephone billing and use the internet to carry telephone conversations. Because of this, the telephone built in to a standard EADL may not meet the needs of a disabled client in an office or other setting. Before recommending an EADL as an access solution for a client, therapists should check that the EADL communication system is compatible with the telecommunications systems in that location.

Because the target consumer for an EADL will have severe restrictions in mobility, the manufacturers of many of these systems consider that a significant portion of the customer's day will be spent in bed, and so include some sort of control system for standard hospital beds. These systems commonly allow the user to adjust head and foot height independently, extending the time the individual can be independent of assistance for positioning. As with telephone systems, different brands of hospital bed use different styles of control. It is essential that the clinician match the controls provided by the EADL with the inputs required by the bed to be controlled.

CONTROLLING EADLs

For an EADL to provide improved function to the individual with a disability, it must provide a control interface that is more usable than that of the devices it controls. The common strategies for control found in EADLs are scanning and voice command.

Scanning Control

While scanning control is not particularly useful for computer control (17), it may be quite satisfactory for control of an EADL. When controlling a computer, the user must select between hundreds of options, and perform thousands of selections per day. When controlling the environment, the user may select among dozens of options, but will probably not be making more than a hundred selections during the day. While the frequent waits for the desired action to be offered in a computer scanning input system can have a crippling effect on productivity, the difference between turning on a light *now* versus a *few*

seconds from now is of little consequence. Because of this, many EADL systems provide a scanning control system as the primary means of controlling the immediate environment.

As with computer scanning, EADL scanning may be arranged in a hierarchical pattern. At the topmost level, the system may scan between lights, telephone, bed, entertainment, and appliances. When "lights" are selected, the system might scan between the various lights that are under EADL control. When a single light is selected, the system may scan between "Off", "Dimmer", "Brighter", and "On". As the number of devices to be controlled increases, the number of selections required to control a particular feature will increase, but the overall complexity of the device need not.

Voice Command

Like scanning, voice command may be significantly more functional in controlling an EADL than a computer. If the user is willing to learn specific voice commands for control that are selected to be highly differentiable, a voice command EADL system can be highly reliable. As the potential vocabulary increases, the likelihood of misrecognitions increases and the quality of control goes down.

As with all voice-command systems, background noise can impair the accuracy of recognition. An EADL may easily recognize the commands to turn on a radio in a quiet room, for example, but may be unable to recognize the command to turn off the radio when it is playing in the background. This problem will be exacerbated if the voice commands are to be received by a remote microphone, and when the user is not looking directly at the microphone. On the other hand, the use of a headset microphone can improve control accuracy, but at the cost of encumbering the user.

In addition to microphone type and quality, voice quality can affect the reliability of the system. For individuals with high level spinal cord injuries or other neurological deficits, voice quality can change significantly during the course of a day. A command that is easily recognized in the morning when the person is well rested may be ignored later in the day, after the user has fatigued. At this later time, the user's tolerance for frustration is also likely to be lessened. The combined effect of vocal changes and frustration may result in the user abandoning the device if an alternative control is not provided. While voice command of EADLs has a "magical" quality of control, for the person whose disability affects vocal quality or endurance, it can be a sporadic and limiting magic.

Other Control Strategies

For those systems that are controlled by a computer, any access method that can be used to control the computer can provide control over the environment as well. This includes mouse emulators and expanded keyboards as well as scanning and voice input. A recent study (18) has also demonstrated the utility of switch-encoding as a means of controlling the environment for children as young as 4, or, by extension, for individuals with significant cognitive limitations. In this strategy, the user closes one or two switches in coded patterns (similar to Morse code) to operate the features of devices in the environment. The devices can be labeled with the control patterns for those users who cannot remember the controls, or during the training phase.

The technology now exists, though it has not been applied to EADL systems, to combine head and eye tracking technologies so that a person with a disability could control devices in the environment by simply looking at them. Head-tracking technology could determine, from the user's location, what objects were in the field of view. Eye tracking could determine the distance and fine direction of the intended object. A "heads-up" display might show the features of the device available for control, and provide the control interface. With such a system, a user wearing a headset might be able to turn on a light simply by looking at it and blinking. This type of control strategy, combined with the V2 control protocol, would allow a person with a disability truly "magical" control of the environment.

THE FUTURE OF EADLs

Recognizing that the EADL provides a bridge between the individual with a disability and the environment in which they live, EADL development is likely to occur in two directions. The interface between the individual and the EADL will be enhanced to provide better input to the EADL, and remote controls will become more pervasive so that more of the world can be controlled by the EADL.

The user's ability to control the EADL is a function of the ability of the person to emit controlled behavior. Scanning requires the presence or absence of an action, and ignores any grading. As assistive technologists develops sensing technologies to identify gradients of movement or neural activity, the number of selections that can be made directly increases. For example, current EADLs may be controlled by eye-blink for severely disabled individuals. But in the future, EADLs could use eye-tracking technology to allow a person to look at the device in the environment to be controlled, and then blink to activate it. Electroencephalography might, eventually, allow the control of devices in the environment by simply *thinking at* them (19,20).

The continued development of remote control technologies will, in the future, allow EADLs to control more of the environment that is currently available. Currently, EADLs can switch power to any device that plugs into the wall or that uses batteries. Feature control, however, is very limited. Cross-walk controls, elevators, and ATM machines do not have the means of remote control today, and are often beyond the reach of a person with a significant disability. Microwave ranges, ovens, and air conditioners also have feature controls that might be, but are not, remotely controllable. If interoperability standards like V2 are accepted, the controls that are provided for sedentary, able-bodied users may provide control for the EADLs of the future. It is generally recognized that the inclusion of remote control for EADL access is not cost-effective, but if the cost of providing remote control for able-bodied users becomes low enough, such options might be made available, which will allow EADLs of the future to control them as well.

BIBLIOGRAPHY

Cited References

1. Assistive Technology Partners. 2002. Direct Access Electronic Aids. Available at http://www.uchsc.edu/atp/library/fast-facts/Direct%20Access%20Electronic%20Aids.htm. Accessed July 26 2004.
2. MacNeil V. 1998. Electronic Aids to Daily Living. Team Rehabilitation Report, 53–56.
3. Center for Assistive Technology. Environmental Control Units. 2004. Available at http://cat.buffalo.edu/newsletters/ecu.php. Accessed 26 July 2004.
4. Cook AM, Hussey SM. Assistive technologies: Principles and practice. 2nd ed. Philadelphia: Mosby International; 2002.
5. Prentke Romich Company, 1022 Heyl Road, Wooster, OH 44691. Phone: (800) 262-1984. Available at http://www.prentrom.com/index.html.
6. Tash Inc., 3512 Mayland Ct., Richmond VA 23233 Phone: 1-(800) 463-5685 or (804) 747-5020. Available at http://www.tashinc.com/index.html.
7. DU-IT Control Systems Group, Inc., 8765 Township Road #513, Shreve, OH 44676, Phone: (216) 567-2906.
8. SmarthHome, Inc., 16542 Millikan Avenue, Irvine, CA 92606-5027, Phone: (949) 221-9200 x109.
9. Butterfield T. 2004. Environmental Control Units. Available at http://www.birf.info/artman/publish/article_418.shtml. Accessed 26 July 2004.
10. Quartet Technology. 2004. Quartet Technology, Inc.-News. Available at http://www.qtiusa.com/ProdOverview.asp?ProdTypeID =1. Accessed at 29 July 2004.
11. asiHome, 36 Gumbletown Rd., CS1, Paupack, PA 18451, Phone: 800-263-8608. Available at http://www.asihome.com/cgi-bin/ASIstore.pl?user_action=detail&catalogno= X10-PEX01.
12. HAVi, Inc., 40994 Encyclopedia Circle, Fremont, CA 94538 USA. Phone: (510) 979-1394. Available at http://www.havi.org/.
13. IrDA Corporate Office, P.O. Box 3883, Walnut Creek, CA 94598. Phone: (925) 943-6546. Available at http://www.irda.org/index.cfm.
14. InterNational Committee for Information Technology Standards. What exactly is V2 - and How Does It Work? Available at http://www.myurc.com/whatis.htm. Accessed at 30 July 2004.
15. Quartet Technology, Inc., 87 Progress Avenue, Tyngsboro, Massachusetts 01879, phone: 1-(978) 649-4328.
16. Madentec, Ltd., 4664 99 St., Edmonton, Alberta, Canada T6E 5H5, phone: (877) 623-3682. Available at http://madentec.com.
17. Anson DK. Alternative Computer Access: A Guide to Selection. Philadelphia: F. A. Davis; 1997.
18. Anson D, Ames C, Fulton L, Margolis M, Miller M. 2004. Patterns For Life: A Study of Young Children's Ability to Use Patterned Switch Closures for Environmental Control. Available at http://atri.misericordia.edu/Papers/Patterns.php. Accessed 4 Oct 2004.
19. Howard T. Beyond the Big Barrier. Available at http://www.cs.man.ac.uk/aig/staff/toby/writing/PCW/bci.html. Accessed 11 Oct 2004.
20. Wolpaw JR. 2004. Brain-Computer Interfaces For Communication And Control. Available at http://www.nichd.nih.gov/about/ncmrr/symposium/wolpaw_abstract.htm. Accessed 11 Oct 2004.

See also COMMUNICATIVE DISORDERS, COMPUTER APPLICATIONS FOR; MOBILITY AIDS.

EQUIPMENT ACQUISITION

ROBERT STIEFEL
University of Maryland
Baltimore, Maryland

INTRODUCTION

Equipment acquisition is the process by which a hospital introduces new technology into its operations. The process involves determining the hospital's needs and goals with respect to new technology and equipment, how best to meet those needs, and instituting the decisions. The process involves virtually every clinical and support department of the hospital. This is consistent with the Joint Commission on Accreditation of Healthcare Organizations (JCAHO) (1) medical equipment management standards which require hospitals to have a process for medical equipment acquisition.

Unfortunately, in many hospitals it is a ritual, the control and details of which are jealously guarded. In fact, the needs of the hospital's operation would be much better served if all departments knew how the process worked. If the rules of the process were known and based upon the stated goals and priorities of the institution, then the people who must attempt to justify requests for new equipment would be able to do their jobs better, and with subsequently better results. If a new technology can improve the hospital's finances and/or clinical or support functions, then the methods by which this can be used to justify requests should be clearly explained. If the hospital's reimbursement structure is such that the introduction of new technology is difficult, but the improvement of current functions is more easily funded, then this should be explained.

In short, there should be a policy and procedure for the method by which the hospital acquires new equipment. It should define what the hospital means by a capital expenditure, reflect the hospital's overall goals and objectives, clearly state how to prepare a justification, and explain, at least in general terms, how a decision is to be made about the funding of a proposal.

The scope of this article is limited to the justification, selection, and implementation of new medical technology and equipment (Table 1). It is not intended to provide cookbook methods to be followed exactly, but instead to explain principles that can be applied to different hospitals and their differing needs. This seems to be particularly appropriate because needs vary not only between hospitals, but also with time.

JUSTIFICATION PROCESS

The justification process lays the groundwork for the acquisition of medical equipment. The better the justification, the more dependable the results will be. If the rules of the justification process are carefully planned, and just as carefully followed, the equipment acquisition function of the hospital will be respected and adhered to by other components of the hospital system. It is via the justification process that the hospital's needs are recognized, proposals

Table 1. Equipment Acquisition: Outline of Process

Justification	Clinical testing
Needs assessment	Use in expected application
Proposal	Questionnaire or interview
Clinical	Assessment
Financial	Ranking
Environmental	Requests for quotations
Budget request	Final choice
Selection	Negotiate
Literature review	Contract
Library	**Implementation**
Subscriptions	Purchase order
Standards	Installation
Manufacturer's literature	Acceptance testing
Vendor list	Training
Request for proposal	Operator
Preliminary review	Service
Engineering testing	**Conclusion**
Safety	Report
Performance	Follow-up
Contact other users	

are created to meet these needs, and sufficient funds are budgeted to fulfill the most acceptable proposals.

Needs assessment is the first step in the justification process. The requirement for new equipment can be based upon a variety of disparate requirements. There can be a need for new technology or expansion of an existing service. Need for equipment can be based upon cost effectiveness of new technology, safety, maintenance costs, or simply a need to replace old equipment. The justification for acquiring equipment based upon any of these reasons must be supported by facts.

The age of equipment by itself is not sufficient justification for replacing equipment. If the age of equipment exceeds accepted guidelines, and maintenance costs, for example, are also exceeding accepted guidelines, then the replacement of equipment is adequately justified. What this points out, however, is that there must be guidelines for replacement of equipment based upon both age and maintenance costs.

A general guideline is that when equipment is over 7 years old, it is time to start considering its replacement. The age when equipment should be considered for replacement can also be based on its depreciation life, which varies depending on the type of device. Wear and tear, along with the advancement of the state of the art in equipment design, will start catching up with the equipment at about this time. When total maintenance costs exceed approximately one and one-half times replacement cost or when an individual repair will cost more than one-half the replacement cost, it is probably more appropriate to consider replacing the equipment. Age and/or maintenance costs by themselves do not necessarily require equipment replacement. There can be mitigating circumstances, such as the fact that the older equipment might be impossible to replace. If an item is one of a standardized group, or part of a larger system, it might be unrealistic to consider replacing the entire group or system.

Safety considerations should be relatively easy to document. If the performance of equipment puts it out of compliance with accepted standards, the standards and the performance of the equipment can be documented. Again, however, judgment can mitigate these standards. If a piece predates current safety or performance standards, it is perfectly acceptable to continue using it if clinical and engineering personnel believe it still performs safely and as intended. This judgment should also be documented.

Cost effectiveness is, by itself, adequate justification for equipment acquisition. If it can be shown that within 3–5 years, the acquisition of new equipment will save more than the cost of the acquisition, it will be money well spent. The justification must be done very carefully, however, to recognize all costs that will be incurred by the acquisition of new equipment, and any additional operating costs.

If an existing clinical service is covering its costs and it can be shown that there is a need to expand the service, the equipment necessary to support the expansion can be cost justified. Again, the justification must be done carefully to ascertain that there are sufficient trained people, or people who can be trained to use the equipment, as well as a sufficient population of patients requiring the service.

The acquisition of new technology requires the most difficult and demanding justification. The technology itself must be assessed. Determine whether the technology is viable, provides a needed service, and will be accepted. There must be clinical professionals capable of utilizing the technology or in a position to be trained to do so. Cost justification of new technology is equally difficult. Very careful thought will have to be given to identifying all costs and revenues. The cost of the equipment, the cost of the supplies necessary to operate it, and the cost of the personnel to operate it must all be determined.

In addition, the space requirements and special installation requirements, such as electrical power, computer networking, air conditioning, plumbing, or medical gases, will all have to be determined. Maintenance costs will have to be identified and provisions made to cover them. It must be determined that there is an adequate population base that will provide patients to take advantage of the new technology. State and local approval (certificate of need) may also have to be met. The entire process is very time consuming, and should be carefully planned to meet the scheduling of the hospital.

Once needs have been justified, primarily by the people intending to apply the new equipment, it will be necessary to create a formal proposal. The formal proposal should be prepared by a select group of people (medical, nursing, purchasing, administration, and clinical engineering) most closely associated with the management and use of the equipment. Physicians are concerned with the function performed by the equipment, nursing with the equipment's operations, and clinical engineering with the design, safety and performance, and dependability of the equipment. Administration and purchasing are involved with managing costs. Information technology staff may need to be involved if equipment is networked, or facilities staff if there are special installation requirements.

The proposal must provide a precise definition of the clinical needs, the intended usage of the equipment, any restrictions of a clinical nature, and a thorough financial plan. The financial planning, in particular, should include accepted planning techniques. For example, life-cycle cost analysis is perhaps the most thorough method for determining the financial viability of a new project. (Life-cycle analysis is the method of calculating the total cost of a project by including the initial capital cost, the operating costs for the expected lifetime of the equipment, and the time cost of money.)

At this stage, it is appropriate to consider a variety of physical or environmental factors that affect or are affected by the proposed new equipment. The equipment will require space for installation, use, and maintenance. Its power requirements might call for a special electrical source, medical gases or vacuum, or a water supply and drain. Its size might prevent if from passing through doors. The equipment might require special air-handling consideration if it generates heat or must be operated in a temperature-and/or humidity-controlled environment. Its weight might preclude transport in elevators or require modification of floors, or its sensitivity to vibration might require a special installation. If the equipment is either susceptible to or generates electrical or magnetic fields, special shielding may be required. There might be special standards that restrict the selection or installation of the type of equipment.

After the staff intending to use and manage the equipment have completed their proposal, it will be necessary to present this to the committee responsible for budgeting new equipment or new projects. This "capital equipment budget committee" will typically be chaired by an individual from the hospital's budget office and should include representatives from central administration, nursing, and clinical engineering. It is their responsibility to review proposals for completeness and accuracy as well as feasibility with respect to the hospital's long-range plans and patient population. Their judgment to approve or disapprove will be the necessary step before providing funds.

SELECTION PROCESS

Once the acquisition of new equipment has been justified, planned, and budgeted, the next step is to select the equipment that will actually be purchased. Again, this is a formal process, with sequential steps that are necessary to achieve the most appropriate equipment selection. There are commercial software products available that help hospitals establish priorities, develop proposals, and select capital equipment (e.g., Strata Decision Technology).

For most equipment, a standing capital equipment committee can oversee the selection and acquisition process. This committee should at least include representatives from clinical engineering, finance, and purchasing. It might also include representatives from nursing and information technology.

For major equipment or new technology, a selection committee should be formed specifically for each acquisition. Such a committee should include physician, nursing, and administrative personnel from the area for which the equipment is intended, and a representative from clinical engineering. Since the representative from clinical engineering will serve on virtually all of these ad hoc selection committees, this person's experience would make him/her the best choice for chairperson. Realistically, however, it might be more politically expedient to allow one of the representatives from the area to chair the committee.

The first step involves a literature review. A library search should be conducted, and the clinical engineering department's professional subscriptions reviewed, for example, *Health Devices* and *Biomedical Instrumentation and Technology*. Look for applicable standards from AAMI or the American National Standards Institute (ANSI). Obtain product literature from the manufacturers being considered. Research the information maintained by the FDA Center for Devices and Radiological Health (CDRH) at http://www.fda.gov/cdrh. The FDA-CDRH Manufacturer and User Facility Device Experience Database (MAUDE) contains reports of adverse events involving medical devices, and their "Safety Alerts, Public Health Advisories, and Notices" page contains safety-related information on medical devices.

A list of proposed vendors to be contacted can be created by consulting the *Health Devices Sourcebook* (ECRI) (2). These vendors should be contacted and their literature on the type of equipment being evaluated requested. Part of the evaluation includes evaluating the manufacturers and vendors. Commencing with this initial contact, notes should be kept on the responsiveness and usefulness of the representatives contacted.

A request for proposal (RFP) should be written on the basis of the needs determined during the justification process and the information acquired from the literature review. When the selection criteria are straightforward, the RFP and the request for quotation (RFQ) can be combined. For more complex selections, such as equipment systems or new technology, the RFP and RFQ should be separate processes.

The RFP should be carefully written, well organized, and thoroughly detailed. It should contain useful background information about the institution and the specific user area. Applicable documents, such as drawings, should be either included or explicitly made available for review by the vendors. The equipment requirements should include a statement regarding the major objectives to be fulfilled by the equipment. The description of the specific requirements must include *what* is to be done, but should avoid as much as possible restrictions on *how* it is to be done. It is likely that, given reasonable latitude in addressing the equipment requirements, manufacturers can make useful suggestions based upon their experience and the unique characteristics of their equipment.

The RFP should contain a description of the acceptance testing that will be conducted before payment is approved. It should also contain a request for a variety of ancillary information: available operator and service documentation; training materials and programs; warranties; and maintenance options and facilities. There should also be a request for the names of at least three users of comparable equipment, located as close as possible to the hospital so that site visits can be conveniently arranged. The RFP should be reviewed by the entire in-house evaluation committee.

A cover letter should accompany the RFP to explain the general instructions for submitting proposals: who to call for answers to questions, deadline for submission, format,

and so on. When appropriate, there can also be such information as to how the proposals will be evaluated, how much latitude the vendors have in making their proposals, and the conditions under which proposals can be rejected. It is not necessary to send the RFP to all known vendors of the type of equipment being evaluated. If there is any reason why it would be undesirable to purchase from particular vendors, for example, poor reputation in the area or unsatisfactory dealings in the past, it would be best to eliminate them from consideration before the RFP process.

The response of the vendors to the RFP will allow the selection committee to narrow the field to equipment that is likely to meet the defined needs. The full evaluation committee should review the proposals. There will have been a deadline for submission of proposals, but it might not be appropriate to strictly enforce it. It is more important to consider the long-term advantages to the hospital and not to discount an otherwise acceptable proposal for being a few days late. The proposals should be reviewed for completeness. Has all of the requested information been provided? Are there any misinterpretations? Are there exceptions? The vendors should be contacted and additional information requested or clarifications discussed as necessary. It will now also be possible to determine the type of acquisition required, that is, whether the equipment can be purchased from a single vendor, whether it will have to be purchased from more than one vendor and assembled, or whether the equipment will require some special development effort to meet the clinical needs.

Evaluate the quality of each proposal. A simple form can be used to record the results. The form should include the elements of the review. Score the responses according to the relative importance of each element, and whether there was a complete, partial, or no response (Table 2). Include comments to explain the reason for each score. Based upon a review of the proposals submitted, the evaluation

Table 2. Proposal Evaluation Form

Evaluation of Response to Request for Proposal
Date:
Reviewer:

Manufacturer:
Equipment included in proposal (models, options, quantity):
Response submitted on time:

Score (0,1):__

Response followed specified format:

Score (0,1,2):__

Response included all information requested:

Score (0,2,4):__

Response includes exceptions:

Score (−2,−1,0):__

Response included additional, useful suggestions:

Score (0,1,2):__

Response included reasonable alternatives:

Score (0,1,2):__

Total Score:__

Percent Score {[*total* score/total possible score (11)] × 100}:_____

committee should agree on which vendors and equipment to consider further (i.e., the proposals that offer equipment that meets the established criteria).

The next step will be to request equipment for an in-house evaluation. The equipment requested for testing should be the exact type that would be ordered. If preproduction prototypes or engineering models are accepted for evaluation, it should be realized that there are likely to be changes in the operation and performance of the final product. The dependability of prototype equipment cannot be judged. In short, the evaluation will not be complete, and, to some extent, the ability to judge the equipment or compare it with other equipment will be hampered.

The most important aspect of the entire equipment acquisition process, especially for equipment types not already in use, is the in-house, comparative evaluation. This evaluation has two phases: engineering and clinical. The equipment requested for comparative evaluation has been selected from the proposals submitted. The field should be narrowed to include only vendors and equipment worthy of serious consideration. Therefore, it will be worthwhile to commit significant effort to this evaluation.

The engineering phase of the evaluation will need test procedures, test equipment, and forms for documenting the results. The clinical phase will need representative areas, users, protocols, a schedule for training as well as use, and a method for collecting results: an interview, a review meeting, or a questionnaire. All of these details should be settled before the arrival of the equipment. For example, it might be determined that sequential testing would be preferable to simultaneous testing. Neither the hospital nor the vendor would want to store equipment while waiting for its evaluation.

The first step in testing equipment is the engineering evaluation, done in the laboratory. The tests include safety and performance aspects. Mechanical safety criteria include consideration of the ruggedness or structural integrity of the equipment as well as the potential for causing injury to patients or personnel. Electrical safety tests are conducted per the requirements in the AAMI/ANSI *Electrical* Safety *Standard* and NFPA electrical standards as appropriate. Performance testing is done as described by any published standards, according to the manufacturer's own service literature, and per the needs determined in the justification process. The results of the engineering tests should be summarized in a chart to facilitate comparison (Table 3). Differences and especially flaws should be highlighted.

Table 3. Engineering Evaluation Form

Engineering Evaluation

Date:
Evaluator:

Manufacturer:
Equipment included in evaluation (models, options):
Safety
  Mechanical: Score (0,1,2):__
  Electrical: Score (0,1,2):__
  Safety Subtotal (weight 0.2 × average score):_____

Performance
  Controls: Score (0,1,2):__
  (List performance features; score according to test results; Score (0,1,2):__
  weight according to importance):
                                        or Score (0,2,4):__
  Performance Subtotal (weight 0.2 × average score):_____

Manufacturer's Specifications
  (List important specifications; score according to test results;
  weight according to importance): Score (0,1,2):__
                                        or Score (0,2,4):__
  Manufacturer's Specifications Subtotal (weight 0.1 × average score):_____

Technical Standards
  (List applicable technical standards; score according to test results;
  weight according to importance): Score (0,1,2):__
                                        or Score (0,2,4):__
  Technical Standards Subtotal (weight 0.2 × average score):_____

Human Engineering
  Design: Score (0,1,2):__
  Size: Score (0,1,2):__
  Weight: Score (0,1,2):__
  Ease of use: Score (0,1,2):__
  Reliability: Score (0,1,2):__
  Serviceability: Score (0,1,2):__
  Operator's manual: Score (0,1,2):__
  Service manual: Score (0,1,2):__
  Human Engineering Subtotal (weight 0.1 × average score):_____
                                        Total Score:_____
  Percent Score [(*total* score/*total* possible score) × 100]:_____

Table 4. User Evaluation Form

User Interview

Date:
Interviewer:

Institution:
Name and title of person(s) interviewed:
Manufacturer:
Equipment at site (models, options):
Years equipment in use:
Safety (any incidents involving patients or personnel)

Score (0,1,2):__
Safety Score (weight 0.2 × score):_____

Performance (does equipment meet user needs)

Score (0,1,2):__
Performance Subtotal (weight 0.2 × average score):_____

Reliability (frequency of equipment failures)

Score (0,1,2):__
Reliability Subtotal (weight 0.1 × average score):_____

Ease of Use (satisfaction with ease of use)

Score (0,1,2):)__
Ease of Use Subtotal (weight 0.1 × average score):_____

Ease of Service (satisfaction with ability to inspect and repair)

Score (0,1,2):__
Ease of Service Subtotal (weight 0.1 × average score):_____

Manufacturer's Support (quality of training and service)

Score (0,1,2):__
Manufacturer's Support Subtotal (weight 0.1 × average score):_____

Overall Satisfaction (would user buy equipment again)

Score (0,1,2):__
Overall Satisfaction Subtotal (weight 0.2 × average score):_____

*Total Score:*_____

Percent Score {[*total* score/*total* possible score (2.0)] × 100}:_____

In addition to the straightforward safety and performance tests, a number of characteristics should be evaluated based upon engineering judgment. The physical construction should be judged, especially if there are constraints imposed by the intended application or installation. A study of the construction and assembly can also allow a judgment regarding reliability. This judgment should be based upon the quality of hardware and components, the method of heat dissipation (fans suggest an exceptional requirement for heat dissipation and require periodic cleaning), the method of construction (the more wiring and connectors, the more likelihood of related failure), and whether the design has had to be modified by such means as alterations on circuit boards or "piggybacked" components.

The maintainability of the equipment is reflected both in the methods of assembly and in the maintenance instructions in the operator and service manuals. The manuals should explain how and how often preventive maintenance or inspection or calibration should be performed. Finally, a clinical engineer should be able to judge human engineering factors. For example, ease of use, how logical and self-explanatory is the front panel, self-test features, and the chances or likelihood of misuse all affect the safety and efficacy of the equipment.

Design a form to record the engineering evaluation results. The form will vary in the specific features and tests that are included, according to the type of equipment that is being evaluated, but the basic format will remain the same. Include comments to explain the reason for the score of each item. Table 3 is an example of an engineering evaluation form.

The physicians, nurses, and clinical engineers on the evaluation committee should contact their counterparts at the institutions named as users of their equipment by the vendors. Site visits can be particularly useful in cases where the equipment or technology being considered is new to the hospital. The committee can see the application firsthand, and talk to the actual users face to face. Record and score the results of interviews (Table 4).

Clinical testing is performed after engineering testing. Equipment must satisfactorily pass the engineering testing to be included in the clinical testing; there is no point in wasting people's time or taking a chance on injuring a patient. Clinical testing should not be taken lightly; it is usually more important than the technical testing. The clinical testing is done only on equipment that has survived all of the previous tests and by the people who will actually be using it.

The clinical testing should be designed so that the equipment is used for the intended application. The equipment included in the clinical testing should be production equipment unless special arrangements are likely to be made with the manufacturer. Users should be trained just as they would be for the actual equipment to be purchased. In fact, the quality of the training should be included in the evaluation. Users should also be given questionnaires or be interviewed after the equipment has been used (Table 5).

Table 5. Clinical Testing Questionnaire

Clinical Trial

Date:
Clinician:

Manufacturer:
Equipment used (models, options):
Number of patients on whom equipment used:
Safety (any incidents involving patients or personnel)

Score (0,1,2):__
Safety Score (weight 0.2 × score):_____

Performance (does equipment meet user needs)

Score (0,1,2):__
Performance Subtotal (weight 0.2 × average score):_____

Reliability (number of equipment failures)

Score (0,1,2):__
Reliability Subtotal (weight 0.1 × average score):_____

Ease of Use (satisfaction with ease of use)

Score (0,1,2):__
Ease of Use Subtotal (weight 0.1 × average score):_____

Manufacturer's Support (quality of training and support)

Score (0,1,2):__
Manufacturer's Support Subtotal (weight 0.1 × average score):_____

Overall Satisfaction (would user recommend equipment)

Score (0,1,2):__
Overall Satisfaction Subtotal (weight 0.2 × average score):__
Total Score:_____
Percent Score {[*total* score/*total* possible score (1.8)] × 100}:_____

One further consideration to be judged by the full evaluation committee is that of standardization. There are numerous valid reasons for standardizing on equipment. Repairs are easier to accomplish because of technicians' familiarity with the equipment. Repair parts are easier and less expensive to keep in inventory. Standardization allows exchange of equipment between clinical areas to help meet varying demands. Training is also more easily provided.

There are also valid drawbacks, however. Standardization makes the hospital dependent upon a limited number of vendors. It can interfere with user acceptance if users feel they have little or no say in the selection process. It can also delay the introduction of new technology. It is important that the evaluation committee consider the relative importance of the pros and cons of standardization in each case.

Assessment of the equipment should result in a ranking of the equipment that has successfully completed the engineering and clinical testing. The advantages and disadvantages of each item should be determined and ranked in order of importance. Alternatively, the criteria can be listed in order of importance. If possible, the list should be divided between criteria that are mandatory and those that are desirable. Then a judgment on whether the equipment does or does not satisfy the criteria can be made. From this charting, it is likely that clear preferences will become obvious.

Ideally, there will be two or more finalists who are close to equal in overall performance at this point. These finalists should be sent a request for quotation (RFQ). The responses to the RFQ will allow a cost comparison. A life-cycle cost analysis can be accomplished with the aid of the hospital's financial officer. This will give a more accurate depiction of the total cost of the equipment for its useful lifetime. While it is beyond the scope of this article to describe life-cycle cost

analysis, it takes into account the initial costs (capital cost of equipment plus installation), operating costs over the anticipated life of the equipment (supplies, service, fees), and the time cost of money (or present value). It may or may not take personnel costs into account, since these will likely be the same or similar for different models. To calculate a percent score for purposes of evaluation, divide the cost of the least expensive equipment by the cost of the equipment being evaluated, and multiply by 100.

With the completion of the evaluation, a comparison of the results should lead to a final selection. This should be as objective as possible. If additional information is needed, there should be no hesitation in contacting vendors. In fact, it may be useful to have a final presentation by the vendors. Develop a point system for the individual elements of the proposal reviews, engineering evaluations, user interviews, clinical testing, and cost comparison. Weight these elements according to their relative importance (Table 6).

Once the hospital has made its final choice of vendor and equipment, it is not necessary to end negotiations. Negotiations should be conducted before a vendor knows that they are the preferred provider. Before a vendor is certain of an order, they are much more likely to make concessions. Requests for extra features, such as spare equipment, spare parts, special tools, and test equipment, may be included in the final quote. Trade-in of old equipment and compromises on special features or warranties, for example, can help reduce the cost of equipment. Consider negotiating hardware and software upgrades for a specified period.

For large orders (e.g., dozens of infusion pumps), or for installations (e.g., monitoring systems), request that the vendor unpack, set up, inspect, distribute or install, and document (using the hospital's system) the equipment at no additional cost. In most cases, the hospital will want

Table 6. Evaluation Scoring Form

Overall Evaluation Score

Date:

Reviewer:

Manufacturer:

Equipment (models, options):

Proposal	Percent Score:__
	Proposal Score (weight 0.1 × score):_____
Engineering Evaluation	Percent Score:__
	Engineering Evaluation Score (weight 0.2 × score):_____
User Evaluation	Percent Score:__
	User Evaluation Score (weight 0.1 × score):_____
Clinical Testing	Percent Score:__
	Clinical Testing Score (weight 0.3 × score):_____
Cost Comparison	Percent Score:__
	Cost Comparison Score (weight 0.1 × score):_____
	Total Score:_____

the vendor to provide user training. This training should meet the hospital's specified needs (e.g., all users, all shifts, train the trainer, videotapes or computer programs, etc.). The hospital may also need service training for its clinical engineering department. The hospital can negotiate not only the service school tuition, but also room, board, and travel. The negotiated quote should be reviewed by the end-users, clinical engineering, finance, purchasing, and a contracts lawyer.

IMPLEMENTATION PROCESS

The successful conclusion of the equipment acquisition process cannot be realized until the equipment is satisfactorily put into use. As with all the previous steps in the equipment acquisition process, the implementation process requires planning and monitoring.

The purchase order with which the equipment is ordered is a legal document: a contract. As such, it can protect the rights of the hospital. Therefore, it is important to include not only the equipment being ordered, but also all agreed-upon terms and conditions. These terms and conditions should include delivery schedules, the work to be performed by the vendor, warranty conditions, service agreements, operator and service manuals, acceptance criteria and testing, and operator and service training.

Before the equipment is delivered, arrangements should be made for installation, for personnel to transport the equipment, for storage, and for personnel to test the equipment. If any of these are to be the responsibility of the vendor, they should be included in the purchase order. If the hospital has to perform any modifications or fabrications, all necessary parts and preparations should be in place.

Acceptance testing should be done upon the completion of installation. The test procedures for acceptance testing should come from the initial specifications and from the manufacturer's data sheets and service manuals. Acceptance testing should be thorough, because this is the ideal time to obtain satisfaction from the manufacturer. It is also appropriate to initiate the documentation per the standard system of the hospital at this time. The invoice for the equipment should not be approved until the equipment has been satisfactorily tested.

After the equipment is in place and working properly, training should be scheduled. The manufacturer or their representative, or the selected hospital personnel, should have the training program completely prepared. If ongoing training will be necessary over the lifetime of the equipment, in-service instructors should be involved in the initial training as well. The clinical engineering personnel responsible for inspection and maintenance should also receive operator and service training.

An often overlooked, but useful, adjunct to the implementation process is follow-up on the installation. Users should be polled for the acceptance of the equipment and their perception of its usefulness. Engineering personnel should review the dependability of the equipment from their service records. All of the people involved in the equipment acquisition process should learn from every acquisition, and what they have learned should be reviewed during this follow-up.

CONCLUSION

Table 1 can be used as a list of tasks that should be accomplished (or at least considered) in the equipment acquisition process. A schedule and checklist can be generated from this list.

The equipment acquisition should be fully documented, ideally by a write-up of the entire process. Table 1 can be used to create an outline for the final report. The results of the equipment acquisition process should be shared with manufacturers and other interested parties. It should always be the intention of the hospital personnel to improve the situation with respect to the manufacturer. Perhaps the most dependable way for medical equipment manufacturers to learn what is important to hospitals is to review what hospitals have said about the evaluation of new equipment during their acquisition process.

BIBLIOGRAPHY

Cited References

1. Comprehensive Accreditation Manual for Hospitals, 2005. Oakbrook Terrace (IL): Joint Commission on Accreditation of Healthcare Organizations; 2004.

Reference List

Health Devices Sourcebook, 2005. Plymouth Meeting (PA): ECRI; 2004.

Larson E, Maciorowski L. Rational Product Evaluation. JONA 16(7, 8):31–36.

Stiefel R, Rizkalla E. The Elements of a Complete Product Evaluation. Biomed Instrum Technol. Nov/Dec 1995. p 482–488.

Stiefel RH. Medical Equipment Management Manual, 2004 Edition. Arlington (VA): Association for the Advancement of Medical Instrumentation; 2004.

Staewen WS. The Clinical Engineer's Role in Selecting Equipment. Med Instrum 18(1):81–82.

See also EQUIPMENT MAINTENANCE, BIOMEDICAL; MEDICAL RECORDS, COMPUTERS IN; OFFICE AUTOMATION SYSTEMS; RADIOLOGY INFORMATION SYSTEMS.

EQUIPMENT MAINTENANCE, BIOMEDICAL

Arif Subhan
Masterplan Technology
Management
Chatsworth, California

INTRODUCTION

Preventive maintenance (PM) is one of the many functions of a clinical engineering department. The other functions include incoming inspection/testing, prepurchase evaluation, coordination of outside equipment service, hazard and recall notification, equipment installation, equipment repair and upgrade, purchase request review, equipment replacement planning, device incident review, and regulatory compliance maintainance. The primary objective of a PM program for biomedical equipment is to prevent failure, which is achieved through (1) detecting the degradation of any non-durable parts of the device and restoring them to like-new condition; (2) identifying any significant degradation of the performance of the device and restoring it to its proper functional level; and (3) detecting and repairing any partial degradation that might create a direct threat to the safety of the patient or operator.

The term preventive maintenance has its origins with mechanical equipment that has parts that are subject to wear and need to be restored or replaced sometime during the useful lifetime of the device. PM refers to the work performed on equipment on a periodic basis and should be distinguished from "repair" work that is performed in response to a complaint from the equipment user that the device has failed completely or is not working properly. The term "corrective maintenance" is often used instead of the term "repair."

Today, medical equipment uses electronic components that fail in an unpredictable manner, and their failure cannot be anticipated through measurements and checks performed during a PM (1). Also, equipment failures that are attributable to incorrect set up or improper use of the device or the use of wrong or defective disposable accessory cannot be prevented by doing PM (2). It has been argued that current medical devices are virtually error-free in terms of engineering performance. Device reliability is an intrinsic function of the design of the device and cannot be improved by PM. In modern equipment, the need for performance and safety testing is greatly reduced because of the design and self-testing capability of the equipment (3,4). For example, the Philips HeartStart FR2+ defibrillator performs many maintenance activities itself. These activities include daily and weekly self-tests to verify readiness for use and more elaborate monthly self-tests that verify the shock waveform delivery system, battery capacity, and internal circuitry. The manufacturer also states that the FR2+ requires no calibration or verification of energy delivery. If the unit detects a problem during one of the periodic self-tests, the unit beeps and displays a flashing red or a solid red warning signal on the status indicator (5).

The term "scheduled (planned) maintenance" was introduced in 1999 by a joint AAMI/Industry Task Force on Servicing and Remarketing while developing a document for submission to the FDA entitled "Joint Medical Device Industry Proposed Alternative to the Regulation of Servicers, Refurbishers, and Remarketers" (6). However, many still use the traditional term preventive maintenance rather than this new, more carefully defined term. According to the document developed by the AAMI/Industry Task Force, the term scheduled (planned) maintenance "consists of some or all of the following activities: cleaning; decontamination; preventive maintenance; calibration; performance verification; and safety testing." Among these activities, the three key activities are (1) preventive maintenance (PM); (2) performance verification (PV) or calibration; and (3) safety testing (ST).

These three terms are defined by the AAMI/Industry Task Force as follows. "Preventive maintenance is the inspection, cleaning, lubricating, adjustment or replacement of a device's nondurable parts. Nondurable parts are those components of the device that have been identified either by the device manufacturer or by general industry experience as needing periodic attention, or being subject to functional deterioration and having a useful lifetime less than that of the complete device. Examples include filters, batteries, cables, bearings, gaskets and flexible tubing." PM performed on a medical device is similar to the oil, filter, and spark plug changes for automobiles.

"Performance Verification is testing conducted to verify that the device functions properly and meets the performance specifications; such testing is normally conducted during the device's initial acceptance testing." This testing is important for detecting performance deterioration that could cause a patient injury. For example, if the output of the device (such as temperature, volume, or some form of energy) is not within specifications, it could result in an adverse patient outcome (7). If the performance deterioration can be detected by visual inspection or by a simple user test, then periodic performance verification becomes less critical. The data obtained during maintenance will also assist in determining the level and intensity of performance verification.

"Safety testing is testing conducted to verify that the device meets the safety specifications, such testing is normally conducted during the device's initial acceptance testing."

HISTORICAL BACKGROUND

The concept of a facility-wide medical equipment management program first emerged in the early 1970s. At that time, there was little management of a facility's medical equipment on a centralized basis. Examples of the lack of proper management that affected the quality of care are described in detail in the literature (8) and include poor frequency response of ECG monitors, heavy accumulation of dirt inside equipment, delay in equipment repair, little attention to the replacement of parts that wear over time, and the consequential high cost of repairs.

The Joint Commission on Accreditation of Healthcare Organizations (JCAHO) has played an important role in promoting equipment maintenance programs in hospitals in the United States. Almost all hospitals in the United States are accredited by the JCAHO and are required to comply with their standards.

Poor or nonexistent equipment maintenance programs (8) coupled with the "great electrical safety scare" of the early 1970s (9) created the impetus that has popularized electrical safety programs in hospitals. Since then, equipment maintenance and electrical safety testing has been an important part of clinical and biomedical engineering programs in hospitals. The JCAHO standards in the mid-1970s required that all electrically powered equipment be tested for leakage current four times a year (10).

In 1983, the Medicare reimbursement program for hospitals changed to a much tighter fixed-cost approach based on so-called Diagnostic Related Groups (DRGs). In 1979, JCAHO increased the maximum testing interval for patient care equipment to six months. This change was made in response to pressure from hospitals to reduce the cost of complying with their standards. It has been reported that there was no other rationale for changing the testing intervals (10).

In 2001, JCAHO removed the annual PM requirement for medical equipment (11). The reason cited for the change was that the safety and reliability of medical equipment had improved significantly during the prior decade. It was also stated that some equipment could benefit from maintenance strategies other than the traditional "interval-based" PM. Other PM strategies suggested included predictive maintenance, metered maintenance (e.g., hours of usage for ventilators), and data-driven PM intervals.

INVENTORY

A critical component of an effective PM program is an accurate inventory, which has been a key requirement in the JCAHO equipment standards since they were introduced in the early 1970s. The 2005 JCAHO standard (EC.6.20) requires a current, accurate, and separate inventory of medical equipment regardless of ownership (12). The inventory should include all medical equipment used in the hospital (owned, leased, rented, physician-owned, patient-owned, etc.). A complete and accurate inventory is also helpful for other equipment management functions including tracking of manufacturer's recalls, documenting the cost of maintenance, replacement planning, and tracking model-specific and device-specific issues. The equipment list should not be limited to those devices that are included in the PM program (13).

INCLUSION CRITERIA

In 1989, JCAHO recognized that not all medical devices are equally critical with respect to patient safety and introduced the concept of risk-based inclusion criteria for the equipment management program. Fennigkoh and Smith developed a method for implementing a risk-based equipment management program by attributing numerical values to three variables; "equipment function," "physical risks associated with clinical application," and "mainte-

nance requirements" (14). They compounded these values into a single variable; the equipment management (EM) number. The EM was calculated as follows: EM = Function rating + Risk rating + Required Maintenance rating. The three variables were not weighted equally. Equipment function constituted 50% of the EM number whereas risk and maintenance each constituted 25%. The authors acknowledge the somewhat arbitrary nature of weighting the equipment function rating at 50% of the EM number; however, they viewed this variable as the device's most significant attribute. The authors also arbitrarily set a level for the EM number at greater than or equal to 12 to determine whether the device will be included in the EM program. Devices included in the EM program are assigned a unique control number. Devices with an EM number less than 12 were excluded from the EM program and are not assigned a control number.

Although risk-based calculators using the original Fennigkoh and Smith model are still widely used, criticism of this method exists on the basis that this particular risk-based approach is arbitrary. The factors used in the calculation, their weighting, the calculated value above which one decides to include the device in the management program, and what PM interval to use are criticized as all being somewhat subjective or arbitrary (13).

SELECTING THE PM INTERVAL

Selecting an appropriate PM interval is a very important element of an effective maintenance program. Several organizations have offered guidelines, requirements, and standards for the selection of an appropriate PM interval for the various devices. This list includes accreditation organizations (e.g., JCAHO), state authorities (e.g., California Code of Regulations, Title 22), device manufacturers, and national safety organizations (e.g., NFPA). JCAHO is the primary authority that most people look to for minimum requirements for the PM intervals for medical devices. However, the JCAHO 2005 Standards [EC.6.10 (4) and EC.6.20 (5)] put responsibility back on the hospital to define PM intervals for devices based on manufacturer recommendations, risk levels, and hospital experience. JCAHO also requires that an appropriate maintenance strategy be selected for all of the devices in the inventory (12).

According to one source (7) two approaches exist to determining the proper PM interval. One is fixed and the other is evidence-based. Other sources present some contradictory views on how to select a PM interval for a device. Some (13,15) argue that PMs should not necessarily follow the manufacturer's recommended intervals but should be adjusted according to the actual PM failure rate. Ridgway and Dinsmore (16,17) argue that if a device is involved in an incident where a patient or staff member is injured, a maintainer who has not followed the manufacturer's recommendations will be deemed to have some legal liability for the injury, irrespective of how the device contributed to the injury. Dinsmore further argues that the manufacturer recommendations should be followed

Table 1. Conflicting Recommendations for the PM Intervals of Some Devices

Device Type	ASHE PM Interval (months)	ECRI PM Interval (months)
Defibrillator	3	6
Apnea monitor	6	12
Pulse oximeter	6	12

because the manufacturer developed the device and proved its safety to the FDA. He states that if no recommendations exist from the manufacturer, then it is up to the clinical engineering department to decide on the maintenance regimen. National organizations such as the American Society of Hospital Engineering (18) and ECRI (2) have published recommendations on PM intervals for a wide range of devices. The recommendations by ASHE were published in 1996 and ECRI in 1995. However, conflicting recommendations exist for certain devices in the list (see Table 1).

The ANSI/AAMI standard EQ56 (19) "Recommended Practice for a Medical Equipment Management Program" does not attempt to make specific recommendations for PM intervals for devices. And, in many cases, PM interval recommendations from the manufacturers are not readily available. Those that have been made available are often vague with little or no rationale provided. In some cases, their recommendations simply state that testing should be conducted periodically "per hospital procedures" or "per JCAHO requirements."

Intervals for the electrical safety testing of patient care equipment are discussed in the latest edition of NFPA 99-2005, Health Care Facilities, Section 8.5.2.1.2.2. Although this document is a voluntary consensus standard and not a regulation, many private and government agencies reference and enforce NFPA standards. The following electrical safety testing intervals are recommended. The actual interval is determined by the device's normal location or area in which the device is used. For general care areas, the recommended interval is 12 months; for critical care areas and wet locations, the recommended interval is 6 months. The standard does allow facilities to use either longer or shorter intervals if they have a documented justification from previous safety testing records or evidence of unusually light or heavy use (20).

In a similar fashion, the evidence-based method allows adjustment of the interval up or down depending on the documented finding of prior testing (7). This method is based on the concepts of reliability-centered maintenance (21). Ridgway proposes an evidence-based method using a PM optimization program that is based on periodic analyses of the results of the PM inspections. His approach takes into consideration the severity of the problems found during the PMs. It classifies the problems found into one of the four PM problem severity levels, level 1 through 4. This method incorporates the concepts found in Failure Modes and Effects Analysis (FMEA)—a discipline that has been recently embraced by the JCAHO. It requires the classification of the problems discovered during the PM testing

into four levels of criticality. Level 1 problems are potentially life-threatening, whereas the other levels are progressively less severe. Examples of level 1 problems include defibrillator output energy being significantly out of specification or an infusion pump significantly under- or overinfusing. For a more detailed explanation of this method, see Ref. (22). Others (23,24) have used maintenance data in a similar way to rationalize longer maintenance intervals.

It should be noted that some regulations, codes, or guidelines and some manufacturer's recommendations may advance more stringent requirements than other standards for the same type of device or an interval that the facility has used successfully in the past. In these instances, the facility needs to balance the risks of noncompliance, which may include injury to the patient, liability, and penalties for the organization, against the cost of compliance. Many factors exist that need to be evaluated before modifying the generally recommended intervals, including equipment maintenance history and experience, component wear, manufacturer recommendations, device condition, and the level of technology used in the device (2).

Note also that the manufacturer's recommendations for their newer models of medical devices are generally less stringent than those for their older models. In many instances, the recommended PM intervals are greater than 12 months.

PM PROCEDURES

Selecting an appropriate PM procedure is another important element of an effective maintenance program. Both ECRI (2) and ASHE (18) have published PM procedures for a broad range of devices. It should be noted that the manufacturers usually provide more detailed PM procedures for their specific devices than those published by ECRI and ASHE. It is recommended that manufacturers' recommendations that appear to be too complex be evaluated to see if they can be simplified (2). See Fig. 1 for a sample PM procedure for an infusion pump.

EVALUATION OF A PM PROGRAM

The facility's PM completion or completed on-time rates are parameters that have been favored by both facility managers and external accreditation agencies (7). However, the value of PM completion rate as an indicator of program quality has been debated for many years (25). Until 2003, to pass the maintenance portion of the medical equipment program, the JCAHO required a consistent 95% PM completion rate. The shortcoming of this standard is that the 5% incomplete PMs could, and sometimes did, include critical devices such as life-support equipment. To address this undesirable loophole, the JCAHO, in 2004, discontinued the 95% requirement. The 2005 standard EC.6.20 now segregates medical equipment into two categories: life-support and nonlife-support devices. Life-support equipment is defined by JCAHO as those devices that are intended to sustain life and whose failure to perform their primary function is expected to result in death. Examples include ventilators, anesthesia

For: **INFUSION PUMP**   PM Procedure No. **IP001**

Interval: **xx months**   Estimated <u>annual</u> m-hrs: **x.0**

Check the <u>AC</u> box if **Action Completed** and the <u>AMR</u> box if Adjustment or Minor Repair was required <u>to bring the device into conformance with the performance or safety specifications</u>. In this case provide a **note*** on the <u>nature of the problem found</u>, in sufficient detail to identify the *level of potential severity* of the problem.

<u>AC</u>/<u>AMR</u>   **Preventive Maintenance**

SM1.   Inspect/clean the chassis/housing especially any moving parts including any user-accessible areas under covers (if applicable). Examine all cable connections. Replace any damaged parts, as required.

SM2.   Confirm that all markings and labeling are legible. Clean or replace, as required.

SM3.   Replace or recondition the battery, as required.

<u>AC</u>/<u>AMR</u>   **Performance Verification**

PV1.   Verify that the flow rate or drip rate is within specification. Perform self-test, if applicable.

PV2.   Verify that all <u>alarms</u> (including occlusion and maximum pressure) and interlocks operate correctly.

PV3.   Verify that the battery charging system is operating within specification.

PV4.   Verify the functional performance of all controls, switches, latches, clamps, soft touch keys, etc.

PV5.   Verify the functional performance of all indicators and displays, in all modes.

PV6.   Verify that the time/date indication is correct, if applicable.

<u>AC</u>/<u>AMR</u>   **Safety Testing**

ST1.   Check that the physical condition of the power cord and plug, including the strain relief, is OK.

ST2.   Check the ground wire resistance. (< 0.5 ohm).

ST3.   Check chassis leakage to ground. (< 300 micro amps).

---

* <u>Notes</u>:

Figure 1. Sample PM procedure for an infusion pump.

machines, and heart–lung bypass machines (12). ECRI suggests that the following additional devices be included in the life-support category: anesthesia ventilators, external pacemakers, intra-aortic balloon pumps, and ventricular assist devices (26). The completion of PMs for life-support equipment is scored more stringently than the completion rate performance for nonlife-support equipment. It is theoretically possible that if the PM of a single life-support device is missed, an adverse, noncompliance "finding" could be generated by the survey team. However, it has been reported that the surveyors will probably be more focused on investigating gaps in the process that led to the missed PM (27).

The 2005 JCAHO standards do not contain an explicit PM completion requirement. However, many organizations appear to have set the goal for on-time PM completion rate for life-support equipment at 100% and the goal for the nonlife-support equipment at better than 90%. An important aspect of PM completion is calculating the on-time PM completion rate. The JCAHO does not state how the PM completion rate should be calculated. Hospitals are free to specify in their management plan that they have allowed themselves either 1 or 2 months of extra time to complete the scheduled maintenance. It is important for those devices that have maintenance intervals of three months or less, or are identified as life-support equipment, that the PMs be completed within the month they are scheduled for PM. Sometimes, additional time may be needed for devices that are unavailable because they are continuously in use. In this case, the appropriate department and the safety/environment of care committee should be informed about the delay.

The real issue to be addressed is the effectiveness of the program. However, "effectiveness" of a PM program is difficult to measure. If the purpose of the restoration of any nondurable parts is to reduce device failures, then the effectiveness of this element can be measured by the resulting reduction in the device failure rate. In principle, repair rates and incident analyses can provide a relationship between PM and equipment failures (7).

Two challenges associated with achieving an acceptable PM completion rate exist. The first is to find the missing equipment listed on the monthly PM schedule. Multiple documented attempts should be made to locate any devices that are not found, which should be followed with a written communication to the clinical users seeking their assistance in locating the device. It is important to get a written acknowledgment from the users that they have attempted to locate the devices in question. This acknowledgment will help when explaining to the surveyors that efforts were made to locate the device. However, if the devices cannot be located after this extended search it must be assumed that they are no longer in use in the facility. The question now is, are they still on site but deliberately or accidentally hidden away? Or have they really been removed from the facility? Until these questions can be answered, no satisfactory way to deal with accounting for the failure to complete the overdue PM exists. One strategy to reduce the potential administrative paperwork is to classify the missing equipment as "unable to locate." This action (classifying the devices as unable to locate) should be commu-

nicated to the appropriate departments including the user departments, the safety/environment of care committee, materials management, and security. Consistent loss or unable to locate device(s) should be viewed negatively by the facility and should serve as an indicator that the facility needs to re-evaluate its security plan. The users should be reminded not to use the missing device(s) if they reappear. These device(s) should be readily identifiable with the "out-of-date" PM sticker. The service provider should be notified so that they can give the re-emerging device(s) an incoming inspection and restart its PM sequence.

The second challenge is to gain access to equipment that is continually in use for patient care. It is difficult to complete the PM if the device is in constant use for monitoring or treatment of patients. Examples of devices that often fall into this category include ventilators, patient monitors, and certain types of laboratory or imaging equipment. Ideally, to alleviate this problem, a back-up unit should be available to substitute while maintenance is performed on the primary unit. This spare unit can also serve as a back-up for emergency failures as well. A good working relationship with the users is very helpful in these circumstances.

DOCUMENTATION

Documentation of maintenance work is another important element of an effective maintenance program. The 2005 JCAHO standards require the hospital to document performance, safety testing, and maintenance of medical equipment (12).

Complete scheduled maintenance and repair documentation is required when a device is involved in an incident, or when the reliability of the device is questioned, and also to determine the cost of maintenance. Most accrediting and regulatory organizations accept a system of exception reporting, which is recording only the results of steps failed during performance verification as part of scheduled maintenance or after repair (18). It has been generally accepted that it is reasonable to record only what went wrong during the PM procedure (13). Having extensive documentation is considered a safe practice. ECRI supports the use of exception reporting and recommends that before deciding on using exception reporting the hospital take into account that exception reporting, with its lack of affirmative detail, may be less than convincing evidence in the event of a liability suit (2).

Two ways to document scheduled (planned) maintenance exist. One is to record the maintenance by hand on a standard form or preprinted PM procedure and file the records manually. The other is to use a computerized maintenance management system (CMMS). Computerized records should help in reducing the time and space required for maintaining manual documentation. Even with CMMS, the department may need to have a way to store or transpose manual service reports from other service providers. Other technologies like laptops and personal data assistants (PDAs) can further assist in implementing the maintenance program. A comprehensive review of CMMSs, that are currently in use can be found in the literature (28). The CMMS helps the clinical engineering staff to manage the medical equipment program effectively. The core of the CMMS consists of equipment inventory, repair and maintenance record, work order subsystem, parts management subsystem, reporting capabilities, and utilities. A CMMS can be classified broadly as internally developed (typically using commercial off-the-shelf personal computer hardware and database software) and commercially available applications (desktop and web-based) (29).

STAFFING REQUIREMENTS

The number of full-time equivalents (FTEs) required to do scheduled maintenance varies based on the experience of the staff, the inventory mix, and the inventory inclusion criteria. See Ref. 30 for the method to determine the number of FTEs required for maintenance. Based on a clinical engineering practice survey, an average of one FTE is required to support 590 medical devices (31). The generally cited relationship between the support required for scheduled maintenance and repair for a typical inventory mix of biomedical devices is 750–1250 devices per FTE. Cost of the maintenance program includes salaries, benefits, overtime and on-call pay, cost of test equipment and tools, and training and education expenses. Salaries for the clinical engineering staff can be obtained from the annual survey published by the Journal of Clinical Engineering and the 24 × 7 magazine.

PM STICKERS

PM stickers or tags are placed on a device to indicate that maintenance has been completed and the device has been found safe to use. They also convey to the user a warning not to use the device when the sticker is out-of-date. Color-coded stickers (see Fig. 2) or tags are not required by the JCAHO, but they can be very helpful in identifying devices that need maintenance (2,32).

ENVIRONMENTAL ROUNDS

Conducting environmental rounds is another important aspect of the medical equipment maintenance program. The intent of these rounds is to discover situations, which are, or could lead to, an equipment-related hazard or safety problem. The clinical engineering staff should conduct

Figure 2. Color coded sticker.

regular inspections of all clinical areas that are considered to be "medical equipment-intensive" areas (e.g., ICU, CCU, NICU, ER, and surgery). The interval between these equipment-related environmental safety rounds should be adjusted according to the frequency with which hazards are found. Other areas that are considered less equipment-intensive such as medical/surgical and other patient care floors, laboratory areas, and outpatient clinics should also be surveyed for electrical safety hazards but less frequently. The equipment items in these areas are usually line-powered items that do not have nondurable parts requiring periodic attention and they usually do not require periodic calibration or performance checking. These items need periodic visual inspections to ensure that they are still electrically safe (i.e., that no obvious defects exist such as a damaged power cord or evidence of physical damage that might electrify the case or controls). The search should be focused on the physical integrity of the equipment, particularly the power cord and the associated connectors and other equipment defects such as dim displays, torn membrane switches, taped lead wires, and so on. Items with power cords that are exposed to possible abuse from foot traffic or the wheels of other equipment should receive closer attention than those items with less exposed cords. Damaged wall outlets should be noted, as should the use of extension cords or similar "jury-rigged" arrangements. Other indicators that should be noted and investigated further are equipment enclosures (cases) that are obviously distorted or damaged from being dropped. In addition to these potential electrical safety hazards, other targets of this survey are devices with past-due PM stickers or with no sticker at all. When a safety hazard is identified, or a past due or unstickered item is found, appropriate corrective actions should be taken immediately and documented as required (33).

BIBLIOGRAPHY

Cited References

1. Wang B, Levenson A. Are you ready? 24 × 7 2004; Jan: 41.
2. ECRI. Health Devices Inspection and Preventive Maintenance System, 3rd ed. Plymouth Meeting (PA): ECRI; 1995.
3. Keil OR. Evolution of the clinical engineer. Biomed Technol Manag 1994; July–Aug: 34.
4. Keil OR. The buggy whip of PM. Biomed Technol Manag 1995; Mar–Apr: 38.
5. Philips. HeartStart FR2+ defibrillator. Instructions for use. M3860A, M3861A, Edition 8, 4-1, 4-9. Seattle (WA): Philips; 2002.
6. Hatem MB. From regulation to registration. Biomed Instrument Technol 1999;33:393–398.
7. Hyman WA. The theory and practice of preventive maintenance. J Clin Eng 2003;28:31–36.
8. JCAHO. Chapter 1. The Case for Clinical Engineering. The Development of Clinical Engineering. Plant, Technology & Safety Management Series Update Number 3. Chicago (IL): JCAHO; 1986. pp 9–11.
9. Nader R. Ralph Nader's most shocking expose. Ladies Home J 1971;88: 98 176–178.
10. Keil OR. Is preventive maintenance still a core element of clinical engineering? Biomed Instrument Technol 1997;31: 408–409.

11. JCAHO. EC Revisions approved, annual equipment PM dropped. Environ Care News 2001;4:1, 3, 9.

12. JCAHO. Hospital Accreditation Standards. Oakbrook Terrace, IL: Joint Commission Resources; 2005.

13. Stiefel RO. Developing an effective inspection and preventive maintenance program. Biomed Instrument Technol 2002;36: 405–408.

14. Fennigkoh L, Smith B. Clinical equipment management. Plant, Technology & Safety Management Series Number 2, Chicago (IL): JCAHO; 1989.

15. Maxwell J. Prioritizing verification checks and preventive maintenance. Biomed Instrument Technol 2005;39:275–277.

16. Ridgway MG. Personal communication. 2005.

17. Baker T. Journal of Clinical Engineering Roundtable: Debating the medical device preventive maintenance dilemma and what is the safest and most cost-effective remedy. J Clin Eng 2003;28:183–190.

18. ASHE. Maintenance Management for Medical Equipment. Chicago, IL: American Hospital Association; 1996.

19. AAMI. Recommended Practice for Medical Equipment Management Program. ANSI/AAMI EQ 56:1999. Arlington, VA: AAMI; 1999.

20. NFPA. NFPA 99, Standard for Health Care Facilities. Quincy (MA): NFPA; 2005.

21. Moubray J. Reliability-Centered Maintenance. New York: Industrial Press; 1997.

22. Ridgway MG. Analyzing planned maintenance (PM) inspection data by failure mode and effect analysis methodology. Biomed Instrument Technol 2003;37:167–179.

23. Acosta J. Data-driven PM Intervals. Biomed Instrument Technol 2000;34:439–441.

24. JCAHO. Data-driven medical equipment maintenance. Environ Care News 2000;3:6–7.

25. Ridgway MG. Making peace with the PM police. Healthcare Technol Manag 1997; Apr: 44.

26. ECRI. Maintaining life support and non-life support equipment. Health Devices 2004;33(7):244–250.

27. AAMI. JCAHO connection. Questions on life support equipment and NPSG # 6. Biomed Instrument Technol 2005;39: 284.

28. Cohen T. Computerized Maintenance Management Systems for Clinical Engineering. Arlington (VA): AAMI; 2003.

29. Cohen T, Cram N. Chapter 36. Computerized maintenance management systems. In: Dyro JF, editor. Clinical Engineering Handbook Burlington (MA): Elsevier Academic Press; 2004;.

30. AHA. Maintenance Management for Medical Equipment. Chicago, IL: American Hospital Association; 1988.

31. Glouhove M, Kolitsi Z, Pallikarakis N. International survey on the practice of clinical engineering: Mission, structure, personnel, and resources. J Clin Eng 2000;25:269–276.

32. Guerrant S. A sticker is worth a thousand words. 24 × 7 44, March 2003.

33. Ridgway MG. Masterplan's Medical Equipment Management Program. Chatsworth (CA): Masterplan; 2005.

See also CODES AND REGULATIONS: MEDICAL DEVICES; SAFETY PROGRAM, HOSPITAL.

ERG. See ELECTRORETINOGRAPHY.

ERGONOMICS. See HUMAN FACTORS IN MEDICAL DEVICES.

ESOPHAGEAL MANOMETRY

VIC VELANOVICH
Henry Ford Hospital
Detroit, Michigan

INTRODUCTION

The purpose of the esophagus is to act as a conduit for food from the mouth to the stomach. This is an active process that is initially a conscious act with chewing and swallowing, then becomes unconscious when the bolus of food enters the esophagus. As part of this process, the esophageal musculature acts to propagate the food bolus to the stomach and to prevent reflux of gastric contents to the esophagus to protect the esophagus itself. There are several disease processes of the esophagus for which the assessment the esophageal musculature function would contribute to the diagnosis and management. This assessment is done indirectly through the measurement of intraesophageal pressures. This pressure measurement is accomplished with esophageal manometry (1).

ESOPHAGEAL ANATOMY AND PHYSIOLOGY

The esophagus is, in essence, a muscular tube that connects the mouth to the stomach. The esophagus anatomically starts in the neck, traverses the thorax, and enters into the abdomen to join the stomach. It is ~20–25 cm in length, with the start of the esophagus being the upper esophageal sphincter and the end being the gastroesophageal junction. The upper esophageal sphincter is primarily made up of the cricopharyngeus muscle attached to the cricoid cartilage anteriorly. The musculature of the upper esophagus is made of striated muscle, and this transitions to smooth muscle in the lower esophagus. The esophagus is made up of three primarily layers: the inner-mucosa, the middle-submucosa, and the outer-muscle layer. The muscle layer is made up of an inner-circular muscle layer and an outer-longitudinal layer. At the inferior end of the esophagus is the lower esophageal sphincter. This is not a true anatomical sphincter, but rather a physiological high pressure zone that is the cumulation of thickening of the distal 5 cm of the esophageal musculature, the interaction of the esophagus with the diaphragmatic hiatus, and at 2 cm of intraabdominal esophagus. The lower esophageal sphincter should not be confused with the gastroesophageal junction, which is the entrance of the esophagus to the stomach, nor with the Z line, which is the transition of the esophageal squamous mucosa to the glandular gastric mucosa.

The lower esophageal sphincter is tonically contracted during rest to prevent reflux of gastric contents into the esophagus. It relaxes with a swallow. Swallowing is divided into an oral stage, pharyngeal stage, and esophageal stage. The oral and pharyngeal stage prepare food by chewing, the tongue pushing the food bolus to the pharynx, the soft palate is pulled upward, the vocal cords are closed and the epiglottis covers the larynx, the upper esophageal sphincter relaxes, and the contraction of the pharyngeal muscles initiate the primary peristaltic wave. The esophageal stage consists of a peristaltic wave that pushes the food bolus to the stomach. There are primary peristaltic waves, which are initiated by the pharynx with a swallow. Secondary peristaltic waves are initiated within the esophagus due to distention of the esophagus with food. As the bolus reaches the lower esophagus, the lower esophageal sphincter relaxes.

INDICATIONS FOR ESOPHAGEAL MANOMETRY

Indications for esophageal manometry include the following five problems. (1) Dysphagia, to assess for an esophageal motility disorder, such as achalasia or nutcracker esophagus. It is important that mechanical causes of dysphagia, such as cancer, have been excluded. (2) Gastroesophageal reflux disease, not for primary diagnosis, but for possible surgical planning. (3) Noncardiac chest pain, which may be of esophageal origin. (4) Exclusion of generalized gastrointestinal disease (e.g., scleroderma or chronic idiopathic pseudoobstruction), and exclusion of an esophageal etiology for anorexia nervosa. (5) Determination of lower esophageal sphincter location for proper placement of an esophageal pH probe (2).

EQUIPMENT

The basic equipment used for manometry is the manometry catheter, infusion system (for water perfused systems), transducers, Polygraf or A/D converter, and computer with appropriate software (Fig. 1).

Esophageal manometry catheters come in two basic types: water perfused and solid state. The water perfused catheter contains several hollow tubes. Each tube has one side opening at a specific site on the catheter. The openings are at various points around the circumference of the catheter. There is a 4 cm catheter with side holes placed 5 cm apart, and a 8 cm catheter with 4 lumens placed 1 cm apart in the most distal end of the catheter, then 5 cm apart for the next proximal 4 lumens. The radial spacing helps to accurately measure upper and lower esophageal sphincter pressures that are asymmetrical. The water perfused catheters require an infusion system. The manometric pump uses regulated compressed nitrogen gas to deliver distilled water through the channels of the catheter. The pressurized water from each channel is connected to a single opening in the catheter within the patient's esophagus. The pressure changes are transmitted to the transducer, and these pressure changes are recorded and charted by a computer with the appropriate software.

The solid-state catheter has internal microtransducers. These directly measure intraesophageal pressures and contractions. Some advantages are that the sensors respond faster, do not require that the patient lie down, and record pressures circumferentially. An additional advantage is that as these catheters do not require fluid containers, potential pitfalls with fluid disinfection are avoided. Some drawbacks are that they are delicate, more expensive to purchase and repair, and are prone to baseline drift. As with the water-perfused system, the

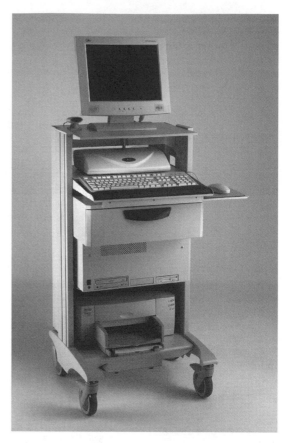

Figure 1. The esophageal manometry console with computer. (Courtesy of Medtronic Corp., Minneapolis, MN.)

solid-state system requires that the catheter be connected to a computer console for recording and storing pressure data.

Additional items are needed. These include a manometer calibration tube for calibrating the solid-state catheter, a viscous lidocaine, a lubricating jelly, tissues, tape, an emesis basin, a syringe, a cup of water, a penlight, and tongue blades.

CONDUCT OF THE TESTING

Prior to the procedure, the pressure channels require calibration and connection to the computer console. The patient sits upright and one of the nares is anesthetized. Lubricating jelly is applied to the catheter and inserted through one the nares into the pharynx. The patient is asked to swallow and the catheter advance into the esophagus to the stomach. The catheter is advanced for ~65 cm to insure all the sensor ports are within the stomach. If using water-perfused catheters, the patient is moved to the supine position. If using solid-state catheters, the patient is kept in the semi-Fowler position, which is the head and torso of the patient at a 45° angle bent at the waist.

The esophageal motility study consists of (1) the lower esophageal sphincter study, (2) esophageal body study, and (3) the upper esophageal sphincter study. Before beginning the study, insure that all sensor ports are within the stomach by having the patient take a deep breath and watching the motility recording. It should show a smooth tracing with a pressure increase during inspiration. An alternative method of determining that all side holes are within the stomach is to apply gentle pressure to the epigastrium to confirm that there is a simultaneous increase in the recorded pressure. When all channels are within the stomach, a gastric baseline is set to establish a reference for pressure changes.

The lower esophageal sphincter study measures sphincter pressure, length, location, and relaxation. This is done using either the "station pull-through" or "slow continuous pull through" methods. With the station pull-through technique, the catheter is slowly withdrawn through the sphincter at 1 cm increments while looking for changes in pressure. When the first channel enters the sphincter, pressure will increase from baseline by at least 2 mmHg (0.266 kPa). This identifies the lower border of the sphincter. As the catheter is continued to be pulled back, the "respiratory inversion point" is reached. This is the transition from the intraabdominal to the intrathoracic esophagus. With inspiration, the catheter will record positive pressures within the abdomen, but negative pressures within the thorax. Inferior to the respiratory inversion point is the lower esophageal sphincter high pressure zone. This is the physiologic landmark used to perform pressure measurements and relaxation studies. When the distal channel passes through the upper border of the lower esophageal sphincter, the pressure should decrease from baseline. The length traveled from the lower to the upper

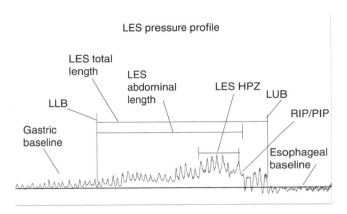

Figure 2. Lower esophageal sphincter profile as determined by esophageal manometry. (Courtesy of Medtronic Corp., Minneapolis, MN.)

border of the sphincter measures the sphincter length. The slow continuous pull-through method is done while the catheter is pulled back continuously, while pressures are being recorded. The catheter is pulled back 1 cm every 10 s. These methods lead to identifying the distal and proximal borders of the sphincter, overall sphincter length, abdominal sphincter length, and resting sphincter pressure (Fig. 2). Sphincter relaxation involves observing the response of the lower esophageal sphincter to swallowing (Fig. 3). This is done by asking the patient to swallow 5 mL of water with the catheter positioned in the high

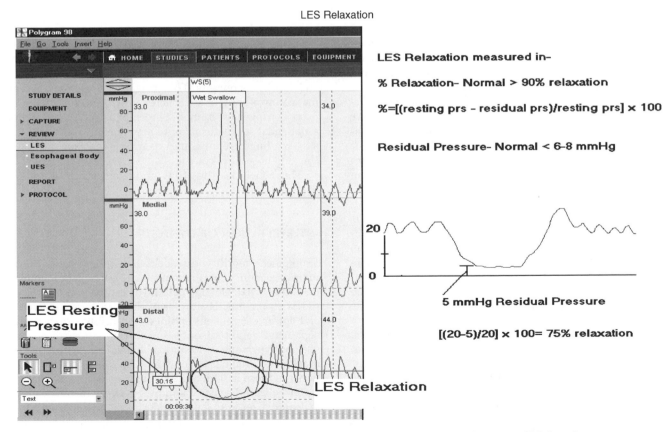

Figure 3. Manometric tracings of lower esophageal sphincter relaxation. (Courtesy of Medtronic Corp., Minneapolis, MN.)

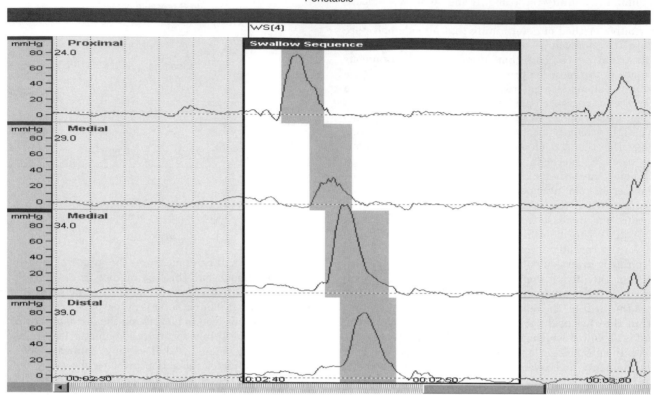

Figure 4. Manometric tracings of esophageal body peristalsis. (Courtesy of Medtronic Corp., Minneapolis, MN.)

pressure zone. The pressures are recorded with the swallow. Accurate determination of lower esophageal relaxation requires a Dent sleeve. Although this can be measured using side holes, artifact can be created.

The study of the esophageal body determines the muscular activity of the esophagus during swallowing. There are four components of the study: (1) peristalsis, (2) amplitude of contractions, (3) duration of contraction, and (4) contraction morphology (Fig. 4). These measurements are made with the distal pressure sensor positioned 3 cm superior to the upper border of the lower esophageal sphincter. The patient takes 10 wet swallows with 5 mL of room temperature water. Esophageal body amplitude is the force with which the esophageal musculature contracts. The amplitude is measured from baseline to the peak of the contraction wave. Duration is the length of time that the esophageal muscle remains contracted. It is measured from the point at which the major upstroke of the contraction begins to the point at which it ends. Velocity is a measurement of the time it takes for a contraction to migrate down the esophagus (unit of measure is cm s^{-1}). These measurements are used to determine esophageal body motility function.

The study of the upper esophageal sphincter includes (1) resting pressure, (2) relaxation, and (3) cricopharyngeal coordination (Fig. 5). The study is done by withdrawing the catheter in 1 cm increments until the upper esophageal sphincter is reached. This is determined when the pressure measured rises above the esophageal baseline. The catheter is positioned so that the first sensor is just superior to the sphincter and the second sensor is at the proximal border of the sphincter. The remaining channels are in the body of the esophagus. The patient is given 5 mL of water for wet swallows. The catheter is withdrawn during this process. However, it should be emphasized that the upper esophageal sphincter is quite asymmetric; therefore, pressure readings are meaningless unless the exact position of the side holes are known.

This concludes the study and the catheter is removed from the patient.

INTERPRETATION OF THE TEST

Esophageal motility disorders are categorized into primary, secondary, and nonspecific (3). Primary esophageal disorders are those in which the dysfunction is limited only to the esophagus. Examples of these include the hypotensive lower esophageal sphincter associated with gastroesophageal reflux disease, achalasia, diffuse esophageal spasm, hypertensive lower esophageal sphincter, and nutcracker esophagus. Secondary esophageal motility disorders are those in which the swallowing occurs as a result of a generalized disease. Examples of these include collagen-vascular disease (e.g., scleroderma), endocrine and metabolic disorders (diabetes mellitus), neuromuscular diseases (myasthenia gravis, multiple sclerosis, and Parkinson's disease), chronic idiopathic intestinal pseudo-obstruction,

UES

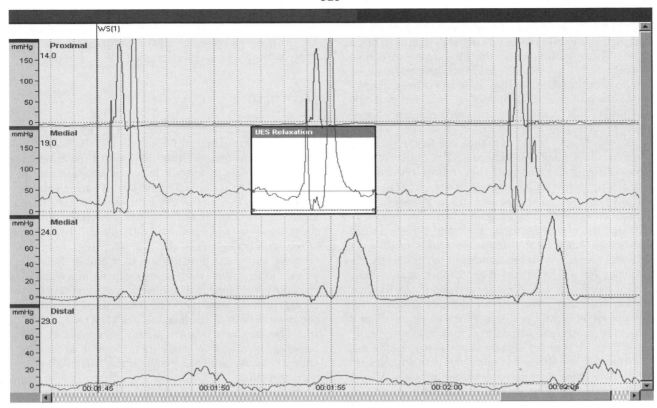

Figure 5. Manometric tracings of the upper esophageal sphincter. (Courtesy of Medtronic Corp., Minneapolis, MN.)

and Chagas' disease. Nonspecific esophageal motility disorders are those that are associated with the patient's symptoms, but the pattern of manometric dysmotility does not fit into the primary or secondary categories. Another category that is becoming better recognized and characterized is ineffective esophageal motility, which occurs when esophageal motility appears normal, but does not result in effective propagation of the food bolus down the esophagus.

BIBLIOGRAPHY

Cited References

1. Castell DO et al. Esophageal Motility Testing, 2nd ed. New York: Elsevier; 1993.
2. Castell JD, Dalton CB. Esophageal manometry. In: Castell DO, editor. The Esophagus. Boston:Little, Brown and Co.; 1992.
3. Stein HJ, Demeester TR, Hinder RA. Outpatient physiologic testing and surgical management of foregut motility disorders. Curr Prob Surg 1992;29:415–555.

See also ENDOSCOPES; GASTROINTESTINAL HEMORRHAGE.

ESU.   See ELECTROSURGICAL UNIT (ESU).

EVENT-RELATED POTENTIALS.   See EVOKED POTENTIALS.

EVOKED POTENTIALS

RODRIGO QUIAN QUIROGA
University of Leicester
Leicester, United Kingdom

INTRODUCTION

Our knowledge about the brain has increased dramatically in the last decades due to the incorporation of new and extraordinary techniques. In particular, fast computers enable more realistic and complex simulations and boosted the emergence of computational neuroscience. With modern acquisition systems we can record simultaneously up to few hundred neurons and deal with issues like population coding and neural synchrony. Imaging techniques such as magnetic resonance imaging (MRI) allow an incredible visualization of the locus of different brain functions. On the other extreme of the spectrum, molecular neurobiology has been striking the field with extraordinary achievements. In contrast to the progress and excitement generated by these fields of neuroscience, electroencephalography (EEG) and evoked potentials (EPs) have clearly decreased in popularity. What can we learn from scalp electrodes recordings, when one can use sophisticated devices like MRI, or record from dozens of intracranial electrodes? Still a lot.

There are mainly three advantages of the EEG: (*1*) it is relatively inexpensive; (*2*) it is noninvasive, and therefore it can be used in humans, (*3*) it has a very high temporal resolution, thus enabling the study of the dynamics of brain processes. These features make the EEG a very accessible and useful tool. It is particularly interesting for the analysis of high level brain processes that arise from the activity of large cell assemblies and may be poorly reflected by single neuron properties. Moreover, such processes can be well localized in time and even be reflected in time varying patterns (e.g., brain oscillations) that are faster than the time resolution of imaging techniques. The caveat of non-invasive EEGs is the fact that they reflect the average activity of sources far from the recording sites, and therefore do not have an optimal spatial resolution. Moreover, they are largely contaminated by noise and artifacts.

Although the way of recording EEG and EP signals did not change as much as multiunit recordings or imaging techniques, there have been significant advances in the methodology for analyzing the data. In fact, due to their high complexity, low signal/noise ratio, nonlinearity, and nonstationarity, they have been an ultimate challenge for most methods of signal analysis. The development and implementation of new algorithms that are specifically designed for such complex signals allow us to get information beyond the one accessible with previous approaches. These methods open a new gateway to the study of high level cognitive processes in humans with noninvasive techniques and at no great expense. Here, we review some of the most common paradigms to elicit evoked potentials and describe basic and more advanced methods of analysis with special emphasis on the information that can be gained from their use. Although we focus on EEG recordings, these ideas also apply to magnetoencephalograpic (MEG) recordings.

RECORDING

The electroencephalogram measures the average electrical activity of the brain at different sites of the head. Typical recordings are done at the scalp with high conductance electrodes placed at specific locations according to the so-called 10–20 system (1). The activity of each electrode can be referenced to a common passive electrode (or to a pair of linked electrodes placed at the earlobes)—monopolar recordings—or can be recorded differentially between pairs of contiguous electrodes—bipolar recordings—. In the latter case, there are several ways of choosing the electrode pairs. Furthermore, there are specific montages of bipolar recordings designed to visualize the propagation of activity across different directions (1). Intracranial recordings are common in animal studies and are very rare in humans. Intracranial electrodes are mainly implanted in epileptic patients refractory to medication in order to localize the epileptic focus, and then evaluate the feasibility of a surgical resection.

Figure 1 shows the 10–20 electrode distribution (a) and a typical monopolar recording of a normal subject with eyes

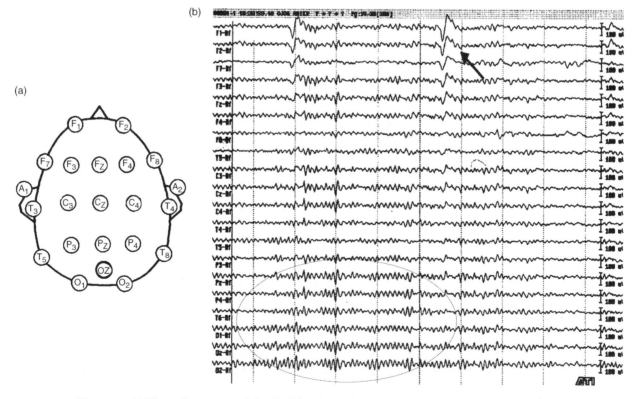

Figure 1. (a) Electrode montage of the 10–20 system and (b) an exemplary EEG recording with this montage. All electrodes are referenced to a linked earlobes reference (A1 and A2). F = frontal, C = central, P = parietal, T = temporal, and O = occipital. Note the presence of blinking artifacts (marked with an arrow) and of posterior alpha oscillations (marked with an oval).

open (b). Note the high amplitude deflections in the anterior recordings due to blinking artifacts. In fact, one of the main problems in EEG analysis is the very low signal/noise ratio. Note also the presence of ongoing oscillations in the posterior sites. These oscillations are ~10 Hz and are known as the alpha rhythm. The EEG brain oscillations of different frequencies and localizations have been correlated with functions, stages and pathologies of the brain (2–4).

In many scientific fields, especially in physics, one very useful way to learn about a system is by studying its reactions to perturbations. In brain research, it is also a common strategy to see how single neurons or large neuronal assemblies, as measured by the EEG, react to different types of stimuli. Evoked potentials are the changes in the ongoing EEG activity due to stimulation. They are time locked to the stimulus and they have a characteristic pattern of response that is more or less reproducible under similar experimental conditions. They are characterized by their polarity and latency, for example, P100 meaning a positive deflection (P for positive) occurring 100 ms after stimulation. The recording of evoked potentials is done in the same way as the EEGs. The stimulus delivery system sends triggers to identify the stimuli onsets and offsets.

GENERATION OF EVOKED POTENTIALS

Evoked potentials are usually considered as the time-locked and synchronized activity of a group of neurons that add to the background EEG. A different approach explains the evoked responses as a reorganization of the ongoing EEG (3,5). According to this view, evoked potentials can be generated by a selective and time-locked enhancement of a particular frequency band or by a phase resetting of ongoing frequencies. In particular, the study of the EPs in the frequency domain attracted the attention of several researchers (see section on Event-Related Oscillations). A few of these works focus on correlations between prestimulus EEG and the evoked responses (4).

SENSORY EVOKED POTENTIALS

There are mainly three modalities of stimulation: visual, auditory, and somatosensory. Visual evoked potentials are usually evoked by light flashes or visual patterns such as a checkerboard or a patch. Figure 2 shows the grand average visual evoked potentials of 10 subjects. Scalp electrodes were placed according to the 10–20 system, with linked earlobes reference. The stimuli were a color reversal of the (black/white) checks in a checkerboard pattern (sidelength of the checks: 50'). There is a positive deflection at ~100 ms after stimulus presentation (P100) followed by a negative rebound at 200 ms (N200). These peaks are best defined at the occipital electrodes, which are the closest to the primary visual area. The P100 is also observed in the central and frontal electrodes, but not as well defined and appearing later than in the posterior sites. The P100–N200 complex can be seen as part of an ~10 Hz event-related oscillation as it will be described in the following sections. Visual EPs can be used clinically to identify lesions in the

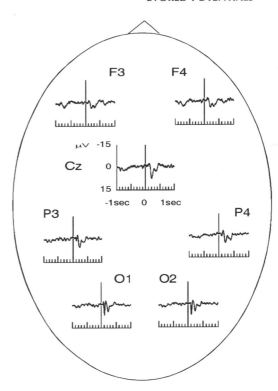

Figure 2. Grand average visual evoked potential. There is mainly one positive response at 100 ms after stimulation (P100) followed by a negative one at 200 ms (N200). These responses are best localized in the posterior electrodes.

visual pathway, such as the ones caused by optic neuritis and multiple sclerosis (6–9).

Auditory evoked potentials are usually elicited by tones or clicks. According to their latency they are further subdivided into early, middle, and late latency EPs. Early EPs comprise: (1) the electrococheleogram, which reflects responses in the first 2.5 ms from the cochlea and the auditory nerve, and (2) brain stem auditory evoked potentials (BSAEP), which reflect responses from the brain stem in the first 12 ms after stimulation and are recorded from the vertex. The BSAEP are seen at the scalp due to volume conduction. Early auditory EPs are mainly used clinically to study the integrity of the auditory pathway (10–12). They are also useful for detecting hearing impairments in children and in subjects that cannot cooperate in behavioral audiometry studies. Moreover, the presence of early auditory EPs may be a sign of recovery from coma.

Middle latency auditory EPs are a series of positive and negative waves occurring between 12 and 50 ms after stimulation. Clinical applications of these EPs are very limited due to the fact that the location of their sources is still controversial (10,11). Late auditory EPs occur between 50 and 250 ms after stimulation and consist of four main peaks labeled P50, N100, P150, and N200 according to their polarity and latency. They are of cortical origin and have a maximum amplitude at vertex locations. Auditory stimulation can also elicit potentials with latencies of > 200 ms. These are, however, responses to the context of the stimulus rather than to its physical characteristics and will be further described in the next section.

Somatosensory EPs are obtained by applying short lasting currents to sensory and motor peripheral nerves and are mainly used to identify lesions in the somatosensory pathway (13). In particular, they are used for the diagnosis of diseases affecting the white matter like multiple sclerosis, for noninvasive studies of spinal cord traumas and for peripheral nerve disorders (13). They are also used for monitoring the spinal cord during surgery, giving an early warning of a potential neurological damage in anesthetized patients (13).

Evoked potentials can be further classified as exogenous and endogenous. Exogenous EPs are elicited by the physical characteristics of the external stimulus, such as intensity, duration, frequency, and so on. In contrast, endogenous EPs are elicited by internal brain processes and respond to the significance of the stimulus. Endogenous EPs can be used to study cognitive processes as discussed in the next section.

EVOKED POTENTIALS AND COGNITION

Usually, the term evoked potentials refers to EEG responses to sensory stimulation. Sequences of stimuli can be organized in paradigms and subjects can be asked to perform different tasks. Event-related potentials (ERPs) constitute a broader category of responses that are elicited by "events", such as the recognition of a "target" stimulus or the lack of a stimulus in a sequence.

Oddball Paradigm and P300

The most common method to elicit ERPs is by using the oddball paradigm. Two different stimuli are distributed pseudorandomly in a sequence; one of them appearing frequently (standard stimulus), the other one being a target stimulus appearing less often and unexpectedly. Standard and target stimuli can be tones of different frequencies, figures of different colors, shapes, and so on. Subjects are usually asked to count the number of target appearances in a session, or to press a button whenever a target stimulus appears.

Figure 3 shows grand-average (10 subjects) visual evoked potentials elicited with an oddball paradigm. Figure 3 a shows the average responses to the frequent (non target) stimuli and (b) shows one to the targets. The experiment was the same as the one described in Fig. 2, but in this case target stimuli were pseudorandomly distributed within the frequent ones. Frequent stimuli (75%) were color reversals of the checks, as in the previous experiment, and target stimuli (25%) were also color reversals but with a small displacement of the checkerboard pattern (see Ref. (14) for details). Subjects had to pay attention to the appearance of the target stimuli.

The responses to the nontarget stimuli are qualitatively similar to the responses to visual EPs (without a task) shown in Fig. 2. As in the case of pattern visual EPs, the P100–N200 complex can be observed upon nontarget and target stimulation. These peaks are mainly related with primary sensory processing due to the fact that they do not depend on the task, they have a relatively short latency

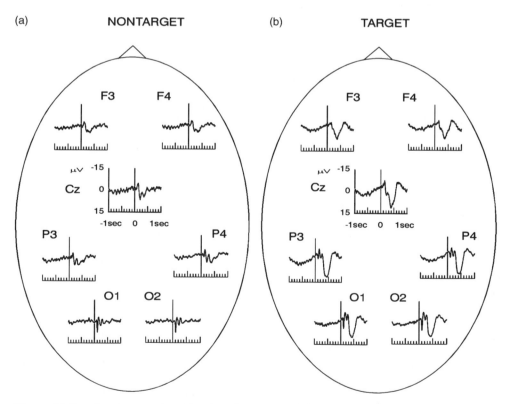

Figure 3. Grand-average pattern visual evoked potentials with an oddball paradigm. (a) Average responses for the nontarget stimuli and (b) average responses for the target stimuli. Note the appearance of a positive deflection at ∼ 400 ms after stimulation (P300) only upon the target stimuli.

(100 ms) and they are best defined in the primary visual area (occipital lobe). Note, however, that these components can also modulate their amplitude in tasks with different attention loads (15,16). Target stimulation led to a marked positive component, the P300, occurring between 400 and 500 ms. The P300 is larger in the central and posterior locations.

While the localization of the P300 in the scalp is well known, the localization of the sources of the P300 in the brain are still controversial (for a review see Ref. (17). Since the P300 is task dependent and since it has a relatively long latency, it is traditionally related to cognitive processes such as signal matching, recognition, decision making, attention and memory updating (6,18,19). There have been many works using the P300 to study cognitive processes [for reviews see Refs. (18–20)]. In various pathologies cognition is impaired and this is reflected in abnormal P300 responses, as shown in depression, schizophrenia, dementia and others [for reviews see (18,21)].

The P300 can be also elicited by a passive oddball paradigm (i.e., an oddball sequence without any task). In this case, a P300 like response appears upon target stimulation, reflecting the novelty of the stimulus rather than the execution of a certain task. This response has been named P3a. It is earlier than the classic P300 (also named P3b), it is largest in frontal and central areas and it habituates quickly (22,23).

Mismatch Negativity

Mismatch negativity (MMN) is a negative potential elicited by auditory stimulation. It appears along with any change in some repetitive pattern and peaks between 100 and 200 ms after stimulation (24). It is generally elicited by the passive (i.e., no task) auditory oddball paradigm and it is visualized by subtracting the frequent stimuli from the deviant one. The MMN is generated in the auditory cortex. It is known to reflect auditory memory (i.e., the memory trace of preceding stimuli) and can be elicited even in the absence of attention (25). It provides an index of sound discrimination and has therefore being used to study dyslexia (25). Since MMN reflects a pre-attentive state, it can be also elicited during sleep (26). Moreover, it has been proposed as an index for coma prognosis (27,28).

Omitted Evoked Potentials

Omitted evoked potentials (OEPs) are similar in nature to the P300 and MMN, but they are evoked by the omission of a stimulus in a sequence (29–31). The nice feature of these potentials is that they are elicited without external stimulation, thus being purely endogenous components. Omitted evoked potentials mainly reflect expectancy (32) and are modulated by attention (31,33). The main problem in recording OEPs is the lack of a stimulus trigger. This results in large latency variations from trial to trial, and therefore OEPs may not be visible after ensemble averaging. Note that trained musicians were shown to have less variability in the latency of the OEP responses (latency jitter) in comparison to non-musicians due to their better time-accuracy (34).

Contingent Negative Variation

Contingent negative variation (CNV) is a slowly rising negative shift appearing before stimulus onset during periods of expectancy and response preparation (35). It is usually elicited by tasks resembling conditioned learning experiments. A first stimulus gives a preparatory signal for a motor response to be carried out at the time of a second stimulus. The CNV reflects the contingency or association between the two stimuli. It has been useful for the study of aging and different psychopathologies, such as depression and schizophrenia (for reviews see Refs. (36,37)). Similar in nature to the CNVs are the "Bereitschaft" or "Readiness" potentials (38), which are negative potential shifts preceeding voluntary movements [for a review see Ref. (36)].

N400

Of particular interest are ERPs showing signs of language processing. Kutas and Hillyard (39,40) described a negative deflection between 300 and 500 ms after stimulation (N400), correlated with the appearance of semantically anomalous words in otherwise meaningful sentences. It reflects "semantic memory"; that is, the predictability of a word based on the semantic content of the preceding sentence (16).

Error Related Negativity

The error related negativity (ERN) is a negative component that appears after negative feedback (41,42). It can be elicited with a wide variety of reaction time tasks and it peaks within 100 ms of an error response. It reaches its maximum over frontal and central areas and convergent evidence from source localization analyses and imaging studies point toward a generation in the anterior cingulated cortex (41).

BASIC ANALYSIS

Figure 4a shows 16 single-trial visual ERPs from the left occipital electrode of a typical subject. These are responses to target stimuli using the oddball paradigm described in the previous section. Note that it is very difficult to distinguish the single-trial ERPs due to their low amplitude and due to their similarity to spontaneous fluctuations in the EEG. The usual way to improve the visualization of the ERPs is by averaging the responses of several trials. Since evoked potentials are locked to the stimulus onset, their contribution will add, whereas one of the ongoing EEG will cancel. Figure 4b shows the average evoked potential. Here it is possible to identify the P100, N200, and P300 responses described in the previous section.

The main quantification of the average ERPs is by means of their amplitudes and latencies. Most research using ERPs compare statistically the distribution of peak amplitudes and latencies of a certain group (e.g., subjects in some particular state or doing some task) with a matched control group. Such comparisons also can be used clinically and, in general, pathological cases show peaks with long latencies and small amplitudes (2,6).

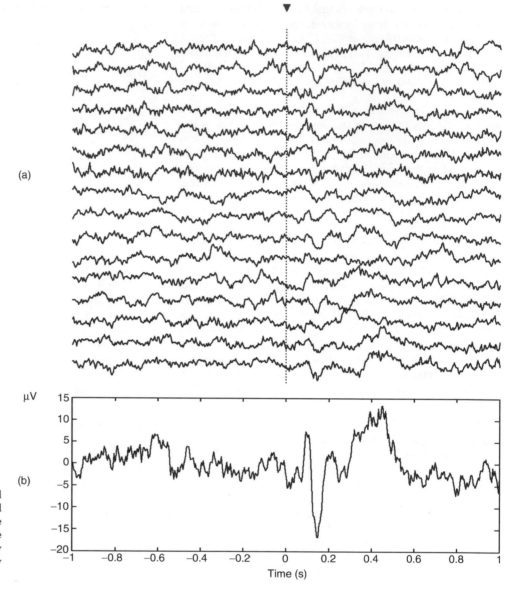

Figure 4. (a) Sixteen single-trial responses of a typical subject and (b) the average response. The triangle marks the time of stimulation. Note that the evoked responses are clearly seen after averaging, but are hardly identified in the single trials.

Another important aspect of ERPs is their topography. In fact, the abnormal localization of evoked responses can have clinical relevance. The usual way to visualize the topography of the EPs is via contour plots (43–47). These are obtained from the interpolation of the EP amplitudes at fixed times. There are several issues to consider when analyzing topographic plots: (1) the way the 3D head is projected into two dimensions, (2) the choice of the reference, (3) the type of interpolation used, and (4) the number of electrodes and their separation (46). These choices can indeed bias the topographic maps obtained.

SOURCE LOCALIZATION

In the previous section, we briefly discussed the use of topographic representations of the EEG and EPs. Besides the merit of the topographic representation given by these maps, the final goal is to get a hint on the sources of the activity seen at the scalp. In other words, given a certain distribution of voltages at the scalp one would like to estimate the location and magnitude of their sources of generation. This is known as the inverse problem and it has no unique solution. The generating sources are usually assumed to be dipoles, each one having six parameters to be estimated, three for its position and three for its magnitude. Clearly, the complexity of the calculation increases rapidly with the number of dipoles and, in practice, no more than two or three dipoles are considered. Dipole sources are usually estimated using spherical head models. These models consider the fact that the electromagnetic signal has to cross layers of different impedances, such as the dura mater and the scull. A drawback of spherical head models is the fact that different subjects have different head shapes. This led to the introduction of realistic head models, which are obtained by modeling the head shape using MRI scans and computer simulations. Besides all these issues, there are already some impressive results in the literature [see Refs. (49,86)] and references cited therein describing the use and applications of the LORETA software; Ref. (50) and an extensive list of publications using the BESA software at http://www.besa.de). Since a reasonable estimation of the EEG and EP sources critically depends on the number of electrodes, dipole location has been quite popular for the analysis of magnetoencephalograms, which have more recording sites.

EVENT-RELATED OSCILLATIONS

Evoked responses appear as single peaks or as oscillations generated by the synchronous activation of a large network. The presence of oscillatory activity induced by different type of stimuli has been largely reported in animal studies. Bullock (51) gives an excellent review of the subject going from earlier studies by Adrian (52) to more recent results in the 1990s (some of the later studies are included in Ref. (53). Examples are event-related oscillations of 15–25 Hz in the retina of fishes in response to flashes (54), gamma oscillations in the olfactory bulb of cats and rabbits after odor presentation (55,56) and beta oscillations in the olfactory system of insects (57,58). Moreover, it has been proposed that these brain oscillations play a role in information processing (55). This idea became very popular after the report of gamma activity correlated to the binding of perceptual information in anesthetized cats (59).

Event-related oscillations in animals are quite robust and in many cases visible by the naked eye. In humans, this activity is more noisy and localized in time. Consequently, more sophisticated time–frequency representations, like the one given by the wavelet transform, are needed in order to precisely localize event-related oscillations both in time and frequency. We finish this section with a cautionary note about event-related oscillations, particularly important for human studies. Since oscillations are usually not clear in the raw data, digital filters are used in order to visualize them. However, one should be aware that digital filters can introduce "ringing effects" and single peaks in the original signal can look like oscillations after filtering. In Fig. 5, we exemplify this effect by showing a delta function (a) filtered with a broad and a narrow band elliptic filter (b,c, respectively). Note that the original delta function can be mistaken for an oscillation after filtering, especially with the narrow band filter [see also Ref. (51)].

WAVELET TRANSFORM AND EVENT-RELATED OSCILLATIONS

Signals are usually represented either in the time or in the frequency domain. The best time representation is given by the signal itself and the best frequency representation is given by its Fourier transform (FT). With the FT it is possible to estimate the power spectrum of the signal, which quantifies the amount of activity for each frequency. The power spectrum has been the most successful method for the analysis of EEGs (2), but it lacks time resolution. Since event-related oscillations appear in a short time range, a simultaneous representation in time and frequency is more appropriate.

The Wavelet transform (WT) gives a time–frequency representation that has two main advantages: (1) optimal

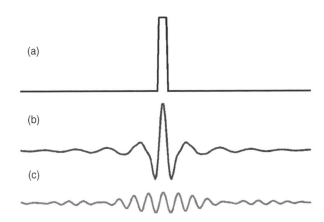

Figure 5. A delta function (a) after broad (b) and narrow (c) band pass filtering. Note that single peaks can look like oscillations due to filtering.

resolution in the time and frequency domains; (2) no requirement of stationarity. It is defined as the correlation between the signal $x(t)$ and the wavelet functions $\psi_{a,b}(t)$

$$W_\psi X(a,b) = \langle x(t) | \psi_{\alpha,\beta}(\tau) \rangle \tag{1}$$

where $\psi_{a,b}(t)$ are dilated (contracted) and shifted versions of a unique *wavelet function* $\psi(t)$

$$\psi_{a,b} = |a|^{-1/2} \psi\left(\frac{t-b}{a}\right) \tag{2}$$

(a, b are the scale and translation parameters, respectively). The WT gives a decomposition of $x(t)$ in different scales, tending to be maximum at those scales and time locations where the wavelet best resembles $x(t)$. Moreover, Eq. 1 can be inverted, thus giving the reconstruction of $x(t)$.

The WT maps a signal of one independent variable t onto a function of two independent variables a, b. This procedure is redundant and not efficient for algorithm implementations. In consequence, it is more practical to define the WT only at discrete scales a and discrete times b by choosing the set of parameters $\{a_j = 2^{-j};\ b_j,\ k = 2^{-j}k\}$, with integers j, k.

Contracted versions of the wavelet function match the high frequency components of the original signal and the dilated versions match low frequency oscillations. Then, by correlating the original signal with wavelet functions of different sizes we can obtain the details of the signal at different scales. The correlations with the different wavelet functions can be arranged in a hierarchical scheme called multiresolution decomposition (60). The multiresolution decomposition separates the signal into "details" at different scales and the remaining part is a coarser representation named "approximation".

Figure 6 shows the multiresolution decomposition of the average ERP shown in Fig. 4. The left part of the figure shows the wavelet coefficients and the right part shows the corresponding reconstructed waveforms. After a five octave wavelet decomposition using B-Spline wavelets (see Refs. 14,61 for details) the coefficients in the following bands were obtained (in brackets the EEG frequency bands that approximately correspond to these values): D1: 63–125 Hz, D2: 31–62 Hz (gamma), D3: 16–30 Hz (beta), D4:

(a)   (b)

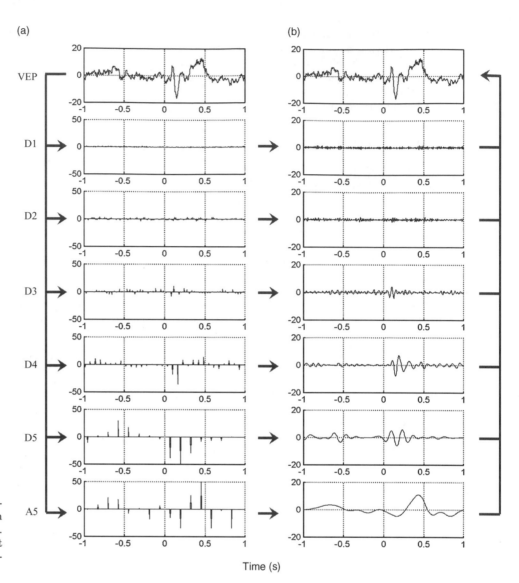

Figure 6. Multiresolution decomposition (a) and reconstruction (b) of an average evoked potential. D1–D5 and A5 are the different scales in which the signal is decomposed.

Time (s)

8–15 Hz (alpha), D5: 4–7 Hz (theta), and A5: 0.5–4 Hz (delta). Note that the addition of the reconstructed waveforms of all frequency bands returns the original signal. In the first 0.5 s after stimulation there is an increase in the alpha and theta bands (D4, D5) correlated with P100–N200 complex and later there is an increase in the delta band (A5) correlated with the P300. As an example of the use of wavelets for the analysis of event related oscillations, in the following we focus on the responses in the alpha band.

The grand average (across subjects) ERP is shown on left side of Fig. 7. Upper plots correspond to the responses to NT stimuli and lower plots to T stimuli. Only left electrodes and Cz are shown, the responses of the right electrodes being qualitatively similar. For both stimulus types we observe the P100–N200 complex and the P300 appears only upon target stimulation. Center and right plots of Fig. 7 show the alpha band wavelet coefficients and the filtered ERPs reconstructed from these coefficients, respectively. Amplitude increases are distributed over the entire scalp for the two stimulus types, best defined in the occipital electrodes. They appear first in the occipital electrodes, with an increasing delay in the parietal, cen-

tral, and frontal locations. The fact that alpha responses are not modulated by the task and the fact that their maximal and earliest appearance is in occipital locations (the primary visual sensory area) point toward a distributed generation and a correlation with sensory processing (14,61). Note that these responses are localized in time, thus stressing the use of wavelets.

In recent years, there have been an increasing number of works applying the WT to the study of event-related oscillations. Several of these studies dealt with gamma oscillations, encouraged by the first results by Gray and coworkers (59). In particular, induced gamma activity has been correlated to face perception (62), coherent visual perception (63), visual search tasks (64) cross-modal integration (64,65), and so on.

Another interesting approach to study event-related oscillations is the one given by the concepts of event-related synchronization (ERS) and event-related desynchronization (ERD), which characterize increases and decreases of the power in a given frequency band (66,67). Briefly, the band limited power is calculated for each single trial and then averaged across trials. Since ERS and ERD are defined as an

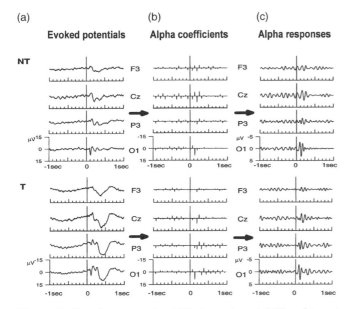

(a) Evoked potentials (b) Alpha coefficients (c) Alpha responses

Figure 7. Grand average visual EPs to nontarget (NT) and target (T) stimuli in an oddball paradigm (a). Both b+ and c plots shows the wavelet coefficients in the alpha band and the corresponding reconstruction of the signal from them, respectively.

SINGLE–TRIAL ANALYSIS

As shown in Fig. 4, averaging several trials increases the signal/noise ratio of the EPs. However, it relies on the basic assumption that EPs are an invariant pattern perfectly locked to the stimulus that lays on an independent stationary and stochastic background EEG signal. This assumption is in a strict sense not valid. In fact, averaging implies a loss of information related to systematic or unsystematic variations between the single trials. Furthermore, these variations (e.g., latency jitters) can affect the validity of the average EP as a representation of the single trial responses.

Several techniques have been proposed to improve the visualization of the single-trial EPs. Some of these approaches involve the filtering of single-trial traces by using techniques that are based on the Wiener formalism. This provides an optimal filtering in the mean-square error sense (69,70). However, these approaches assume that the signal is a stationary process and, since the EPs are compositions of transient responses with different time and frequency localizations, they are not likely to give optimal results. A obvious advantage is to implement time-varying strategies. In the following, we describe a recently proposed denoising implementation based on the WT to obtain the EPs at the single trial level (71,72). Other works also reported the use of wavelets for filtering average EPs or for visualizing the EPs in the single trials (73–77) see a brief discussion of these methods in Ref. 72.

In Fig. 6, we already showed the wavelet decomposition and reconstruction of an average visual EP. Note that the P100–N200 response is mainly correlated with the first poststimulus coefficient in the details D4–D5. The P300 is mainly correlated with the coefficients at ~400–500 ms in A5. This correspondence is easily identified because: (1) the coefficients appear in the same time (and frequency) range as the EPs and (2) they are relatively larger than the rest due to phase-locking between trials (coefficients related with background oscillations are diminished in the average). A straightforward way to avoid the fluctuations related with the ongoing EEG is by equaling to zero those coefficients that are not correlated with the EPs. However, the choice of these coefficients should not be solely based on the average EP and it should also consider the time ranges in which the single-trial EPs are expected to occur (i.e., some neighbor coefficients should be included in order to allow for latency jitters).

Figure 8a shows the coefficients kept for the reconstruction of the P100–N200 and P300 responses. Figure 8b shows the contributions of each level obtained by eliminating all the other coefficients. Note that in the final reconstruction of the average response (uppermost right plot) background EEG oscillations are filtered. We should remark that this is usually difficult to be achieved with a Fourier filtering approach due to the different time and frequency localizations of the P100–N200 and P300 responses, and also due to the overlapping frequency components of these peaks and the ongoing EEG. In this context, the main advantage of Wavelet denoising over conventional filtering is that one can select different time windows for the different scales. Once the coefficients of interest are identified from the average ERP, we can apply the same procedure to each single trial, thus filtering the contribution of background EEG activity.

Figure 9 shows the first 15 single trials and the average ERP for the recording shown in the previous figure. The raw single trials have been already shown in Fig. 4. Note that with denoising (red curves) we can distinguish the P100–N200 and the P300 in most of the trials. Note also that these responses are not easily identified in the original signal (gray traces) due to their similarity with the ongoing EEG. We can also observe some variability between trials. For an easier visualization Fig. 10 shows a contour plot of the single trial ERPs after denoising. This figure is the output of a software package for denoising EPs (EP_den) available at www.vis.caltech.edu/~rodri. In the denoised plot, we observe a gray pattern followed by a black one between 100 and 200 ms, corresponding to the P100–N200 peaks. The more unstable and wider gray pattern at ~400–600 ms corresponds to the P300. In particular, it has been shown that wavelet denoising improves the visualization of the single trial EPs (and the estimation of their amplitudes and latencies) in comparison with the original data and in comparison with previous approaches, such as Wiener filtering (72).

APPLICATIONS OF SINGLE-TRIAL ANALYSIS

The single-trial analysis of EPs has a wide variety of applications. By using correlations between the average EP and the single-trial responses, it is possible to calculate

(a)   (b)

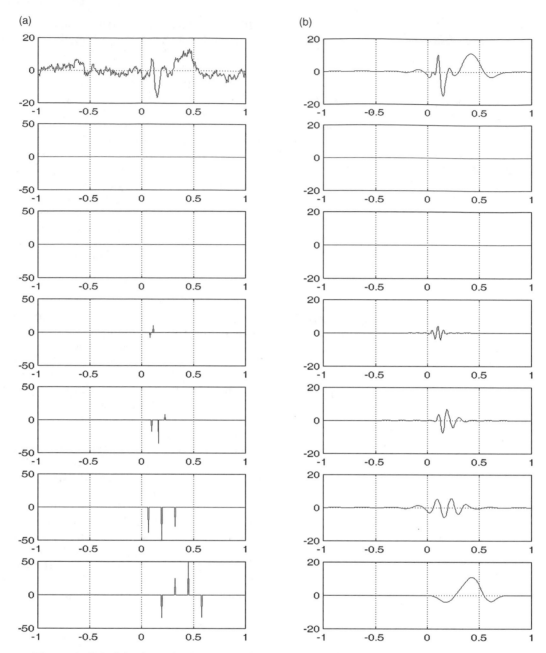

Figure 8. Principle of wavelet denoising. The reconstruction of the signal is done using only those coefficients correlated with the EPs. See text for details.

selective averages including only trials with good responses (71,72). Moreover, it is possible to eliminate effects of latency jitters by aligning trials according to the latency of the single-trial peaks (71,72). The use of selective averages as well as jitter corrected averages had been proposed long ago (78,79). Wavelet denoising improves the identification of the single-trials responses, thus facilitating the construction of these averages.

Some of the most interesting features to study in single-trial EPs are the changes in amplitude and latency of the peaks from trial to trial. It is possible to calculate amplitude and latency jitters: information that is not avail-

able in the average EPs. For example, trained musicians showed smaller latency jitters of omitted evoked potentials in comparison with nonmusicians (34). Variations in amplitude and latency can be also systematic. Exponential decreases in different EP components have been related to habituation processes both in humans and in rats (80–82). Furthermore, the appearance of a P3-like component in the rat entorhinal cortex has been correlated to the learning of a go/no-go task (83). In humans, it has recently been shown that precise timing of the single-trial evoked responses accounts for a sleep-dependent automatization of perceptual learning (84).

(a)

(b)

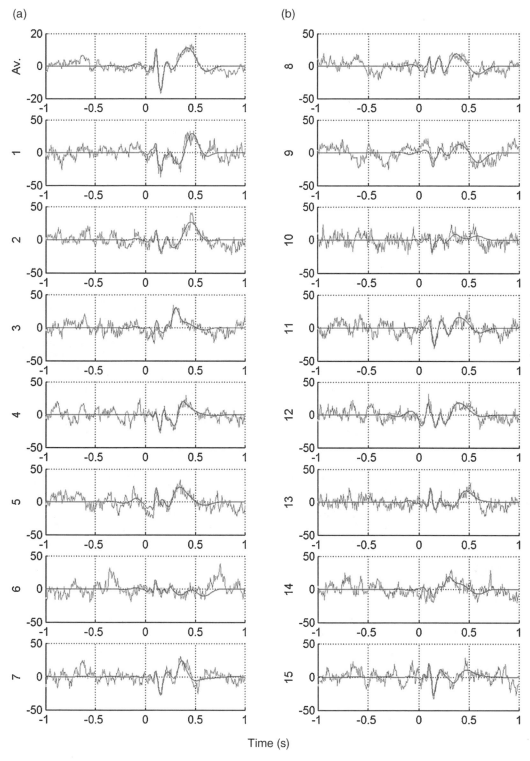

Time (s)

Figure 9. Average EP and the single-trial responses corresponding to the data shown in the previous figure, with (black) and without (gray) denoising. Note that after denoising it is possible to identify the single-trial responses.

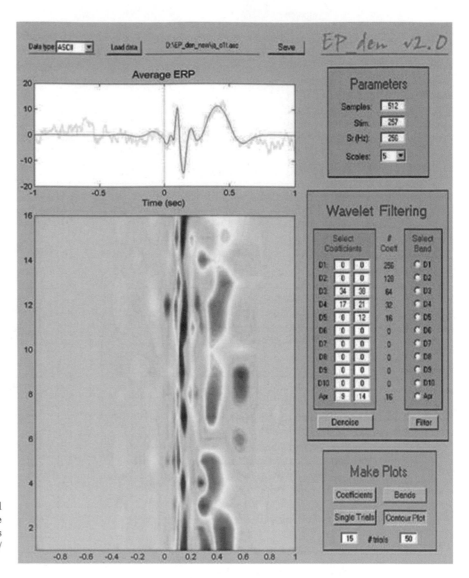

Figure 10. Contour plot of the single-trial responses shown in the previous figure. The graphic user interface used to generate this plot is available at www.vis.caltech.edu/~rodri.

CONCLUDING COMMENT

In addition to clinical applications, EPs are very useful to study high level cognitive processes. Their main advantages over other techniques are their low cost, their non-invasiveness and their good temporal resolution. Of particular interest is the study of trial-to-trial variations during recording sessions. Supported by the use of new and powerful methods of signal analysis, the study of single trial EPs and their correlation to different behavioral processes seems one of the most interesting directions of future research. In conclusion, the good and old EEG and its cousin, the EP, have a lot to offer, especially when new and powerful methods of analysis are applied.

BIBLIOGRAPHY

Cited References

1. Reilly EL. EEG recording and operation of the apparatus. In: Niedermeyer E, Lopes da Silva F, editors. Electroencephalography: Basic principles, clinical applications and related fields. Baltimore: Williams and Wilkins; 1993.

2. Niedermeyer E, Lopes da Silva F, editors. Electroencephalography: Basic principles, clinical applications and related fields. Baltimore: Williams and Wilkins; 1993, p 1097–1123.

3. Başar E. EEG-Brain dynamics. Relation between EEG and brain evoked potentials. Amsterdam: Elsevier; 1980.

4. Başar E. Brain function and oscillations. Vol. I: Brain oscillations, principles and approaches. Vol. II: Integrative brain function: Neurophysiology and cognitive processes. Berlin-Heidelberg-New York: Springer; 1999.

5. Sayers B, Beagley HA, Hanshall WR. The mechanisms of auditory evoked EEG responses. Nature (London) 1974;247: 481–483.

6. Regan D. Human brain electrophysiology. Evoked potentials and evoked magnetic fields in science and medicine. Amsterdam: Elsevier; 1989.

7. Celesia GG. Visual evoked potentials and electroretinograms. In: Niedermeyer E, Lopes da Silva F, editors. Electroencephalography: Basic principles, clinical applications and related fields. Baltimore: Williams and Wilkins; 1993.

8. Desmetdt JE, editor. Visual evoked potentials in man: new developments. Oxford: Clarendon Press; 1977a.

9. Epstein CM. Visual evoked potentials. In: Daly DD, Pedley TA, editor. Current practice of clinical electroencephalography. New York: Raven Press; 1990.

10. Celesia GG, Grigg MM. Auditory evoked potentials. In: Niedermeyer E, Lopes da Silva F, editors. Electroencephalography: Basic principles, clinical applications and related fields. Baltimore: Williams and Wilkins; 1993.

11. Picton TW. Auditory evoked potentials. In: Daly DD, Pedley TA, editors. Current practice of clinical electroencephalography. New York: Raven Press; 1990.

12. Desmetdt JE, editor. Auditory evoked potentials in man. Basel: S. Karger; 1977b.

13. Erwin CW, Rozear MP, Radtke RA, Erwin AC. Somatosensory evoked potentials. In: Niedermeyer E, Lopes da Silva F, editors. Electroencephalography: Basic principles, clinical applications and related fields. Baltimore: Williams and Wilkins; 1993.

14. Quian Quiroga R, Schürmann M. Functions and sources of event-related EEG alpha oscillations studied with the Wavelet Transform. Clin Neurophysiol 1999;110:643–655.

15. Hillyard SA, Hink RF, Schwent VL, Picton TW. Electrical signs of selective attention in the human brain. Science 1973;182:177–179.

16. Hillyard SA, Kutas M. Electrophysiology of cognitive processing. Ann Rev Psychol 1983;34:33–61.

17. Molnar M. On the origin of the P3 event-related potential component. Int J Psychophysiol 1994;17:129–144.

18. Picton TW. The P300 wave of the human event-related potential. J Clin Neurophysiol 1992;9:456–479.

19. Polich J, Kok A. Cognitive and biological determinants of P300: an integrative review. Biol Psychol 1995;41:103–146.

20. Pritchard WS. Psychophysiology of P300. Psychol Bull 1984;89:506–540.

21. Polich J. P300 in clinical applications: Meaning, method and measurement. Am J EEG Technol 1991;31:201–231.

22. Polich J. Neuropsychology of P3a and P3b: A theoretical overrier. In: Arikan K, Moore N, editors. Advances in Electrophysiology in Clinical Practice and Research. Wheaton (IL): Kjellberg; 2002.

23. Polich J, Comerchero MD. P3a from visual stimuli: Typicality, task, and topography. Brain Topogr 2003;15:141–152.

24. Naatanen R, Tervaniemi M, Sussman E, Paavilainen P, Winkler I. 'Primitive intelligence' in the auditory cortex. Trends Neurosci 2001;24:283–288.

25. Naatanen R. Mismatch negativity: clinical research and possible applications. Int J Psychophysiol 2003;48:179–188.

26. Atienza M, Cantero JL. On-line processing of complex sounds during human REM sleep by recovering information from long-term memory as revealed by the mismatch negativity. Brain Res 2001;901:151–160.

27. Kane NM, Curry SH, Butler SR, Gummins BH. Electrophysiological indicator of awakening from coma. Lancet 1993; 341:688.

28. Fischer C, Morlet D, Bouchet P, Luante J, Jourdan C, Salford F. Mismatch negativity and late auditory evoked potentials in comatose patients. Clin Neurophysiol 1999;11:1601–1610.

29. Simson R, Vaughan HG Jr, Ritter W. The scalp topography of potentials associated with missing visual or auditory stimuli. Electr Clin Neurophysiol 1976;40:33–42.

30. Ruchkin DS, Sutton S, Munson R, Silver K, Macar F. P300 and feedback provided by absence of the stimulus. Psychophysiology 1981;18:271–282.

31. Bullock TH, Karamursel S, Achimowics JZ, McClune MC, Basar-Eroglu C. Dynamic properties of human visual evoked and omitted stimulus potentials. Electr Clin Neurophysiol 1994;91:42–53.

32. Jongsma MLA, Eichele T, Quian Quiroga R, Jenks KM, Desain P, Honing H, VanRijn CM. The effect of expectancy on omission evoked potentials (OEPs) in musicians and nonmusicians. Psychophysiology 2005;42:191–201.

33. Besson M, Faita F. An event-related potential (ERP) study of musical expectancy:Comparisons of musicians with nonmusicians. J Exp Psychol, Human Perception and Performance 1995;21:1278–1296.

34. Jongsma MLA, Quian Quiroga R, VanRijn CM. Rhythmic training decreases latency-jitter of omission evoked potentials (OEPs). Neurosci Lett 2004;355:189–192.

35. Walter WG, Cooper R, Aldridge VJ, McCallum WC, Winter AL. Contingent negative variation. An electric sign of sensorimotor association and expectancy in the human brain. Nature (London) 1964;203:380–384.

36. Birbaumer N, Elbert T, Canavan AGM, Rockstroh B. Slow potentials of the cerebral cortex and behavior. Physiol Rev 1990;70:1–41.

37. Tecce JJ, Cattanach L. Contingent negative variation (CNV). In: Niedermeyer E, Lopes da Silva F, editors. Electroencephalography: Basic principles, clinical applications and related fields. Baltimore: Williams and Wilkins; 1993. p 1097–1123.

38. Kornhuber HH, Deeke L. Hirnpotentialanderungen bei Willkurbewegungen und passiven Bewegungen des Menschen. Bereitschaftspotential und reafferente Potentiale. Pfluegers Arch 1965;248:1–17.

39. Kutas M, Hillyard SA. Event-related brain potentials to semantically inappropriate and surprisingly large words. Biol Psychol 1980a;11:99–116.

40. Kutas M, Hillyard SA. Reading senseless sentences: Brain potentials reflect semantic incongruity. Science 1980b;207: 203–205.

41. Holroyd CB, Coles GH. The neural basis of human error processing: Reinforcement learning, dopamine, and the error-related negativity. Psychol Rev 2002;109:679–709.

42. Nieuwenhuis S, Holroyd CB, Mol N, Coles MGH. Neurosci Biobehav Rev 2004;28:441–448.

43. Vaughan HG Jr, Costa D, Ritter W. Topography of the human motor potential. Electr Clin Neurophysiol 1968;27:(Suppl.) 61–70.

44. Duffy FH, Burchfiel JL, Lombroso CT. Brain electrical activity mapping (BEAM): a method for extending the clinical utility of EEG and evoked potential data. Ann Neurol 1979;5:309–321.

45. Lehman D. Principles of spatial analysis. In: Gevins AS, Remond A, editors. Methods of analysis of brain electrical and magnetic signals. Amsterdam: Elsevier; 1987.

46. Gevins AS. Overview of computer analysis. In: Gevins AS, Remond A, editors. Methods of analysis of brain electrical and magnetic signals. Amsterdam: Elsevier; 1982.

47. Lopes da Silva F. EEG Analysis: Theory and Practice. In: Niedermeyer E, Lopes da Silva F, editors. Electroencephalography: Basic principles, clinical applications and related fields. Baltimore: Williams and Wilkins; 1993.

48. Fender DH. Source localization of brain electrical activity. In: Gevins AS, Remond A, editors. Methods of analysis of brain electrical and magnetic signals. Amsterdam: Elsevier; 1987.

49. Pascual-Marqui RD, Esslen M, Kochi K, Lehmann D. Functional imaging with low resolution brain electromagnetic tomography (LORETA): a review. Methods Findings Exp Clin Pharmacol 2002;24:91–95.

50. Scherg M, Berg P. New concepts of brain source imaging and localization. Electr Clin Neurophysiol 1996;46: (Suppl.) 127–137.

51. Bullock TH. Introduction to induced rhythms: A widespread, hererogeneous class of oscillations. In: Başar E, Bullock T, editors. Induced rhythms in the brain. Boston: Birkhauser; 1992.

52. Adrian ED. Olfactory reactions in the brain of the hedgehog. J Physiol 1942;100:459–473.

53. Başar E, Bullock T, editors. Induced rhythms in the brain. Boston: Birkhauser; 1992.

54. Bullock TH, Hofmann MH, New JG, Nahm FK. Dynamics properties of visual evoked potentials in the tectum of cartilaginous and bony fishes, with neuroethological implications. J Exp Zool 1991;5: (Suppl.) 142–155.

55. Freeman WJ. Mass action in the nervous system. New York: Academic Press; 1975.

56. Freeman WJ, Skarda CA. Spatial EEG-patterns, non-linear dynamics and perception: the neo-Sherringtonian view. Brain Res Rev 1981;10:147–175.

57. Laurent G, Naraghi M. Odorant-induced oscillations in the mushroom bodies of the locust. J Neurosci 1994;14:2993–3004.

58. Laurent G, Wehr M, Davidowitz H. Odour encoding by temporal sequences of firing in oscillating neural assemblies. J Neurosci 1996;16:3837–3847.

59. Gray CM, Koenig P, Engel AK, Singer W. Oscillatory responses in cat visual cortex exhibit inter-columnar synchronization which reflects global stimulus properties. Nature (London) 1989;338:334–337.

60. Mallat S. A theory for multiresolution signal decomposition: the wavelet representation. IEEE Trans Pattern Anal Machine Intell 1989;2:674–693.

61. Quian Quiroga R, Sakowicz O, Basar E, Schurmann M. Wavelet transform in the analysis of the frequency composition of evoked potentials. Brain Res Protocols 2001;8:16–24.

62. Rodriguez E, George N, Lachaux JP, Martinerie J, Renault B, Varela FJ. Perception's shadow: Long-distance synchronization of human brain activity. Nature (London) 1999;397:430–433.

63. Tallon-Baudry C, Bertrand O, Delpuech C, Pernier J. Simulus specificity of phase-locked and non-phase-locked 40 Hz visual responses in human. J Neurosc 1996;16:4240–4249.

64. Sakowicz O, Quian Quiroga R, Başar E, Schürmann M. Bisensory stimulation increases gamma-responses over multiple cortical regions. Cogn Brain Res 2001;11:267–279. Tallon-Baudry C, Bertrand O, Delpuech C, Pernier J. Activity induced by a visual search task in humans. J Neurosc 1997;15:722–734.

65. Sakowicz O, Quian Quiroga R, Schürmann M, Basar E. Spatio-temporal frequency characteristics of intersensory components in audio-visually evoked Potentials. Cog Brain Res 2005;23:87–99.

66. Pfurtscheller G, Lopes da Silva RH. Event-related EEG/MEG synchronization and desynchronization: basic principles. Clin Neurophysiol 1999a;110:1842–1857.

67. Pfurtscheller G, Lopes da Silva RH, editors. Event-related desynchronization. Amsterdam: Elsevier; 1999b.

68. Quian Quiroga R, Basar E, Schürmann M. Phase locking of event-related alpha oscillations. In: Lehnertz K, Elger CE, Arnhold J, Grassberger P, editors. Chaos in Brain? World Scientific; 2000.

69. Walter DO. A posteriori "Wiener filtering" of average evoked responses. Electr Clin Neurophysiol 1969;27: (Suppl.) 61–70.

70. Doyle DJ. Some comments on the use of Wiener filtering for the estimation of evoked potentials. Electr Clin Neurophysiol 1975;38:533–534.

71. Quian Quiroga R. Obtaining single stimulus evoked potentials with wavelet denoising. Phys D 2000;145:278–192.

72. Quian Quiroga R, Garcia H. Single-trial event-related potentials with wavelet denoising. Clin Neurophysiol 2003;114:376–390.

73. Bartnik EA, Blinowska KJ, Durka PJ. Single evoked potential reconstruction by means of wavelet transform. Biol Cybern 1992;67:175–181.

74. Bertrand O, Bohorquez J, Pernier J. Time-frequency digital filtering based on an invertible wavelet transform: an application to evoked potentials. IEEE Trans Biomed Eng 1994;41:77–88.

75. Thakor N, Xin-rong G, Yi-Chun S, Hanley D. Multiresolution wavelet analysis of evoked potentials. IEEE Trans Biomed Eng 1993;40:1085–1094.

76. Effern A, Lehnertz K, Fernandez G, Grunwald T, David P, Elger CE. Single trial analysis of event related potentials: non-linear de-noising with wavelets. Clin Neurophysiol 2000a;111:2255–2263.

77. Effern A, Lehnertz K, Schreiber T, David P, Elger CE. Non-linear denoising of transient signal with application to event related potentials. Physica D 2000b;140:257–266.

78. Pfurtscheller G, Cooper R. Selective averaging of the intra-cerebral click evoked responses in man: An improved method of measuring latencies and amplitudes. Electr Clin Neurophysiol 1975;38:187–190.

79. Woody CD. Characterization of an adaptive filter for the analysis of variable latency neuroelectric signals. Med Biol Eng 1967;5:539–553.

80. Quian Quiroga R, van Luijtelaar ELJM. Habituation and sensitization in rat auditory evoked potentials: a single-trial analysis with wavleet denoising. Int J Psychophysiol 2002;43:141–153.

81. Sambeth A, Maes JHR, Quian Quiroga R, Coenen AML. Effects of stimulus repetitions on the event-related potentials of humans and rats. Int J Psychophysiol 2004a;53:197–205.

82. Sambeth A, Maes JHR, Quian Quiroga R, van Rijn CM, Coenen AML. Enhanced re-habituation of the orienting response of the human event related potential. Neurosci Lett 2004b;356:103–106.

83. Talnov A, Quian Quiroga R, Meier M, Matsumoto G, Brankack J. Enthorinal imputs to dentate gyrus are activated mainly by conditioned events with long time intervals. Hippocampus 2003;13:755–765.

84. Atienza M, Cantero JL, Quian Quiroga R. Precise timing accounts for posttraining sleepdependent enhancements of the auditory mismatch negativity. Neuroimage 2005;26:628–634.

See also ELECTROENCEPHALOGRAPHY; ELECTROPHYSIOLOGY; SLEEP STUDIES, COMPUTER ANALYSIS OF.

EXERCISE FITNESS, BIOMECHANICS OF.   See BIOMECHANICS OF EXERCISE FITNESS.

EXERCISE, THERAPEUTIC.   See REHABILITATION AND MUSCLE TESTING.

EXERCISE STRESS TESTING

HENRY CHEN
Stanford University
Palo Alto, California

VICTOR FROEHLICHER
VA Medical Center
Palo Alto, California

INTRODUCTION

Exercise is the body's most common physiologic stress, and while it affects several systems, it places major demands on the cardiopulmonary system. Because of this interaction, exercise can be considered the most practical test of cardiac perfusion and function. Exercise testing is a noninvasive tool to evaluate the cardiovascular system's response to exercise under carefully controlled conditions.

THEORY

The major cardiopulmonary adaptations that are required during acute exercise make exercise testing a practical evaluation of cardiac perfusion and function. Exercise testing is not only useful in clinical evaluation of coronary status, but also serves as a valuable research tool in the study of cardiovascular disease, evaluating physical performance in athletes, and studying the normal and abnormal physiology of other organ systems.

A major increase and redistribution of cardiac output underlies a series of adjustments that allow the body to increase its resting metabolic rate as much as 10–20 times with exercise. The capacity of the body to deliver and utilize oxygen is expressed as the maximal oxygen uptake, which is defined as the product of maximal cardiac output and maximal arteriovenous oxygen difference. Thus, the cardiopulmonary limits are defined by (1) a central component (cardiac output) that describes the capacity of the heart to function as a pump, and (2) peripheral factors (arteriovenous oxygen difference) that describe the capacity of the lung to oxygenate the blood delivered to it as well as the capacity of the working muscle to extract this oxygen from the blood. Hemodynamic responses to exercise are greatly affected by several parameters, such as presence or absence of disease and type of exercise being performed, as well as age, gender, and fitness of the individual.

Coronary artery disease is characterized by reduced myocardial oxygen supply, which, in the presence of an increased myocardial oxygen demand, can result in myocardial ischemia and reduced cardiac performance. Despite years of study, a number of challenges remain regarding the response to exercise clinically. Although myocardial perfusion and function are intuitively linked, it is often difficult to separate the impact of ischemia from that of left ventricular dysfunction on exercise responses. Indexes of ventricular function and exercise capacity are poorly related. Cardiac output is considered the most important determinant of exercise capacity in normal subjects and in most patients with cardiovascular or pulmonary disease. However, among patients with disease, abnormalities in one or several of the links in the chain that defines oxygen uptake contribute to the determination of exercise capacity.

The transport of oxygen from the air to the mitochondria of the working muscle cell requires the coupling of blood flow and ventilation to cellular metabolism. Energy for muscular contraction is provided by three sources: stored phosphates [adenosine triphosphate (ATP) and creatine phosphate]; oxygen-independent glycolysis, and oxidative metabolism. Oxidative metabolism provides the greatest source of ATP for muscular contraction. Muscular contraction is accomplished by three fiber types that differ in their contraction speed, color, and mitochondrial content. The duration and intensity of activity determine the extent to which these fuel sources and fiber types are called upon.

All of the physiologic responses to exercise are mediated by the autonomic nervous system. As such, the exercise test and the response in recovery after exercise are increasingly recognized as important surrogate measures of autonomic function. The balance between sympathetic and parasympathetic influences to the cardiovascular sytem is critical, as they determine heart rate, blood pressure, cardiac output redistribution, and vascular resistance during exercise. Furthermore, indirect measures of autonomic function, such as heart rate variability and the rate in which heart rate recovers after exercise, are important prognostic markers in patients with cardiovascular disease (1).

There are several advantages to using the exercise test. These include the test's widespread availability, multiple uses, and high yield of clinically useful information. These factors make it an important "gatekeeper" for more expensive and invasive procedures. However, a major drawback has been the non-uniform application in clinical practices. To approach this problem, excellent guidelines have been developed based on expert consensus and research performed over the last 20 years. These include an update of the AHA/ACC guidelines on exercise testing, the American Thoracic Society/American College of Chest Physicians Statement on Cardiopulmonary Exercise Testing, an AHA Scientific Statement on Exercise and Heart Failure, new editions of the American Association of Cardiovascular and Pulmonary Rehabilitation Guidelines, and American College of Sports Medicine Guidelines on Exercise Testing and Prescription (cardiologyonline.com/guidelines.htm, cardiology.org). These have made substantial contributions to the understanding and greater uniformity of application of exercise testing.

EQUIPMENT

Current technology, although adding both convenience and sophistication, has raised new questions about methodology. For example, all commercially available systems today use computers. Do computer-averaged exercise electrocardiograms (ECGs) improve test performance, and what should the practitioner be cautious of when considering computer measurements? What about the many computer-generated exercise scores? When should ventilatory gas exchange responses be measured during testing and what special considerations are important when using them? The following text is intended to help answer such questions.

Blood Pressure Measurement

While there have been a number of automated devices developed for blood pressure measurement during exercise, our clinical impression is that they are not superior to manual testing. The time-proven method of the physician holding the patient's arm with a stethoscope placed over the brachial artery remains most reliable. An anesthesiologist's auscultatory piece or an electronic microphone can be fastened to the arm. A device that inflates and deflates the cuff with the push of a button can be helpful.

A dropping or a flat systolic response during exercise are ominous and can be the most important indicator of adverse events occurring during testing. If systolic blood pressure fails to increase or appears to be decreasing, it should be taken again immediately. The exercise test should be stopped if the systolic blood pressure drops by

10 mmHg (1.33 kPa) or more or falls below the value of the blood pressure taken in a standing position before testing. Additionally, if the systolic blood pressure exceeds 250 mmHg (33.33 kPa) or the diastolic blood pressure reaches 115 mmHg (15.33 kPa), the test should be stopped.

ECG Recording

Electrodes and Cables. Several disposable electrodes perform adequately for exercise testing. Disposable electrodes can be quickly applied and do not have to be cleaned for reuse. A disposable electrode that has an abrasive center spun by an applicator after the electrode is attached to the skin (Quickprep) is available from Quinton Instrument Co. This approach does not require skin preparation. The applicator of this system has a built-in impedance meter that stops the spinning when the skin impedance has been appropriately lowered.

Previously used buffer amplifiers and digitizers carried by the patient are no longer advantageous. Cables develop continuity problems with use and require replacement rather than repair. It is often found that replacement is necessary after roughly 500 tests. Some systems have used analog-to-digital converters in the electrode junction box carried by the patient. Because digital signals are relatively impervious to noise, the patient cable can be unshielded and is therefore very light.

Careful skin preparation and attention to the electrode–cable interface are important for a safe and successful exercise test and are necessary no matter how elaborate or expensive the ECG recording device used.

Lead Systems

Bipolar Lead Systems. Bipolar leads have been traditionally used owing to the relatively short time required for placement, the relative freedom from motion artifact, and the ease with which noise problems can be located. The usual positive reference is an electrode placed the same as the positive reference for V_5 (2). The negative reference for V_5 is Wilson's central terminal, which consists of connecting the limb electrodes to the right arm, left arm, and left leg. Virtually all current ECG systems, however, use the modified 12-lead system first described by Mason and Likar (3).

Mason–Likar Electrode Placement. Because a 12-lead ECG cannot be obtained accurately during exercise with electrodes placed on the wrists and ankles, the electrodes are placed at the base of the limbs for exercise testing. In addition to lessening noise for exercise testing, the Mason–LIkar modified placement has been demonstrated to exhibit no differences in ECG configuration when compared to the standard limb lead placement (3). However, this finding has been disputed by others who have found that the Mason–Likar placement causes amplitude changes and axis shifts when compared with standard placement. Because these could lead to diagnostic changes, it has been recommended that the modified exercise electrode placement not be used for recording a resting ECG. The preexercise ECG has been further complicated by the recommendation that it should be obtained standing, since

that is the same position maintained during exercise. This situation is worsened by the common practice of moving the limb electrodes onto the chest to minimize motion artifact.

In the Mason–Likar torso-mounted limb lead system, the conventional ankle and wrist electrodes are replaced by electrodes mounted on the torso at the base of the limbs. This placement avoids artifact caused by movement of the limbs. The standard precordial leads use Wilson's central terminal as their negative reference, which is formed by connecting the right arm, left arm, and left leg. This triangular configuration around the heart results in a zero voltage reference through the cardiac cycle. The use of Wilson's central terminal for the precordial leads (V leads) requires the negative reference to be a combination of three additional electrodes rather than the single electrode used as the negative reference for bipolar leads.

The modified exercise electrode placement should not be used for routine resting ECGs. However, the changes caused by the exercise electrode placement can be kept to a minimum by keeping the arm electrodes off the chest and putting them on the anterior deltoid and by having the patient supine. In this situation, the modified exercise limb lead placement of Mason–Likar can serve well as the resting ECG reference before an exercise test.

For exercise testing, limb electrodes should be placed as far from the heart as possible, but not on the limbs; the ground electrode (right leg) can be on the back out of the cardiac field and the left leg electrode should be below the umbilicus. The precordial electrodes should be placed in their respective interspaces.

Inferior Lead ST-Segment Depression

One potential area of confusion in interpretation lies in ST-segment depression in the inferior leads. Miranda et al. (4) studied 178 men who had undergone exercise testing and coronary angiography to evaluate the diagnostic value of ST-segment depression occurring in the inferior leads. The area under the curve in lead II was not significantly > 0.50, suggesting that for the identification of coronary artery disease, isolated ST-segment depression in lead II appears unreliable. The ST depression occurring in the inferior leads alone (II, AVF) can sometimes represent a false positive response. ST elevation in these leads, however, suggests transmural ischemia in the area of RCA blood distribution.

Number of Leads to Record

In patients with normal resting ECGs, a V_5 or similar bipolar lead along the long axis of the heart is usually adequate. The ECG evidence of myocardial infarction or history suggestive of coronary spasm necessitate use of additional leads. As a minimal overall approach, it is advisable to record three leads: a V_5-type of lead, an anterior V_2-type of lead, and an inferior lead, such as a V_F. Alternatively, Frank X, Y, and Z leads may be used. Either of these approaches can be helpful additionally for the detection and identification of dysrhythmias. It is also advisable to record a second three-lead grouping consisting of V_4, V_5, and V_6. Occassionally abnormalities seen as borderline in V_5 can be better defined in neighboring V_4 or V_6.

Because most meaningful ST-segment depression occurs in the lateral leads (V_4, V_5, and V_6) when the resting ECG is normal, other leads are only necessary in patients who had a myocardial infarction, those with a history consistent with coronary spasm or variant angina, or those who have exercise- induced dysrhythmias of an uncertain type.

Although as much as 90% of abnormal ST depression occurs in V_5 (or the two adjacent precordial leads), other leads should also be used, particularly in patients with history of known myocardial infarction or variant angina, and especially since ST elevation can localize ischemia to the area beneath the electrodes, and multiple-vector leads can be useful for studying arrhythmias (best diagnosed with inferior and anterior leads where the P waves are best seen).

ECG Recording Instruments

There have been several advances in ECG recorder technology. The medical instrumentation industry has promptly complied with specifications set forth by various professional groups. Machines with a high input impedance ensure that the voltage recorded graphically is equivalent to that on the surface of the body despite the high natural impedance of the skin. Optically isolated buffer amplifiers have ensured patient safety, and machines with a frequency response up to 100 Hz are commercially available. The lower end is possible because direct current (dc) coupling is technically feasible.

Waveform Processing

Analog and digital averaging techniques have made it possible to filter and average ECG signals to remove noise. There is a need for consumer awareness in these areas because most manufacturers do not specify how the use of such procedures modifies the ECG. Both filtering and signal averaging can, in fact, distort the ECG signal. Averaging techniques are nevertheless attractive because they can produce a clean tracing when the raw data is noisy. However, the clean-looking ECG signal produced may not be a true representation of the waveform and in fact may be dangerously misleading. Also, the instruments that make computer ST-segment measurements are not entirely reliable because they are based on imperfect algorithms. While useful in reducing noise, filtering and averaging can cause false ST depression due to distortion of the raw data.

Computerization

There are a host of advantages of digital over analog data processing. These include more accurate measurements, less distortion in recording, and direct accessibility to digital computer analysis and storage techniques. Other advantages are rapid mathematical manipulation (averaging), avoidance of the drift inherent in analog components, digital algorithm control permitting changes in analysis schema with ease (software rather than hardware changes), and no degradation with repetitive playback. When outputting data, digital processing also offers higher plotting resolution and easy repetitive manipulation.

Computerization also helps meet the two critical needs of exercise ECG testing: the reduction of the amount of ECG data collected during testing and the elimination of electrical noise and movement artifact associated with exercise. Because an exercise test can exceed 30 min (including data acquisition during rest and recovery) and many physicians want to analyze all 12 leads during and after testing, the resulting quantity of ECG data and measurements can quickly become excessive. The three-lead vectorcardiographic [or three-dimensional (3D) (i.e., aV_F, V_2, V_5)] approach would reduce the amount of data; however, clinicians favor the 12-lead ECG. The exercise ECG often includes random and periodic noise of high and low frequency that can be caused by respiration, muscle artifact, electrical interference, wire continuity, and electrode–skin contact problems. In addition to reducing noise and facilitating data collection, computer processing techniques may also make precise and accurate measurements, separate and capture dysrhythmic beats, perform spatial analysis, and apply optimal diagnostic criteria for ischemia.

Although most clinicians agree that computerized analysis simplifies the evaluation of the exercise ECG, there has been disagreement about whether accuracy is enhanced (5). A comparison of computerized resting ECG analysis programs with each other and with the analyses of expert readers led to the conclusion that physician review of any reading is necessary (6). Although computers can record very clean representative ECG complexes and neatly print a wide variety of measurements, the algorithms they use are far from perfect and can result in serious differences from the raw signal. The physician who uses commercially available computer-aided systems to analyze the results of exercise tests should be aware of the problems and always review the raw analog recordings to see whether they are consistent with the processed output.

Even if computerization of the original raw analog ECG data could be accomplished without distortion, the problem of interpretation still remains. Numerous algorithms have been recommended for obtaining the optimal diagnostic value from the exercise ECG. These algorithms have been shown to provide improved sensitivity and specificity compared with standard visual interpretation. Often, however, this improvement has been demonstrated only by the investigator who proposed the new measurement. Furthermore, the ST measurements made by a computer can be erroneous. It is advisable to have the devices mark both the isoelectric level and the point of ST0. Even if the latter is chosen correctly, misplacement of the isoelectric line outside of the PR segment can result in incorrect ST level measurements. Computerized ST measurements require physician over reading; errors can be made both in the choice of the isoelectric line and the beginning of the ST segment.

Causes of Noise

Many of the causes of noise in the exercise ECG signal cannot be corrected, even by meticulous skin preparation. Noise is defined in this context as any electrical signal that is foreign to or distorts the true ECG waveform. Based on this definition, the types of noise that may be present can be caused by any combination of line-frequency (60 Hz), muscle, motion, respiration, contact, or continuity artifact. Line-frequency noise is generated by the interference of the 60 Hz electrical energy with the ECG. This noise can be reduced by using shielded patient cables. If in spite of these precautions this noise is still present, the simplest way to remove it is to design a 60-Hz notch filter and apply it in series with the ECG amplifier. A notch filter removes only the line frequency; that is, it attenuates all frequencies in a narrow band ~ 60 Hz. This noise can also be removed by attenuating all frequencies > 59 Hz; however, this method of removing line-frequency noise is not recommended because it causes waveform distortion and results in a system that does not meet AHA specifications. The most obvious manifestation of distortion caused by such filters is a decrease in R-wave amplitude; therefore a true notch filter is advisable.

Muscle noise is generated by the activation of muscle groups and is usually of high frequency. This noise, with other types of high frequency noise, can be reduced by signal averaging. Motion noise, another form of high frequency noise, is caused by the movement of skin and the electrodes, which causes a change in the contact resistance. Respiration causes an undulation of the waveform amplitude, so the baseline varies with the respiratory cycle. Baseline wander can be reduced by low-frequency filtering; however, this results in distortion of the ST segment and can cause artifactual ST-segment depression and slope changes. Other baseline removal approaches have been used, including linear interpolation between isoelectric regions, high order polynomial estimates, and cubic-spline techniques, which can each smooth the baseline to various degrees.

Contact noise appears as low frequency noise or sometimes as step discontinuity baseline drift. It can be caused by poor skin preparation resulting in high skin impedance or by air bubble entrapment in the electrode gel. It is reduced by meticulous skin preparation and by rejecting beats that show large baseline drift. Also, by using the median rather than the mean for signal averaging, this type of drift can be reduced. Continuity noise caused by intermittent breaks in the cables is rarely a problem because of technological advances in cable construction, except, of course, when cables are abused or overused.

Most of the sources of noise can be effectively reduced by beat averaging. However, two types of artifact can actually be caused in the signal-averaging process by the introduction of beats that are morphologically different from others in the average and the misalignment of beats during averaging. As the number of beats included in the average increases, the level of noise reduction is greater. The averaging time and the number of beats to be included in the average have to be compromised, though, because the morphology of ECG waveforms changes over time.

For exercise testing, the raw ECG data should be considered first, and then the averages and filtered data may be used to aid interpretation if no distortion is obvious.

ECG Paper Recording

For some patients, it is advantageous to have a recorder with a slow paper speed option of $5\,mm \cdot s^{-1}$. This speed is optimal for recording an entire exercise test and reduces the likelihood of missing any dysrhythmias when specifically evaluating such patients. Some exercise systems allow for a total disclosure print out option similar to that provided and many holter systems. In rare instances, a faster paper speed of $50\,mm \cdot s^{-1}$ can be helpful for making particular evaluations, such as accurate ST-segment slope measurements.

Thermal head printers have effectively become the industry standard. These recorders are remarkable in that they can use blank thermal paper and write out the grid and ECG, vector loops, and alpha-numerics. They can record graphs, figures, tables, and typed reports. They are digitally driven and can produce very high resolution records. The paper price is comparable with that of other paper, and these devices have a reasonable cost and are very durable, particularly because a stylus is not needed.

Z-fold paper has the advantage over roll paper in that it is easily folded, and the study can be read in a manner similar to paging through a book. Exercise ECGs can be microfilmed on rolls, cartridges, or fiche cards for storage. They can also be stored in digital or analog format on magnetic media or optical disks. The latest technology involves magnetic optical disks that are erasable and have fast access and transfer times. These devices can be easily interfaced with microcomputers and can store megabytes of digital information. Lasers or ink jet printers have a delay making them unsuitable for medical emergencies but offer the advantage of the inexpensiveness of standard paper and long lived images.

Many available recording systems have both thermal head and laser or inkjet printers and use the cheaper, slower printers for final reports and summaries while the thermal head printers are used for live ECG tracings (i.e., real time). The standard 3 lead × 4 lead groups print out leaves only 2.5 s to assess ST changes or arrhythmias.

Exercise Test Modalities

Types of Exercise. Two types of exercise can be used to stress the cardiovascular system: isometric and dynamic, though most activities are a combination of the two. Isometric exercise, which involves constant muscular contraction with minimal movement (e.g., a handgrip), imposes a disproportionate pressure load on the left ventricle relative to the body's ability to supply oxygen. Dynamic exercise, or rhythmic muscular activity resulting in movement, initiates a more appropriate balance between cardiac output, blood supply, and oxygen exchange. Because a delivered workload can be accurately calibrated and the physiological response easily measured, dynamic exercise is preferred for clinical testing. Dynamic exercise is also superior because it can be more easily graduated and controlled. Using gradual, progressive workloads, patients with coronary artery disease can be protected from rapidly increasing myocardial oxygen demand. Although bicycling is also a dynamic exercise, most individuals perform more work on a treadmill because of greater muscle mass involved and generally more familiarity with walking than cycling.

Numerous modalities have been used to provide dynamic exercise for exercise testing, including steps, escalators, ladder mills, and arm ergometers. Today, however, the bicycle ergometer and the treadmill are the most commonly used dynamic exercise devices. The bicycle ergometer is usually cheaper, takes up less space, and makes less noise. Upper body motion is usually reduced, but care must be taken so that isometric exercise is not performed by the arms. The workload administered by the simple, manually braked cycle ergometers is not well calibrated and depends on pedaling speed. It can be easy for a patient to slow pedaling speed during exercise testing and decrease the administered workload, making the estimation of exercise capacity unreliable. More expensive electronically braked bicycle ergometers keep the workload at a specified level over a wide range of pedaling speeds, and have become the standard for cycle ergometer testing today.

Dynamic exercise, using a treadmill or a cycle ergometer, is a better measure of cardiovascular function and better method of testing than isometric exercise.

Arm Ergometry. Alternative methods of exercise testing are needed for patients with vascular, orthopedic, or neurological conditions who cannot perform leg exercise. Arm ergometry can be used in such patients (7). However, non-exercise techniques (such as pharmacologic stress testing) are currently more popular for these patients.

Bicycle Ergometer. The bicycle ergometer is usually cheaper, takes up less space, and makes less noise than a treadmill. Upper body motion is usually reduced, but care must be taken so that the arms do not perform isometric exercise. The workload administered by the simple bicycle ergometers is not well calibrated and is dependent on pedaling speed. It can be easy for a patient to slow pedaling speed during exercise testing and decrease the administered workload. More modern electronically braked bicycle ergometers keep the workload at a specified level over a wide range of pedaling speeds and are recommended. Since supine bicycle exercise is so rarely used, we will not address it here except to say that maximal responses are usually lower than with the erect position.

Treadmill. Treadmills should have front and side rails to allow patients to steady themselves. Some patients may benefit from the help of the person administering the test. Patients should not grasp the front or side rails because this decreases the work performed and the oxygen uptake and, in turn, increases exercise time, resulting in an overestimation of exercise capacity. Gripping the handrails also increases ECG muscle artifact. When patients have difficulty maintaining balance while walking, it helps to have them take their hands off the rails, close their fists, and extend one finger to touch the rails after they are accustomed to the treadmill. Some patients may require a few moments to feel comfortable enough to let go of the handrails, but grasping the handrails after the first minute of exercise should be strongly discouraged.

Table 1. Two Basic Principles of Exercise Physiology

Myocardial oxygen consumption	≈ Heart rate × systolic blood pressure (determinants include wall tension ≅ left ventricular pressure × volume; contractility; and heart rate)
Ventilatory oxygen consumption (VO_2)	≈ External work performed, or cardiac output × a-VO_2 difference[a]

[a]The arteriovenous O_2 difference is ~ 15–17 vol% at maximal exercise in most individuals; therefore, the VO_2 max generally reflects the extent to which cardiac output increases.

A small platform or stepping area at the level of the belt is advisable so that the patient can start the test by pedaling the belt with one foot before stepping on. The treadmill should be calibrated at least monthly. Some models can be greatly affected by the weight of the patient and will not deliver the appropriate workload to heavy patients. An emergency stop button should be readily available to the staff only.

Bicycle Ergometer versus Treadmill

Bicycle ergometry has been found to elicit similar maximum heart rate values to treadmill exercise in most studies comparing the methods. However, maximum oxygen uptake has been 6–25% greater during treadmill exercise (8). Some studies have reported similar ECG changes with treadmill testing as compared with bicycle testing (9), whereas others have reported more significant ischemic changes with treadmill testing (10). Nonetheless, the treadmill is the most commonly used dynamic testing modality in the United States, and the treadmill may be advantageous because patients are more familiar with walking than they are with bicycling. Patients are more likely to give the muscular effort necessary to adequately increase myocardial oxygen demand by walking than by bicycling.

Treadmills usually result in a higher MET values, but maximal heart rate is usually the same as with a bike. Thus, bike testing can result in a lower prognosis estimate but has similar ability to predict ischemic disease.

Exercise Protocols

The many different exercise protocols in use have led to some confusion regarding how physicians compare tests between patients and serial tests in the same patient. A recent survey performed among VA exercise laboratories confirmed that the Bruce protocol remains the most commonly used; 83% of laboratories reported using the Bruce test for routine testing. This protocol uses relatively large and unequal increments in work (2–3 MET) every 3 min. Large and uneven work increments, such as these have been shown to result in a tendency to overestimate exercise capacity and the lack of uniform increases in work rate, can complicate the interpretation of some ST segment measurements and ventilatory gas exchange responses (11,12). (Table 1). Thus, exercise testing guidelines have recommended protocols with smaller and more equal increments. It is also important to individualize the test to target duration in the range of 8–12 min.

Ramp Testing

An approach to exercise testing that has gained interest in recent years is the ramp protocol, in which work increases constantly and continuously.

Questionnaires

The key to appropriately targeting a ramp is accurately predicting the individual's maximal work capacity. If a previous test is not available, a pretest estimation of an individual's exercise capacity is quite helpful to set the appropriate ramp rate. Functional classifications are too limited and poorly reproducible. One problem is that usual activities can decrease, so an individual can become greatly limited without having a change in functional class. A better approach is to use the specific activity scale of Goldman et al. (13) (Table 2), the DASI (Table 3), or the VSAQ (Table 4). Alternatively, the patient may be questioned regarding usual activities that have a known MET cost (Table 5) (14).

Borg Scale

Instead of simply tracking heart rate to clinically determine the intensity of exercise, it is preferable to use the 6–20 Borg scale or the nonlinear 1–10 scale of perceived exertion (Table 6) (15). The Borg scale is a simple, valuable way of assessing the relative effort a patient exerts during exercise.

Table 2. Specific Activity Scale of Goldman

Class I ($\geq$ 7 METs)	A patient can perform any of the following activities: Carrying 24 lb (10.88 kg) up eight steps Carrying an 80 lb (16.28 kg) object Shoveling snow Skiing Playing basketball, touch football, squash, or handball Jogging/walking 5 mph
Class II ($\geq$ 5 METs)	A patient does not meet Class I criteria, but can perform any of the following activities to completion without stopping: Carrying anything up eight steps Having sexual intercourse Gardening, raking, weeding Walking 4 mph
Class III ($\geq$ 2 METs)	A patient does not meet Class I or Class II criteria but can perform any of the following activities to completion without stopping: Walking down eight steps Taking a shower Changing bedsheets Mopping floors, cleaning windows Walking 2.5 mph Pushing a power lawnmower Bowling Dressing without stopping
Class IV ($\leq$ 2 METs)	None of the above

Table 3. Duke Activity Scale Index[a]

Activity	Weight
Can you?	
1. Take care of yourself, that is, eating, dressing, bathing, and using the toilet?	2.75
2. Walk indoors, such as around your house?	1.75
3. Walk a block or two on level ground?	2.75
4. Climb a flight of stairs or walk up a hill?	5.50
5. Run a short distance?	8.00
6. Do light work around the house like dusting or washing dishes?	2.7
7. Do moderate work around the house like vacuuming, sweeping floors, or carrying in groceries?	3.50
8. Do heavy work around the house like scrubbing floors or lifting and moving heavy furniture?	8.00
9. Do yard work like raking leaves, weeding, or pushing a power mower?	4.5
10. Have sexual relations?	5.25
11. Participate in moderate recreational activities like golf, bowling, dancing, doubles tennis, or throwing a basketball or football?	6.00
12. Participate in strenuous sports like swimming, singles tennis, football, basketball, or skiing?	7.50

[a]Duke activity scale index = DASI = sum of weights for "yes" replies. $VO_2 = 0.43 \times DASI + 9.6$.

Postexercise Period

The patient should assume a supine position in the postexercise period to achieve greatest sensitivity in exercise testing. It is advisable to record $\sim$ 10 s of ECG data while the patient is motionless, but still experiencing near-maximal heart rate before having the patient lie down. Some patients must be allowed to lie down immediately to avoid hypotension. Letting the patient have a cool-down walk after the test is discouraged, as it can delay or eliminate the appearance of ST-segment depression (16). According to the law of La Place, the increase in venous return and thus ventricular volume in the supine position increases myocardial oxygen demand. Data from our laboratory (17) suggests that having patients lie down may enhance ST-segment abnormalities in recovery. However, a cool-down walk has been suggested to minimize the postexercise chances of dysrrhythmic events in this high risk time when catecholamine levels are high. The supine position after exercise is not as important when the test is not being performed for diagnostic purposes, for example, fitness testing. When testing is not performed for diagnostic purposes, it may be preferable to walk slowly (1.0–1.5 mph) or continue cycling against zero or minimal resistance (up to 25 W when testing with a cycle ergometer) for several minutes after the test.

Monitoring should continue for at least 6–8 min after exercise or until changes stabilize. An abnormal response occurring only in the recovery period is not unusual. Such responses are not false positives. Experiments confirm mechanical dysfunction and electrophysiological abnormalities in the ischemic ventricle after exercise. A cool down walk can be helpful when testing patients with an

Table 4. Veterans Specific Activity Questionnaire[a,b]

1 MET:	Eating; getting dressed; working at a desk
2 METs:	Taking a shower; shopping; cooking; walking down eight steps
3 METs:	Walking slowly on a flat surface for one or two blocks; doing moderate amounts of work around the house like vacuuming, sweeping the floors, or carrying in groceries
4 METs:	Doing light yard work, i.e., raking leaves, weeding, sweeping, or pushing a power mower; painting; light carpentry
5 METs:	Walking briskly; social dancing; washing the car
6 METs:	Playing nine holes of golf, carrying your own clubs; heavy carpentry; mowing lawn with a push mower
7 METs:	Carrying 60 lb; performing heavy outdoor work, that is, digging, spading soil; walking uphill
8 METs:	Carrying groceries upstairs; moving heavy furniture; jogging slowly on flat surface; climbing stairs quickly
9 METs:	Bicycling at a moderate pace; sawing wood; jumping rope (slowly)
10 METs:	Briskly swimming; bicycling up a hill; jogging 6 mph
11 METs:	Carrying a heavy load (i.e., a child or firewood) up two flights of stairs; cross-country skiing; bicycling briskly and continuously
12 METs:	Running briskly and continuously (level ground, 8 mph)
13 METs:	Performing any competitive activity, including those that involve intermittent sprinting; running competitively; rowing competitively; bicycle racing.

[a]Veterans Specific Activity Questionnaire = VSAQ

[b]Before beginning your treadmill test today, we need to estimate what your usual limits are during daily activities. Following is a list of activities that increase in difficulty as you read down the page. Think carefully, then underline the first activity that, if you performed it for a period of time, would typically cause fatigue, shortness of breath, chest discomfort, or otherwise cause you to want to stop. If you do not normally perform a particular activity, try to imagine what it would be like if you did.

established diagnosis undergoing testing for other than diagnostic reasons, when testing athletes, or when testing patients with dangerous dysrhythmias.

The recovery period is extremely important for observing ST shifts and should not be interrupted by a cool down walk or failure to monitor for at least 5 min. Changes isolated to the recovery period are not more likely to be false positives.

The recovery period, particularly between the second and fourth minute, are critical for ST analysis. Noise should not be a problem and ST depression at that time has important implications regarding the presence and severity of coronary artery disease (CAD).

Additional Techniques

Several ancillary imaging techniques have been shown to provide a valuable complement to exercise electrocardiography for the evaluation of patients with known or suspected CAD. They can localize ischemia and thus guide interventions. These techniques are particularly helpful among patients with equivocal exercise electrocardiograms or those likely to exhibit false-positive or false-negative responses. The guidelines call for their use when testing patients with more than 1.0 mm of ST depression at rest, LBBB, WPW, and paced rhythms. They are frequently used to clarify abnormal ST segment responses in asymptomatic people or those in whom the cause of chest discomfort remains uncertain, often avoiding angiography. When exercise electrocardiography and an imaging technique are combined, the diagnostic and prognostic accuracy is enhanced. The major imaging procedures are myocardial perfusion and ventricular function studies using radionuclide techniques, and exercise echocardiography. Some of the newer add-ons or substitutes for the exercise test have the advantage of being able to localize ischemia as well as diagnose coronary disease when the baseline ECG exhibits the above-mentioned abnormalities, which negate the usefulness of ST analysis. While the newer technologies are often suggested to have better diagnostic characteristics, this is not always the case particularly when more than the ST segments from the exercise test are used in scores. Pharmacologic stress testing is used in place of the standard exercise test for patients unable to walk or cycle or unable to give a good effort. These nonexercise stress techniques (persantine or adenosine with nuclear perfusion, dobutamine or arbutamine with echocardiography) permit diagnostic assessment of patients unable to exercise.

The ancillary imaging techniques are indicated when the ECG exhibits more than a millimeter of ST depression at rest, LBBB, WPW, and paced rhythms or when localization of ischemia is important.

Ventilatory Gas Exchange Responses

Because of the inaccuracies associated with estimating METs (ventilatory oxygen uptake) from workload (i.e., treadmill speed and grade), it can be important for many patients to measure physiologic work directly using ventilatory gas exchange responses, commonly referred to as cardiopulmonary exercise testing. Although this requires metabolic equipment, a facemask or mouthpiece and other equipment, advances in technology have made these measurements widely available. Cardiopulmonary exercise testing adds precision to the measurement of work and also permits the assessment of other parameters, including the respiratory exchange ratio, efficiency of ventilation, and the ventilatory anaerobic threshold. The latter measurement is helpful because it usually represents a comfortable sub maximal exercise limit and can be used for setting an optimal exercise prescription or an upper limit for daily activities. Clinically, this technology is often used to more precisely evaluate therapy, for the assessment of disability, and to help determine whether the heart or lungs limit exercise. Computerization of equipment has also led to the widespread use of cardiopulmonary exercise testing in sports medicine. Gas exchange measurements

Table 5. MET Demands for Common Daily Activities[a]

Activity	METs
Mild	
Baking	2.0
Billiards	2.4
Bookbinding	2.2
Canoeing (leisurely)	2.5
Conducting an orchestra	2.2
Dancing, ballroom (slow)	2.9
Golfing (with cart)	2.5
Horseback riding (walking)	2.3
Playing a musical instrument	2.0
Volleyball (noncompetitive)	2.9
Walking (2 mph)	2.5
Writing	1.7
Moderate	
Calisthenics (no weights)	4.0
Croquet	3.0
Cycling (leisurely)	3.5
Gardening (no lifting)	4.4
Golfing (without cart)	4.9
Mowing lawn (power mower)	3.0
Playing drums	3.8
Sailing	3.0
Swimming (slowly)	4.5
Walking (3 mph)	3.3
Walking (4 mph)	4.5
Vigorous	
Badminton	5.5
Chopping wood	4.9
Climbing hills	7.0
Cycling (moderate)	5.7
Dancing	6.0
Field hockey	7.7
Ice skating	5.5
Jogging (10 min mile)	10.0
Karate or judo	6.5
Roller skating	6.5
Rope skipping	12.0
Skiing (water or downhill)	6.8
Squash	12.0
Surfing	6.0
Swimming (fast)	7.0
Tennis (doubles)	6.0

[a]These activities can often be done at variable intensities if one assumes that the intensity is not excessive and that the courses are flat (no hills) unless so specified.

Table 6. Borg Scales of Perceived Exertion[a]

Borg 20-Point Scale of Perceived Exertion

6	
7	Very, very light
8	
9	Very light
10	
11	Fairly light
12	
13	Somewhat hard
14	
15	Hard
16	
17	Very hard
18	
19	Very, very hard
20	

Borg Nonlinear 10-Point Scale of Perceived Exertion

0	Nothing at all
0.5	Extremely light (just noticeable)
1	Very light
2	Light (Weak)
3	Moderate
4	Somewhat heavy
5	Heavy (Strong)
6	
7	Very heavy
8	
9	
10	Extremely heavy (almost maximal)
•	Maximal

[a]Top: Borg 20-point scale; bottom: Borg nonlinear 10-point scale.

camera, images of the blood circulating within the LV chamber could be obtained. While regurgitant blood flow from valvular lesions could not be identified, ejection fraction and ventricular volumes could be estimated. The resting values could be compared to those obtained during supine exercise and criteria were established for abnormal. The most common criteria involved a drop in ejection fraction. This procedure is now rarely performed because its test characteristics have not fulfilled their promise.

Nuclear Perfusion Imaging. After initial popularity, the blood volume techniques have become surpassed by nuclear perfusion techniques. The first agent used was thallium, an isotopic analog of potassium that is taken up at variable rates by metabolically active tissue. When taken up at rest, images of metabolically active muscle such as the heart are possible. With the nuclear camera placed over the heart after intravenous injection of this isotope, images were initially viewed using X-ray film. The normal complete donut shaped images gathered in multiple views would be broken by cold spots where scar was present. Defects viewed after exercise could be due to either scar or ischemia. Follow up imaging confirmed that the cold spots were due to ischemia if they filled in later. As computer imaging techniques were developed, 3D imaging (SPECT) and subtle differences could be plotted and scored. In recent years, ventriculograms based on the imaged wall as apposed to the blood in the chambers (as with RNV) could be constructed.

can supplement the exercise test by increasing precision and providing additional information concerning cardiopulmonary function during exercise. It is particularly needed to evaluate therapies using serial tests, since workload changes and estimated METs can be misleading. Because of their application for assessing prognosis in patients with heart failure, their use has become a standard part of the work-up for these patients.

Nuclear Techniques

Nuclear Ventricular Function Assessment. One of the first imaging modalities added to exercise testing was radionuclear ventriculography (RNV). This involved the intravenous injection of technetium tagged red blood cells. Using ECG gating of images obtained from a scintillation

Because of the technical limitations of thallium (i.e., source and lifespan), it has largely been replaced by chemical compounds called isonitriles which could be tagged with technetium, which has many practical advantages over thallium as an imaging agent. The isonitriles are trapped in the microcirculation permitting imaging of the heart with a scintillation camera. Rather than a single injection as for thallium, these compounds require an injection at maximal exercise then later in recovery. The differences in technology over the years and the differences in expertise and software at different facilities can complicate the comparisons of the results and actual application of this technology. The ventriculograms obtained with gated perfusion scans do not permit the assessment of valvular lesions, or as accurate an assessment of wall motion abnormalities or ejection fraction as echocardiography.

Nuclear perfusion scans can now permit an estimation of ventricular function and wall motion abnormalities.

Echocardiography

Echocardiography has made a significant and impressive impact on the field of cardiology. This imaging technique comes second only to contrast ventriculography via cardiac catheterization for measuring ventricular volumes, wall motion, and ejection fraction. With Doppler added, regurgitant flows can be estimated as well. With such information available, this imaging modality was quickly added by echocardiographers to exercise testing. Most studies showed that supine, posttreadmill assessments were adequate and the more difficult imaging during exercise was not necessary. The patient must be placed supine as soon as possible after treadmill or bicycle exercise and imaging begun. A problem can occur when the imaging requires removal or displacement of the important V_5 electrode where as much as 90% of the important ST changes are observed.

Biomarkers

The latest ancillary measures added to exercise testing in an attempt to improve diagnostic accuracy are biomarkers. The first and most logical biomarker evaluated to detect ischemia brought out by exercise was troponin. Unfortunately, it has been shown that even in patients who develop ischemia during exercise testing, serum elevations in cardiac specific troponin do not occur, demonstrating that myocardial damage does not occur (18,19). B-type natriuretic peptide (BNP) is a hormone produced by the heart that is released by both myocardial stretching and by myocardial hypoxia. Armed with this knowledge, investigators have reported several studies suggesting improvement in exercise test characteristics with BNP and its isomers (20,21). BNP is also used to assess the presence severity of (CHF) coronary heart failure, and has been shown to be a powerful prognostic marker (22,23). The point of contact analysis techniques available for these assays involves a hand held battery powered unit that uses a replaceable cartridge. Finger stick blood samples are adequate for these analyses and the results are available immediately. If validated using appropriate study design (similar to QUEXTA), biomarker measurements could greatly improve the diagnostic characteristics of the standard office–clinic exercise test.

In summary, use of proper methodology is critical for patient safety and accurate results. Preparing the patient physically and emotionally for testing is necessary. Good skin preparation will cause some discomfort but is necessary for providing good conductance and for avoiding artifact. The use of specific criteria for exclusion and termination, physician interaction with the patient, and appropriate emergency equipment are essential. A brief physical examination is always necessary to rule out important obstructive cardiomyopathy and aortic valve disease. Pretest standard 12-lead ECGs are needed in the supine and standing positions. The changes caused by exercise electrode placement can be kept to a minimum by keeping the arm electrodes off the chest, placing them on the shoulders, placing the leg electrodes below the umbilicus, and recording the baseline ECG supine. In this situation, the Mason–Likar modified exercise limb lead placement, if recorded supine, can serve as the resting ECG reference before an exercise test.

Few studies have correctly evaluated the relative yield or sensitivity and specificity of different electrode placements for exercise-induced ST-segment shifts. Using other leads in addition to V_5 will increase the sensitivity; however, the specificity is decreased. The ST-segment changes isolated to the inferior leads can on occasion be false-positive responses. For clinical purposes, vectorcardiographic and body surface mapping lead systems do not appear to offer any advantage over simpler approaches.

The exercise protocol should be progressive with even increments in speed and grade whenever possible. Smaller, even, and more frequent work increments are preferable to larger, uneven, and less frequent increases, because the former yield a more accurate estimation of exercise capacity. The value of individualizing the exercise protocol rather than using the same protocol for every patient has been emphasized by many investigators. The optimum test duration is from 8 to 12 min; therefore the protocol workloads should be adjusted to permit this duration. Because ramp testing uses small and even increments, it permits a more accurate estimation of exercise capacity and can be individualized to yield targeted test duration. An increasing number of equipment companies manufacture a controller that performs such tests using a treadmill.

Target heart rates based on age is inferior because the relationship between maximum heart rate and age is poor and scatters widely around many different recommended regression lines. Such heart rate targets result in a submaximal test for some individuals, a maximal test for others, and an unrealistic goal for some patients. Blood pressure should be measured with a standard stethoscope and sphygmomanometer; the available automated devices cannot be relied on, particularly for detection of exertional hypotension. Borg scales are an excellent means of quantifying an individual's effort. Exercise capacity should not be reported in total time but rather as the oxygen uptake or MET equivalent of the workload achieved. This method permits the comparison of the results of many different exercise testing protocols. Hyperventilation should be avoided before testing. Subjects with and without disease may exhibit ST-segment changes with hyperventilation; thus, hyperventilation to identify false-positive responders is no longer considered useful by most researchers. The postexercise period is a critical period diagnostically; therefore the patient should be placed in the supine position immediately after testing.

EVALUATION

Hemodynamics involves studying the body's adaptations to exercise, commonly evaluated in heart rate and blood pressure. However, it also includes changes in cardiac output and its determinants, as well as the influence of cardiovascular disease on cardiac output. Exercise capacity is an important clinical measurement and is also influenced strongly by exercise hemodynamics. There are several factors that affect exercise capacity, and there is an important issue of how normal standards for exercise capacity are expressed.

When interpreting the exercise test, it is important to consider each of its responses separately. Each type of response has a different impact on making a diagnostic or prognostic decision and must be considered along with an individual patient's clinical information. A test should not be called abnormal (or positive) or normal (or negative), but rather the interpretation should specify which responses were abnormal or normal, and each particular response should be recorded. The final report should be directed to the physician who ordered the test and who will receive the report. It should contain clear information that helps in patient management rather than vague "med-speak". Interpretation of the test is highly dependent upon the application for which the test is used and on the population tested.

Predicting Severe Angiographic Disease

Exercise Test Responses. Studies have long attempted to assess for disease in the left main coronary artery using exercise testing (24–26). Different criteria have been used with varying results. Predictive value here refers to the percentage of those with the abnormal criteria that actually had left main disease. Naturally, most of the false positives actually had coronary artery disease, but less severe forms. Sensitivity here refers to the percentage of those with left main disease only that are detected. These criteria have been refined over time and the last study by Weiner using the CASS data deserves further mention (27). A markedly positive exercise test (Table 7) defined as 0.2 mV or more of downsloping ST-segment depression beginning at 4 METs, persisting for at least six minutes into recovery, and involving at least five ECG leads had the

Table 7. Markedly Positive Treadmill Test Responses

Markedly Positive Treadmill Test Responses
More than 0.2 mV downsloping ST-segment depression
   Involving five or more leads
   Occurring at < 5 METs
   Prolonged ST depression late into recovery

greatest sensitivity (74%) and predictive value (32%) for left main coronary disease. This abnormal pattern identified either left main or three-vessel disease with a sensitivity of 49%, a specificity of 92% and a predictive value of 74%.

It appears that individual clinical or exercise test variables are unable to detect left main coronary disease because of their low sensitivity or predictive value. However, a combination of the amount, pattern, and duration of ST-segment response was highly predictive and reasonably sensitive for left main or three-vessel coronary disease. The question still remains of how to identify those with abnormal resting ejection fractions, those that will benefit the most with prolonged survival after coronary artery bypass surgery. Perhaps those with a normal resting ECG will not need surgery for increased longevity because of the associated high probability of normal ventricular function.

Blumenthal et al. (28) validated the ability of a strongly positive exercise test to predict left main coronary disease even in patients with minimal or no angina. The criteria for a markedly positive test included (1) early ST-segment depression, (2) 0.2 mV or more of depression, (3) downsloping ST depression, (4) exercise-induced hypotension, (5) prolonged ST changes after the test, and (6) multiple areas of ST depression. While Lee et al. (29) included many clinical and exercise test variables, only three variables were found to help predict left main disease: angina type, age, and the amount of exercise-induced ST segment depression.

Meta Analysis of Studies Predicting Angiographic Severity

To evaluate the variability in the reported accuracy of the exercise ECG for predicting severe coronary disease, Detrano et al. (30) applied meta analysis to 60 consecutively published reports comparing exercise-induced ST depression with coronary angiographic findings. The 60 reports included 62 distinct study groups comprising 12,030 patients who underwent both tests. Both technical and methodologic factors were analyzed. Wide variability in sensitivity and specificity was found [mean sensitivity 86% (range 40–100%); mean specificity 53% (range 17–100%)] for left main or triple vessel disease. All three variables found to be significantly and independently related to test performance were methodological. Exclusion of patients with right bundle branch block and receiving digoxin improved the prediction of triple vessel or left main coronary artery disease and comparison with a better exercise test decreased test performance.

Hartz et al. (31) compiled results from the literature on the use of the exercise test to identify patients with severe coronary artery disease. Pooled estimates of sensitivity and specificity were derived for the ability of the exercise test to identify three-vessel or left main coronary artery disease. One millimeter criteria averaged a sensitivity of 75% and a specificity of 66% while two millimeters criteria averaged a sensitivity of 52% and a specificity of 86%. There was great variability among the studies examined in the estimated sensitivity and specificity for severe coronary artery disease that could not be explained by their analysis.

Table 8. Summary of Studies Assessing Maximal Heart Rate

Investigator	No. Subjects	Population[a] Studied	Mean Age ±SD (Range)	Mean HR Max (SD)	Regression Line	Correlation[a] Coefficient	Standard Error of the Estimate, beats/min[a]
Astrand[b]	100	Asymptomatic men	50 (20–69)	166 ± 22	$y = 211 - 0.922$ (age)	NA	NA
Bruce	2091	Asymptomatic men	44 ± 8	181 ± 12	$y = 210 - 0.662$ (age)	−0.44	14
Cooper	2535	Asymptomatic men	43 (11–79)	181 ± 16	$y = 217 - 0.845$ (age)	NA	NA
Ellestad[c]	2583	Asymptomatic men	42 ± 7 (10–60)	173 ± 11	$y = 197 - 0.556$ (age)	NA	NA
Froelicher	1317	Asymptomatic men	38 ± 8 (28–54)	183	$y = 207 - 0.64$ (age)	−0.43	10
Lester	148	Asymptomatic men	43 (15–75)	187	$y = 205 - 0.411$ (age)	−0.58	NA
Robinson	92	Asymptomatic men	30 (6–76)	189	$y = 212 - 0.775$ (age)	NA	NA
Sheffield	95	Men with CHD	39 (19–69)	176 ± 14	$y \pm 216 - 0.88$ (age)	−0.58	11^d
Bruce	1295	Men with CHD	52 ± 8	148 ± 23	$y = 204 - 1.07$ (age)	−0.36	25^d
Hammond	156	Men with CHD	53 ± 9	157 ± 20	$y = 209 - 1.0$ (age)	−0.30	19
Morris	244	Asymptomatic men	45 (20–72)	167 ± 19	$y = 200 - 0.72$ (age)	−0.55	15
Graettinger	114	Asymptomatic men	46 ± 13 (19–73)	168 ± 18	$y = 199 - 0.63$ (age)	−0.47	NA
Morris	1388	Men referred for evaluation for CHD, normals only	57 (21–89)	144 ± 20	$y = 196 - 0.9$ (age)	−0.43	21

[a]CHD = coronary heart disease; HR max = maximal heart rate; NA = not able to calculate from available data.
[b]Astrand used bicycle ergometry; all other studies were performed on a treadmill.
[c]Data compiled from graphs in reference cited.
[d]Calculated from available data.

Studies Using Multivariate Techniques to Predict Severe Angiographic CAD

Multiple studies have reported combining the patient's medical history, symptoms of chest pain, hemodynamic data, exercise capacity and exercise test responses to calculate the probability of severe angiographic coronary artery disease (32–43). The results are summarized in Table 8. Of the 13 studies, 9 excluded patients with previous coronary artery bypass surgery or prior percutaneous coronary intervention (PCI) and in the remaining 4 studies, exclusions were unclear. The percentage of patients with one vessel, two vessels and three vessels was described in 10 of the 13 studies. The definition of severe disease or disease extent also differed. In 5 of the 13 studies disease extent was defined as multivessel disease. In the remaining 8 studies, it was defined as three-vessel or left main disease and in one of them as only left main artery disease and in another the impact of disease in the right coronary artery disease on left main disease was considered. The prevalence of severe disease ranged from 16 to 48% in the studies defining disease extent as multivessel disease and from 10 to 28% in the studies using the more strict criterion of three-vessel or left main disease.

Chosen Predictors

Interestingly, some of the variables chosen for predicting disease severity are different than those for predicting disease presence. While gender and chest pain were chosen to be significant in more than half of the severity studies, age was less important and resting ECG abnormalities and diabetes were the only other variables chosen in more than half the studies. In contrast, the most consistent clinical variables chosen for diagnosis were age, gender, chest pain type, and hypercholesterolemia. ST depression and slope were frequently chosen for severity, but METs and heart rate were less consistently chosen than for diagnosis. Double product and delta SBP were chosen as independent predictors in more than half of the studies predicting severity.

Consensus to Improve Prediction

So far, only two studies [Detrano et al. (41) and Morise et al. (38)] have published equations that have been validated in large patient samples. Even though validated, however, the equations from these studies must be calibrated before they can be applied clinically. For example, a score can be discriminating but provide an estimated probability that is higher or lower than the actual probability. The scores can be calibrated by adjusting them according to disease prevalence; most clinical sites however, do not know their disease prevalence and even so, it could change from month to month.

In NASA trajectories of spacecraft are often determined by agreement between three or more equations calculating the vehicle path. With this in mind, we developed an agreement method to classify patients into high, no agreement, or low risk groups for probability of severe disease by requiring agreement in all three equations [Detrano, Morise, and ours (LB–PA)] (44). This approach adjusts the calibration and makes the equations applicable in clinical populations with varying prevalence of coronary artery disease.

It was demonstrated that using simple clinical and exercise test variables could improve the standard application of

ECG criteria for predicting severe coronary artery disease. By setting probability thresholds for severe disease of < 20% and > 40% for the three prediction equations, the agreement approach divided the test set into populations with low, no agreement, and high risk for severe coronary artery disease. Since the patients in the no agreement group would be sent for further testing and would eventually be correctly classified, the sensitivity of the agreement approach is 89% and the specificity is 96%. The agreement approach appeared to be unaffected by disease prevalence, missing data, variable definitions, or even by angiographic criterion. Cost analysis of the competing strategies revealed that the agreement approach compares favorably with other tests of equivalent predictive value, such as nuclear perfusion imaging, reducing costs by 28% or $504 per patient in the test set.

Requiring agreement of these three equations to make diagnosis of severe coronary disease has made them widely applicable. Excellent predictive characteristics can be obtained using simple clinical data entered into a computer. Cost analysis suggests that the agreement approach is an efficient method for the evaluation of populations with varying prevalence of coronary artery disease, limiting the use of more expensive noninvasive and invasive testing to patients with a higher probability of left main or three vessel coronary artery disease. This approach provides a strategy for assisting the practitioner in deciding when further evaluation is appropriate or interventions indicated.

Predicting Improved Survival with Coronary Artery Bypass Surgery

Which exercise test variables indicate those patients who would have an improved prognosis if they underwent coronary artery bypass surgery (CABS)? Research in this area is limited by the fact the available studies did not randomize patients to surgery based on the results of their exercise test results and the retrospective nature of the studies.

Bruce et al. (45) demonstrated noninvasive screening criteria for patients who had improved 4 year survival after coronary artery bypass surgery. Their data have come from 2000 men with coronary heart disease enrolled in the Seattle Heart Watch who had a symptom-limited maximal treadmill test; these subjects received usual community care, which resulted in 16% of them having coronary artery bypass surgery in nonrandomized fashion. Patients with cardiomegaly, < 5 MET exercise capacity and/or a maximal systolic blood pressure of < 130 would have a better outcome if treated with surgery. Two or more of the above parameters present the highest risk and the greater differential for improved survival with bypass. Four year survival in this group would be 94% for those that had surgery versus 67% for those who received medical management (in those who had two or more of the above factors). In the European surgery trial (46), patients who had an exercise test response of 1.5 mm of ST segment depression had improved survival with surgery. This also extended to those with baseline ST segment depression and those with claudication.

From the CASS study group (47), in > 5000 nonrandomized patients, though there were definite differences between the surgical and nonsurgical groups, this could be accounted for by stratification in subsets. The surgical benefit regarding mortality was greatest in the 789 patients with 1 mm ST segment depression at < 5 METs. Among the 398 patients with triple vessel disease with this exercise test response, the 7 year survival was 50% in those medically managed versus 81% in those who underwent coronary artery bypass surgery. There was no difference in mortality in randomized patients able to exceed 10 METs. In the VA surgery randomized trial (48), there was a 79% survival rate with CABS versus 42% for medical management in patients with two or more of the following: 2 mm or more of ST depression, heart rate of 140 or greater at 6 METs, and/or exercise-induced PVCs.

Evaluation of Percutaneous Coronary Interventions

One important application of the exercise test is to assess the effects of percutaneous coronary intervention (PCI) on physical function, ischemic responses, and symptoms in the immediate and longer period following the various interventions that now fall under the general term PCI. The exercise test has been used for this purpose in numerous trials of PCI, and a few notable examples are described in the following. Berger et al. (49) reported follow-up data in 183 patients who had undergone PCI at least 1 year earlier. PCI was initially successful in 141 patients (79%). Of the 42 patients in whom PCI was unsuccessful, 26 underwent CABG, while 16 were maintained on medical therapy. When compared to the medical patients at time of follow-up, successful PCI patients experienced less angina (13 vs. 47%), used less nitroglycerin (25 vs. 73%), were hospitalized less often for chest pain (8 vs. 31%), and subjectively felt their condition had improved (96 vs. 20%).

Vandormael and colleagues reported the safety and short-term benefit of multi-lesion PCI in 135 patients (50). Primary success, defined as successful dilation of the most critical lesion or all lesions attempted, occurred in 87% of the 135 patients. Exercise-induced angina occurred in 11 (12%) and an abnormal exercise ECG in 30 (32%) of the 95 patients with post-PCI exercise test data. Of 57 patients who had paired exercise test data before and after angioplasty, exercise-induced angina occurred in 56% of patients before the procedure, compared with only 11% of patients after angioplasty. Exercise-induced ST-segment depression of > 0.1 mV occurred in 75% of patients before PCI versus 32% after the procedure.

Rosing et al. (51) reported that exercise testing after successful PCI exhibited improved ECG and symptomatic responses, as well as improved myocardial perfusion and global and regional left ventricular function.

Prediction of Restenosis with the Exercise Test

To determine whether a treadmill test could predict restenosis after angioplasty, Honan et al. (52) studied 289 patients six months after a successful emergency angioplasty of the infarct-related artery for acute myocardial infarction (MI). After excluding those with interim interventions, medical events, or medical contraindications to follow-up testing, both a treadmill test and a cardiac catheterization were completed in 144 patients; 88% of those eligible for this assessment. Of six clinical and treadmill variables examined by multivariable logistic analysis, only exercise ST deviation was independently correlated with restenosis. The sensitivity of ST deviation of 0.10 mV or greater for detecting restenosis was only 24% (13 of 55 patients), and the specificity was 88% (75 of 85 patients). Extent or severity of wall motion abnormalities at follow-up did not affect the sensitivity of exercise-induced ST deviation for detection of restenosis, by the timing of thrombolytic therapy or of angioplasty, or by the presence of collateral blood flow at the time of acute angiography. A second multivariable analysis evaluating the association of the same variables with number of vessels with significant coronary disease at the 6 month catheterization found an association with both exercise ST deviation and exercise duration. Angina symptoms and exercise test results in this population had limited value for predicting anatomic restenosis six months after emergency angioplasty for acute myocardial infarction.

Bengtson et al. (53) studied 303 consecutive patients with successful PCI and without a recent myocardial infarction. Among the 228 patients without interval cardiac events, early repeat revascularization or contraindications to treadmill testing, 209 (92%) underwent follow-up angiography, and 200 also had a follow-up treadmill test and formed the study population. Restenosis occurred in 50 patients (25%). Five variables were individually associated with a higher risk of restenosis: recurrent angina, exercise-induced angina, a positive treadmill test, greater exercise ST deviation, and a lower maximum exercise heart rate. However, only exercise-induced angina, recurrent angina, and a positive treadmill test were independent predictors of restenosis. Using these three variables, patient subsets could be identified with restenosis rates ranging from 11 to 83%. The exercise test added independent information to symptom status regarding the risk of restenosis after elective PCI. Nevertheless, 20% of patients with restenosis had neither recurrent angina nor exercise-induced ischemia at follow-up.

The ROSETTA registry was studied to demonstrate the effects of routine post-PCI functional testing on the use of follow-up cardiac procedures and clinical events (54). The ROSETTA (Routine versus Selective Exercise Treadmill Testing after Angioplasty) registry is a prospective multicenter observational study examining the use of functional testing after PCI. A total of 788 patients were enrolled in the registry at 13 clinical centers in 5 countries. The frequencies of exercise testing, cardiac procedures and clinical events were examined during the first 6 months following a successful PCI. Patients were predominantly elderly men (mean age, 61 ± 11 years; 76% male) who underwent single-vessel PCI (85%) with stent implantation (58%). During the 6 month follow-up, a total of 237 patients underwent a routine exercise testing strategy (100% having exercise testing for routine follow-up), while 551 pts underwent a selective (or clinically driven) strategy (73% having no exercise testing and 27% having exercise testing for a clinical indication). Patients in the routine testing group underwent a total of 344 exercise tests compared with 165 tests performed in the selective testing group (mean, 1.45 tests/patient vs. 0.3 tests/patient). However, clinical events were less common among those who underwent routine exercise testing, for example, unstable angina (6% vs. 14%), myocardial infarction (0.4% vs. 1.6%), death (0% vs. 2%) and composite clinical events (6% vs. 16%). After controlling for baseline clinical and procedural differences, routine exercise testing had a persistent independent association with a reduction in the composite clinical event rate. This association may be attributable to the early identification and treatment of patients at risk for follow-up events, or it may be due to clinical differences between patients who are referred for routine and selective exercise testing.

The ACC/AHA Guidelines for the Prognostic Use of the Standard Exercise Test

The task force to establish guidelines for the use of exercise testing has met and produced guidelines in 1986, 1997, and 2002. The following is a synopsis of these evidence-based guidelines.

Indications for Exercise Testing to Assess Risk and prognosis in patients with symptoms or a prior history of CAD:

Class I (Definitely Appropriate). Conditions for which there is evidence and/or general agreement that the standard exercise test is useful and helpful to assess risk and prognosis in patients with symptoms or a prior history of CAD.

  Patients undergoing initial evaluation with suspected or known CAD. Specific exceptions are noted below in Class IIb.
  Patients with suspected or known CAD previously evaluated with significant change in clinical status.

Class IIb (Maybe Appropriate). Conditions for which there is conflicting evidence and/or a divergence of opinion that the standard exercise test is useful and helpful to assess risk and prognosis in patients with symptoms or a prior history of coronary artery disease but the usefulness/efficacy is less well established.

  Patients who demonstrate the following ECG abnormalities: pre excitation (Wolff–Parkinson White) syndrome; electronically paced ventricular rhythm; >1 mm of resting ST depression; and complete left bundle branch block.
  Patients with a stable clinical course who undergo periodic monitoring to guide management

Class III (Not Appropriate). Conditions for which there is evidence and/or general agreement that the standard exercise test is not useful and helpful to assess risk and prognosis in patients with symptoms or a prior history of CAD and in some cases may be harmful.

  Patients with severe comorbidity likely to limit life expectancy and/or candidacy for revascularization.

In summary, the two principal reasons for estimating prognosis are to provide accurate answers to patient's questions regarding the probable outcome of their illness and to identify those patients in whom interventions might improve outcome. There is a lack of consistency in the available studies because patients die along a pathophysiological spectrum ranging from those that die due to CHF with little myocardium remaining to those that die from an ischemic related event with ample myocardium remaining. Clinical and exercise test variables most likely associated with CHF deaths (CHF markers) include a history or symptoms of CHF, prior MI, Q waves, and other indicators of LV dysfunction. Variables most likely associated with ischemic deaths (ischemic markers) are angina, and rest and exercise ST depression. Some variables can be associated with either extremes of the type of CV death; these include exercise capacity, maximal heart rate, and maximal systolic blood pressure that may explain why they are reported most consistently in the available studies. A problem exists that ischemic deaths occur later in follow up and are more likely to occur in those lost to follow up whereas CHF deaths are more likely to occur early (within 2 years) and are more likely to be classified. Work-up bias probably explains why exercise-induced ST depression fails to be a predictor in most of the angiographic studies. Ischemic markers are associated with a later and lesser risk, whereas CHF or left ventricular dysfunction markers are associated with a sooner and greater risk of death.

Recent studies of prognosis have actually not been superior to the earlier studies that considered CV endpoints and removed patients from observation who had interventions. This is because death data is now relatively easy to obtain while previously investigators had to follow the patients and contact them or review their records. CV mortality can be determined by death certificates. While death certificates have their limitations, in general they classify those with accidental, GI, Pulmonary and cancer deaths so that those remaining are most likely to have died of CV causes. This endpoint is more appropriate for a test for CV disease. While all-cause mortality is a more important endpoint for intervention studies, CV mortality is more appropriate for evaluating a CV test (i.e., the exercise test). Identifying those at risk of death of any cause does not make it possible to identify those who might benefit from CV interventions, one of the main goals of prognostication.

The consistencies actually overshadow the differences. Considering simple clinical variables can assess risk. A good exercise capacity, no evidence or history of CHF or ventricular damage (Q waves, history of CHF), no ST depression or only one of these clinical findings are associated with a very low risk. These patients are low risk in exercise programs and need not be considered for interventions to prolong their life. High risk patients can be identified by groupings of the clinical markers; that is, two or more. Exertional hypotension is particularly ominous. Identification of high risk implies that such patients in exercise training programs should have lower goals and should be monitored. Such patients should also be considered for coronary interventions to improve their longevity. Furthermore, with each drop in METs there is a 10–20%

increase in mortality so simple exercise capacity has consistent importance in all patient groups.

The mathematical models for determining prognosis are usually more complex than those used for identifying severe angiographic disease. Diagnostic testing can utilize multivariate discriminant function analysis to determine the probability of severe angiographic disease being present or not. Prognostic testing must utilize survival analysis which includes censoring for patients with uneven follow-up due to "lost to follow up" or other cardiac events (i.e., CABS, PCI) and must account for time-person units of exposure. Survival curves must be developed and the Cox proportional hazards model is often preferred.

From this perspective, it is obvious that there is substantial support for the use of the exercise test as a first noninvasive step after the history, physical exam, and resting ECG in the prognostic evaluation of coronary artery disease patients. It accomplishes both of the purposes of prognostic testing: to provide information regarding the patient's status and to help make recommendations for optimal management. The exercise test results help us make reasonable decisions for selection of patients who should undergo coronary angiography. Since the exercise test can be performed in the doctor's office and provides valuable information for clinical management in regard to activity levels, response to therapy, and disability, the exercise test is the reasonable first choice for prognostic assessment. This assessment should always include calculation of the estimated annual mortality using the Duke treadmill score though its ischemic elements have less power in the elderly.

There has been considerable debate in screening asymptomatic patients. Screening has become a controversial topic because of the incredible efficacy of the statins (drugs that lower cholesterol) even in asymptomatic individuals (55). There are now agents that can cut the risk of cardiac events almost in half. The first step in screening asymptomatic individuals for preclinical coronary disease should be using global risk factor equations, such as the Framingham score. This is available as nomograms that are easily applied by healthcare professionals or it can be calculated as part of a computerized patient record. Additional testing procedures with promise include the simple ankle-brachial index (particularly in the elderly), CRP, carotid ultrasound measurements of intimal thickening, and the resting ECG (particularly spatial QRS-T wave angle). Despite the promotional concept of atherosclerotic burden, EBCT does not have test characteristics superior to the standard exercise test. If any screening test could be used to decide regarding statin therapy and not affect insurance or occupational status, this would be helpful. However, the screening test should not lead to more procedures.

True demonstration of the effectiveness of a screening technique requires randomizing the target population, applying the screening technique to half, taking standardized action in response to the screening test results, and then assessing outcomes. Efficacy of the screening test necessitates that the screened group has lower mortality and/or morbidity. Such a study has been completed for mammography, but not for any cardiac testing modalities. The next best validation of efficacy is to demonstrate that the technique improves the discrimination of those asymptomatic individuals with higher risk for events over that possible with the available risk factors. Mathematical modeling makes it possible to determine how well a population will be classified if the characteristics of the testing method are known.

Additional follow-up studies and one angiographic study from the CASS population (where 195 individuals with abnormal exercise-induced ST depression and normal coronary angiograms were followed for 7 years) improve our understanding of the application of exercise testing as a screening tool. No increased incidence of cardiac events was found, and so the concerns raised by Erikssen's findings in 36 subjects that they were still at increased risk have not been substantiated.

The later follow-up studies (MRFIT, Seattle Heart Watch, Lipid Research Clinics, and Indiana State Police) have shown different results compared to prior studies, mainly because hard cardiac end points and not angina were required. The first ten prospective studies of exercise testing in asymptomatic individuals included angina as a cardiac disease end point. This led to a bias for individuals with abnormal tests to subsequently report angina or to be diagnosed as having angina. When only hard end points (death or MI) were used, as in the MRFIT, Lipid Research Clinics, Indiana State Police or the Seattle Heart Watch studies, the results were less encouraging. The test could only identify one-third of the patients with hard events and 95% of abnormal responders were false positives; that is, they did not die or have a MI. The predictive value of the abnormal maximal exercise electrocardiogram ranged from 5 to 46% in the studies reviewed. However, in the studies using appropriate endpoints (other than angina pectoris) only 5% of the abnormal responders developed coronary heart disease over the follow-up period. Thus, >90% of the abnormal responders were false positives. However, the exercise test's characteristics as a screening test probably lie in between the results with hard or soft endpoints since some of the subjects who develop chest pain really have angina and coronary disease. The sensitivity is probably between 30 and 50% (at a specificity of 90%) but the critical limitation is the predictive value (and risk ratio) which depends upon the prevalence of disease (which is low in the asymptomatic population).

Although some of these individuals have indolent coronary disease yet to be manifest, angiographic studies have supported this high false positive rate when using the exercise test in asymptomatic populations. Moreover, the CASS study indicates that such individuals have a good prognosis. In a second Lipid Research Clinics study, only patients with elevated cholesterol's were considered, and yet only a 6% positive prediction value was found. If the test is to be used to screen it should be done in groups with a higher estimated prevalence of disease using the Framingham score and not just one risk factor. The iatrogenic problems resulting from screening must be considered. Hopefully, using a threshold from the Framingham score would be more successful in identifying asymptomatic individuals that should be tested.

Some individuals who eventually develop coronary disease will change on retesting from a normal to an abnormal response. However, McHenry and Fleg have reported that a change from a negative to a positive test is no more predictive than is an initially abnormal test. One individual has even been reported who changed from a normal to an abnormal test, but was free of angiographically significant disease (56). In most circumstances an added imaging modality (echocardiographic or nuclear) should be the first choice in evaluating asymptomatic individuals with an abnormal exercise test.

The motivational impact of screening for CAD is not evidence based with one positive study for exercise testing and one negative study for EBCT. Further research in this area certainly is needed.

While the risk of an abnormal exercise test is apparent from these studies, the iatrogenic problems resulting from screening must be considered (i.e., employment, insurance). The recent U.S. Preventive Services Task Force statement states that "false positive tests are common among asymptomatic adults, especially women, and may lead to unnecessary diagnostic testing, over treatment and labeling". This statement summarizes the current U.S. Preventive Services Task Force (USPSTF) recommendations on screening for coronary heart disease and the supporting scientific evidence and updates the 1996 recommendations on this topic. The complete information on which this statement is based, including evidence tables and references, is available in the background article and the systematic evidence review, available through the USPSTF Web site (http://www.preventiveservices.ahrq.gov) and through the National Guideline Clearinghouse (http://www.guidclinc.gov) (57). In the majority of asymptomatic people, screening with any test or test add-on, is more likely to yield false positives than true positives. This is the mathematical reality associated with all of the available tests.

There are reasons to include exercise testing in the preventative health recommendations for screening healthy, asymptomatic individuals along with risk factor assessment. The additional risk classification power documented by the data from Norway (2000 men, 26 year follow up), the Cooper Clinic (26,000 men, 8 year follow up), and Framingham (3000 men, 18 year follow up) provide convincing evidence that the exercise test should be added to the screening process. Furthermore, exercise capacity itself has substantial prognostic predictive power. Given the emerging epidemic of physical inactivity, including the exercise test in the screening process sends a strong message to our patients that we consider their exercise status as important.

However, if screening could be performed in a logical way with test results helping to decide on therapies rather than leading to invasive interventions, insurance or occupational problems, then the recent results summarized above should be applied to preventive medicine policy.

Because of the inherent difficulties, few preventive medicine recommendations are based on randomized trials demonstrating improved outcomes but rely on reasonable assumptions from available evidence. There is now enough evidence to consider recommending a routine exercise test every five years for men > 40 and women > 50 years of age, especially if one of the potential benefits is the adoption of an active lifestyle (58).

CONCLUSION

While there are important technological considerations and the need to be very knowledgeable of the guidelines to insure its proper application, exercise testing remains one of the most widely used and valuable noninvasive tools to assess cardiovascular status.

The following precepts regarding methodology are important to follow:

The treadmill protocol should be adjusted to the patient and one protocol is not appropriate for all patients.

Exercise capacity should be reported in METs not minutes of exercise.

Hyperventilation prior to testing is not indicated, but can be utilized at another time, if a false positive is suspected

ST measurements should be made at ST0 (J-junction) and ST depression should only be considered abnormal if horizontal or downsloping; 95% of the clinically important ST depression occurs in V_5 particularly in patients with a normal resting ECG.

Patients should be placed supine as soon as possible postexercise with a cool down walk avoided in order for the test to have its greatest diagnostic value.

The 2–4 min recovery period is critical to include in analysis of the ST response.

Measurement of systolic blood pressure during exercise is extremely important and exertional hypotension is ominous; at this point, only manual blood pressure measurement techniques are valid.

Age-predicted heart rate targets are largely useless because of the wide scatter for any age; a relatively low heart rate can be maximal for a given patient and submaximal for another.

The Duke Treadmill Score should be calculated automatically on every test except for the elderly.

Other predictive equations and heart rate recovery should be considered a standard part of the treadmill report.

BIBLIOGRAPHY

Cited References

1. Freeman J, Dewey R, Hadley D, Froelicher V. Evaluation of the Autonomic Nervous System with Exercise Testing. Prog Cardiovasc Dis 2005 Jan-Feb;47(4):285–305.
2. Froelicher VF, et al. A comparison of two-bipolar electrocardiographic leads to lead V_5. Chest 1976;70:611.
3. Gamble P, et al. A comparison of the standard 12-lead electrocardiogram to exercise electrode placements. Chest 1984;85: 616–622.
4. Miranda CP, et al. Usefulness of exercise-induced ST-segment depression in the inferior leads during exercise testing as a marker for coronary artery disease. Am J Cardiol 1992;69: 303–307.
5. Milliken JA, Abdollah H, Burggraf GW. False-positive treadmill exercise tests due to computer signal averaging. Am J Cardiol 1990;65:946–948.
6. Willems J, et al. The diagnostic performance of computer programs for the interpretation of ECGs. N Engl J Med 1991;325:1767–1773.
7. Balady GJ, et al. Value of arm exercise testing in detecting coronary artery disease. Am J Cardiol 1985;55:37–39.
8. Myers J, et al. Comparison of the ramp versus standard exercise protocols. J Am Coll Cardiol 1991;17:1334–1342.
9. Wickes JR, et al. Comparison of the Electrocardiographic changes induced by maximum exercise testing with treadmill and cycle ergometer. Circulation 1978;57:1066–1069.
10. Hambrecht RP, et al. Greater diagnostic sensitivity of treadmill versus cycle exercise testing of asymptomatic men with coronary artery disease. Am J Cardiol 1992 Jul 15;70(2): 141–146.
11. Sullivan M, McKirnan MD. Errors in predicting functional capacity for post myocardial infarction patients using a modified Bruce protocol. Am Heart J 1984;107:486–491.
12. Webster MWI, Sharpe DN. Exercise testing in angina pectoris: the importance of protocol design in clinical trials. Am Heart J 1989;117:505–508.
13. Goldman L, et al. Comparative reproducibility and validity of systems for assessing cardiovascular function class: advantages of a new specific activity scale. Circulation 1981;64: 1227–1234.
14. Fletcher GF, et al. Exercise standards for testing and training: a statement for healthcare professionals from the American Heart Association. Circulation 2001;104:1694–1740.
15. Borg G, Holmgren A, Lindblad I. Quantitative evaluation of chest pain. Acta Med Scand 1981;644:43–45.
16. Gutman RA, et al. Delay of ST depression after maximal exercise by walking for two minutes. Circulation 1970;42: 229–233.
17. Lachterman B, et al. "Recovery only" ST segment depression and the predictive accuracy of the exercise test. Ann Intern Med 1990;112:11–16.
18. Ashmaig ME, et al. Changes in serum concentrations of markers of myocardial injury following treadmill exercise testing in patients with suspected ischaemic heart disease. Med Sci Monit 2001;7:54–57.
19. Akdemir I, et al. Does exercise-induced severe ischaemia result in elevation of plasma troponin-T level in patients with chronic coronary artery disease? Acta Cardiol 2002;57: 13–18.
20. Sabatine MS, et al. TIMI Study Group. Acute changes in circulating natriuretic peptide levels in relation to myocardial ischemia. J Am Coll Cardiol 2004;44(10):1988–1995.
21. Foote RS, Pearlman JD, Siegel AH, Yeo KT. Detection of exercise-induced ischemia by changes in B-type natriuretic peptides. J Am Coll Cardiol 2004;44(10):1980–1987.
22. Wang TJ, et al. Plasma natriuretic peptide levels and the risk of cardiovascular events and death. N Engl J Med 2004 Feb 12; 350(7):655–663.
23. Kragelund C, Gronning B, Kober L, Hildebrandt P, Steffensen R. N-terminal pro-B-type natriuretic peptide and long-term mortality in stable coronary heart disease. N Engl J Med 2005 Feb 17;352(7):666–75.
24. Cheitlin MD, et al. Correlation of "critical" left coronary artery lesions with positive submaximal exercise tests in patients with chest pain. Am Heart J 1975;89(3):305–310.
25. Goldschlager N, Selzer A, Cohn K. Treadmill stress tests as indicators of presence and severity of coronary artery disease. Ann Int Med 1976;85:277–286.
26. NcNeer JF, et al. The role of the exercise test in the evaluation of patients for ischemic heart disease. Circulation 1978; 57:64–70.

27. Weiner DA, McCabe CH, Ryan TJ. Identification of Patients with left main and three vessel coronary disease with clinical and exercise test variables. Am J Cardiol 1980;46:21–27.

28. Blumenthal DS, Weiss JL, Mellits ED, Gerstenblith G. The predictive value of a strongly positive stress test in patients with minimal symptoms. Am J Med 1981;70:1005–1010.

29. Lee TH, Cook EF, Goldman L. Prospective evaluation of a clinical and exercise-test model for the prediction of left main coronary artery disease. Med Decis Making 1986;6: 136–144.

30. Detrano R, et al. Exercise-induced ST segment depression in the diagnosis of multivessel coronary disease: A meta analysis. J Am Coll Cardiol 1989;14:1501–1508.

31. Hartz A, Gammaitoni C, Young M. Quantitative analysis of the exercise tolerance test for determining the severity of coronary artery disease. Int J Cardiol 1989;24:63–71.

32. Cohn K, et al. Use of treadmill score to quantify ischemic response and predict extent of coronary disease. Circulation 1979;59:286–296.

33. Fisher L, et al. Diagnostic quantification of CASS (Coronary artery surgery study) clinical and exercise test results in determining presence and extent of coronary artery disease. Circulation 1981;63:987–1000.

34. McCarthy D, Sciacca R, Blood D, Cannon P. Discriminant function analysis using thallium 201 scintiscans and exercise stress test variables to predict the presence and extent of coronary artery disease. Am J Cardiol 1982;49:1917–1926.

35. Lee T, Cook E, Goldman L. Prospective evaluation of a clinical and exercise test model for the prediction of left main coronary artery disease. Med Decis Making 1986;6:136–144.

36. Hung J, et al. A logistic regression analysis of multiple noninvasive tests for the prediction of the presence and extent of coronary artery disease in men. Am Heart J 1985;110:460–469.

37. Christian T, Miller T, Bailey K, Gibbons R. Exercise tomographic thallium-201 imaging in patients with severe coronary artery disease and normal electrocardiograms. Ann Intern Med 1994;121:825–832.

38. Morise A, Bobbio M, Detrano R, Duval R. Incremental evaluation of exercise capacity as an independent predictor of coronary artery disease presence and extent. Am Heart J 1994;127:32–38.

39. Morise A, Diamond G, Detrano R, Bobbio M. Incremental value of exercise electrocardiography and thallium-201 testing in men and women for the presence and extent of coronary artery disease. Am Heart J 1995;130:267–276.

40. Moussa I, Rodriguez M, Froning J, Froelicher VF. Prediction of severe coronary artery disease using computerized ECG measurements and discriminant function analysis. J Electrocardiol 1992;25:49–58.

41. Detrano R, et al. Algorithm to predict triple-vessel/left main coronary artery disease in patients without myocardial infarction. Circulation 1991;83(3):89–96.

42. Christian TF, Miller TD, Bailley KR, Gibbons RJ. Noninvasive identification of severe coronary artery disease using exercise tomographic thallium-201 imaging. Am J Cardiol 1992;70:14–20.

43. Hung J, et al. Noninvasive diagnostic test choices for the evaluation of coronary artery disease in women:a multivariate comparison of cardiac fluoroscopy, exercise electrocardiography and exercise thallium myocardial perfusion scintigraphy. J Am Coll Cardiol 1984;4:8–16.

44. Do D, West JA, Morise A, Froelicher VF. Agreement Predicting Severe Angiographic Coronary Artery Disease Using Clinical and Exercise Test Data. Am Heart J 1997;134: 672–679.

45. Bruce RA, Hossack KF, DeRouen TA, Hofer V. Enhanced risk assessment for primary coronary heart disease events by maximal exercise testing: 10 years' experience of Seattle Heart Watch. J Am Coll Cardiol 1983;2:565–73.

46. European Cooperative Group. Long-term results of prospective randomized study of coronary artery bypass surgery in stable angina pectoris. Lancet 1982; 1173–1180.

47. Weiner DA, et al. The role of exercise testing in identifying patients with improved survival after coronary artery bypass surgery. J Am Coll Cardiol 1986;8(4):741–748.

48. Hultgren HN, Peduzzi P, Detre K, Takaro T. The 5 year effect of bypass surgery on relief of angina and exercise performance. Circulation 1985;72:V79–V83.

49. Berger E, Williams DO, Reinert S, Most AS. Sustained efficacy of percutaneous transluminal coronary angioplasty. Am Heart J 1986;111:233–236.

50. Vandormael MG, et al. Immediate and short-term benefit of multilesion coronary angioplasty: Influence of degree of revascularization. J Am Coll Cardiol 1985;6:983–991.

51. Rosing DR, et al. Exercise, electrocardiographic and functional responses after percutaneous transluminal coronary angioplasty. Am J Cardiol 1984;53:36C–41C.

52. Honan MB, et al. Exercise treadmill testing is a poor predictor of anatomic restenosis after angioplasty for acute myocardial infarction. Circulation 1989;80:1585–1594.

53. Bengtson JR, et al. Detection of restenosis after elective percutaneous transluminal coronary angioplasty using the exercise treadmill test. Am J Cardiol 1990;65:28–34.

54. Eisenberg MJ, et al. ROSETTA Investigators. Utility of routine functional testing after percutaneous transluminal coronary angioplasty: results from the ROSETTA registry. J Invasive Cardiol 2004;16:318–322.

55. Downs JR, et al. Primary prevention of acute coronary events with lovastatin in men and women with average cholesterol levels: Results of AFCAPS/TexCAPS. Air Force/Texas Coronary Atherosclerosis Prevention Study. JAMA 1998;279: 1615–1622.

56. Thompson AJ, Froelicher VF. Normal coronary angiography in an aircrewman with serial test changes. Aviat Space Environ Med 1975;46:69–73.

57. U.S. Preventive Services Task Force. Screening for coronary heart disease: recommendation statement. Ann Intern Med 2004 Apr 6; 140(7):569–572.

58. DiPietro L, Kohl HW 3rd, Barlow CE, Blair SN. Improvements in cardiorespiratory fitness attenuate age-related weight gain in healthy men and women: the Aerobics Center Longitudinal Study. Int J Obes Relat Metab Disord 1998 Jan; 22(1):55–62.

See also BIOMECHANICS OF EXERCISE FITNESS; BLOOD PRESSURE MEASUREMENT; ELECTROCARDIOGRAPHY, COMPUTERS IN.

EXTRACORPOREAL SHOCK WAVE LITHOTRIPSY.   See LITHOTRIPSY.

EYE MOVEMENT, MEASUREMENT TECHNIQUES FOR

JOSHUA BORAH
Applied Science Laboratories
Bedford, Massachusetts

INTRODUCTION

The terms eye movement measurement, eye tracking, and oculogragphy refer to measurement of the orientation and motion of the eye, either with respect to the head, or with respect to the visual environment. This may include not only rotations of the eye that cause changes in gaze direction, but also rotations of the eyeball about the line of sight, called ocular torsion. Point-of-gaze is the point in the visual environment whose image forms on the small, high acuity area of the retina, called the fovea. Line-of-gaze is the imaginary line connecting the eye to the point-of-gaze. Sometimes the term gaze tracker is used to describe a system whose primary function is to determine a subject's fixation point or line of gaze with respect to the visual environment, rather than the dynamics of eyeball motion with respect to the head.

Eye movement measurement devices have long been used for research in reading, various aspects of visual perception and cognition, neurology, instrument panel layout, and advertising. Technological advances, especially in the areas of digital processing and solid-state sensor technology, have made eye tracking possible under progressively less and less restrictive conditions. In recent years, uses have expanded to include computer application usability research, communication devices for the disabled, sports and gait research, Lasik surgery instrumentation, and research requiring simultaneous fMRI (functional magnetic resonance imaging) measurement. In the past decade it has also become practical to measure ocular torsion with optical, noncontacting methods.

Figure 1 shows some of the structures and dimensions of the eye that are important in eye movement measurement (1,2). In an idealized model, the optical axis of the eye is the line that passes through the centers of curvature of the cornea, and lens, and the center of rotation of the eyeball. The visual axis (or line of sight) is the ray that passes from the fovea, through the nodal points of the lens and inter-

sects the point-of-gaze. It is important to note that the fovea is not centered on the retina, but rather is located 5–7° toward the temporal side. The visual and optical axes are therefore, not identical. An idealized model of the eye usually assumes the pupil and iris to be centered on the optical axis, and assumes that the eyeball and eye socket operate as a perfect ball and socket joint, with the eyeball rotating about a single point within the eye socket.

The idealized model is often perfectly adequate for making good measurements of gaze direction and eye movement dynamics, but is not precisely accurate. For example, the eye does not rotate about a single center of rotation within the eye socket (3). The pupil is not precisely centered with respect to the optical axis, visual axis, or iris, and its center moves as the iris opens and closes (4).

Eye position with respect to the head can be described by a three element rotation vector, by a four element rotation specification called a quarternion, or by a set of three angles that describe the positions of an imaginary set of nested gimbals, with the outer gimbal fastened to the head, and the eye attached to the inner most gimbal. In the latter case, the three angles are usually referred to as Fick or Euler angles, and consist of an azimuth (or horizontal) angle, an elevation (or vertical) angle, and a roll (or torsion) angle. In all cases rotations are measured from a somewhat arbitrary reference position that loosely corresponds to looking straight ahead when the head is upright. A complete description of the methods and underlying mathematics for specifying eye rotation is available in an article by Haslwanter (5).

Gaze tracking devices usually report point-of-gaze in terms of a coordinate system defined on a surface in the

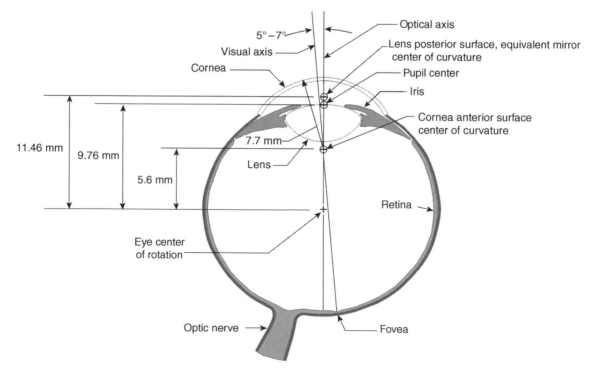

Figure 1. Schematic diagram of the eye showing typical values of dimensions that are important in eye movement measurement. The dimension values, which do vary between individuals, are derived from Refs. 1 and 2.

environment; or an eye location in space plus a gaze direction vector, with respect to an environment coordinate frame.

Normal human eye movements fall into the broad categories of conjugate and nonconjugate movements. Conjugate movements, in which both eyes move together, include the rapid, ballistic jumps between fixation points, called saccades; smooth compensatory movements to hold the gaze steady in the presence of head motion; smooth movements to track objects that are moving across the visual field; and the saw tooth pattern of movement called nystagmus that occurs in response to an inertial rotation of the body or a rotation of the visual field. There are also miniature motions during the relatively stationary fixation periods, which are $< 1°$ and are not perceived. Vergence is a nonconjugate motion used to keep a visual target at the same position on both retinas. As a visual target moves closer, the visual axes of the two eyes rotate toward each other. Ocular torsion occurs in response to inertial rotation or lateral acceleration, or rotation of the visual field about a horizontal axis. It is associated with perceptions of tilt. A thorough review of eye movement behavior can be found in Hallett (6).

The eye movement measurement techniques currently in most frequent use fall into the major categories of magnetic search coil, a technique that measures magnetically induced current in a tiny wire coil fastened to the eye; electrooculography, which uses surface electrodes to measure the direction of an electrical potential between the cornea and retina; and optical techniques that rely on optical sensors to detect the position or motion of features on the eye. Their optical technique category includes many subcategories, and has the largest variety of different systems in current use. Background theory and system descriptions for eye movement measurement devices, in all of these categories, are presented in the following sections.

Eye tracker performance is usually described by some subset of the following parameters. *Accuracy* is the expected difference between the measured value and the true value. *Resolution* is the smallest change that can be reported by the device. *Precision* is the expected difference in repeated measurements of the same true value. *Range* describes the span of values that can be measured by the device. *Linearity* is the degree to which a given change in the real quantity results in a proportional change in the measured value, usually expressed as percent of the measurement range. *Update rate* is the frequency with which data is output (samples per second). *Bandwidth* is the range of sinusoidal input frequencies that can be measured without significant distortion or attenuation. *Transport delay* is the time required for data to pass through the system and become available for use.

SCLERAL SEARCH COIL

The scleral search coil technique, first described by Robinson (7), requires that a sensing element be placed on the eye. The technique is based on the principle that a changing electric field can induce a current in a coil of wire. If the coil lies in a plane parallel to a uniform, alternating current (ac) magnetic field, no current is induced. If the plane of the coil is not parallel to the field lines, an ac current *will* be induced in the coil. Current amplitude will be proportional to the coil area projected onto the plane that is perpendicular to the magnetic field lines. For example, if the plane of the coil is tilted about an axis perpendicular to the magnetic field lines, the induced current will be proportional to the sine of the tilt angle. A tilt in the opposite direction results in an induced current with the opposite phase (180° phase shift). The sign of the tilt angle can, therefore, be deduced from the phase of the induced current.

As shown in Fig. 2, a pair of Helmholz coils, which set up uniform ac magnetic fields in both vertical and horizontal axes, surrounds the subject's head. The driving circuitry ensures that the two fields are exactly 90° out of phase. An induction coil, made of very fine wire, is held on the eye so

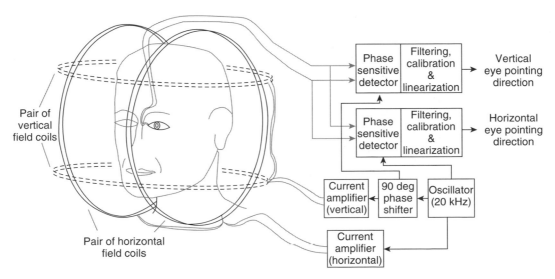

Figure 2. Schematic illustrating the scleral search coil method of measuring eye movement. (Adapted from a diagram in Ref. 8.)

that it forms a circle about the limbus (iris–sclera boundary). Robinson embedded the coil in a scleral contact lens, which was held to the limbus by the action of a tiny suction tube. Collewijn et al. (9) developed a technique for using an annular ring made of silicone rubber and having a slightly hollow inner surface. The ring adheres to the limbic area by capillary action and does not require the suction tube. Fine wire leads, from the induction coil, extend out of the eye at the canthus (the corner of the eye). A drop of anesthetic is generally administered prior to insertion, but the devise is usually tolerated well once the anesthetic wears off (9).

The induction coil encloses an area that is approximately in the plane of the pupil. Horizontal eye movement varies the current induced by the horizontal ac magnetic field, and vertical motion varies the current induced by the vertical field. By detecting the phase, as well as the amplitude of the induced current, it is possible to obtain separate analog signals proportional to the sine of vertical and horizontal eye rotations. A simple calibration is required to find the initial reference orientation of the eye.

It is possible to embed, in the annular ring, a second induction coil that encloses an area having a component parallel to the optical axis of the eye. The second coil is shown in Fig. 3. Torsional eye movement can be computed from the current induced in this second coil.

When used with nonhuman primates, a magnetic induction coil is often implanted surgically, using a method described by Judge et al. (10).

Scleral search coil systems can be expected to measure with a resolution > 1 arc·min over a range of ±15–20°, and accuracy of ~ 1–2% of the range. Slippage of the annular ring on the eyeball is possible, and can produces significant additional error. Temporal bandwidth is a function of the coil excitation frequency and filtering, and depends on the specific implementation, but 0–200 Hz or better is probably achievable. The accuracy of the torsional measurement is generally assumed to be a fraction of a degree, but may be affected by slippage of the annulus as well as variation in eyeball curvature, and is not well documented.

The method is distinctly invasive, and requires the head to be confined within the Helmholz coil assembly, but does not require the head to be rigidly fixed. It offers good measurement performance, measures all three axes of rotation simultaneously, and is not affected by eyelid closure, or ambient illumination. In the past, complete systems for use with humans have been commercially available from C-N-C Engineering, Seattle, WA; and Skalar Medical BV (11), The Netherlands. Current commercial availability is uncertain. Systems for use with animals are available from Riverbend Instruments, Inc., Birmingham, AL (12).

ELECTRO-OCULOGRAPHY

Electro-oculography (EOG) has a relatively long history, dating from the 1920s and 1930s (13–17). The retina of the eye carries a slightly negative electrical charge, varying from ~ 0.4 to 1.0 mV, with respect to the cornea, probably because the retina has a higher metabolic rate. This charge difference constitutes an electrical dipole, which is approximately, although not exactly, aligned with the optical axis of the eye. Electro oculography refers to the use of surface skin electrodes to measure the position of the cornea-retinal dipole. When used with ac recording techniques to measure nystagmus, or when used in a neurological test setting to measure any type of eye movement, it is often called electronystagmography (ENG).

Ideally, when the electrical dipole is midway between two electrodes that have been placed near the eye, the differential voltage between the electrodes would be zero, and as the eye rotates from this position, the differential voltage would increase with the sine of the angle. Although this is indeed the qualitative result, in practice there is a great deal of direct current (dc) drift. Skin conductance varies over time, and the corneo-retinal potential changes with light adaptation, alertness, and the diurnal cycle. In fact, EOG is sometimes used explicitly to measure the changes in corneo-retinal potential as a function of light stimuli, rather than eye movement (18,19). Electromyographic activity from facial muscles can also interfere with EOG measurement.

After cleaning the skin with an alcohol swab, electrodes are often placed as shown in Fig. 4. In this case, the differential voltage between the electrodes placed near the outer canthi (junction of upper and lower eyelid) of each eye is used to measure horizontal motion of the two eyes together. It is also possible to use additional electrodes near the nose, or on the bridge of the nose, to measure the two horizontal eye positions independently. The vertically positioned electrode pairs measure both eyes together when wired as shown in the diagram, but can also be used to measure vertical position of each eye, independently. The electrode at the center of the forehead is used as a reference. Other placement patterns can be used as well.

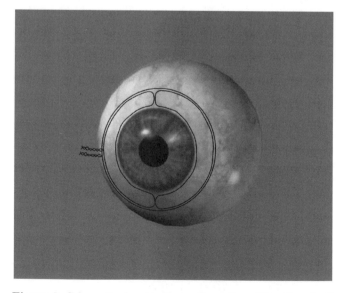

Figure 3. Schematic showing the configuration of the second coil in a dual induction coil system. The second coil forms a loop whose enclosed area has a component parallel to the optical axis of the eye (perpendicular to the plane of the pupil), and is used to measure torsional movement. Although the diagram shows only one winding for each coil, the actual device uses multiple windings.

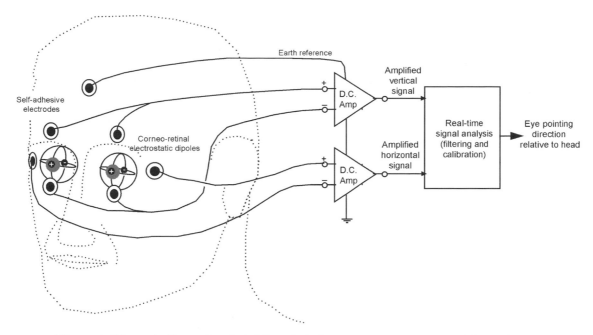

Figure 4. Schematic illustrating the EOG method of measuring eye movement. (Adapted from a diagram in Ref. 8.)

The electrodes most often used are silver–silver chloride, self-adhesive models, designed as infant heart monitor electrodes. They are connected to high gain, low impedance, low noise, differential amplifiers. The output from the differential amplifiers is now most commonly digitized and input to digital processors for linearization, scaling, and other processing. The ac coupling or frequent rezeroing is required to keep the analog signal in range of the analog to digital converters.

The high and unpredictable rate of dc drift makes this technique less suitable than others for point of gaze measurement. However, the drift is usually slower than eye movement velocities, and EOG provides an excellent means for measuring eye velocity profiles, slow and fast phases of nystagmus, and patterns of fixations and saccades, as long as the precise point of regard is unimportant. Research laboratories often assemble their own EOG devices from commercially available amplifiers, electrodes, and digital data processing software packages. Both EOG and ENG devices are sometimes used for neurological testing in clinical settings, and commercially available EOG or ENG devices are most often packaged as part of neurological testing suites.

The EOG and ENG systems are commercially available from: Cambridge Research Systems Ltd., UK (20); GN Otometrics, Denmark (21); Guymark UK Ltd., UK (22); Metrovision, Pérenchies, France (23), and Neuro Kinetics, Inc., Pittsburgh, PA (24).

OPTICAL TECHNIQUES

Noncontacting optical sensors can be used to deduce the orientation of the eyeball from the position of optical features, optically detectable geometry, or the pattern of reflectivity on the eye and the facial area surrounding the eye. The sensors may range from a small number of individual photodiodes to CCD or CMOS arrays, which provide two-dimensional (2D) gray scale image data. In most cases the eye area is illuminated with a near infrared (IR) light that is within the sensitive spectral region for solid-state light sensors, but minimally visible to the human eye. Optics may be mounted to head gear and move with the head, the head may be restrained to prevent or limit motion with respect to optics that are not head mounted, or movement with respect to non-head-mounted optics may be allowed. In the latter case the sensor field of view must either be large enough to accommodate the expected head movement, or some component of the optical assembly must automatically move to keep the sensor aimed at the eye being measured.

OPTICAL SENSORS

Sensors used by optical eye trackers include the following: *Quadrant and bicell photodetectors*: The disk shaped detector surface is divided into two (bicell) or four (quadrant) discrete photosensitive areas. The devices are configured to produce analog signals (one for bicell detectors, and two for quadrant detectors) proportional to the difference in light intensity sensed on adjacent areas. The signals are monotonically related to small displacements of a light spot from the center of the detector either in one (bicell) or two (quadrant) dimensions. The light spot must remain completely on the detector surface, and displacements must be smaller than the diameter of the spot. *Lateral effect photo diodes (position sensitive detectors)*: A solid-state detector provides analog information proportional to the one (1D) or two dimensional location of the incident light center of gravity. *Small arrays of discreet solid state photosensors*: The array provides a small number of analog light intensity signals. *Large, linear, photo-sensor arrays*: The array provides gray scale image data in a single dimension. *Large, 2D, solid-state, photosensor arrays (CCD and CMOS)*: The array provides two dimensional gray scale image data. Commercially available video cameras, based on CCD and CMOS sensors provide analog and digital signals in standard formats, usually at 50 or 60 fields/ second. Using a CCD chip that supports double the normal pixel output rate, a small number of cameras are available that output 120 fields/second. By using a subset of the video lines for each field, these devices can also deliver field update rates that are higher than 120 Hz. Some CMOS sensor chips allow even more flexibility to receive data from a dynamically determined subset of the pixels and to vary the update rate. Higher update rates always mean that each pixel has less time to accumulate charge, resulting in less effective sensitivity and lower signal to noise ratios.

Quadrant detectors, lateral effect photo diodes, and small arrays of discrete photo sensors provide information with low spatial bandwidth content. The information can usually be processed at high temporal bandwidth with relatively little digital processing requirement. Large linear and 2D arrays offer much richer spatial information, but require more processing power to interpret the information, often leading to reduced temporal bandwidth.

FEATURES OF THE EYE

The eye image features most often used for eye movement measurement are *Limbus:* Boundary between the colored iris and white sclera. *Iris*: "Colored" ring that opens and closes to adjust pupil size. *Pupil*: Circular opening defined by inner boundary of iris. *First Purkinje image (corneal reflection)*: Reflection of a light source from the outer surface of the cornea. *Fourth Purkinje image*: Reflection of a light source from the inner surface of the lens.

These features are shown schematically in Fig. 5.

The pattern of blood vessels on the retina, if imaged with an opthalmoscope or fundus camera, also constitute markings that can be used to track eye movement, but this is a less commonly used technique.

In some cases facial landmarks, such as the canthus (corner of the eye, where the upper and lower eyelid meet), another facial feature, or a dot placed on the skin, may be used to compare with the location of features on the eye. As discussed later on, facial features can also be used to guide remote camera based trackers.

The iris has a distinctive pattern of radial markings that can be used to track ocular torsion. Its inner boundary, defining the pupil, is nominally circular and centered with respect to the limbus. Detailed examination, however, reveals a slightly noncircular shape that is off center (usually toward the nasal side) with respect to the limbus. Furthermore, both the shape and position of the pupil change slightly with pupil diameter and vary across the population. The characteristics of the pupil form are described and quantified in considerable detail by Wyatt (4). He found that pupil position tends to shift in the nasal and superior directions (with respect to the limbus) as the pupil contracts, and that pupil diameter and circularity tend to decrease with age.

Light rays that enter the eye through the pupil are reflected by the retina and directed back toward their

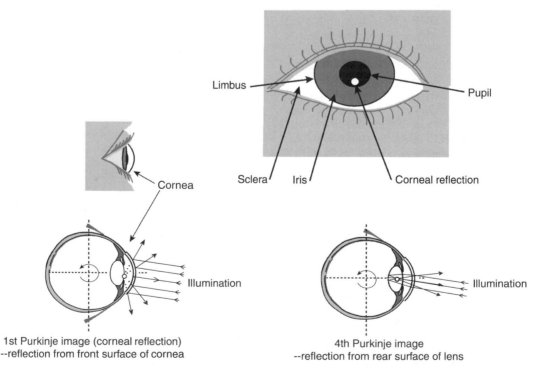

Figure 5. Schematic illustrating the various features of the eye often used by optical eye movement measurement techniques. (Adapted from a diagram in Ref. 8.)

"Bright pupil" optics

"Dark pupil" optics

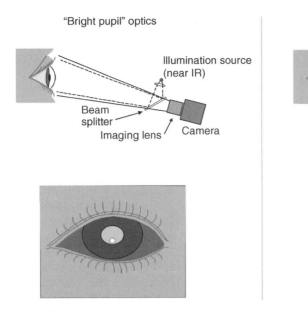
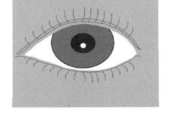

Figure 6. Illustration of bright and dark pupil optics. In the bright pupil example, retroreflected light from the retina will beam back into the camera lens, resulting in a bright, back lit, pupil image.

source. The eye therefore acts as a retroreflector. If the eye is viewed by a detector that is coaxial with an illumination beam, as shown in Fig. 6, the retroreflected light from the retina makes the pupil appear to be a bright, back lit circle. Some of the retinal reflection can be received slightly off axis from the illumination beam (25). Although the bright pupil effect falls off sharply as the detector moves off axis, this accounts for the "red eye" effect in flash photography. In an idealized case, the apparent brightness of the pupil retroreflection will vary inversely with the square of pupil diameter (26). As shown by Nguyen et al. (25), brightness of the retinal reflection also varies between individuals and as a function of gaze direction with respect to the illumination source, but these variations are small compared to the effect of pupil diameter.

If the detector is off axis from the illumination beam, the retroreflected light does not enter the detector and the pupil appears as the familiar dark circle.

The corneal reflection is a virtual image that appears to be just behind the plane of the pupil. If the cornea is assumed to be spherical, the corneal reflection image will form at point half way between the surface of the cornea and its center of curvature, along a ray that is parallel to the illumination beam and passing through the corneal center of curvature. If the light source is far away compared to the radius of the eyeball, then as the eye rotates, the corneal reflection always appears to move the same amount as the corneal center of curvature. A similar analysis will show that the fourth Purkinje image forms in almost the same plane as the corneal reflection, but appears to move the same amount as the posterior lens surface center of curvature.

FEATURE RECOGNITION

Systems that use 2D detector arrays typically perform a pattern recognition task to identify features in the 2D

image. Digital processing power has always been a critical limitation. When video cameras were first used as sensors, analog preprocessing was often used to find edge points, or other pixel subgroups, and thereby reduce the amount of data that needed to be processed digitally. Algorithms requiring relatively few computational steps were used to process the reduced digital information in order to recognize features and find their centers.

Increased digital processing capability has very significantly eased, although by no means eliminated, this limitation. It is now practical to digitally process 2D, gray scale image buffers while maintaining reasonable, real-time update rates. It has become possible to use, in real-time, elements of classical digital image processing such as Sobel or other convolution based edge detection algorithms, and circle or ellipse best-fit algorithms. Less computationally intensive algorithms are still often used, however, to maximize update rates.

Digital processing components in current use range from commercially available PCs and frame grabbers to custom processing boards that include field programmable gate arrays (FPGAs) and digital signal processors (DSPs), and microcontrollers.

When using an array sensor, the location of individual boundary points on an image feature (e. g., the pupil) are often identified with only single pixel resolution. If the object covers multiple pixels, knowledge of the object shape allows its position to be computed with subpixel resolution. For example, if a group of pixels are thought to define the edge of circular object, a least mean squared error circle fit will define the circle center location with sub pixel resolution. If sufficient computation time is available, it is also possible to use gray scale information to define edge points with subpixel accuracy.

Pupil Recognition

The pupil is generally recognized as a circular or elliptical area that is darker (in the case of a dark pupil image) or brighter (in the case of a bright pupil image) than surrounding features. The pupil is often partially occluded by eyelids and corneal reflections, and algorithms must therefore recognize the circular or elliptical shape even when occluded to some degree. Furthermore, it is important that only the real pupil boundaries be used to determine the center of the object, rather than the occlusion boundaries. Examples of pupil detection algorithms can be found in Zhu (27), Mulligan (28), Ohno et al. (29), Charlier et al. (30), and Sheena (31).

The retroreflective property of the eye makes possible a signal enhancement technique that can be exploited to help identify the pupil (32–35). If a camera is equipped with both a coaxial and off-axis illuminator, a bright pupil image will be produced when only the coaxial illuminator is on, and a dark pupil image will be produced when only the off-axis source is on. If the illuminators are alternately activated for sequential camera images, the result will be alternating bright and dark pupil images. Assuming elements in the camera field of view have not moved between images, all other image features, which are not retroreflectors, will remain essentially unchanged.

If two such images are subtracted, one from the other, the result should leave only the pupil image. In practice, there will still be small differences in all parts of the image due to the different illumination angles, but contrast between the retroreflective pupil and the rest of the image is still greatly enhanced. There are some drawbacks to the technique. If the eye moves significantly between the sequential images, subtraction results in a distorted pupil image. The need to digitize and subtract two images increases memory and processing requirements, and the fact that two sequential images are required to create one data sample limits the temporal bandwidth of the measurement. Note that a similar result can be obtained with images from two cameras that are carefully aligned with respect to the same image plane, and an illumination source that is coaxial with only one of them.

Corneal Reflection Recognition

The corneal reflection (first Purkinje image) is usually recognized, within a larger 2D image, by its intense brightness, predictable size and shape, and proximity to the pupil. It can, however, be confused with small reflections from tear ducts or eyeglass frames, and reflections from external sources emitting light in same spectral band as the intended source. Its center can be identified with subpixel accuracy only if it is large enough to cover multiple pixels. Eizenman et al. (36) describe a technique for using knowledge of brightness pattern across the corneal reflection (first Purkinje image) to find its position, on a linear array, with subpixel resolution. To accomplish this they used a precisely designed illumination source to insure a known pattern of luminance.

Fourth Purkinje Image Recognition

The fourth Purkinje image is very dim and very difficult to reliably identify in a larger image. The one system in common use that requires the fourth Purkinje image relies on careful initial alignment to focus the image on a quadrant detector. The detector is not much larger than the Purkinje image; and, in this case, automatic recognition in a larger field is not necessary.

Face Recognition

In some cases wide-angle images are now used to recognize the presence of a face, and to find the location of one or both eyes. If the eye is identified in a wide angle image, and assuming the camera is well calibrated to account for lens or sensor distortions, it is reasonably straight forward to compute the direction of the line extending from the camera to the eye. Finding the distance to the eye is more difficult. If the eye is identified on the images from two cameras, it is possible to triangulate. If both eyes are identified on a single camera image, knowledge of the true interpupillary distance can be used to compute distance to the face. However, head rotation with respect to the camera will cause some error if not taken into account. Some face recognition systems are able to determine head orientation. For example, Xiao et al. (37) and Matthew and Baker (38) describe a method, based on a technique known as active appearance modeling (39), to make real-time measurements of the position and 3D orientation of a face.

Information about eye location can be used to direct a separate sensor to obtain a more magnified view of the eye, or the wide-angle image itself may also be used to find the position of features within the eye. In the latter case, there is a clear trade off between spatial resolution and wide-angle coverage. Recognition of facial features can also be exploited in order to use a facial landmark, such as the canthus or the center of the eyeball as one of the elements in a dual feature-tracking algorithm (40,41). However, the plasticity of facial features makes it difficult to determine their position with the same precision as the pupil and Purkinjie images.

EYE ORIENTATION AS A FUNCTION OF SINGLE OR DUAL FEATURE POSITION

If we have a sphere of radius r, with a mark on its outer surface as shown in Fig. 7, and if we assume the center of the sphere is fixed with respect to an observer, then the observer can compute the rotation (θ) of the sphere about its center by noting the displacement (d) of the surface mark.

$$\theta = \arcsin(d/r)$$

This is the principle behind single feature eye tracking. However, if the center of the sphere moves with respect to the observer, observation of a single mark provides no way to distinguish such motion from rotation.

If there are two visible marks on the sphere, which are fixed to the sphere at different distances from its center, then observing the relative position of these marks does allow rotation of the sphere to be distinguished from translations. So long as the distance between the sphere and observer remains the same or is independently known, translation with respect to the observer can be unambiguously distinguished from rotation.

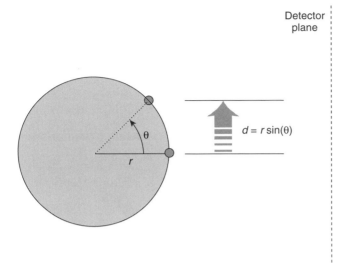

Figure 7. Schematic illustrating the principle behind single feature tracking. Observation of a single land mark on the sphere allows determination of its orientation, so long as the center of the sphere does not move.

If the two marks are located along the same radius line, at distances r_1 and r_2 from the center of the sphere, as shown in Fig. 8, then the rotation angle (θ) of this radius line, with respect to the line connecting the sphere and observer, is

$$\theta = \arcsin(\,\Delta d/(r_1 r_2))$$

where Δd is the observed separation between the marks. This is the basic principle behind dual feature tracking techniques. The underlying sine function has a steep slope at small angles, maximizing the sensitivity of the technique. Note that if the distance between the observer and sphere changes, the dual feature relation still leaves some

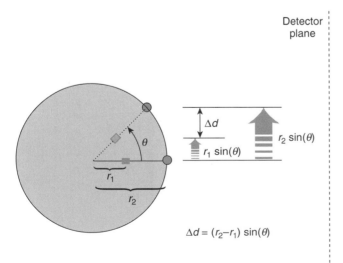

Figure 8. Schematic illustrating the principle behind dual feature tracking. The relative position of the two landmarks defines the orientation of the sphere, even if its center moves in a direction parallel to the plane of the detector.

ambiguity since the apparent separation between the marks will change as a function of distance.

In the case of the eye, referring to Fig. 1, the center of the pupil or center of the iris is a mark that is $\sim 9.8\,\text{mm}$ from the center of the eyeball. Although both images appear to be just behind the plane of the pupil, the first Purkinje image moves the same amount, with eye rotation, as would a mark $\sim 5.6\,\text{mm}$ from the center of the eyeball (the position of the anterior corneal surface center of curvature); and the fourth Purkinje image moves the same amount as would a mark $\sim 11.5\,\text{mm}$ from the center of the eyeball (the position of the posterior lens surface center of curvature).

Eye Orientation as a Function of Just Pupil or Corneal Reflection Position

The pupil, and corneal reflection (first Purkinje image) are the features most commonly used for single-feature tracking. If a sensor and light source are mounted so that they do not move with respect to a person's head, either of these features can be used as a marker to compute eye rotation in the eye socket. Either the head must be stabilized with some type of head restraint mechanism, or the sensor must be mounted to head gear that moves with the subject.

When measuring only a single feature, any translation of the head with respect to the sensor will be erroneously interpreted as eye rotation. For example, if using pupil position, a 1 mm motion of the head parallel to the detector image plane may be mistaken for about a $5°$ eye rotation. If the corneal reflection is the feature being tracked, a 1 mm slippage will be indistinguishable from an $\sim 12°$ eye rotation.

Eye Orientation as a Function of Relative Pupil and Corneal Reflection Positions (CR/Pupil)

The pupil and first Purkinje image (corneal reflection), or the first and fourth Purkinje image are the most commonly used feature pairs for dual feature tracking. The pupil to corneal reflection technique (CR/Pupil) was first described by Merchant et al. (42). As shown in Fig. 9, if the sensor is close to the light source that produces the corneal reflection, the angle (θ) of the eye optical axis with respect to the sensor is described by

$$\theta = \arcsin(\,d/k)$$

where d is the apparent distance between the pupil and corneal reflection (from the point of view of the sensor), and k is the distance from the pupil to the corneal center of curvature. If the sensor is not close to the illumination source, the relation changes only slightly so long as the sensor does not move with respect to the illumination source.

$$\theta = \arcsin((d - k_d)/k)$$
$$k_d = k_{cr}\sin(\gamma)$$

where k_{cr} is half the cornea radius of curvature and γ is the angle between the illumination beam and the sensor line of sight (see Fig. 10). If the sensor and illumination source are very far away, compared to the radius of the eyeball, then γ,

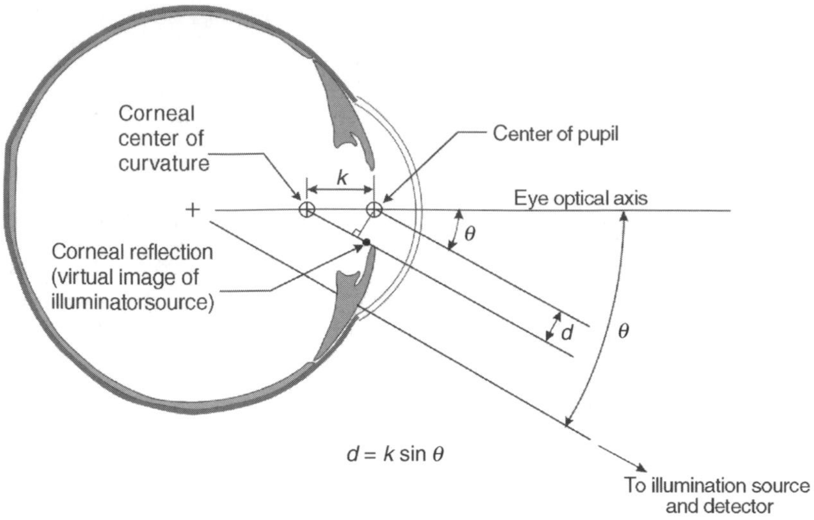

Figure 9. Schematic illustrating the basic relationship behind the pupil-to-corneal-reflection technique for measurement eye movement. The diagram assumes that the detector and illumination source are coaxial, or very close together, and are very far from the eye, compared to the radius of the eyeball. The optical axis of the eye is rotated away from the detector by an angle θ. From the vantage point of the detector, the pupil and corneal reflection appear to be separated by distance d.

and hence k_d, are constants. One drawback to the pupil to corneal reflection technique is that the pupil center is not completely stable with respect to the optical or visual axis, but moves slightly as the pupil size changes (4). Theoretically, this effect can be measured in an individual and accounted for (43,44), but would add to the time and effort required to calibrate an individual. Note also that the pupil-to-corneal-reflection vector is less sensitive to eye rotation than either of the individual features. A 5° eye rotation, from an initial position in which the pupil and corneal reflections are aligned, causes the pupil and corneal reflection images to separate by only $\sim 0.4\,\mathrm{mm}$, whereas the pupil center moves $\sim 1\,\mathrm{mm}$.

The equations given above describe the major effect, but not all secondary effects, and are not precise. For example, the cornea is not perfectly spherical, the eyeball does not rotate about a perfectly stable central point, and the pupil image is slightly magnified by the refractive power of the cornea. To the extent that secondary effects are large enough to be detected, they are often accounted for by the results of an empirical calibration procedure, as discussed later.

The range of gaze angles that can be measured by the pupil to corneal reflection technique is limited by the range over which the corneal reflection remains visible to the detector. In the horizontal axis, this range is usually $\sim 70°$ visual angle for a given illumination source and sensor pair. It is usually less in the vertical axis due to occlusion of either the pupil or corneal reflection by the eyelids. The range can be extended by using multiple illumination sources, at different positions, to create multiple corneal reflections, but the system must be able to uniquely recognize each reflection even when not all are visible.

Eye Orientation as a Function of Relative First and Fourth Purkinje Image Positions (CR/4PI)

As described by Cornsweet and Crane (45), the first and fourth Purkinje images can also be used for dual feature tracking. The same type of arcsine relation applies, but with d the apparent distance between the two Purkinje images, and with k equal to the distance between the corneal center of curvature and the posterior lens surface center of curvature. The technique has the advantage that the Purkinje image positions can be more precisely defined than the pupil center. In addition, the separation of the posterior lens surface and corneal centers of curvature ($\sim 6\,\mathrm{mm}$) is greater than that between the pupil and corneal centers of curvature ($\sim 4.2\,\mathrm{mm}$), yielding greater

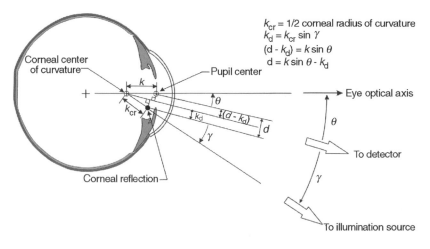

Figure 10. Schematic illustrating the detected pupil to corneal reflection separation (d) when the detector and illuminator optical paths are not coaxial. The diagram assumes that the detector and illumination source are far from the eye, so that lines from the detector to various points on the eye are essentially parallel, as are rays from the illumination source to various points on the eye. The angle θ is between the eye and detector optical axes, and γ is the angle between the detector optical axis and the illumination beam.

sensitivity to eye rotation than the pupil to corneal reflection technique. Drawbacks are that the fourth Purkinje image is relatively dim and difficult to find, and is only visible within the iris opening.

Eye Orientation as a Function of Relative Pupil or Iris and Facial Landmark Positions

The relative position of the pupil or iris and a facial landmark can also be used to measure eye orientation. This is sometimes done when image magnification and resolution is such that the smaller Purkinje images cannot be reliably detected, or the relative motion of the pupil and corneal reflection cannot be sufficiently resolved. Since the facial landmark does not move with eye rotation, the governing relationship is described by

$$\theta = \arcsin(\ (d-d_i)/r)$$

where d is the distance (in the camera image plane) from the facial landmark to the pupil or iris center, d_i is the distance (also in the camera image plane) from the facial landmark to the eye center of rotation, r is the distance from the pupil or iris center to the eyeball center of rotation, and θ is the angle of the eye optical axis with respect to the detector. If the head rotates with respect to the detector, d_i will appear to shorten, in the direction perpendicular to the rotation axis, by an amount proportional to the cosine of the rotation angle. Advantages over other dual feature techniques are greater sensitivity (larger change of the measured quantity for a given eye rotation), and the possibility of identifying the features in wider, less magnified images. Disadvantages are that facial image features are not stable, but usually move at least slightly as a function of facial muscle activity, and head orientation must also be measured or must remain stable with respect to the detector. Zhu and Yang (40) describe a dual feature technique using the relative position of the canthus and iris; and Tomono et al. (41) describe an algorithm for using the relative position of the pupil, and a computed position of the center of the eyeball.

EYE ORIENTATION AS A FUNCTION OF FEATURE SHAPE

It is also possible to extract eye orientation information from feature shape. A circle (e. g., the pupil outline or outer boundary of the iris) appears elliptical if viewed from an angle. As the circle tilts, the minor axis of the ellipse, which is perpendicular to the axis of rotation, appears shortened by the cosine of the rotation angle. A major drawback to using pupil or iris ellipticity as an orientation measure is that, due to symmetry considerations, the direction of rotation about the rotation axis remains ambiguous. A second major limitation is that the underlying cosine function has a shallow slope at small angles resulting in poor sensitivity.

EYE ORIENTATION AS A FUNCTION OF REFLECTIVITY PATTERN MOVEMENT

Different structures on the eye, primarily the pupil, iris, and sclera have different reflectivity properties. As the eye rotates this reflectivity pattern moves, and that property can be exploited to measure eye movement. Movement of the reflectivity pattern can be detected with small numbers of individual photodetectors, and this, in turn, makes it possible to achieve relatively high temporal bandwidth. The technique was pioneered by Torok et al. (46) and Smith and Warter (47), using a photomultiplier as the detector, and further developed by Stark and Sandberg (48), Wheeless et al. (49), Young (50), Findlay (51) and Reulen et al. (52). The most prominent contrast feature in the pattern is generally produced by the boundary between the iris and sclera. Therefore, devices that use small arrays of photodetectors to measure motion of this pattern are often called limbus trackers.

The iris sclera boundary is easily visible along the horizontal axis, but along the vertical axis it is usually obscured by the eyelids. In fact the boundary between the eyelids and iris are often the most prominent reflectivity boundaries along the vertical axis. The eyelids do tend to move in proportion to vertical eye motion, and are useful as measures of vertical eye position; but motion of the reflectivity pattern remains a much less dependable function of vertical (as opposed to horizontal) eye rotation.

In principle, reflectivity pattern tracking is similar to single feature tracking. As with single feature tracking, any movement of the sensors with respect to the eye produces erroneous measurements.

MEASURING POINT-OF-GAZE IN THE PRESENCE OF HEAD MOTION

Dual feature techniques, such as the pupil to corneal reflection method, permit computation of gaze direction with respect to a detector. However, head motion may still need to be measured in order to accurately determine the point of gaze on other objects in the environment.

First, consider an example in which the head is free to move with respect to a stationary detector, and the task is to measure point of gaze on other stationary objects in the environment. This is illustrated by the two dimensional example in Fig. 11. The point of gaze, defined by x, is dependent not only on θ, which can be measured by one of the dual feature tracking methods previously described, but also on ϕ and d_1, which define head position. If head motion is small compared to distance to the detector and scene surface, then changes in head position will have little effect and can be ignored.

If the location of the eye in the environment space can be independently measured, and if the detector and scene surface positions are known with respect to the same environment space, the following general algorithm can be used to determine point of gaze on a surface. Use a dual feature technique to determine direction of the gaze vector with respect to the detector, and knowledge of the detector orientation to express this as a direction in the environment space. Use the gaze direction and known start point (the location of the eye in space) to write the parametric equation for a line in the environment coordinate space. Use knowledge of the scene surface position to solve for the intersection of a line and a plane. Ohno et al. (29) describe a version of this strategy.

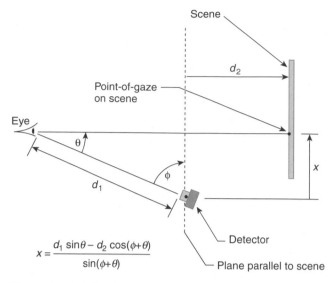

$$x = \frac{d_1 \sin\theta - d_2 \cos(\phi+\theta)}{\sin(\phi+\theta)}$$

Figure 11. Relationship, in one plane, between point-of-gaze on a flat scene and relative eye, detector, and scene positions. (From Ref. 26.)

Next, consider an example in which the detector is fastened to headgear worn by the subject. Single as well as dual feature techniques may be adequate to measure gaze direction with respect to the head (although single feature methods will have larger errors if there is any slippage of the headgear). In order to find point-of-gaze on objects in the environment it is clearly necessary to know the position and orientation of the head with respect to those objects. The following strategies can be used.

1. A second camera can be mounted to the head gear so that it shares the same reference frame as the eye sensor, but points toward the subject's field of view. Point-of-gaze can be indicated as a cursor superimposed on the image from this scene camera.
2. Light-emitting diodes (LEDs) or other special emitters can be fastened to an object in the environment, such as the bezel of a computer monitor, and detected by head mounted sensors to locate the object in the head reference frame. Measurement of gaze in the

head reference frame can then be related to position on that object.

3. A separate head tracking system can be used to measure head position and orientation with respect to the environment. This information can then be used to compute the location and direction of the gaze vector in the environment coordinate frame. If the locations of surfaces are also known in the same environment reference frame, it is possible to solve for the intersection of a line (line-of-gaze) with a surface, to find point-of-gaze on various surfaces in the environment. A general method for doing this computation is described in Appendix E of Leger et al. (8). Duchowski (53) also describes specific algorithms for handling this type of task.

In the case of the head mounted scene camera described above, it is important to be aware of possible parallax error. The scene camera is usually not viewing the scene from exactly the same vantage point as the eye being tracked. As shown in Fig. 12, eye rotation angle data can be mapped to the scene camera image plane at a particular image plane distance, but the relation changes as the image plane distance changes. The resulting parallax error can be easily corrected if there is knowledge of the distance, from he subject to the gaze point. The parallax error may be negligible if the distance to the gaze point is large compared to the distance of the scene camera from the eye. It is also possible, as shown in Fig. 13, to minimize parallax by bending the scene camera optical path with a beam splitter, such that the scene camera has the same vantage point as the eye being measured (54).

If a person with normal ocular function is fixating a point not infinitely far away, the lines of gaze from the two eyes should converge. If the gaze angles of both eyes are measured with head mounted optics, the intersection of the two lines-of-gaze theoretically indicates the three-dimensional (3D) point-of-gaze in space, with respect to a head fixed coordinate system. If head position and orientation are known, this can be transformed to a position in environment space. Duchowski (53) describes an algorithm for this computation. It should be noted that, in practice, the measured lines-of-gaze from the two eyes will almost

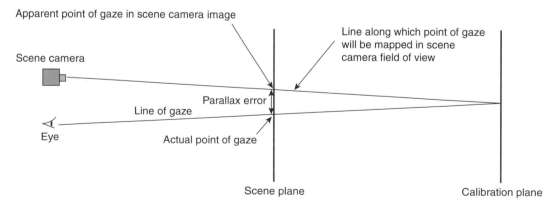

Figure 12. Parallax error when gaze direction is mapped (calibrated) to a scene camera image plane that is different from the plane being fixated.

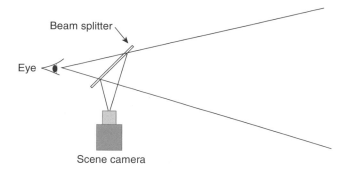

Figure 13. Use of a beam splitter to minimize scene camera parallax.

never intersect. Neither the measurement nor the system being measured is infinitely precise. The discrepancy can be resolved as follows. Consider the two planes that are parallel to the "vertical" axis of the head and which contain the gaze vector from each of the two eyes. If the two lines-of-gaze converge, these two planes will intersect along a line, which is also intersected by the line-of-gaze from each eye. The 3D point-of-gaze can be chosen to be either of these intersection points or a point half way between the two.

As the point-of-gaze moves farther from the head, the vergence angle (the angle formed by the gaze vector from each eye) diminishes. The relation is reasonably sensitive at close distances. As distances become longer, a very small change in vergence corresponds to an increasingly large change in distance, and moderate measurement noise or error in the eye tracker may result in a very noisy or inaccurate point of gaze computation.

MEASUREMENT OF TORSIONAL EYE MOVEMENT

Optical measurement of torsional eye movement was first accomplished by offline photographic analysis techniques and later by tracking artificial marks placed on the eye. For example, Edelmann (55) created a visible mark by sandwiching a human hair between two contact lenses. Use of standard surgical markers, applied just outside the limbus, has also been reported (56), although this requires that a local anesthetic be applied to the cornea.

Over the last decade or so it has become practical to make real time measurements using the patterns that are naturally visible on the iris, as captured by CCD or CMOS cameras. This has generally been done either by using a cross-correlation technique, or a template-matching scheme to compare the iris over sequential video frames. The first method, first described by Hatamian and Anderson (57), and further developed and automated by Clarke et al. (58) and Bucher et al. (59), cross-correlates the pixel sequence from a 1 pixel wide path around the pupil, sampled from each video frame, with that from an initial reference frame.

It is important that the same strip of iris be sampled each time, so unless the eye is stationary in the two nontorsional degrees of freedom, the pupil center must be accurately tracked. Even so, at eccentric eye positions, geometric image distortions can cause errors. The points on

the iris that form a circle on the camera image plane when the subject looks directly at the camera, begin to form a more elliptical shape on the camera image plane as the subject looks farther away from the camera. Moore et al. (60), and Peterka et al. (61) describe a method for avoiding this type of error by correctly computing the projection of the eye onto the camera image plane and empirically solving for parameters that correspond to physical characteristics of the eye or eye-to-camera geometry. A template-matching scheme described by Groen (62) and further developed by Zhu et al. (63) is also designed to minimize geometrical perspective errors by tracking distinctive landmarks on the iris. Changes in pupil diameter can affect the pattern of radial markings on the iris and lead to some torsion measurement error. Guillemant et al. (64) describes a technique using neural network software to identify the pupil and iral patterns for torsion measurement.

CALIBRATION

Most eye tracking systems require a practical method to relate a measured quantity, such as the relative position of the pupil and corneal reflection, to a desired quantity, such as point of gaze on a particular scene space. The underlying relationships behind several techniques for measuring eye orientation have been presented in preceding sections. In practice, however, eye tracking systems often rely on completely empirical techniques to map the measured quantity to gaze points on a scene space. The measured quantity is recorded as a subject looks at several known points in the scene space and either a polynomial curve fit, an interpolation scheme, or some combination is used to map (transform) one to the other.

The process of gathering data to compute the transform is referred to as the calibration. In this way, the precise physical dimensions, such as the corneal radius of curvature or angle between the optical and visual axis of the eye, and precise geometrical relationships between detectors, illumination sources and scene surfaces do not have to be explicitly determined. Rather, these relations are automatically incorporated in the implicitly determined function.

Theoretically, the calibration transformation can remove any systematic error that is a function of the measured variables. More calibration data points allow higher order polynomial transforms or more interpolation points, and usually improve the result, but with diminishing returns. Too many calibration points also result in a time consuming and onerous procedure. Systems often require subjects to look at either five or nine target points, and rarely > 20. In some cases, precise knowledge of the geometrical relation between components, along with knowledge of the underlying mechanism, can be used to reduce the number or calibration points required while preserving accuracy. Ohno et al. (29) describes a scheme using this type of strategy.

A cascaded polynomial curve fit scheme, used with pupil to corneal reflection method eye trackers, is described by Sheena and Borah (43). The same paper describes a method to account for changes in pupil position associated with change in pupil diameter. A 2D interpolation scheme is described by McConkie (65), and Kliegle and Olson (66). Possible variations are unlimited, and available systems employ a wide variety of calibration schemes. Sometimes, in order to further reduce systematic error, the users of commercially produced eye trackers add their own calibration and transform process onto data that has already been processed by the manufacturer's calibration and transform scheme. Jacob (67) and Duchowsky (53) describe specific examples.

Amir et al. (68) describe a method for using two cameras and illumination sources to compute the location of the eye optical axis with no calibration requirement. The pupil and corneal reflection images, on each camera, can be used to determine a plane that must contain the eye optical axis. The intersection of the two planes, one computed from each camera image, defines the optical axis. The orientation of the visual axis relative to the optical axis of the eye varies across the population, however, and this uncertainty cannot be removed without requiring at least one calibration point.

If eye tracker optics are head mounted, the calibration transform is often designed to map gaze to the image plane of a head mounted scene camera or a similar imaginary plane that travels with the head. Head position and orientation measurements can then be combined with this result, as described in the previous section, to derive the line-of-gaze in space, and to compute its intersection with known surfaces.

COMPATIBILITY WITH EYEGLASSES AND CONTACT LENSES

Eye glasses may present mechanical problems for systems that require sensors to be very close to the eye. Systems having sensors that are farther away and that view the eye through the spectacle lens must contend with mirror reflections from the spectacle lens and frame, or obstruction by elements of the frame. Distortion of the image by the spectacle lens usually does not present a significant problem since such effects are removed by the calibration scheme. The biggest reflection problem is often posed by the illumination source that is part of the eye tracking system, especially since the sensor must be sensitive in the spectral region of this light. Antireflective coatings are usually not good enough to eliminate the problem.

The position of the specular reflection (mirror image of the illumination source) is determined by the incidence angle of the illumination beam with the spectacle lens surface, and it is often possible to position the source so that the specular reflection does not cover a feature of interest, although this can become more difficult in the case of very high power (high curvature) spectacle lenses. If the specular reflection is not occluding an important feature, but is still in the sensor field of view, the system must be able to distinguish the features of interest from the specular reflection without confusion. This ability varies among systems, but video based systems can often be used successfully with eyeglasses.

Contact lenses also are often tolerated well by optical eyetracking devices, but may sometimes present the following problems. An edge of the contact lens may sometimes be visible or generate a bright reflection, and may confuse feature recognition, especially if the edge intersects a feature of interest.

When a light source reflection from the outer surface of the contact lens is visible to the detector, it usually appears to replace the first Purkinje image (corneal reflection). This is not a problem for systems that use the corneal reflection, so long as the visible corneal reflection is always the reflection from the contact lens. Although the contact lens surface will have a slightly different position and curvature than the cornea, the difference in the motion of this reflection from that of the real corneal reflection is easily accounted for by whatever calibration scheme is used. However, if the contact lens moves so that the reflection appears to fall off the edge of the contact lens and onto the cornea, there will be a shift in the computed gaze position. Hard contact lenses, which tend to be relatively small and float about on the tear film, are more likely to cause this problem than the larger soft lenses.

The contact lens surface may be less reflective than the cornea, resulting in a dimmer first Purkinje image, and making detection more difficult.

ILLUMINATION SAFETY

Most eye trackers that use optical sensors also include a means to illuminate the eye, usually with nonlaser light at the lower end of the near-infrared (IR-A) spectral region. The IR-A region spans the wavelengths between 770 and 1400 nm. To prevent harming the eye, it is important to avoid excessively heating the cornea and lens, and to avoid focusing too much energy on too small a spot on the retina.

The American Conference of Governmental Industrial Hygienists (ACGIH) suggests the following safety criteria for extended exposure to nonlaser, near-IR light (69). To protect the cornea and lens from thermal injury, irradiance at the eye should be no $> 10\,\mathrm{mW \cdot cm^{-2}}$. To protect the retina, near-IR radiance, expressed in units of $\mathrm{W \cdot (cm^2 \cdot sr)^{-1}}$, should be limited to no $> 0.6/\alpha$, where α is the angular subtense, in radians, of the source as seen by the subject.

Various safety standards are specified by many other organizations, including the American National Standards Institute (ANSI), The U. S. Food and Drug Administration (FDA), the International Electrotechnical Commission (IEC), and others, although some of these are intended specifically for laser sources. A comprehensive review of light safety issues can be found in a book by Sliney and Wolbarsht (70).

SPECIFIC IMPLEMENTATIONS OF OPTICAL TECHNIQUES

Photo Electric, Reflectivity Pattern (Limbus) Trackers

There are a small number of commercially available systems that use a photo electric reflectivity pattern (limbus) tracking technique to measure eye movements. Figure 14 shows a schematic for a basic system measuring horizontal eye position. The difference in the signal received by the two photodetectors is roughly proportional to the

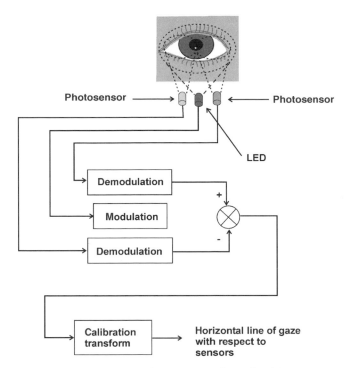

Figure 14. Schematic showing simple reflectivity pattern (limbus) tracker for horizontal measurement. (Adapted from a diagram in Ref. 8.)

horizontal position of the reflectivity pattern across the eye, with the most prominent feature of the pattern being the contrast between the white sclera and darker iris (limbus). Modulation of the LED, typically at 2 kHz or higher, and corresponding demodulation of photosensor signals diminishes the effect of ambient light. The signal is low pass filtered to remove modulation artifacts; and is either scaled and linearized with analog controls, or sampled and processed digitally, in order to scale and linearize the values.

Vertical position can be measured by orienting a similar LED and sensor array vertically, instead of horizontally. The results are less dependable because the high contrast iris to sclera boundary is often obscured by the eyelids. Alternately, vertical position is measured by aiming the horizontally oriented LED and sensor array at the boundary between the eye and the lower eyelid, and summing (instead of differencing) the photodetector signals. The result is really a measure of lower eyelid position, and takes advantage of the fact that the lower eyelid moves roughly in proportion to vertical eye motion. In either case, the vertical measurement is less accurate and less repeatable than the horizontal measure.

The LED and photosensors are positioned within ~ 2 cm of the eye, and are mounted to a head band, goggles, or spectacle frames. The detector assembly inevitably obscures some of the visual field. Since the reflectivity pattern moves more or less as would a landmark on the eyeball surface, a 1 mm shift of the optics on the head is expected to produce an error of $\sim 5°$. There is often a distinct cross-talk effect. The horizontal measurement values are affected by vertical eye position, and visa versa.

If vertical eye position is also measured, the calibration transform can attempt to correct cross-talk.

System bandwidth is limited, primarily, by the low pass filtering needed due to the modulation scheme, and is typically 50–100 Hz. If the signal is processed digitally, sample rates are often at least 1000 Hz. It is possible to achieve resolutions of $> 0.05°$ visual angle. While bandwidth and resolution are very good, accuracy is somewhat undependable because of headgear slippage affects, cross-talk effects, and, especially in the vertical axis, eye lid effects. Accuracy of $1°$ along a horizontal axis and $2°$ along a vertical axis may be achievable over a short period, but errors of several degrees would not be unusual, especially over longer periods or in the presence of vigorous head motion. Devices that use one LED and two photo sensors for a given axis tend to become very nonlinear, and difficult to calibrate over > 30 or $40°$ in either axis. The range can be extended somewhat by using multiple LEDs and sensors for each axis. Neither torsional eye movements, nor pupil diameter is measured.

Photoelectric, reflectivity pattern (limbus) trackers are best suited to measure dynamics of horizontal eye movements as opposed to point of regard measurement, although they are sometimes used for point of regard measurement as well. Systems in this category are commercially available from Applied Science Laboratories Bedford, MA (71), Cambridge Research Systems Ltd, UK (20), and Optomotor Laboratory, Freiburg, Germany (72).

Cambridge Research Systems offers a version of their device designed for use in fMRI environments. In this case, the LEDs and photodiodes are located outside of the magnet bore, and connected to the eye piece, within the magnet bore, via fiber optic cables (20).

Dual Purkinje Image Measurement (CR/4PI)

A technique was described by Cornsweet and Crane (45) and further developed by Crane and Steele (73,74) in which the eye is illuminated with a modulated IR source, and servo controlled mirrors are used to image the first and fourth Purkinje images onto solid state quadrant detectors. Demodulated, analog signals from the quadrant detectors are used, in separate feed back loops, to move the mirrors and keep the images centered on the detectors. The resulting mirror positions constitute measures of the feature positions, and can be used to compute eye rotation with respect to the optics.

A schematic representation of the system is shown in Fig. 15. The entire optics platform is mounted to a servo controlled *XYZ* stage, which automatically moves to optimize overall system alignment. A hot mirror beam splitter is used to direct light to and from the optical unit, which is positioned off to one side of the subject. By looking through the beam splitter, the subject has an unimpeded view of the forward visual field. Not shown in the simplified schematic is an autofocus mechanism, implemented with a beam splitter, off axis aperture, and bicell detector, in the first Purkinje image optical path.

The subject's motion is usually restricted with a head rest or bite bar assembly. Because the fourth Purkinje image is visible only through the iris opening, measurement range is

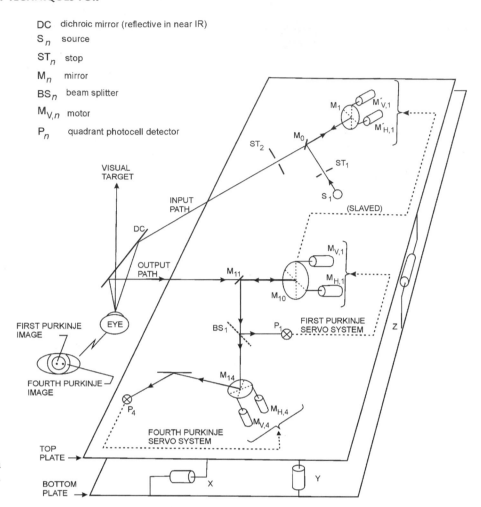

DC   dichroic mirror (reflective in near IR)

S_n source

ST_n stop

M_n mirror

BS_n beam splitter

$M_{V,n}$ motor

P_n quadrant photocell detector

Figure 15. Schematic representation of Double Purkinje image eye tracker. (Redrawn from Ref. 73.)

restricted to about $\pm\,10^\circ$ visual angle, although this can be extended to about $\pm\,15^\circ$ by using drops to dilate the pupil. Accuracy is ~ 1 min of arc, and the sampling rate is 1000 Hz. Using a model eye, Crane and Steele (74) empirically measured a bandwidth of ~ 500 Hz (for small eye movements of several degrees), noise levels of 20 arc·s rms, response delay of 0.25 ms, and linear tracking of slew rates up to $\sim 2000^\circ$·s. The manufacturer of a current version of the device specifies a bandwidth of 400 Hz and a 1 ms response time (75).

Although the subject's motion is restricted and the measurement range is small, the accuracy, precision and temporal bandwidth are exceptionally good. The Dual Purkinje image eye-tracking device can provide a precise enough and fast enough measurement, for example, to allow another device to effectively stabilize an image on the retina (74). Neither torsional eye movement nor pupil diameter is measured, but it is possible to attach an infrared optometer, which provides a real-time, analog measure of changes in eye accommodation (75). An updated version of the device described by Crane and Steele (74) is offered commercially by Fourward Technologies, Inc, Buena Vista, VA. (75).

Systems Using Two-Dimensional Video Sensor Arrays

Eye tracking devices that use 2D CCD or CMOS sensor arrays exist in wide variety, and are commercially available from numerous sources. The term video oculography (VOG) is often used to describe this type of system. Most current systems in this category use the pupil to corneal reflection technique (CR/Pupil) to measure gaze direction, but there are exceptions. Some systems designed primarily to measure ocular torsion might rely on a fixed position of the head with respect to the sensor, and use only pupil position to correct for gaze direction changes. Some systems that normally use the pupil-to-corneal reflection technique have options to measure with only pupil position or only corneal reflection position under certain circumstances. Other systems are designed to measure the position of the head in space, and use a camera (or cameras) with a wide-angle view on which the corneal reflection can not easily be resolved. These systems may use the pupil position relative to a facial feature to compute gaze direction.

Note that a pupil-only or corneal-reflection-only measurement offers some advantages if it can be assured that the head does not move with respect to the camera. Pupil and corneal reflection motion have greater sensitivity to eye rotation than the pupil-to-corneal reflection vector, and the measurement is therefore more precise (less noisy).

Figure 16 is a functional diagram consistent with most current video-based eye tracking systems. The functions may be distributed among physical components in various ways, and a separate head tracking system may or may not be included.

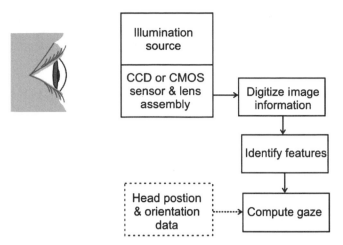

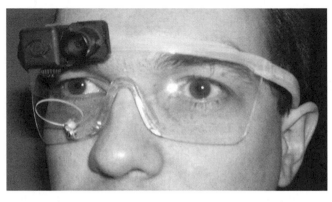

Figure 17. The head mounted module directly over the subject's right eye contains both a camera and illuminator, which are aimed down at the tilted, hot mirror beam splitter. The camera views a reflection of the eye from the hot mirror. The module just to the nasal side of the eye camera module is a scene camera, aimed toward the subject's field of view. (Courtesy of Applied Science Laboratories, Bedford, MA.)

Figure 16. Schematic showing the basic functional architecture of most video-based eye tracking systems.

Some systems input video from the eye camera to PC memory via a USB or firewire connection, either directly from a camera with compatible output, or via a video–firewire converter. The PC CPU then does all the processing. Alternately, a frame grabber installed on the PC may be used to digitize a composite video signal. In some cases, the resulting digitized signal is processed by the PC, or in other cases, a frame grabber that also has image processing capabilities may do some portion of the processing before making a reduced information set available to the PC. In still other cases, the eye video is input to a completely custom processing board or board set, often populated with some combination of field programmable gate array (FPGA) and digital signal processing (DSP) components. The custom processor may do virtually all of the processing or may send a reduced data set to a PC. Other workstations may be used in place of the PCs referred to above. In short, there is enormous variety possible. Systems that support > 60 Hz update rates often make some use of custom processing components to help achieve the high speed. Although not yet capable of the same temporal bandwidth as several other types of system, camera based systems are now available with high enough update rates to make them useful for study of eye movement dynamics. Video-based systems almost always measure pupil diameter as a byproduct of computations for eye rotation measurement, and these systems are sometimes used as pupillometers.

The two main subcategories within this group are head mounted systems, having the sensor and small optics packages mounted to headgear; and remote systems, relying on optics mounted to the environment (table, instrument panel, etc.). In both cases these are often used as real time point of regard systems. The following sections describe video based, head mounted and remote eye tracking systems, respectively; followed by a section describing systems designed to handle the special problem of operation in MRI devices.

Head Mounted, Video Based Systems. Head mounted sensors are often designed to view a reflection of the eye from a beam splitter, as shown in Fig. 17, so that the

sensor, lens, and illumination source need not be in the subject's field of view. The beam splitter is coated to be a hot mirror, very reflective in the near-IR, but not in the visible spectrum. In some cases, however, the module containing the sensor is positioned on a stalk (Fig. 18), so that it can obtain a direct view of the eye. Illumination is most often provided by one or more near infrared LEDs. Both dark-pupil and bright-pupil type optics are represented among available systems.

The image of the eye produced by a head mounted sensor yields information about eye orientation with respect to the head. In other words, line of gaze is measured in the head reference frame. In some cases, this is the

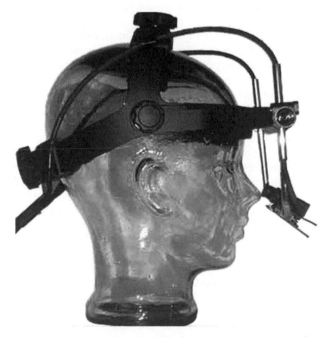

Figure 18. Head mounted camera and illuminator module is mounted on a stalk, and is aimed directly at the eye. (Courtesy of SR Research, Inc. Ontario.)

desired quantity, but in other cases the desired measurement is point of gaze in the environment. In order to find point of gaze on objects in the environment, it is necessary either to detect the positions of those objects with respect to the head, or to detect the position and orientation of the head with respect to the environment containing the objects. Techniques for accomplishing this were discussed in the previous section titled "Measuring point of gaze in the presence of head motion". All of these methods have been used by commercially available, pupil–corneal reflection-type systems.

Head tracking devices commonly used with head mounted eye trackers include magnetic systems that measure the position and orientation of a small, head gear mounted, set of coils with respect to a larger set of stationary coils; optical systems that use head mounted sensors to detect moving laser beams from a stationary source; and a system which uses a head mounted inertial package for high frequency measurements, corrected by lower frequency measurements from ultrasonic emitter–receiver pairs (8).

Head tracking systems capable of sufficient 6 degree of freedom accuracy usually restrict subject to 1 or 2 m of movement. Applications requiring a subject to walk or run a significant distance usually rely on a head mounted scene camera to capture information about point of gaze on the environment. There are available systems that allow subjects to move about with no physical connection to stationary components while either recording data on a recording device attached to the subject, or using wireless transmission to send data to a base station.

Systems intended for neurological testing are usually concerned with eye movement with respect to the head rather than point of gaze in the environment. Some of these systems measure ocular torsion as well as measuring gaze direction with either pupil-to-corneal-reflection or pupil position techniques.

Measurement accuracy, of head mounted, video-based, eye trackers, tends to be from ~ 0.5–$1.0°$ visual angle, with resolutions varying from 0.01 to 0.1° visual angle. Measurement range is 60–70° visual angle in the horizontal axis and usually $\sim 20°$ less in the vertical axis due to eye lid interference. These figures refer to measurement of gaze direction with respect to the head. The measurable field of view with respect to the environment is unrestricted because subjects are free to turn their heads and bodies. Measurement update rates range from 25 to 360 Hz, with at least one commercially available system offering 500 Hz with pupil-only (as opposed to pupil-to-corneal-reflection) measurements.

When a head mounted eye tracker is used in combination with a head tracking device, in order to measure point of gaze in the environment, error in the final measurement is increased due to noise and error in the head position and orientation measurement. The result can vary widely depending on the particular combination of equipment being used and the conditions under which measurements are being made. It is probably not unreasonable to expect additional errors corresponding to ~ 0.2–$1.0°$ visual when using a head mounted eye tracker to measure point of gaze on elements in the environment. Conditions that create

difficulties for either the eye or head tracking components, for example a heavily metallic environment when using a magnetic head tracker, or occlusion of the pupil image by reflection from some external light source, may cause errors to increase dramatically or make the measurement impossible.

Manufacturers of systems that include ocular torsion measurement usually specify torsion measurement range of 18–20°, resolution of $\sim 0.1°$, and linearity between 1 and 4% of full scale. Video-based, head mounted systems, using the pupil to corneal reflection measurement technique, currently allow point of gaze measurement in the most diverse environments, and with the greatest freedom of subject motion and measurable field of view. They do not have as high a temporal bandwidth as scleral search coil, double Purkinje image, EOG, or reflectivity (limbus) tracking systems, and are not as accurate or precise as scleral search coil, or double Purkinje image systems. They do require that the subject wear head gear, but the size and obtrusiveness of the headgear has been steadily reduced over the past decade. It is becoming practical to use this type of system in ever more flexible and varied settings. The amount and severity of subject motion, the amount of variation in ambient light environment, and the length of time over which continuous measurements must be made, all tend to trade off somewhat against the accuracy and dependability of the measurements.

Video-based eye trackers, with head mounted optics, are offered commercially by: Alphabio, France (76); Applied Science Laboratories, Bedford, MA (71); Arrington Research, Inc., Scottsdale, AZ (77); Chronos Vision GmbH, Berlin, Germany (78); Guymark UK Ltd, UK (22); EL-MAR Inc, Downsview, ON (79); ISCAN, Inc., Burlington, MA (80); Neuro Kinetics Inc. Pittsburgh, PA (24); SensoMotoric Insturments GmbH, Berlin, Germany (81); and SR Research Ltd, Osgoode, ON (82).

Head mounted systems that measure ocular torsion are available from: Alphabio, France (72); Arrington Research, Inc, Scottsdale, AZ (77); Chronos Vision GmbH, Berlin, Germany (78); Neuro Kinetics Inc. Pittsburgh, PA (24); and SensoMotoric Insturments GmbH, Berlin, Germany (81).

Remote, Video-Based Systems. Remote (nonhead mounted) systems have one or more videosensors and illumination sources that are fixed to the environment and aimed toward the subject. The optics may sit on a table next to or underneath a monitor being viewed by the subject, may be mounted on a vehicle (or vehicle simulator) instrument panel, or mounted to the environment in some other way. Both bright and dark pupil optics are used. In the case of bright pupil optics, the illuminator may be a ring of LEDs placed close enough to the lens to produce the bright pupil effect, an LED actually mounted over the front of the lens so that it blocks only a small portion of the lens surface, or a beam splitter arrangement like that shown in Fig. 6.

Some system configurations require the subject's head to be stabilized. This maximizes achievable accuracy and dependability. A chin and forehead rest, or cheek and forehead rest are commonly used, although these do allow a small amount of head motion. A bite bar with a dental impression can be used to completely stabilize the head. In the case of nonhuman primate research the head is often stabilized by other means appropriate for neural recording. The camera, lens, and illumination sources can all be stationary (after manual positioning) and may safely be designed with a field of view just large enough to comfortably accommodate pupil and corneal reflection motion due to eye rotation. If the subject's head is sufficiently stabilized there is no need to differentiate between eye rotation and head translation, and, as previously discussed, there is a precision advantage to using a pupil-only (rather than pupil-to-corneal-reflection) measurement. Some systems allow a choice between pupil-to-corneal-reflection and pupil-only or corneal-reflection-only measurement.

Alternately, the eye camera field of view may be wide enough to accommodate the motions of a seated subject who is not restrained, but is voluntarily remaining still. If the subjects are provided with feedback, so that they know when they are moving out of the legal space, it is only necessary to allow ∼5 cm of head motion.

Many remote systems are designed to allow enough head motion to accommodate normal motions of a person working at a computer terminal or driving a car, and so on. This is accomplished either by dynamically rotating the camera optical axis so that it always points toward the eye being tracked, by using a camera field of view wide enough to accommodate the desired head motion, using multiple cameras, or some combination of these. Head motion toward and away from the eye camera must either be small enough to remain within the lens system depth-of-field, or must be accommodated by an autofocus mechanism. Figure 19 shows an example of an eye camera that automatically moves in azimuth (pan) and elevation (tilt) to follow the eye as the head moves about. Figure 20 is an example of a camera with a moving mirror used to direct

the optical path, and Fig. 21 shows a system with stationary sensor and illumination components in an enclosure.

Systems with a moving camera or moving mirrors often use a closed loop control, based on the detected position of the pupil in the camera field of view. The moving element is driven to move the image toward center. If the pupil is completely lost from the camera field of view, the system may execute a search pattern, use information from a separate head tracking system to reacquire the eye, or require that a human operator intervene to reacquire the eye image. Such systems can have a relatively narrow eye camera field of view, thus maximizing image resolution for the features of interest. The disadvantage is the need for moving parts, and the possibility of failing to maintain the eye image in the camera field of view.

Systems that do not require moving parts have a significant advantage in terms of system simplicity and dependability. Furthermore, there is the possibility of using the same image to measure head position and to track the movement of both eyes. The trade-off is that as the field of view is increased, resolution is reduced (the pupil and corneal reflection are imaged onto fewer pixels) and the features of interest must be identified within a larger, more complex image.

The pupil-to-corneal-reflection method alone can be used to determine gaze angle with respect to the eye camera. However, in the presence of significant head motion, this is not sufficient to accurately determine point of gaze on other surfaces. Some remote systems use head trackers to find the position of the eye in space, and use the information to more accurately compute point of gaze on other stationary surfaces. Head position information can also be used to help aim moving cameras or mirrors. The same types of head tracker mentioned in the previous section, as being appropriate for use with head mounted eye trackers, are sometimes used in conjunction with remote eye trackers.

Figure 19. Example of a remote eye tracker optics module that moves in azimuth (pan) and elevation (tilt) to keep a telephoto eye image within the camera field of view. (Courtesy of Applied Science Laboratories, Bedford, MA.)

Figure 20. Example of a remote eye tracker optics module that uses a moving mirror to keep a telephoto eye image within the camera field of view. The optics module is shown with its cover removed. (Courtesy of Applied Science Laboratories, Bedford, MA.)

In the case of remote eye trackers, it is also possible to use one or more wide angle cameras to locate the head and eyes. The head tracking camera (or cameras) may be separate from the eye tracking camera, or alternately, the same cameras (or cameras) may be used for both.

At present, remote systems that offer > 60 Hz update rates either require the head to be stabilized, or use a narrow field image that is dynamically directed to follow head motions. Systems that use stationary wide-angle optics typically update at 60 Hz or less, probably because of the extra processing required for feature recognition in the more complex wide-angle image. Systems that either require the head to be stabilized or direct a narrow field of camera to follow head motion are available with update rates of up to 360 Hz. As with head mounted systems, higher update rates result in lower signal/noise ratios for the sensor.

Figure 21. Example of a remote eye tracker optics module using stationary wide angle optics. Sensor, lens, and illumination components are within the enclosure. (Courtesy of Tobii Tehnology AB, Stockholm, Sweden.)

For configurations that stabilize the subject's head, accuracy of remote, video-based eye tracking systems tends to be ~ 0.5–1.0° visual angle, with resolutions varying from 0.01 to 0.1° visual angle. When a remote eye tracker is used to measure point of gaze on a stationary surface, and head motion is allowed, some additional error can be expected. If no attempt is made to account for head motion, the equation in Fig. 11 can be used to estimate the amount of error expected from a given change in head position. If an attempt is made to correct for head movement, the amount of additional error depends on the accuracy of the head position measurements, and the way the information is used in the point-of-gain computation, as well as the range of head motion. It is probably not unreasonable to expect an additional 0.5–1.5° of error when there is significant head motion. Data may also have additional error, or even be briefly lost, during fast head motion, due to instability of the eye image on the eye-camera field of view.

Remote systems using one eye-camera and one illumination source, and the pupil-to-corneal-reflection method, can generally measure gaze directions that are within ~ 25–35° visual angle from the eye-camera lens. At more eccentric angles the corneal reflection is either not visible to the eye camera, or easily confused with multiple reflections from the sclera. The exception is that when line of gaze is below the eye-camera, the upper eyelid often begins to occlude part of the pupil. The amount of eye lid occlusion under this condition varies from subject to subject. Furthermore, different eye tracking systems, using different recognition and center computation algorithms, are tolerant of different amounts of occlusion. This often limits measurement range to 5–10° visual angle below the eye-camera, and for this reason, the environment is usually arranged with the eye-camera at the bottom of the scene space that is of interest. The result is a horizontal gaze measurement range of ~ 50–70° visual angle, and a vertical range of ~ 35–45°.

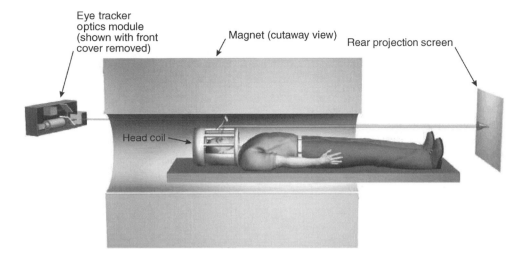

Figure 22. Example of eye tracker optical path in an fMRI system. The drawing shows the magnet in cutaway view. One mirror, just above the subject's face, reflects the eye image to the eye tracker camera, while another allows the subject to view a stimulus on a rear projection screen. Many variations to this arrangement are used, including configurations in which both the projection screen and eye tracker optics module are at the same end of the magnet. (Courtesy of Applied Science Laboratories, Bedford, MA.)

The range of measurable gaze directions, using the pupil-to-corneal-reflection method, can be increased by using multiple illumination sources (and thus multiple corneal reflections). The horizontal range can be increased to as much as 90° by widely spacing illumination sources. Only a more modest range expansion is possible in the vertical axis because of eyelid interference.

Remote, video-based, eye trackers, offer the possibility of measuring eye movement very unobtrusively. Unlike head mounted systems, nothing need be worn by the subject (unless using a type of head tracker that requires a head mounted sensor), and the optics can be hidden from obvious view by a filter that is transmissive in the near-IR. The range of subject motion and measurable field of view is significantly more restricted than for systems with head mounted optics, although less so than for most other techniques. Measurement accuracy, resolution, and temporal bandwidth are not as good as that available with the much more restrictive and obtrusive double Purkinje image method, or scleral search coil method.

Remote, video-based, eye movement measurement systems are commercially available from: Applied Science Laboratories, Bedford, MA (71); Arrington Research, Inc., Scottsdale, AZ (77); Cambridge Research Systems Ltd, UK (20), ISCAN, Inc., Burlington, MA (80); LC Technologies Inc., Fairfax, VA (83); Metrovision, Pérenchies, France (23), Seeing Machines, Canberra, Australia (84); SensoMotoric Insturments GmbH, Berlin, Germany (81); SR Research Ltd, Osgoode, ON (82); and Tobii Technology AB, Stokholm, Sweeden (85).

Video-Based Systems for use in Conjunction with fMRI.
There are commercially available, video-based eye tracker systems, which use the pupil-to-corneal-reflection method, and are specifically configured to be used in conjunction with fMRI measurements of brain activity.

During fMRI measurements, the subject's head and torso are inside the main MRI magnet bore, and a smaller head coil is placed fairly closely around the subject's head. The head coil always has an opening over the subject's eyes, and there is usually some provision for the subject to look at a visual display, often with the help of a mirror or set of mirrors. The environment is challenging because there is

not much room in which to arrange an optical path from the subject's eye to a camera, and also because any electronic equipment used must operate in a high magnetic field, without creating even small amounts of magnetic field noise that would interfere with the MRI measurement.

One approach is to use a camera equipped with a telephoto lens, and placed outside of the magnet. The camera is aimed down the magnet bore, and views a reflection of the eye on a small mirror placed inside of the magnet, near the subject's head. The optical path length from the camera lens to the eye is typically at least 2.5–3 m, necessitating a fairly large telephoto lens, and powerful illumination source. An example is shown in Fig. 22, and is similar to a setup described by Gitleman et al. (86). The system shown is using a bright pupil technique, so there is only one coaxial path for both the camera and illumination beam. In this case, a second mirror allows the subject to look at a display screen. Many variations of both the eye camera and display screen optical paths are possible.

Other systems bring the eye image out of the magnet, to the sensor, with a coherent, fiber optic bundle. This has the advantage of not requiring large, telephoto optics, but the resolution of fiber bundles are limited and light is attenuated somewhat unevenly by the individual fibers that comprise the bundle. Fiber optics may also be used to relay the output from an illumination source into the magnet. It is possible to place a small camera inside the magnet, but it can be difficult to avoid some interference with the magnetic field and resulting degradation of the fMRI measurement.

Video-based, eye tracking systems designed for use during fMRI measurement, are commercially available from Applied Science Laboratories, Bedford, MA (71); Arrington Research, Inc., Scottsdale, AZ (77); ISCAN, Inc., Burlington, MA (80); and SensoMotoric Insturments GmbH, Berlin, Germany (81).

COMPARISON OF EYE MOVEMENT MEASUREMENT TECHNIQUES

Table 1 is an abbreviated comparison of the various techniques discussed.

Table 1. Summary of Most Prevalent Eye Tracking Techniques[a]

Method	Typical Applications	Typical Attributes	Typical Reference Frame(s)	Typical Performance
EOG	Dynamics of saccades smooth pursuit nystagmus	High bandwidth Eyes can be closed In expensive Drift problem (poor position accuracy) Requires skin electrodes - otherwise unobtrusive	head	static accuracy: $\sim$3-7° resolution: with low pass filtering and periodic rezero, virtually infinite bandwidth: $\sim$100 Hz
Scleral Coil	Dynamics of saccades, smooth pursuit, nystagmus Miniature eye movements Point of gaze Scan path Torsion	Very high accuracy and precision Invasive Very obtrusive High bandwidth	Room	accuracy:$\sim$0.2° resolution: $\sim$1 arc min. range: $\sim$ 30° bandwidth: $\sim$200 Hz
Limbus (using small number of photo sensors)	Dynamics of saccades, smooth pursuit, nystagmus Point of gaze Scan path	High bandwidth Inexpensive Poor vertical accuracy Obtrusive (sensors close to eye) Head gear slip errors	head gear	accuracy: varies resolution: 0.1° (much better res. possible) range: $\sim$30° update rate: 1000 samples/s
CR/Pupil	Point of gaze Line of gaze Scan path Dynamics of eye movements (only with subset of systems offering high update rates) Torsion (available only on some systems)	Minimal head gear slip error Unobtrusive Low to medium bandwidth Problems with sunlight Wide variety of configurations and environments	Head gear Room Cockpit	accuracy: $\sim$1° resolution: $\sim$ 0.2° hor. range:$\sim$50° vert. range: $\sim$40° update rate: 50 or 60 samples/s. Update rates up to 360 samples/s available on some systems (higher for pupil only measurement);
CR/4PI	Dynamics of saccades, smooth pursuit, nystagmus Miniature eye movements Point of gaze Scan path Image stabilization on retina Accommodation	Very high accuracy and precision High bandwidth Obtrusive (large optics package, restricted head motion) Limited range	Room	accuracy: $\sim$1 arc min. resolution: $<$ 1 arc min range: $\sim$20° bandwidth: $\sim$400 Hz

[a]Adapted from Ref. 8.

An earlier, but comprehensive and still pertinent review of eye movement measurement techniques, can be found in Young and Sheena (87). A book by Duchowski (53) presents a detailed treatment of many aspects of eye tracking methodology. The journal *Computer Vision and Image Understanding* has devoted an entire special edition to developments in optical eye tracking, with an emphasis on real-time, nonintrusive techniques (88). A database of commercially available eye movement measurement equipment is available at a web site hosted by the University of Derby (89).

BIBLIOGRAPHY

Cited References

1. Roth EM, Finkelstein S. Light environment. In: Roth EM, editor. Compendium of Human responses to the Aerospace Environment. NASA CR-1205(1); 1968.

2. Westheimer G. The eye, In: Mountcastle VB, editor, Medical Physiology, Volume One. Saint Louis: The C. V. Mosby Co.; 1974.

3. Fry GA, Hill WW. The center of rotation of the eye. Am J Optom Arch Am Acad Optom 1962;390:581–595.

4. Wyatt HJ. The form of the human pupil. Vision Res 1995;35(14):2021–2036.

5. Haslwanter T. Mathematics of three-dimensional eye rotations. Vision Res 1995;35:1727–1739.

6. Hallett P. Eye movements. In: Boff KR, Kaufman L, Thomas JP, editors. Handbook of Perception and Human Performance. New York: John Wiley & Sons, Inc.; 1986.

7. Robinson DA. A method of measuring eye movement using a scleral search coil in a magnetic field, IEEE Transactions on Bio-Medical Electronics 1963;BMI-10:137–145.

8. Leger A et al. Alternative Control Technologies, North Atlantic Treaty Organization, RTO-TR-7, 1998.

9. Collewijn F, van der Marx S, Jansen TC. Precise recording of human eye movements. Vision Res 1975;15:447–450.

10. Judge SJ, Richmond BJ, Chu FC. Implantation of magnetic search coils for measurement of eye position: an improved method. Vision Res 1980;20:535–538.

11. Skalar Medical BV. (2002, December 20). Home. (Online). Available at http://www.skalar.nl, Accessed Feb. 9, 2005.

12. Riverbend Instruments. (No date). Home. (Online). Available at http://www.riverbendinst.com/index.htm. Accessed 13 April 2005.

13. Mowrer OH, Ruch RC, Miller NE. The corneo-retinal potential difference as the basis of the galvanometric method of recording eye movements. Am J Physiol 1936;114:423.

14. Marg E. Development of electro-oculography—standing potential of the eye in registration of eye movement. AMA Arch Ophthalmol 1951;45:169.

15. Kris C. Vision: electro-oculography. Glasser O, editor. Medical Physics: Chicago: Chicago Yearbook Publishers; 1960.

16. Shackel B. Review of the past and present in oculography. Medical Electronics Proceedings of the Second International Conference. London: Hiffe; 57, 1960.

17. Shackel B. Eye movement recording by electro-oculography. In: Venables PH, Martion I, editors, A manual of Psychophysiological Methods. Amsterdam: North-Holland Publishing Co.; pp 300–334, 1967.

18. Arden GB, Barrada A, Kelsey JH. New clinical test of retinal function based upon the standing potential of the eye. Br J Ophthalmol 1962;46:467.

19. De Rouck A, Kayembe D. A clinical procedure of the simultaneous recording of fast and slow EOG oscillations. Int Ophthalmol 1981;3:179–189.

20. Cambridge Research Systems. (2005, February 9). Home. (Online). Available at http://www.crsltd.com Accessed 2005, Feb. 10.

21. GN Otometrics. (2004). Products. (Online). Available at http://www.gnotometrics.com/products.htm. Accessed 2005 Feb. 9.

22. Guymark UK Ltd. (No date). Vestibular Assessment. (Online). Available at http://www.guymark.com/vestibul.html. Accessed 2005, Feb. 9.

23. Metrovision. (No date). Homepage (Online). Available at http://www.metrovision.fr. Accessed 2005, Feb. 9.

24. Neuro Kinetics Inc. (No date). Products and Services. (Online). Available at http://neuro-kinetics.com/products-research.htm. Accessed 2005, Feb. 9.

25. Nguyen K, Wagner C, Koons D, Flickner M. Differences in the infrared bright pupil response of human eyes. Proceedings ETRA 2002 Eye Tracking Research & Applications Symposium. New Orleans: March 25–27, 2002.

26. Borah J. Helmet Mounted EyeTtracking for Virtual Panoramic Display Systems–Volume I: Review of Current Eye Movement Measurement Technology. US Air Force report AAMRL-TR-89019, 1989.

27. Zhu S, Moore T, Raphan T. Robust pupil center detection using a curvature algorithm. Computer Methods Programs Biomed 1999;59:145–157.

28. Mulligan JB. A software-based eye tracking system for the study of air-traffic displays. Proceedings ETRA 2002 Eye Tracking Research & Applications Symposium. New Orleans: March 25–27, 2002.

29. Ohno T, Mukawa N, Yoshikawa A. FreeGaze: a gaze tracking system for everyday gaze interaction. Proceedings ETRA 2002 Eye Tracking Research & Applications Symposium. New Orleans: March 25–27, 2002.

30. Charlier J et al. Real time pattern recognition and feature analysis from video signals applied to eye movement and pupillary reflex monitoring. Proceedings of the 6th Int Visual Field Symposium. In: Heijl A, Greve EL, editors. Dordrecht, The Netherlands: Dr. W Junk Publisheres; 1985.

31. Sheena D. Pattern-recognition techniques for extraction of features of the eye from a conventional television scan. In: Monty RA, Senders JW, editors. Eye Movements and Psychological Processes. Hillsdale (NJ): Lawrence Erlbaum; 1976.

32. U.S. Pat. 5,302,819 1994. Kassies M. Method of and Apparatus for, Detecting an Object.

33. Zhai S, Morimoto C, Ihde S. Manual and Ggaze input cascaded (MAGIC) pointing. Proceedings CHI'99: ACM Conference on Human Factors in Computing Systems. Pittsburgh: ACM: 246–253, 1999.

34. Tomono I, Muneo X, Lobayashi Y. A TV camera system that extracts feature points for non-contact eye movement detection. iSPIE vol. 1194 Optics, Illumination, and Image Sensing for Machine Vision IV; 1989.

35. Ebisawa Y, Satoh S. Effectiveness of pupil area detection technique using two light sources and image difference method. 15th Annual International Conference of the IEEE Engineering in Medicine and Biology Society. San Diego: 1993.

36. Eizenman M, Frecker RC, Hallett PE. Precise non-contacting measurement of eye movements using the corneal reflex. Vision Res 1984;24:167–174.

37. Xiao J, Baker S, Matthews I, Kanade T. Real-Time Combined 2D+3D Active Appearance Models. Proceedings of the IEEE Conference on Computer Vision and Pattern Recognition. June, 2004.

38. Matthew X, Baker S. (none) Real-Time AAM fitting algorithms. [Online]. Carnegie Mellon University. Available at http:// www.ri.cmu.edu/projects/project_448.html. Accessed 9 Feb. 2005.

39. Cootes TF, Edwards GJ, Taylor CJ. Active appearance models. IEEE Transactions on Pattern Analysis and Machine Intelligence 2001;23:681–685.

40. Zhu J, Yang J. Subpixel Eye-Gaze Tracking. Proceedings of the fifth IEEE International Conference on Face and Gesture Recognition 2002; 131–136.

41. U.S. Pat. 5,818,954 1998. Tomono M, Iida K, Ohmura X. Method of Detecting Eye Fixation using Image Processing.

42. Merchant J, Morrissette R, Perterfield JI. Remote measurement of eye direction allowing subject motion over one cubic foot of space. IEEE Transactions on Biomedical Engineering 1974; BME-21:309–317.

43. Sheena D, Borah J. Compensation for some second order effects to improve eye position measurements. In: Fisher DF, Monty RA, Senders JW, editors. Eye Movements: Cognition and Visual Perception. Hillsdale (NJ): Lawrence Erlbaum Associates; 1981.

44. U.S. Pat. 6,5987,971, 2003. Cleveland D. Method and System for Accommodating Pupil Non-concentricity in Eyetracker Systems.

45. Cornsweet TN, Crane HD. Accurate two-dimensional eye tracker using first and fourth Purkinje images. J Opt Soc Am 1973;63:921.

46. Torok N, Guillemin B, Barnothy JM. Photoelectric nystagmography. Ann Otol Rhinol Laryngol 1951;60:917.

47. Smith WM, Warter PJ. Eye movement and stimulus movement: New Photoelectric electromechanical system for recording and measuring tracking motions of the eye. J Opt Soc Am 1960;50:245.

48. Stark L, Sandberg A. A simple instrument for measuring eye movements, Quarterly Progresss Report 62, Research Laboratory of Electronics, Massachusetts Insstitute of Technology; 1961. p. 268.

49. Wheeless LL, Boynton RM, Cohen GH. Eye movement responses to step and pulse-step stimuli. J Opt Soc Am 1966;56:956–960.

50. Young LR. Recording eye position. In: Clynes M, Milsum JH, editors. Biomedical Engineering Systems. New York: McGraw-Hill; 1970.

51. Findlay JM. A simple apparatus for recording micorsaccades during visual fixation. Quart J Exper Psychol 1974;26:167–170.

52. Reulen JP et al. Precise recording of eye movement: the IRIS technique. part 1. Med Biol Eng Comput 1988;26:20–26.

53. Duchowski AT. Eye Tracking Methodology Theory and Practice. London: Springer-Verlag; 2003.

54. U.S. Pat. 4,852,988, 1989. Velez J, Borah J. Visor and camera providing a parallax-free field-of-view image for a head-mounted eye movement measurement system.

55. Edelman ER. Video based monitoring of torsional eye movements. S.M. dissertation, Massachusetts Institute of Technology, Cambridge (MA). 1979.

56. Proceedings of the 3rd VOG workshop, Tubingen, Germany, Nov. 30–Dec. 2, 1999.

57. Hatamian M, Anderson DJ. Design considerations for a real-time ocular counterroll instrument. IEEE Trans Biomed Eng 1983;30:278–288.

58. Clarke AH, Teiwes W, Scherer H. Videooculography—an alternative method for measurement of three-dimensional eye movements. In: Schmid R, Zambarbieri D, editors. Oculomotor Control and Cognitive Processes, Amsterdam: Elsevier; 1991.

59. Bucher U, Heitger F, Mast F. A novel automatic procedure for measuring ocular counterrolling. Behav Res Methods Instr Comp 1990;22:433–439.

60. Moore ST, Haslwanter T. Curthoys IS, Smith ST. A geometric basis for measurement of three-dimensional eye poisition using image processing. Vision Res 1996;36: 445–459.

61. Petrka RJ, Merfeld DM. Calibration techniques for video-oculography, Poster Presentation at Barany Society Meeting. Sydney Australia; 1996.

62. Groen E, Nacken PFM, Bos JE, De Graaf B. Determination of ocular torsion by means of automatic pattern recognition. IEEE Trans Biomed Eng 1996;43(5): 471–479.

63. Zhu D, Moore ST, Raphan T. Teal-time torsional eye position calculation from video images. Soc Neurosci Abstr 1999;25:1650.

64. Guillemant P, Ulmer E, Freyss G. 3-D eye movement measurements on four Comex's drivers using video CCD cameras, during high pressure diving. Acta Otolaryngol (Suppl) (Stockh) 1995;520:288–292.

65. McConkie GW. Evaluating and reporting data quality in eye movement research. Behav Res Methods Instrum 1981;13: 97–106.

66. Kliegle R, Olson RK. reduction and calibration of eye monitor data. Behav Res Methods Instrum 1981;13:107–111.

67. Jacob RK. Eye tracking in advanced interface design. In: Barfield W, Furness T, editors. Virtual Environments and Advanced Interface Design. New York: Oxford University Press; 1995.

68. U.S. Pat. 6,578,962, 2003. Amir MD, Flickner DB, Koons X, Russell GR. (to Calibration-Free Eye Gaze Tracking).

69. American Conference of Governmental Industrial Hygienists, 2001 TLVs and BEIs, ACGIH, 1330 Kemper Meadow Drive, Cincinnati, OH; 2001.

70. Sliney DH, Wolbarsht M. Safety with Lasers and Other Optical Sources: A Comprehensive Handbook. New York: Plenum Press; 1980.

71. U. S. Pat. (to Applied Science Laboratories). (2005, January 5) Home. (Online). Available at http:http://www.a-s-l.com. Accessed 9 Feb. 2005.

72. U. S. Pat. (to Applied Science Laboratories). (No date). Express Eye. (Online), Available at http://www.optom.de/english/exe. htm. Accessed 9 Feb. 2005.

73. Crane HD, Steele CM. Accurate three-dimensional eye-tracker. Appl Opt 1978;17:691.

74. Crane HD, Steele CM. Generation-V dual-Purkinje image eyetracker. Appl Opt 1985;24:527–537.

75. Fourward Technologies Inc. (2004, November 11). Home (Online). Available at http://www.fourward.com. Accessed 9 Feb. 2005.

76. Alphabio. (No date). Home. (Online). Available at http://www.electronica.fr/alphabio. Accessed 9 Feb. 2005.

77. Arrington Research, Inc. (2004, September 9). Home. (Online). Available at http://www.arringtonresearch.com. Accessed 9 Feb. 2005.

78. Chronos Vision. (no date). Eye Tracking. (Online). Available at http://www.chronos-vision.de/eyetracking/default_start_eyetracking.htm. Accessed 9 Feb. 2005.

79. EL-MAR, Inc. (No date). Menu. (Online). Available at http://www.interlog.com/~elmarinc/menu.htm. Accessed 9 Feb. 2005.

80. ISCAN Inc., (no date). Home. (Online). Available at http://iscaninc.com. Accessed 9 Feb. 2005.

81. SensoMotoric Instruments. (No date). Home. (Online). Available at http://www.smi.de. Accessed 9 Feb. 2005.

82. SR Research Ltd. (2004, Nov. 24). Home. (Online). Available at http://www.eyelinkinfo.com/. Accessed 9 Feb. 2005.

83. LC Technologies Inc. (2003)., Home. (Online). Available at http://www.eyegaze.com. Accessed 9 Feb. 2005.

84. Seeing Machines. (2004). Home. (Online). Available at http://www.seeingmachines.com. Accessed 9 Feb. 2005.

85. Tobii Technology. (2005). Home. (Online). Available at http://www.tobii.se. Accessed 9 Feb. 2005.

86. Gitelman DR, Parrixh TB, LaBar KS, Mesulam MM. Real-time monitoring of eye movements using infrared video-oculography during functional magnetic resonance imaging of the frontal eye fields. NeuroImage 2000;11:58–65.

87. Young LR, Sheena D. Survey of eye movement recording methods. Behav Res Methods Instrumen 1975;7:397–429.

88. Wechsler H, Duchowski A, Flickner M. Editorial, special issue: eye detection and tracking. Computer Vision Image Understanding 2005;98:1–3.

89. Eye Movement Equipment Data Base (EMED). (2002, May 2). Home. (Online). Available at http://ibs.derby.ac.uk/emed. Accessed 9 Feb. 2005.

See also ELECTRORETINOGRAPHY; FIBER OPTICS IN MEDICINE; OCULAR MOTILITY RECORDING AND NYSTAGMUS.

F

FES. See FUNCTIONAL ELECTRICAL STIMULATION.

FETAL MONITORING

MICHAEL R. NEUMAN
Michigan Technological
University
Houghton, Michigan

INTRODUCTION

Fetal monitoring is a special type of electronic patient monitoring aimed at obtaining a record of vital physiologic functions during pregnancy and birth. Such monitoring is applied in assessing the progress of pregnancy and labor, and it can identify conditions that concern the clinician caring for the patient. These nonreassuring recordings can lead to special considerations in caring for the pregnant patient and in managing her labor. Although these recordings are no longer considered to be definitive in identifying most forms of fetal distress, they can help to reassure patient and clinician that the fetus is able to withstand the physiologic stress of labor and delivery. The technology is also useful in assessing high risk pregnancies, which, in most cases, is only reassuring as opposed to giving a definitive diagnosis. Although this technology is now recognized to have diagnostic limitations, it is still frequently used in the hospital and clinics as an adjunct to other diagnostic evaluations.

PHYSIOLOGIC VARIABLES MONITORED

The goal of fetal monitoring is to ensure that vital fetus organs receive adequate perfusion and oxygen so that metabolic processes can proceed without compromise and these organs can carry out their functions. Thus, an ideal situation for monitoring from the physiologic standpoint would be to monitor the perfusion and oxygen tension in the fetal central nervous system, heart, kidneys, and brain, with the brain being by far the most important. It is also important to know that the fetus is receiving adequate oxygen and nutrients from the mother through the placenta. Unfortunately, it is not possible to directly or even indirectly measure these variables in the fetus *in utero* using currently available technology. One, therefore, must look for related secondary variables that are practical for monitoring and are related to these critical variables. In the following paragraphs, some of these variables and the methods used to obtain them are described.

METHODS OF MONITORING BY A HUMAN OBSERVER

Any discussion of fetal monitoring must begin by pointing out an obvious, but often overlooked, fact that fetal monitoring does not always require expensive electronic equipment. Basic fetal monitoring can be carried out by a trained clinician using his or her hands, ears, and brain. A fetoscope is a stethoscope especially designed for listening to the fetal heart sounds through the maternal abdomen, which can be used to follow the fetal heart rate (FHR), and a hand placed on the abdomen over the uterus can be used to detect the relative strength, frequency, and duration of uterine contractions during the third trimester of pregnancy and labor. Any woman who has experienced labor will point out that the patient is also able to detect the occurrence of uterine contractions during labor. Although these techniques are only qualitative, they can be quite effective in providing information on the patient in labor and frequently represent the only fetal monitoring that is necessary in following a patient.

The main problems with this type of fetal monitoring are associated with convenience, fatigue, data storage and retrieval, and the difficulty of simultaneously processing multiple inputs. Electronic instrumentation can help to overcome these types of problems. Although electronic devices are less flexible and, at the present time, unable to interpret data as well as their human counterparts, the electronic devices can provide quantitative data, continuously monitor patients with minimal interruption of hospital routines, monitor for extended periods of time without fatigue, store data in forms that can be reevaluated at a later time, and, in some circumstances, make elementary logical decisions and calculations based on the data. Thus, the electronic monitor can serve as an extension of the clinician's data-gathering senses and provide a convenient method of recording and summarizing these data. Such a monitoring apparatus has the potential of allowing the clinician to optimize his or her limited available time.

FETAL HEART RATE MONITORING

The widespread use of electronic fetal monitoring was the result of the development of a practical method of sensing the fetal electrocardiogram and determining the instantaneous fetal heart rate from it. Much of the early work in this area was carried out by Dr. Edward Hon and associates who demonstrated a practical technique for directly obtaining the fetal electrocardiogram during labor (1). Techniques for obtaining the fetal heart rate can be classified as direct or indirect. The former involves invasive procedures in which a sensor must come into contact with the fetus to pick up the fetal electrocardiogram; the latter techniques are relatively noninvasive procedures where the mother's body serves as an intermediary between the fetus and the electronic instrumentation. In this case, the maternal tissue conducts a signal (electrical or mechanical) between the fetus and the surface of the mother's abdomen.

Direct Determination of Fetal Heart Rate

Direct FHR determinations are made from the fetal electrocardiogram (FECG). This signal is obtained by placing an electrode on the fetus and a second electrode in the maternal vaginal fluids as a reference point. These electrodes are connected to a high input impedance, high common-mode rejection ratio bioelectric amplifier. Such a direct connection to the fetus can be made only when the mother is committed to labor, the uterine cervix has dilated at least 2 cm, and the chorioamniotic membranes have been ruptured. In principle, it is possible to pass a wire through the maternal abdominal wall into the uterine cavity and beneath the fetal skin to obtain the FECG, and this technique was experimentally reported in the past (2). The method, however, places the mother and fetus at risk and is not used or suitable for routine clinical application. Indirect methods of determining the FHR that are available today make the application of such a technique unnecessary. Thus, the method that is directly used to obtain the FECG is to attach an electrode to the fetal presenting part through the cervix once the mother is committed to labor and the fetal membranes can be ruptured.

Although many different types of electrodes for obtaining the FECG have been described, best results are obtained when the electrode actually penetrates the fetal skin. The reason is illustrated in Fig. 1. The fetus lines in a bath of a amniotic fluid that is electrically conductive due to its electrolyte content. This amniotic fluid tends to short-out the fetal electrocardiogram on the skin surface, therefore, those potentials that are seen on the surface are relatively weak and affected by noise. Even if it were physically possible to place conventional chest electrodes on the fetus for picking up the electrocardiogram, a poor-quality signal would be obtained because of this shunting effect. The amniotic fluid does, however, provide a good central terminal voltage for the fetal electrocardiogram because it contacts most of the fetal body surface.

The fetal head is normally positioned against the dilating cervix when the mother is in labor, but it is possible for the fetal buttocks or other parts to present first. As the cervix dilates, the skin on the presenting part can be observed through the vagina, and it is possible to place an electrode on or within this skin. If this electrode penetrates the fetal scalp (or other exposed skin surface), it contacts the subcutaneous tissue. As an electrical resistance associated with the surface layers of the fetal skin exists, as indicated in Fig. 1, placing the electrode subcutaneously bypasses this resistance and gives a stronger, more reilable signal. Penetrating the skin also helps to physically keep the electrode in place on the fetus during movement associated with labor.

Various types of penetrating fetal electrodes ranging from fish hooks (3) to wound chips (4) have been developed over the years. Today, the most frequently applied electrode in the helical electrode originally described by Hon et al. (5). This electrode, as illustrated in Fig. 2, consists of a section of a helix of stainless-steel wire on an electrically insulating support. The tip of the wire is sharpened to a point that can penetrate the fetal skin when pressed against it and rotated to advance the helix. Typical dimensions of the wire helix are 5 mm in diameter with 1.25 turn of the wire exposed so that the tip of the helix is 2 mm from the surface of the insulator. A second stainless-steel electrode consisting of a metal strip is located on the opposite end of the insulator from the helix and is used to establish contact with the amniotic fluid through the fluid in the vagina. Lead wires connect the two electrodes to the external monitor.

The electrode is attached to the fetal presenting part by means of a special applicator device, which allows the electrode helix to be pressed against the fetal head to penetrate the skin and be twisted so that the entire wire is advanced beneath the surface of the skin until the insulating portion of the electrode contacts the skin. The flexible lead wires then exit through the vagina and can be connected to the monitoring electronics.

Signal Processing

In fetal heart monitoring, it is desired to have a continuous recording of the instantaneous heart rate. A fetal monitor must, therefore, process the electrocardiogram sensed by

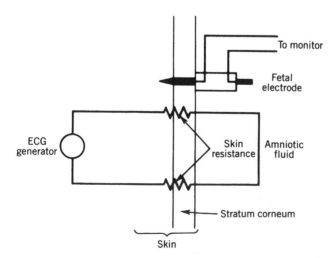

Figure 1. Schematic view and equivalent circuit of a direct fetal ECG electrode penetrating the fetal skin.

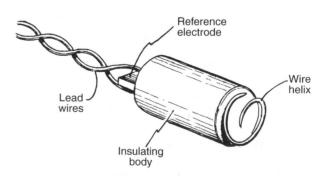

Figure 2. A helical direct fetal ECG scalp electrode of the type described by Hon et al. (5).

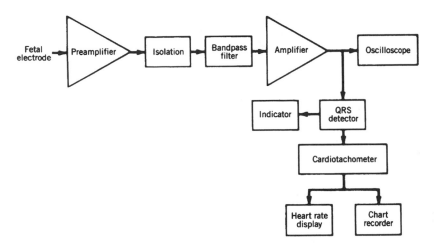

Figure 3. Block diagram of the electronic circuit of a direct fetal heart rate monitor.

the electrode and present the results on a computer monitor or paper printout. A typical electronic system for doing this recording is illustrated in Fig. 3. The signal from the fetal electrode has an amplitude ranging from $50\,\mu V$ to $1.2\,mV$, which is amplified to a more suitable level for processing by an amplifier stage. The input of this aimplifier is electrically isolated and must have a very high input impedance and low lekage current because of the polarizable nature of most fetal electrodes. A high common-mode rejection ratio is also important, because a relatively strong maternal electrocardiogram signal is present on both electrodes. Another characteristic of the amplifier system is that it includes filtering to minimize the amplification of noise and motion artifact from the fetal electrode. As the purpose of the electronics is primarily to display the instantaneous heart rate, the filtering can distort the configuration of the fetal electrocardiogram as long as it does not affect the time at which the QRS complex appears, as it is used to determine the heart rate. For this reason, a relatively narrow band-pass filter is often used. The QRS complex contains higher frequencies than the rest of the electrocardiogram, and noise frequently has a predominance of the lower frequencies. For this reason, the band-pass filter can be centered at frequencies as high as $40\,Hz$.

Many ways exist for the QRS complex can be detected. The simplest of these is a threshold detector that indicates whenever the output voltage from the amplifier exceeds a preset threshold. The level of this threshold is adjusted such that it is usually greater than the noise level but less than the minimum amplitude of a typical QRS complex. The majro limitation of this method lies in the fact that wide variation exists in fetal QRS complex amplitudes. If the threshold level were fixed such that the minimum fetal QRS complex would cross it, this would mean that, for stronger signals, the threshold would not be optimal and interference from noise exceeding the threshold level would be quite possible. One way to get around this problem is to use some type of adaptive threshold. In this case, the threshold level is adjusted based on the amplitude of the electrocardiogram. A simple example of how this can be done is illustrated in Fig. 3. An automatic gain control circuit determines the amplitude of the fetal electrocardiogram at the output of the amplifier, and uses this amplitude to set the gain of that amplifier. This closed-loop

control system, therefore, results in a constant-amplitude electrocardiogram appearing at the output of the amplifier even though the actual signal from the fetal electrode at the input might vary in amplitude from one patient to the next. Using a simple threshold detector with this automatic gain control will greatly improve the reliability of the fetal monitor in detecting true fetal heartbeats. Often, instead of using a simple threshold detector, a detector with hysteresis is used to minimize multiple triggers in the presence of noise. One can also use matched filters in the amplifier to recognize only true QRS complexes. A peak detector may be used to locate the true peak of the QRS complex (the R wave) for better timing, and pattern-recognition algorithms can be used to confirm that the detected pulse is most likely to be a fetal heartbeat. Of course, the best consideration for an accurate determination of the instantaneous fetal heart rate is to have a good signal at the input to the electronic instrumentation. Thus, care should always be taken to have the fetal electrode well positioned on the fetal presenting part so that one has the best possible input to the electronic system.

The cardiotachometer block of the fetal monitor determines the time interval between successive fetal QRS complexes and calculates the heart rate for that interval by taking the reciprocal of that time. Although it is obvious that such a cardiotachometer can introduce errors when it erroneously detects a noise pulse rather than a fetal QRS complex, other errors resulting from the method of heartbeat detection can exist. For a cardiotachometer to accurately determine the heart rate, it must measure the time interval over one complete cycle of the electrocardiogram. In other words, it must detect each QRS complex at the same point on the complex to ensure that the complete cycle period has been recorded. If one beat is detected near the peak of the R wave and the next beat is detected lower on the QR segment, the beat-to-beat interval measured in that case will be too short and the heart rate determined from it will be slightly greater than it should be. Normally, such a concern would be of only minimal significance, because the Q-R interval of the fetal electrocardiogram is short. However, because the variability in fetal heart rate from one interval to the next may be important in interpreting the fetal heart rate pattern, detection problems of this type can affect the apparent variability of the signal and, perhaps, influence the interpretation of the pattern. The output from the cardiotachometer is recorded on one channel of a strip chart recorder and is also often indicated on a digital display. In both cases, the output is presented in the units of beats per minute, and standard chart speeds of 1 and $3\,cm\cdot mm^{-1}$ are used.

Indirect Sensors of Fetal Heart Rate

Indirect methods of sensing the fetal heart rate involve measurement of a physiologic variable related to the fetal heartbeat from the surface of the maternal abdomen. Unlike the fetal scalp ECG electrode, these methods are noninvasive and can be used prior to committing the patient to labor. The most frequently applied method is transabdominal Doppler ultrasound. Lesser used techniques involve transabdominal phonocardiography and electrodiography. Each of these techniques will be described in the paragraphs that follow.

Transabdominal Doppler Ultrasound. Ultrasonic energy propagates relatively easily through soft tissue, and a portion of it is reflected at surfaces where the acoustic impedance of the tissue changes such as at interfaces between different tissues. If such an interface is in motion relative to the source of the ultrasound, the frequency of the reflected signal radiation will be shifted from that of the incident signal according to the Doppler effect. This principle can be used to detect the fetal heartbeat from the maternal abdominal surface. A beam of ultrasound is passed through the abdomen from a transducer acoustically coupled to the abdominal surface. Frequencies around 2 MHz are generally used, because ultrasound of moderate source energy at this frequency can penetrate deep enough into the abdomen to sufficiently illuminate the fetus. Wherever this ultrasound beam encounters an abrupt change in tissue acoustical impedance, some of it is reflected back toward the transducer. If the incident ultrasound illuminates the fetal heart, some of it will be reflected from the various heart-blood interfaces in this organ. Many of these interfaces, such as the valve leaflets, experience periodic movement at rhe rate of the cardiac cycle. In the case of the valve leaflets, relatively high velocities can be obtained during portions of the cardiac cycle. Ultrasound reflected from these interfaces can, therefore, be significantly shifted in frequency so that the reflected wave can be identified at the maternal abdominal surface because of its frequency shift. This frequency shift will be related to the velocity of the reflecting surface and, hence, will be able to indicate each fetal heartbeat. Thus, by detecting and processing this reflected Doppler-shifted ultrasonic wave, it is possible to determine each heartbeat and, hence, the fetal heart rate.

A block diagram of a typical indirect fetal heart rate monitoring system using Doppler ultrasound is shown in Fig. 4(b). As continuous wave ultrasound is used, separate adjacent transducers are employed to establish the ultrasonic beam and detect the Doppler-shifted reflected waves. The reflected ultrasound signal is mixed with the transmitted wave, and beat frequencies are produced when a Doppler shift in frequency occurs for the reflected wave.

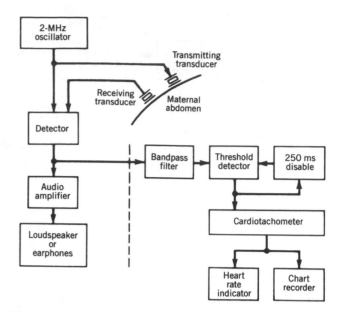

Figure 4. Block diagram of a Doppler ultrasound indirect fetal heart rate monitor. [Reprinted with permission from CRC Press (6).]

This beat frequency is amplified and used to indicate the occurrence of a heartbeat to a cardiotachometer. Many monitors also provide this signal to a loudspeaker to assist the clinical personnel in positioning the transducers for optimal signal pickup or for auditory monitoring.

The reflected ultrasound signal is different from an electrocardiogram, although it can also be used to identify various events in the cardiac cycle. A typical signal is illustrated in Fig. 5. Here one sees two principal peaks per heartbeat, one corresponding to valve opening and the other to valve closing. Actually, such signals can be quite useful in measuring fetal systolic time intervals, but from the standpoint of the cardiotachometer for determining heart rate, they can create problems. If the cardiotachometer is set to trigger at the peak of each wave it sees, as it is for the electrocardiogram, it could measure two beats per cardiac cycle and would give an erroneously high fetal heart rate. One way to avoid this problem is to detect only the first peak of the signal, and, once it is detected, to disable the detection circuit for a period of time that is less than the shortest expected beat-to-beat interval but longer than the time necessary for the second Doppler-shifted signal to occur. In this way, only one peak per cardiac cycle will be registered.

A second, more sophisticated method for detecting the fetal heartbeat involves the use of short-range autocorrelation techniques. The monitor recognizes the beat signal from the reflected wave for a given cardiac cycle and looks for a signal that most closely correlates with this signal over the period of time in which the next heartbeat is likely to occur. The time interval between that time when the initial wave was measured and the point of best correlation corresponds to a beat-to-beat interval of the fetal heart. Thus, instead of relying only on the peaks of the ultrasound signal, this method looks at the entire signal and, therefore, is more accurate. Some manufacturers of commercial fetal monitors claim their ultrasonic systems using this

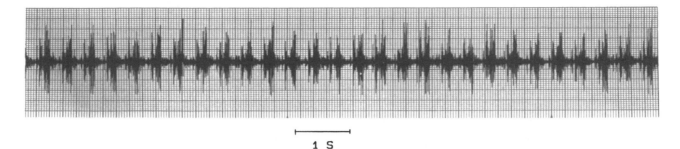

1 S

Figure 5. Illustration of the raw reflected ultrasound signal from the beating fetal heart *in utero*. Note that two ultrasonic bursts occur per cardiac cycle.

type of autocorrelation technique can detect fetal heart-beats as well as can be done by the direct electrocardiographic technique.

The major limitations of the Doppler ultrasound technique are related to its sensitivity to movement. As the Doppler effect will respond to movements of any tissue interfaces illuminated by the ultrasound beam with respect to the signal source, movement of the mother or fetus can result in Doppler-shifted reflected waves that are stronger than the cardiac signal, which, with artifact, can completely obliterate the signal of interest. Thus, this technique is really only reliable when the patient is resting quietly, and it often fails to provide reliable information in the active phase of labor. The other movement-related problem is that the fetus can move *in utero* so that the heart is no longer illuminated by the ultrasound beam or the orientation of the heart with respect to the ultrasonic beam is such that it produces only a minimum. Doppler shift in the reflected ultrasonic wave. Thus, while monitoring a patient, it sometimes necessary to reposition the ultrasonic sensors on the maternal abdomen from time to time because of the movement of the fetus.

Acoustic Pickup of the Fetal Heart. Until the advent of electronic fetal monitoring, the standard method of detecting the fetal heartbeat to measure the fetal heart rate was to use a fetoscope. When the bell of this instrument was firmly pressed against the maternal abdomen, fetal heart sounds could be heard and the heart rate could be determined from them. The acoustic method of indirect fetal heart monitoring follows the fetal heartbeat by a similar technique (7). A sensitive contact microphone is placed on the maternal abdomen over the point where the loudest fetal heart sounds are heard with a fetoscope. The signal picked up by this microphone is filtered to improve the signal-to-noise ratio, and the resulting signal drives a cardiotachometer to give the instantaneous fetal heart rate. The acoustic signal from the fetal heart is similar to the Doppler ultrasound signal in that it generally has two components per heartbeat. The cardiotachometer is set to trigger when the peak signal comes from the acoustic transducer, so it is possible that two apparent fetal heartbeats can exist for each cardiac cycle. Thus, as was the case for the Doppler ultrasound, it is wise to have a processing circuit that selects only the first of the two heart sounds to trigger the cardiotachometer. Unlike the ultrasonic Doppler signal, the fetal heart sounds produce

sharp pulses that are narrower so that the detection of the time of the peak can be more precise. Thus, it is generally more accurate to measure the beat-to-beat interval using a peak detector with the acoustic signal than it is with the Doppler ultrasound. The use of the electrocardiogram still represents the best way to measure beat-to-beat cardiac intervals.

The major limitation of the acoustic method of detecting fetal heart sounds is the poor selectivity of the acoustic transducer. It not only is sensitive to the fetal heart sounds, but it will also respond to any other intraabdominal sounds in its vicinity. Also, a finite sensitivity to environmental sounds, exists which is an especially severe limitation for patients in active labor on a busy, noisy delivery service. For this reason, the acoustic method is limited primarily to patients who can lie quietly in a quiet environment to be monitored. The advent and use of the home-like labor/delivery rooms has helped to create an atmosphere that is more conducive to acoustic fetal heart monitoring, yet it is still not a widely applied approach.

The acoustic technique also has the limitation that when used for antepartum (before labor and delivery) monitoring the fetus can move such that the microphone is no longer ideally positioned to pick up the fetal heart sounds. Thus, it is frequently necessary to relocate the microphone on the maternal abdomen with this monitoring approach.

The major advantages of the acoustic method lie in the fact that not only is there better accuracy in determining the instantaneous fetal heart rate, but unlike the ultrasound method, which must illuminate the fetus with ultrasonic energy, the acoustic method derives its energy entirely from the fetus, and no possibility exists of placing the fetus at risk due to exogenous energy. As a result, investigators have considered the possibility of using the acoustic method for monitoring the high-risk fetus at home (8).

Abdominal Electrocardiogram. Although the fetus is bathed in amniotic fluid located within the electrically conductive uterus and maternal abdomen, one can still see small potentials on the surface of the maternal abdomen that correspond to the fetal electrocardiogram. These signals are generally very weak, ranging in amplitude from 50 to 300 μV. Methods of obtaining the abdominal fetal electrocardiogram and clinical application of the information have been known for many years as described by

Larks (9), yet signal quality remains a major problem. Nevertheless, some methods of improving the quality of the signal have been developed. These methods can allow a much more detailed fetal electrocardiogram to be obtained from the maternal abdomen under ideal conditions, and such electrocardiograms can be used in some cases for more detailed diagnosis than from just looking at heart rate. One of these methods involves applying signal-averaging techniques to several subsequent fetal heartbeats using the fetal R wave as the time reference (10). In this way, the full P-QRS-T wave configuration can be shown, but heart rate information and its variability will be lost.

As the fetal electrocardiogram at the maternal abdominal surface is very weak, it is easy for other signals and noise to provide sufficient interference to completely obliterate the fetal signal. Having the subject rest quietly during the examination and removing the stratum corneum of the skin at the electrode sites can reduce noise due to motion artifact and electromyograms from the abdominal muscles. Nevertheless, one major interference source exists that requires other types of signal processing to eliminate. This source is the component of the maternal electrocardiogram seen on the abdominal leads. This signal is generally considerably higher in amplitude than the fetal signal. Thus, observation of the fetal electrocardiogram could be greatly improved by the elimination or at least the reduction of the maternal signal. One method of reducing this signal involves simultaneously recording the maternal electrocardiogram from chest electrodes and subtracting an appropriate component of this signal from the abdominal lead so that only the fetal signal remains. Under idealized circumstances, this process can give a greatly improved abdominal fetal electrocardiogram, but the conditions for subtraction of the maternal signal are likely to vary during a recording session so that frequent adjustments may be necessary to maintain the absence of the maternal signal (11).

The abdominal fetal electrocardiogram can be used for antepartum fetal heart monitoring. In this case, the goal of the instrumentation is to collect fetal R-R intervals as done with the direct monitoring of the fetal electrocardiogram and to determine the instantaneous heart rate from these intervals, which strong maternal component in the abdominal fetal electrocardiogram can make a very difficult task electronically, and so most abdominal fetal electrocardiogram fetal heart rate monitors need to eliminate the maternal component of the abdominal signal. The substraction method described in the previous paragraph would be ideal for this purpose because if fetal and maternal heartbeats occur in approximately the same time, subtracting the maternal component should leave the fetal component unaffected. Unfortunately, because the conditions under which the maternal component is added to the fetal signal change from one minute to the next, it is not always practical to use this subtraction technique. Thus, a simpler technique that loses more information is used.

A typical abdominal fetal electrocardiogram is shown in Fig. 6 (lower panel) along with a direct fetal electrocardiogram taken from a scalp electrode and the maternal electrocardiogram taken from a chest lead. In the abdominal

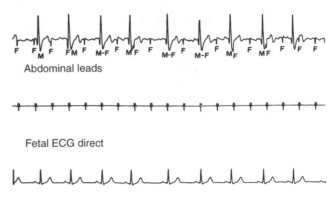

Abdominal leads

Fetal ECG direct

Maternal ECG

Figure 6. An example of a fetal electrocardiogram as obtained from the maternal abdomen. F, fetal QRS complexes; M, maternal QRS complexes. The direct fetal electrocardiogram and maternal electrocardiogram are recorded simultaneously for comparison. [Reprinted with permission from CRC Press (6).]

fetal electrocardiogram, fetal heartbeats are indicated by F and maternal heartbeats by M. Note that some beats exist where the fetal and maternal heartbeats occur at the same time. The strategy of the abdominal fetal electrocardiogram/fetal heart rate monitor is to monitor two signals, the maternal electrocardiogram from a chest lead and the fetal and maternal electrocardiograms from an abdominal lead. As shown in the block diagram in Fig. 7, the maternal electrocardiogram triggers a gate such that the input from the abdominal lead is interrupted every time a maternal beat occurs. Thus, this process eliminates the maternal component from the abdominal signal, but it can also eliminate a fetal QRS complex if it occurs at a time close to or during the maternal QRS complex. Thus, the cardiotachometer estimates the intervals where one or more fetal beats is missing. Due to the random relationship between maternal and fetal heartbeats, it is most likely that only one fetal beat would be missing at a time because of this mechanism, and so when maternal and fetal beats coincide, the fetal R-R interval should be approximately double the previous interval. Some monitors look for this condition and imply that it is the result of simultaneous maternal and fetal beats. The monitor, therefore, artificially introduces a fetal beat at the time of the maternal beat so that an abrupt (and presumably incorrect) change in the fetal heart rate will not occur.

Although such processing of the fetal signal makes the resulting heart rate recordings appear to have less artifact, this technique can loose some important information. For example, if the fetus suffers from a cardiac arrhythmia such as second-degree heart block, in which the fetal heart can miss a beat every so often, the monitor would reintroduce the missing beat, and this arrhythmia would not be detected. The principal advantage of the abdominal electrocardiogram method of fetal heart rate monitoring is that it can, under optimal conditions, provide the closest indirect observation of the fetal heart rate as compared with direct observations. No risk to the patient exists from this procedure, and inexpensive disposable electrodes can be used as the sensors.

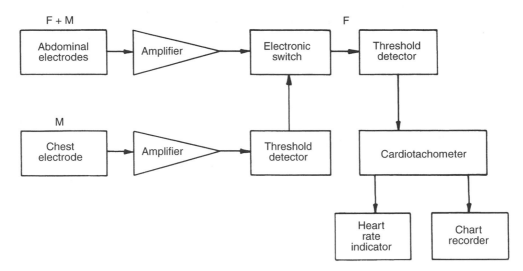

Figure 7. Block diagram of a monitor for processing the abdominal electrocardiogram shown in Fig. 6 using the anticoincidence detector method. [Reprinted with permission from CRC Press (6).]

The limitations of this method include its being based on a very weak signal in an environment that can contain a great amount of artifact. Thus, low signal-to-noise ratios are frequently encountered. Patients must be resting quietly for the method to work. Furthermore, electrodes must be optimally placed for good results, which requires some experimentation with different electrode sites, and skill is required on the part of the user in finding optimal electrode positions for a particular patient. Various signal processing techniques have been used over the years to get a more reliable fetal signal from the abdominal surface, but most of these techniques only improve signal quality under very special circumstances (12,13).

UTERINE CONTRACTIONS

Although the fetal heart rate is an important fetal variable for clinical monitoring, an equally important maternal variable is uterine activity. In fetal monitoring, one must detect the occurrence of uterine contractions, their frequency, their duration, and their intensity. As was the case with the fetal heart rate, it is possible to monitor uterine contractions by both direct methods and indirect methods.

Direct Monitoring of Uterine Contractions

Uterine contractions are periodic coordinated contractions of the myometrium, the muscle of the uterine wall. In an ideal method of direct measurement of uterine contractions, the tension and displacement of the myometrium would be measured, but this measurement cannot be done for routine fetal monitoring as only invasive methods of making this measurement exist. Uterine contractions, however, are reflected in increases in hydrostatic pressure of the amniotic fluid within the pregnant uterus. If this fluid is continuous and in a closed system, pressure increases resulting from uterine contractions should be seen throughout the amniotic fluid and should be related to the overall strength of the contraction but not necessa-

rily to the tension at any one particular location in the myometrium.

The pressure change in the amniotic fluid during a contraction can be measured directly by coupling the amniotic fluid to a manometer, which consists of an electrical pressure sensor and the appropriate electronic circuitry for processing and indicating or recording the measured pressure. Intrauterine pressure can be measured by placing the pressure sensor directly in the amniotic fluid or by using a fluid-filled catheter to couple the amniotic fluid to an external pressure sensor. This latter method is the method most frequently employed in clinical fetal monitoring. The catheter used for coupling the amniotic fluid to an external pressure sensor can be placed only when the membranes surrounding the fetus have been ruptured, which should only be done if the patient is in labor. Unfortunately, rupture of the fetal membranes sometimes occurs spontaneously before the patient goes into labor or when the patient is in premature labor. It is unwise to place a catheter under these circumstances unless labor will be induced and the patient will deliver within 24 h. The reason is that the catheter can serve as a conduit for introducing infectious agents into the uterus or such agents can be introduced during the process of placing the catheter. When the distal tip of the catheter is within the amniotic fluid and its proximal end is connected to a pressure sensor at the same elevation as the distal end, the pressure seen at the sensor will, according to Pascal's law, be the same as that in the amniotic fluid. Thus, when a contraction occurs, the pressure increase will be transmitted along the catheter to the external pressure sensor.

Although the fluid-filled catheter provides a direct conduit from the amniotic fluid to the externally located pressure sensor, it can also be responsible for measurement errors. As was pointed out earlier, the proximal and distal ends of the catheter must be at the same level if one is to avoid the gravitational hydrostatic errors that give incorrect baseline pressure readings. Pascal's law applies only to the static solution where no fluid movement exists in the system. Once fluid movement occurs in the catheter, pressure drops along the length of the catheter can result. Such fluid movement can occur when a small leak in the plumbing system exists at the sensor end of the catheter. Movement of the catheter itself or of the patient with respect to the pressure sensor can also produce alterations in the observed dynamic pressure. The most serious violation of Pascal's law is that once the fetal membranes have been ruptured, a truly closed system no longer exists. The fetal head or other presenting part approximated against the cervix does, indeed, isolate the intrauterine amniotic fluid from the outside world, but amniotic fluid can leak through the cervix, thus, no longer providing a static situation. Furthermore, after membranes have been ruptured, the total amount of amniotic fluid in the uterine cavity is reduced. It is possible that there might be local non-communicating pools of amniotic fluid delineated by fetal parts on one side and by the uterine wall on the other. The pressure in one of these isolated pools possibly can be different from that of another. The measured pressure will, therefore, be dependent on which pool contains the distal tip of the catheter. A statistical study by Knoke et al. has shown that when three identical catheters are placed in the pregnant uterus, the pressure measured by each can be considerably different, and differences of more than 10 mmHg (1.3 kPa) can be seen between different sensors (14), which is probably due to the fact that the distal tip of each catheter is located in a different part of the uterus and is coupled to a pocket of amniotic fluid at a different pressure.

Other problems exist that can affect the quality of intrauterine pressure measurement with the catheter-external sensor method. Poor recordings are obtained when catheter when catheter placement is not optimal and when limited communication exists between the fluid in the catheter and the intrauterine amniotic fluid. Many catheters in use today have a single hole either in the end or at the side of the catheter that communicates with the amniotic fluid. Mucus or vernix caseosa (a substance of a consistency similar to soft cheese that is found on the fetus) can obstruct or partially obstruct this opening resulting in poor quality recordings. When the distal tip of an open-ended intrauterine catheter becomes obstructed, the obstruction can frequently be "blown" off by forcing fluid through the catheter. In practice, this procedure is done by attaching a syringe filled with normal physiologic saline at the proximal end of the catheter near the pressure sensor and introducing fluid into the catheter when an obstruction is suspected.

It is possible to minimize these obstruction problems by modifying the catheter (15). Increasing the number of openings at the catheter tip is one of the simplest ways of minimizing obstructions. By placing an open-celled sponge on the catheter tip, it is possible to obtain a greater surface area in contact with the amniotic fluid because of the multiple openings of the sponge, which also tends to keep the tip of the catheter away from fetal or uterine structures, minimizing the possibility of complete obstruction or injury to the fetus. A small balloon placed at the distal tip of the catheter will prevent the fluid in the catheter from making actual contact with the amniotic fluid, and so this interface cannot be obstructed. As the

wall of the balloon is flexible, the pressure in the fluid within the balloon will be equal to the pressure in the fluid surrounding the balloon plus the pressure resulting from the tension in the balloon wall itself. This system, however, has the disadvantage that it will respond to a direct force on the balloon as well as to hydrostatic pressure in the fluid outside of the balloon; thus, fetal movements or the entrapment of the balloon between the fetus and the uterine wall during a contraction can lead to erroneous pressure measurements.

Interuterine pressure can be directly measured using a miniature pressure sensor that can be placed in the intrauterine cavity (16). These devices are based on a miniature silicon pressure sensor that can be placed on the tip of a probe that has a similar appearance to an intrauterine catheter. In some cases, the probe at the catheter tip is no longer than the catheter itself, so the method of placement is the same as that used for the catheter. The advantage of the intrauterine pressure sensor is its location within the uterine cavity, which aviods artifact introduced by the fluid-filled catheter, and the problem of zeroing the pressure measurement system due to elevation differences is avoided because the sensor is at the pressure source. Investigators have compared the performance of intrauterine sensors with that of intrauterine catheters and have found the newer devices to provide equivalent data to the previously accepted technology (17).

Indirect Monitoring of Uterine Contractions

The clinician is able to sense uterine contractions by palpating (feeling) the maternal abdomen. Indirect uterine contraction sensors known as tocodynamometers are electrical sensors for doing the same thing. The basic principle of operation of these sensors is to press against the abdomen to measure the firmness of the underlying tissues. A contracting muscle will feel much more firm than a relaxed one. Most tocodynamometers carry out this function by pressing a probe against the abdomen and measuring its displacement.

The construction of a typical tocodynamometer is shown in Fig. 8. The sensor is held in place against the surface of

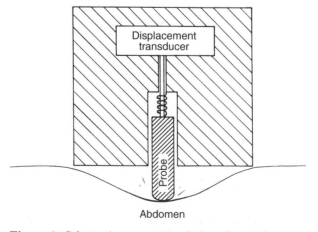

Figure 8. Schematic cross-sectional view of a tocodynamometer. [Reprinted with permission from CRC Press (6).]

the abdomen with an elastic strap. A movable probe protrudes beyond the surface of the sensor so that it causes a slight indentation in the abdominal wall. It is loaded by a spring, which makes it somewhat compliant; it can be either extended further into the abdominal wall or retracted into the body of the sensor depending on the firmness of the tissue of the abdominal wall under the probe. In some tocodynamometers, the spring tension, and hence the force that the probe exerts on the abdomen wall, can be adjusted by means of a small knob so that optimal operation of the sensor can be achieved. What a uterine contraction occurs, the abdominal wall will become tense, and it tends to push the probe back into the housing of the tocodynamometer. Following the contraction, the spring is again able to push the probe deeper into the abdomen. In some tocodynamometers this actual movement is very slight, whereas in others it can be as great as a few millimeters.

A displacement sensor inside the tocodynamometer provides an electrical signal proportional to the position of the probe. This displacement reflects myometrial activity. Different types of displacement sensors can be used in tocodynamometers. Including a strain gage on a cantilever arm, mutual inductance coils, a linear variable differential transformer, or a piezoelectric crystal.

The principal advantage of the tocodynamometer is the noninvasive way in which it measures uterine contractions. It is the only method that can be safely used before the patient is in active labor. It has serious limitations, however, in the quantitative assessment of labor. The method can be used only to quantitatively determine the frequency and duration of uterine contractions. Its output is only qualitative with respect to the strength of the contractions. Signal levels seen are a function of the position of the sensor on the maternal abdomen and the tension of the belt holding it in place. Signal amplitudes are also strongly related to maternal anatomy, and the method is virtually useless in obese patients. Many patients in active labor complain that the use of the tocodynamometer with a tight belt is uncomfortable and irritating.

Electronic Signal Processing

A block diagram for the uterine contraction channel of an electronic fetal monitor is illustrated in Fig. 9. The sensor can be either an internal or external pressure transducer or a tocodynamometer. Signals are sometimes filtered in the amplifier stages of the monitor because the uterine contraction information includes only dc and very low ac frequencies. Nevertheless, filtering is generally not necessary for high quality signals, and often the presence of artifact due to breathing movements of the patient is useful in demonstrating that the pressure measuring system is functional.

In some cases, it is necessary to adjust the baseline pressure to establish a zero reference pressure when using the monitor. In the case of direct uterine contraction monitoring when a single pressure sensor is always used with the same monitor, this adjustment should be made by the manufacturer, and additional adjustment should not be necessary. As a matter of fact, making such a zero-level adjustment control available to the operator of the monitor runs the risk of having significantly altered baseline pressures that can affect the interpretation of uterine basal tone. The adjustment of a zero-level control should not replace the requirement of having the proximal and distal end of the fluid-filled catheter at the same level. It is far better to adjust zero levels in uterine pressure monitoring by raising or lowering the external pressure transducer than by adjusting the electrical zero. On the other hand, when the tocodynamometer is used, no physiologically significant zero level exists. It is not possible to establish uterine basal tone with a tocodynamometer. Baseline levels are frequently dependent on how the tocodynamometer is attached to the patient and the structure of the sensor itself. In this case, it is reasonable to adjust the baseline level between uterine contractions so that the tracing conveniently fits on the chart. When doing so, it is important that the chart indicates that the uterine contractions were measured using a tocodynamometer so that the individual reading the chart does not ascribe inappropriate information to the baseline.

Uterine Electromyogram

The uterus is a muscle, and electrical signals are associated with its contraction as they are for any kind of muscle. These signals can be detected from electrodes on the maternal abdomen or the uterine cervix during uterine contractions. Garfield and others have studied these signals and suggested that they might be useful in managing

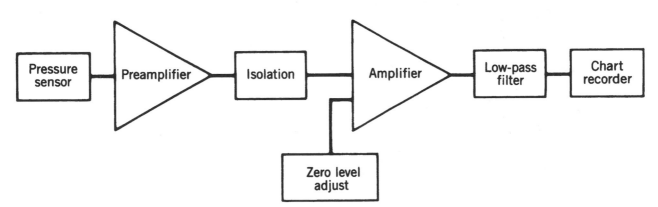

Figure 9. Block diagram of the signal processing electronics for intrauterine pressure measurement.

patients during pregnancy and, perhaps, even during labor (18–20). These techniques are still experimental and not yet ready for clinical application. Nevertheless, they offer a new approach to assessing uterine activity and the possibility of differentiating contractions leading to cervical dilatation from those that are nonprogressive.

THE FETAL CARDIOTOCOGRAPH

Electronic fetal monitoring is accomplished using a fetal cardiotocograph, such as illustrated in Fig. 10, which is basically a two-channel instrument with a two-channel chart recorder as the output indicator. One of the channels records the fetal heart rate, whereas the second channel records the uterine contractions. Most cardiotocographs are capable of accepting direct or indirect signals as inputs for each channel, although specialized monitors for antepartum assessment have only the indirect signal input capability. To aid clinicians in interpreting monitored patterns, most instruments use chart paper that is 70 mm wide for the fetal heart rate channel and calibrated from 30 to 240 beats·min^{-1}. The uterine contraction channel is 40 mm wide and calibrated with a scale from 0 to 100. The scale is only qualitative when a tocodynamometer is the input source, but corresponds to the pressure in millimeters of mercury when direct sensors of uterine contractions are used. The standard speeds for the chart paper are 1 or 3 cm/min. The use of a standardized chart and chart speed results in fetal heart rate—uterine contraction patterns that appear the same no matter what monitoring device is used—which is important because cardiotocograms are read by visually recognizing patterns on the chart. Changing the chart speed and scale significantly changes the appearance of the patterns even though the data remain unchanged. Thus, a clinician would have to learn to interpret patterns from each of the different types of monitors used if they each had different chart speeds and scales, because the same pattern can appear quite different when the chart speed or signal amplitude is changed.

Information Obtained from Fetal Cardiotocography

In interpreting a cardiotocogram, a clinician considers the heart rate and uterine contraction information separately as well as the interaction between the two signals. The frequency, duration, and, in the case of direct monitoring, amplitude and baseline information have already been discussed. Similar types of information can be obtained from the directly and indirectly monitored fetal heart rate recordings. Specifically in the fetal heart rate channel, one looks for the average baseline value of the fetal heart rate, which should generally be in the range 120–160 beats·min^{-1} and when outside of this range can be cause for concern. The beat-to-beat variability of the fetal heart rate can also be an important indicator of fetal condition, and so the use of an instantaneous cardiotachometer in a cardiotocograph is mandatory. Certain recurring patterns in the fetal heart rate recording can also be important indicators of fetal condition. Sinusoidally varying fetal heart rate has been described as an ominous sign (21), and sometimes fetal cardiac arrhythmias can be detected by observing the heart rate pattern.

The information that is most frequently obtained from the cardiotocogram and applied clinically comes from both the heart rate and uterine contraction channels and is concerned with the relationship between these two signals. One can consider a uterine contraction as a stress applied to the fetus and the resulting changes in the fetal heart rate as the response to this stress. When the changes occur in direct relationship to the uterine contractions, they are referred to as periodic changes in the fetal heart rate. Several possibilities exist for fetal heart rate changes during and following a uterine contraction. One can see no change, an acceleration, or a deceleration in the fetal heart rate. In the case of decelerations, three basic patterns are seen, and representative examples of these are shown in Fig. 11. The different patterns are characterized by the shape of the deceleration curve and the temporal relationship of its onset and conclusion with the uterine contraction.

Early decelerations begin during the rising phase of the uterine contraction and return to baseline during the falling phase. They frequently appear to be almost the inverse of the uterine contraction waveform. Periodic decelerations of this type are thought to not represent a serious clinical problems.

Late decelerations refer to fetal heart rate decelerations that begin during a uterine contraction but late in the duration of that contraction. The rate of heart rate descent is not rapid, and the deceleration lasts beyond the end of the contraction and then slowly returns to baseline. Such patterns sometimes can be associated with fetal distress, although they should not be considered definitive of fetal distress.

The third type of periodic deceleration of the fetal heart rate is known as a variable deceleration. In this pattern, the deceleration of heart rate is sharp and can occur either early or late in the duration of the uterine contraction. Following the contraction, a rapid return to baseline values occurs. Sometimes one sees rapid return to baseline while the uterus is still contracting and then a rpaid fall back to the reduced heart rate. Variable decelerations have a flat "U" shape, whereas early and late decelerations represent a more smooth curve that could be characterized as shaped as the letter "V" with the negative peak rounded. Variable

Figure 10. A commercially available fetal cardiotocograph. (Courtesy of Portage Health System, Hancock, Michigan.)

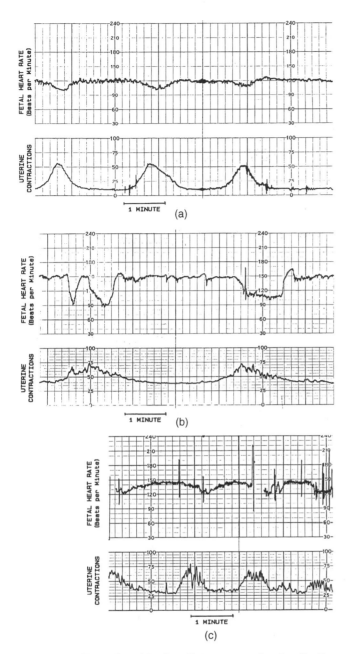

Figure 11. Examples of fetal cardiotocograms showing the three basic patterns: (a) early deceleration, (b) late deceleration, and (c) variable deceleration.

decelerations can be sometimes associated with involvement of the umbilical cord, and, in some cases, they can indicate the presence of fetal distress. As more recent clinical studies have shown that a simple relationship between late and variable decelerations and fetal compromise, does not exist these patterns are not considered to indicate fetal distress as they once were. Now clinicians refer to them as being "nonreassuring," and their presence should encourage the application of other clinical measures to evaluate the fetus.

These basic thoughts for interpreting the fetal cardiotocogram are very elementary and should not be used for diagnostic purposes. The reader is referred to the current obstetrical literature for more detailed descriptions of fetal cardiotocogram and their clinical significance.

Clinical Applications of Fetal Cardiotocography

Electronic fetal monitoring can be applied during the antepartum (before labor and delivery) and intraparturn (during labor and delivery) periods of pregnancy. In the anterpartum period, only indirect methods of fetal monitoring can be used. A primary application of fetal cardiotocography in this period is in nonstress testing. In this test, a cardiotocograph is applied to a patient who is resting quietly. In the United States, the ultrasonic Doppler method of detecting the fetal heart rate and the tocodynamometer are the sensors of choice. The patient is monitored for 1–2 h, and the cardiotocogram is examined for spontaneously occurring uterine contractions of fetal movements, which can also be indicated by the tocodynamometer. In some cases, the mother is asked to activate an event marker on the chart when she feels a fetal movement. The response of the fetal heart rate to these stimuli is noted in interpreting the cardiotocogram. In a reactive nonstress test, a response to these stimuli occurs, which is usually in the form of a brief fetal heart rate acceleration following the uterine contraction or fetal movement. Although nonstress testing is not routinely applied to apparently normal pregnancies, it is indicated for complications of pregnancy such as maternal diabetes, Rh sensitization, intrauterine growth retardation , decreased fetal movement, known fetal anomalies, oligohydrainnios or poly-hydramnios (too little or too much amniotic fluid), pregnancy-induced hypertension, pregnancy lasting beyond the normal 40 weeks, and other maternal and fetal complications.

A second antepartum test involving fetal cardiotocography is the oxytocin challenge test, which is usually applied when the nonstress test yields positive results, such as when fetal heart rate decelerations follow spontaneous uterine contractions or fetal movements. In this test, the patient is given intravenous oxytocin, a hormone that stimulates uterine contractions. The response of the fetal heart rate to the induced contractions is then examined, looking for the periodic changes described before.

Intrapartum monitoring of the fetal heart and uterine contractions can be carried out using the indirect techniques in early labor with the direct techniques applied during active labor. The indications for intrapartum fetal monitoring are controversial. Some obstetricians feel that all labors should be monitored whether they are complicated or not, whereas others feel that only those patients considered being at risk should have monitors. As monitoring is no longer considered to give a definitive diagnosis of fetal distress, some clinicians find it of little value and to not make use of the technology. As internal monitoring gives the most efficacious results, this modality is recommended in cases when it can be applied and the indirect methods do not give satisfactory results. Otherwise, indirect methods can be used as long as they give readable results.

The preceding paragraphs describe fetal cardiotocography as clinically applied in most major medical centers. Although this technology has the advantage of providing continuous surveillance of the mother and fetus, it also has some limitations that prevent if from providing optimal information to obtain the earliest indications of fetal or maternal problems. The major limitation is in the data. Although uterine contractions provide a good indication of the intensity of labor, they do not necessarily indicate its effectiveness in dilating the cervix and expelling the fetus. If, in addition to uterine contractions, one should monitor whether labor is progressing, better information about some maternal aspects of labor and delivery could be obtained.

A similar argument can be made for the use of the fetal heart rate as the primary variable for evaluating the status of the fetus. Heart rate is a very non specific variable, and, in same cases, the fetus must be seriously compromised before any problem is detected by the heart rate. The goal of fetal monitoring as mentioned at the beginning of this article is to make certain that vital organs such as the fetal brain are adequately perfused so as to receive necessary nutrients and oxygen. Although the heart rate is related, it is not the principal variable for determining this perfusion.

Accepting these principal limitations for the variables measured, limitations still exist to the practical application of the cardiotocograph. Sensor placement, especially for indirect monitoring, is important for optimal recordings. The operator of the instrumentation, therefore, must be skilled in determining the best placement for the sensors. Most cardiotocographs are connected to the sensors on the patients by wires and catheters. Although this method is quite adequate while the patient is in bed, it can become quite inconvenient when it is necessary to transfer the patient to another location or to have the patient stand up and walk around. Many of these problems have been overcome by the use of biotelemetry for fetal monitoring (see Biotelemetry).

A final limitation of fetal cardiotocography is associated with the fact that some of the monitored patterns are not easily recognized and interpreted, which means that different clinicians looking at the data can see different things, lead to uncertain diagnoses. Periodic decelerations are usually not as clear, as illustrated in Fig. 11. Again, experience is an important factor here. Even when patterns can be readily determined, the relationship between certain patterns and pathology is not completely clear. As is so often the case medicine, one can only suggest from monitored tracets that certain problems might be present, and other tests need to be performed for confirmation.

OTHER METHODS OF FETAL MONITORING

Although the cardiotocogram is the usual method used to monitor the fetus, other techniques have been developed and experimentally employed to more accurately assess fetal status during the antepartum and intrapartum periods. One of these techniques, fetal microblood analysis is routinely used at major medical centers that care for patients deemed to have high risk pregnancies; the other techniques are still experimental or relatively new and have not enjoyed routine application at the time of this writing.

Fetal Microblood Analysis

About the time when electronic fetal monitoring was developed, Saling (22) was working on a new technique for taking a small sample of fetal capillary blood during active labor and measuring its hydrogen ion activity. This technique, known as fetal microblood analysis, made it possible to determine whether acidosis that could be associated with fetal distress was present during the labor. The technique involves observing a portion of the fetal presenting part (usually the scalp) through the cervix using a vaginal endoscope. By cleaning this portion of fetal skin and even, in some cases, shaving a small amount of hair from the scalp, the obstetrician is able to make a small superifical incision in the skin using a scalpel blade. A droplet of capillary blood will form at this site, and it can be collected in a miniature heparinized glass pipet. Generally, 100–300 μL of blood can be collected in this way. The blood sample is transferred to a special instrument designed to measure the pH of very small blood specimens. This instrument can be a part of a more extensive blood gas analysis instruments in a blood gas laboratory or it can be a relatively simple bedside device that uses disposable pH sensor cartrnidges. In either case, it is possible to measure the pH of this small sample and get the results back to the clinician within a few minutes of collecting the sample.

Chronic hypoxia can cause tissue and, hence, blood pH to drop as a result of the formation of acidic products of anaerobic metabolism such as lactic acid. Thus, if a blood sample is found to have a low pH (most clinical guidelines say lower than 7.2 or in some cases 7.15), it is possible that the fetus is experiencing some form of distress. Often, this technique is used in conjunction with fetal cardiotocography. When the cardiotocograph indicates possible fetal distress, such as when late decelerations are seen, the clinician can get a better idea as to whether distress is indeed present by performing a fetal microblood analysis. If the results indicate acidosis, the probability of actual fetal distress is higher, and appropriate actions can be taken.

A major limitation of the Saling technique is that it gives only an indication of the fetal acid-base status at the time the blood sample was taken. It would be far better to have a continuous or quasi-continuous measure of fetal tissue pH. Stamm et al. (23) have described a technique in which a miniature glass pH sensor is placed in the fetal scalp during active labor. This sensor can continuously record the pH of the fetal scalp. Clinical studies of this technique have shown that a drop in fetal tissue pH can occur along with a cardiotocographic indication of fetal distress (24). The major limitation of this as yet experimental technique is technical. The sensor is fragile, and it is not always possible to obtain efficacious recordings from it. Other sensor are under development in an attempt to overcome some of these limitations (25), yet this technique remains experimental due to the lack of practical devices.

Monitoring of Fetal Blood Gases

Many investigators have been interested in developing technology to continuously monitor fetal oxygenation during active labor and delivery. A review of some of the earlier techniques showed different types of oxygen sensors that could be placed in the fetal scalp using structures similar to electrodes for directly obtaining the fetal electrocardiogram. Investigators also have used transcutaneous oxygen sensors on the fetus (26), and the most recent approach has been the uses of fetal pulse oximetry (27–29). In the transcutaneous oxygen case (see Blood Gas Measurement, Transcutaneous), a miniature sensor is attached to the fetal scalp once the cervix has dilated enough to make this physically possible, and fetal membranes have been ruptured. The technique is considerably more difficult than that for neonates, and it is important to have a preparation where the sensor surface is well approximated to the feal skin so no chance exists for environmental air to enter the electrode, as fetal Po_2 is much lower than that of the air. Most investigators who use this technique experimentally find that gluing the sensor to a shaved region of fetal scalp is the best technique to maintain contact (26).

Fetal pulse oximetry is performed in a similar way, but the sensor probe does not have to be physically fixed to the fetus as was the case for the transcutaneous oxygen tension measurement described above (27–29). Instead, the probe is a flat, flexible structure that contains ligh-emitting diodes at two different wavelengths and photodetector for sensing the reflected light. It is slid between the fetal head and the cervix once the head is engaged and membranes have been ruptured and is oriented so that the light sources and detector are pressed against the fetal skin by the uterine wall. The reflected light at each wavelength will vary in intensity as the blood volume in the fetal tissue changes over the cardiac cycle. As with the routine clinical pulse oximeter, the ratio of amplitudes of the reflected light at the different wavelengths is used to determine the oxygen saturation of the fetal arterial blood.

Recent improvements in the technology of making transcutaneous carbon dioxide sensors have allowed miniature transcutaneous sensors to be built in the laboratory. These have been applied to the fetus during active labor to continuously measure carbon dioxide tensions (30). All of these transcutaneous methods of measuring fetal blood gases are experimental at the time of this writing and have limitations regarding the technique of application and the quality of recorded information. Nevertheless, they present an interesting new approach to monitoring the fetus using variables more closely related to fetal metabolism and, hence, with greater potential for accurately detecting fetal distress.

Fetal Activity and Movements

The amount of time that the fetus spends in different activity states may be an important indicator of fetal condition. The fetus, as does the neonate, spends time in different activity states. Part of the time it may be awake and active, moving around in the uterus; at other times, it may be quiet and resting or sleeping. By establishing norms for the percentage of time that the fetus spends in these states, one can measure the activity of a particular fetus over a period of time and determine whether it falls within the normal classifications as a means of evaluating fetal condition.

One of the simplest ways to measure fetal activity is to have the mother indicate whether she feels fetal movements over a period of time, which can be done and recorded for several days as an assessment of fetal wellbeing. Fetal movements can also be detected by tocodynamometers. If the fetus is located under the probe of a tocodynamometer and moves or kicks, it can be detected as a short-duration pulse of activity on the chart recording from the sensor. Maternal movements can appear on this sensor as well, and so it is not easy to differentiate between the two. Timor-Trich et al. have developed a technique using two tocodynamometers to minimize this problem (31). By placing one over the fundus of the uterus and the second at a lower level, and recording the signals on adjacent channels of a chart recorder, fetal movements very often either are seen only on one sensor or produce pulses of opposite sign on the two sensors. Maternal movements, on the other hand, are usually seen on both sensors and are similar in shape and sign.

One of the most elegant methods of measuring fetal movements is to directly observe these movements using real-time ultrasonic imaging (see Ultrasonic Imaging). The main limitation of this technique is that an ultrasonographer must continuously operate the apparatus and reposition the ultrasonic transducer to maintain the best image. It also requires the subject to rest quietly during the examination. Although not believed to be a problem, no definite evidence currently exists that long-term exposure of the fetus to ultrasonic energy is completely safe.

One special type of fetal movement that is of interest to obstetricians is fetal breathing movement. The fetus goes through periods of *in utero* movement that are very similar to breathing movements. The relative percentage of these movements during a period of time may be indicative of fetal condition (32). Such movements can be observed using real-time ultrasound as described above. One can also select specific points on the chest and abdomen and use the ultrasonic instrument to record movements of these points as a function of time as one does for echocardiography (see Echocardiography). Measurement of fetal breathing movements by this technique also requires an experienced ultrasonographer to operate and position the instrumentation during examinations. For this reason, it is not a very practical technique for routine clinical application.

Fetal Electroencephalography

As one of the principal objectives of fetal monitoring is to determine if conditions are adequate to maintain fetal brain function, it is logical to consider a measure of this function as an appropriate measurement variable. The electroencephalogram (EEG) is one such measure that is routinely used in the neurological evaluation of patients. The EEG from the fetus during labor has been measured and shown to undergo changes commensurate with other indicators of fetal distress during labor and delivery (33,34). The monitoring of fetal EEG involves placement of two electrodes on the fetal scalp and measurement of the differential signal between them. These electrodes can be similar to the electrodes used for detecting the fetal electrocardiogram, or they can be electrodes especially designed for EEG. Of course, when either of these electrodes is used in the unipolar mode, the fetal electrocardiogram can be obtained.

Rosen et al. obtained good-quality fetal EEG recordings using a specially designed suction electrode (34). By observation of configurational or power spectrum changes in the EEG, it may be possible to indicate conditions of fetal distress.

Continuous Monitoring of Cervical Dilatation

In the routine method used to assess the progress of labor, the examiner places his or her fingers in the vagina and feels the uterine cervix to determine its length, position, and dilatation. Although this technique is simple and quick and requires no special apparatus, it has some limitations as well. It is an infrequent sampling method, and each time a measurement is made there can be discomfort for the patients as well as risk of intrauterine infection. The technique is also very subjective and depends on the experience of the examiner. A more reliable and reproducible technique that is capable of giving continuous records could be useful in the care of high-risk patients and patients with increased risk of intrauterine infection. Mechanical, caliper-like devices attached to opposite sides of the cervix have been described by Friedman (35) and others. These devices measure a cervical diameter with an electrical angular displacement transducer attached to the calipers. These devices are somewhat big and awkward, and Richardson et al. have optimised the mechanical structure by reducing its size (36). Other investigators have eliminated the mechanical calipers and used a magnetic field to measure the distance between two points on diametrically opposed sides of the cervix (37). In another technique for continuously monitoring cervical dilatation reported by Zador et al., ultrasound is used to measure the cervical diameter (38). A brief pulse of ultrasound is generated at a transducer on one side of the cervix and is detected, after propagating across the cervical canal, by a similar transducer on the opposite side. By measuring the transit time of the ultrasonic pulse between the two transducers, one can determine the distance between them, because ultrasound propagates through soft tissue a nearly constant known velocity. By generating an ultrasonic pulse once a second, a continuous recording of cervical dilatation as a function of time can be produced, which can be recorded either on an adjacent channel with the fetal cardiotocogram or on a separate display that generates a curve of cervical dilatation as a function of time known as a labor graph. Many clinicians plot such a curve as a result of their digital examinations of the cervix.

SUMMARY

As seen from this article, the use of biomedical instrumentation in obstetrical monitoring is fairly extensive, but the variables measured are not optimal in achieving the goals of fetal monitoring. Some of the newer and yet experimental techniques offer promise of getting closer to the question of whether vital structures in the fetus are being adequately perfused, but at the present time, none of these techniques are ready for general widespread application. Fetal monitoring is important if it can detect correctable fetal distress, as the results of such distress can remain with the newborn for life. It is important that the fetal monitoring techniques used will eventually benefit this patient. Some critics of currently applied fetal cardiotocography claim that the only result of fetal monitoring has been increase in the number of cesarean sections performed, and this might have a negative rather than positive effect on patient care. It is important that as this area of biomedical instrumentation progresses, biomedical engineers, clinicians, and device manufacturers are not only concerned with the technology. Instead , true progress will be seen when measured variables and their analysis are more closely and more specifically related to fetal status, and measurements can be made in a less invasive way without disturbance or discomfort. The application of this technology must be a benefit to the patients and to society.

BIBLIOGRAPHY

Cited References

1. Hon EH. Apparatus for continuous monitoring of the fetal heart rate. Yale J,Biol Med 1960;32:397.
2. Shenker L. Fetal electrocardiography. Obstet,Gynecol Surv 1966;21:367.
3. LaCroix GE. Fetal electrocardiography in labor: A new scalp electrode. Mich Med 1968;67:976.
4. Hon EH. Instrumentation of fetal heart rate and fetal electrocardiography. II. A vaginal electrode. Am J Obstet Gynecol 1963;86:772.
5. Hon EH, Paul RH, Hon RW. Electrode evaluation of the fetal heart rate. XI. Description of a spiral electrode. Obstet Gynecol 1972;40:362.
6. Roux JF, Neuman MR, Goodlin RC. Monitoring intrapartum phenomena. CRC Crit Rev Bioeng 1975;2:119.
7. Hammacher K. Neue methode zur selectiven registrierung der fetalen herzschlagfrequenz. Geburteh Frauenkeilk 1962;22:1542.
8. Talbert DO, Davies WL, Johnson F, Abraham N, Colley N, Southall DP. Wide bandwidth fetal phonography using a sensor matched to the compliance of the mother's abdominal wall. IEEE Trans Biomed Eng 1986;BME-33:175.
9. Larks SD. Normal fetal electrocardiogram, statistical data and representative waveforms. Am J Obstet Gynecol 1964;90:1350.
10. Cox JR. An algorithmic approach to signal estimation useful in electrocardiography. IEEE Trans Biomed Eng 1969;BME-16:3.
11. Nagel J, Schaldach M. Processing the abdominal fetal ECG using a new method. In: Rolfe P, editor. Fetal and Neonatal Physiological Measurements. London: Pitman; 1980. p. 9.
12. Tal Y, Akselrod S. A new method for fetal ECG detection. Comput Biomed Res 1991;24(3):296–306.
13. Assaleh K, Al-Nashash H. A novel technique for the extraction of fetal ECG using polynomial networks. IEEE Trans Biomed Eng 2005;52(6):1148–1152.
14. Knoke JD, Tsao LL, Neuman MR, Roux JF. The accuracy of measurements of intrauterine pressure during labor: A statistical analysis. Comput Biomed Res 1976;9:177.
15. Csapo A. The diagnostic significance of the intrauterine pressure. Obstet Gynecol Surv 1970;25:403–515.
16. Neuman MR, Picconnatto J, Roux JF. A wireless radiotelemetry system for monitoring fetal heart rate and intrauterine pressure during labor and delivery. Gynecol Invest 1970;1(2):92–104.
17. Devoe LD, Gardner P, Dear C, Searle N. Monitoring intrauterine pressure during active labor. A prospective comparison of two methods. J Reprod Med 1989;34(10):811–814.

18. Garfield RE, Maner WL, Maul H, Saade GR. Use of uterine EMG and cervical LIF in monitoring pregnant patients. Brit J Obstet Gynecol 2005;112(Suppl 1):103–108.

19. Garfield RE, Maul H, Shi L, Maner W, Fittkow C, Olsen G, Saade GR. Methods and devices for the management of term and preterm labor. Ann N Y Acad Sci 2001;943:203–24.

20. Devedeux D, Marque C, Mansour S, Germain G, Duchêne J. Uterine electromyography: A critical review. Am J Obstet Gynecol 1993;169(6):1636–1653.

21. Egley C. The clinical significance of intermittent sinusoidal fetal heart rate. Am J Obstet Gynecol 1999;181(4):1041–1047.

22. Saling E. A new method of safeguarding the life of the fetus before and during labor. J Int Fed Gynaecol Obstet 1965; 3:100.

23. Stamm O, Latscha U, Janecek P, Campana A. Development of a special electrode for continuous subcutaneous pH measurement in the infant scalp. Am J Obstet Gynecol 1976;24:193.

24. Lauersen NH, Hochberg HM. Clinical Perinatal Biochemical Monitoring. Baltimore: Williams & Wilkins; 1981.

25. Hochberg HM, Hetzel FW, Green B. Tissue pH monitoring in obstetrics. Proc 19th Annual Meeting Association Advanced Medical Instrumentation. Boston: 1984. 36.

26. Huch A, Huch R, Schneider H, Rooth G. Continuous transcutaneous monitoring of fetal oxygen tension during labour. Br J Obstet Gynaecol 1977;84(Suppl 1):1.

27. Johnson N, Johnson VA, Fisher J, Jobbings B, Bannister J, Lilford RJ. Fetal monitoring with pulse oximetry. Br J Obstet Gynaecol 1991;98(1):36–41.

28. König V, Huch R, Huch A. Reflectance pulse oximetry–Principles and obstetric application in the Zurich system. J Clin Monit Comput 1998;14(6):403–412.

29. Yam J, Chua S, Arulkumaran S. Intrapartum fetal pulse oximetry. Part I: Principles and technical issues. Obstet Gynecol Surv 2000;55(3):163–172.

30. Lysikiewicz A, Vetter K, Huch R, Huch A. Fetal transcutaneous Pco_2 during labor. In: Huch R, Huch A, editors. Conenuous Transcutaneous Blood Gas Monitoring. New York: Dekker; 1983. p. 641.

31. Timor-Trich I, Zador I, Hertz RH, Roser MG. Classification of human fetal movement. Am J Obstet Gynecol 1976;126: 70.

32. Boddy K, Dawes GS. Fetal breathing. Br Med Bull 1975;31:3.

33. Mann LI, Prichard J, Symms D. EEG, EKG and acid-base observations during acute fetal hypoxia. Am J Obstet Gynecol 1970;106:39.

34. Rose MO, Scibetta J, Chik L, Borgstedt AD. An approach to the study of brain damage. Am J Obstet Gynecol 1973;115:37.

35. Friedman EA, Micsky L. Electronic cervimeter: A research instrument for the study of cervical dilatation in labor. Am J Obstet Gynecol 1963;87:789.

36. Richardson JA, Sutherland IA, Allen DW. A cervimeter for continuous measurement of cervical dilatation in labor-preliminary results. Br J Obstet Gynaecol 1978;85:178.

37. Kriewall TJ, Work BA. Measuring cervical dilatation in human parturition using the Hall effect. Med Instrum 1977;11:26.

38. Zador I, Neuman MR, Wolfson RN. Continuous monitoring of cervical dilatation during labor by ultrasonic transit time measurement. Med Biol Eng 1976;14:299.

Reading List

Hon EH. An Atlas of Fetal Heart Rate Patterns. New Haven, (CT): Marty Press; 1968.

See also ELECTROCARDIOGRAPHY, COMPUTERS IN; ELECTROENCEPHALOGRAPHY; INTRAUTERINE SURGICAL TECHNIQUES.

FETAL SURGERY.   See INTRAUTERINE SURGICAL TECHNIQUES.

FEVER THERAPY.   See HYPERTHERMIA, SYSTEMIC.

FIBER OPTICS IN MEDICINE

MARK D. MODELL
LEV T. PERELMAN
Harvard Medical School
Beth Israel Deaconess Medical Center
Boston, Massachusetts

INTRODUCTION

In the first edition of the Wiley *Encyclopedia of Medical Devices and Instrumentation*, our friend and colleague Max Epstein, Professor Emeritus at Northwestern University, wrote an excellent article on Fiber Optics in Medicine. Now, almost 20 years later, applications of fiberoptics in medicine underwent dramatic changes and expansions. Thus on Max's recommendation, this article has been updated and rewritten for the application of fiber optics in medicine for the second edition while keeping, where it was appropriate, the original text.

For a long time optical fibers in medicine have been primarily used in endoscopy, where they have been employed for transmission of illumination to the distal end of the fiberoptic endoscope and for conveying images for the visualization of otherwise inaccessible organs and tissues. However, in the past 20 years, the science of biomedical optics of the light–tissue interaction has been dramatically advanced. The new methods of imaging, often based on substantial utilization of the optical fibers, for example, optical coherence tomography (OCT) and fiber-based confocal microscopy have been introduced. Also, the new methods of the diagnostics employing various spectroscopic techniques, for example reflectance spectroscopy, light scattering spectroscopy (LSS), fluorescence spectroscopy, and Raman spectroscopy have been developed. To be useful in the diagnosis of tissue in the lumens of the human body, these methods utilize fiber-based catheters. Phototherapy and diagnoses of internal organs also require optical fiber catheters.

The goal of this article is to give the reader basic tools necessary to understand principles of biomedical fiber optics and its applications. In addition to diagnostic, imaging, and therapeutic applications that are described in this article, optical fibers have been employed in a number of biomedical applications, for example, laser surgery and fiber-based transducers for monitoring physiologically important parameters (temperature, pressure, oxygen saturation, blood flow). All those subjects have been covered in detail in the dedicated articles of this encyclopedia.

The structure of this article is the following. The first section provides general physical and engineering principles of fiber optics needed to understand the rest of the article. It discusses the physics of total internal reflection and throughput, fiber propagation modes, optical fiber construction, and types of fibers. The next section provides the reader with the review of illumination applications of fibers in medicine. The third section discusses the diagnostic applications of the biomedical fibers, including imaging and spectroscopy. The last section reviews therapeutic applications of fibers.

GENERAL PRINCIPLES OF FIBER OPTICS

The Physics of Fiber Optics: Total Internal Reflection

For almost 400 years, it has been well known from classical optics that when light enters a medium with a lower refractive index it bends away from the imaginary line perpendicular to the surface of the medium. However, if the angle of incidence is sufficiently large, the angle in the medium with lower refractive index can reach 90°. Since the maximum possible angle of refraction is 90°, the light with higher angles of incidence would not enter the second medium and will be reflected entirely back in the medium from which it was coming. This particular angle is called the critical angle and the effect is called total internal reflection.

The effect of total internal reflection that makes an optical fiber possible is depicted in Fig. 1. The light in an optical fiber propagates through the medium with high refractive index, which is called core (usually the silica glass). The core is surrounded by another medium with lower refractive index, which is called cladding (usually another type of the silica glass). If light reaches the core–cladding interface with an incident angle higher than the critical angle it will be entirely reflected back into the optical fiber. However, if light reaches the core–cladding in terface with an incident angle lower than the critical angle, it will leave the core and will be lost. Thus, optical fibers can propagate light only at a certain angular composition, which depends on the critical angle of the core–cladding interface and thus on the refractive indexes of the core and the cladding.

Light, being continuously reflected from the core–cladding interface, can propagate very far through an optical fiber, even if the fiber is bent or is placed in a highly absorptive medium. However, if the fiber is bent too much some of the light can escape the core of the fiber. Other sources of losses are impurities in the glass. Typical optical fiber has 50–60% losses per kilometer of its length.

Throughput

There is a limit on how much light an optical fiber can transmit. Intuitively, it should be limited by an acceptance angle of a fiber and its area. This rule is often formulated as a conservation of throughput principle, which is a very general principle in optics.

For a given aperture in an optical system, the throughput T is defined by

$$T = S(\text{NA})^2$$

where S is the area of the aperture, and NA is numerical aperture of the optical element equal to the sine of the maximum divergence angle of radiation passing through the aperture (1). Conservation of throughput says that it can be no greater than the lowest throughput of any aperture in the system (2). It is very important to take the throughput of the fiber into consideration when one calculates the power, which can pass through the fiber.

Propagation Modes

By solving Maxwell's equations for an optical fiber one can find various patterns of the electromagnetic field inside the fiber. Those patterns are modes of the fiber. There are two main types of an optical fiber. The fiber can be either single mode or multimode (see Fig. 2). The difference between these types is the number of modes that the fiber can propagate.

A single-mode fiber is a fiber through which only one mode can propagate. Usually, a single-mode fiber has a very small core, ~5–10 μm in diameter. Due to their size and also because of their small NA, these fibers have a very small throughput. In medicine, such fibers are used, for example, in confocal microscopy because of the requirement for the small core diameter of the fiber tip (see the section Fiber-Based Confocal Microscopy) and OCT. However, in OCT they are used not for their size, but because the coherence of the light pulse is critical for OCT to work (this is described in detail in the section Optical Coherence Tomography characterization of flexible imaging fiber Bundles).

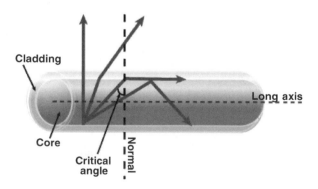

Figure 1. Total internal reflection confines light within optical fibers. (From The Basics Of Fiber Optic Cable, ARC Electronics.)

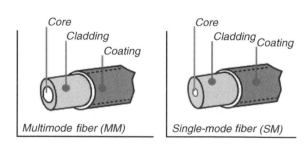

Figure 2. Types of fibers. (From Basic Principles of Fiber Optics, Corning Incorporated © 2005.)

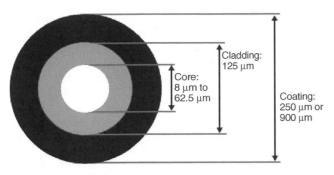

Figure 3. Fiber components. (From Basic Principles of Fiber Optics, Corning Incorporated © 2005.)

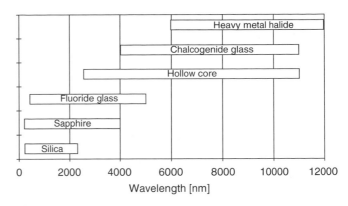

Figure 4. Transmission ranges of materials used for fibers (From Ref. 3)

The core of the multimode fiber can be much larger, somewhere between 50 and 1500 μm in diameter. These fibers are ideal for light delivery and collection, and are used in medicine when the throughput is important (see the Section Spectroscopy).

In addition to the single-mode and multimode fibers, there is another important type, that is, a polarization-maintaining fiber. The polarization-maintaining capability of a fiber is provided by the induced birefringence in the core of the fiber. This birefringence causes the polarization in these fibers to remain in the axis that it was launched into the fiber and is not changing randomly as in the regular fiber.

There are two main types of the polarization-maintaining fibers: one with the geometrical birefringence and another with the stress-induced birefringence. Geometrical birefringence is created by the elliptically shaped core. The stress-induced birefringence is created by using two stress rods as a core.

In the application where fibers are exposed to the physical stress and temperature changes, the former type is used mostly since it maintains its polarization.

Optical Fiber Construction

Optical fiber consists of three main components: core, cladding, and coating as shown in Fig. 3. The fiber is constructed by drawing a solid glass rod in a high purity graphite furnace. The rod consists of a core with high refractive index inside a low refractive index cladding. Thus both core and cladding are produced from a single piece of glass.

After core and cladding are formed, the protective coating is applied to the fiber. This protective coating is called a jacket and it guarantees that the fiber is protected from the outside environment.

Types of Fibers

Transmission ranges of materials used for fibers are shown in Fig. 4 (3). Most lasers operate in the range from 300 to 2500 nm, where silica fibers have the best overall properties and thus are commonly used.

Ultraviolet, Visible, and Near-Infrared Fibers. Ultraviolet(UV), visible, and near-infrared (NIR) light spans the range from 200 nm to 2.5 μm. Visible and near-IR light

propagates in the silica-based fibers with practically no losses due to the low absorption of the order of tenths of percent per meter. In the IR range (wavelength >2.4 μm) and UV (wavelength <400 nm) absorption is higher. The silica-based fiber absorption is caused primarily by hydroxyl radicals (OH), and thus is determined by OH concentration resulting from the presence of free water during the fiber production. Low OH concentration determines excellent transmission of these fibers in the NIR range up to 2.4 μm. At wavelengths longer than 2.5 μm, the absorption of silica limits the use of silica fibers. In the UV range, most of the silica fibers are usable down to 300 nm, particular the fibers with a high OH concentration. For shorter wavelengths fibers with both core and cladding made of silica, silica–silica fibers are used.

For applications in wavelengths <230 nm, special attention should be paid to the solarization effect caused by the exposure to the deep UV light. The solarization effect is induced by the formation of "color centers" with an absorbance at the wavelength of 214 nm. These color centers are formed when impurities (like Cl) exist in the core and cladding fiber materials, and form unbound electron pairs in the Si atom, which are affected by the deep UV radiation. Recently, solarization resistant fibers have been developed. It consist of a silica core, surrounded by silica cladding that is coated in aluminum, which prevents the optical fiber from solarizing. The fiber preform (a high grade silica rod used to make the fiber) is hydrogen loaded in a hydrogen-rich environment that helps to heal the silicone–oxygen bonds broken down by UV radiation.

As far as power-handling capability is concerned, the typical glass optical fiber is quite adequate in applications where the laser beam energy is delivered continuously or in relatively long pulses such that the peak power in the optical fiber does not exceed power densities of several megawatts per square millimeter. When the laser energy is delivered in very short pulses, however, even a moderate energy per pulse may result in unacceptable levels of peak power. Such may be the case of Nd-YAG lasers, operating in mode-locked or Q-switched configurations, which produce laser beam energy in the form of pulses of nanosecond duration or less. On the other hand, excimer lasers, which are attractive in a number of applications (4), generate energy in the UV range of the spectrum (200–400 nm) in very short pulses; they, therefore, require solarization-resistant silica–silica fibers, which can transmit light en ergy at such short wavelengths and, at the same time, carry the high power densities.

The limitation on power-handling capability of a glass optical fiber is due to several nonlinear effects, for example, Raman and Brillouin scattering (5), avalanche breakdown (6), and self-focusing (7). Stimulated Raman scattering, which occurs when, because of molecular vibrations, a photon of one wavelength, say that of the laser, is absorbed and a photon of another wavelength, known as a Stoke's photon, is emitted, has been observed at power densities of $6\,MW\cdot mm^{-2}$. The time varying electric field of the laser beam generates, by electrostriction, an acoustic wave, which in turn modulates the refractive index of the medium and gives rise to Brillouin scattering. Thus, Brillouin scattering is analogous to stimulated Raman scattering wherein the acoustic waves play the same role as the molecular vibrations. Although the Brillouin gain is higher than the one measured for the stimulated Raman scattering, the latter is usually the dominant process in multimode fibers (8).

Under the influence of an intense electromagnetic field, free electrons, which may exist in the optical fiber as a result of ionized impurities, metallic inclusions, background radiation, or multiphoton ionization, are accelerated to energies high enough to cause impact ionization within the medium. If the rate of electron production due to ionization exceeds the electron loss by diffusion out of the region, by trapping, or by recombination, then an avalanche breakdown may occur, resulting in material damage. If high enough power densities ($>100\,MW\cdot mm^{-2}$) are applied to the fiber core, avalanche breakdown is the main mechanism of permanent damage to the optical fiber. The fiber surface should be polished and chemically processed with great care to avoid reduction in the damage threshold level of the fiber surfaces. The latter is usually lower by two orders of magnitude than that of the bulk material as a result of the presence of foreign materials embedded during improper polishing or because of mechanical defects. The threshold of induced Raman and Brillouin scattering and avalanche breakdown can be further substantially reduced by self-focusing of the laser beam. Self-focusing may occur when the refractive index of the nonlinear medium increases with beam intensity. The possible physical mechanisms involved are vibration, reorientation, and redistribution of molecules, electrostrictive deformation of electronic clouds, heating, and so on. Thus, a laser beam with a transverse Gaussian profile causes an increase in the refraction index in the central portion of its path of propagation, and becomes focused toward the center. Self-focusing is counteracted by the diffraction of the beam and the balancing effects of the two determine the threshold of power that causes self-focusing; for glass it was found to be ~4 MW. Damage to optical fibers can also occur if a pulsed-laser beam is not properly aligned with the entrance face of the fiber (8,9).

IR Fibers. For the IR region beyond 2500 nm, materials other than silica are being used. These IR fibers can be classified into three categories: IR glasses fibers, crystal-

line fibers, and hollow fibers (10,11). Fluorozirconate and fluoroaluminate glass fibers can be used in the 0.5–4.5 μm region. Commercially, they are produced in diameters from 100 to 400 μm. Because of their low melting point, they cannot be used $>150\,°C$; however, they have a high damage threshold. The refractive index is similar to silica (~1.5) and the transmission is $>95\%$ for several meters. For longer wavelengths (4–11 μm) chalcogenide glass fibers are available in diameters of 150–500 μm. The disadvantage of these fibers is that they are mechanically inferior to silica fibers and are toxic. They have high Fresnel reflection because of the high refractive index (2.8) and relatively high absorption; as a result they have low transmission [losses are several tens of percent per meter (12)].

The crystalline fibers can be better candidates for the mid-IR range. Sapphire can be grown to a single crystal with a diameter of 200–400 μm, which is strong, hard, and flexible (13). It can be used up to a wave length of 4 μm. Sapphire, however, has a high index of refraction (1.75), which produces rather high reflection losses at each surface. Silver halide and thallium halide polycrystalline alloys (e.g., KRS-13), in contrast, can successfully transmit even high power CO_2 light (14). From these alloys, good quality fibers are manufactured with high transmission of a few $dB\cdot m^{-1}$ and that are insoluble in water, are nontoxic, and are fairly flexible.

Hollow fibers are built as flexible tubes, which are hollow inside, that is with air. They transmit light in the whole IR range with high efficiency (15–17). One type of these fibers comprises metallic, plastic, or glass tubing that is coated on the inside with a metallic or dielectric film with a refractive index $n>1$ (18). Another type has the tubing coated with a dielectric coating of $n<1$ for 10.6 μm on the inside of hollow glass (15) or crystalline tubes (19). The losses are 0.4–7 $dB\cdot m^{-1}$ depending on the core size. The losses due to bending are inversely proportional to the core radius. Power transmissions $>100\,W$ have been achieved. The damage threshold for high power densities is comparable with that of solid core fibers. It has been reported that the dielectric-coated metallic hollow fiber is the most promising candidate for IR laser light transmission. The standard hollow fiber is 2 m in length with an inner diameter of 0.7 mm and has transmission $>75\%$ of Er-YAG or CO_2 laser light under practical usage conditions (20).

ILLUMINATION APPLICATIONS

Introduction and Applications

Optical fibers are used for various illumination needs. The use of fiber optics bundles allows illuminating the desired area without the problem associated with the presence of the lamp-based light source.

For example, the fiber bundle brings the visible light to illuminate the area under examination with the colposcope and surgical microscope while the light source is placed in the area not interfering with the physician's activities.

In endoscopy, the fiber optic bundles are incorporated into the small diameter flexible cylindrical body of the endoscope, which is dictated by the necessity to pass through narrow (<2 cm at the most) pathway of lumens

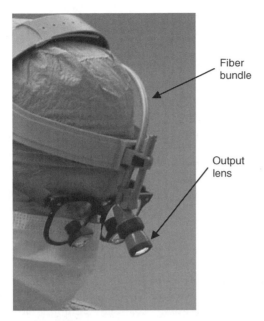

Figure 5. Headlight attached to the head bend on the surgeon head. The light from the lamp is transmitted through the fiber bundle to the output lens. (From www.luxtec.com.)

of the human body. The fiber optic bundles are used to bring the light down to the distal end and illuminate the target tissue, which the physician is examining through the imaging channel of the endoscopes.

In surgery, especially in microsurgery, dentistry, and so on fiber optic bundles are used to build a headlight, which creates bright illumination of the surgery area where the

surgeon eyes are pointed. Here, the fiberoptic illumination allows mounting the output lens on the headband of the surgeon and leaving their hands free for the operation (see Fig. 5). Recent development in the fiber optics flat and flexible illumination panels (see below) allows bringing the visible light inside the deep cavities of the human body for illumination during surgery.

In ophthalmology, early application of fiber optic illumination has included its use as a light source in the indirect ophthalmoscope. The resulting small light spot at the distal end of the optical fibers allows for the use of variable apertures and provides sharply focused and uniformly illuminated circles of light upon the retina (21). Fiber optic illuminators are also used in conjunction with intraocular surgery. Thus a variety of miniature devices are available for the visualization of the interior of the eye to provide improved illumination in microsurgery of the retina and vitreous.

Currently, a number of companies are developing the "solid-state" light based on the light emitting diodes (LED). The advantages of this type of illumination is that it produces much less heat and emits in the visible spectral range thus making the illumination fiber bundle less useful for some a pplications. However, it will take another decade before this type of the light will become commercially viable.

In addition to transmitting the light through the fiber bundle to the target for illumination, the fiber optics often serves to shape the light in the form most advantages for the application. For example, in the form of the rigid or flexible flat panel that can be used during almost any deep cavity surgery. These fiberoptic panels are made of woven plastic optical fibers as shown in Fig. 6.

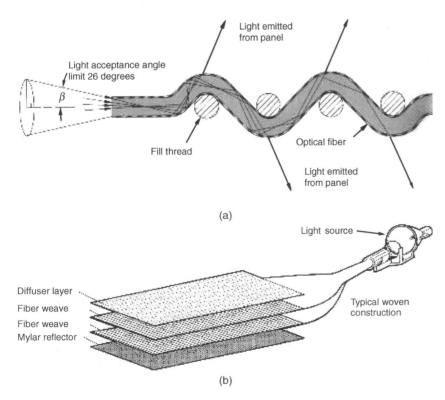

Figure 6. Principle of the fiber optics panels with the side emission. (From www.lumitex.com.) (a) All fiber optics illumination bundles, li ght enters the panel through each highly polished fiber end. Here, the computer controlled "macrobends" in the fibers cause the transmitted light to be emitted from the sides of the fibers through the cladding. Precisely engineered construction causes all light to be emitted uniformly along the length of the panel. (b) Layers of fiber optic weave are assembled together with double-sided adhesive into as many as eight layers. A mylar reflector is laminated to the back and a clear vinyl top layer is added for extra durability. For some applications (e.g., LCD backlighting), a semitransparent diffuser layer is placed between the top weave layer and the clear vinyl. The optical fibers extend from the panel in cable form and are bundled into a brass ferrule and highly polished. These ferrules are then connected to a remote light source.

Requirements for Illumination Fibers

A thin optical fiber is obtained by drawing in a furnace a glass rod inside glass tubing, which together form the high refractive index core and the low refractive index cladding, respectively. The core material usually used for the illumination fibers is silica, which, as was discussed earlier, is the best material for the visible spectral range. When drawn to a core diameter of 25 μm or less, the optical fiber is quite flexible, allowing for bending radii of <1 cm.

Light Sources Used with Illumination Fibers.

The light sources for the fiberoptic illuminators are tungsten or quartz halogen projection lamps, mercury or xenon high pressure arc in quartz glass enclosures, and metal halide lamps. Color temperatures of tungsten sources vary between 2800 and 3500 K, which is a low color temperature with the illumination appearing yellow. Arc and halide lamps are used in the commercially available illuminators and provide light at the color temperature of ~5400 K.

These light sources are a black body type radiator, thus they radiate into the sphere, that is, solid angle of 4π steradian and the amount of the output light they emit is almost proportional to the emitting body (filament, for the tungsten lamps, and arc, for other types). Thus they have the emitting body of >1 mm^2 for the arc lamp and much greater for the tungsten and quartz halogen lamps.

Coupling Efficiency.

The fact that the light source radiates in 4π steradian from the emitting body lead to requirements that the optical fibers used for the illumination have a high aperture and sufficient cross-section. To transmit the required illumination through the bundle and keep the fibers flexible, the conventional practice is to gather a lot of thin flexible fibers into a bundle, with the ends bound and polished. This bundle is almost as flexible as the single fiber. The fibers in the ends of the illumination fiber bundles are arranged at random and these bundles are called incoherent . This organization of the fibers in a bundle, in addition to being inexpensive to assemble, is also sometimes useful to provide uniform illumination at the output end.

Various schemes have been employed to maximize the light intensity delivered from the source to the fiber or fiber bundle and to obtain uniform illumination at the distal end of the fiber. For example, a special reflecting mirror, in the form of an ellipsoid of revolution, has the light source and the fiber input at the focal points with the major axis of the ellipsoid at an angle to the receiving fiber. However, note that light lamp sources, as opposed to lasers, could not be focused or concentrated onto areas smaller than their emitting body without considerable loss of the luminous flux. This follows from the fundamental relationship in radiometry, which states that the radiance of an image cannot exceed that of an object for the case when both lie in a medium with the same index of refraction (22). Consequently, the optimum illumination from an optical fiber is obtained when the image of the source completely fills the entry face of the fiber and the cone of light forming this image is equal to the numerical aperture of the fiber. In some cases, when the light source consists of a filament, the use of reflectors increases the effective area of the source by redirecting light rays through the voids in the filament.

Removal of Heat, UV, and IR Radiation.

All lamps produce heat directly or by absorption of light in the visible to IR spectral ranges. Also, the UV radiation portion of the spectrum may be hazardous to the eyes and illuminated tissue. The high intensity light sources needed to deliver adequate illumination to and through the optical fiber cause the latter to become very hot. In the case of fiber bundles, the heat generated at the proximal end of the fibers requires that they not be bonded with epoxy, but instead be fused together. In some cases this may also apply to the distal end if enough heat is conducted through the fiber or is generated by absorption of light in the fiber. Of course, this is highly undesirable and should be avoided. After all, one of the main objectives of the use of illumination fibers is to keep the heat of the light source away from the area being illuminated. Indeed, the early applications of optical fibers were referred to as "cold-light" illumination. Special provisions must, therefore, be made to dissipate and divert large amounts of heat. Most light sources utilize dichroic reflectors and/or heat filters to block the heat from reaching the fiber.

Transmission.

The light propagating through the optical fiber is attenuated by bulk absorption and scattering losses. These losses vary with the wavelength of light and are usually higher at short wavelengths, that is, the blue end of the visible spectrum of 400–700 nm. In most illumination applications, the typical length of the fibers does not exceed 2 m and thus the attenuation of light due to absorption is of no consequence. Antireflection coating can reduce Fresnel losses due to reflection at the end faces of the fiber.

The faithful rendition of color in viewing depends on the spectral content of the light illumination. This is of particular significance in medicine where the diagnosis of disease is often determined by the appearance and color of organs and tissue. Hence, optical fibers made of plastic materials, for example, polystyrene for the core and Lucite, polymethylene methacrylate (PMMA), for the cladding, despite their flexibility and low cost, do not find extensive use as illumination fibers.

DIAGNOSTIC APPLICATIONS

Introduction

One of the earliest applications of optical fiber in medicine were in imaging. The bundle of the optical fibers has been used to transmit image from the distal to the proximal end, where the physician could see the image of the target tissue in real time. The device using this technique is known as an endoscope. The endoscopes are used widely in current medical practice for imaging of lumens in the human body.

In the past 20 years, many new diagnostic applications of fiber optics have appeared as a result of the developments in biomedical optics (23). Most all of them utilize a single or a few fibers. Some of them, for example OCT (24) and confocal imaging (CI) (25) use scanning to produce the images of the lateral or axial cross-sections of the body. Others are utilizing the spectroscopic differentiation of the tissue, using natural (23) or man-made markers (26). Among these techniques, fluorescence (23), reflectance (23), light scattering, and Raman spectroscopic methods (27) are most promising and thus most developed.

Another new area for the application of fiber optics in medicine is the fiberoptic biosensor (FOBS). This is a sensor consisting of optical fiber with a light source and detector and an integrated biological component that provides the selective reaction to the biochemical conditions of the tissue and body fluids. These sensors have been applied for glucose and blood gas measurements, catheter-based oximetry, billirubin, and so on. The detailed discussion about the principle and applications of these sensors is in "Optical Sensors" article of this encyclopedia.

Imaging

Endoscopy. The word endoscopy derives from two Greek words meaning "inside" and "viewing". Its use is limited to applications in medicine and is concerned with visualization of organs and tissue by passing the instrument through natural openings and cavities in the human body or through the skin, that is percutaneously. The endoscope has become an important and versatile tool in medicine. It provides a greater flexibility than it is possible with instruments that consist of a train of optical lenses, and transmits illumination to the distal end of the probe. The greater flexibility of the flexible endoscope enables the visualization around corners and the eliminati on of "blind areas" obtained with the rigid instrument. It should be noted that optical fibers are used to deliver illumination light in rigid endoscopes that in some cases may employ lenses for image transmission.

Endoscopes, which utilize optical fibers to transmit the image from the distal to the proximal end, are often called fiberscopes to differentiate them from the video or electronic endoscopes where a semiconductor imager is placed at the distal end and the image is transmitted electronically. Progress in the production of the relatively inexpensive high quality miniature semiconductor imagers based on Charge Coupled Device (CCD) technology and Complementary Metal Oxide Semiconductor (CMOS) led to the development of the high quality electronic (video) endoscopes. Currently, most of the endoscope vendors produce such endoscopes. It appears that these video endoscopes produce higher quality images with considerably higher magnification (28) and are replacing the fiberscopes where it is prac tical (29–31). However, a large number of fiberscopes are still used in the clinics and new fiberscopes are being sold (e.g., see www.olympusamerica.com). Moreover, in the areas that require thin and ultrathin endoscopes of <2–4 mm (32), fiberscopes are still the only practical solutions. The general discussion on endoscopy, its features, and applications is presented in the Endoscopy article. This article will primarily discuss fiberscopes.

In addition to the imaging and illumination channels, a typical endoscope includes channels to accommodate tools for biopsy to aspirate liquids from the region being inspected, and to inflate the cavity or to inject clear fluids to allow for better visualization. The overall dimensions of such instruments vary between 5 and 16 mm in diameter, the thinner versions being smaller and more versatile than the corresponding rigid systems. The fiberscope can be made as long as nece ssary, since the light losses in most fibers made of glass cores and cladding are tolerable over distances of up to several meters. These images can be recorded using film, analog and digital still, and video cameras.

Most of the fiberscopes use similar optical structures and vary mainly in length, total diameter, maneuverability, and accessories, for example, biopsy forceps. The diameter of the individual glass fibers in the image-conveying aligned bundle are made as small as possible limited only by the wavelength of the light to be transmitted. In practical applications, the diameter is ranging from 2 to 15 μm. Moreover, if the fibers are densely packed, cross-talk problems may arise due to the evanescent electromagnetic field (light waves) in each individual fiber (33).

A variety of fiberscopes have been developed, each with features that are best suited for specific applications.

Transmission of Images Through Optical Fibers. As shown in the section General Principles of Fiber Optics, an optical fiber cannot usually transmit images. However, a flexible bundle of thin optical fibers (obviously, with silica core) can be constructed in a manner that does allow for the transmission of images. If the individual fibers in the bundle are aligned with respect to each other, each optical fiber can transmit the intensity and color of one object point. This type of fiber bundles is usually called a "coherent" or "aligned" bundle. The resulting array of aligned fibers then conveys a halftone image of the viewed object, which is in contact with the entrance face of the fiber array. To obtain the image of objects that are at a distance from and larger than the imaging bundle, or imaging guide, it is necessary to use a distal lens or lens system that images the distal object onto the entrance face of the aligned fiberoptic bundle. The halftone screen-like image formed on the proximal or exit face of a bundle of aligned fibers can be viewed through magnifying systems or on the video monitor if this exit face is projected onto the video camera.

Fabrication of Flexible Imaging Bundles. The fabrication of flexible imaging bundles involves winding of the optical fibers on a highly polished and uniform cylinder or drum, with the circumference of the latter determining the length of the imaging structure. The aligned fibers can be wound directly from the fiber-drawing furnace or a separate spool of individual fibers. When the entire bundle is obtained in a single winding process, similar to a coil, an overhang of fibers wound on the outer layers develops after the structure is removed from the winding cylinder. Such overhang can be eliminated by winding single or a small number of layers and cutting them into strips to form the aligned fiber bundle. This process, although more laborious, usually renders better uniformity in the imaging bundles. Some users find the evenness of the fiber arrangement distracting and prefer the irregular structure. This may be compared with the viewing of a television image wherein the horizontal line scan has accentuated by imperfect interlacing. In either method, the diameter of the fibers is on the order of 10 μm, which is thinner than hair or about the size of a strand in a cottonball, and must therefore be soaked in binding epoxy during the winding process. When the completed imaging bundle is cut and remove d from the drum, its ends are bound to secure the alignment and the epoxy is removed along the structure to let the individual fibers remain loose, and thus flexible. A flexible imaging bundle can also be obtained by first drawing a rigid preform made up of individual optical fibers, each with a double coating, where the outer cladding is chosen to be selectively etched afterwards. Such a structure has nearly perfect alignment, sin ce the preform can be constructed of fibers thick enough to allow parallel arrangement and the individual fibers in the drawn bundle are fused together without any voids be tween them. Moreover, such fabrication technique, as compared with the winding method, is far less expensive since it permits batch processing.

The size of the imaging bundle varies from 1 to 2 mm in diameter for miniature endoscopes, for example, angioscopes or pediatric bronchoscopes, where the fiberscopes are currently primarily utilized, to 6 mm in diameter for large colonoscopes, and lengths exceeding 2 m. Large imaging bundles may consist of as many as 100,000 individual fibers (28), each fiber providing one point of resolution or pixel. For such large numbers of fibers, it is often necessary to fabricate the final device by draw ing first fiber bundles containing a smaller number of fibers and then joining them together to form a larger structure.

Characterization of Flexible Imaging Fiber Bundles. The resolution attainable with a perfectly aligned fiberoptic imaging structure is determined by the fiber diameter. For a hexagonal configuration, the resolution in optical line pairs per millimeter is, at best, equal to, and usually slightly $<1/2d$, where d is the fiber diameter in millimeters. Figure 7 shows a cross-section of an imaging bundle of fibers aligned in a closely packed hexagonal pattern, i.e., each fiber has six closest neighbors. In Fig. 7a are shown two opaque stripes separated by a distance equal to their width, which are placed at the distal end of an entrance face of the imaging bundle. When this end is illuminated, the corresponding fibers at the output or proximal exit end of the bundle will be either partially or totally darkened, depending on the relative location of the opaque stripes with respect to the fiber pattern. In Fig. 7b, the partially illuminated fibers indicate that the smallest resolvable separation of the two stripes is equal to twice the fiber diameter. In practice, the packing of the fiber bundle is often not perfect, leaving spaces between the fibers; thus, the line resolution is further reduced and is usually $<1/2d$. For a typical imaging bundle in endoscopy, the diameter of the individual fiber is ~10 μm and, therefore, the resolution is better than 50 line-pairs·mm^{-1}. This resolution is considerably poorer than in most lens systems.

The line resolution for the hexagonally aligned imaging bundle does depend on the orientation of the stripes (Fig. 7). Thus, for an orientation different from that shown in Fig. 7, the line resolution may be somewhat different from that

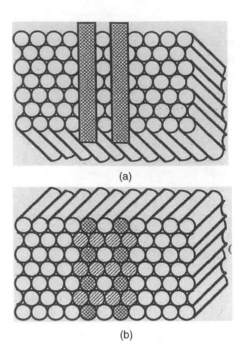

(a)

(b)

Figure 7. Image transmission of two opaque stripes from (a) distal end to (b) proximal face of an imaging fiber optics bundle.

obtained. The spatial variance of the image transfer properties of imaging bundle has led to the use of averaging techniques in the evaluation of their limits of image resolution (34). The modulation-transfer function (MTF) of an imaging bundle is the contrast response at its exit face to a sinusoidal test pattern of varying spatial periodicity imaged onto the input or entrance face. The fabrication of an imaging bundle may result in the misalignment or deviation from perfect alignment of the adjacent fibers. A method of evaluation of the image-conveying properties of practical imaging bundles has been developed (35), which takes into account the fact that the arrangement of the fibers may differ at the input and output faces of the imaging bundle. It also uses a statistical approach in determining an average modulation-transfer function. The aligned fiber bundle exhibits a mosaic pattern that represents the boundaries of the individual fibers and that appears superimposed on the viewed image. Hence, the viewer sees the image as if through a screen or mesh. Any broken fiber in the imaging bundle appears as a dark spot. In some applications, these two features can be very annoying and distracting, although most medical practitioners have become accustomed to these peculiarities and have learned to discount them. In some sense, it is comparable to viewing through a window with a wire screen, which in most cases is unnoticed.

Rigid Imaging Bundles. Fiberoptics bundles composed of individual optical fibers far smaller than those described earlier can be obtained by drawing the entire assembly in the furnace in a manner that preserves their alignment. This method is similar to that employed in the preparation of flexible imaging structures by etching a double-clad rigid imaging bundle. However, since the fibers become fused together during the drawing process, the structure

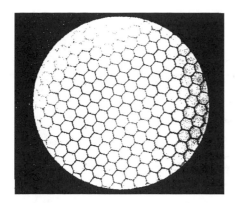

Figure 8. Cross-section of the imaging bundle.

remains a rigid solid-glass rod. A segment of a cross-section of such an imaging bundle, which is 0.5 mm in diameter and contains 11,000 fibers, each 4.2 μm in diameter, is shown in Fig. 8. The honeycomb pattern represents the boundaries of the individual fibers and is superimposed on the image, similar to the flexible fiberoptic imaging bundle.

Fiber-Based Confocal Microscopy. Confocal imaging (CI) is based on illuminating a single point on the sample and collecting scattered light from the same single point of the sample (Fig. 9a). The illumination point on the sample is the image of the illumination pinhole and is imaged into the detector pinhole, thus making both pinholes and the illuminated spot on the sample optically conjugated. In this way, stray light from points outside the collection volume is effectively filtered, improving contrast, particularly for scattering media, for examples, biological tissues. Scanning the position of this point on the sample and acquiring the signal from each position creates a lateral image of the sample. For axial cross-sectional imaging, the focal point on the sample is axially moved while acquiring the signal from each position. Combining both lateral and axial scans provides a perpendicular cross-sectional image of the tissue. The nature of the signal from the illuminated

point depends on the specific embodiment. Either back-scattered or fluorescent signals are most often used for CI; however, other signals have been collected and proven clinically useful.

Both pinholes contribute to depth sectioning ability. The source pinhole causes the illuminating irradiance to be strongly peaked at the image formed by the objective and to fall off rapidly both transversely and along the line of sight (depth). There is, therefore, very little light available to illuminate out-of-focus volumes. This implies that these volumes will not significantly mask the image of the more brightly illuminated focal region. The detector pinhole acts similarly. It is most efficient in gathering the radiation from the volume at the focal point, and its efficiency falls off rapidly with distance, both transversely and in depth. Together they assure that the totality of the out-of-focus signal is strongly rejected.

Confocal imaging provides enhanced lateral and axial resolutions and improved rejection of light from the out-of-focus tissue. As a result, the confocal imaging (CI) can achieve a resolution sufficient for imaging cells to depths of several hundreds of microns. At these depths, CI has been especially helpful in imaging the epithelium of many organs, including internal organs via endoscopic access. Images displayed by the CI system are similar to images of histopathology slides under high resolution conventional microscope, thus, are familiar to the physicians.

The fiberoptical confocal microscope has been introduced in late 1980s and early 1990s (see, e.g., Ref. 37). In this kind of the CI microscope, the light source is not a point source, but the tip of an optical fiber, and the signal is collected by another optical fiber that delivers the signal to a detector. This makes the CI system compact and thus convenient in numerous biomedical applications.

There are a variety of optical systems and scanning arrangements, which allows for optimization of the CI device for the required image and particular clinical application (25). For example, the scanning can be organized by moving the fiber tips, especially, when the common fiber is used for the illumination and detection as shown in Fig. 10b, or by moving the objective lens or by placing in

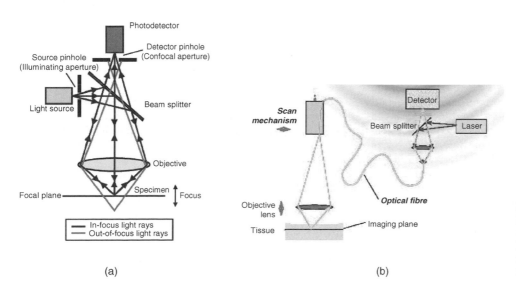

(a)   (b)

Figure 9. Confocal microscopy. (a) Principle of confocal arrangement. (From MPE Tutorial, Coherent Incorporated © 2000.) (b) Possible implementation of fiber-based confocal microscope (36).

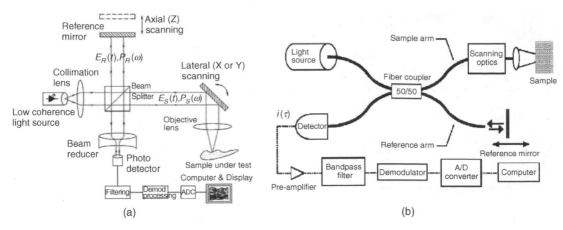

Figure 10. Optical coherence tomography (OCT). (a) Principle of OCT (From Wikipedia.org). (b) Possible implementation of fiber-based OCT (38).

the pinhole plane an output face of coherent fiber bundle and scanning an mage of a pinhole at the entrance end of it.

The CI system can provide a cellular level resolution: A lateral integrated image of tissue (similar to C-scan imaging), a lateral cross-sectional image at the desirable depth of tissue, a perpendicular-to-the-surface cross-sectional image of the tissue (similar to B-scan imaging), and a combination of the above images, thus a three-dimensional (3D) image of tissue. In each case, the CI system could be optimized to achieve the best overall performance while utilizing the same basic platform. The application of the CI imaging has been successfully demonstrated in diagnostics of intraepithelial neoplasia and cancer of the colon (39). It is reasonable to envision that the CI-based endoscope will be used for the screening of Barrett's esophagus and cancer and other areas of the upper GI tract. It appears that there is potential for application of the CI device combined with one of the scattering-based spectroscopies for the vulnerable plaque screening and triage.

The optical fibers used for the CI application are usually a single mode type because of the requirement for the small core diameter of the fiber tip as discussed above.

Optical Coherence Tomography. Optical coherence tomography is a relatively new technology and has been developed and investigated for the past 15 years (38). It uses the wave property of light called coherence to perform ranging and cross-sectional imaging. In OCT systems, a light beam from a light source is split into a reference light beam and a sample light beam (see Fig. 10a). The sample light beam is direct ed onto a point on the sample and the light scattered from the sample is combined w ith the reference light beam. The combined reference and sample light beams interfere if the difference of their optical paths is less than the coherence length. The reference and the collected sample beam are mixed in a photodetector, which detects the interference signal. The light source used in OCT has a low coherence length so that only the light s cattered from the sample within the close proximity around a certain depth will satisfy this condition. The strength of the interference signal corresponds to the scattering around this depth. To acquire the signal from another depth in the sample the optical pa th of one of the beams

is changed so that the same condition is now satisfied by the li ght scattered from another depth of the sample. The component of the OCT system providing this change is called an optical delay line. By sequentially changing the optical path of one of the beams and processing the photodetector output, a cross sectional image of the sample is generated. By laterally moving the sample beam along the sample provides a perpendicular cross-sectional image of the sample. The OCT image is similar to high frequency ultrasound B-scan images.

Usually a moving mirror in the optical path of one of the beams performs the continuous scan of the optical path. The shortest coherence length of available light sources allows the OCT systems to achieve a depth resolution higher than in high frequency ultrasound imagers, but lower than in confocal imaging. As a direct correlate, the depth penetration of OCT systems is lower than for the high frequency ultrasound imagers and higher than the confocal imaging. These parameters make OCT a useful technology in the biological and medical examinations and procedures that require good resolutions to 2 mm depths.

The OTC systems utilizing the fiberoptic components are most often used (Fig. 10b). These systems are compact, portable, and modular in design. The sample arm of the OCT can contain a variety of beam-delivery options including fiberoptic radial- and linear-scanning catheter-probes for clinical endoscopic imaging. The aiming beam is used so that the clinicians could see the location on the tissue where the OCT catheter is acquiring image. The optical fiber based OCT systems require using single mode fibers in both reference and sample arms to keep the coherence of the light guided by the fiber. In some cases, when the OCT system is using the polarization properties of the light, it must utilize the polarization-maintaining fibers.

There is a variety of OCT optical systems and scanning arrangements, which provide room for optimization of the OCT device for the specific clinical application. Recently, several system modifications of the basic OCT have been reported; for example, Fourier transform OCT (40,41), spectroscopic OCT (42), and polarization OCT (43). Some of them appear to promise practical OCT systems with higher data acquisition rates, higher resolution (comparable with the resolution of CI), better noise immunity, or capabilities to acquire additional information on tissue.

In the endoscopic applications, the sample arm fiber is designed as a probe that goes through the endoscope to the target tissue. There are a number of OCT probes developed and suggested designs (44–54) that provide room for optimization of the OCT device for the specific clinical application. A number of successful clinical studies have been carried out demonstrating the clinical applicability of the endoscopic fiber OCT technique for clinical imaging, for example, for imaging of Barrett's esophagus (54) and esophageal cancer (55), bile duct (56), and colon cancer (55).

Spectroscopy

Reflectance and Fluorescence Spectroscopy. Diffuse reflectance spectroscopy is one of the simplest spectroscopic techniques for studying biological tissue. Light delivered to the tissue surface undergoes multiple elastic scattering and absorption, and part of it returns as diffuse reflectance carrying quantitative information about tissue structure and composition.

This technique can serve as a valuable supplement to standard histological techniques. Histology entails the removal, fixation, sectioning, staining, and visual examination of a tissue sample under the microscope. Tissue removal is subject to sampling errors, particularly when the lesions are not visible to the eye. Also, the multiple-stage sample preparation process is time-consuming, labor intensive, and can introduce artifacts that are due to cutting, freezing, and staining of the tissue. Most importantly, the result is largely qualitative in nature, even though quantitative information is available through techniques, for example, morphometry and DNA multiploidy analysis.

Spectroscopy, in contrast, can provide information in real time, is not greatly affected by artifacts or sampling errors, and can provide quantitative information that is largely free of subjective interpretation. Because it does not require tissue removal, it can be conveniently used to examine extended tissue areas.

Usually, light is delivered and collected using an optical fiber probe that can be advanced through the accessory channel of the endoscope and brought into contact with the tissue. The probe can consist of several delivery and collection fibers. Probably the simplest geometry is a central optical fiber for light delivery and six fibers for light collection arranged in a circle around the central fiber. In Zonios et al. (57), all fibers had a 200 μm core diameter and a NA of 0.22, and were packed tightly with no gap between them. The probe tip was fitted with a quartz shield ~1.5 mm in length and in diameter, which provided a fixed delivery and collection geometry with uniform circular delivery and collection spots in the form of overlapping cones. The tip was beveled at an angle of 17° to eliminate unwanted specular reflections from the shield–tissue interface (see Fig. 11).

To extract quantitative properties of tissue collected with the above probe the model of light transport in tissue should take into account probe geometry. To find the total light collected by the probe, diffuse reflectance from a point source must be integrated over the spatial extent of the light delivery and collection areas, characterized by

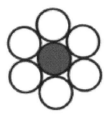

Figure 11. Configuration of fibers in a typical reflectance and fluorescence optical probe. Probe contains six fibers for light collection arranged in a circle around the central fiber.

radii r_d and r_c, respectively. Assuming the incident light intensity to be uniform over the entire delivery area, the diffuse reflectance $R_p(\lambda)$ collected by the probe is given by (57).

$$R_p(\lambda) = \frac{1}{r_d^2} \int_0^{r_c} r\,dr \int_0^{2\pi} d\phi \int_0^{r_d} R(\lambda, |\mathbf{r} - \mathbf{r}'|)r'\,dr'$$

where r_d and r_c are radii of the delivery and collection spots of the fiber optics diffuse reflectance probe and $R(\lambda, |\mathbf{r}-\mathbf{r}'|)$ is diffuse reflectance predicted by the physical model, which depends on tissue morphological and biochemical composition.

A similar probe (Fig. 11) can be used for fluorescence spectroscopy measurements. For fluorescence measurements, it is especially important that delivery and collection signals are delivered over the separate optical fibers. This is because intense illumination light can easily induce certain amount of fluorescence in the delivery fiber. This fluorescence of the fiber is likely to be weak, however, tissue fluorescence is also very weak. Thus if the delivery and collection fibers coincide, the fluorescence from the probe itself would significantly perturb the tissue fluorescence observed by the instrument. Hence, fluorescence fiber probes should always have separate delivery and collection fibers.

Light Scattering Spectroscopy. In addition to the multiply scattered light described by the diffuse reflectance, there is a single scattering component of the returned light that contains information about the structure of the uppermost epithelial cells (58). It has been shown that light scattering spectroscopy (LSS) (Fig. 12) enables

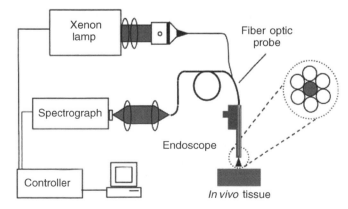

Figure 12. Schematic diagram of the LSS system (59).

quantitative characterization of some of the most important changes in tissues associated with precancerous and early cancerous transformations, namely, enlargement and crowding of epithelial cell nuclei (58,60). Typical nondysplastic epithelial cell nuclei range in size from 4 to 10 μm. In contrast, dysplastic and malignant cell nuclei can be as large as 20 μm. Single scattering events from such particles, which are large compared to the wavelength of visible light (0.5–1 μm), can be described by the Mie theory. This theory predicts that the scattered light undergoes small but significant spectral variations. In particular, the spectrum of scattered light contains a component that oscillates as a function of wavelength. The frequency of these oscillations is proportional to the particle size. Typically, normal nuclei undergo one such oscillation cycle as the wavelength varies from blue to red, whereas dysplastic/malignant nuclei exhibit up to two such oscillatory cycles. Such spectral features were observed in the white light directly backscattered from the uppermost epithelial cell nuclei in human mucosae (60,61).

When the epithelial nuclei are distributed in size, the resulting signal is a superposition of these single frequency oscillations, with amplitudes proportional to the number of particles of each size. Thus, the nuclear size distribution can be obtained from the amplitude of the inverse Fourier transform of the oscillatory component of light scattered from the nuclei. Once the nuclear size distribution is known, quantitative measures of nuclear enlargement (shift of the distribution toward larger sizes) and crowding (increase in area under the distribution) can be obtained. This information quantifies the key features used by pathologists in the histological diagnosis of dysplasia and Carcinoma *in situ* (CIS), and can be important in assessing premalignant and noninvasive malignant changes in biological tissue *in situ*.

However, single scattering events cannot be directly observed in tissue *in vivo*. Only a small portion of the light incident on the tissue is directly backscattered. The rest enters the tissue and undergoes multiple scattering from a variety of tissue constituents, where it becomes randomized in direction, producing a large background of diffusely scattered light. Light returned after a single scattering event must be distinguished from this diffuse background. This requires special techniques because the diffusive background itself exhibits prominent spectral features dominated by the characteristic absorption bands of hemoglobin and scattering of collagen fibers (there is abundance of them in the connective tissue laying below the epithelium). The first technique of diffusive background removal uses a standard reflectance probe described in the section References and Fluorescence Spectroscopy. This technique is based on observation that the diffuse background is typically responsible for >95–98% of the total reflectance signal. Therefore, the diffusive background is responsible for the coarse features of the reflectance spectra. The diffusion approximation-based model may account for this component by fitting to its coarse features. After the model fit is subtracted, the single backscattering component becomes apparent and can be further analyzed to obtain nuclear size distribution (58).

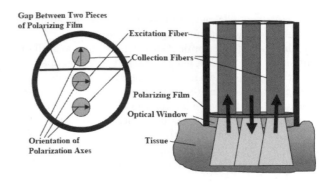

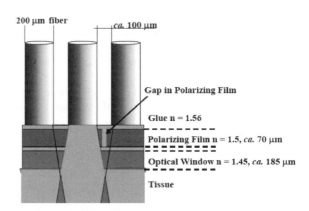

Figure 13. Design of the polarized fiber probe (62).

Another technique would use a special polarized probe. One possible implementation of the polarized probe is described by Utzinger and Richards-Kortum (62) (see Fig. 13). Recently, similar polarized fiber probe was developed by Optimum Technologies, Inc. The working principle of this probe is based on the fact that initially polarized light loses its polarization when traversing a turbid medium such as biological tissue. Consider a mucosal tissue illuminated by linearly polarized light. A small portion of the incident light will be backscattered by the epithelial cell nuclei. The rest of the signal diffuses into the underlying tissue and is depolarized by multiple scattering. In contrast, the polarization of the light scattered backward after a single scattering event is preserved. Thus, by subtracting the unpolarized component of the reflected light, the contribution due to the backscattering from epithelial cell nuclei can be readily distinguished. The residual spectrum can then be analyzed to extract the size distribution of the nuclei, their population density, and their refractive index.

Raman Spectroscopy. Raman spectroscopy is a very powerful technique, which should be capable of performing detailed analysis of tissue biochemical composition. However, in biological tissues the intensity of the Raman signal is only 10^{-9}–10^{-11} of the intensity of the incident light. What make it even worse is the fact that fluorescence signal excited in tissue by the same incident light is much higher, $\sim 10^{-3}$–10^{-5} of the incident light. And if the signal is detected by the regular optical fiber probe it will also overlap with the fluorescence signal originated in the fiber

Figure 14. Design of the fiber tip of a typical fiber optic probe for Raman spectroscopy (62).

itself. Hence, development of a reliable biomedical Raman fiber probe is still a challenge.

In Fig. 14 one can see the design of a Raman fiber optic probe developed by Visionex Inc. It consists of a central delivery fiber and seven collection fibers. A bandpass filter is located in front of the delivery fiber and a longpass filter in front of the collection fibers. Those filters ensure that no laser light and fluorescence light originated in the delivery fiber can enter the collection fiber.

THERAPEUTIC APPLICATIONS

Introduction

In therapeutic application, fibers are used primarily as a part of light delivery system. Utilization of fibers allows for flexible and convenient methods of delivering light from bulky high energy light sources to the diseased location. Often, using the fiber optics delivery system makes otherwise inaccessible locations accessible. Variety of applications, light sources and energy requirements require different fibers with different features.

Fiber Surgical Applications

The monochromatic nature of laser beam energy, temporal coherence, and the ability to focus it onto a very small spot because of its spatial coherence, allow for efficient surgical methods, for example, tissue cutting or ablation, as well for the transmission of large amounts of power through a single optical fiber. This localized energy can be used to cauterize tissue by evaporation while reducing bleeding by coagulation. The use of lasers in surgical and other therapeutic applications is most effective when employed in conjunction with the flexible delivery systems and, in particular, with laser-beam delivery which is compatible with the endoscopy. Lasers utilized in medical applications range in wavelength from vacuum UV (excimer laser at 193 nm) to infrared (IR) (CO_2 laser at 10.6/11.1 μm).

Laser-beam-delivery systems depend on the wavelength laser energy. At the present time there is no single delivery system that can be used over the entire range of medical lasers. The primary design considerations in such systems are efficiency, maximum power delivered, preservation of beam quality, and mechanical properties (flexibility, degrees of freedom, range, size, and weight).

Low efficiency (output power/input power) results in losses in the delivery system, which, in turn, requires higher power and thus more expensive lasers. Moreover, the power lost in the delivery system is generally dissipated as heat, resulting in a temperature rise that causes damage to the device, leading to a further deterioration of efficiency or catastrophic failure, Hence, efficiency, together with heat dissipation, can be considered to be the limiting factors in maximum power delivery.

A well-designed laser oscillator emits a highly collimated beam of radiation, which can then be focused to a spot size of just a few wavelengths, yielding power densities not achievable with conventional light sources. A useful delivery system should, therefore, preserve this quality of the beam as much as possible; otherwise the main advantages of using a laser are lost.

The most desirable flexible system should be easy to handle mechanically, perform a large variety of tasks, and, of course, still satisfy the above properties. From a mechanical point of view, the ideal laser-beam guide would be an optical fiber of small cross section and long enough to reach any site, at any orientation, over a wide range of curvatures, through openings and inside complex structures.

The choice of fibers for such system is mainly determined by the wavelength of the laser (see Fig. 4) and discussion in the section Types of Fibers.

Most lasers operate in the range of 300–2500 nm, where silica fibers have the best overall properties and, thus, are commonly used.

Outside this range, there are few wavelengths used for laser surgery, in IR, 2.94 μm of the erbium:YAG laser, 5–6 μm of the CO laser, 10.6/11.1 μm of the CO_2 laser and in the UV, the 193 and 248 nm of the ArF and KF excimer lasers, respectively. In the infrared range, silver halide and sapphire are being used for the fiber core. Also, hollow fibers have become available for efficient guidance of infrared light with minimal losses (20,63). In the UV range, the solarization resistant silica/silica fibers are available (64) and hollow fibers have been reported recently (16,17). For pulsed lasers, utilization of fibers can be limited by the high peak powers, which could damage the fiber; for example, the peak power of 106 kW·cm^{-2} is a threshold for silica. In this case, the flexible arms with reflective surfaces have to be used.

As far as power-handling capability is concerned, the typical glass optical fiber is quite adequate in applications where the laser beam energy is delivered continuously or in relatively long pulses such that the peak power in the optical fiber does not exceed power densities of several megawatts per square millimeter. When the laser energy is delivered in very short pulses, however, even a moderate energy per pulse may result in unacceptable levels of-peak power. Such may be the case of Nd-YAG lasers, operating in a mode-locked or Q-switched configuration, which produces laser beam energy in the form of pulses of nanosecond duration or less. On the other hand, excimer lasers, which are attractive in a number of applications (4), generate energy in the UV range of the spectrum (200–400 nm) in very short pulses; they, therefore, require solarization-resistant silica/silica fibers, which can transmit light energy at such short wavelengths and, at the same time, carry the high power densities.

For lasers in the IR region beyond 2500 nm, fibers made of materials other than silica are being used as shown in the section Types of Fibers.

Photodynamic Therapy

Photodynamic therapy utilizes the unique properties of a substance known as photosensitizer (65). When administered systemically it is retained selectively by cancerous tissue. The substance is photosensitive and produces two effects: When excited by light at a photosensitizer-specific wavelengt hit fluoresces. This effect is used in photodynamic diagnostics. When irradiated with light, the photosensitizer undergoes a photochemical reaction, which results in the formation of a singlet oxygen and the subsequent destruction of the cell (usually malignant cell) that retained the substance.

In photodynamic diagnostics, since the fluorescence efficiency of the photosensitizer is low, high intensity illumination at the photosensitizer-specific wavelength and high gain imaging systems are required to detect very small tumors. The excitation is in the spectral range, where silica fibers have excellent properties thus glass (silica) fibers are used for delivery of the excitation light.

In order to obtain an effective cure rate in photodynamic therapy, it is essential that the optical fibers, which deliver the light energy to the tumor site, provide uniform light distribution. Since an optical fiber, even if uniformly irradiated, yields a nonuniform output light distribution, it is necessary to employ special beam shapers at the exit face of the fiber (66). The specifics of the fiber used for the delivery is determined by the absorption wavelength of the selected photosensitizer, the laser power and mode of operation, and the site being treated. These are the same consideration as in the fiber delivery for the laser surgery as discussed in the section Surgical Application of Fibers.

It is noteworthy that the illumination in photodynamic therapy does not require that it be obtained with coherent light. The only reason that lasers are used is that unlike conventional light sources, spatially coherent radiation can be efficiently focused onto and transmitted through a small diameter fiber. Light emitting diodes are often used as a compromise between efficient but expensive lasers and inexpensive and inefficient conventional light sources.

The application of photodynamic therapy in oncology has been investigated over the past 25 years. It appears that this modality is being used more often now (67) for variety of cutaneous and subcutaneous tumors.

BIBLIOGRAPHY

Cited References

1. Webb MJ. Practical considerations when using fiber optics with spectrometers. Spectroscopy 1989;4:26.
2. Basic Principles of Fiber Optics, Corning Incorporated.
3. Verdaasdonk RM, van Swol CFP. Laser light delivery systems for medical applications. Phys Med Biol 1997;42:869–894.
4. Parrish JA, Deutsch TF. Laser photomedicine. IEEE J Quantum Electron, 1984; QE- 20:1386.
5. Stolen RH. Nonlinearity in fiber transmission. Proc IEEE- 68: 1980; 1232.
6. Bass M, Barrett HH. Avalanche breakdown and the probabilistic nature of laser induced damage. IEEE J. Quantum Electron 1972; QE- 8:338.
7. Chiao RY, Garmire E, Townes CH. Self-trapping of optical beams. Phys Rev Lett 1964;13:479.
8. Smith RG. Optical power handling capacity of low loss optical fibers as determined by stimulated Raman and Brillouin scattering. Appl Opt 1972;11:2489.
9. Allison W, Gillies GT, Magnuson DW, Pagano TS. Pulsed laser damage to optical fibers. Appl Opt 1985;4:3140.
10. Harrington JA, ed. Selected Papers on Infrared Fiber Optics (SPIE Milestone Series MS-9). Bellingham (WA): SPIE; 1990.
11. Merberg GN. Current status of infrared fiber optics for medical laser power delivery. Lasers Surg Med 1993;13:572–576.
12. Harrington JA. Laser power delivery in infrared fiber optics. Proc SPIE Eng Comput 1992;1649:14–22.
13. Barnes AE, May RG, Gollapudi S, Claus RO. Sapphire fibers: optical attenuation and splicing techniques. Appl Opt 1995;34:6855–6858.
14. Shenfeld O, Ophir E, Goldwasser B, Katzir A. Silver halide fiber optic radiometric temperature measurement and control of CO_2 laser-irradiated tissues and application to tissue welding. Lasers Surg Med 1994;14:323–328.
15. Abel T, Harrington JA, Foy PR. Optical properties of hollow calcium aluminate glass waveguides. Appl Opt 1994;33:3919–3922.
16. Matsuuga Y, Yamamoyo T, Miyagi M. Delevirey of F2-excimer laser light by aluminum hollow fibers. Opt Exp 2000;6:257–261.
17. Matsuuga Y, Miyagi M. Hollow Optical Fibers for Ultraviolet Light. IEEE J Quantum Electron 2004; QE- 10:1430–1439.
18. Cossmann PH, et al. Plastic hollow waveguides: properties and possibilities as a flexible radiation delivery system for CO_2-laser radiation. Lasers Surg Med 1995;16:66–75.
19. Matsuura Y, Abel T, Harrington JA. Optical properties of small-bore hollow glass waveguides. Appl Opt 1995;34:6842–6847.
20. Hongo A, Koike T, Suzuki T. Infrared hollow fibers FOR medical applications. Hitachi Cable Rev 2004;23:1–5.
21. Havener WH. The fiber optics indirect ophthalmoscope. Eye Ear Nose Throat Mon 1970;49:26.
22. Bass M, editor in chief, Handbook of Optics. Vol 1 New York: McGraw-Hill; 1995. Chapt. 1.
23. Tuchin VV, ed., Handbook of Optical Biomedical Diagnostics. Bellingham (WA): SPIE Press; 2002.
24. Bouma BE, Tearney GJ. Handbook of Optical Coherence Tomography. New York: Marcel Dekker; 2001.
25. MacAulay C, Lane P, Richards-Kortum R, In vivo pathology: microendoscopic imaging modality. Gastroint Endoscopy Clin N Am 2004;14:595–620.
26. Wagnieres GS, Star WM, Wilson BC. In vivo fluorescence spectroscopy and imaging for oncology applications. Photochem Photobiol 1998;68:603–632.
27. Perelman LT, Modell MD, Vitkin E, Hanlon EB. Scattering spectroscopy: from elastic to inelastic. In: Tuchin VV, ed., Coherent Domain Optical Method: Biomedical Diagnostics, Environmental and Material Science. Vol. 1. Boston: Kluwer Academic; 2004. pp 355–396.
28. Niederer P, et al. Image quality of endoscope. Proc SPIE 2001;4158:1–9.
29. Nelson DB. High resolution and high magnification endoscopy. Gastrointes Endosc 2000;52:864–866.
30. Kourambas J, Preminger GM. Advances in camera, video, and imaging technologies in laparoscopy. Urolog. Clinics N. Am 2001;28:5–14.

31. Korman LY. Digital imaging in endoscopy. Gastrointest Endosc 1998;48:318-326.

32. Nelson DB. Ultrathin endoscopes esophagogastroduodeno-scopy. Gastrointest Endosc 2000;51:786–789.

33. Lipson SG, Lipson HS, Tannhauser DS. Optical Physics. New York: Cambridge University Press; 1995.

34. Sawatari T, Kapany NS. Statistical evaluation of transfer properties in fiber assemblies. SPIE Proc 1970;21:23.

35. Marhie ME, Schacham SE, Epstein M. Misalignment of imaging multifibers. Appl Opt 1978;17:3503.

36. OptiScan Pty, Ltd. [Online], Investor Presentation, October 2003. Available at http://www.optiscan.com.

37. Gu M, Sheppard CJR, Gan X. Image formation in a fiber-optical confocal scanning microscope. J Opt Soc Am A 8(11): 1755 (November 1991).

38. Huang D, et al. Optical coherence tomography. Science 1991;254:1178–1181.

39. Kiesslich R, et al. Confocal laser endoscopy for diagnosing intraepithelial neoplasias and colorectal cancer in vivo. Gastroenterology 2004;127:706–713.

40. Morgner U, et al. Spectroscopic optical coherence tomography. Opt Lett 2000;25:111–113.

41. Wax A, Yang C, Izatt JA. Fourier-domain low-coherence interferometry for light-scattering spectroscopy. Opt Lett 2003;28:1230–1232.

42. Vakhtin AB, Peterson KA, Wood WR, Kane DJ. Differential spectral interferometry: an imaging technique for biomedical applications. Opt Lett 2003;28:1332–1334.

43. Jiao S, Yu W, Stoica G, Wang LV. Optical -fiber-based Mueller optical coherence tomography. Opt Lett 2003;28:1206–1208.

44. Tearney GJ, et al. Scanning single- mode fiber optic catheter-endoscope for optical coherence tomography. Opt Lett 1996;21:543–545.

45. Gelikonov FI, Gelikonov VM. Design of OCT Scanners. Bouma BE, Tearney GJ, editors. Handbook of Optical Coherence Tomography. New York: Marcel Dekker; 2001. pp 125–142.

46. Liu X, Cobb MJ, Chen Y, Li X. Miniature lateral priority scanning endoscope for real-time forward-imaging optical coherence tomography. OSA Biomed Top Meeting Tech Dig SE6 2004.

47. Zara JM, et al. Electrostatic micromachine scanning mirror for optical coherence tomography. Opt Lett 2003;28:628–630.

48. Zara JM, Smith SW. Optical scanner using a MEMS actuator. Sens Actuators A 2002;102:176–184.

49. Piyawattanametha W, et al. Two-dimensional endoscopic MEMS scanner for high resolution optical coherence tomography. Tech Digest Ser Conf Lasers Electro-Optics (CLEO) CWS 2 2004.

50. Pan Y, Xie H, Fedder GK. Endoscopic optical coherence tomography based on a microelectromechanical mirror. Opt Lett 2001;26:1966–1968.

51. Xie H, Pan Y, Fedder GK. Endoscopic optical coherence tomographic imaging with a CMOS-MEMS micromirror. Sens Actuators A 2003;103:237–241.

52. Pan Y, Fedder GK, Xie H. Endoscopic imaging system. U.S. Pat. Appl. US2003/0142934, 2003.

53. Tran PH, Mukai DS, Brenner M, Chen Z. In vivo endoscopic optical coherence tomography by use of a rotational microelectromechanical system probe. Opt Lett 2004;29:1236–1238.

54. Qi B, et al. Dynamic focus control in high-speed optical coherence tomography based on a microelectromechanical mirror. Opt Commun 2004;232:123–128.

55. Brand S, et al. Optical coherence tomography in the gastrointestinal tract. Endoscopy 2000;32:796–803.

56. Seitz U, et al. First in vivo optical coherence tomography in the human bile duct. Endoscopy 2001;33:1018–1021.

57. Zonios G, et al. Diffuse reflectance spectroscopy of human adenomatous colon polyps in vivo. Appl Opt 1999;38:6628–6637.

58. Perelman LT, et al. Observation of Periodic Fine Structure in Reflectance from Biological Tissue: A New Technique for Measuring Nuclear Size Distribution. Phys Rev Lett 1998;80:627–630.

59. Wallace MB, et al. Endoscopic Detection of Dysplasia in Patients with Barrett's Esophagus using Light Scattering Spectroscopy. Gastroenterology 2000;119:677–682.

60. Perelman LT, Backman V. Light scattering spectroscopy of epithelial tissues: principles and applications. In: Tuchin VV, editor. Handbook on Optical Biomedical Diagnostics. Bellingham: SPIE Press; 2002.

61. Backman V, et al. Diagnosing cancers using spectroscopy. Nature (London) 2000;405:35–36.

62. Utzinger U, Richards-Kortum RR. Fiber optic probes for biomedical optical spectroscopy. J Biomed Opt 2003;8:121–147.

63. Matsuura Y, Miyagi M. Er:YAG, CO, and CO^2 laser delivery by ZnS-coated Ag hollow waveguides. Appl Opt 1993;32:6598–6601.

64. Solarization Resistant Optical Fiber, SolarGuide 193 [online], Fiberguide Industries, Inc, Stirling, N.J. Available at www.fiberguide.com.

65. Dougherty TJ, et al. Photodynamic Therapy. J Nat Cancer Inst 1998;90:889–905.

66. Panjehpour M, Overholt DF, Haydek JM. Light sources and delivery devices for photodynamic therapy in the gastrointestinal tract. Gastrointest Endosc Clin N Am 2000;10:513–532.

67. Brown SB, Brown EA, Walker I. The present and future role of photodynamic therapy in cancer treatment. The Lancet 2004;5:497–508.

See also ENDOSCOPES; OPTICAL SENSORS.

FICK TECHNIQUE.   See CARDIAC OUTPUT, FICK TECHNIQUE FOR.

FIELD-EFFECT TRANSISTORS, ION-SENSITIVE.   See ION-SENSITIVE FIELD-EFFECT TRANSISTORS.

FITNESS TECHNOLOGY.   See BIOMECHANICS OF EXERCISE FITNESS.

FIXATION OF ORTHOPEDIC PROSTHESES.   See ORTHOPEDICS, PROSTHESIS FIXATION FOR.

FLAME ATOMIC EMISSON SPECTROMETRY AND ATOMIC ABSORPTION SPECTROMETRY

ANDREW W. LYON
MARTHA E. LYON
University of Calgary
Calgary, Canada

INTRODUCTION

The observation that atoms of each element can emit and absorb light at specific wavelengths is a fundamental property of matter that fostered the study of chemistry and development of structural models of atoms during the past century. The interaction of light and matter can be traced to use of the lens of Aristophanes ∼ 423 BC, and to studies of mirrors by Euclid (300 BC) and Hero (100 BC). Seneca (40 AD) observed the ability of prisms to scatter

Table 1. Selected Wavelengths and Excitation Energies for Alkali and Alkaline Earth Elements

Element	Wavelength, nm Light Emission	Excitation Energy, eV
Lithium	670.7	1.85
Sodium	589.0/589.6	2.1
Magnesium	285.2	4.34
Potassium	766.5/769.9	1.61
Calcium	422.7	2.93
Manganese	403.1/403.3/403.4	3.07
Rubidium	780.0/794.8/	1.56/1.59/
	420.2/421.6	2.95/2.94
Strontium	460.7	2.69
Cesium	894.3/852.1/	1.39/1.45/
	455.5/459.3	2.72/2.69
Barium	553.6	2.24

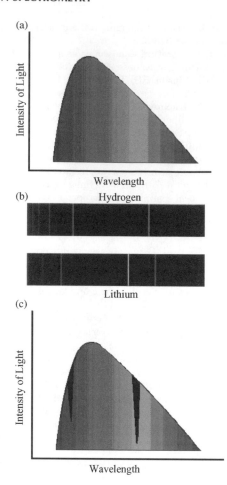

Figure 1. Emission and absorption spectra: (a) A continuous spectrum emitted from a hot solid, liquid, or gas under pressure. (b) A line spectrum emitted from a low pressure hot gas. (c) A dark line or absorption spectrum observed when a continuous spectrum is viewed behind a cool gas at low pressure.

light and Ptolemy (100 AD) studied angles of incidence and refraction. Sir Isaac Newton (1642–1727) performed many experiments to separate light into its component spectrum described in the 1718 volume Opticks: or, a treatise of the reflections, refractions, inflections and colours of light (sic).

The scientific basis of flame emission spectrometry (also called flame photometry) arose from studies of the light emitting and absorbing behaviors of matter when introduced into a flame. Observations can be traced to Thomas Melvill (1752), who observed the change of color and intensity of light when different materials were introduced into flames (1). Wollaston (1802) separated candlelight by a prism and noted that it contained a bright yellow line (2). The term line spectra is used to describe spectra composed of discontinuous narrow bands of light wavelengths or lines of different colors. Herschel (1822) observed different spectral diagrams for different salts and Talbot suggested that the yellow light indicated the presence of soda (various forms of sodium carbonate) (3,4). In 1860, Kirchhoff and Bunsen reported the correlation of spectral lines with specific elements by introducing solutions into a Bunsen burner using a platinum wire or hydrogen spray (5). This body of work is Table 1 is depicted in Fig. 1 and summarized by Kirchhoff's laws of spectroscopy that describe emission and absorption of light by matter. Kirchhoff's laws are a hot solid, liquid, or gas, under high pressure, gives off a continuous spectrum; a hot gas, under low pressure, produces a bright line or emission line spectrum; a dark line or absorption line spectrum is seen when a source of continuous spectrum is viewed behind a cool gas at low pressure.

In 1930, Lundegardh developed the first practical method for quantitative spectrochemical analysis by showing that the intensity of light emitted at specific wavelengths represented the concentration of an element being introduced into a flame (6,7). Flame emission methods were rapidly developed to determine alkali and alkaline earth metals. With the introduction of interference filters by Leyton and Ivanov in the 1950s to select narrow bands of wavelengths of light, multichannel flame emission photometers were developed that allow quantification of several elements simultaneously (potassium, sodium, lithium) from a single flame (8,9).

Following the development of commercial flame emission photometers, clinical laboratories used the analyzers to measure sodium and potassium concentrations in body fluids from the mid-1950s until the mid-1970s, when potentiometric technologies largely replaced this technology. Flame emission photometers were initially retained in clinical laboratories to allow monitoring of serum lithium in patients treated with oral lithium salts as a therapy for manic depressive disorders. The rugged and reliable flame emission photometers are still used in specialized clinical laboratories to analyze fluid matrices that can render potentiometric analyses unstable (e.g., the analysis of stool electrolyte concentrations).

In addition to flame, other energy sources can be used to produce atomic emission spectra in the ultraviolet (UV) and visible regions including electric arc, electric spark, or inductively coupled plasma. The use of higher temperature energy sources to induce emission allows this technology to assess quantitative elemental composition of materials. These high energy instruments are used to support analysis of specimens for geology, manufacturing, and clinical toxicology.

Kirchhoff's third law describes the absorption of specific wavelengths of light by cool gases, the basis of atomic absorption spectrometry. In the 1950s, Walsh (10) and Alkemade and Milatz (11) independently reported the analytical potential of light absorption by atoms in a flame, extending the observation that Woodson (12) reported for determination of mercury in air. The temperature of the flame is sufficient to disperse atoms in a low density gaseous phase, enabling the atoms of individual elements to absorb light at specific wavelengths. An intermittent external light source is used to assess light absorption by atoms in the flame while the continuous light emission from the flame is measured and subtracted. Hollow cathode lamps were subsequently developed that enabled atomic absorption methods to achieve both greater sensitivity and selectivity than flame emission photometry and many more analytical applications. The development of electrothermal atomic absorption spectrometers and preanalytical derivitization of chemicals into compounds that minimize interference during analysis make this technology a valuable part of analytical laboratories. Clinical laboratories continue to use atomic absorption spectrometric methods to evaluate the concentration of lead, copper, zinc, and other trace metal elements in body fluids.

THEORY: FLAME ATOMIC EMISSION SPECTROMETRY

A theoretical basis for flame emission spectrometry can be described using a model of the atom, where the atom is composed of a nucleus surrounded by a cloud of electrons that fluctuate between different energy levels. The temperature of a solution of salt is rapidly elevated when the solution is introduced into a flame. The solvent evaporates and most metallic ions are reduced to the elemental neutral atom state by free electrons in the flame. A small propor-

tion of monatomic ions are also formed. The temperature of the flame can confer sufficient quantum energy to excite electrons in a low energy level to a higher energy state, however, the higher energy state is inherently unstable. When the unstable high energy electron returns to the ground state, energy is released from the atom in the form of monochromatic light; a spectral line that is characteristic of that atom and energy transition. The analysis of the intensity and wavelengths of the line spectra emitted from materials in high temperature sources is the basis of flame atomic emission spectrometry, Fig. 2.

The intensity of light in the line spectra is related to the number of atoms with electrons in the higher energy level. The relationship between number of atoms in each energy state (N_0: ground state or N_1 elevated energy state) is dependent on the temperature: T, and the size of the energy interval between the ground state and elevated energy state: $E_1 - E_0$. This relationship is depicted by the Boltzmann equation:

$$\frac{N_1}{N_0} = \frac{P_1}{P_0} e^{-(E_1 - E_0)/kT}$$

The important features of this equation are that the ratio of atoms in each energy state (N_1/N_0) is dependent on the magnitude of the quantity $-(E_1 - E_0)/kT$. As temperature increases, the number of atoms in the high energy state increases in an exponential manner. The values P_1 and P_0 are the quantum mechanical properties called statistical weights and represent the number of ways in which an atom can be arranged in a given state.

The Boltzmann equation describes the proportion of atoms with electrons in an excited or high energy state at specific temperatures. As electrons decay from excited states to lower energy states, light energy is released in discrete energy intervals associated with the line spectrum for that element. The relationship between the light energy

Absorption of energy (light or heat) as an

electron moves to a higher energy level

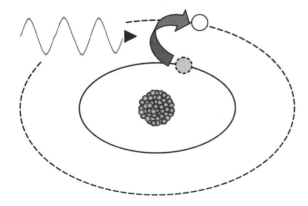

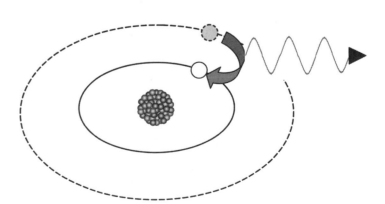

Emission of light energy as an electron

moves to a lower energy level

Figure 2. A model of the atom depicting (left panel) the absorption of energy associated with an electron moving to a higher energy state or (right panel) the emission of light energy associated with an electron moving to a lower energy state.

released and the wavelength or color of light emitted is described by Planck's law: E: Energy of electron transition (J), h: Planck's constant (J·s), c: speed of light (m/s), λ: wavelength of light (m^{-1}).

$$E = hc/\lambda$$

Atoms can have electrons at many different energy levels and this results in line spectra emissions at many different wavelengths. When a flame has limited energy and the atoms are at low density, only a small fraction of atoms are excited and only a simple line spectrum of low energy wavelengths is observed. Line spectra are characteristic of each element allowing analysts to measure emitted light to determine the chemical composition of solutions in a flame or the chemical composition of a star, based on starlight Table 1 (13). For example, the line spectrum of sodium is characterized by an intense doublet of yellow lines at 589 and 589.6 nm. The intensity of light at 588–590 nm can be measured in a flame photometer to measure the concentration of sodium in a fluid. Other alkali metals have low excitation energy that results in emission of light at longer wavelengths. Lithium and rubidium produce a red color and potassium a red-violet color when introduced into a flame. Higher energy associated with higher temperatures is required for many other elements to emit light in the visible or UV regions. To achieve the higher temperatures, electrical arc, spark, ionized gases or plasmas are used to induce light emission instead of a cooler conventional flame. The inductively coupled plasma atomic emission spectrometer (ICP–AES) (also called an inductively coupled plasma optical emission spectrometer, ICP–OES) devices are versatile instruments capable of measuring elemental composition with high sensitivity and accuracy and are currently used to detect trace elements and heavy metals in human urine or serum. The ICP–AES devices use temperatures in excess of 5000 °C and can achieve resolution of 0.0075–0.024 nm.

The line spectra of pure elements often contains doublets, triplets, or multiplets of emission lines. Planck's law implies there is very little difference in energy level associated with spectral lines that have similar wavelengths. Spectral line splitting is attributed to the quality of electron spin that can have two quantum values of slightly different energy. Doublets are observed from atoms with a single outer electron (with two possible electron spin values), triplets from atoms with two outer electrons and multiplets from atoms with several outer electrons. High resolution instruments are required to distinguish some of the multiplet lines. With use of high temperature flames or ICP source and high resolution instruments, the spectra of atoms, ions, and molecules can be observed. Alkaline earth metals can become ionized and the resulting cations can generate emission spectra. Oxides and hydroxides of alkaline earth metals can be present in samples or generated in the flame and can also generate emission spectra. In practice, an emission wavelength used to measure an element is selected when the line provides sufficient intensity to provide adequate sensitivity and the line is free from interference from other lines near the selected wavelength. The amount of light emitted at a specific emission wavelength may not be sufficient for analysis. For this reason, flame

emission spectrometry methods may not be applicable and alternate methods, such as atomic absorption spectrometry or inductively coupled plasma mass spectrometry, may be preferred.

THEORY: ATOMIC ABSORPTION SPECTROMETRY

Atomic absorption spectrometry developed from the observations used to establish Kirchhoff's third law: 'That a dark line or absorption line spectrum is seen when a source of continuous spectrum is viewed behind a cool gas under pressure' (Fig. 1). When a sample is introduced into a flame, only a small fraction of the atoms are in an excited or high energy state (according to the Boltzmann equation) and most atoms of the element remain in the unexcited or ground state and are capable of absorbing energy. If light energy is provided as a continuous spectrum, ground-state atoms will selectively absorb wavelengths of light that correspond with the energy intervals required to excite electrons from low energy to higher energy levels. The wavelengths of light that are absorbed as electrons are excited from ground state to higher energy levels are analogous to the wavelengths of light emitted as high energy electrons return to the ground state. The advantage of atomic absorption spectrometry is that most of the atoms in a flame remain in the unexcited ground state and are capable of light absorption, allowing this method to be ~ 100-fold more sensitive than flame emission spectrometry.

Atomic absorption of light is described by an equation analogous to the Lambert–Beer law. The absorbance of light, A, is defined as the logarithm of the ratio of initial light beam intensity I_0 to the intensity of the beam after light absorption, I. The absorbance of light is directly proportional to the concentration of absorbing atoms c and the path length of the light beam in the atoms d. The value k is a constant referred to as the absorptivity constant.

$$A = \log\left(\frac{I_0}{I}\right) = kcd$$

The emission of characteristic wavelengths or colors of light from materials in a flame was initially observed by the unaided human eye and preceded the measurement of atomic absorption that required instrumentation. To observe atomic absorption of light, a beam of light was passed through atoms introduced to a flame and the ratio of initial and postabsorption light beams was determined. However, the light beam emitted from the flame also contains light derived from the combustion of materials in the flame that was not present in the initial light beam. To measure the atomic absorption of light, the intensity of background light emitted from the flame itself must be subtracted from the postabsorption light beam intensity prior to calculating the absorbance. Electrothermal atomic absorption (also called flameless atomic absorption or graphite furnace atomic absorption) uses an electrically heated furnace to atomize the sample. By design, the electrothermal instruments have greater sensitivity and lower detection limits because analyte atoms can achieve higher gaseous

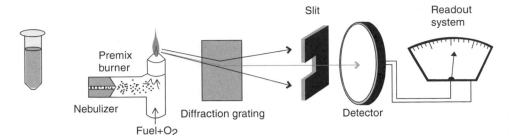

Figure 3. Components of a flame atomic emission spectrometer.

concentrations and residence time in the light beam and there is less background light emission.

The sensitivity of atomic absorption methods was improved by using light sources that generated wavelengths of light that exactly matched the wavelength of maximum absorption by the element undergoing analysis. The light sources are referred to as hollow cathode lamps. These lamps heat a small amount of the element of interest and generate line spectra emission for the light beam at wavelengths that exactly match the wavelengths maximally absorbed by atoms of that element in the ground state in the flame. Often hollow cathode lamps contain a single element, however, multielement lamp sources with six or more elements are commercially available.

EQUIPMENT

Flame Atomic Emission Spectrometry

The components of a flame photometer are illustrated schematically in Fig. 3 and consist of a nebulizer within a spray chamber, a premix burner, flame, monochromator, detector, and readout system. The monochromator, detector, and readout systems are similar to those used in a spectrophotometer: monochromatic light is obtained by means of an interference filter or diffraction grating, a detector consisting of a photomultiplier tube, result display (visual meter, digital display, or data capture a computer system), and the flame serves as a light source and sample compartment.

Each of the basic components of the instrument contributes to the analytical process. Various combinations of combustible gases and oxidants can be used to create flames with different temperatures (e.g., acetylene, propane, natural gas, oxygen, compressed air). The nebulizer is an important component responsible for mixing the analyte liquid with compressed air and converting it into a fine mist that enters the flame. For precise analytical measurements, the nebulizer must create a mist of consistent composition (10–15% of the aspirated sample) and mix the mist into the gases that combust to create a consistent flame. Within the spray chamber, large droplets settle out and are drained to waste, and only a fine mist enters the flame. A wetting agent (e.g., a nonionic detergent) may be added to standards and samples to minimize changes in nebulizer flow rates that result from differences in the viscosity or matrix of samples.

The intensity of emitted line spectra from atoms excited in the flame must be discriminated from other light in the flame. Less sophisticated instruments (e.g., flame photo-

meters) rely on filters with narrow bandpass to eliminate background light emitted from the flame. Diffraction gratings are used as monochromators in more sophisticated instruments with a slit to limit the bandwidth of light being detected.

To improve the reliability, an internal standard may be used when designing a method. Usually an element not present in the sample is introduced into the calibration standards, quality control, and unknown solutions. In biological applications, lithium or cesium is frequently used for this purpose. By measuring the ratio of emission light intensity of the target element to the internal standard, there is compensation for small variation in nebulization rate, flame stability and solution viscosity, and methods perform with greater precision. Addition of either lithium or cesium to all solutions can also prevent interference by acting as an ionization buffer (previously referred to as a radiation buffer). A relatively high concentration of an alkali metal ion in the solutions creates a buffer mixture of its ionic and atomic species and free electrons in the flame. The elevated free-electron concentration in the flame represses and stabilizes the degree of ionization of analyte atoms and improves the atomic emission signal.

Inductively coupled plasma emission spectrometers have similar components. The sample is nebulized into argon gas that is heated to extreme temperatures ($>5000\,°C$) that efficiently breaks molecular bonds to generate gaseous atoms. Because few molecules exist at these temperatures, ICP–AES has few chemical interferences compared to flame techniques, however, because of the high temperature and large proportion of atoms and ions that emit light, ICP–AES methods are more prone to spectral interferences associated with inadequate resolution of spectral lines.

Atomic Absorption Spectrometry

The components of a flame atomic absorption spectrometer are illustrated schematically in Fig. 4 and consist of a nebulizer, premix burner, hollow cathode lamp, modulation system or beam chopper, flame, monochromator, detector, and readout system. The monochromator, detector, and readout systems are similar to those used in a spectrophotometer. Monochromatic light is obtained by isolating a single spectral line emitted from a hollow cathode lamp using an interference filter. The detector is a photomultiplier tube and the results can be displayed by a visual meter, digital display, or captured by a computer system. The flame serves as a sample compartment and

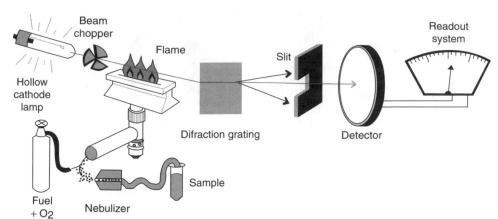

Figure 4. Components of a flame atomic absorption spectrometer.

light from the hollow cathode lamp is passed through the flame to measure absorption. Electronic modulation or a mechanical beam chopper is used to interrupt the light from the hollow cathode lamp into pulses so the extraneous light emitted from the flame alone can be measured and subtracted prior to measuring atomic absorption.

Electrothermal atomic absorption instruments typically use a graphite furnace to rapidly heat a sample in a uniform manner to ~2600 °C. Samples are typically introduced into the furnace on a carbon rod. As the temperature is raised, the samples are dried, charred, rendered to ash, and atomized (e.g., Table 2). Hollow cathode lamps are used as a source of monochromatic light to determine atomic absorption. In contrast to flame-based methods that rapidly dilute and disperse analyte atoms, electrothermal methods retain atoms at higher concentrations with longer residence time in the light beam. In addition, there is less extraneous background light. These advantages allow electrothermal atomic absorption instruments to measure many trace elements and heavy metals not capable of measurement with flame atomic absorption instruments (15).

There are four categories of interference that can occur with atomic absorption spectrometry: chemical, spectral, ionization, and matrix. The element undergoing analysis may exist or form stable molecules within the flame. The atoms in the stable molecules do not generate line spectra at the characteristic wavelengths of the free atoms and this chemical interference results in a negative analytic bias for the method. Spectral interference can occur when there is nonspecific absorption of light by material in the sample. Nonspecific absorption of light can be caused by solids or

particles in the flame. Spectral interference can often be reduced by performing maintenance on the burner head to remove residues and assure a steady flame is maintained. Emission of light from other components in the flame can interfere with analysis and can often be removed with a background correction method. Ionization interference occurs when atoms in the flame are converted to ions that will not absorb the line spectra at the characteristic wavelengths for the atoms. Ionization interference can be reduced by an ionization buffer or by lowering flame temperature to reduce the extent of ionization. Matrix interferences occur when the calibration standards have a different composition than the samples (e.g., viscosity, salt concentration or composition, pH, or presence of colloids, e.g., protein). Matrix interferences can be avoided in the method design by assuring calibrator solutions have similar composition to the samples or that samples are highly diluted so they acquire a matrix similar to the calibrator solutions.

Several methods have been developed to improve the sensitivity of atomic absorption by reducing background radiation spectral interference. Electronically modulated light beams or mechanical beam choppers have been used to generate pulsed light so the background emission of light from the flame can be distinguished, measured and subtracted. Analogous to molecular absorption spectrophotometers, atomic absorption spectrometers can have a single beam design (Fig. 4) or a split double-beam design where the beam modulator acts as a beam splitter to create the light beam that passes through the sample and a second beam that acts as a reference. Split double-beam instrumentation has an advantage of compensating for variation in light intensity from the lamp or variation of detector sensitivity.

In electrothermal atomic absorption methods, the rapid heating of samples can cause the release of smoke or nonspecific vapor that blocks some light from reaching the detector. To correct for this nonspecific light scatter, in addition to the beam from a hollow cathode lamp, a beam from a deuterium continuum lamp can be direct through the sample chamber and light intensity measured at a wavelength other than that used for the assay to assess the extent of light scatter.

A third method of background correction is based on the Zeeman effect. When atoms are placed in a strong magnetic

Table 2. Graphite Furnace Settings for Blood Lead Determination[a]

Stage	Temperature, °C	Ramp Time, s	Hold Time, s	Gas flow, $L \cdot min^{-1}$
Dry	120	5	15	300
Preash	260	1	5	300
Ash	800	5	27	300
Cool	20	1	4	300
Atomize	1600	0	5	0
Clean	2700	1	2	300

[a]See Ref. 14.

field, the emission and absorption line spectra are dramatically altered as the single line spectra are split into triplets or more complex patterns. Splitting of a spectral line into two different wavelengths implies that the electrons involved have slightly different energy levels, according to Planck's law. The small difference in the energy levels is attributed to slightly different quantum numbers for the electrons involved and this difference in energy level can only be detected in the presence of the magnetic field. In the absence of the magnetic field, slight differences in energy levels are not apparent and a single wavelength of absorption or emission is observed. By turning the magnetic field off and on during atomic absorption spectrometry, the population of atoms being tested is modulated from a uniform population to discrete populations with different quantum numbers. By measuring the element-specific difference in atomic absorption as the magnetic field is modulated, nonspecific background absorption or emission can be accurately subtracted. Implementation of instruments with Zeeman correction is complex, expensive, and requires optimization of the magnetic field for each element. Consequently, Zeeman correction approaches are usually reserved for methods involving strong background interferences.

MEDICAL APPLICATIONS

Flame emission spectrometry and atomic absorption spectrometry are sophisticated methods of elemental analyses that are robust and flexible techniques that have been applied to the analysis of many different types of biological or clinical samples. In the 1950s and 1960s, use of flame photometers provided state-of-the-art accurate and precise determinations of serum, sodium, and potassium. Stable, reliable, and inexpensive potentiometric methods for determination of electrolyte concentrations were developed and replaced flame photometric methods for most medical applications. The hardy reliable nature of flame photometers is still used for some analysis (e.g., the determination of stool electrolytes). While flame emission photometric methods can be applied to determine the concentration of trace elements or heavy metals in biological specimens, many clinical laboratories use electrothermal atomic absorption spectrometry or inductively coupled plasma mass spectrometry that offer greater specificity, sensitivity, precision, and adaptability.

Flame emission spectrometric determinations are prone to negative bias when biological samples undergoing analysis have high concentrations of lipid or protein (16). Biological samples are mixed with a diluent to create a mist in the nebulizer and the mist is created by mixing constant proportions of each fluid. However, when a biological fluid has a high concentration of lipid or protein, the constant volume of sample fluid has a reduced amount of aqueous dissolved ions because lipid particles suspended in the fluid occupy volume and reduce the volume of water available to dissolve the ions. In a similar manner, proteins bind water molecules creating a hydration shell and at high concentration in an aqueous solution, the proteins occupy volume in the sample and reduce the volume of water available to dissolve ions. This negative bias is known as

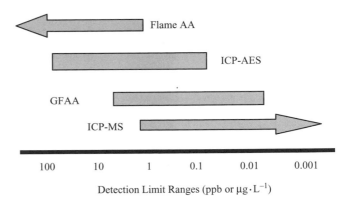

Figure 5. Typical detection limit ranges for major spectrometry techniques. Flame atomic absorption, inductively coupled plasma-atomic emission spectrometry, graphite furnace atomic absorption, inductively coupled plasma–mass spectrometry (19).

the volume exclusion effect. The term pseudohyponatremia is used to describe the misleading low serum sodium concentrations attributed to this bias. This volume exclusion effect was initially characterized with flame photometers and it remains a concern with potentiometric instruments that use indirect electrodes and rely on sample dilution (17).

The use of atomic absorption spectrometers in clinical laboratories for the determination of calcium, magnesium, and other metal cations reached a peak in the 1970s. In the following decades, potentiometric methods became commercially available for the most common tests (e.g., calcium, magnesium, sodium, and potassium) and atomic absorption spectrometers were not commonly used in each hospital laboratory. A concise review of the history of atomic absorption spectrometry application in clinical chemistry was published in 1993 (18). The atomic absorption spectrometer remained an important platform for the determination of serum zinc, copper, and lead. Medical laboratory services have often been consolidated, and infrequent tests (e.g., serum zinc, copper, lead, trace elements, and heavy metals) are shipped to reference laboratories for analyses. While inductively coupled plasma mass spectrometry offers greater sensitivity and flexibility of analysis, flame and electrothermal atomic absorption spectroscopic methods remain cost-effective, precise analytic methods, widely used in clinical reference laboratories to support the evaluation of serum or blood concentrations of zinc, copper, aluminum, chromium, cadmium, mercury and lead (Fig. 5).

BIBLIOGRAPHY

Cited References

1. Melvill T. Observations on light and colors, Essays Observ. Phys Lit Edinburgh 1756;2:12–90.
2. Wollaston WH. A method of examining refractive and dispersive powers. Philos Trans R Soc London 1802;92:365–380.
3. Herschel JFW. On the absorption of light by coloured media. Trans R Soc Edinburgh 1823;9:445–460.
4. Talbot HF. Some experiments on colored flames. Edinburgh J Sci 1826;5:77–81.
5. Kirchhoff G, Bunsen R. Chemical analyses by means of spectral observations. Ann Phy (Leipzig) 1860;110(2):160–189.

6. Lundegardh H. New contributions to the technique of quantitative chemical spectral analysis. Z Phys 1930;66: 109–118.

7. Lundegardh H. Investigations into the quantitative emission spectral analysis of inorganic elements in solutions. Lantbrukshoegsk Ann 1936;3:49–97.

8. Leyton L. An improved flame photometer. Analyst 1951;76: 723–728.

9. Ivanov DN. The use of interference filters in the determination of sodium and potassium in soils. Pochvovedenie [N.S.] 1953;1: 61–66.

10. Walsh A. The application of atomic absorption spectra to chemical analysis. Spectrochim Acta 1955;7:108–117.

11. Alkemade CTJ, Milatz JMW. Double beam method of spectral selection and flames. J Opt Soc Am 1955;45:583–584.

12. Woodson TT. A new mercury vapor detector. Rev Sci Instrum 1939;10:308–311.

13. Lutz RA, Stojanov M. Flame Photometry. In: Webster JG, editor. Encyclopedia of Medical Devices and Instrumentation. New York: John Wiley & Sons Inc; 1988.

14. Bannon DI, et al. Graphite furnace atomic absorption spectroscopic measurement of blood lead in matrix-matched standards. Clin Chem 1994;40(9):1730–1734.

15. Butcher DJ. Joseph Sneddon. A Practical Guide to Graphite Furnace Atomic Absorption Spectrometry. New York: John Wiley & Sons Inc; 1998.

16. Waugh WH. Utility of expressing serum sodium per unit of water in assessing hyponatremia. Metabolism 1969;18: 706–712.

17. Lyon AW, Baskin LB. Pseudohyponatremia in a myeloma patient: Direct electrode potentiometry is a method worth its salt. Lab Med 2003;34(5):357–360.

18. Willis JB. The birth of the atomic absorption spectrometer and its early applications in clinical chemistry. Clin Chem 1993; 39(1):155–160.

19. Perkin Elmer Life Sciences (2004). Guide to Inorganic Analysis. Available at http://las.perkinelmer.com/Content/Related Materials/005139C_01_Inorganic_Guide_web.pdf Accessed 1 Aug, 2005.

See also ANALYTICAL METHODS, AUTOMATED; FLUORESCENCE MEASUREMENTS.

FLAME PHOTOMETRY.   See FLAME ATOMIC EMISSION SPECTROMETRY AND ATOMIC ABSORPTION SPECTROMETRY.

FLOWMETERS

ARNOLD A. FONTAINE
STEVEN DEUTSCH
KEEFE B. MANNING
Pennsylvania State University
University Park, Pennsylvania

INTRODUCTION

Fluid flow occurs throughout biomedical engineering in areas as different as air flow in the lungs (1,2) to diffusion of nutrients or wastes through membranes (3–5). Flow-related problems can involve fluid media in the form of gas, liquid, or multiphase flows of both liquids and gas together or in combination with solid matter. Biomedical flows occur in both *in vivo* (6) and *in vitro* (7,8). They can involve relatively benign flows like that of saline through an intravenous tube to a biochemically active flow of a non-Newtonian fluid such as blood. Many biomedical or bioengineering processes require the quantification of some flow field that may be directly or indirectly related to the process. Such quantification can involve the measurement of volume or mass flow, the static and dynamic pressures, the local velocity of the fluid, the motion (speed and direction) of particles such as cells, the flow-related shear, or the diffusion of a chemical species.

Interest in understanding fluid flows and attempts to measure flow-related phenomena has had a long history in the scientific and medical communities with early work by Newton, DaVinci, and others. Early studies often involved observations of flow-related phenomena that can be characterized as simple flow visualization or particle tracking (9) or the estimation of a pressure by the displacement of a fluid. Rouse and Ince (10) provide a historical review of these early works. Throughout the years, flow measurement techniques have advanced significantly in capability (what is measured and how), versatility, and in many ways, complexity. Some techniques, such as photographic flow visualization, have changed little in over 100 years, whereas others are only possible because of advances in electronics, optics, and physics. Improved capability and versatility are evidenced through the increased ease of use in some systems and the ability to measure more quantities with increased accuracy and resolution. However, this improved capability and versatility has, in some cases, come at the cost of increased complexity in system hardware, calibration requirements, and application complexity.

Measurement techniques can be characterized as invasive or noninvasive and direct or indirect. Invasive measurement techniques require the insertion of a sensing or a sampling element directly into the flow field. As a result of this direct insertion, invasive probes may alter the flow field characteristics or induce bias errors associated with the presence of the probe in the flow field or by the operation of the probe (11,12). Invasive probes are often designed to minimize flow disturbance by miniaturizing the sensing elements or by displacing the sensing elements some distance from the hardware holding the probe in the flow, as illustrated in Fig. 1. Invasive probes also require some type of closure at the penetration site through the boundary of the flow, which must be accounted for in the test design and can be particularly important in *in vivo* applications to prevent fluid loss or infection. White et al. (13) measured wall shear stress in the abdominal aorta of dogs using an invasive, flush-mounted hot-film probe and described how the probe tip is modified to provide an effective entry mechanism through the arterial wall with an adequate seal.

Noninvasive techniques do not involve direct insertion of a sensor into the flow but provide sensing capability through either access to the flow at the flow boundary or through the use of some form of electromagnetic radiation (EMR) transfer or propagation. Wall-mounted thermal sensors, static pressure taps and transducers, or surface

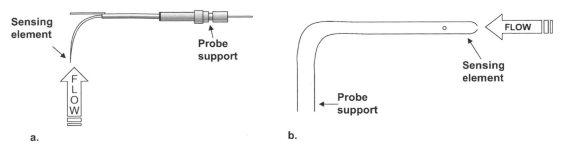

Figure 1. Examples of invasive velocity measurement sensors with displaced sensing elements relative to their probe supports. (a) a boundary layer style hot-wire probe for thermal anemometry. Picture from TSI Inc. catalog, probe catalog 1261A. (b) Pitot static probe for velocity measurement.

sampling probes are examples of noninvasive techniques that require access to the boundary of the flow field through a wall penetration (13). Ultrasound (14) magnetic resonance (MR), X-ray, and optical techniques are all examples of electromagnetic radiation that can be used to probe flow fields (15). Unlike the wall-mounted invasive probes describe above, EMR-based measurement systems do not require physical penetration into the flow field or access through the flow field boundary. They do, however, require a "window" into the flow through the boundary enclosing the flow field of interest. This "window" depends on the type of technique being used. Optical-based techniques require an optically clear window that may not be suitable for *in vivo* applications, whereas ultrasound and X-ray techniques require that the material properties of the flow boundaries are transparent to these forms of EMR waves. For example, lead will shield X-ray penetration and metal objects on a surface or in the flow may create local artifacts (noise or error) in MR measurements.

Direct and indirect measurements are defined by how quantities of interest are measured. The displacement of a particle or cell in a flow can be directly measured by photographing the movement of the particle over a finite time interval (16). Most flow-related measurement systems, however, are indirect. In general, velocity or flow is indirectly calculated from the direct measurement of a quantity and the application of a calibration that relates the magnitude of the measured quantity to the parameter of interest. This calibration may involve not only a conversion of a measured quantity like a voltage to a physical quantity such as a pressure, but may also involve the application of a functional relationship (i.e., Bernoulli's equation), which requires assumptions about the flow. For example, volume flow probes often assume a characteristic velocity profile at the location of the probe (17). Blood flow in the microcirculation can be estimated by indirectly measuring the cell velocity using a time-of-flight optical technique where the time a cell takes to move a known distance is measured and the cell velocity is calculated by the ratio of the distance divided by the transit time (18).

Indirect measurement can also impact the uncertainty in the estimated quantity. The measurement of velocity using a Pitot probe is one example of an indirect measurement (19). The Pitot probe measures the local dynamic and static pressures in the flow. These pressures are most often measured using a pressure transducer that provides a voltage output in response to an applied pressure. A cali-

bration is then applied to the measured voltage to convert it to pressure. Velocity is indirectly calculated from the estimated pressures using the Bernoulli equation. The error in Pitot probe velocity measurements includes the pressure transducer calibration uncertainty, noise and statistical uncertainty during the pressure measurement, electronic noise in the acquisition of the transducer output voltage, and transducer drift and potential bias due to the physical size of the probe relative to the flow scales being measured. These errors are nonlinearly propagated into the estimate of the velocity uncertainty.

Measurement accuracy is also a function of the physical and operating characteristics of the probe itself. Many flows exhibit a range of spatial and temporal scales. The physical size and the frequency response of the sensing element must be taken into account when choosing a measurement system for a particular application. A large sensing element or an element with poor frequency response has the effect of low pass filtering the measured signal (20). This low pass filtering will cause a bias in the measured quantity. The total measurement uncertainty must also take into account statistical errors associated with random processes, cycle-to-cycle variability in pulsatile systems, and noise. The reader is referred to the texts by Coleman and Steele (21) and Montgomery (22) for a detailed approach to experimental uncertainty analysis. The focus of this chapter will be on measurement techniques, their fundamentals of operation, their advantages and disadvantages, and examples of their use.

The name flow measurement is a broad term that can encompass the measurement of many different flow-related parameters. In this chapter, the authors will focus on the measurement of those parameters that are most often desired in a biomedical/bioengineering application, volume flow rate, and velocity. Imaging, Doppler echocardiography, and MR techniques are addressed in other chapters within the encyclopedia and, thus, will only be briefly introduced in this article when applicable. This chapter is subdivided into sections that will address volume flow and velocity separately, with a detailed presentation of systems that are available for the measurement of each. Although ultrasound and MR techniques are often used to measure flow-related parameters, a detailed discussion of the principles of operation will not be presented here as these topics are covered in depth in other chapters of this Encyclopedia.

FLOW MEASUREMENT APPLICATIONS

Volume Flow Measurement

In both the clinical environment and the laboratory environment, the measurement of the volume flow rate of a fluid as a function of time can be an important parameter. In internal flow applications, which comprise most biomedical flows of interest, the volume flow of a fluid (Q) is related to the local fluid velocity (V) through the integration of the velocity over the cross sectional area of the duct or vessel (19,23).

$$Q = \int V \, dA \tag{1}$$

The flow rate Q, velocity V, and area A have dimensions of volume per time, length per time, and length squared, respectively. The SI units are typically used in the bioengineering field with mass units of grams or kilograms, length units of meters (millimeter and centimeters), and time units of seconds. The mass flow (M) is directly related to the flow volume through the fluid density, ρ, with units of mass/volume.

$$M = \rho \cdot Q \tag{2}$$

Fluid pressure and velocity are related through the Navier–Stokes equations, which govern the flow of fluids in internal and external flows [see White (23)].

The volume flow rate of blood is often measured in many cardiovascular applications (24). For example, cardiac output (CO) is the integrated average of the instantaneous volume flow rate of blood (Q_b) exiting the aortic valve over one cardiac cycle (T_c):

$$CO = \left[\int Q_b \, dt \right] / T_c \tag{3}$$

The cardiac output has units of volume flow rate, volume per unit time. The following subsections will provide an overview of measurement techniques typically used for both volume flow and velocity measurement in *in vivo* and *in vitro* studies. This article will be limited to those flow measurement techniques most often used in the biomedical and bioengineering fields. Specialized measurement techniques, such as concentration or species measurement through laser-induced fluorescence (LIF) or mass spectrometry, will not be addressed.

Electromagnetic Flow Probes. Carolina Medical Inc. developed the first commercially available electromagnetic flow meter in 1955. The design provided scientists and clinicians with a noninvasive tool that could directly measure the volume flow rate (17). Clinical probes were developed that could be attached to a vessel for extravascular measurement of blood flow without the need for cannulation of a surgically exposed vessel.

Electromagnetic flow meters are volumetric flow measuring devices designed to measure the flow of an electrically conducting liquid in a closed vessel or pipe. Commercial meters, used in the biomedical and general engineering fields, come in a variety of sizes and designs that can measure flow rates from $\sim 1 \, \text{mL·min}^{-1}$ to $>100,000 \, \text{L·min}^{-1}$.

Reported uncertainties are typically on the order of a few percent, but can be as low as 0.5% of reading in some specialized meter designs. Low uncertainties are dependent on proper use and installation of the meter. Improper use or installation (mounting the meter in a location with a complex flow profile) will increase measurement uncertainty. Most clinical quality meters that are mounted externally to a vessel exhibit uncertainties that can approach 15% in some applications (Carolina Medical Inc., product support literature). In cardiovascular applications, these meters may also be susceptible to electrical interference by the heart or by measurement anomalies due to motion of the probe.

The principle governing the operation of an electromagnetic flow meter is Faraday's law of electromagnetic induction. This law states that an induced electrode voltage is proportional to the velocity of flow of a conductor through a magnetic field of a known density. Mathematically, this law is represented as:

$$E_e = K[V \cdot B \cdot L_e] \tag{4}$$

Here, E_e is the induced voltage between two electrodes (with units of volts) separated by a known conductor length L_e (provided by the conducting fluid between the electrodes) in units of millimeters or centimeters, B is the magnetic field strength in units of Tesla's, and V is the conducting fluid average velocity in units of length per time. The parameter K is a dimensionless constant. Meter output is linear and proportional to flow velocity.

Fluid properties such as viscosity and density are absent from Eq. 4. Thus, the output of an electromagnetic flow meter is independent of these properties and its calibration is independent of the type of fluid, provided the fluid meets minimum conductivity levels. This meter can then be used for highly viscous fluids, Newtonian fluids, and nonNewtonian fluids such as blood. The requirement of an electrically conductive fluid can disqualify an electromagnetic meter in some applications. Typical meters require a minimum fluid conductivity of $\sim 1 \, \mu\text{S/cm}^{-1}$. However, low conductivity designs are capable of operating with fluid conductivities as low as $0.1 \, \mu\text{S/cm}^{-1}$. The presence of gas bubbles in the fluid can cause erratic behavior in the meter output.

A typical meter design has electromagnetic coils mounted on opposing sides of an electrically insulated duct with two opposing electrodes mounted 90° relative to the electromagnets. The two electrodes are mounted such that they are in contact with the conducting fluid. Figure 2 illustrates the typical configuration.

The meter is designed to generate a magnetic field that is perpendicular to the axis of motion of the flowing fluid. A voltage is generated when the conducting fluid flows through the magnetic field. This voltage is sensed by the two opposing electrodes. The supporting material around the meter is made of a nonconducting material to prevent leakage of the voltage generated in the moving fluid into the surrounding material. In practice, the conductor length, L_e, is not the simple path illustrated, but is rather the integral of all possible path lengths between the two electrodes across the cross section of the duct or vessel. The

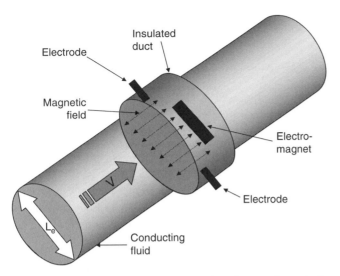

Figure 2. Illustration of the principle of operation of the electromagnetic flow meter. Note, the magnetic field, electrodes, and flow direction are all mutually perpendicular to one another.

signal generated along each path length is proportional to the fluid velocity across that path. Thus, the two electrodes measure the integrated sum of all velocities across every possible path in the vessel cross section. This signal is then directly proportional to the volume flow rate of the fluid passing through the magnetic field.

The magnetic field generated in commercial meters may be anything from a uniform field to a specifically designed field with prescribed characteristics. Meters with uniform magnetic fields can exhibit some sensitivity to the conducting liquid's velocity profile. Fluid flowing through a vessel does not have the same velocity at all locations across the vessel. The no-slip condition at the walls of the duct ensures that the fluid velocity at the wall is zero (19). Viscosity then generates a gradient between the flowing fluid in the vessel and the stationary fluid at the wall. In complex geometries, secondary flows may occur and velocity gradients in other directions may also develop (25,26). This variation in fluid velocity across the magnetic field coupled with variations in the conductor length generates a variation in the magnitude of the voltage measured across the duct. As a result, installation of these meters must be carefully performed to ensure that the velocity profile of the liquid in the tube is close to that used during calibration.

A number of commercial meters shape the magnetic field coils to generate a magnetic field that exhibits a field strength with a prescribed pattern across the duct. This field shaping compensates for some velocity variations in the duct and provides a meter with a reduced sensitivity to flow profile. As a result, this type of meter is better able to measure in vessels with upstream and downstream characteristics, such as curvature and nonuniformity in the vessel cross section, that generate asymmetric flow profiles with secondary flow velocity components.

Commercial meters generate magnetic fields using either an ac excitation or a pulsed dc excitation. The ac excitation generates a magnetic field with a strength that varies with the frequency of the applied ac voltage. This

configuration produces a meter with a relatively high frequency response but with the disadvantage that the output signal not only varies with the flow velocity but also with the magnitude of the alternating excitation voltage. Thus, the output of the meter for a flow with constant velocity across the vessel will exhibit a sinusoidal pattern. In addition, zero flow will produce an offset output due to the presence of the nonmoving conductor in a moving magnetic field. Quadrature signal rejection techniques can be used to filter the unwanted signal generated by the ac excitation from the desired signal generated by the flowing liquid, but this correction requires careful compensation and zeroing of the meter in the flow field before data acquisition.

The pulsed dc excitation was developed to reduce or eliminate the zero shift encountered with ac excitation. This improvement has the cost of reduced frequency response and increased sensitivity to the presence of particulates in a fluid. Particles that impact the electrodes in a pulsed dc operated meter produce output fluctuations that can be characterized as noise. The accuracy in each system is comparable.

A low sensing voltage at the electrodes requires a signal conditioning unit to provide a measurable output with a good signal-to-noise ratio. Meter calibrations typically involve one of two calibration techniques. The meter and signal conditioning unit are calibrated separately or the meter and signal conditioning unit are calibrated as a system. The latter provides the most accurate calibration with accuracies that can approach 0.5% of reading in certain applications. The reader is referred to literature by various manufacturers of electromagnetic flow meters for a more comprehensive discussion of the techniques, operation, and use for specific applications.

Ultrasound Techniques – Transit Time Volume Flow Meters. Ultrasonic transit time flow meters provide a direct measure of volume flow by correlating the change in the transit time of sound waves across a pipe or vessel with the average velocity of the liquid flowing through the pipe (17,27,28). Transit time ultrasonic flow meters are widely used in clinical cardiovascular applications. In recent years, a number of studies have been performed to evaluate and compare transit time ultrasonic flow measurement techniques with other techniques used clinically (29). The typical configuration for an ultrasonic flow probe involves one or two ultrasonic transducers (transmitters/receivers) and possibly an ultrasonic reflector. Transducers and reflectors are positioned in opposing configurations across a tube or vessel as illustrated in Fig. 3. The time it takes an ultrasound wave to propagate across a fluid depends on the distance of propagation and the acoustic velocity in the fluid. If the fluid is moving, the motion of the fluid will positively or negatively add a phase shift to the time of propagation (transit time) through the fluid, which can be written mathematically as:

$$T_t = D_p / (c \pm V \cdot \cos \theta) \qquad (5)$$

Here, T_t is the measured transit time (s), D_p is the total propagation distance of the wave (length units), c is the acoustic speed of the fluid (units of length per time), V is

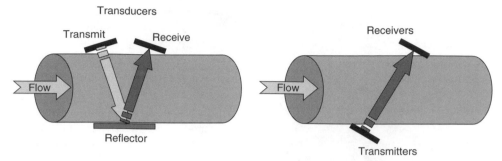

Figure 3. Illustration of principal of transit time ultrasonic flow probe operation.

the average velocity of the fluid (units of length per time) over the propagation length, and θ is the angle between the flow direction and the propagation direction of the wave. The configurations illustrated in Fig. 3 have an inherent dependency of the measured transit time on the coupling of the transducer with the vessel. Acoustic impedance characteristics of the vessel wall, and mismatches in impedance at the vessel wall–transducer and wall–fluid interfaces will affect the accuracy of the flow rate measurement.

The approach to using equation 5 in a metering device is to incorporate bidirectional wave propagation in opposing directions, as shown in Fig. 4, which will produce two independent transit time measurements (T_{t1} and T_{t2}), one from each direction of propagation. The forward direction transit time T_{t1} is defined by Eq. 5 with a plus sign before V, and T_{t2} is defined by the minus sign. The fluid velocity can then be obtained by taking the difference of the transit times ($T_{t1}-T_{t2}$). It can be shown that, for fluid velocities small relative to the acoustic velocity of the fluid ($V^2 << c^2$), this difference reduces to:

$$(T_{t1} - T_{t2}) = 2 D_{p}(V \cdot \cos\theta)/c^2 \qquad (6)$$

With the probe geometry defined (the propagation distance and propagation angle relative to the vessel or flow direction) and with the fluid properties known (acoustic speed of the fluid), the average speed of the fluid along the propagation path of a narrow beam can be calculated from the transit times. Wide beam illumination, where the beam width is wider than the vessel diameter, effectively integrates the measured transit time phase shift over the cross section of the vessel. The wide beam transmission can be approximated by the summation of many infinitesimally narrow beams adjacent to one another. Thus, the measured

wide beam phase shift is proportional to the sum of these narrow beams propagating through the vessel. As the phase shift encountered in a narrow beam transmission is proportional to the average fluid velocity times the beam path length, integrating or summing over all narrow beams across the vessel results in a measured total transit time phase shift that is proportional to the average fluid velocity times the area of the vessel sliced by the ultrasound beam, or the volume flow rate.

The popular Transonic Systems Inc. flow meter uses bidirectional transmission with two transducers operating in transmit and receive modes, and a single reflector configured as illustrated in the left schematic of Fig. 4. This approach increases the propagation length while effectively reducing sensitivity to wall coupling or misalignment of the probe with the wall. The increased path length improves uncertainty and provides a probe body with a relatively small footprint, an advantage in *in vivo* or surgical applications.

The basic operation of a bidirectional transit time meter involves the transmission of an ultrasound plane wave at a specific frequency. This wave propagates through the vessel wall and fluid where it is either received at the opposite wall or is reflected to another transducer operating as an acoustic receiver. This received signal is recorded, processed, and digitized before the transducer is reconfigured to transmit a second pulse in the opposite direction. The overall frequency response of such a probe is dependent on the pulse time, the time delay between the forward and reverse pulses, the acoustic speed through the medium, the propagation distance, and the signal conditioning electronics, which can include analog signal acquisition and filtering. The meter size governs the propagation distance and, thus, the size of the vessel that the meter can be mounted on.

The frequency response of commercial probes varies from approximately 100 Hz to more the 1 kHz, where the highest frequency responses are obtained in the smaller probes. As a result, commercial probes have sufficient frequency response for most clinically or biomedically relevant flow regimes. Velocity and flow resolution is governed, in part, by the propagation length over which the flow is integrated and the resolution of the transit time measurement. The reader is referred to the meter manufacturers for detailed information about the operating specifications of particular meters. Reported uncertainties in transit time meters can be better than 15%. Actual uncertainties will depend on meter use, the experience of the operator, the meter calibration, and the acoustic properties of the

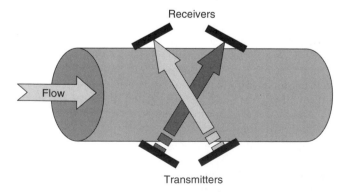

Figure 4. Bidirectional wave propagation.

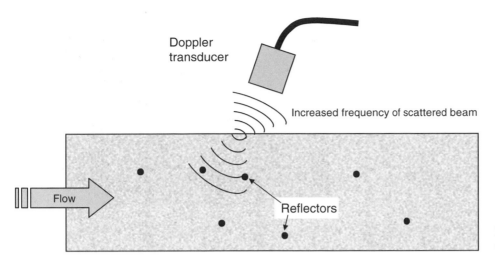

Figure 5. Schematic of the operation of a Doppler ultrasound probe.

fluid measured and how these properties differ from those of the calibration fluid.

Ultrasound Techniques – Doppler Volume Flow Meters. Flow can also be measured by ultrasound using the Doppler shift in a propagating sound wave generated by moving objects in a fluid flow (17,30). The primary difference is in the principal of operation. Devices using the Doppler approach measure the Doppler frequency shift of the transmitted beam due to the motion of particles encountered along the beam path, as illustrated in Fig. 5. The Doppler shift due to reflection of an incident wave by a moving particle is given by

$$F_D = 2F_o V \cos\theta / c \qquad (7)$$

The shift frequency F_D (units of $L \cdot s^{-1}$) is linearly related to the component of the speed of a particle, V, in the direction of the wave propagation, the initial transmission frequency of the wave, F_o, and the speed of sound in the fluid, c. The reader is referred elsewhere in this Encyclopedic series and to the text by Weyman (30) for a detailed presentation of the Doppler technique.

The Doppler meter provides a direct measure of velocity that can be used to calculate the volume flow rate indirectly. Most biomedical applications involving volume flow measurement are performed on flow through a duct or vessel of given shape and size. Thus, the volume flow is the integral of the measured velocity profile across the vessel cross sectional area as defined in equation 1. The integral in equation 1 can be related to the average velocity $\bar{U}$ across the duct multiplied by the cross-sectional area of the duct (23). The Doppler technique then requires not only an estimate of the average velocity in the vessel but knowledge of the vessel area as well.

Commercially available Doppler volume flow meters, although not commonly used in biomedical applications, can be attached to a pipe or duct wall as with transit time meters. The commercial Doppler flow meters measure volume flow by integrating the measured Doppler shift frequency generated by particles throughout the flow. This integration is performed over a predefined path length in the flow, and is dependent on the number and type of particles, their size and distribution. The meter accuracy

is also dependent on the velocity profile in the flow. Careful *in situ* calibrations are often needed to obtain accuracies of less than 10%. The Doppler meter has several disadvantages when compared with the transit time meter. It requires a fluid that contains a sufficient concentration of suspended particles to act as scattering sites on the incident ultrasound wave. These particles must be large enough to scatter the incident beam with a high intensity level but small enough to ensure that they follow the fluid flow (31). As a result of their dependence on flow profile, Doppler flow meters are not well-suited for measurement of flow in vessels with curvature or branching. Doppler flow meter measurements in blood rely on blood cells to act as scatterers. Clinical Doppler ultrasound machines, commonly used in echocardiography, can also be used to indirectly infer volume flow through the direct measure of the fluid velocity, and will be discussed later in the subsection on velocity measurements.

Invasive or Inline Volume Flow Measurement. Invasive or inline flow meters must be installed inline as a part of the piping or vessel network and involve hardware that is in contact with the fluid. These meters often have a nonnegligible pressure drop and may adversely interact with the flowing fluid. As a result, these meters are not often used in *in vivo* applications. Meters that fall in this category are variable area rotameters, turbine/paddle wheel meters, and vortex shedding meters. The primary advantage of these meters, is low cost and ease of use. However, these meters typically exhibit sensitivity both to fluid properties, which can be dependent on temperature and pressure, and to flow profile, White (23).

Variable area rotameters are simple measurement devices that can be used with a variety of liquids and gases. The flow of fluid through the meter raises a float in a tapered tube, as shown in Fig 6. The higher the float is raised, the larger the diameter of the tapered tube, increasing the cross-sectional area of the tube for passage of the fluid. As the flow rate increases, the float is elevated higher in the tube. The height of the float is directly proportional to the fluid flow rate. In liquid flow, the float is raised by the combination of the buoyancy of the liquid and the fluid drag on the float. Buoyancy is negligible in gaseous flows

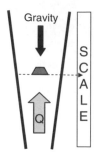

Figure 6. Schematic of a variable area flowmeter.

and the float moves in response to the drag by the gas flow. For constant flow, the float reaches a stable position in the tube when the upward force generated by the flowing fluid equals the downward force of gravity. The float will move to a new equilibrium position in response to a change in flow. These meters must be installed vertically to operate properly, although spring-loaded meters have been designed to eliminate the need for gravity and permit installation in other orientations.

Rotameters are designed and calibrated for the type of fluid (fluid properties such as viscosity and density) and flow range expected. They do not function properly in nonNewtonian fluids. Use of a meter with a fluid different from that the meter was calibrated for, or with, a fluid at a different temperature or pressure requires a correction to the meter reading. Meter uncertainty and repeatability will vary with operation of the meter, but can approach a few percent with proper operation.

Turbine and paddle wheel meters measure volume flow rate through the rotation of a vaned rotor in contact with the flowing fluid. These meters are intrusive flow measurement devices that produce higher pressure drops than others in the class of invasive flow probes. The turbine meter has a turbine mounted across the pipe or duct in full contact with the flow, whereas the paddle wheel meter has a vaned wheel mounted on the side of the duct with half of the wheel in contact with the flow. Accuracy and repeatability is best with the turbine meter, but the pressure drop is also higher. An ac voltage is induced in a magnetic pickup coil as the turbine or paddle wheel rotates. Each pulse in the ac signal represents the passage of one blade of the turbine. As the turbine fills the flow path, a pulse represents a distinct volume of fluid being displaced between two turbine blades. This design provides an accurate direct measure of the volume flow rate.

Flowmeter selection must take into account the type of fluid, the flow rate range under study, and the acceptablepressure drop for a given flow application. In general, these meters have a sensitivity to flow profile, and the pressure drop is dependent on the fluid properties. The meters incorporate moving parts within the flow and thus use bearings to ensure smooth operation. Bearing wear will affect the meter accuracy and must be monitored for the life of the meter. The paddle wheel meter operates in a similar manner as the turbine meter. The primary difference is that only part of the rotor is in contact with the fluid and, thus, as the paddle wheel meter is more sensitive to flow profile it has a smaller pressure drop. Installation of these

Figure 7. Vortex shedding from a circular cylinder. Picture from White (23), courtesy of the U.S. Naval Research Laboratory.

meters often involves a specified number of straight pipe sections upstream and downstream of the meter and may also require installation of a flow straightener inline upstream of the meter.

Vortex meters operate on the principal of Strouhal shedding. Separating flow over an obstruction such as a cylinder or sharp-edged bar results in a pulsatile or oscillatory pattern as shown in Fig. 7. The shedding frequency, ω (units of $L \cdot s^{-1}$), is related to the fluid velocity by

$$\omega = V \cdot S_t / L \tag{8}$$

where S_t is the Strouhal number, which is a nondimensional number that is a function of the flow Reynolds number and geometry of the obstruction, and L is a characteristic length scale (23). For a cylinder, L is the diameter of the cylinder. The Reynolds number is a dimensionless number that is the ratio of inertial to viscous forces in the flow, and is defined as

$$Re = V \cdot L / \upsilon \tag{9}$$

Here, V and L are defined as in Eq. 8 and υ is the kinematic viscosity of the fluid with units of length squared per time.

The vortex meter is an intrusive meter that has a "shedder bar" installed across the diameter of the duct. The flow separates off this bar and generates a shedding frequency that is transmitted through the bar to a piezoelectric sensor attached to the bar. The meter is sensitive to flow and fluid properties, and rated accuracy and pressure drop depend on application.

Volume flow rate can also be estimated through an indirect measure of the velocity profile in the flow and the use of Eq. 1. A number of instruments are available that measure fluid velocity in biomedical engineering applications. Doppler ultrasound and MR phase velocity encoding are standard clinical techniques used to measure velocity of a flowing fluid noninvasively. *In vitro* systems that are commonly used to measure fluid velocity, in addition to Doppler and MR, are laser Doppler velocimetry (LDV), particle image velocimetry (PIV), and thermal anemometry. Besides an estimate of volume flow rate, fluid velocity measurement can provide quantification of flow profiles, fluid wall shear, pressure gradient, and flow mixing. The following section summarizes velocity measurement techniques commonly used in biomedical/bioengineering applications.

Velocity Measurements

Thermal Anemometry. Thermal anemometry is an invasive technique used to measure fluid velocity or wall shear. A heated element is inserted into the flow, and specialized electronic circuitry is used to measure the rate of change in heat input into the element in response to changes in the flow field (32). Thermal anemometry, when used properly, is characterized by high accuracy, low noise, and high spatial and temporal resolution. Its main disadvantages are sensitivity to changes in fluid temperature and properties, particulates and bubbles suspended in a fluid, nonlinear response to velocity, and its invasive characteristics (geometry, size, vibration, etc.).

Hot-film anemometry has been used to measure blood velocity and wall shear in biomedical and bioengineering applications both *in vitro* and *in vivo*. Arterial blood flow velocity measurements were performed by Nerem et al. (33,34) in horses and by Falsetti et al. (35) in dogs. Tarbell et al. (36) used flush-mounted hot films to measure wall shear in the abdominal aorta of a dog. *In vitro* applications of hot-film anemometry include the measurement of wall shear in an LVAD device (37), and the *in vitro* measurement of flow in rigid models of arterial bifurcations by Batten and Nerem (38). Although rarely used in biomedical applications now, we will briefly present hot-film anemometry here for completeness. The reader is referred to the text by Bruun (39) and the symposium proceedings by Stock (40) for a detailed presentation of thermal anemometry.

Thermal anemometry operates on the principal of convective cooling of a heated element. A thin wire or coated quartz element, mounted between supports, is heated and exposed to a fluid flow, as shown in Fig. 8. The element is heated by passing a current through the wire or element. The amount of heat generated is proportional to the current, I (A), and the resistance of the element, R (Ω), by I^2R. The element is convectively cooled by flow until an equilibrium state is reached between electrical heating of the

element and flow-induced convective cooling, $DE/Dt = W + H$, where E is the thermal energy stored in the element, W is the heat added by joule heating, and H is the heat loss to the environment by cooling. At equilibrium, $DE/Dt = 0$ and $W = H$.

Changes in flow velocity will increase or decrease the amount of convective cooling and produce changes in the equilibrium state of the element or the temperature of the element. Commercial anemometers employ a four-arm electronic bridge circuit to maintain constant element temperature, current, or voltage in response to changes in convective cooling. As convective cooling changes, the anemometer output, current, or voltage changes in response to maintaining the desired set condition. This equilibrium condition assumes that radiation losses are small, conduction to supports is small, temperature is uniform over length of sensor, velocity impinges normally on the sensor, velocity is uniform over sensor length and is small compared with the sonic speed, and finally, the fluid temperature and density are constant.

An energy balance between convective heat cooling and joule heating can be performed to derive a set of governing equations that relate input current, I, to convective velocity, V. The "King's law" is the classic result of this energy balance:

$$I^2R^2 = V_0^2 = (T_w - T_a)(A + B \cdot V^n) \qquad (10)$$

where V_0 is the measured voltage drop in response to a velocity, V, and T_w and T_a are the wire and ambient fluid temperatures ($^\circ$C), respectively. The coefficients, A and B, and power, n, are determined through careful calibration over the velocity and temperature range that will be observed experimentally. In the event of a three-component flow, the probe must be calibrated for yaw and pitch angles between the probe and the flow velocity vector, and the velocity, V, in Eq. 10 must be replace by a term related to the velocity vector magnitude. Bridge-type circuits are also prone to stable and unstable performance under unsteady operation. Thus, the overall calibration of a hot-wire/film system must involve the element and electronics as a system and must also involve dynamic calibrations to characterize the frequency response of the system.

Hot-wire/film probes come in a variety of sizes, shapes, and configurations. Probes are manufactured from platinum, gold-plated tungsten, or nickel-plated quartz, and come in single-element or multielement configurations for measurement in a variety of flow conditions. The reader is referred to the hot-wire/film manufacturers for a complete summary of probe types and conditions for use. In general, wire probes are used when possible due to lower cost, improved frequency response, and ease of repair. However, wire probes are more fragile compared with film-type probes and are usually restricted to air flows. Film probes are used in rough environments, such as liquid flows.

The following considerations should be addressed to ensure accurate measurements when using thermal anemometry. The type of flow should be assessed for velocity range, flow scales, and fluid properties (clean gas or particle contaminated liquid, etc.). The flow characteristics will define the right probe, anemometer configuration, and

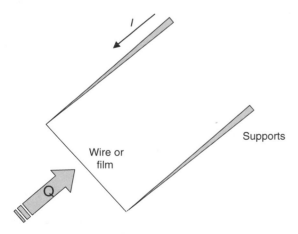

Figure 8. Illustration of a hot wire or film element.

A/D setup to use. Perform appropriate calibrations with complete hardware setup. Perform the experiment and post calibrations to ensure that the anemometer/probe calibration has not changed.

Doppler Ultrasound And Magnetic Resonance Flow Mapping. The focus of this subsection is to introduce the concept of Doppler ultrasound and MR flow mapping for local velocity measurement. Flow measurement using clinical Doppler can suffer from the same limitations as the small Doppler meter, but has several advantages over these small meters. Most ultrasound machines can operate in continuous wave or pulsed Doppler modes of operation; see Weyman (30) for a more detailed discussion of the modes of operation.

Pulsed Doppler ultrasound offers the advantage of localizing a velocity measurement within a flow and can be used to measure the velocity profile across a vessel or lesion. This information coupled with echocardiographic imaging of the geometry can be used to calculate the flow rate from Eq. 1. Unfortunately, the implementation of this technique is not straightforward due to limitations in resolution, velocity aliasing, and the need to know the relative angle between the transmitted ultrasound beam and the local flow.

Velocity aliasing in pulsed-mode Doppler occurs because the signal can only be sampled once per pulse transmission (e.g., the pulse repetition frequency). Frequency aliasing, or the ambiguous characterization of a waveform, occurs in signal processing when a waveform is sampled at less than one-half of its fundamental frequency, referred to as the Nyquist condition in signal processing. In a pulsed Doppler system, velocity aliasing will occur when the Doppler shift of the moving particles exceeds half of the pulse repetition frequency. As the pulse repetition frequency is a function of the depth at which a sample is measured, the alias velocity will vary with imaging depth. Increasing the imaging depth will lower the velocity at which aliasing will occur. Continuous wave Doppler signals are typically digitized at higher sampling frequencies, limited by the Nyquist frequency associated with the frequency of the transmission wave. Velocities observed clinically produce Doppler shifts that are generally lower than sampling frequencies used in continuous wave Doppler. As a result, velocity alias is not usually observed with continuous mode Doppler in clinical applications.

Aliased signals that are processed by the measuring system are not directly proportional to the Doppler shift generated by the velocity of the particle but can be related to the Doppler shift. Undersampling a wave underestimates the frequency and produces a phase shift. Most clinical Doppler machines use the frequency and phase of the sampled wave to infer velocity magnitude and direction. Velocity alias will produce a lower velocity magnitude with increasing Doppler shift above the alias limit. As frequency is, by definition, positive and Doppler machines use the signal phase to determine direction, the measured frequency is usually reported as a negative velocity above the alias limit, which is often displayed as an increasing positive velocity magnitude with increasing Doppler shift up to the alias limit. Further increases in the Doppler shift

(particle velocity) result in a sign change at the velocity magnitude of the alias velocity with a continued decrease in velocity with increasing Doppler shift. Velocity alias can be reduced or eliminated by frequency unwrapping and baseline shifting, or through the careful selection of machine settings during data acquisition.

Frequency unwrapping is simply correcting the reported aliased velocity by a factor that is related to the alias velocity limit and the magnitude of the reported negative velocity. This correction is, roughly speaking, adding the relative difference in magnitude between the measured aliased velocity and the velocity alias limit to the velocity alias limit. This method of addressing velocity alias is often accomplished by baseline shifting in commercial Doppler machines. In baseline shifting, the phase angle at which a negative velocity is defined is shifted with the effect of a relative shift in the reported alias velocity. Baseline shifting or frequency unwrapping does not eliminate velocity alias but provides a correction to extend the measurement to higher frequencies.

Velocity alias can be "eliminated" by reducing the Doppler frequency of moving particles and thereby shifting the measurable range below the alias limit, which can be accomplished by reducing the carrier frequency of the ultrasound wave that will, in turn, reduce the Doppler frequency shift induced by a moving particle and increase the maximum velocity that can be recorded before reaching the Nyquist limit. Alternatively, the Doppler shift frequency can be reduced by increasing the angle between the propagation of the ultrasound wave and the velocity vector, which reduces the magnitude of the Doppler shift frequency by the cosine of this included angle. Angle correction has limitations in that the flow direction must be known and the uncertainty in the correction increases with increasing included angle. As color flow mappers operate in the pulsed Doppler mode, they are subject to velocity alias. Color flow mappers indicate velocity direction by a color series (for example, shades of red or blue). Velocity alias is displayed as a change in a color series from red-to-blue or blue-to-red.

The clinical measurement of many velocity ensembles across a vessel and the integration across the vessel geometry can be time-consuming and problematic in pulsatile flow through a compliant duct. Furthermore, lesions are often complex in shape and cannot be adequately defined by echo. Doppler echocardiographers and scientists have exploited the physics of fluid flow to develop diagnostic tools that complement these capabilities of commercial Doppler systems. The text by Otto (41) provides an excellent review of these diagnostic tools. Techniques such as the PISA or proximal flow convergence use the capability of color Doppler flow mapping machines to estimate volume flow through an orifice, such as a heart valve. Figure 9 illustrates the concept of the proximal flow convergence method.

The flow accelerating toward a small circular orifice will increase in velocity V_a, until a maximum velocity at the orifice V_j is reached. This acceleration occurs in a symmetric pattern around the orifice and is characterized by hemispheres of constant velocity. As the orifice is approached, the velocity increases and the radius of the

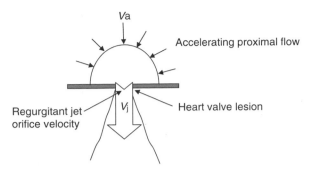

Figure 9. Illustration of the proximal isovelocity surface area (PISA) concept.

hemisphere decreases. The regurgitant flow through the orifice can then be calculated as:

$$Q = (2\pi r^2)V_a \tag{11}$$

The combined imaging and Doppler characteristics of color Doppler flow mapping are exploited in the PISA approach. The location of the alias velocity in the flow map provides a measure of V_a, which is then coupled with a measure of the radial location from the orifice using the imaging capability of color flow mapping. As flow is velocity times area, the hemispheric assumption provides a shell with a surface area of $2\pi r^2$ that velocity V_a is passing through. Figure 10 illustrates the concept of PISA with a color Doppler flow image of valvular regurgitation in a patient.

The PISA approach assumes a round orifice with a hemispherical acceleration zone. In clinical applications, orifices are rarely round and the acceleration zone is not hemispherical with the result of under or over estimation of the flow rate depending on what radial velocity contour is used in Eq. 11. The semielliptic method is one approach at considering nonhemispheric geometries in an attempt to correct for errors associated with the PISA technique.

The combination of continuous wave and pulsed Doppler ultrasound is exploited in the turbulent jet decay method of measuring flow through a stenotic lesion or a regurgitant valve. While continuous wave Doppler does not suffer from velocity alias as does pulsed Doppler, it cannot provide spatial localization of the velocity. The turbulent

jet decay method uses continuous wave Doppler to measure the high velocity at the lesion orifice and then uses pulsed Doppler to measure the velocity decay at specified location downstream of the orifice. Turbulent jet theory can be used to relate the flow rate of the turbulent jet to the decay of the jet velocity downstream of the orifice, as in Eq. 12

$$Q = (\pi V_m^2 x^2)/160\,V_j \tag{12}$$

The velocity V_m is measured by pulsed Doppler at location x measured from the jet orifice, whereas the orifice velocity, V_j, is measured by continuous wave Doppler; Fig. 10b illustrates this decay phenomenon. This equation is valid for round jets and has been extended to jets with other geometries by Cape et al. (42) with the resulting change to Eq. 12:

$$Q = (V_m^2 H_x)/5.78\,V_j \tag{13}$$

where H is the width of the jet measured by color Doppler.

Doppler velocity measurements are also used to estimate pressure gradients in various cardiovascular applications. The Bernoulli equation can be used to estimate the pressure drop across a stenotic lesion or through a valve by measuring the velocity both upstream and downstream of the lesion or valve. The Bernoulli equation is

$$\Delta P = (P_1 - P_2) = 1/2 \cdot \rho(V_2^2 - V_1^2)$$

where position 1 is often measured upstream of the lesion and position 2 is at the lesion or downstream. In this equation, the pressure drop ΔP has units of Pascal's (Pa). A Pascal is a Newton (N) per square meter, where a Newton has units of mass (kg) times length (meter) per time squared. Bioengineering and biomedical applications often use the units of millimeters of mercury (mmHg) in defining a pressure value. A mmHg is related to a Pa by the conversion; 1 mmHg = 133.32 Pa.

Magnetic resonance flow mapping has the advantage over Doppler that it can measure the full three-component velocity field over a volume region (43–45), which eliminates the uncertainty in flow direction and enables the use of standard fluid dynamic control volume analysis. The advantages of MR flow mapping come at the cost of long imaging times and increased sensitivity to motion artifacts in *in vivo* applications, where phase locking to the heart rate or breathing cycle can increase complexity.

The velocity of moving tissue can be detected by a time-of-flight technique (46) and by phase velocity encoding (47,48). The time-of-flight method tracks a selected number of protons in a plane and measures the displacement of the protons over a time interval defined by the imaging rate. *In vivo* (49) and phantom (50) studies have shown that the time-of-flight technique is capable of accurate velocity measurement up to velocities at least as high as 0.5 m·s^{-1}. However, the time-of-flight method requires a straight length of vessel on the order of several centimeters for accurate velocity estimation. This requirement reduces its usability in most *in vivo* applications. The phase velocity encoding method has become the preferred technique in most clinical applications.

Phase velocity encoding directly relates the local velocity of nuclei to the induced phase shift in an imaging voxel.

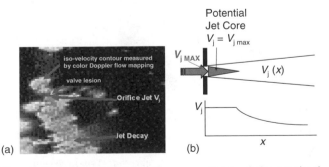

Figure 10. (a) Color Doppler flow map image of the proximal isovelocity surface area (PISA) in valvular regurgitation. (b) Illustration of the jet decay downstream of an orifice. The parameter V is the jet velocity and x is measured from the orifice.

Properly defined bipolar magnetic field gradients are produced in the direction of interest for velocity measurement. The velocity of hydrogen nuclei are then encoded into the phase of the detected signal (51). Chatzimavroudis et al. (52) and Firmin et al. (53) provide a discussion of the phase encoding technique with an assessment of its accuracy and limitations for flow measurement.

The technique uses two image acquisitions to provide velocity compensation and velocity encoding. Velocity information in each voxel is obtained by a voxel-by-voxel subtraction of the two images with respect to the phase of the signal. Like Doppler ultrasound, phase velocity encoding can suffer from aliasing effects, alignment error, and limits in spatial and temporal resolution. Velocity estimation using phase shift measurement is limited to a maximum range of phase of 2π radians without ambiguity or aliasing. However, the estimation of the phase shift using phase subtraction between two images reduces that sensitivity to this problem. Studies have been conducted that show MR phase velocity encoding can measure velocities covering the complete physiologic range up to several meters per second (54). Misalignment of the flow direction with the encoding direction will produce a negative bias in the measured flow where the measured velocity will be lower than the true velocity. Like Doppler, this bias follows a cosine behavior where $V_{meas} = V_{act} \cos(\theta)$, where V_{meas} is the measured velocity, V_{act} is the actual velocity, and θ is the misalignment angle. This error is typically less than 1% in most applications.

The size of a voxel and the sampling capabilities of the hardware characterize the spatial and temporal resolution of the system. Spatial resolution affects the size of a flow structure that can be measured without spatially filtering or averaging the structure or velocity measurement. Spatial velocity gradients that are small relative to the voxel size will not be adequately resolved and will be averaged over the voxel volume (55). In addition, rapidly varying velocity fluctuations in time will produce a similar low pass frequency filtering effect if these fluctuations occur with a time scale that is much smaller than the imaging time scale of the measurements. Turbulent flow can produce spatial and temporal scales that could be small relative to the imaging characteristics and can result in what is referred to as signal loss in the image (56). Stenotic lesions and valvular regurgitation are clinical examples where turbulent flow can occur with spatial and temporal scales that could compromise measurement accuracy.

Phase velocity encoding has the drawback of fairly long imaging or magnet residence times, which is particularly true for three-component velocity mapping. Although this may be acceptable for *in vitro* testing with flow loop phantoms, it can present problems and concerns with clinical measurements. Patients can be exposed to long time intervals in the magnetic with the associated problems of patient comfort and care. *In vivo* velocity measurements are often phase-locked with cardiac cycle or breathing rhythm. Long imaging times can increase potential for measurement error due to patient movement and variability in the cardiac cycle or breathing rhythm, which can cause noise in the phase-averaged, three-component velocity measurements. Research, in recent years, has focused on hardware and software improvements to increase spatial resolution and reduce imaging time [see, e.g., Zhang et al. (57)].

Magnetic resonance phase velocity encoding provides coupled 3D geometric imaging using traditional MR imaging methods with three-component velocity information. This coupled database provides a powerful diagnostic tool for blood flow analysis and has been used extensively in *in vitro* and clinical applications. Jin et al. (6) used this coupled imaging flow mapping capability to study the effects of wall motion and compliance on flow patterns in the ascending aorta. Standard imaging was used to measure aortic wall movement and the range of lumen area and diameter change over the cardiac cycle. This aortic wall motion was phase-matched with phase velocity encoded axial velocity distributions in the ascending aorta. Similar to the PISA approach in Doppler ultrasound, a control volume approach using phase velocity encoded MR velocities can be applied to the assessment of valvular regurgitation (58,59). The control volume approach is illustrated in Fig. 11.

Laser Doppler Velocimetry. The Doppler shift of laser light scattered by particles or cells in a fluid is the basis of laser Doppler velocimetry (LDV). Detailed presentations of the LDV technique are provided in the works by Drain (60) and Durst et al. (61). The scattered radiation, from a laser beam directed at moving particles in a fluid, has a Doppler-shifted frequency defined as:

$$f_D \sim (1 - V/C_1) f' \tag{14}$$

where C_1 is the speed of light in a vacuum, V is the particle velocity, and f' is incident light frequency. The Doppler-shifted frequency is very small relative to the frequency of

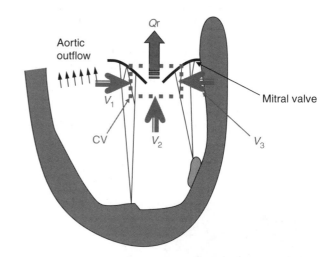

Figure 11. Illustration of the control volume method in MR phase velocity assessment of valvular regurgitation. The control volume (CV) is the heavy dotted line box around the mitral regurgitant orifice. The box edges are usually selected to correspond with rows and columns in the MR image. V_i represents the three-component velocities measured with MR through the i faces of the box. Faces 4 and 5 are in the plane of the image at $\pm Z$ offsets from the plane of the image. The regurgitant flow Q_r is the sum of the V_iA_i on each face.

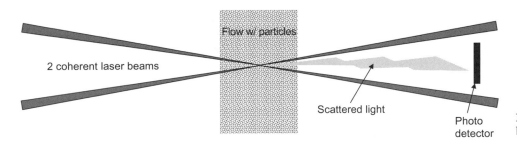

Figure 12. Illustration of the dual-beam or fringe mode LDV setup.

light and, thus, dual-beam or fringe mode LDV is the system configuration of choice. The dual-beam mode of operation is schematically shown in Fig. 12. In fringe mode LDV, two coherent laser beams of the same wavelength or frequency are focused to a common point (control volume) in the flow field. The scattered light from a particle moving through the control volume is received by a photodetector.

The crossing of two coherent, collimated laser beams forms interference fringes as the propagating light waves constructively and destructively interfere with one another. This interference creates a series of light and dark bands with spacing, d_f, of

$$d_f = \lambda / 2 \sin(\kappa) \quad (15)$$

The number of fringes, N_{FR}, in the measurement volume is given by

$$N_{FR} = 1.27 d / D_e^{-2} \quad (16)$$

where d is the spacing between the two parallel laser beams before the focusing lens and D_e^{-2} is the beam diameter before the lens. Figure 13 illustrates the probe geometry generated by the intersection of two focused coherent laser beams with a common wavelength.

The spatial resolution of a dual-beam system is affected by the distribution of the light intensity at the intersection of the two focused beams, referred to as the probe or measurement volume. When the laser is operating in the TEM_{oo} mode, the laser cavity sustains a purely longitudinal standing wave oscillation along its axis with no transverse modes. The laser output has an axisymmetric intensity profile that is approximately a Gaussian function

of radial distance from the axis. In the far field, the beam divergence is small enough to appear as a spherical wave from a point source located at the front of the lens. A lens is used to focus the beam into a converging spherical wave. The radius of this wave decreases until the focal point of the lens is reached. At the focal point, the beam has almost a constant radius and planar behavior. The beam is focused to its minimum diameter or focal waist, d_e^{-2}, and is defined as:

$$d_e^{-2} = (4\lambda f)/(\pi D_e^{-2})$$

where λ is the wavelength of the laser beam and f is the focal length of the lens. A single pair of laser beams generates an ellipsoidal geometry having dimensions of major axis l_m and minor axis d_m given by

$$l_m = d_e^{-2}/\sin(\kappa) \text{ and } d_m = d_e^{-2}/\cos(\kappa) \quad (17)$$

where κ is the half angle between the two laser beams, as illustrated in Fig. 13.

The particle velocity is calculated by the fluctuating light intensity collected by the receiver as the particle passes through the measurement volume and scatters light from the fringes. The intensity change of the scattered light from the light and dark fringes is converted into an electrical signal by a photomultiplier tube (PMT). The electrical signal represents an amplitude-modulated sine wave, with frequency proportional to the Doppler frequency shift (f_D) of the particle traveling through the measurement volume. The particle velocity is then equal to the Doppler frequency multiplied by the fringe spacing. In a two-beam LDV system, the measured velocity component is in the plane of the two laser beams and in the direction perpendicular to the long axis of the measurement volume.

Coherent laser beams with the same frequency produce stationary fringes. A particle moving in either direction across the fringes will produce the same frequency independent of sign, such that a stationary fringe system can only determine the magnitude of the velocity, not the direction. To avoid this directional ambiguity, one of the laser beams of a beam pair is shifted to a different frequency, using a Bragg cell, to provide a moving fringe pattern. One laser beam from each beam pair passes through a transparent medium such as glass, in which acoustic waves, generated by a piezoelectric transducer, are traveling. If the angle between the laser beam and the acoustic waves satisfies the Bragg condition, reflections from successive acoustic wave fronts reinforce the laser beam. The beam exits at a higher frequency and a prism

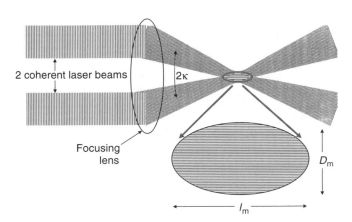

Figure 13. Illustration of the measurement volume generated in fringe mode LDV.

directs the beam to its original direction. The Bragg shift causes the fringes in the probe volume to move at a constant speed in either the positive or negative direction relative to the flow. The measured frequency by the PMT and processor is then the sum or difference of the Bragg cell frequency (typically 40 MHz) and the Doppler shift frequency. This measured frequency is then downmixed with a frequency that is a percent of the Bragg frequency (called the shift frequency) producing a frequency that has a zero shifted to a higher baseline frequency (usually on the order of several MHz). This zero shift eliminates directional ambiguity in LDV signal processing.

Laser Doppler velocimetry has excellent spatial and temporal frequency response compared with most other measurement systems. It is considered a gold standard measurement technique in biomedical applications and is the noninvasive measurement system of choice for turbulence measurements. Two disadvantages of LDV worth noting are (1) LDV noise and (2) velocity bias. The LDV is noisy when compared with other turbulence measurement systems, such as thermal anemometry, due to the use of photomultiplier tubes. These optical detectors, used for their sensitivity and high frequency response, suffer from higher noise floors than other photo detectors.

Velocity bias is a result of the random sampling characteristics of LDV. As a velocity ensemble is randomly recorded when a particle passes through a probe volume, the statistics of the measured velocity ensembles are not independent of the particle velocity. A greater number of higher speed particles will cross the measurement volume over a specified time than will slower speed particles. Standard ensemble averaging will produce mean velocity estimates that are biased toward higher velocities. This velocity bias can have a significant impact on the velocity statistics, particularly in turbulent flow. In addition to velocity bias, two other biases may occur, fringe bias and gradient bias.

Fringe bias is an error that is minimized by frequency shifting. This type of bias is created by not having enough fringe crossings to satisfy processor validation criteria when calculating a velocity, which occurs when a particle crosses the edge of the probe volume or if the particle velocity is nearly parallel to the fringes. Thus, velocity ensemble averages weight velocities from particles traveling near the center of the measurement volume or those particles that cross more fringes then others. By frequency shifting with a fringe velocity at least two times greater than the flow velocity, particles moving parallel to the fringes can cross the minimum number of fringes for validation by a processor.

Gradient bias results from a nonnegligible mean gradient across the probe volume. This bias depends on the fluid flow and the measurement volume dimensions. The mean velocity and the odd order moments are the only statistics affected by gradient bias. In general, LDV-transmitting optics are chosen to provide as small a measurement volume as possible to increase spatial resolution and reduce gradient bias. As the LDV measurement volume is longer than it is wide, experiments should be designed to ensure that the LDV optical setup is oriented to position the measurement volume diameter in the direction of the maximum gradients in the flow.

Several post processing techniques have been developed to reduce velocity bias. The recommended technique is to use a transit time weighting when computing the velocity statistics. This transit time weighting approximates the ensemble average as a time average. The reader is referred to Edwards (62) for a detailed discussion of the transit time technique and its implementation in LDV data processing.

Multiple pairs of laser beams with different wavelength (color) or polarization can be used to produce a multicomponent velocity measuring system. Two or three laser beam pairs can be focused to the same point in the flow. Each beam pair can then be used to independently measure a different component of the velocity vector. As more than one particle can pass through a measurement volume at one time, it is possible to get valid velocity component estimates from different particles. The ellipsoidal geometry of the measurement volumes exaggerates this problem. As a result, LDV data are often processed in one of two modes, random and coincident.

Random mode processing records every valid velocity ensemble as it arrives at the measurement volume, which can generate uneven sample distributions in the different velocity components or LDV channels being measured. Random mode processing has a negligible impact on mean velocity statistics but can be detrimental to turbulence estimates. Coincident mode processing uses hardware or software filters to ensure that each velocity component ensemble is measured from the same particle. Filters are used to ensure that the Doppler bursts measured on the different LDV channels overlap in time. Bursts generated by one particle should be measured on each channel with perfect overlap. Time window filters are used to reject bursts that do not occur within a window defined by a percentage of the burst length. The effect of coincident mode processing is usually a reduction in the overall data rate by a factor of at least two but provides the necessary data quality for turbulence measurements.

Laser Doppler velocimetry is primarily an *in vitro* tool, although systems have been developed for blood flow measurement (17,63). Blood is a multiphase fluid composed of a carrier liquid, plasma, and a variety of cells and other biochemical components. Plasma is optically clear to the wavelengths of light used in LDV. The optical opacity of blood is due to the high concentration of cells, in particular red cells. On the microscopic level, however, blood can transmit light over a short distance due to the plasma carrier fluid. Clinical-style probes have been developed to measure the velocity of blood cells in blood using catheter-type insertion into vessels of suitable size or through transcutaneous measurement of capillary flow below the skin. These *in vivo* systems are designed with very short focal length transmitting lenses providing a small measurement volume located a very short distance from the transmitting lens. Laser light is propagated through the plasma and focused a few millimeters from the probe tip. Blood cells act as particles in the fluid and scatter light that is collected by the transmitting lens and directed to a PMT system for recording of the Doppler bursts. Manning et al. (64) and Ellis et al. (65) have used LDV to measure the velocity fields around mechanical heart valves in *in vitro* studies. Figure 14 shows the measured velocity

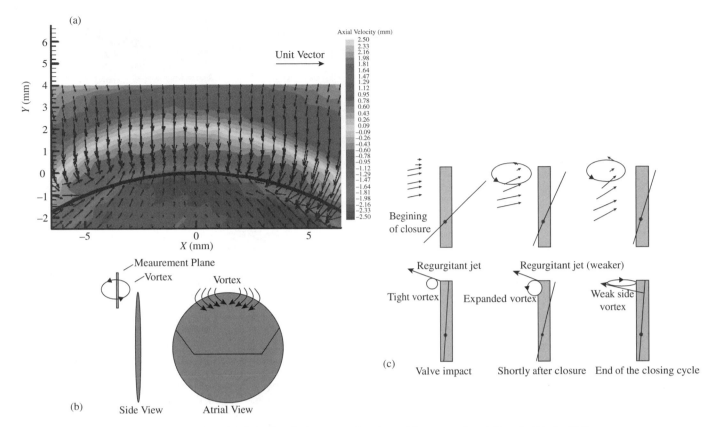

Figure 14. 3D phase-averaged velocity map of major orifice regurgitant flow in Bjork–Shiley monostrut valve. (a) 3 mm from valve housing, 4 ms after impact. (b) Illustration of measurement plane and vortex flow pattern. (c) Flow field schematic during valve closure determined from multicomponent LDV measurements (64).

distributions associated with impact of a Bjork–Shiley monostrut valve (64).

Particle Image Velocimetry and Particle Tracking Velocimetry. Particle image velocimetry (PIV) and particle tracking velocimetry (PTV) have been applied in fluid flow applications for over a decade. They are noninvasive experimental techniques that provides a quantitative, instantaneous velocity vector field with good spatial resolution, an appealing feature when studying complex, time-dependent fluid flows that can occur in biomedical applications. The method allows the quantitative visualization of instantaneous flow structures over a spatial region, as opposed to a point measurement like LDV, which can provide insight into the flow physics. The two techniques, PIV and PTV, differ in the way particle displacements are measured. Particle tracking follows and tracks the motion of individual particles whereas PIV measures the displacement of groups of particles within the flow. Particle tracking velocimetry, although not commonly used, is a subset of the more common PIV technique and is still used in specific applications of PIV. Raffel et al. (66) provide a comprehensive discussion of the PIV and PTV technique with a detailed presentation of the physics, processing tools, capabilities, and errors.

The instantaneous velocity field is computed by measuring an instantaneous particle displacement field over a specified, finite time interval. Laser-based PIV and PTV are noninvasive velocity measurement tools that require good optical access to the flow field. As a result, they are essentially *in vitro* tools (8,67) that are of limited use *in vivo*. Figure 15 shows an example of the use of PIV in a bioengineering application of flow through one chamber of an artificial heart (68). X-ray-based PTV systems are being developed and will be capable of *in vivo* use. In this section, the authors will focus on PIV; however, system concepts (seeding, acquisition, processing, noise, and errors) would be applicable to some degree to systems like X-ray PTV (69).

Particle image velocimetry uses a double-pulsed light source, usually a laser, to illuminate a thin sheet in the flow field. Particles suspended in the fluid scatter light during each pulse, and this scattered light is recorded on a digital camera. The optimal setup has the recording device located 90° to the illumination sheet. Figure 16 illustrates the typical PIV setup. (66)

Two lasers with coincident beam paths are used to illuminate a desired plane of the flow by incorporating optics to produce thin laser sheets. During image acquisition, the two lasers are pulsed with the specified time separation (typically between 1 and 1000 μs). A trigger system, referred to as a synchronizer, controls the firing of the two lasers relative to the shuttering of a CCD camera. The camera, usually placed orthogonal to the laser sheet, collects the light scattered by tracer particles in the flow and records an image. The synchronizer, used in cross-correlation-based PIV systems, delays the firing of the first

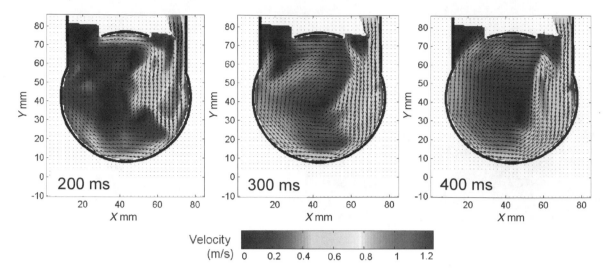

Figure 15. Phase-average velocity maps from mid to late diastole for a prototype artificial heart ventricular chamber (time reference is from the onset of diastole, 4.7 L·min^{-1} CO, 75 bpm) (68).

laser such that the camera shutter is centered between the firing of the two lasers. This synchronization technique is called frame straddling and produces two sequential images of each laser beam pulse. Although the time between successive camera frames may be much larger than the time duration between laser pulses, the two images of the particle field created are separated by the specified time interval between the two laser pulses.

A discussion of PIV must begin with a brief introduction of the terminology commonly used. Figure 17 provides a schematic representation of geometric imaging (66). The "light plane" is the "volume" of the fluid illuminated by the light sheet. The "image plane" is the image from the light plane captured on the CCD sensor. It is important to note that the light plane is a 3D space or volume, whereas the image plane is a 2D space or "surface." The subvolume

selected from the light plane for cross correlation is called the interrogation subvolume. The corresponding location of this interrogation volume captured on the image plane is called the interrogation subregion. Please note that the displacement vectors in an interrogation volume are three-component vectors, whereas those in an interrogation area are two-component vectors. "Particle" is the physical particle suspended in the fluid volume. "Particle image" is the image of each particle in the image plane. Particle density, intensity, and pair refer to particle properties, whereas image density, intensity, and pair refer to particle image properties.

Most commercial systems use a cross-correlation-based image processing technique to compute the particle displacement. Images are subdivided into small interrogation regions, typically a fraction of the overall image size, and

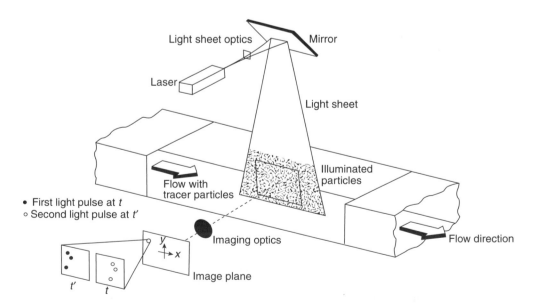

Figure 16. Schematic of a PIV setup (66). (With kind permission of Springer Science and Business Media.)

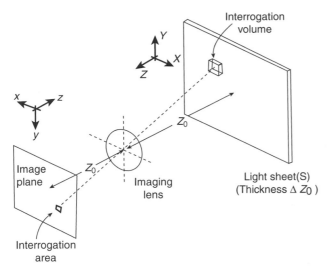

Figure 17. Schematic representation of geometric imaging (66). (With kind permission of Springer Science and Business Media.)

the same two subregions are cross-correlated between the two images to obtain the average displacement of the particles within that subregion. From this displacement and the known time delay, the velocities within the interrogation region are obtained.

Statistical PIV assumes all particles in an interrogation subregion move a similar distance and direction. The processing algorithm then computes the mean displacement vector for the particles in the interrogation volume. Therefore, the local particle distribution pattern captured on each exposure should be similar; but the group of local particles is displaced from image to image. Statistical PIV is then "pattern recognition" of the particle distribution within an interrogation subregion, instead of the averaged

particle displacements. Sophisticated pattern recognition schemes have been developed by a number of researchers; however, the cross-correlation tends to be the algorithm of choice. The use of a cross-correlation as opposed to an autocorrelation eliminates directional ambiguity in the velocity measurement. Most commercial systems use advanced cross-correlation algorithms, such as the Hart-Correlation, developed to improve signal to noise in the correlation estimate and enhance resolution (70,71).

The cross-correlation function for two interrogation subregions of frames A and B is defined by:

$$R_{II}(s, \Gamma, D) = \langle I(x, \Gamma) I'(x + s, \Gamma', D) \rangle \quad (18)$$

where s is the shifting vector of the second interrogation window, Γ is the series of location vectors for each particle in the interrogation volume, D is the displacement vector for each particle, x is the spatial domain vectors within the interrogation area, I is the intensity matrix for the interrogation area from frame A, and I' is the intensity matrix for the interrogation area from frame B. A detailed mathematical derivation of the cross-correlation for (group) particle tracking is beyond the scope of this presentation. The location of the maximum value of cross-correlation function represents the mean particle displacement for the interrogation image. Figure 18 is an example of a cross-correlation function between two images.

The location of the cross-correlation peak should be determined with subpixel accuracy. Several curve-fitting algorithms have been developed to identify the peak in the cross-correlation. Gaussian, parabolic, and centroid are three commonly used methods in commercial software, although others exist. A Gaussian peak fit is most commonly used because the cross-correlation function of two Gaussian functions also yields a Gaussian function, which

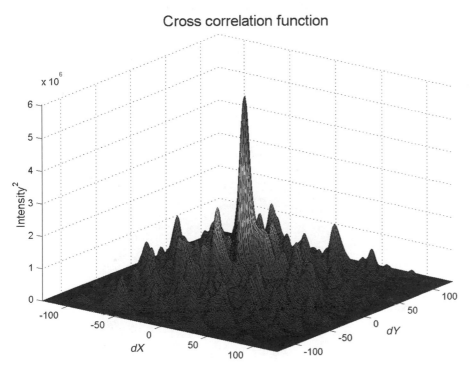

Figure 18. Representative cross-correlation map between frames A and B.

means that if the intensities of individual particle images in the interrogation area fit a Gaussian shape, the cross-correlation function can be accurately approximated by the Gaussian function, which occurs only under the condition of low displacement gradient, so that the particle distribution pattern is preserved in windows A and B. The cross-correlation function in distorted, particle image intensity distribution patterns are less accurately approximated by a Gaussian distribution. A three-point estimator for Gaussian fit works better with a narrow and symmetric peak. Centroid peak finding should be considered when the cross-correlation peak is expected to have asymmetric and irregular patterns. Such cases occur for particle images larger than 2–3 pixels in diameter, for a low intensity ratio of particle image to background noise, or for a high gradient displacement field. For "correlation-based correction," the centroid peak finding might be more suitable than Gaussian because the multiplications could distort the cross-correlation peak.

The use of a digital CCD camera presents an error source known as "peak locking." This error impacts the accuracy of the subpixel estimation in the correlation peak and thus impacts the velocity measurement. This error will be discussed later.

Like LDV and Doppler, PIV and PTV require that the fluid be seeded with tracer particles that follow the fluid motion. The particle density, the number of particles per unit volume of fluid, determines what technique should be used, PIV or PTV. Flows with a low particle density are more suited to PTV, wheras PIV works best in high particle density flows. It is assumed that the tracer particles follow the flow accurately to give an exact representation of the flow field at various times and locations. The particle density, however, should be sufficiently low to preserve the original flow dynamics. Such a dilute condition is expressed by the inequality:

$$(\rho_\mathrm{p} \pi d_\mathrm{p}^4 \upsilon_\mathrm{r})/(18\mu\,\delta_\mathrm{p}^3) < 1 \qquad (19)$$

where d_p and ρ_p are the particle diameter and density, respectively, μ is fluid viscosity, υ_r is the averaged particle velocity relative to neighboring particles, and δ_p is the average distance between particles.

Particles must be small enough to follow the fluid flow but large enough to be seen. The particle relaxation time, τ_s, should be small relative to the flow time scales.

$$\tau_\mathrm{s} = d_\mathrm{p}^2(\rho_\mathrm{p}/18\mu) \qquad (20)$$

In practice, τ_s is made small by using very small particles. The Stokes number for PIV experiments, St_PIV, can be defined as $\tau_\mathrm{s}/\tau_\mathrm{PIV}$, where τ_PIV is the small finite separation time between two observations (pulse separation time). St_PIV should be much less than 1 to assure negligible particle-fluid velocity differences over the pulse separation. However, particles must be large enough to scatter sufficient light energy to be visualized on the recording device (e.g., a CCD camera) with the goal that a particle image is at least several pixels in size. Increasing the light source energy can improve visibility, but a saturation point is reached where increasing light source energy does not help. Furthermore, high energy can damage windows and plastic test models.

The time separation of the two laser pulses must be small enough to minimize particle loss through too large a particle displacement between the first and second frames of the interrogation window. However, the time separation must be long enough both to permit adequate displacement of particles at the lowest measurable velocities in each velocity component and to minimize the impact of pixel peak locking (72). Complex and highly 3D flows can be biased in a 2D PIV system. PIV can provide very high spatial resolution, but suffers from low temporal resolution. Furthermore, high magnification imaging used in high resolution PIV introduces additional constraints and limitations that must be overcome to achieve high quality vector maps.

The challenge for PIV is to correctly track particle motion. Figure 19 shows an example of a PIV particle image. The statistical cross-correlation approach is used to track the displacement of a "group" of particles within an indicated small volume or subregion. The location of a velocity vector is at the center of the subregion spot. The spatial resolution for a "velocity vector" in PIV is the size of the interrogation subregion. Overlapping adjacent interrogation subregions is commonly used to reduce the distance between adjacent vectors and provide additional vectors in the overall vector map. However, this overlapping does not increase the spatial resolution of the velocity field, which is limited to the size of the interrogation subregion.

Commercial PIV systems use a multigrid recursive method to reduce interrogation subregion size. In the hierarchical approach, a PIV measurement with large interrogation subregions is first computed. Subsequently, the initial interrogation area is evenly divided into smaller areas for processing. Each smaller interrogation area is offset by the displacement obtained from its parent interrogation area. This process is repeated until the smallest possible interrogation size is reached (e.g., $128 \times 128 \rightarrow 64 \times 64 \rightarrow 32 \times 32 \rightarrow 16 \times 16$) for the given flow field. An iterative method can be applied at the final grid level.

Figure 19. Example of a PIV particle image.

Similar to the multigrid method, the iterative method uses the obtained displacement from the first cross correlation to offset the second window for the next cross correlation. The difference is that it does not break down the interrogation areas into smaller windows, which is repeated until the difference between the displacements from successive cross correlations is less than one pixel. If the window B is shifted by the converged displacement, windows A and B should be virtually the same, as long as the gradient is sufficiently low. Iterative cross correlation is another way to increase accuracy.

A minimum number of particle pairs (on the order of 10) are required in PIV processing. The particle density in the flow will determine the minimum size subregion that can be used to obtain adequate vector maps. Thus, the spatial resolution is governed by the particle density. Near solid surfaces, the particle density is often lower in flows with strong wall gradients. Reducing the interrogation window size increases spatial resolution. However, an overly small window causes in-plane particle loss due to particles moving out of the interrogation spot. Several techniques exist to capture the particles moving out of the window without enlarging the interrogation spot. The first is simply to enlarge the second window to cover the expected displaced particle. The original interrogation window (frame A) is enlarged to the same size as window B and zero-padded at the extended part. The second technique is to offset the second window to the location of anticipated displacement.

Errors in PIV processing can occur from several sources. The spatial resolution for a velocity vector is the dimension of the interrogation volume. If the particles are evenly distributed, the center of an interrogation volume can be used as the vector location. The accuracy of the 'displacement depends on both the subpixel accuracy of the peak finding algorithm and the image quality. A one-tenth pixel is the most accurate (66). Time resolution for the velocity is the separation time between two pulses, as the information during this period is not recorded. The velocity error is composed of systematic and residual error. Systematic errors come from the algorithm and experiment setting or image quality, which can be minimized and uncertainty in the time separation. Residual errors are inherent in the processing, such as errors due to the peak finding algorithm. The residual errors are usually not a function of Δt. Therefore, a too small separation time increases the velocity error as this error is proportional to $1/\Delta t$.

The following discussion is relevant to the effect of image quality on PIV accuracy. Large particle images can result in wide cross-correlation peaks, which can reduce the accuracy of the peak finding algorithm. In addition, large particles require larger interrogation spots to contain an appropriate number of particles, which leads to a reduction in spatial resolution. Particle images smaller than 2 pixels in diameter, or particle displacements that are less than 2 pixels, can introduce a displacement bias, called "peak locking." The displacement peaks tend to be biased toward integer values. Peak locking presents itself as a "staircased" velocity pattern, in a region with a velocity gradient where the velocity distribution should be smooth. The calculation of spatial derivatives of this vector map then produces a mosaic pattern in the gradient map.

Figure 20 illustrates these patterns. Techniques, such as multigrid or continuous window-shifting or iterative image deformation, have been proposed to overcome peak locking. Image preconditioning, such as filtering, or defocusing can optimize the image diameter. The resolution of the CCD sensor, therefore, also limits the use of smaller particles to increase the velocity resolution.

The methods developed to increase displacement accuracy rely on the assumption of low displacement gradient. High gradient tends to bias the displacement toward low values because the particles with smaller displacements are more likely to remain in the interrogation volume longer than those with higher displacements. This bias can be minimized by reducing the size of the interrogation volume and separation time. For high distortion of the particle pattern in a high gradient spot, the centroid peak finding algorithm is more suitable than the Gaussian. However, PTV, as it follows an individual particle, is not affected by high gradient. Several research groups use the displacement results from PIV to guide the local search areas for PTV in a coupled PIV/PTV processing algorithm. These coupled techniques relieve the gradient limit in PIV and increases resolution of both velocity vectors and velocity fields.

The motion across the light sheet in highly 3D flows can bias the local velocity estimation due to perspective projection, if the particle has not left the light sheet. The effect of perspective projection velocity bias, illustrated in Fig. 21 (68), is usually more severe at the edges of the image, where the projection angle increases. At high image magnification, the focal length becomes shorter and the projection angle increases, which worsens the perspective projection. Strong perspective projection could vary the magnification factor through the image plane resulting in an image distortion.

In general, the light sheet thickness is smaller than the depth of focus, δ_z. The light sheet thickness, therefore, determines the thickness of the effective interrogation volume: All illuminated particles are well focused. Most commercial systems using standard-grade Yag lasers with appropriate optics can generate light sheets that have a thickness on the order of 100–200 μm, although light sheets can be as thick as a 1 mm. In high magnification imaging, the depth of focus can become smaller than the light sheet thickness. The thickness of the effective interrogation volume is then constrained to the depth of focus. In this case, particles located beyond the focal plane but within the illuminated plane are out-of-focus and appear as highly nonuniform background image noise and can affect the cross correlation. In addition, the thickness of the effective interrogation volume determines the tolerance to out-of-plane motion. A smaller effective volume thickness increases the probability for out-of-plane particle loss. In general, the estimated maximum out-of-plane motion should be less than one-fourth of the effective volume thickness.

Data validation is another source of uncertainty in PIV. Bad velocity vectors will ultimately appear within a vector map due to noise in the displacement estimation. Several filtering algorithms have been developed to remove these bad vectors. These filter routines operate on thresholding velocity at a particular location and use either the

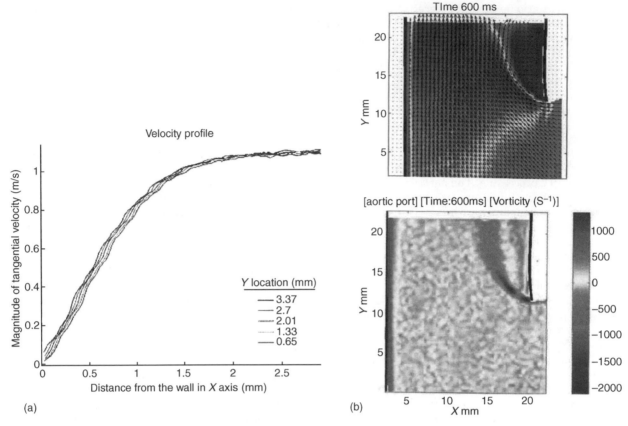

Figure 20. (a) Staircased pattern in a velocity profile at a wall. (b) Gradient field calculation (bottom image) of a measured velocity field (top image) showing mosaic pattern (68).

magnitude of the velocity estimate, the mean, the median or the rms within a predefined subregion, or other more complicated thresholds to low pass filter the estimates. Improper application of a validation scheme can overfilter the velocity map and throw away good data. For example, rms validation techniques should be carefully used in turbulent shear layers where high rms values are normally encountered. It is possible to inadvertently filter good instantaneous turbulent velocity ensembles with a tight rms filter setting. In general, some knowledge of the flow under study is needed to accurately perform vector validation.

Commercial PIV systems can be two-component or three-component, planar or volume systems. A two-component, planar PIV system provides information on two components of the velocity vector. In two-component PIV, the measured displacement vector is the projection of the three-component velocity vector on the 2D plane of the light sheet. Flow information for highly three-component flows can be inaccurately represented by planar PIV images. Stereographic and holographic PIV systems have been developed for three-component measurement in a plane or volume, respectively. Although the instantaneous velocity field obtained by PIV has an advantage over LDV (or Doppler), two-exposure PIV only provides information on the particle motion during the two exposures and also suffers from poor temporal frequency response in the measurement of adjacent vector maps in time. Particle acceleration cannot be measured by direct two-exposure PIV. Four-exposure systems have been developed to permit calculation of the particle acceleration by Hassan and Phillip (73) and Lui and Katz (74), although the temporal resolution for the acceleration is not yet comparable with that of LDV.

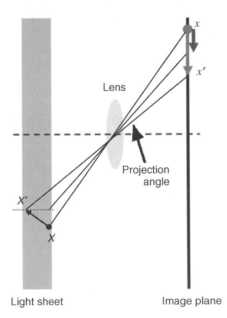

Figure 21. Illustration of the perspective projection; the red arrow is the vector obtained from perspective projection, the blue is the correct projection of displacement vector on the XY plane (68).

BIBLIOGRAPHY

Cited References

1. Fung YC. Respiratory gas flow. In: Biomechanics: Motion, Flow, Stress and Growth. New York: Springer-Verlag; 1990.
2. Primiano FP. "Measurements of the respiratory system. In: Webster JG, editor. Medical Instrumentation: Application and Design. New York: John Wiley & Sons; 1998.
3. Hampers CL, Schuback E, Lowrie EG, Lazarus JM. Clinical engineering in hemodialysis and anatomy of an artificial kidney unit. In: Long Term Hemodialysis. New York: Grune and Stratten; 1973.
4. Neuman MR. Therapeutic and prosthetic devices. In: Webster JG, ed., Medical Instrumentation: Application and Design. 3rd ed. New York: John Wiley & Sons; 1998.
5. Tedgui A, Lever MJ. Filtration through damaged and undamaged rabbit thoracic aorta. Am J Physiol 1984;247:784.
6. Jin S, Oshinski J, Giddens DP. Effects of wall motion and compliance on flow patterns in the ascending aorta. J Biomech Eng 2003;125:347–354.
7. Hochareon P, Manning KB, Fontaine AA, Tarbell JM, Deutsch S. Wall shear-rate estimation within the 50cc Penn State artificial heart using particle image velocimetry. J Biomech Eng 2004;126:430–437.
8. Hochareon P, Manning KB, Fontaine AA, Tarbell JM, Deutsch S. Correlation of in vivo clot deposition with the flow characteristics in the 50cc Penn State artificial heart: A preliminary study J ASAIO 2004;50(6):537–542.
9. Merzkirch W. Flow Visualization. New York: Academic Press; 1974.
10. Rouse H, Ince S. History of hydraulics. Iowa Institute of Hydraulics Research Report. Ames, (IA); State University of Iowa; 1957.
11. Latto B, El Riedy O, Vlachopoulos J. Effect of sampling rate on concentration measurements in nonhomogeneous dilute polymer solution flow. J Rheol 1981;25:583–590.
12. Lauchle GC, Billet ML, Deutsch S. Hydrodynamic measurements in high speed liquid flow facilities. Gad-el-Hak M. In: Lecture Notes in Engineering, 46, Experimental Fluid Mechanics. New York: Springer Verlag, 1989.
13. White KC, Kavanaugh JF, Wang DM, Tarbell JM. Hemodynamics and wall shear rate in the abdominal aorta of dogs. Circ Res 1994;75(4):637–649.
14. Povey MJW. Ultrasonic Techniques for Fluids Characterization. New York: Academic Press; 1997.
15. Siedband MP. Medical imaging systems. In: Webster J, editor. Medical Instrumentation: Application and Design. New York: John Wiley & Sons; 1998.
16. Adrian RJ. Particle imaging techniques for experimental fluid mechanics. Ann Rev Fluid Mech 1991;23:261–304.
17. Webster JG. Measurement of flow and volume of blood. In: Webster JG, editor. Medical Instrumentation: Application and Design. 3rd ed. New York: John Wiley & Sons; 1998.
18. Lipowski HH, McKay CB, Seki J. Transit time distributions of blood flow in the microcirculation. In: Lee, Skalak, editors. Microvascular Mechanics. New York: Springer Verlag; 1989. p 13–27.
19. Shaughnessy EJ, Katz IM, Schaffer JP. Introduction to Fluid Mechanics. New York: Oxford University Press; 2005.
20. Bendat JS, Peirsol AG. Random Data: Analysis and Measurement Procedures. New York: John Wiley and Sons; 1986.
21. Coleman HW, Steele WG. Experimentation and Uncertainty Analysis for Engineers. New York: John Wiley and Sons; 1999.
22. Montgomery DC. Design and Analysis of Experiments, 3rd ed. New York: John Wiley and Sons; 1991.
23. White FM. Fluid Mechanics. New York: McGraw-Hill; 1979.
24. Caro CG, Pedley TJ, Schroter RC and Seed WA. The Mechanics of the Circulation. New York: Oxford University Press; 1978.
25. Pedley TJ, Schroter RC, Sudlow MF. Flow and pressure drop in systems of repeatedly branching tubes. J Fluid Mech 1971; 46(2):365–383.
26. Mori Y, Nakayama W. Study on forced convective heat transfer in curved pipes. Int J Heat Mass Transfer 1965;8:67–82.
27. Drost CJ. Vessel diameter-independent volume flow measurements using ultrasound. Proceedings of the San Diego Biomedical Symposium, 17, 1978: p 299–302.
28. Lynnworth LC. Ultrasonic flow meters. In: Mason WP, Thurston RN eds. Physical Acoustics. Academic Press; 1979.
29. Beldi G, Bosshard A, Hess OM, Althaus U and Walpoth BH. Transit time flow measurement: Experimental validation and comparison of three different systems. Ann Thorac Surg 2000;70:212–217.
30. Weyman AE. Principles and Practice of Echocardiography, 2nd ed., New York: Lea & Febiger; 1994.
31. Crowe CT, Sommerfeld M, Tsuji Y. Multiphase Flows with Droplets and Particles. Boca Raton (FL): CRC Press; 1998.
32. Comte-Bellot G. Hot-wire anemometry. Ann Rev Fluid Mech 1976;8:209–232.
33. Nerem RM, Rumberger JA, Gross DR, Muir WW, Geiger GL. Hot film coronary artery velocity measurements in horses. Cardiovasc Res 1976;10(3):301–313.
34. Nerem RM, Rumberger JA, Gross DR, Hamlin RL, Geiger GL. Hot film anemometer velocity measurements of arterial blood flow in horses. Circ Res 1974;34(2):193–203.
35. Falsetti HL, Carroll RJ, Swope RD, Chen CJ. Turbulent blood flow in the ascending aorta of dogs. Cardiovasc Res 1983;17(7):427–436.
36. Baldwin, JT, Tarbell, JM, Deutsch S, Geselowitz DB, Rosenberg G, Hot-film wall shear probe measurements inside a ventricular assist device," J Biomech Eng, 1988;110(4):326–333.
37. Baldwin JT, Tarbell KM, Deutsch S, Geselowitz DB. Wall shear stress measurements in a ventricular assist device. J BioMech Eng 1988;V110:326–333.
38. Batten JR, Nerem RM. Model study of flow in curved and planar arterial bifurcations. Cardiovasc Res 1982;16(4):178–186.
39. Bruun HH. Hot Wire Anemometry: Principles and Signal Analysis. New York: Oxford University Press; 1995.
40. Stock DE Ed. Thermal anemometry 1993. Proceedings of the 3rd Int. Symposium on Thermal Anemometry – ASME Fluids Engineering Conference, Washington, (DC), FED 167 1993.
41. Otto, CM, The Practice of Clinical Echocardiography, Philadelphia, PA., WB Saunders Co.; 1997.
42. Cape EG, Nanda NC, Yoganathan AP. Quantification of regurgitant flow through Bileaflet heart valves: theoretical and in vitro studies. Ultrasound Med Biol 1993;19:461–468.
43. Pettigrew RI. Magnetic resonance in cardiovascular imaging. In Zaret BL et al., editors. Frontiers in Cardiovascular Imaging. Baltimoc, MD: Raven Press; 1993.
44. Ku DN, Biancheri CL, Pettigrew RI, et al. Evaluation of magnetic resonance velocimetry for steady flow. J Biomech Eng 1990;112:464–472.
45. Hahn EL. Detection of sea-water motion by nuclear preemission. J Geophys Res 1960;65:776–777.
46. Singer JR, Crooks LE. Nuclear magnetic resonance blood flow measurements in the human brain. Science 1983;221:654–656.
47. Moran PR, Moran RA, Karstaedt N. "Verification and evaluation of internal flow and motion. True magnetic resonance imaging by the phase gradient modulation method." Radiology. 1985 Feb; 154(2):433–41.
48. Bryant DJ, Payne JA, Firmin DN, Longmore DB. Measurement of flow with NMR imaging using a gradient pulse and phase difference technique. J Comp Assist Tomogr 1984;8: 588–593.

49. Matsuda T, Shimizu K, Sakurai T, et al. Measurement of aortic blood flow with MR imaging: Comparative study with Doppler ultrasound. Radiology 1987;162:857–861.

50. Edelman RR, Heinrich PM, Kleefield J, Silver MS. Quantification of blood flow with dynamic MR imaging and presaturation bolus tracking. Radiology 1989;171:551–556.

51. Moran, PR, "A flow velocity zeugmatographic interlace for NMR imaging in humans," Magn. Res. Imaging, 1982, 1:197–203.

52. Chatzimavroudis GP, Oshinski JN, Franch RH, et al. Evaluation of the precision of magnetic resonance phase velocity mapping for blood flow measurements. J Card Mag Res 2001;3:11–19.

53. Firmin DN, Nayler GL, Kilner PJ, Longmore DB. The application of phase shifts in NMR for flow measurement. Mag Res Med 1990;14:230–241.

54. Zhang H, Halliburton SS, White RD, Chatzimavroudis GP. Fast measurements of flow through mitral regurgitant orifices with magnetic resonance phase velocity mapping. Ann Biomed Eng 2004;32(12):1618–1627.

55. Kraft KA, Fei DY, Fatouros PP. Quantitative phase-velocity MR imaging of in-plane laminar flow: Effect of fluid velocity, vessel diameter, and slice thickness. Med Phys 1992;19: 79–85.

56. Suzuki J, Caputo GR, Kondo C, Higgins CB. Cine MR imaging of valvular heart disease: Display and imaging parameters affect the size of the signal void caused by valvular regurgitation. Am J Roentgenol 1990;155:723–727.

57. Zhang H, Halliburton SS, Moore JR, Simonetti OP, et al. Ultrafast flow quantification with segmented k-space magnetic resonance phase velocity mapping. Ann Biomed Eng 2002;30:120–128.

58. Walker PG, Oyre S, Pedersen EM, Houlind K, Guenet FS, Yoganathan AP. A new control volume method for calculating valvular regurgitation. Circ 1995;92:579–586.

59. Chatzimavroudis GP, Oshinski JN, Pettigrew RI et al. Quantification of mitral regurgitation with magnetic resonance phase velocity mapping using a control volume method. J Mag Reson Imag 1998;8:577–582.

60. Drain LE. The Laser Doppler Technique. New York: John Wiley and Sons; 1980.

61. Durst F, Melling A, Whitelaw JH. Principles and Practice of Laser Doppler Anemometry. San Diago, (CA): Academic Press; 1976.

62. Edwards, Robert V, "Report of the special panel on statistical particle bias problems in laser anemometry," J Fluids Eng, Transactions of the ASME, V 109, n 2, Jun, 1987, p 89–93.

63. Tomonaga G, Mitake H, Hoki N, Kajiya F. Measurement of point velocity in the canine coronary artery by laser doppler velocimeter with optical fiber. Jap J Surg 1981;11(4):226–231.

64. Manning KB, Przybysz TM, Fontaine AA, Tarbell JM, Deutsch S. Near field flow characteristics of the Bjork-Shiley monostrut valve in a modified single shot valve chamber. ASAIO J 2005; 51(2):133–138.

65. Ellis JT, Healy TM, Fontaine AA, Westin MW, Jarret CA, Saxena R, Yoganathan AP. An in vitro investigation of the retrograde flow fields of two bileaflet mechanical heart valves. J Heart Valve Disease 1996;5:600–606.

66. Raffel M, Willert CE, Kompenhans J. Particle Image Velocimetry: A Practical Guide. New York: Springer; 1998.

67. Oley LA, Manning KB, Fontaine AA, Deutsch S. "Off design considerations of the 50cc Penn State ventricular assist device. Art Organs 2005;29(5):378–386.

68. Hochareon P. Development of particle image velocimetry (PIV) for wall shear stress estimation within a 50cc Penn State artificial heart ventricular chamber. Ph.D. dissertation Bioengineering Department, Penn State University, 2003.

69. Lee SJ, Kim GB. X-ray particle image velocimetry for measuring quantitative flow information inside opaque objects. J Appl Phys 2003;94:3620–3623.

70. Hart DP. Super-resolution PIV by recursive local correlation. J Visualization 1999;10:1–10.

71. Hart DP. PIV error correction. Exp Fluids 2000;29(1):13–22.

72. Christensen, KT, "The influence of peak-locking errors on turbulence statistics computed from PIV ensembles," Experiments in Fluids, Vol. 36, n 3, March, 2004, p 484–497.

73. Hassan YA, Phillip OG. A new artificial neural network tracking technique for particle image velocimetry. Exp Fluids 1997;23(2):145–154.

74. Lui X, Katz J. Measurements of pressure distribution in a cavity flow by integrating the material acceleration. Proceedings of 2004 Heat Transfer and Fluids Engineering Conference, ASME HT-FED04-56373, July, 2004.

Reading List

Adrian RJ. Laser velocimetry. In: Fluid Mechanics Measurements. *New York*: 1983.

See also BLOOD RHEOLOGY; HEMODYNAMICS; PERIPHERAL VASCULAR NONINVASIVE MEASUREMENTS.

FLOWMETERS, RESPIRATORY.   See PNEUMOTACHOMETERS.

FLUORESCENCE MEASUREMENTS

ROBERT J. KLEBE
GUSTAVO ZARDENETA
PAUL M. HOROWITZ
University of Texas
San Antonio, Texas

INTRODUCTION

Following the absorption of light, fluorescent compounds release light at a less energetic wavelength. This phenomenon can be used to detect the presence of many compounds with exceedingly high sensitivity and selectivity.

The objective of this article is to acquaint the reader with the basic principles of fluorescence spectroscopy. For a more detailed analysis of this area, several reviews exist (1–5) that should be useful to readers with different levels of sophistication.

Most readers are probably familiar with the basic phenomenon involved in fluorescence measurements. For example, when certain minerals are irradiated with ultraviolet light, light in the visible region of the electromagnetic spectrum is released. In this case, absorption of high energy ultraviolet (UV) light excites electrons to leave their lowest energy state (the ground state); upon return to the ground state, energy is emitted as either light or heat. The energy emitted during fluorescence must be of a lower energy than the light that initially excited a compound and, hence, high energy ultraviolet light is emitted from a mineral as lower energy visible light.

When one deals with biological samples, one is always confronted with the problem of detecting extraordinarily small amounts of a compound mixed in an array of other compounds. As this article will show, fluorescence analysis provides a rather simple means to detect the existence of a compound within a complex mixture. For obvious reasons, the ability of fluorescence spectroscopy to detect exceptionally small amounts of a compound in biological specimens has many applications of biomedical interest.

The analytical power of fluorescence measurements may be appreciated by consideration of the two following practical analogies. The sensitivity of fluorescence measurements is similar to the increased sensitivity with which one could measure the light output of a flashlight in a completely darkened room versus a sunlit room. In the case of the sunlit room, the light output of the flashlight would represent < 1% of the total light in the room; indeed, it would be difficult for an observer in a sunlit room to detect whether the flashlight was on or off. In contrast, in a darkened room, an observer would readily be able to sense whether a flashlight was turned on. In fluorescence measurements, light strikes a sensor only if the compound of interest is present and, thus, as in the case of the darkened room, the sensitivity of the fluorescence measurements is quite high.

The ability of fluorescence measurements to detect particular compounds in complex mixtures is due to the fact that several conditions must be met before a fluorescence signal is detected. Just as the location of a person would be precisely defined if one knew their street address, which consists of a city, state, street name, and house address, the selective ability of fluorescence measurements to detect the presence of only desired compounds arises from the fact that a compound will fluoresce only under a series of quite restrictive conditions. In the case of fluorescence spectroscopy, one observes light emanating from a compound if and only if the following conditions are met: (a) the compound must absorb light; (b) the compound must be capable of emitting light following excitation (the compound must be fluorescent); (c) the compound must be irradiated at a particular excitation wavelength; (d) a detector must be set to sense light at a different, less energetic emission wavelength; (e) a discrete time (in nanoseconds) must elapse between the time of excitation and emission; and (f) other conditions, such as type of solvent and pH, must be satisfactory. As a street address defines the location of a person, the parameters involved in the fluorescence of a compound can be used to determine the presence or absence of just a single compound. For example, the presence of the compound, fluorescein, can be accurately detected in a biological specimen because (a) fluorescein is a fluorescent compound; (b) fluorescein is excited to fluoresce only at wavelengths near 488 nm (nanometer) in aqueous solvents in the neutral to alkaline range; and (c) following excitation, fluorescein gives off light maximally at 514 nm (see Fig. 1). Hence, since only a very few compounds have fluorescence properties similar to that of fluorescein, selecting light with a wavelength of 488 nm to irradiate the sample and setting a detector to only respond to light with a 514 nm wavelength allows one to detect fluorescein in a complex mixture.

In addition to an overview of the theory involved in this area, a brief introduction into the instrumentation involved in fluorescence determinations will be presented. While there are numerous applications of fluorescence that are outside of the scope of this article (1–5), a few specific examples of the use of fluorescence measurements in biomedical studies will be presented and technical problems in interpretation of fluorescence data will be pointed out.

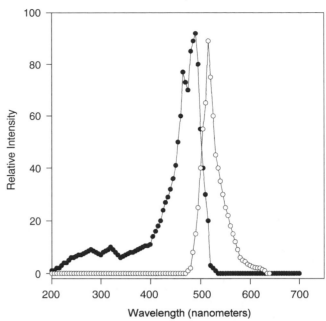

Figure 1. Excitation and emission spectra of the fluorescent compound, fluorescein. As described in the text, excitation and emission maxima are determined empirically and, then, spectra are obtained for the excitation and emission of fluorescein. Excitation of fluorescein at any wavelength in the excitation spectrum will produce emission at all wavelengths in the emission spectrum, with the exception of wavelengths of higher energy. Thus, fluorescence involves the conversion of high energy photons into photons of lower energy plus heat.

THEORY AND INSTRUMENTATION

After absorption of light energy, all compounds release some energy as heat; fluorescent compounds represent a special case in which some energy is also given off as light. Spectrofluorimeters are optical devices that measure the amount of light emitted by fluorescent compounds.

Spectrofluorimeters

While a complete description of the theory of fluorescence is quite involved (see Ref. 3, for a review), one can grasp the basic principles of the phenomenon from the following examples. If one were to aim a flashlight at a mirror, one would find that light would reflect off the surface of the mirror; in contrast, if one were to direct a flashlight beam at a black wall, one would see no light reflected from the wall. In this example, molecules in the black wall stop the transmission of light by absorbing the energy of the light and then converting the energy of light into heat energy. (Very precise measurements would reveal that the black wall was slightly warmer at the site at which the light beam struck it.) In the case of fluorescent compounds, light absorbed by a compound is converted into light (of a lower energy) as well as heat. The brighter the light that strikes a fluorescent compound, the stronger the fluorescent light emitted by the compound. The particular wavelength absorbed by a compound and the wavelength that is later emitted are characteristics of each fluorescent compound (Fig. 1).

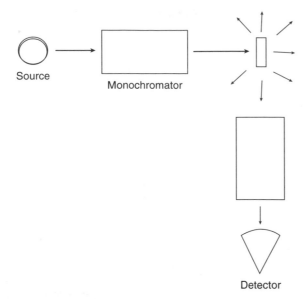

Figure 2. Design of a typical spectrofluorimeter. As described in the text, a desired wavelength is selected with the aid of a monochromator and is used to irradiate a sample (large arrows); light emitted by the fluorescent sample is released in all directions (small arrows). The detector is positioned at a right angle with respect to the excitation beam in order to avoid light from the exciting beam entering the detector.

While light reflection from a mirror occurs at an angle determined by the angle of the incident light, light emission by a fluorescent compound occurs in all directions. For example, if one observed a beam of light striking a mirror in a smoke filled room, one would see a beam of light reflected at a unique angle from the mirror. In contrast, if one observed a beam of light striking a solution of fluorescent compound, one would see the emitted light appear as if it originated from a glowing sphere (and, compared to the original beam of light, the emitted light would have a color that was shifted toward the red end of the spectrum, i.e., lower energy).

The above features of fluorescence dictate the design of instruments used in the measurement of fluorescence (see Fig. 2). First, an intense beam of light is passed through a filter (or monochromator) that permits only light of a desired band of wavelengths to proceed toward the sample. As indicated in Fig. 2, light not absorbed by the sample passes straight through the sample without change in angle while the light emitted by the fluorescent sample is released in all directions. By positioning the detector at a 90° angle with respect to the exciting beam, instrument designers can minimize the amount of stray light from the exciting beam that reaches the detector. As indicated above, the light emitted by the fluorescent sample is given off in all directions and, thus, the positioning of the detector with respect to the exciting beam does not affect the relative amount of the emitted fluorescent light received by the detector. Hence, as we have shown above, the design of a fluorimeter is predicated upon maximizing the signal/noise ratio.

The sensitivity of fluorescence measurements is inherent in the design of instruments used to make such measurements. Since only light that is emitted from a fluorescent compound reaches the detector, the sensitivity

of the measurement is equivalent to our analogy of measuring the light from a flashlight in a darkened room. The ability to selectively measure the presence of a given compound is determined by the appropriate selection of filters (or settings on monochomators) used to make the measurement. In contrast, analytical measurements based on absorption are inherently less sensitive (due to signal/noise problems).

Fluorescence Microscopy

The design of a fluorescence microscope is based on the same principles used in the construction of a spectrofluorimeter. In the technique of reflected fluorescence microscopy, light of a desired wavelength is used to irradiate a sample and, then, a dichroic mirror is used to separate light emitted by the fluorescent compounds in the sample from light in the original exciting beam (Fig. 3). A dichroic mirror is an optical device that reflects light of certain

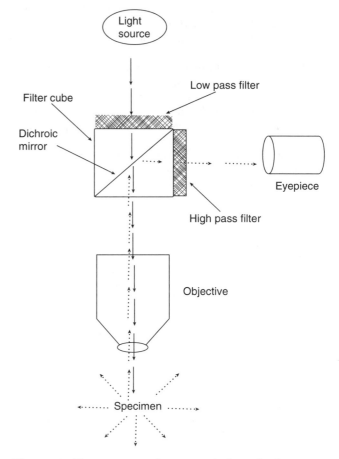

Figure 3. Fluorescence microscope design. A fluorescence microscope employes optical components that allow one to irradiate a specimen at one wave lenght and then observe the fluorescence of a microscopic object at a second wavelength. The excitation beam (dashed line) is directed toward the specimen by the dichroic mirror and objective lens. Due to the fact that light emitted by a fluorescent object is released in all directions, one can observe the fluorescence of the specimen with the same objective lens that was used to irradiate the specimen. The emitted light (dotted line) is collected by the objective, passes through the dichroic mirror, and is observed via the eyepiece.

desired wavelengths and transmits light of other wavelengths.

Fluorescence microscopes can be used to detect fluorescence compounds in exceedingly small specimens. Via the technique of immunofluorescence, one can visualize the distribution of virtually any biological compound in living or fixed tissues by using antibodies tagged with fluorescent compounds.

PRACTICAL APPLICATIONS

The high sensitivity and selectivity of fluorescence methods permits the detection of exceedingly low amounts of many compounds of biological and biomedical interest.

In the preceding section, the basic methods that are used to detect fluorescent compounds were described. In the following section, the criteria that are used in establishing the presence or absence of particular chemical species are presented.

Detection of Fluorescent Compounds

It should be pointed out initially that only a small percentage of the known compounds are fluorescent. Thus, while most compounds cannot be studied by fluorescence techniques, the mere fact that an unknown compound is fluorescent can be used as a diagnostic means to positively identify it. In a following section, literature references are provided that list many fluorescent compounds and their properties (6,7).

Once it is established that a sample contains a fluorescent compound, one can use several criteria to establish the identity of the unknown compound. First, by trial-and-error, one establishes the wavelength(s) at which the compound is maximally stimulated to fluoresce (the excitation maxima). Second, again by trial-and-error, one determines the wavelength(s) at which the emission of light is highest (the emission maxima). Fluorescence emission spectra are then generated by holding the excitation monochromator at the excitation maximum wavelength and recording the intensity of light released from the sample as the wavelength of the emission monochromator is varied (Fig. 2). In this manner, a spectrum is produced that (a) describes the wavelengths at which a compound emits light and (b) establishes the relative efficiency with which light of different wavelengths is emitted (Fig. 1). The shape of the emission spectrum and the number of major and minor peaks are characteristics of each compound and are important parameters in establishing the identity of an unknown compound. The excitation spectrum, when appropriately corrected, is often identical to the absorption spectrum of the fluorescent compound. (In a similar fashion, the fluorescence excitation spectra are established by holding the emission monochromator at the emission maximum and varying the settings of the excitation monochromator.)

Strong evidence for the identity of an unknown compound is provided by (a) the establishment that a compound is fluorescent and (b) the shapes of the excitation and emission spectra. In addition, one could use other parameters involved in fluorescence to characterize a com-

pound, namely, (a) the fluorescent lifetime of a compound, (b) the quantum yield, and (c) the perturbation of fluorescence by various solvents. It is possible that two compounds could have identical excitation and emission spectra just as it is possible that two individuals could have the same blood groups. Other analytical methods may be required to establish the identity of a compound with certitude. If the sample under study is quite impure, it is possible to use high performance liquid chromatography (HPLC) in conjunction with a fluorescence detector to gain yet another parameter that is characteristic of a compound, namely, its elution time from a particular HPLC column. The nature of the sample and the reasons for interest in it will determine the rigor required in its characterization.

Biomedical Applications

As described above, fluorescence methods provide means to identify compounds in complex mixtures. The sensitivity of fluorescence methods coupled with their inherent selectivity provide valuable means for biomedical analysis; several examples are described below.

Identification of Compounds in Biological Samples. If a compound of interest proves to be fluorescent, several analytical methods for the detection and quantitation of this compound immediately become available. Many drugs and biological compounds can be detected by fluorescence analysis (6,7). Fluorescence can be used to identify these and many other biologically active compounds in complex pathological specimens.

Due to the sensitivity of fluorescence methods to detect compounds at the level of one part per billion, such methods can be used to determine the presence of environmental pollutants with great sensitivity. The presence of specific pesticides and potentially carcinogenic aromatic hydrocarbons from cigarette smoke can be easily detected. Leakage of industrial wastes into the environment can be proven by placing fluorescent compounds into waste containers and later detecting the fluorescent compound in streams and lakes.

Fluorescence spectra of many chemicals and pharmaceuticals have been obtained under well controlled conditions and are published as the Sadtler Standard Fluorescence Spectra (6). Passwater (7) published a three volume series that presents literature citations for the fluorescence properties of a wide variety of compounds.

Fluorescence Microscopy. Following staining with fluorescent dyes, the use of fluorescence at the microscopic level can be used to determine the sex of an individual from single hair follicle cells, teeth, or from blood smears (8). Sex determination is based upon the fact that male cells have a highly condensed Y chromosome that appears as an intense fluorescent spot in the nucleus of each male cell. Fluorescent dyes have found many applications in the biomedical sciences (1,5).

Immunofluorescence. Antibodies can be prepared that specifically recognize a wide variety of molecules and microbial organisms. By labeling such antibodies with fluorescein or other fluorescent probes, one can visualize the presence of antigens at the subcellular level; this approach has been widely used in molecular biology (9). In addition, one can visualize specific molecules and organisms in pathological specimens. The identification of disease causing microorganisms in pathological specimens by immunofluorescence (10) can be used. Immunofluorescence microscopy can also be employed in the identification of bacteria in food products (11).

Fluorescence Imaging. In addition to localizing molecules in histological sections, fluorescence has found numerous novel applications in cell biology. The availability of highly sensitive optical detection systems has permitted the localization of specific molecules in living cells. By tagging recombinant proteins with the green fluorescent protein (GFP) of jellyfish origin, one can track the expression and translocation of specific proteins in living cells during hormonal responses (12). The rate at which molecules move within living cells can be determined by fluorescence recovery after photobleaching (FRAP), which involves the photobleaching of molecules in a small region of a cell and then monitoring of the recovery of fluorescence with time as unbleached molecules return to the bleached area (13). Fluorescence energy transfer and fluorescence polarization methods (14–16) can also be used to study the interaction of molecules within living cells (15,16).

Fluorescence Polarization. Fluorescence polarization is perhaps the only method in biology that is directly attributable to Albert Einstein. The principle of fluorescence polarization involves the fact that emission of a photon is delayed by a few nanoseconds following absorption of light. During the delay in emission of light, Brownian motion will result in the movement of a molecule and smaller molecules will move more than larger molecules. Thus, molecules excited by polarized light will emit progressively depolarized light as the size of the molecule increases. Hence, if a small fluorescent molecule binds to a large nonfluorescent molecule, the light emitted by the bound small fluorescent molecule will become more polarized. Thus, the method of fluorescence polarization permits one to measure the binding of ligands to large molecules in real time (14).

Forster Resonance Energy Transfer (FRET). This method involves the transfer of energy from a fluorophore, which absorbs light to a second molecule that emits light at a different wavelength. Since the efficiency of FRET depends on the distance between the absorbing and emitting molecules, FRET permits one to obtain information about the distance between two molecules. Thus, if the molecule that absorbs light binds to a molecule that emits light via FRET, one can measure the binding event via release of light.

Molecular Biology Applications of Fluorescence. The high sensitivity of Fluorescence has been used as the basis of several important methods in molecular biology. For example, real-time polymerase chain reaction (PCR) methods can be used to quantify the amount of a specific ribonucleic acid (RNA) species in a small sample of cells. In this method, fluorescence energy transfer is used to detect the increase in PCR product with time (16).

The jellyfish green fluorescent protein (GFP) has become an important tool in molecular biology because it is fluorescent without the necessity of binding or reacting with a second molecule. Mutant GFP and GFP-like molecules from various species have been described (17) that emit light at a variety of wavelengths. Thus, one can engineer the sequence of the GFP into a recombinant protein and have a fluorescently tagged protein (12). In contrast, the firefly luciferase protein must react with ATP and luciferin in order to release a photon (18).

The high sensitivity and specificity of fluorescence should find many new applications in the future.

GLOSSARY

Dichroic Mirror. An optical device that reflects light of a desired band of wavelengths yet permits the transmission of light of another band of wavelengths. Dichroic mirrors are used in fluorescence microscopes to separate the light that excites a sample from the light emitted from the sample.

Emission (Fluorescent). The release of light by a compound that follows the absorption of a photon.

Emission Wavelengths. Following absorption of light at a wavelength capable of inducing fluorescence, light is released by the compound at less energetic wavelengths, termed the emission wavelengths (Fig. 1).

Excitation (Fluorescent). Following the absorption of a photon, one or more electrons of a fluorescent compound are promoted to a more energetic, "excited" state of the compound.

Excitation Wavelengths. While light can be absorbed by fluorescent compounds at many wavelengths, only certain wavelengths, termed excitation wavelengths, are capable of inducing fluorescent emission of light. The wavelengths are often the same as those absorbed by the compound of interest.

Filter (Optical). Generally a colored piece of glass that transmits only certain wavelengths of light. Interference filters reflect light that is not transmitted while color glass filters absorb light and convert the absorbed energy into heat.

Fluorescence. The emission of light by certain compounds that is induced by the absorption of light at a more energetic wavelength.

Fluorescent Lifetime. The amount of time, generally in nanoseconds, that expires between the absorption of a photon and the emission of a photon.

Monochromator. An optical device that is used (a) to separate light into its component wavelengths and, then, (b) to isolate a desired group of wavelengths. Such devices employ either a prism or a diffraction grating.

Nanosecond. $0.000000001 \text{ s} = 10^{-9} \text{ s}$. (There are 1000 million ns in 1 s).

Quantum Yield. The percentage of excited compounds that release a photon of light during their return to the ground state. In most cases, the absorption of light by a compound is followed by the liberation of heat, rather than light.

Spectrofluorimeter. An optical device that is used to measure the amount of light that is emitted by fluorescent compounds.

BIBLIOGRAPHY

Cited References

1. Hof M, Hutterer R, Fidler V. Fluorescence Spectroscopy in Biology. New York: Springer; 2005.
2. Lakowicz JR. Principles of Fluorescence Spectroscopy. New York: Plenum; 1983.
3. Valeur B. Molecular Fluorescence: Principles and Applications. New York: Wiley-VCH; 2004.
4. Albani JR. Structure and Dynamics of Macromolecules: Absorption and Fluorescence Studies. New York: Elsevier; 2004.
5. Kohen E, Hirschberg JG, Santus R. Fluorescence Probes in Oncology. Imperial College Press; 2002.
6. Sadtler Standard Fluorescence Spectra. Philadelphia: Sadtler Research Laboratories;
7. Passwater RA. Guide to Fluorescence Literature. Vols. 1–3. New York: Plenum; 1967.
8. Kringsholm B, Thomsen JL, Henningsen K. Fluorescent Y-chromosomes in hairs and blood stains. Forensic Sci 1977;9:117.
9. Giambernardi TA, et al. Neutrophil collagenase (MMP-8) is expressed during early development in neural crest cells as well as in adult melanoma cells. Matrix Biol 2001;20:577–587.
10. Stenfors LE, Raisanen S. Quantification of bacteria in middle ear effusions. Acta Otolaryngol 1988;106:435–440.
11. Rodrigues UM, Kroll RG. Rapid and selective enumeration of bacteria in foods using a microcolony epifluorescence microscopy technique. J Appl Bacteriol 1988;64:65–78.
12. Rivera OJ, et al. Role of promyelocytic leukemia body in the dynamic interaction between the androgen receptor and steroid receptor coactivator-1 in living cells. Mol Endocrinol 2003;17:128–140.
13. Snickers YH, van Donkelaar CC. Determining diffusion coefficients in inhomogeneous tissues using fluorescence recovery after photobleaching. Biophys J 2005.
14. Bentley KL, Thompson L, Klebe RJ, Horowitz P. Florescence polarization: A general method for studing ligand interactions. Bio Techniques 1985;3:356–366.
15. Rizzo MA, Piston DW. Hight-contrast imaging of fluorescent protein FRET by fluorescence polarization microscopy. Biophys J 2005;88:14–16.
16. Peng XH, et al. Real-time detection of gene expression in cancer cells using molecular beacon imaging: New Strategies for cancer research. Cancer Res 2005;65: 1909–1917.
17. Shagin DA, et al. GFP-like proteins as ubiquitous metazoan superfamily: Evolution of functional features and structural complexity. Mol Biol Evol 2004;21:841–850.
18. Branchini BR, et al. An alternative mechanism of bioluminescence color determination in firefly luciferase. Biochemistry 2004;43:7255–7262.

See also COLORIMETRY; MICROSCOPY, FLUORESCENCE; ULTRAVIOLET RADIATION IN MEDICINE.

FLUORESCENCE MICROSCOPY. See MICROSCOPY, FLUORESCENCE.

FLUORESCENCE SPECTROSCOPY. See FLUORESCENCE MEASUREMENTS.

FLUORIMETRY. See FLUORESCENCE MEASUREMENTS.

FRACTURE, ELECTRICAL TREATMENT OF. See BONE UNUNITED FRACTURE AND SPINAL FUSION, ELECTRICAL TREATMENT OF.

FUNCTIONAL ELECTRICAL STIMULATION

GANAPRIYA VENKATASUBRAMANIAN
RANU JUNG
JAMES D. SWEENEY
Arizona State University
Tempe, Arizona

INTRODUCTION

Functional electrical stimulation (FES) is a rehabilitative technique where low level electrical voltages and currents are applied to an individual in order to improve or restore function lost to injury or disease. In its broadest definition, FES includes electrical stimulation technologies that, for example, are aimed at restoration of a sense of hearing for the deaf, vision for the blind, or suppression of seizures in epilepsy or tremors for people with Parkinson's disease. Most FES devices and systems are known then as "neuroprostheses" because through electrical stimulation they artificially modulate the excitability of neural tissue in order to restore function. While sometimes used synonymously with FES, the term functional neuromuscular stimulation (FNS) is most commonly used to describe only those FES technologies that are applied to the neuromuscular system in order to improve quality of life for people disabled by stroke, spinal cord injury, or other neurological conditions that result in impaired motor function (e.g., the abilities to move or breathe). Another technology closely related to FES is that of therapeutic electrical stimulation (TES), wherein electrical stimulation is applied to provide healing or recovery of tissues (e.g., muscle conditioning and strengthening, wound healing). As will be seen, some FES and FNS technologies concurrently provide or rely upon such therapeutic effects in order to successfully restore lost function. For illustrative purposes, much of this article is centered on FNS and related TES devices and technologies. For a wider exposure to additional FES approaches and neural prosthetic devices, the reader is referred to this article's *Reading List*, which contains references to a number of general books, journal articles, and on-line resources.

An important consideration in most all FNS technologies is that significant neural tissue remains intact and functional below the level of injury or disease so that electrical stimulation can be applied effectively. Individuals exhibiting hemiplegia (i.e., paralysis on one side of the body) due to stroke, for example, will exhibit paralysis in an impaired limb due to loss of control from the central nervous system (CNS), not because the peripheral nervous system (PNS) innervation of skeletal muscles in the limb has been lost. Similarly, while spinal cord injury (SCI) destroys motor neurons at the level of injury either partially or completely, many motor neurons below the level of injury may be spared and remain intact. Therefore, in stroke or SCI the axons of these intact motor neurons can be artificially excited by introducing an appropriate electrical field into the body using electrodes located on the skin surface, or implanted within the body. Artificial excitation of motor nerves by electrical excitation can generate action potentials (propagating excitation waves) along axons that, when they arrive at synaptic motor-endplate connections to skeletal muscle fibers, act to generate muscle force much as the intact nervous system would. Thus, lower extremity FNS systems often have the objective of restoring or improving mobility for stroke or SCI individuals. Upper extremity FNS systems often are designed to restore or augment reaching and grasping movements for SCI subjects. Both FNS and TES technologies are of course not a cure for stroke, spinal cord injury or diseases (e.g., cerebral palsy or multiple sclerosis where FNS also has been used). They are also not universally beneficial, and must be carefully matched by a clinician to an individual and their medical condition (1). On the other hand, as will be seen in the remainder of this article, FES and TES systems can provide greatly improved quality of life for many people who use them.

THEORY AND APPLICATION

In 1961, Liberson and co-workers proposed the usage of electrical stimulation in what was called functional electrotherapy to restore or augment movement capability that has been lost or compromised due to injury or disease (2). Specifically, Liberson's group developed the first electrical stimulation system for correction of hemiplegic drop foot: a gait disability occurring in some stroke survivors (for an excellent review of the history of development of neural orthoses for the correction of drop foot see Ref. 3). Moe and Post subsequently coined the term functional electrical stimulation to describe such techniques (4).

Electrical stimulation devices and systems now have been developed to activate paralyzed muscles in human subjects for a variety of applications in both the research lab and the clinic. Both FES and FNS systems have seen their greatest use as a tool for long-term rehabilitation of persons with neurological disorders (e.g., spinal cord injury, head injury, stroke) (5–10). For example, implanted electrical stimulation devices have been developed that can restore hand-grasp function to people with tetraplegia (11). Stimulation devices that utilize percutaneous electrodes (thin wires that cross the skin) have been developed to provide individuals with thoracic-level spinal cord injury with the ability to stand and step (12–14). Other devices that utilize electrodes placed on the surface of the skin can restore standing and locomotor function to individuals with spinal cord injury or other neuromuscular disorders (6,8,15,16). One system that uses surface electrodes (Parastep, Sigmedics Inc.) is FDA approved for use by people with thoracic level spinal cord injury and has been used at several rehabilitation centers worldwide. These efforts have clearly demonstrated that neuromuscular stimulation can be effectively used to activate paralyzed muscles for performing motor activities of daily living.

The basis by which all neuromuscular stimulation systems function is artificial electrical activation of muscle force, usually through excitation of the nerve fibers that innervate the skeletal muscle(s) of interest.

Excitation, Recruitment, and Rate Modulation

The nerve fibers that innervate skeletal muscle fibers are myelinated in nature, which means that they are regularly along their lengths ensheathed within layers of Schwann-cell derived myelin separating exposed axonal membrane at nodes of Ranvier. Myelination enables increased propagation velocities via saltatory conduction in such nerve fibers. The cell bodies of these alpha motor neurons lie within the ventral horn of the spinal cord. The efferent axons of these cells (~ 9–$20\,\mu$m in diameter) pass out from the spinal cord via the ventral roots and project then to muscle fibers within peripheral nerve trunks. When spared during damage or disease of the nervous system, alpha motor neurons and their axons usually form the substrate of electrical activation of skeletal muscle force in FNS applications. This may come as something of a surprise to the reader, in that skeletal muscle cells are themselves also excitable. Why then is indirect stimulation of the innervating nerve fiber generally the mechanism by which force is generated rather than direct stimulation of the muscle cells themselves? The reason is that large myelinated nerve fibers are usually excited at lower stimulus amplitudes (voltage or current) and with shorter stimulus pulse widths than are skeletal muscle cells (assuming similar spatial separations of electrodes to cells) (17). Electrical stimulation of myelinated nerves to threshold occurs when a critical extracellular potential distribution is created along or near the cell. At threshold, outward transmembrane currents are sufficient to depolarize the nerve cell membrane voltage to the level where an action potential is generated.

In normal physiology, there exist two natural control mechanisms to regulate the force a single muscle produces—recruitment and rate coding. Motor units are recruited naturally according to the Size Principle (18,19). Small alpha motor neurons innervating slow motor units have a low synaptic threshold for activation, and therefore are recruited first. As more force is demanded by an activity, progressively larger alpha motor neurons that innervate fast motor units are recruited. The second method of natural force regulation is called rate coding. Within a given motor unit there is a range of firing frequencies. Alpha motor neurons innervating fast-twitch motor units have firing rates that are higher than those that innervate slow-twitch units (20,21). Within that range, the force generated by a motor unit increases with increasing firing frequency. If an action potential reaches a muscle fiber before it has completely relaxed from a previous impulse, then force summation occurs. Twitches generated by the slow motor units have a fusion frequency of 5–10 Hz and reach a tetanic state at 25–30 Hz. The fast motor units may achieve fusion at 80–100 Hz (21,22).

The contractile properties of the muscle are largely dependent on the composition of the skeletal muscle (i.e., the muscle fiber types). The composition of muscle fibers varies across species. The composition of muscle fibers in the hindlimbs of the rat are predominantly fast fibers (23) whereas, human skeletal muscle is composed of a heterogenous collection of muscle fiber types (24). This is also indicated in the differences in fusion frequencies observed

Table 1. Skeletal Muscle Fiber Types and Their Characteristics

Fiber type	Skeletal Muscle Fiber Types and Characteristics		
	Type I	Type IIa	Type IIb
Other names	Slow red Slow oxidative (SO) Slow (S)	Fast red Fast oxidative (FOG) Fast resistant (FR)	Fast white Fast glycolytic (FG) Fast fatigable (FF)
Motor unit size	Smallest	Moderate	Largest
Firing order	1	2	3
Stimulation threshold	Lowest	Moderate	Highest
Force production	Lowest	Moderate	Highest
Resistance to fatigue	Highest	Moderate	Lowest
Contraction time	Slowest	Fast	Fastest
Mitochondrial density	High	High	Low
Capillary density	Highest	Moderate	Lowest

in the two species. The fusion frequency for muscles in the human is 25 Hz (25) and those for the muscles in the rat are higher ($\sim$ 75 Hz) (26). As summarized in Table 1, from various mammalian studies, skeletal muscle fibers have been grouped into many different types according to physiological, ultrastructural, and metabolic properties. Based on histochemical measurements of adenosinetriphosphatase (ATPase) reactivities, muscles were classified into type I, type IIA, and type IIB (27). A differentiation based on combination of physiological and metabolic properties categorized muscle fibers as SO-, FOG-, FG- (28). Based on twitch speed and fatigue resistance, muscle fiber types were identified as S, FR, and FF (29). There is also an intermediate type of fast muscle fiber in certain muscles denoted type IIAB or FI (Fast Intermediate resistance to fatigue). The different muscle fiber types vary in the amount of force generated, speed of contraction, and fatigability. The slow fiber types (SO, Type I, S) generate lower force, but for a prolonged duration. They are very fatigue resistant. The fast fiber types (FG, IIB, and FF) are on the other end of the spectrum with greater force generating capacity, but briefer intervals of time. Also, these fatigue very quickly compared to slow fibers. Therefore, there is a trade off between the ability to produce force quickly and powerfully or slowly and steadily. Though slow fibers are able to generate a steady force for long periods of time, their force output is less. Fast fibers on the other hand can generate quicker, greater forces, but they fatigue very fast. Some fibers are classified in between the two extremes of slow and fast and are termed intermediate fibers. These are fast fibers, but with fatigue resistant capability (FOG, IIA, FR, IIAB, FI). The properties of these intermediate fibers lie between those of slow fibers and fast fibers. The force generated by these fibers is less that those generated by fast fibers and greater than the force produced by slow fibers.

The heterogeneity of muscle fibers within the muscle is in part due to the hierarchy of motor unit recruitment order (the Size Principle, described above) (30) indicating the influence of motor neuron activity upon muscle fiber phenotypes. The fiber-type composition within a muscle can be altered by altering the excitation patterns delivered to the muscle (induced by various exercise regimes). The best documented effects of such transformations are those that

occur after chronic, low frequency stimulation (CLFS) of a predominantly fast muscle using implanted electrode systems. The fast skeletal muscles of a number of mammalian species have been shown to change to the slower phenotype in response to chronic electrical stimulation (31–39). The muscle phenotype can be manipulated to enhance fatigue resistance at the expense of contractile power and speed (40–45). Changes in metabolic activity, and muscle mass have been documented too (38,46). These transformations are also dose dependent. A continuous stimulation of rabbit fast muscle at 10 Hz completely transform the muscle fibers to the slow phenotype, but lower frequencies of stimulation produce an intermediate state of conversion. However, stimulation at 2.5 Hz for 12 weeks (47,48) or 10 months (49) results in a whole muscle consisting mainly of the fast phenotype.

CLFS has been shown to affect human muscle in a manner similar to that in animals (50–57). Electrical stimulation has shown to increase strength–force and build fatigue resistance in muscles in both healthy and SCI individuals (56,58–63). An increase in passive range of motion has also been observed (64). Electrical stimulation has been shown to prevent the shift and loss of fibers in patients with paralyzed muscles thereby increasing fatigue resistance (60,65–67). A well-defined progression of changes is observed, whereby the muscle changes first its metabolic and then its contractile properties to become slow muscle (68). This has been documented in different species and muscles suggesting that probably the effects observed are not species or muscle specific. Following transformation, the new slow fibers are indistinguishable from normal slow skeletal muscle fibers. Also, from time series studies (69) and single fiber biochemistry (70,71) it is clear that the changes that occur result from transformation at the level of the single fiber and not from fast-fiber degeneration with subsequent slow-fiber regeneration.

From the above sections, it is clear that skeletal muscle is very adaptive, and therefore provides an opportunity for conditioning and therapy after an injury. Electrical stimulation based exercise has gained much significance in toning and conditioning muscles. Even though electrical stimulation techniques are being used increasingly for rehabilitation and therapy, note that in general electrical stimulation systems generate activation patterns and

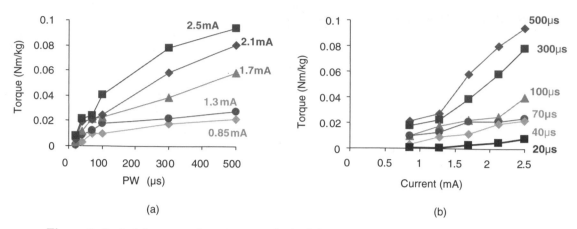

Figure 1. Typical force recruitment curves obtained from the ankle dorsiflexor muscle (Tibialis anterialis) of a rat through intramuscular stimulation. The recruitment curves indicate two techniques of force–torque modulation (a) pulse width modulation (PWM) and (b) pulse amplitude modulation (PAM). Single, symmetric, charge balanced, biphasic (cathodic first) pulses at an interval of 60 s were delivered. The currents were chosen as multiples of the twitch threshold current at 40 μs.

recruitment characteristics quite different from the normal physiological mechanisms. With electrical stimulation, physiological muscle force regulation is controlled either by spatial summation or by temporal summation (72). Spatial summation (or electrical recruitment) is achieved by increasing the pulse width (Fig. 1a) and/or the pulse amplitude (Fig. 1b) of the electrical stimulus—extending the excitatory extracellular potential distribution further out from the stimulating electrode(s) to greater numbers of nerve fibers, and/or longer in time. Force recruitment curves are in general quite nonlinear. The isometric recruitment curve (IRC) of a muscle can be defined as the static gain relation between stimulus level and output force/torque when the muscle is held at a fixed length. The features of a typical IRC are an initial dead-zone region, a high slope, monotonically increasing region, and a saturation region (73,74). These features can be explained by recognizing that the slope of the IRC is primarily a function of the electrode–nerve interface. The shape is dictated by the location and size distributions of the individual motor unit axons within the nerve with large diameter axons having a lower stimulus activation threshold than small diameter axons. The IRC depends on the past history of muscle activation and location of the electrode relative to the motor point. The motor point functionally is defined as the location (on the skin surface, or for implanted electrodes on the muscle overlying its innervation) where stimulation thresholds are lowest for the desired motor response. There is a drop in the maximum magnitude and slope of the monotonic region of the IRC on muscle fatigue (73,75). The IRC is also influenced by the muscle length tension curve (76) and, if muscle force is estimated by measuring joint torque, by the muscle nonlinear moment arm as it crosses the joint. Because of these factors, the IRC shape will be different for each muscle and set of experimental configurations and will also vary between subjects.

Temporal summation (also called rate modulation) varies the stimulus frequency or the rate of action potential firing on the nerve fiber(s). When electrodes are located

closer to the motor point for stimulation, enhanced spatial selectivity can be achieved because the electric field introduced can be focused closer to the α motor neuron fibers of interest. Another aspect of recruitment selectivity is fiber diameter, which relates to the tendency to stimulate subpopulations of nerve fibers based on their size. In electrical stimulation of myelinated fibers, there will be a tendency to recruit large axons at small stimulus magnitudes and then smaller axons with increased stimulus levels unlike during normal physiological recruitment—this is often dubbed reverse recruitment (77–79). Such reversed recruitment of motor units will inappropriately utilize fast, more readily fatigued muscle fibers for low force tasks. Slower fatigue resistant muscle fibers will only be recruited at higher stimulus levels. This also results in an undesirable steep relation between force output and stimulus magnitude. After injuries causing paralysis and disuse of muscle, many fatigue resistant muscle fibers tend to shift their metabolism toward less oxidative and more anaerobic, more readily fatigued mechanisms. Electrical stimulation therapy in such instances will recruit the faster muscle fibers first thereby inducing fatigue at a very early stage in the therapy.

FES DEVICES AND SYSTEMS

As illustrated in Fig. 2, all modern FES and FNS devices and systems incorporate (1) surface or implanted electrodes to generate an excitatory electric field within the body, (2) a regulated-current or regulated-voltage output stage that delivers stimulus pulses to the electrodes, (3) the stimulator pulse conditioning circuitry that creates the desired pulse shape, amplitude, timing, and pulse delivery (often within trains of pulses at set frequencies and for intended intervals), and (4) an open- or closed-loop stimulator controller unit. Systems may be completely or partially implanted and often incorporate a microcontroller or computer interface. Smith and colleagues at the Cleveland FES Center, for example, have developed an externally

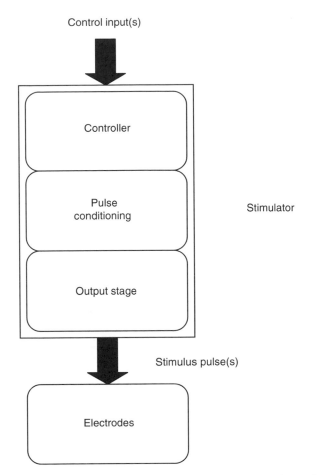

Control input(s)

Controller

Pulse
conditioning

Stimulator

Output stage

Stimulus pulse(s)

Electrodes

Figure 2. The FES systems typically incorporate control signals from the user that a Controller stage acts upon. Patterns of stimulation pulses are shaped with a pulse conditioning module that in turn feeds pulse information to an output stage that delivers regulated-current or regulated-voltage pulses of the desired amplitudes and timing to one or more channels of electrodes which are in contact with, or implanted within, the body.

powered, multichannel, implanted stimulator with telemetry for control of grasp and release functions in individuals with cervical level (C5 and C6) spinal cord injuries (80). Wu et al. designed a PC-based LabView controlled multichannel FES system with regulated-current or regulated-voltage arbitrary stimulation waveform pattern capability (81).

Commercialized FES systems include, for example, the Bioness, Inc. H200/Handmaster. This U.S. Food and Drug Administration (FDA) approved device incorporates microprocessor controlled surface stimulation into a portable, noninvasive hand–wrist orthosis for poststroke rehabilitation [see, e.g., (82)]. The FreeHand System, commercialized by NeuroControl Corporation in Cleveland, implements implanted receiver-stimulator, external controller, electrode, and sensor technologies (Fig. 3) developed through the Cleveland FES Center into a system for restoration of control of hand grasp and release for C5/C6 level spinal cord injured individuals. Compex Motion (Fig. 4), a programmable transcutaneous electrical stimulation product of Compex SA, is designed as a multipurpose FES system for incorporation into rehabilitation therapies (83). The Parastep System developed by Sigmedics, Inc. is designed

to enable independent, unbraced standing and walking for spinal cord injured people. Parastep is a noninvasive system that incorporates a battery-powered, microcomputer controlled stimulator unit (Fig. 5), surface electrodes, and a control and stability walker with finger activated control switches.

Electrode Designs for Electrical Stimulation

In the implementation of FES and FNS techniques, surface or implanted electrodes are used to create an excitatory electric field distribution within the targeted tissues. Researchers over the years have identified a number of important criteria for stimulation electrode selection and have developed a variety of electrode designs in order to meet specific application requirements (for an excellent recent review see Ref. 84).

Criteria for Electrode Selection. A few of the important factors identified for long-term applications are anatomical and surgical factors, mechanical and electrochemical characteristics, biocompatibility, long-term stability, and economics. Anatomical and surgical factors include ease of identification of stimulation site, either on the skin surface or through implantation. In the event of damage to the electrode, any implanted region should be easily accessible for retrieval and replacement. The mechanical properties of electrodes are important particularly with respect to implants whose lifetime is measured in years. Electrodes that are flexible, and consequently smaller in diameter, induce less trauma to muscles during movement. Instead of straight wires, coiled electrode wires provide for greater tension, and reduce the stress. The use of multistranded wires reduces breakage or provides redundancy if some wires should fail.

The electrical stability of the electrode is usually judged based upon reproducibility of muscle force recruitment curves. These depict some stimulation parameter (e.g., pulse width or current) against muscle force or torque output. As we have seen, the normal order of recruitment is generally reversed (larger motor units are activated before smaller ones). The threshold and the steepness of the curve are important properties that vary with electrode design, fiber size, and strength duration relations.

Another important criterion of consideration for choice of electrodes that are chronically implanted and tested over time is biocompatibility. The charge carriers in the electrode material (metal) are electrons unlike in our body wherein the charge carriers are ions. This results in a change of charge carriers when currents cross the metal–body interface. A capacitive double layer of charge arises at the metal–electrolyte interface; the single layer in the metal arises because of its connection to the battery, whereas that in the electrolyte is due to the attraction of ions in the electric field (85,86). These layers are separated by the molecular dimensions of the water molecule so the effective capacitance (being inversely proportional to charge separation) is quite high. At sufficiently low levels, the current will be primarily capacitive. But for high currents that exceed the capabilities of the capacitance channel, irreversible chemical reactions will take place

Functional electrical stimulation hand grasp system

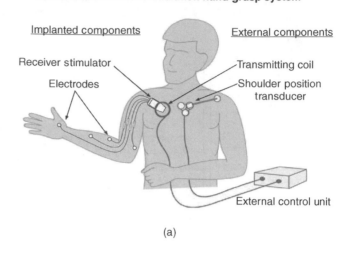

(a)

Figure 3. (a) Diagram of components for the implanted stimulation system developed at the Cleveland FES Center and commercialized as the Freehand neuroprosthesis by NeuroControl Corp. In the hand-grasp example shown, shoulder position is transduced for use as the command input. (b) The external control unit (ECU) provides the transducer interface, user control algorithm, multichannel stimulus coordination, and power for the implanted receiver-stimulator system. (c) The implanted receiver-stimulator provides multiple channels of stimulus output via the leads seen in the figure. It also transmits implantable sensor data to the ECU, and is powered through an inductive link that forms a coreless bidirectional transformer. Intramuscular or epimysial electrodes implanted in the forearm or hand are attached to the stimulator leads (not shown). (Courtesy of the Cleveland FES Center.)

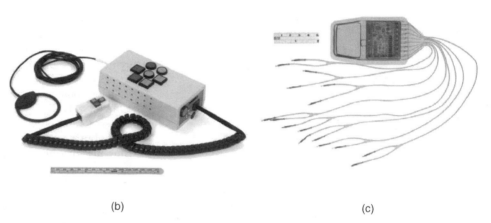

(b)   (c)

that are undesirable since they are detrimental to the tissue or electrode or both. Therefore, the electrode material must have little impact on the electrochemistry at the electrode–tissue interface. For biocompatibility and to avoid local tissue damage induced by high current levels, the electrode materials used are essentially inert (e.g., platinum, platinum–iridium, and 316LVM stainless steel).

The above mentioned criteria for electrode selection are a general guideline for either skin surface or chronically implanted electrode systems. However, the choice of electrode is also application dependent. For example, during stimulation of the brain, of particular concern is prevention of breakdown of the blood–brain barrier. For nerve stimulation circular (82) electrodes can be placed within an insulating cuff; consequently, smaller amounts of current are required because the field is greatly confined. Also, lower current tends to minimize unwanted excitation of surrounding tissue. Finally, intramuscular electrodes, because of the implant flexing that must be withstood, are usually of the coiled-wire variety discussed above.

Electrode Classification. In general, electrodes designed to deliver electrical pulses to excitable tissue are classified based on the site of stimulation or placement of electrodes. Motor nerves can be stimulated through electrodes

placed on the surface of the skin (surface electrodes) or implanted within the body. Implanted electrodes include those placed on or in the muscle (epimysial or intramuscular electrodes, respectively); as well as within or adjacent to a motor nerve (intraneural or extraneural electrodes). Electrodes that stimulate the spinal cord and BIONs (electrodes integrated with sensing and processing and packaged into a capsule) are recent additions to the family of implanted electrode technologies. The above classification of electrodes is further described below and summarized in Table 2.

Surface Electrodes. Surface electrodes as the name implies are placed on the surface of the skin and are the earliest of the electrodes to be used for applications in electrotherapy. These consist of conductive plates and are available in many types including conductive rubber patches coated with electrolyte gel, metal plates contacting the skin via thin, moist sponges and flexible, disposable, stainless steel mesh or rubber electrodes with self-adhesive conductive polymers (98–100). They do not need any implantation and are therefore noninvasive and relatively easy to apply and replace. An excellent description on the placements of these electrodes can be found in the Rancho Los Amigos Medical Center's practical guide to neuromuscular electrical stimulation (101). Surface electrodes

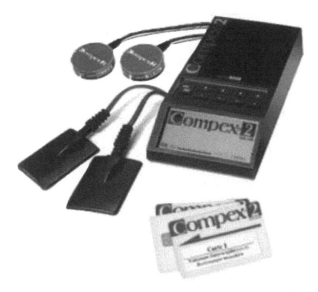

Figure 4. The Compex Motion FES system, manufactured by the Swiss based company Compex SA, is a general purpose programmable transcutaneous electrical stimulation device. Seen are the stimulator unit, three memory chip-cards that are inserted into the stimulator and used to store all pertinent information for a specific protocol, two EMG sensors, and two surface electrodes. (Reprinted from Ref. 83 with permission from the Institute of Physics and Engineering in Medicine.)

Figure 5. The neuromuscular stimulation unit for the Parastep system manufactured by Sigmedics, Inc. is battery-powered and microcomputer controlled. Cables connect the unit to surface electrodes, as well as to finger activated control switches on a walker. (Courtesy of Sigmedics, Inc.)

do have some disadvantages. They offer relatively poor selectivity for stimulation, have elevated threshold levels, may activate skin pain receptors, and do not have highly reproducible positioning capability. When higher currents are delivered to stimulate deeper muscles, spill over of charge to the nontargeted superficial muscles occurs. It is sometimes difficult to anchor surface electrodes in moving limbs and electrical properties at the skin–electrode interface can be variable.

Surface electrodes have been used for both lower limb and upper limb motor prosthesis, including the aforementioned Parastep system for ambulation (Fig. 6). WalkAid was designed for the management of foot drop to help toe clearance during the swing phase of walking (102). A single channel stimulator, the Odstock Dropped Foot Stimulator (ODFS) and later a two channel stimulator (O2CHS) designed for foot drop correction, used self-adhesive skin surface electrodes placed on the side of the leg (103,104). MikroFES was another orthotic stimulator for correction of foot drop in paralyzed patients (9). The Hybrid Assist System (HAS) (105) and the RGO system (106) use surface stimulation along with braces. Upper extremity applications include the Handmaster (107), the Belgrade Grasp System (BGS) (108), and the Bionic Glove (109) which focus on improving hand grasp.

Implanted Electrodes. Implanted electrodes can either be in direct contact with a muscle or peripheral nerve, within a muscle and only separated by muscle tissue from the motor nerves innervating the muscles, or within the spinal cord. Since peripheral electrodes are closer to the motor nerves than surface electrodes, they allow for better selectivity and more repeatable excitation. Their positioning and implantation is more permanent. Implanted electrodes have the advantage of place and forget by comparison to surface electrodes. That is, once the system is implanted, the user potentially can forget it is there. The chances of spill over are reduced since the electrodes can be placed close to the target muscle or nerve. The sensation to the user is usually much more comfortable as the implantation is away from the cutaneous pain receptors and the threshold current amplitude is lower. However, the implant procedure is invasive and in case of implant failure an invasive revision procedure can be required. Improper design and implantation can lead to tissue damage and infection. Insufficient tensile strength, high threshold levels, highly nonlinear recruitment curves, poor selectivity of activation and repeatability and adverse pain sensation (110–112) indicate failure. Excess encapsulation and infection (113); mechanical failures of electrode lead breakage and corrosion of electrodes and the insulator (114,115) can also impair the system.

Electrodes in or on the Muscle: Intramuscular and Epimysial Electrodes. Implanted electrodes that are placed on or in the muscle consist of intramuscular (87,88,116–121) and epimysial electrodes (89,122–125). Intramuscular electrodes (88,126) can, for example, be fabricated from multistranded Teflon coated stainless steel wires. This configuration provides good tensile strength and flexibility. They are implanted by injecting a hypodermic needle

Table 2. Electrical Stimulation Electrode Classifications and Types

Location/Type	Features and Advantages	Example	References
Surface	Metal plate with electrolyte gel, noninvasive	WalkAid, ODFS, MikroFES, HAS, RGO, Handmaster, BGS, Bionic Glove	
In/On Muscle	lower thresholds and better selectivity compared to surface electrodes		
Intramuscular	Implanted in the muscle, multistranded Teflon coated stainless steel wire, monopolar and bipolar configurations, good tensile strength, and flexibility		87,88
Epimysial	Implanted under the skin: on the muscle, monopolar and bipolar configurations, less prone to mechanical failure		89
BIONs	Injected into or near the muscle, hermetically sealed glass/ceramic capsule integrated with electronics		90
Near/On Nerve	Lower threshold levels and better selectivity than the above mentioned electrodes		
Nerve Cuffs	Monopolar, bipolar and tripolar configurations, good power efficiency, improved selectivity, comparatively stable		91,92
FINE	Reshape or maintain nerve geometry		93
Intrafascicular	Penetrate the epineurium and into the fascicle, selective stimulation, lower current and charge levels		
LIFE	Stable, suitable for stimulating and recording		94
SPINE	Reduced nerve damage		95
Intraspinal			
Microwires	Near to normal recruitment, reduced fatigue, highly selective stimulation		96,97

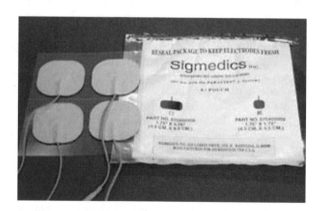

Figure 6. Examples of self-adhesive, reusable surface electrodes. The electrodes shown are used in the Parastep neuromuscular stimulation system. (Courtesy of Sigmedics, Inc.)

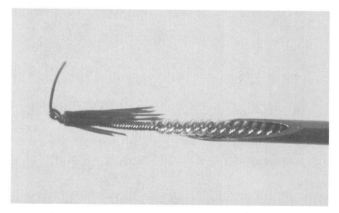

Figure 7. A "Peterson" type intramuscular electrode design. This is a helically wound PFS insulated multistranded 316LVM stainless steel wire design that is attached to a barb-like anchoring structure constructed of polypropylene suture material. The wound section of the electrode is $\sim 800\,\mu m$ in diameter and is partially loaded into a hypodermic needle. (Courtesy of J.T. Mortimer and reproduced by permission of World Scientific Publishing Co.)

either nonsurgically or through an open incision. A fine needle probe used by itself or in conjunction with a surface probe is used to detect the motor point; the motor point for an intramuscular electrode is usually just below the muscle surface beneath the motor point position as defined by surface electrode. These electrodes can elicit a maximal muscular contraction with only $\sim 10\%$ of the stimulus charge required by equivalent surface electrodes (25). Figure 7 depicts a Peterson type intramuscular electrode developed at Case Western Reserve University (121).

Both monopolar and bipolar intramuscular electrodes have been used. Bipolar intramuscular electrodes that straddle the nerve entry point can be as effective at activating the muscles as a nerve cuff. If bipolar electrodes do not straddle the nerve entry point, full recruitment of the muscle can require large stimulation charge and stimulation cannot be achieved without activating the surrounding muscles. In contrast, monopolar stimulation is less position dependent, though it cannot match the selectivity obtained with good bipolar placement (127). The size of the muscle will determine the limit of electrode size, although large electrodes are more efficacious.

A recent development in the intramuscular stimulating electrode world are BIONs (for BIOnic Neurons), that can potentially provide precise and inexpensive interfaces between electronic controllers and muscles (90). The BIONs consist of a hermetically sealed glass–ceramic capsule with integral capacitor electrodes for safety and reliability (128). The internal electronics include an antenna coil wrapped around a sandwich of hemicylindrical ferrites over a ceramic microprinted circuit board carrying a custom integrated circuit chip. In animal studies, these electrodes have demonstrated long-term biocompatibility (129) and ability to achieve selective muscle stimulation (130). The first generation of BIONs, BION1, generates stimulation pulses of 0.2–30 mA at 4–512 µs duration. This system is now in clinical trials to provide therapeutic electrical stimulation to patients with disabilities (131–135). The second generation BION, BION2, is under development. BION2s are expected to sense muscle length, limb acceleration and bioelectrical potentials for feedback control in FES (136–138).

Intramuscular electrodes have been used to activate paralyzed muscles that retain a functional motor neuron in the muscles of the upper extremity (139,140), lower extremity (118,140,141) and the diaphragm (142). Muscles also have been stimulated to correct spinal deformities in the treatment of scoliosis (143).

Epimysial electrodes (89,110) are positioned on the surface of a muscle below the skin but not within the muscle. They have a smooth circular disk on one side and a flat, insulating backing, reinforced with mesh. The motor point is usually identified by moving a stimulating electrode across the muscle surface to locate the surface position that requires the least amplitude to fully excite the muscle. Replacing this electrode in the event of failure is comparatively easier. The stimulation levels and impedance are also similar to that of intramuscular electrodes. A perceived advantage of epimysial electrodes over intramuscular electrodes is that they are less prone to mechanical failure and less likely to move in the hours and days immediately after implantation.

Epimysial electrodes also can be used either in the monopolar mode or the bipolar mode (89,108,119,120,123). Use of a monopolar epimysial electrode close to the motor nerves results in reduced threshold stimulus amplitude, higher gain and selectivity, and decrease in length dependent recruitment. When a bipolar epimysial electrode is used, the stimulus current is constrained to regions closer to the two electrodes. Compared to the results with monopolar electrodes, the threshold is increased, relative gain decreased, and though greater selectivity is found with stimulation current levels close to twitch threshold poorer selectivity is present in the stimulus range needed for maximum activation of the muscle (108).

Epimysial electrodes have been used for a number of years in the implementation of upper extremity assist devices for C5 or C6 adult subjects with tetraplegia (Fig. 8), including incorporation into the FDA approved FreeHand System (144) and more recently for providing the capability of standing after paraplegia (117).

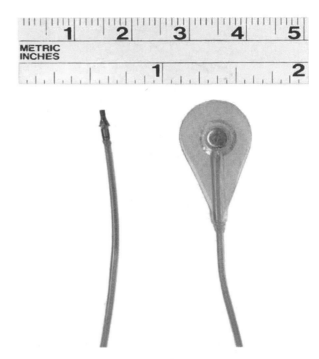

Figure 8. An example implantable epimysial electrode (right) with intramuscular electrode (left), typical of those used with the Cleveland FES Center's implanted hand-grasp system. (Courtesy of the Cleveland FES Center.)

Implanted Nerve Electrodes. Electrodes that are placed in contact with the nerve include extraneural and intraneural electrodes. Extraneural electrodes do not penetrate the epineurium and include varying designs of nerve cuffs (91,92,145–149) and the recently investigated flat interface nerve electrodes (FINE) (93,150,152). Intraneural electrodes penetrate the epineurium and include intrafascicular and interfascicular electrodes (94,95,153–157). Nerve electrodes have several potential advantages over intramuscular electrodes—including, lower power requirements, the ability to control several muscles with a single implant, and the ability to place the electrodes far from contracting muscles (158).

Electrodes placed on the surface of the nerve, and housed in an insulative carrier that encompasses the nerve trunk, are cuff electrodes (91,151,159,160). The cuff material is often silicone rubber and sometimes reinforced with Dacron. Cuff-type electrodes hold the stimulating contacts in close proximity to the nerve trunk. Holding the target tissues close to the stimulating contacts offers opportunities for power efficiency and improved selectivity. Less power is spent on electrical conduction through space between the electrode and target tissues. Improved selectivity is possible because the electric potential gradient is larger when the spacing between the stimulating contact and the target tissue is least. Further, these electrodes are less likely to move in relationship to the target tissues after implantation (161–164). However, while nerve cuffs stimulate effectively and selectively they require invasive surgery for implantation. They may also damage the nerves they enclose unless carefully designed, sized, and implanted.

Figure 9. A self-sizing cuff electrode design fabricated using PMP (polymer–metal–polymer) technology and laser machining. (Courtesy of J.T. Mortimer and M. Tarler.)

0.8mm

Figure 10. The FINE nerve cuff design, intended to flatten peripheral nerve trunks into a layering of nerve fascicles. Electrode contacts are seen as small dots within the overall structure. (Courtesy of D. Durand.)

To overcome potential problems with a fixed cuff-size, nerve cuff electrodes have been designed with different configuration. The Huntington nerve cuff (165), is a helix-type nerve electrode system that has exposed metal sections as stimulating contacts along the internal diameter of the helix. The open helix design can accommodate some swelling. Other self-sizing cuff electrode designs sometimes have a spiral configuration that enables opening or closing to accommodate a range of different diameter nerves (91). Figure 9, for example, is a photo of a self-sizing nerve cuff fabricated at Case Western Reserve University using PMP technology and laser machining. Both cuff and spiral electrode configurations can be used in various monopolar, bipolar or tripolar configurations (91,164). Cuff electrodes with multiple electrical contacts can produce selective activation of two antagonistic muscle groups innervated by that nerve trunk (166). Increased function and additional control of muscles with minimum number of electrodes can be achieved. Self-sizing nerve-cuff electrodes, with multiple contacts in a tripolar configuration, have been shown to produce controlled and selective recruitment of some motor nerves in a nerve trunk (145,158,167–170). A monopolar electrode with four radially placed contacts can work as well as a tripolar electrode with four radially placed tripoles (171,172). A four contact self-sizing spiral cuff electrode has been described as a tunable electrode that is capable of steering the excitation from an undesirable location to a preferred location (92).

The flat interface nerve electrode, or FINE system as seen in Fig. 10, has been introduced in an attempt to improve the stimulation selectivity of extraneural electrodes (151). The goal with the FINE is to create a geometry that optimizes stimulation selectivity. In contrast to cylindrical electrodes, the FINE either reshapes the nerve into, or maintains the nerve in, an ovoid geometry. Chronic studies in rats have demonstrated that nerves and fascicles can be safely reshaped (150,173). Also, acute experiments and finite element models have demonstrated that it is possible to selectively activate individual fascicles in the cat sciatic nerve using this electrode (151,152,174). This could be important in both reducing fatigue and selectively activating individual muscles (153,175). A potential disadvantage

is that a fibrous capsule with electrical properties different from the surrounding tissues will envelope the electrode (176,177), potentially rendering the recruitment properties unstable, although a recent study has shown that both selectivity measurements and the recruitment curve characteristics can remain stable for a prolonged implant period (93).

Intraneural electrodes are positioned to penetrate the epineurium around the nerve trunks. Intraneural electrodes utilize a conductor that invades the epineurium. Maximal contraction is elicited at stimulation levels an order of magnitude lower than with nerve cuff electrodes (200 μA, pulse duration 300 μs). However, connectors, fixation, and neural damage are still not completely resolved to allow routine clinical usage. Intraneural multipolar sword type electrodes have been made out of solid silicon with golden contacts and can be very selective (178). Such electrodes could minimize the needs for using many electrodes for activation of different muscles that are innervated from a single nerve (179).

A subset of intraneural electrodes are meant to enter the perineurium around the fascicles and go between the nerve fibers: These are so-called intrafascicular electrodes. Intrafascicular electrodes place stimulating elements inside the fascicles, in close proximity to axons (126,153,160,175,178,180,181). They have been shown to produce axonal recruitment with almost no excitation of muscles that are not targeted (181). A variation of the intrafascicular electrode is the longitudinal intrafascicular electrode (LIFE) (94,153). Compared with extraneural electrodes, LIFEs have many advantages and can be implanted into any of the fascicles of peripheral nerves to selectively stimulate a single fascicle thereby offering highly selective stimulation. Also they serve as excellent recording electrodes. When LIFEs are used as recording electrodes, the amplitudes of motor evoked potentials (MEPs) recorded by LIFEs implanted in fascicles are much larger than those of EMGs recorded from the skin by surface electrodes and the signals recorded are not affected by external electrical fields (155,182). Therefore, the signals recorded by LIFE can be used to control a prosthetic limb more accurately than those controlled by EMGs (183). In addition, LIFEs have excellent biocompatibility with peripheral fascicles (156,184,185).

While intrafascicular electrodes can provide high degrees of selectivity, it remains unclear whether penetrating the perineurium will lead to long-term nerve injury (126,186). Interestingly, an intraneural electrode system dubbed the slowly penetrating interfascicular electrode (SPINE) has been developed, which has been reported to penetrate a peripheral nerve within 24 h without evidence of edema or damage of the perineurium and showed functional selectivity (95).

In general, compared to externally placed electrodes, the current and charge stimulation requirements for intraneural electrodes are low since they are positioned inside the nerve trunk to be excited. Also, the stimulation selectivity is high compared to extraneural electrodes where stimulation selectivity suffers from the relatively large amount of tissue interposed between the stimulating contacts and the target axons.

Micro wires: Electrodes for Intraspinal Stimulation
Spinal circuits that are shown to have the capacity of generating complex behaviors with coordinated muscle activity can be activated by intraspinal electrical stimulation (187–190). Microwires that are finer than a human hair have been used to stimulate the spinal cord neurons to control single muscles or small group of synergists (96,97,191–193). Stimulation through single wires in a few sites has been shown to have the ability to elicit whole-limb activation sufficient to support the animal's weight (191,192,194–196). The stimuli were not perceived but were able to produce strong coordinated movements. Near normal recruitment order, minimal changes in kinematics and little fatigue and functional, synergistic movements induced by stimulation in the lumbosacral cord (97,194,196) are some of the promising advantages of stimulating the spinal cord with microwires. However, the clinical and long-term feasibility of implanting many fine microwires into the spinal cord remains questionable. In addition, stimulating the spinal cord results in steep recruitment curves compared to muscle and nerve stimulation thereby limiting the degree of control achievable.

Controllers and Control Strategies

Besides stimulating the paralyzed muscles, it is also important to control and regulate the artificial movements produced. The control task refers to specification of the temporal patterns of muscle stimulation to produce the desired movements; and the regulation task is the modification of these patterns during use to correct for unanticipated changes (disturbances) in the stimulated muscles or in the environment. A major impediment to the development of satisfactory control systems for functional neuromuscular stimulation has been the nonlinear, time varying properties of electrically activated skeletal muscle that make control difficult to achieve (7,76,197). With FNS, the larger, fatigable muscle fibers are recruited at low levels of stimulation before the more fatigue-resistant fibers are activated thereby inducing rapid fatigue (56). It is important that the output of any FNS control system results in stable, repeatable, regulated muscle input–output properties over a wide range of conditions of muscle length, electrode movement, potentiation, and fatigue. To improve control strategies to provide near physiological control, inherent muscle characteristics (force-activation, force-length, and force-velocity), muscle modeling studies, studies on understanding how to model the patterns of neural prostheses and how neural prostheses respond to disturbances have been performed (197–200).

As depicted in Fig. 11 (201), FNS control methods include feedforward (open-loop), feedback, and adaptive control. Feedforward control requires a great deal of information about the biomechanical behavior of the limb. The control algorithms specify the stimulus parameters (musculoskeletal system inputs) that are expected to be needed to produce the desired movement (system outputs). In an open-loop control system these parameters are often identified by trial and error (6,13,202–205). The same stimulation pattern, which is often stored in the form of a lookup table, is delivered for each cycle of movement.

Three major problems exist with this form of fixed-parameters, open-loop control (204–206). First, the process of specifying the parameters for a single stimulation pattern for a single user often requires several extensive sessions involving the user, therapist, physician, and engineer. This process is often expensive, time consuming, and often only minimally successful in achieving adequate performance. Second, the fixed parameter stimulation pattern may not be suitable after muscles fatigue that is exacerbated by the stimulation paradigm itself. The third problem is that the open-loop stimulation pattern does not respond to changing environments (e.g., slope of walking surface) and external perturbations (e.g., muscle spasms).

To address the limitations of open-loop control systems feedback control was implemented (12,14,207,208). In a feedback control system, sensors monitor the output and corrections are made if the output does not behave as

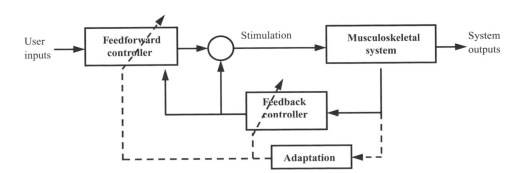

Figure 11. A representation of FNS control system components and strategies (feedforward, feedback and adaptive). (Reproduced by permission from Neuromodulation 2001;4: 187–195.)

desired. The corrections are made based on a control law, which is a mathematical prescription for how to change the input to reduce the difference (error) between the desired output and the actual output. Feedback control requires output sensors, and compensation is generally slower than in feedforward control since an output error must be present to generate a controller response. Thus feedback control might best be used for slow movements and for maintaining a steady posture. Since the output of the feedback controller is highly dependent on sensor signals, the quality of the control that is achieved will be compromised by the relatively low quality of sensors that are available. Feedback control has been successful in regulating hand grasp (209) and standing posture (12), but it appears that another strategy, adaptive feedforward control, is likely to be required for dynamic activities such as locomotion.

To improve performance of feedback control systems, adaptive control strategies were developed that automatically adjusted the overall system behavior (i.e., the combined response of the controller and the system) so that it is more linear, repeatable, and therefore predictable (75,210–213). These techniques adjust the parameters of the control system and attempt to self-fit the system to the user in order to make it easier to use and learn to use (206,212,214). The control system developed by Abbas and Chizeck has a pattern generator (PG) and a pattern shaper (PS) (211,215). The PG generates the basic rhythm for controlling a given movement. The PS adaptively filters those signals and sends its output to the muscles. The adaptive properties of the PS provide the control system with the ability to customize stimulation parameters for a particular individual and to adjust them on-line to account for fatigue. In some of the computer simulation experiments a proportional-derivative feedback controller was also active. Studies have shown that the pattern generator/pattern shaper (PG/PS) adaptive neural network controller is able to account for nonlinear and dynamic system properties and muscle fatigue (73,75,213). To summarize, adaptive control systems have replaced other developed control system strategies because this strategy can (1) provide the ability to automatically customize the stimulation pattern for a given user, (2) automatically adjust stimulation parameters to account for fatigue, and (3) automatically adjust to allow the voluntary motor commands to recover control of the movement pattern (in the case of partial recovery in a person with an incomplete spinal cord lesion).

Apart from the above other strategies, such as fuzzy logic (216) and proportional–integral–derivative (PID) controllers (217) have also been implemented to investigate automatic fatigue compensation. However, fatigue remains one of the major factors limiting utility of FES/FNS because such adaptive systems can adjust for fatigue only up to the contractile limits of the muscle.

Rather than initiating and modulating control of FES systems indirectly through residual motor function (e.g., as in the Freehand system for grasping, where paralyzed hand closure and opening were command controlled through sensing of opposite shoulder position), future FES devices might be controlled directly through thought—by tapping into the subject's remaining cortical intent to move via a brain–machine interface (BMI) [or sometimes brain–computer interface (BCI)]. So-called direct brain–machine interfaces utilize arrays of intracortical recording electrodes to sense action potentials from a host of individual neurons in regions of the brain where cells code for movement and its intent. A number of research teams have in recent years demonstrated the feasibility of recording and processing movement related signals from cortex (in both animals and in humans), and then enabling the subject to control computers or devices directly through such processed thought (218–220). Ultimately, BMI technologies hold promise that paralyzed individuals might one day be able to control FES devices for movement restoration with little or no effort or learning other than forming the simple intent to move (221).

THERAPEUTIC EFFECTS OF ELECTRICAL STIMULATION

While this article is focused mainly on electrical stimulation therapies for restoring lost function, it is important to recognize that electrical stimulation techniques are used also for therapeutic reasons. A recent review summarizes the current state of therapeutic and neuroprosthetic applications of electrical stimulation after spinal cord injury and identifies some future directions of research and clinical and commercial development (222). Functional electrical stimulation therapy individually and in combination with other rehabilitation therapies also is being utilized after incomplete spinal cord injury to influence the plasticity within the nervous system for improved recovery (9,223–228).

Therapeutic electric stimulation (TES) can affect the restoration of muscle strength (229). Therapeutic electric stimulation in humans has been shown to prevent muscle atrophy thereby increasing muscle cross-sectional area, torque, and force (230–234). Such electrical therapy has been effective in reversing the increased fatigability associated with the change in fiber type in both animals (31–37) and humans (56,59–61,65–67) after spinal cord injury. Electrical stimulation has also been able to reduce spasticity among patients with neurological disorders (reference).

While osteoporosis has been prevented in the limbs of paralyzed individuals, in menopausal women, and in the elderly and fracture patients through electrical stimulation therapy (235–240), certain other studies have shown little or no change in bone density (235,241–244). These contradictory results suggest the importance of other characteristics, such as the stimulation patterns, specifications for training (intensity, duration, loading), and the time postinjury. Enhancing fracture–wound healing is another therapeutic application of electrical stimulation (245–249). The theory here is to attract negatively or positively charged cells into the wound area, such as neutrophils, macrophages, epidermal cells, and fibroblasts that in turn will contribute to wound healing processes by way of their individual cellular activities (250). Electrical stimulation may also play a role in wound healing through improved blood flow (251,252), prevent occurrence of pressure sores thereby improving general tissue health (253). A recent review details all the theories suggested and experimental studies and clinical trials performed on wound healing through electrical stimulation (254).

Recent applications of electrical stimulation have also been successful in altering neural function. For example, deep brain stimulation (DBS) is being used to treat a variety of disabling neurological symptoms, most commonly the debilitating symptoms of Parkinson's disease (PD), such as tremor, rigidity, stiffness, slowed movement, and walking problems [for a review, see (255,256)]. Deep brain stimulation uses a surgically implanted, neurostimulator approximately the size of a stopwatch. The implanted device delivers electrical stimulation to targeted areas in the brain that control movement, blocking the abnormal nerve signals that cause tremor and PD symptoms. Vagal nerve stimulator (VNS), approved by the FDA in 1997 are used to treat patients with intractable epilepsy. These devices controls seizures by sending electrical pulses to the vagus nerve (257,258). Transcutaneous electrical nerve stimulation (TENS), wherein electrical signals are sent to underlying nerves, can relieve a wide range of chronic and acute pain (259). The TENS devices are small battery-powered stimulators that produce low intensity electrical signals through electrodes on or near a painful area, producing a tingling sensation that reduces pain. Chronic electrical stimulation of the GI tract has been found to be a potential therapy for the treatment of obesity (260–262). It is clear that in future development of electrical stimulation technologies many devices will be designed to achieve both therapeutic and functional outcomes.

ACKNOWLEDGMENT

This work was in part supported by NIH (NCMRR)–HD-40335.

BIBLIOGRAPHY

Cited References

1. Kilgore KL, Kirsch RF. Upper and lower extremity motor neuroprostheses. In: Horch KW, Dhillon GS, editors. Neuroprosthetics: Theory and Practice, New Jersey: World Scientific; 2003. pp 844–877.
2. Liberson WT, Holmquest HJ, Scot D, Dow M. Functional electrotherapy: stimulation of the peroneal nerve synchronized with the swing phase of the gait of hemiplegic patients. Arch Phys Med Rehabil 1961;42:101–105.
3. Lyons GM, Sinkjaer T, Burridge JH, Wilcox DJ. A review of portable FES-based neural orthoses for the correction of drop foot. IEEE Trans Neural Syst Rehabil Eng 2002;10(4):260–279.
4. Moe JH, Post HW. Functional electrical stimulation for ambulation in hemiplegia. J Lancet 1962;82:285–288.
5. Bajd T, Andrews BJ, Kralj A, Katakis J. Restoration of walking in patients with incomplete spinal cord injuries by use of surface electrical stimulation-preliminary results. Prosthet Orthot Int 1985;9(2):109–111.
6. Krajl A, Bajd T. Functional Electrical Stimulation: Standing and Walking After Spinal Cord Injury. Boca Raton (FL): CRC Press; 1989.
7. Yarkony GM, Roth EJ, Cybulski G, Jaeger RJ. Neuromuscular stimulation in spinal cord injury: I: Restoration of functional movement of the extremities. Arch Phys Med Rehabil 1992;73(1):78–86.
8. Stein RB, et al. Electrical systems for improving locomotion after incomplete spinal cord injury: an assessment. Arch Phys Med Rehabil 1993;74(9):954–959.
9. Bajd T, Kralj A, Stefancic M, Lavrac N. Use of functional electrical stimulation in the lower extremities of incomplete spinal cord injured patients. Artif Organs 1999;23(5):403–409.
10. Stein RB. Functional electrical stimulation after spinal cord injury. J Neurotrauma 1999;16(8):713–717.
11. Peckham PH, Keith MW. Motor prostheses for restoration of upper extremity function., in Neural prostheses: Replacing motor function after disease or disability. New York: Oxford University Press; 1992. pp 162–190.
12. Chizeck HJ, et al. Control of functional neuromscular stimulation systems for standing and locomotion in paraplegics. Proc IEEE 1988;1155–1165.
13. Marsolais EB, Kobetic R. Development of a practical electrical stimulation system for restoring gait in the paralyzed patient. Clin Orthop 1988;233:64–74.
14. Abbas JJ, Chizeck HJ. Feedback control of coronal plane hip angle in paraplegic subjects using functional neuromuscular stimulation. IEEE Trans Biomed Eng 1991;38(7):687–698.
15. Solomonow M. Biomechanics and physiology of a practical functional neuromuscular stimulation walking orthosis for paraplegics. In: Stein RB, Popovic DP, editors. Neural Prostheses: Replacing motor function after disease or disability. New York: Oxford University Press; pp 202–232.
16. Graupe D, Kohn KH. Functional electrical stimulation for ambulation by paraplegics, in Functional electrical stimulation for ambulation by paraplegics. Krieger; 1994. p 194.
17. Mortimer JT. Motor Prostheses. In: Brookhart JM, Mountcastle VB, Brooks VB, Geiger SR, editors. Handbook of Physiology, Section 1: The Nervous System, Vol. II Motor Control, Part I. Bethesda (MD): American Physiological Society; 1981.
18. Henneman E, Somjen G, Carpenter DO. Functional Significance of Cell Size in Spinal Motoneurons. J Neurophysiol 1965;28:560–580.
19. Henneman E, Somjen G, Carpenter DO. Excitability and inhibitability of motoneurons of different sizes. J Neurophysiol 1965;28(3):599–620.
20. Burke RE. Firing patterns of gastrocnemius motor units in the decerebrate cat. J Physiol 1968;196(3):631–654.
21. Burke RE. Motor units: Anatomy, physiology and functional organization. In: Brooks VB, editor. Handbook of Physiology Section 1: The Nervous System. Vol. III. Motor Systems. Bethesda (MD): American Physiology Society; 1981. pp 345–422.
22. McPhedran AM, Wuerker RB, Henneman E. Properties of Motor Units in a Heterogeneous Pale Muscle. J Neurophysiol 1965;28:85–99.
23. Armstrong RB, Phelps RO. Muscle Fiber Type Composition of the Rat Hindlimb. Am J Anat 1984;171:256–272.
24. Staron RS. Human skeletal muscle fiber types: delineation, development, and distribution. Can J Appl Physiol 1997;22(4):307–327.
25. Popovic D, Sinkjaer T. Control of Movement for the Physically Disabled. London: Springer-Verlag; 2003.
26. Ichihara K, et al. Muscle stimulation in a rodent model: electrode design, implantation and assessment. 9th Annual Conference of the International FES Society. Bournemouth (UK): 2004.

27. Brooke MH, Kaiser KK. Muscle fiber types:How many and what kind? Arch Neurol 1970;23:369–379.

28. Peter JB, et al. Metabolic profiles of three fiber types of skeletal muscle in guinea pigs and rabbits. Biochemistry 1972;11:2627–2633.

29. Burke RE, Levine DN, Tsairis P, Zajac FE. Physiological types of histochemical profiles in motor units of the cat gastrocnemius. J Physiol 1973;234:723–748.

30. Pette D, Staron RS. Cellular and molecular diversities of mammalian skeletal muscle fibers. Rev Physiol Biochem Pharmacol 1990;116:1–76.

31. Brown WE, Salmons S, Whalen RG. The sequential replacement of myosin subunit isoforms during muscle type transformation induced by long term electrical stimulation. J Biol Chem 1983;258(23):14686–14692.

32. Brownson C, et al. Changes in skeletal muscle gene transcription induced by chronic stimulation. Muscle Nerve 1988; 11(11):1183–1189.

33. Brownson C, Little P, Jarvis JC, Salmons S. Reciprocal changes in myosin isoform mRNAs of rabbit skeletal muscle in response to the initiation and cessation of chronic electrical stimulation. Muscle Nerve 1992;15(6): 694–700.

34. Carraro U. Contractile proteins of fatigue-resistant muscle. Semin Thorac Cardiovasc Surg 1991;3(2):111–115.

35. Kirschbaum BJ, Heilig A, Hartner KT, Pette D. Electrostimulation-induced fast-to-slow transitions of myosin light and heavy chains in rabbit fast-twitch muscle at the mRNA level. FEBS Lett 1989;243(2):123–126.

36. Pette D, Muller W, Leisner E, Vrbova G. Time dependent effects on contractile properties, fibre population, myosin light chains and enzymes of energy metabolism in intermittently and continuously stimulated fast twitch muscles of the rabbit. Pflugers Arch 1976;364(2):103–112.

37. Sreter FA, Gergely J, Salmons S, Romanul F. Synthesis by fast muscle of myosin light chains characteristic of slow muscle in response to long-term stimulation. Nat New Biol 1973;241(105): 17–19.

38. Pette D, et al. Partial fast-to-slow conversion of regenerating rat fast-twitch muscle by chronic low-frequency stimulation. J Muscle Res Cell Motil 2002;23(3):215–221.

39. Putman CT, et al. Fiber-type transitions and satellite cell activation in low-frequency-stimulated muscles of young and aging rats. J Gerontol A Biol Sci Med Sci 2001;56(12):B510–B519.

40. Jarvis JC. Power production and working capacity of rabbit tibialis anterior muscles after chronic electrical stimulation at 10 Hz. J Physiol 1993;470:157–169.

41. Mannion JD, et al. Histochemical and fatigue characteristics of conditioned canine latissimus dorsi muscle. Circ Res 1986;58(2):298–304.

42. Trumble DR, LaFramboise WA, Duan C, Magovern JA. Functional properties of conditioned skeletal muscle: implications for muscle-powered cardiac assist. Am J Physiol 1997;273(2 Pt. 1):C588–C597.

43. Salmons S, Vrbova G. The influence of activity on some contractile characteristics of mammalian fast and slow muscles. J Physiol 1969;201(3):535–549.

44. al-Amood WS, Buller AJ, Pope R. Long-term stimulation of cat fast-twitch skeletal muscle. Nature (London) 1973; 244(5413): 225–257.

45. Glatz JF, et al. Differences in metabolic response of dog and goat latissimus dorsi muscle to chronic stimulation. J Appl Physiol 1992;73(3):806–811.

46. Ferguson AS, et al. Muscle plasticity: comparison of a 30-Hz burst with 10-Hz continuous stimulation. J Appl Physiol 1989;66(3):1143–1151.

47. Jarvis JC, et al. Fast-to-slow transformation in stimulated rat muscle. Muscle Nerve 1996;19(11):1469–1475.

48. Mayne CN, et al. Induction of a fast-oxidative phenotype by chronic muscle stimulation: histochemical and metabolic studies. Am J Physiol 1996;270(1 Pt 1):C313–C320.

49. Sutherland H, et al. The dose-related response of rabbit fast muscle to long-term low-frequency stimulation. Muscle Nerve 1998;21(12):1632–1646.

50. Andersen JL, et al. Myosin heavy chain isoform transformation in single fibres from m. vastus lateralis in spinal cord injured individuals: effects of long-term functional electrical stimulation (FES). Pflugers Arch 1996;431(4):513–518.

51. Theriault R, Theriault G, Simoneau JA. Human skeletal muscle adaptation in response to chronic low-frequency electrical stimulation. J Appl Physiol 1994;77(4):1885–1889.

52. Gordon T, Pattullo MC. Plasticity of muscle fiber and motor unit types. Exerc Sport Sci Rev 1993;21:331–362.

53. Lenman AJ, et al. Muscle fatigue in some neurological disorders. Muscle Nerve 1989;12(11):938–942.

54. Rutherford OM, Jones DA. Contractile properties and fatigability of the human adductor pollicis and first dorsal interosseus: a comparison of the effects of two chronic stimulation patterns. J Neurol Sci 1988;85(3):319–331.

55. Scott OM, Vrbova G, Hyde SA, Dubowitz V. Effects of chronic low frequency electrical stimulation on normal human tibialis anterior muscle. J Neurol Neurosurg Psychiat 1985; 48(8): 774–781.

56. Stein RB, et al. Optimal stimulation of paralyzed muscle after human spinal cord injury. J Appl Physiol 1992;72(4): 1393–1400.

57. Theriault R, Boulay MR, Theriault G, Simoneau JA. Electrical stimulation-induced changes in performance and fiber type proportion of human knee extensor muscles. Eur J Appl Physiol Occup Physiol 1996;74(4):311–317.

58. Currier DP, Mann R. Muscular strength development by electrical stimulation in healthy individuals. Phys Ther 1983;63(6):915–921.

59. Hartkopp A, et al. Effect of training on contractile and metabolic properties of wrist extensors in spinal cord-injured individuals. Muscle Nerve 2003;27(1):72–80.

60. Mohr T, et al. Long-term adaptation to electrically induced cycle training in severe spinal cord injured individuals. Spinal Cord 1997;35(1):1–16.

61. Gerrits HL, et al. Variability in fibre properties in paralysed human quadriceps muscles and effects of training. Pflugers Arch 2003;445(6):734–740.

62. Ragnarsson KT, et al. Clinical evaluation of computerized functional electrical stimulation after spinal cord injury: a multicenter pilot study. Arch Phys Med Rehabil 1988;69(9): 672–677.

63. Sloan KE, et al. Musculoskeletal effects of an electrical stimulation induced cycling programme in the spinal injured. Paraplegia 1994;32(6):407–415.

64. Baker LL, Yeh C, Wilson D, Waters RL. Electrical stimulation of wrist and fingers for hemiplegic patients. Phys Ther 1979;59(12):1495–1499.

65. Martin TP, Stein RB, Hoeppner PH, Reid DC. Influence of electrical stimulation on the morphological and metabolic properties of paralyzed muscle. J Appl Physiol 1992;72(4): 1401–1406.

66. Crameri RM, et al. Effects of electrical stimulation leg training during the acute phase of spinal cord injury: a pilot study. Eur J Appl Physiol 2000;83(4–5):409–415.

67. Munsat TL, McNeal D, Waters R. Effects of nerve stimulation on human muscle. Arch Neurol 1976;33(9):608–617.

68. Salmons S, Henriksson J. The adaptive response of skeletal muscle to increased activity. Muscle Nerve 1981;4: 94–105.

69. Eisenberg BR, Salmons S. The reorganization of subcellular structure in muscle undergoing fast-to-slow type transformation. A stereological study. Cell Tissue Res 1981; 220(3):449–471.

70. Nemeth PM. Electrical stimulation of denervated muscle prevents decreases in oxidative enzymes. Muscle Nerve 1982;5(2): 134–139.

71. Sreter FA, Pinter K, Jolesz F, Mabuchi K. Fast to slow transformation of fast muscles in response to long-term phasic stimulation. Exp Neurol 1982;75(1):95–102.

72. Peckham PH. Principles of electrical stimulation. Top spinal cord injury rehabilitation 1999;5(1):1–5.

73. Abbas JJ, Triolo RJ. Experimental evaluation of an adaptive feedforward controller for use in functional neuromuscular stimulation systems. IEEE Trans Rehabil Eng 1997; 5(1):12–22.

74. Durfee WK, MacLean KE. Methods for estimating isometric recruitment curves of electrically stimulated muscle. IEEE Trans Biomed Eng 1989;36(7):654–667.

75. Riess J, Abbas JJ. Adaptive control of cyclic movements as muscles fatigue using functional neuromuscular stimulation. IEEE Trans Neural Syst Rehabil Eng 2001;9(3):326–330.

76. Crago PE, Peckham PH, Thrope GB. Modulation of muscle force by recruitment during intramuscular stimulation. IEEE Trans Biomed Eng 1980;27(12):679–684.

77. Fang ZP, AJTM. A method of attaining natural recruitment order in artificially activated muscles. Proceedings 9th IEEE-EMBS Conference; 1987. pp 657–658.

78. Blair EA, Erlanger J. A comparison of the characteristics of axons through their individual electrical responses. Am J Physiol 1933;106:565–570.

79. Petrofsky JS. Control of the recruitment and firing frequencies of motor units in electrically stimulated muscles in the cat. Med Biol Eng Comput 1978;16(3):302–308.

80. Smith B, et al. An externally powered, multichannel, implantable stimulator-telemeter for control of paralyzed muscle. IEEE Trans Biomed Eng 1998;45(4):463–475.

81. Han-Chang Wu, Young S-T, Kuo T-S. A versatile multichannel direct-synthesized electrical stimulator for FES applications. IEEE Trans Instrum Meas 2002;51(1):2–9.

82. Ring H, Rosenthal N. Controlled study of neuroprosthetic functional electrical stimulation in sub-acute post-stroke rehabilitation. J Rehabil Med 2005;37(1):32–36.

83. Popovic MR, Keller T. Modular transcutaneous functional electrical stimulation system. Med Eng Phys 2005;27(1):81–92.

84. Mortimer JT, Bhadra N. Peripheral Nerve and Muscle Stimulation. In: Horch KW, Dhillon GS, editors. Neuroprosthetics: Theory and Practice. New Jersey: World Scientific (Series on Bioengineering & Biomedical Engineering; 2004.

85. Conway B. Theory and Principles of Electrode Processes. New York: Ronald Press; 1965.

86. Dymond AM. Characteristics of the metal-tissue interface of stimulation electrodes. IEEE Trans Biomed Eng 1976;23(4): 274–280.

87. Scheiner Λ, Polando G, Marsolais EB. Design and clinical application of a double helix electrode for functional electrical stimulation. IEEE Trans Biomed Eng 1994;41(5):425–431.

88. Daly JJ, et al. Performance of an intramuscular electrode during functional neuromuscular stimulation for gait training post stroke. J Rehabil Res Dev 2001;38(5):513–526.

89. Uhlir JP, Triolo RJ, Davis JA Jr, Bieri C. Performance of epimysial stimulating electrodes in the lower extremities of individuals with spinal cord injury. IEEE Trans Neural Syst Rehabil Eng 2004;12(2):279–287.

90. Loeb GE, Peck RA, Moore WH, Hood K. BION system for distributed neural prosthetic interfaces. Med Eng Phys 2001;23(1):9–18.

91. Naples GG, Mortimer JT, Scheiner A, Sweeney JD. A spiral nerve cuff electrode for peripheral nerve stimulation. IEEE Trans Biomed Eng 1988;35(11):905–916.

92. Tarler MD, Mortimer JT. Selective and independent activation of four motor fascicles using a four contact nerve-cuff electrode. IEEE Trans Neural Syst Rehabil Eng 2004; 12(2):251–257.

93. Leventhal DK, Durand DM. Chronic measurement of the stimulation selectivity of the flat interface nerve electrode. IEEE Trans Biomed Eng 2004;51(9):1649–1658.

94. Lawrence SM, Dhillon GS, Horch KW. Fabrication and characteristics of an implantable, polymer-based, intrafascicular electrode. J Neurosci Methods 2003;131(1–2): 9–26.

95. Tyler DJ, Durand DM. A slowly penetrating interfascicular nerve electrode for selective activation of peripheral nerves. IEEE Trans Rehabil Eng 1997;5(1):51–61.

96. Mushahwar VK, Gillard DM, Gauthier MJ, Prochazka A. Intraspinal micro stimulation generates locomotor-like and feedback-controlled movements. IEEE Trans Neural Syst Rehabil Eng 2002;10(1):68–81.

97. Saigal R, Renzi C, Mushahwar VK. Intraspinal microstimulation generates functional movements after spinal-cord injury. IEEE Trans Neural Syst Rehabil Eng 2004;12(4): 430–440.

98. McNeal DR, Baker LL. Effects of joint angle, electrodes and waveform on electrical stimulation of the quadriceps and hamstrings. Ann Biomed Eng 1988;16(3):299–310.

99. Bowman BR, Baker LL. Effects of waveform parameters on comfort during transcutaneous neuromuscular electrical stimulation. Ann Biomed Eng 1985;13(1):59–74.

100. Bajd T, Kralj A, Turk R. Standing-up of a healthy subject and a paraplegic patient. J Biomech 1982;15(1):1–10.

101. Baker L, et al. NeuroMuscular Electrical Stimulation: A Practical Guide. 4th ed. Los Amigos Research & Education Institute; 2000.

102. Wieler M, SN, Stein RB. WalkAid: An improved functional electrical stimulator for correcting foot-drop. Proceeding of the 1st Anuual Conference IFES; Cleveland (OH): 1996.

103. Burridge J, Taylor P, Hagan S, Swain I. Experience of clinical use of the Odstock dropped foot stimulator. Artif Organs 1997;21(3):254–260.

104. Taylor PN, et al. Clinical use of the Odstock dropped foot stimulator: its effect on the speed and effort of walking. Arch Phys Med Rehabil 1999;80(12):1577–1583.

105. Popovic D, Tomovic R, Schwirtlich L. Hybrid assistive system–the motor neuroprosthesis. IEEE Trans Biomed Eng 1989;36(7):729–737.

106. Solomonow M, et al. Reciprocating gait orthosis powered with electrical muscle stimulation (RGO II). Part II: Medical evaluation of 70 paraplegic patients. Orthopedics 1997; 20(5):411–418.

107. Snoek GJ, et al. Use of the NESS handmaster to restore handfunction in tetraplegia: clinical experiences in ten patients. Spinal Cord 2000;38(4):244–249.

108. Popovic MR, Popovic DB, Keller T. Neuroprostheses for grasping. Neurol Res 2002;24(5):443–452.

109. Popovic D, et al. Clinical evaluation of the bionic glove. Arch Phys Med Rehabil 1999;80(3):299–304.

110. Grandjean PA, Mortimer JT. Recruitment properties of monopolar and bipolar epimysial electrodes. Ann Biomed Eng 1986;14(1):53–66.

111. Gruner JA, Mason CP. Nonlinear muscle recruitment during intramuscular and nerve stimulation. J Rehabil Res Dev 1989;26(2):1–16.

112. Crago PE, Peckham PH, Mortimer JT, Van der Meulen JP. The choice of pulse duration for chronic electrical stimulation via surface, nerve, and intramuscular electrodes. Ann Biomed Eng 1974;2(3):252–264.

113. Mortimer T. Motor prosthesis. In: B VB, editor. Handbook of Physiology. Bethesda (MD): American Physiologer Society; 1981.

114. Smith BT, Betz RR, Mulcahey MJ, Triolo RJ. Reliability of percutaneous intramuscular electrodes for upper extremity functional neuromuscular stimulation in adolescents with C5 tetraplegia. Arch Phys Med Rehabil 1994;75(9):939–945.

115. Scheiner A, Mortimer JT, Roessmann U. Imbalanced biphasic electrical stimulation: muscle tissue damage. Ann Biomed Eng 1990;18(4):407–425.

116. Daly JJ, Ruff RL. Feasibility of combining multi-channel functional neuromuscular stimulation with weight-supported treadmill training. J Neurol Sci 2004;225(1-2):105–115.

117. Uhlir JP, Triolo RJ, Kobetic R. The use of selective electrical stimulation of the quadriceps to improve standing function in paraplegia. IEEE Trans Rehabil Eng 2000;8(4):514–522.

118. Prochazka A, Davis LA. Clinical experience with reinforced, anchored intramuscular electrodes for functional neuromuscular stimulation. J Neurosci Methods 1992;42(3): 175–184.

119. Marsolais EB, Kobetic R. Functional walking in paralyzed patients by means of electrical stimulation. Clin Orthop Relat Res 1983;175:30–36.

120. Handa Y, Hoshimiya N, Iguchi Y, Oda T. Development of percutaneous intramuscular electrode for multichannel FES system. IEEE Trans Biomed Eng 1989;36(7):705–710.

121. Peterson DK, et al. Electrical activation of respiratory muscles by methods other than phrenic nerve cuff electrodes. Pacing Clin Electrophysiol 1989;12(5):854–860.

122. Degnan GG, Wind TC, Jones EV, Edlich RF. Functional electrical stimulation in tetraplegic patients to restore hand function. J Long Term Eff Med Implants 2002;12(3):175–188.

123. von Wild K, et al. Computer added locomotion by implanted electrical stimulation in paraplegic patients (SUAW). Acta Neurochir Suppl 2002;79:99–104.

124. Davis JA Jr, et al. Preliminary performance of a surgically implanted neuroprosthesis for standing and transfers—where do we stand? J Rehabil Res Dev 2001; 38(6):609–617.

125. Sharma M, et al. Implantation of a 16-channel functional electrical stimulation walking system. Clin Orthop Relat Res 1998;347:236–242.

126. Bowman BR, Erickson RC. 2nd, Acute and chronic implantation of coiled wire intraneural electrodes during cyclical electrical stimulation. Ann Biomed Eng 1985;13(1):75–93.

127. Popovic D, Gordon T, Rafuse VF, Prochazka A. Properties of implanted electrodes for functional electrical stimulation. Ann Biomed Eng 1991;19(3):303–316.

128. Singh J, Peck RA, Loeb GE. Development of BION Technology for functional electrical stimulation: Hermetic Packaging. Proceedings of the 23rd Annual EMBS International Conference; Istanbul, Turkey: 2001. pp 1313–1316.

129. Cameron T, Liinamaa TL, Loeb GE, Richmond FJ. Long-term biocompatibility of a miniature stimulator implanted in feline hind limb muscles. IEEE Trans Biomed Eng 1998; 45(8):1024–1035.

130. Cameron T, et al. Micromodular implants to provide electrical stimulation of paralyzed muscles and limbs. IEEE Trans Biomed Eng 1997;44(9):781–790.

131. Richmond FJ, et al. Therapeutic electrical stimulation with BIONs to rehabilitate shoulder and knee dysfunction. 2002. Ljubljana, Slovenia:IFESS.

132. Dupont A, et al. Therapeutical electrical stimulation with BIONS: Clinical trial report. in 2nd Joint Conference of the IEEE Engineering in Medicine and Biology Society and the Biomedical Engineering Society; Huston (TX): 2002.

133. Dupont AC, et al. Clinical Trials of BION Injectable Neuromuscular Stimulators. Reno (NV): RESNA; 2001.

134. Baker L. Rehabilitation of the Arm and Hand Following Stroke - A Clinical Trial with BIONsTM. Proceeding of the 26th Annual International Conference IEEE Engineering in Medicine and Biology Society; San Francisco: 2004.

135. Dupont AC, et al. First patients with BION implants for therapeutic electrical stimulation Neuromodulation. Neuromodulation 2004;7:38–47.

136. Arcos I, et al. Second-generation microstimulator. Artif Organs 2002;26(3):228–231.

137. Troyk PR, Brown IE, Moore WH, Loeb GE. Development of BION Technology for functional electrical stimulation: Bidirectional Telemetry. Istanbul, Turkey: IEEE-EMBS; 2001.

138. Zou Q, Kim ES, Loeb GE. Implantable Bimorph Piezoelectric Accelerometer for Feedback Control of Functional Neuromuscular Stimulation. The 12th International Conference on Solid State Sensors, Actuators and Microsystems; Boston: 2003.

139. Peckham PH, Mortimer JT, Marsolais EB. Controlled prehension and release in the C5 quadriplegic elicited by functional electrical stimulation of the paralyzed forearm musculature. Ann Biomed Eng 1980;8(4–6):369–388.

140. Triolo RJ, et al. Implanted Functional Neuromuscular Stimulation systems for individuals with cervical spinal cord injuries: clinical case reports. Arch Phys Med Rehabil 1996;77(11):1119–1128.

141. Marsolais B, RK. Experience with a helical percutaneous electrode in the human lower extremity. Proceedings of the RESNA 8th Annual Conference; 1985. pp 243–245.

142. Peterson DK, Nochomovitz ML, Stellato TA, Mortimer JT. Long-term intramuscular electrical activation of the phrenic nerve: efficacy as a ventilatory prosthesis. IEEE Trans Biomed Eng 1994;41(12):1127–1135.

143. Herbert MA, Bobechko WP. Paraspinal muscle stimulation for the treatment of idiopathic scoliosis in children. Orthopedics 1987;10(8):1125–1132.

144. Peckham PH, et al. Efficacy of an implanted neuroprosthesis for restoring hand grasp in tetraplegia: a multicenter study. Arch Phys Med Rehabil 2001;82(10):1380–1388.

145. Veraart C, Grill WM, Mortimer JT. Selective control of muscle activation with a multipolar nerve cuff electrode. IEEE Trans Biomed Eng 1993;40(7):640–653.

146. Walter JS, et al. Multielectrode nerve cuff stimulation of the median nerve produces selective movements in a raccoon animal model. J Spinal Cord Med 1997;20(2):233–243.

147. Crampon MA, Brailovski V, Sawan M, Trochu F. Nerve cuff electrode with shape memory alloy armature: design and fabrication. Biomed Mater Eng 2002;12(4):397–410.

148. Navarro X, Valderrama E, Stieglitz T, Schuttler M. Selective fascicular stimulation of the rat sciatic nerve with multipolar polyimide cuff electrodes. Restor Neurol Neurosci 2001; 18(1):9–21.

149. Loeb GE, Peck RA. Cuff electrodes for chronic stimulation and recording of peripheral nerve activity. J Neurosci Methods 1996;64(1):95–103.

150. Tyler DJ, Durand DM. Chronic response of the rat sciatic nerve to the flat interface nerve electrode. Ann Biomed Eng 2003;31(6):633–642.

151. Tyler DJ, Durand DM. Functionally selective peripheral nerve stimulation with a flat interface nerve electrode. IEEE Trans Neural Syst Rehabil Eng 2002;10(4):294–303.

152. Leventhal DK, Durand DM. Subfascicle stimulation selectivity with the flat interface nerve electrode. Ann Biomed Eng 2003;31(6):643–652.

153. Yoshida K, Horch K. Selective stimulation of peripheral nerve fibers using dual intrafascicular electrodes. IEEE Trans Biomed Eng 1993;40(5):492–494.

154. McDonnall D, Clark GA, Normann RA. Selective motor unit recruitment via intrafascicular multielectrode stimulation. Can J Physiol Pharmacol 2004;82(8–9):599–609.

155. Zheng X, Zhang J, Chen T, Chen Z. Longitudinally implanted intrafascicular electrodes for stimulating and recording fascicular physioelectrical signals in the sciatic nerve of rabbits. Microsurgery 2003;23(3):268–273.

156. Lawrence SM, et al. Long-term biocompatibility of implanted polymer-based intrafascicular electrodes. J Biomed Mater Res 2002;63(5):501–506.

157. Yoshida K, Jovanovic K, Stein RB. Intrafascicular electrodes for stimulation and recording from mudpuppy spinal roots. J Neurosci Methods 2000;96(1):47–55.

158. Grill WM Jr, Mortimer JT. Quantification of recruitment properties of multiple contact cuff electrodes. IEEE Trans Rehabil Eng 1996;4(2):49–62.

159. Goodall EV, de Breij JF, Holsheimer J. Position-selective activation of peripheral nerve fibers with a cuff electrode. IEEE Trans Biomed Eng 1996;43(8):851–856.

160. Veltink PH, van Alste JA, Boom HB. Multielectrode intrafascicular and extraneural stimulation. Med Biol Eng Comput 1989;27(1):19–24.

161. Hoffer JA, Loeb GE. Implantable electrical and mechanical interfaces with nerve and muscle. Ann Biomed Eng 1980; 8(4–6):351–360.

162. Juch PJ, Minkels RF. The strap-electrode: a stimulating and recording electrode for small nerves. Brain Res Bull 1989; 22(5):917–918.

163. Stein RB, et al. Stable long-term recordings from cat peripheral nerves. Brain Res 1977;128(1):21–38.

164. Sweeney JD, Mortimer JT. An asymmetric two electrode cuff for generation of unidirectionally propagated action potentials. IEEE Trans Biomed Eng 1986;33(6): 541–549.

165. Agnew WF, McCreery DB, Yuen TG, Bullara LA. Histologic and physiologic evaluation of electrically stimulated peripheral nerve: considerations for the selection of parameters. Ann Biomed Eng 1989;17(1):39–60.

166. McNeal DR, Bowman BR. Selective activation of muscles using peripheral nerve electrodes. Med Biol Eng Comput 1985;23(3):249–253.

167. Sweeney JD, Ksienski DA, Mortimer JT. A nerve cuff technique for selective excitation of peripheral nerve trunk regions. IEEE Trans Biomed Eng 1990;37(7):706–715.

168. Sweeney JD, Crawford NR, Brandon TA. Neuromuscular stimulation selectivity of multiple-contact nerve cuff electrode arrays. Med Biol Eng Comput 1995;33(3 Spec No): 418–425.

169. Rozman J, Sovinec B, Trlep M, Zorko B. Multielectrode spiral cuff for ordered and reversed activation of nerve fibres. J Biomed Eng 1993;15(2):113–120.

170. Rozman J, Trlep M. Multielectrode spiral cuff for selective stimulation of nerve fibres. J Med Eng Technol 1992;16(5): 194–203.

171. Deurloo KE, Holsheimer J, Boom HB. Transverse tripolar stimulation of peripheral nerve: a modelling study of spatial selectivity. Med Biol Eng Comput 1998;36(1):66–74.

172. Tarler MD, Mortimer JT. Comparison of joint torque evoked with monopolar and tripolar-cuff electrodes. IEEE Trans Neural Syst Rehabil Eng 2003;11(3):227–235.

173. Tyler DJ. Functionally Selective Stimulation of Peripheral Nerves: Electrodes That Alter Nerve Geometry. Cleveland (OH): Case Western Reserve University; 1999.

174. Choi AQ, Cavanaugh JK, Durand DM. Selectivity of multiple-contact nerve cuff electrodes: a simulation analysis. IEEE Trans Biomed Eng 2001;48(2):165–172.

175. Branner A, Stein RB, Normann RA. Selective stimulation of cat sciatic nerve using an array of varying-length microelectrodes. J Neurophysiol 2001;85(4):1585–1594.

176. Anderson JM. Inflammatory response to implants. ASAIO Trans 1988;34(2):101–107.

177. Grill WM, Mortimer JT. Neural and connective tissue response to long-term implantation of multiple contact nerve cuff electrodes. J Biomed Mater Res 2000;50(2):215–226.

178. Rutten WL, van Wier HJ, Put JH. Sensitivity and selectivity of intraneural stimulation using a silicon electrode array. IEEE Trans Biomed Eng 1991;38(2):192–198.

179. Rutten WL, Meier JH. Selectivity of intraneural prosthetic interfaces for muscular control. Med Biol Eng Comput 1991;29(6):3–7.

180. Meier JH, Rutten WL, Boom HB. Force recruitment during electrical nerve stimulation with multipolar intrafascicular electrodes. Med Biol Eng Comput 1995;33(3 Spec No): 409–417.

181. Nannini N, Horch K. Muscle recruitment with intrafascicular electrodes. IEEE Trans Biomed Eng 1991;38(8):769–776.

182. Lawrence SM, et al. Acute peripheral nerve recording characteristics of polymer-based longitudinal intrafascicular electrodes. IEEE Trans Neural Syst Rehabil Eng 2004; 12(3):345–348.

183. Dhillon GS, Lawrence SM, Hutchinson DT, Horch KW. Residual function in peripheral nerve stumps of amputees: implications for neural control of artificial limbs. J Hand Surg [Am] 2004;29(4):605–615; discussion 616–618.

184. Zheng XJ, et al. [Experimental study of biocompatibility of LIFEs in peripheral fascicles]. Zhonghua Yi Xue Za Zhi 2003;83(24):2152–2157.

185. Malmstrom JA, McNaughton TG, Horch KW. Recording properties and biocompatibility of chronically implanted polymer-based intrafascicular electrodes. Ann Biomed Eng 1998;26(6):1055–1064.

186. Lundborg G, Richard P. Bunge memorial lecture. Nerve injury and repair—a challenge to the plastic brain. J Peripher Nerv Syst 2003;8(4):209–226.

187. Herman R, He J, D'Luzansky S, Willis W, Dilli S. Spinal cord stimulation facilitates functional walking in a chronic, incomplete spinal cord injured. Spinal Cord 2002;40(2): 65–68.

188. Pinter MM, Dimitrijevic MR. Gait after spinal cord injury and the central pattern generator for locomotion. Spinal Cord 1999;37(8):531–537.

189. Field-Fote E. Spinal cord stimulation facilitates functional walking in a chronic, incomplete spinal cord injured subject. Spinal Cord 2002;40(8):428.

190. Prochazka A, Mushahwar V, Yakovenko S. Activation and coordination of spinal motoneuron pools after spinal cord injury. Prog Brain Res 2002;137:109–124.

191. Mushahwar VK, Horch KW. Selective activation of muscle groups in the feline hindlimb through electrical microstimulation of the ventral lumbo-sacral spinal cord. IEEE Trans Rehabil Eng 2000;8(1):11–21.

192. Mushahwar VK, Horch KW. Proposed specifications for a lumbar spinal cord electrode array for control of lower extremities in paraplegia. IEEE Trans Rehabil Eng 1997; 5(3):237–243.

193. Tai C, et al. Multi-joint movement of the cat hindlimb evoked by microstimulation of the lumbosacral spinal cord. Exp Neurol 2003;183(2):620–627.

194. Mushahwar VK, Horch KW. Selective activation and graded recruitment of functional muscle groups through spinal cord stimulation. Ann NY Acad Sci 1998;860:531–535.

195. Mushahwar VK, Collins DF, Prochazka A. Spinal cord microstimulation generates functional limb movements in chronically implanted cats. Exp Neurol 2000;163(2):422–429.

196. Mushahwar VK, Horch KW. Muscle recruitment through electrical stimulation of the lumbo-sacral spinal cord. IEEE Trans Rehabil Eng 2000;8(1):22–29.

197. Durfee WK, Palmer KI. Estimation of force-activation, force-length, and force-velocity properties in isolated, electrically stimulated muscle. IEEE Trans Biomed Eng 1994;41(3):205–216.

198. Durfee W. Muscle model identification in neural prosthesis systems. In: Stein R, Peckham H, editors. Neural Prostheses: Replacing Motor Function After Disease or Disability. Oxford University Press; 1992.

199. Durfee WK. Control of standing and gait using electrical stimulation: influence of muscle model complexity on control strategy. Prog Brain Res 1993;97:369–381.

200. Veltink PH, Chizeck HJ, Crago PE, el-Bialy A. Nonlinear joint angle control for artificially stimulated muscle. IEEE Trans Biomed Eng 1992;39(4):368–380.

201. Wame X. Newromoclulation 2001;4:187–195.

202. Marsolais EB, Kobetic R. Functional electrical stimulation for walking in paraplegia. J Bone Joint Surg Am 1987;69(5): 728–733.

203. Yamaguchi GT, Zajac FE. Restoring unassisted natural gait to paraplegics via functional neuromuscular stimulation: a computer simulation study. IEEE Trans Biomed Eng 1990; 37(9):886–902.

204. Quintern J, Minwegen P, Mauritz KH. Control mechanisms for restoring posture and movements in paraplegics. Prog Brain Res 1989;80:489–502; discussion 479–480.

205. Nathan RH. Control strategies in FNS systems for the upper extremities. Crit Rev Biomed Eng 1993;21(6):485–568.

206. Crago PE, et al. New control strategies for neuroprosthetic systems. J Rehabil Res Dev 1996;33(2):158–172.

207. Bajzek TJ, Jaeger RJ. Characterization and control of muscle response to electrical stimulation. Ann Biomed Eng 1987; 15(5):485–501.

208. Lan N, Crago PE, Chizeck HJ. Feedback control methods for task regulation by electrical stimulation of muscles. IEEE Trans Biomed Eng 1991;38(12):1213–1223.

209. Crago PE, Nakai RJ, Chizeck HJ. Feedback regulation of hand grasp opening and contact force during stimulation of paralyzed muscle. IEEE Trans Biomed Eng 1991;38(1): 17–28.

210. Kataria P, Abass J. Adaptive user-specific control of movements with functional neuromuscular stimulation. Proceedings of the IEEE/BMES Conference; Atlanta (GA): 1999. p. 604.

211. Abbas JJ, Chizeck HJ. Neural network control of functional neuromuscular stimulation systems: computer simulation studies. IEEE Trans Biomed Eng 1995;42(11):1117–1127.

212. Chang GC, et al. A neuro-control system for the knee joint position control with quadriceps stimulation. IEEE Trans Rehabil Eng 1997;5(1):2–11.

213. Riess J, Abbas JJ. Adaptive neural network control of cyclic movements using functional neuromuscular stimulation. IEEE Trans Rehabil Eng 2000;8(1):42–52.

214. Chizeck HJ. Adaptive and nonlinear control methods for neural prostheses. In: Stein PP, RB, Popovic DB, editor. Neural prostheses: replacing motor function after disease or disability. New York: Oxford University Press; 1992. pp 298–328.

215. Abbas J, Chizeck H. A neural network controller for functional neuromuscular stimulation systems. Proceedings IEEE/EMBS Conference. Orlando (FL): 1991. p 1456–1457.

216. Chen JJ, et al. Applying fuzzy logic to control cycling movement induced by functional electrical stimulation. IEEE Trans Rehabil Eng 1997;5(2):158–169.

217. Veltink PH. Control of FES-induced cyclical movements of the lower leg. Med Biol Eng Comput 1991;29(6):NS8–NS12.

218. Friehs GM, et al. Brain-machine and brain-computer interfaces. Stroke 2004;35(11 Suppl. 1):2702–2705.

219. Patil P, Carmena J, Nicolelis MA, DA T. Ensemble recordings of human subcortical neurons as a source of motor control signals for a brain-machine interface. Neurosurgery 2004; 55(1):27–35.

220. Schwartz AB. Cortical neural prosthetics. Annu Rev Neurosci 2004;27:487–507.

221. Donoghue JP. Connecting cortex to machines: recent advances in brain interfaces. Nat Neurosci 2002;5(Suppl.): 1085–1088.

222. Creasey GH, et al. Clinical applications of electrical stimulation after spinal cord injury. J Spinal Cord Med 2004;27(4): 365–375.

223. Postans NJ, Hasler JP, Granat MH, Maxwell DJ. Functional electric stimulation to augment partial weight-bearing supported treadmill training for patients with acute incomplete spinal cord injury: A pilot study. Arch Phys Med Rehabil 2004;85(4):604–610.

224. Field-Fote EC. Combined use of body weight support, functional electric stimulation, and treadmill training to improve walking ability in individuals with chronic incomplete spinal cord injury. Arch Phys Med Rehabil 2001;82(6): 818–824.

225. Field-Fote EC, Tepavac D. Improved intralimb coordination in people with incomplete spinal cord injury following training with body weight support and electrical stimulation. Phys Ther 2002;82(7):707–715.

226. Barbeau H, Ladouceur M, Mirbagheri MM, Kearney RE. The effect of locomotor training combined with functional electrical stimulation in chronic spinal cord injured subjects: walking and reflex studies. Brain Res Brain Res Rev 2002; 40(1–3):274–291.

227. Barbeau H, et al. Tapping into spinal circuits to restore motor function. Brain Res Brain Res Rev 1999;30(1):27–51.

228. Field-Fote EC. Electrical stimulation modifies spinal and cortical neural circuitry. Exerc Sport Sci Rev 2004;32(4): 155–160.

229. Waters RL, Campbell JM, Nakai R. Therapeutic electrical stimulation of the lower limb by epimysial electrodes. Clin Orthop 1988;233:44–52.

230. Kagaya H, Shimada Y, Sato K, Sato M. Changes in muscle force following therapeutic electrical stimulation in patients with complete paraplegia. Paraplegia 1996;34(1): 24–29.

231. Baldi J, Jackson RD, Moraille R, Mysiw WJ. Muscle atrophy is prevented in patients with acute spinal cord injury using functional electrical stimulation. Spinal Cord 1998;36(7): 463–469.

232. Scremin AM, et al. Increasing muscle mass in spinal cord injured persons with a functional electrical stimulation exercise program. Arch Phys Med Rehabil 1999;80(12): 1531–1536.

233. Crameri RM, et al. Effects of electrical stimulation-induced leg training on skeletal muscle adaptability in spinal cord injury. Scand J Med Sci Sports 2002;12(5):316–322.

234. Kern H, et al. Long-term denervation in humans causes degeneration of both contractile and excitation-contraction coupling apparatus, which is reversible by functional electrical stimulation (FES): a role for myofiber regeneration? J Neuropathol Exp Neurol 2004;63(9):919–931.

235. Hangartner TN, Rodgers MM, Glaser RM, Barre PS. Tibial bone density loss in spinal cord injured patients: effects of FES exercise. J Rehabil Res Dev 1994;31(1):50–61.

236. Mohr T, et al. Increased bone mineral density after prolonged electrically induced cycle training of paralyzed limbs in spinal cord injured man. Calcif Tissue Int 1997;61(1):22–25.

237. Bloomfield SA, Mysiw WJ, Jackson RD. Bone mass and endocrine adaptations to training in spinal cord injured individuals. Bone 1996;19(1):61–68.

238. Lee YH, Rah JH, Park RW, Park CI. The effect of early therapeutic electrical stimulation on bone mineral density in the paralyzed limbs of the rabbit. Yonsei Med J 2001;42(2):194–198.

239. Pettersson U, Nordstrom P, Lorentzon R. A comparison of bone mineral density and muscle strength in young male adults with different exercise level. Calcif Tissue Int 1999;64(6):490–498.

240. Belanger M, et al. Electrical stimulation: can it increase muscle strength and reverse osteopenia in spinal cord injured individuals? Arch Phys Med Rehabil 2000;81(8):1090–1098.

241. Pacy PJ, et al. Muscle and bone in paraplegic patients, and the effect of functional electrical stimulation. Clin Sci (London) 1988;75(5):481–487.

242. Leeds EM, et al. Bone mineral density after bicycle ergometry training. Arch Phys Med Rehabil 1990;71(3): 207–209.

243. Eser P, et al. Effect of electrical stimulation-induced cycling on bone mineral density in spinal cord-injured patients. Eur J Clin Invest 2003;33(5):412–419.

244. BeDell KK, Scremin AM, Perell KL, Kunkel CF. Effects of functional electrical stimulation-induced lower extremity cycling on bone density of spinal cord-injured patients. Am J Phys Med Rehabil 1996;75(1):29–34.

245. Weiss DS, Kirsner R, Eaglstein WH. Electrical stimulation and wound healing. Arch Dermatol 1990;126(2):222–225.

246. Castillo E, Sumano H, Fortoul TI, Zepeda A. The influence of pulsed electrical stimulation on the wound healing of burned rat skin. Arch Med Res 1995;26(2):185–189.

247. Reich JD, Tarjan PP. Electrical stimulation of skin. Int J Dermatol 1990;29(6):395–400.

248. Thawer HA, Houghton PE. Effects of electrical stimulation on the histological properties of wounds in diabetic mice. Wound Repair Regen 2001;9(2):107–115.

249. Reger SI, et al. Experimental wound healing with electrical stimulation. Artif Organs 1999;23(5):460–462.

250. Kloth LC. Physical modalities in wound management: UVC, therapeutic heating and electrical stimulation. Ostomy Wound Manage 1995;41(5):18–20, 22–24, 26–27.

251. Gentzkow GD. Electrical stimulation to heal dermal wounds. J Dermatol Surg Oncol 1993;19(8):753–758.

252. Kloth LC, McCulloch JM. Promotion of wound healing with electrical stimulation. Adv Wound Care 1996;9(5):42–45.

253. Agarwal S, et al. Long-term user perceptions of an implanted neuroprosthesis for exercise, standing, and transfers after spinal cord injury. J Rehabil Res Dev 2003;40(3):241–252.

254. Kloth LC. Electrical stimulation for wound healing: a review of evidence from in vitro studies, animal experiments, and clinical trials. Int J Low Extrem Wounds 2005;4(1): 23–44.

255. Stewart RM, Desaloms JM, Sanghera MK. Stimulation of the subthalamic nucleus for the treatment of Parkinson's disease: postoperative management, programming, and rehabilitation. J Neurosci Nurs 2005;37(2):108–114.

256. Garcia L, D'Alessandro G, Bioulac B, Hammond C. High-frequency stimulation in Parkinson's disease: more or less? Trends Neurosci 2005;28(4):209–216.

257. Uthman BM. Vagus nerve stimulation for seizures. Arch Med Res 2000;31(3):300–303.

258. McLachlan RS. Vagus nerve stimulation for intractable epilepsy: a review. J Clin Neurophysiol 1997;14(5):358–368.

259. Carroll D, et al. Transcutaneous electrical nerve stimulation (TENS) for chronic pain. Cochrane Database Syst Rev 2001;3:CD003222.

260. Chen JD, Lin HC. Electrical pacing accelerates intestinal transit slowed by fat-induced ileal brake. Dig Dis Sci 2003;48(2):251–256.

261. Sun Y, Chen J. Intestinal electric stimulation decreases fat absorption in rats: therapeutic potential for obesity. Obes Res 2004;12(8):1235–1242.

262. Xing J, et al. Gastric electrical-stimulation effects on canine gastric emptying, food intake, and body weight. Obes Res 2003;11(1):41–47.

Reading List

Baker L, et al. NeuroMuscular Electrical Stimulation: A Practical Guide 4th ed., California: Los Amigos Research & Education Institute, 2000, Available at www.ranchorep.org/Publications. htm, An excellent resource book on the basics of electrical stimulation, emphasizing the clinical uses and outcomes of neuromuscular stimulation.

Horch KW, Dhillon GS, editors. Neuroprosthetics: Theory and Practice. New Jersey: World Scientific (Series on Bioengineering & Biomedical Engineering—Vol. 2, J K-J Li, Series Editor); 2004. An extensive and comprehensive text covering a wide range of neuroprosthetic subjects.

McNeal DR. 2000 Years of Electrical Stimulation. In: Hambrecht FT, Reswick JB, editors. Functional Electrical Stimulation: Applications in Neural Prostheses. New York: Marcel Dekker; 1977. pp 3–35. A now classic historical perspective on the usage of electrical stimulation for medical purposes from the time of the ancient Greeks up until the modern advent of functional electrical stimulation techniques in the 1970s.

Popovic D, Sinkjaer T. Control of Movement for the Physically Disabled. London: Springer Verlag; 2003. Reviews the state of the art of rehabilitation systems and methods used to restore movement, including the combined use of electrical stimulation systems and orthotics.

Reilly JP. Applied Bioelectricity: From Electrical Stimulation to Electropathology. New York: Springer-Verlag; 1998. A detailed text covering the fundamental principles of electrical stimulation, including sensory, motor and cardiac responses.

The web-site of the Cleveland FES Center at Case Western Reserve University contains an excellent "Resource Guide" as well as an extensive glossary of FES terms. Available at fescenter.case.edu/.

The International Functional Electrical Stimulation Society (IFESS) acts to promote the research, application and understanding of electrical stimulation as it is utilized in the field of medicine. The official journal of IFESS along with the International Neuromodulation Society (INS) is Neuromodulation. For a number of general resources on FES technologies see the IFESS web-site available at www.ifess.org/index.htm.

See also ELECTROPHYSIOLOGY; REHABILITATION AND MUSCLE TESTING; SPINAL CORD STIMULATION.

FUNCTIONAL NEUROMUSCULAR STIMULATION.   See FUNCTIONAL ELECTRICAL STIMULATION.

G

GAMMA CAMERA. See Anger camera.

GAMMA KNIFE

Steven J. Goetsch
San Diego Gamma Knife Center
La Jolla, California

INTRODUCTION

The Leksell Gamma Knife is one of the most massive and costliest medical products ever created (see Fig. 1). It is also one of the most clinically and commercially successful medical products in history, with > 180 units installed worldwide at this writing. The device is used exclusively for the treatment of brain tumors and other brain abnormalities. The Gamma Knife, also known as the Leksell Gamma Unit, contains 201 sources of radioactive cobalt-60, each of which emits an intense beam of highly penetrating gamma radiation (see Cobalt-60 units for radiotherapy). Due to the penetrating nature of the gamma rays emitted by these radiation sources, the device must be heavily shielded, and therefore it weighs ~ 22 tons. The Gamma Knife must also be placed in a vault with concrete shielding walls 2–4-ft thick.

This remarkable device is used in the following way: A patient known from prior medical diagnosis to have a brain tumor or other treatable brain lesion, is brought to a Gamma Knife Center on the selected day of treatment. Gamma Knife treatment is thus intended for elective surgery and is never used for emergency purposes. The patient is prepared for treatment, which normally occurs with the patient alert and awake, by a nurse. Then, a neurosurgeon injects local anesthetic under the skin of

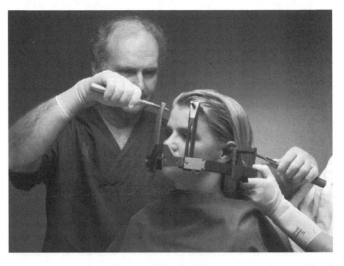

Figure 2. Patient with Leksell Model G stererotactic frame affixed to their head. This frame restricts patient motion during imaging and treatment and also allows placement of fiducial markers to localize the volume to be treated.

the forehead and posterior of the skull. He/she then affixes a stereotactic head frame (see Fig. 2) with sharp pins to the patient's head (much like a halo fixation device for patients with a broken neck). The patient is transported by wheelchair or gurney to a nearby imaging center where a Computed Tomography (CT) X-ray scan or a Magnetic Resonance Imaging (MRI) scan of the brain (with the stereotactic head frame on) is obtained (see articles on Computed Tomography and Magnetic Resonance Imaging). Specially constructed boxes consisting of panels containing geometric localization markers are attached to the stereotactic frame and surround the patient's head during imaging. The markers contained in the localization boxes are visible on the brain scan, just outside the skull (see Fig. 3). All imaging studies are then exported via a PACS computer network or DAT tape (see the article on Picture Archiving and Communication Systems) into a powerful computer, where a treatment plan is created. A neurosurgeon, a radiation oncologist, and a medical physicist participate in the planning process. When the plan is satisfactorily completed, the patient (still wearing the stereotactic frame) is brought into the treatment room. The patient is then placed on the couch of the treatment unit and the stereotactic frame is docked with the trunnions affixed to the helmet (see Fig. 4). After the staff members leave the room and the room shielding doors are closed, the Gamma Knife vault door automatically opens and the patient couch is pushed forward into the body of the device, so that the holes in the collimating helmet line up with the radiation source pattern inside the central body of the device. The treatment begins at that point. Any given patient may be treated in this manner with a single "shot" (e.g., treatment) or with

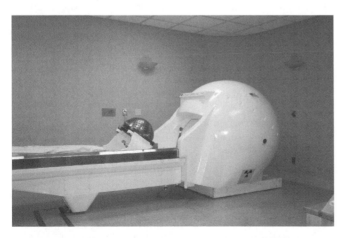

Figure 1. The Leksell Gamma Unit Model U for treatment of patients with brain tumors and other brain abnormalities.

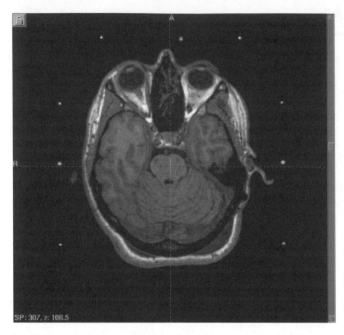

Figure 3. Axial MRI scan of patients brain, with external fiducial markers filled with copper sulfate solution to enable localization of target volumes.

many shots (30 or more in some cases). The collimating helmet may be changed to use one or more of the available helmet sizes, corresponding to a roughly spherical volume 4, 8, 14, or 18 mm in diameter. At the conclusion of treatment, the stereotactic frame is removed and most patients are then discharged. Thus Gamma Knife radiosurgery is most commonly performed on an outpatient basis.

Gamma Knife radiosurgery has shown rapidly increasing acceptance, since the first commercial unit was introduced at the University of Pittsburgh in 1987 (1). Despite the high purchase price (a little >$3 million) and single purpose, Gamma Knife units are widely available in the

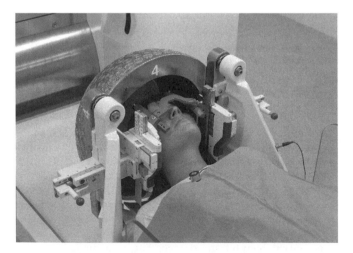

Figure 4. Supine patient in position for treatment in Gamma Knife. Stereotactic head frame is carefully docked with trunnions, which slide in precise channels in the secondary collimator helmet.

United States, Europe, Asia, and other parts of the world. All units are manufactured by Elekta Instruments, of Stockholm, Sweden. Use of this device can eliminate the need for open surgery of the brain. Modern surgical techniques and nursing follow-up have reduced the death rate due to brain surgery from as much as 50% in the 1930s to <1% in the United States in 2002. However, Gamma Knife patients most commonly do not have to remain overnight in the hospital at all (an important consideration in a very cost conscious healthcare environment), while craniotomy patients typically have a 2–5 day stay. Conventional brain surgery patients sometimes require 30 days or more of hospitalization if extremely adverse effects occur. Thus, the cost of the Gamma Knife outpatient procedure is typically far less than that for inpatient open brain surgery. Recovery of the patient is much more rapid for Gamma Knife patients, with most patients going home immediately and returning to work or other normal routines in 1–2 days.

EARLY HISTORY OF THE DEVICE

Gamma Knife radiosurgery was an outgrowth of several prior inventions. Dr. Lars Leksell, a Swedish neurosurgeon, was one of the pioneers in the field of stereotaxis (see the article on Stereotactic Surgery). Dr. Leksell was motivated to find minimally invasive ways to treat brain abnormalities by the appalling death rate for early twentieth century brain surgery, which could be as high as 50% (2). Leksell was one of the first surgeons to create a workable stereotactic frame (in 1949) that could be affixed to a patients skull, together with a set of indexed external markers (called fiducials) that were visible on an X-ray of the patient's head. Only primitive brain imaging procedures were available in 1950 to the late 1970s, so stereotactic surgery patients had to undergo a painful procedure called pneumoencephalography. A lumbar puncture was used to introduce air into the spinal canal while the patient (strapped into a special harness) was manipulated upside down, back and forth, while air was injected under positive pressure to displace the cerebro-spinal fluid in the ventricles of the brain. A pair of plane orthogonal X-ray images (anterior–posterior and lateral) were then taken. Since the air-filled ventricles were well imaged by this technique, standard atlases of the human brain such as Schaltenbrand and Wahren (3) were then used to compute the location of the desired target relative to these landmarks. The imaging procedure alone was considered extremely painful and typically required hospitalization. The early stereotactic frames were applied by drilling into the patient's skull and driving screws into the calvarium (outer table of the skull), which was then topped with a Plaster of Paris cap that could be rigidly fixed. A twist drill could then be guided to create a small hole (a few millimeters in diameter) in the patient's skull, through which a catheter could be passed to a designated target, such as the globus pallidum for treatment of Parkinsons disease. A radio frequency transmitter was passed through the catheter and a small volume of tissue was heated to high temperature, creating a deliberate, controlled brain lesion. This procedure, though rigorous, was far less invasive and dangerous than open brain surgery, called craniotomy.

Leksell then attached an X-ray treatment unit to his stereotactic frame and used it in 1951 to treat brain structures without opening of the skull. He called this procedure "radiosurgery", which he defined as "a single high dose of radiation stereotactically directed to an intracranial region of interest" (4). Leksell was successfully in treating previously intractable cases of trigeminal neuralgia, an extremely painful facial nerve disease, by stereotactically irradiating the very narrow ($\sim$ 2–4 mm diameter) nerve as it enters the brainstem. Only a few patients were treated with this X-ray unit.

Leksell then collaborated with physicist Borge Larsson in treating patients at a cyclotron located at Uppsala University near Stockholm beginning in 1957. Tobias and others had just begun treating patients with proton therapy (see article on Proton Beam Radiotherapy) at the University of California Berkeley in 1954. The proton is a positively charged subatomic particle, a basic building block of matter, which has extremely useful properties for treatment of human patients. The charged particles, accelerated to very high energies by a massive cyclotron (typically located at a high energy physics research laboratory) are directed at a patient, where they begin to interact through the Coulomb force while passing through tissue. At the end of the proton range, however, the particles give off a large burst of energy (the Bragg peak) and then stop abruptly. Leksell and Larsson utilized these properties with well-collimated beams of protons directed at intracranial targets. A few other centers in the United States and Russia also began proton therapy in the 1950s and 1960s.

The Gamma Knife was invented in Stockholm, Sweden by Leksell and Larsson and was manufactured (as a prototype) by the Swedish shipbuilding firm Mottola. The first unit had 179 radioactive cobalt-60 sources and was installed in 1968 at Sophiahemmet Hospital in Stockholm, Sweden (5). This original unit had three interchangeable helmets with elliptically shaped collimators of maximum diameter 4, 8, or 14 mm. Despite the lack of good brain imaging techniques at that time, the Gamma Knife was used to successfully treat Parkinson's disease (a movement disorder), trigeminal neuralgia (extreme facial pain), and arteriovenous malformations (AVMs), which are a tangle of congenitally malformed arteries and veins inside the brain. Several years later a second nearly identical unit was manufactured for Leksell when he became a faculty member at Karolinska Hospital in Stockholm. The original unit lay idle for a number of years, until it was donated to UCLA Medical Center in Los Angeles, where it was moved in 1982 (Fig. 5). It was used in animal research and treated a limited number of human patients before it was retired permanently in 1988 (6). The two original, custom-made Gamma Knife units were unique in the world and did not immediately enjoy widespread acceptance or gain much notice. A large number of patients with AVMs began to be treated at the Gamma Knife Center in Karolinska, by Dr. Ladislau Steiner, a neurosurgical colleague of Lars

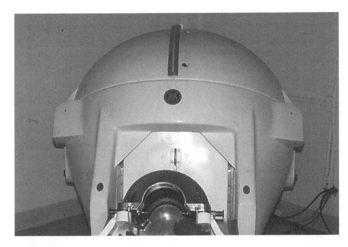

Figure 5. Original Gamma Knife, after being moved from Stockholm to UCLA Medical Center in Los Angeles.

Leksell. Arteriovenous malformations are prone to spontaneous hemorrhage that can cause sudden coma or death. Open surgical techniques for removal of these life-threatening vascular anomalies were extremely difficult and dangerous in the 1970s. Patients came from all over the world to be treated for these AVMs at the Gamma Knife Center in Stockholm.

In 1984 and 1985, two new Gamma Knife units were manufactured using Dr. Leksell's specifications by Nucletec SA of Switzerland (a subsidiary of Scanditronix AB, Sweden) for installation in hospitals in Buenos Aires, Argentina and Sheffield, England, respectively (7,8). These units also had three sets of collimators, which were now circular in shape, of 4, 8, and 14 mm diameter, but the number of cobalt-60 sources was increased to 201. The mechanical tolerance was exquisite: The convergence of all 201 beams at the focal point was specified as $\pm$ 0.1 mm. The total radioactivity was 209 TBq (5500 Ci) and the sources were distributed evenly over a $160 \times 60°$ sector of the hemispherical secondary collimators (Fig. 6). An ionization chamber (a type of radiation detector) placed at the center of a spherical phantom 16 cm in diameter was used to measure an absorbed dose rate of $\sim$ 2.5 gray$\cdot$min^{-1} for the Sheffield unit. This was adequate to treat patients to a large radiation dose in a reasonable period of time. Both Gamma Knife units were successfully used fo many years to treat patients in their respective countries.

A new corporation, called Elekta Instruments, AB, of Stockholm was created in 1972 by Laurent and Dan Leksell, sons of Lars Leksell, to manufacture neurosurgical products, including the Gamma Knife, which is now a trademark of this firm. Elekta created the first commercial Gamma Knife product, the Model U, and has manufactured all Gamma Knife units worldwide since that time. This new 201 source unit was installed at the University of Pittsburgh in 1987 and expanded the available beam diameters to include a fourth secondary collimator with a nominal diameter of 18 mm (1). The trunnions, which connect the secondary collimator helmets to the patient, were now configured to dock with connecting points located

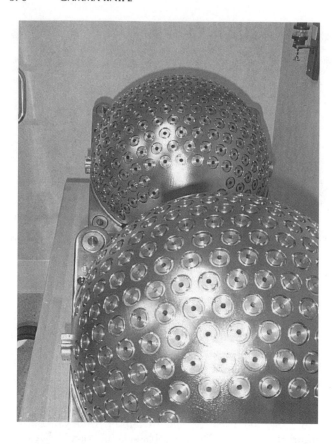

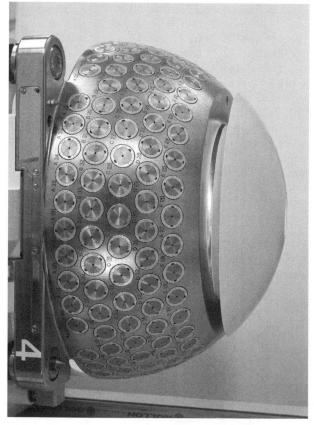

Figure 6. (Upper panel) Collimator helmets for Leksell Gamma Unit Model U. (Lower panel) Collimator helmet for Leksell Gamma Unit Models B and C.

on removable *Y–Z* positioning bars on the patients headframe. The earliest versions of the Gamma Knife had required implantation of screws into the patients skull and covering with Plaster of Paris to achieve this docking. The unit (like the two previous units) utilized a hydraulic drive to open the shielding door in the central body and propel the patient couch into treatment position.

Elekta also introduced a radiation therapy treatment plan called KULA to calculate the size and shape of the radiation volume to be treated for each patient and compute the necessary duration of the treatment. The Sophia-hemmet and Karolinska Gamma Knife Centers had relied on manual calculations until this time. The KULA plan could calculate isodose lines (lines of equal radiation dose) in a two-dimensional (2D) plane, which could then be manually traced onto an axial brain image. The advent of stereotactic imaging with computed tomography also eliminated the need for the difficult and painful pneumoencephalograms and was capable of localizing brain tumors as well as anatomical targets. The University of Pittsburgh Gamma Knife Center enjoyed a relatively high degree of acceptance from the time of installation, and was soon joined by other Leksell Gamma Units in the United States, Europe, and Asia.

One drawback of the Leksell Gamma Unit Model U, which is no longer manufactured, is that it was shipped to the hospital unloaded and then loaded with cobalt-60 sources on site. This necessitated the shipment of many tons of shielding materials to create a temporary hot cell, complete with remote manipulating arms (Fig. 7). A further difficulty with Gamma Units is that the radioactive cobalt-60 is constantly being depleted by radioactive decay. The half-life is cobalt-60 is ∼ 5.26 years, which means that the radiation dose rate decreases ∼ 1%/month. The Sheffield Gamma Unit (manufactured by Nucletec) was reloaded after a number of years of use by British contractors who had not been involved in designing or building the unit and it therefore took ∼ 12 months to complete the task. The University of Virginia Leksell Model U Gamma Unit was the first to be reloaded (in 1995) and it was out of service for only 3 weeks. Nevertheless, the necessity of having the treatment unit down for a period of weeks after 5–6 years of operation, at a cost approaching $500,000 with a very elaborate construction scenario inhibited the early acceptance of these units. A compensating advantage of the Gamma Unit Model U was the extremely high reliability of these devices. Routine maintenance is required once every 6 months and mechanical or electrical breakdowns preventing use of the device are very rare.

Leksell introduced the Gamma Unit Model B in Europe in 1988, although it was not licensed in the United States until 5 years later. The new unit, which strongly resembles the later Model C (Fig. 8), departed from the unique spherical shape of the earlier unit and more closely resembled the appearance of a CT scanner. The source configuration was changed to five concentric rings (Fig. 6b), although the number and activity of the cobalt-60 sources remained the same as in the Model U. The new Gamma Unit Model B was designed so that the radioactive cobalt-60 sources could be loaded and unloaded by means of a special

Figure 7. Loading cobalt-60 sources into Gamma Unit with remote manipulating arms.

Figure 9. Special loading device for insertion and removal of cobalt-60 sources with the Leksell Gamma Units Models B and C.

11 ton loading device (Fig. 9), without the necessity for creating a large and costly hot cell. This significantly reduced installation and source replenishment costs and speeded up these operations as well. The hydraulic operating system used to open the shielding doors and to move the patient treatment couch was replaced with a very quiet electrically powered system.

Extensive innovations were introduced with the Leksell Gamma Unit Model C with optional Automatic Positioning System (APS) in calendar year 2000. This unit was awarded three American patents and one Swedish patent. The new unit provided several upgrades at once: a new computer control system operates the shielding door and patient transport mechanism and is networked via RS232C protocol with the Leksell GammaPlan treatment planning computer. All previous models required manual setting (and verification) of helmet size, stereotactic coordinates and treatment time for each shot. An optional APS system (Fig. 10) has motorized trunnions that permit the patient to be moved from one treatment isocenter to another without human intervention. This automated system can only be utilized if the secondary helmet, gamma angle (angle of patients stereotactic frame Z axis with respect to the longitudinal axis of symmetry of the Gamma Unit helmet) and patient position (prone or supine) are identical to the values for these respective treatment parameters as provided in the final approved treatment plan. In addition, the isocenters (or shots) are grouped into "runs" having stereotactic coordinates within a predefined distance of each other (typically ±2 cm) so as not to introduce unacceptable strain on the patient's neck while their head is being moved

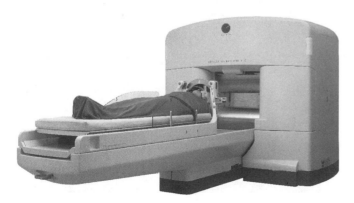

Figure 8. Leksell Gamma Unit Model C, which strongly resembles the previous Leksell Gamma Unit Model B.

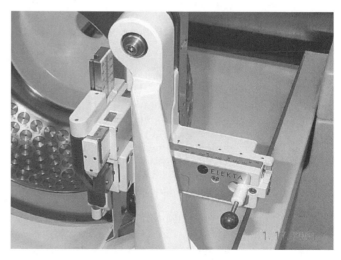

Figure 10. Close-up view of trunnions and helmet of Leksell Gamma Unit model C with Automatic Positioning System in place.

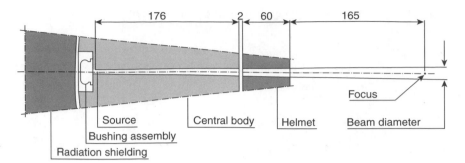

Figure 11. Geometry of sources installed in Leksell Gamma Units.

to a new position relative to the immobile body. Within these limitations the efficiency of a complex treatment plan can be greatly increased. Additionally, two independent electromechanical devices verify the positioning of the patient's stereotactic coordinates to within 50 μm (below the resolution of the unaided human eye).

THEORY

The invention of the Gamma Knife built on seven decades of previous experience with radiation therapy (see related articles). Early external beam radiation treatments used X-ray sets with low energies, in the range of 100–300 kV, which have the disadvantage of depositing a maximum dose at the surface of the skin. This physical characteristic makes it difficult to treat deep seated tumors without causing unacceptable damage to the overlying skin and tissue. Lars Leksell used a 200 kV X-ray set to treat his first radiosurgery patient in 1951, but abandoned that technique after only a few patients to work with far more penetrating proton beam radiotherapy (see article Proton beam radiotherapy). The disadvantage of proton beam therapy was that the patient had to be brought to a high energy physics laboratory, which was not otherwise equipped to treat sick or infirm patients and was often located at a great distance from the surgeon's hospital. This made treatments somewhat difficult and awkward, and the cyclotron was not always available. An important breakthrough came when two Canadian cancer centers introduced cobalt-60 teletherapy (see article Cobalt-60 units for radiotherapy) in the early 1950s. Leksell and Larsson realized that this new, more powerful radiation source could be utilized in a hospital setting. They also realized that rotational therapy, where a radiation source is revolved around a patient's tumor to spread out the surface dose, could be mimicked in this new device by creating a static hemispherical array of smaller radiation sources. Since Leksell was interested only in treating intracranial disease, where the maximum patient dimension is only ~ 16 cm, the device could place radiation sources relatively close to the center of focus. The Leksell Gamma Knife uses a 40 cm Source to Surface Distance (SSD), far shorter than modern linear accelerators (see article Medical linear accelerator), which typically rotate around an isocenter at a distance of 100 cm (see Fig. 11). This short SSD allows the manufacturer to take advantage of the inverse square principle, which implies that a nominal 30-curie source at 40 cm achieves the same dose rate at the focus as a 187.5

curie source would achieve at 100 cm. This makes loading and shielding a Gamma Knife practical.

The Gamma Knife treats intracranial tumors or other targets by linear superposition of 201 radiation beams. The convergence accuracy of these sources is remarkable: The radiation focus of the beams converge at the center point of stereotactic space (e.g., 100, 100, 100 in Leksell coordinates) to within < 0.3 mm. Thus the targeting accuracy of treatment of brain tumors is essentially not limited at all by mechanical factors, and is primarily limited by the inaccuracy of imaging techniques and by target definition. Each cobalt-60 beam interacts by ionization and excitation (primarily by Compton scattering) as it passes through the skull of the patient. The intensity of each beam is diminished by ~ 65% while passing through 16 cm of tissue (a typical skull width, approximated as water for purposes of calculation). At the mechanical intersection of all 201 radiation sources, which are collimated to be 18, 14, 8, or 4 mm in diameter, the useful treatment volume is formed (see Fig. 12). Outside this volume the radiation dose rate drops off precipitously (90% of Full Maximum to 50% in 1 mm for the 4 mm beam) thereby mimicking the behavior of protons at the end of their range. The mathematics of this 3D convergent therapeutic beam irradiation

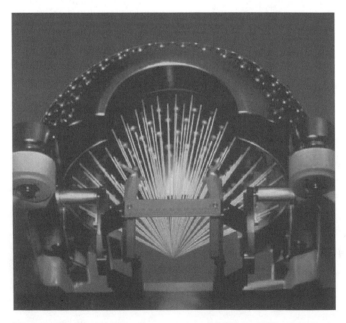

Figure 12. Illustration of convergence of 201 radiation beams to create treatment volume.

is relatively simple: The radiation absorbed dose adds up in linear superposition.

$$D(P) = D_{\mathrm{fi}} / \sum D_{\mathrm{fi}} \times [d_{\mathrm{fs}}/(d_{\mathrm{fs}} - dz)]^2 \times \mu dz$$

where $D(P)$ is the total dose at arbitrary point P, D_{fi} is the relative contribution from source i to the total dose at the point of focus, d_{fs} is the distance from the source to the focus (40 cm), dz is the distance along the beam axis from the focal point to intersection with the perpendicular from point P, and μ is the linear attenuation coefficient for Co-60 gamma radiation in tissue.

A radiation therapy treatment planning code (see article Radiation Therapy Treatment Planning) called Leksell GammaPlan is provided by the manufacturer for the purpose of preparing treatment plans for the Leksell Gamma Unit for use with human patients. An early treatment plan called KULA calculated treatment time and created a 2D plot of lines of equal radiation dose (or isodose lines), but with severe limitations. The early code could only calculate plans with a single center of irradiation (called an isocenter in general radiosurgery applications, or a "shot" in Gamma Knife usage). Calculated isodose lines had to be transferred by hand from a plot to a single CT slice in the axial plane. In 1991 the Leksell GammaPlan software was introduced (and premarket clearance by the FDA was obtained), which permitted on-screen visualization of isodose lines in multiple CT slices. The improved code could calculate and display the results of multiple shots and could model the effect of "plugging" some of the 201 source channels with thick steel plugs to "turn off" certain radiation sources. The software was written for UNIX workstations and has rapidly become increasingly powerful and much more rapid as processing speed and computer memory increased in the last decade. Leksell GammaPlan utilizes a matrix of $> 27,000$ equally spaced points (in the shape of a cube), which can be varied from 2.5 cm on a side to 7.5 cm on a side. Within this cube a maximum radiation dose is computed from the linear superposition of all 201 radiation beams (some of which may be plugged), from collimator helmets of 18, 14, 8, or 4 mm diameter, and this calculation is integrated over each "shot". More than 30 different shots (each with a discrete X, Y, and Z stereotactic coordinate, in 0.1 mm increments) can be computed and integrated, with user selectable relative weighting of each shot. Whereas the original KULA plan required ~ 15 min for one single shot calculation, modern workstations with Leksell GammaPlan can now compute 30 shot plans in up to 36 axial slices in < 1 min. Leksell GammaPlan can now utilize stereotactic CT, MRI, and Angiographic studies in the planning process. Each study must be acquired with the stereotactic frame in place and registered separately. Image fusion is now available. Figures 13 and 14 give two of the many possible screen presentations possible with a very sophisticated Graphical User Interface.

CLINICAL USE OF GAMMA KNIFE

The Gamma Knife has gained widespread acceptance in the neurosurgical and radiation oncology community as an

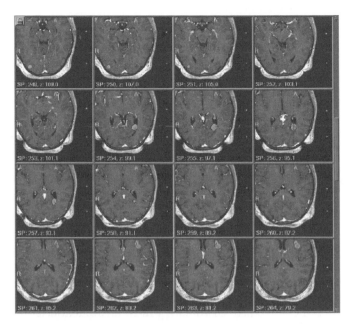

Figure 13. Screen from Leksell GammaPlan illustrating multiple MRI slices with superimposed isodose display.

effective treatment for many different pathologies of brain tumors, neurovascular abnormalities and functional disorders. Gamma Knife radiosurgery may in some cases be used as an alternative to open craniotomy while for other patients it may be used after previous surgeries have been attempted. Since Gamma Knife radiosurgery infrequently requires overnight hospitalization, and generally has a very low probability of adverse side effects, it may in many cases be much less costly, with lower chance of complication and much less arduous recovery.

A typical Gamma Knife procedure is performed after a patient has been carefully screened by a neurosurgeon, a radiation oncologist, and a neuroradiologist. The procedure is scheduled as an elective outpatient procedure and typically lasts from 3 to 8 h. The first critical step is placement of a stereotactic frame (see article Stereotactic Surgery) to provide rigid patient fixation and to allow stereotactic imaging to be performed with special fiducial attachments. The exact position of the frame (offset to left or right, anterior or posterior) is crucial, since the tumor must be well centered in stereotactic space to permit

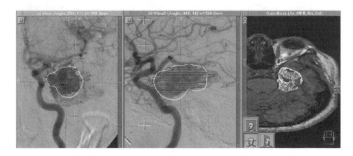

Figure 14. Screen from Leksell GammaPlan illustrating angiographic study and three-dimensional display of AVM nidus.

accommodation of the patient's skull (with the frame attached) inside the small volume of the secondary collimator helmet. The Leksell Model G frame encloses $\sim 2900 \text{ cm}^3$ of accessible stereotactic space, significantly less than other stereotactic frames which do not have to fit inside the Gamma Knife helmet. A stereotactic localization study is immediately performed, using one or more modalities such as computed tomography, and magnetic resonance imaging. Patients with vascular abnormalities may also undergo an angiographic study: A radiologist inserts a thin catheter (wire) into a vein in the patient's groin and then carefully advances the wire up through one of the major arteries leading to the brain, then into the area of interest. The catheter is then used to inject X-ray opaque dye, which reveals the extent of the vascular lesion (see Fig. 14). These imaging studies must then be promptly networked (via a hospital PACS system) to the planning computer. There the images are registered from couch coordinates (left-right, in-out, up-down) to stereotactic space (X, Y, and Z). At that point, each individual point in the brain corresponds to a specific stereotactic coordinate, which can be identified from outside the brain.

Gamma Knife radiosurgery, both in the United States and worldwide, has enjoyed a very rapid acceptance since the first commercial unit was produced in 1987. The number of procedures performed annually, subdivided by indication, is compiled by the nonprofit Leksell Society. Only results voluntarily reported by Gamma Knife centers are tallied, with no allowance for nonreporting centers, so their statistics are conservative. The growth in use of this device has been explosive, with < 7000 patients treated worldwide by 1991 and $> 297,000$ patients reported treated through December, 2004 (see Fig. 15). This parallels the increase in number of installed Leksell Gamma units, going from 17 in 1994 in the United States to 96 centers by the end of 2004. The number of Gamma Knife cases reported performed in the United States has increased by an average of 17%/year for the last 10 years, a remarkable increase. Table 1 indicates the cumulative number of patients treated with Gamma Knife radiosurgery in the western hemisphere and worldwide through December, 2004, subdivided by diagnosis.

Treatment objectives for Gamma Knife patients vary with the diagnosis. The most common indication for treatment is metastatic cancer to the brain. An estimated

Table 1. Cumulative Reported Gamma Knife Radiosurgery Procedures through December, 2004

Indication	Western Hemisphere Procedures	Worldwide Procedures
AVM and other vascular	9,793	43,789
Acoustic neuroma	7,719	28,306
Meningioma	11,016	36,602
Pituitary adenoma	3,577	24,604
Other benign tumors	3,137	14,884
Metastatic brain tumors	29,285	100,098
Glial tumors	7,727	20,614
Other malignant tumors	1,501	6,492
Trigeminal neuralgia	11,609	17,799
Other functional disease	1,135	4,441
TOTAL INDICATIONS:	**67,336**	**297,529**

1,334,000 cancers (not including skin cancers) were diagnosed in the United States in calendar year 2004. Approximately 20–30% of those patients will ultimately develop metastatic tumors in the brain, which spread from the primary site. These patients have a survival time (if not treated) of 6–8 weeks. The treatment objective with such patients is to palliate their symptoms and stop the growth of known brain tumors, thereby extending lifespan. A recent analysis (9) reported a median survival of patients treated with radiosurgery of 10.7 months, a substantial improvement. Approximately 18,000 patients were diagnosed with primary malignant brain tumors in the United States in calendar year 2004, with 13,000 deaths from this cause. Patients with primary malignant brain tumors (i.e., those originating in the brain) have a lifespan prognosis varying from 6 months to many years, depending on the grade of the pathology. Many glioma patients are offered cytoreductive brain surgery to debulk the tumor and may have an extended period of recovery and significant loss of quality of life afterwards. At time of tumor recurrence for these patients, the noninvasive Gamma Knife procedure may accomplish as much as a second surgery, while sparing the patient the debilitation of such a procedure. Recent reports in the clinical literature indicate that Gamma Knife radiosurgery is effective in improving survival for glioma patients.

Many patients with nonmalignant brain tumors are also treated with Gamma Knife radiosurgery. Meningiomas are the most common nonmalignant brain tumor, arising from the meninges (lining of the brain) as pathologically altered cells and causing neurological impairment or even death. Approximately 7000 new meningiomas are diagnosed in the United States each year. Most grow very slowly ($\sim 1 \text{ mm} \cdot \text{year}^{-1}$) while the most aggressive tumors may grow rapidly to as much as 12–15 cm in length and may even invade the bone. Gamma Knife radiosurgery has been reported for treatment of meningioma as far back as 1987 and is considered a well-established treatment for this extremely persistent disease, with > 1000 Gamma Knife treatments reported for meningioma in the United States during calendar year 2001. Another common nonmalignant tumor is the acoustic neuroma (also called vestibular schwannoma), which arises from the auditory nerve

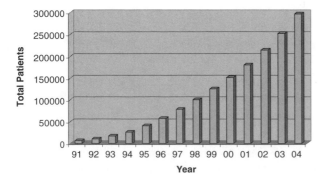

Figure 15. Cumulative number of Gamma Knife patients treated worldwide from 1991 through December, 2004.

(cranial nerve VIII). It can cause deafness and imbalance, and in severe cases motor impairment as it compresses the brainstem. The incidence of newly diagnosed acoustic neuromas is 2500–3000/year in the United States. Craniotomy for acoustic neuroma is among the most challenging brain operations, typically requiring 8–24 h on the operating table. Potential complications range from loss of residual hearing to devastating facial palsy to cerebrospinal fluid leak requiring as much as 30 days of hospitalization. Extremely high control rates of up to 97% (no additional tumor growth or moderate shrinkage) have been reported for Gamma Knife radiosurgery of these tumors with extremely low complication rates (10). This may explain why >1000 acoustic neuromas were treated with Gamma Knife radiosurgery in the United States during Calendar Year 2003, nearly one-third of all such tumors diagnosed that year.

Arteriovenous malformations are a rare genetic disorder of the vascular system of the brain and spinal cord. Estimates of incidence ranges from 5 to > 600/100,000 people. The lesion consists of a tangle of abnormal arteries and veins that may not be detected until late in life. The AVMs can cause debilitating headaches, epileptic seizures, coma, and even sudden death due to cerebral hemorrhage. Arteriovenous malformations were first described in the 1800s and the first surgical resection was credited to Olivecrona in Stockholm in 1932. The AVMs were categorized by Spetzler and Martin into five distinct surgical categories in order of increasing surgical risk and one additional category for inoperable lesions (11). Surgery for these lesions remained extremely challenging until late in the twentieth Century. Therefore, when angiography became available in the 1960s, Ladislau Steiner (a neurosurgical colleague of Lars Leksell at Karolinska Hospital) began to treat AVMs with the Gamma Knife as early as 1970 (12). A large number of AVMs were treated in the early days of Gamma Knife radiosurgery both because of the extreme risk of open surgery and the early success with this technique in obliterating these lesions. Recent clinical studies report an obliteration rate for these lesions of 75–85% within 3 years of Gamma Knife radiosurgery, with similar obliteration rates if a second Gamma Knife treatment is necessary. Over 33,000 AVMs have been treated with Gamma Knife radiosurgery worldwide, making it the second most common indication after metastatic brain tumors.

Trigeminal neuralgia is a neurological condition marked by excruciating pain of the fifth cranial nerve that enervates the face in three branches between the eyebrows and the jawline. The pain may be caused by a blood vessel pressing on a nerve, by a tumor, by multiple sclerosis, or for unknown reasons. This is the first condition ever treated by Lars Leksell, using a 200 kVp X-ray unit affixed to a stereotactic frame in a treatment performed in 1951. The root entry zone of the nerve as it enters the brainstem is the target volume. The nerve diameter at that point is only 2–4 mm and the consequences of a geometric miss with the radiosurgery treatment volume accidentally being directed to the brainstem could be quite severe. Alternative treatments include injection of glycerol into the cistern under radiographic guidance, radio frequency "burn" of

the nerve under radiographic guidance and microvascular decompression which is a fairly major brain surgery. Physicians at the University of Pittsburgh recently reviewed their first 10 years of treatments on 202 trigeminal neuralgia patients and found that > 85% had complete or partial relief of pain at 12 months after Gamma Knife radiosurgery (13). Over 12,500 patients with trigeminal neuralgia have been treated with Gamma Knife radiosurgery at this writing.

QUALITY CONTROL/QUALITY ASSURANCE

Quality Control and Quality Assurance for Gamma Knife radiosurgery is of critical importance. Unlike fractionated radiation therapy, Gamma Knife treatments are administered at one time, with the full therapeutic effect expected to occur in weeks, months, or years. Errors in any part of the radiosurgery process, from imaging to planning to the treatment itself could potentially have severe or even fatal consequences to the patient. An international group of medical physicists published a special task group report on Quality Assurance in stereotactic radiosurgery in 1995 (14) and the American Association of Physicists in Medicine discussed Quality Assurance for Gamma Knife radiosurgery in a task group report in that same year (15). Each group stressed the need for both routine Quality Control on a monthly basis, examining all physical aspects of the device, and calibration of radiation absorbed dose measurements with traceability to national standards. Both groups also emphasized detailed documentation and independent verification of all treatment parameters for each proposed isocenter before the patient is treated. An Information Notice was published by the U.S. Nuclear Regulatory Commission (NRC) on December 18, 2000 that documented 16 misadministrations in Leksell Gamma Knife radiosurgery cases in the United States over a 10-year period (16). The Nuclear Regulatory Commission defines a misadministration as "A gamma stereotactic radiosurgery radiation dose: (1) Involving the wrong individual, or wrong treatment site; or (2) When the calculated total administered dose differs from the total prescribed dose by > 10% of the total prescribed dose." Fifteen of the 16 incidences were ascribed to human error while utilizing the Leksell Gamma Knife models U, B, and B2. Six of the reported errors involved setting incorrect stereotactic coordinates (often interchanging Y and Z coordinates). Two errors occurred when the same shot was inadvertently treated twice. One error involved interchanging left and right side of the brain. One error involved using the wrong helmet. No consequences to patients were reported, but would be expected to be minor in most of the reported cases.

It is important to note in this respect that the new Leksell Gamma Unit Model C with (optional) Automatic Positioning System has the potential to eliminate many of the reported misadministrations. The older Leksell Gamma Unit Models U and B are manual systems in which the treatment plan is printed out and hand carried to the treatment unit. Stereotactic coordinates for each of the isocenters (shots) are set manually by one clinician and checked by a second person. It is thus possible to treat the patient with the wrong helmet, prone instead of supine, wrong gamma angle, incorrect plugged shot pattern, wrong time, or to repeat or omit shots. The operation of the new Model C Gamma Unit is computer controlled. The Leksell GammaPlan workstation is networked via an RS232C protocol with full error checking, thus transferring the treatment plan electronically. The shots may be treated in any order, but no shot may be repeated and the screen will indicate shots remaining to be treated. The helmet size is remotely sensed and treatment cannot commence if an incorrect helmet is selected. If the optional Automatic Positioning System is used, the X, Y, and Z stereotactic coordinates are automatically sensed to within 0.05 mm. The device will not permit treatment until the X, Y, and Z coordinates sensed by the APS system match those indicated on the treatment plan. Thus, it appears that all of the 15 misadministrations due to human error as reported by the Nuclear Regulatory Commission would have been prevented by use of the Model C with APS.

RISK ANALYSIS

The concept of misadministration should be placed in the larger concept of risk analysis. All medical procedures have potential adverse effects and, under state laws, patients must be counseled about potential consequences and sign an informed consent before a medical procedure (even a very minor procedure) may be performed. The relative risk of misadministration of Gamma Knife misadministration may be computed, utilizing the NRC report and data from the Leksell society on number of patients treated per year in the United States. Since $\sim$ 28,000 patients received Gamma Knife radiosurgery between 1987 and 1999, while 16 misadministrations were reported during the same interval, a relative risk of misadministration of 0.00057 per treatment may be computed for that period. Using the most recent year (1999) for which both NRC and patient treatment data are available, the relative risk drops to 0.0001/patient treatment.

These risks may be compared with other risks for patients undergoing an alternative procedure to Gamma Knife radiosurgery, namely, open craniotomy with hospital stay (17). A report by the National Institute of Medicine estimates that medical errors kill between 44,000 and 98,000 patients/year in the United States (18). These deaths reportedly occur in hospitals, day-surgery centers, outpatient clinics, retail pharmacies, nursing homes, and home care settings. The committee report states that the majority of medical errors do not result from individual recklessness, but from basic flaws in the way the health system is organized. A total of 33.6 million hospital admissions occur in the United States each year, which yields a crude risk estimate range of 0.0013–0.0029 death per admission to hospital or outpatient facility.

A second source of inadvertent risk of injury or death must also be considered. The National Center for Infectious Diseases reported in December, 2000 that an estimated 2.1 million nosocomial (hospital based) infections occur in the United States annually (19). These infections are often drug resistant and require extremely powerful antibiotics with additional adverse effects. Given that there are 31 million acute care hospital admissions annually in the United States, the relative risk of a hospital based infection may be computed as 0.063/patient admission, or roughly one chance in 16. The risk of infection from craniotomy was given by the same report as 0.82/100 operations for the time period January, 1992–April, 2000.

The Leksell Gamma Knife Model C system is one example of a computer-controlled irradiation device. The rapidly developing field of Intensity Modulated Radiation Therapy (IMRT) is the subject of a separate article in this work. These complex treatments require extraordinary care on the part of treatment personnel to minimize the possibility of misadministration. Only rigorous Quality Assurance and Continuing Quality Improvement in radiation oncology can make such treatments safe, reliable and effective. Leveson has studied the use of computers to control machinery which could potentially cause human death or injury, such as linear accelerators, nuclear reactors, modern jet aircraft and the space shuttle (20).

EVALUATION

The Leksell Gamma Knife, after a long period as a unique invention of limited applicability, has enjoyed explosive growth in medical application in the last 10 years. It is one of a number of medical instruments specifically created to promote minimally invasive surgery. Such instruments subject human patients to less invasive, painful, and risky procedures, while often enhancing the probability of success or in fact treating surgically inoperable patients. Over 297,000 Gamma Knife treatments have been performed worldwide as of the last tally. Most treatments are successful in achieving treatment objectives in 85–90% of patients treated. Although the Gamma Knife is the most massive and probably the costliest single medical product ever introduced, it has enjoyed widespread commercial and clinical success in 31 countries. The simplicity and reliability of operation of the unit make its use an effective treatment strategy in lesser developed countries where difficult craniotomies may not be as successful as in more developed countries. The newest version of the unit addresses the issues of automation, efficiency, treatment verification, and increased accuracy. The instrument appears to be well established as an important neurosurgical and radiological tool.

BIBLIOGRAPHY

Cited References

1. Wu A, et al. Physics of Gamma Knife Approach on Convergent Beams in Stereotactic Radiosurgery. Int J Rad Oncol Biol Phys 1990;18:941–949.
2. Greenblatt SH. The crucial decade: modern neurosurgery's definitive development in Harvey Cushing's early research and practice. 1900 to 1910, J Neurosurg 1997;87:964–971.
3. Schaltenbrand G, Wahren W. Atlas for stereotaxy of the human brain. 2nd ed. Stuttgart: Thieme; 1977.
4. Leksell L. The Stereotactic Method and Radiosurgery of the Brain. Acta Chir Scand 1951;102:316–319.

5. Leksell L. Stereotaxis and radiosurgery: an operative system. Springfield: Thomas Publishers; 1971.

6. Rand RW, Khonsary A, Brown WJ. Leksell stereotactic radiosurgery in the treatment of eye melanoma. Neurol Res 1987; 9:142–146.

7. Walton L, Bomford CK, Ramsden D. The Sheffield stereotactic radiosurgery unit: physical characteristics and principles of operation. Br J Rad 1987;60:897–906.

8. Bunge HJ, Guevara JA, Chinela AB. Stereotactic brain radiosurgery with Gamma Unit III RBS 5000. Barcelona Proceeding 8th European Congress Neurology Surgery; 1987.

9. Sanghavi SN, Miranpuri BS, Chapell R. Multi-Institutional Analysis of Survival Outcome for Radiosurgery Treated Brain Metastases, Stratified by RTOG RPA Classification. Int J Rad Oncol Biol Phys October, 2001;51(2):426–434.

10. Flickinger JC, et al. Results of acoustic neuroma radiosurgery: an analysis of 5 years' experience using current methods. J Neurosurg Jan, 2001;94(1):141–142.

11. Spetzler RF, Martin NA. A proposed grading system for arteriovenous malformations. J Neurosurg 1986;65(4):476–83.

12. Steiner L, Lindquist C, Adler JR, Torner JC, Alves W, Steiner M. Clinical outcome of radiosurgery for cerebral arteriovenous malformations. J Neurosurg 1992;77(1):1–8.

13. Kondziolka D, Lunsford LD, Flickinger JC. Stereotactic radiosurgery for the treatment of trigeminal neuralgia. Clin J Pain 2002;18(1):42–7.

14. Lutz W, Arndt J, Ermakov I, Podgorsak EB, Schad L, Serago C, Vatnitsky SM. Quality Assurance Program on Stereotactic Radiosurgery. Berlin: Springer; 1995.

15. Schell MC, Bova FJ, Larson DA, Leavitt DD, Lutz WR, Podgorsak EB, Wu A. Stereotactic Radiosurgery. AAPM Report No. 54, College Park: American Association of Physicists in Medicine; 1995.

16. Medical Misadministrations Caused by Human Errors Involving Gamma Stereotactic Radiosurgery, Bethesda. U.S.: Nuclear Regulatory Commission; 2000.

17. Goetsch SJ. Risk Analysis of Leksell Gamma Knife Model C with Automatic Positioning System. Int J Rad Oncol Biol Phys March, 2002;52(3):869–877.

18. To Err is Human: Building a Safer Health System. Bethesda: National Academy Press; 2000.

19. Geberding JL. Semiannual Report. National Nosocomial Infections Surveillance (NNIS) System. Rockville; U.S.: Public Health Service; June, 2000.

20. Leveson N. Safeware: System Safety and Computers. Reading: Addison-Wesley; 1995.

Reading List

De Salles AAF, Lufkin R, Minimally Invasive Therapy of the Brain. New York: Thieme Medical Publishers, Inc.; 1997.

Pollock B. Contemporary Stereotactic Radiosurgery: Technique and Evaluation. Oxford: Futura Publishing Company; 2002.

Coffey R, Nichols D. A Neuroimaging Atlas for Surgery of the Brain: Including Radiosurgery and Stereotaxis. Philadelphia: Lippincott-Raven; 1998.

Ganz J. Gamma Knife Radiosurgery. 2nd ed. New York: Springer-Verlag; 1997.

Webb S. Physics of Three-Dimensional Radiation Therapy: Conformal Radiotherapy, Radiosurgery and Treatment Planning. London: Institute of Physics Publishing; 1993.

Lunsford LD, Kondziolka D, Flickinger JC. Gamma Knife Brain Surgery. S. Karger Publishing; March, 1998.

See also COBALT 60 UNITS FOR RADIOTHERAPY; RADIOSURGERY, STEREOTACTIC.

GAS AND VACUUM SYSTEMS, CENTRALLY PIPED MEDICAL

BURTON KLEIN
Burton Klein Associates
Newton, Massachusetts

INTRODUCTION

Terminal gas outlets or vacuum inlets are as common a fixture today in hospital rooms as stethoscopes. Even clinics, outpatient surgery facilities, and some nursing homes utilize them. But how did they get there? And have they helped medical and nursing staff give better patient care?

This article is intended to give readers a brief look at how and why these systems were developed, how they operate, what hazards they pose, what standards have been developed to mitigate hazards as well as to standardize operation, and why maintenance of these systems is very important. In a sense, medical gas and vacuum systems are a reflection, in part, of how the practice of medicine has changed over the past 60–70 years: Both systems have become more complex and sophisticated in order to meet and treat more serious illnesses.

The systems discussed below are those involving the distribution of pressurized gases (or suctioning of air)or the creation of a vacuum via rigid metal pipes, with the source of gas or suction *not* in the same room as the end-use terminals of the system. Further, the description of these systems is a generalized one; specific systems may have different operating characteristics to meet a particular need. The authority(ies) having jurisdiction (AHJ) should be consulted for specific locations (e.g., hospital, clinic, nursing home) and application purpose (medical surgical, dental, laboratory, veterinary).

Finally, the limited bibliography provided at the end of this article has been included (1) for readers who wish to pursue this subject further, and (2) to show the various standards organizations involved in setting standards that are used in designing, installing and using these systems.

GAS SYSTEMS (PRESSURIZED)

To understand how and why the piping of medical gases to operating rooms and other patient care areas came into practice, it is necessary to briefly review how the practice of medicine, and in particular the practice of anesthesiology, changed from the mid-1800s to the early 1900s, for it was advances in administering anesthesia that led to the piping of gases into operating rooms, and from there to many other patient care areas.

Some History

The first public demonstration of inhalation anesthetics took place on October 16, 1846 at the Massachusetts General Hospital in Boston. There had been some experimentation prior to this date, but this publicized demonstration by Dr. John W. Collins clearly showed that patients could be kept unconscious as long as necessary and have surgery performed without their sensing pain. A giant step forward in the practice of medicine had been achieved.

These first years of anesthesiology were relatively simple in that only a sponge soaked in ether and placed over the nose and mouth of patients was used to induce anesthesia. In 1868, Andrews introduced oxygen mixed with nitrous oxide as an adjunct to inhalation anesthesia. In 1871, cylinders of nitrous oxide became available. In 1882, cyclopropane was discovered, though it was not until the 1930s that it was found useful for anesthesia. And in 1887, Hewitt developed the first gas anesthesia machine that used compressed gases in cylinders.

This controlled unconsciousness sparked a dramatic increase and change in medical practice and hospital activities. No longer did patients enter a hospital only for terminal care or to feel the cutting edge of a scalpel. Problems occurring inside the body could now be exposed for examination and possible correction. And, as anesthesia systems became more available and sophisticated, the volume and type of operations increased dramatically. Finally, the discovery that oxygen enrichment helped patients during anesthesia and operations increased the use of oxygen in operating rooms tremendously.

By the 1920s, cylinders of oxygen and nitrous oxide were constantly in motion about hospitals, from loading docks to storage rooms to operating rooms and back again. But, occasionally, cylinders did not make the entire circuit in one piece. Thus, the question occurred to healthcare staff: Was there another, better way to provide gas in operating rooms?

Some sources credit the late Albert E. McKee, who was working with a Dr. Waters at the University of Wisconsin Medical Center in the 1920s, with installing the first medical piped gas system that used high pressure cylinders of oxygen connected by pipes to outlets in nearby operating rooms. He (and his counterparts) saw this as a better method of providing gases to operating rooms. Their installation had some very positive effects (it also had some negative ones that will be discussed shortly):

1. There was an immediate reduction in operating costs. Instead of many small cylinders, fewer but larger cylinders could be utilized, with concurrent reduction in the unit cost per cubic foot of gas. (It has been reported to this author that the amount saved after McKee installed his system was sufficient to pay the salaries of the University of Wisconsin anesthesiology departmental staff.) Fewer cylinders also meant less loss of residual gas that remained in empty cylinders. When individual cylinders were used, they would be replaced when the pressure inside the cylinder dropped down to $\sim$500 psi (lb $\cdot$ in^{-2} or 3448 kPa); when two or more cylinders were manifolded together as a source, however, individual cylinders could be allowed to go down to $\sim$ 40 psi (276 kPa), since there were other cylinders in the system from which gas could be drawn.

2. This method provided immediate access to gases. Operating room staff only needed to connect hoses to gas outlets to obtain gas. The large supply at the central dispersion point could be monitored by one person (instead of each anesthesiologist worrying about their own individual small cylinders). Since several large cylinders were grouped together, when one became empty, or nearly empty, others could be switched on line and the empty one replaced. Thus, operating room staff were assured of a constant supply of gas.

3. Safety was improved. No longer were cylinders, with their inherent hazards, inside the operating room. Cylinder movement around the hospital was dramatically reduced.

Industry had been using gas under pressure in pipes since the late 1800s (e.g., street lamps). Piping gases around a hospital was, thus, a natural extension of this methodology, though components had to be modified to meet medical needs. These new installations were not without problems, however. The system had to be leak-free, since an escape and buildup of gases (flammable or oxidizing) within a building was dangerous. Also, having these gases carried in pipes around a healthcare facility meant that an incident in one place now had a means of becoming an incident in another place. Finally, if more than one gas were piped, the possibility of cross-connection and mixing of gases existed (and cross-connecting of some gases can create explosive possibilities).

This last problem was of particular concern since initially there was no restriction on the piping of flammable anesthetic gases. Several institutions, including the University of Wisconsin, installed systems to pipe ethylene gas. Even though the standardization of terminal connectors began in the late 1940s, explosions in operating rooms continued to occur. While the number of such incidents was not large, the occurrence was always devastating, almost always killing the patient, and sometimes maiming medical–surgical–nursing staff. In 1950, the National Fire Protection Association (NFPA) Committee on Hospital Operating Rooms proposed a number of changes, including prohibiting the piping of flammable anesthetic gases. The proposal, adopted by the NFPA membership, eliminated one possible source of explosions and fire.

The relatively recent introduction (late-1940s) of storing a large volume of oxygen on-site in a liquid state presented a new host of concerns. While large-volume storage replaced the use of many cylinders, it vastly increased the amount of oxygen in one location and introduced the hazard associated with gas in a cryogenic state (i.e., gas at an extremely low temperature).

System Components

The following is a general description of components used in piped gas systems today (Fig. 1). An actual system may not utilize all these components. However, all systems have certain minimum safety features, as discussed below. In addition, standardization of some components (e.g., threaded station outlet connections) and practices (e.g., operating pressures) has evolved over the years, which will be discussed later as well.

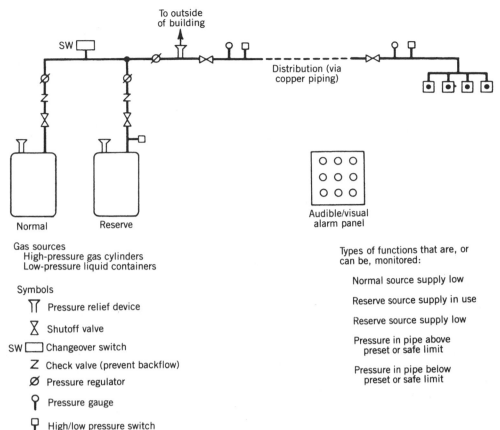

Figure 1. Components of a medical gas (pressurized) central piping system (simplified). Standards have been developed for component and total-system performance and safety.

Gases. The most common nonflammable medical gases piped under pressure today include medical air, oxygen, nitrogen, and nitrous oxide. These gases are available from manufacturers in cylinders into which a large volume of the gas has been compressed. The pressure of the gas in these cylinders can be > 2000 psig (13.8 GPa). Some of these gases are also available in a liquefied state, through a refrigeration process, and are supplied in portable containers or in large stationary bulk units (tanks). (When the gas is used, it is allowed to evaporate and return to its gaseous state.) The gas in the liquefied state is placed under relatively low pressure [~ 75 psig (520 kPa)]. One gas (air) can also be obtained on-site using compressors. Whichever method is used to obtain a specific gas, it must interface with the piping portion of the system; that is, the mechanical parts must interconnect. It also means that the pressure of the source gas needs to be regulated to pressure at which the system is operating. Gas in the liquid state must be transformed to the gaseous state. For all gases, except nitrogen, a pressure between 50 and 55 psig (344 and 379 kPa) at station outlets has become the standard. For nitrogen, which is used to power nonelectric surgical tools, such as drills, bone saws, and dermatomes, a pressure between 160 and 185 psig (1103 and 1379 kPa) is used. This regulation can be likened to electricity and the use of transformers that are installed between the power generators of utility companies (where voltages upward of 10,000 V are generated) and buildings where the voltage

is regulated down to 208 or 110 V. In gas systems, these transformers are called pressure regulators.

In the last few years, other nonpatient medical gases (called support gases in NFPA 99, *Standard for Health Care Facilities*) have begun to be piped. These gases are used for powering equipment that use pressurized gas in order to function (e.g., pneumatically operated utility columns). Gases in this category include nitrogen and instrument air.

Source Equipment

Other devices used at the source portion of the piped gas system include (1) shutoff valves at prescribed locations so that a complete or partial shutdown of a source can be accomplished; (2) check valves to control the direction of gas flow (i.e., one direction only); (3) pressure-relief valves, which are preset to vent gas to the atmosphere if the pressure in a cylinder, container, or pipeline becomes excessive enough to cause a rupture or explosion if allowed to continue to increase; and (4) signals to alarm panels to indicate such conditions as low and high pressure.

A separate reserve supply of the gas is also included in some piped systems. This reserve serves as a backup if the main (normal) source is interrupted or requires repair. This reserve can be adjacent to, or remote from, the main source. Its remote location precludes both sources from damage should an accident occur to one source. Such separation, however, may not always be possible.

A requirement added in NFPA 99 in the early 1990s called for a piped bulk-oxygen system that has its source sources located outside the building to have a separate connection to the piping system, also located outside of the building and accessible to a bulk oxygen delivery truck (ODT). Thus, if both the main and reserve supplies of oxygen were to fail or become damaged or depleted, the ODT could be used as the source. This emergency connection is required only for the oxygen supply because it is a life-support gas.

Finally, if extra cylinders or containers of gases are kept stored within a facility or within close proximity to a healthcare facility, a safe means of storing the gas must be provided. These storage requirements are intended to provide safety for occupants should an incident occur outside the storage room (i.e., in order to protect the cylinders from adding to the incident), or should an incident occur inside the storage room (i.e., in order to protect occupants in the building from the incident in the storage room).

Piping (Distribution) System. From the source piping is installed to distribute the gas to patient care areas. (Standards require gases piped to laboratory areas to be supplied from a separate system from gases piped to patient care areas. This is to prevent any backfeeding of gas from laboratory systems into patient care systems, and to allow for different pressures where required or desired for laboratory purposes.) Sizes and locations of main, riser, and lateral pipes should take into consideration both present and future needs or plans. As with the source, shutoff valves and flow-control devices (check valves) are required by standards at certain locations in the piping (distribution) system.

Terminal Units (Station Outlets). The endpoints (called outlets) of the piped gas system are very important since it must be very clear what gas is flowing to each outlet. To eliminate any chance of mix-up, noninterchangeable mechanical connectors have been designed for each type of gas. These different connectors are similar to the different configurations of electrical outlets for 110, 220–208 (single-phase), 220–208 V (three-phase), and so on. Labeling of gas outlets and piping is also required. Color coding of new piping became a requirement in NFPA 99 in 2005. However, it requires staff to remember the color coding scheme. It also poses problems for persons who are color-blind.

Alarm Panels/Monitoring. Because gases are relied upon for life support, system monitoring is essential and has become standard practice. Sensors and alarms are required to be installed in all critical care areas to detect if the pressure decreases or increases beyond specified limits (e.g., should someone inadvertently or deliberately close a shutoff valve). Other sensors are required to detect when the normal source and/or reserve supply are low and when the reserve supply has been switched in.

All signals are fed to one or more master alarm panels, one of which is required to be constantly monitored by facility staff. The electrical power for these alarms is to be connected to the facility's emergency power system so that alarms will continue to function if normal electrical power is interrupted. This constant surveillance is required because of fire hazards that could develop should something in the system malfunction, and for patient safety should gas delivery be interrupted. Immediate action (corrective, responsive) is necessary in either situation.

Installation of Systems. In the early 1990s, concern about the quality of the installation of medical piped gas (and vacuum) systems resulted in the technical committee responsible for piping system requirements listed in NFPA 99, *Standard for Health Care Facilities*, to revise and expand requirements for their installation. To assure the system has been installed according to the design drawings, extensive requirements were included not only for the installer, but also for a verifier who is to be totally independent of the installer, and who tested the system after everything was connected and readied for operation (i.e., for patient use).

Performance Criteria and Standards

When first installed, medical piped gas systems generally followed the practices then in use for the piping of nonmedical gases. These practices were considered adequate at the time. In 1932, the subject came to the attention of the NFPA Committee on Gases, which noted the following hazards that the installation of these systems posed for hospitals:

1. Pipes, running through an extensive portion of a building into operating rooms, carried gases that were of the flammable type (those that burn or explode if ignited) or of the oxidizing type (those that support and intensify the burning of combustibles that have been ignited).
2. A large quantity of gas in cylinders was being concentrated and stored in one area.
3. The possible buildup of potentially hazardous gas concentrations existed should the pipes leak.
4. A possible explosion in an operating room was possible if a hose on an anesthesia machine were connected to the wrong gas.
5. A compromising of patient safety existed in that a mix-up of gases could be injurious or even fatal.

This notification came in the form of identification of hazards resulted in a request by the National Board of Fire Underwriters that this Committee develop a set of guidelines on the subject. The Committee studied the subject and, in 1933, proposed "Recommended Good Practice Requirements for the Construction and Installation of Piping Systems for the Distribution of Anesthetic Gases and Oxygen in Hospitals and Similar Occupancies, and for the Construction and Operation of Oxygen Chambers." The proposed document contained guidance on acceptable types of piping, the length of pipe runs, the identification of piping, the kind of manifolds that were acceptable, and the number and location of shutoff valves. As noted in the title, it permitted the distribution of anesthetic gases, which were flammable. The NFPA did not formally adopt the proposed Recommended Good Practice until 1934. Over the years, as more knowledge was gained from the hazards, installation, and use of piped gas systems, the NFPA standard also changed. In addition, other organizations prepared standards addressing other aspects of piped gas systems (1–10). A brief summary of these efforts follows.

National Fire Protection Association Standards. The original NFPA document, first designated NFPA 565 and later, NFPA 56F, which in turn was incorporated into NFPA 99 (11) remained unchanged until 1950 when the NFPA Hospital Operating Room Committee, working with the NFPA Gas Committee, recommended that piping of flammable anesthetic gases be prohibited. Later, specific safety requirements were added, such as for the storage of gases, shutoff valve locations, check valves, line-pressure gages, pressure switches and alarm panels, and installation and testing criteria. Performance criteria, in the form of operating pressure limits for different gases, were added because no other organization had included them in their documents, and uniformity of systems operations was helpful to both medical staff, designers, and the industry producing the equipment for these piped systems.

Other NFPA documents have been developed over the years that impact on medical piped gas systems. These include documents on the subjects of emergency electric power, bulk oxygen supplies, and building construction (11–16).

Compressed Gas Association (CGA) Standards. The CGA, an organization of manufacturers of gases and gas equipment, publishes many documents on the subject of gases. Some of these apply directly to medical gas piping systems; others are generic and affect any closed gas system. Topics addressed include gas cylinder criteria; noninterchangeable connectors for cylinders and terminal outlets; liquefied gas transfer connections; compressed gas-transfer connections; and commodity specifications for nitrogen, air, nitrous oxide, and oxygen (1–6).

Other Organizations. (7–10).

VACUUM SYSTEMS

Some History

The development of vacuum central piped vacuum systems, in place of portable suction machines, occurred over a period of time from the late-1940s to the early-1950s. These systems did not have to face the same unknowns and problems that the development of piped gases faced 20 years earlier. While they did not pose as great a threat as piped gases [i.e., they were not carrying oxidizing gases at 50 psig (344.8 kPa)], they did have their own hazards (e.g., they carried flammable and/or nonflammable oxidizing gases around a facility; created patient risks should the system stop; created possible restrictive contamination of

orifices and low vacuum levels if orifices became clogged or contaminated; and created excecssive loading on emergency electrical power systems if not properly provided for in the system). Since many vacuum systems were and still are installed simultaneously with central piping systems for gases, they added to the problems associated with installing two or more systems simultaneously (e.g., cross-connections, incorrect labeling).

Until vacuum central piping systems were installed, medicine utilized small portable suction pumps that created a vacuum much the same way an ordinary vacuum cleaner creates suction (Fig. 2). A major difference, however, is the type of collection container used. For a home vacuum cleaner, a semiporous bag collects dirt; for a medical vacuum machine, a nonporous "trap" is necessary because of the products collected (e.g., body fluids of all kinds, semiliquid bulk material). In addition, a major problem with portable suction machines is the airborne bacteria it can spread as it operates. Since vacuum pumps operate on the principle of moving air from one place to another, this movement can be unhealthy in a healthcare setting where airborne bacteria can be infectious. Another problem with individual suction pumps was their need to be safe when flammable anesthetics were in use. (This ceased to be a problem as flammable anesthetics were replaced by nonflammable anesthetics in the 1960s and 1970s.) A central vacuum system eliminated these two problems, since it exhausted contaminated air outdoors and no electric motor was needed in the patient area in order to provide the vacuum. (It should *not* be concluded that portable suction pumps are no longer used. With bacteria filters now available and used on suction pumps,

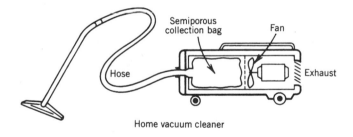

Home vacuum cleaner

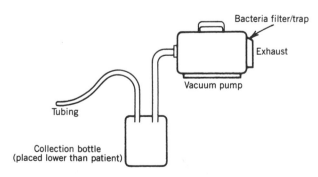

Portable medical vacuum pump

Figure 2. Home vacuum cleaner (high volume, low degree of vacuum) versus portable medical vacuum pump [low volume, low-to-high (adjustable) level of vacuum].

these devices are still quite suitable, in the same fashion that individual gas cylinders are still used.) It is necessary, however, that a trap unit be used between the patient and the vacuum control regulator and station inlet, so that nothing but air is drawn into the piping system.

The other reason central (add) vacuum systems began to be installed was the result of studies by the late David A. McWhinnie, Jr., in the early 1950s that showed the economic viability of these systems. Initially, vacuum central piped vacuum systems served only operating rooms and specialty areas, such as postanesthesia recovery rooms and emergency rooms. General patient care areas were added as demand for suction increased and economics made their installation viable. The reduction in the spread of airborne bacteria that central piped vacuum systems provided also contributed to their installation in general patient care areas as hospitals become more aware and concerned about this hazard. Pediatric and neonatal areas were last to install piped vacuum systems because of concern over what high degrees of vacuum and flow rates might do to babies (e.g., damage to very delicate tissues, possible collapse of newly functioning lungs). With improvements in the regulation of the degree of vacuum and more staff education, this concern abated, and piped vacuum systems were installed in these areas as well.

System Components

A medical piped vacuum system can be diagrammed, as shown in Fig. 3, in a fashion similar to the piped gas system described above. However, remember that the flow of subatmospheric air is opposite to the flow of pressurized gases in centrally piped gas systems. Note that piped vacuum systems require much larger orifices at

inlet terminals than those at outlet terminals for gas systems because of (1) the pressures involved [i.e., 12 in. (30.5 cm) of Hg (40.6 kPa) (negative pressure) as opposed to 50 psi (344.8 kPa)]; and (2) the need for high flow. As noted previously for piped gas systems, the following description for piped vacuum systems includes the major components of a large system. Of course, individual systems will vary.

Sources for Vacuum. Pumps provide the means by which suction is created. They draw in the air that exists within the piped system, and exhaust it via a vent discharge located on the outside of the building (generally on a roof) and away from any intake vents. This configuration allows exhausted air, which may be infectious, to dissipate into the atmosphere.

At least two pumps are required to be installed, each with either one capable of providing adequate vacuum to the entire system. This redundancy is necessary to keep the vacuum system functioning in case one pump fails or needs maintenance. To smooth out pump impulses and provide a constant vacuum, a receiver (surge tank) is required to be installed at the source site between the pumps and the rest of the system. Shutoff valves and check valves are to be installed for maintenance, and efficiency, and to shut down the system (or portions of the system) in the event of an emergency.

Piping (Distribution) System. Like piped gas systems, the first standard on piped vacuum systems required metal pipes to be used to connect the various patient care areas to the receiver. And like gas systems, there were and still are prescribed locations for shutoff valves, check valves, vacuum switches, and vacuum-level gages.

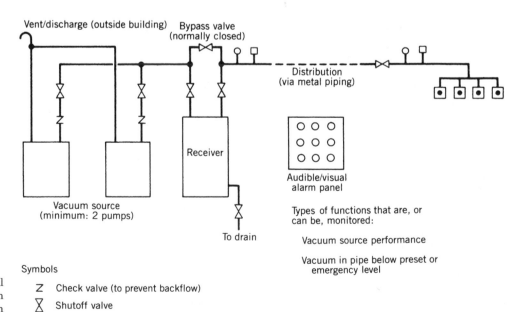

Figure 3. Components of a medical piped vacuum central piping system (simplified). Standards have been developed for component and total-system performance and safety. A collection unit and a trap are required between the inlet terminal and the patient.

Symbols

Z   Check valve (to prevent backflow)

X   Shutoff valve

⚲   Vacuum gauge

⚲   Low-vacuum switch (to alarm)

⊡   Station (vacuum) inlet terminal

However, because of the subatmospheric operating pressures and lower concentration of oxidizing gases in a piped vacuum system as opposed to a piped gas system, more types of metal pipes are allowed in the first standard on these vacuum systems. Piping for vacuum systems may have to be larger than piping for gas systems because of the level of airflow (vacuum) required by medical staff. Also, originally, the melting point allowed for joints can be lower for piped vacuum systems was permitted to be lower than the 1000° withstand-temperature required for piped gas systems. However, it is recognized that vacuum systems are sometimes installed at the same time as gas systems; as such, it may be prudent in those situations to use one type of piping throughout in order to reduce the chance of using the wrong piping and/or brazing on the piped gas system. In recent years, the committee responsible for piped vacuum system requirements has gradually required the type of piping for vacuum systems to be closer to that required for piped gas systems.

A significant difference of piped vacuum systems from piped gas systems permits connection of medical laboratories into patient care vacuum systems, though with the stipulation that the connection be made directly into the receiver and not via the pipes serving patient areas, so that a fluid trap and manual shut off valve are included. Separate systems, however, are encouraged.

Terminal Units (Station Inlets). The terminals for vacuum systems (called inlets), resemble the outlets of gas systems. Thus, it is required that they be clearly labeled vacuum or suction. To preclude problems (since piped vacuum systems sometimes are installed along at the same time with piped gas systems), the connector used for vacuum inlets is to be mechanically different from all gas outlet connectors, thereby reducing the chance of interconnection of gas and vacuum equipment.

Alarm Panels/Monitoring. Because vacuum is now a critical tool in the practice of medicine, it, too, requires constant monitoring. An audible/visual alarm panel (integrated with one for a piped gas system if also installed) alerts staff to problems similar to those of gas systems (e.g., pump malfunction, a drop in vacuum below a prescribed level).

Performance Criteria and Standards

With no vacuum standards in existence, the first piped vacuum systems installed were based on prevailing engineering expertise. While vacuum systems may seem similar to gas systems (e.g., piping, the movement of gas, although in the opposite direction), the design criteria for them are very different technically. With a piped gas system, after the source gas has been connected, the whole system reaches and stabilizes at a narrow range of pressure. In a piped vacuum system, a pump is trying to evacuate a space and provide a degree of vacuum [measured in inches of Hg (negative) *and* in volume displacement (flow)] at each inlet. In the former, the gas itself within the system provides a positive pressure and flow; in the latter, a pump is required to create a subatmospheric pressure and flow.

In the early 1950s, ineffective performance plagued many piped vacuum systems. Staff techniques, the lack of appropriate check valves, and widely divergent pump sizing contributed to the problems. One city known to have been investigating the problem was Detroit. During the 1950s, the city attempted to establish a municipal standard for the piped vacuum systems in city hospitals. Several of the major manufacturers of vacuum pumps became involved in the effort. Because general agreement could not be reached, the manufacturers suggested that industry try to develop a standard. This led to Compressed Gas Association (CGA) involvement, since many of its members were by the late 1950s supplying vacuum pumps and inlet connectors. In 1961, the CGA released a document (designated P-2.1) that included recommendations on pumps, warning systems, piping, installation, and labeling. It also included recommendations on pump sizing.

During the 1960s, staff practices were improved or standardized. This included the location of collection bottles (below patient level) and the use of regulator bypasses. This helped system performance as well. Because there continued to be differences of opinion in the engineering world regarding piped vacuum system design, the CGA approached the NFPA in the early-1970s about the NFPA developing a medical–surgical vacuum system standard. The NFPA agreed to the idea and a subcommittee of the then Committee on Hospitals was established. After tests of various pumps and suction-therapy equipment, and surveys of actual systems in hospitals, a recommended practice (designated NFPA 56K) was adopted by NFPA in 1980. After 3 years, it was revised and changed to a standard (being incorporated into NFPA 99, Standard for Health Care Facilities, at the same time) (11). The NFPA recommended practice (and then standard) generally contained the same topics as the CGA document. Other standards that impact piped vacuum systems have been developed. Most have already been mentioned or listed for piped gas systems and cover such subjects as cleaning and purging, pressure testing, and connection for emergency electrical power.

As noted, the initial criteria for installing vacuum central piped vacuum systems differed from piped gas systems. Of late, the major document on the subject (NFPA 99) has gradually revised piped vacuum system requirements, particularly on piping material, to that required for piped gas systems. But if they a piped vacuum system is are installed alongside a piped gas system at the same time the gas system is installed, the installation standards of the gas system should be considered to avoid possible degradation of the gas system, which requires more stringent standards.

Requirements on piped vacuum system design have also been deleted from the NFPA 99 document as it was seen to be outside the scope of the document (NFPA 99 is a minimum performance safety standard), as well as not changed in the document for > 20 years.

MAINTENANCE OF SYSTEMS

A separate note on maintenance is deemed warranted because of the inherent hazards posed by piped gas and vacuum systems, as well as the high reliance now placed on these systems by medical–surgical–nursing staff. In the former, everyone is affected by the hazards; in the latter, failure of these systems can place patients at considerable medical risk.

Like any system, periodic maintenance is necessary in order to assure continuous, and optimum and safe level of operation. For piped gas or vacuum systems, this includes visual inspection of exposed pipes and outlets–inlets, sampling of gases (gas systems), measurement of pressures (gas systems), measurement of flow rates (vacuum systems), and testing of alarms. Guidance on this subject is included in such documents as NFPA 99, *Standard for Health Care Facilities* (11).

While today's standards assure a high probability of a safe and reliable system, mechanical failures can and do occur, and human error or abuse still remain. Thus, should a fault of some kind occur, or a wrong connection be made, periodic maintenance should detect the condition so that corrective action can be taken before a serious incident occurs. This maintenance is particularly necessary whenever either system is breached for upgrading, component maintenance occurs, or system expansion purposesis made. The value of these systems in the treatment of patients demands no less.

Originial manuscript for this article was reviewed for technical accuracy by John M.R. Bruner, M.D., W.E. Doering, William H.L. Dornette, M.D., James F. Ferguson, Edwin P. Knox, (the late) David A. McWhinnie, Jr., Ralph Milliken, M.D., and (the late) Carl Walter, M.D.

BIBLIOGRAPHY

Cited References

1. Compressed Gas Association. Commodity Specification of Air. G-7.1, Arlington Chantilly (VA): CGA; 1973–2004.
2. Compressed Gas Association. Standard for the Installation of Nitrous Oxide System at Consumer Sites. G-8.1. Chantilly Arlington (VA): CGA; 1979–1990.
3. Compressed Gas Association. Standard Color-Marking of Compressed Gas Cylinders Intended for Medical Use in the OR, C-9. ChantillyArlington (VA): CGA; 1982–2004.
4. Compressed Gas Association. Commodity Specification for Nitrogen, G-10.1. ChantillyArlington (VA): CGA; 1985–2004.
5. Compressed Gas Association. Compressed Gas Cylinder Valve Outlet and Inlet Connections. V-1. ChantillyArlington (VA): CGA; 1977–2003.
6. Compressed Gas Association. Diameter Index Safety System, V-5. ChantillyArlington (VA): CGA; 1978–2005.
7. American Society of Sanitary Engineering. Professional Qualifications Standard for Medical Gas Systems, Installer, Inspectors, Verifiers, Maintenance Personnel and Instructors. ASSE, Series 6000: Westlake (OH); 2004.
8. American Society for Testing and Materials. Specification for Seamless Copper Water Tube. B-88, Philadelphia: ASTM; 1986–2003.
9. American Society for Testing and Materials. Specifications for Seamless Copper tube for Medical Gas Systems. B-819, Philadelphia: ASTM; 1992.
10. American Society for Testing and Materials. Standard Test Method for Behavior of Materials in a Vertical Tube Furnace at 750°C, E-136, Philadelphia: ASTM; 1982–2004.
11. National Fire Protection Association. Standard for Health Care Facilities (which includes criteria on piped medical gas systems, piped medical–surgical vacuum systems, and emergency electrical power), NFPA 99. Quincy (MA): NFPA; 1987–2005.
12. National Fire Protection Association. Standard for Bulk Oxygen System at Consumer Sites, NFPA 50. Quincy (MA): NFPA; 1985 (now included in NFPA 55, Standard for the Storage, Use, and Handling of Compressed Gases and Cryogenic Fluids in Portable and Stationary Containers, Cylinders, and Tanks; 2005).
13. National Fire Protection Association. National Electrical Code. NFPA 70, Quincy (MA): NFPA; 1987–2005.
14. National Fire Protection Association. Life Safety Code. NFPA 101, Quincy (MA): NFPA; 1985–2003.
15. National Fire Protection Association. Standard on Types of Building Construction. NFPA 220, Quincy (MA): NFPA; 1985–1999.
16. National Fire Protection Association. Standard Test Method for Potential Heat of Building Materials. NFPA 259, Quincy (MA): NFPA; 1987–2003.

Reading List

National Fire Protection Association. Historical Proceedings. Annual Meeting, Quincy (MA): NFPA; 1933.
National Fire Protection Association. Historical Proceedings. Annual Meeting, Quincy (MA): NFPA; 1934.
National Fire Protection Association. Historical Proceedings. Annual Meeting, Quincy (MA): NFPA; 1950.
National Fire Protection Association. Historical Proceedings. Annual Meeting, Quincy (MA): NFPA; 1951.
American Welding Society. Specification for Brazing Metal. A5.8, Miami (FL): AWS; 1981–2003.
American Society of Mechanical Engineers. Boiler and Pressure Vessel Code. New York: ASME; 1986–2001.

See also CODES AND REGULATIONS: MEDICAL DEVICES; EQUIPMENT MAINTENANCE, BIOMEDICAL; SAFETY PROGRAM, HOSPITAL.

GAS EXCHANGE.   See RESPIRATORY MECHANICS AND GAS EXCHANGE.

GASTROINTESTINAL HEMORRHAGE

R.C. BRITT
L.D. BRITT
Eastern Virginia Medical School
Norfolk, Virginia

INTRODUCTION

Gastrointestinal (GI) hemorrhage is a common medical problem, with significant morbidity and mortality. Traditionally, GI hemorrhage was managed by medically supporting the patient until the bleeding stopped or surgical intervention was undertaken. The modern day management of GI hemorrhage involves a multidisciplinary approach, including gastroenterologists, surgeons, interventional radiologists, primary care physicians, and intensivists. Despite the evolution in management of GI hemorrhage, the mortality rate has remained fairly constant, concentrated in the elderly with greater comorbidity (1). Additionally, medical advances, e.g., proton pump inhibitors, H$_2$ blockers, antimicrobial treatment of *Helicobacter pylori*, and endoscopic management have led to a decrease in the number of operations for hemorrhage, but not in the actual number of hemorrhages (2). The incidence of upper GI bleeding has remained relatively constant at 100–150/ 100,000 people (3), with an estimated 300,000–350,000 admissions annually and a mortality rate of 7–10% (4). Lower GI hemorrhage accounts for 20–30% of GI hemorrhage, and typically has a lower mortality rate than upper GI bleeding.

There are three major categories of GI hemorrhage, including esophageal variceal bleeding, nonvariceal upper GI bleeding, and lower GI bleeding. Typically, upper GI bleeding is classified as that bleeding occurring from a source proximal to the ligament of Treitz, with lower GI bleeding occurring distally. When bleeding occurs in the upper GI tract, it can be vomited as bright red blood, referred to as hematemesis. Slower bleeding from the upper GI tract is often referred to as "coffee-ground emesis", which refers to the vomiting of partially digested blood. Black, tarry stool is referred to as melena, and usually originates from an upper GI source, with the black color due to the action of acid on hemoglobin. Visible blood in the stool, or bright red blood per rectum, is referred to as hematochezia. Hematochezia is usually indicative of lower GI bleeding, although brisk upper GI bleeding may also present as hematochezia. The stool may also be maroon, suggesting the blood has mixed with liquid feces, usually in the right colon.

INITIAL EVALUATION AND RESUSCITATION

Upon presentation with GI hemorrhage, two large-bore (16 gauge or larger) peripheral IVs should be placed and intravascular volume resuscitation initiated with an isotonic solution. Lactated Ringers is frequently preferred to 0.9% Normal Saline because the sodium and chloride concentrations more closely approximate whole blood. The ABCs of resuscitation are a priority in the initial evaluation of the massive GI bleed, with careful attention given to the airway because of the high incidence of aspiration. The patient must be carefully monitored to ensure the adequacy of resuscitation. In the presence of continued rapid bleeding or failure of the vital signs to improve following 2 L of crystalloid solution, the patient should also begin receiving blood. If type-specific blood is not yet available, the patient may receive O negative blood.

On presentation, blood is drawn for hematocrit, platelets, coagulation profile, electrolytes, liver function tests, and a type and cross. Caution must be used when evaluating the initial hematocrit, as this does not accurately reflect the true blood volume with ongoing hemorrhage. A foley catheter should be inserted to monitor for adequate urine output as a marker for adequate resuscitation. An NG tube should be inserted to evaluate for the presence of upper GI bleeding, as bright red blood per NG tube indicates recent or active bleeding. While clear, bilious aspirate usually indicates that the source of bleeding is not upper GI, this is not a definite as absence of blood on

nasogastric aspirate is associated with a 16% rate of actively bleeding lesions found on upper endoscopy (5).

A thorough history is paramount when evaluating a patient presenting with GI hemorrhage. The clinical history may suggest the etiology of hemorrhage, as well as offer prognostic indicators. Important features in the history include a history of previous bleeding, history of peptic ulcer disease, history of cirrhosis or hepatitis, and a history of alcohol abuse. Also important is a history of medication use, particularly aspirin, nonsteroidals, and anticoagulants. Symptoms the patient experiences prior to the onset of bleeding, such as, the presence or absence of abdominal pain, can also be useful in the diagnosis.

A comprehensive physical exam must be done to evaluate the severity of the hemorrhage, as well as to assess for potential etiology. Massive hemorrhage is associated with signs and symptoms of shock, including tachycardia, narrow pulse pressure, hypotension, and cool, clammy extremities. The rectal exam may reveal the presence of bright red blood or melena, as well as evidence of bleeding hemorrhoids in a patient with bright red blood per rectum. Physical exam is also useful to evaluate for stigmata of liver failure and portal hypertension, such as, jaundice, ascites, telangiectasia, hepatosplenomegaly, dilated abdominal wall veins, and large hemorrhoidal veins.

When faced with a patient presenting with GI hemorrhage, the complete history and physical exam will help direct further management by assessing the likely source of bleed. The initial questions that must be answered to determine management priorities include whether the likely source of hemorrhage is from the upper or lower GI tract, and if the bleeding is from an upper source, whether the bleed is likely variceal or nonvariceal (Table 1).

UPPER GASTROINTESTINAL BLEEDING

Upper gastrointestinal bleeding is shown in Table 2.

Gastroesophageal Varices

Portal hypertension, defined as an increase in pressure > 5 mmHg (0.666 kPa) in the portal venous system (6), can lead to acute variceal hemorrhage. Cirrhosis, related to either chronic alcohol abuse or hepatitis, is the most common cause of portal hypertension, and leads to an increased outflow resistance, which results in the formation of a collateral portosystemic circulation. Collaterals form most commonly in the gastroesophageal junction and form submucosal variceal veins. Patients with isolated splenic vein thrombosis often form submucosal varices in the fundus of the stomach. Some 30–60% of all cirrhotic patients will have varices at the time of diagnosis, and 5–8% develop new varices each year (7). One-third of patients with varices will experience variceal hemorrhage, with mortality from the first variceal bleed as high as 50% (8). Rebleeding occurs frequently, especially in the first 6 weeks. Risk factors for early rebleeding, within the first 6 weeks, include renal failure, large varices, and severe initial bleeding with hemoglobin $< 8\,\mathrm{g\cdot dL^{-1}}$ at admission (6). The risk of late rebleeding is related to the severity of liver

Table 1. Localization of Gastrointestinal Hemorrhage

Diagnosis	History	Physical Examination
Esophagus		
Nasopharyngeal bleeding	Epistaxis	Blood in nares, blood dripping down pharynx, evidence for telangiectasias
Esophagogastric varices	Alcoholism, lived in area where schistosomiasis is endemic, history of blood transfusions or hepatitis B	Stigmata of chronic liver disease, (e.g., gynecomastia, testicular atrophy, parotid enlargement Cachexia, Kaposi's sarcoma, oral candidiasis)
Esophagitis	Dysphagia, odynophagia; immunosuppressed, (e.g., AIDS); diabetes mellitus, lymphoma, elderly	
Esophageal neoplasm	Progressive dysphagia for solids	Cachexia
Mallory–Weiss tear	Retching or vomiting prior to hematemesis	Not specific
Stomach		
Acute gastric ulcer	Intensive care unit setting	Comatose, multiple burns, on respirator
Chronic gastric ulcer	Peak between 55 and 65 years old	Not specific
Acute hemorrhagic gastritis	History of aspirin use, intensive care unit setting	Similar to acute gastric ulcer
Gastric neoplasm	Weight loss, early satiety; obstructive symptoms	Cachexia, Virchow's node; abdominal mass
Gastric angiodysplasia	Elderly	Aortic stenosis murmur
Gastric telangiectasia	Epistaxis, family history of Osler-Weber-Rendu disease or history of renal failure	Telangiectasias on lips, buccal mucosa, palate
Duodenum		
Duodenal ulcer	Epigastric pain	Not specific
Aortoenteric fistula	History of abdominal aortic aneurysm repair	Laparotomy scar
Colon		
Colonic neoplasm	Often occult; if located in rectosigmoid then may have obstructive symptoms	Mass on rectal examination
Cecal angiodysplasia	Elderly, recurrent bleeding, low grade	Aortic stenosis murmur
Colonic diverticuloses	Severe, single episode of bright red blood per rectum	Not specific

failure, ongoing alcohol abuse, variceal size, and renal failure (6).

Variceal bleeding classically presents as massive, painless upper GI hemorrhage in a patient with known cirrhosis. The management of acute variceal bleeding requires attention to adequate resuscitation as well as control of the active bleeding and minimization of complications related to the bleed (Fig. 1). Early endoscopy is imperative for the successful management of variceal bleeding. Frequently, endoscopy is performed in conjunction with pharmacologic therapy. Endoscopy is essential to confirm the diagnosis of bleeding varices, as many patients with cirrhosis bleed from a source other than varices. Endoscopic sclerotherapy is effective in managing active variceal hemorrhage 70–90% of the time, and is superior to balloon tamponade or vasopressin (6). Intravariceal and paravariceal injections are equally efficacious. Sclerotherapy should be repeated at 1 week, and then at 1–3 week intervals until the varices are obliterated. Endoscopic variceal band ligation achieves hemostasis 90% of the time, and is felt to have a lower rebleeding and complication rate than sclerotherapy (9).

Pharmacologic therapy is used in conjunction with early endoscopy in massive variceal bleeding. Vasopressin, which works to cause splanchnic and systemic vasoconstriction and thus decrease portal venous flow, was traditionally used to control hemorrhage, but its use is limited by systemic side effects in 20–30% of patients (10). Vasopressin causes systemic vasoconstriction, which is particularly problematic in patients with coronary artery disease, in which vasoconstriction may induce myocardial infarction. Simultaneous administration of intravenous nitroglycerine will minimize the cardiac complications

Table 2. Upper Gastrointestinal Bleeding

Differential Diagnosis of Upper GI Hemorrhage

Gastroesophageal varices
Mallory–Weiss tear
Esophagitis
Neoplasm esophagus, stomach, small bowel
Gastritis: stress, alcoholic, drug-induced
Angiodysplasia of stomach, small bowel
Peptic ulcer disease: stomach, duodenum
Dieulafoy ulcer
Aortoenteric fistula
Hemobilia

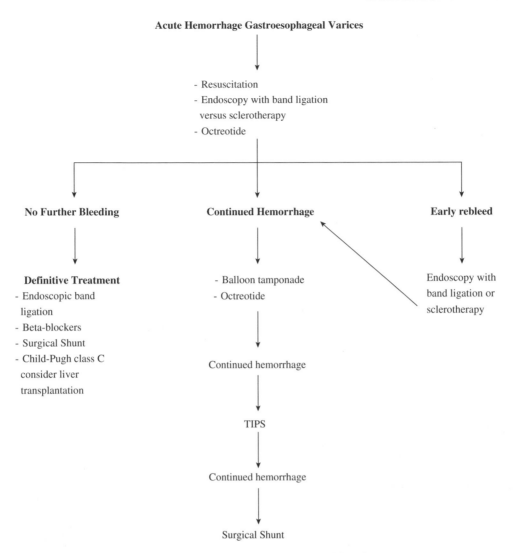

Figure 1. Management of acute variceal bleeding.

associated with vasopressin. Somatostatin, and its long-acting analog Octreotide, work via inhibition of various vasodilatory hormones, and therefore inhibit splanchnic vasodilatation and decrease portal pressure. Somatostatin is as effective as vasopressin, but without the systemic side effects (11), and is currently the medication of choice to reduce portal pressure.

Occasionally, a patient presents with massive variceal hemorrhage not amenable to endoscopic or pharmacologic

Table 3. Lower Gastrointestinal Bleeding

Differential Diagnosis of Lower GI Hemorrhage

Colonic diverticular disease
Colonic arteriovenous malformations
Neoplasm: colon, small bowel
Meckel's diverticulum
Ulcerative colitis
Crohn's disease
Colitis: infectious, ischemic, radiation-induced
Internal hemorrhoidal disease

therapy. Balloon tamponade, generally done with the Sengstaken–Blakemore tube, can be used to achieve short-term hemostasis, which is successful 60–90% of the time (12). Caution must be taken to secure the airway with endotracheal intubation prior to placement of the tamponade balloon because of the high risk of aspiration. Care must be used to ensure that the gastric balloon is in the stomach prior to full insufflation, as migration or inflation of the gastric balloon in the esophagus can lead to esophageal rupture. The balloon can be left in place for 24 h, at which time endoscopic band ligation or sclerotherapy can be performed.

Bleeding that cannot be controlled by endoscopic therapy or that recurs should be managed with portal pressure reduction. The initial approach currently used is the transjugular intrahepatic portosystemic shunt (TIPS), which can be done with or without general anesthesia. Potential benefits to the use of general anesthesia include advanced management of fluid dynamics by the anesthesiologist and pain management for the patient. The TIPS method works by creating a channel between the hepatic and portal veins, which is kept patent by a metal stent, which achieves

Acute Nonvariceal Upper GI Hemorrhage

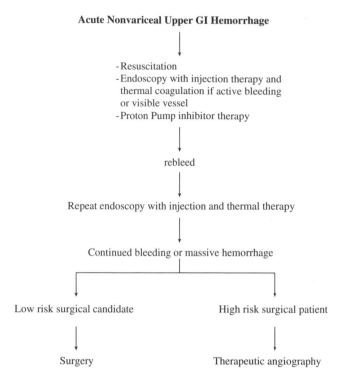

Figure 2. Management of acute nonvaricea bleeding.

hemostasis in 90% of patients (13). The downside of TIPS is related to shunt thrombosis, which can occur early or late and may result in recurrent variceal bleeding. Approximately 20% of patients by 1 year and 30% by 2 years experience recurrent bleeding related to shunt thrombosis (14,15). Complications related to TIPS procedures include a 30% rate of encephalopathy, procedure related complications including inadvertent puncture of the portal vein leading to massive hemorrhage, stent stenosis and malfunction, and TIPS-associated hemolysis.

Traditionally, reduction of portal pressure was achieved by surgical shunt procedures or devascularization. Surgical shunt procedures include nonselective shunts, which divert all the portal flow away from the liver, and selective shunts. Nonselective shunts include the portacaval end-to-side and side-to-side shunts, the central spleno-renal shunt, and the interposition portacaval shunt. Nonselective shunts are successful in achieving hemostasis in the actively bleeding patient, but frequently lead to hepatic encephalopathy as well as acceleration of liver failure. Selective shunts include the distal splenorenal shunt and the small-diameter mesocaval shunt. The selective shunts have a lower rate of encephalopathy, but are frequently complicated by uncontrollable ascites given the continued portal flow. Nonshunt operations, including esophageal transection and devascularization of the gastroesophageal junction are rarely used today. In the setting of emergent operation for ongoing hemorrhage, a nonselective portacaval shunt is most frequently employed. The distal splenorenal shunt is the most common shunt used for elective control.

Once control of active bleeding is achieved, the focus shifts to prevention of future bleeding. Endoscopic band

ligation is the treatment of choice for long-term management of variceal hemorrhage (16). Beta-blockers in combination with nitrates have been shown to synergistically lower portal pressures and thus decrease the risk of rebleeding. Surgical shunting is an option in patients refractory to endoscopic or pharmacologic therapy, with the distal splenorenal shunt the most frequently used for this indication. For patients with liver failure, liver transplantation is effective for both longterm prevention of bleeding as well as hepatic decompensation and death.

NONVARICEAL UPPER GI BLEEDS

Peptic Ulcer Disease

Peptic ulcer disease is the most common cause of upper GI hemorrhage; accounting for between 40 and 50% of all acute upper GI bleeds. Major complications related to peptic ulcer disease include perforation, obstruction, and hemorrhage and occur in ~25% of patients, with hemorrhage the most common complication. Risk factors for peptic ulcer disease include infection with *H. pylori*, nonsteroidal antiinflammatory use, and physiologic stress related to critical illness. Medical advances including proton pump inhibitors and H2 blockers have led to a decreased need for operation for hemorrhage, but no decrease in the actual number of hemorrhages (17).

Hemorrhage related to peptic ulcer will classically present as hematemesis. In the setting of massive bleeding, the patient may also present with hematochezia. The patient may give a history of midepigastric abdominal pain preceding the bleeding. Important elements in the history include a history of peptic ulcer disease and recent usage of aspirin or nonsteroidal medications. Adverse clinical prognostic factors include age >60 years, comorbid medical conditions, hemodynamic instability, hematemesis, or hematochezia, the need for emergency surgical interventions, and continued or recurrent bleeding (18).

The initial diagnostic test on all patients presenting with an upper GI bleed should be endoscopy. Endoscopy is the best test for determining the location and nature of the bleeding lesion, provides information regarding the risk of further bleeding, and allows for therapeutic interventions. Endoscopy should be performed urgently in all high-risk patients, and within 12–24 h for patients with acute, self-limited episodes of bleeding. The goal of endoscopy is to stop the active hemorrhage and reduce the risk of recurrent bleeding. Stigmata of recent hemorrhage (SRH) are endoscopically identified features that help determine which patients should receive endoscopic therapy. The SRH include active bleeding visible on endoscopy, visualization of a nonbleeding visible vessel, adherent clot, and a flat, pigmented spot (18). Certainly, patients with the major SRH including active bleeding or a visible vessel should undergo endoscopic therapy, as meta-analysis has shown a significant reduction in rates of continued or recurrent bleeding, emergency surgery, and mortality in those who received endoscopic therapy versus those who did not (19).

A variety of modalities exist for endoscopic therapy, including injection, thermal, and mechanical therapy. Epinephrine diluted 1:10,000 is the most frequently used

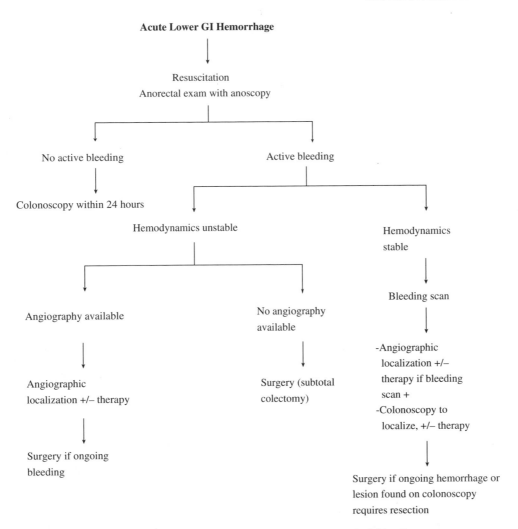

Acute Lower GI Hemorrhage

Resuscitation
Anorectal exam with anoscopy

No active bleeding

Colonoscopy within 24 hours

Active bleeding

Hemodynamics unstable

Hemodynamics stable

Angiography available

Angiographic localization +/– therapy

Surgery if ongoing bleeding

No angiography available

Surgery (subtotal colectomy)

Bleeding scan

-Angiographic localization +/– therapy if bleeding scan +
-Colonoscopy to localize, +/– therapy

Surgery if ongoing hemorrhage or lesion found on colonoscopy requires resection

Figure 3. Management of acute lower gastrointestinal bleeding.

injection therapy, with injection into and adjacent to the bleeding point until hemostasis is achieved. Other agents used for injection therapy include ethanol, ethanolamine, thrombin, and fibrin. Thermal therapy is generally delivered by coaptive techniques, including bipolar electrocoagulation or heater probe. With coaptive coagulation, the probe is used to physically compress the vessel prior to delivery of heat to seal the vessel. Laser photocoagulation and argon beam plasma coagulation are noncoaptive techniques that are used less frequently. Mechanical therapy with hemoclips can also be used in bleeding, although the efficacy may be limited by ulcer location or a firm, scarred ulcer base preventing adequate application of the clips. A combination of injection therapy with epinephrine and bipolar thermal therapy is the most common endoscopic management of an acute bleed.

Despite initial success with endoscopic therapy, 15–20% of patients will rebleed, generally within the initial 72 h (18). While this was traditionally considered a surgical indication, endoscopic retreatment is now recommended for most patients. Repeat endoscopy rather than surgery was found in a prospective, randomized study to be associated with less complications and similar mortality (20). Surgical indications include massive hemorrhage

unresponsive to resuscitation and continued bleeding unresponsive to nonoperative management. For bleeding gastric ulcers, the operation of choice is a wedge resection to include the portion of the stomach containing the ulcer with or without vagotomy. For duodenal ulcers, truncal vagotomy, pyloroplasty, and direct oversewing of the bleeding ulcer via duodenotomy is the most common operation. Care must be taken to incorporate the proximal and distal gastroduodenal artery as well as the transverse pancreatic artery.

Therapeutic angiography is an option when therapeutic endoscopy is unsuccessful and may be performed prior to surgical intervention, as is effective, less invasive than surgery, and does not impact on the ability to surgically manage the bleeding if the intervention is unsuccessful. Angiographic options include selective intra-arterial vasopressin infusion or embolotherapy with microcoils, poly-(vinyl alcohol) (PVA) particles, or gelatin sponge. Embolization is considered the first line angiographic therapy, with success rates as high as 88% (21). Vasopressin is selectively infused after bleeding has been identified by contrast extravasation at an initial rate of 0.2 units per minute, with an increase to 0.4 units per minute then 0.6 units per minute if hemostasis is not achieved. The infusion is continued for 12–24 h, and then gradually tapered. Efficacy of vasopressin infusion ranges from 60 to 90% (22). Side effects related to selective infusion of vasopressin include abdominal cramping, fluid retention, hyponatremia, cardiac arrhythmias, and systemic hypertension. Vasopressin should not be used in patients with coronary artery disease because of the risk for myocardial ischemia.

Pharmacologic therapy to reduce gastric acidity is generally started as an adjunct to endoscopic therapy. The H2 blockers were found in meta-analysis to reduce the rate of continued bleeding, surgery, and death (23); however, a subsequent multicenter randomized trial found no difference in rebleeding rates in patients randomized to famotidine infusion versus placebo (24). Intravenous proton pump inhibitors have been shown to reduce rebleeding rates, length of hospital stay, and need for blood transfusion (25). Treatment with a proton pump inhibitor is generally started on admission for upper GI bleed, and continued as an adjunct to endoscopic therapy.

Mallory–Weiss Tear

The Mallory–Weiss syndrome describes acute upper GI bleeding that occurs after retching or vomiting, and was first described by Kenneth Mallory and Soma Weiss in 1929 (26). The increased intragastric pressure caused by vomiting causes mucosal lacerations, which are usually longitudinal. The typical presentation is a patient who initially vomits gastric material, followed by hematemesis and melena. Mallory–Weiss tears can also occur after anything that raises intragastric pressure, such as blunt abdominal trauma, severe coughing, childbirth, seizures, and closed chest cardiac compression. Mallory–Weiss tears classically occur at the gastroesophageal junction, but can occur in the distal esophagus. The lesion is common, occurring in 5–15% of patients presenting with upper GI bleeding. The majority of Mallory–Weiss tears will stop bleeding spontaneously, although some patients will require emergency treatment for ongoing hemorrhage. Endoscopic options for Mallory–Weiss bleeding include band ligation, epinephrine injection, and hemoclip application. In cases not amenable to endoscopic management, operative therapy involves oversewing the laceration via a gastrotomy.

Gastritis

Stress gastritis is associated with multiple superficial gastric ulcerations and is typically seen in the critically ill patient. Mechanical ventilation and coagulopathy increase the risk for hemorrhage in the critically ill. Prophylaxis with H2 blockers and proton pump inhibitors has led to a decrease in the incidence of stress gastritis in the critically ill. Bleeding from gastritis usually is self-limited, not requiring intervention. Early endoscopy is essential to establish the diagnosis and rule out other sources of upper GI bleeding. The patient should be started on pharmacologic therapy with either proton pump inhibitors or H2 blockers at a therapeutic dose. Endoscopic laser anticoagulation has been used for bleeding gastritis. Intraarterial infusion of vasopressin or selective embolization may also be used to arrest hemorrhage in gastritis. Ongoing hemorrhage not amenable to nonsurgical management is operatively managed with vagotomy, pyloroplasty, and oversewing of the bleeding sites versus total gastrectomy. The mortality for a patient requiring surgical management of bleeding gastritis remains quite high.

Esophagitis

Esophagitis is an unusual cause of acute gastrointestinal bleeding, and only rarely occurs in association with severe reflux esophagitis. The history would be suggestive of gastroesophageal reflux, with symptoms, such as, heartburn, cough, and hoarseness. Rare causes of esophagitis associated with bleeding in the immunocompromised patient include infection with candida, herpes, or cytomegalovirus (27).

Neoplasm

Acute upper GI bleeding is a rare manifestation of esophageal neoplasms, with <5% of esophageal malignancies presenting with an acute bleed. Occult, slow GI bleeding is much more common with esophageal neoplasms. Benign tumors of the esophagus include leiomyomas and polyps, and are very unlikely to present with GI bleeding. Esophageal hemangiomas, which constitute only 2–3% of benign esophageal tumors, may present with potentially massive GI hemorrhage. Leiomyosarcomas are more likely to bleed than benign leiomyomas. When brisk bleeding occurs in the setting of esophageal cancer, one also must consider the erosion of the tumor into a major thoracic vessel.

Dieulafoy Vascular Malformation

Dieulafoy lesions are the result of arterioles of large diameter (1–3 mm) running through the submucosa, with erosion of the overlying mucosa resulting in bleeding. The mucosal defect is usually small, without evidence of chronic inflammation. Dieulafoy lesions generally present with brisk hemorrhage, due to their arterial nature. Diagnosis is made by upper endoscopy, with visualization of a small mucosal defect with brisk bleeding. Management is initially endoscopic with epinephrine injection and bipolar thermal therapy. Catheter embolization is generally successful in patients who fail endoscopic management. For patients requiring surgical management, the operation involves a wedge resection of the lesser curve of the stomach at the site of the lesion.

AORTOENTERIC FISTULA

Aortoenteric fistula classically occurs as a communication between a prosthetic aortic graft and the distal duodenum, and the diagnosis should be entertained in any patient presenting with an upper GI bleed who has undergone aortic reconstruction. The time period from aortic surgery to presentation is varied, and many patients present years down the road. The patient will frequently present initially with a sentinel bleed, which may be followed by massive upper GI hemorrhage. Upper endoscopy is paramount to making the diagnosis, as well as ruling out other sources of upper GI bleeding. Upon making the diagnosis of aortoenteric fistula, the optimal management is surgical, with removal of the aortic prosthesis, extra-anatomic bypass, and repair of the duodenum.

HEMOBILIA

Hemobilia classically presents as upper GI bleeding, melena, and biliary colic. Diagnosis is established by upper endoscopy, with visualization of blood from the ampulla. Endoscopic retrograde cholangiopancreatography can more clearly delineate the source of hemobilia. A variety of disease processes can lead to hemobilia, including hepatobiliary trauma, chronic cholelithiasis, pancreatic cancer, cholangiocarcinoma, and manipulation of the hepatobiliary tree. While hemobilia remains a rare cause of upper GI bleeding, the frequency is increasing related to increased manipulation of the hepatobiliary system and improved diagnostic modalities. Many cases of hemobilia will resolve without intervention. In the setting of ongoing hemorrhage, angiography with selective embolization of the bleeding vessel is the primary treatment modality. Surgery is reserved for failure of angiographic management.

LOWER GI BLEEDING

The passage of bright red or maroon blood via the rectum suggests a bleeding source distal to the ligament of Treitz, although blood can originate from any portion of the GI tract, depending on the rate of bleeding. Some 80–90% of lower GI bleeding will stop spontaneously. Initial resuscitation is similar to the patient presenting with upper GI bleeding, with hemodynamic assessment, establishment of appropriate access, and thorough history and physical exam. Visual inspection of the anorectal region, followed by anoscopy is essential to rule out a local anorectal condition such as hemorrhoids as the source of bleeding. A variety of modalities are available to further define the etiology of the bleeding, including endoscopy, nuclear medicine, angiography, and intraoperative localization.

The timing of colonoscopy for acute lower GI bleeding is controversial, with early (within 24 h of admission) colonoscopy increasingly advocated. Certainly, visualization may be difficult in a massively bleeding patient. Aggressive bowel prep can be given for 6–12 h prior to endoscopy, with the benefit of improved visualization. The benefit of early colonoscopy, similar to early upper endoscopy for upper GI bleed, is the opportunity for endoscopic diagnosis and therapy, using injection therapy and thermal modalities. Colonoscopy can directly visualize the bleeding source, which is beneficial in directing the surgeon in resection if the patient has continued or recurrent hemorrhage. Additionally, early colonoscopy may shorten hospital length of stay (28).

Nuclear scans can localize the site of lower GI bleeding and confirm active bleeding, with sensitivity to a rate of bleeding as low as $0.05-0.1 \, \text{mL·min}^{-1}$. Bleeding scans use either ^{99m}Tc sulfur colloid or ^{99m}Tc-labeled erythrocytes, with radioactivity detected by a gamma camera, analyzed by computer, and recorded onto photographic film. The ^{99m}Tc sulfur colloid has the advantage of detection of bleeding as slow as $0.05 \, \text{mL·min}^{-1}$, is inexpensive, and easy to prepare, but only will detect bleeding within 10 min of injection as it disappears quickly from the bloodstream (21). ^{99m}Tc-labeled erythrocytes detect bleeding as slow as $0.1 \, \text{mL·min}^{-1}$, and circulate within the bloodstream for 24 h. The ^{99m}Tc-labeled erythrocyte technique is generally considered the test of choice because of an increased sensitivity and specificity when compared with the ^{99m}Tc sulfur colloid (21). When a bleeding scan is positive, angiography or endoscopy is recommended to confirm the location of bleeding, to diagnose the specific cause, and to possible apply either endoscopic or angiographic therapy.

Angiography is advantageous because of the potential for both localization and treatment. Angiographic control can permit elective rather than emergent surgery in patients who are good surgical candidates, and can provide definitive treatment for poor surgical candidates. Bleeding can be detected at a rate as low as $0.5 \, \text{mL·min}^{-1}$. The SMA is cannulated initially, with cannulation of the IMA if the SMA study is nondiagnostic. When bleeding is localized, either vasopressin infusion or superselective catheter embolization may be used. Vasopressin is used in a method similar to upper GI bleeding, with an infusion rate of 0.2–0.4 units·min^{-1}. Efficacy varies from 47–92%, with rebleeding in up to 40% of patients (21). Vasopressin is particularly effective for bleeding from diverticula.

Angiographic embolization may be done with a variety of agents, including coil springs, gelatin sponge, cellulose, and (PVA). There is less collateral blood supply in the lower G tract than in the upper, so embolization was initially thought to be a salvage therapy for those patients who would not tolerate an operation. Recent innovations in catheter and guidewire design, however, have enabled the interventional radiologist to superselectively embolize the bleeding vasa recta, sparing the collateral vessels and thus minimizing ischemia. Several small studies have reported successful embolization without intestinal infarction (21), with combined results showing successful hemostasis in 34 of 37 patients. Superselective embolization with coaxial microcatheters is currently considered the optimal angiographic therapy.

Traditionally, emergency operations for lower GI bleeding were required in 10–25% of patients presenting with bleeding (29). Surgical indications traditionally include hemodynamic instability, ongoing transfusion requirements, and persistent or recurrence of hemorrhage. If the bleeding has not been localized, a total abdominal colectomy with ileorectal anastomosis or end ileostomy is performed. If the lesion has been localized to either the right or left side of the colon, a hemicolectomy may be performed. With the advances in angiography available, the surgical indications are evolving. If an angiographer is readily available, angiographic localization and therapy is a viable option even for the hemodynamically unstable or persistently bleeding patient, thus avoiding the high morbidity and mortality associated with emergent total colectomy in this patient population.

Diverticulosis

The most common cause of lower GI bleeding is diverticular disease. Diverticular disease increases with age and is present in 50% of people > 80 years. Less than 5% of these patients, however, will hemorrhage. While most diverticula are found distal to the splenic flexure, bleeding diverticula more frequently occur proximal to the splenic flexure. Classically, the patient will present with sudden onset of mild lower abdominal pain and the passage of maroon or bright red bloody stool per rectum. The majority of diverticular bleeds will stop spontaneously, with a recent study showing spontaneous resolution in 76% of patients (1). About 20–30% of patients will have a recurrent bleeding episode, of which the majority will again stop without intervention. Patients that have persistent or recurrent bleeding should be considered for surgical therapy, particularly if the site of bleeding has been localized. High risk surgical patients can be treated with angiographic or endoscopic therapy.

Angiodysplasia

Angiodysplasia arises from age-related degeneration of submucosal veins and overlying mucosal capillaries, with frequency increasing with age. The bleeding tends to be less severe than with diverticular bleeds, and frequently resolves spontaneously, although recurrence is common. Diagnosis can be made by colonoscopy, with electrocoagulation as definitive therapy. Angiography may also be used for diagnosis, with the angiographic hallmarks a vascular tuft from an irregular vessel, an early and intensely filling vein resulting from arteriovenous communication, and persistent venous filling (30). Angiographic therapy with vasopressin can be used for treatment.

Neoplasm

While polyps and cancers frequently present with blood per rectum, they rarely cause massive hemorrhage as the presenting symptom. Diagnosis is made with colonoscopy. Management of a polyp is via colonoscopic polypectomy, while cancer requires surgical resection. Occasionally, a patient will present up to 1 month following a polypectomy with lower GI bleeding, which should prompt colonoscopy and thermal or injection therapy to the bleeding polypectomy site.

Meckel's Diverticulum

Meckel's diverticulum is an unusual cause of GI bleeding, and usually occurs in the first decade of life. The etiology of the bleeding is ectopic gastric mucosa in the diverticulum with resultant ulceration of adjacent bowel. Diagnosis is usually demonstrated by nuclear scanning demonstrating the ectopic gastric mucosa. Management is with surgical resection of the diverticulum as well as the adjacent bowel.

Ischemic Colitis

Ischemic colitis generally presents with bloody diarrhea, and massive lower GI bleeding is rare in this population. The bloody diarrhea is due to mucosal sloughing. Ischemic colitis should be suspected in patients with a history of vascular disease and in critically ill, hypotensive patients with a low flow state. Diagnosis is made by flexible endoscopy showing evidence of ischemia. Management for early ischemia is resuscitation and improvement of blood flow. Advanced ischemia requires surgical resection of the necrotic portion of the bowel.

Inflammatory Bowel Disease

GI bleeding characterizes both Crohn's disease and ulcerative colitis; however, massive bleeding is quite uncommon. The bleeding from inflammatory bowel disease is usually self-limited, and rarely acutely requires surgical attention. Diagnosis is made by colonoscopy, with identification of features unique to these entities and biopsy for pathology. Occasionally, ulcerative colitis will present fulminantly with massive hemorrhage and require surgical resection, consisting of total colectomy, end ileostomy, and Hartman's pouch, leaving the possibility for future conversion to an ileo-pouch anal anastomosis. Both entities are managed with immunosuppressive medications.

BIBLIOGRAPHY

Cited References

1. Hamoui N, Docherty DO, Crookes PH. Gastrointestinal hemorrhage: is the surgeon obsolete? Emerg Med Clin N Am 2003;21:1017–1056.
2. Bardhan KD, et al. Changing patterns of admission and operations for duodenal ulcer. Br J Surg 1989;76:230–236.
3. Longstreth GF. Epidemiology of hospitalization for acute upper gastrointestinal hemorrhage: a population based study. Am J Gastroenterol 1995;90:206–210.
4. Yavorski R, et al. Analysis of 3294 cases of upper gastrointestinal bleeding in military medical facilities. Am J Gastroenterol 1995;90:568–573.
5. Gilbert DA, et al. The national ASGE survey on upper gastrointestinal bleeding III. Endoscopy in upper gastrointestinal bleeding. Gastrointest Endosc 1982;27:94
6. Comar KM, Sanyal AJ. Portal hypertensive bleeding. Gastroenterol Clin N Am 2003;32:1079–1105.
7. Lebrec D, et al. Portal hypertension, size of esophageal varices, and risk of gastrointestinal bleeding in alcoholic cirrhosis. Gastroenterology 1980;79:1139–1144.
8. Pagliaro L, et al. Prevention of the first bleed in cirrhosis. A metaanalysis of randomized trials of non-surgical treatment. Ann Intern Med 1992;117:59–70.
9. Stiegmann GV, Goff GS, Sun JH, Wilborn S. Endoscopic elastic band ligation for active variceal hemorrhage. Am Surg 1989;55:124–128.
10. Conn HO. Vasopressin and nitroglycerine in the treatment of bleeding varices: the bottom line. Hepatology 1986;6:523–525.
11. Inperiale TF, Teran JC, McCullough AJ. A meta-analysis of somatostatin versus vasopressin in the management of acute esophageal variceal hemorrhage. Gastroenterology 1995;109:1289–1294.
12. Feneyrou B, Hanana J, Daures JP, Prioton JB. Initial control of bleeding from esophageal varices with the Sengstaken-Blakemore tube: experience in 82 patients. Am J Surg 1988;155:509–511.
13. LaBerge JM, et al. Creation of transjugular intrahepatic portosystemic shunts with the wallstent endoprothesis: results in 100 patients. Radiology 1993;187:413–420.
14. Sanyal AJ, et al. Transjugular intrahepatic portosystemic shunts compared with endoscopic sclerotherapy for the prevention of recurrent variceal hemorrhage: a randomized, controlled trial. Ann Intern Med 1997;126:849–857.
15. LaBerge JM, et al. Two-year outcome following transjugular intrahepatic portosystemic shunt for variceal bleeding: results in 90 patients. Gastroenterology 1995;108:1143–1151.

16. Grace ND. Diagnosis and treatment of gastrointestinal bleeding secondary to portal hypertension. American College of Gastroenterology Practice Parameters Committee. Am J Gastroenterol 1997;92:1081–1091.

17. Bardhan KD, et al. Changing patterns of admissions and operations for duodenal ulcer. Br J Surg 1989;76:230–236.

18. Huang CS, Lichtenstein DR. Nonvariceal upper gastrointestinal bleeding. Gastroenterol Clin N Am 2003;32:1053–1078.

19. Cook DJ, Guyatt GH, Salena BJ, Laine LA. Endoscopic therapy for acute nonvariceal upper gastrointestinal hemorrhage:a meta-analysis. Gastroenterology 1992;102:139–148.

20. Lau JY, et al. Endoscopic retreatment compared with surgery in patients with recurrent bleeding after initial endoscopic control of bleeding ulcers. N Engl J Med 1999;340:751–756.

21. Gomes AS, Lois JF, McCoy RD. Angiographic treatment of gastrointestinal hemorrhage: comparison of vasopressin infusion and embolization. Am J Roentgenol 1986;146:1031–1037.

22. Lefkovitz Z, et al. Radiologic diagnosis and treatment of gastrointestinal hemorrhage and ischemia. Med Clin N Am 2002;86:1357–1399.

23. Collins R, Langman M. Treatment with histamine H2 antagonists I acute upper gastrointestinal hemorrhage: implications of randomized trials. N Engl J Med 1985;313: 660–666.

24. Walt RP, et al. Continuous infusion of famotidine for hemorrhage from peptic ulcer. Lancet 1992;340:1058–1062.

25. Javid G, et al. Omeprazole as adjuvant therapy to endoscopic combination injection sclerotherapy for treating bleeding peptic ulcer. Am J Med 2001;111:280–284.

26. Mallory GK, Weiss S. Hemorrhage from lacerations of the cardiac orifice of the stomach due to vomiting. Am J Med Sci 1929;178:506.

27. Onge GS, Bezahler GH. Giant esophageal ulcer associated with cytomegalovirus. Gastroenterology 1982;83:127–130.

28. Schmulewitz N, Fisher DA, Rockey DC. Early colonoscopy for acute lower GI bleeding predicts shorter hospital stay: A retrospective study of experience in a single center. Gastrointest Endosc 2003;58:841–846.

29. Colacchio TA, Forde KA, Patsos TJ, Nunez D. Impact of modern diagnostic methods on the management of rectal bleeding. Am J Surg 1982;143:607–610.

30. Boley SJ, et al. The pathophysiologic basis for the angiographic signs of vascular ectasias of the colon. Radiology 1977;125:615–621.

See also ELECTROGASTROGRAM; ENDOSCOPES.

GEL FILTRATION CHROMATOGRAPHY.   See CHROMATOGRAPHY.

GLUCOSE SENSORS

FARBOD N. RAHAGHI
DAVID A. GOUGH
University of California,
La Jolla, California

INTRODUCTION

Glucose assay is arguably the most common of all medical measurements. Billions of glucose determinations are performed each year by laypeople with diabetes based on "fingersticking" and by healthcare professionals based on blood samples. In fingersticking, sample collection involves the use of a lancet to puncture the skin of the fingertip or forearm to produce a small volume of blood and tissue fluid, followed by collection of the fluid on a reagent-containing strip and analysis by a handheld meter. Glucose measurements coupled to discrete sample collection continue to be the most common method of glucose monitoring. However, new types of sensors capable of continuous glucose monitoring are nearing clinical introduction. Continuous or near-continuous glucose sensors may make possible new and fundamentally different approaches to the therapy of the disease. This article reviews recent progress in the development of new glucose sensors and describes the potential roles for these sensors in the improved treatment of diabetes.

THE CASE FOR NEW GLUCOSE SENSORS

The objective of all forms of therapy for diabetes is the maintenance of blood glucose near normal levels (1). The Diabetes Control and Complications Trial (or DCCT) and counterpart studies such as the United Kingdom Prevention of Diabetes Study (UKPDS) have clearly demonstrated (Fig. 1) that lower mean blood glucose levels resulting from aggressive treatment can lead to a reduced incidence and progression of retinopathy, nephropathy, and other complications of the disease (2,3). These prospective studies showed definitively that there exists a cause-and-effect relationship between poor blood glucose control and the complications of diabetes. As convenient means for frequent glucose assay were not available at the time, glucose control was assessed in these trials by glycosylated hemoglobin levels (Hb_{A1c}), which indicate blood glucose concentrations averaged over the previous 3 month period. Although Hb_{A1c} levels are useful for assessment of longitudinal blood glucose control, the values indicate only *averaged* blood glucose, rather than blood glucose *dynamics* (i.e., how blood

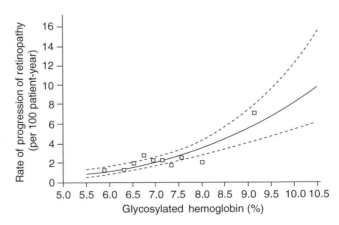

Figure 1. The results of the DCCT (2). Results show that improved glucose control, measured by a reduction in the fraction of glycosylated hemoglobin, leads to reduced long-term complications of diabetes. (Copyright © 1993, Massachusetts Medical Society.)

glucose changes with time), and cannot be used for immediate adjustment of therapy (4). There is general agreement that frequent determination of glucose by a sensing method that is convenient and widely acceptable to people with diabetes would allow a finer degree of control. Normalization of blood glucose dynamics may be of equal or greater importance than normalization of average blood glucose. The results of the DCCT and related studies point to the need for practical new approaches to achieve control.

The primary need for a new type of glucose sensor is to facilitate improved treatment of type 1 diabetes. In this case, the insulin producing ability of the pancreas has been partially or fully destroyed due to a misdirected autoimmune process, making insulin replacement essential. The sensor would help avoid the long-term complications associated with hyperglycemia (i.e., above-normal blood glucose) by providing information to specify more timely and appropriate insulin administration. It is now becoming widely appreciated that a new sensor could also be beneficial for people with the more common type 2 diabetes, where a progressive resistance of peripheral tissues to insulin develops, leading to glucose imbalances that can eventually produce long-term clinical consequences similar to type 1 diabetes. Type 2 diabetes is related to obesity, lifestyle, and inherited traits. In recent years, the incidence of type 2 diabetes has increased at extraordinary rates in many populations, to the point of becoming a worldwide epidemic (5). It is estimated that within 10 years, the prevalence of diabetes may approach 210 million cases worldwide (6). This places increased urgency on developing new approaches to managing or preventing the disease where possible, and a meliorating its consequences.

In addition, an automatic or continuous sensor may also have an important role in preventing hypoglycemia (i.e., below-normal blood glucose). Hypoglycemia is caused primarily by a mismatch between the insulin dosage used and the amount of insulin actually needed to return the blood glucose level to normal. Many people with diabetes can reduce the mean blood glucose by adjustment of diet, insulin, and exercise, but when aggressively attempted, this has led to a documented increase in the incidence of hypoglycemia (7). Below-normal glucose values can rapidly lead to cognitive lapses, loss of consiousness, and life-threatening metabolic crises. In children, there is concern that severe hypoglycemic events may lead to neurologic sequelea (8). A significant percentage of deaths of people under 40 with type 1 diabetes is due to the "dead-in-bed" syndrome (9), which may be linked to nocturnal hypoglycemia. Some experts claim that "... the threat of severe hypoglycemia remains the single most important barrier to maintaining normal mean blood glucose" (10). A continuous glucose sensor that does not depend on user initiative could be part of an automatic alarm system to warn of hypoglycemia and provide more confidence to the user to lower mean blood glucose, in addition to preventing hypoglycemia by providing improved insulin dosages. Hypoglycemia detection may be the most important application of a continuous glucose sensor. Ultimately, a glucose sensor may also be useful in the prediabetic state to indicate behavior modification for reduction of metabolic stress on the pancreas.

Beyond applications in diabetes, it has recently been shown that stricter glycemic control during surgery and intensive care can reduce mortality in non-diabetic patients and significantly shorten the hospital stay (11). The exact mechanism of this effect has not been elucidated, but the benefit is closely tied to the extent of glucose control and not simply insulin dosage (12). This is another important application for new glucose sensors.

Alternatives to sensor-based therapies for diabetes are more distant. Several biological approaches to diabetes treatment have been proposed, including pancreatic transplantation, islet transplantation, genetic therapies, stem cell-based therapies, new pharmaceutical strategies, islet preservation, and others. Whole or partial organ and islet transplantation requires discovery of methods for assuring immuno-tolerance that do not rely on anti-rejection drugs and approaches for overcoming the shortage of transplantable pancreatic tissue. Potential therapies based on stem cells, if feasible, require basic research on growth, regulation, and implementation of the cells, and share the immuno-intolerance problem. Genetic therapies are limited by incomplete understanding of the complex genetic basis of diabetes, as well as progress in developing site-specific gene delivery, activation, and inactivation. Although transplantation, stem cell, and genetic approaches are based wholly on biological materials, it is not certain that the glucose and insulin dynamics resulting from their use will necessarily be near-normal or readily adjustable. Immunotherapeutic approaches for *in situ* preservation of islets are also being studied but, if eventually feasible, are far off and may require lifetime immune system modulation. The possibility of prevention of type 1 diabetes relies on development of timely methods for early detection of the disease and discovery of an acceptable approach to avoid or interrupt the islet destruction process. Furthermore, prevention will have little value for people who already have diabetes. These alternatives require substantial basic research and discovery, and while often highly publicized, are not likely to be available until far into the future, if eventually feasible.

Although new glucose sensors have the advantage of being closer to clinical introduction, there are certain other advantages as well. First, no anti-rejection medication will be needed. Second, the sensor will provide real-time information about blood glucose dynamics that is not available from other technologies. Third, in addition to real-time monitoring, continuous sensor information may be useful to *predict* blood glucose ahead of the present (13), a capability not feasible with the other approaches. Real-time monitoring and predictive capabilities may lead to entirely new applications of present therapies. Fourth, the sensor could operate in parallel with various other therapies, should they become available. The glucose sensor will likely have broad application, regardless of whether or when other technologies are introduced.

The sensor is also key to the implementation of the mechanical artificial beta cell. In the ideal configuration, this device would have an automatic glucose sensor, a refillable insulin pump, and a controller containing an algorithm to direct automatic pumping of insulin based on information provided by the sensor. There has been progress on development of several of the components of this system, including: (1) external insulin pumps, which operate in a substantially preprogrammed mode with minor adjustments by the user based on fingerstick glucose information; (2) long-term implantable insulin pumps that operate in a similar way; (3) models of glucose and insulin distribution in the body that may eventually be useful in conjunction with control systems; and (4) controllers to direct insulin pumping based on sensor information. In contrast to other approaches to insulin delivery, the mechanical artificial beta cell has the advantage that the insulin response can be reprogrammed to meet the changing needs of the user. Development of an acceptable glucose sensor has thus far been the most difficult obstacle to implementation of the mechanical artificial beta cell.

THE IDEAL GLUCOSE SENSOR

The likelihood that glucose monitoring will reach its full potential as a tool for the therapy of diabetes depends on the technical capabilities of candidate sensors and the general acceptance of sensors by people with diabetes. Technical requirements of the sensor system include: specificity for glucose in the presence of interfering biochemicals or physiological phenomena that may affect the signal; sensitivity to glucose and adequate concentration resolution over the relevant range; accuracy as compared to a "gold standard" blood glucose assay; a sufficiently short response lag to follow the full dynamic range of blood glucose variations; reliability to detect mild hypoglycemia without false positives or negatives; and sufficient stability

that recalibration is rarely needed. The specific criteria for sensor performance remain a matter of consensus and may become better defined as sensors are introduced. The general acceptance of new sensors by people with diabetes will be based on such factors as safety, convenience, reliability, automatic or initiative-independent operation, infrequent need for recalibration, and independence from fingersticking.

For the glucose sensor to be a widely accepted innovation, the user must have full confidence in its accuracy and reliability, yet remain uninvolved in its operation and maintenance. Sensor systems under development have yet to reach this ideal, but some promising aspirants are described below. Short of the ideal, several intermediate sensing technologies with limited capabilities may find some degree of clinical application and, if used effectively, may lead to substantial improvements in blood glucose control. Nevertheless, the most complete capabilities will lead to the broadest adoption by users.

GLUCOSE SENSORS AND SENSING METHODOLOGIES

Several hundred physical principles for monitoring glucose have been proposed since the 1960s. Many are capable of glucose measurement in simple solutions, but have encountered limitations when used with blood, employed as implants, or tested in clinically relevant applications. Certain others have progressed toward clinical application. A brief summary of the history of events related to glucose sensor development is shown in Figure 2.

Present Home Glucose Monitoring

A major innovation leading to improved blood glucose management was the widespread use of home glucose

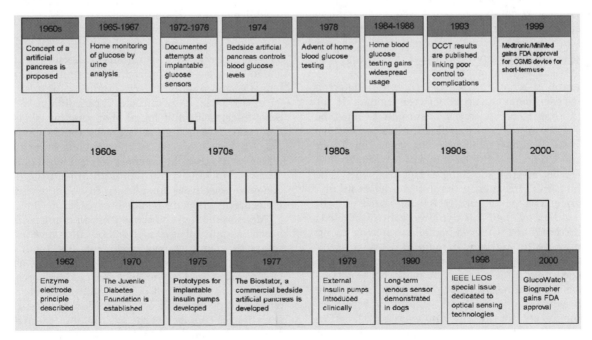

Figure 2. A time-line of some important developments relating to glucose sensors (2,14–24).

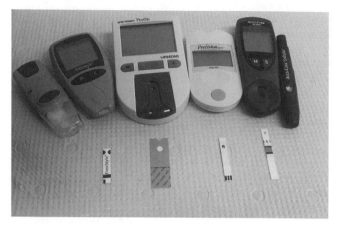

Figure 3. A small collection of home glucose monitoring equipment developed over the past decade. At either end (above) are devices used to puncture the skin for sample collection. Examples of commercial glucose meter (above, center) are also shown. Strips (below) contain immobilized glucose oxidase and are discarded after a single measurement.

monitoring in the 1980s (22). Present commercial versions of this technology are available with respective methods for glucose assay, data presentation and storage, sample volume requirements, and various convenience features (Fig. 3). These devices employ single-use strips based on enzyme methods discussed below. The widespread application of home glucose monitoring has permitted laypeople with diabetes to assume a newfound role in the management of their disease. The present standard-of-care recommends glucose measurement three or more times a day for insulin-dependent individuals (25), but a small number of individuals samples 10 or more times daily. It is generally suspected that the average sampling rate is inadequate and a recent publication noted that only 56% of diabetic individuals sampled their blood glucose once or more daily (26). The general resistance to more frequent sampling may be related to several factors, including: the pain associated with finger puncture, the requirement for user initiative, the general inconvenience of the assay, and unwillingness to carry out nocturnal testing (27).

When sampling is not sufficiently frequent, undetected blood glucose excursions can occur between samples. It has been shown that blood glucose measurements must be obtained every 10 min to detect all blood glucose excursions in the most severe diabetic subjects (28), although slower blood glucose excursions in the majority of people with diabetes may not require sampling at this frequency. The fact that the sample frequency required to detect all glycemic excursions is not clinically feasible with present technology indicates that the dynamic control of blood glucose is currently not practiced in diabetes management.

To compensate for infrequent monitoring, users typically adopt various strategies to estimate blood glucose concentration using subjective *ad hoc* models. These strategies rely on the most recent reported values of glucose, in conjunction with the timing and content of recent or upcoming meals, insulin therapy, and exercise. The effectiveness of these strategies is limited and the constant attention required to make such estimates represent a

substantial intrusion in lifestyle. Although glucose monitoring by fingersticking is likely to become more acceptable as the sample volume and the pain associated with sample collection are reduced, the problem of infrequent sampling and the requirement for user initiative will continue to be the major obstacles to the improvement of glucose control based on this technology.

Noninvasive Optical Sensing Concepts

Noninvasive optical methods are based on directing a beam of light onto the skin or through superficial tissues, and recording the reflected, transmitted, polarized, or absorbed components of the light (29). A key requirement for success of these methods is a specific spectral region that is sufficiently sensitive to glucose, but insensitive to other similar optically active interfering molecules and tissue structures. Several optical methods allow straightforward glucose measurement in simple aqueous solutions, but are ineffective at detecting glucose in tissue fluid, plasma, or blood. If an optical approach can be validated, a noninvasive sensor might be possible. For this reason, an intensive research effort and substantial industrial investment over the past two decades have gone into investigation of these concepts.

Infrared (IR) absorption spectroscopy is based on excitation of molecular motions that are characteristic of the molecular structure. The near-infrared (NIR) region of the spectrum (750–2500 nm) is relatively insensitive to water content so that the beam penetration depth in tissues can be substantial (30). Trials to identify a clinical correlation between NIR signals and blood glucose have employed various computational methods for analyzing the absorption spectrum. Basic studies have focused on identifying the absorbing species and tissue structures responsible for optical signals. However, after much effort the operating conditions that provide selectivity for glucose have yet to be established, leading one investigator to conclude that "... signals can be attributed to chance" (31).

Raman spectroscopy relies on detecting scattered emissions associated with vibrational molecular energy of the chemical species (as opposed to transmitted, rotational, or translational energy). Early studies compared the measurement in water of three different analytes (urea, glucose, lactic acid) and found that glucose levels could be determined with limited accuracy (32). Raman spectroscopy has been applied in the aqueous humor of the eye (33), which is thought to reflect delayed blood glucose levels over certain ranges (34). As with other optical methods, adequate specificity for glucose in the presence of other molecules remains to be demonstrated.

Measurement of the concentration using *polarimetry* is based on ability of asymmetric molecules such as glucose to rotate the plane of polarized light (35). This method is limited by the presence of other interfering asymmetric molecules, as well as the thickness and light scattering by tissues in the region of interest (30). Considerable development of polarimetry has centered on measurements in the anterior chamber of the eye (36), but there is yet to be a demonstration of sufficient selectivity under biological conditions.

Attempts at validation of optical sensor concepts have involved two general approaches. One approach endeavors to establish selectivity for glucose by identification of the components of tissues besides glucose that contribute to the optical signal, and determine if the effects of these interfering substances can be eliminated or the observed signals can be reconstructed based on all contributions. This has not yet been successful, in spite of intensive efforts. The impediment is the large number of optically active components in tissues, many of which produce much stronger effects than glucose. A second approach to validation involves identifying an empirical relationship between the observed optical signal *in vivo* and simultaneously recorded blood glucose concentration. Noninvasive optical approaches have been the premise of several human clinical trials, all of which have been unsuccessful. The prospects for a non-invasive optical glucose sensor are distant.

Implantable Optical Sensor Concepts

Implanted optical sensors offer the prospect of a less congested optical path, at the expense of requiring a more complicated device and confronting the foreign body response. One promising optical concept is based on chemical interaction between glucose and an optically active chemical species that is immobilized in an implanted, glucose-permeable chamber. The interaction creates a change in the optical signal which, under ideal conditions, may indicate glucose concentration. An example is the "affinity sensor" (37), in which glucose competes with a fluorescent substrate for binding with a macromolecule, Con A, resulting in a change in the optical signature. A similar detection strategy has been proposed as part of an implantable intraocular lens (38). There are difficulties with biocompatibility of the implant, design of the chamber, specificity of the optical detector, as well as toxicity and photobleaching of the indicator molecules (38). These systems have yet to be extensively tested.

Tissue Fluid Extraction Techniques

The interstitial fluid that irrigates tissues contains glucose derived from the blood in local capillaries. Several strategies have been devised to extract this fluid for glucose assay.

Microdialysis is based on a probe consisting of a fine hairpin loop of glucose-permeable dialysis tubing in a probe that is inserted into subcutaneous tissues (39). A fluid perfusate is continuously circulated through this tubing by a pump contained in an external apparatus (40), collected, and assayed for glucose concentration using an enzyme electrode sensor (Fig. 4). This methodology relies on the exchange of glucose between the microvascular circulation and the local interstitial compartment, transfer into the dialysis tube, and appropriate adjustment of the pumping pressure and perfusion rate (41). The advantage of the system is that a foreign body response of the tissue and local mass transfer resistance are slow to develop due to the sustained tissue irrigation, but drawbacks include the requirement for percutaneous access, the need for frequent relocation of the probe to minimize the chance

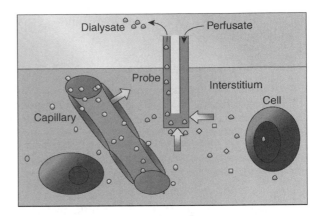

Figure 4. Diagram of a microdialysis probe (39). The semipermeable membrane at the probe tip allows exchange of soluble molecules between the probe and surrounding tissue. Samples are continuously collected and analyzed. (Used with permission of BMJ Publishing Group.)

of infection, and management of the external apparatus by the user. This device may find clinical applications for short-term monitoring.

Reverse iontophoresis employs passage of electrical current between two electrodes placed on the surface of the body to extract tissue fluid directly through the intact skin (42). Glucose in the fluid has been measured by an enzyme electrode-type sensor as part of a wristwatch-like apparatus (43) (Fig. 5). With a 2 h equilibration process after placing the device and a fingerstick calibration, the sensor can take measurements as often as every 10 min for 12 h, at which time sensor components must be replaced and the sensor recalibrated (44). This sensor was approved by the Food and Drug Administration (FDA) for indicating glucose trends, but users are instructed to revert to more reliable conventional assays for insulin dosing decisions. Minor skin irritation has been reported as a side effect (45). Although this sensor was briefly available commercially, it

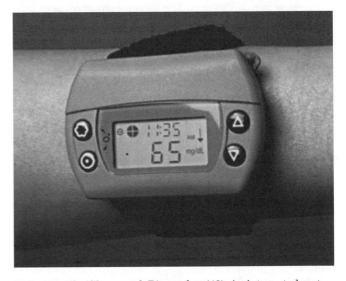

Figure 5. The Glucowatch Biographer (43). An integrated system for sample extraction by reverse iontophoresis and glucose sensing. (Used with permission from John Wiley & Sons, Inc.)

was not successful as a product due to its limited capabilities and inconvenience.

Implantable Enzyme Electrode Sensors

The most promising approaches have been various configurations of the enzyme electrode sensor based on immobilized glucose oxidase coupled to electrochemical detectors. The enzyme catalyzes the reaction:

$$\text{glucose} + O_2 + H_2O \rightarrow \text{gluconic acid} + H_2O_2 \qquad (1)$$

Monitoring of glucose can be based on detection of hydrogen peroxide production, oxygen depletion, or electron transfer via a conductive polymer link, as described below. Enzyme electrode sensors must contact the sample fluid to be assayed, and therefore require either sensor implantation or sample extraction (as in the case of reverse iontophoresis, microdialysis sensors and fingerstick devices). By employing the enzyme, sensors can have a significant advantage over non-enzymatic sensors of being *specific* for glucose rather than just selective. However, the benefits of enzyme specificity may not be fully realized unless the sensor is properly designed. To achieve the best performance, enzyme electrode sensors must include design features to address enzyme inactivation, biological oxygen variability, mass transfer dependance, generation of peroxide, electrochemical interference, and other effects.

From the perspective of biocompatibility, sensors can be implanted either in direct contact with blood or with tissues. Biocompatibility in contact with blood depends on the surface properties of the sensor as well flow characteristics at the implant site. Implantation in an arterial site, where the pressure and fluid shear rates are high, poses the threat of blood clotting and embolization, and is rarely justified. Central venous implantation is considerably safer, and there are several examples of successfull long-term implants in this site.

Implantation of the sensor in a tissue site is safer, but involves other challenges. The sensing objective is to infer blood glucose concentration from the tissue sensor signal, and factors that affect glucose mass transfer from nearby capillaries to the implanted sensor must be taken into account. These factors include: the pattern and extent of perfusion of the local microvasculature; regional perfusion of the implant site, the heterogeneous distribution of substrates within tissues, and the availability of oxygen. There are also substantial differences in performance between short- and long-term implant applications. In the short term, a dominant wound healing response prevails, whereas in the long term, encapsulation may occur. Definitive studies are needed to establish the real-time accuracy of implanted sensors and determine when recalibration is necessary. Studies should be designed to ascertain whether signal decay is due to enzyme inactivation, electrochemical interference, or tissue encapsulation. More information is needed about the effect of these processes on the sensor signals.

There are $>10^4$ technical publications and several thousand patents related to glucose measurement by glucose oxidase-based enzyme electrodes, although only a fraction of these address implant applications. Rather than an attempt to be comprehensive, comments here are limited to examples of the most advanced approaches intended for implant applications.

Enzyme Electrode Sensors Based on Peroxide Detection. Detection of hydrogen peroxide, the enzyme reaction product, is achieved by electrochemical oxidation of peroxide at a metal anode resulting in a signal current that passes between the anode and a counterelectrode (46). A membrane containing immobilized glucose oxidase is attached to the anode and, in the presence of glucose and oxygen under certain conditions, the current can reflect glucose concentration.

The peroxide-based sensor design is used in several home glucose monitoring devices and has been highly successful for glucose assay on an individual sample basis. However, it is not easily adapted as an implant, especially for long-term applications. The peroxide-based sensor is subject to electrochemical interference by oxidation of small molecules due to its requirement of a porous membrane and an aqueous pathway to the electrode surface for transport of the peroxide molecule. This factor partially accounts for a documented decay in sensitivity to glucose during sensor use. In addition, this sensor design can incorporate only a limited excess of immobilized glucose oxidase to counter enzyme inactivation, as high enzyme loading reduces peroxide transport to the electrode (47). Coimmobilization of catalase to avoid peroxide-mediated enzyme inactivation is not an option because it would prevent peroxide from reacting with the anode. There are also no means to account for the effects of physiologic variation in oxygen concentration and local tissue perfusion on the sensor response.

There have, nevertheless, been proposals to address some of these challenges. Composite membranes with reduced pore size have markedly reduced electrochemical interference from a variety of species over the short-term (48). A "rechargeable" enzyme system has been devised for periodically replenishing enzyme activity (49), in which a slurry of carbon particles with immobilized glucose oxidase is pumped between membrane layers of a peroxide electrode from a refillable reservoir. A gas-containing chamber has been proposed (50) to address the "oxygen deficit" (51), or stoichiometric limitation of the enzyme reaction by the relativley low tissue oxygen concnetration. Certain other challenges of the peroxide sensor principle remain to be addressed. As a result of the inherent features of this sensor principle, the peroxide-based sensor may be best suited to short-term implant applications and where frequent sensor recalibration is acceptable.

Small, needle-like *short-term peroxide-based sensors* connected by wire to a belt-mounted monitor have been developed for percutaneous implantation (52) (Fig. 6). The sensor was ultimately intended for insertion by the user for operation up to 3 days at a given tissue site before relocation. Sensors based on peroxide detection have been tested extensively in animals and humans (52–55) and, in some cases have functioned remarkably well, although frequent recalibration was required. In

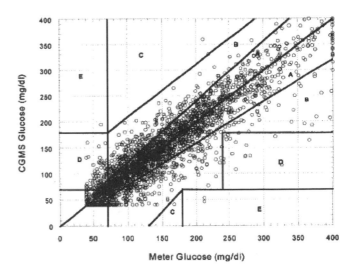

Figure 7. An approach to sensor validation. Comparison of 2477 glucose values determined by a CGMS sensor system and a standard meter (63). Data pairs were collected during home use from 135 patients. The plot, known as the Clarke Error Grid, has zones with differing clinical implications. This type of plot is widely used to describe glucose sensor performance, but has limited ability to discriminate ineffective sensors (64). (Copyright © 1999, Elsevier.)

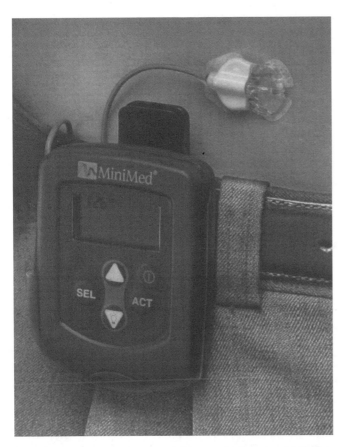

Figure 6. Components of the MiniMed CGMS system (62). This sensor, based on the peroxide-based sensing principle, is FDA approved for short-term monitoring. (Copyright © 1999, Elsevier.)

some cases, difficulties of two types have been identified (56–61). First, the sensors themselves have not been specific for glucose, sufficiently sensitive, or stable. In some cases where the sensor response in simple buffer solutions was acceptable, ambiguous signals have sometimes resulted when used as an implant. Examples of such responses are: sensor signals that decay over the short term while blood concentrations remain constant, signals that apparently follow blood concentrations during some periods, but not at other times; and identical sensors implanted in different tissue sites in a given subject that sometimes produce opposite signals (55). The peroxide-based subcutaneous sensor was the first commercially available near continuous sensor. In small controlled studies, use of the sensor was shown to lower Hb_{A1c} levels (62). Latter versions of this sensor have been approved by the FDA to be used as monitors to alarm for hyper- and hypoglycemia in real time. Although there are reservations about its accuracy, the needle sensor has been used in clinical research settings (Figs. 7, 8). A recent study found substantial error in values produced by a prominent commercial needle sensor and concluded that this sensor "... cannot be recommended in the workup of hypoglycemia in nondiabetic youth" (66) and, by extension, to other diabetic subjects. Reasons for these signal deviations are not fully understood.

Although the needle sensor may be acceptable only to a relatively small group of the most motivated individuals, it represents an advance in glucose sensor technology. Perhaps the most important contribution of the short-term needle sensor has been the revelation to users and clinicians that blood glucose excursions generally occur much more frequently and in a greater number of people than previously thought. This heightened awareness of blood glucose dynamics may lead to a greater appreciation of the need for dynamic control in improved metabolic management.

Long-term peroxide-based sensors have been implanted in the peritoneal cavity of dogs and in humans in conjunction with battery-operated telemetry units (67,68). Although the sensors remained implanted in humans for up to 160 days, the sensitivity to glucose decayed during the study and frequent recalibration was required.

Peroxide-based sensors with telemetry systems have also been implanted in the subcutaneous tissues of human type 1 diabetic subjects to determine if the mere presence of a nonreporting sensor can improve metabolic control (69). Glucose was monitored in parallel by fingerstick throughout the study as a basis for insulin dosage. Study subjects were able to reduce the time spent in hyperglycemia, increase the time spent in normoglycemia and modest hypoglycemia, and markedly reduce the time spent in severe hypoglycemia, but reductions in Hb_{A1c} values were not observed. The study was not specifically designed to validate sensor function and a more straightforward and informative study design is needed.

Short-Term Enzyme Electrodes Based on Conductive Polymers. Another principle for glucose monitoring is based on

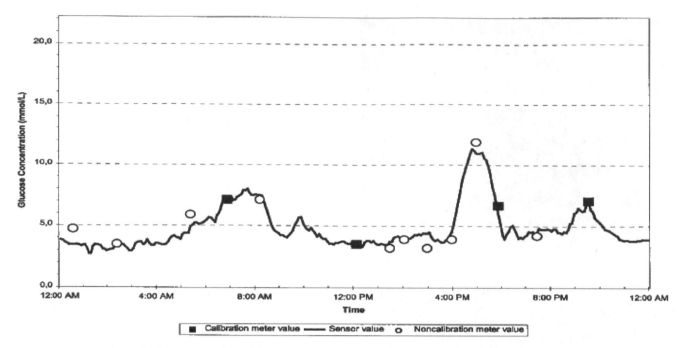

Figure 8. An example of the CGMS sensor response (65). Squares are reference values utilized in sensor calibration. Circles are additional reference blood glucose values. Reference values were obtained from a standard glucose meter. Values are in mmol · L^{-1}. (Used with permission.)

immobilization of glucose oxidase to electron-conducting polymers that can act as "chemical wires", providing a means for direct electron transport between glucose oxidase and the electrode (70). This priniciple eliminates the need for oxygen as a coreactant and, although a porous membrane that can allow passage of ionic current and interferants is still required, the electrode can be operated at lower anodic potentials to reduce electrochemical interference (71). A short-term needle-like version of this sensor for 3 day operation is under development.

Long-Term Enzyme Electrode Sensors Based on Oxygen Detection. Glucose can also be monitored by detecting differential oxygen consumption from the glucose oxidase reaction. In this case, the process is based either on glucose oxidase alone (reaction 1), or a two-enzyme reaction including catalase in excess, which produces the following overall process:

$$\text{Glucose} + 0.5\ O_2 \rightarrow \text{gluconic acid} \qquad (2)$$

The enzymes are immobilized within a gel membrane in contact with the electrochemical oxygen sensor. Excess oxygen not consumed by the enzymatic process is detected by an oxygen sensor and, after comparison with a similar background oxygen sensor without enzymes, produces a differential signal current that is related to glucose concentration.

This approach has several unique features (23). Electrochemical interference and electrode poisoning from endogenous biochemicals are prevented by a pore-free silicone rubber membrane between the electrode and the enzyme layer. This material is permeable to oxygen but completely impermeable to polar molecules

that cause electrochemical interference. Appropriate design of the sensor results in sufficient supply of oxygen to the enzyme region to avoid a stoichiometric oxygen deficit (51), a problem that has not been addressed in the peroxide-based sensor system. The differential oxygen measurement system can also readily account for variations in oxygen concentration and local perfusion, which may be particularly important for accurate function of the implant in tissues. Vast excesses of immobilized glucose oxidase can be incorporated to extend the effective enzyme lifetime of this sensor, a feature not feasible with peroxide- and conductive polymer-based sensors. Co-immobilization of catalase can further prolong the lifetime of glucose oxidase by preventing peroxide-mediated enzyme inactivation, the main cause of reduced enzyme lifetime (72). This sensor design also avoids current passage through the body and hydrogen peroxide release into the tissues.

A long-term oxygen-based sensor has been developed as a central venous implant (23) (Fig. 9). The sensor functioned with implanted telemetry (73) in dogs for >100 days and did not require recalibration during this period (Fig. 10). The central venous site permitted direct exposure of the sensor to blood, which allowed simple verification of the sensor function without mass transfer complications. This was particularly beneficial for assessing sensor stability. In clinical trials, this system has been reported (74) to function continuously for >500 days in humans with <25% change in sensitivity to glucose over that period. This achievement represents a world record for long-term, stable, implanted glucose sensor operation, although there may still exist hurdles to commercialization.

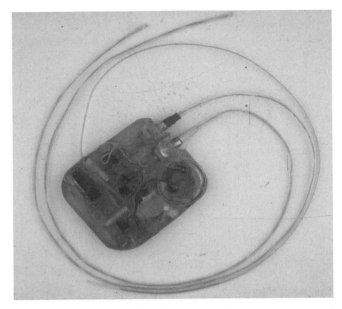

Figure 9. Animal prototype long-term central venous glucose sensor with implanted telemetry (73). Glucose and oxygen sensors are at the end of the catheters. Telemetry antenna emerges from the top, left. The telemetry body is 2 × 2.5 in.

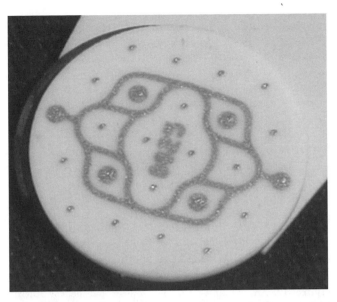

Figure 11. Close-up view of tissue glucose and oxygen sensor array (77). Sensor array with small (125 μm diameter) independent platinum working electrodes, large (875 μm diameter) common platinum counterelectrodes, and a curved common Ag/AgCl reference electrode. The membrane is not present. (Copyright 2003, American Physiological Society.)

These results have lead to several unanticipated conclusions. Although native glucose oxidase is intrinsically unstable, with appropriate sensor design the apparent catalytic lifetime of the immobilized enzyme can be substantially extended (75). The potentiostatic oxygen sensor is remarkably stable (76) and the oxygen deficit, once thought to be an insurmountable barrier, can be easily overcome (51). The central venous implant site, which is uniquely characterized by slow, steady flow of blood, allows for sufficient long-term biocompatibility with blood that the sensor stability can be documented (23). Nevertheless, the

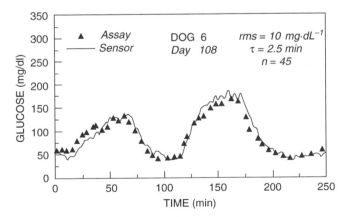

Figure 10. Response of an implanted intravenous sensor to glucose challenges on day 108 after implantation in a dog (23). The solid line is the sensor signal and triangles are venous blood glucose assays. Blood glucose excursions with initial rates of 0.2-8 $mM \cdot min^{-1}$ were produced by infusions of sterile glucose solutions through an intravenous catheter in a foreleg vein. (*Note*: 90 mg·dL^{-1} glucose = 5.0 mM.) (Copyright © 1990, American Diabetes Association.)

potential for thromoembolic events, anticipated to be rare but potentially significant over many years of sensor use, suggests reservations that may limit clinical acceptance and provides motivation for development of a potentially safer long-term sensor implant in tissues.

Long-term oxygen-based sensors have also been implanted in tissues. The successful central venous sensor cannot be simply adopted for use in the safer tissue site, but certain design features of that sensor which promote long-term function, such as immobilized enzyme design, the stable potentiostatic oxygen sensor, and membrane design to eliminate the oxygen deficit, can be incorporated (Fig. 11).

A systematic approach is required to validate sensor function, based on quantitative experimentation, mass transfer analysis, and accounting for properties of tissues that modulate glucose signals. Several new tools and methods have been developed. A tissue window chamber has been developed that allows direct optical visualization of implanted sensors in rodents, with surrounding tissue and microvasculature, while recording sensor signals (77) (Fig. 12). This facilitates determination of the effects of microvascular architecture and perfusion on the sensor signal. A method has been devised for sensor characterization in the absence of mass transfer boundary layers (78) that can be carried out before implantation and after explantation to infer stability of the implanted sensor. This allows quantitative assessment of mass transfer resistance within the tissue and the effects of long-term tissue changes. A sensor array having multiple glucose and oxygen sensors has also been developed that shows the range of variation of sensor responses within a given tissue (77). This provides a basis for averaging

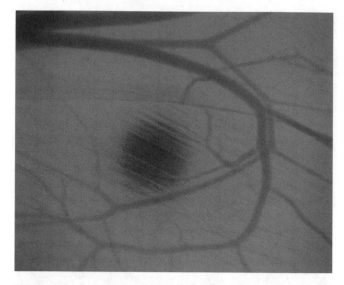

Figure 12. An implanted glucose sensor and nearby micro-vasculature (79). Optical image taken in a hamster window chamber. Sensor diameter is 125 μm.

sensor signals for quantitative correlation to blood glucose concentration.

REMAINING CHALLENGES FOR SENSOR DEVELOPMENT

Although there has been recent progress in sensor development and implementation, certain challenges remain. In many cases, there is need for improvement in data presentation and sensor validation. Standard glucose measurements for validation of sensor signals are often either not reported or are not obtained frequently enough to validate sensor responses. Requirements for sensor recalibration are not always given. Published results are often selected to show what may be ideally possible for a particular sensor rather than what is typical, sometimes conveying the impression that sensors are more accurate than may be the case.

There is a need to understand the effects of physiologic phenomena such as local perfusion, tissue variability, temperature and movement, that modulate sensor responses to glucose and affect measurement accuracy. A detailed understanding of these effects and their dynamics is needed for a full description of the glucose sensing mechanism. Robust sensor designs and modes of operation are required that assure reliable determination of glucose during exercise, sleeping and other daily life conditions.

A complete explanation for the response of every sensor should be sought, whether it is producing "good" or "bad" results, as more can often be learned for sensor improvement from sensors that produce equivocal results than from those that produce highly predictable signals (56). Present definitions as to what constitutes an acceptable sensor are based on narrow technical criteria proposed by sponsors of individual sensors that apply under specific conditions and lead to limited-

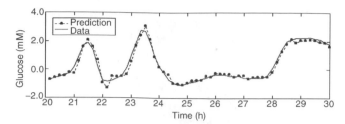

Figure 13. Blood glucose prediction based on recently sampled values (13). A 10 min prediction in a non-diabetic, average rms error = 0.2 m*M*. (Copyright © 1999, American Diabetes Association.)

use approvals by the FDA. There is a need to establish rational criteria for sensor validation and performance specific for the intended use (80). As sensors must be useful for hypoglycemia detection, sensor function must be validated in the hypoglycemic state. Correlation with Hb_{A1c} levels may not be useful for sensor validation, as the detailed information from sensors is likely to supplant Hb_{A1c} as a more comprehensive index of control.

BLOOD GLUCOSE PREDICTION

The ability to monitor blood glucose in real-time has major advantages over present methods based on sample collection that provide only sparse, historical information. There exists, however an additional possibility of using sensor information to *predict* future blood glucose values. It been demonstrated that blood glucose dynamics are not random and that blood glucose values can be predicted using autoregressive moving average (ARMA) methods, at least for the near future, from frequently sampled previous values (13) (Fig. 13). Prediction based only on recent blood glucose history is particularly advantageous because there is no need to involve models of glucose and insulin distribution, with their inherent requirements for detailed accounting of glucose loads and vascular insulin availability. This capability may be especially beneficial to children. Glucose prediction can potentially amplify the considerable benefits of continuous glucose sensing, and may represent an even further substantial advance in blood glucose management.

CLOSING THE LOOP

Glucose control is an example of a classical control system (Fig. 14). To fully implement this system, there is a need to establish a programmable controller based on continuous glucose sensing, having control laws or algorithms to counter hyper- and hypoglycemic excursions, identify performance targets for optimal insulin administration, and employ insulin pumps. The objective is restore optimal blood glucose control while avoiding over-insulinization by adjusting the program, a goal that may not be possible to achieve with alternative cell- or tissue-based insulin replacement strategies.

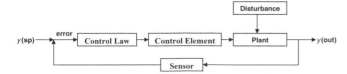

Figure 14. A simple control system for blood glucose. y(out) is the blood glucose concentration, y(sp) is the desired blood glucose, the natural sensor is in the pancreatic beta cell, the plant is the body over which glucose is distributed, and the disturbance is absorption of glucose from the gut via digestion. The control element can be an insulin pump. The control law is an algorithm that directs the pump in response to the difference between measured and target blood glucose.

Programmable external pumps that deliver insulin to the subcutaneous tissue are now widely used and implanted insulin pumps may soon become similarly available. At present, these devices operate mainly in a preprogrammed or *open-loop* mode, with occasional adjustment of the delivery rate based on fingerstick glucose information. However, experimental studies in humans have been reported utilizing *closed-loop* systems based on implanted central venous sensors and intra-abdominal insulin pumps in which automatic control strategies were employed over periods of several hundred days (81) (Fig. 15). Initial inpatient trials using subcutaneous peroxide sensors to close the loop with an external insulin pump are also underway. There is a need to expand development of such systems for broad acceptance. Extensive reviews of pump development can be found elsewhere (82–84).

These results demonstrate that an implantable artificial beta cell is potentially feasible, but more effort is required to incorporate a generally acceptable glucose sensor, validate the system extensively, and demonstrate its robust response.

CONCLUSIONS

The need for new glucose sensors in diabetes is now greater than ever. Although development of an acceptable, continuous and automatic glucose sensor has proven to be a substantial challenge, progress over the past several decades has defined sensor performance requirements and has focused development efforts on a limited group of promising candidates. The advent of new glucose sensing technologies could facilitate fundamentally new approaches to the therapy of diabetes.

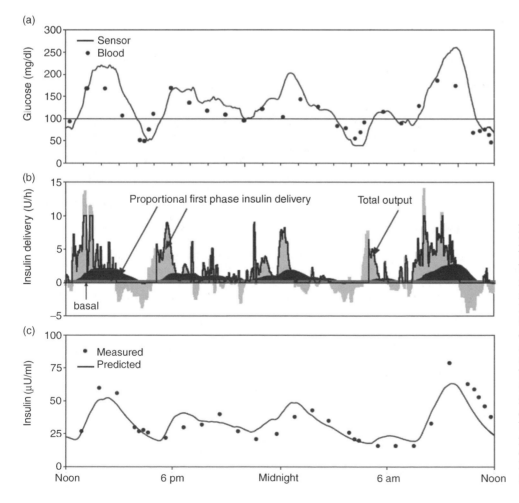

Figure 15. Blood glucose control in humans by an implanted artificial beta cell. A chronic, central venous blood glucose sensor and implanted insulin pump (Medtronic/MiniMed) implanted in a human subject. (a) Plasma (solid circles) and sensor (line) glucose following initiation of closed-loop control (noon). Solid line at 100 mg · dL^{-1} indicates setpoint. (b) Proportional (medium shading), basal (light shading), and derivative (dark shading) insulin delivery during the closed-loop (solid line indicates total, which is not allowed to go below zero). (c) Plasma (circles) and predicted insulin (solid line) concentrations. (Study performed by Medical Research Group. Copyright 2004, Elsevier.)

ACKNOWLEDGMENTS

This work was supported by grants from the National Institutes of Health and the Technology for Metabolic Monitoring Initiative of the Department of Defense. D.G. holds equity interest in GlySens, Inc., a company dedicated to the development of a new glucose sensor, and is a scientific advisor. This arrangement that has been approved by the University of California, San Diego in accordance with its conflict of interest policies.

BIBLIOGRAPHY

Cited References

1. Cahill Jr GF, Etzwiler LD, Freinkel N. Editorial: "Control" and diabetes. N Engl J Med 1976;294(18): 1004–1005.
2. The effect of intensive treatment of diabetes on the development and progression of long–term complications in insulin-dependent diabetes mellitus. The Diabetes Control and Complications Trial Research Group. N Engl J Med 1993;329(14): 977–986.
3. Stratton IM, et al. Association of glycaemia with macrovascular and microvascular complications of type 2 diabetes (UKPDS 35): prospective observational study. Bmj 2000; 321(7258):405–412.
4. Nathan DM, et al. The clinical information value of the glycosylated hemoglobin assay. N Engl J Med 1984;310(6): 341–346.
5. Skyler JS, Oddo C. Diabetes trends in the USA. Diabetes Metab Res Rev 2002;18(3 Suppl):S21–S26.
6. Zimmet P, Alberti KG, Shaw J. Global and societal implications of the diabetes epidemic. Nature (London) 2001;414 (6865):782–787.
7. Egger M, et al. Risk of adverse effects of intensified treatment in insulin-dependent diabetes mellitus: a meta-analysis. Diabet Med 1997;14(11):919–928.
8. Rovet JF, Ehrlich RM. The effect of hypoglycemic seizures on cognitive function in children with diabetes: a 7-year prospective study. J Pediatr 1999;134(4):503–506.
9. Sovik O, Thordarson H. Dead-in-bed syndrome in young diabetic patients. Diabetes Care 1999;22(2 Suppl):B40–B42.
10. Santiago JV. Lessons from the Diabetes Control and Complications Trial. Diabetes 1993;42(11):1549–1554.
11. van den Berghe G, et al. Intensive insulin therapy in the critically ill patients. N Engl J Med 2001;345(19):1359–1367.
12. Van den Berghe G, et al. Outcome benefit of intensive insulin therapy in the critically ill: Insulin dose versus glycemic control. Crit Care Med 2003;31(2):359–366.
13. Bremer T, Gough DA. Is blood glucose predictable from previous values? A solicitation for data. Diabetes 1999;48 (3):445–451.
14. Clark Jr LC, Lyons C. Electrode systems for continuous monitoring in cardiovascular surgery. Ann NY Acad Sci 1962; 102:29–45.
15. Chang KW, et al. Validation and bioengineering aspects of an implantable glucose sensor. Trans Am Soc Artif Intern Org 1973;19:352–360.
16. Slama G, Bessman SP. [Results of in vitro and in vivo use of a prototype implantable glucose sensor]. J Annu Diabetol Hotel Dieu 1976; 297–302.
17. Albisser AM, et al. An artificial endocrine pancreas. Diabetes 1974;23(5):389–396.
18. Thomas LJ, Bessman SP. Prototype for an implantable insulin delivery pump. Proc West Pharmacol Soc 1975;18:393–398.
19. Clemens AH, Chang PH, Myers RW. The development of Biostator, a Glucose Controlled Insulin Infusion System (GCIIS). Horm Metab Res 1977; (7 Suppl):23–33.
20. Skyler JS, et al. Home blood glucose monitoring as an aid in diabetes management. Diabetes Care 1978;1(3):150–157.
21. Pickup JC, et al. Clinical application of pre-programmed insulin infusion: continuous subcutaneous insulin therapy with a portable infusion system. Horm Metab Res 1979; (8) (Suppl):202–204.
22. McCall AL, Mullin CJ. Home monitoring of diabetes mellitus–a quiet revolution. Clin Lab Med 1986;6(2):215–239.
23. Armour JC, et al. Application of chronic intravascular blood glucose sensor in dogs. Diabetes 1990;39(12):1519–1526.
24. Sage Jr BH. FDA panel approves Cygnus's non-invasive GlucoWatch. Diabetes Technol Ther 2000;2(1): 115–116.
25. Clinical Practice Recommendations 2005. Diabetes Care 2005;28(1 Suppl):S1–S79.
26. Lojo J, et al. Prevantive Care Practices Among Persons with Diabetes, United States, 1995 and 2001. Morbidity Mortality Weekly Rep, Center Disease Control Prevention 2002;51(43): 965–967.
27. Bennion N, Christensen NK, McGarraugh G. Alternate site glucose testing: a crossover design. Diabetes Technol Ther 2002;4(1):25–33; discussion 45–47.
28. Gough DA, Kreutz-Delgado K, Bremer TM. Frequency characterization of blood glucose dynamics. Ann Biomed Eng 2003;31(1):91–97.
29. McNichols RJ, Cote GL. Optical glucose sensing in biological fluids: an overview. J Biomed Opt 2000;5(1):5–16.
30. Cote GL. Noninvasive and minimally-invasive optical monitoring technologies. J Nutr 2001;131(5):1596S–604S.
31. Arnold MA, Burmeister JJ, Small GW. Phantom glucose calibration models from simulated noninvasive human near-infrared spectra. Anal Chem 1998;70(9):1773–1781.
32. Goetz Jr MJ, et al. Application of a multivariate technique to Raman spectra for quantification of body chemicals. IEEE Trans Biomed Eng 1995;42(7):728–731.
33. Steffes PG. Laser-based measurement of glucose in the ocular aqueous humor: an efficacious portal for determination of serum glucose levels. Diabetes Technol Ther 1999;1(2):129–133.
34. Cameron BD, Baba JS, Cote GL. Measurement of the glucose transport time delay between the blood and aqueous humor of the eye for the eventual development of a noninvasive glucose sensor. Diabetes Technol Ther 2001;3(2):201–207.
35. Blass DA, Adams E. Polarimetry as a general method for enzyme assays. Anal Biochem 1976;71(2):405–414.
36. Baba JS, et al. Effect of temperature, pH, and corneal birefringence on polarimetric glucose monitoring in the eye. J Biomed Opt 2002;7(3):321–328.
37. Schultz JS, Mansouri S, Goldstein IJ. Affinity sensor: a new technique for developing implantable sensors for glucose and other metabolites. Diabetes Care 1982;5(3): 245–253.
38. March WF, Ochsner K, Horna J. Intraocular lens glucose sensor. Diabetes Technol Ther 2000;2(1):27–30.
39. Muller M. Science, medicine, and the future: Microdialysis. Bmj 2002;324(7337):588–591.

40. Meyerhoff C, et al. On line continuous monitoring of sub-cutaneous tissue glucose in men by combining portable glucosensor with microdialysis. Diabetologia 1992;35(11): 1087–1092.

41. Hoss U, et al. A novel method for continuous online glucose monitoring in humans: the comparative microdialysis technique. Diabetes Technol Ther 2001;3(2):237–243.

42. Tamada JA, Bohannon NJ, Potts RO. Measurement of glucose in diabetic subjects using noninvasive transdermal extraction. Nat Med 1995;1(11):1198–1201.

43. Potts RO, Tamada JA, Tierney MJ. Glucose monitoring by reverse iontophoresis. Diabetes Metab Res Rev 2002;18 (1 Suppl):S49–S53.

44. Lenzen H, et al. A non-invasive frequent home blood glucose monitor. Practical Diabetes Int 2002;19(4):101–103.

45. Tierney MJ, et al. The GlucoWatch biographer: a frequent automatic and noninvasive glucose monitor. Ann Med 2000; 32(9):632–641.

46. Bindra DS, et al. Design and *In vitro* studies of a needle-type glucose sensor for subcutaneous monitoring. Anal Chem 1991;63(17):1692–1696.

47. Jablecki M, Gough DA. Simulations of the frequency response of implantable glucose sensors. Anal Chem 2000;72(8):1853–1859.

48. Ward WK, et al. A new amperometric glucose microsensor: *in vitro* and short-term *In vivo* evaluation. Biosens Bioelectron 2002;17(3):181–189.

49. Xie SL, Wilkins E. Rechargeable glucose electrodes for long-term implantation. J Biomed Eng 1991;13(5):375–378.

50. Clark Jr LC. Membrane Polarographic Electrode System and Method with Electrochemical Compensation. US pat 3,539,455. 1970.

51. Gough DA, Lucisano JY, Tse PH. Two-dimensional enzyme electrode sensor for glucose. Anal Chem 1985;57(12):2351–2357.

52. Mastrototaro JJ. The MiniMed continuous glucose monitoring system. Diabetes Technol Ther 2000;2(1 Suppl): S13–S18.

53. Rebrin K, et al. Subcutaneous glucose predicts plasma glucose independent of insulin: implications for continuous monitoring. Am J Physiol 1999;277(3 Pt. 1): E561–E571.

54. Gross TM, Mastrototaro JJ. Efficacy and reliability of the continuous glucose monitoring system. Diabetes Technol Ther 2000;2(1 Suppl):S19–S26.

55. Metzger M, et al. Reproducibility of glucose measurements using the glucose sensor. Diabetes Care 2002;25(7):1185–1191.

56. Gough DA, Armour JC. Development of the implantable glucose sensor. What are the prospects and why is it taking so long? Diabetes 1995;44(9):1005–1009.

57. Shichiri M, et al. Telemetry glucose monitoring device with needle-type glucose sensor: a useful tool for blood glucose monitoring in diabetic individuals. Diabetes Care 1986;9 (3):298–301.

58. Abel P, Muller A, Fischer U. Experience with an implantable glucose sensor as a prerequisite of an artificial beta cell. Biomed Biochim Acta 1984;43(5):577–584.

59. Moatti-Sirat D, et al. Towards continuous glucose monitoring: *In vivo* evaluation of a miniaturized glucose sensor implanted for several days in rat subcutaneous tissue. Diabetologia 1992;35(3):224–230.

60. Johnson KW, et al. *In vivo* evaluation of an electroenzymatic glucose sensor implanted in subcutaneous tissue. Biosens Bioelectron 1992;7(10):709–714.

61. Kerner W, et al. The function of a hydrogen peroxide-detecting electroenzymatic glucose electrode is markedly impaired in human sub-cutaneous tissue and plasma. Biosens Bioelectron 1993;8(9–10):473–482.

62. Bode BW, et al. Continuous glucose monitoring used to adjust diabetes therapy improves glycosylated hemoglobin: a pilot study. Diabetes Res Clin Pract 1999;46(3):183–190.

63. Gross TM, et al. Performance evaluation of the MiniMed continuous glucose monitoring system during patient home use. Diabetes Technol Ther 2000;2(1): p. 49–56.

64. Gough DA, Botvinick EL. Reservations on the use of error grid analysis for the validation of blood glucose assays. Diabetes Care 1997;20(6): p. 1034–1036.

65. Kerssen A, de Valk HW, Visser GH. The Continuous Glucose Monitoring System during pregnancy of women with type 1 diabetes mellitus: accuracy assessment. Diabetes Technol Ther 2004;6(5): p. 645–51.

66. Mauras N, et al. Lack of accuracy of continuous glucose sensors in healthy, nondiabetic children: results of the Diabetes Research in Children Network (DirecNet) accuracy study. J Pediatr 2004;144(6):770–775.

67. Gilligan BJ, et al. Evaluation of a subcutaneous glucose sensor out to 3 months in a dog model. Diabetes Care 1994;17(8): 882–887.

68. Updike SJ, et al. A subcutaneous glucose sensor with improved longevity, dynamic range, and stability of calibration. Diabetes Care 2000;23(2):208–214.

69. Garg SK, Schwartz S, Edelman SV. Improved glucose excursions using an implantable real-time continuous glucose sensor in adults with type 1 diabetes. Diabetes Care 2004; 27(3):734–738.

70. Csoregi E, Schmidtke DW, Heller A. Design and optimization of a selective subcutaneously implantable glucose electrode based on "wired" glucose oxidase. Anal Chem 1995;67(7): 1240–1244.

71. Heller A. Implanted electrochemical glucose sensors for the management of diabetes. Annu Rev Biomed Eng 1999;1:153–175.

72. Tse PHS, Leypoldt JK, Gough DA. "Determination of the Intrinsic Kinetic Constants of Immobilized Glucose Oxidase and Catalase". Biotechnol Bioeng 1987;29:696–704.

73. McKean BD, Gough DA. A telemetry-instrumentation system for chronically implanted glucose and oxygen sensors. IEEE Trans Biomed Eng 1988;35(7):526–532.

74. Medtronic Minimed talk at the Diabetes Technology and Therapeutics Conference, S.F., CA. 2003.

75. Gough DA, Bremer T. Immobilized glucose oxidase in implantable glucose sensor technology. Diabetes Technol Ther 2000; 2(3):377–380.

76. Lucisano JY, Armour JC, Gough DA. *In vitro* stability of an oxygen sensor. Anal Chem 1987;59(5):736–739.

77. Makale MT, et al. Tissue window chamber system for validation of implanted oxygen sensors. Am J Physiol Heart Circ Physiol 2003;284(6):H2288–H2294.

78. Makale MT, Jablecki MC, Gough DA. Mass transfer and gas-phase calibration of implanted oxygen sensors. Anal Chem 2004;76(6):1773–1777.

79. Makale MT, Chen PC, Gough DA. Variants of the tissue/sensor array chamber. Am J Physiol Heart Circ Physiol 2005;286, in press.

80. Bremer TM, Edelman SV, Gough DA. Benchmark data from the literature for evaluation of new glucose sensing technologies. Diabetes Technol Ther 2001;3(3):409–418.

81. Steil GM, Panteleon AE, Rebrin K. Closed-loop insulin delivery-the path to physiological glucose control. Adv Drug Deliv Rev 2004;56(2):125–144.

82. Saudek CD. Implantable Pumps. 3rd ed. International Textbook of Diabetes Mellitus; 2004.

83. Selam JL. External and implantable insulin pumps: current place in the treatment of diabetes. Exp Clin Endocrinol Diabetes 2001;109(2 Suppl):S333–S340.

84. Vague P, et al. The implantable insulin pump in the treatment of diabetes. Hopes and reality?. Bull Acad Natl Med 1996; 180(4):831–41. discussion 841–843.

See also FIBER OPTICS IN MEDICINE; OPTICAL SENSORS; OXYGEN SENSORS; PANCREAS, ARTIFICIAL.

H

HBO THERAPY. See HYPERBARIC OXYGENATION.

HEARING IMPAIRMENT. See AUDIOMETRY.

HEART RATE, FETAL, MONITORING OF. See FETAL MONITORING.

HEART VALVE PROSTHESES

K. B. CHANDRAN
University of Iowa
Iowa City, Iowa

INTRODUCTION

The human circulatory system provides adequate blood flow without interruption to the various organs and tissues and regulates blood supply to the demands of the body. The contracting heart supplies the energy required to maintain the blood flow through the vessels. The human heart consists of two pumps in series. The right side of the heart, a low pressure pump, consisting of the right atrium and the right ventricle supplies blood to the pulmonary circulation. The left side consisting of the left atrium and the left ventricle is the high pressure pump circulating blood through the systemic circulation. Figure 1 is a schematic representation of the four chambers of the heart and the arrows indicate the direction of blood flow. The pressure gradients developed between the main arteries supplying blood to the systemic and pulmonary circulation and the respective venous ends are the driving forces causing the blood flow and the energy is dissipated in the form of heat due to frictional resistance.

The four valves in the heart ensure that the blood flows only in one direction. The blood from the systemic circula-tion supplies nutrients and oxygen to the cells for the various tissues and organs and removes carbon dioxide at the level of capillaries. The oxygen depleted blood returns through the systemic veins to the right atrium. During the ventricular relaxation or diastole, the blood passes through the tricuspid valve into the right ventricle. In the ventricular contraction phase of the cardiac cycle or systole, the tricuspid valve closes and the pulmonic valve opens to pump the blood to the lungs through the pulmonary arteries. Carbon dioxide is removed and oxygen is absorbed by the blood in the capillaries of the lungs that is surrounded by the alveolar sac with the air we breathe. The oxygen-rich blood returns to the left atrium via the pulmonary veins and passes through the mitral (bicuspid) valve into the left ventricle during the ventricular diastole. During the ventricular contraction, the mitral valve closes and the aortic valve opens to pump the blood through the systemic circulation. The main function of the heart valves is to control the direction of blood flow permitting flow in the forward direction and preventing regurgitation or back flow through the closed valves.

Anatomy of the Native Valves

The aortic valve (Fig. 2) consists of three semicircular (semilunar) leaflets or cusps within a connective tissue sleeve (1) attached to a fibrous ring. The cusps meet at three commissures that are equally spaced along the circumference at the supraaortic ridge. This ridge is thickening of the aorta at which the cusps insert and there is no continuity of tissue from one cusp to the other across the commissure. The leaflet consists of three layers as shown in Fig. 3: the aortic side layer is termed the fibrosa and is the

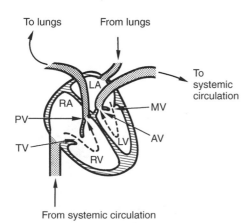

Figure 1. Schematic of blood flow in the human heart. LA-Left atrium; RA-Right atrium; LV-Left ventricle; RV-Right ventricle; PV-pulmonary valve; TV-Tricuspid valve; AV-Aortic valve; and MV-Mitral valve.

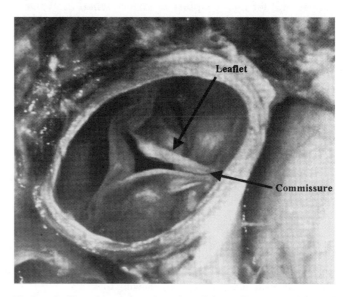

Figure 2. Human aortic valve viewed from the aorta. (Adapted with permission from Otto, C. M. Valvular Heart Disease, Second Edition, 2004, Saunders/Elsevier, Inc., Philadelphia, PA.)

Figure 3. A histologic section of an aortic valve leaflet depicting the three layers along the thickness of the leaflet: F = Fibrosa; S = Spongiosa; and V = Ventricularis. (Courtesy of Prof. Michael Sacks of the University of Pittsburgh, Pittsburgh, PA.)

major fibrous layer within the body of the leaflet; layer on the ventricular side termed ventricularis is composed of both collagen and elastin; and the central portion of the valve termed the spongiosa consisting of loose connective tissue, proteins, and glycosaminoglycans (GAG). The leaflet length is larger than the radius of the annulus, and hence a small overlap of the tissue from each leaflet protrudes and forms a coaptation surface when the valve is closed to ensure that the valve is sealed in the closed position. The sinus of Valsalva is attached to the fibrous annular ring on the aortic side and is comprised of three

bulges at the root of the aorta. Each bulge is aligned with the belly or the central part of the valve leaflet. The left and the right sinuses contain the coronary ostia (openings) giving rise to the left and right coronary arteries, respectively, providing blood flow and nutrients to the cardiac muscles. When the valve is fully open, the leaflets extend to the upper edges of the sinuses. The anatomy of the pulmonic valve is similar to that of the aortic valve, but the sinuses in the pulmonary artery are smaller than the aortic sinuses, and the pulmonic valve orifice is slightly larger. The average aortic valve orifice area is $\sim$4.6 cm^2 and is $\sim$4.7 cm^2 for the pulmonic valve (2). In the closed position, the pulmonic valve is subject to a pressure of $\sim$30 mmHg (3.99 kPa) while the load on the aortic valve is $\sim$100 mmHg (13.30 kPa).

The mitral and tricuspid valves are also anatomically similar with the mitral valve consisting of two main leaflets (cusps) compared to three for the valve in the right side of the heart. The valves consist of the annulus, leaflets, papillary muscles, and the chordae tendinae (Fig. 4). The average mitral and tricuspid valve orifice areas are 7.8 and 10.6 cm^2, respectively (2). The atrial and ventricular walls are attached to the mitral annulus, consisting of dense collagenous tissue surrounded by muscle, at the base of the leaflets. The chordae tendinae are attached to the free edge of the leaflets at multiple locations and extend to the tip of the papillary muscles. Anterior and posterior leaflets of the mitral valve are actually one continuous tissue with two regularly spaced indentations called the commissures. The combined surface area of both the leaflets is approximately twice the area of the valve orifice and thus the leaflets coaptate during the valve closure. The posterior leaflet encircles two-thirds of the annulus and is quadrangular shaped, while the anterior leaflet is semilunar shaped. The left ventricle has two papillary muscles that attach to the ventricular free wall and tether the mitral valve in place via the chordae tendinae. This tethering prevents the leaflets from prolapsing into the left atrium during ventricular ejection. Improper tethering will result in the leaflets extending into the atrium and incomplete apposition of the leaflets will permit blood to

Figure 4. Schematic of the human mitral (bicuspid) valve and a photograph showing the anterior leaflet with the chordae tendinae attachment with papillary muscles. (Courtesy of Prof. Ajit Yoganathan from Georgia Institute of Technology.)

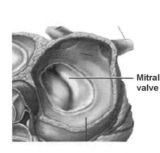

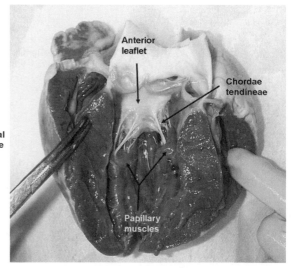

regurgitate back to the atrium. The tricuspid valve has three leaflets, a septal leaflet along with the anterior and posterior leaflets, and is larger and structurally more complicated than the mitral valve.

Valve Dynamics

At the beginning of systole, the left ventricle starts to contract and with the increase in pressure, the mitral valve closes preventing regurgitation of blood to the left atrium. During the isovolumic contraction with both the mitral and aortic valves closed, the ventricular pressure rises rapidly. The aortic valve opens when the left ventricular pressure exceeds the aortic pressure. The blood accelerates rapidly through the open valve and peak velocity of flow occurs during the first third of systole. The pressure difference between the left ventricle and the aorta required to open the valve is of the order of 1–2 mmHg (0.13–0.26 kPa). During the forward flow phase, vortices develop in the three sinuses behind the open leaflets and the formation of such vortices was first described by Leonardo da Vinci in 1513. Several studies have suggested that the vortices and the difference in pressure between the sinuses and the center of the aortic orifice pushes the leaflets toward closure even during the second third of systole when forward flow of blood continues. With the ventricular relaxation and rapid drop in the ventricular pressure, an adverse pressure gradient between the ventricle and the aorta moves the leaflets toward full closure with negligible regurgitation of blood from the aorta to the ventricle. Systole lasts for about one-third of the cardiac cycle and the peak pressure reached in the aorta during systole in healthy humans is ~120 mmHg (15.96 kPa) and the diastolic pressure in the aorta with the aortic valve closed is ~80 mmHg (10.64 kPa).

At the beginning of diastole, the aortic valve closes and the ventricular pressure decreases rapidly during the isovolumic relaxation. As the ventricular pressure falls below the atrial pressure, the mitral valve opens and the blood flows from the atrium to the ventricle. The pressure difference between the left atrium and the ventricle required to open the mitral valve and drive the blood to fill the ventricle is smaller than that required with the aortic valve (< 1 mmHg or 0.13 kPa). As the blood fills the ventricle, vortical flow is established in the ventricle, and it has been suggested that the leaflets move toward closure due to the same. The atrial contraction induces additional flow of blood from the atrium into the ventricle during the second half of diastole and the adverse pressure gradient at the beginning of ventricular contraction forces the mitral valve to close and isovolumic contraction takes place. The chordae tendinae prevents the prolapse of the leaflets into the left atrium when the mitral valve is in the closed position. The dynamics of the mitral valve opening and closing is a complex process involving the leaflets, mitral orifice, chordae tendinae, and the papillary muscles. During systole, the closed mitral valve is subjected to pressures of ~120 mmHg (15.96 kPa).

The dynamics of opening and closing of the pulmonic and the tricuspid valves are similar to the aortic and mitral valve, respectively, even though the pressures generated in the right ventricle and the pulmonary artery are generally about a one-third of the corresponding magnitudes in the left side of the heart. From the description of the valve dynamics, one can observe several important features on the normal functioning of the heart valves. These include opening efficiently with minimal difference in pressure between the upstream and downstream sides of the valve, and efficient closure to ensure minimal regurgitation. In addition, the flow past the valves are laminar with minimal disturbances in flow and the fluid induced stresses do not activate or destroy the formed elements in blood such as the platelets and red blood cells. As the valves open and close, the leaflets undergo complex motion that includes large deformation, as well as bending. The leaflet material is also subjected to relatively high normal and shear stresses during these motions. The valves open and close at about once every second, and hence functions over several million cycles during the normal life of a human. These are some of the important considerations in the design and functional evaluation of heart valve prostheses that we consider in detail below.

Valvular Diseases and Replacement of the Diseased Valves

Valvular diseases are more common on the left heart due to the high pressure environment for the aortic and mitral valves and also with the tricuspid valve on the right side of the heart. Valvular diseases include stenosis and incompetence. Stenosis of the leaflets is due to calcification resulting in stiffer leaflets that will require higher pressures to open the valves. Rheumatic fever is known to affect the leaflets resulting in stenosed valves (3). Premature calcification of the bicuspid valve, as well as significant obstruction of the left ventricular outflow in congenital aortic valve stenosis, also affects the valves of the left heart (3). Aortic sclerosis due to aging can also advance to valvular stenosis in some patients. Mitral stenosis may be the result of commissural fusion in younger patients and may also be due to cusp fibrosis. In the case of valvular stenosis, higher pressure needs to be generated to force the stiffer leaflets to open and the valvular orifice area in the fully open position may be significantly reduced. Effective orifice area (EOA) can be computed by the application of the Gorlin equation (4) based on the fluid mechanic principles and is given by the following relationship:

$$\mathrm{EOA(cm^2)} = \frac{Q_{\mathrm{rms}}}{C\sqrt{\Delta\bar{p}}} \qquad (1)$$

In this equation, Q_{rms} is the root-mean-square (rms) flow rate (mL/s) during the forward flow through the valve and $\Delta\bar{p}$ is the mean pressure drop (mmHg) across the open valve. The measurement of mean pressure drop *in vivo* is described later, and the flow rate across the valve during the forward flow phase is computed from the measurement of cardiac output and the heart rate. The parameter C represents a constant that is based on the discharge coefficient used for the aortic or mitral valve, and the unit conversion factors to result in the computed area in terms of square centimeter. A more direct technique to estimate the effective orifice area is the application of conservation of mass principle. The systolic volume flow through the left ventricular outflow tract is determined as the product of the outflow tract cross-sectional area and the flow velocity–time integral. Since the same blood volume must also pass through the valve orifice, the valve orifice area is computed by dividing the volume flow with the measured aortic valve velocity–time integral (5). Replacement of the aortic valve is generally considered when the measured valvular orifice area is < 0.4 $\mathrm{cm}^2 \cdot \mathrm{m}^{-2}$ of body surface area (6). The corresponding value for the mitral stenosis is 1.0 $\mathrm{cm}^2 \cdot \mathrm{m}^{-2}$.

Valvular incompetence results from incomplete closure of the leaflets resulting in significant regurgitation of blood. Incompetence could be the result of decrease in leaflet area due to rheumatic disease or perforations in the leaflets due to bacterial endocarditis. Structural alterations due to loss of commissural support or aortic root dilatation can result in aortic valve incompetence. Rupture of chordae tendinae, leaflet perforation, and papillary muscle abnormality may also affect the mitral valve closure and increase in regurgitation. Optimal timing for valvular replacement in the case of native valve incompetence is not clearly defined.

Various methods for valvular reconstruction or repair are also being developed instead of replacement with prostheses since these techniques are associated with lower risk of mortality and lower risk of recurrence (7). Valvular repair rather than replacement is generally preferred for regurgitation due to segmental prolapse of the posterior mitral leaflet. Implantation of a prosthetic ring to reduce the size of the mitral orifice and improve leaflet coaptation is performed in the case of mitral regurgitation due to ring dilatation. Mitral endocarditis with valvular or chordal lesions is also repaired rather than replacing the whole valve. Dilatation of the root and prolapse of the cusps are also the most important causes for regurgitation of the aortic valves and several techniques have also been developed to correct these pathologies (7).

The cardiopulmonary bypass technique to reroute the blood from the vena cava to the ascending aorta, and the introduction of cold potassium cardioplegia to arrest the heart to perform open heart surgery introduced in the 1950s enabled the replacement of diseased valves. Replacement of severely stenosed and/or incompetent valves with prostheses is a common treatment modality today and patients with prosthetic valves lead a relatively normal life. Yet, significant problems are also encountered with implanted prosthetic valves and efforts are continuing to improve the design of the valves for enhanced functionality and minimizing the problems encountered with implantation.

PROSTHETIC HEART VALVES

Ideal Valve Design

An ideal valve to be used as replacement for a diseased valve should mimic the functional characteristics of the native human heart valves with the following characteristics (adapted from Ref. 8). The prosthetic valve should open efficiently with a minimal transvalvular pressure drop. The blood should flow through the orifice with central and undisturbed flow as is observed with healthy native heart valves. The valve should close efficiently with minimal amount of regurgitation. The material used for the valve should be biocompatible, durable, nontoxic, and nonthrombogenic. The valve will be anticipated to open and close for > 40 million cycles/year for many years and must maintain the structural integrity throughout the lifetime of the implant. Blood is a corrosive and chemically unstable fluid that tends to form thrombus in the presence of foreign bodies. To avoid thrombus formation, the valve surfaces must be smooth, and the flow past the valve should avoid regions of stagnation and recirculation as well as minimize flow-induced stresses that are factors related to thrombus initiation. The prosthetic valve should be surgically implantable with ease and should not interfere with the normal anatomy and function of the cardiac structures and the aorta. The valve should be easily manufactured in a range of sizes, inexpensive, and sterilizable.

Transplanting freshly explanted human heart valves from donors who died of noncardiovascular diseases is probably the most ideal replacement and such homograft valves have been successfully used as replacements. These are the only valves entirely consisting of fresh biological tissue and sewn into place resulting in an unobstructed central flow. The transplanted homograft valves are no longer living tissue, and hence lack the cellular regeneration capability of the normal valve leaflets. Thus, the transplanted valves are vulnerable to deterioration on a long-term use. Moreover, homograft valves are difficult to obtain except in trauma centers in large population areas, and hence not a viable option generally.

Numerous prosthetic valves have been developed over the past 40 years and most of the design and development of valvular prostheses have been empirical. The currently available heart valve prostheses can be broadly classified into two categories: mechanical heart valves (MHV) and bioprosthetic heart valves (BHV). Even though valves from both categories are routinely implanted in patients with valvular disease and the patients with prosthetic implants lead a relatively normal life, several major problems are encountered with the mechanical and biological prosthetic valves (9). These problems include: (1) thromboembolic complications; (2) mechanical failure due to fatigue or chemical changes; (3) mechanical damage to the formed elements in blood including hemolysis, activation, and destruction of platelets and protein denaturation; (4) perivalvular leak due to healing defects; (5) infection; and (6) tissue overgrowth. The first three problems with implanted valves can be directly attributed to the design of the mechanical and biological prostheses and the fluid and solid mechanics during valve function. Thrombus deposition on the valvular surface and subsequent breakage of the thrombus to form emboli that can result in stroke or heart failure is still a major problem with MHV implants. Hence, patients with MHV implants need a long-term anticoagulant therapy that can lead to complications with bleeding. On the other hand, patients implanted with bioprostheses do not generally require anticoagulant therapy except immediately after surgery. Yet, leaflet tearing with structural disintegration results in the need for BHV implants to be replaced at about 10–12 years after implantation on the average. Due to the necessity of multiple surgeries during a lifetime, the tissue valves are generally not implanted in younger patients. Patients who cannot tolerate or cannot be on long-term anticoagulant therapy are also candidates for the BHV.

Mortality is higher among patients after prosthetic valve replacement than among age-matched controls. Mortality rate is not significantly different between MHV and BHV implantation. In addition to the mortality rate and valve related morbidity, quality of life for the patient must be an important consideration in the choice of valve implantation and the quality of life is difficult to quantify. Individual patients may place different emphasis on mortality, freedom from reoperation, risk of thromboembolism and stroke, risk of anticoagulation-related hemorrhage, and lifestyle modification required with chronic anticoagulation. Patients may choose to accept the high probability of reoperation within 10–12 years with BHV in order to avoid long-term anticoagulation with MHV, whereas others may want to avoid the likelihood of reoperation (10).

We will review the historical development of heart valve prostheses, functional evaluation of these valves in order to understand the relationship between the dynamics of the valve and the problems associated with the valve implants and our continuing efforts on the understanding of the problem and improvements in design.

Mechanical Heart Valves

Mechanical valves are made of blood compatible, nonbiological material, such as metals, ceramics, or polymers. The initial designs of mechanical valve prostheses were of the centrally occluding type with either a ball or disk employed as the moving occluder. The occluder passively responds to the difference in pressure in opening and closing of the valves. Starr-Edwards caged ball valve (Fig. 5a) was the first mechanical valve to be implanted in 1960 in the correct anatomical position as replacement for the diseased native mitral valve (11–14). The caged disk prostheses (Fig. 5b), in which a flat disk was employed as the occluder, were of a lower profile than the ball valves, and hence were thought to be advantageous especially as a replacement in the mitral position in order to minimize interference with the cardiac structures. However, increased flow separation and turbulence in the flow past the flat disk compared to that for the spherical ball occluder in the caged ball prostheses resulted in larger pressure drop across the valve with the caged disk design. An increased propensity for thrombus deposition in the recirculation region was also observed, and hence this design was not successful in spite of the low profile. The cage in the caged ball prostheses is made of a polished cobalt–chromium alloy and the ball is made of a silicone rubber that contains 2% by weight barium sulfate for radiopacity. The sewing ring contains silicone rubber insert under knitted composite polytetrafluoroethylene (PTFE: Teflon) and polypropylene cloth. With the centrally occluding design, the flow dynamics past the caged ball prostheses is vastly different from that of flow past native aortic or mitral valves.

Within the next two decades of 1970s and 1980s, valve prostheses with a tilting disk or bileaflet designs were introduced with significantly improved flow characteris-

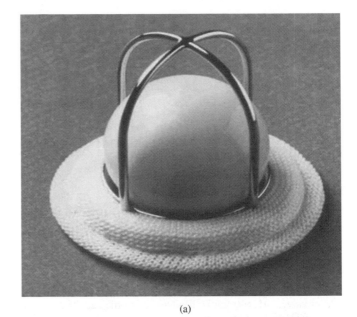

(a)

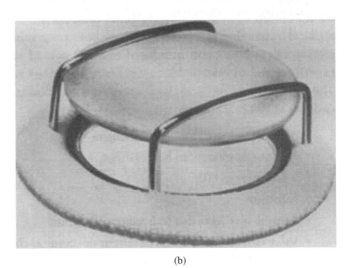

(b)

Figure 5. Photographs of (a) Starr-Edwards caged ball valve (Edwards Lifesciences, LLC, Irvine, CA); and (b) a caged disk valve (Courtesy of Prof. Ajit Yoganathan of Georgia Institute of Technology, Atlanta, GA) as examples of early mechanical valve designs.

tics. The first tilting disk valve that was clinically used was a notched Teflon occluder that engaged in another pair of notches in the housing (15). The stepped occluder with the notches was not free to rotate. Clinical data soon indicated severe thrombus formation around the notches and wear of the Teflon disk leading to severe valvular regurgitation or disk embolization (16). A major improvement to this design was the introduction of hinge-less free-floating tilting disk valves in the Bjork–Shiley (17) and the Lillehei–Kaster valves. The Bjork–Shiley valve had a depression in the disk and two welded wire struts in the valve housing to retain the disk. The occluder tilted to the open and closed position and it was free to rotate around its center. The Bjork–Shiley valve housing was made from Stellite-21 with a Teflon sewing ring and a Delrin disk. Compared to the

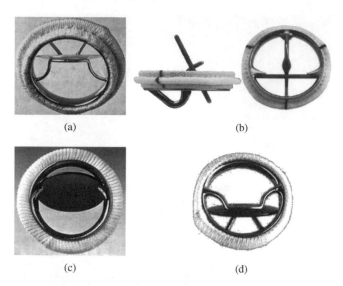

(a)

(b)

(c)

(d)

Figure 6. Tilting disk mechanical valve prostheses: (a) Bjork–Shiley valve; (b) Medtronic Hall valve (Medtronic, Inc., Minneapolis, MN; (c) Omni Carbon valve (MedicalCV Inc., Minneapolis, MN); (d) Sorin valve (Sorin Biomedica Cardio S.p.A., Via Crescentino, Italy).

centrally occluding caged ball valves, a large annulus diameter compared to the tissue annulus diameter in the tilting disk valve resulted in a very low pressure drop and thus energy loss in the flow across the valve in the open position. The disk opened to an angle of 60° or more, and hence the flow was more central. The free floating disk that rotated during the opening and closing phases prevented any build up of thrombus. However, the Delrin disk had a propensity for swelling during autoclaving that may compromise the proper functioning of the leaflets (18,19). The disk for the Lillehei–Kaster valve consisted of a graphite substrate coated with a 250 μm thick layer of a carbon–silicon alloy (Pyrolite). The pyrolytic carbon has proven to be a very durable and blood-compatible material for use in prosthetic heart valves and is the preferred material for the MHVs currently available for implants. The Bjork–Shiley valve also had the Delrin disk replaced with pyrolytic carbon disk shortly thereafter (Fig. 6a). The Medtronic Hall tilting disk valve (Fig. 6b) has a central, disk control strut. An aperture in the flat pyrolytic carbon disk affixes it to this central guiding strut and allows it to move downstream by ~2.0 mm. This translation improves the flow velocity between the orifice ring and the rim of the disk. The ring and strut combination is machined from a single piece of titanium for durability and the housing can be rotated within the knitted Teflon sewing ring for optimal orientation of the valve within the tissue annulus. The Omniscience valve was an evolution of the Lillehei–Kaster valve and the Omnicarbon valve (Fig. 6c) is the only tilting disk valve with the occluder and housing made of pyrolytic carbon. Sorin Carbocast tilting disk valve (Fig. 6d), made in Italy and available in countries outside United States, has the struts and the housing made in a single piece by a microcast process and thus eliminates the need for welding the struts to the housing. The cage for this valve is made of a chrome–cobalt alloy and coated with a carbon film. The

tilting disk valves of the various manufacturers open to a range of angles varying from 60 to 85° and in the fully open position, the flow passes through the major and minor orifices. Some of the valve designs, such as the Bjork–Shiley valve, encountered unforeseen problems with structural failure due to further design modifications, and hence are currently not used for replacement of diseased native heart valves. However, some of these designs are still being used in the development of artificial heart and positive displacement left ventricular assist devices.

Another major change in the MHV design was the introduction of a bileaflet valve in the late 1970s. The St. Jude Medical bileaflet valve (Fig. 7a) incorporates two semicircular hinged pyrolytic carbon leaflets that open to an angle of 85° and the design is intended to provide minimal disturbance to flow. The housing and the leaflets of the bileaflet valve is made of pyrolytic carbon. Numerous other bileaflet designs have since been introduced into the market. Several design improvements have also been incorporated in the bileaflet valve models in order to improve their hemodynamic performance. The design improvements have included a decrease in thickness of the sewing cuff that allows the placement of a larger

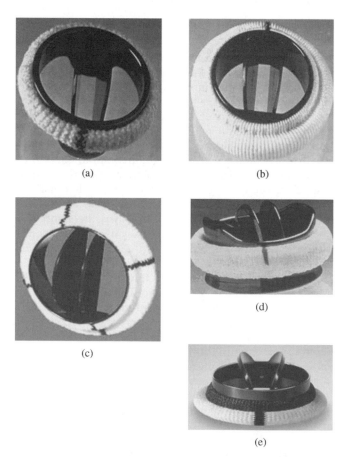

(a)

(b)

(c)

(d)

(e)

Figure 7. Bileaflet mechanical valve prostheses: (a) St. Jude valve (St. Jude Medical, Inc., St. Paul, MN); (b) Carbomedics valve (Carbomedics Inc., Austin Texas); (c) ATS valve (ATS Medical Inc., Minneapolis, MN); (d) On-X valve (Medical Carbon Research Institute, LLC, Austin, Texas); and (e) Sorin valve (Sorin Biomedica Cardio S.p.A., Via Crescentino, Italy).

housing within the cuff for a given tissue annulus diameter with the resulting hemodynamic improvement. Structural reinforcement of the housing has also allowed reducing its thickness that increases the internal orifice area for improved hemodynamics. The Carbomedics bileaflet valve (Fig. 7b) was introduced into the market in the 1990s with a recessed pivot design. The aortic version of this valve is designed to be implanted in the supraannular position enabling a larger size valve to be implanted with respect to the aortic annulus. More recent bileaflet valve designs available in the market include the ATS valve (Fig. 7c) with an open leaflet stop rather than the recessed hinges and the On-X bileaflet valve (Fig. 7d) that has a length to diameter ratio close to the native heart valves, a smoothed pivot recess allowing for the leaflet to open to 90°, a flared inlet for reducing flow disturbances, and a two point landing mechanism for smoother closing of the leaflets. The Sorin Bicarbon valve (Fig. 7e), marketed outside the United States, has curved leaflets, and hence increases the area of the central orifice. The pivots of this valve with two spherical surfaces enable the leaflet projections to roll against the surfaces rather than with the sliding action between the leaflet and the housing at the hinges.

Studies have shown that the bileaflet valves generally have a smaller pressure drop compared to the tilting disk valves, especially in the smaller sizes. However, there are several characteristic differences between the bileaflet and tilting disk valve designs that must be noted. The bileaflet valve designs include a hinge mechanism generally by introducing a recess in the housing in which a protrusion from the leaflets interacts during the opening and closing of the leaflets, or open pivots for the retention of the leaflets. On the other hand, the tilting disk valve designs do not have a hinge mechanism for retaining the occluder and the occluder is freefloating. The free-floating disk rotates as the valve opens and closes, and hence the stresses are distributed around the leaflets as opposed to the bileaflet designs. In spite of the advances in the MHV valves by the introduction of the tilting disk and bileaflet designs, design improvements aimed at enhancing the flow dynamics, and material selection, problems with thromboembolic complications and associated problems with bleeding (20) are still significant with the implanted valves and the possible relationship between the flow dynamics and initiation of thrombus will be discussed in detail later. An example of thrombus deposition and tissue ingrowth with an explanted MHV is shown in Fig. 8.

Bioprosthetic Valves

With the lack of availability of homograft valves as replacement of diseased valves, and as alternative to MHV that required long-term anticoagulant therapy, numerous attempts have been made in the use of various biological tissues as valvular replacement material. BHV made out of fascia lata (a layer of membrane that encases the thigh muscles) as well as human duramater tissue has been attempted. Fascia lata tissue was prone to deterioration, and hence unsuitable while the duramater tissue suffered from lack of availability for commercial manufacture in sufficient quantities. Harvested and preserved porcine

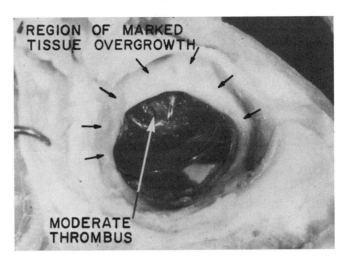

Figure 8. Photograph of an explanted mechanical heart valve prosthesis with thrombus deposition on the leaflet surface and tissue ingrowth (Courtesy of Prof. Ajit Yoganathan of Georgia Institute of Technology, Atlanta, GA.)

aortic valve as well as BHV made of bovine pericardial tissue have been employed as replacement and have been available commercially for > 30 years. The early clinical use of a xenograft (valve made from animal tissue) employed treatment of the leaflets with organic mercurial salts (21) or formaldehyde (22) to overcome the problems of rejection of foreign tissue by the body. Formaldehyde is used to fix and preserve the tissue in the excised state by histologists and results in shrinkage as well as stiffening of the tissue. Formaldehyde treated valves suffered from durability problems with 60% failure rates at 2 years after implantation. Subsequently, it was determined that the durability of the tissue cross-links was important in maintaining the structural integrity of the leaflets and gluteraldehyde was employed as the preservation fluid (23). Glutaraldehyde also reduces the antigenicity of the foreign tissue, and hence can be implanted without significant immunological reaction.

Porcine aortic valves are excised from pigs with the aortic root and shipped to the manufacturer in chilled saline solution. Support stents, configured as three upright wide posts with a circular base, is manufactured out of metal or plastic material in various sizes and covered in a fabric. A sewing flange or ring is attached to the base of the covered stent and used to suture the prostheses in place during implantation. The valve is cleaned, trimmed, fitted, and sewn to the appropriate size cloth covered stent. The stented valve is fixed in gluteraldehyde with the valve in the closed position. Glutaraldehyde solution with concentrations ranging from 0.2 to 0.625% is used in the fixation process at pressures of < 4 mmHg (0.53 kPa) to maintain the valve in the closed position. The low pressure fixation maintains the microstructure of collagen. The first glutaraldehyde-treated porcine aortic valve prosthesis mounted on metallic stent was implanted in 1969 (24). The metallic frame was soon replaced by a flexible stent on a rigid base ring; the Hancock Porcine Xenograft was commercially introduced in 1970. The stent in this valve is made of polypropylene with stainless steel radiopaque marker,

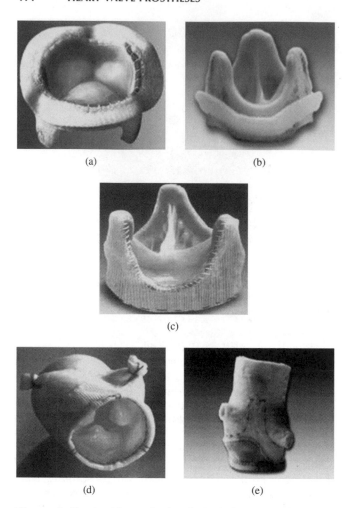

(a)

(b)

(c)

(d)

(e)

Figure 9. Porcine bioprosthetic valve prostheses: (a) Hancock II valve (Medtronic Inc., Minneapolis, MN); (b) Carpentier–Edwards valve (Edwards Lifesciences, LLC, Irvine, CA); (c) Toronto stentless valve (St. Jude Medical, Inc., St. Paul, MN); (d) Medtronic Freestyle stentless valve (Medtronic Inc., Minneapolis, MN); and (e) Edwards Prima Plus stentless valve (Edwards Lifesciences, LLC, Irvine, CA).

solution such as sodium dodecyl sulfate (SDS) in order to reduce calcification.

Another major innovation in the bioprosthetic valve design was the introduction of stentless bioprostheses. In the stented bioprostheses, the regions of stress concentration are observed at the leaflet–stent junction and the stentless valve design is intended to avoid such regions prone to failure. The absence of the supporting stents also results in less obstruction to flow, and hence should improve the hemodynamics across the valves. Due to the lack of stents, larger size valve can be implanted for a given aortic orifice to improve the hemodynamics. Stentless porcine bioprostheses are only approved for aortic valve replacement in the United States. *In vitro* studies have shown improved hemodynamic performance with the stentless designs in the mitral position, but questions remain about the durability of these valves, and the implantation techniques are also complex in the mitral position. Examples of stentless bioprostheses currently available in the United States include: St. Jude Medical Toronto SPV (Fig. 9c); Medtronic Freestyle (Fig. 9d); and Edwards Prima (Fig. 9e) valves. The Edwards Prima prosthesis is the pig's aortic valve with a preserved aortic root, with a woven polyester cloth sewn around the inflow opening to provide additional support and with features that make it easier to implant. The other stentless valve designs are also porcine aortic valves with the intact aortic root and specific preservation techniques in order to improve the hemodynamic characteristics and durability after implantation. In the Toronto SPV valve, a polyester cloth covering around the prosthesis separates the xenograft from the aortic wall of the host, making it easier for handling and suturing, and also promotes tissue ingrowth.

Both stented and stentless porcine tissue valves are from the pig's native aortic valve, and hence individual leaflets need not be manufactured. In order to have sufficient quantities of these valves in various sizes available for implant, a facility to harvest adequate quantities of these valves become necessary. As an alternative, pericardial valves are made by forming the three leaflets from the bovine pericardial tissue, and hence valves of various sizes can be made. In the pericardial valves, bovine pericardial sac is harvested and shipped in chilled saline solution. At the manufacturing site, the tissue is debrided of fatty deposits and trimmed to remove nonusable areas before the tissue is fixed in glutaraldehyde. After fixation, leaflets are cut out from the selected areas of the pericardial tissue and sewn to the cloth-covered stent in such a fashion to obtain coapting and fully sealing cusps. Since the valve is made in the shape of the native human aortic valve, the hemodynamic characteristics were also superior to the porcine valves in comparable sizes. The Ionescu-Shiley pericardial valve introduced into the market in the 1970s was discontinued within a decade due to problems associated with calcification and decreased durability. For this reason, pericardial valve were not marketed by the valve companies for several years. With advances in tissue processing and valve manufacturing technology, pericardial valves were reintroduced into the commercial market in the 1990s. The Edwards Lifesciences introduced the Carpentier-Edwards PERIMOUNT Bioprostheses

sewing ring made of silicone rubber foam fiber and polyester used as the cloth covering. Hancock Modified Orifice valve (Fig. 9a) was introduced in 1977 as a refinement of the earlier valve. The right coronary cusp of the pig's aortic valve is a continuation of the septal muscle, and hence stiffer. In the modified orifice valve, the right coronary cusp is replaced with a non-coronary cusp of comparable size from another valve. The Carpentier–Edwards Porcine valve (Fig. 9b) also employs a totally flexible support frame. The Hancock II and the Carpentier–Edwards supra-annular porcine bioprostheses employed modified preservation techniques in which the porcine tissue is initially fixed at 1.5 mmHg (0.2 kPa), and then at high pressure in order to improve the preservation of the valve geometry. The supra-annular valve is designed to be implanted on top of the aortic annulus while aligning the internal diameter of the valve to the patient's annulus and this technique allows implantation of a larger valve for any given annular size. These valves are also treated with antimineralization

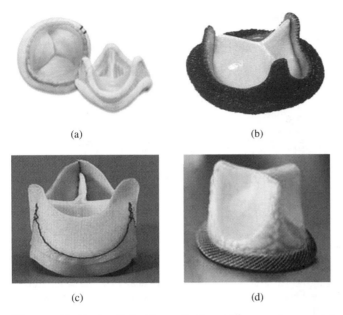

(a)   (b)

(c)   (d)

Figure 10. Pericardial bioprosthetic valve prostheses: (a) Carpentier-Edwards valve (Edwards Lifesciences, LLC, Irvine, CA), (b) Sorin Pericarbon valve (Sorin Biomedica Cardio S.p.A., Via Crescentino, Italy); (c) Sorin Pericarbon stentless valve (Sorin Biomedica Cardio S.p.A., Via Crescentino, Italy); and (d) Mitroflow Aortic pericardial valve (Sorin Group Canada, Inc., Mitroflow Division, Burnaby, B.C., Canada.)

(Fig. 10a) in the early 1990s. The pericardial tissue in this valve is mounted on a lightweight frame that is covered with porous, knitted PTFE material. A sewing ring made of molded silicone rubber covered with PTFE cloth is incorporated to suture the valve in place. Sorin pericardial valve (Fig. 10b), Sorin stentless pericardial valve (Fig. 10c) and the Mitroflow aortic pericardial valve (Fig. 10d) are available in markets outside the United States.

FUNCTIONAL CHARACTERISTICS

From the functional point of view, the implanted valve prostheses should open with minimal transvalvular pressure drop and have minimal energy loss in blood flow across the valve, and the valve must close efficiently with minimal regurgitation. The valve should mimic the central flow characteristics that are observed with native human heart valves with minimally induced fluid dynamic stresses on the formed elements in blood. The flow characteristics across the implants should also avoid regions of stasis or flow stagnation where thrombus deposition and growth can be enhanced. *In vivo* measurements, *in vitro* experimental studies, and computational simulations have been used over the past 40 years in order to assess the functional characteristics of the mechanical and bioprosthetic heart valves. The information gained from such studies have been exploited to improve the design of the valves in order to improve the performance characteristics of the valves and also increase the durability in order to provide a "normal" life style for the patient with prosthetic valve implants.

Pressure Drop and Effective Orifice Area

In vivo measurement of pressure drop requires placing pressure transducers inserted via a catheter both on the inflow and outflow side of the valve, and computing the pressure drop during the phase when the valve is open. For the aortic valve, the pressure transducers are placed in the left ventricular outflow tract and in the ascending aorta. The peak or the average pressure drop during the forward flow phase is computed from the recorded data. To avoid the invasive technique of catheterization, the fluid mechanics principle can also be applied to estimate the pressure drop across the valve in the aortic position. The average velocity, V in $m \cdot s^{-1}$, in the ascending aortic cross-section is measured noninvasively and the pressure drop, Δp expressed in millimeters of mercury can be computed using the equation

$$\Delta p \cong 4V^2 \qquad (2)$$

Using this simplified equation and noninvasive measurement of the aortic root velocity using the Doppler technique, the pressure drop across the aortic valve can be computed. With the availability of several designs of MHV and BHV, it is desirable to compare the pressure drop and regurgitation for the various valve designs. In the case of development of a new valve design, United States Federal Drug Administration (FDA) requires that these quantities measured *in vitro* for the new designs are compared with currently approved valves in the market. *In vitro* comparisons are performed in pulse duplicators that mimic the physiological pulsatile flow in the human circulation. One of the initial designs of a pulse duplicator for valve testing was that of Wicting (25) that consisted of a closed-loop flow system that is actuated by a pneumatic pump to initiate pulsatile flow through the mitral and aortic valves in their respective flow chambers. Pressure transducers were inserted through taps in the flow chambers on the inflow and outflow sides to measure the pressure drop across the valves. The fluid used in such *in vitro* experimental studies, referred to as the blood-analogue fluid, is designed to replicate the density (1060 $kg \cdot m^{-3}$) and viscosity coefficient (0.035 P or 35×10^{-4} Pa·s) of whole human blood. A glycerol solution (35–40% glycerin in water) has been generally used as the blood analog fluid in these studies. Prosthetic valves are made in various sizes (specified in sewing ring diameter magnitude). In comparing the pressure drop data for the various valve designs, proper comparison can be made on data only with comparable valve sizes. Since the flow across the valve during the forward flow phase is not laminar, the pressure drop has a nonlinear relationship with flow rate. Hence, pressure drop is measured for a range of flow rates and the pressure drop data is presented as a function of flow rate. Numerous studies comparing the pressure drop data for the various mechanical and bioprosthetic valves have been reported in the literature. Typical pressure drop comparisons for MHV and BHV (of nominal size of 25 mm) are shown in Fig. 11 (14). As can be observed, stented porcine tissue valves have the higher pressure drop, and hence are considered to be stenotic, especially in smaller valve sizes. The advantage of the stentless bioprostheses design is

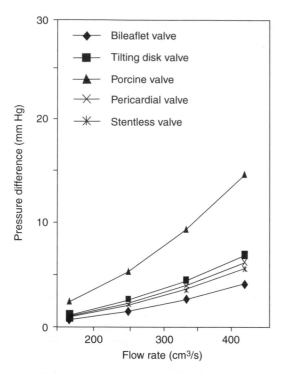

Figure 11. Typical plots for the pressure drop as a function of flow rate for the various mechanical and bioprosthetic valve prostheses (Courtesy of Prof. Ajit Yoganathan of Georgia Institute of Technology, Atlanta, GA.)

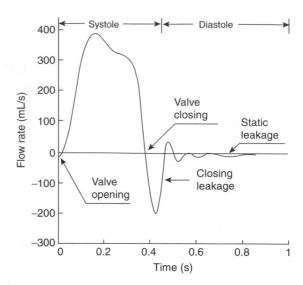

Figure 12. Flow rate across mechanical valve prosthesis in the aortic position obtained in a pulse duplicator *in vitro* measured with an electromagnetic flow meter.

obvious especially in smaller sizes since the flow orifice will be larger due to the absence of supporting stents. Supra-annular design also permits the implantation of a larger sized valve for a given annulus orifice, thus providing a smaller pressure drop and energy loss. Smaller pressure drops across the valve prostheses will result in a reduced workload of the left ventricle as the pump. Gorlin equation, described in Eq. 1, has also been employed to compute the effective orifice area for the various valve designs (14). Generally, the pericardial and bileaflet valve designs have the largest effective orifice area, followed by the tilting disk, and porcine valves, with the caged ball valve exhibiting the smallest effective orifice area for a given sewing ring diameter. Valves with the larger EOA correspond to a smaller pressure drop and energy loss in flow across the valve. Performance index (PI), computed as the ratio of effective orifice area to the sewing ring area is also used for comparison of the various valve designs. Table 1 includes data on the EOA and PI for the various MHV and BHV with a 27 mm tissue annulus diameter from *in vitro* experimental studies. It can be observed that the centrally occluding caged ball valve and stented porcine valves have lower values of PI where as the tilting disk, bileaflet, pericardial valves, and stentless tissue valves have higher values indicating improved hemodynamics for the same values of the tissue annulus diameter.

Regurgitation

In discussing the flow dynamics with native heart valves, it was observed that the anatomy and the fluid dynamics enable the leaflets to close efficiently with minimal amount

of regurgitation. On the other hand, the adverse pressure gradient at the end of the forward flow motion induces the occluders to move toward closure and all prosthetic valves exhibit a finite amount of regurgitation. Figure 12 shows a typical flow rate versus time curve obtained from an electromagnetic flow meter recording obtained *in vitro* in a pulse duplicator with a mechanical valve in the aortic position. As can be observed, a certain volume of reverse flow is observed as the valve closes and is termed as closing leakage. The closing leakage is related to the geometry of the valve and the closing dynamics. The rigid occluders in the mechanical valves also prevent the formation of a tight seal between the occluder and the seating ring when the valve is closed. With the tilting disk and bileaflet valves, a small gap between the leaflet edge and the valve housing is also introduced in order to provide a continuous wash-out of blood in the hope of preventing any thrombus deposition. Hence, even when the valve is fully closed, a small volume of blood is continuously leaking and is termed the static leakage. Percent regurgitation is defined as the ratio of the leakage volume over the net forward flow volume expressed as a percentage. Percent regurgitation can be computed by recording the flow rate versus time curve in an *in vitro* experimental set up and measuring the area under the forward and reverse flow phases from the data. These can be compared for the various size valves of the same model and also for comparison across the various valve models. Table 1 shows typical data of regurgitant volumes measured *in vitro* under physiological pulsatile flow in a pulse duplicator. The BHV designs result in more efficient leaflet closure with relatively small regurgitant volumes followed by the caged ball valve design. The magnitudes of the regurgitant volumes for the tilting disk and bileaflet valves are relatively larger and comparable to each other.

Quantitative measurement of percent regurgitation *in vivo* with both incompetent native valves or with prostheses has not been successful, even though attempts have

Table 1. Comparison of Effective Orifice Area (EOA)[a], Performance Index (PI), Regurgitation Volume, and Peak Turbulent Shear Stresses for the Various Models of Commercially Available Valve Prostheses

Valve Type	EOA[b] cm^2	PI	Reg. Vol., cm^3/beat	Peak Turb. SS, Pa[c]
Caged ball	1.75	0.30	5.5	185
Tilting Disk[d]	3.49	0.61	9.4	180
Bileaflet[d]	3.92	0.68	9.15	194
Porcine (Stented)[d]	2.30	0.40	<2	298
Pericardial (Stented)[d]	3.70	0.64	<3	100
Stentless BHV	3.75	0.65	<4	NA[e]

[a]EOA = Effective orifice area computed by the application of Gorlin's equation (4).
[b]Values compiled from Yoganathan (14).
[c]Turbulent stresses were measured at variable distances from the valve seat.
[d]Values reported are mean values from several valve models of the same type. Data reported are for 27-mm tissue annulus diameter size of the valves with measurements obtained *in vitro* in a pulse duplicator with the heart rate of 70 bpm and a cardiac output of 5.0 L·min^{-1}.
[e]Not available = NA.

been made to employ fluid mechanical theories for regurgitant flow in order to estimate the leakage volume. Turbulent jet theory and proximal flow convergence theory have been employed in an attempt to measure the regurgitant glow volume quantitatively (26,27). However, *in vivo* application has not been successful due to the restrictive assumptions of steady flow and alterations due to impingement of the jet on the ventricular wall in the theoretical considerations as well as lack of *in vivo* validation.

Dynamics of Valve Function

As discussed earlier, significant problems still exist with the implantation of heart valve prostheses in patients with disease of native heart valves. These include thrombus initiation and subsequent embolic complications with MHV implantation. The thromboembolic rates with MHV have been estimated at 2%/patient year (28). Structural disintegration and tearing of leaflets are the major complications with BHV requiring reoperation in ~10–12 years after implantation. Flow past healthy native valves are central with minimal flow disturbances and fluid induced stresses and it can be anticipated that the fluid dynamics past the mechanical valve prostheses will be drastically different from those of native heart valves. Flow induced stresses with MHV function have long been implicated with hemolysis and activation of platelets that may trigger thrombus initiation. Regions of stress concentration on the leaflets during the opening and closing phases have been implicated on structural alterations of collagen fibers resulting in leaflet tears with BHV. Detailed functional analysis of implanted valve prostheses *in vivo* is impractical. Limited attempts have been made in the measurement of velocity profiles distal to the valve prostheses in the aortic position with hot film anemometry (29). Doppler and MR phased velocity mapping techniques have also been used to measure the velocity profiles distal to heart valves (30–32). However, detailed velocity measurements very close to the leaflets and housing of prosthetic valves are not possible *in vivo*, and hence *in vitro* experimental studies and computational fluid dynamic simulation are necessary for the same. Limited *in vivo* studies in animal models have also been employed to describe the complex leaflet motion with native aortic valves (33–37). *In vitro* studies and computer simulations are also necessary for a

detailed analysis of stress distribution in the leaflets of native valves and bioprostheses during the opening and closing phases of the valve function and to determine its relationship with failure of the leaflets.

Flow Dynamics Past Mechanical Valves

With the assumption that the deposition of thrombi in MHV implants is related to the flow induced stresses, studies are continuing to date on the deterministic relationship between fluid induced stresses and damage to formed elements in blood. Subjecting blood cells to precise flow fields and assessing the destruction or activation (of platelets), magnitudes of turbulent stresses beyond which damage can be expected has been established. In addition to the magnitude of the flow induced stresses, the time for which the blood elements are exposed to the stresses also need to be considered in assessing the destruction or activation of the platelets. Nevaril et al. (38) reported that blood cells can be hemolyzed with shear stresses of the order of 150–400 Pa. In the presence of foreign surfaces, the threshold for red blood cell damage reduces to ~1–10 Pa (39). Sublethal damage to red blood cells have also been reported at turbulent shear stress levels of about 50 Pa (40). Shear induced platelet activation and aggregation is observed to be a function of both magnitude and duration of shear stresses. The larger the magnitude of the shear stress, the shorter is the duration to which platelets are subjected to the shear before they get activated. Platelets have been shown to be activated with 10–50 Pa of shear stresses with a duration of the order of 300 ms (41). Platelet damage also increases linearly with time of exposure when subjected to constant magnitudes of shear (42).

Hence, it is of interest to determine the level of wall shear and turbulent shear stresses in flow past valve prostheses as factors causing initiation and deposition of thrombus.

For the first two to three decades after the implantation of the first mechanical valve, investigations concentrated on the flow dynamics past the valves during the forward flow phase and measurements of velocity profiles, regions of stasis and recirculation, high wall shear and bulk turbulent shear stresses. Wieting (25) employed a pulse duplicator and flow visualization studies using illuminated neutrally buoyant particles in order to qualitatively describe the nature of flow past the prosthetic valves.

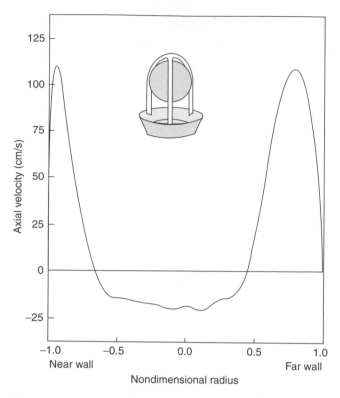

Figure 13. Velocity profile measured distal to a caged ball valve in the aortic position *in vitro* in a pulse duplicator using laser Doppler anemometry technique.

Yoganathan (43) employed laser Doppler velocimetry (LDV) technique to measure the velocity profiles and turbulent shear stresses under steady flow past the valve prostheses. Since then numerous detailed studies have

been reported in the literature on the detailed measurement of velocity profiles and turbulent shear stresses distal to the prostheses under physiological pulsatile flow (13,14).

Figure 13 shows the velocity profile distal to a caged ball valve during the peak forward flow phase measured under physiological pulsatile flow *in vitro* (44). Jet-like flow is observed around the circumference that is separated by the ball and high turbulent stresses were measured at the edge of the jet. A wake is observed behind the ball with slow moving fluid. With the caged ball valve, higher incidences of thrombus deposition have been observed at the top of the cage and correspond to the slow moving fluid in this region behind the wake of the ball. With the tilting disk valves in the fully open position, the blood flows through the major and minor orifices as shown in Fig. 14 where the velocity profile during peak forward flow phase is once again depicted (44). Two jets are formed corresponding to the two orifices with the major orifice jet having larger velocity magnitudes. The amount of blood flow through the major and minor orifices will depend on the angle of opening of the occluder as well as the geometry. A region of reverse flow is also observed adjacent to the valve housing in the minor orifice. The velocity profile measured distal to the leaflet along the major flow orifice in the perpendicular orientation is also included in the figure.

Velocity profiles with three jets corresponding to the central orifice and two peripheral orifices are observed with the bileaflet valve as shown in Fig. 15 (45). The velocity profile along the central orifice in the perpendicular orientation is also included in this figure. Regions of flow reversals near the valve housing are also observed in the figure. Typical magnitudes of turbulent shear stresses measured in the various positions distal to the MHV under pulsatile flow conditions are also included in Table 1. It can be

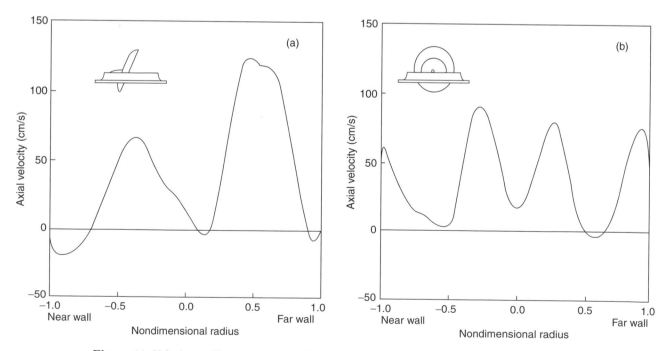

Figure 14. Velocity profile measured distal to a tilting disk valve in the aortic position *in vitro* in a pulse duplicator using laser Doppler anemometry technique.

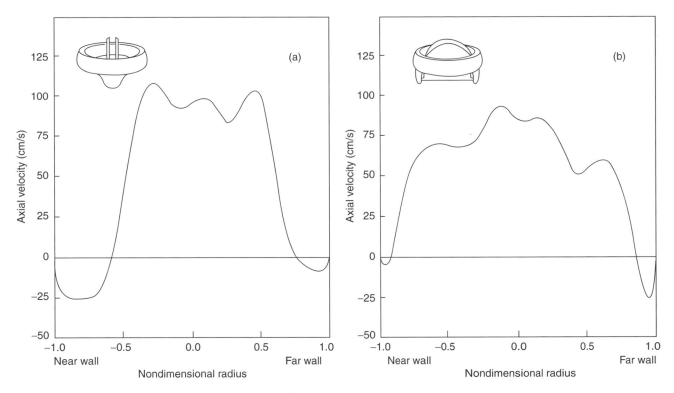

Figure 15. Velocity profile measured distal to a bileaflet valve in the aortic position *in vitro* in a pulse duplicator using laser Doppler anemometry technique.

observed that the measured bulk turbulent stresses are large enough to cause hemolysis and platelet activation that can be related to thrombus deposition with MHV. Thrombus deposition is generally observed on the leaflets and the valve housing with the tilting disk valves and also in the hinge region in the case of bileaflet valves.

More recently, it has been suggested that the relatively high turbulent stresses observed during the forward flow phase may not necessarily be the only reason for problems associated with MHV implantation. High turbulent stresses that may damage the formed elements occur in bulk flow distal to and moving away from the valve during the forward flow phase. The activated platelets will need to go through the systemic and pulmonic circulation before it will get deposited once again in the vicinity of the housing in the case of tilting disk and bileaflet valves. Several other experiences with mechanical valve prostheses designs have also indicated the importance of the valve dynamics during the closing phase to be more important for structural integrity and also in the initiation of thrombus. Medtronic Parallel valve design was introduced in the European market in the 1990s with the two leaflets opening to 90° in the fully open position. *In vitro* studies suggested that the fluid dynamics past this valve in the forward flow phase is superior or at least comparable to the currently available bileaflet valves. However, soon after human implantation trials in Europe, increased incidences of thrombus deposition was observed with this valve model, and hence it was withdrawn from clinical trials.

Another example is a design change in the tilting disk valve resulting in major changes in valve dynamics that

resulted in structural failure in a small percentage of implanted valves. In an effort to increase the flow through the minor orifice with an aim of preventing thrombus deposition, the flat disk geometry of the original Bjork–Shiley valve was changed to a curved geometry in the Bjork–Shiley convexo-concave valve. Even though this design change resulted in improved forward flow hemodynamics, this change resulted in alterations in the dynamics of valve closure with the leaflet overrotating and subjecting the outlet strut to additional loading (46). In a small percentage of valves particularly in the mitral position, single leg separation followed by outlet strut fracture resulted in leaflet escape, and hence this valve was withdrawn from the market. These developments also suggest the importance of understanding the mechanics of valve function throughout the cardiac cycle with any mechanical valve designs. In addition, structural failure and leaflet escape was reported with the implantation of a newly introduced bileaflet valve (Edwards-Duromedics) that resulted in the withdrawal of the valve from the market (47,48). The structural failure was thought to be due to pitting and erosion of the valve structures due to cavitation damage on the pyrolytic carbon (49). These reports also spurred a number of investigations on the closing dynamics and the potential for the mechanical valves to cavitate during the closing phase.

The occluders in the mechanical valves move toward closure with the onset of adverse pressure gradients, and the time taken to move from the fully open to the fully closed position is ~30 ms. Toward the end of the leaflet closure, the leaflet edge moves with a velocity of $\sim 3\text{--}4\,\mathrm{m}\cdot\mathrm{s}^{-1}$

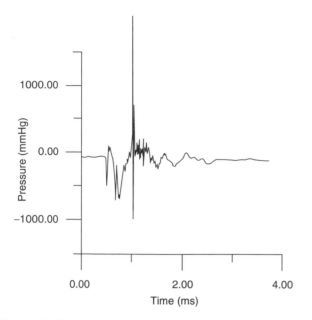

Figure 16. Typical negative pressure transients recorded at the instant of mechanical heart valve closure (in the mitral position) with the pressure transducer placed very close to the leaflet on the inflow side (atrial side) from *in vitro* experiments.

(50–53) and comes to a sudden stop as it impacts the seating lip. This produces a water hammer effect with large positive pressure transient on the outflow side (left ventricle in the mitral position and aorta in the aortic position) and a large negative pressure transient on the inflow side (left atrium in the mitral position and the left ventricular outflow tract in the aortic position). Several *in vitro* studies have recorded negative pressure transients (54) with magnitudes below that of the vapor pressure for blood (ca. −713 mmHg or −94.8 kPa) and cavitation bubbles have also been visualized at the edge of the leaflets (53–58) in the region corresponding to large negative pressure transients and where the linear velocity of the leaflet edge will be the largest. Figure 16 depicts measured negative pressure transients with the pressure transducer placed near the leaflet edge on the inflow side (atrial side) of the leaflet of a tilting disk valve at the instant of valve closure from *in vitro* experiments. Note that structural failure due to cavitation type of damage has been reported with only one model of the bileaflet valve and there are no other reports of pitting and erosion on the valve material reported with implanted mechanical valves. It is also not possible to visualize cavitation bubbles *in vivo* with implanted mechanical valves. However, *potential* for mechanical valves to cavitate has been demonstrated in an animal model with the recording of negative pressure transients, similar to those measured *in vitro*, in the left atrium in the vicinity of the implanted mechanical valves in the mitral position (59,60). The actual mechanism of cavitation bubble formation, whether due to the negative pressure magnitudes below the vapor pressure for blood or due to strong vortices forming in the atrial chamber providing additional pressure reductions, is still being debated. It has also been suggested that vortex cavitation

bubbles forming away from the valve surfaces, can trap the dissolved gas from blood and form stable gas bubbles that travel with blood to the circulation and induce neurological deficit due to the gas emboli (61). Number of attempts has also been reported on the detection of the induced cavitation *in vivo* with implanted mechanical valves from acoustic signals (62–64). Another aspect of MHV cavitation that has been neither fully understood nor fully investigated is the development of stable bubbles, found by microembolic signals (MES) or high intensity transient signals (HITS) during and post-MHV implantation. *In vitro* studies have shown the development of stable bubbles (HITS) in an artificial heart and closing dynamics experimental models, and are affected by the concentration of CO_2(65–67). *In vivo*, HITS have been visualized during and post-MHV implantation through transcranial Doppler ultrasound (68,69). These events have been implicated as a cause of strokes and neurological deficits. Further evidence has shown that these HITS are in fact, gaseous, and not solid. Patients placed on pure O_2 after MHV implantation showed a large decrease in the number of HITS recorded, when compared to patients on normal air (70). These stable bubbles are believed to develop when gaseous nuclei that are present in blood, flow into low pressure regions associated with valve closure. As the valve closes and rebounds inducing vaporous cavitation, gas diffuses into the nuclei enlarging the bubble. When the pressure recovers and the vapor collapses, the bubble dynamics and local fluid mechanics prevent the gas from diffusing back into solution causing the bubble to stabilize and allowing it to flow freely in the vasculature. There is some discussion as to which gas stabilizes the nuclei. Both N_2 and CO_2 have been suggested as the link to MES/HITS/stable bubble formation (71), but there has yet to be concrete proof indicating which one does.

Large negative pressure transients occur due to the rigidity of the occluder and negative pressures do not occur at the instant of valve closure in the case of bioprosthetic valves (59). *In vitro* measurements with a tilting disk valve design employing a flexible occluder being implanted in India has also demonstrated that large negative pressures do not develop in such designs because the leaflets deform at the instant of valve closure and absorb part of the energy (12,59).

Irrespective of the formation of cavitation bubbles and subsequent collapse with implanted mechanical valves, the flow induced stresses during the valve closing phase has been suggested as of sufficient magnitude to induce platelet activation and initiation of thrombus. Even if the negative pressure transients do not reach magnitudes below the vapor pressure for blood, the large positive and negative pressure transients on the outflow and inflow sides of the valve at the instant of valve closure can induce high velocity flows through the gap between the leaflet and the housing, in the central gap between the two leaflets in the bileaflet valve, and also through the hinge region. The wall shear stress in the clearance region have been computed to be relatively high, even though present only for a fraction of a second. They induce platelet activation in the region where thrombus deposition is observed with mechanical valves (72). Relatively high turbulent shear stresses have also been reported from *in vitro* studies distal to the hinge region of bileaflet valves during the valve closing phase (73,74) indicating the presence of high fluid induced regions near the leaflet edges during the valve closing phase that may be a significant contributor for thrombus initiation.

Most of the studies described above for the measurement of velocity profiles and turbulent stresses employed the laser Doppler velocimetry (LDV) technique. This is a point-velocity measurement technique, and hence measurement of the complete three-dimensional (3D) velocity profile distal to the valve under unsteady flows is tedious and time consuming. On the other hand, particle image velocimetry (PIV) technique has the ability to capture the whole flow field information in a relatively shorter time. Lim et al.employed the PIV technique to study the flow field distal to prosthetic heart valves in steady (75) and pulsatile (76) flow conditions. Along with the description of the flow field at the aortic root that was employed to identify the regions of flow disturbances, they also presented the results of turbulent Reynolds stress and turbulent intensity distributions. Under pulsatile flow conditions distal to a BHV, the velocity vector fields and Reynolds stress mappings at different time steps were used to estimate the damage of shear induced damage to formed elements of blood (76). Browne et al. (77) and Castellini et al. (78) compared the LDV and PIV techniques for the measurement of flow past MHV. Both these works conclude that PIV has the advantage of describing the detailed flow field distal to the valve prostheses in a relatively short time, but LDV technique affords more accurate results in the measurement of turbulent stresses. Regurgitant flow fields and the details of the vortical flow, a potential low pressure field for cavitation initiation have also been measured employing PIV techniques (79,80) and with a combination of PIV and LDV techniques (81).

Bioprosthetic Valve Dynamics

Measurement of velocity profiles and turbulent stresses from *in vitro* tests in pulse duplicators past BHV have also been reported in the literature (82). Figure 17 depicts typical velocity profiles past a porcine bioprosthesis under normal physiological flow simulation (82). With the porcine valve, a jet-like flow is observed during the peak forward flow phase with high turbulent shear stresses at the edge of the jet. With the pericardial valves (82), the peak velocity magnitudes in the jet-like flow during the peak forward flow phase were smaller than those for the porcine valves in comparable sizes (Fig. 18). It can be observed from Table 1 that the peak turbulent stresses are also smaller in the pericardial valves with geometry closer to the native aortic valves compared to that of the porcine prostheses. It should be noted that the magnitudes of turbulent stresses with the BHV also exceed those values suggested for activation of platelets. However, the leaflets of the BHV are treated biological tissue rather than artificial surfaces. Long-term anticoagulant therapy is generally not required with the implantation of bioprostheses since thrombus initiation is not a significant problem with these valves.

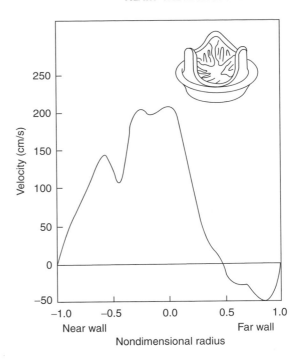

Figure 17. Velocity profile measured distal to a porcine bioprosthetic valve in the aortic position *in vitro* in a pulse duplicator using laser Doppler anemometry technique.

On the other hand, these valves fail after an average of 10–12 years of implantation and replacement surgery is required with leaflet failure. A number of studies have been reported on the analysis of the complex leaflet motion during the valve function in order to determine a causative relationship between regions of high stress concentration

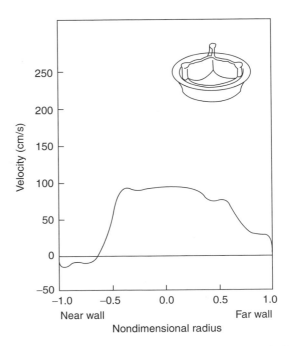

Figure 18. Velocity profile measured distal to a pericardial bioprosthetic valve in the aortic position *in vitro* in a pulse duplicator using laser Doppler anemometry technique.

on the leaflets and its attachment sites and structural failure. Based on the analysis of native aortic leaflet motion *in vivo* with the use of radiopaque markets, it has been reported that the design of the native aortic leaflets affords minimal stresses on the leaflets (33,35). On the other hand, stress analysis on the BHV valve leaflets *in vivo* suggests that mechanical stresses play a role in calcification of the leaflets (33,34,83). A number of studies has been reported on a finite element stress analysis on the BHV valve leaflets in the closed position in order to correlate regions of high stresses with calcification and tissue failure (84,85). Recent studies have indicated that damage to the valvular structural matrix occurs as the result of mechanical stresses and that the structural damage occurs in spatially distinct sites from those of cuspal mineralization (86). Hence, there is a renewed interest in analyzing the stresses on the leaflets during the opening and closing phases of the leaflet function since loss in flexural rigidity has been demonstrated after the valves undergo 50 million cycles of loading *in vitro*(87). Detailed analysis on the mechanism of structural failure under cyclic loading requires experimental quantification of the complex motion and also detailed description of the nonlinear anisotropic material property description of the BHV leaflets (88). Leaflet motion quantification by the application of ink markers on the leaflet surface (89), laser profiling technique (90), and noncontacting structured laser light projection technique (91) have been employed for the BHV leaflets. Several studies have been reported on the material characterization of chemically treated BHV leaflet tissue (92–94). Even though these studies have yielded valuable information about the nonlinear material characterization of the leaflets, true physiological characterization requires biaxial loading tests. Recently, constitutive models for the BHV leaflets under biaxial loading have been reported (95,96), which takes into consideration the local architecture of the valve fibers. Since the treated aortic valve leaflet consists of three layers, it can also be anticipated that the behavior of each layer will be different under the physiological loading during the valve opening and closing phases. Recent studies have included separating the fibrosa and ventricularis layers by dissecting microscope and each layer being subjected to biaxial testing (97). Incorporating such detailed material description in the computational analysis in order to determine the flexural and shear stresses on the multilayered tissue leaflets may yield valuable information on the nature of the effect of mechanical stresses on structural disintegration and limited durability with implanted BHV.

Computational Simulation of Heart Valve Function

With the advent of high speed computing capabilities, advances in computational fluid dynamic and finite element numerical analyses algorithms, simulations to determine the mechanical stresses on the blood elements and leaflets during the valve function, is being increasingly employed to understand the mechanics of valve function. In the case of MHV, recent studies have focused on the mechanical stresses developed during the closing phase of valve function. Wall shear stresses of the order of 1000 Pa

have been reported through a numerical simulation in the central clearance of a bileaflet valve with the leaflets in the fully closed position (98). Since the simulation was with the leaflets in the closed position, this simulation did not incorporate the effects of a large pressure gradient across the leaflet at the instant of valve closure (54). A quasistatic simulation in which the leaflets were in the fully closed position and with the application of the measured transient pressures at the instant of valve closure indicated the presence of shear stresses of ~ 2200 Pa (99,100). Employing moving boundaries for the mechanical valve leaflet very near the time of valve closure, simulations of the flow dynamics in the clearance region between the edge of the leaflet and the seat stop have demonstrated fluid velocities of the order of $28 \text{ m} \cdot \text{s}^{-1}$ with large negative pressure regions on the inflow side of the occluder (101,102). The computed resulting wall shear stress magnitudes at the edge of the leaflet exceeds 17 kPa for a fraction of a second at the instant of valve closure and first rebound after impact with the seat stop (72).

In the case of BHV, finite element analysis has been the popular technique employed to determine the stress distribution on the leaflets with the blood pressure applied as the load on the leaflets. Such analyses have also been employed to perform stress analysis with native aortic and mitral valves (103,104). Stress analysis with BHV geometry have incorporated the nonlinear material property of the leaflets and employed both the rigid and flexible stent support (84,85). These results have generally suggested a correlation between regions of high stresses with the leaflets in the fully closed position and calcification observed with implanted valves.

Numerical simulation of native and prosthetic heart valve function is quite challenging with the necessity of addressing several difficult issues. In the case of MHV simulation, the mesh generation for the detailed 3D geometry of the tilting disk and bileaflet valves, including the complex geometry of the hinge mechanism, is required. The simulation must also have the ability to deal with the moving leaflets. In the case of BHV simulation, it is necessary to incorporate the nonlinear anisotropic material property of the leaflets and compute the complex motion of the leaflets due to the external load imparted on the same by the surrounding fluid. In modeling the valve function, the fluid domain is most conveniently described using the Eulerian reference frame in which the fluid moves through a fixed mesh. A Lagrangian formulation is more appropriate for the leaflet motion in which the mesh moves together with the leaflet. The complete simulation of the heart valve function requires a fluid-structure interaction simulation and the two formulations are incompatible for such an analysis. Two methods have generally been employed to circumvent this problem. An Eulerian method used in the simulation of heart valve function is the fictitious domain method employed by de Hart et al. (105,106) and Baaijens (107). This simulation requires careful attention to accurate mesh representation for flows near the leaflet due to the use of fixed grid and numerical stability. It is also computationally very intensive. The second approach is the arbitrary Lagrangian–Eulerian (ALE) method in which the computational mesh is allowed to deform (108) and move in space arbitrarily to conform to the moving boundaries. It has also been successfully employed in the heart valve simulation. The disadvantage with this method is the need for a mechanism to adapt or deform the mesh to conform to the boundary motion at each new time step. Large and complex deformation of the BHV leaflets within the computational domain makes the mesh adaptation very difficult when structured mesh is used and the mesh topology has to be maintained. Mesh regeneration has been employed to avoid the problems with the maintenance of mesh topology, and requires reinterpolation of the flow variables that can be expensive to perform and may result in large artificial errors. The ALE method recently has been employed in the simulation of prosthetic valve function by several investigators (102,109–111). More recently, a fluid-structure interaction simulation for the mechanical valve closing dynamics employing the ALE method has been presented for both two-dimensional (2D) (112) and 3D (72) geometry of a bileaflet valve. In this analysis in which the details of the hinge geometry was not included, the leaflet was specified to rotate around an axis and the motion of the leaflets was computed by solving the governing equation of motion for the leaflet with fluid-induced stresses specified as the external load. The pressure and velocity field is calculated employing the CFD solver employing the ALE method. The grid points on the leaflets are rotated, based on the solution of the equation of motion of the leaflet. An elliptic solver is employed to the entire mesh using the initial methods and the computed displacement of the leaflet grid points. This simulation has also clearly demonstrated the presence of abnormally high wall shear stresses in the clearance region of the leaflet and the valve housing at the instant of valve closure and leaflet rebound. These studies indicate that the shear stress-time magnitude present in this region far exceeds magnitudes suggested for platelet activation, and hence may be the critical factor for thrombus initiation with MHV implants.

SUMMARY

Since the introduction of the first prosthetic valve replacement for a diseased native heart valve > 40 years ago, numerous developments in design and manufacturing process have resulted in improved performance of these implants and patients are leading a relatively normal life. Continued efforts are underway to minimize the effect of thrombus deposition with MHV and to improve the durability of implanted BHV. State-of-the-art experimental techniques and computational simulations are being applied to improve our understanding on the relationship between the complex solid and fluid mechanics during the prosthetic valve function and its relationship with the problems still continuing to be observed with the implants. With the advent of high performance computers and advances in computational flow dynamics algorithms, more detailed 3D unsteady laminar and disturbed flow simulations are becoming a reality today. Development of fluid–structure interaction simulations, inclusion of detailed structural analysis of biological leaflet valves during the valve function, and the behavior of platelets

and red blood cells in the unsteady 3D flow field to simulate the platelet and red blood cell motion in the crevices are crucial for our further understanding of the mechanical valve function. However, complementary experimental studies to validate the simulations are also essential to gain confidence in the results of the complicated numerical simulations. A deterministic study of the effect of such stresses on the leaflet structures as well as formed elements in blood will require an analysis that includes computational algorithms that span multiple length and time scales. Efforts are underway to develop such simulations. These studies will also provide valuable information toward the development of tissue engineered valve replacements.

BIBLIOGRAPHY

Cited References

1. Yoganathan A, Lemmon JD, Ellis JT. Heart Valve Dynamics. In: Bronzino JD, editor. The Biomedical Engineering Handbook. 2nd ed. Boca Raton: CRC Press and IEEE Press; 2000. p 29.1–29.15.
2. Westaby S, Karp RB, Blackstone EH, Bishop SP. Adult human valve dimensions and their surgical significance. Am J Cardiol 1984;53(4):552–556.
3. Davies MJ. Pathology of Cardiac Valves. London: Butterworth; 1980.
4. Gorlin R, Gorlin SG. Hydraulic formula for calculation of the area of the stenotic mitral valve, other cardiac valves, and central circulatory shunts. I Am Heart J 1951;41(1):1–29.
5. Schlant RC, Alexander RW, editors. Technique of Doppler and color flow Doppler in the evaluation of cardiac disorders and function. in The Heart: Arteries and Veins. The Heart: Arteries and Veins. 6th ed. New York: McGraw-Hill; 1994.
6. Braunwald E. Heart Disease: A Textbook of Cardiovascular Medicine. 2nd ed. Philadelphia: W. B. Saunders; 1984.
7. Aazami M, Schafers HJ. Advances in heart valve surgery. J Interv Cardiol 2003;16(6):535–541.
8. Harken DE, Taylor WJ, Lefemine AA, Lunzer S, Low HB, Cohen ML, Jacobey JA. Aortic valve replacement with a caged ball valve. Am J Cardiol 1962;9:292–299.
9. Hammermeister K, Sethi GK, Henderson WG, Grover FL, Oprian C, Rahimtoola SH. Outcomes 15 years after valve replacement with a mechanical versus a bioprosthetic valve: final report of the Veterans Affairs randomized trial. J Am Coll Cardiol 2000;36(4):1152–1158.
10. Bach DS. Choice of prosthetic heart valves: update for the next generation. J Am Coll Cardiol 2003;42(10):1717–1719.
11. Starr A, Edwards ML. Mitral replacement: Clinical experience with a ball valve prosthesis. Ann Surg 1961;154:726.
12. Bhuvaneshwar G, Ramani AV, Chandran KB. Polymeric occluders in tilting disk heart valve prostheses. In: Dumitriu S, editor. Polymeric Biomaterials. 2nd ed. New York: Marcel Dekker; 2002. p 589–610.
13. Chandran KB. Dynamic behavior analysis of mechanical heart valve prostheses. In: Leondes C, editor. Cardiovascular Techniques. Boca Raton: CRC Press; 2001. p 3.1–3.31.
14. Yoganathan A. Cardiac valve prostheses. In: Bronzino JD, editor. Biomedical Engineering Handbook. Boca Raton: CRC Press and IEEE Press; 2000. p 127.1–127.23.
15. Wada J. Knotless suture method and Wada hingeless valve. Jpn J Thor Sur 1967;15:88.
16. Bjork VO. Experience with the Wada-Cutter valve prostheses in the aortic area- One year follow-up. J Thorac Cardiovasc Surg 1970;60:26.

17. Bjork VO. A new tilting disk valve prostheses. Scand J Cardiovasc Surg 1969;3:1.

18. Bjork VO. Delrin as implant material for valve occluders. Scand J Thorac Cardiovasc Surg 1972;6:103.

19. Bjork VO. The pyrolytic carbon occluder for the Bjork–Shiley tilting disk valve prostheses. Scand J Thorac Cardiovasc Surg 1972;6:109–113.

20. Butchart EG, Bodnar E. Thrombosis, Embolism, and Bleeding. Marlow, Bucks (UK): ICR Publishing; 1992.

21. Carpentier A, Lamaigre RL, Carpentier S, Dubost C. Biological factors affecting long-term results of valvular heterografts. J Thorac Cardiovasc Surg 1969;58:467.

22. O'Brien M, Clarebrough JK. Heterograft aortic valve transplant for human valve disease. Aust Med J 1966;2(228).

23. Carpentier A, Dubost C. From xenograft to bioprosthesis: Evolution of concepts and techniques of valvular xenografts. In: Ionescu MI, Ross DN, Wooler GH, editors. Biological Tissue in Heart Valve Replacement. London: Butterworth; 1971. p 515–541.

24. Yoganathan AP. Cardiac Valve Prostheses. In: Bronzino JD, editor. The Biomedical Engineering Handbook. 2nd ed. Boca Raton (FL): CRC Press; 2000. p 127.1–127.23.

25. Wieting DW. *Dynamic flow characteristics of heart valves*, Ph.D., dissertation. University of Texas, Austin, TX.

26. Cape EG, Nanda NC, Yoganathan AP. Quantification of regurgitant flow through bileaflet heart valve prostheses: theoretical and *in vitro* studies. Ultrasound Med Biol 1993;19(6):461–468.

27. Cape EG, Kim YH, Heinrich RS, Grimes RY, Muralidharan E, Broder JD, Schwammenthal E, Yoganathan AP, Levine RA. Cardiac motion can alter proximal isovelocity surface area calculations of regurgitant flow. J Am Coll Cardiol 1993;22(6):1730–1737.

28. Edmunds LH Jr. Thrombotic and bleeding complications of prosthetic heart valves. Ann Thorac Surg 1987;44(4):430–445.

29. Paulsen PK, Nygaard H, Hasenkam JM, Gormsen J, Stodkilde-Jorgensen H, Albrechtsen O. Analysis of velocity in the ascending aorta in humans. A comparative study among normal aortic valves, St. Jude Medical and Starr-Edwards Silastic Ball valves. Int J Artif Organs 1988;11(4):293–302.

30. Farthing S, Peronneau P. Flow in the thoracic aorta. Cardiovasc Res 1979;13(11):607–620.

31. Rossvoll O, Samstad S, Torp HG, Linker DT, Skjaerpe T, Angelsen BA, Hatle L. The velocity distribution in the aortic anulus in normal subjects: a quantitative analysis of two-dimensional Doppler flow maps. J Am Soc Echocardiogr 1991;4(4):367–378.

32. Kilner PJ, Yang GZ, Mohiaddin RH, Firmin DN, Longmore DB. Helical and retrograde secondary flow patterns in the aortic arch studied by three-directional magnetic resonance velocity mapping. Circulation 1993;88(5 Pt 1):2235–2247.

33. Thubrikar M, Piepgrass WC, Shaner TW, Nolan SP. The design of the normal aortic valve. Am J Physiol 1981;241(6):H795–801.

34. Thubrikar MJ, Skinner JR, Eppink RT, Nolan SP. Stress analysis of porcine bioprosthetic heart valves *in vivo*. J Biomed Mater Res 1982;16(6):811–826.

35. Thubrikar M, Skinner JR, Aouad J, Finkelmeier BA, Nolan SP. Analysis of the design and dynamics of aortic bioprostheses in vivo. J Thorac Cardiovasc Surg 1982;84(2):282–290.

36. Thubrikar M, Eppink RT. A method for analysis of bending and shearing deformations in biological tissue. J Biomech 1982;15(7):529–535.

37. Thubrikar M, Carabello BA, Aouad J, Nolan SP. Interpretation of aortic root angiography in dogs and in humans. Cardiovasc Res 1982;16(1):16–21.

38. Nevaril C, Hellums J, Alfrey C Jr. Physical effects in red blood cell trauma. J Am Inst Chem Eng 1969;15:707.

39. Mohandas N, Hochmuth RM, Spaeth EE. Adhesion of red cells to foreign surfaces in the presence of flow. J Biomed Mater Res 1974;8(2):119–136.

40. Sutera SP, Mehrjardi MH. Deformation and fragmentation of human red blood cells in turbulent shear flow. Biophys J 1975;15(1):1–10.

41. Ramstack JM, Zuckerman L, Mockros LF. Shear-induced activation of platelets. J Biomech 1979;12(2):113–125.

42. Anderson GH, Hellums JD, Moake JL, Alfrey CP Jr. Platelet lysis and aggregation in shear fields. Blood Cells 1978;4(3):499–511.

43. Yoganathan AP, Corcoran WH, Harrison EC. In vitro velocity measurements in the vicinity of aortic prostheses. J Biomech 1979;12(2):135–152.

44. Chandran KB, Cabell GN, Khalighi B, Chen CJ. Laser anemometry measurements of pulsatile flow past aortic valve prostheses. J Biomech 1983;16(10):865–873.

45. Chandran KB. Pulsatile flow past St. Jude Medical bileaflet valve. An in vitro study. J Thorac Cardiovasc Surg 1985;89(5):743–749.

46. Chandran KB, Lee CS, Aluri S, Dellsperger KC, Schreck S, Wieting DW. Pressure distribution near the occluders and impact forces on the outlet struts of Bjork–Shiley convexo-concave valves during closing. J Heart Valve Dis 1996;5(2):199–206.

47. Dimitri W, Williams BT. Fracture of a Duromedics mitral valve housing with leaflet escape. J Cardiovas Sur 1990;31:41–46.

48. Deuvaert FE, Devriendt J, Massaut J. Leaflet escape of a mitral Duromedics prosthesis (case report). Acta Chirur 1989;89:15–18.

49. Kafesjian R, Howanec M, Ward GD, Diep L, Wagstaff LS, Rhee R. Cavitation damage of pyrolytic carbon in mechanical heart valves. J Heart Valve Dis 1994;3:S2–S7.

50. Guo GX, Chiang TH, Quijano RC, Hwang NH. The closing velocity of mechanical heart valve leaflets. Med Eng Phys 1994;16(6):458–464.

51. Wu ZJ, Shu MC, Scott DR, Hwang NH. The closing behavior of Medtronic Hall mechanical heart valves. ASAIO J 1994;40(3):M702–6.

52. Wu ZJ, Wang Y, Hwang NH. Occluder closing behavior: a key factor in mechanical heart valve cavitation. J Heart Valve Dis 1994;3(Suppl 1):S25–33; discussion S33–4.

53. Chandran KB, Aluri S. Mechanical valve closing dynamics: relationship between velocity of closing, pressure transients, and cavitation initiation. Ann Biomed Eng 1997;25 (6):926–938.

54. Chandran KB, Lee CS, Chen LD. Pressure field in the vicinity of mechanical valve occluders at the instant of valve closure: correlation with cavitation initiation. J Heart Valve Dis 1994;3(Suppl 1):S65–75; discussion S75–6.

55. Graf T, Reul H, Dietz W, Wilmes R, Rau G. Cavitation of mechanical heart valves under physiologic conditions. J Heart Valve Dis 1992;1(1):131–141.

56. Graf T, Fischer H, Reul H, Rau G. Cavitation potential of mechanical heart valve prostheses. Int J Artif Organs 1991;14(3):169–174.

57. Zapanta CM, Liszka EG Jr., Lamson TC, Stinebring DR, Deutsch S, Geselowitz DB, Tarbell JM. A method for real-time in vitro observation of cavitation on prosthetic heart valves. J Biomech Eng 1994;116(4):460–468.

58. Garrison LA, Lamson TC, Deutsch S, Geselowitz DB, Gaumond RP, Tarbell JM. An *in vitro* investigation of prosthetic heart valve cavitation in blood. J Heart Valve Dis 1994;3(Suppl 1):S8–22; discussion S22–4.

59. Chandran KB, Dexter EU, Aluri S, Richenbacher WE. Negative pressure transients with mechanical heart-valve closure: correlation between in vitro and in vivo results. Ann Biomed Eng 1998;26(4):546–556.

60. Dexter EU, Aluri S, Radcliffe RR, Zhu H, Carlson DD, Heilman TE, Chandran KB, Richenbacher WE. *In vivo* demonstration of cavitation potential of a mechanical heart valve. ASAIO J 1999;45(5):436–441.

61. Kleine P, Perthel M, Hasenkam JM, Nygaard H, Hansen SB, Laas J. High-intensity transient signals (HITS) as a parameter for optimum orientation of mechanical aortic valves. Thorac Cardiovasc Surg 2000;48(6):360–363.

62. Johansen P, Andersen TS, Hasenkam JM, Nygaard H. In-vivo prediction of cavitation near a Medtronic Hall valve. J Heart Valve Dis 2004;13(4):651–658.

63. Johansen P, Manning KB, Tarbell JM, Fontaine AA, Deutsch S, Nygaard H. A new method for evaluation of cavitation near mechanical heart valves. J Biomech Eng 2003;125(5):663–670.

64. Johansen P, Travis BR, Paulsen PK, Nygaard H, Hasenkam JM. Cavitation caused by mechanical heart valve prostheses—a review. APMIS 2003;(Suppl)(109):108–112.

65. Biancucci BA, Deutsch S, Geselowitz DB, Tarbell JM. *In vitro* studies of gas bubble formation by mechanical heart valves. J Heart Valve Dis 1999;8(2):186–196.

66. Lin HY, Bianccucci BA, Deutsch S, Fontaine AA, Tarbell JM. Observation and quantification of gas bubble formation on a mechanical heart valve. J Biomech Eng 2000;122 (4):304–309.

67. Bachmann C, Kini V, Deutsch S, Fontaine AA, Tarbell JM. Mechanisms of cavitation and the formation of stable bubbles on the Bjork–Shiley Monostrut prosthetic heart valve. J Heart Valve Dis 2002;11(1):105–113.

68. Dauzat M, Deklunder G, Aldis A, Rabinovitch M, Burte F, Bret PM. Gas bubble emboli detected by transcranial Doppler sonography in patients with prosthetic heart valves: a preliminary report. J Ultrasound Med 1994;13(2):129–35.

69. Reisner SA, Rinkevich D, Markiewicz W, Adler Z, Milo S. Spontaneous echocardiographic contrast with the carbomedics mitral valve prosthesis. Am J Cardiol 1992;70(18): 1497–1500.

70. Droste DW, Hansberg T, Kemeny V, Hammel D, Schulte-Altedorneburg G, Nabavi DG, Kaps M, Scheld HH, Ringelstein EB. Oxygen inhalation can differentiate gaseous from nongaseous microemboli detected by transcranial Doppler ultrasound. Stroke 1997;28(12):2453–2456.

71. Georgiadis D, Wenzel A, Lehmann D, Lindner A, Zerkowski HR, Zierz S, Spencer MP. Influence of oxygen ventilation on Doppler microemboli signals in patients with artificial heart valves. Stroke 1997;28(11):2189–2194.

72. Cheng R, Lai YG, Chandran KB. Three-dimensional fluid-structure interaction simulation of bileaflet mechanical heart valve flow dynamics. Ann Biomed Eng 2004;32(11): 1469–1481.

73. Ellis JT, Yoganathan AP. A comparison of the hinge and near-hinge flow fields of the St Jude medical hemodynamic plus and regent bileaflet mechanical heart valves. J Thorac Cardiovasc Surg 2000;119(1):83–93.

74. Saxena R, Lemmon J, Ellis J, Yoganathan A. An in vitro assessment by means of laser Doppler velocimetry of the Medtronic advantage bileaflet mechanical heart valve hinge flow. J Thorac Cardiovasc Surg 2003;126(1):90–98.

75. Lim WL, Chew YT, Chew TC, Low HT. Steady flow dynamics of prosthetic aortic heart valves: a comparative evaluation with PIV techniques. J Biomech 1998;31(5):411–421.

76. Lim WL, Chew YT, Chew TC, Low HT. Pulsatile flow studies of a porcine bioprosthetic aortic valve in vitro: PIV measurements and shear-induced blood damage. J Biomech 2001; 34(11):1417–1427.

77. Browne P, Ramuzat A, Saxena R, Yoganathan AP. Experimental investigation of the steady flow downstream of the St. Jude bileaflet heart valve: a comparison between laser Doppler velocimetry and particle image velocimetry techniques. Ann Biomed Eng 2000;28(1):39–47.

78. Castellini P, Pinotti M, Scalise L. Particle image velocimetry for flow analysis in longitudinal planes across a mechanical artificial heart valve. Artif Organs 2004;28(5):507–513.

79. Manning KB, Kini V, Fontaine AA, Deutsch S, Tarbell JM. Regurgitant flow field characteristics of the St. Jude bileaflet mechanical heart valve under physiologic pulsatile flow using particle image velocimetry. Artif Organs 2003;27(9):840–846.

80. Kini V, Bachmann C, Fontaine A, Deutsch S, Tarbell JM. Flow visualization in mechanical heart valves: occluder rebound and cavitation potential. Ann Biomed Eng 2000; 28(4):431–441.

81. Kini V, Bachmann C, Fontaine A, Deutsch S, Tarbell JM. Integrating particle image velocimetry and laser Doppler velocimetry measurements of the regurgitant flow field past mechanical heart valves. Artif Organs 2001;25(2):136–145.

82. Chandran KB, Cabell GN, Khalighi B, Chen CJ. Pulsatile flow past aortic valve bioprostheses in a model human aorta. J Biomech 1984;17(8):609–619.

83. Thubrikar MJ, Deck JD, Aouad J, Nolan SP. Role of mechanical stress in calcification of aortic bioprosthetic valves. J Thorac Cardiovasc Surg 1983;86(1):115–125.

84. Hamid MS, Sabbah HN, Stein PD. Influence of stent height upon stresses on the cusps of closed bioprosthetic valves. J Biomech 1986;19(9):759–769.

85. Rousseau EP, van Steenhoven AA, Janssen JD. A mechanical analysis of the closed Hancock heart valve prosthesis. J Biomech 1988;21(7):545–562.

86. Sacks MS, Schoen FJ. Collagen fiber disruption occurs independent of calcification in clinically explanted bioprosthetic heart valves. J Biomed Mat Res 2002;62(3):359–371.

87. Gloeckner DC, Billiar KL, Sacks MS. Effects of mechanical fatigue on the bending properties of the porcine bioprosthetic heart valve. ASAIO J 1999;45(1):59–63.

88. Sacks MS. The biomechanical effects of fatigue on the porcine bioprosthetic heart valve. J Long-Term Effects Med Implants 2001;11(3–4):231–247.

89. Lo D, Vesely I. Biaxial strain analysis of the porcine aortic valve. Ann Thorac Surg 1995;60(2 Suppl):S374–8.

90. Donn AW, Bernacca GM, Mackay TG, Gulbransen MJ, Wheatley DJ. Laser profiling of bovine pericardial heart valves. Int J Artif Organs 1997;20(8):436–439.

91. Iyengar AKS, Sugimoto H, Smith DB, Sacks MS. Dynamic in vitro quantification of bioprosthetic heart valve leaflet motion using structured light projection. Ann Biomed Eng 2001;29(11):963–973.

92. Lee JM, Boughner DR, Courtman DW. The gluteraldehyde-stabilized porcine aortic valve xenograft. II. Effect of fixation with or without pressure on the tensile viscoelastic properties of the leaflet material. J Biomed Mater Res 1984;18 (1):79–98.

93. Lee JM, Courtman DW, Boughner DR. The gluteraldehyde-stabilized porcine aortic valve xenograft. I. Tensile viscoelastic properties of the fresh leaflet material. J Biomed Mater Res 1984;18(1):61–77.

94. Vesely I, Noseworthy R. Micromechanics of the fibrosa and the ventricularis in aortic valve leaflets. J Biomech 1992;25(1): 101–113.

95. Billiar KL, Sacks MS. Biaxial mechanical properties of the native and gluteraldehyde-treated aortic valve cusp: Part II—A structural constitutive model. J Biomech Eng 2000;122(4):327–335.

96. Billiar KL, Sacks MS. Biaxial mechanical properties of the natural and gluteraldehyde treated aortic valve cusp—Part I: Experimental results. J Biomech Eng 2000;122(1):23–30.

97. Sacks M, Stella J, Chandran KB. *A bi-layer structural constitutive model for the aortic valve leaflet*. Philadelphia: BMES Annu Conf, Abstract No. 1164, 2004.

98. Reif TH. A numerical analysis of the backflow between the leaflets of a St. Jude Medical cardiac valve prosthesis. J Biomech 1991;24(8):733–741.

99. Lee CS, Chandran KB. Instantaneous back flow through peripheral clearance of Medtronic Hall tilting disk valve at the moment of closure. Ann Biomed Eng 1994;22(4):371–380.

100. Lee CS, Chandran KB. Numerical simulation of instantaneous backflow through central clearance of bileaflet mechanical heart valves at closure: shear stress and pressure fields within clearance. Med Biol Eng Comput 1995; 33(3):257–263.

101. Bluestein D, Einav S, Hwang NH. A squeeze flow phenomenon at the closing of a bileaflet mechanical heart valve prosthesis. J Biomech 1994;27(11):1369–1378.

102. Makhijani VB, Yang HQ, Singhal AK, Hwang NH. An experimental-computational analysis of MHV cavitation: effects of leaflet squeezing and rebound. J Heart Valve Dis 1994;3(Suppl 1):S35–44; discussion S44–8.

103. Grande KJ, Cochran RP, Reinhall PG, Kunzelman KS. Stress variations in the human aortic root and valve: the role of anatomic asymmetry. Ann Biomed Eng 1998;26(4): 534–545.

104. Kunzelman KS, Quick DW, Cochran RP. Altered collagen concentration in mitral valve leaflets: biochemical and finite element analysis. Ann Thorac Surg 1998;66(6 Suppl):S198–205.

105. De Hart J, Peters GW, Schreurs PJ, Baaijens FP. A two-dimensional fluid-structure interaction model of the aortic valve [correction of value]. J Biomech 2000;33(9):1079–1088.

106. De Hart J, Peters GW, Schreurs PJ, Baaijens FP. A three-dimensional computational analysis of fluid-structure interaction in the aortic valve. J Biomech 2003;36(1):103–112.

107. Baaijens FPT. A fictitious domain/mortar element method for fluid-structure interaction. Int J Numerical Met Fluids 2001;35(7):743–761.

108. Lai YG. Unstructured grid arbitrarily shaped element method for fluid flow simulation. AIAA J 2000;38 (12):2246–2252.

109. Makhijani VB, Yang HQ, Dionne PJ, Thubrikar MJ. Three-dimensional coupled fluid-structure simulation of pericardial bioprosthetic aortic valve function. ASAIO J 1997;43(5):M387–M392.

110. Aluri S, Chandran KB. Numerical simulation of mechanical mitral heart valve closure. Ann Biomed Eng 2001; 29(8): 665–676.

111. Lai YG, Chandran KB, Lemmon J. A numerical simulation of mechanical heart valve closure fluid dynamics. J Biomech 2002;35(7):881–892.

112. Cheng R, Lai YG, Chandran KB. Two-dimensional fluid-structure interaction simulation of bi-leaflet mechanical heart valve flow dynamics. Heart Valve Dis 2003;12: 772–780.

See also Biocompatibility of materials; biomaterials, corrosion and wear of; tissue engineering.

HEART VALVE PROSTHESES, IN VITRO FLOW DYNAMICS OF

Steven Ceccio
University of Michigan
Ann Arbor, Michigan

INTRODUCTION

The cardiac cycle begins with venous blood passing through the tricuspid valve in response to relaxation of the right ventricle. Then, during ventricular contraction, the tricuspid valve closes and blood flows through the open pulmonary valve to the lungs. Similarly, oxygenated blood leaving the lungs crosses the mitral valve during filling of the left ventricle and is then ejected through the aortic valve when the left ventricle contracts. A normally functioning heart valve must open and close approximately once every second without posing significant resistance to flow during opening and without allowing significant leakage during closure.

Due to the critical role that heart valves play in controlling pressures and flows in the heart and throughout the body, valve disease is a serious health risk and can be fatal if not treated. In the United States almost 20,000 people die annually as a result of heart valve disease (1). Although the causes and mechanisms of heart valve diseases are varied, their effect can usually be reduced to either failure of the valve to open fully (stenosis) or failure to prevent leakage of blood (regurgitation). Patients with stenosis or regurgitation may experience chest pain, labored breathing, lightheadedness, and a reduced tolerance for exercise.

Because of the mechanical nature of valve dysfunction, treatments for severe valve disease usually involve surgical intervention to restore the flow control function of the valve. Early surgical treatments consisted of a surgeon using a tool, or his fingers, to reach into the beating heart and forcefully open a stenotic mitral valve. With the advent of cardiopulmonary bypass in the 1950s, the notion of fabricating and implanting a prosthetic valve became more feasible and by the early 1960s the first successful and repeatable prosthetic valve implants were performed by Starr (2,3). The valve he developed with Edwards, an engineer, consisted of a ball trapped in a rigid, dome-shaped cage. At the inflow edge of the valve was a cloth flange, the sewing ring, which enabled the surgeon to sew the valve into the patient's heart (see Fig. 1). Although crude in comparison to the native valve structure, this

Figure 1. Starr–Edwards ball–cage valve. (Courtesy of Edwards Lifesciences, Irvine, CA).

Starr–Edwards valve and subsequent models had been used successfully in >175,000 people by 1991 (4).

Over the past four decades, numerous valve designs have been developed and used clinically. There were ~80,000 heart valve-related surgeries in the United States in 1999, ~50,000 of which were implants of prosthetic aortic valves (5). But despite the success of prosthetic valve technology, currently no valve is optimal, so surgeons and engineers continue their collaborative efforts in pursuit of improved designs.

Due to the lower pressures in the right ventricle, the tricuspid and pulmonary valves are implicated far less in heart disease than the valves of the left heart (1). Consequently, prosthetic heart valve technology is focused almost entirely on mitral and aortic valves. The following discussion will focus on the primary tools, techniques, and data that are of interest when evaluating these valves *in vitro*.

NATIVE VALVE STRUCTURE AND HEMODYNAMICS

Although it has been shown that a prosthetic heart valve need not look like a native heart valve to have adequate *in vivo* function, it must have geometry suitable for the intended implantation site and must function without impeding other aspects of cardiac function. It is therefore important to understand the anatomy and physiology of native heart valves as well as the process of surgical implantation of prosthetic valves. These will be reviewed briefly here; more detailed reviews can be found elsewhere (6–10). Figure 2 shows schematic drawings of a cross-section of the left ventricle in systole and diastole.

The base, or inflow perimeter, of the aortic valve is contiguous with the left ventricular outflow tract and the anterior leaflet of the mitral valve. The valve is comprised of three flexible, triangular leaflets, each of which attaches along ~120° of the circumference of the aorta. The inflow attachment line curves upward from the annulus at both ends, giving the leaflet its curved, three-dimensional (3D) geometry. At the outflow aspect of the valve, each adjacent leaflet pair meet at a commissure, a junction point on the aortic wall, and the aorta surrounding the valve leaflets is comprised of three bulbous regions, the sinuses of Valsalva.

Contraction of the left ventricle causes ventricular pressure to increase until it exceeds that in the aorta at which point the aortic valve opens rapidly and allows flow into the aorta. The flow reaches its peak amplitude about one-third of the way through the flow cycle. As the left ventricle relaxes, ventricular pressure falls, which reduces the pressure gradient and causes flow in the aorta to decelerate. Eventually, the pressure in the aorta exceeds that in the left ventricle and the aortic valve closes. During forward flow, a vortex forms in each sinus, which may play a role in the subsequent closure of the leaflets (11).

Aortic flow continues forward for a short period of time after the reversal of the pressure gradient due to the momentum of the blood (12), and a small volume of blood, the closing volume, is pushed back into the ventricle at closure due to the motion of the closing leaflets. During diastole, the closed aortic valve is under a back pressure of 80–100 mmHg (10.66–13.33 kPa). Representative pressure and flow waveforms of the aortic valve are shown in Fig. 3. At a heart rate of 70 beats · min^{-1}, systole will typically last ~35% of the whole cardiac cycle, or ~300 ms. At an exercise heart rate of 120 beats · min^{-1}, the systolic ratio will increase to near 50%.

Although there are many factors involved in deciding whether a patient needs surgery, general guidelines suggest that a diseased aortic valve may be considered for replacement with a prosthesis when the area of the valve has been reduced to 0.5–1.0 cm^2 (compared to a normal range of 3–4 cm^2), or when regurgitation has caused an ejection fraction of <50% (13). When a diseased aortic valve is replaced with a prosthetic valve, the aorta is cut open, all three leaflets are cut out, and any calcium is removed from the valve annulus. The diameter of the annulus is then measured to determine the appropriate sized prosthetic valve to use. The prosthetic valve is then implanted by stitching the sewing ring to the tissue of the native annulus, although the exact implantation process as well as the positioning of the valve will vary based on valve type.

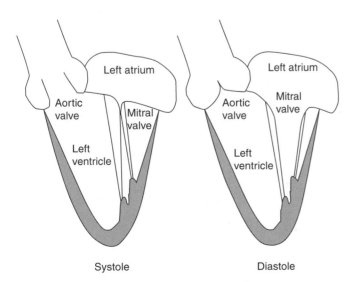

Figure 2. Schematic of the left ventricle and valves during systole and diastole.

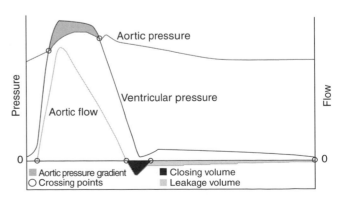

Figure 3. Idealized waveforms of pressure and flow through the aortic valve.

The mitral valve is structurally and functionally distinct from the aortic valve. It has two primary, unequally sized leaflets, each of which is made from several segments. The longer anterior leaflet, adjacent to the aortic valve, is attached along one-third of the annulus, while the shorter posterior leaflet connects to about two-thirds of the annulus. The leaflets contain many furrows or wrinkles that allows the valve to occupy minimal space in the heart while still having a large surface area for preventing leakage and supporting the stress of closure. The annulus itself is not circular, but D-shaped, and has a 3D, saddle-shaped curvature (14).

Many tendonous chords emanate from the underside and edge of the leaflets and attach to the papillary muscles, which are part of the wall of the left ventricle. The chords and papillary muscles comprise the tensioning component of the mitral apparatus, helping the valve to support and balance the stresses on the leaflets during closure. The entire mitral apparatus is a dynamic and active structure: the annulus, leaflets, and papillary muscles all move in coordination throughout the cardiac cycle in support of proper valve function (15,16). Currently, no prosthetic mitral valve can replicate or synchronize with the complicated force and motion dynamics of the native mitral apparatus.

During left ventricular filling, diastolic flow through the mitral valve is equal to the subsequent flow out of the aortic valve, assuming there are no leaks through either valve. Although the same volume passes through the mitral valve, the flow profile is very different than that of aortic flow. First, diastolic flow occupies ~65% of the cardiac cycle, lasting ~557 ms at a heart rate of 70 beats · min^{-1}. Due to the longer flow period, peak flow rates and pressure gradients are usually lower through the mitral valve than the aortic valve. Second, diastolic flow occurs in two phases. As the left ventricle relaxes, pressure falls to near zero and the mitral valve opens to allow ventricular filling from the left atrium, which acts as a compliant filling chamber and maintains a fairly constant blood pressure of ~15 mmHg (1.99 kPa). The pressure gradient between the atrium and the ventricle lessens as the ventricle fills, causing the flow to approach zero and the leaflets to nearly close. The left atrium then contracts, opening the leaflets again and sending a second bolus of blood, less than the first, into the ventricle. The two waves of the biphasic diastolic flow pattern are referred to as the E and A waves. The valve closes fully at the end of the A wave and is under a back pressure of >100 mmHg (13.33 kPa) during systole. Figure 4 shows a schematic representation of pressure and flow waveforms through the mitral valve.

The decision to surgically replace the mitral valve with a prosthesis, as with the aortic valve, is based on many factors. But general functional criteria include an effective orifice area <1.0 cm^2 (depending on body size) or regurgitation causing an ejection fraction <50% (13). During surgical replacement, the left atrium is opened, the native leaflets are cut out, and any calcium is removed. As with aortic replacement, the diameter of the annulus is measured to determine the prosthetic valve size needed. For valves with stent posts, care must be taken that the posts do not impinge on the wall of the left ventricle. The chords

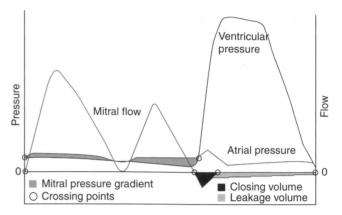

Figure 4. Idealized waveforms of pressure and flow through the mitral valve.

are generally removed, but may be left in the heart in some cases to provide structural support to the left ventricle. For some patients, the mechanical functionality of the mitral valve can be restored with surgical repair techniques, obviating the need for a prosthesis.

PROSTHETIC HEART VALVE TECHNOLOGY

A prosthetic heart valve must meet several basic functional requirements. In addition to having adequate opening and closing characteristics, it must also be durable, and biocompatible. All prosthetic valves compromise at least one of these features in favor of one of the others. The primary distinction between valve types is materials: Most valve designs use either synthetic, rigid components or flexible, biologically derived tissue. Based on these material features, a prosthetic valve is generally categorized as either a mechanical valve or a tissue valve.

Most mechanical heart valves (MHVs) are made from a rigid ring of metal or pyrolytic carbon, the outer perimeter of which is covered with a cloth sewing ring. The ring, or housing, contains one or more rigid occluders that are free to swivel on struts or hinges in response to a pressure gradient. The occluders are constrained within the housing, but are not mechanically coupled to it, allowing blood to flow completely around them, which helps to avoid flow stagnation.

Although they can adequately prevent backflow, MHVs do not create a seal during closure, and thus allow some regurgitation. The volume of blood regurgitated is tolerable for the patient, but the squeezing of blood through the closed valve creates high velocity, high shear jets that may be responsible for blood damage that leads to thrombosis (17). Whatever the mechanism, all MHV patients must take a daily dose of anticoagulant to counteract the thrombogenic effects of the valve. Without anticoagulation MHVs will develop clots that can impede valve function or become embolized into the bloodstream. One advantage of MHVs is that they are highly durable and can usually function for the duration of the recipient's life.

Although widely used, the Starr–Edwards valve was eventually surpassed by MHVs with better flow characteristics. Bileaflet valves and tilting disk valves are the two

incubated in glutaraldehyde, and then mounted on a cloth-covered frame with three commissure posts and a sewing ring. The frame, made of metal wire or plastic, provides structural support and allows ease of handling and implantation. The other main type of THV is the pericardial valve. Pericardium, the tissue that surrounds the heart, is separated from the heart of a cow or horse and then flattened and incubated in glutaraldehyde, producing a sheet of material that may be cut and assembled as desired to form a valve. Commercial pericardial valves typically incorporate three separate leaflets onto a three-pronged support structure similar to those used for porcine valves. Although structurally similar to the aortic valve, both porcine and pericardial valves are also implanted in the mitral position. Examples of a porcine valve and a pericardial valve are shown in Fig. 6. The Edwards Perimount pericardial valve is currently the most widely used THV, comprising >70% of all tissue valves used in the

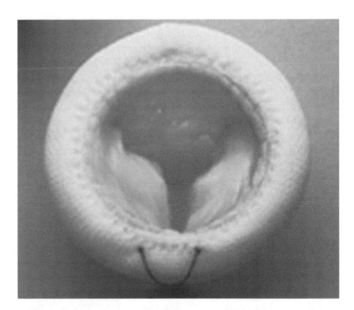

Figure 5. Examples of two mechanical heart valves: A bileaflet St. Jude Mechanical valve, and a Medtronic tilting disc valve. (Courtesy of Medtronic, Inc.)

mostly popular types in use clinically (see Fig. 5). The St. Jude bileaflet mechanical valve is by far the most widely used mechanical valve. It accounted for >70% of all mechanical valves sold in the United States in 2003 (18). Most innovations in MHVs in the last 10 years have focused on optimizing this type of bileaflet design, either through improved materials, geometries, or sewing rings. Other manufacturers of mechanical heart valves include Carbomedics, Medtronic, Sorin, and Medical Carbon Research Institute.

Tissue heart valves (THVs), also called bioprosthetic valves, were developed based on the idea that valves made of biologic tissue and with a structure similar to the native heart valve would function better *in vivo* than rigid mechanical valves. The leaflets of these valves are made from animal tissue that has been treated with a dilute solution of glutaraldehyde. Glutaraldehyde cross-links the collagen in the tissue, which prevents its breakdown and reduces its antigenicity *in vivo*. The cross-linked tissue is slightly stiffer than fresh tissue, but it still retains a functional amount of flexibility.

Porcine THVs are made from the aortic valve of a pig. To construct the prosthesis, the aortic valve is first excised from the pig heart, trimmed to remove excess tissue,

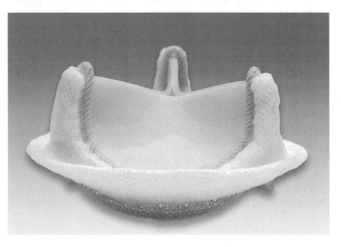

Figure 6. Examples of two tissue heart valves: A Mosaic porcine tissue valve. (Courtesy of Medtronic, Inc.) A Perimount pericardial tissue valve. (Courtesy of Edwards Lifesciences, Irvine, California.)

United States (18). Other manufacturers of tissue valves include Medtronic, St. Jude, Sorin, and 3F Therapeutics.

In contrast to MHVs, THVs generally have an unobstructed flow orifice, seal during closure, and do not typically require long-term anticoagulation. However, THVs have limited structural durability compared to MHVs, which usually last the duration of the recipient's lifetime. Although some pericardial valves have performed well for as long as 17 years in patients (19), the THVs in general are expected to degenerate or calcify within 10–15 years, at which time they must be surgically replaced with a new prosthesis. Due to this limitation, THVs are typically only implanted in patients older than 65 years of age (13). Over the past several years there has been an increased use of tissue valves in the United States, while mechanical valve usage has declined, a trend that is expected to continue (18).

Another type of THV that was widely pursued in the 1990s is the stentless valve. This valve type, intended for aortic valve replacement only, is made from porcine aortic roots that are fixed in glutaraldehyde, but do not have any rigid support materials added. The surgeon must attach both ends of the device into the patient's aorta without the aid of a sewing ring or support structures. The intended benefit of these valves is better hemodynamics because of their flexibility and lack of a sewing ring, and greater durability due to lower stresses in the tissue. However, they are not used as extensively as other THVs because they are more difficult to implant, which results in extended surgery time. Examples of two stentless valves used clinically are shown in Fig. 7. Similarly, human aortic roots can be removed from cadavers and processed with preservative techniques to make an implantable replacement. These valves, called homografts or allografts, have all the perceived benefits of stentless valves, but are just as difficult to implant, and long-term clinical results have been mixed (5). The primary commercial source of homografts is Cryolife, which cryogenically preserves the aortic roots.

There have been numerous attempts at fabricating prosthetic valves from polymers but, to date, none have achieved clinical success. Polymer valves are conceptually attractive because they would be flexible, with a reproducible geometry and relatively economical and straightforward to manufacture, and ideally would be more durable than tissue valves, while not requiring chronic anticoagulation like mechanical valves. But design difficulties and calcification have prevented these valves from realizing their full potential (20). Like polymer valves, tissue-engineered valves have many theoretical advantages over current mechanical and tissue valves. As a result, processes for growing a valve *in vitro* from human cells seeded on a scaffold has been an area of active research (21–23), but has also not yet produced a clinically viable product.

IN VITRO TESTING

In vitro evaluations of prosthetic heart valves are performed to understand the detailed flow characteristics of a given design. The flow area, the amount of leakage, the ultrasound compatibility, the presence of deleterious flow

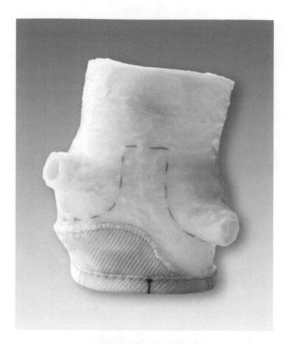

Figure 7. Examples of two stentless porcine heart valves: A Prima stentless valve. (Courtesy of Edwards Lifesciences, Irvine, Ca.) A Freestyle stentless valve. (Courtesy of Medtronic, Inc.)

patterns, the velocity magnitude of regurgitant jets, the motion of the leaflets during forward flow, and the position of the leaflets during closure can all be assessed *in vitro* and be used to improve valve designs and assess the appropriateness for human implantation.

Equipment

To evaluate the hemodynamic performance of prosthetic valves, a pulse duplicator or mock flow loop is implemented to act as an artificial left ventricle, and therefore must be able to simulate the pressure and flow waveforms shown in

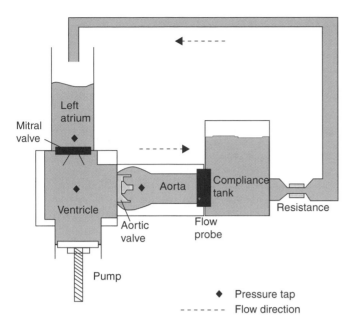

Figure 8. Schematic of a pulse duplicator for *in vitro* heart valve testing.

Figs. 2 and 3 over a range of physiologic hemodynamic conditions.

Figure 8 is a schematic representation of a pulse duplicator for *in vitro* heart valve testing. Generally, the left atrium is simulated with an open or compliant reservoir that maintains a static fluid height above the mitral valve so as to provide a diastolic filling pressure of ~10–15 mmHg (1.33–1.99 kPa). The mitral valve should open directly into the left ventricle as it does anatomically. The left ventricle can be simulated with a rigid chamber with a volume similar to the human left ventricle (~70–100 mL), although some pulse duplicators utilize a flexible, ventricular-shaped sac that can be hydraulically compressed to simulate the squeezing action of the heart (24,25). The aortic valve should be mounted at the outflow of the ventricular chamber so that flow enters the valve directly from the chamber. Tubing between the ventricle and the valve should be avoided as it will cause higher than physiologic velocities upstream of the valve. The flow exiting the aortic valve enters a tubular chamber with dimensions similar to the native aorta, including three sinuses and a diameter appropriate for the aortic valve being tested. Although not necessary for valve closure, the presence of sinuses allows for more realistic flow patterns behind the leaflets during systole.

The system should be instrumented to allow instantaneous pressure measurements in the left atrium, left ventricle, and aorta and flow measurements through both valves. The flowmeter must be able to accurately measure both forward flow, which can reach peak valves of $30 \, \text{L} \cdot \text{min}^{-1}$, and leakage flows, which may be on the order of $1 \, \text{mL} \cdot \text{s}$. Test fluids can be either saline or a blood analog fluid with a density and viscosity similar to that of blood, $\sim 1.1 \, \text{g} \cdot \text{mL}^{-1}$ and 3.5 cp, respectively. A mixture of water and glycerin is the most common blood analog fluid. Because blood is a non-Newtonian fluid (i.e., its viscosity

varies with shear rate), polymeric solutions, which more closely mimic this property, have also been used for *in vitro* valve testing (26).

In order to produce physiologic waveforms, compliant and resistive elements must be present downstream of the aorta. Compliant elements are expansive chambers that may utilize air, springs, flexible beams, or compliant tubing to absorb fluid volume during the systolic stroke. The compliant element should be located as close as possible to the aortic valve to minimize "ringing" in the pressure and flow waveforms. The impedance of the arterial system can be simulated with simple pinch clamp on flexible tubing, or a variably restrictive orifice.

The flow is driven with a pulsatile pumping system that interfaces with the left ventricular chamber. The pump must be able to create physiologic beat rates, flow rates, and systolic/diastolic ratios. One type of pumping system utilizes positive and negative air pressure, cycled with solenoid valves and timers, to drive a flexible diaphragm (27). Alternatively, a large, cam-driven syringe (piston–cylinder) pump can be employed with controlled piston displacement designed to produce the desired systolic and diastolic timing. Syringe pumps driven by computer-controlled servo motors are the optimal pumping systems because they allow the greatest amount of control over the motion of the piston and can best simulate complicated physiologic flow patterns (e.g., biphasic diastolic flow).

Valve mounting in a pulse duplicator can be accomplished by mechanical compression of the sewing ring, or by suturing the ring to a gasket. However, any valve that is flexible and mechanically coupled to the native anatomy during implantation (e.g., a stentless aortic valve) must be tested in a valve chamber that mimics the dynamic motion of the valve *in vivo*, because this may affect hemodynamic performance. Standardized testing guidelines suggest that stentless valves be tested in flexible chambers that undergo a 4 and 16% change in diameter per 40 mmHg (5.33 kPa) pressure change (28).

Test Conditions and Data Analysis

Hemodynamic conditions considered typical for an adult would be the following: cardiac output: $5 \, \text{L} \cdot \text{min}^{-1}$; left atrial pressure: 10–15 mmHg (1.33–1.99 kPa); left ventricular pressure: 120/0 mmHg (systolic/diastolic) (15.99/0 g); aortic pressure: 120/80 mmHg (systolic/diastolic) (15.99/10.66 kPa); heart rate: 70 beats $\cdot \text{min}^{-1}$.

Although flow and pressure conditions are often quantified according to mean, peak, and minimum values this way, the shape of the pressure and flow waveforms are also important in determining the appropriateness and quality of the test condition. Excess "ringing", large pressure spikes, and square waveforms are examples of unphysiologic waveform features that may be seen *in vitro*.

A thorough *in vitro* investigation of heart valve performance should also include different flow, heart rate, and pressure conditions that correspond to the range of physiologic hemodynamic conditions that the valve will experience *in vivo*. At rest, the heart rate may decrease to 45 beats $\cdot \text{min}^{-1}$ or lower, while during exercise it may increase to >120 beats$\cdot \text{min}^{-1}$; cardiac output will vary accordingly at these conditions, typically ranging from 2 to $7 \, \text{L} \cdot \text{min}^{-1}$. It is at these extremes that leaflet dynamics, for example, may change and reveal performance limitations of the valve.

Typically, a minimum of 10 cycles, or heart beats of flow, inflow pressure and outflow pressure are collected for analysis at each test condition. The waveforms are analyzed by first identifying the key crossing points that define the start and end of forward flow, regurgitant flow, and positive (forward flow) pressure gradient (see Figs. 3 and 4). These pressure and flow data are used to calculate the key variables that define the hemodynamic performance of a valve: the pressure gradient (ΔP), the effective orifice area (EOA), and regurgitant volume.

The pressure gradient is the most basic measure of valve resistance to flow. *In vivo*, a prosthetic aortic valve with a high pressure gradient will force the left ventricle to expend more energy than one with a low pressure gradient. The regions of positive pressure gradient for aortic and mitral valves are shown in Figs. 3 and 4. These pressures can be measured in a pulse duplicator through wall taps in the testing chambers or through a catheter inserted into the flow. The pressure gradient will increase with decreasing valve size and, for a given valve size, increasing flow rate. It has also been shown that aortic pressure gradient measurements can be dependent on left ventricular ejection time (29) and aortic pressure (30).

The EOA is not a physical valve dimension, but a calculation of the minimal flow area allowed by the open valve. Under pulsatile flow conditions, it is calculated as $Q_{\text{rms}}/[51.6(\Delta P)^{1/2}]$, where Q_{rms} is the root-mean square of the average flow rate through the valve. The EOA is a more useful measure than geometric orifice dimensions because mechanical valves, with occluders that occupy the orifice, restrict the orifice flow in variable ways based on the degree of opening and flow rate. Similarly, the extent of tissue valve leaflet opening will vary based on leaflet stiffness and flow rate. A method of dynamically measuring the actual valve area in an *in vitro* tester by measuring the amount of light that passes through it has also been proposed and tested (31).

All valves move some volume of fluid back into the inflow chamber during closure. The closing volume of a prosthetic valve is calculated from the area under the flow curve immediately after valve closure as shown in Figs. 3 and 4. This volume is a function of the valve area and the travel distance of the leaflets during closure. Although it will vary based on valve design, the closing volume is relatively small and typically does not have a significant hemodynamic effect. Leakage volume, by contrast, is the volume of blood that passes through the valve during closure, and is a critical measure of valve performance. Excessive leakage reduces the efficiency of the heart and can cause progressive deterioration of ventricular function. Ideally, the leakage volume through a valve is zero, although most mechanical valves allow some leakage of blood in order to provide "washout" of mechanical junctions. Tissue valves do not allow any leakage as long as there is adequate apposition, or coaptation, of the leaflets.

The total regurgitant volume is the sum of the closing and leakage volumes and represents the total fluid loss during one valve closure. Regurgition can also be expressed as a percentage of forward volume. Clinically, the ejection fraction, the ratio of the ejected blood volume to the left ventricular volume, is used to assess the hemodynamic severity of regurgitation. As with other hemodynamic variables, regurgitant volumes will vary with valve design, valve size, flow rate, pressure, and heart rate.

When assessing prosthetic valve regurgitation it is important to discriminate between transvalvular regurgitation and paravalvular regurgitation. Transvalvular regurgitation occurs through the valve mechanism, such as past the hinges or occluder gaps in mechanical valves or through regions of inadequate leaflet coaptation in tissue valves. Because prosthetic valves are sewn in place with a porous cloth sewing ring, leakage can also occur through the sewing ring or spaces created by inadequate fixation to the annulus. This is paravalvular leakage, occurring around or adjacent to the valve, and should be differentiated from transvalvular leakage *In vitro*, test valves should be sealed to eliminate paravalvular leakage so that transvalvular leakage can be measured independently.

In vitro performance comparisons of different prosthetic valve designs are complicated by variations between labeled valve size and actual valve size (32). Most commercial prosthetic valves are sized according to diameter in 2 mm increments and typically range from 19 to 29 mm for aortic valves, and 25 to 31 mm for mitral valves. However, the diameter measurement is made at different locations based on manufacturer and valve type. The performance index (PI), which is the ratio of EOA to actual sewing ring area (33), is calculated in order to control for these variations and allow for more reliable valve-to-valve comparisons. Clinically, an EOA Index (EOAI) is calculated by dividing the EOA by the patient's body surface area, in order to normalize EOA values based on patient size and, indirectly, hemodynamic requirements.

Due to the variability of pulse duplicator systems and the sensitivity of test results on hemodynamic conditions, comparing valves evaluated in different test systems can also be problematic (34). Differences in pressure and flow waveform shapes, drive types, location of pressure taps, chamber geometries, and other system configuration characteristics can cause variations in measured valve performance. Comparative *in vitro* studies of different mechanical valves (35,36) and tissue valves (37) tested in the same pulse duplicator under the same conditions can be found in the literature. Figures 9–10 show *in vitro* pressure drop results for various tissue valves from one of these studies.

Doppler Ultrasound

Doppler echocardiography is the most widely used modality for assessing prosthetic valve performance *in vivo* and can be useful for *in vitro* studies as well. *In vitro* Doppler and imaging assessments are an important part of valve evaluations prior to human implants because they may reveal signature ultrasound features (e.g., acoustic shadowing, eccentric flow profiles, jet morphology), which are unique to the valve design and different from a native valve. It is important for clinical sonographers to understand these

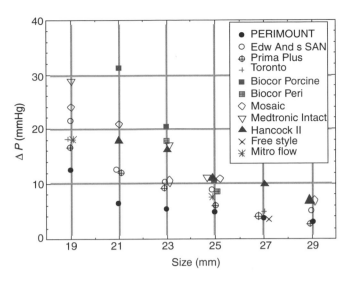

Figure 9. *In Vitro* pressure drop data of various prosthetic aortic valves. (Reprinted with permission from *The Journal of Heart Valve Disease*.)

features so they can distinguish between normal and abnormal valve function. In addition, two-dimensional (2D) echo images of the valve can be used to obtain an axial cross-sectional view of the valve during cycling that can be useful in assessing the motion and coaptation morphology of tissue leaflets. And although a flowmeter can be used to quantify the amount of regurgitation through a valve, Color Doppler ultrasound allows for visualization of the size and shape of any jets and detecting their point of origin.

Doppler measurements and 2D imaging can be performed *in vitro* providing there is an acoustic window for the ultrasound beam that allows the transducer to be aligned roughly perpendicular to the axis of valve flow and does not attenuate the acoustic signal. It is also

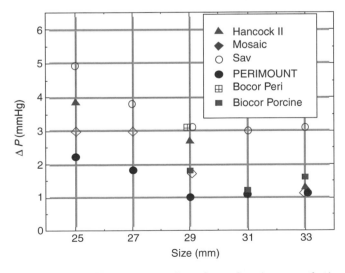

Figure 10. *In Vitro* pressure drop data of various prosthetic mitral valves. (Reprinted with permission from *The Journal of Heart Valve Disease*.)

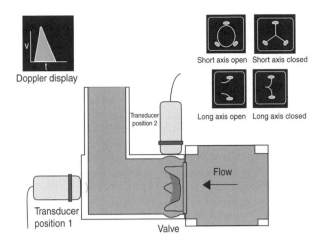

Figure 11. Ultrasound measurements in a pulse duplicator. In transducer position 1, the transducer is aligned with the flow and emits a sound wave that reflects off particles in the fluid; the return signal is measured by the transducer and the Doppler equation is used to calculate velocity. Transducer position 2 is used to obtain short-axis (cross-sectional) and long axis images of the valve as it opens and closes.

necessary to add small, insoluble particles to the test fluid to provide acoustic scatter for Doppler measurements and Color Doppler imaging. Figure 11 shows how an ultrasound transducer would be positioned on a pulse duplicator and the types of images that can be obtained.

Doppler measurements of pressure gradients can be performed *in vitro* to assess the hemodynamic performance of a valve as it would be done in a patient. During forward flow, continuous wave Doppler measures a real-time spectrum of velocities across the open valve. Standard echocardiographic software allows the user to acquire and analyze the spectrum and determine an average velocity.

Clinically, the average velocity value is used to compute pressure gradient using a simplified version of the Bernoulli equation. Ignoring viscous losses, acceleration effects, and changes in height, the Bernoulli relationship between two points, 1 and 2, on a streamline is:

$$P_1 + \frac{1}{2}\rho v_1^2 = P_2 + \frac{1}{2}\rho v_2^2 \qquad (1)$$

where P is the pressure, is the fluid density, and v is the velocity.

Rearranging Eq. 1 and expressing the upstream, or proximal valve velocity as v_p, and the downstream, or distal velocity as v_d, the pressure gradient ($P_1 - P_2 = \Delta P$) across a valve can be expressed as:

$$\Delta P = \frac{1}{2}\rho(v_d^2 - v_p^2) \qquad (2)$$

Doppler ultrasound machines typically report velocity in units of meter per second ($m \cdot s^{-1}$) and pressure in units of millimeter of mercury (mmHg). Assuming a blood density of $\sim 1.1\,g \cdot mL^{-1}$ and applying the appropriate unit conversions, the 1/2 term is approximated as $4\,mmHg \cdot m^{-2} \cdot s^{-2}$. A further simplification can be made if the proximal velocity term (the velocity at the left ventricular outflow tract for aortic valves) is small compared to the distal velocity.

Neglecting the proximal velocity term yields the equation used by ultrasound software programs to compute pressure gradient from continuous wave velocity measurements:

$$\Delta P = 4(v^2) \tag{3}$$

If the proximal velocity term is needed, it can be measured with pulsed wave Doppler ultrasound, which, unlike continuous wave Doppler, provides velocity spectra at a user-defined location in space (e.g., the left ventricular outflow tract).

Prior to the use of Doppler, hemodynamic assessment of prosthetic valves *in vivo* was performed by placing a pressure-measuring catheter into the heart. Comparisons between pressure gradients measured by catheter and by Doppler typically show some discrepancy between the two methods, causing some concern as to which method is more accurate. Several groups have performed *in vitro* and *in vivo* studies to compare and contrast catheter-based measurements and Doppler measurements (38–41). These comparisons are complicated by the fact that the two techniques use fundamentally different methods to arrive at the pressure gradient and they each have individual sources of approximation and error.

The primary valve-related reason for differences in Doppler and catheter measurements of pressure gradient is pressure recovery. Pressure recovery occurs when some portion of the downstream kinetic energy is converted back to potential energy (i.e., pressure). Like flow through a Venturi nozzle, the lowest pressure and highest velocity through a heart valve will occur at the narrowest point, which is typically the region immediately downstream of the valve. If the flow out of the valve is allowed to gradually expand and remains attached to the walls, there will be an increase, or recovery, of pressure further downstream relative to the pressure at the valve exit. Due to this local variation in pressure, the location of pressure measurement becomes critical. Because continuous wave Doppler measures velocities all along the beam path, it will detect the highest velocity (regardless of its location), which will then be used to calculate the pressure gradient. In contrast, catheters take direct pressure measurement at a specific location; if that location is not at the point of lowest pressure, catheter measurements of pressure gradient will tend to be lower than those derived from Doppler.

Pressure measurements across bileaflet MHVs (e.g., the St. Jude Mechanical) are further complicated because, during opening, the two side orifices are larger than the central orifice, resulting in a nonuniform velocity profile. This may be one reason why better agreement between *in vitro* Doppler and catheter measurements has been found for a tissue valve, which has an unobstructed central orifice, than for a bileaflet mechanical valve (39).

Flow Visualization

Clear chambers that provide visual access to the valve as well as the flow fields in its vicinity are an important feature of *in vitro* test systems. The use of video to asses leaflet dynamics can provide important information about valve performance that cannot be obtained from hemodynamic measurements or *in vivo* studies. Tissue contact with a stent, poor coaptation, asymmetric or asynchronous leaflet closure, and leaflet rebound are all visual clues that either the valve design is inadequate or the test conditions are inappropriate.

Flow field visualization and measurements are equally important because thrombotic complications *in vivo* can be caused by unphysiologic, valve-induced flow patterns. Both flow stasis and high velocity leakage jets (and associated turbulent stresses) can trigger thrombosis and subsequent thromboemboli. *In vitro* pulsatile flow studies using blood are not a practical means of determining the thrombogenicity of a valve design because of the complexity and sensitivity of the coagulation process. Although the use of milk as an enzymatically coagulable test fluid has been reported (42–44), most *in vitro* assessments of a valve's thrombogenicity are made indirectly based on its flow characteristics.

Illuminating the flow field of interest with a sheet of laser light and seeding the fluid with neutrally buoyant particles will allow for qualitative visual assessment of flow patterns (i.e, uniformity and direction of forward flow, presence of vortices or recirculation zones, areas of washout during closure.) The motion of the particles can be analyzed quantitatively with digital particle image velocimetry (DPIV), which uses the translocation of seeded particles in consecutive, laser-illuminated video frames to compute the velocity of many particles at one point in time. With these numerous velocity measurements, a velocity map of the flow field can be created from any point in the cardiac cycle.

Velocity measurements can also be performed with laser Doppler velocimetry (LDV), which allows good temporal resolution of velocity, but only at a single point. Multiple point measurements at different locations can be made sequentially and then compiled in order to construct a phase averaged velocity profile of the entire flow field. Simultaneous LDV and DPIV measurements have been performed on a MHV in an attempt to integrate the relative benefits of each method (45). Thorough flow field measurements of several prosthetic heart valves, both tissue and mechanical, have been published by Yoganathan and co-workers (46,47). During forward flow, MHVs are seen to have disrupted or eccentric velocity profiles reflective of the open occluder position, while THVs tend to have more uniform, unobstructed flow profiles.

Since blood cell damage will likely occur above some critical shear stress threshold, *in vitro* velocity data is used to calculate shear stresses created throughout the flow cycle, and indirectly assess the potential for hemolysis and platelet activation *in vivo*. The MHV leakage jets, in particular, have the potential to create high shear stresses in blood (48). It is difficult, however, for *in vitro* flow characterization studies to be conclusive with regard to blood damage because of the myriad of variables that interact to trigger coagulation. In addition to the magnitude of shear stresses, the exposure time to those stresses, as well as flow stagnation, material surface interactions, and patient factors can contribute to prosthetic valve thrombosis.

Design Specific Testing

Guidelines for *in vitro* testing heart valves prior to human use were originally introduced by the U.S. Food and Drug

Administration (FDA) in 1982. A revised document, that included guidelines for testing stentless valves, was introduced in 1994 (28). The International Standards Organization publishes a similar set of guidelines for heart valve evaluation, including the equipment and data requirements for *in vitro* studies (49). Many of the testing techniques in use today are motivated by and in response to these guidelines. However, standardized guidelines are often insufficient for testing new and innovative designs, since each new valve design will have unique features that may require special testing methods to evaluate. A risk analysis or failure mode analysis can be used to assess the need for *in vitro* testing beyond that described in the standards.

In addition to predictive studies of *in vivo* performance, *in vitro* studies may be conducted retrospectively, in order to elucidate a particular failure mode seen after a valve has been used clinically. These types of studies typically require the development of new or improved testing methodologies to investigate a particular phenomenon, as shown in the following examples.

In 1988, structural failures of the Baxter Duromedics MHV were reported in several patients (50). Surface pitting was observed on the pyrolytic carbon leaflets of the explanted valves, suggestive of cavitation-induced erosion. Cavitation occurs in a fluid when the pressure drops rapidly below the vapor pressure of the fluid, causing the formation of small vaporous cavities, which collapse violently when the pressure increases. Many *in vitro* studies were conducted, employing novel measurement techniques, to assess the propensity for cavitation to occur during leaflet closure of mechanical heart valves (51,52) and cavitation testing is now part of U.S. Food and Drug Administration (FDA) required preclinical testing for all new MHVs.

In the mid-1990s, the Medtronic parallel bileaflet valve experienced an unanticipated number of thrombosed valves in early clinical trials (53). Explanted valves were observed to have clot formation in the hinge region of the valve, indicating a flow-related problem in this vicinity (54). These failures occurred despite the full set of required tests having been conducted with apparently satisfactory results. Prototype valves with clear housings were constructed to allow flow visualization studies in the region of the hinge mechanism (55). Results of these studies suggested the geometry of the hinge created stagnant flow regions which may have been responsible for the clinical failures.

Clinical studies of patients with stentless valves revealed that some developed aortic regurgitation several years after implantation (56,57). Because these valves lack a support structure, it was believed that age-related dilation of the aorta strained the outflow edge of the valve, which lead to insufficient leaflet coaptation. Although FDA testing requirements for stentless valves included flow testing at elevated aortic pressures in compliant chambers, testing of mechanical dilation without an increase in pressure was not required. An *in vitro* study using canine hearts showed that simply pulling the aorta outward at the commisssures with sutures prevented the aortic leaflets from closing in the center (58). This study helped confirm the need to band the aorta during implantation of some stentless valves in order to prevent later dilation.

In some cases, *in vitro* testing may also be employed to test the actual valve that was explanted from a patient. Depending on the age and condition of the valve, it may be mounted in a pulse duplicator and studied for signs of performance, structural or manufacturing abnormalities responsible for an adverse event in a patient.

Finally, *in vitro* flow studies may be conducted using valves excised from animal hearts in order to study the flow or mechanical dynamics of the native valve that in turn may be used to design improved prosthetic devices. Fresh porcine aortic roots have been installed in pulse duplicators to study the motion of the aortic valve as well to serve as a testing chamber for prosthetic valves (59–61). The porcine mitral apparatus, including chords and papillary muscles, has also been mounted and tested *in vitro* (62,63).

FUTURE DIRECTIONS

After more than four decades of prosthetic heart valve development, surgeons have come to rely on just a few valve designs, both mechanical and tissue, for the large majority of implants. These valves have achieved widespread use because their performance is reliable and their limitations and failure modes are known and reasonably predictable. None of these valves are ideal, however, and new designs that try to improve upon the state of the art will continue to emerge.

Although polymer valves and tissue-engineered valves still hold promise, the greatest change to be expected in the near future is not the valve itself but the way it is implanted in the heart. The success of catheter-based technologies in treating other heart diseases (e.g., coronary stents, septal defect closure devices) has inspired the pursuit and clinical evaluation of a prosthetic heart valve that can be placed in the beating heart with a catheter (64–66). This technology is attractive because it would not require opening the chest and stopping the heart, nor the use of cardiopulmonary bypass equipment.

Catheter-delivered valves will still require all the testing and analysis described above, but will also necessitate new equipment and methodologies that take into account their unique delivery and implantation method. For example, the actual delivery and deployment of these valves may first need to be simulated under *in vitro* pulsatile conditions, requiring testers that simulate the entry point and anatomy of the delivery path. Once deployed, the ability of the valve to seal around its perimeter and remain in place without migrating must be evaluated, which will require deployment chambers with the appropriate surface properties, mechanical properties, and dynamic motion. Also, current *in vitro* valve tests rarely simulate coronary flow out of the sinuses. But when testing valves that are placed in the aortic annulus under image guidance, rather than direct visualization, it will be important to evaluate valve designs in terms of their propensity to block coronary flow.

With aging populations in the United States and Europe, heart valve disease is likely to remain a significant and prevalent problem. Better, more efficient prosthetic heart valve technology will surely emerge to address this need and *in vitro* testing technologies will need to keep pace to continue helping to improve these devices and ensure their safety and efficacy for human use.

BIBLIOGRAPHY

Cited References

1. Heart Disease and Stroke Statistics—2004 Update. American Heart Association, 2004.
2. Starr A, Edwards ML. Mitral replacement: clinical experience with a ball-cage prosthesis. Ann Surg 1961;154:726–740.
3. Starr A, Edwards ML. Mitral replacement: late results with a ball valve prosthesis. J Cardiovasc Surg 1963;45:435–447.
4. Pluth JR. The Starr valve revisited. Ann Thor Surg 1991;51: 333–334.
5. Otto CM. Valvular Heart Disease, 2nd ed. Philadelphia: Saunders; 2004.
6. Anderson RH. Clinical anatomy of the aortic root. Heart 2000;84:670–673.
7. Thubrikar M. The Aortic Valve. Boca Raton (FL): CRC Press; 1990.
8. Antunes MJ. Functional Anatomy of the Mitral Valve. In: Barlow JB, editor. Perspectives on the Mitral Valve. Philadelphia: FA Davis; 1987.
9. Kalmanson D. The mitral valve: a pluridisciplinary approach. London: Arnold; 1976.
10. Zipes DP. Braunwald's Heart Disease. 7th ed. Philadelphia: WB Saunders; 2005.
11. Thubrikar MJ, Heckman JL, Nolan SP. High speed cineradiographic study of aortic valve leaflet motion. J Heart Valve Dis 1993;2(6):653–661.
12. Nichols WW, O'Rourke MF. McDonald's Blood Flow in Arteries, 4th ed. London: Arnold; 1998.
13. Bonow RO, et al. ACC/AHA guidelines for the management of patients with valvular heart disease: a report of the American College of Cardiology/American Heart Association Task Force on Practice Guidelines. J Am College Cardiol 1998;32: 1486–1588.
14. Levine RA, et al. Three-Dimensional echocardiographic reconstruction of the mitral valve, with implications for the diagnosis of mitral valve prolapse. Circulation 1989;80: 589–598.
15. Yellin EL, et al. Mechanisms of mitral valve motion during diastole. Am J Physiol 1981;241:H389–H400.
16. Komeda M, et al. Papillary muscle-left ventricular wall "complex". J Thor Cardiovasc Surg, 1997;113:292–301.
17. Ellis JT, et al. An in vitro investigation of the retrograde flow fields of two bileaflet mechanical heart valves. J Heart Valve Dis 1996;5(6):600–606.
18. Anonymous Focus on Heart Valves, Medical Device and Diagnostic Industry, vol. 126, March 2004.
19. Banbury MK, et al., Hemodynamic stability during 17 years of the Carpentier-Edwards aortic pericardial bioprosthesis. Ann Thor Surg 2002;73(5):1460–1465.
20. Hyde JA, Chinn JA, Phillips RE., Jr., Polymer Heart Valves. J Heart Valve Dis 1999;8(3):331–339.
21. Hoerstup SP, Kadner A, Melnitchouk S. Tissue engineering of a functional trileaflet heart valve from human marrow stromal cells. Circulation 2002;106 (Suppl. I):143–150.
22. Cebatori S, Mertsching H, Kallenbach K. Construction of autologous human heart valves based on an acellular allograft matrix. Circulation 2002;106 (Suppl. I):63–68.
23. Bertipaglia B, Ortolani F, Petrelli L. Cell cellularization of porcine aortic valve and decellularized leaflets repopulated

24. Reul H, Minamitani H, Runge J. A hydraulic analog of the systemic and pulmonary circulation for testing artificial hearts. Proc ESAO 1975;2:120.
25. Scotten LN, et al. New tilting disc cardiac valve prostheses. J Thor Cardiovasc Surg 1986;82:136–146.
26. Pohl M, et al. *In vitro* testing of artificial heart valves:comparison between Newtonian and non-Newtonian fluids. Arti Organs 1996;20(1):37–46.
27. Weiting DW. Dynamic flow characteristics of heart valves dissertation. University of Texas, Austin, 1969.
28. U.S. Department of Health and Human Services, Public Health Service, Food and Drug Administration, Center for Devices and Radiological Health, Replacement Heart Valve Guidance, 1994.
29. Kadem L, et al. Independent contribution of left ventricular ejection time to the mean gradient in aortic stenosis. J. Heart Valve Dis 2002;11(5):615–623.
30. Razzolini R, et al. Transaortic gradient is pressure-dependent in a pulsatile model of the circulation. J Heart Valve Dis 1999;8(3):279–283.
31. Scotten LN, Walker DK. New laboratory technique measures projected dynamic area of prosthetic heart valves. J Heart Valve Dis 2004;13(1):120–132.
32. Cochran RP, Kunzelman KS. Discrepancies between labeled and actual dimensions of prosthetic valves and sizers. J Cardiovasc Surg 1996;11:318–324.
33. Stewart SFC, Bushar HF. Improved statistical characterization of prosthetic heart valve hemodynamics using a performance index and regression analysis. J Heart Valve Dis 2002;11:270–274.
34. Van Auker MD, Strom JA. Inter-Laboratory comparisons: approaching a new standard for prosthetic heart valve testing *in vitro*. J Heart Valve Dis 1999;8(4):384–391.
35. Walker DK, Scotten LN. A database obtained from *in vitro* function testing of mechanical heart valves. J Heart Valve Dis 1994;3(5):561–570.
36. Fisher J. Comparative study of the hydrodynamic function of six size 19 mm bileaflet heart valves. Eur J Cardiothora Surg 1995;9(12):692–695.
37. Marquez S, Hon RT, Yoganathan AP. Comparative hemodynamic evaluation of bioprosthetic heart valves. J Heart Valve Dis 2001;10(6):802–811.
38. Vandervoort PM, et al. Pressure recovery in bileaflet heart valve prostheses. Circulation 1995;92:3464–3472.
39. Baumgartner H, et al. Discrepancies between Doppler and catheter gradients in aortic prosthetic valves *in vitro*. A manifestation of localized gradients and pressure recovery. Circulation 1990;82:1467–1475.
40. Burstow DJ, et al. Continuous wave Doppler echocardiographic measurement of prosthetic valve gradients. A simultaneous Doppler-catheter correlative study. Circulation 1989; 80(3):504–514.
41. Bech-Hanssen O, et al. Assessment of effective orifice area of prosthetic valves with Doppler echocardiography: and *in vivo* and *in vitro* study. J Thorac Cardiovasc Surg 2001;122(2): 287–295.
42. Lewis JM, Macleod N. A blood analogue for the experimental study of flow-related thrombosis at prosthetic heart valves. Cardiovasc Res 1983;17(8):466–475.
43. Keggen LA, et al. The use of enzyme activated milk for *in vitro* simulation of prosthetic valve thrombosis. J Heart Valve Dis 1996;5(1):74–83.
44. Martin AJ, Christy JR. An *in vitro* technique for assessment of thrombogenicity in mechanical prosthetic cardiac valves: evaluation with a range of valve types. J Heart Valve Dis 2004;13(3):509–520.

with aortic valve interstitial cells: the VESALIO project. Ann Thor Surg 2003;75:1274–1282.

45. Kini V, et al. Integrating Particle Image Velocimetry and Laser Doppler Velocimetry Measurements of the Regurgitant Flow Field Past Mechanical Heart Valves. Art Organs 2001;25(2):136.

46. Woo YR, Yoganathan AP. *In vitro* pulsatile flow velocity and shear stress measurements in the vicinity of mechanical mitral heart valve prostheses. J Biomechan 1986;19(1):39–51.

47. Yoganathan AP, Woo YR, Sung HW. Turbulent shear stress measurements in the vicinity of aortic heart valve prostheses. J Biomechan 1986;19(6):433–442.

48. Ellis JT, et al. An in vitro investigation of the retrograde flow fields of two bileaflet mechanical heart valves. J Heart Valves Dis 1996;5:600–606.

49. International Standard 5840:2004(E), Cardiovascular Implants—Cardiac Valve Prostheses, International Standards Organization, June 22, 2004.

50. Quijano RC. Edwards-Duromedic dysfunctional analysis, Proceedings of Cardiostimulation; 1988.

51. Hwang NH. Cavitation potential of pyrolytic carbon heart valve prostheses: a review and current status. J Heart Valve Dis 1998;7:140–150.

52. Hwang NHC, editor. Cavitation in mechanical heart valves, Proceedings of the First International Symposium. J Heart Valve Dis 3(Suppl. I), 1994.

53. Bodnar E. The Medtronic Parallel valve and the lessons learned. J Heart Valve Dis 1996;5:572–673.

54. Ellis JT, et al. Velocity measurements and flow patterns within the hinge region of a Medtronic Parallel bileaflet mechanical valve with clear housing. J Heart Valve Dis 1996;5:591–599.

55. Gross JM, et al. A microstructural flow analysis within a bileaflet mechanical heart valve hinge. J Heart Valve Dis 1996;5:581–590.

56. Jin XY, Westaby S. Aortic root geometry and stentless porcine valve competence. Seminars Thorac Cardiovasc Surg 1999;11(Suppl I):145–150.

57. David TE, et al. Dilation of the sinotubular junction causes aortic insufficiency after aortic valve replacement with the Toronto SPV bioprosthesis. J Thorac Cardiovascu Surg 2001;122(5):929–934.

58. Furukawa K, et al. Does dilation of the sinotubular junction cause aortic regurgitation. Ann Thorac Surg 1999;68:949–954.

59. Thubrikar MJ, et al. The influence of sizing on the dynamic function of the free-hand implanted porcine aortic homograft: an in vitro study. J Heart Valve Dis 1999;8(3):242–253.

60. Revanna P, Fisher J, Watterson KG. The influence of free hand suturing technique and zero pressure fixation on the hydrodynamic function of aortic root and aortic valve leaflets. Eur J Cardiothor Surg 1997;11(2):280–286.

61. Jennings LM, et al. Hydrodynamic function of the second-generation mitroflow pericardial bioprosthesis. Ann Thorac Surg 2002;74:63–68.

62. Jensen MO, et al. Harvested porcine mitral xenograft fixation: impact on fluid dynamic performance. J Heart Valve Dis 2001;10(1):111–124.

63. Jensen MO, Fontaine AA, Yoganathan AP. Improved *in vitro* quantification of the force exerted by the papillary muscle on the left ventricular wall: three-dimensional force vector measurement system. Ann Biomed Eng 2001;29(5):406–413.

64. Cribier A, et al. Early experience with percutaneous transcatheter implantation of heart valve prosthesis for the treatment of end-stage inoperable patients with calcific aortic stenosis. J Am College Cardiol 2004;43(4):698–703.

65. Boudjemline Y, Bonhoffer P. Steps toward percutaneous aortic valve replacement. Circulation 2002;105(6):775–778.

66. Lutter G, et al. Percutaneous aortic valve replacement: an experimental study. J Thorac Cardiovasc Surg 2002; 123(4):768–776.

Reading List

Thubrikar M. The Aortic Valve. Boca Raton (FL): CRC Press; 1990.

Antunes MJ. Functional Anatomy of the Mitral Valve. In: Barlow JB, editor. Perspectives on the Mitral Valve, Philadelphia: FA Davis; 1987.

Kalmanson D. The mitral valve: a pluridisciplinary approach. London: Arnold; 1976.

Zipes DP. Braunwald's Heart Disease. 7th ed. Philadelphia: WB Saunders; 2005.

Otto CM, Valvular Heart Disease. 2nd ed. Philadelphia: WB Saunders; 2004.

Nichols WW, O'Rourke MF. McDonald's Blood Flow in Arteries. 4th ed. London: Arnold; 1998.

Reul H, Talukder N. Heart Valve Mechanics. In: Hwang NHC, et al., editors. Quantitative Cardiovascular Studies: Clinical and Research Applications of Engineering Principles. Vol. 12. Baltimore: University Park Press; 1979. p 527–564.

Nanda N. Doppler Echocardiography. 2nd ed. Philadelphia: Lee & Febiger; 1993.

See also BLOOD PRESSURE MEASUREMENT; MONITORING, HEMODYNAMIC.

HEART VALVES, PROSTHETIC

ROBERT MORE
Austin, Texas

INTRODUCTION

Blood flow through the four chambers of the normal human heart is controlled by four one-way valves. These valves open and close in response to local pressure gradients and flow during the cardiac cycle. The atrioventricular mitral and tricuspid valves open to admit blood flow from the atria into the ventricles during diastole and then close during systolic ventricular contraction to prevent backflow into the atria. The semilunar aortic and pulmonary valves open during systole to eject blood from the ventricles and then close during diastole to prevent backflow. Various pathological states, either congenital or acquired, may result in the failure of one or more of these valves. A valve may become stenotic, in which case forward flow through the valve is impaired, or a valve may become incompetent or regurgitant, closing improperly, which allows excessive backflow losses. Loss of valve function has a profound degenerative effect on quality of life and is ultimately life-threatening.

THE CHALLENGE

The goal for valve prosthesis design is to provide a functional substitute for a dysfunctional native valve that will endure for a patients' lifetime while requiring minimal chronic management. The development of open-heart surgery opened the technical possibility of valve replacement in the late 1950s and early 1960s. However, the practicality of an effective, durable artificial heart valve was to prove elusive in the early 1960s because of problems with valve construction materials causing blood clotting, material fatigue, and degradation. Materials, related problems were greatly diminished when Carpentier (1) introduced the use of aldehyde preserved biological valves in the early 1960s. In 1963, Bokros discovered isotropic pyrolytic carbon (PyC), and shortly thereafter Gott and Bokros discovered the remarkable blood compatibility of PyC (2). These events led to two separate avenues of approach to the development of successful heart valve replacements, namely *biological valves*, derived from biological valves themselves or biological tissues, and *mechanical valves*, manufactured from synthetic materials. Significant process has occurred toward the goal of a lifelong valve substitute, last 30 years to the extent that valve replacement surgery is now a commonplace procedure worldwide. However, todays valve prostheses are nonideal and a tradeoff remains between long-term durability and the need for chronic anticoagulant management.

Although the history of early development of replacement heart valves is fascinating in it's own right, it is well chronicled and, thus, will not be addressed here (3–5). See http://members.evansville.net/ict/prostheticvalveimage-gallery.htm for a chronology of valve designs. Rather, the focus here is to contemporary replacement heart valve designs, which is a story of interplay between advances in materials technology, design concepts, methods for evaluation, and regulatory issues.

EVOLUTION OF REGULATION

Prior to 1976, heart valve prostheses were unregulated devices—clinical studies, use, and introduction to the marketplace could be initiated without a formalized performance evaluation conducted under regulatory oversight. Regulation in the United States by the Food and Drug Administration (FDA), and by similar agencies in other countries, became necessary because not all valve prostheses were clinically successful. With the passage of the Medical Device Amendments in 1976, heart manufacturers were required to register with the FDA and to follow specific control procedures. Heart valve replacements were required to have premarket approval (PMA) by the FDA before they could be sold. To quote from the FDA website, http://www.FDA.gov, (the) "FDA's mission is to promote and protect the public health by helping safe and effective products reach the market in a timely way, and monitoring products for continued safety after they are in use".

The establishment of FDA oversight was a watershed event because it formalized the approval process and led to the development of criteria for evaluating new valve prosthesis designs. The preclinical performance tests and clinical performance evaluation criteria that have been developed are given in the FDA Document *Replacement Heart Valve Guidance - Draft Document, October 14, 1994* (6). This document may be downloaded from the FDA website at: http://www.fda.gov/cdrh/ode/3751.html. The document is deliberately labeled "Draft" because it is a "living" document that is periodically updated as heart valve technology evolves. Furthermore, requirements for

postmarket monitoring exist to continually verify valve safety and effectiveness.

A parallel international effort gave rise to the International Standards Organization (ISO) Standard 5840, *Cardiovascular Implants-Cardiac Valve* (7), which is currently being harmonized with the European Committee for Standardization (CEN) to produce a European Standard EN prEN12006. A major benefit of the European Union (EU) is that member countries agree to recognize a single set of standards, regulatory processes and acceptance criteria. Prior to the establishment of the European Union, each individual country imposed their own standards and regulatory requirements. The ISO committee hopes to ultimately merge with or at least to philosophically harmonize the ISO, EN requirements with the FDA Guidance Document requirements in order to produce a single international standard.

The regulatory approval process for a new heart valve design consists of two general stages, a preclinical study and a subsequent clinical study. The preclinical study consists of *in vitro* tests, orthotopic animal implants, and a survey of manufacturing quality assurance systems. When the preclinical requirements are satisfied, the FDA or EU agency may issue an Investigational Device Exemption (IDE) that allows human clinical trials. Clinical trials are conducted according to specified study designs, patient numbers, and evaluation methods. Upon the successful completion of clinical trials, the regulatory agency such as the FDA may issue a PMA, or equivalent EU approval, that allows the open marketing of the valve design. As a result of the importance and complexity of regulatory requirements in heart valve design, evaluation, manufacture, marketing, and monitoring, major heart valve companies now have dedicated "Regulatory Affairs" departments.

DEMOGRAPHICS AND ETIOLOGY

Patients require heart valve replacement for a number of reasons, which vary by sex, valve position, and the prevailing levels of socioeconomic development. Valves become dysfunctional because of stenosis, inadequate forward flow when open or insufficient and excessive backflow when closed. A summary of typical patient data by sex, lesion, and disease state that led to dysfunction is given in Table 1. This data is taken from PMA studies of the On-X mechanical valve conducted in North America and Western Europe (8,9).

Patients in developing countries tend to require valves at a younger age (mean 33 years) and have a significantly higher incidence of rheumatic disease and endocarditis. The proportion of aortic implants relative to mitral implants is higher in North America and Western Europe, whereas the proportion of mitral implants and double-valve implants is higher in the developing countries.

CLINICAL PERFORMANCE

Regulatory agencies (6,7) gage the results of observational clinical (IDE) trials in terms of Objective Performance

Table 1. Summary of Typical Patient Data

Category	Aortic	Mitral
Mean age years	60	59
Sex	%	%
Male	66	38
Female	34	62
Lesion	%	%
Stenosis	47	13
Insufficiency	21	48
Mixed/other	32	39
Etiology (can be > one per patient)	%	%
Calcific	50	16
Degenerative	28	27
Rheumatic	13	38
Congenital	10	2
Endocarditis	4	7
Previous prosthetic valve dysfunction	3	3
Other	3	16

Criteria (OPC). Ultimately, the performance of a new investigational device is compared with existing devices. The OPCs are linearized complication rate levels. An IDE study uses statistical methods to demonstrate that observed complication rates for the investigational device are less than two times the OPC rate. The linearized rates are given in units of percent per patient year (6,7) in Table 2. A patient year is simply the total post implant follow-up duration in years for patients enrolled in the study. For example, 100 patients followed one year each post implant is 100 patient years (pt-yr). For the occurrence of complication events, the linearized rate is 100% × (number of events/number of patient years). Currently, a followup of 400 patient years is the minimum FDA regulatory requirement for each valve position, aortic and mitral each.

Thromboembolism (blood clotting-related strokes) and hemmohrage (excessive bleeding) are measures of the stability of the valve design relative to hemostasis and chronic anticoagulation therapy. Perivalvular leaks occur at the interface between the valve sewing cuff attachment mechanism and the cardiac tissue and are a measure of how well the valve design seals. Endocarditis is any infection involving the valve. A number of other complications, or morbid events, including hemolysis, unacceptable hemodynamics, explant, reoperation, and death are also monitored and compared on the basis of percent per patient

Table 2. Linearized Rates in Units of Percent Patient Year

Complication	Mechanical Valve (%/pt-yr)	Biological Valve (%/pt-yr)
Thromboembolism	3.0	2.5
Valve thrombosis	0.8	0.2
All hemmhorage	3.5	1.4
Major hemmhorage	1.5	0.9
All perivalvular leak	1.2	1.2
Major perivalvular leak	0.6	0.6
Endocarditis	1.2	1.2

year. Other presentations of data often used are Kaplan Meier "Survival or Freedom from complications" plots or tables in which the percentage of the study population that has survived, or has not experienced a given complication, is plotted vs. implant duration in years (10).

HEMODYNAMIC EVALUATION

Methods of evaluating heart valve hemodynamic function, both natural and prosthetic, have advanced considerably in the past 30 years. With today's technology, most evaluations of valve performance can be performed noninvasively, and more information regarding valve design performance is available than ever before, both *in vivo* and *in vitro*. Early methods, such as ascultation, listening to valve sounds with a stethoscope, have been greatly improved on, and invasive catheter techniques are no longer needed except in special cases.

Cine fluoroscopy involves recording a fluoroscopic examination with film or digital recording media. It provides a direct dynamic visualization of radio-opaque prosthetic valve features and, thus, can be used to identify a valve type and to assess function. Cine fluoroscopy used in combination with a catheter delivered radio-opaque contrast media, cineangiography, provides a means for basic flow visualization. Noninvasive echocardiography, however, has become the most common and important method for assessing valve function.

The physics of echocardiography are simple; a beam of ultrasound (1–5 MHz) is aimed into the patient's chest and, as the propagating sound wave encounters the different tissue layers, some of the energy is reflected back. The existence and depth of the reflecting tissue layers are then inferred from the time history of the reflections relative to the incident beam and the strength of the reflection. The reflection time history and strength from an angular sweep are then used to construct a 2D tomographic image of the underlying tissue structures. Thus, cardiac structures and their motions are readily visualized.

Echocardiography also produces an angiographic image for visualizing flow by using the moving red blood cells as a contrast medium. A Doppler shift occurs between the frequency of the incident sound and the sound reflected by the moving red blood cells, which is in proportion to the cells' velocity. This information is used to map flow direction and velocity throughout the cardiac cycle. Furthermore, the Doppler frequency shifts can be rendered audible to produce sound signatures. Thus, echocardiography and Doppler echocardiography provide powerful visual and audible diagnostic tools to interrogate cardiac dynamic structure and function.

Although the above description of echocardiography is admittedly brief, the real complexity of echocardiography lies in transducer mechanics, sophisticated electronics, and computational techniques required to produce near real-time dynamic 2D images and flow maps. Further details are readily available; for example, a very accessible tutorial can be found at the http://www.echoincontext.com/basicEcho.asp "Introduction to Echocardiography," prepared by Duke University, 2000.

For the purposes of heart valve evaluation, echocardiography and Doppler echocardiography give dynamic visualizations of valve component motion and reasonable measurements of flow velocities, flow distributions valve effective orifice areas, and transvalvular pressure differences. Definitions and technical details of valve evaluations can be found in Appendix M of the FDA Document *Replacement Heart Valve Guidance - Draft Document, October 14, 1994* (6,7). A summary of the important echocardiographic parameters for valve evaluation are listed below.

1. Visualizations of valve motion and structure: Provides a diagnostic tool for both the native valve and valve prostheses and can detect the presence of thrombosis.

2. Doppler measurements of transvalvular pressure difference (or gradient): A stenotic valve, with impaired forward flow, either native or prosthetic, will have an elevated transvalvular pressure difference. The transvalvular gradient (peak or mean) is determined by measuring the forward flow velocity through the valve. If the proximal velocity is greater than $1 \text{ m} \cdot \text{s}^{-1}$, then it is also measured. Velocity values are diagnostic and can be used to derive an estimate of the transvalvular pressure difference using the complete (long) Bernoulli equation,

$$\Delta P = 4(V_2^2 - V_1^2),$$

where ΔP is the peak pressure difference mmHg, V_2 is the continuous-wave Doppler peak transvalvular velocity $(\text{m} \cdot \text{s}^{-1})$, and V_1 is the proximal pulse-wave Doppler peak velocity $(\text{m} \cdot \text{s}^{-1})$. If the proximal velocity is less than $1 \text{ m} \cdot \text{s}^{-1}$, as is usual in the mitral position, the V_1 term can be dropped and, as such, the single-term equation is called the short Bernoulli equation. Forward flow velocity-time envelopes may be integrated to give mean velocity values or mean pressure values depending on the echo system's processing capabilities. The mean velocity values are substituted into the Bernoulli equation, or the mean transvalvular and proximal pressures are subtracted to give the mean transvalvular pressure difference.

3. Effective orifice area: A stenotic valve, either native or prosthetic, will have decreased effective orifice area. Effective orifice area is determined using the classical continuity equation:

$$\text{EOV}(\text{cm}^2) = \text{CSA} \, vti_1/vti_2$$

CSA is the cross-sectional area of the left ventricular outflow tract (LVOT) for the aortic position. Variables vti_1 and vti_2 are the velocity-time integrals at the pulse-wave proximal LVOT vti_1 and transvalvular continuous wave for the aortic position vti_2. For the mitral position, CSA is the aortic cross-sectional area; vti_1 and vti_2 are the velocity-time integrals for the continuous-wave transvalvular velocities for the aortic valve, vti_1, and vti_2 for the mitral valve.

4. Regurgitation: Regurgitation or incompetence is graded as trivial, mild, moderate, or severe according to the regurgitant flow jet size expressed as a height or area relative to the LVOT area in the aortic position or left atrium in the mitral position. An incompetent, leaky valve will have severe regurgitation.

As a result of variations in patient body size and because the valve must fit the native valve annulus, prosthetic valve designs are prepared in annulus sizes ranging from an approximate 19 mm annulus diameter up to 33 mm diameter. However, valve sizes are not scalar in the sense that a size 19 valve or its sizer will often not be 19 mm. The valve and its associated sizer are supposed to fit a 19 mm annulus. For this reason, it is critical to use the sizers and associated instrumentation provided by the manufacturer for a specific valve design.

Typically, a design will have six sizes in 2 mm diameter increments. The size differences can consist of larger valve components or larger sewing cuffs. In general, it is desired to present the largest internal diameter possible for flow. Thus, it is important to recognize that hemodynamic quantities vary strongly with valve size in the aortic position and, for this reason, aortic hemodynamics comparisons must be interpreted according to size. One can find studies in the literature that report single values for a valve design that lump together and ignore the effects of valve size, the utility of such information is limited to the engineer, but meaningful to the cardiologist (11).

Patient body surface area (BSA) is often used to normalize valve effective orifice area, to give an orifice area $\text{cm}^2 \cdot \text{m}^{-2}$. EOA is measured using echocardiography and BSA is calculated using the patient height and weight. The ratio of valve EOA to patient BSA is known as the indexed orifice area (IEOA)

Patient prosthesis mismatch has recently been recognized as a potential error in selecting valves for patients. It is possible to replace a diseased aortic valve with a valve prosthesis that is too small for the patient body size. A generally accepted criteria for sizing is that the indexed effective orifice area should not fall below a value of 0.85 (12).

Left ventricular mass regression is another contemporary measure of valve effectiveness. Ventricular dimensions are measured from echocardiography for patients preoperatively and postoperatively. When overloaded by a dysfunctional valve, the heart responds with an increase in mass. An effective prosthetic valve lowers the work load on the heart and thus causes a regression back toward normal size (13).

The information presented above provides a basic framework for evaluating and comparing valve design clinical performance: complication rates and hemodynamics. Complication rates are assessed using statistics presented as linearized rates, percent per patient year. However, the linearized rates tend to assume that the risk of a complication is constant with time (14), which may not hold for all complications. Furthermore, it is necessary to compare linearized rates determined for comparable population demographics, implant experience, and complication definition. Short-term clinical experience study rates do not compare well with long-term clinical experience. For this reason, it is convenient to compare results from PMA pre-clinical studies. Such studies have comparable patient numbers, demographics, duration, statistics, and study methods.

Results from PMA preclinical studies can be found by searching a database for "Summary of Safety and Effectiveness" documents on the FDA website (http://www.fda.gov/cdrh/). A Summary of Safety and Effectiveness is prepared for each device that has received a PMA and contains a summary of the clinical study, the preclinical studies, and the manufacturing facilities. Furthermore, most valves in the United States are packaged with "Instructions for Use" that include the clinical complication and hemodynamics data. Most valve manufacturers have websites that either contain the pertinent data or provide relevant references to the literature because of the importance of this data for marketing purposes. In general, high survival, freedom from complications or low complication rates, low pressure gradients, high effective orifice areas, and low regurgitation are essential.

Much of the same type of information is also available in the open literature; however, because of differences in study design, methods, and size that occur, it is more difficult to compare results from different studies and sources. An excellent review of heart valve complication comparison problems can be found in a paper by G. Grunkmeier and Y. Wu, "Our complication rates are lower than theirs: Statistical critique of heart valve comparisons" (14). An excellent review of hemodynamics for a number of valves may also be found in a paper by Zang and Chambers (15). As with complication rates, hemodynamics comparisons based on studies in the open literature can be difficult because of differences in technique and quantity definitions.

Examples of valve comparisons with the above information will be presented in subsequent sections of this text.

VALVE DESIGN ELEMENTS

The challenge is to provide valve prostheses that can endure the aggressive biological and mechanical environment for a patients' lifetime without structural or functional degradation. A myriad of disciplines, including medicine, biology, engineering design, materials engineering, structural engineering, and fluid mechanics, are required to meet this challenge. Advances in each of these disciplines over the past 30 years have made possible significant improvements in the safety and efficacy of valve prostheses. At the current state of development, common structural features exist among the various valve designs required to meet the goals of function and durability. Conceptually, a valve prosthesis is a simple one-way valve that opens easily to allow forward flow and closes adequately to prevent regurgitant losses during the cardiac cycle. Each design has:

- An orifice body that defines the open forward flow channel.
- Leaflets or occluders that reversibly open to permit forward flow and close (or occlude) to prevent reverse flow.

- Struts or pivots that capture and guide leaflet motion during opening and closing.
- A sewing cuff or sewing ring that is used to attach the valve to the cardiac anatomy.

In addition, bioprosthetic valves may have stents, which are metallic or polymeric structures that define and support the soft tissue valve shape.

Many of the design elements of valves are radio-opaque and present a unique appearance upon examination by X-ray or cine fluoroscopy. Guides to identifying valves by radiographic appearance have recently been published by Butany et al. (16,17).

The size, shape, strength, durability, and biocompatibility of these common design elements define the performance of the valve. Each element is, in turn, defined by the materials used in its construction, hence, the importance of materials technology.

MATERIALS TECHNOLOGY

The materials available in many ways dictate design possibilities. Our repertoire of manmade materials for mechanical valves includes certain polymers, metals, and pyrolytic carbons in a class generally known as biomaterials. For heart valve applications, the materials used must be blood- and tissue-compatible. The presence of the material itself, wear particles, or the material in combination with others cannot provoke adverse biological reactions. Some of the more important adverse biological reactions include intense inflammatory response, carcinogenicity, mutagenicity, or the inception of thrombosis.

The materials must be biodurable: The mechanical integrity must resist degradation by the *in situ* physiochemical environment. Assuming that the human heart beats approximately 40 million times a year, valve prostheses must endure on the order of 600 million cycles over a 15 year period without functional degradation due to exposure to the aggressive biological reactions and repeated cyclic stresses in the cardiac environment. Early valve designs failed within a short period because of thrombosis (blood clotting), and damage due to exposure to the rigorous biological environment manifest as distortion, wear, and fatigue. Additionally, the materials must tolerate sterilization without adverse affects. A more current requirement is that a material cannot interact adversely with high strength electromagnetic radiation, as may be encountered with modern diagnostic techniques such MRI (magnetic resonance imaging).

The short list of materials that meet these requirements for heart valve structural components includes isotropic pyrolytic carbons, certain cobalt-chrome alloys, and certain titanium alloys. The polymers most often used today in the fabric sewing cuffs are polyethylene terephthalate (PET, tradename Dacron) and polytetrafluoroethylene (PTFE, tradename Dacron). Polyacetyl (polyformaldehyde, tradename Delrin) and poly dimethyl siloxane (Silicone rubber) are used as occluders in some of the older designs. Some important physical and mechanical characteristics of these materials are given in Table 3.

Table 3. Representative Material Properties

Property	Unit	PyC	CoCr	Delrin
Density	$g \cdot cm^{-3}$	1.93	8.52	1.41
Bend strength	MPa	494	690 (uts)	90
Young's modulus	GPa	29.4	226	3.1
Hardness	HV	236[a]	496	86 Shore D
Fracture toughness K_{1c}	$MN \cdot m^{3/2}$	1.68		
Elongation at failure	%	2		30
Poisson's ratio		0.28	0.3	0.35

[a]The hardness value for PyC is a hybrid definition that represents the indentation length at a 500 g load with a diamond penetrant indentor.

As the mechanical properties of biological materials are more complex than for manmade materials, they will not be addressed here, rather references are suggested in the recommended reading section.

Each material used in a prosthetic heart valve must meet a battery of tests for biocompatibility as currently defined in the FDA Guidance document (8) ISO 5840 and ISO 10993. The interested reader can find the details of these tests in the reference documents.

DESIGN CONCEPTS

Two broad categories of valve prostheses have evolved, mechanical and biological, each with advantages and disadvantages. The selection of valve type for a given recipient is made by balancing the individual's needs relative to the valve type advantages and disadvantages. Factors for consideration in valve type selection include:

- Thromboembolism/Stroke,
- Bleeding Risk,
- Reoperation,
- Survival.
- Guidelines from The American Heart Association/ American College of Cardiology (AHA/ACC) are listed below in Table 4 (18).

Table 4. AHA/ACC Guidelines[a]

AHA/ACC Indications for Valve types	
Mechanical Valve	Tissue Valve
Expected long lifespan	AVR >65 years age, No risk factors for TE
Mechanical valve in another position	MVR >70 years age
AVR ≤ 65 years age	Cannot take warfarin
MVR ≤70 years age	
Requires warfarin due to risk factors	
Atrial fibrillation	
LV Dysfunction	
Prior thromboembolism	
Hypercoagulable condition	

[a]AVR –aortic valve replacement MVR – mitral valve replacement.

In the absence of concomitant disease, aortic mechanical valves are recommended for patients less than 65 years of age and mitral mechanical valves for patients less than 70 years of age.

Biological valves offer the advantage that chronic anticoagulation therapy (warfarin) may not be required beyond an aspirin taken daily. Many patients benefit from biological valves because of an inability to tolerate or to comply with chronic anticoagulation. However, many patients with biological valves may require anticoagulation for other concomitant conditions, such as atrial fibrillation that pose a coagulation risk, and some physicians may recommend anticoagulation for mitral replacements.

The disadvantage is that biological valves have limited lifetimes for reasons inherent to host responses that degenerate the implant tissue and the relatively low fatigue resistance of the devitalized biological tissues. Expected patient survival (median) in the United States equals expected biological valve survival at 18 years for aortic valves and 69 years for mitral valves (19).

The risk of biological valve dysfunction increases inversely with age, with failure being more rapid in younger patients. Degenerative mechanisms such as calcification are accelerated in young growing patients, and an active lifestyle may more severely stress the valve. As a mitigating factor, biological valve degeneration tends to be gradual, hence readily detected so that a re-operation to replace the valve can be scheduled before the patients' health becomes compromised. Here, the tradeoff is the risk of re-operation vs. the risk of anticoagulant-related complications. Therefore, the ideal biological valve candidate patient tends to be an elderly person with a relatively sedentary lifestyle and an expected lifetime that is not likely to exceed the useful lifetime of the valve.

A mechanical valve has the advantage of unlimited durability, but has the disadvantage of requiring chronic anticoagulant therapy.

Many excellent resources are available on the Internet providing illustrations, some design, details, and performance details for valve prostheses. A general overview of valves and other devices approved for marketing in the United States can be found at the Cardiovascular and Thoracic Surgery website (http://www.ctsnet.org/sections/industryforum/products/index.cfm). Other very accessible resources include the Evansville Heart Center prosthetic heart valve gallery (http://members.evansville.net/ict/prostheticvalveimagegallery.htm) and the Cedar Sinai Medical Center (http://www.csmc.edu/pdf/Heart Valves.pdf). Heart valve manufacturers also have websites that provide information regarding valves and performance. A partial list follows:

- ATS Medical, Inc. http://www.atsmedical.com/
- Carbomedics, A Sorin Group Company, http://www.carbomedics.com/, and Sorin Biomedica, http://www.sorin-cid.com/intro_valves.htm
- Cryolife, Inc., http://www.cryolife.com/products/cryo-valvenew.htm.
- Edwards Lifesciences, http://www.edwards.com/Products/HeartValves

- Medical Carbon Research Institute, LLC., http://www.onxvalves.com/
- Medical CV, http://www.vitalmed.com/products/medicalcv/omnicarbon_heart_valve.htm
- Medtronic, Inc., http://www.edwards.com/Products/HeartValves
- St. Jude Medical, Inc., http://www.sjm.com/conditions/condition.aspx?name=Heart+Valve+Disease&section=RelatedFeaturesProducts

Many other resources are available, other manufacturers, a number of professional journals and archives of news stories, and press releases available on the web. A web search for "artificial heart valves" will yield more than 300,000 sites. The sites listed above are given as a simple quick start for the reader, without implied endorsement.

As a point of interest, the valve manufacturers' websites are highly colored by competitive marketing and tend to present a rather focused view. However, they often provide copies of studies in the literature along with supplemental interpretative information. As an example, aspects of the selection of type, biological and mechanical, for a given patient are controversial. To address this controversy, the Edwards Lifesciences and St. Jude Medical sites have detailed, authoritative panel discussions of issues related to valve selection that are well worth reviewing.

BIOLOGICAL VALVES

Bioprosthetic valve replacements tend to mimic the semilunar aortic valve directly by using a natural aortic valve from a donor (allograft) or an aortic valve from another species (xenograft). Alternatively, the natural aortic valve may be imitated by design using a combination of nonvalvular natural tissues and synthetic materials (heterograft). Typically, the semilunar aortic-type valve is used in all positions for valve replacement.

The native aortic valve can be visualized as a short flexible tube within a longer tube. The outer tube, the aorta, has three symmetric scalloped bulbous protuberances (sinus of Valsalva) at its root. Within the sinus of Valsalva, there is the short inner tubular aortic valve, with a length approximately the same as the diameter. The aortic valve tube is constrained at three equidistant points along its upper circumference at the commissures. When pressurized from above, the aortic valve tube wall collapses onto itself into three symmetric crescent-shaped leaflet cusps. Along the lower circumference, the leaflets attach to the aorta wall following the scalloped curve of the aortic root. When pressurized from below, the valve opens, the cusps rise together and separate to form the short open internal tube, which is the flow region of the valve. When closed, the internal tube walls, cusps, collapse, inflate, and form a seal (coaptation) at the center of the valve (nodus Aranti) along lines 120° apart (see Fig. 1). A good dynamic, multimedia visualization of this collapsing tube concept of the aortic valve can be found at the 3F Technologies website (http://www.3ftherapeutics.com/us/products.html).

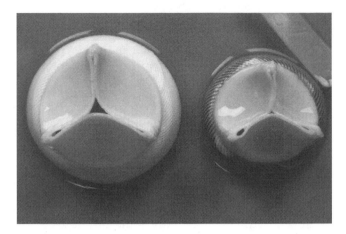

Figure 1. Mitroflow bovine pericardial prostheses heterograft shown in two sizes. This figure illustrates the collapsed tube concept of aortic valve structure.

Valves constructed from biological materials have several categories, including:

- Homograft (also Allograft), native valves taken from members of the same species,
- Xenograft (also Heterograft), valves taken from other species,
- Autograft, valves transposed from one position to another in the same patient. Valves may also be constructed from nonvalvular tissues such as pericardium, dura mater, or facia lata and the tissues may be autogenic, allogenic, or xenogenic. Valves may be stented (with added rigid support structures) or unstented.

Valves transplanted from one individual to another do not remain vital. Although the valve may be initially coated with endothelium, the valve eventually becomes acellular with only the extracellular matrix collagenous structures remaining. A biological tissue originating from another species (xenograft) must be killed and stabilized by aldehyde cross-linking or some other method. Fixation masks autoimmune rejection processes and adds stability and durability. However, the tissue again consists of an acellular, but now cross-linked, extracellular matrix collagenous structure. Biological valves lifetimes are limited due to degradation of the relatively fragile nonvital, acellular material, which results in structural and functional failure. Degradation processes are exeraberated by calcification and elevated applied stress levels in the cardiovascular environment.

Valve Homografts (Also Allograft)

Use of an orthotopic implanted homologous cardiac valve (e.g., a valve transplant from the same species) as a valve substitute was first successfully accomplished in humans by Donald Ross in 1962 (20). Harvest must occur as rapidly as possible following the donor's death, if within 48 h, the valve may still be vital. However, with human or animal origin, a risk of disease transmission exists, particularly if the valve is not sterilized. Furthermore, competition exists with the need for intact donor hearts for heart transplantation, so the supply of homografts is limited.

During the period between 1962 and 1972, homografts were widely used because other valve substitutes were not yet satisfactory. As other valve substitutes became effective and more readily available during the 1970s, homograft use decreased. With the development of cryo-preservation technology in the mid-1980s and commercialization by Cryolife, homograft use has increased again. However, unless the homograft is an autograft (from the same patient), it becomes a devitalized acellular, collagenous structure within the first year of implantation (21). Homograft lifetime in the aortic position tends to be limited to approximately 10 years due to primary tissue failure because of the inherent fragility of the devitalized, acellular homograft.

The most successful use of homografts is for a pulmonary valve replacement in the Ross procedure in which the patient's own viable pulmonary valve (autograft) is transplanted into the aortic position as a replacement for diseased aortic valve and a heterograft used to replace the patient's pulmonary valve (22). In principal, the autograft pulmonary valve in the aortic position remains vital and grows with the patient. The heterograft in the pulmonary position endures well because the pressure loads and the consequences of regurgitation are much less severe on the right side of the heart.

Xenografts

Aldehyde-fixed porcine xenograft valves were initially developed in 1969 by Carpentier et al. (1), which consisted of porcine aortic valves mounted on a stent frame to maintain shape and then fixed with dilute gluteraldehyde (see Fig. 2). Aldehyde fixation masks autogenic host reactions, cross-links the extracellular matrix collagen, and kills the native valve cells. Infectous cells and viri are also cross-linked and killed, so fixation provides a means of disinfection. During the same time period, Hancock et al.

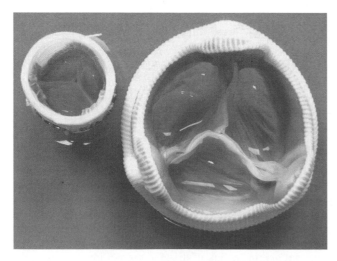

Figure 2. A porcine valve xenograft design (Xenomedica) shown in two sizes. Here, a porcine valve is mounted in a stented structure.

independently developed another porcine prosthesis (23). Yet another design, constructed from bovine pericardium, was developed by Ionescu in 1971 (24). These valves provided readily available alternatives to homografts and were widely used during the early 1970s. Although these valves offered freedom from chronic anticoagulation, lifetimes were limited to around 10 years. The valves failed because of calcification, leaflet tears, and perforations. Freedom from reoperation 15 years post implant was about 50% for the porcine valves and slightly worse for the bovine pericardial valves. Other important lessons were learned the Hancock valve polymeric Delrin stents would creep, leading to valve dysfunction. With the Ionescu valve, a suture located at the valve commissures initiated tears in the leaflets. The attachment of leaflets at the commisures tends to be the most highly stressed point, because of pressure and reversed bending stresses, which leads to commissural tearing.

Xenograft biological valve replacements have gone through three generations of evolution. The first generation and some of its problems are described above by Hancock and Ionescu. Tissues used in these valves were fixed with aldehydes at high pressure in the closed position. As the collagen fibers within the porcine leaflets tend to have a wavy or crimped structure, the high pressure fixation tended to straighten the fibers out, removing the crimp, which was found to have an adverse effect on fatigue endurance.

For the second generation valves, low and zero pressure aldehyde fixation were used, flexible stent materials were employed, and assembly strategies improved. These improvements retained the natural collagen crimp and lowered stresses at the commissural attachment. Also, additional fixation treatments and chemical posttreatments were employed in hopes of reducing calcification.

Contemporary third-generation-valves tend to use fixation at physiological pressures, flexible stents materials, and include stentless designs. Some of these newer aortic valve designs are challenging the age limits given above for valve-type selection (25). A number of stentless designs are available that typically consist of a xenograft or homograft valve mounted in fabric or tissue to enable sewing (see Fig. 3). Some designs include the portions of the aortic arch along with the valve. Stenting and annular support structures tend to decrease the annular flow area. Removal of the stents should enhance hemodynamic performance.

A unique design using equine pericardium and an external expandable stent for use with a transcatheter delivery system is under development by 3F Theraputics. Furthermore, tissue engineering concepts are being explored as a means of developing valve scaffolding structures in expectation that the host cells will infiltrate, populate, and ultimately render the scaffolding into an autologous tissue structure. One such example is a chemically decellularized valve in development by Cryogenics. Carbomedics has also developed a method for nonaldehyde cross-linking and decellularization of collageneous cardiac tissues (26).

Hemodynamics for some of the contemporary biological valves cited from FDA PMA Summaries of Safety and

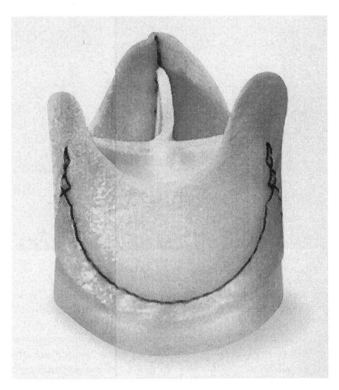

Figure 3. A stentless pericardial aortic valve. Sorin Biomedica Pericarbon.

Effectiveness and the literature for uniform echocardiographic protocols and comparable patient experience are listed in Tables 5 and 6. All of the contemporary biological valves exhibit satisfactory hemodynamics and acceptably low risk of valve-related complications.

MECHANICAL VALVES

Contemporary mechanical valves are primarily bileaflet valve designs constructed from pyrolytic carbon (PyC).

Some monoleaflet valves and ball in cage valves, such as the Medtronic Hall valve and the Starr-Edwards valve, are still used. Some examples of bileaflet and monoleaflet designs are shown in Fig. 4. The bileaflet designs consist of two flat half circular leaflets inserted into a cylindrical annulus. The orifice may be stiffened by external metallic rings and is encircled by a fabric sewing cuff. Small depressions, or extrusions, within the orifice accept mating features on the leaflet and provide a pivot mechanism to capture and guide the leaflets during opening and closing excusions. Figure 5 provides a cut-away illustration of general bileaflet valve features. Figure 6 provides a high detail view of a hinge pivot.

PyC is used almost exclusively in bileaflet design structural components because of excellent blood compatibility, biostability, strength, and fatigue endurance. This biomaterial is an isotropic, fine-grained turbostratic carbon that is typically prepared as a coating on an inner graphite substrate. Most PyC valve components are on the order of slightly less than 1 mm in thickness and have a low enough density so as to be easily moved by cardiovascular flow and pressure. Two general types are available, silicon-alloyed and pure carbon; both have roughly equivalent and adequate properties for heart valve application. The history of PyC applications in heart valves and its material science is another story in itself and will not be addressed here. Rather, sources about PyC can be found in the recommended reference list.

Supraannular placement has been the most prominent design evolution in contemporary mechanical valves. Originally, the sewing cuff girdled the mechanical valve orifice at approximately mid-height and was intended to fit entirely within the native valve annulus. However, inserting the bulk of the sewing cuff and valve into the native annulus reduces the available flow area. Supraannular designs are modifications to the sewing cuff and valve orifice exterior wall that allow the sewing cuff to be positioned above the annulus, which removes the sewing cuff bulk from within the native annulus and, thus, increases the available flow area.

Table 5. Bioprosthetic Aortic Valve Pressure Gradients (mmHg)

Valve/Size		19	21	23	25	27	29	PMA
Toronto	Stentless	577	10 ± 9.0	7.3 ± 4.8	6.4 ± 5.1	5.1 ± 3.1	3.8 ± 2.3	P970030
Freestyle	Stentless	631	11.7 ± 4.7	9.8 ± 7.4	8.8 ± 6.8	5.1 ± 3.3	4.4 ± 2.9	P970031
Prima Plus	Stentless	366	15.9 ± 7.0	9.9 ± 5.7	8.8 ± 4.9	6.5 ± 3.9		P000007
SAV	Stented	337	12.7 ± 4.2	10.5 ± 4.3	11.3 ± 5.5	8.3 ± 3.3		P010041
Mosaic	Stented	1252	14.5 ± 5.3	12.8 ± 5.0	11.8 ± 5.2	10.0 ± 4.0	10.3 ± 2.6	P990064
Hancock II	Stented	205	12.9 ± 4.2	13.2 ± 4.6	11.3 ± 4.4	11.7 ± 4.8	10.5 ± 3.6	P980043

Table 6. Bioprosthetic Aortic Valve Effective Orifice Areas (cm²)

Valve/Size		19	21	23	25	27	29	PMA
Toronto	Stentless	577	1.3 ± 0.7	1.5 ± 0.6	1.7 ± 05	2.0 ± 0.6	2.5 ± 0.8	P970030
Freestyle	Stentless	631	1.1 ± 0.3	1.4 ± 0.4	1.7 ± 0.5	2.0 ± 0.5	2.5 ± 0.7	P970031
Prima Plus	Stentless	366	1.1 ± 0.4	1.6 ± 0.5	1.9 ± 0.4	2.1 ± 0.6		P000007
SAV	Stented	337	1.3 ± 0.4	1.5 ± 0.4	1.7 ± 0.5	1.8 ± 0.6		P010041
Mosaic	Stented	1252	1.3 ± 0.4	1.5 ± 0.4	1.8 ± 0.5	1.9 ± 0.6	2.2 ± 0.7	P990064
Hancock II	Stented	205	1.4 ± 0.5	1.3 ± 0.2	1.4 ± 0.4	1.6 ± 0.4	1.4 ± 0.3	P980043

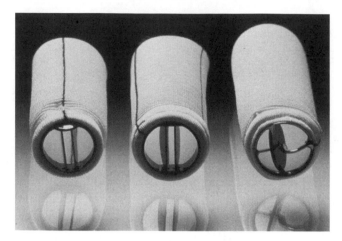

Figure 4. Three mechanical valve designs shown attached to aortic grafts. Left to right: St. Jude Medical bileaflet, Carbomedics bileaflet, and Medtronic-Hall monodisk.

Effectively, a supraannular sewing cuff allows the placement of an upsized valve relative to the intraannular cuff. Most often, the valve orifice and leaflet components are unchanged, only the sewing cuff is modified. Thus, supraannular cuffs are a subtle modification in that the renamed supraannular design is not really a new design, rather it is just an upsized version of the original valve. Virtually every bileaflet valve manufacturer now has supraannular cuffs. Renaming the supraannular versions has caused a bit of confusion because it is not always obvious that improved hemodynamic performance is due solely to the placement of an upsized valve of the same design. However, two recent designs, the Medical Carbon Research Institute On-X valve and the SJM Regent, actually incorporate significant design features as supraannular valves beyond a modified sewing cuff.

Hemodynamics for some contempory mechanical valves are given in Tables 7 and 8 (27–38). All of the contemporary mechanical valves exhibit satisfactory hemodynamics and acceptably low risk of valve-related complications. As a

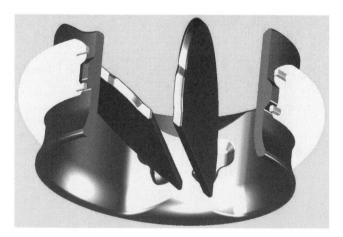

Figure 5. A cut-away view of the On-X valve showing leaflets in the closed (left) and open (right) positions. The valve orifice or annulus ring has pivot depressions in the inside diameter wall and is encased in a sewing cuff. The orifice wall has a bulge and two external metal wires for stiffening.

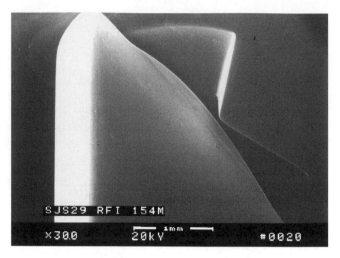

Figure 6. Detail of the St. Jude Medical bileaflet valve pivot mechanism. A tab on the leaflet inserts into a butterfly shaped depression in the valve annulus ring. The pivot depression retains and guides the leaflet motion during opening and closing excusions.

point of interest, only aortic hemodynamics are presented for both the biological and mechanical valve designs, because clinical data for aortic valves demonstrates a strong dependence on valve size. Effective orifice areas increase and pressure gradients decrease with increasing valve annulus size. Clinical hemodynamic data for mitral valves typically does not depend as strongly on annulus size (15). Mitral sizes range from 25 to 33 mm annulus diameters. In many designs, the larger annulus diameter sizes, above 27–29 mm, are attained by increasing the sewing cuff bulk. However, the difference between effective orifice area and pressure gradients with size is minimal. For mechanical mitral valves, mean pressure gradients are on the order of 4 mmHg and effective orifice areas of 2.5–3 cm^2.

COMPARISON OF BIOLOGICAL AND MECHANICAL

Detailed comparisons of biological and mechanical valve experience in large multicenter randomized studies have been recently published (39–41). Freedom from complications were found to be roughly comparable. Rates for thromboembolism and stroke have been found to be the same, and strongly related to nonvalve factors. Bleeding is more common for mechanical valves, but depends on the anticoagulant control and the use of aspirin. Freedom from reoperation is significantly better for mechanical valves. At 15 years, the freedom from reoperation for mechanical valve patients exceeded 90%, whereas while for biological valves, the percent free from reoperation was 67% for the aortic position and 50% for the mitral position. Survival at 15 years was slightly better with aortic mechanical valves, mortality of 66 versus 79% for biological valves. No difference exizted in survival for mitral valves. Overall, the survival of biological and mechanical valve patients was similar over extended periods, with the tradeoff being an increased risk of hemorrhage with mechanical valves vs. and increased risk of reoperation for biological valves. This risk of reoperation for biological valves increases in time. Thus, valve type selection is a balance

Table 7. Mechanical Aortic Valve Pressure Gradients (mmHg)

Valve/Size		19	21	23	25	27	29	PMA
On-X	SA	8.9 ± 3.0	7.7 ± 2.9	6.7 ± 3.1	4.3 ± 2.4	5.6 ± 0.3	5.6 ± 0.3	P000037
SJM Regent	SA	9.7 ± 5.3	7.6 ± 5.2	6.3 ± 3.7	5.8 ± 3.4	4.0 ± 2.6		[a]
SJM HP	SA	13.6 ± 5	12.6 ± 6.5	13 ± 6				[a]
SJM		17 ± 7	15.1 ± 3.2	16 ± 6	13 ± 6	11.5 ± 5	7 ± 1	[a]
ATS		20.2 ± 2.8	18.0 ± 1.6	13.1 ± 0.8	11.1 ± 0.8	8.0 ± 0.8	7.8 ± 1.1	P990046
CMI		21.7 ± 9.1	16.2 ± 7.9	9.9 ± 4.2	10.5 ± 2.8	7.2 ± 3.9	5.1 ± 2.8	P00060
Sorin Bicarbon		14.1 ± 2.9	10.1 ± 3.3	7.7 ± 3.3	5.6 ± 1.6			[b]
Omnicarbon				19 ± 8	16 ± 8	12 ± 4		P8300039

[a]See Refs. 27–36.
[b]See Refs. 37,38.

Table 8. Mechanical Aortic Valve Effective Orifice Areas (cm^2)

Valve/Size		19	21	23	25	27	29	PMA
On-X	SA	1.5 ± 0.3	1.9 ± 0.5	2.4 ± 0.7	2.7 ± 0.7	2.9 ± 0.7	2.9 ± 0.7	P000037
SJM Regent	SA	1.6 ± 0.4	2.0 ± 0.7	2.2 ± 0.9	2.5 ± 0.9	3.6 ± 1.3		[a]
SJM HP	SA	1.25 ± 0.2	1.3 ± 0.3	1.8 ± 0.4				[a]
SJM		0.99 ± 0.2	1.3 ± 0.2	1.3 ± 0.3	1.8 ± 0.4	2.4 ± 0.6	2.7 ± 0.3	[a]
ATS		1.2 ± 0.3	1.5 ± 0.1	1.7 ± 0.1	2.1 ± 0.1	2.5 ± 0.2	3.1 ± 0.4	P990046
CMI		0.9 ± 0.3	1.3 ± 0.4	1.4 ± 0.4	1.5 ± 0.3	2.2 ± 0.7	3.2 ± 1.5	P00060
Sorin Bicarbon		0.8 ± 0.2	1.1 ± 0.2	1.6 ± 0.2	2.4 ± 0.3			[b]
Omnicarbon				1.8 ± 0.9	1.9 ± 0.8	2.5 ± 1.4		P8300039

[a]See Refs. 27–36.
[b]See Refs. 37,38.

between the patient ability to tolerate chronic anticoagulation therapy and the risk of re-operation. To reiterate the AHA/ACC criteria, in the absence of concomitant disease, aortic mechanical valves are recommended for patients less than 65 years of age and mitral mechanical valves for patients less than 70 years of age.

IMPROVEMENTS IN ANTICOAGULATION AND DURABILITY

As mentioned earlier, although contemporary PyC mechanical valves have virtually unlimited durability and extremely low incidences of structural failure, the biocompatibility is not perfect. Hence, because of imperfect biocompatibility, chronic anticoagulation therapy is required, which is the disadvantage of mechanical valves. Anticoagulation therapy carries with it the risk of hemorrhage, thromboembolism, and thrombosis.

Valve thrombogenicity can be thought of in terms of an extrapolation of Virchow's triad from veins to valves. Here, a predisposition to thrombogenicity is attributed to three factors, (1) the blood compatibility of the artificial valve material, (2) the hemodynamics of blood flow through the valve, and (3) patient specific hemostasis. In effect, resistance to valve thrombogenicity occurs through a balance of all three factors. Conversely, valve thrombosis could be provoked by a poor material, poor hemodynamics, or a patient-specific hypercoagulable condition.

Improvements are being made in the need for and control of anticoagulant therapy for mechanical valve patients. Anticoagulant-related complications have been shown to be reduced by the use of INR (international normalized ratio) self-monitor units (42). The success of

INR self-monitoring leads to the possibility of reduced anticoagulant levels, which affects the patient-specific hypercoagulable condition of Virchow's triad. Valve manufacturers have also made improvements in material quality and hemodynamics to the extent that aspirin-only and reduced warfarin level studies are planned and in place for certain low risk aortic mechanical valve patients (43).

On the other hand, concurrent improvements in biological valve design and tissue treatments can lead to extended durability, which reduces the risk of re-operation. However, comparison of Tables 5 and 8 for effective orifice areas shows that biological valves tend to be more stenotic than mechanical valves. In the interest of avoiding patient-prosthesis mismatch, smaller patients tend to be better candidates for biological valves. Valve selection should ultimately be a matter of patient and surgeon preference.

FUTURE DIRECTIONS

Advances in tissue engineering probably hold the brightest promise, because a viable, durable, autologous valve is the best possible valve replacement. However, until realized, improvements in materials technology, anticoagulant therapy, and valve hemodynamics can all be expected to improve the outcome of valve replacements. Developments in techniques for valve annuloplasty and valve repair (44) have been highly successful in allowing the surgical restoration of the native valve, which often eliminates the need for a valve replacement. In fact, mitral valve implant rates have decreased with the popularity of valve repair. Strides are also being made in minimally invasive techniques, robotic surgery, and transcatheter surgery so as to minimize the risk and trauma of valve replacement surgery.

CONCLUSION

Valve replacements are a commonplace occurrence worldwide, and many advances in design and patient management have been made over the past 30 years. Current replacements are certainly safe and effective. Surgical outcomes and patient quality of life can only be expected to improve.

BIBLIOGRAPHY

Cited References

1. Carpentier A, et al. Biological factors affecting long-term results of valvular heterografts. J Thorac Cardiovas Surg 1969;58:467–482.
2. LaGrange LD, Gott VL, Bokros JC, Ramos MD. In: Johnson Hegyeli R, editor. Compatibility of Carbon and Blood, Chapter 5. Artificial Heart Program Conference, Washington, DC: National Heart Institute Artificial Heart Program; June 9–13,1969; 47–58.
3. Roe B. Chapter 13. Extinct. In: Bodnar E, Frater R, editors. Cardiac Valve Prostheses, Replacement Cardiac Valves, New York: Pergamon Press; 1969; 307–332.
4. Dewall RA, Qasim N, Carr L. Evolution of mechanical heart valves. Ann Thorac Surg 2000;69:1612–1621.
5. Brewer L, Prosthetic Heart Valves, Springfield (IL): C.C. Thomas; 1969.
6. Replacement Heart Valve Guidance—Draft Document, October 14, 1994. Available at http://www.fda.gov/cdrh/ode/3751.html.
7. ISO Standard 5840, Cardiovascular Implants—Cardiac Valve.
8. Summary of Safety and Effectiveness Data, On-X Prosthetic Heart Valve Model ONXA, P000037, May 30, 2001.
9. Summary of Safety and Effectiveness Data, On-X Prosthetic Heart Valve Model ONXM and ONXMC, P000037/S1, Marxh 6, 2002.
10. Hammermeister K, Sethi GK, Henderson WG, Grover FL, Oprian C, Rahimtoola SH. Outcomes 15 years after valve replacement with a mechanical versus a bioprosthetic valve: Final report of the Veterans Affairs randomized trial. J Am Coll Cardiol 2000 Oct;36(4):1152–1158.
11. Walther T, Lehmann S, Falk V, Kohler J, Doll N, Bucerius J, Gummert J, Mohr FW. Experimental evaluation and early clinical results of a new low-profile bileaflet aortic valve. Artif Organs. 2002 May;26(5):416–419.
12. Pibarot P, et al. Impact of prosthesis-patient mismatch on hemodynamics and asymptomatic status, morbidity after aortic valve replacement with a bioprosthetic heart valve. J Heart Valve Dis 1998;7:211–218.
13. Walther T, et al. Left ventricular remodeling after stentless aortic valve replacement. In: Huysmans H, David T, Westaby S, editors. Stentless Bioprosthesis. Oxford (UK): Isis Medical Medica Ltd; 1999; p 161–165.
14. Grunkemeir GL, Wu Y. Our complication rates are lower than theirs: Statistical critique of heart valve comparisons. J Thorac Cardiovas Surg 2003;125(2):290–300.
15. Wang Z, Grainger N, Chambers J. Doppler echocardiography in normally functioning replacement heart valves: A literature review. J Heart Valve Dis 1995;4:591–614.
16. Butany J, Fayet C, Ahluwalia MS, Blit P, Ahn C, Munroe C, Israel N, Cusimano RJ, Leask RL. Biological replacement heart valves. Identification and evaluation. Cardiovasc Pathol 2003 May–Jun; 12(3):119–39.
17. Butany J, Ahluwalia MS, Munroe C, Fayet C, Ahn C, Blit P, Kepron C, Cusimano RJ, Leask RL. Mechanical heart valve prostheses: Identification and evaluation. Cardiovasc Pathol 2003 Jan–Feb;12(1):1–22.
18. Bonow R, et al. ACC/AHA Guidelines for the Management of Patients With Valvular Heart Disease. Executive Summary. A report of the American College of Cardiology/American Heart Association Task Force on Practice Guidelines (Committee on Management of Patients With Valvular Heart Disease). J Heart Valve Dis 1998 Nov;7(6):672–707.
19. Khan S. Long-term outcomes with mechanical and tissue valves. J Heart Valve Dis 2002 Jan; 11 (Suppl 1): S8–S14.
20. Ross DN. Homograft replacement of the aortic valve. Lancet 1962 2:447.
21. Koolbergen DR, Hazenkamp MG, de Heer E, Bruggemans EF, Huysmans HA, Dion RA, Bruijn JA. The pathology of fresh and cryopreserved homograft heart valves: An analysis of forty explanted homograft valves. J Thoracic Cardiovas Surg; 124;4:689–697.
22. Ross DN. Replacement of the aortic and mitral valves with a pulmonary autograft. Lancet 1967;2:956–958.
23. Reis RL, Hancock WD, Yarbrough JW, Glancy DL, Morrow AG. The flexible stent: A new concept in the fabrication of tissue heart valve prostheses. J Thoracic Cardiovas Surg 1971;62:683–689.
24. Masters RG, Walley VM, Pipe AL, Keon WJ. Long-term experience with the Ionescu-Shiley pericardial valve. Ann Thorac Surg 1995;60(Suppl):S288–S291.
25. Glower DD, et al. Determinants of 15-year outcome with 1,119 standard Carpentier-Edwards porcine valves. Ann Thorac Surg 1998;66(Suppl):S44–S48.
26. Moore M. PhotoFix: Unraveling the mystery. J Long Term Eff Med Implants 2001;11(3–4):185–197.
27. Bach DS, Goldbach M, Sakwa MP, Petracek M, Errett L, Mohr F. Hemodynamic and early performance of the St. Jude Medical regent aortic valve prosthesis. J Heart Valve Dis 2001;10:436–442.
28. Lercher AJ, Mehlhorn U, Muller-Riemschneider, Rainer de Vivie E. In vivo evaluation of the SJM regent valve at one-year follow up after aortic valve replacement. The Society For Heart Valve Disease FIRST BIENNIAL MEETING Queen Elizabeth II Conference Centre, London June 15–18; 2001; Abstract 101.
29. Prasad SU, Prendergast B, Codispoti M, Mankad PS. Evaluation of a new generation mechanical prosthesis: Preliminary results of the St. Jude regent aortic prosthesis. The Society For Heart Valve Disease FIRST BIENNIAL MEETING Queen Elizabeth II Conference Centre, London June 15–18;2001; Abstract 104.
30. Chafizadeh E, Zoghbi W. Doppler echocardiographic assessment of the St. Jude Medical prosthetic valve in the aortic position using the continuity equation. Circulation 1991; 83:213–223.
31. Flameng W, Vandeplas A, Narine K, Daenen W, Herijgers P, Herregods M. Postoperative hemodynamics of two bileaflet valves in the aortic position. J Heart Valve Dis 1997;6:269–273.
32. Kadir I, Izzat M, Birdi I, Wilde P, Reeves B, Bryan A, Angelini G. Hemodynamics of St. Jude Medical prostheses in the small aortic root: In vivo studies using dobutamine Doppler echocardiography. J Heart Valve Dis 1997;6:123–129.
33. Zingg U, Aeschbacher B, Seiler C, Althaus U, Carrel T. Early experience with the new masters series of St Jude Medical heart valve: In vivo hemodynamics and clinical results in patients with narrowed aortic annulus. J Heart Valve Dis 1997(6):535–541.
34. De Paulis R, Sommariva L, De Matteis G, Polisca P, Tomai F, Bassano C, Penta de Peppo A, Chiariello L. Hemodynamic performance of small diameter carbomedics and St. Jude valves. J Heart Valve Dis 1996;(5: SIII):S339–S343.

35. Carrel T, Zingg U, Jenni R, Aeschbacher B, Turina M. Early in vivo experience with the hemodynamic plus St. Jude Medical valve in patients with narrowed aortic annulus. Ann Thorac Surg 1996;61:1418–1422.

36. Vitale N, et al. Clinical evaluation of St. Jude Medical hemodynamic plus versus standard aortic valve prostheses: The Italian multicenter, prospective, randomized study. J Thorac Cardiovas Surg 2001;122(4):691–698.

37. Flameng W, et al. Postoperative hemodynamics of two bileaflet heart valves in the aortic position. J Heart Valve Dis 1997; 6:269–273.

38. Kadir I, Wan IY, Walsh C, Dip-Rad, Wilde P, Byran AJ, Angelini GD. Hemodynamic performance of the 21-mm sorin bicarbon mechanical aortic prosthesis using dobutamine Doppler echocardiography. Ann Thorac Surg 2001; 72:49–53.

39. Hammermeister K, Sethi GK, Henderson WG, Grover FL, Oprian C, Rahimtoola SH. Outcomes 15 years after valve replacement with a mechanical versus a bioprosthetic valve: Final report of the Veterans Affairs randomized trial. J Am Coll Cardiol 2000 Oct; 36(4):1152–1158.

40. Khan S, et al. Twenty-year comparison of tissue and mechanical valve replacement. J Throrac Cardiovas Surg 2001; 122:257–269.

41. Bloomfield P. Choice of prosthetic heart valves: 20-year results of the Edinburgh Heart Valve Trial. J Am Coll Cardiol 2004 Aug 4;44(3):667.

42. Koertke H, Minami K, Boethig D, Breymann T, Seifert D, Wagner O, Atmacha N, Krian A, Ennker J, Taborski U, Klovekorn WP, Moosdorf R, Saggau W, Koerfer R. INR self-management permits lower anticoagulation levels after mechanical heart valve replacement. Circulation 2003; 108(Suppl 1):I175–178

43. On-X Aspirin-Only Study with Selected Isolated Aortic Valve Replacements. Available at http://www.onxvalves.com/Med_Aspirin_Study.asp.

44. Cohn L, Soltesz E. The evolution of mitral valve surgery: 1902-2002. Am Heart Hosp J 2003 Winter; 1(1):40–46.

Reading List

Standards

FDA Document *Replacement Heart Valve Guidance-Draft Document, October* 14, 1994.
ISO 5840, *Cardiovascular Implants—Cardiac valve.*
EN prEN12006 *Cardiovascular Implants—Cardiac valve.*

Pathology

Schoen FJ. Cardiovascular Pathology: Pathology of Heart Valve Substitution With Mechanical and Tissue Prostheses. New York: Churchill Livingston; 2001.

Biological Tissue Properties

Sacks MS, Schoen FJ. Mechanical damage to collagen independent of calcification limits bioprosthetic heart valve durability. Biomed Mater Res 2002;62:359–371.

Billiar KL, Sacks MS. Biaxial mechanical properties of the fresh and glutaraldehyde treated porcine aortic valve: Part I - Experimental results. J Biomechan Eng 2000;122:23–30.

Wells SM, Sacks MS. Effects of fixation pressure on the biaxial mechanical behavior of porcine bioprosthetic heart valves with long-term cyclic loading. Biomaterials 2002;23(11): 2389–2399.

Sacks MS. The biomechanical effects of fatigue on the porcine bioprosthetic heart valve. Long-term Effects Med Implants 2001;11(3&4):231–247.

Pyrolytic Carbon

More R, Bokros J. Carbon Biomaterials, Encyclopedia of Medical Devices and Instrumentation EMD 023. Heart Valve Prostheses In Vitro Flow Dynamics, Encyclopedia of Medical Devices and Instrumentation.

HEART VIBRATION.   See PHONOCARDIOGRAPHY.

HEART, ARTIFICIAL

CONRAD M. ZAPANTA
Penn State College of Medicine
Hershey, Pennsylvania

INTRODUCTION

Artificial hearts are broadly defined as devices that either supplement or replace the native (natural) heart. These devices can be classified into two groups: ventricular assist devices and total artificial hearts. This article will define the clinical need, review the native heart anatomy and function, describe design considerations for ventricular assist devices and total artificial hearts, review selected designs, and recommend areas for future development.

CLINICAL NEED

Cardiovascular disease accounted for 38% of all deaths (almost 1.4 million people) in the United States in 2002 (1). Coronary heart disease (53%) represented the majority of these deaths, followed by stroke (18%), and congestive heart failure (6%). Almost 4.9 million Americans suffer from congestive heart failure, with ~550,000 new cases diagnosed each year. Over 80% of men and 70% of women with congestive heart failure under the age of 65 will die within 8 years. In people diagnosed with congestive heart failure, sudden cardiac death occurs at six to nine times the rate of the general population.

One treatment for congestive heart failure is heart transplantation. It is estimated that 40,000 Americans could benefit from a heart transplant each year (1,2). However, only ~2100 donor hearts were available each year from 1999 to 2004. The number of donor hearts dropped during this period, from a high of 2316 in 1999 to 1939 in 2004. Over 3300 patients were on the waiting list for a donor heart at any time during this period, with >65% of these patients on the waiting list for >1 year. From 1998 to 2004, ~630 patients died each year waiting for transplant.

These numbers clearly demonstrate the clinical need for ventricular assist devices and total artificial hearts that support the patient until transplant (bridge to transplant) or permanently assist or replace the natural heart (destination therapy).

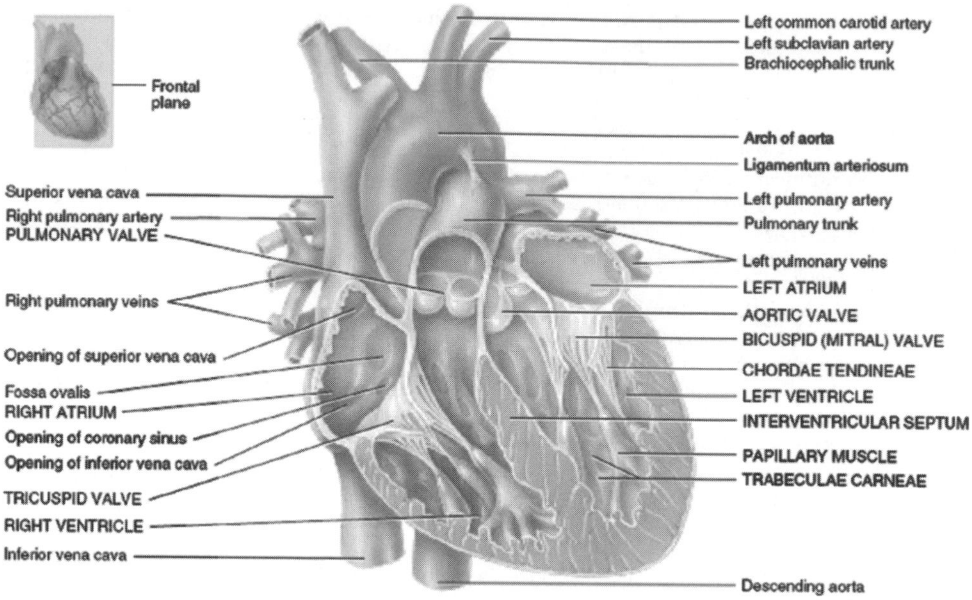

Figure 1. Anatomy of the Native Heart. The heart is composed of two pumps (right and left side) that work simultaneously. The right pump delivers blood to the pulmonary circulation (the lungs), while the left pump delivers blood to the systemic circulation (the body). Each pump consists of an atrium and ventricle. [Reprinted with permission from Gerard J. Tortora and Bryan H. Derrickson, *Principles of Anatomy and Physiology, 11th ed.*, Hoboken (NJ): John Wiley & Sons; 2006.]

NATIVE HEART ANATOMY AND FUNCTION

The anatomy of the native (or natural) heart is shown in Fig. 1. The heart is composed of two pumps (right and left side) that work simultaneously. The right pump delivers blood to the pulmonary circulation (the lungs) while the left pump delivers blood to the systemic circulation (the body). Each pump consists of an atrium and ventricle that make up the heart's four distinct chambers: right atrium, right ventricle, left atrium, and left ventricle. The atria act as priming chambers for the ventricles. The ventricles pump blood out of the heart to either the pulmonary or systemic circulation. Heart valves located between each atrium and ventricle and at the outlet of each ventricle maintain flow direction during pulsatile flow.

Blood from the systemic circulation enters the right atrium through the superior vena cava (from the head and upper extremities) and inferior vena cava (from the trunk and lower extremities). The blood is then pumped to the right ventricle, which pumps blood to the pulmonary circulation via the pulmonary arteries. Oxygenated blood returns to the left atrium heart from the lungs via the pulmonary vein and is then pumped to the left ventricle. The left ventricle pumps blood to the systemic circulation via the aorta.

Table 1 lists the nominal pressures and flows in the native heart (3). A ventricular assist device or total artificial heart must be able to generate these pressures and flows in order to meet the needs of the recipient.

DESIGN CONSIDRATIONS FOR VENTRICULAR ASSIST DEVICES AND TOTAL ARTIFICIAL HEARTS

Several design considerations must be taken into account when developing a ventricular assist device or total artificial heart. These considerations are detailed below:

1. *Size of the Intended Patient*: The size of the patient will determine the amount of blood flow required to adequately support the patient. This then determines the size of the ventricular assist device or total artificial heart. For example, a total artificial heart designed for adults would most likely be too large to be implanted within small children. A larger ventricular assist device may be placed externally, while a smaller ventricular assist device could be placed within the native heart. In addition, the size of the patient may dictate the location of some of the components. For example, the power sources may be located either internally (in the abdominal cavity) or externally depending on the size and type of the power source.

2. *Pump Performance:* A ventricular assist device or total artificial heart can be used to support or replace the native heart. Each of these support modes requires a different cardiac output. For example, a ventricular assist device can provide

Table 1. Nominal Pressures and Flows in the Native (Natural) Heart

Pressures	
Left ventricle	120 mmHg (16.0 kPa) peak systolic normal (into aorta)
	10 mmHg (1.33 kPa) mean diastolic (from left atrium)
Right ventricle	25 mmHg (3.33 kPa) peak systolic (into pulmonary artery)
	5 mmHg (0.667 kPa) mean diastolic (from right atrium)
Flows	
Normal healthy adult at rest: 5 L·min^{-1}	
Maximum flow: 25 L·min^{-1}	

either a portion of the blood flow required by the patient (partial support) or the entire blood flow (total support). In addition, the decision must be made whether to include a controller that will either passively or actively vary the cardiac output of the ventricular assist device or total artificial heart based on the patient demand.

3. *Reliability:* The National Institutes of Health (NIH) proposed a reliability goal for ventricular assist devices and total artificial hearts of 80% for a 2 year operation with an 80% confidence level before an artificial heart can begin clinical trials. However, the desired reliability may need to be more demanding for long-term clinical use, such as 95% reliability with 95% confidence for a 5 year operation. The design and components of ventricular assist devices and total artificial hearts must be carefully selected to achieve this reliability goal.

4. *Quality of Life:* The patient's quality of life can have a significant impact on the design of a ventricular assist device or total artificial heart. It is important to clearly define what constitutes an acceptable quality of life. For example, if a patient desires to be ambulatory following the implantation of a ventricular assist device or total artificial heart, the power supply must be easily transportable. The environment of the patient (home, work, car, etc.) should also be considered to insure proper function in these different environments. The patient should be able to monitor basic pump operation without the need for a physician or other trained medical personnel. The ventricular assist device or total artificial heart should be designed to clearly provide information through displays and provide alarms to warn the patient of potentially dangerous situations, such as a battery that is running low on power.

VENTRICULAR ASSIST DEVICES

A ventricular assist device (VAD) is designed to assist or replace the function of either the left or right ventricle. These devices are intended to provide either temporary support until a donor heart has been located or the native heart has recovered function, or as a permanent device.

As shown in Table 1, the left ventricle pumps against a higher pressure system than the right ventricle. Therefore, the left ventricle is typically more in need of assistance. Consequently, left ventricular assist devices (LVADs) are more prevalent than right ventricular assist devices (RVADs).

Ventricular assist devices can generate either pulsatile or continuous (nonpulsatile) flow.

Pulsatile

Pulsatile flow ventricular assist devices are composed of a single ventricle that mimics the native ventricle. The ventricle is placed either outside the patient's body or within the abdominal cavity. There are two types of pulsatile flow ventricular assist devices: pneumatic and electric.

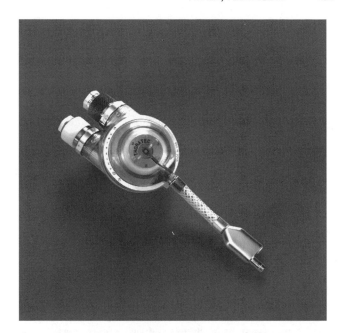

Figure 2. Penn State/Thoratec Pneumatic Ventricular Assist Device. The ventricle contains a flexible blood sac made of segmented polyurethane that is housed within a rigid polysulfone case. Mechanical heart valves are located in the inlet and outlet positions of the ventricle to control flow direction. Air pulses that are generated by an external drive unit are used to periodically compress the flexible blood sac. (Reprinted with permission from Thoratec Corporation.)

Pneumatic Ventricular Assist Devices. Figure 2 shows a pneumatic ventricular assist device that was originally developed by the Pennsylvania State University and later purchased by Thoratec Corporation (Pleasanton, CA) (4). The ventricle contains a flexible blood sac made of segmented polyurethane that is housed within a rigid polysulfone case. This blood sac is extremely smooth to prevent the formation of clots (or thrombi). Mechanical heart valves are located in the inlet and outlet positions of the ventricle to control flow direction. Air pulses that are generated by an external drive unit are used to periodically compress the flexible blood sac. An automatic control system varies the cardiac output by adjusting the heart rate and the time for ventricular filling in response to an increase in filling pressure.

The device is placed outside the patient on the patient's abdomen (paracorporeal). The device can be used to assist a single ventricle, or simultaneously with an additional device that assists the both ventricles, as shown in Fig. 3. For the LVAD configuration (right pump in Fig. 3), the inlet cannula is inserted into the apex of the left ventricle and connected to the inlet port of the ventricular assist device. The outflow cannula is attached between the outflow port of the ventricular assist device and the ascending aorta. For the RVAD configuration (left pump in Fig. 3), the inlet cannula is connected to the right atrium and the outlet cannula to the main pulmonary artery. For both types of configurations, the inflow and outflow cannulae pass through the skin below the rib cage. Over 2850 implants have occurred worldwide with the longest duration of 566 days (5). An implantable version of this pump (with a titanium pump casing) was approved by the FDA in August of 2004 (6).

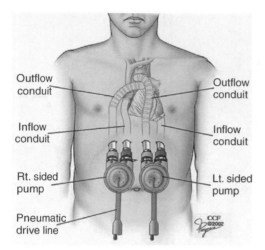

Figure 3. Implant Location of Penn State/Thoratec Pneumatic Ventricular Assist Device in the RVAD (left) and LVAD (right) Configuration. For the LVAD configuration, the inlet cannula is inserted into the apex of the left ventricle and connected to the inlet port of the ventricular assist device. The outflow cannula is attached between the outflow port of the ventricular assist device and the ascending aorta. For the RVAD configuration, the inlet cannula is connected to the right atrium and the outlet cannula to the main pulmonary artery. For both types of configurations, the inflow and outflow cannulae pass through the skin below the rib cage. (Reprinted with permission from The Cleveland Clinic Foundation.)

Another type of pneumatic ventricular assist device is the Thoratec HeartMate IP (intraperitoneal). The HeartMate IP is an implantable blood pump that is connected to an external drive unit via a percutaenous air drive line (7). The interior of this device is shown in Fig. 4. A unique feature of this device is the use of a textured blood surface in the ventricle that promotes the formation of a cell layer. The cell layer is believed to decrease thrombus formation because the layer mimics the blood contacting surface of a blood vessel. Bioprosthetic heart valves are used to regulate the direction of flow. The HeartMate IP has been implanted in >1300 patients worldwide with the longest duration of 805 days.

Two other types of pneumatic devices include the BVS5000 and AB5000 (both made by ABIOMED, Danvers, MA). Both devices are intended to provide cardiac support as a bridge to transplant or until the native heart recovers. The BVS5000, illustrated in Fig. 5, is an external dual-chamber device that can provide support to one or both ventricles as a bridge to transplant (8). The chambers utilize polyurethane valves to regulate the flow direction. More than 6000 implants have been performed worldwide (9). The AB5000 is a pneumatically driven, paracorporeal device that is similar to a single ventricle of the AbioCor total artificial heart (described in a later section) (10). This device was approved by the FDA in October of 2003 as has been used in >88 patients. The average duration of support is 15 days with the longest duration of 149 days.

Additional types of pneumatic devices include the Berlin Heart Excor (Berlin Heart AG, Berlin, Germany), the

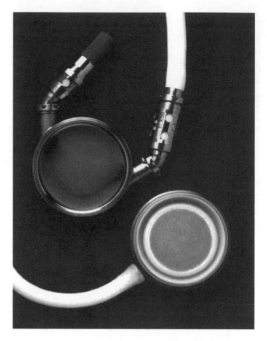

Figure 4. Interior of Thoratec HeartMate IP Pneumatic Ventricular Assist Device. A unique feature of this device is the use of a textured blood surface in the ventricle that promotes the formation of a cell layer. The cell layer is believed to decrease thrombus formation because the layer mimics the blood contacting surface of a blood vessel. (Reprinted with permission from Thoratec Corporation.)

MEDOS/HIA (Aachen, Germany), and the Toyobo Heart (National Cardiovascular Center, Osaka, Japan). The Berlin Heart Excor is available in a range of sizes (10–80 mL stroke volume) with either tilting disk or polyurethane valves, and has been implanted in >500 patients (11). The MEDOS/HIA system is also available in a range of sizes and has been implanted in >200 patients (12). The Toyobo LVAS has been implanted in >120 patients (13).

ELECTRIC VENTRICULAR ASSIST DEVICES

Electric ventricular assist devices mainly differ from their pneumatic counterparts in their source of power. Electric ventricular assist devices are typically placed within the

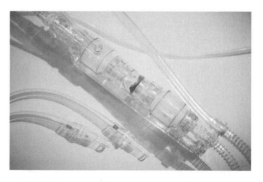

Figure 5. ABIOMED BVS 5000 Pneumatic Ventricular Assist Device. The BVS5000 is an external dual-chamber device that can provide support to one or both ventricles as a bridge to transplant (8). The chambers utilize polyurethane valves to regulate the flow direction. (Reprinted with permission from ABIOMED, Inc.)

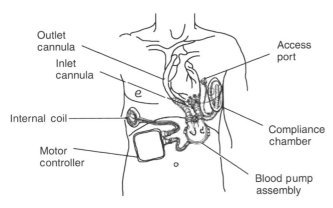

Figure 6. Penn State/Arrow Electric Ventricular Assist Device (LionHeart). The blood pump assembly utilizes a rollerscrew energy converter with a pusher plate. The motion of the pusher plate compresses the blood sac and ejects blood from the ventricle. Mechanical heart valves are used to control the direction of flow into and out of the pump. Energy passes from the external power coil to the subcutaneous (internal) coil by inductive coupling via the transcutaneous energy transmission system (TETS). (Reprinted with permission from Arrow International.)

abdominal cavity. The inlet cannula is inserted into the apex of the native left ventricle and connected to the inlet port of the device. The outlet cannula is attached between the outflow port of the device and the ascending aorta via an end-to-side anastomosis. These types of devices can be used as either bridge-to-transplant or as permanent implants (destination therapy).

Figure 6 illustrates an electric ventricular assist device (LionHeart) developed by the Pennsylvania State University in conjunction with Arrow International (Reading, PA) (14). The blood pump assembly utilizes a rollerscrew energy converter with a pusher plate. The motion of the pusher plate compresses the blood sac and ejects blood from the ventricle. Mechanical heart valves are used to control the direction of flow into and out of the pump. Energy passes from the external power coil to the subcutaneous (internal) coil by inductive coupling via the transcutaneous energy transmission system (TETS). The controller and internal battery supply are also implanted in the abdomen. The internal battery permits operation without the external power coil for ~ 20 min. Air displaced by the blood pump housing enters the polyurethane compliance chamber. Because the air in the compliance chamber can slowly diffuse across the wall of the compliance chamber, the air in the chamber is periodically replenished via the subcutaneous access port. The LionHeart is intended to be used as destination therapy. This device was approved for use in Europe in 2003.

Another type of electric ventricular assist devices is the Novacor LVAS (left ventricular assist system), produced by WorldHeart Corporation (Ottawa, ON). The Novacor LVAS, illustrated in Fig. 7, contains a polyurethane blood sac that is compressed between dual pusher plates (15). The pusher plates are actuated by a solenoid that is coupled to the plates via springs. Bioprosthetic heart valves are utilized to control the direction of flow. A percutaneous power line connects the pump to an external battery pack and controller. The Novacor LVAS has been implanted in

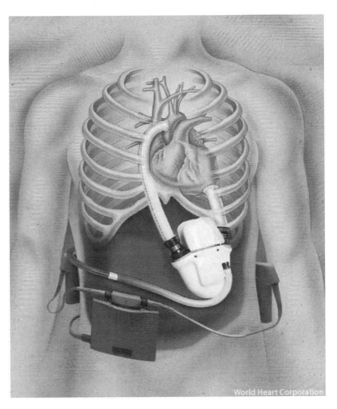

Figure 7. WorldHeart Novacor Electric Ventricular Assist Device. The Novacor contains a polyurethane blood sac that is compressed between dual pusher plates. The pusher plates are actuated by a solenoid that is coupled to the plates via springs. A percutaneous power line connects the pump to an external battery pack and controller. (Reprinted with permission from World Health Corporation, Inc.)

over 1500 patients worldwide. The longest implant duration is > 6 years. No deaths have been attributed to device failure with only 1.4% of the devices needing replacement. The Novacor LVAS is approved as a bridge-to-transplant in the United States and Europe and is in clinical trials for destination therapy in the United States.

The HeartMate XVE (illustrated in Fig. 8) is a derivative to the HeartMate IP (7). The HeartMate XVE uses an electric motor and pusher plate system to pump blood. A percutaneous power line is used to connect the pump to an external battery pack and controller. The HeartMate VE (an earlier version of the XVE) and XVE have been implanted in > 2800 patients worldwide with the longest duration of 1854 days. The HeartMate SNAP-VE was recently approved by the FDA as destination therapy.

The Randomized Evaluation of Mechanical Assistance for the Treatment of Congestive Heart Failure (REMATCH) study examined the clinical utility of ventricular assist devices (16). Patients with end-stage heart failure who were ineligible for cardiac transplantation were split into two groups. The first group ($n = 68$) received the HeartMate VE LVAS while the second group ($n = 61$) received optimal medical management. The results showed a reduction of 48% in the risk of death from any cause in the LVAD group versus the medical-therapy group ($p = 0.001$). The 1 year survival was 52% for the VAD group and 25% for the medical-therapy group ($p = 0.002$). The 2 year survival

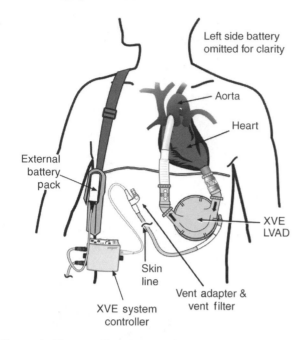

Figure 8. Thoratec HeartMate XVE Electric Ventricular Assist Device. The HeartMate XVE uses an electric motor and pusher plate system to pump blood. A percutaneous power line is used to connect the pump to an external battery pack and controller. (Reprinted with permission from Thoratec Corporation.)

was 23% for the VAD group and 8% for the medical-therapy group ($p = 0.09$). Finally, the median survival was 408 days for the VAD group and 150 days for the medical-therapy group. This study clearly showed the clinical utility of ventricular assist devices.

Continuous Flow

Continuous flow ventricular assist devices deliver nonpulsatile flow. Consequently, they do not require heart valves to regulate the direction of blood flow. Continuous flow ventricular assist devices are classified as either centrifugal flow or axial flow pumps based on the direction of the flow as it passes through the pump. These types of pumps are implanted in a similar fashion as their pulsatile counterparts.

Continuous flow assist devices have several potential advantages over pulsatile systems. First, these devices are typically smaller than their pulsatile counterparts and can be used in smaller patients (such as small adults and children). In addition, these pumps have fewer moving parts and are simpler devices than pulsatile systems. These types of devices typically require less energy to operate than the pulsatile pumps.

However, continuous flow pumps have several potential disadvantages. The main disadvantage is that the long-term effects of continuous flow in patients are unknown. Some studies suggest that continuous flow results in lower tissue perfusion (17,18). In addition, these types of devices typically have higher fluid stresses than their pulsatile counterparts, potentially exposing blood components to stress levels that may damage or destroy the cells. However, due to the short residence time of the blood compo-

nents within these pumps, the potential for damage or destruction is reduced (19). Finally, feedback control mechanisms for altering pump speed and flow in response to patient demand are complex and unproven.

Centrifugal Flow Ventricular Assist Device. In a centrifugal flow ventricular assist device, the direction of the outlet port is orthogonal (at a right angle) to the direction of the inlet port. Blood flowing into a centrifugal pump moves onto a spinning impeller. This causes the blood to be propelled away from the impeller due to centrifugal forces. The blood is then channeled to the outlet port by a circular casing (known as the volute) around the impeller. Finally, the blood is discharged through the outlet at a higher pressure than the inlet pressure.

The impeller typically consists of various numbers and geometric configurations of blades, cones, or disks. Typical motor speeds (or rotation rates) for centrifugal flow pumps range vary from 1500 to 5000 rpm (revolutions per minute). This results in flow rates of 2–10 $L \cdot min^{-1}$. Many centrifugal flow pumps utilize electromagnetic impellers that do not make any contact with the interior of the pump when the impeller is spinning. The inlet and outlet ports are connected to the native ventricle and the aorta, respectively, as described previously for pulsatile electric ventricular assist devices.

A major drawback with centrifugal flow pumps is that they are outlet pressure sensitive and may not produce flow if the outflow pressure (the pressure that the pump is working against) becomes greater than the outlet pressure. When this happens, the impeller will continue to spin without producing any flow. In order for the pump to produce flow, either the outflow pressure must be reduced or the impeller speed must be increased (to increase the outlet pressure).

The Bio-Pump (Medtronic BioMedicus, Inc., Minneapolis, MN), shown in Fig. 9, is an extracorporeal, centrifugal flow pump that was originally developed for cardiopulmonary bypass (20). It has been used to provide support for one or both ventricles as a bridge to transplant for short periods

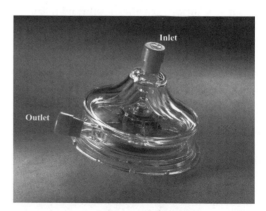

Figure 9. Medtronic BioMedicus Bio-Pump Centrifugal Flow Ventricular Assist Device. The Bio-Pump is an extracorporeal, centrifugal flow pump that was originally developed for cardiopulmonary bypass. It has been used to provide support for one or both ventricles as a bridge to transplant for short periods. (Reprinted with permission from Medtronic, Inc.)

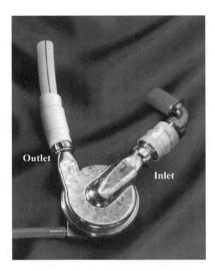

Figure 10. Thoratec HeartMate III Centrifugal Flow Ventricular Assist Device. The HeartMate III is a centrifugal pump that features a magnetically levitated impeller. (Reprinted with permission from Thoratec Corporation.)

(5 days or less). The pump consists of an acrylic pump head with inlet port and outlet ports placed at right angles to each other. The impeller consists of a stack of parallel cones within a conical acrylic housing. A single magnetic drive unit is coupled with a magnet in the impeller. The pump is driven by an external motor and power console. Two different sizes are available to provide support for both adults and children. Recipients of the Bio-Pump have had mixed results (21). The Sarns/3M Centrifugal system (Terumo, Ann Arbor, MI) is another centrifugal pump that is used primarily for cardiopulmonary bypass (22).

The HeartMate III (Thoratec), shown in Fig. 10, is a centrifugal pump that features a magnetically levitated impeller (23,24). The entire pump is fabricated from titanium. The interior of the pump uses the same type of textured blood contacting surfaces utilized in the Heart-Mate VE. In addition, the HeartMate III incorporates a TETS that permits it to be fully implantable as a permanent device for destination therapy. The controller is designed to respond automatically to the patient's needs and to permit both pulsatile and continuous flow. This pump is currently under development. Other centrifugal pumps that utilize a magnetically levitated impeller include the HeartQuest (MedQuest, Salt Lake City, UT) (25) and the Duraheart (Terumo) (26). The Duraheart was first implanted in 2004.

Two centrifugal flow pumps utilize hydrodynamic forces, rather than magnetic levitation, to suspend the impeller: the CorAide (Arrow International) (27) and the VentrAssist (Ventrcor Limited, Chatswood, Australia) (28). The CorAide (shown in Fig. 11) began clinical trials in Europe in 2003 (29), while the VentrAssist (shown in Fig. 12) began clinical trials in Europe in 2004 (30).

Axial Flow Ventricular Assist Devices. An axial flow ventricular assist device is also composed of an impeller spinning in a stationary housing. However, the blood that flows into and out of the device travels in the same direction

Figure 11. Arrow CorAide Centrifugal Flow Ventricular Assist Device. The CorAide utilizes a hydrodynamic bearing, rather than magnetic levitation, to suspend the impeller. (Reprinted with permission from Arrow International, Inc.)

as the axis of rotation of the impeller. The impeller transfers energy to the blood by the propelling, or lifting, action of the vanes on the blood. Stators (stationary flow straighteners) stabilize the blood flow as it enters and exits the impeller. Magnets are embedded within the impeller and are coupled with a rotating magnetic field on the housing. The pumps are typically constructed of titanium.

Axial flow pumps run at speeds of 10,000–20,000 rpm, generating flow rates of up to 10 L·min^{-1}. These high motor speeds are not expected to cause excessive hemolysis (damage to blood components) because of the limited exposure of blood within the axial flow pump (19). Like centrifugal pumps, axial flow pumps are also outlet pressure sensitive and may not produce flow in cases when the outflow pressure exceeds the outlet pressure. Mechanical bearings are typically used to support the impeller within the stator.

Figure 12. Ventracor VentrAssist Centrifugal Flow Ventricular Assist Device. The VentrAssist utilizes a hydrodynamic bearing, rather than magnetic levitation, to suspend the impeller. (Reprinted with permission from Ventracor, Inc.)

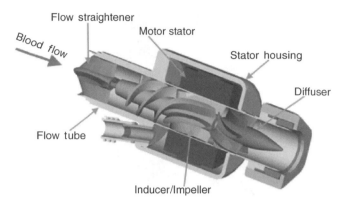

Figure 13. MicroMed Debakey Axial Flow Ventricular Assist Device. The MicroMed is connected to an external controller and power unit. The pump speed is varied manually to meet the needs of the patient. (Reprinted with permission from Micromed Technology, Inc.)

Figure 13 shows the MicroMed Debakey VAD (Micro-Med Technology, Houston, TX) axial flow pump. This device operates from 7500 to 12,000 rpm and can provide flows up to 10 L·min^{-1} (31). The flow curves, speed, current and power are displayed in a bedside monitor unit. A pump motor cable along with the flow probe wire exit transcutaneously from the implanted device and connect to the external controller and power unit. The pump speed is varied manually to meet the needs of the patient. The pump can be actuated by two 12 V dc batteries for 4–6 h. This device was approved in Europe in 2001 (32). Clinical trials in the United States began in 2000. Over 280 patients have received the MicroMed Debakey VAD as of January 2005 worldwide. Although this device was originally approved as a bridge to transplant, clinical trials are underway to use the device for destination therapy.

Figure 14 shows the HeartMate II (Thoratec) axial flow ventricular assist device. The rotating impeller is surrounded by a static pump housing with an integral motor (33). The pump's speed can be controlled either manually or by an automatic controller that relies on an algorithm based on pump speed, the pulsatility of the native heart, and motor current. The HeartMate II is designed to operate between 6000 and 15,000 rpm and deliver as much as 10 L·min^{-1}. The initial version of this device is powered through a percutaneous small-diameter electrical cable connected to the system's external electrical controller. A fully implantable system utilizing a TETS is under development. The first implant HeartMate II implant occurred in 2000 (34). Clinical trials in Europe and the United States are ongoing. This device is intended for both bridge to transplant and destination therapy.

Figure 15 illustrates the Jarvik 2000 (Jarvik Heart, New York). The Jarvik 2000 is intraventricular axial flow pump. The impeller is a neodymium–iron–boron magnet, which is housed inside a welded titanium shell and supported by ceramic bearings (35). A small, percutaneous cable delivers power to the impeller. All of the blood-contacting surfaces are made of highly polished titanium. The normal operating range for the control system is 8000–12,000 rpm, which generates an average pump flow rate of 5 L·min^{-1}. The pump is placed within the left ventricle with

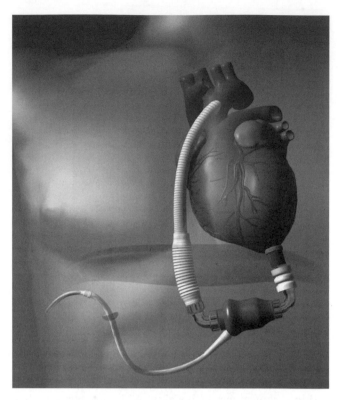

Figure 14. Thoratec HeartMate II Axial Flow Ventricular Assist Device. The rotating impeller is surrounded by a static pump housing with an integral motor. The pump's speed can be controlled either manually or by an automatic controller that relies on an algorithm based on pump speed, the pulsatility of the native heart, and motor current. (Reprinted with permission from Thoratec Corporation.)

a sewing cuff sutured to the ventricle, eliminating the need for an inflow cannula. Over 100 patients have received the Jarvik 2000 as a bridge to transplant or destination therapy, with the longest implant duration of > 4 years (36).

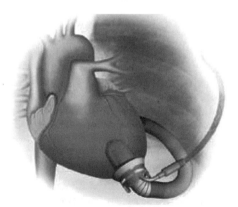

Figure 15. Jarvik 2000 Axial Flow Ventricular Assist Device. Unlike most other axial flow devices, the Jarvik 2000 is intraventricular axial flow pump. The impeller is a neodymium-iron-boron magnet, which is housed inside a welded titanium shell and supported by ceramic bearings. (Reprinted with permission from Jarvik Heart, Inc.)

TOTAL ARTIFICIAL HEARTS

The total artificial heart (TAH) is designed to support both the pulmonary and systemic circulatory systems by replacing the native heart. Two types of artificial hearts have been developed: pneumatic and electric.

Pneumatic Total Artificial Heart

A pneumatic total artificial heart is composed of two ventricles that replace the native left and right ventricle. Each ventricle is of similar design to the Penn State/Thoratec pneumatic ventricular assist device (as described in a previous section) (4). Both ventricles are implanted within the chest. The air pulses are delivered to the ventricles via percutaneous drivelines. An automatic control system varies cardiac output by adjusting the heart rate and the time for ventricular filling in response to an increase in filling pressure.

Pneumatic total artificial hearts are currently used as a bridge to transplant. Several different pneumatic artificial hearts have been used clinically around the world. The only pneumatic TAH approved as a bridge to transplant in the United States is the CardioWest (SynCardia, Tucson, AZ) TAH, illustrated in Fig. 16 (37). The CardioWest (with a stroke volume of 70 mL) is based on the Jarvik-7, which has a stroke volume of 100 mL. A study of 81 recipients of the CardioWest revealed a survival rate to transplant of 79% and a 1 year survival rate of 70%.

Pneumatic total artificial hearts have also been used as a permanent replacement device. The Jarvik-7 pneumatic TAH was permanently implanted in five patients (38).

Although the longest survivor lived for 620 days, all five patients had hematologic, thromboembolic, and infectious complications. The pneumatic artificial heart is no longer considered for permanent use because of infections associated with the percutaneous pneumatic drive lines and quality of life issues related to the bulky external pneumatic drive units.

Electric Total Artificial Heart

The electric TAH is completely implantable and is designed for permanent use. The Penn State/3M Electric TAH is shown in Fig. 17. The artificial heart is composed of two blood pumps that are of similar design to the Penn State/Arrow electric ventricular assist device (39). However, the electric TAH uses a single implantable energy converter that alternately drives each ventricle. The implantable controller adjusts the heart rate in response to ventricular filling and maintains left–right balance. The design for this system was completed in 1990 and was the first to incorporate the controller, transcutaneous energy transmission system (TETS), telemetry, and internal power (via rechargeable batteries) into a completely implantable system. The Penn State electric TAH has been successfully implanted in animals for >1 year without thromboembolic complications. In 2000, ABIOMED acquired the rights to the Penn State/3M Electric TAH.

The ABIOMED AbioCor TAH, illustrated in Fig. 18, uses an electrohydraulic energy converter to alternately compress each blood sac (40,41). In addition, the AbioCor uses polymer valves to control flow into and out of each ventricle. The AbioCor is currently undergoing clinical

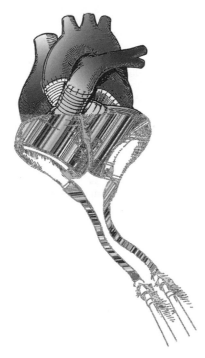

Figure 16. SynCardia CardioWest Pneumatic Total Artificial Heart. The CardioWest is based on the Jarvik-7 and is the only pneumatic TAH approved as a bridge to transplant in the United States. (Reprinted with permission from SynCardia Systems, Inc.)

Figure 17. Penn State/3M Electric Total Artificial Heart. This artificial heart is composed of two blood pumps that are of similar design to the Penn State/Arrow electric ventricular assist device. However, the electric TAH uses a single implantable energy converter that alternately drives each ventricle.

Figure 18. ABIOMED AbioCor Electric Total Artificial Heart. The AbioCor TAH uses an electro hydraulic energy converter to alternately compress each blood sac. The AbioCor is currently undergoing clinical trials in the United States. Two patients were discharged from the hospital, with one patient surviving for >1 year. (Reprinted with permission from ABIOMED, Inc.)

trials in the United States. Fourteen critically ill patients (with an 80% chance of surviving < 30 days) have been implanted. Two patients were discharged from the hospital (one to home), with one patient surviving for >1 year. The causes of death were typically end organ failure and strokes. One pump membrane wore out at 512 days. Smaller, improved totally implantable artificial hearts are currently under development.

FUTURE DIRECTIONS OF RESEARCH

The ventricular assist devices and total artificial hearts presented in this article successfully provide viable cardiac support by either assisting or replacing the native heart. However, there are several areas for future research on artificial hearts. These include the following: Power sources to permit longer intervals between battery changes; Improved control schemes for both pulsatile and nonpulsatile devices that enhance the response of the cardiac assist device to meet physiologic demands; Decrease thromboembolic events associated by modifying the device geometry and/or blood-contacting materials; Determine the long-term effects of continuous, nonpulsatile flow; Decrease incidence of infection by the elimination of all percutaneous lines and creating smaller implantable electronic components; Reduced pump sizes to fit smaller adults, children, and infants; Increased reliability for 5 or more years to 95% (with a 95% confidence level).

Significant progress has been made in the last 20 years. One can only imagine what the next 20 years will bring!

ACKNOWLEDGMENTS

The author would like to acknowledge the support of William S. Pierce, M.D., Gerson Rosenberg, Ph.D., David B. Geselowitz, Ph.D., and the past and present faculty, staff, and graduate students at the Division of Artificial Organs at the Penn State College of Medicine and the Department of Bioengineering at the Pennsylvania State University. The financial support from the National Institutes of Health is also recognized.

BIBLIOGRAPHY

Cited References

1. American Heart Association. Heart Disease and Stroke Statistics—2005 Update. Dallas (TX): American Heart Association; 2005.
2. Organ Procurement and Transplantation Network Data as of May 29, 2005. Available at http://www.optn.org. 2005.
3. Guyton A, Hall J. Textbook of Medical Physiology. Philadelphia: Saunders; 2000.
4. Richenbacher WE, Pierce WS. In: Braunwald HE, editor. Assisted Circulation and the Mechanical Heart Disease: A Textbook of Cardiovascular Medicine, 6th ed. Philadelphia: Saunders; 2001. p 534–547.
5. Thoratec VAD Clinical Results, Available at http://www.thoratec.com. Accessed Nov 2004.
6. Thoratec Corporation Press Release, Aug. 5, 2004 [Online]. Thoratec. Available at http://www.thoratec.com/index.htm. [5/19/2005]. Accessed 2005.
7. HeartMate LVAS Clinical Results, Nov. 2004. Available at http://www.thoratec.com/index.htm. Accessed 2004.
8. Berger EE. ABIOMED's BVS 5000 biventricular support system. J Heart Lung Transplant 2004;23(5):653.
9. Clinical Information, BVS Clinical Update 2004 [online] ABIOMED. Available at http://www.abiomed.com/clinical information/BVS5000Update.cfm. [5/19/2005]. Accessed 2004.
10. Clinical Information, AB5000 Clinical Update 2004 [online] ABIOMED. Available at http://www.abiomed.com/clinical information/AB5000Update.cfm. [5/19/2005]. Accessed 2004.
11. Berlin Heart AG- The VAD System [online] Berlin Heart. http://www.berlinheart.com/download/system.pdf. [5/17/2005]. 2003.
12. Reul H. The MEDOS/HIA system: development, results, perspectives. Thorac Cardiovasc Surg 1999;47(Suppl 2):311–315.
13. Takano H, Nakatani T. Ventricular assist systems: experience in Japan with Toyobo pump and Zeon pump. Ann Thorac Surg 1996;61(1):317–322.
14. El-Banayosy A, et al. Preliminary experience with the LionHeart left ventricular assist device in patients with end-stage heart failure. Ann Thorac Surg 2003;75(5):1469–1475.
15. Novacor LVAS-Products-WorldHeart [Online] WorldHeart. Available at http://www.worldheart.com/products/novacor lvas.cfm. [5/17/2005]. Accessed 2005.
16. Rose EA, et al. Long-term mechanical left ventricular assistance for end-stage heart failure. N Engl J Med 2001;345(20): 1435–1443.
17. Baba A, et al. Microcirculation of the bulbar conjunctiva in the goat implanted with a total artificial heart: effects of pulsatile and nonpulsatile flow. ASAIO J 2004;50(4):321–327.
18. Undar A. Myths and truths of pulsatile and nonpulsatile perfusion during acute and chronic cardiac support. Artif Organs 2004;28(5):439–443.
19. Arora D, Behr M, Pasquali M. A tensor-based measure for estimating blood damage. Artif Organs 2004;28(11):1002–1015.

20. Noon GP, BallJr JW, Papaconstantinou HT. Clinical experience with BioMedicus centrifugal ventricular support in 172 patients. Artif Organs 1995;19(7):756–760.
21. Noon GP, Lafuente JA, Irwin S. Acute and temporary ventricular support with BioMedicus centrifugal pump. Ann Thorac Surg 1999;68(2):650–654.
22. Williams M, Oz M, Mancini D. Cardiac assist devices for end-stage heart failure. Heart Dis 2001;3(2):109–115.
23. Bourque K, et al. HeartMate III: pump design for a centrifugal LVAD with a magnetically levitated rotor. ASAIO J 2001;47(4):401–405.
24. Bourque K, et al. Incorporation of electronics within a compact, fully implanted left ventricular assist device. Artif Organs 2002;26(11):939–942.
25. Chen C, et al. A magnetic suspension theory and its application to the HeartQuest ventricular assist device. Artif Organs 2002;26(11):947–951.
26. Terumo Heart, Inc. Press Release, Jan. 19, 2004 [Online]: Available at http://www.terumocvs. com/newsandevents/rendernews.asp?newsId=5. Terumo [5/17/2005]. Accessed 2005.
27. Doi K, et al. Preclinical readiness testing of the Arrow International CorAide left ventricular assist system. Ann Thorac Surg 2004;77(6):2103–2110.
28. James NL, et al. Implantation of the VentrAssist Implantable Rotary Blood Pump in sheep. ASAIO J 2003;49(4):454–458.
29. Arrow International-Cardiac Assist-CorAide [online] Arrow International. Available at http://www.arrowintl.com/products/cardassist/. [2/28/05]. Accessed 2003.
30. Ventracor Half-Year Report Summary, Feb 16 2005 [online] Ventracor. Available at http://www.ventracor.com/default.asp?cp=/news/newsitem.asp%3FnewsID%3D316. [5/19/2005]. Accessed 2005.
31. Goldstein DJ. Worldwide experience with the MicroMed DeBakey Ventricular Assist Device as a bridge to transplantation. Circulation 2003;108(Suppl 1):II272–277.
32. MicroMed Technology Press Release, Jan. 20 2005 [Online] MicroMed. Available at http://www.micromedtech.com/news/01-20-05.htm. [5/19/2005]. Accessed 2005.
33. Burke DJ, et al. The Heartmate II: design and development of a fully sealed axial flow left ventricular assist system. Artif Organs 2001;25(5):380–385.
34. Frazier OH, et al. First clinical use of the redesigned HeartMate II left ventricular assist system in the United States: a case report. Tex Heart Inst J 2004;31(2):157–159.
35. Westaby S, et al. The Jarvik 2000 Heart. Clinical validation of the intraventricular position. Eur J Cardiothorac Surg 2002;22(2):228–232.
36. Frazier OH, et al. Use of the Flowmaker (Jarvik 2000) left ventricular assist device for destination therapy and bridging to transplantation. Cardiology 2004;101(1–3):111–116.
37. Copeland JG, et al. Cardiac replacement with a total artificial heart as a bridge to transplantation. N Engl J Med 2004; 351(9):859–867.
38. DeVries WC. The permanent artificial heart. Four case reports. JAMA 1988;259(6):849–859.
39. Weiss WJ, et al. Steady state hemodynamic and energetic characterization of the Penn State/3M Health Care Total Artificial Heart. ASAIO J 1999;45(3):189–193.
40. Dowling RD, et al. The AbioCor implantable replacement heart. Ann Thorac Surg 2003;75(6 Suppl):S93–S99.
41. ABIOMED AbioCor Press Release, Nov. 4, 2004 [Online] ABIOMED. Available at http://www.abiomed.com/news/Fourteenth-AbioCor-Patient.cfm. [5/19/2005]. Accessed 2004.

Reading List

Lee J, et al. Reliability model from the in vitro durability tests of a left ventricular assist system. ASAIO J 45(6):595–601 1999.

Raman J, Jeevanadam V. Destination therapy with ventricular assist devices. Cardiology 101(1–3):104–110 2004.
Rosenberg G. Artificial Heart and Circulatory Assist Devices. In: Bronzino J, editor. The Biomedical Engineering Handbook, Boca Raton (FL): CRC Press; 1995: 1839–1846.
Song X, et al. Axial flow blood pumps. ASAIO J 49(4):355–364 2003.
Stevenson LW, Kormos RL. Mechanical Cardiac Support 2000: Current applications and future trial design. J Thorac Cardiovasc Surg 121(3):418–424 2001.
Zapanta CM, et al. Durability testing of a completely implantable electric total artificial heart. ASAIO J 51(3):214–223 2005.

See also HEART VALVE PROSTHESES; HEMODYNAMICS; QUALITY OF LIFE MEASURES, CLINICAL SIGNIFICANCE OF.

HEART–LUNG MACHINES

DAVID D'ALESSANDRO
ROBERT MICHLER
Montefiore Medical Center
Bronx, New York

INTRODUCTION

The heart–lung machine is perhaps the most important contribution to the advancement of surgery in the last century. This apparatus was designed to perform the functions of both the human heart and the lungs allowing surgeons to suspend normal circulation to repair defects in the heart. The development of a clinically safe and useful machine was the rate-limiting step to the development of modern cardiac surgery. Since its inception, the heart–lung machine has enabled the surgical treatment of congenital heart defects, coronary heart disease, valvular heart disease, and end-stage heart disease with heart transplantation and mechanical assist devices or artificial hearts.

The heart–lung machine consists of several components that together make up a circuit that diverts blood away from the heart and lungs and returns oxygenated blood to the body. Commercial investment and production of these components has resulted in wide variability in the design of each, but the overall concept is preserved. During an operation, a medical specialist known as a perfusionist operates the heart–lung machine. The role of the perfusionist is to maintain the circuit, adjust the flow as necessary, prevent air and particulate emboli from entering the circulation, and maintain the various components of the blood within physiologic parameters.

HISTORY OF CARDIAC SURGERY AND ASSISTED CIRCULATION

The emergence of modern heart surgery and the ability to correct congenital and acquired diseases of the heart were dependent on the work of innovative pioneers who developed a means to stop the heart while preserving blood flow to the remainder of the body. This new technology was aptly named *cardiopulmonary bypass* (CPB), which simply means circulating blood around the heart and lungs. Since the inception of the heart–lung machine, continued refinements and widespread adoption of CPB led to the rapid growth of congenital and adult heart surgery.

Lessons learned during World War II and corresponding advances in critical care emboldened surgeons to consider surgical solutions for diseases of the heart, an organ long considered to be inoperable. One such surgeon was Dr. Dwight Harken who pioneered closed heart surgery, which he initially used to remove foreign bodies from the heart such as bullets and shrapnel. Having achieved success with this approach, he and others modified the techniques, developing a blinded technique for performing closed valvuloplasty primarily used to ameliorate rheumatic valvular disease. Although this method proved safe and reproducible, its applicability to other diseases of the heart was limited.

The next leap came when surgeons attempted open-heart procedures using brief periods of total circulatory arrest. As the brain is highly sensitive to hypoxic injury, very few patients were successfully treated with his approach. The safety of circulatory arrest was greatly increased when Dr. Bill Bigelow at the University of Minnesota introduced the concept of induced hypothermia. This method of lowering body temperature to reduce metabolic demand provided protection for the oxygen-starved organs allowing for modestly longer periods of circulatory arrest. Inspired by Bigelow's work, Lewis and Taufic first used this approach clinically on September 2, 1952, at the University of Minnesota (1). Under moderate total body hypothermia, Lewis and Taufic used a short period of circulatory arrest to repair a congenital defect in a 5 year-old girl. This was a landmark achievement in surgery and marks the true beginning of open-heart surgery. For the first time, surgeons had the ability to open the heart to repair defects under direct vision. Despite this great achievement, however, the relatively brief periods of circulatory arrest that this technique provided were sufficient only for the repair of simple defects and did little to broaden the scope of surgically treatable cardiac diseases.

The development of assisted circulation was a quantum leap in the field of cardiac surgery. Although investigation into mechanical means of circulation began during the early part of the twentieth century, an effective and safe design would not emerge for several years. An alternative approach named *cross-circulation* was used for several years in the interim. Dr. C. Walt Lillehei, again at the University of Minnesota, was the first to use this technique clinically when on March 26, 1954, when he repaired a VSD in a 12 month-old infant. During this operation, the child's mother acted as a blood reservoir, a pump, and an oxygenator, allowing a safe period of extended circulatory arrest to repair a complicated congenital defect. The child's circulation was connected in series to her mother's diverting oxygen poor blood away from the patient's heart and lungs and returning oxygen-saturated blood to her arterial system. Although many were amazed by and congratulatory of Dr. Lillehei's efforts, critics were outspoken of their disapproval of a technique that risked the death of two patients for the benefit of one. Nevertheless, these early successes provided proof of the concept and a mechanical substitute for cross-circulation soon followed.

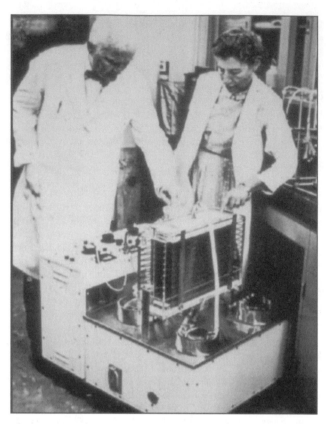

Figure 1. John and Mary Gibbon with their heart–lung machine. (Reprinted with permission from Thomas Jefferson University Archives.)

John and Mary Gibbon are credited with building the first usable CPB machine, a prototype of which John Gibbon employed in 1935 to support a cat for 26 min (2) (Fig. 1). This pump, which used two roller pumps and other similar early designs, were traumatic to blood cells and platelets and allowed for easy air entry into the circulation, which often proved catastrophic. Later, in 1946, Dr. Gibbon in collaboration with Thomas Watson, then chairman of IBM, made further refinements. Together they were able to successfully perform open-heart surgery in dogs, supporting the animals for period exceeding 1 h (3). Finally on May 6, 1953, Dr. Gibbon used his heart–lung machine to successfully repair an atrial septal defect in an 18 year-old girl, marking the first successful clinical use of a heart–lung machine to perform open-heart surgery. Gibbon's first attempt in 1952 followed two unsuccessful attempts by Clarence Dennis et al. in 1951 (4), and it too ended in failure. Sadly, subsequent failures led Gibbon to finally abandon this new technology, finally giving up heart surgery completely.

Kirklin's group at the Mayo Clinic modified the Gibbon pump oxygenator and reported a series of eight patients in 1955 with a 50% survival (5). Around the same time, Frederick Cross and Earl Kay developed a rotating disk oxygenator that had a similar effectiveness (6,7). This apparatus was similar in design to that used by Dennis et al. several years earlier (8). While this Kay-Cross unit became commercially available, it shared many of the

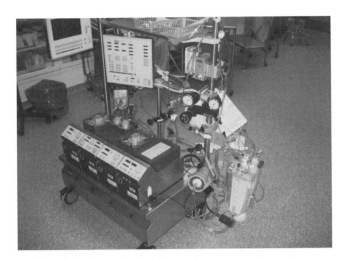

Figure 2. Typical cardiopulmonary bypass circuit demonstrating the various components. This circuit contains four accessory pumps for suction/venting and the administration of cardioplegia. (Reprinted with permission from Ref. 13.)

limitations of the Gibbon system. These pumps were cumbersome, difficult to clean and sterilize, and were inefficient and thus wasteful of blood.

Collaboration between Richard DeWall and Lillehei provided the next advance in pump oxygenators, and in 1956, they reported their first clinical experience using a bubble oxygenator in seven patients, five of whom survived (9). Further refinements by Gott et al. (10) resulted in a reproducible, disposable plastic sheet oxygenator, a system that was readily accepted by pioneering centers around the world. The bubble oxygenator became the predominant pump oxygenator for the next three decades. More recently, the membrane oxygenator, an early design of which was described by Kolff et al. in 1956 (11), has replaced it. Lande et al. described the first commercially available disposable membrane oxygenator (12), but its advantages to the bubble oxygenator were not readily apparent. By the mid-to-late 1980s, advanced microporous membrane oxygenators began to supplant the bubble oxygenator in the clinical arena and they remain the predominant design in use today (Fig. 2).

THE CIRCUIT

The first requirement of a CPB circuit is a means to evacuate and return blood to the circulation. Special drainage and return catheters known as cannulae have been devised for this purpose. One or more venous cannulae are placed in a central vein or in the heart to empty the heart. The venous cannulae are connected to a series of poly(vinyl chloride) tubing that deliver the blood to the remainder of the circuit.

The next component is a reservoir that collects and holds the blood. The reservoir is often used to collect shed blood from the operative field and to store excess blood volume while the heart and lungs are stopped. Drainage of the blood is accomplished by gravitational forces and is sometimes assisted with vacuum. The movement of blood

through the remainder of the circuit, however, requires the use of a pump. Many pump designs have been used over the years, but the most common in current use are the roller pump and the centrifugal pump. A roller pump is a positive displacement pump and has been the most prolific design to date for this purpose. These pumps use rollers mounted on the ends of a rotating arm that displace blood within the compressed tubing, propelling it forward with faint pulsatility. Both the rate and direction of rotation and the occlusion pressure can be adjusted to adjust the flow. In contrast, centrifugal pumps use a magnetically driven impeller design that create a pressure differential between the inlet and the outlet portion of the pump housing propelling blood. Less traumatic to the blood elements, the centrifugal pump has largely replaced the roller pump for central circulation at most large centers.

The blood is actively pumped through a series of components before it is returned to the body. A heat exchange system is used to first lower the blood and body temperature during heart surgery. Controlled hypothermia reduces the body's energy demands, safely allowing reduced or even complete cessation of blood flow, which is necessary during certain times in the course of an operation. Reversing the process later rewarms the blood and body.

The pump oxygenator is by far the most sophisticated component of the circuit. As discussed, the membrane oxygenator is the predominant design in use today. Membrane oxygenators employ a microporous membrane that facilitates gas exchange in the circulating blood. Pores less than 1 μm prevent the leakage of serum across the membrane yet allow efficient gas exchange. An integrated gas mixer or blender enables the perfusionist to adjust the oxygen and carbon dioxide content in the blood and thus regulate the acid–base balance. In addition, an inhalational gas vaporizer provides anesthetic to the patient for the period during CPB. In line oxygen saturation monitors supplemented with frequent blood gas analysis ensure physiologic requirements are met during CPB.

Finally, an in line filter (typically 40 μm) prevents air and thromboembolic particles from returning to the arterial circulation. In parallel with this circuit, several additional roller pumps are used to provide suction that returns shed blood form the operative field. These are critical in open-heart procedures where a dry operative field is imperative for good visualization. An additional pump is often needed to deliver a cardioplegia solution, a mixture of blood and hyperkalemic solution used to stop the heart. Various monitors, hemofiltration units, blood sampling manifolds, gauges, and safety valves are usually present.

The main arterio-venous circuit is generally primed with a balanced electrolyte solution, which obviates the need for the blood prime used in the earlier, higher volume circuits. Other prime constituents include buffers such as sodium bicarbonate, oncotics such as albumin and mannitol, and anticoagulation in the form of heparin. Some protocols also call for the addition of an antifibrinolytic such as aminocaproic acid or aprotinin to the prime.

The overall construction of the CPB circuit is highly variable among institutions depending on the preferences of the surgeons and perfusionists. Although the principals of perfusion are universal, individual and group practices are not. Perfusionists are a highly trained group of specialists dedicated to the safe operation of the heart–lung machine. Perfusion education and training has evolved concurrently with improvements in the bypass circuit, and accredited perfusion programs are included in many allied health curriculums at select universities. Perfusionists are required to pass a board examination, and many government organizations are now enacting licensure legislation. These persons must be able to adapt to the ever-changing technology present in cardiac surgery as well as to protocols and circuitry that vary among surgeons and institutions.

FUTURE DIRECTIONS

There are many commercial designs currently available that incorporate several of these components into a single disposable unit. Some units, for example, combine a reservoir with a membrane oxygenator and a heat exchanger. Continued innovation has resulted in more efficient, compact designs that limit the blood contacting surface area and decrease the need for large priming volumes and thus the need for patient transfusions. Improving biocompatibility of materials has lessened the inflammatory response that is associated with extracorporeal (outside the body) circulation. Sophisticated cannula designs are less traumatic to the vessels and have improved flow dynamics, causing less sheer stress on circulating red blood cells. New safety mechanisms such as alarm systems, pop-off valves, and automatic air evacuation devices will further increase the efficacy of these lifesaving machines.

The field of cardiac surgery is similarly evolving. Minimally invasive approaches to many heart operations are developing driven by patient demand as well as efforts to reduce postoperative hospital stay and patient morbidity. Recent debate has also focused on the possible adverse neurologic sequelae associated with the use of CPB. A growing number of coronary bypass procedures are now performed without the use of a heart–lung machine. Some centers have demonstrated success with this approach, decreasing the need for postoperative blood transfusions and end-organ dysfunction. To date, however, there is no conclusive evidence that avoidance of the heart lung machine results in improved neurologic outcomes or patient survival.

The creation of the heart–lung machine was the rate-limiting step to the development of the field of cardiac surgery. The ability to stop the heart while preserving flow the remainder of the heart has given surgeons the ability to repair defects in the heart and great vessels in patients of all ages. The pioneers in this field demonstrated remarkable courage and conviction in persevering in the face of overwhelming odds. Their collaborative efforts during one of the most prolific periods in the history of medicine have had a remarkable impact on human health. The future of cardiac surgery is largely dependent on continued advances in perfusion techniques and in the components of the heart–lung machine. In this endeavor, industry plays a pivotal role in developing and manufacturing improved products tailored to meet the needs of an everchanging field. Ultimately, evidence-based outcomes research will

help ensure these innovations result in improved outcome for patients.

BIBLIOGRAPHY

Cited References

1. Lewis FJ, Taufic M. Closure of atrial septal defects with the aid of hypothermia; experimental accomplishments and the report of one successful case. Surgery 1953;33:52–59.
2. Gibbon JH Jr. Artificial maintainance of circulation during experimental occlusion of the pulmonary artery. Arch Surg 1937;34:1105–1131.
3. Gibbon JH Jr., et al. the closure of interventricular septal defects in dogs during open cardiotomy with the maintainance of the cardiorespiratory functions by a pump oxygenator. J Thor Surg 1954;28:235–240.
4. Dennis C, et al. Development of a pump oxygenator to replace the heart and lungs: an apparatus applicable to human patients and application to one case. Ann Surg 1951;134: 709–721.
5. Kirklin JW, et al. Intracardiac surgery with the aid of a mechanical pump oxygenator system (Gibbon type): Report of eight cases. Proc Mayo Clin 1955;30:201–206.
6. Kay EB, et al. Certain clinical aspects in the use of the pump oxygentor. JAMA 1956;162:639–641.
7. Cross FS, et al. Description and evaluation of a rotating disc type reservoir oxygenator. Surg Forum 1956;7:274–278.
8. Dennis C, Karleson KE, Nelson WP, Eddy FD, Sanderson D. Pump-oxygenator to supplant the heart and lunf for brief periods. Surgery 1951;29:697–713.
9. Lillehei CW, et al. Direct vision intracardiac surgery in man using a simple, disposable artificial oxygenator. Dis Chest 1956;29:1–8.
10. Gott VL, et al. A self-contained, disposable oxygenator of plastic sheet for intracardiac surgery. Thorax 1957;12:1–9.
11. Kolff WJ, et al. Disposable membrane oxygenator (heart-lung machine) and its use in experimental surgery. Clev Clin Q 1956;23:69–97.
12. Lande AJ, et al. A new membrane oxygenator-dialyzer. Surg Clin North Am 1967;47:1461–1470.
13. Gravlee GP, Davis RF, Kurusz M, Utley JR, editors. Cardiopulmonary Bypass; Principles and Practice, Philadelphia: Lippincott Williams and Wilkins; 2000.

Reading List

Edmunds LH. Cardiopulmonary bypass after 50 years. New Engl J Med 2004;351:1603–1606.
Iwahashi H, Yuri K, Nosé Y. Development of the oxygenator: Past, present and future. J Artif Organs 2004;7:111–120.

See also BLOOD GAS MEASUREMENTS; PULMONARY PHYSIOLOGY; RESPIRATORY MECHANICS AND GAS EXCHANGE.

HEAT AND COLD, THERAPEUTIC

MARY DYSON
United Kingdom

INTRODUCTION

Therapeutic levels of heating accelerate the resolution of inflammation, relieve pain, and promote tissue repair and regeneration. Therapeutic levels of cooling reduce inflammation, relieve pain, and can reduce the damage caused by some injurious agents. To use therapeutic heating (thermotherapy) and therapeutic cooling (cryotherapy) effectively, the clinician should know:

- The physiological effects of heat and cold
- What devices are available to produce temperature changes

  ○ How they work
  ○ When and when not to use them
  ○ Their advantages and disadvantages.

This information assists the clinician in selecting the most suitable therapy and device for each patient.

The aim of this article is to provide this information, following a brief description of the history of these therapies, (this being included because history informs current and future usage).

HISTORY OF THERMOTHERAPY AND CRYOTHERAPY

Heat and cold have been used to treat diseases, aid healing, and reduce pain for many millennia. Exposure to the warmth of sunlight and thermal mineral springs continues to be used to this day. The ancient Greeks and Romans also advocated heated sand and oils for the treatment of injuries (1). Heated aromatic oils were rubbed on the body before massage, a form of therapy still in use today. In contrast, lesions accompanied by burning sensations (e.g., abscesses) were commonly treated by the application of cooling substances such as cold water and ice packs, another form of therapy still in use today. Fever was treated with cold drinks and baths; eating snow alleviated heartburn.

Thermotherapy

In the seventeenth century, one treatment for arthritis and obesity was to bury patients up to their necks in the sun-warmed sand of beaches. Warm poultices made from vegetable products such as leaves and cereals were used to treat musculoskeletal and dermal ailments. Molten wax was used to treat bruises around the eyes and infected eyelids, a technique not recommended because of the danger of getting hot wax into the orbit and damaging the eye.

In the eighteenth century, hot air produced by ovens and furnaces was used to induce perspiration and improve the circulation (2).

In the nineteenth century, Guynot found that wounds healed faster when the ambient temperature was 30 °C than when the ambient temperature was lower than this. After the invention of electric light bulbs, these were used to produce light baths and heating cabinets that were used to treat neuralgia, arthritis, and other conditions. Different wavelengths of light were also produced, initially by the use of prisms and passing light through media of different colors, and most recently by the use of lasers. There is evidence that different wavelengths of the electromagnetic spectrum including light and infrared radiation have different biomedical effects (3). When light is used to produce heat, both phototherapeutic and thermotherapeutic effects are produced that may reinforce each other.

In the late nineteenth and early twentieth century, the availability of electricity as a power supply led to the development of a host of novel thermotherapeutic devices including whirlpool baths, paraffin baths, and diathermy machines. In diathermy, high frequency electric currents are used to heat deeply located muscles. This effect was discovered in the 1890s when d'Arsonval passed a 1 A current at high frequency through himself and an assistant. He experienced a sensation of warmth (4). d'Arsonval worked on the medical application of high-frequency currents throughout the 1890s, work that led to the design of a prototype medical device by Nagelschmidt in 1906 and the coining of the term "diathermy" for what had previously been known as "darsonvalization." The first diathermy machines were long wave, and in the second decade of the twentieth century, these were used to treat a wide range of diseases, including arthritis, poliomyelitis, and unspecified pelvic disorders. In 1928, short-wave diathermy was invented, superceding long-wave diathermy in Europe after the Second World War.

Therapeutic ultrasound also became popular in the post-war period, initially as a method of heating tissues deep within the body to relieve pain and assist in tissue repair. It was demonstrated experimentally that ultrasonic heating was accompanied by primarily nonthermal events such as micromassage, acoustic streaming, and stable cavitation. These events, which also occur at intensities less than are required to produce physiologically significant heating of soft tissues, produce cellular events that accelerate the healing process (3,5). Therapeutic ultrasound is generally used at frequencies in the 0.75–3 MHz range. The higher the frequency, the shorter the wavelength and the lower the intensity needed to produce physiologically significant heating of soft tissues. However, higher frequencies, because they are more readily absorbed than lower frequencies, are less penetrative and are therefore less suitable for heating deep soft tissues. Since the 1990s, low kilohertz ultrasound, also known as long-wave ultrasound, has been used to initiate and stimulate the healing of chronic wounds (6). Although long-wave ultrasound is not primarily a heating modality (7), as with all therapies in which energy is absorbed by the body, it is inevitably transduced into heat, although in this instance insufficient to be clinically significant (8).

Cryotherapy

Until the invention in 1755 of artificial snow, which was made by placing a container of water over nitrous ether as the latter vaporized, only cold water and naturally occurring ice and snow were available as means of cooling the body. In the nineteenth century, ice packs were used over the abdomen to reduce the pain of appendicitis and over the thorax to reduce the pain of angina. Ice was also used as a local anesthetic. In 1850, evaporative cooling methods were introduced; for example, ether applied to the warm forehead had a cooling effect as it evaporated. Since the early days of physical therapy, the local application of ice has been used to treat acute musculoskeletal injuries; usually a combination of rest, ice, compression, and elevation (RICE) are recommended.

Contrast Bath Hydrotherapy

Hydrotherapy is the use of water as a therapeutic agent. Hot or cold water, usually administered externally, has been used for many centuries to treat a wide range of conditions, including stress, pain, and infection. Submersion in hot water is soothing and relaxing, whereas cold water may be anti-inflammatory and limit the extent of tissue damage. In contrast bath hydrotherapy, the patient is exposed to hot and cold water alternately to stimulate the circulation. The blood vessels dilate in the heat and constrict in the cold. Pain is also reduced. Contrast baths have been used to treat overuse injuries such as carpal tunnel syndrome and tendonitis of the hand and forearm (http://www.ithaca.edu/faculty/nquarrie/contrast.html). Although there has been little research on the efficacy of contrast baths, they remain in use and may be of clinical value (9).

THE PHYSIOLOGICAL EFFECTS OF HEAT AND COLD

When healthy we keep a fairly constant body temperature by means of a very efficient thermoregulatory system. We are homoeothermic organisms. Homoeothermic is a pattern of temperature regulation in which cyclic variation of the deep body (core) temperature is maintained within arbitrary limits of ±2 °C despite much larger variations in ambient temperature. In health and at rest, our core temperature can be maintained by the thermoregulatory system within ±0.3 °C of 37 °C in accordance with the body's intrinsic diurnal temperature cycle. Superimposed on the diurnal temperature cycles are monthly and seasonal temperature cycles. Hyperthermia is a core temperature greater than 39 °C. Hypothermia is a core temperature less than 35 °C.

The physiological effects of heat and cold are generally independent of the agent producing the temperature change, although in some cases, for example, ultrasound therapy, primarily nonthermally induced events accompany those induced thermally (3).

Physiological Effects of Heat

Heating of tissues occurs when these tissues absorb energy. The different effects of heating are due to many factors, including the following (10):

- The volume of tissue absorbing the energy
- The composition of the absorbing tissue
- The capacity of the tissue to dissipate heat, a factor largely dependent on its blood supply
- The rate of temperature rise
- The amount by which the temperature is raised

Cell Metabolism. Cell metabolism increases by about 13% for each 1 °C increase in temperature up to the temperature at which the proteins of the cell, many of which are vital enzymes, coagulate. Enzyme activity first increases as the temperature increases, then peaks, then declines as the enzymes denature, and is finally abolished when the temperature reaches about 45 °C when heat kills the cells. Only temperature increases less than those producing enzyme denaturation are therapeutic. The cells' membranes are particularly sensitive to heat, which increases the fluidity of their lipoproteinaceous components, producing changes in permeability (11). At sublethal temperatures, heat-shock proteins, which give some protection to cells reexposed to heat, accumulate in the cells.

Abnormal and normal cells are affected differently by mild hyperthermia (~ 40 °C). Synthesis of deoxyribonucleic acid (DNA), ribonucleic acid (RNA), and proteins can all be inhibited by mild hyperthermia in abnormal cells, irreversibly damaging their membranes and killing the cells; this does not occur in normal cells subjected to mild hyperthermia (12). The technique can therefore be used to kill abnormal cells selectively.

Collagen. The extensibility of collagen is increased by thermotherapy. Raising the temperature of a highly collagenous structure such as a tendon increases the extent of elongation produced by stretch of a given intensity (13). Joint capsules are also highly collagenous, so thermotherapy decreases the resistance of a joint to movement. Cryotherapy, however, increases this resistance. Changes in resistance to movement are also due to changes in the viscosity of the synovial fluid.

Laboratory experiments suggest that heat should be applied to tendons with caution because *in vitro* exposure of excised tendons to temperatures in the range of 41–45 °C is accompanied by a reduction in their tensile strength (13). It is unlikely, however, that the stresses produced by passive stretch during physical therapy and normal exercise will reach the levels at which rupture occurs, although overvigorous exercise could be damaging, particularly in those who ignore protective pain.

Blood Flow. Thermotherapy causes blood vessels to widen (vasodilatation), increasing local blood flow and causing the skin to redden (erythema). Heat-induced vasodiltation has several causes, including the following:

- The direct effect of heat on the smooth muscle cells of the arterioles and venules.
- If there is local damage, then the damage-induced release of vasodilators such as bradykinin will cause further vasodiltation.

Bradykinin, histamine, and other chemical mediators released in response to injury and to heating increase capillary and venule permeability that, together with an increase in capillary hydrostatic pressure, can produce local swelling of the tissues (edema). The application of local heat immediately after injury should therefore be avoided (14).

Heating also induces changes in blood flow in subcutaneous tissues and organs. These changes depend on the amount of heating. First blood flow increases in these structures, but then it decreases if heating is sufficient for the core temperature to rise, as blood is diverted to the skin where heat is lost from it to the external environment as part of the thermoregulatory process.

Different local heating techniques can have different effects on blood flow, due to differences in their depth of penetration. Methods producing superficial heating include infrared radiation and contact with a heated material; those producing deep heating include short-wave and microwave diathermy. Infrared (IR) radiation increases cutaneous blood flow (15) but not in the underlying skeletal muscle (16). In contrast, diathermy increases blood flow and temperature in both skin and skeletal muscle in humans (17), hyperemia being sustained for at least 20 min after the cessation of treatment, probably because of an increase in the metabolic rate of the heated tissues.

Neurological Effects.   Therapeutic heat produces changes in muscle tone and pain relief.

Muscle Tone.   Increased muscle tone can sometimes be reduced by the application of either superficial or deep heat. In 1990, Lehmann and de Lateur (18) showed that heating skin and skeletal muscle of the neck relieved muscle spasm secondary to underlying pathology. The Ia afferents of muscle spindles increase their firing rate on receipt of heat within the therapeutic range, as do tendon organs (19). Most secondary afferents decrease their firing rate (20). Collectively these neurological changes reduce muscle tone locally.

Pain Relief.   People in pain generally consider heat to be beneficial, even on the intense pain experienced by patients with cancer (21).

Therapeutic heat produces vasodiltation and may therefore relieve pain related to ischemia. Pain related to muscle spasm may also be relieved by therapeutic heat; this reduces muscle spasm secondary to underlying pathology (18). Heat may also act as a counterirritant and relieve pain via the pain gate mechanism (22), in that the thermal sensations take precedence in the central nervous system over nociceptive sensations.

Increasing the temperature within the therapeutic range increases the velocity of nerve conduction. An increase in sensory conduction increases the secretion of pain-relieving endorphins.

Tissue Injury and Repair.   The initial response of vascularised tissues to injury is acute inflammation. This is characterized by:

- Heat
- Redness
- Swelling (edema)
- Pain
- Loss of function

During it a host of chemical mediators and growth factors are secreted; these collectively limit the extent of tissue damage and lead to healing. The application of therapeutic levels of heat can accelerate the resolution of acute inflammation leading to faster healing. Although uncomfortable, acute inflammation is not a disease but a vital component of tissue repair and regeneration. The pain associated with it is generally of a short duration and is of survival value in eliciting protective actions.

The management of acute trauma by thermotherapy and cryotherapy is based on a pathophysiological model often referred to as the secondary injury model (23). According to this model, secondary injury is the damage that occurs as a consequence of the primary injury in previously unaffected cells. The mechanisms initially hypothesized as producing this damage were classified as being either enzymatic or hypoxic (24). Since then, knowledge of the mechanisms involved in cell death from trauma has increased dramatically and a third mechanism is now postulated, namely the delayed death of primary injured cells. A review and update by Merrick in 2002 (25) attempts to reconcile the secondary injury model with current knowledge of pathophysiology. Secondary hypoxic injury has been reclassified as secondary ischemic injury, and specific mechanisms for ischemic injury have been identified. In addition to changes of vascular origin, there is now evidence that apparently secondary injury may be due, in part, to the delayed death of cells subjected, for example, to mitochondrial damage during the primary trauma. A better understanding of secondary injury should inform methods of treatment, some of which, although traditional, may need to be altered to improve their effectiveness. For example, the rationale that short-term cryotherapy of acute injuries was effective because it limited edema through vasoconstriction has been replaced by the currently accepted theory that it also retards secondary injury, regardless of the cellular mechanisms by which this occurs, be they lysosomal mechanisms, protein denaturation mechanisms, membrane permeability mechanisms, mitochondrial mechanisms, and/or apoptotic mechanisms (25). By reducing metabolic demand, the application of cryotherapy as soon as possible after an acute injury should reduce the speed and, possibly the extent, of secondary injury. There is evidence that continuous cryotherapy for 5 hours after a crush injury inhibited the loss of mitochondrial oxidative function that follows such injuries (25). The effect of continuous and intermittent cryotherapy for other durations on this and other pathophysiological events remains to be examined. This must be done if clinical treatments are to be improved.

Systemic and local therapeutic heating can reduce post-operative wound infection (26). The use of a warm-up dressing (Augustine Medical), which radiates heat and provides a moist wound environment, eradicated methicillin-resistant *Staphylococcus aureus* (MRSA) infection from pressure ulcers within 2 weeks (27). It should be appreciated that patients with MRSA-infected pressure ulcers, also known as pressure sores and bed sores, have increased morbidity and mortality; these ulcers can kill. Warming the tissues induces vasodiltation, giving rise to the high oxygen tension required for the production of oxygen-free radicals; these are an important part of the body's defense against bacterial infection (28) and initiate collagen synthesis and re-epithelialization. Vasodiltation also aids healing by increasing the efficiency with which the cells and growth factors needed for this are transported to the injured region and the efficiency with which waste products are removed from it.

Warming has been shown to increase the proliferation of fibroblasts (29), the cells that synthesize collagen, and of endothelial cells (30), the cells lining blood vessels, *in vitro*. If this also occurs *in vivo*, then warming will accelerate the formation of granulation tissue. More research is required to confirm whether therapeutic heating reduces the risk of wound infection and accelerates healing in a clinically significant fashion. The data gathered to date suggest that it may. If so, then "there is a real prospect of reducing complications, improving patient outcomes, shortening hospital stays and minimizing costs" (31).

Physiological Effects of Cold

The physiological effects of therapeutic cold (9) depend on a range of factors, including

- The volume of tissue to which cooling is applied
- The composition of the tissues cooled
- The capacity of the tissues to moderate the effects of cooling
- The rate of temperature fall
- The amount by which the temperature of the tissues is lowered

Cell Metabolism.   The vital chemical processes occurring in cells generally slow as the temperature is lowered. These processes are catalyzed by enzymes, many of which are associated with the cells' membranes. Cell viability relies on passive and active membrane transport systems, the latter involving ionic pumps activated by enzymes. These transport systems are necessary to maintain the intracellular ionic composition required for cell viability. Below a threshold temperature, the pumps fail, and consequently, the membranes lose their selective permeability; the intracellular concentration of Na^+ and Ca^{2+} increases whereas that of K^+ decreases. Between normal body temperature and this threshold, cooling is therapeutic. The application of therapeutic levels of cold can reduce cell degeneration and therefore limit the extent of tissue damage (31). The induction of mild hypothermia in the brain of a baby starved of oxygen at birth can interrupt the cascade of chemical processes that cause the neurons of the brain to die after being deprived of oxygen (32), reducing disability.

Collagen.   Collagen becomes stiffer when cooled. People with rheumatoid arthritis generally experience a loss of mobility of their affected joints at low temperatures, due in part to increased stiffness of the collagen of their joint capsules (9).

Blood Flow.   Lowering the skin temperature is detected by dermal thermoreceptors that initiate an autonomic reflex narrowing of the blood vessels of the skin (vasoconstriction). This results in reduction of the flow of blood to the dermis. Cold also has a direct constrictor effect on the smooth muscle of the blood vessels, and the arteriovenous anastomoses that shunt blood to the skin close. The resulting reduction in the flow of blood to the dermis diminishes heat loss through it. Countercurrent heat exchange between adjacent arteries and veins reduces heat loss.

These processes collectively reduce the rate at which the core temperature of the body falls.

Dermal vasoconstriction induced by lowering the temperature of the skin to approximately $10\,°C$ is followed after a few minutes by cold-induced vasodiltation (CIVD) followed by cycles of alternating vasoconstriction and vasodiltation, resulting in alternating decrease and increase on dermal blood flow. Originally thought to be either a local neurogenic axon reflex or due to the local release of vasodilator materials into the tissues, CIVD is now believed to be due to paralysis of vascular smooth muscle contraction in direct response to cold (33). This reaction may provide some protection to the skin from damage caused by prolonged cooling and ischemia. However, in CIVD of the skin, the erythema produced is a brighter red than in erythema produced by heating because at low temperatures, the bright red oxyhemoglobin dissociates less readily than at higher temperatures. Therefore, although in CIVD the skin receives oxygen-rich blood, it is still starved of oxygen, suggesting that cryotherapy may not aid wound healing (9).

In contrast to skin, the blood flow in the skeletal muscles is determined more by local muscle metabolic rate than by temperature changes at the skin because muscles are insulated from these by subcutaneous fat.

Neurological Effects.   Cold can be used therapeutically to affect muscle tone and pain.

Muscle Tone.   Although cooling generally decreases muscle tone, this can be preceded by a temporary increase in tone (34), possibly related to tactile stimulation accompanying the application of cryotherapy by means of ice massage. Decrease in muscle tone in response to cryotherapy is likely to be due to a decrease in muscle spindle sensitivity as the temperature falls, together with slowing of conduction in motor nerves and skeletal muscle fibers.

Pain Relief.   The immediate response of the skin to cold is a stimulation of the sensations of cold and pain. However, if the cold is sufficiently intense, nerve conduction is inhibited, causing both sensations to be suppressed. Less intense cold can be used as a counterirritant, its effects being explicably by the pain gate theory of Wall and Melzack (35). Enkephalins and endorphins may also be involved (36).

Tissue Injury and Repair.   Cryotherapy has beneficial effects on the acute inflammatory phase of healing in that it reduces bleeding, decreases edema at the injury site, gives pain relief, and reduces local muscle spasm, as described above. Once these events have occurred, it should be replaced by thermotherapy because, as previously described, this accelerates the resolution of acute inflammation leading to faster healing.

DEVICES AVAILABLE FOR PRODUCING TEMPERATURE CHANGE

Thermotherapy Devices

These can be classified into those producing superficial heating and those producing deep heating. All should be

Figure 1. Hydrocollator pack for moist heating by conduction. Each pack consists of silica gel covered by cotton fabric. After submersion in hot water, the pack is applied to the area to be heated over layers of terry cloth.

used with caution in patients unable to detect heat-induced pain or over tissues with a poor blood supply because these cannot dissipate heat efficiently.

Superficial Heating. This is provided by devices that transfer energy to the body by conduction, convection, or radiation (37).

Conduction. Devices that heat by conduction include hydrocollator packs, hot water bottles, electric heating pads, and baths of heated paraffin wax.

Hydrocollator packs Figs. 1 and 2 consist of cotton fabric bags containing a silica gel paste that absorbs water equal to 10 times its weight. They are placed in thermostatically controlled tanks of hot water, and after they have absorbed this, one or more packs are placed on towcling-covered skin, into which heat is conducted until the pack has cooled, usually or 20 30 min, depending on the ambient temperature. Heat is also conducted into the subcutaneous tissues to a depth of about 5 mm. The transportation of blood warmed in the superficial tissues to deeper tissues may result in the latter also becoming warmer. Hot water bottles and thermostatically controlled electric heating pads act in a similar fashion.

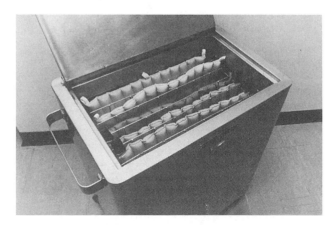

Figure 2. Hydrocollator packs being heated in a thermostatically controlled tank containing hot water.

When and When Not to Use These Devices. They are useful in the relief of pain and muscle spasm. Abdominal cramps can be treated effectively through the induction of smooth muscle relaxation. Superficial thrombophlebitis and localized skin infections such as boils or furuncles may also be treated by heat conduction with these devices. It has been advised that these and other primarily heating devices should not be used over injured tissue within 36 h of the injury. Nor should they be used where there is decreased circulation, decreased sensation, deep vein thrombophlebitis, impaired cognitive function, malignant tumors, a tendency toward hemorrhage or swelling, an allergic rash, an open cut, skin infection, or a skin graft (http://www.colonialpt.com/CPTFree.pdf). A search of English–language textbook and peer–reviewed sources and computerized databases from January 1992 to July 2002 (38) has revealed "generally good agreement among contra-indication sources for superficial heating devices"; however, agreement ranged from 11% to 95% and was lower for pregnancy, metal implants, edema, skin integrity, and cognitive/communicative concerns.

Advantages and Disadvantages. The moist heat produced by the hydrocollator packs has a greater sedative effect than the dry heat produced by the other conduction devices listed above. The hydrocollator packs are, however, heavy because of the amount of water they absorb and should not be applied to very frail patients, for example those with advanced osteoporosis. Hot water bottles occasionally leak, producing scalding and should therefore be inspected after filling and before each use. Electric pads should be thermostatically controlled and used with an automatic timer to reduce the risk of burning the skin.

Paraffin Wax Baths. Molten paraffin wax is mixed with liquid paraffin wax and kept molten in a thermostatically controlled bath (Fig. 3) at between 51.7 and 59.9 °C. Typically used for heating hands or feet, these are either dipped into the bath several times so that a thick layer of wax forms on them, or immersed in the bath for 20–30 min. After dipping, the hand or foot is wrapped in a cotton towel to retain the heat; after 20–30 min, the towel and wax are peeled off.

When and When Not to Use These Devices. Paraffin wax baths are suitable for treating small areas with irregular surfaces. The heat provides temporary relief of joint stiffness and pain in patients with chronic osteoarthritis and rheumatoid arthritis. By increasing the elasticity of collagen, it helps to increase soft-tissue mobilization in the early stages of Dupuytren's contracture and after traumatic hand and foot injuries provided that the lesions have closed. Paraffin wax baths should not be used to open or infected wounds.

Advantages and Disadvantages. Molten paraffin wax is suitable for applying heat to small irregular surfaces such as those of the hands and feet. The baths can be used several times without changing the wax, and several patients can be treated at the same time if the baths are sufficiently large.

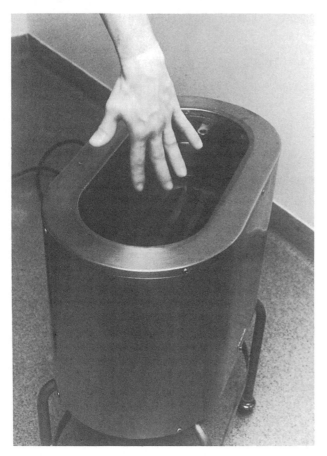

Figure 3. Paraffin wax bath used for heating hands and feet by the dip method.

A disadvantage is that the wax collects debris from the surface of the skin and other particles that settle at the bottom of the bath and are difficult to remove without emptying the bath. Treatments must be supervised and the temperature of the wax monitored to avoid the risk of burning.

Convection.   This involves immersion of either the whole of the body or part of the body in heated water. Hubbard tanks (Fig. 4), hydrotherapy pools (Fig. 5), akin

Figure 4. Hubbard tank used for partial or total body submersion.

to heated swimming pools, and baths can be used for either total or partial immersion. The Hubbard tanks can be fitted with agitation devices. Some baths incorporate low-frequency ultrasound transducers. It is recommended that the water temperature should not exceed 40.6 °C when used for total immersion and 46.1 °C when used for partial immersion (1). Treatments typically last for 20–30 min.

Hydrotherapy is useful for treating multiple arthritic joints simultaneously and for treating extensive soft-tissue injuries. Although total immersion can raise the core body temperature, the main use of hydrotherapy is for the relief of pain, muscle spasm, and joint stiffness. Exercise regimes can be incorporated into hydrotherapy treatment to take advantage of the increase in joint movement made possible by the decrease in joint stiffness and the pain associated with movement. The addition of an agitation device to a Hubbard tank allows it to be used for the gentle debridement of burns, pressure ulcers, and other skin conditions, provided that thorough cleaning and disinfection of the tank are carried out between treatments. Baths incorporating low frequency (kHz) ultrasound transducers have the additional advantage that they can stimulate the repair process as well as assist in debridement.

The greater the surface of the body that is immersed, the greater the number of joints and the greater the area of skin and subcutaneous tissue that can be heated. Immersion also provides buoyancy, helping the patient to exercise more easily.

Disadvantages are that

- Each patient needs one-to-one supervision to ensure that the mouth and nose are kept free from water.
- The equipment is relatively expensive and is time-consuming to maintain.

Radiation.   Heat transmission by radiation is generally provided by infrared lamps (Fig. 6). These are photon producers, emitting wavelengths from the red end of the visible part of the electromagnetic spectrum, together with IR. The production of visible light provides a visual indication when the lamps are active. The IR component provides heating.

Infrared emitters used for heating are either nonlaser or laser devices, the former being the most commonly used.

Nonlaser.

- Luminous
- Nonluminous

Luminous emitters are effectively light bulbs, each mounted in the center of a parabolic reflector that projects a bright beam like a floodlight. About 70% of the energy emitted consists of IR rays in the wavelength range of 700–4000 nm.

Luminous emitters can be classified into those with large, high-wattage (600–1500 W) bulbs, and those with smaller, lower wattage (250–500 W) bulbs.

   1. The large luminous emitters are used to treat large regions of the body such as the lumbar spine. They

Figure 5. Hydrotherapy pool used for partial or total body submersion.

are positioned about 0.6 m from the patient and used for 20 min.

2. The small luminous emitters are used to treat smaller regions of the body such as the hands and feet.

Figure 6. Luminous generator infrared lamp used for superficial heating by radiation. The height and angle of the lamp can be adjusted.

They are positioned closer to the patient, about 0.45–0.5 m away, and they are typically used for 20 min.

The former produce a greater sensation of warmth than the latter. The skin becomes temporarily erythematous, resembling blushing. If the heat is too great because, for example, the lamp is too close to the skin, *erythema ab igne*, a mottled appearance indicative of damage, may be produced.

Nonluminous emitters are usually cylinders of either metal or an insulating material, the radiation source, around which a resistance wire, the heat source, is coiled. They emit longer wavelengths of IR that penetrate deeper than the shorter IR wavelengths of the luminous emitters. About 90% of the emission of the nonluminous IR device is in the wavelength range of 3500–4000 nm.

When and When Not to Use These Devices. IR radiation when used as a heating modality can alleviate pain and muscle spasm. It also accelerates the resolution of acute inflammation. It should not be applied to skin that is photosensitive. It should only be applied to skin free from creams that, if heated, could burn the skin. Goggles should be worn to protect the eyes from radiation damage to either the lens or retina. Patients with abnormally low blood pressure should not receive any form of treatment that increases the temperature of large areas of skin and subcutaneous tissue because of the redistribution of blood to these areas and away from vital organs such as the brain, heart, and lungs.

Advantages and Disadvantages. IR heating is suitable for use in the home as well as in clinics and hospitals. Because the devices described in this section are noncontact, they can be used to treat open and infected wounds, but only for short periods because they cause fluid evaporation and wounds need to be kept moist if they are to heal efficiently.

The main disadvantage is that of burning if the IR heater is placed too close to the patient. Only mechanically stable devices should be used to ensure that they do not fall on the patient.

Laser.   Lasers emitting IR energy have been used to treat injured tissue in Japan and Europe for over 25 years (39). The term "LASER" is an acronym for light amplification by the stimulated emission of radiation. Laser devices are a very efficient means of delivering energy of specific wavelengths to injured tissue. Laser radiation is coherent; that is the waves are in phase, their peaks and troughs coinciding in time and space. Some semiconductor diodes produce coherent and others noncoherent IR radiation. Both can accelerate repair and relieve pain, as does red light (39). The technique of using low intensity lasers therapeutically is usually referred to as LILT, an acronym for low intensity laser therapy, or low level laser therapy (LLLT) to distinguish it from the surgical use of high intensity lasers. Surgical CO_2, ruby, and Nd-YAG surgical lasers can be used for thermal biostimulation provided that the amount of energy absorbed by the tissues is reduced sufficiently. Manufacturers achieve this either by defocusing and spreading the beam over an area large enough to reduce the power density below the threshold for burning or by scanning rapidly over the area to be treated with a narrow beam. The patient will then feel only a mild sensation of heat (39).

Nonsurgical lasers are typically used at power densities below those producing a sensation of heating; although their bioeffects are primarily nonthermal, absorption of light inevitably involves some heating. LILT can be used in either continuous or pulsed mode, the latter further reducing heating.

Single probes, each containing a single diode, are used to treat small areas of skin and small joints. Some are suitable for home use; their output is generally too low to produce the sensation of heat although the radiation they emit is transduced into heat after absorption. The effects of these devices on pain and tissue repair are essentially nonthermal. Clusters of diodes are used in clinics and hospitals to treat large areas of skin, skeletal muscle, and larger joints. These cluster probes generally include, in the interest of cost-effectiveness, nonlaser diodes. Treatment times vary with the type of probe and the area to be treated. Typically they range from about 3 to 10 min. They are calculated in terms of the time taken to deliver an effective amount of energy, generally 4 or more $J \cdot cm^{-2}$, joules being calculated by multiplying the power density of the probe (in $W \cdot cm^{-2}$) by the irradiation time in seconds. The larger cluster probes have sufficient output to produce a sensation of warmth. They also have essentially nonthermal effects that assist in the relief of pain and the stimulation of tissue repair (39).

When and When Not to Use These Devices.   LILT devices are effective on acute and chronic soft-tissue injuries of patients of all ages. Most are designed for use in clinics and hospitals, but small, handheld devices are also available for home use as part of a first aid kit. The radiation they produce is absorbed by all living cells and transduced into chemical and thermal energy within these cells, which are activated by it. Temporary dilatation of superficial blood vessels occurs, aiding oxygenation.

Contraindications for LILT have been described in some detail by Tuner and Hode (39). They include the following:

- As a matter of prudence, LILT should not be applied over the abdomen during pregnancy to ensure that no harm comes to the fetus although it can be applied to other parts of the body.
- In patients with epilepsy who may be sensitive to pulses of light in the 5–10 Hz range, it is advisable to use either unpulsed (continuous) LILT or much higher pulsing frequencies.
- Treatment over the thyroid gland is contraindicated because this gland may be sensitive to absorbed electromagnetic radiation in the red and IR parts of the spectrum.
- Patients with cancer or suspected cancer should only be treated by a cancer specialist.

Advantages and Disadvantages.   LILT is easy to apply directly to the skin. The treatment is rapid and painless, the only sensations being of those of contact and, if the power is sufficiently high, of warmth. If used over an open wound, this should first be covered with a transparent dressing through which the radiation can pass unimpeded. The use of goggles is recommended as a precaution. The larger devices require clinical supervision, but some of the smaller predominantly nonthermal devices are suitable for home use.

Deep Heating.   Deep heating devices produce heat within the tissues via electrostatic, electromagnetic, and acoustic fields. The term diathermy is used to describe the conversion or transduction of any form of energy into heat within the tissues. The devices used for deep heating include short-wave, microwave, and ultrasound generators. They are relatively expensive and should only be used by trained operators.

Short-Wave diathermy (SWD).   SWD equipment Figs. 7 and 8 produces nonionizing radiation in the form of radio waves in the frequency range of 10–100 MHz. The most commonly used frequency is 27.12 MHz; this has a wavelength of 11 m.

SWD machines consist of:

- An oscillating circuit that produces the high frequency current
- A patient circuit connected to the oscillating circuit through which electrical energy is transferred to the patient
- A power supply

The patient's electrical impedance is a component of the patient circuit. Because the patient's electrical impedance is variable, the patient circuit must be tuned to be in resonance with the oscillating circuit to ensure maximum flow of current through the patient.

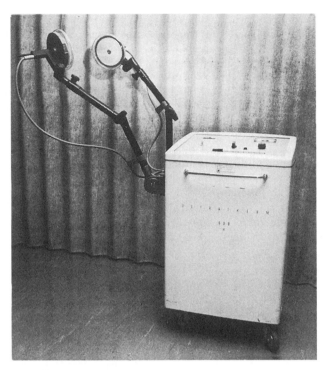

Figure 7. SWD equipment. The SW energy leaving the insulated disk-shaped applicators produces small eddy currents in the tissues, thus heating them.

The most commonly used means of application of SWD are as follows:

- The capacitor plate technique
- The inductive method

The capacitor plate technique entails the placement of two capacitor plates near to the part of the patient's body that is to be heated, with a spacer between each plate and the skin. The spacer allows the electric radiation to diverge just before entering, preventing thermal damage to the surface of the skin. The current density depends on the resistance of the tissues and on capacitance factors. Both subcutaneous fat and superficial skeletal muscle can be heated with this technique. The patient feels a sensation of warmth. Treatment is usually for $20 \, \text{min} \cdot \text{day}^{-1}$.

In the inductive method, a high-frequency alternating current is applied to a coiled cable generally incorporated into an insulated drum that is placed close to the part of the body requiring treatment. Alternatively, but these days rarely, an insulated cable is wrapped around the limb to be treated. The passing of an electric current through the cable sets up a magnetic field producing small eddy currents within the tissues, increasing tissue temperature. The patient feels a sensation of warmth. Treatment is usually for $20 \, \text{min} \cdot \text{day}^{-1}$.

Pulsed Shortwave. Some SWD devices allow the energy to be applied to the patient in short pulses. This form of application is termed pulsed short–wave diathermy (PSWD). The only physical difference between SWD and PSWD is that in the latter the electromagnetic field is interrupted at regular intervals. Pulsing reduces the amount of energy available for absorption by the tissues and therefore reduces the thermal load, allowing its nonthermal effects to be exploited without the risk of a potentially damaging thermal overload. It has been suggested that the applied energy produces ionic and molecular vibration affecting cellular metabolism (40).

With some PSWD machines, the therapist can vary the:

- Pulse repetition rate (PRR)
- Pulse duration (PD)
- Peak pulse power (PPP).

The mean power applied is the product of these three variables.

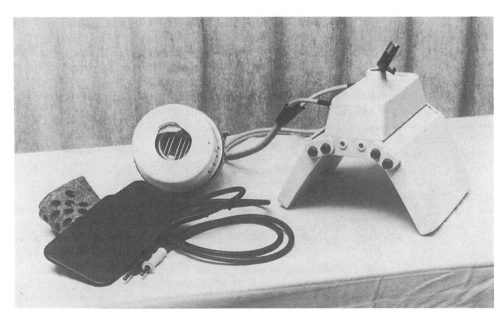

Figure 8. Applicator heads for administration of SWD to produce deep heating of tissues.

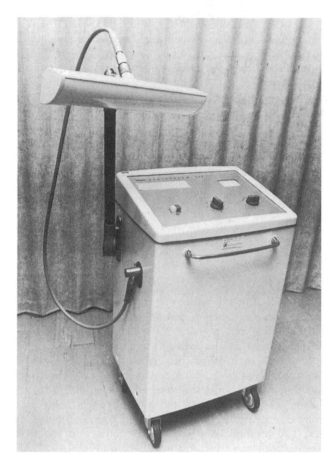

Figure 9. Microwave diathermy equipment. The microwave generator contains a magnetron that produces high frequency alternating current. This current is carried by a coaxial cable shown on the left to a transducer (upper left) containing an antenna and a reflector. The antenna converts the electric current into microwaves that are collected by the reflector. This focuses them and beams them into the body, where they are transduced into heat.

Tissues with good conductivity, such as muscle and blood, i.e., with a high proportion of ions, should absorb energy preferentially from the SWD field. However, there is still debate about which tissues are heated the most during SWD and PSWD treatments.

Microwave Diathermy. Microwaves are that part of the electromagnetic spectrum within the frequency range of 300 MHz–300 GHz and, therefore, with wavelengths between 1 m and 1 mm. Microwave diathermy, although deeper than superficial (surface) heating, is not as deep as capacitative short-wave or ultrasonic heating (40).

Microwave generators (Fig. 9) contain a magnetron, which produces a high frequency alternating current that is carried to a transducer by a coaxial cable. The transducer consists of an antenna and a reflector. The electric current is transduced into electromagnetic energy on passing through the antenna. The reflector focuses this energy and beams it into the tissues to be heated (1).

On entering the body the microwave energy is absorbed, reflected, refracted, or transmitted according to the physical properties of the tissues in its path. When microwaves are absorbed, their energy is transduced into heat. Tissues with a low water content (e.g., superficial adipose tissue) absorb little microwave energy, transmitting it into those with a high water content (e.g., skeletal muscle) that are very absorptive and therefore readily heated, warming adjacent tissues by conduction. As with SWD, there is no objective dosimetry, the intensity of treatment being judged by the patient's sensation of warmth (41).

When and When Not to Use These Devices. As with other forms of heating, SWD and microwave diathermy are used to relieve pain and muscle spasm. They are also used to stimulate repair.

According to the Medical Devices Agency of the United Kingdom, there are several groups of patients on whom these devices must not be used:

1. Those with implanted drug infusion pumps because the energy provided during therapy may affect the pump's electronic control mechanisms causing temporary alteration in drug delivery.
2. Women in the first trimester of pregnancy should not be treated with SWD because heating may be teratogenic.
3. Patients with metallic implants because metals are heated preferentially and may burn the surrounding tissue.
4. Patients fitted with active (i.e., powered) implants such as neurostimulators and cardiac pacemakers/defibrillators because there have been reports of tissue damage, including nerve damage adjacent to stimulation electrodes on implanted lead systems.
5. Patients with pacemakers are also unsuitable for SWD because the frequency of the short-wave may interfere with cardiac pacing.

Patients with rheumatoid arthritis have had their joint pain reduced and walking time increased after treatment with microwave diathermy, and it has been suggested that the heat produced in the joints may have potentiated the effects of concurrent anti-inflammatory medication (42).

In the interests of safety, microwave diathermy should be restricted to patients in whom skin pain and temperature sensation are normal. It should not be used near the eyes, sinuses, and moist open wounds, all of which could be heated excessively because of their high water contact. Nor should it be used on any of the groups of patients in whom SWD is contraindicated.

Advantages and Disadvantages. The devices do not need to be in direct contact with the body; however, the body part being treated must be immobile during treatment because movement interferes with the field, resulting in too little heating in some regions and too much in other. Its disadvantages include its potential for burning tissue exposed to it due, for example, to treatment over regions in contact with metal. As with SWD, the therapist must be careful to avoid being exposed to the radiation, bearing in mind that some of the microwaves will be reflected from the patient.

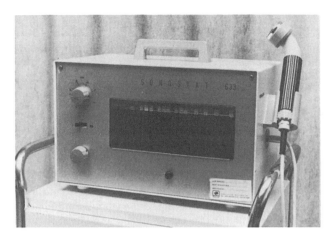

Figure 10. Portable ultrasound therapy equipment. The ultrasound transducer is housed in the head of the applicator shown on the right.

Ultrasound. Ultrasound therapy devices (Fig. 10 designed to produce clinically significant tissue heating operate at 0.75–3.0 MHz, the lower frequencies mechanical waves being more penetrative and affecting deeper tissues than the higher frequency waves. The ultrasound (US) is produced by the reverse piezoelectric effect when a high frequency alternating current is applied to a ceramic crystal causing it to vibrate at the same frequency. The crystal is located inside an applicator fitted with a metal cap into which these vibrations pass. A coupling medium such as either water or a gel with a high water content transmits the ultrasonic vibration into the tissues. US is transmitted readily by water and by tissue fluids. As US is more readily absorbed by protein than by fat, it can be used to heat muscle and collagenous tissue selectively without heating subcutaneous fat to a clinically significant level. Heating occurs where the US energy is absorbed. If the US enters the tissue at an angle other than 90°, it is refracted; the transducer should therefore be held perpendicular to the surface being treated.

Absorption, and therefore heating, is greatest at higher frequencies. As a consequence, it is generally accepted that higher frequency US (e.g., 3 MHz) penetrates less deeply into the body than lower frequency US (e.g., 1 MHz). However, in some circumstances, 3 MHz may heat deeper tissues than originally theorized, when for example, it is transmitted into muscle over a bony prominence via a gel pad coated on both sides with an ultrasound transmitting gel (43). In this clinical investigation, it was found that 3 MHz ultrasound applied over the lateral malleolus of the ankle heated the musculotendinous tissue deep to the peroneal groove 0.5 cm deeper than suggested by others. The 3 MHz ultrasound produced heating deeper into muscle in healthy volunteers than did the 1 MHz ultrasound. This interesting but surprising observation may be an artifact due to variations in coupling of the US transducers to the skin. The authors point out that air trapped at the gel/pad interfaces might result not only in nonuniform heating but also in hot spots on the US transducer faceplate and possibly the skin.

In recent years, US therapy devices producing kilohertz US have been developed (6); these are more penetrative than the MHz devices but produce little heat, using the nonthermal effects of US to produce their effects (7).

When US enters tissue, its intensity gradually decreases as energy is lost from the US beam by absorption, scattering, and reflection. This lessening of the force of the US is termed attenuation. US is scattered by structures smaller than its wavelength. It is reflected at the interfaces between materials with different acoustic impedances, e.g., air and skin, collagen and ground substance, periosteum, and bone. Muscle and bone are preferentially heated at their interfaces with other tissues, e.g., tendon and periosteum. This is because mode conversion occurs when US waves are reflected from these interfaces, the longitudinal incident waves being changed into reflected transverse waves, creating additional heating. Therapists can take advantage of this enhanced thermal effect to target inflamed joints. The thickness of tissue through which the US must pass for its intensity to be reduced to half the level applied to its surface is termed the half-value thickness. The half-value thickness of 1 MHz US is theoretically three times more than that of 3 MHz US although artifacts during clinical use may alter this (43).

The intensity range of US necessary to elevate tissue temperature to between 40 and 45 °C is 1.0–2.0 $W \cdot cm^{-2}$ applied in continuous mode for 5–10 min (44). This increase is in the generally accepted thermal therapeutic range. Temperatures exceeding this can cause thermal necrosis and must be avoided. In poorly vascularized tissues, even the upper end of the therapeutic range can produce thermal damage. When US is absorbed or when mode conversion occurs, heat is generated in the tissues. This produces local increases in blood flow and softens collagenous deposits such as those in scars.

If the temperature increase is less than 1 °C, this is not considered to be clinically significant. Therapeutic effects produced by US used in a manner that produces a temperature increase of less than 1 °C are considered to be predominantly nonthermal. These effects, which include stable cavitation and acoustic streaming (3,44), occur at intensities lower than those necessary to produce therapeutic increases in temperature. The amount of acoustic energy entering the tissues can be reduced by pulsing the US. A commonly used pulsing sequence is 2 ms ON, 8 ms OFF. By reducing the total amount of energy supplied to the tissues, the thermal load on these tissues is reduced. The intensity during the pulse is sufficient to permit the predominantly nonthermal therapeutic events to occur. This is of particular value when ultrasound is used to treat tissue with a blood supply too poor to ensure adequate heat dissipation via the circulation.

The nonthermal therapeutic effects of US include stable cavitation and acoustic streaming. Cavitation is the production of bubbles of gas a few microns in diameter in fluid media such as tissue fluid and blood. The bubbles increase in size during the negative pressure or rarefaction part of the US wave and decrease during the positive pressure part. If the intensity is sufficiently great, the bubbles collapse, damaging the tissues. At lower intensities, within the therapeutic range, the cavities are stable and acoustic streaming is increased around them. Acoustic streaming has been shown to cause reversible changes in membrane permeability to calcium and other ions, stimulating cell division, collagen synthesis, growth factor secretion, myofibroblast contraction, and other cellular events that collectively accelerate the resolution of acute inflammation leading to more rapid tissue repair (3).

When and When Not to Use These Devices. US therapy is used as a thermotherapeutic modality to treat acute and chronic injuries of the

- Skin, e.g., pressure ulcers and venous ulcers
- Musculoskeletal system, e.g., arthritis, bursitis, muscle spasms, and traumatic injuries to both soft tissue and bone.

In addition to superficial injuries, US can be used to heat tissues at a considerable depth from the surface of the skin. Sites of pathology such as damaged skeletal muscle, tendon, and ligaments within 5 cm of the surface can be heated preferentially; any adipose tissue superficial to these lesions is heated less because it absorbs less US than highly proteinaceous muscle, tendons, and ligaments. It is the most suitable form of thermotherapy to use on deeply located areas of damage in obese patients. At least 3 treatments per week are generally recommended, daily treatment being preferable. The treatment head should be moved throughout treatment to reduce the possibility of thermal tissue damage due to local hot spots and mechanical damage due to standing wave formation.

In the interests of safety, continuous US should only be used to heat tissues in patients sensitive to heat-induced pain. In patients lacking this sensitivity, pulsed ultrasound can be used. Although this may not heat the tissues significantly, nonthermal events will occur that can accelerate healing (3,44).

It should not be used either over tumors because it can increase cell division or over the eye because of the risk of collapse cavitation. Nor should it be used on any of the groups of patients in whom SWD is contraindicated.

Advantages and Disadvantages. Its differential absorption makes it the therapy of choice for treating muscle, tendons, ligaments, and bone without heating superficial adipose tissue.

The effects of ultrasound are, however, local and only small regions can be treated because of the small size of the treatment heads, rarely greater than 5 cm^2. Another disadvantage is that a coupling medium such as water or a gel with a high water content must be used to transit the US from the applicator to the tissues. If there is an open wound, this must be covered with an ultrasound transmitting dressing such as Opsite to the surface of which the coupling medium can be applied. Alternatively the intact tissue adjacent to the wound and from which reparative tissue grows into the wound can be treated (3). Care must be taken to ensure that there are no bubbles in the coupling medium because reflection from these

can reduce the efficiency to energy transfer from the treatment head to the tissue. If the US energy is reflected back into the treatment head because of inadequate coupling, this will increase in temperature and could burn a patient, lacking adequate sensitivity to heat-induced pain.

Cryotherapy

Cryotherapy produces a rapid fall in the temperature of the skin and a slower fall in the temperature of the subcutaneous tissues, skeletal muscle, and the core temperature of the body. The rate of fall in skeletal muscle and the core temperature is dependent, in part, on the amount and distribution of adipose tissue. In a slender person with less than 1 cm of adipose tissue in the hypodermis, cooling extends almost 2.5 cm into the skeletal muscle after 10 min of applying cryotherapy to the skin surface. In an obese person with more than 2.5 cm of adipose tissue in the hypodermis, cooling extends only 1 cm into the muscle in the same time (1). To get deeper cooling in an obese person, the cooling agent must be applied for longer than in a slender person, for example, 30 min compared with 10 min to get adequate cooling extending 2.5 cm into the muscle. This concept has been confirmed recently by Otte et al. (45) who found that although 25 min of treatment may be adequate for a slender patient with a skinfold of 20 mm or less, 40 min is needed if the skinfold is between 21 and 30 mm, and 60 min if the skinfold is between 30 and 40 mm. The subcutaneous adipose tissue thickness is an important determinant of the time required for cooling in cryotherapy.

Cryotherapy is used to relieve pain, retard the progression of secondary injury, and hasten return to participation in sport and work. In a literature review by Hubbard et al. in 2004 (46), the conclusion drawn was that cryotherapy may have a positive effect on these aims, but attention was drawn to the relatively poor quality of the studies reviewed. There is a clear need for randomized, controlled clinical studies of the effect of cryotherapy on acute injury and return to participation in sport or work.

The main equipment used in cryotherapy is a refrigerator/freezer necessary for cooling gels and for producing ice. The ice is mixed with water, reducing the temperature of the water to just above its freezing point. The temperature of an injured limb can be reduced by immersing it in this ice/water mixture, an effective but initially uncomfortable experience for the patient. Alternatively, a cold compress containing the mixture can be applied to the region to be cooled. Also, a terry cloth can be soaked in the mixture, wrung out, and then applied. Blocks of ice can be used to massage an injured area if the skin over the injury is intact, an initially uncomfortable experience for both patient and therapist.

Another technique is to spray a vapor coolant on the skin. As the coolant evaporates, the temperature of the skin is reduced, but there is no clinically significant cooling of subcutaneous tissues. Ethylene chloride, chlorofluoromethane, or preferably a non-ozone-depleting vapor coolant, is sprayed over the area to be treated in a stroking fashion at a rate of about $1 \, cm \cdot s^{-1}$ (1). Concern over the ozone-depleting properties of chlorofluorocarbons has led to the development of vapor coolant sprays that are not ozone-depleting (http://www.gebauerco.com/Default.asp) and that may be substituted for those noted above.

When and When Not to Use Cooling. The main use of cryotherapy is after acute skin and musculoskeletal injury to reduce swelling, bleeding, and pain. It helps give temporary relief to painful joints. It also reduces muscle spasms and spasticity, again temporarily. Trigger points, myofascial pain, and fibrositis may be treated with vapor coolant sprays.

Cooling is an excellent way of minimizing the effect of burns and scalds by reducing the temperature at the injury site provided that it is done as soon as possible after the incident. It can also slow brain damage after the deprival of oxygen in, for example, babies for whom delivery has been prolonged and difficult. A cool cap filled with chilled water is placed on the baby's head within a few hours of birth and kept there for several days while the baby is in intensive care. A computerized controller circulates cold water through the cap, reducing the temperature of the brain by several degrees, minimizing cell death within the brain. The results of a clinical trial showed a significantly lower disability and death rate in children at risk of post-natal brain damage due to oxygen deprivation if they were given this treatment than in those whose brains had not been cooled in this way (47). It has been suggested that a similar technique may help patients with hemorrhagic strokes where bleeding has occurred on to the surface of the brain.

Mild systemic hypothermia has also been reported to reduce brain damage after severe head trauma. Patients in whom the core temperature was reduced to and maintained at 33–35 °C with a cooling blanket for 4 days had reduced extradural pressure, an increase in the free radical scavenger superoxide dismutase, and consequently, less neuron loss and improved neurological outcomes (48).

Retarding secondary injury is an important benefit of cryotherapy. Secondary tissue death occurring after the initial trauma has been attributed to subsequent enzymatic injury and hypoxic injury (23). Cryotherapy reduces tissue temperature, slowing the rate of chemical reactions and therefore the demand for adenosine triphosphate (ATP), which in turn decreases the demand for oxygen, leading to longer tissue survival during hypoxia. By decreasing the amount of damaged and necrotic tissue, the time taken to heal may be reduced. In an extensive review of the secondary injury model and the role of cryotherapy, Merrick et al. (24) addressed the following question: "Is the efficacy of short-term cryotherapy explained by rescuing or delaying the death of the cells that were primarily injured but not initially destroyed?" He recommended the replacement of the term "secondary hypoxic injury" by "secondary ischemic injury" because hypoxia presents tissue with the challenge of reduced oxygen only, whereas ischemia presents inadequacies in not only oxygen but also fuel and anabolic substrates and in waste removal, all of which may contribute to secondary injury. Short-term cryotherapy may lessen the demand for these necessities.

Cooling should not be used:

- In people who are hypersensitive to cold-induced pain

- Over infected open wounds
- In people with a poor circulation

Advantages and Disadvantages. Disadvantages are that many people find prolonged exposure to cold therapy uncomfortable. Frostbite can occur if the skin freezes, as is possible if vapor coolant sprays are overused. Furthermore, its effects on chronic inflammatory conditions are generally temporary.

SUMMARY

Changing the temperature of tissues can reduce pain, muscle spasms, spasticity, stiffness, and inflammation. Heating the tissues by a few degrees centigrade increases tissue metabolism and accelerates the healing process. Cooling tissue by a few degrees centigrade can limit secondary damage to soft tissues, nerves, and the brain after trauma.

The correct selection of either thermotherapy or cryotherapy depends on an understanding of the physiological effects of heat and cold, and on knowledge of the patient's medical condition. It is suggested that acute skin lesions and musculoskeletal injuries be treated:

- First by the application of cold as soon as possible after the injury to limit its extent
- Followed a few hours later by the application of moderate heat to accelerate the resolution of inflammation enabling healing to progress more rapidly.

It is recommended that the progress of healing be monitored noninvasively so that treatment can be matched to the response of the injury. In recent years this has become possible by means of high resolution diagnostic ultrasound or ultrasound biomicroscopy (49).

Having decided which is required, heat or cold, the next question is which modality to use. The following should be considered:

- Size and location of the injury
- Type of injury
- Depth of penetration of the modality
- Ease of application
- Duration of application
- Affordability
- Medical condition of the patient
- Contraindications

Patients heal their own injuries if they can to. This can be facilitated by the timely and correct application of therapeutic heat or cold. Their pain can also be relieved, improving their quality of life.

BIBLIOGRAPHY

Cited References

1. Tepperman PS, Kerosz V. In: Webster JG, ed., Encyclopedia of Medical Devices and Instrumentation. 1st ed. New York: John Wiley & Sons; 1988. p 14811–1493.

2. Lehmann JF. Therapeutic Heat and Cold. 3rd ed. Baltimore: Williams & Wilkins; 1982.
3. Dyson M. Adjuvant therapies: ultrasound, laser therapy, electrical stimulation, hyperbaric oxygen and negative pressure therapy. In: Morison MJ, Ovington LG, Wilkie LK, editors. Chronic Wound Care: A Problem-Based Learning Approach. Edinburgh: Mosby; 2004. 129–159.
4. Guy AW, Chou CK, Neuhaus B. Average SAR and distribution in man exposed to 450 mMHz radiofrequency radiation. IEEE Trans Microw Theory Tech 1984;MTT–32:752–762.
5. Dyson M, Suckling J. Stimulation of tissue repair by ultrasound. A survey of the mechanisms involved. Physiotherapy 1978;64:105–108.
6. Peschen M, Weichenthal MM, Schopf E, Vanscheidt W. Low-frequency ultrasound treatment of chronic venous ulcers in an outpatient therapy. Acta Dermatovenereol 1997;77:311–314.
7. Ward AR, Robertson VJ. Comparison of heating of nonliving soft tissue produced by 45 kHz and 1 MHz frequency ultrasound machines. J Org Sports Physiotherapists 1996;23:258–266.
8. Kitchen S, Dyson M. Low-energy treatments: non-thermal or thermal? In: Kitchen S, editor. Electrotherapy. Evidence-Based Practice. Edinburgh: Churchill Livingstone; 2002. 107–112.
9. Stanton DB, Bear-Lehman J, Graziano M, Ryan C. Contrast baths: What do we know about their use? J Hand Ther 2003;16:343–346.
10. Kitchen S. Thermal effects. In: Kitchen S, editor. Electrotherapy. Evidence-Based Practice. Edinburgh: Churchill Livingstone; 2002. 89–105.
11. Bowler K. Cellular heat injury: Are membranes involved? In: Bowler K, Fuller BJ, editors. Temperature and animal cells. Cambridge: Company of Biologists; 1987. 157–185.
12. Westerhof W, Siddiqui AH, Cormane RH, Scholten A. Infrared hyperthermia and psoriasis. Arch Dermatol Res 1987;279:209–210.
13. Lehmann JF, Masock AJ, Warren CG, Koblanski JN. Effects of therapeutic temperatures on tendon extensibility. Arch Phys Med Rehab 1970;51:481–487.
14. Feebel H, Fast H. Deep heating of joints: A reconsideration. Arch Phys Med Rehab 1976;57:513–514.
15. Millard JB. Effects of high frequency currents and infrared rays on the circulation of the lower limb in man. Ann Phys Med 1961;6:45–65.
16. Wyper DJ, McNiven DR. Effects of some physiotherapeutic agents on skeletal muscle blood flow. Physiotherapy 1976;62:83–85.
17. McMeeken JM, Bell C. Microwave irradiation of the human forearm and hand. Physiother Theory Practice 1990;6:171–177.
18. Lehmann JF, de Lateur BJ. Therapeutic heat. In: Lehmann JF, editor. Therapeutic Heat and Cold. 4th ed. Baltimore: Williams & Watkins; 1990. 444.
19. Mense S. Effects of temperature on the discharges of muscle spindles and tendon organs. Pflug Arch 1978;374:159–166.
20. Lehmann JF, de Lateur BJ. Ultrasound, shortwave, microwave, laser, superficial heat and cold in the treatment of pain. In: Wall PD, Melzack R, editors. Textbook of Pain. 4th ed. New York: Churchill Livingstone; 1999. 1383–1397.
21. Barbour LA, McGuire DS, Kirchott KT. Nonanalgesic methods of pain control used by cancer outpatients. Oncol Nurs For 1986;13:56–60.
22. Doubell P, Mannon J, Woolf CJ. The dorsal horn: state dependent sensory processing, plasticity and the generation of pain. In: Wall PD, Melzack R, editors. Textbook of Pain. 4th ed. New York: Churchill Livingstone; 1999. 165–182.
23. Knight KL. Cryotherapy in Sports Injury Management. Champaign (IL): Human Kinetics; 1995.
24. Merrick MA, Rankin JM, Andrea FA, Hinman CL. A preliminary examination of cryotherapy and secondary injury in skeletal muscle. Med Sci Sports Exerc 1999;31:1516–1521.
25. Merrick MA. Secondary injury after musculoskeletal trauma: A review and update. J Athl Train 2002;37:209–217.
26. Melling DAC, Baqar A, Scott EM, Leaper DJ. Effects of preoperative warming on the incidence of wound infection after clean surgery: a randomized controlled trial. Lancet 2001;358:876–880.
27. Ellis SL, Finn P, Noone M, Leaper DJ. Eradication of methicillin-resistant Staphylococcus aureus from pressure sores using warming therapy. Surg Infect 2003;4:53–55.
28. MacFie CC, Melling AC, Leaper DJ. Effects of Warming on Healing. J Wound Care 2005;14:133–136.
29. Xia Z, Sato A, Hughes MA, Cherry GW. Stimulation of fibroblast growth in vitro by intermittent radiant warming. Wound Rep Regen 2000;8:138–144.
30. Hughes MA, Tang C, Cherry GW. Effect of intermittent radiant warming on proliferation of human dermal endothelial cells in vitro. J Wound Care 2003;12:135–137.
31. Zarro V. Mechanisms of inflammation and repair. In: Michlovitz SL, editor. Thermal Agents in Rehabilitation. Philadelphia: Davis; 1986. 3–17.
32. Gluckman PD, Wyatt JS, Azzopardi D, Ballard R, Edwards AD, Ferriero DM, Polin RA, Robertson CM, Thoresen M, Whitelaw A, Gunn AJ. Selective head cooling with mild systemic hypothermia after neonatal encephalopathy: multicentre randomized trial. Lancet 2005;365:632–634.
33. Keatinge WR. Survival in Cold Water. Oxford: Blackwell; 1978. 39–50.
34. Price R, Lehmann JF, Boswell-Bessett S, Burling S, de Lateur R. Influence of cryotherapy on spasticity at the human ankle. Arch Phys Med Rehab 1993;74:300–304.
35. Wall PD, Melzack R, editors. Textbook of Pain. 4th ed. New York: Churchill Livingstone; 1999.
36. Fields HL, Basbaum AI. Central nervous system mechanisms of pain. In: Wall PD, Melzack R, editors. Textbook of Pain. 4th ed. New York: Churchill Livingstone; 1999. 309–330.
37. Tepperman PS, Devlin M. Therapeutic heat and cold. Postgrad Med 1983;73:69–76.
38. Batavia M. Contraindications for superficial heat and therapeutic ultrasound: Do sources agree? Arch Phys Med Rehabil 2004;85:1006–1012.
39. Tuner J, Hode L. The Laser Therapy Handbook. Grangesberg: Prima Books AB; 2004.
40. Scott S, McMeeken J, Stillman B. Diathermy. In: Kitchen S, editor. Electrotherapy. Evidence-Based Practice. Edinburgh: Churchill Livingstone; 2002. 145–170.
41. Low J. Dosage of some pulsed shortwave clinical trials. Physiotherapy 1995;81:611–616.
42. Weinberger A, Fadilah R, Lev A, et al. Treatment or articular effusions with local deep hyperthermia. Clin Rheumatol 1989;8:461–466.
43. Hayes BT, Merrick MA, Sandrey MA, Cordova ML. Three-MHz ultrasound heats deeper into tissue than originally theorized. J Athl Train 2004;39:230–243.
44. Sussman C, Dyson M. Therapeutic and diagnostic ultrasound. In: Sussman C, Bates-Jensen BM, editors. Wound Care. A Collaborative Practice Manual for Physical Therapists and Nurses. 2nd ed. Gaithersburg: Aspen Publishers; 2001. 596–620.
45. Otte JW, Merick MA, Ingersoll CD, Cordova ML. Subcutaneous adipose tissue thickness alters cooling time during cryotherapy. Arch Phys Med Rehabil 2002;83:1501–1505.
46. Hubbard TJ, Aronson SL, Denegar CR. Does cryotherapy hasten return to participation? A systematic review. J Athl Train 2004;39:88–94.

47. Gluckman PD, Wyatt JS, Azzopardi D, Ballard R, Edwards AD, et al. Selective head cooling with mild systemic hypothermia after neonatal encephalopathy: A multicentre randomised trial. Lancet 2005;365:663–670.

48. Qui WS, Liu WG, Shen H, Wang WM, Hang ZL, Jiang SJ, Yang XZ. Therapeutic effect of mild hypothermia on severe traumatic head injury. Chin J Traumatol 2005;8:27–32.

49. Dyson M, Moodley S, Verjee L, Verling W, Weinman J, Wilson P. Wound healing assessment using 20 MHz ultrasound and photography. Skin Res Technol 2003;9:116–121.

HEAVY ION RADIOTHERAPY.   See Radiotherapy, heavy ion.

HEMODYNAMICS

Patrick Segers
Pascal Verdonck
Ghent University
Belgium

INTRODUCTION

"Arterial blood pressure and flow result from the interaction of the heart and the arterial system. Both subsystems should be considered for a complete hemodynamic profile and a better diagnosis of the patient's disease". This statement seems common sense, and a natural engineering approach of the cardiovascular system, but is hardly applied in clinical practice, where clinicians have to deal with limitations imposed by the clinical environment and ethical and economical considerations. The result is that the interpretation of arterial blood pressure is (too) often restricted to the interpretation of systolic and diastolic blood pressure measured using the traditional cuff around the upper arm (cuff sphygmomanometry). Blood flow, if even measured, is usually limited to an estimate of cardiac output.

The purpose of this article is to provide the reader with an overview of both established and newer methods and techniques that allow us to gain more insight into the dynamics of blood flow in the cardiovascular system (the hemodynamics), based on both invasive and noninvasive measurements. The emphasis is that hemodynamics results from the interaction between the action of the heart and the arterial system, and can be analyzed as the interplay between a (complex) pump and a (complex) tube network. This article, has been divided into three main sections. First the (mechanical function of the) heart is considered, followed by a major section on arterial function analysis. The final section deals with cardiovascular interaction.

THE HEART AS A PUMP...

The heart is a hollow muscle, consisting of four chambers, whose function is to maintain blood flow in two circulations: the systemic (or large) and the pulmonary circulation. The left atrium receives oxygenized blood from the lungs via the pulmonary veins. Blood flows (through the mitral valve) into the left ventricle, where it is pumped into the aorta (through the aortic valve) and distributed toward the organs, tissue, and muscle for exchange of O_2 and CO_2, nutrients, and waste products. Deoxygenated blood is collected via the systemic veins (ultimately the inferior and superior vena cava) into the right atrium and flows, via the tricuspid valve, into the right ventricle, where it is pumped (through the pulmonary valve) into the pulmonary artery toward the lungs. Functionally, the pulmonary and systemic circulation are placed in series, and there is a "serial interaction" between the left and right heart. Anatomically, however, the left and right heart are embedded within the pericardium (the thin membrane surrounding the whole heart) and are located next to each other. The part of the cardiac muscle (myocardium) that they have in common is called the myocardial septum. Due to these constraints, the pumping action of one chamber has an effect on the other, a form of "parallel interaction". In steady-state conditions, the left and right heart generate the same flow (cardiac output), on average ~ 6 L/min in an adult at rest.

The Cardiac Cycle and Pressure–Volume Loops

The most heavily loaded chamber is the left ventricle (LV), pumping ~ 80 mL of blood with each contraction (70 beats $\cdot$ min^{-1}), with intraventricular pressure increasing from ~ 5–10 mmHg (1 mmHg = 133.3 Pa) at the onset of contraction (i.e., at the end of the filling period or diastole) to 120 mmHg (6.0 kPa) in systole (ejection period) (Fig. 1). In heart physiology research, it is common to study the function of the ventricle using pressure–volume loops (PV loops; Fig. 1), with volume on the x axis and pressure on the y axis. Considering the heart at the end of diastole, it has reached its maximal volume (EDV; end-diastolic volume). Specialized pacemaker cells within the heart generate the (electrical) stimulus for the contraction, initiating the depolarization of cardiac muscle cells (myocytes). Electrical depolarization causes the muscle to contract and ventricular pressure increases. With this, the mitral valve closes, and the ventricle contracts at closed volume, with a rapidly increasing pressure (isovolumic contraction). When LV pressure becomes higher than the pressure in the aorta, the aortic valve opens, and the ventricle ejects blood. The ventricle then starts its relaxation, slowing down the ejection, with a decrease in LV pressure. At the end of ejection, the LV has reached its end-systolic volume (ESV), and LV pressure drops below the pressure in the aorta, closing the aortic valve. Relaxation then (rapidly) takes place at closed volume (isovolumic relaxation), until LV pressure drops below LA pressure and LV early filling begins (E-wave). After complete relaxation of the ventricle, contraction of the LA is responsible for an extra (late) filling wave (A-wave). The difference between EDV and ESV is the stroke volume, SV. Multiplied with heart rate, one obtains cardiac output (CO), the flow generated by the heart, commonly expressed in L $\cdot$ min^{-1}. The time course of cardiac and arterial pressure and flow is shown in Fig. 1.

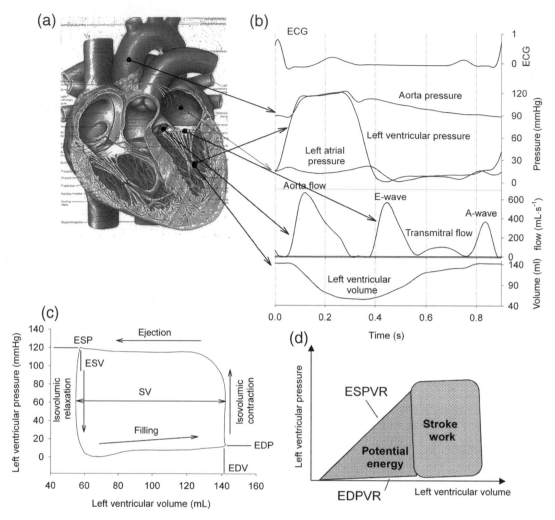

Figure 1. The heart (a), and the variation in time of pressure, flow, and volume within the left ventricle (b). Plotting left ventricular pressure as a function of volume (c), a pressure–volume loop is obtained. (d) Illustrates the association between area's defined within the pressure–volume plane and the mechanical energy.

Time Varying Elastance and Cardiac Contractility

The intrinsic properties of cardiac muscle are responsible for making the functional pumping performance of the ventricle determined by different factors (1): the degree to which the cardiac muscle is prestretched prior to contraction (preload), the intrinsic properties of the muscle (the contractility or inotropy), the load against which the heart ejects (the afterload), and the speed with which the contraction takes place (reflected by the heart rate; chronotropy). In *muscle physiology*, preload is muscle length and is related to the overlap distance of the contractile proteins (actin and myosin) of the sarcomere (the basic contractile unit of a muscle cell), while afterload is the load against which a muscle strip or fiber contracts. In *pump physiology*, ventricular end-diastolic volume is often considered as the best approximation of preload (when unavailable, ventricular end-diastolic pressure can be used as a surrogate). To characterize afterload, one can estimate maximal ventricular wall stress (e.g., using Laplace formula), but most often, mean or systolic arterial blood pressure is taken as a measure of afterload.

Most difficult to characterize is the intrinsic contractility of the heart, which is important to know in diagnosing the severity of cardiac disease. At present, the gold standard is still considered to be the slope of the end-systolic pressure–volume relation (2,3). To fully comprehend this measure, the time varying elastance concept has to be introduced.

Throughout a cardiac cycle, cardiac muscle contracts and relaxes. The functional properties of fully relaxed muscle-at the end of diastole-can be studied in the pressure–volume plane. This relation, sometimes called the (end-) diastolic pressure–volume relation (EDPVR), is nonlinear, that is, for higher volumes, a higher increase in pressure (ΔP) is required to realize a given increase in volume (ΔV). With the volume/pressure ratio defined as compliance, the ventricle behaves less compliant (stiffer) at high volume. The EDPVR represents the passive properties of the ventricular chamber.

Similarly, if one could "freeze" the ventricle at its maximal contraction (as is reached at the end of systole), and measure the pressure–volume relation of the chamber in this maximally contracted state, one could assess the (end-) systolic pressure–volume relation (ESPVR) and the maximal stiffness of the ventricle. The ESPVR represents the active, contractile function of the ventricle. Throughout the cycle, the stiffness (or elastance) of the ventricle varies in between its diastolic and end-systolic value, hence the conceptual model of the time-varying elastance (Fig. 2).

The basis of this time-varying elastance model was laid by Suga et al. in the 1970s. They performed experiments in isolated hearts, and found that by increasing the initial volume of the heart (increasing preload), the maximally developed pressure in an isovolumic beat increased linearly with preload (2,3) (Frank–Starling mechanism). Obviously, the minimal pressure, determined by the passive properties, also increased. When they allowed these ventricles to eject against a quasiconstant afterload pressure, PV loops were obtained, with wider loops (higher stroke volume) being obtained for the more filled ventricles. When connecting all end-systolic points of the PV loops, it was found that the slope of this line, the end-systolic ventricular stiffness, was not different from the line obtained with the isovolumic experiments, demonstrating that it is independent of the load against which the ventricle ejects. Moreover, connecting data points on the PV loops occurring at the same instant in the cardiac cycle (isochrones), these points were also found to line up (Fig. 2). The slopes of these lines have the dimension of stiffness ($\Delta P/\Delta V$; mmHg·mL^{-1} or Pa·mL^{-1}) or elastance (E). In addition, it is often assumed that these isochrones all have the same intercept with the volume axis (which is, however, most often not the case). This volume is called V_0 and represents the volume for which the ventricle no longer develops any pressure.

The slope of the isochronic lines, E, is given by $E = P/(V-V_0)$ and can be plotted as a function of time, yielding the time varying elastance curve, $E(t)$ (Fig. 2). The experiments of Suga and co-workers further pointed out that the maximal slope of the ESPVR, also called end-systolic (E_{es}) elastance, is sensitive to inotropic stimulation. The parameter E_{es} is, at present, still considered as the gold standard measurement of ventricular contractility.

Since these experiments, it has been shown that the ESPVR is not truly linear (4,5), especially not in conditions of high contractility or in small mammals. Since V_0 is a value derived from linear extrapolation of the ESPVR, one often finds negative values, which clearly have no physiological meaning at all. Nevertheless, the time varying elastance remains an attractive concept to concisely describe ventricular function. In practice, PV loops with altered loading conditions are obtained via inflation of a balloon in one of the caval veins, reducing venous return, or with a Valsalva maneuver. The PV loops can be measured invasively with a conductance catheter (6), or by combining intraventricular pressure (measured with a catheter) with volumes measured with a medical imaging technique that is fast enough to measure instantaneous volumes during the load manipulating operations (e.g., echocardiography).

The area enclosed within the PV loop is the work performed by the heart per stroke (stroke work, SW). Furthermore, when the heart contracts, it pressurizes the volume within the ventricle, giving it a potential energy (PE). In the PV plane, PE is represented by the area enclosed within the triangle formed by V_0 on the volume axis, the end-systolic point, and the left bottom corner of the PV loop (Fig. 1). The sum of SW and PE is also called the total pressure–volume area (PVA) and it has been shown

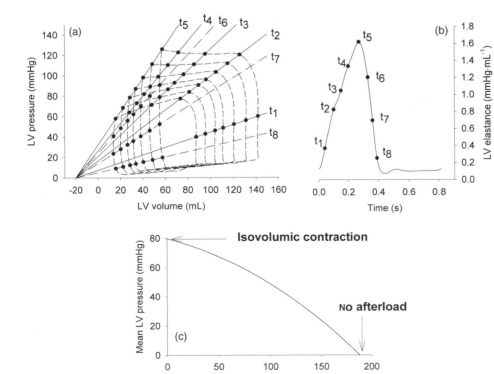

Figure 2. Pressure–volume loops recorded in the left ventricle during transient loading conditions (a), and the concept of the time-varying elastance (b). (c) Illustrates an alternative representation of ventricular function through the pump function graph.

that the consumption of oxygen by the myocardium, VO_2, is proportional to PVA: $VO_2 = c_1 PVA + c_2 E_{es} + c_3$, with c_{1-3} constants to be determined from experiments. The constant c_1 represents the O_2 cost of contraction, c_2 is the O_2 cost of Ca handling related to the inotropic state, and c_3 is the O_2 cost of basal metabolism. Mechanical efficiency can then be expressed as the ratio of SW and VO_2. Recent reviews of the relation between pressure–volume area and ventricular energetics can be found in Refs. 7,8.

Another measure of ventricular function, also derived from PV loop analysis, is the so-called preload recruitable stroke work (PRSW) (9). Due to the Frank–Starling effect (10), a ventricle filled up to a higher EDV will generate a higher pressure and/or stroke volume, and hence a higher SW. Plotting SW as a function of EDV yields a quasilinear relation, of which the slope is sensitive to the contractile state of the ventricle (9).

Alternative Ways of Characterizing LV Systolic Function

Pump function of the ventricle may also be approached in a way similar to hydraulic pumps through its pump function graph (11,12), where the pressure generated by the pump (e.g., mean LV pressure) is plotted as a function of its generated flow (cardiac output). With no outflow, the ventricle contracts isovolumically, and the highest possible pressure is generated. Pumping against zero load, no pressure is built up, but outflow is maximal. The ventricle operates at some intermediate stage, in between these two extreme cases (Fig. 2). One such pump function curve is obtained by keeping heart rate, inotropy, and preload constant, while changing afterload. Although the principle is attractive, it appears to be difficult to measure pump function curves *in vivo*, even in experimental conditions.

Assessing Cardiac Function in Real Life

Although pressure–volume loop-based cardiac analysis still has the gold standard status in experimental work, the applicability in clinical conditions is rather limited. First, the method requires intraventricular pressure and volume. While volumes can, more and more, be measured noninvasively with magnetic resonance imaging (MRI) and even real-time three-dimensional (3D) echocardiography, the pressure requirement still implies invasive measurements. Combined pressure–volume conductance catheters are available (6), but these require calibration to convert conductance into volume data. This requires knowledge of the conductance of the cardiac structures and blood outside the cardiac chamber under study (offset correction), and an independent measurement of stroke volume for scaling of the amplitude of the conductance signal. Second, and perhaps even more important, measuring preload-recruitable stroke work or the end-systolic pressure–volume relation requires that PV loops are recorded during transient loading conditions, which is experimentally obtained via inflation of a balloon in the caval vein to reduce the venous return and cardiac preload. It is difficult to (ethically) justify an extra puncture and the insertion of an additional catheter in patients, knowing also that these maneuvers induce secondary changes in the overall autonomic state of the patient and the release of cathechola-

mines, making this method limited to assess the pump function in a "steady state". To avoid the necessity of the caval vein balloon, so-called "single beat" methods have been proposed, where cardiac contractility is estimated from data measured at baseline steady-state conditions (13,14). The accuracy, sensitivity, and specificity of these methods, however, remains a matter of debate (15,16).

Since it is easier to measure aortic flow than ventricular volume, indicies based on the concept of "hydraulic power" have been proposed. As the ventricle ejects blood, it generates hydraulic power (Pwr), which is calculated as the instantaneous product of aorta pressure and flow. The peak value of power (Pwr_{max}) has been proposed as a measure of ventricular performance. However, due to the Frank–Starling mechanism, ventricular performance in general, and hydraulic power in particular, is highly dependent on the filling state of the ventricle, and correction of the index for EDV is mandatory. Preload-adjusted maximal power, defined as Pwr_{max}/EDV^2, has been proposed as a "single beat" (i.e., measurable during steady-state conditions) index of ventricular performance (17). It has, however, been suggested that the factor 2 used in the denominator is not a constant, but depends on ventricular size (18). It has been demonstrated that the most correct approach is to correct Pwr_{max} for $(EDV - V_0)^2$, V_0 being the intercept of the end-systolic pressure–volume relation (19,20). Obviously, the index then loses its main feature, that is, the fact that it can be deduced from steady-state measurements.

In clinical practice, cardiac function is commonly assessed with ultrasound echocardiography in its different modalities. Imaging the heart in two-dimensional (2D) planar views (2 and 4 chamber long axis views, short axis views), allows us to visually inspect ventricular wall motion and to identify noncontracting zones. With (onboard) image processing software, parameters such as ejection fraction or the velocity of circumferential fiber shortening can be derived. Echocardiography has played a major role in quantifying "diastolic function", that is, the filling of the heart (21,22). Traditionally, this was based on the interpretation of flow velocity patterns at the mitral valve (23) and the pulmonary veins (24). With the advent of more recent ultrasound processing tools, the arsenal has been extended. Color M-mode Doppler, where velocities are measured along a base-to-apex directed scanline, allows us to measure the propagation velocity of the mitral filling wave (25,26). More recent, much attention has been and is being paid to the velocity of the myocardial tissue (in particular the motion of the mitral annulus) (27). Further processing of tissue velocity permits us to estimate the local strain and strain rate within sample volumes positioned within the tissue. Strain and strain rate imaging are new promising tools to quantify local cardiac contractile performance (28,29). Further advances are directed toward real time 3D imaging with ultrasound and quantification of function. Some of the aforementioned ultrasound modalities are illustrated in Fig. 3.

An important domain where assessment of cardiac function is important, is in the catheterization laboratory (cath-lab), where patients are "catheterized" to diagnose and/or treat cardiovascular disease. Access to the vasculature is gained via a large vein (the catheter then ends in the right

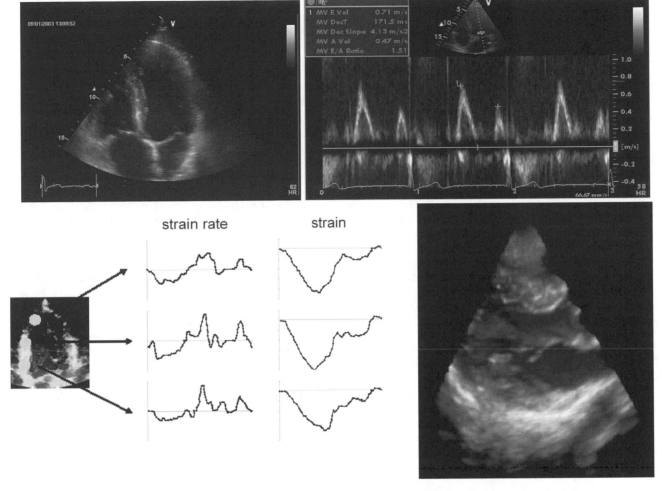

Figure 3. Echocardiography of the heart. (a) Classical four-chamber view of the heart, with visualization of the four cardiac chamber; (b) transmitral flow velocity pattern; (c) strain and strain rate imaging in an apical, mid and basal segment of the left ventricle [adapted from D'hooge et al. (29)]; (d) real-time 3D visualization of the mitral valve with the GE Dimension. (Courtesy of GE Vingmed ultrasound, Horten, Norway.)

atrium/ventricle/pulmonary artery: right heart catheterization) or large artery, typically the femoral artery in the groin (the catheter then resides in the aorta/left ventricle/left atrium: left heart catheterization). Cath-labs are equipped with X-ray scanners, allowing us to visualize cardiovascular structures in one or more planes. With injection of contrast medium, vessel structures (e.g., the coronary arteries) can be visualized (angiography) as well ventricular cavity (ventriculography). The technique, through medical image processing, allows us to estimate ventricular volumes at high temporal resolution. At the same time, arterial pressures can be monitored. Although one cannot deduce intrinsic ventricular function from pressure measurements alone, it is common to use the peak positive (dp/dt_{max}) and peak negative value (dp/dt_{min}) of the time derivative of ventricular pressure as surrogate markers of ventricular contractility and relaxation, respectively. These values are, however, highly dependent on ventricular preload, afterload, and heart rate, and only give a rough estimate of ventricular function. Measurement of dp/dt_{max} or dp/dt_{min} at different loading states, and

assessing the relation between changes in volume and changes in dp/dt, may compensate for the preload dependency of these parameters.

There are also clinical conditions where the physician is only interested in a global picture of cardiac function, for example, in intensive care units, where the principal question is whether the heart is able to deliver a sufficient cardiac output. In these conditions, indicator-dilution methods are still frequently applied to assess cardiac output. In this method, a known bolus of indicator substance (tracer) is injected into the venous system, and the concentration of the substance is measured on the arterial side (the dilution curve), after the blood has passed through the heart/cardiac output = [amount of injected indicator]/[area under the dilution curve]. A commonly used indicator is cold saline, injected into the systemic veins, and the change in blood temperature is then measured with a catheter equipped with a thermistor, positioned in the pulmonary artery. This methodology is commonly referred to as "thermodilution". With valvular pathologies, as tricuspid or pulmonary valve regurgitation, the method becomes less accurate. The variability of the method is quite high, and the method should be repeated (three times or more) so that results can be averaged to provide a reliable estimate. Obviously, the method only yields intermittent estimates of cardiac output. Cardiac output monitors based on other measuring principles are in use, but their accuracy and/or responsiveness is still not optimal and they often require a catheter in the circulation (30,31).

Finally, it is also worth mentioning that cardiac MRI is an emerging technology, able to provide full 3D cardiac morphology and function data (e.g., with MRI tagging) (32,33). Especially for volume estimation, MRI is considered the gold standard method, but new modalities also allow us to measure intracardiac flow velocities. The high cost, the longer procedure, the fact that some materials are still banned from the scanner (e.g., metal containing pacemakers) or cause image artifacts and the limited availability of MRI scanner time in hospitals, however, make that ultrasound is still the first method of choice. For reasons of completeness, computed tomography (CT) and nuclear imaging are mentioned as medical imaging techniques that provide morphological and/or functional information on cardiac function.

The Coronary Circulation

The coronary arteries branch off the aorta immediately distal to the aortic valve (Fig. 4), and supply blood to the heart muscle itself. With their specific anatomical position, they are often considered as part of the heart, although they could as well be considered as being part of the arterial circulation. This ambiguity is also reflected in the medical specialism: coronary artery disease is the territory of the (interventional) cardiologist, and not of the angiologist. The right coronary artery mainly supplies the right heart, while the left coronary artery, which bifurcates into the left anterior descending (LAD) and left circumflex (LCX) branch, mainly supplies the left heart.

As they have to perfuse the cardiac muscle, the coronaries protrude the ventricular wall, which has a profound effect on coronary hemodynamics (34,35). Upon ventricular contraction, blood flow in the coronaries is impeded, leading to a typical biphasic flow pattern, with systolic flow impediment, and predominantly flow during the diastolic phase (Fig. 4). This pattern is most obvious in the LAD, which supplies oxygenized blood to the left ventricle. The resistance of the coronary arteries is thus not constant in time, and contains an active component. When coronary flow is plotted as a function of coronary pressure, other typical features for the coronary circulation are observed. Under normal conditions, coronary flow is highly regulated, so that blood flow is constant for a wide range of coronary perfusion pressures (35,36). The level up to which the flow is regulated is a function of the activity of the ventricle, and hence of the metabolic demands. This seems to suggest that at least two different mechanisms are involved: metabolic autoregulation (flow is determined by the metabolic demand) and myogenic autoregulation. Myogenic autoregulation is the response of a muscular vessel on an increase in pressure: the vessel contracts, reducing its diameter and increasing its wall thickness, which tends to normalize the wall stress. It is only after maximal dilatation of the coronary vessels [e.g., through infusion of vasodilating pharmacological substances such as adenosine or papaverine, or immediately following a period of oxygen deficiency (ischemia)] that autoregulation

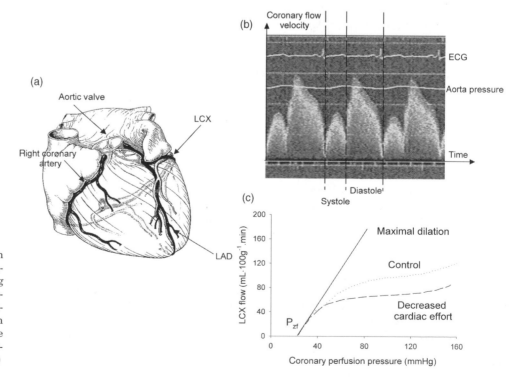

Figure 4. (a) Anatomical situation of the coronary arteries. (b) Demonstration of flow impediment during systole. (c) Coronary flow as a function of perfusion pressure, demonstrating the aspect of autoregulation and the non-zero intercept with the pressure axis (P_{zf}, zero-flow pressure). (Reconstructed after Ref. 36.)

can be "switched off", and that the pressure-flow relation becomes linear (Fig. 4). Note, however, that the pressure-flow relation does not pass through the origin: it requires a certain pressure value (zero-flow pressure P_{zf}) to generate any flow, an observation first reported by Bellamy et al. (37). The origin of P_{zf} is still not fully understood and has been attributed to (partial) vessel collapse by vascular tone and extravascular compression. This is a conceptual model, also known as the "waterfall" model, as flow is determined by the pressure difference between inflow and surrounding pressure (instead of outflow pressure) and pressure changes distal to the point of collapse (the waterfall) have no influence on flow (38,39). Note, however, that P_{zf} is often an extrapolation of pressure-flow data measured in a higher pressure range. It is plausible that the relation becomes nonlinear in the lower perfusion pressure range, due to the distensibility of the coronary vessels and compression of vessels due to intramyocardial volume shifts (35). Another model, based on the concept of an intramyocardial pump and capable of explaining phasic patterns of pressure and flow waves, was developed by Spaan et al. (35,40), and is mentioned here for reasons of completeness.

An important clinical aspect of the coronary circulation, which we will only briefly touch here, is the assessment of the severity of coronary artery stenosis (34). The stenosis forms an obstruction to flow, causes an extra pressure drop, and may result in coronary perfusion pressures too low to provide oxygen to the myocardial tissues perfused by that coronary artery. Imaging of the coronary vessels and the stenosis with angiography (in the cath-lab) is still an important tool for clinical decision making, but more and more attention has been attributed to quantification of functional severity of the stenosis. One of the most known indicies in use are the coronary flow reserve (CFR), that is, the ratio of maximal blood flow (velocity) through a coronary artery (after induction of maximal vasodilation) and baseline blood flow (velocity), with values > 2 indicating sufficient reserve and thus a subcritical stenosis. Coronary flow reserve has the drawback that flow or velocity measurements are required, which are not common in the cath-lab. From that perspective, the fractional flow reserve (FFR) is more attractive, as it requires only pressure measurements. It can be shown that $FFR = P_d/P_a$ (with P_d the pressure distal to the stenosis and P_a the aorta pressure) is the ratio of actual flow through the coronary, and the hypothetical flow that would pass through the coronary in the absence of the stenosis (34,41). A FFR value >0.75 indicates nonsignificant stenosis (34). Measurement of FFR is done in conditions of maximal vasodilation, and requires ultrathin pressure catheters (pressure wires) that can pass through the stenosis without causing (too much) extra pressure drop. Other indicies combine pressure and flow velocity (42) and are, from fluid dynamic perspective, probably the best characterization of the extra resistance created by the stenosis. It deserves to be mentioned that intravascular ultrasound (IVUS) is, currently, increasingly being applied, especially when treating complex coronary lesions. In addition to quantifying stenosis severity, much research is also focused on assessing the histology of the lesions and their vulnerability and risk of rupture via IVUS or other techniques.

THE ARTERIAL SYSTEM: DAMPING RESERVOIR AND/OR A BLOOD DISTRIBUTING NETWORK...

Basically, there are two ways of approaching the arterial system (43): (1) one can look at it in a "lumped" way, where abstraction is being made of the fact that the vasculature is a network system with properties distributed in space; or (2) one can take into account the network topology, and analyze the system in terms of pressure and flow waves propagating along the arteries in a forward and backward direction.

Irrespective of the conceptual framework within which one works, the analysis of the arterial system requires (simultaneously measured) pressure and flow, preferably measured at the upstream end of the arterial tree (i.e., immediately distal to the aortic valve). The analysis of the arterial system is largely analogous to the analysis of electrical network systems, where pressure is equivalent to voltage and flow to current. Also steming from this analogy is the fact that the arterial system is often analyzed in terms of impedance. Therefore, before continuing, the concept of impedance is introduced.

Impedance Analysis

While electrical waves are sine waves, with a zero time-average value, arterial pressure and flow waves are (approximately) periodical, but certainly nonsinusoidal, and their average value is different from zero (in humans, mean systemic arterial blood pressure is ~ 100 mmHg (13.3 kPa), while mean flow is ~ 100 mL·s^{-1}). To bypass this limitation, one can use the Fourier theorem, which states that any periodic signal, such as arterial pressure and flow, can be decomposed into a constant (the mean value of the signal) and a series of sinusoidal waves (harmonics). The frequency of the first harmonic is cardiac frequency (the fundamental frequency), while the frequency of the nth harmonic is n times the fundamental frequency.

Fourier decomposition is applicable if two conditions are fulfilled: (1) the cardiovascular system operates in steady-state conditions (constant heart rate; no respiratory effects); (2) the mechanical properties of the arterial system are sufficiently linear so that the superposition principle applies, meaning that the individual sine waves do not interact and that the sum of the effects of individual harmonics (e.g., the flow generated by a pressure harmonic) is equal to the effect caused by the original wave that is the sum of all individual harmonics.

Harmonics can be represented using a complex formalism. For the nth harmonic, the pressure (P) and flow (Q) component can be written as

$$P_n = |P_n|e^{i(n\omega t + \Phi_{P_n})} \qquad Q_n = |Q_n|e^{i(n\omega t + \Phi_{Q_n})}$$

where $|P_n|$ and $|Q_n|$ are the amplitudes (or moduli) of the pressure and flow sine waves, having phase angles Φ_{P_n} and Φ_{Q_n} (to allow for a phase lag in the harmonic), respectively. Time is indicated by t, and ω is the fundamental angular frequency, given by $2\pi/T$ with T the duration of a heart cycle (RR-interval). For a heart rate of 75 beats·min^{-1}, T is 0.8 s. The fundamental frequency is 1.25 Hz, and ω becomes 7.85 rad·s^{-1}.

In general, 10–15 harmonics are sufficient to describe hemodynamic variables, such as pressure, flow, or volume

(44,45). Also, to avoid aliasing, the sampling frequency should be twice as high as the frequency of the signal one is measuring (Nyquist limit). Thus, when measuring hemodynamic data in humans, the frequency response of the equipment should be $> 2 \times 15 \times 1.25$ Hz, or > 37.5 Hz. These requirements are commonly met by Hi–Fi pressure tip catheters (e.g., Millar catheters) with a measuring sensor embedded within the tip of the sensor, but not by the fluid-filled measuring systems that are frequently used in the clinical setting. Here, the measuring sensor is outside the body (directly or via extra fluid lines) connected to a catheter. Although the frequency response of the sensor itself is often adequate, the pressure signal is being distorted by the transmission via the catheter (and fluid lines and eventual connector pieces and three-way valves). The use of short, rigid, and large bore catheters is recommended, but it is advised to assess the actual frequency response of the system (as it is applied *in vivo*) if the pressure data is being used for purposes other than patient monitoring. Note that in small rodents like the mouse, where heart rate is as high as 600 beats min^{-1}, the fundamental frequency is 10 Hz, posing much higher measuring equipment requirements with a frequency response flat up to 300 Hz.

Impedance Z is generally defined as the ratio of pressure and flow: $Z = P/Q$, and thus has the dimensions of mmHg·mL^{-1} s [kg·m^{-4}·s^{-1} in SI units; dyn·cm^{-5} · s in older units] if pressure is expressed in mmHg and flow in mL·s^{-1}. The parameter Z is usually calculated for each individual harmonic, and displayed as a function of frequency. Since both P and Q are complex numbers, Z is complex as well, and also has a modulus and a phase angle, except for the steady (dc) component at 0 Hz, which is nothing but the ratio of mean pressure and mean flow (i.e., the value of vascular resistance). For all higher harmonics, the modulus of the nth harmonic is given as $|Z_n| = |P_n|/|Q_n|$, and its phase, Φ_z, is given as $\Phi_{Pn} - \Phi_{Qn}$.

Arterial impedance requires simultaneous measurement of pressure and flow at the same location. Although it can be calculated all over the arterial tree, it is most commonly measured at the entrance of the systemic or pulmonary circulation, and is called 'input' impedance, often denoted as Z_{in}. The parameter Z_{in} fully captures the relation between pressure and flow, and is determined by all downstream factors influencing this relation (arterial network topology, branching patterns, stiffness of the vessel, vasomotor tone, ...). In a way, it is a powerful description of the arterial circulation, since it captures all effects, but this is at the same time its greatest weakness, as it is not very sensitive to local changes in arterial system properties, such as focal atherosclerotic lesions.

The interpretation of (input) impedance is facilitated by studying the impedance of basic electrical or mechanical "building blocks" (43–45): (1) When a system behaves strictly resistive, there is a linear relation between the pressure (difference) and flow. Pressure and flow are always in phase, Z_n is a real number and Φ_z is zero. (2) In the case where there is only inertia in a system, pressure is ahead of flow; for sine waves, the phase difference between both is a quarter of a wavelength, or $+90°$ in terms of phase angle. (3) In the case where the system behaves like a capacitor, flow is leading pressure, again 90° out of phase, so that Φ_z is $-90°$.

Figure 5 displays the input impedance of these fundamental building blocks, as well as Z_{in} calculated from the

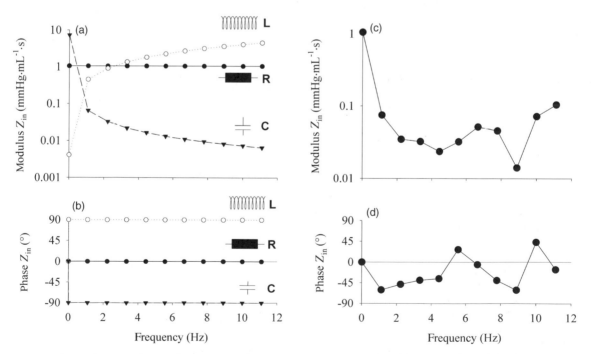

Figure 5. (a and b) Impedance modulus (a) and phase angle (b) of fundamental electrical/mechanical building blocks of the arterial system: resistance (R; 1.028 mmHg·mL^{-1}·s (137.0×10^6 Pa·m^{-3}·s)), inertia [L; 0.065 mmHg·mL^{-1}·s^2 (8.7×10^6 Pa·m^{-3}·s^2)] and compliance [C; 2.25 mL·mmHg^{-1} (16.9×10^{-9} m^3·Pa^{-1})]. (c and d) input impedance modulus (c) and phase (d) calculated from aortic pressure and flow given in Fig. 1.

aorta pressure and flow shown in Fig. 1. The impedance modulus drops from the value of total systemic vascular resistance at 0 Hz to much lower values at higher frequencies. The phase angle is negative up until the fourth harmonic, showing that capacitive effects dominate at these low frequencies, although the phase angle never reaches $-90°$, indicating that inertial effects are present as well. For higher harmonics, the phase angle is close to zero, or at least oscillating around the zero value. At the same time, the modulus of Z_{in} is nearly constant. For these high frequencies, the system seems to act as a pure resistance, and the impedance value, averaged over the higher harmonics, has been termed the characteristic impedance (Z_0).

The Arterial System as a "Windkessel" Model

The most simple approximation of the arterial system is based on observations of reverend Stephen Hales (1733), who drew the parallel between the heart ejecting in the arterial tree, and the working principle of a fire hose (46). The pulsatile action of the pump is damped and transformed into a quasicontinuous outflow at the downstream

end of the system, that is, the outflow nozzle for the fire hose and the capillaries for the cardiovascular system. In mechanical terms, this type of system behavior can be simulated with two mechanical components: a buffer reservoir (compliant system) and a downstream resistance. In 1899, Otto Frank translated this into a mathematical formulation (47,48), and it was Frank who introduced the terminology "windkessel models", windkessel being the German word for air chamber, as the buffer chamber used in the historical fire hose. The windkessel models are often described as their electrical analogue (Fig. 6).

The Two-Element Windkessel Model. While there are many different "windkessel" models in use (49,50), the two basic components contained within each model are a compliance element, C (mL·mmHg^{-1} or m^3·Pa^{-1}), and a resistor element, R (mmHg·mL^{-1}·s or Pa·m^{-3}·s in SI units). The compliance element represents the volume change associated with a unit change in pressure; R is the pressure drop over the resistor associated with a unit flow. In diastole, when there is no new inflow of blood into the compliance, the arterial pressure decays exponentially following $P(t) = P_0 e^{-t/RC}$. The parameter RC is the product

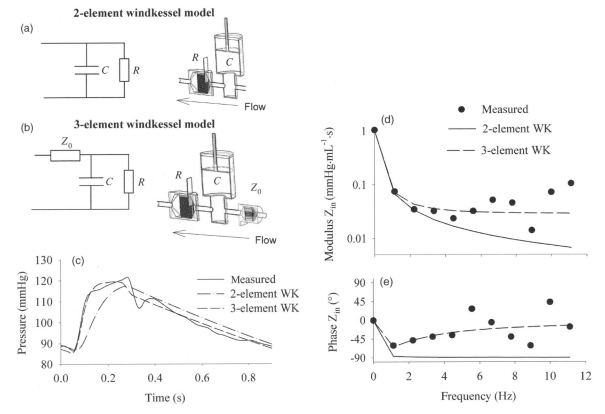

Figure 6. (a and b) Electrical analog and mechanical representation of a two- and three-element windkessel model. (c) Agreement between measured pressure, and the pressure obtained after fitting a two- and three-element windkessel model to the pressure (1 mmHg = 133.3 Pa) and flow data from Fig. 1. Model parameters are $R = 1.056$ mmHg·mL^{-1}·s (140.8×10^6 Pa·m^{-3}·s) and $C = 2.13$ mL·mmHg^{-1} (16.0×10^{-9} m^3·Pa^{-1}) for the two-element windkessel model; $R = 1.028$ mmHg·mL^{-1}·s (137.0×10^6 Pa·m^{-3}·s). $C = 2.25$ mL·mmHg^{-1} (16.9×10^{-9} m^3·Pa^{-1}) and $Z_0 = 0.028$ mmHg·mL^{-1}·s (3.7×10^6 Pa·m^{-3}·s) for the three-element windkessel model. (d and e) Input impedance modulus (d) and phase angle (e) of these lumped parameter models and their match to the *in vivo* measured input impedance (1 mmHg·mL^{-1}·s = 133.3×10^6 Pa·m^{-3}·s).

of R and C and is called the arterial decay time. The higher the RC time, the slower the pressure decay. It is the time required to reduce P_0 to 37% of its initial value (note that the 37% is a theoretical value, usually not reached *in vivo* because the next beat impedes a full pressure decay). One can make use of this property to estimate the arterial compliance: By fitting an exponential curve to the diastolic decaying pressure, RC is obtained and thus, when R is known we also know C (49,51–53). This method is known as the decay time method.

For typical hemodynamic conditions in humans at rest (70 beats $\cdot$ min^{-1}, systolic/diastolic, and mean pressure of 120 (16.0 kPa), 80 (10.7 kPa), and 93 (12.4 kPa) mmHg respectively, stroke volume 80 mL), mean flow is 93 mL$\cdot$s^{-1} $(0.93 \times 10^{-4}$ m$^3\cdot$s$^{-1})$, and R is 1 mmHg (mL$\cdot$s$^{-1})$ $(133.3 \times 10^6$ Pa$\cdot$m$^{-3}\cdot$s). Assuming that the whole stroke volume is buffered in systole, C can be estimated as the ratio of stroke volume and pulse pressure (systolic–diastolic pressure difference), ~2 mL$\cdot$mmHg^{-1} $(1.50 \times 10^{-8}$ m$^3\cdot$Pa$^{-1})$. This value is considered as an overestimation of the actual arterial compliance (49,54). In humans, RC time is thus of the order of 1.5–2 s.

The question, How well does a windkessel model represents the actual arterial system?, can be answered by studying the input impedance of both. In complex formulation, the input impedance of a two-element windkessel model is given as

$$Z_{i\text{-WK2}} = \frac{R}{1 + i\omega RC}$$

with i the complex constant, and $\omega = 2\pi f$, f is the frequency. The dc value (0 Hz) of Z_{in} is thus R; at high frequencies, it becomes zero. The phase angle is 0 at 0 Hz, and $-90°$ for all other frequencies. Compared to input impedance as measured in mammals, the behavior of a two-element windkessel model reasonably represents the behavior of the arterial system for the low frequencies (up to third harmonic), but not for higher frequencies (43,49,54) (Fig. 6). This means that it is justified to use the model for predicting the low frequency behavior of the arterial system, that is, the low frequency response to a flow input. This property is used in the so-called "pulse pressure method", an iterative method to estimate arterial compliance: with R assumed known, the pulse pressure response of the two-element windkessel model to a (measured) flow stimulus is calculated with varying values of C. The value of C yielding the pulse pressure response matching the one measured *in vivo*, is considered to be the correct one (55). Compared to the decay time method, the advantage is that the pulse pressure method is insensitive to deviations of the decaying pressure from the true exponential decay (53).

The Three-Element and Higher Order Windkessel Models. The major shortcoming of the two-element windkessel model is the inadequate high frequency behavior (43,49,56). Westerhof et al.resolved this problem by adding a third resistive element proximal to the windkessel, accounting for the resistive-*like* behavior of the arterial system in the high frequency range (56). The third element represents the characteristic impedance of the proximal

part of the ascending aorta, and integrates the effects of inertia and compliance. Adding the element, the input impedance of the three-element windkessel model becomes

$$Z_{i\text{-WK3}} = Z_0 + \frac{R}{1 + i\omega RC}$$

it making the phase angle negative for lower harmonics, but returns to zero for higher harmonics, where the impedance modulus asymptotically reaches the value of Z_0. For the systemic circulation, the ratio of Z_0 and R is 0.05–0.1 (57).

The major disadvantage of the three-element windkessel model is the fact that Z_0, which should represent the high frequency behavior of the arterial system, plays a role at all frequencies, including at 0 Hz. This has the effect that, when the three-element windkessel model is used to fit data measured in the arterial system, the compliance is systematically overestimated (53,58): the only way to "neutralize" the contribution of Z_0 at the low frequencies, is to artificially increase the compliance of the model. The "ideal" model would incorporate both the low frequency behavior of the two-element windkessel model, and the high frequency Z_0, though without interference of the latter at all frequencies. This can be achieved by adding an inertial element in parallel to the characteristic impedance, as demonstrated by Stergiopulos et al. (59), elaborating on a model first introduced by Burattini et al. (60). For the DC component and low frequencies, Z_0 is bypassed through the inertial element. For the high frequencies, Z_0 takes over. It has been demonstrated, fitting the four-element windkessel model to data generated using an extended arterial network model, that L effectively represents the total inertia present in the model (59).

Obviously, by adding more elements, it is possible to develop models that are able to further enhance the matching between model and arterial system behavior (49,50), but the uniqueness of the model may not be guaranteed, and the physiological interpretation of the model elements is not always clear.

Although lumped parameter models cannot explain all aspects of hemodynamics, they are very useful as a concise representation of the arterial system. Mechanical versions are frequently used in hydraulic bench experiments, or as highly controllable afterload systems for *in vivo* experiments. An important field of application of the mathematical version is for parameter identification purposes: fitting arterial pressure and flow data measured *in vivo* to these models, the arterial system can be characterized and quantified (e.g., the total arterial compliance) through the model parameter values (43,49).

It is important to stress that these lumped models represent the behavior of the arterial system as a whole, and that there is no relation between model components and anatomical parts of the arterial tree (43). For example, although Z_0 represents the properties of the proximal aorta, there is no drop in mean pressure along the aorta, which one would expect if the three-element windkessel model were to be interpreted in a strict anatomical way. The combination of elements simply yields a model that represents the behavior of the arterial system as it is seen by the heart.

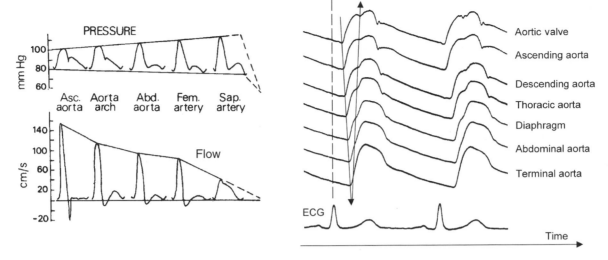

Figure 7. (a) Shows both pressure (1 mmHg = 133.3 Pa) and flow velocity measured along the arterial tree in a dog [after Ref. (44).] (b) Pressure wave forms measured between the aortic valve (AoV) and the terminal aorta (Term Ao) in one patient. [Modified from Ref. 61.]

Wave Propagation and Reflection in the Arterial Tree

As already stressed, lumped parameter models simply represent the behavior of the arterial system as a whole, and as seen by the heart. As soon as the model is subjected to flow or pressure, there is an instantaneous effect throughout the whole model. This is not the case in the arterial tree: when measuring pressure along the aorta, it can be observed that there is a finite time delay between the onset of pressure rise and flow in the ascending and distal aorta (Fig. 7). The pressure pulse travels with a given speed from the heart toward the periphery.

When carefully analyzing pressure wave profiles measured along the aorta, several observations can be made: (1) there is a gradual increase in the steepness of the wave front; (2) there is an increase in peak systolic pressure (at least in the large-to-middle sized arteries); (3) diastolic blood pressure is nearly constant, and the drop in mean blood pressure (due to viscous losses) is negligible in the large arteries. The flow (or velocity) wave profiles exhibit the same time delay in between measuring locations, but their amplitude decreases. Also, by comparing the pressure and flow wave morphology, one can observe that these are fundamentally different.

In the absence of wave reflection (assuming the aorta to be a uniform, infinitely long elastic tube), dissipation would only lead to less steep wave fronts, and damping of maximal pressure. Also, in these conditions, one would expect similarity of pressure and flow wave morphology. Thus, the above observations can only be explained by wave reflection. This is, of course, not surprising given the complex anatomical structure of the arterial tree, with geometric and elastic tapering (the further away from the heart, the stiffer the vessel), its numerous bifurcations, and the arterioles and capillaries making the distal terminations.

The Arterial Tree as a Tube. Despite the complexity described above, arterial wave reflection is often approached in a simple way, conceptually considering the arterial tree

as a single tube [or T-tube (62)], with one (or 2) discrete reflection site(s) at some distance from the heart. Within the single tube concept, the arterial tree is seen as a uniform or tapered (visco-)elastic tube (63,64), with an "effective length" (65,66), and a single distal reflection site. The input impedance of such a system can be calculated, as demonstrated in Fig. 8 for a uniform tube of length 50 cm, diameter 1.5 cm, Z_0 of 0.275 mmHg·mL^{-1}·s (36.7 × 10^6 Pa·m^{-3}·s) and wave propagation speed of 6.2 m·s^{-1}. The tube is ended

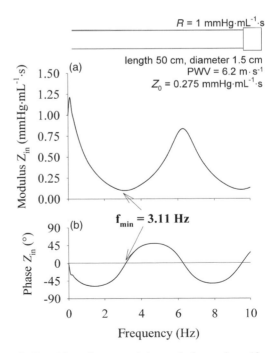

Figure 8. Input impedance modulus and phase of a uniform tube with length 50 cm, diameter 1.5 cm, characteristic impedance of 0.275 mmHg·mL^{-1}·s (36.7 × 10^6 Pa·m^{-3}·s) and wave propagation speed of 6.2 m·s^{-1}. The tube is ended by a (linear) resistance of 1 mmHg·mL^{-1}·s (133.3 × 10^6 Pa·m^{-3}·s).

by a (linear) resistance of 1 mmHg·mL^{-1}·s (133.3×10^6 Pa·m^{-3}·s). The impedance mismatch between the tube and its terminal resistance gives rise to wave reflections and oscillations in the input impedance pattern, and many features, observed *in vivo* (67), can be explained on the basis of this model.

Assume there is a sinusoidal wave running in the system with a wavelength λ being four times the length of the tube. This means that the phase angle of the reflected wave, when arriving back at the entrance at the tube, will be 180° out of phase with respect to the forward wave. The sum of the incident and reflected wave will be zero, since they interfere in a maximally destructive way. If this wave is a pressure wave, measured pressure at the entrance of the tube will be minimal for waves with this particular wave length. There is a relation between pulse wave velocity (PWV), λ, and frequency f (i.e., $λ = \mathrm{PWV}/f$). Thus, in an input impedance spectrum, the frequency f_{min}, where input impedance is minimal, corresponds to a wave with a wavelength that is equal to four times the distance to the reflection site, L_s: $4L = \mathrm{PWV}/f_{min}$, or $L = \mathrm{PWV}/4f_{min}$ and $f_{min} = \mathrm{PWV}/4L$. Applied to the example of the tube, f_{min} is expected at $6.2/2 = 3.1$ Hz. This equation is known as the "quarter wavelength" formula, and is used to estimate the effective length of the arterial system.

Although the wave reflection pattern in the arterial system is complex (68), pressure (P) and flow (Q) are mostly considered to be composed of only one forward running component, P_f (Q_f) and one backward running component, P_b (Q_b), where the single forward and backward running components are the resultant of all forward and backward traveling waves, including the forward waves that result from rereflection at the aortic valve of backward running waves (69).

At all times,

$$P = P_f + P_b \quad \text{and} \quad Q = Q_f + Q_b$$

Furthermore, if the arterial tree is considered as a tube, defined by its characteristic impedance Z_0, the following relations also apply:

$$Z_0 = P_f/Q_f = -P_b/Q_b$$

since Z_0 is the ratio of pressure and flow in the absence of wave reflection (43–45), which is the case when only forward or backward running components are taken into consideration. The negative sign in the equation above appears because the flow is directional, considered positive in the direction away from the heart, and negative toward the heart, while the value of the pressure is insensitive to direction.

Combining these equations, $P = P_f - Z_0 Q_b = P_f - Z_0 (Q - Q_f) = 2P_f - Z_0 Q$, so that

$$P_f = (P + Z_0 Q)/2$$

Similarly, it can be deduced that

$$P_b = (P - Z_0 Q)/2$$

These equations were first derived by Westerhof et al., and are known as the linear wave separation equations (67). In principle, the separation should be calculated on individual

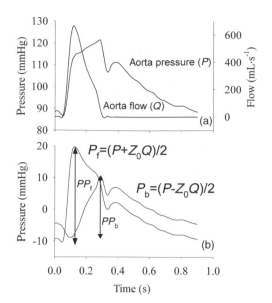

Figure 9. Application of linear wave separation analysis to the pressure (1 mmHg = 133.3 Pa) and flow data of Fig. 1. The ratio of PP_b and PP_f can be used as a measure of wave reflection magnitude.

harmonics, and the net P_f and P_b wave then follows from summation of all forward and backward harmonics. In practice, however, the equations are often used in the time domain, using measured pressure and flow as input. Note, however, that wave reflection only applies to the pulsatile part of pressure and flow, and mean pressure and flow should be subtracted from measured pressure and flow before applying the equations. An example of wave separation is given in Fig. 9.

Note also that wave separation requires knowledge of characteristic impedance, which can be estimated both in the frequency and time domain (70). When discussing input impedance, it was already noted that for the higher harmonics (> fifth harmonic), Z_{in} fluctuates around a constant value, Z_0, and with a phase angle approaching zero. Assuming wave speed to be ~5 m·s^{-1}, the wave length λ for these higher harmonics (e.g., the fifth harmonic for a heart rate of 60 beats·min^{-1}), being the product of the wave speed and wave period (0.2 s) becomes shorter (1 m) than the average arterial pathlength. Waves reflect at distant locations throughout the arterial tree (with distal ends of vascular beds < 50 cm to ~2 m away from the heart in humans), return back to the heart with different phase angles and destructively interfere with each other, so that the net effect of the reflected waves appears inexistent. For these higher harmonics, the arterial system thus appears reflectionless, and under these conditions, the ratio of pressure and flow is, by definition, the characteristic impedance. Therefore, averaging the input impedance modulus of the higher harmonics, where the phase angle is about zero, yields an estimate of the characteristic impedance (43–45).

Characteristic impedance can also be estimated in the time domain. In early systole, the reflected waves did not yet reach the ascending aorta, and in the early systolic ejection period, the relation between pressure and flow is

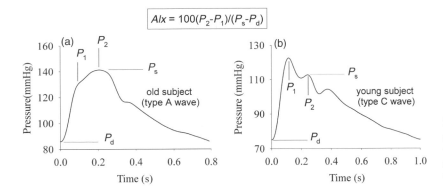

$$AIx = 100(P_2-P_1)/(P_s-P_d)$$

Figure 10. Typical pressure wave contours measured noninvasively at the common carotid artery with applanation tonometry in a young and old subject, and definition of the "augmentation index" (AIx) (1 mmHg = 133.3 Pa). See text for details.

linear, as can be observed when plotting P as a function of Q (70,71). The slope of the Q–P relationship during early systole is the time domain estimate of Z_0, and is in good agreement with the frequency domain estimate (70).

Both the time and frequency domain approach, however, are sensitive to subjective criteria, such as the selection of the early systolic ejection period, or the selection of the harmonic range that is used for averaging.

Pathophysiological Consequences of Arterial Wave Reflection. Figure 10 displays representative carotid artery pressure waves ($\sim$ aorta pressure) for a young subject, (b), and for an older, healthy subject, (a). It is directly observed that the morphology of the pressure wave is fundamentally different. In the young subject, pressure first reaches its maximal systolic pressure, and an "inflection point" is visible in late systole, generating a late shoulder (P_2). In the older subject, this inflection point appears in early systole, generating an early shoulder (P_1).

Measurements of pressure along different locations in the aorta, conducted by Latham et al. (67), showed that, when the foot of the wave on one hand, and the inflection point on the other, are interconnected (Fig. 7): (1) these points seem to be aligned on two lines; (2) the lines connecting these characteristic marks intersect. This pattern is consistent with the concept of a pressure wave being generated by the heart, traveling down the aorta, reflect, and superimpose on the forward going wave. The inflection point is then a visual landmark, designating the moment in time where the backward wave becomes dominant over the forward wave (61).

In young subjects, the arteries are most elastic (deformable), and pulse wave velocity (see below) is much lower than in older subjects, or in patients with hypertension or diabetes. In the young, it takes more time for the forward wave to travel to the reflection site, and for the reflected wave to arrive at the ascending aorta. Both interact only in late systole (late systolic inflection point), causing little extra load on the heart. In older subjects, on the other hand, PWV is much higher, and the inflection point shifts to early systole. The early arrival of the reflected wave literally boosts systolic pressure, causing pressure augmentation, and augmenting the load on the heart (72). At the same time, the early return of the reflected wave impedes ventricular ejection, and thus may have a negative impact on stroke volume (72,73). An increased load on the heart increases the energetic cost to maintain stroke

volume, and will initiate cardiac compensatory mechanisms (remodeling), which may progress into cardiac pathology (74–76).

Wave Intensity Analysis. With its origin in electrical network theory, much of the arterial function analysis, including wave reflection, is done in the frequency domain. Besides the fact that this analysis is, strictly speaking, only applicable in linear systems with periodic signal changes, the analysis is quite complex due to the necessity of Fourier decomposition, and it is not intuitively comprehensible. An alternative method of analysis, performed in the time domain and not requiring linearity and periodicity is the analysis of the wave intensity, elaborated by Parker and Jones in the late 1980s (77).

Disturbances to the flow lead to changes in pressure (dP) and flow velocity (dU), "wavelets", which propagate along the vessels with a wave speed (PWV), as defined above. By accounting for conservation of mass and momentum, it can be shown that

$$dP_\pm = \pm PWV\rho dU_\pm$$

where the "+" denotes a forward traveling wave (for a defined positive direction), while "−" denotes a backward traveling wave. This equation is also known as the waterhammer equation. Waves characterized by a $dP > 0$, that is, a rise in pressure, are called compression waves, while waves with $dP < 0$ are expansion waves. Note that this terminology still reflects the origin of the theory in gas dynamics.

Basically, considering a tube, with left-to-right as the positive direction, there are four possible types of waves: (1) Blowing on the left side of the tube, pressure rises ($dP > 0$), and velocity increases ($dU > 0$). This is a *forward compression wave*. (2) Blowing on the right side of the tube, pressure increases ($dP > 0$), but velocity decreases ($dU < 0$) with our convention. This is a *backward compression wave*. (3) Sucking on the left side of the tube, pressure decreases ($dP < 0$), as well as the velocity ($dU < 0$), but the wavefront propagates from left to right. This wave type is a *forward expansion wave*. (4) Finally, one can also suck on the right side of the tube, causing a decrease in pressure ($dP < 0$) but an increase in velocity ($dU > 0$). This is a *backward expansion wave*.

The nature of a wave is most easily comprehended by analysing the wave intensity, dI, which is defined as the product of dU and dP, and is the energy flux carried by the

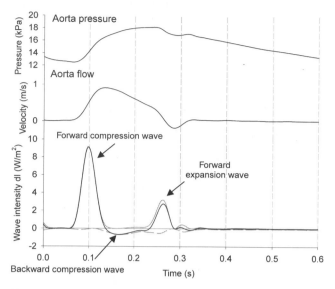

Figure 11. The concept of wave intensity analysis, applied to pressure and flow measured in the ascending aorta of a dog.

wavelet. It can be deduced from the above that dI is always positive for forward running waves, and always negative for a backward wave. When dI is positive, forward waves are dominant; otherwise, backward waves are dominating. Analysis of dP reveals whether the wave is a compression or an expansion wave. Figure 11 shows the wave intensity calculated from pressure and flow velocity measured in the ascending aorta of a dog (71). A typical wave intensity pattern is characterized by three major peaks. The first one is a forward compression wave, associated with the ejection of blood from the ventricle. The second positive peak is associated with a forward running wave, but $dP < 0$, and this second peak is thus a forward running expansion wave, due to ventricular relaxation, slowing down the ejection from the heart. During systole, reflected waves are dominant, resulting in a negative wave intensity, but with, in this case, positive dP. The negative peak is thus a backward compression wave, resulting from the peripheral wave reflections.

Further, note that similar to Westerhof's work (69), the wavelets dP and dU also can be decomposed in the forward and backward components (71). It can be derived that

$$dP_{\pm} = \frac{1}{2}(dP \pm \rho \text{PWV}\, dU) \quad \text{and} \quad dU_{\pm} = \pm\frac{1}{2}\left(\frac{dP}{\rho\text{PWV}} \pm dU\right)$$

The total forward and backward pressure and flow wave can be obtained as

$$P_+ = P_\text{d} + \sum_{t=0}^{t} dP_+$$

with P_d the diastolic blood pressure, which is added to the forward wave, and $P_- = \sum_{t=0}^{t} dP_-$ (71). Similarly, for the forward and backward velocity wave, it applies that

$$U_+ = \sum_{t=0}^{t} dU_+ \quad \text{and} \quad U_- = \sum_{t=0}^{t} dU_-$$

Wave Intensity in itself, dI, can also be separated in a net forward and backward wave intensity: $dI_+ = dP_+ dU_+$ and $dI_- = dP_- dU_-$ with $dI = dI_+ + dI_-$.

Wave intensity is certainly an appealing method to gain insight into complex wave (reflection) patterns, as in the arterial system, and the use of the method is growing (78–80). The drawback of the method is the fact that dI is calculated as the product of two derivatives dP and dU, and thus is highly sensitive to noise in the signal. Adequate filtering of basic signals and derivatives is mandatory. As for the more "classic" impedance analysis, it is also required that pressure and flow be measured at the exact same location, and preferably at the same time.

Assessing Arterial Hemodynamics in Real Life

Analyzing Wave Propagation and Reflection. The easiest approach of analyzing wave reflection is to simply quantify the global effect of wave reflection on the pressure wave morphology. This can be done by formally quantifying the observation of Murgo et al. (61), and is now commonly known as the augmentation index (AIx). This index was first defined in the late 1980s by Kelly et al. (81). Although different formulations are used, AIx is nowadays most commonly defined as the ratio of the difference between the "secondary" and "primary peak", and pulse pressure: AIx $= 100 \, (P_2 - P_1)/(P_\text{s} - P_\text{d})$, and expressed as a percentage (Fig. 10). For A-type waves (Fig. 10) with a shoulder preceding systolic pressure, P_1 is a pressure value characteristic for the shoulder, while P_2 is systolic pressure. These values are positive. For C-type waves, the shoulder follows systolic pressure, and P_1 is systolic pressure, while P_2 is a pressure value characteristic for the shoulder, thus yielding negative values for AIx (Fig. 10). There are also alternative formulations in use, for example, $100(P_2 - P_\text{d})/(P_1 - P_\text{d})$, which always yields a positive value ($< 100\%$ for C-type waves, and $> 100\%$ for A-type waves). Both are, of course, mathematically related to each other. Since AIx is an index based on pressure differences and ratios, it can be calculated from noncalibrated pressure waveforms, which can be obtained noninvasively using techniques such as applanation tonometry (see below). In the past 5 years, numerous studies have been published using AIx as a quantification of the contribution of wave reflection to arterial load.

It is, however, important to stress that AIx is an integrated measure and that its value depends on all factors influencing the magnitude and timing of the reflected wave: pulse wave velocity, magnitude of wave reflection, and the distance to the reflection site. This AIx is thus strongly dependent on body size (82). In women, the earlier return of the reflected wave leads to higher AIx. The relative timing of arrival of the reflected wave is also important, so that AIx is also dependent on heart rate (83). For higher heart rates, systolic ejection time shortens, so that the reflected wave arrives relatively later in the cycle, leading to an inverse relation between heart rate and AIx.

Instead of studying the net effect of wave reflection, one can also measure pressure and flow (at the same location) and separate the forward and backward wave using the aforementioned formulas. An example is given in Fig. 9, using aorta pressure and flow from Fig. 1. The ratio of the amplitude of P_b (PP_b) and P_f (PP_f) then yields an easy measure of wave reflection, which can be considered as a wave reflection coefficient, although it is to be emphasized that it is not a reflection coefficient in a strict sense. It is not the reflection coefficient at the site of reflection, but at the upstream end of the arterial tree and thus also incorporates the effects of wave dissipation, playing a role along the pathlength for the forward and backward component.

Another important determinant of the augmentation index is pulse wave velocity, which is an interesting measure of arterial stiffness in itself. This is easily demonstrated via the theoretical Moens–Korteweg equation for a uniform 1D tube, stating that $PWV = \sqrt{Eh/\rho D}$ with E the Young elasticity modulus (400–1200 kPa for arteries), ρ the density of blood ($\sim 1060 \ kg \cdot m^{-3}$), h the wall thickness, and D the diameter of the vessel. Another formula, sometimes used, is the so-called Bramwell–Hill (84) equation: $PWV = \sqrt{A \partial P / \rho \partial A} = \sqrt{A/\rho \, 1/C_A}$ with P intra-arterial pressure, A cross-sectional area and C_A the area compliance, $\partial A/\partial P$. For unaltered vessel dimensions, an increase in stiffness (increase in E, decrease in C_A) yields an increase in PWV. The most common technique to estimate PWV is to measure the time delay between the passage of the pressure, flow, or diameter distension pulse at two distinct locations, for example, between the ascending aorta (or carotid artery) and the femoral artery (85). These signals can be measured noninvasively, for example, with tonometry (86,87) (pressure pulse) or ultrasound [flow velocity, vessel diameter distension (88,89)]. Reference work was done by Avolio et al., who measured PWV is large cohorts in Chinese urban and rural communities (90,91).

When the input impedance is known, which implies that pressure and flow are known, the "effective length" of the arterial system, that is, the distance between the location where the input impedance is measured and an "apparent" reflection site at a distance L, can be estimated using the earlier mentioned quarter wavelength formula. Murgo et al. found the effective length to be ~ 44 cm in humans (61), which would suggest a reflection site near the diaphragm, in the vicinity of where the renal arteries branch off the abdominal aorta. Latham et al. demonstrated that local reflection is higher at this site (67), but one should keep in mind that the quarter wavelength formula is based on a conceptual model of the arterial system, and the apparent length does not correspond to a physical obstruction causing the reflection. Nevertheless, there is still no clear picture of the major reflection sites, which is complex due to the continuous branching, the dispersed distal reflection sites, and the continuous reflection caused by the geometric and elastic tapering of the vessels (45,64,67,68,92–96).

Central and Peripheral Blood Pressure: On the Use of Transfer Functions. Wave travel and reflection generally result in an amplification of the pressure pulse from the aorta (central) toward the periphery (brachial, radial, femoral artery) (44,45). Since clinicians usually measure blood pressure at the brachial artery (cuff sphygmomanometry), this implies that the pressure measured at this location is an overestimation of central pressure (97,98). It is the latter against which the heart ejects in systole, and which is therefore of primary interest.

In the past few years, much attention has been attributed to the relation between radial artery and central pressure (99–102). The reason for this is that radial artery pressure pulse is measurable with applanation tonometry, a noninvasive method (86,87,103). The relation between central and radial artery pressure can be expressed with a "transfer function", most commonly displayed in the frequency domain (Fig. 12). The transfer function expresses

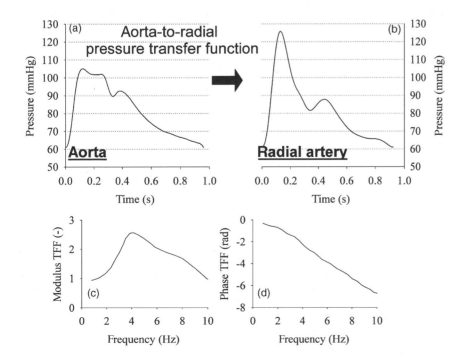

Figure 12. Demonstration of the aorta-to-radial pressure pulse amplification: (a and b) are carotid (as substitute for aorta pressure) and radial artery pressure measured in the same subject with applanation tonometry (1 mmHg = 133.3 Pa). (c and d) Display the modulus and phase angle of the radial-to-aorta transfer function as published by Ref. (99).

the relation between the individual harmonics at both locations, with a modulus (damping or amplification of the harmonic) and a phase angle (the time delay). In humans, the aorta-to-radial transfer function shows a peak ~ 4 Hz, so that amplification is maximal for a harmonic of that frequency (99,101). It has been demonstrated that the transfer function is surprisingly constant, with albeit little variation among individuals (99). It is this observation that led to the use of generalized transfer functions, embedded in commercial systems, that allow us to calculate central pressure from a peripheral pressure measurement (104). In these systems, the transfer is commonly done using a time domain approach [autoregressive exogenous (ARX) models (100)]. There have been attempts to "individualize" the transfer function, but until now these attempts were unsuccessful (105). The general consensus appears to be that the generalized transfer function can be used to estimate central systolic blood pressure and pulse pressure (104), but that one should be cautious when using synthesized central pressures for assessing parameters based on details in the pressure wave (e.g., the augmentation index) (106). The latter requires high frequency information that is more difficult to guarantee with a synthesized curve, both due to the increase in scatter in the generalized transfer function for higher frequencies, and the absence of high frequency information in the peripherally measured pressure pulse (99,107).

One more issue worth noting is the fact that the radial-to-aorta transfer function peaks at 4 Hz (in humans), which implies that the peripheral pulse amplification is frequency, and thus heart rate dependent (83,108,109). In a pressure signal, most power is embedded within the first two to three harmonics. When heart rate shifts from 60 to 120 beats/min (e.g., during exercise), the highest amplification thus occurs for these most powerful, predominant harmonics, leading to a more excessive pressure amplification. This means, conversely, that the overestimation of central pressure by a peripheral pressure measurement is a function of heart rate. As such, the effect of drugs that alter the heart rate (e.g., beta blockers, slowing down heart rate) may not be fully reflected by the traditional brachial sphygmomanometer pressure measurement (97,98).

Practical Considerations. The major difficulty, transferring experimental results into clinical practice, is the accurate measurement of the data necessary for the hemodynamic analysis. Nevertheless, there are noninvasive tools available that do permit "full" noninvasive hemodynamic assessment in clinical conditions, as possible with central pressure and flow.

Flow is at present rather easy to measure with ultrasound, using pulsed Doppler modalities. Flow velocities can be measured in the left ventricular outflow tract and, when multiplied with outflow tract cross-section, be converted into flow. Also, velocity-encoded MRI can provide aortic blood flow velocities.

As for measuring pressure, applanation tonometry is an appealing technique, since it allows us to measure pressure pulse tracings at superficial arteries such as the radial, brachial, femoral, and carotid artery (86,87,103,110,111). The carotid artery is located close to the heart, and is often used as a surrogate for the ascending aorta pulse contour (87,111). Others advocate the use of a transfer function to obtain central pressure from radial artery pressure measurement (104), but generalized transfer functions do not fully capture all details of a central pressure (106). Nevertheless, applanation tonometry only yields the morphology of the pressure wave and not absolute pressure values (although this is theoretically possible, but virtually impossible to achieve in practice). For the calibration of the tracings, one still relies on brachial cuff sphygmomanometry (yielding systolic and diastolic brachial blood pressure). Van Bortel et al. validated a calibration scheme in which first a brachial pressure waveform is calibrated with sphygmomanometer systolic and diastolic blood pressure (81,112). Averaging of this calibrated curve subsequently yields mean arterial blood pressure. Pulse waveforms at other peripheral locations are then calibrated using diastolic and mean blood pressure. Alternatively, one can use an oscillometric method to obtain mean arterial blood pressure, or estimate mean arterial blood pressure from systolic and diastolic blood pressure using the two-third/one-third rule of thumb (mean blood pressure $= \frac{2}{3}$ diastolic $+ \frac{1}{3}$ systolic pressure). With the oscillometric method, one makes use of the oscillations that can be measured within the cuff (and in the brachial artery) when the cuff pressure is lowered from a value above systolic pressure (full occlusion) to a value below diastolic pressure (no occlusion).

Nowadays, ultrasound vessel wall tracking techniques also allow us to measure the distension of the vessel upon the passage of the waveform (88,113). These waveforms are quasiidentical to the pressure waveforms [but not entirely, since the pressure–diameter relationship of blood vessels is not linear (114,115)] and can potentially be used as an alternative for tonometric waveforms (112,116).

When central pressure and flow are available, the arterial system can be characterized using all described techniques, from impedance analysis, parameter estimation by means of a lumped parameter windkessel or tube model, analysis of wave reflection, and so on. The augmentation index can be derived from pressure data alone. A parameter that is certainly relevant to measure in addition to pressure and flow is pulse wave velocity, as it provides both functional information, and helps to elucidate wave reflection. For measuring the transit time (e.g., from carotid to femoral) of the arterial pressure pulse, the flow velocity wave or the diameter distension wave is the most commonly applied technique (85).

HEART–ARTERIAL COUPLING

The main function of the heart is to maintain the circulation, that is, to provide a sufficient amount of blood at sufficiently high pressures to guarantee the perfusion of vital organs, including the heart itself. Arterial pressure and flow arise from the interaction between the ventricle and the arterial load. In the past few years, the coupling of the heart and the arterial system in cardiovascular pathophysiology has been recognized (74–76,117,118). With ageing or in hypertension, for example, arterial stiffening leads to an increase of the pressure pulse and an early

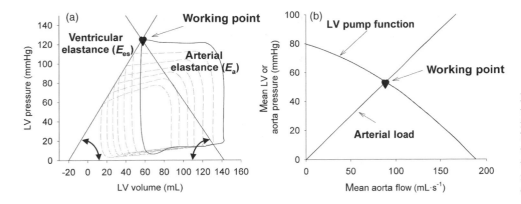

Figure 13. Heart–arterial coupling using the E_a–E_{es} (a) of the pump function graph (b) framework. In both cases, the intersection between the curves characterizing the ventricle and the arterial system determines the working point of the cardiovascular system.

return of reflected waves, increasing the load on the heart (73,119), which will react through adaptation (remodeling) to cope with the increased afterload.

Factors affecting ventricular pump function are preload (venous filling pressure and/or end diastolic volume EDV), heart rate, and the intrinsic contractile function of the heart muscle (here assumed to be best represented by the slope of the end-systolic PV relation, E_{es}). The two main mechanical afterload parameters determining systolic and pulse pressure are total peripheral resistance (R) and total arterial compliance (C) (120,121). Arterial blood pressure and flow are thus determined by a limited number of cardiac and arterial mechanical factors.

The most common framework, however, to study the heart–arterial (or ventricular–vascular) coupling, is the E_a–E_{es} framework (23,122–124), where E_a is effective arterial elastance (125), and where E_a/E_{es} is a parameter reflecting heart–arterial coupling. In the pressure–volume plane, E_a is the slope of the line connecting the end-systolic point (ESV, ESP) with the end-diastolic volume point on the volume axis (EDV, 0) (Fig. 13). As such, $E_a = ESP/SV$, with SV the stroke volume. The parameter E_a can be written as a function of three-element windkessel parameters describing the arterial system (125), but it has been shown that E_a can be approximated as R/T with T the duration of the cardiovc cycle (124,126).

The first major advantage of this framework is that it allows us to graphically study the interaction between the heart and the arterial system. Cardiac function is characterized by the ESPVR, which is a fully defined line in the PV plane. For a given preload (EDV) and arterial stiffness E_a, the line characterizing the arterial system can be drawn as well. The intersection of these two lines determines the end-systolic point (ESV, ESP), and thus the stroke volume and (end-)systolic pressure (Fig. 13).

Second, the analysis in the PV plane also allows us to calculate some energetic parameters, and to relate them to E_a/E_{es}. It can be shown that SW is maximal only when $E_a = E_{es}$. Thus when $E_a/E_{es} = 1$. Mechanical efficiency is maximal when $E_a/E_{es} = 0.5$, when arterial elastance is one-half of the end-systolic elastance (124). In normal hearts, the heart operates in conditions with an E_a/E_{es} ratio in the range 0.5:1, and it has been shown that this ratio is preserved in normal aging (74,75). Arterial stiffening is thus paralleled by an increase in ventricular end-systolic stiffness. This happens even in patients with heart failure,

but with preserved ejection fraction (76). In this patient population, both E_a and E_{es} are higher than expected with normal ageing, but the ratio is more or less preserved.

Although the E_a/E_{es} parameter certainly has its value, it has been shown under experimental conditions that the E_a/E_{es} range, where the heart operates near optimal efficiency, is quite broad (122,127). Thus the parameter is not a very sensitive tool to detect "uncoupling" of the heart and the arterial system, which commonly implies a increase in E_a and a poorer ventricular function (depressed E_{es}) leading to increased E_a/E_{es} ratios (128,129). Also, one should not focus only on the coupling, but also on the absolute values of E_a and E_{es}. Increased stiffening leads to hypertensive response in exercise, where small changes in volume (filling volumes, stroke volumes) have an amplified effect on arterial pressure and workload, and lead to an increased energy cost to increase stroke volume (75,76).

By using dimensional analysis, Stergiopulos et al. (130) and later Segers et al. (131), demonstrated that blood pressure and stroke volume are mainly determined by ventricular preload, and by 2 dimensionless parameters: $E_{es}C$ and RC/T, where T is the heart period. The first is the product of ventricular (end-systolic) stiffness and arterial compliance; the latter is the ratio of the time constant of arterial blood pressure decay (RC) and the heart period.

It is also worth noting the link between the coupling parameters as proposed by Stergiopulos et al., and E_a/E_{es}: $(RC/T)/(E_{es}C) = (R/T)/E_{es} \approx E_a/E_{es}$. E_a/E_{es} thus combines the two coupling parameters following from the dimensional analysis into a single dimensionless parameter. This is, however, at the cost of eliminating the contribution of total arterial compliance from the parameter. Within this perspective, the terminology of E_a as arterial stiffness, is perhaps not entirely justified, as it is a parameter related to peripheral resistance and the heart period, rather than to arterial compliance or stiffness (126).

Finally, it should be mentioned that heart–arterial interaction can also be studied within the pump function framework of Westerhof and co-workers (Fig. 13). While the function of the heart can be described with the pump function graph, the function of the arterial system can be displayed in the same graph. In its simplest approximation, considering only mean pressure and flow, the arterial system is characterized by the total vascular resistance. In a pump function graph, this is represented by a straight line: mean flow is directly proportional to mean arterial pressure.

The intersection with the pump function graph yields the working point of the cardiovascular system. Note, however, that although both the pump and arterial function curve are displayed in the same figure, their y axis is not the same.

BIBLIOGRAPHY

Cited References

1. Strackee EJ, Westerhof N. Physics of Heart and Circulation. Bristol (UK): Institute of Physics; 1993.
2. Suga H, Sagawa K, Shoukas AA. Load independence of the instantaneous pressure–volume ratio of the canine left ventricle and effects of epinephrine and heart rate on the ratio. Circ Res 1973;32:314–322.
3. Suga H, Sagawa K. Instantaneous pressure–volume relationships and their ratio in the exercised, supported canine left ventricle. Circ Res 1974;35:117–126.
4. van der Velde ET, Burkhoff D, Steendijk P, Karsdon J, Sagawa K, Baan J. Nonlinearity and load sensitivity of end-systolic pressure–volume relation of canine left ventricle *in vivo*. Circulation 1991;83:315–327.
5. Kass DA, Beyar R, Lankford E, Heard M, Maughan WL, Sagawa K. Influence of contractile state on curvilinearity of in situ end-systolic pressure–volume relations. Circulation 1989;79:167–178.
6. Baan J, van der Velde ET, de Bruin HG, Smeenk GJ, Koops J, van Dijk AD, Temmerman D, Senden J, Buis B. Continuous measurement of left ventricular volume in animals and humans by conductance catheter. Circulation 1984;70:812–823.
7. Suga H. Cardiac energetics: from E(max) to pressure–volume area. Clin Exp Pharmacol Physiol 2003;30:580–585.
8. Suga H. Global cardiac function: mechano-energetico-informatics. J Biomech 2003;36:713–720.
9. Glower DD, Spratt JA, Snow ND, Kabas JS, Davis JW, Olsen CO, Tyson GS, Sabiston Jr DC, Rankin JS. Linearity of the Frank–Starling relationship in the intact heart: the concept of preload recruitable stroke work. Circulation 1985;71:994–1009.
10. Starling E, Vischer M. The regulation of the output of the heart. J Physiol Cambridge 1927;62:243–261.
11. Elzinga G, Westerhof N. How to quantify pump function of the heart. The value of variables derived from measurements on isolated muscle. Circ Res 1979;44:303–308.
12. Elzinga G, Westerhof N. Pump function of the feline left heart: changes with heart rate and its bearing on the energy balance. Cardiovasc Res 1980;14:81–92.
13. Takeuchi M, Igarashi Y, Tomimoto S, Odake M, Hayashi T, Tsukamoto T, Hata K, Takaoka H, Fukuzaki H. Single-beat estimation of the slope of the end-systolic pressure–volume relation in the human left ventricle. Circulation 1991;83: 202–212.
14. Chen CH, Fetics B, Nevo E, Rochitte CE, Chiou KR, Ding PA, Kawaguchi M, Kass DA. Noninvasive single-beat determination of left ventricular end-systolic elastance in humans. J Am Coll Cardiol 2001;38:2028–2034.
15. Lambermont B, Segers P, Ghuysen A, Tchana-Sato V, Morimont P, Dogne JM, Kolh P, Gerard P, D'Orio V. Comparison between single-beat and multiple-beat methods for estimation of right ventricular contractility. Crit Care Med 2004;32:1886–1890.
16. Kjorstad KE, Korvald C, Myrmel T. Pressure–volume-based single-beat estimations cannot predict left ventricular contractility *in vivo*. Am J Physiol Heart Circ Physiol 2002;282:H1739–H1750.
17. Kass DA, Beyar R. Evaluation of contractile state by maximal ventricular power divided by the square of end-diastolic volume. Circulation 1991;84:1698–1708.
18. Nakayama M, Chen CH, Nevo E, Fetics B, Wong E, Kass DA. Optimal preload adjustment of maximal ventricular power index varies with cardiac chamber size. Am Heart J 1998;136:281–288.
19. Segers P, Leather HA, Verdonck P, Sun Y-Y, Wouters PF. Preload-adjusted maximal power of right ventricle: contribution of end-systolic P-V relation intercept. Am J Physiol 2002;283:H1681–H1687.
20. Segers P, Tchana-Sato V, Leather HA, Lambermont B, Ghuysen A, Dogne JM, Benoit P, Morimont P, Wouters PF, Verdonck P, Kolh P. Determinants of left ventricular preload-adjusted maximal power. Am J Physiol Heart Circ Physiol 2003;284:H2295–H2301.
21. Ommen SR. Echocardiographic assessment of diastolic function. Curr Opin Cardiol 2001;16:240–245.
22. Garcia MJ, Thomas JD, Klein AL. New Doppler echocardiographic applications for the study of diastolic function. J Am Coll Cardiol 1998;32:865–875.
23. Appleton CP, Hatle LK, Popp RL. Relation of transmitral flow velocity patterns to left ventricular diastolic function: new insights from a combined hemodynamic and Doppler echocardiographic study. J Am Coll Cardiol 1988;12:426–440.
24. Thomas JD, Zhou J, Greenberg N, Bibawy G, McCarthy PM, Vandervoort PM. Physical and physiological determinants of pulmonary venous flow: numerical analysis. Am J Physiol 1997;272:H2453–H2465.
25. Garcia MJ, Ares MA, Asher C, Rodriguez L, Vandervoort P, Thomas JD. An index of early left ventricular filling that combined with pulsed Doppler peak E velocity may estimate capillary wedge pressure. J Am Coll Cardiol 1997;29:448–454.
26. De Mey S, De Sutter J, Vierendeels J, Verdonck P. Diastolic filling and pressure imaging: taking advantage of the information in a colour M-mode Doppler image. Eur J Echocardiog 2001;2:219–233.
27. Nagueh SF, Middleton KJ, Kopelen HA, Zoghbi WA, Quinones MA. Doppler tissue imaging: a noninvasive technique for evaluation of left ventricular relaxation and estimation of filling pressures. J Am Coll Cardiol 1997;30:1527–1533.
28. Heimdal A, Stoylen A, Torp H, Skjaerpe T. Real-time strain rate imaging of the left ventricle by ultrasound. J Am Soc Echocardiog 1998;11:1013–1019.
29. D'Hooge J, Heimdal A, Jamal F, Kukulski T, Bijnens B, Rademakers F, Hatle L, Suetens P, Sutherland GR. Regional strain and strain rate measurements by cardiac ultrasound: principles, implementation and limitations. Eur J Echocardiog 2000;1:154–170.
30. Chaney JC, Derdak S. Minimally invasive hemodynamic monitoring for the intensivist: current and emerging technology. Crit Care Med 2002;30:2338–2345.
31. Leather HA, Vuylsteke A, Bert C, M'Fam W, Segers P, Sergeant P, Vandermeersch E, Wouters PF. Evaluation of a new continuous cardiac output monitor in off-pump coronary artery surgery. Anaesthesia 2004;59:385–389.
32. O'Dell WG, McCulloch AD. Imaging three-dimensional cardiac function. Annu Rev Biomed Eng 2000;2:431–456.
33. Matter C, Nagel E, Stuber M, Boesiger P, Hess OM. Assessment of systolic and diastolic LV function by MR myocardial tagging. Basic Res Cardiol 1996;91:23–28.
34. Pijls NHJ, de Bruyne B. Coronary Pressure. Dordrecht: Kluwer Academic Publishers; 1997.
35. Spaan JAE. Coronary Blood Flow: Mechanics, Distribution, and Control. Dordrecht and Boston: Kluwer Academic Publishers; 1991.
36. Mosher P, Ross Jr J, McFate PA, Shaw RF. Control of Coronary Blood Flow by an Autoregulatory Mechanism. Circ Res 1964;14:250–259.

37. Bellamy RF. Diastolic coronary artery pressure-flow relations in the dog. Circ Res 1978;43:92–101.

38. Downey JM, Kirk ES. Inhibition of coronary blood flow by a vascular waterfall mechanism. Circ Res 1975;36:753–760.

39. Permutt S, Riley RL. Hemodynamics of Collapsible Vessels with Tone: The Vascular Waterfall. J Appl Physiol 1963;18:924–932.

40. Spaan JA, Breuls NP, Laird JD. Diastolic-systolic coronary flow differences are caused by intramyocardial pump action in the anesthetized dog. Circ Res 1981;49:584–593.

41. Matthys K, Carlier S, Segers P, Ligthart J, Sianos G, Serrano P, Verdonck PR, Serruys PW. In vitro study of FFR, QCA, and IVUS for the assessment of optimal stent deployment. Catheter Cardiovasc Interv 2001;54:363–375.

42. Siebes M, Verhoeff BJ, Meuwissen M, de Winter RJ, Spaan JA, Piek JJ. Single-wire pressure and flow velocity measurement to quantify coronary stenosis hemodynamics and effects of percutaneous interventions. Circulation 2004; 109:756–762.

43. Westerhof N, Stergiopulos N, Noble M. Snapshots of Hemodynamics. An aid for clinical research and graduate education. New York: Springer Science + Business Media; 2004.

44. Nichols WW, O'Rourke MF. McDonald's Blood Flow in Arteries. 3rd ed. London: Edward Arnold; 1990.

45. Milnor WR. Hemodynamics. 2nd ed. Baltimore (MA): Williams & Wilkins; 1989.

46. Hales S. Statical Essays: Containing Haemostatics (reprint 1964). New York: Hafner Publishing; 1733.

47. Frank O. Die Grundfurm des arteriellen Pulses. Erste Abhandlung. Mathematische Analyse. Z Biol 1899;37: 483–526.

48. Frank O. Der Puls in den Arterien. Z Biol 1905;46:441–553.

49. Segers P, Verdonck P. Principles of Vascular Physiology. In: Lanzer P, Topol EJ, editors. Pan Vascular Medicine. Integrated Clinical Management. Heidelberg: Springer-Verlag; 2002.

50. Toy SM, Melbin J, Noordergraaf A. Reduced models of arterial systems. IEEE Trans Biomed Eng 1985;32:174–176.

51. Liu Z, Brin K, Yin F. Estimation of total arterial compliance: an improved method and evaluation of current methods. Am J Physiol 1986;251:H588–H600.

52. Simon A, Safar L, London G, Levy B, Chau N. An evaluation of large arteries compliance in man. Am J Physiol 1979;237:H550–H554.

53. Stergiopulos N, Meister JJ, Westerhof N. Evaluation of methods for the estimation of total arterial compliance. Am J Physiol 1995;268:H1540–H1548.

54. Chemla D, Hébert J-L, Coirault C, Zamani K, Suard I, Colin P, Lecarpentier Y. Total arterial compliance estimated by stroke volume-to-aortic pulse pressure ratio in humans. Am J Physiol 1998;274:H500–H505.

55. Stergiopulos N, Segers P, Westerhof N. Use of pulse pressure method for estimating total arterial compliance in vivo. Am J Physiol 1999;276:H424–H428.

56. Westerhof N, Elzinga G, Sipkema P. An artificial arterial system for pumping hearts. J Appl Physiol 1971;31:776–781.

57. Westerhof N, Elzinga G. Normalized input impedance and arterial decay time over heart period are independent of animal size. Am J Physiol 1991;261:R126–R133.

58. Segers P, Brimioulle S, Stergiopulos N, Westerhof N, Naeije R, Maggiorini M, Verdonck P. Pulmonary arterial compliance in dogs and pigs: the three-element windkessel model revisited. Am J Physiol 1999;277:H725–H731.

59. Stergiopulos N, Westerhof B, Westerhof N. Total arterial inertance as the fourth element of the windkessel model. Am J Physiol 1999;276:H81–H88.

60. Burattini R, Gnudi G. Computer identification of models for the arterial tree input impedance: comparison between two new simple models and first experimental results. Med Biol Eng Comp 1982;20:134–144.

61. Murgo JP, Westerhof N, Giolma JP, Altobelli SA. Aortic input impedance in normal man: relationship to pressure wave forms. Circulation 1980;62:105–116.

62. Burattini R, Campbell KB. Modified asymmetric T-tube model to infer arterial wave reflection at the aortic root. IEEE Trans Biomed Eng 1989;36:805–814.

63. Chang KC, Tseng YZ, Kuo TS, Chen HI. Impedance and wave reflection in arterial system: simulation with geometrically tapered T-tubes. Med Biol Eng Comput 1995;33:652–660.

64. Segers P, Verdonck P. Role of tapering in aortic wave reflection: hydraulic and mathematical model study. J Biomech 2000;33:299–306.

65. Wang DM, Tarbell JM. Nonlinear analysis of oscillatory flow, with a nonzero mean, in an elastic tube (artery). J Biomech Eng 1995;117:127–135.

66. Campbell K, Lee CL, Frasch HF, Noordergraaf A. Pulse reflection sites and effective length of the arterial system. Am J Physiol 1989;256:H1684–H1689.

67. Latham R, Westerhof N, Sipkema P, Rubal B, Reuderink P, Murgo J. Regional wave travel and reflections along the human aorta: a study with six simultaneous micromanometric pressures. Circulation 1985;72:1257–1269.

68. Berger D, Li J, Laskey W, Noordergraaf A. Repeated reflection of waves in the systemic arterial system. Am J Physiol 1993;264:H269–H281.

69. Westerhof N, Sipkema P, Van Den Bos G, Elzinga G. Forward and backward waves in the arterial system. Cardiovasc Res 1972;6:648–656.

70. Dujardin J, Stone D. Characteristic impedance of the proximal aorta determined in the time and frequency domain: a comparison. Med Biol Eng Comp 1981;19:565–568.

71. Khir AW, O'Brien A, Gibbs JS, Parker KH. Determination of wave speed and wave separation in the arteries. J Biomech 2001;34:1145–1155.

72. O'Rourke MF. Mechanical principles. Arterial stiffness and wave reflection. Pathol Biol (Paris) 1999;47:623–633.

73. Westerhof N, O'Rourke MF. Haemodynamic basis for the development of left ventricular failure in systolic hypertension and for its logical therapy. J Hypertens 1995;13:943–952.

74. Chen C-H, Nakayama M, Nevo E, Fetics BJ, Maughan WL, Kass DA. Coupled systolic-ventricular and vascular stiffening with age. Implications for pressure regulation and cardiac reserve in the elderly. J Am Coll Cardiol 1998;32: 1221–1227.

75. Kass DA. Age-related changes in venticular-arterial coupling: pathophysiologic implications. Heart Fail Rev 2002;7:51–62.

76. Kawaguchi M, Hay I, Fetics B, Kass DA. Combined ventricular systolic and arterial stiffening in patients with heart failure and preserved ejection fraction: implications for systolic and diastolic reserve limitations. Circulation 2003;107: 714–720.

77. Parker KH, Jones CJ. Forward and backward running waves in the arteries: analysis using the method of characteristics. J Biomech Eng 1990;112:322–326.

78. Bleasdale RA, Mumford CE, Campbell RI, Fraser AG, Jones CJ, Frenneaux MP. Wave intensity analysis from the common carotid artery: a new noninvasive index of cerebral vasomotor tone. Heart Vessels 2003;18:202–206.

79. Wang Z, Jalali F, Sun YH, Wang JJ, Parker KH, Tyberg JV. Assessment of Left Ventricular Diastolic Suction in Dogs using Wave-intensity Analysis. Am J Physiol Heart Circ Physiol 2004.?

80. Sun YH, Anderson TJ, Parker KH, Tyberg JV. Effects of left ventricular contractility and coronary vascular resistance on

coronary dynamics. Am J Physiol Heart Circ Physiol 2004;286:H1590–H1595.

81. Kelly R, Hayward C, Avolio A, O'Rourke M. Noninvasive determination of age-related changes in the human arterial pulse. Circulation 1989;80:1652–1659.

82. Hayward CS, Kelly RP. Gender-related differences in the central arterial pressure waveform. J Am Coll Cardiol 1997;30:1863–1871.

83. Wilkinson IB, MacCallum H, Flint L, Cockcroft JR, Newby DE, Webb DJ. The influence of heart rate on augmentation index and central arterial pressure in humans. J Physiol 2000;525:263–270.

84. Bramwell CJ, Hill A. The velocity of the pulse wave in man. Proc R Soc London (Biol) 1922;93:298–306.

85. Lehmann ED. Noninvasive measurements of aortic stiffness: methodological considerations. Pathol Biol (Paris) 1999;47:716–730.

86. Drzewiecki GM, Melbin J, Noordergraaf A. Arterial tonometry: review and analysis. J Biomech 1983;16:141–152.

87. Chen CH, Ting CT, Nussbacher A, Nevo E, Kass D, Pak P, Wang SP, Chang MS, Yin FC. Validation of carotid artery tonometry as a means of estimating augmentation index of ascending aortic pressure. Circulation 1996;27:168–175.

88. Hoeks AP, Brands PJ, Smeets FA, Reneman RS. Assessment of the distensibility of superficial arteries. Ultrasound Med Biol 1990;16:121–128.

89. Segers P, Rabben SI, De Backer J, De Sutter J, Gillebert TC, Van Bortel L, Verdonck P. Functional analysis of the common carotid artery: relative distension differences over the vessel wall measured *in vivo*. J Hypertens 2004;22:973–981.

90. Avolio A, Chen S, Wang R, Zhang C, Li M, O'Rourke M. Effects of aging on changing arterial compliance and left ventricular load in a northern Chinese urban community. Circulation 1983;68:50–58.

91. Avolio A, Fa-Quan D, Wei-Qiang L, Yao-Fei L, Zhen-Dong H, Lian-Fen X, M. OR. Effects of aging on arterial distensibility in populations with high and low prevalence of hypertension: comparison between urban and rural communities in China. Circulation 1985;71:202–210.

92. Karamanoglu M, Gallagher D, Avolio A, O'Rourke M. Functional origin of reflected pressure waves in a multibranched model of the human arterial system. Am J Physiol 1994;267:H1681–H1688.

93. Karamanoglu M, Gallagher D, Avolio A, O'Rourke M. Pressure wave propagation in a multibranched model of the human upper limb. Am J Physiol 1995;269:H1363–H1369.

94. O'Rourke MF. Pressure and flow waves in the systemic arteries and the anatomical design of the arterial system. J Appl Physiol 1967;23:139–149.

95. Avolio A. Multi-branched model of the human arterial system. Med Biol Eng Comp 1980;18:709–718.

96. Stergiopoulos N, Young DF, Rogge TR. Computer simulation of arterial flow with applications to arterial and aortic stenoses. J Biomech 1992;25:1477–1488.

97. Takazawa K, Tanaka N, Takeda K, Kurosu F, Ibukiyama C. Underestimation of Vasodilator Effects of Nitroglycerin by Upper Limb Blood Pressure. Hypertension 1995;26:520–523.

98. Vlachopoulos C, Hirata K, O'Rourke MF. Pressure-altering agents affect central aortic pressures more than is apparent from upper limb measurements in hypertensive patients: the role of arterial wave reflections. Hypertension 2001;38: 1456–1460.

99. Chen C-H, Nevo E, Fetics B, Pak PH, Yin FCP, Maughan L, Kass DA. Estimation of Central Aortic Pressure Waveform by Mathematical Transformation of Radial Tonometry Pressure: Validation of Generalized Transfer Function. Circulation 1997;95:1827–1836.

100. Fetics B, Nevo E, Chen CH, Kass DA. Parametric model derivation of transfer function for noninvasive estimation of aortic pressure by radial tonometry. IEEE Trans Biomed Eng 1999;46:698–706.

101. Karamanoglu M, O'Rourke M, Avolio A, Kelly R. An analysis of the relationship between central aortic and peripheral upper limb pressure waves in man. Eur Heart J 1993; 14:160–167.

102. Karamanoglu M, Fenely M. Derivation of the ascending aorta-carotid pressure transfer function with an arterial model. Am J Physiol 1996;271:H2399–H2404.

103. Matthys K, Verdonck P. Development and modelling of arterial applanation tonometry: a review. Technol Health Care 2002;10:65–76.

104. Pauca AL, O'Rourke MF, Kon ND. Prospective evaluation of a method for estimating ascending aortic pressure from the radial artery pressure waveform. Hypertension 2001;38: 932–937.

105. Hope SA, Tay DB, Meredith IT, Cameron JD. Comparison of generalized and gender-specific transfer functions for the derivation of aortic waveforms. Am J Physiol Heart Circ Physiol 2002;283:H1150–H1156.

106. Segers P, Qasem A, De Backer T, Carlier S, Verdonck P, Avolio A. Peripheral "Oscillatory" Compliance Is Associated With Aortic Augmentation Index. Hypertension 2001;37: 1434–1439.

107. O'Rourke MF, Nichols WW. Use of arterial transfer function for the derivation of aortic waveform characteristics. J Hypertens 2003;21:2195–2197; author reply 2197–2199.

108. Albaladejo P, Copie X, Boutouyrie P, Laloux B, Declere AD, Smulyan H, Benetos A. Heart rate, arterial stiffness, and wave reflections in paced patients. Hypertension 2001;38: 949–52.

109. Wilkinson IB, Mohammad NH, Tyrrell S, Hall IR, Webb DJ, Paul VE, Levy T, Cockcroft JR. Heart rate dependency of pulse pressure amplification and arterial stiffness. Am J Hypertens 2002;15:24–30.

110. Marcus R, Korcarz C, McCray G, Neumann A, Murphy M, Borow K, Weinert L, Bednarz J, Gretler D, Spencer K, Sareli P, Lang R. Noninvasive method for determination of arterial compliance using Doppler echocardiography and subclavian pulse tracings. Circulation 1994;89:2688–2699.

111. Kelly R, Karamanoglu M, Gibbs H, Avolio A, O'Rourke M. Noninvasive carotid pressure wave registration as an indicator of ascending aortic pressure. J Vas Med Biol 1989;1:241–247.

112. Van Bortel LM, Balkestein EJ, van der Heijden-Spek JJ, Vanmolkot FH, Staessen JA, Kragten JA, Vredeveld JW, Safar ME, Struijker Boudier HA, Hoeks AP. Noninvasive assessment of local arterial pulse pressure: comparison of applanation tonometry and echo-tracking. J Hypertens 2001;19:1037–1044.

113. Rabben SI, Baerum S, Sorhus V, Torp H. Ultrasound-based vessel wall tracking: an auto-correlation technique with RF center frequency estimation. Ultrasound Med Biol 2002;28: 507–517.

114. Langewouters G, Wesseling K, Goedhard W. The static elastic properties of 45 human thoracic and 20 abdominal aortas in vitro and the parameters of a new model. J Biomech 1984;17:425–435.

115. Hayashi K. Experimental approaches on measuring the mechanical properties and constitutive laws of arterial walls. J Biomech Eng 1993;115:481–488.

116. Meinders JM, Hoeks AP. Simultaneous assessment of diameter and pressure waveforms in the carotid artery. Ultrasound Med Biol 2004;30:147–154.

117. Lakatta EG, Levy D. Arterial and cardiac aging: major shareholders in cardiovascular disease enterprises:

Part II: the aging heart in health: links to heart disease. Circulation 2003;107:346–354.

118. Lakatta EG, Levy D. Arterial and cardiac aging: major shareholders in cardiovascular disease enterprises: Part I: aging arteries: a "set up" for vascular disease. Circulation 2003;107:139–146.

119. O'Rourke M. Arterial stiffness, systolic blood pressure, and logical treatment of arterial hypertension. Hypertension 1990;15:339–347.

120. Stergiopulos N, Westerhof N. Role of total arterial compliance and peripheral resistance in the determination of systolic and diastolic aortic pressure. Pathol Biol (Paris) 1999;47:641–647.

121. Stergiopulos N, Westerhof N. Determinants of pulse pressure. Hypertension 1998;32:556–559.

122. Sunagawa K, Maughan WL, Burkhoff D, Sagawa K. Left ventricular interaction with arterial load studied in isolated canine ventricle. Am J Physiol 1983;245:H773–H780.

123. Sunagawa K, Maughan WL, Sagawa K. Optimal arterial resistance for the maximal stroke work studied in isolate canine left ventricle. Circ Res 1985;56:586–595.

124. Burkhoff D, Sagawa K. Ventricular efficiency predicted by an analytical model. Am J Physiol 1986;250:R1021–R1027.

125. Kelly R, Ting C, Yang T, Liu C, Lowell W, Chang M, Kass D. Effective arterial elastance as index of arterial vascular load in humans. Circulation 1992;86:513–521.

126. Segers P, Stergiopulos N, Westerhof N. Relation of effective arterial elastance to arterial system properties. Am J Physiol Heart Circ Physiol 2002;282:H1041–H1046.

127. De Tombe PP, Jones S, Burkhoff D, Hunter WC, Kass DA. Ventricular stroke work and efficiency both remain nearly optimal despite altered vascular loading. Am J Physiol 1993;264:H1817–H1824.

128. Asanoi H, Sasayama S, Kameyama T. Ventriculoarterial coupling in normal and failing heart in humans. Circ Res 1989;65:483–493.

129. Sasayama S, Asanoi H. Coupling between the heart and arterial system in heart failure. Am J Med 1991;90: 14S–18S.

130. Stergiopulos N, Meister JJ, Westerhof N. Determinants of stroke volume and systolic and diastolic pressure. Am J Physiol 1996;270:H2050–H2059.

131. Segers P, Steendijk P, Stergiopulos N, Westerhof N. Predicting systolic and diastolic aortic blood pressure and stroke volume in the intact sheep. J Biomech 2001;34:41–50.

See also BIOIMPEDANCE IN CARDIOVASCULAR MEDICINE; BLOOD PRESSURE MEASUREMENT; FLOWMETERS, ELECTROMAGNETIC; MONITORING, HEMODYNAMIC.

HEMODYNAMIC MONITORING.   See MONITORING, HEMODYNAMIC.

HIGH FREQUENCY VENTILATION

J. BERT BUNNELL
Bunnell Inc.
Salt Lake City, Utah

INTRODUCTION

High frequency ventilators (HFVs) were designed to eliminate many of the problems that conventional ventilators create as they try to mimic normal breathing. When we breathe normally, we draw gas into the lungs by creating a negative pressure with our diaphragm. Iron lungs were created to replicate that activity, and they worked very well for thousands of polio patients. However, when patients are sealed off in airtight iron lungs numerous practical problems unrelated to breathing arise.

Positive pressure ventilators made assisting ventilation much easier. Attaching the ventilator to the patient's lungs via an endotracheal (ET) tube greatly simplified patient care. But, lungs, especially premature lungs, are not designed to tolerate much positive pressure.

The lungs of prematurely borne infants have yet to be fully formed, and they lack the surfactant that enables alveoli to expand with very little pressure gradient. Hence, a considerable pressure gradient must be applied to ventilate them. Applying that pressure from the outside in, as *conventional* ventilators (CVs) have been doing since the early 1970s, causes problems. Tiny infant's airways get distended, alveoli are ruptured, and inflammatory sensors are triggered. Even if an infant receives artificial surfactant to lessen the need for assisted ventilation, and they grow new lung fast enough to survive, they may well develop chronic lung disease at a time when most of us were just taking our first breaths. Most premature infants outgrow their chronic lung disease, but many struggle mightily with every virus they encounter in their first few years of life, and they have an increased incidence of neurodevelopmental problems, such as cerebral palsy. Some infants lose those struggles and die of pneumonia.

Acute respiratory distress syndrome (ARDS) is the primary problem for mechanical ventilation of adults. This disease affects ~50 people per 100,000 with a mortality of 30–50%, and there have been few improvements in this mortality rate over the past several decades.

High frequency ventilators were developed in response to problems associated with CVs, but HFVs do not try to replicate normal breathing. They assist ventilation using much smaller tidal volumes delivered at rates ~10 times higher than normal. Animal and clinical studies indicate that smaller tidal volumes cause less lung injury (1,2).

Swedish anesthesiologists in the 1970s turned up the rate of their anesthesia ventilators to enable them to use smaller breaths to assist patients during neurosurgery (3). Their regular ventilators caused pulsations in blood pressure, causing brain movement every time the ventilator pushed in a breath, which was an obvious problem during microsurgery.

Auto accident victims whose heads went through car windshields also pose problems during surgery when access to the lungs has to pass right through the area of the face and neck where major reconstruction is required. So, another anesthesiologist, Dr. Miroslav Klain, began sticking needles into patients' necks to gain access to their tracheas, and he made it work by delivering very tiny breaths at very rapid rates (4).

The HFVs have shown great promise in supporting premature infants where fragile, underdeveloped and surfactant deficient lungs need to be gently ventilated until growth and maturation allow the newborn to catch up both anatomically and physiologically. In newborn infants, where various modes of positive pressure ventilation have produced lung injury, HFVs have been widely accepted. Early studies using HFV to treat adults with severe ARDS have also shown promise in lessening lung injury and improving survival (5,6).

This article will review our current understanding of how HFVs work, the types of HFV equipment that are available for treating infants and adults, the results of several key animal and clinical studies that indicate how to optimize applications of HFVs, and what controversies remain to be resolved before HFVs can be considered as a primary mode of ventilation.

THEORETICAL BASIS FOR HFV: HOW HFVs WORK

High frequency ventilators are different from all other types of mechanical ventilators. They do not mimic normal breathing; rather, they facilitate gas exchange in a manner similar to panting in animals.

There are two elements to explaining how HFVs work with the higher than normal frequencies and smaller than normal tidal volumes. We begin with the assumption that ventilation or CO_2 elimination is proportional to minute volume or frequency times tidal volume, as

$$\dot{V}_{CO_2} \propto \dot{V}_{min} = f \times V_T \qquad (1)$$

where $\dot{V}_{CO_2}$ = rate of carbon dioxide elimination; $\dot{V}_{min}$ = minute volume; f = ventilator frequency; and V_T = tidal volume.

If the practical limits of the high frequency end of this relationship are considered, there must be a lower limit on tidal volume size that will effectively provide alveolar ventilation. That lower limit is related to the effective or physiologic dead space of the lungs by the following equation:

$$\dot{V}_A = f \times (V_T - V_D) \qquad (2)$$

where $\dot{V}_A$ = alveolar ventilation, and V_D = effective or physiologic dead space.

Thus, as tidal volume approaches the size of the effective dead space of the lungs, ventilation of the alveoli becomes nil.

Physiologic Dead Space and the Lower Limit of Tidal Volume

Anatomic dead space of mammalian lungs is generally considered to be $2 \, mL \cdot kg^{-1}$ body weight (7). When one breathes normally, effective or physiologic dead space must be at least as large as anatomic dead space, because one pushes the dead space gas back into the alveoli ahead of the fresh gas during inhalation. What happens in the panting animal is another matter, as Henderson and associates described in 1915 (8).

Henderson et al. (8) demonstrated that panting animals breathe very shallowly as well as rapidly. They hypothesized that physiologic dead space changes at rapid respiratory rates in mammals, and they measured these effects on themselves. They also performed a series of experiments using smoke to demonstrate how inhaled gas penetrates through the anatomic dead space with rapid inhalations in a manner that makes physiologic dead space become less than anatomic dead space (Fig. 1).

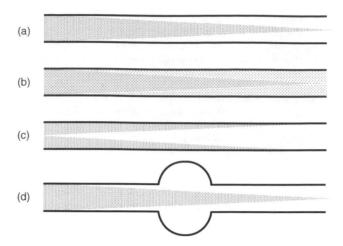

Figure 1. Henderson's smoke experiment. (a) A long thin spike or jet stream of smoke shoots downstream when suddenly blown into a glass tube. (b) The jet stream disappears when flow stops and diffusion takes place. (c) This effect can be duplicated in the opposite direction if fresh gas is drawn back into the smoke filled tube with a sudden inhalation. (d) Imperfections in the tube walls (such as a bulb) have little effect on the shape of the jet stream. (Adapted with permission from Ref. 8, p. 8. © 1915, American Physiology Society.)

Regardless of how much physiologic dead space can be reduced in the panting animal, there is still the matter of providing adequate alveolar ventilation as defined by Eq. 2. Thus, the extent to which smaller tidal volumes can be used to ventilate the alveoli has to be balanced by an increase in breathing frequency, as defined by Eq. 1. Panting animals breathe very rapidly, of course, but humans do not pant as a rule. So, how can the benefits of increasing ventilator frequency for humans be explained?

The Natural Frequency of the Lungs

There is a mechanical advantage to ventilating lungs at frequencies higher than normal breathing frequency. This phenomenon was revealed by a diagnostic technique for measuring airway resistance called forced oscillations (9).

Applying forced oscillations to measure airway resistance requires a person to hold a large bore tube in their mouth and allow small volumes of gas to be oscillated in and out of their lungs by a large loudspeaker. The frequency of oscillations produced by the speaker is varied through a spectrum from low (~1 Hz) to high (~60 Hz), and the pressure amplitudes of the oscillations are measured along with the flow rate of the gas that is passing in and out of the lungs (Fig. 2 depicts the test set up). Although the volume of gas moving in and out of the lungs is a constant, pressure amplitude varies with frequency and is minimized at the resonant or natural frequency of the lungs.

The concept that the lungs have a natural frequency is explained by consideration of lung mechanics. There are three elements to lung impedance (those things that impede the flow of gas in and out of the lungs): airway resistance, lung compliance, and inertance. We normally are not concerned about inertance, since it is concerned with the energy involved in moving the mass in the system,

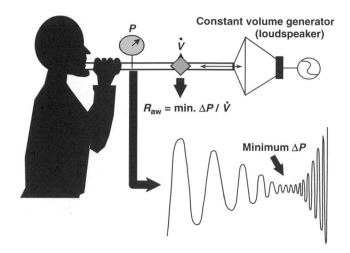

Figure 2. Measuring airway resistance at the resonant or natural frequency of the lungs using forced oscillations. (Used with permission. © 2003, Bunnell Inc.)

most of which is gas, and gas does not have much mass. Therefore, it does not take much energy to overcome inertance when one breathes: unless one is breathing very rapidly.

In the forced oscillations determination of airway resistance, the point of minimum pressure amplitude marks the frequency at which the energy necessary to overcome the elasticity of the lungs is supplied by the energy temporarily stored in the inertial elements of the system (i.e., the gas rushing in). (We normally measure lung elasticity inversely as lung compliance.) As the lungs recoil at the end of the gas-in phase, the elasticity of the lungs imparts its energy to turn the gas around and send it back out to the loudspeaker.

When the natural frequency or resonance is reached, the speaker and lungs exchange the gas being forced in and out of the lungs with ease. The lungs and the speaker accept the gas and recoil at just the right times to keep the gas oscillating back and forth with minimal energy required to keep the gas moving. At this point, the only element impeding gas flow is frictional airway resistance, which works against the gas coming in and going out. Its value can be calculated by dividing pressure amplitude by the gas flow rate when pressure amplitude is minimized.

The smaller the lungs are, the higher the natural frequency. The natural frequency of adult lungs is ~4 Hz, while that of premature infant lungs is closer to 40 Hz.

Putting Two and Two Together: How Can We HFV?

Combining the two concepts that describe the relationships of gas velocity, physiologic dead space, breathing frequency, and lung mechanics led us to HFV. We reported that one can then achieve adequate minute ventilation and compensate for very small tidal volumes in paralyzed animals by increasing ventilatory frequency to several hundred breaths per minute in 1978 (10). Pushing small volumes of gas into the lungs at high velocities reduced effective dead space volume and pushed the lower limit of effective tidal volume below anatomic dead space volume

($\sim$2 mL·kg^{-1}). Increasing frequency to near resonant frequency also allowed us to minimize airway pressure.

As HFVs were developed and clinical use in newborn intensive care units (NICUs) became widespread in the 1980 and 1990s, numerous theories and experiments refined our concepts of how it all works. A number of prominent physiologists and bioengineers tackled the analysis and interpretation of gas exchange within the lungs during HFV while clinicians were seeking to identify appropriate applications of the new technique and all its intricacies. A few notable contributions will be discussed here.

Fredberg (11) and Slutsky et al. (12) analyzed mechanisms affecting gas transport during high frequency oscillation, expanding traditional concepts of convection and diffusion to include their combined effects, and termed the collection: augmented transport. Their analyses and those of Venegas et al. (13) and Permutt et al. (14) revealed that our traditional appreciation of the relationship between minute volume and CO_2 elimination must be modified during HFV to reflect the increased contribution of tidal volume, as

$$V_{CO_2} \propto f^a \times V_T^b \qquad (3)$$

where the exponent b is greater than the exponent a. For practical purposes, most people now accept this relationship as

$$\dot{V}_{CO_2} \propto f \times V_T^2 \qquad (4)$$

Slutsky also explored the limitations of HFV by measuring the effect of bronchial constriction on gas exchange. When the peripheral airways of dogs were constricted by administration of histamine, HFOV was no longer as effective at higher frequencies. (This issue is discussed later when the effectiveness of various types of HFVs for different pathophysiologies are explored.)

Venegas and Fredberg explored the importance of frequency during HFV in their classic paper of 1994, subtitled: "Why does high frequency ventilation work?" (15). They found that the resonant frequency of the smallest prematurely born infant with RDS (respiratory distress syndrome) is approximately 40 Hz. At that frequency, the minimum pressure amplitude is required to ventilate the lungs. However, the shape of theoretical curves of pressure amplitude measured at the carina versus frequency for infants with various lung conditions is most interesting as illustrated in Fig. 3, which was constructed using their concepts.

Figure 3 illustrates four essential considerations concerning the application of HFV for newborn infants.

1. Decreasing lung compliance moves the optimal frequency for HFV to the right (i.e., toward higher frequencies).
2. Increasing airway resistance moves the optimal frequency for HFV to the left (i.e., toward lower frequencies); and
3. There is diminishing value in applying HFV for infants at frequencies above ~ 10 Hz as far as airway pressure is concerned.
4. Choosing to operate at the "corner frequency" is an appropriate choice for HFV since there is little benefit

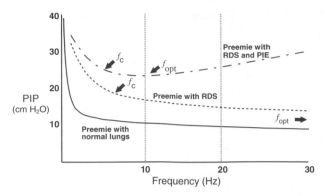

Figure 3. Theoretical peak carinal pressures for infants with normal lungs and lungs with poor compliance (RDS), poor airway resistance (asthma), and both conditions (RDS + PIE). Note how infants with RDS are well served using the corner frequency (f_c) of ~10 Hz (600 breaths per minute, bpm). Larger patients will exhibit curves with nearly identical shapes, but they will all be shifted to the left. (Adapted with permission from Ref. 15.)

above that frequency and more chance for gas trapping. Venegas and Fredberg define corner frequency as that frequency above which airway pressure required to provide adequate ventilation no longer rapidly decreases.

In other words, ventilating premature babies at 10 breaths·s^{-1} is *practically* as efficient as ventilating them at their theoretical resonant frequency of 40 "breaths"·s^{-1}, where the danger of gas trapping is greatly increased. Patients with increased airway resistance require more careful consideration of the decreased benefits of exceeding the corner frequency and are more safely ventilated at lower frequencies.

One can calculate the resonant and corner frequencies if values of lung compliance (C_L), airway resistance (R_{aw}), and inertance (I) are known, but that is rarely the case with patients in intensive care. Venegas and Fredberg provided the following formulas:

$$f_0 = 1/(2\pi\sqrt{\bar{C}_I}) \qquad (5)$$

where f_0 = resonant frequency, and

$$f_c = 1/(2\pi CR) \qquad (6)$$

where f_c = corner frequency. (Plug in typical values for lung compliance, inertance, and airway resistance of a premature infant, $0.5\,\text{mL·cm}^{-1}\ \text{H}_2\text{O}$, $0.025\,\text{cm}\,\text{H}_2\text{O·L·s}^{-2}$, and $50\,\text{cm}\,\text{H}_2\text{O·L}^{-1}\text{·s}^{-1}$, respectively, and $f_0 = 45\cdot\text{s}^{-1}$ and $f_c = 6.4\cdot\text{s}^{-1}$.)

Finally, Venegas and Fredberg illustrated the value of using appropriate levels of positive end-expiratory pressure (PEEP). The PEEPs of 5–10 cm H$_2$O dramatically decrease the pressure amplitude necessary to ventilate premature infants at all frequencies when lung compliance is normal, and at all frequencies above ~6 Hz when lung compliance is reduced.

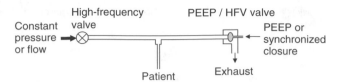

Figure 4. Basic design of HFPPVs. (Used with permission. © 2003, Bunnell Inc.)

HFV EQUIPMENT

Design Classifications

Figures 4–7 illustrate four different ways HFVs have been created. Figure 4 illustrates high frequency positive-pressure ventilation (HFPPV), which is basically a CV that operates at HFV rates. Early devices worked in this manner, but they seldom worked at the very high frequencies used with infants.

Figure 5 illustrates high frequency flow interruption (HFFI), where positive pressure oscillations are created by releasing gas under pressure into the breathing circuit via an HFV valve mechanism. The valve may be a solenoid valve or valves, a spinning ball with a hole in it, and so on. Early HFVs used long, small diameter exhaust tubes to increase impedance to gas flow oscillating at HFV frequencies in the expiratory limb so that the HFV oscillations would preferentially flow in and out of the patient.

High frequency oscillatory ventilators (HFOVs) work in a similar manner to HFFIs, as shown in Fig. 6, except that the pressure oscillations in the patient's breathing circuit are caused by an oscillating piston or diaphragm. Again, the impedance of the expiratory limb of the circuit tubing must be higher than the impedance of the patient and his ET tube when the gas flowing through the circuit is oscillating at HFV frequencies. The major difference between HFOV and HFFI is that pressure in the ventilator circuit during HFOV oscillates below atmospheric pressure in an effort to actively assist the patient's expiration. (This topic is discussed further below when gas trapping is addressed.)

Finally, Fig. 7 illustrates high frequency jet ventilation (HFJV), where inspiratory gas is injected into the patient's ET tube via a jet nozzle. Jet nozzles have been fashioned out of needles or built into special ET tubes or ET tube adapters as discussed below.

Each HFV approach introduces fresh gas into the patient's airway at about 10 times the patient's normal breathing frequency. The last three designs incorporate a separate constant flow of gas that passes by the patient's

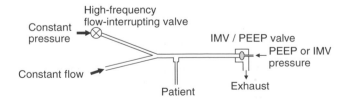

Figure 5. Basic design of HFFIs. (Used with permission. © 2003, Bunnell Inc.)

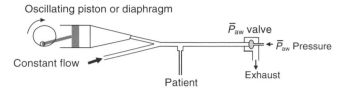

Figure 6. Basic design of HFOVs. (Used with permission. © 2003, Bunnell Inc.)

ET tube and out a large orifice valve to control baseline PEEP and mean airway pressure in the circuit. The patient may also breathe spontaneously from this gas stream, which may be provided by a built-in mechanism or by a separate conventional ventilator. Conventional IMV (intermittent mandatory ventilation) may be combined with HFV in this way. Additional hybrid devices that are more difficult to characterize have also been created, but the currently most common used HFVs are HFOVs, HFJVs, and conventional ventilators with built-in HFOV modules.

In the early 1980s, the FDA (U.S. Food and Drug Administration) decided that 150 breaths·min^{-1} would be the lower limit of what they would define as an HFV, and they placed rigorous Class III restrictions on any ventilator that operates above that frequency. As a result, there have been only six HFVs approved for use in the United States, three for infants and children, two for adults, and one HFV that was granted Class II approval (i.e., not needing proof of safety and efficacy since it was substantially equivalent to devices marketed before 1976, the date U.S. law was amended to require proof of safety and efficacy before new products can be marketed). At least four other HFVs are available outside the United States.

Of the FDA approved devices, two HFVs have been withdrawn from the market by the major corporation that acquired the smaller companies that developed them. Therefore, we are left with one HFJV, one HFOV for infants and children, one HFOV for children and adults, and a Class II HFV hybrid device designed for patients of all sizes. These four devices will be discussed in more detail below.

HFJV: High Frequency Jet Ventilators

The HFJVs inject inspired gas into the endotracheal tube via a jet nozzle. The Bunnell LifePulse High Frequency

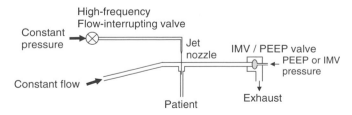

Figure 7. Basic design of HFJVs. (Used with permission. © 2003, Bunnell Inc.)

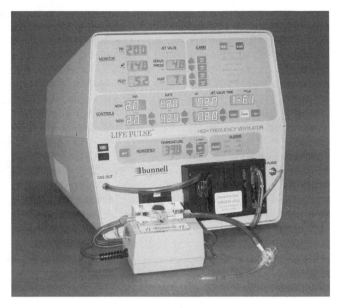

Figure 8. Bunnell life pulse HFV. (Used with permission. © 2003, Bunnell Inc.)

Ventilator is the only HFJV currently available for intensive care in the United States (Fig. 8). It was designed for infants and children up to ~10 years of age and operates at rates between 240 and 660 bpm. It is also used in tandem with a conventional ventilator, which provides for the patient's spontaneous breathing, delivery of occasional sigh breaths, and PEEP.

The LifePulse is a microprocessor controlled, pressure limited, time cycled ventilator that delivers heated and humidified breaths to the ET tube via a LifePort adapter (Fig. 9). A small Patient Box placed close to the patient's head contains an inhalation valve and pressure transducer for monitoring airway pressures in conjunction with the LifePort adapter. Peak inspiratory pressure (PIP) is feedback controlled by regulating the driving pressure (servo pressure) behind the jet nozzle. More detailed information on the device can be found on the manufacturer's website: www.bunl.com.

The theory of operation behind the LifePulse is that pulses of high velocity fresh gas stream down the center of the airways, penetrating through the dead-space gas, while exhaled gas almost simultaneously moves outward in the annular space along the airway walls. This countercurrent action facilitates mucociliary clearance while it minimizes effective dead space volume.

The pressure amplitude (ΔP) of LifePulse HFJV breaths is determined by the difference between PIP and the CV-controlled PEEP. Its value is displayed continuously on the LifePulse front panel along with mean airway pressure, PIP, PEEP, and Servo Pressure.

Servo Pressure on the LifePulse is a direct reflection of the gas flow needed to reach the set PIP, so it varies with the patient's changing lung mechanics. Alarm limits are automatically set around the Servo Pressure to alert the operator to significant changes in the patient's condition as well as the tubing connecting the patient to the HFJV.

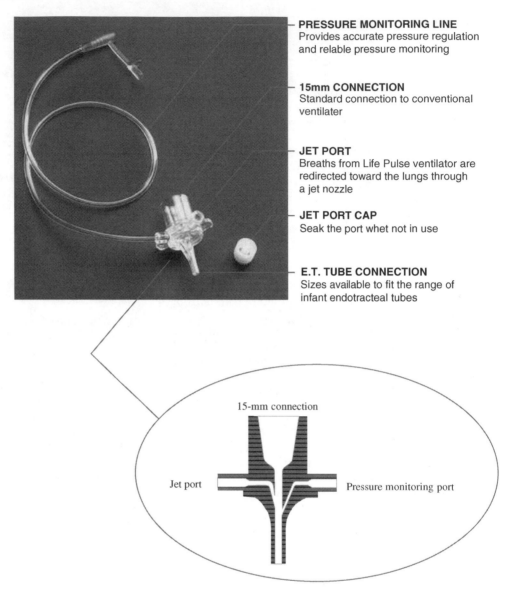

PRESSURE MONITORING LINE
Provides accurate pressure regulation
and reliable pressure monitoring

15mm CONNECTION
Standard connection to conventional
ventilater

JET PORT
Breaths from Life Pulse ventilator are
redirected toward the lungs through
a jet nozzle

JET PORT CAP
Seak the port whet not in use

E.T. TUBE CONNECTION
Sizes available to fit the range of
infant endotracteal tubes

Figure 9. LifePort ET tube adapter
for HFJV. (Used with permission. ©
2003, Bunnell Inc.)

A low limit alarm would infer that the patient's lung compliance or airway resistance has worsened, and the LifePulse is using less gas (i.e., smaller tidal volumes) to reach the set PIP in that circumstance. A high limit alarm would infer that the patient's condition has improved, larger tidal volumes are being delivered, and the operator should consider weaning PIP in order to avoid hyperventilation.

Alarm limits are also automatically set around monitored mean airway pressure.

HFOV: High Frequency Oscillatory Ventilators

The SensorMedics 3100A HFOV for infants and children and its sister model, the 3100B for adults, are the only pure HFOVs currently available in the United States. Sinusoidal oscillatory ventilation is produced by an electromagnetically driven floating piston with adjustable frequency and amplitude. Inspiratory gas is supplied as bias flow, which escapes from the very large diameter (1.5 in. ID, 38 mm) patient breathing circuit via a traditional dome valve that controls mean airway pressure. All pressures are monitored at the connection to the ET tube. The SensorMedics 3100A HFOV is illustrated in Fig. 10 and more information is available at www.sensormedics.com.

The 3100 HFOVs operate on the principle that high frequency oscillations that are in tune with the natural frequency of the patients lungs will preferentially move in and out of the lungs, as opposed to the exhaust system of the patient circuit. Haselton and Scherer illustrated a new gas transport principle that applies to HFOV (16). Differences in the velocity profiles of inspiration and expiration during HFOV created by the branching architecture of the lungs enables inspiratory gas to advance down the center of the airways while exhaled gas moves up along the airway walls as the piston of the HFOV pulls the gas back. The net effect of many oscillations is similar to, but less pronounced than, the flow characteristics of HFJV flow in the airways. Fresh gas tends to flow down the center of the airways while exhaled gas recedes back along the airway walls.

Figure 10. SensorMedics 3100A high frequency oscillatory ventilator. (Used with permission. © 2003, SensorMedics Inc.)

The 3100 HFOVs have six control settings:

1. Frequency, which is adjusted to suit patient size and lung time constants.
2–4. Bias gas flow rate, Mean Pressure Adjust and Mean Pressure Limit, which together set mean airway pressure.
5. Power, which sets ΔP.
6. % Inspiratory Time, which sets I:E (inspiratory to expiratory time ratio; typically set at 33%).

Mean airway pressure is the primary determinant of oxygenation, and ΔP is the primary determinant of tidal volume and ventilation (CO_2 removal). However, all controls are open-loop: increasing frequency decreases tidal volume and visa versa, and changing bias gas flow rate or power may change mean airway pressure. Mean airway pressure is monitored in the HFOV circuit, so changes there are apparent, but changes in tidal volume due to setting changes or changes in a patient's lung mechanics are not apparent. Given the squared relationship between tidal volume and CO_2 removal noted above, changes in frequency move $PaCO_2$ in the opposite direction of what one would anticipate with conventional ventilation. (Increasing frequency increases $PaCO_2$; decreasing frequency decreases $PaCO_2$.) Thus, continuous or frequent monitoring of arterial $PaCO_2$ is recommended during HFOV, as it is with all HFVs and conventional ventilation of premature infants due to the potential for cerebral injury associated with hyperventilation (more on that topic later).

HFFIs and Other Hybrids

Conventional infant ventilators with built-in HFV modules, such as the Dräger Babylog 8000 *plus* (Dräger Medical AG & Co. KGaA) and the popular Infant Star Ventilator, which will no longer be supported by its manufacturer after May 2006, have been widely used in the United States, Canada, Europe, and Japan. In general, these hybrid HFVs are not as powerful as the stand-alone HFVs, so their use is limited to smaller premature infants (<2 kg). The VDR Servolator Percussionator (Percussionaire Corporation, Sand Point ID), however, was designed to ventilate adults as well as premature infants. (Detailed information on these devices can be viewed on the manufacturers' websites: www.draeger-medical.com and www.percussionaire.com.) The mechanical performances of these devices vary widely, as do their complexities of operation and versatilities as infant ventilators. (See the section Equipment Limitations.)

Design Philosophy for Clinical Applications

The philosophy for controlling arterial blood gases with HFVs is similar to that used with pressure-limited conventional ventilation, especially when HFVs are used to treat homogeneous lung disorders such as RDS (respiratory distress syndrome) in prematurely born infants. The alveoli of these surfactant-deficient lungs must be opened with some type of recruitment maneuver and kept open with appropriate mean airway pressure (Paw) or PEEP in order for the lungs to make oxygen available to the blood stream. Ventilation is accomplished at a frequency proportionate to the patient's size and lung mechanics using a peak airway pressure or pressure amplitude above PEEP that creates a tidal volume that produces an appropriate arterial PCO_2. Pulse oximeters, which report the oxygen percent saturation of arterial blood, are great indirect indicators of when lungs have opened up, because oxygenation is highly dependent on the number of alveoli that are open and participating in gas exchange with the blood stream. Chest wall motion is a good indirect indicator of ventilation since it reflects the amount of gas that is passing in and out of the lungs.

The usual approach to initiation of HFV is to choose a frequency that is appropriate for the size of the patient and his lung mechanics, starting with 10 Hz or 600 bpm for the smallest premature infant with RDS and working downward as the size of the patient increases and lung mechanics improve. A good rule of thumb is to choose a frequency 10 times greater that the patient's normal breathing frequency, which would put HFV for adults at rates <200 bpm. Higher rates may be used with HFOV since exhalation is nonpassive, but gas trapping can still result unless mean airway pressure is kept high enough to keep the airways open during the active exhalation phase. Operating an HFOV with a 33% inspiratory time ($I{:}E = 1{:}2$) lessens negative pressure during exhalation compared to longer I-times (e.g., $I{:}E = 1{:}1$) thereby decreasing the potential for causing airway collapse.

With HFJV, the shortest possible inspiratory time (~0.020 s) usually works best; it maximizes inspiratory velocity, which helps reduce effective dead space, and minimizes $I{:}E$, which allows more time for exhalation to avoid gas trapping. These characteristics also minimize mean airway pressure, which is very useful when treating airleaks and for ventilation during and after cardiac surgery. The high velocity inspirations also enable ventilation of patients with upper airway leaks and tracheal tears.

Treatment of obstructive lung disorders absolutely requires longer exhalation times, so HFV must be used at lower frequencies on these patients. HFJV *I:E* varies from 1:3.5 to 1:12 as frequency is reduced from 660 to 240 bpm when inspiratory time is held constant at its shortest value.

The HFV is not intended and may in fact be contraindicated for patients with asthma, unless helium–oxygen mixtures become part of the mix (17).

Once a frequency and duty cycle (% *I*-time or *I:E*) is chosen, airway pressure settings (PIP, PEEP, or ΔP) are set to provide HFV tidal volumes that noticeably move the chest. If chest wall movement is not apparent, ventilation is probably not adequate. Use of transcutaneous CO_2 monitoring is of great benefit here.

Finally, mean airway pressure (*Paw*) or PEEP must be optimized. Too little *Paw* or PEEP will lead to atelectasis and hypoxemia, and too much *Paw* or PEEP will interfere with cardiac output. One of the true benefits of HFV, however, is that higher *Paw* and PEEP can be used without increasing the risk of iatrogenic lung injury. (The small HFV tidal volumes do not create the same potential for creating alveolar "stretch" injury as larger CV tidal volumes do.) Pulse oximeters can be great indirect indicators of appropriate lung volume, but one must be vigilant in detecting signs of decreased cardiac output.

Conventional ventilation is sometimes required or available for tandem use with certain HFVs. The CV breaths are most useful with nonhomogeneous lung disorders and to facilitate alveolar recruitment with atelectatic lungs. The usual strategy is to reduce CV support when starting HFV (assuming the patient is on CV prior to HFV) to 5–10 bpm while optimal *Paw* and PEEP is being sought, and then reduce CV support further.

Now some of the performance differences in HFV equipment and how those differences may affect successful HFV implementation will be examined.

HFV Equipment Limitations

There have been few head-to-head comparisons of HFV equipment. The most recent comparison were by Hatcher et al. and Pillow et al. where they compared several neonatal HFOVs and found wide variations in performance, complexity, and versatility (18,19). Pillow et al. concluded that the clinical effects of manipulating ventilator settings may differ with each HFOV device. In particular, the pressure amplitude required to deliver a particular tidal volume varies with device, and the effect of altering frequency may result in very different effects on tidal volume and $PaCO_2$.

The first rigorous analysis of HFVs was undertaken by Fredberg et al. in preparation for the HiFi Study (20). They bench tested eight HFVs in an effort to provide the clinicians who were to participate in the study comparative data that they could use to select an HFV for use in their study. (They selected the Hummingbird HFOV, manufactured by MERA of Japan.) Despite the wide diversity of ventilator designs tested, certain common features emerged. In almost all devices, delivered tidal volume was sensitive to endotracheal tube size and airway resistance and invariant with respiratory system compliance. These results supported the theoretical basis for why high frequency ventilation may be a better treatment for RDS compared to pressure-limited CV (conventional ventilation), because low lung compliance is its paramount pathophysiologic feature.

These HFV bench tests also found that tidal volume decreased with increasing frequency with all HFOVs where I:E (inspiratory to expiratory time ratio) was held constant and was invariant with HFJV and HFFI devices where I-time was held constant. Peak inspiratory flow rates for a given tidal volume and frequency were significantly higher with the HFJV and HFFI as well. Proximal airway pressure was also a poor indicator of distal pressure with all devices.

Two other studies compared HFJV to HFOV. Boros and associates compared the pressure waveforms measured at the distal tip of the endotracheal tube of the Bunnell LifePulse HFJV and the Gould 4800 HFOV (precursor to the SensorMedics 3100A HFOV) in normal, paralyzed, and anesthetized cats (21). They found that the HFOV required higher PIP, ΔP, and *Paw* to get the same $PaCO_2$, PaO_2, and pH compared to HFJV. Likewise, $PaCO_2$ was higher and pH and PaO_2 were lower with HFOV when the same airway pressures were used. However, different frequencies were used with the two ventilators; 400 bpm with HFJV and 900 bpm (15 Hz) with HFOV.

Zobel and associates also found that HFJV was effective at lower airway pressure compared to HFOV (22). They used a piglet model of acute cardiac failure and respiratory failure and also measured airway pressure at the distal tip of the endotracheal tube. The HFJV used was an Acutronic AMS-1000® (Acutronic Medical Systems AG, Switzerland) operating at 150 bpm with an *I:E* of 1:2. The HFOV was a SensorMedics 3100A operating at 10 Hz and 1:2.

Why do HFOVs (presumably) operate at higher *Paw* compared to HFJV? The answer to this question may be related to gas trapping, HFV rates, and what happens during exhalation. In both of the animal studies just discussed, HFJV rate was considerably lower than HFOV rate. Exhalation is passive during HFJV, so lower rates must be employed to allow sufficient exhalation time to avoid gas trapping. The HFOVs suck the gas back out of the lungs during the expiratory phase, and the physiologic consequence can be, not surprisingly, airway collapse. However, the *Paw* employed during HFOV determines the importance of this effect.

Bryan and Slutsky set the tone for the future of HFVs when they noted that this mode of ventilation is ideally designed for treatment of patients with poor lung compliance (23). The higher *Paw* required to match the pathophysiology of such patients also serves to splint the airways open during HFOV so that the choking effect of active expiration is mitigated.

In conclusion, both modes of HFV can cause gas trapping; they just do it by different mechanisms. The HFOV can choke off airways when *Paw* is insufficient to mitigate the effect of active expiration, and HFJV will trap gas when expiratory time is insufficient to allow complete exhalation of inspired tidal volume. One cannot lower *Paw* during HFOV beyond the point where choking is made evident by a rise in a patient's PCO_2. With HFJV, one should not increase frequency beyond the point where PEEP *monitored* in the endotracheal tube, as it is with the Bunnell LifePulse, begins to rise inadvertently. If the automatically set upper alarm limit on mean airway pressure with the LifePulse is activated, there is a good chance that this rise in *Paw* is due to inadvertent PEEP. The remedy for that circumstance is to decrease HFJV frequency, which lengthens exhalation time and allows the PEEP to fall back to *set* level.

Airway Pressure Monitoring During HFV

While airway pressures were monitored at the distal tip of the ET tube in the animal studies noted above, monitoring at this location is seldom done currently, because the Hi-Lo ET tubes (formerly manufactured by Mallinckrodt, Inc.) are no longer available. Thus, airway pressure monitoring is done either at the standard ET tube adapter connection during HFOV or at the distal tip of the special LifePort adapter during HFJV. In either case, the pressure waveform measured deep in the lungs at the alveolar level is greatly damped (Fig. 11). Gerstmann et al. reported that measurement of pressure amplitude in the alveoli of rabbits during HFOV at 15 Hz was only 10% of that measured proximal to the ET tube (24).

Meaningful monitoring of airway pressure during HFOV is limited to mean airway pressure, and that is only representative of mean alveolar pressure in the absence of gas trapping, as noted above. Relative values of pressure amplitude at the proximal end of the ET tube are indicative of tidal volume size, and they are typically expressed as such by the various HFOVs.

Peak inspiratory pressure (PIP) and PEEP as well as *Paw* are measured during HFJV at the distal tip of the LifePort ET adapter. The PEEP is representative of alveolar PEEP at this location in the absence, again, of gas trapping. However, the PIP at this location is a gross overestimate of peak pressure in the alveoli. Mean airway pressure may slightly overestimate mean alveolar pressure as shown by the study of Perez-Fontan et al. (25).

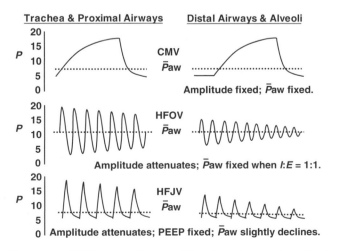

Figure 11. HFV Airway Pressure Waveform Dampening. (Used with permission. © 2003, Bunnell Inc.)

HFV APPLICATIONS IN NEONATES AND CLINICAL OUTCOMES

Homogeneous Atelectatic Lung Disease (e.g., RDS) and Prevention of Lung Injury

Ever since the completion of the first multicenter, randomized, controlled HFV trial was published in 1989, reporting no benefit for premature infants with RDS and an increased risk of severe cerebral injury (26), the choice of HFV to prevent lung injury in preterm infants has been hotly debated. Some recent trails have demonstrated that if HFVs are implemented within hours of a premature infant's birth with the proper strategy, results are positive. Other recent studies have not been positive.

The HiFi Trial, as the first multicenter, randomized, controlled trial was labeled, was criticized for the general lack of clinical experience of the investigators and failure to adhere to the most appropriate strategy for recruiting and maintaining appropriate lung volume (15). Later multicenter, randomized controlled trials conducted in the 1990s using both HFJV and HFOV demonstrated significant reductions in chronic lung disease (CLD) measured at 36 weeks postconceptional age (PCA) in this patient population with practically no difference in adverse effects (27,28). (There was a slightly higher incidence of PIE in the experimental group of the HFOV study.) The demographics and results of these two trials are illustrated in Tables 1 and 2.

The results of the HFJV study were criticized for a lack of well-defined ventilator protocols for the conventionally ventilated control group, whereas protocols for both the HFOV and SIMV control groups in the HFOV study, conducted several years later, were well conceived and monitored during the study. Therefore, it is interesting to note that the major outcome measures of CLD at 36 weeks PCA in the control groups of the two studies were almost identical.

Other HFV studies revealed an increase in severe cerebral injury that appears to be related to hyperventilation and hypocarbia during HFV (29–32). Other criticisms of recent trials with negative or equivocal results include the same strategy issues plus choice of HFV devices, limited time on HFV before weaning back to CV, and so on (33).

Because of these mixed results, HFVs have yet to be generally accepted for early treatment of premature infants with RDS and prevention of lung injury.

Table 1. Demographics of Two Multicenter, Randomized Controlled Trials with HFOV and HFJV

Design/Demographics	HFJV Study[a]		HFOV Study[10]	
Treatment Groups	HFJV	CV	HFOV	SIMV
Number of Patients	65	65	244	254
Mean Birth Weight, kg	1.02	1.02	0.86	0.85
Mean Gestational Age	27.3	27.4	26.0	26.1
Age at Randomization, h	8.1	8.3	2.7	2.7
1 min/5 min Apgar Scores	3.5/7	4/7	5/7	5/7
F_1O_2 at Entry	0.62	0.69	0.57	0.60
Mean Airway Pressure at Entry	10	10	8.2	8.3

[a]See Ref. 9.

Table 2. Significant Respiratory and Clinical Outcomes of HFOV and HFJV Early Application Trials on Premature Infants with RDS

Significant Respiratory and Clinical Outcomes	HFOV Study		HFJV Study	
	HFOV	SIMV	HFJV	CV
Alive w/o CLD at 36 weeks PCA	56%	47%	68%	48%
	$p = 0.046$		$p = 0.037$	
Age at extubation, days	13	21	-	
	$p < 0.001$			
Crossovers or Exits[a]	25/244 (10%)	49/254 (19%)	3/65 (5%)	21/65 (32%)
	$p = 0.07$		$p < 0.01$	
Success after Crossover			14/21 (67%)	0/3 (0%)
			$p = 0.06$	
Supplemental O_2	27%	31%	5.5%	23%
	$p = 0.37$		$p = 0.019$	
PIE	20%	13%		
	$p = 0.05$			
Pulmonary Hemorrhage	2%	7%	6.3%	10%
	$p = 0.02$		$p > 0.05$	

[a]Similar failure criteria were prospectively defined in both studies. Those who met the criteria in the HFJV study were crossed over to the other mode, while those who met the criteria in the HFOV study exited the study and were treated with whatever mode of ventilation the investigators deemed appropriate, including HFJV (personal communication, David Durand, MD). Data on all patients were retained in their originally assigned group in both studies.

Homogeneous Restrictive Lung Disease (e.g., Congenital Diaphragmatic Hernia)

While theories support use of HFV in cases where the lungs are uniformly restricted by acute intra-abdominal disease or postsurgically in infants with congenital diaphragmatic hernia, omphalocele, or gastroschisis, there are no randomized controlled trials due to the rarity of these disorders. Despite this lack of controlled trials, HFV has been widely accepted as an appropriate treatment for this category of lung disease due to the futility of CV treatment in severe cases.

Keszler et al. demonstrated improved gas exchange and better hemodynamics with HFJV in an animal model of chest wall restriction (34) and later reported improved ventilation and hemodynamics in a series of 20 patients with decreased chest wall compliance (35). Fok et al. reported improved gas exchange with HFOV in eight similar patients who were failing CV (36).

Nonhomogeneous Atelectatic and Restrictive Lung Disease (e.g., RDS with Tension PIE)

Pulmonary interstitial emphysema (PIE) in the premature infant creates a non-homogeneous lung disease: parts of the lungs are collapsed as a result of surfactant deficiency while other parts become overexpanded with gas trapped in interstitial areas. Air leaks like PIE originate most commonly in premature infants near the terminal bronchial (37). As gas dissects into interstitial spaces, it invades and dissects airway and vascular walls moving towards the larger airways and vessels and the pleural space where pneumothoraces are formed (38). While positive-pressure CV may successfully penetrate such restricted airways, the consequence may well be accumulation of trapped gas in the alveoli and subsequent alveolar disruption, which produces the classical picture of PIE on X ray.

The HFJV quickly gained a reputation for superior treatment of PIE in the early days of its clinical application. A multicenter randomized trial of HFJV compared to rapid rate (60–100 bpm), short I-time (0.20–0.35 s) CV for the treatment of PIE confirmed anecdotal findings of faster and more frequent resolution of PIE on HFJV. Survival in the stratified group of 1000–1500 g birth weight infants was most evident (79% with HFJV vs. 44% with CV; $p < 0.05$). There was no difference in the incidence of adverse side effects.

There is, as yet, no comparable randomized trial of HFOV treatment for PIE. While anecdotal success has been reported, attempts to show an advantage with HFOV in a randomized controlled trial have so far been unsuccessful. It may be that the physical characteristics of the two types of HFVs coupled with the pathophysiologic characteristics of PIE are the reasons for this lack of success. Recall that one difference between HFV devices reported in the pre-HiFi bench studies by Fredberg et al. was that HFJVs squirt gas into the lungs at much higher flow rates compared to HFOV. That fact may make HFJV more sensitive to airway patency compared to HFOV.

Since CV breath distribution may be more affected by lung compliance while HFV breaths may be more affected by airway resistance, especially HFJV breaths with their high velocity inspirations, the distribution of ventilation in the nonhomogeneous PIE lung may be markedly affected by mode of ventilation. While the path of least resistance for CV breaths may lead to more compliant, injured areas of the lungs, HFJV breaths may automatically avoid injured areas where airway and vascular resistances are increased. Therefore, HFJV breath distribution may favor relatively normal airways in the uninjured parts of the lungs where ventilation/perfusion matching is more favorable.

The CV tidal volumes delivered with higher PEEP and *Paw* may dilate airways enough to help gas get into restricted areas in babies with PIE, but those larger tidal volumes take longer to get back out. Much smaller HFV tidal volumes are more easily expired, especially those that were unable to penetrate the restricted airways where the lungs are injured.

Upper Airway Fistulas and Pneumothoraces

Theoretically, the small tidal volumes, high inspiratory velocities, and short inspiratory times of HFJV are ideally suited for treating pneumothoraces and broncho-pleural and tracheal-esophageal fistulae. Gonzalez et al. found that gas flow in chest tubes, inserted in a series of infants with pneumothoraces, dropped an average of 54% when six infants were switched from CV to HFJV (39). Their mean $PaCO_2$ dropped from 43 to 34 Torr at the same time that their peak and mean airway pressures measured at the distal tip of the ET tube dropped from means of 41–28 and 15 to 9.7 cm H_2O, respectively.

Goldberg et al. (40) and Donn et al. (41) similarly reported improved gas exchange and reduced flow through tracheal–esophageal fistulas.

Homogeneous Obstructive Lung Disease (e.g., Reactive Airway Disease, Asthma)

The HFV should theoretically not be of much benefit in treating lung disorders such as asthma wherein airway resistance is uniformly increased. Low rates and long expiration times should be more effective. However, recent work with HFJV and helium-oxygen mixtures (heliox) demonstrated interesting potential for treating such disorders in patients requiring no more than 80% oxygen.

Tobias and Grueber improved ventilation in a one-year old infant with respiratory syncytial virus and progressive respiratory failure related to bronchospasm with HFJV by substituting a mixture of 80% helium/20% oxygen for compressed air at the air/oxygen blender (42). They hypothesized that the reduced density of helium compared to nitrogen enhanced distal gas exchange. Gupta and associates describe another case where HFJV and heliox rescued a 5 month old infant with acute respiratory failure associated with gas trapping, hypercarbia, respiratory acidosis, and air leak (43). The combination of HFJV with heliox led to rapid improvements in gas exchange, respiratory stabilization, and the ability to wean the patient from mechanical ventilation.

Nonhomogeneous Obstructive Lung Disease (e.g., MAS) and ECMO Candidates

Clinical studies of infants with meconium aspiration syndrome (MAS) provide support for the use of HFV with this type of lung disease. These patients are potential candidates for extracorporeal membrane oxygenation (ECMO), so ability to avoid ECMO is a typical outcome variable in such studies.

Clark et al. randomized 94 full-term infant ECMO candidates to HFOV or CV in a multicenter study (44). Prospectively defined failure criteria were met by 60% of those infants randomized to CV while only 44% of those randomized to HFOV failed. Cross-overs to the alternate mode by those who failed were allowed, and 63% of those

who failed CV were rescued by HFOV, while only 23% of those who failed HFOV were rescued by CV. (The latter comparison was statistically significant.) Overall, 46% of the infants who met ECMO criteria required ECMO.

A similar single-center study of HFJV versus CV involved 24 ECMO candidates with respiratory failure and persistent pulmonary hypertension of the newborn (PPHN) (45). Most of the infants in the HFJV-treated group (8 of 11) and 5 of 13 of the conventionally treated infants had either MAS or sepsis pneumonia. Treatment failure within 12 h of study entry occurred in only two of the HFJV-treated infants versus seven of the conventionally treated infants. The ECMO was used to treat 4 of 11 HFJV infants versus 10 of 13 control infants. Zero of nine surviving HFJV-treated infants developed chronic lung disease compared to four of 10 surviving controls ($p = 0.08$). Survival without ECMO in the HFJV group was 5 of 11 (45%) versus 3 of 13 (23%) in the control group. There was no statistical significance in any of these comparisons due to the small number of patients.

The degree to which pathophysiology predicts positive outcomes with respect to the ability of HFVs to rescue infants that become ECMO candidates has been explored in two additional clinical studies. Baumgart et al. evaluated their success with HFJV prior to instituting an ECMO program in 73 infants with intractable respiratory failure who by age and weight criteria may have been ECMO candidates (46). They found survival after HFJV treatment to be much higher in infants with RDS and pneumonia (32/38, 84%) compared to MAS/PPHN (10/26, 38%) or congenital diaphragmatic hernia (3/9, 33%). All patients initially responded rapidly to HFJV as measured by oxygen index (O.I., calculated as mean airway pressure in cm H_2O multiplied by fraction of inhaled O_2 divided by P_{aO2} in Torr). However, that improvement in survivors was realized and sustained during the first 6 h of HFJV treatment.

Paranka et al. studied 190 potential ECMO candidates treated with HFOV during 1985–1992 (47). All patients were born at 35 weeks gestational age or more and developed severe respiratory failure, as defined by an arterial to alveolar oxygen ratio ($P_{(A-a)O_2}$) < 0.2 or the need for a peak pressure of >35 cm H_2O on CV. Fifty-eight percent (111 patients) responded to HFOV and 42% (79 patients) were placed on ECMO. Gas exchange improved in 88% of the infants with hyaline membrane disease (RDS), 79% of those with pneumonia, 51% with meconium aspiration, and 22% of those with congenital diaphragmatic hernia. They also found failure to demonstrate an improvement in $P_{(A-a)O_2}$ after six hours on HFOV to be predictive of failure.

During and After Cardiac Surgery

The ability of HFJV to hyperventilate while using lower mean airway pressure is a great asset when treating patients with cardiac problems. During surgery, the small tidal volumes and low mean airway pressure allow the surgeon to move the lungs out of the way, in order to visualize and work on the heart. After surgery, HFJV can gently hyperventilate the patient to encourage increased pulmonary blood flow while mean airway pressure is kept down (48–51).

PPHN and Nitric Oxide Therapy

Kinsella et al. demonstrated the potential of HFV to enhance delivery of nitric oxide (NO) for the treatment of PPHN in a large, multicenter, randomized controlled trial (52). Nitric oxide delivered with HFOV to infants with significant parenchymal lung disease was more effective than NO delivered by CV. NO has also been delivered successfully with HFJV (53). However, NO must be administered via the HFJV circuit in order for the patient to realize any beneficial effect from the gas (54). Inhaled NO does not work with HFJV when administered exclusively through the conventional ventilator circuit (55).

HFV APPLICATIONS IN CHILDREN AND ADULTS

While the bulk of the research and application of HFV has been aimed at the benefit of infants to date, the sheer number of potential applications for children and adults is far greater. Unfortunately, the number of HFVs available to treat adults is severely limited. There is only one instrument currently available in the United States specifically designed for ARDS in children and adults, the SensorMedics 3100B. (The Percussionaire VDR4-F00008 ventilator also provides HFV for adults. It was approved as a Class II device by the FDA.)

Acute respiratory distress syndrome is the obvious target for HFV treatment in adult intensive care. This syndrome affects ~50 per 100,000 population with a mortality of 30–50%. It is a clinical syndrome of noncardiogenic pulmonary edema associated with pulmonary infiltrates, stiff lungs, and severe hypoxemia (56). Although the pathology of ARDS involves a number of features similar to RDS in infants, such as hyaline membranes, endothelial and epithelial injury, loss of epithelial integrity, and increased alveolar-capillary permeability, it may have a much greater inflammatory component.

The only treatment shown to positively impact mortality over the past several decades came from the ARDSnet Trial where CVs were used with a low tidal volume ventilatory strategy designed to reduce iatrogenic lung injury (57). Comparative treatments in this multicenter study of 861 patients included an experimental group where mean tidal volumes for the first 3 days of their treatments were $6.2\,mL\cdot kg^{-1}$ body weight and a control group where tidal volumes were $11.8\,mL\cdot kg^{-1}$. The experimental group had lower mortality and fewer days on mechanical ventilators.

With ARDSnet trial pointing in the general direction of smaller tidal volumes, it is not surprising that recent HFV trials appear very promising, especially since HFV investigators focused on NICU patients and worked their way up the learning curve. The most important lesson learned, and one that took many years to learn in the treatment of infants, was the importance of recruiting and maintaining adequate lung volume during HFV. Adult trials of HFV for ARDS now begin with a Paw $5\,cm\,H_2O$ greater than that currently being used with CV. Just as was learned with infants, it is safe to use higher PEEPs and mean airway pressures with HFVs smaller tidal volumes.

HFV Clinical Trails with Children and Adults

The importance of starting early with HFV on adults and children with ARDS was highlighted in several anecdotal and pilot trials. Smith et al. treated 29 children with severe ARDS complicated by pulmonary barotrauma with HFJV (58). Twenty (69%) survived, and the only statistically significant difference between survivors and nonsurvivors was the mean time on CV before initiating HFJV (3.7 days in survivors vs. 9.6 days in nonsurvivors). Fort et al. similarly found that survivors in a pilot study of HFOV for adults with ARDS were on CV 2.5 days before initiation of HFOV, while nonsurvivors were on CV for 7.2 days (59). Expected survival in the pilot study was <20%, actual survival was 47%.

Arnold et al. compared HFOV to CV in children with respiratory failure (60). Optimizing lung volume was emphasized in both the experimental and control groups. The strategy for optimizing lung volume in the CV group was to lengthen inspiratory times and increase PEEP in order to decrease required PIPs. They found significant improvement in oxygenation in the HFOV group as well as a lower need for supplement oxygen at 30 days postenrollment.

A recent prospective trial of HFOV for ARDS had similar results. Mehta et al. treated a series of 24 adults with severe ARDS with HFOV (61). Five of the patients were burn victims. Within 8 h of HFOV initiation, F_IO_2 and $PaCO_2$ were lower and PaO_2/F_IO_2 was higher than baseline values during CV throughout the duration of the trial. An obvious focus was placed on recruiting and maintaining adequate lung volume while on HFOV, since Paw was also significantly higher than that applied during CV throughout the HFOV trial. Unfortunately, this increase in Paw was associated with significant changes in hemodynamic variables including an increase in pulmonary artery occlusion pressure (at 8 and 40 h) and central venous pressure (at 16 and 40 h), and a reduction in cardiac output throughout the study. Thus, Paw may not have been optimized. However, 10 patients were successfully weaned from HFOV and 7 survived. Again, there was a statistically significant difference in the time spent on CV prior to initiation of HFV: 1.6 days for survivors versus 5.8 days for the nonsurvivors.

Noting the importance of early intervention, Derdak et al. designed a multicenter, randomized, controlled trial comparing the safety and effectiveness of HFOV versus CV in adults with less severe ARDS (62). (The authors nicknamed their trial: the MOAT Study.) Inclusion criteria included $PaO_2/F_IO_2 \leq 200\,mmHg$ (26.66 kPa) on $\geq 10\,cm$ H_2O PEEP, and 148 adults were evenly randomized. Applied Paw was significantly higher in the HFOV group compared with the CV group throughout the first 72 h. The HFOV group showed improvement in PaO_2/F_IO_2 at <16 h, but this difference did not persist beyond 24 h. Thirty day mortality was 37% in the HFOV group and 52% in the CV group ($p = 0.102$). At 6 months, mortality was 47% in the HFOV group and 59% in the CV group ($p = 0.143$). There were no significant differences in hemodynamic variables, oxygenation failure, ventilation failure, barotraumas, or mucus plugging between treatment groups.

The MOAT Study indicates that HFOV is safe and effective for ARDS, and the FDA approved the SensorMedics 3100B for ARDS. Outcome data from this study are comparable to those of the ARDSnet Trial. The control group in the MOAT study was not ventilated with tidal volumes as small as those used in the experimental group of the ARDSnet trial (6–10 vs. $6.2\ mL\cdot kg^{-1}$), but they were generally smaller than the ARDSnet control group ($11.8\ mL\cdot kg^{-1}$). Mortality at 30 days in the MOAT Study was not quite as good as that in the ARDSnet Trial (37 vs. 31%, respectively), but sepsis was much more prevalent in the MOAT Study compared to the ARDSnet Trial (47 vs. 27%, respectively).

STATUS OF HFV, RISKS, AND OUTLOOK FOR THE FUTURE

Are HFVs Safe and Effective?

Use of HFVs for newborn infants and adults began in the early 1980s. Fifteen randomized controlled trials with infants and about one-half that many randomized studies with children and adults were conducted over the next 20+ years. Over 1000 articles about HFV have been published. Yet, there are still questions about HFV safety and efficacy.

There are certainly adequate data to suggest that HFVs are effective in lessening chronic lung injury. The fact that not all studies have been successful in this regard is a reflection of differences in infant populations, ventilator strategies, and devices used. There is little argument that use of antenatal steroids, exogenous surfactant, and ventilator strategies using smaller tidal volumes have greatly improved mortality and morbidity of premature infants.

Not surprisingly, as clinicians have become more successful with HFV and other small tidal volume strategies, the age of viability of premature infants has gone down. Thus, the challenge of preventing chronic lung disease in NICU patients never gets easier, because the patients keep getting more premature.

What Are the Risks Associated with HFV in the NICU?

The greatest controversy in consideration of HFVs as a primary mode of ventilation of premature infants is safety, particularly whether HFV use increases the risk of cerebral injury. Clark et al. evaluated the probability of risk of premature infants suffering from intraventricular hemorrhage (IVH) or periventricular leukomalacia (PVL) by conducting a meta-analysis of all prospective randomized controlled trials of HFV published by 1996

(63). The meta-analysis showed that use of HFV was associated with an increased risk of PVL (odds ratio = 1.7 1.7 with a confidence interval of 1.06–2.74), but not IVH or severe ($\geq$grade 3) IVH. In addition, since the largest study in the group by far was the HiFi Trial (14), where implementation strategy was reputed to be less than optimal, they repeated the analysis without that study. When the results of the HIFI study were excluded, there were no differences between HFV and conventional ventilation in the occurrence of IVH or PVL.

Since 1996, seven additional randomized controlled trials of early use of HFV have been conducted on 1726 patients. Only one of the newer studies demonstrated a possible increased risk of cerebral injury (64), and that study included 273 patients or 16% of the total in these 7 studies. Thus, a more current meta-analysis would be even more convincingly positive today, and one could even say that there is *little* evidence of increased risk of cerebral injury during HFV. Why then, is this matter still controversial?

The risk of causing cerebral injury in premature infants is associated with hyperventilation and hypocarbia as noted earlier. There will never be a randomized controlled trial to prove cause and effect here, for obvious reasons. Therefore, all we can do is try to avoid hyperventilation and hypocarbia and see if outcomes get better over time.

Avoiding hyperventilation and hypoxemia first requires proper monitoring. Pulse oximetry, transcutaneous CO_2 monitoring, and continuous or frequent arterial blood gas monitoring are essential during HFV. Control of $PaCO_2$ during HFV often requires optimization of PEEP, Paw, and pressure amplitude (ΔP) as shown in Fig. 12. The HFVs are noted for their ease of blowing off CO_2 at lower airway pressures compared to CV, so PEEP and Paw must often be increased above those used during CV, if hypoxemia is to be avoided.

With HFOV, one often adjusts mean airway pressure without knowing the resulting baseline pressure or PEEP, whereas with HFJV, PEEP is adjusted to get an appropriate Paw. Therefore, one must not be fearful of higher PEEP when higher mean airway pressure is required. PEEP as high as 10 cm H_2O is not unusual when HFJV is being used to treat premature infants.

One must also recognize that raising PEEP will reduce ΔP when PIP is held constant, as shown in Fig. 12, which causes both PaO_2 and $PaCO_2$ to rise.

Other safety concerns with early use of HFVs for preventing lung injury are interference with cardiac output by using too much PEEP or Paw. Since interference with

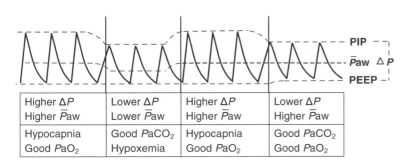

Higher ΔP	Lower ΔP	Higher ΔP	Lower ΔP
Higher Paw	Lower Paw	Higher Paw	Higher Paw
Hypocapnia	Good $PaCO_2$	Hypocapnia	Good $PaCO_2$
Good PaO_2	Hypoxemia	Good PaO_2	Good PaO_2

Figure 12. Adjusting pressure waveforms to correct arterial blood gases. (Used with permission. © 2003, Bunnell Inc.)

venous return by elevated intrathoracic pressure raises intracranial pressure, there is associated fear of causing IVH by this mechanism as well.

A related issue, for those HFVs with that capability, is using too many CV breaths or using overly large CV tidal volumes during HFV. The latter use increases the risk of causing lung injury when HFV is implemented with higher PEEP.

Optimizing PEEP and minimizing the risk of using too many CV breaths during HFV can be achieved at the same time. The following flowchart for finding optimal PEEP during HFJV illustrates this point (Fig. 13).

The flowchart in Fig. 13 is based on the concept that CV breaths will be most effective in opening up collapsed alveoli, while PEEP or baseline pressure will prevent alveoli from collapsing during exhalation. The longer I times and tidal volumes of CV breaths provide a greater opportunity to reach the critical opening pressure of collapsed alveoli, and if PEEP is set above the critical closing pressure of those alveoli, they will remain open throughout the ventilatory cycle. Once PEEP is optimized, there is less value in using CV in tandem with HFV.

Although Fig. 13 was designed for use during HFJV, its principles are equally applicable to HFOV when CV may not be available. In this case, mean airway pressure is raised until an improvement in oxygenation makes it apparent that alveolar recruitment has occurred. At that point, it should be possible to decrease mean airway pressure somewhat without compromising oxygenation. However, the appropriate strategy here would be to set a goal for lowering the fraction of inhaled oxygen (F_IO_2) *before*

attempting to lower *Paw*. In this way, one should avoid inadvertently weaning *Paw* too fast and risking catastrophic collapse of alveoli. An appropriate F_IO_2 goal in this circumstance might be 0.3–0.4 depending on the vulnerability of the patient to high airway pressures and the magnitude of the mean airway pressure present at the time.

The final risk to be mentioned here will be the greatest risk associated with HFV: inadequate humidification and airway damage. In unsuccessful applications of HFV in infants in the early 1980s, necrotizing tracheal bronchitis (NTB) was frequently noted at autopsy (65). First discovered during HFJV, it was subsequently discovered during HFOV as well (66). Fortunately, the development of better humidification systems coupled with earlier implementation of HFV seems to have eradicated this problem as an extraordinary adverse side effect. None of the 14 randomized controlled trials has found an increase in NTB associated with HFV treatment.

Humidification during HFV is challenging, especially for HFOVs that use high gas flow rates and HFJVs. Gas humidified under pressure will not hold as much water as unpressurized gas, so HFJVs must humidify their inspiratory gas at higher than normal temperatures in order to reach anything near 100% relative humidity at body temperature.

Bunnell Incorporated's HFJV addressed this inherent problem by minimizing the driving pressure behind their jet nozzle using an inspiratory pinch valve in a little box placed near the patient's head. The pinch valve reduces the pressure drop through that part of the system because of

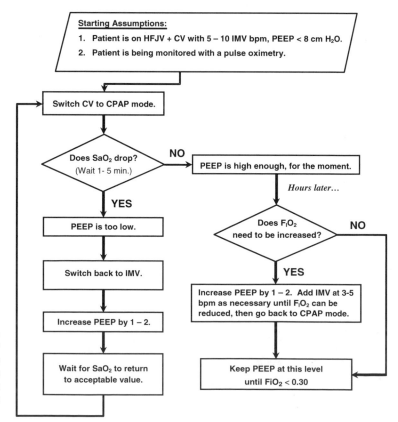

Figure 13. Optimal PEEP flowchart. (Used with permission. © 2005, Bunnell Inc.) (*Warnings*: Lowering PEEP may improve SaO_2 in some cases. Optimal PEEP may be lower in patients with active air leaks or hemodynamic problems. Do not be shocked if optimal PEEP = 8 – 12 cm H_2O. Using IMV PIP with high PEEP is hazardous. Do not assume high PEEP causes overexpansion.)

the relatively large internal diameter (ID) of the tubing in the valve (0.13 in., 3.2 mm). Placing the pinch valve within 35 cm of the patient where inspired gas is delivered via a jet nozzle embedded in a special ET tube adapter also enables the LifePulse Ventilator to deliver its tiny tidal volumes without much driving pressure. (A typical driving pressure needed for HFJV with the LifePulse on a premature infant is between 1.5 and 3.5 psi.) The HFVs that work at higher pressures and gas flow rates sometimes provide humidity with liquid water. (See Acutronic jet ventilation systems on their website: www.acutronic-medical.ch.)

Working Within the Limitations of HFVs in the NICU

The HFVs have been widely accepted for treating newborn infants with lung injury. Whether HFV will ever be widely accepted as a primary mode of ventilation is another matter. There have been many advances in conventional ventilator therapy over the past several years and, to a certain extent, techniques used to optimize HFVs have been applied to CVs (67).

Like most therapies, the skill with which HFV is implemented is probably the most critical determinant of clinical success. Starting HFV on patients sooner rather than waiting for worsening lung disease to become nearly hopeless is also extraordinarily important. Having written protocols defining the patient population and how HFV is optimized for that patient population can also be very helpful.

Early-to-moderately early stage treatment of homogeneously noncompliant lungs is the most obvious choice as an appropriate indication for HFV. If hyperventilation and gas trapping are avoided and appropriate resting lung volume is achieved, better outcomes should result. The ability of all HFVs to ventilate lungs using less pressure, independent of lung compliance, is nearly universal as long as the lungs are not too large for the ventilator's output. Many of the HFV modes built into CVs are not powerful enough to ventilate even a term infant, so operators need to know the relative output capacities of their HFVs.

Using HFVs as rescue ventilators usually means that the underlying lung disorder has become nonhomogeneous or even obstructive. Ironically, most clinicians will only use HFVs as rescue ventilators even though these disorders are much harder to treat with HFV. Gas trapping is a much greater risk with heterogeneous obstructive disorders, and it may be avoided via use of small HFV tidal volumes. However, even HFV tidal volumes can be trapped if expiratory times are not several times greater than inspiratory times. The HFJVs with their very short I:E ratios and very high velocity inspiratory flows have been demonstrated to be more effective with these types of lung disease. Combined CV and HFJV have also been shown to work even better than pure HFJV in severe nonhomogeneous lung disorders (68).

What Are the Risks and Limitations Associated with HFV in Treating Children and Adults?

The only risks unique to HFV for children and adults are those associated with the necessity for delivering larger tidal volumes at higher flow rates compared to HFV for infants. Humidification of HFV gases is crucial as was learned with the early trials in infants, and gas is more difficult to humidify under pressure, as discussed above. The most challenging mode of HFV in this respect is HFJV, since it takes elevated pressure to push gas through a jet nozzle. Perhaps this is one reason for the lack of success of the only HFJV for adults approved by the FDA for use with adults (the APT 1010 Ultrahigh Frequency Ventilator, developed by the Advanced Pulmonary Technologies, Inc., Glastonbury, CT). Humidification with this device was only provided via supersaturated entrained gas. However, the same corporation that pulled this product off the market is also planning to discontinue manufacturing the Infant Star HFV for infants, and that ventilator had no such humidification issues. Thus, it appears likely that these products were discontinued for other (i.e., business) reasons.

There is also danger of operator error when any machine is used by untrained or unskilled operators. Given the tendency for some hospitals to only use HFVs as last resort rescue ventilators, one must consider those types of risks. Since HFOV is most successful in homogeneous lung disorders, it only makes sense for it to be used relatively early in the course of ARDS before significant lung injury results from CV treatment. Once the patient's condition deteriorates into ARDS complicated by airleaks, chances for HFOV success may be significantly decreased.

Treating children and adults with HFVs also has to take optimal frequency into account. The primary determinant of optimal frequency is lung compliance, which is primarily determined by the patient's size. An adult's lung is larger than that of a child, which is larger than that of a term newborn, which is larger than that of a preemie. Thus, HFV frequency should be reduced as patient size increases. Optimal HFV for adults may occur at 150 bpm, whereas operation at that frequency with infants would not even be considered HFV.

Given the evidence that HFJVs have been more successful with nonhomogeneous lung disorders in infants, as described above, it would seem likely that HFJVs would find a role for treating adult patients with ARDS as well. Unfortunately, the application of HFJV for adults has been tarnished by a lack of success in very early studies.

Carlon et al. conducted a randomized controlled trial of HFJV versus volume-cycled CV on adults with acute respiratory failure (69). While they reported patients failing on CV improved more rapidly and in greater number when switched to HFJV compared to those who were failing on HFJV and crossed to CV, there were no advantages with respect to survival and total duration of stay in the ICU. Thus, they concluded that HFJV offered no obvious benefits over CV. The study was published in 1983, long before there was much appreciation of the need to optimize PEEP and maintain adequate lung volume during HFV. One wonders what results would come from a similar trial 20 years later, but such a trial will probably never happen now. The cost, time, and effort of seeking FDA approval for any Class III device is so high now, that HFJV may never find its way back into an adult ICU in the United States.

What Is the Outlook for HFV in the Future?

The HFVs evolved as people discovered problems with trying to replicate breathing in compromised patients. Unlike conventional ventilation, HFV is designed to facilitate gas exchange rather than mimic how people breathe normally. The differences between HFV and CV have led to creative solutions for many of the problems that investigators set out to solve, but they have also created their own problems.

It was discovered how HFVs can ventilate much more effectively than CV, and in the process, it was discovered how hypocapnia can lead to severe cerebral injury in premature infants.

The HFV still uses positive pressure where we create negative pressure to draw gas into the lungs. So, the problems related to use of ET tubes and its bypassing the normal humidification system of the body (i.e., the nose) are still there.

The HFV uses much smaller tidal volumes than CV, so the damage we have come to call volutrauma has lessened. In the process, it was discovered that more mean airway pressure or PEEP is required to keep sick lungs open while they are being ventilated. Then it was discovered that the new problems were associated with too much pressure in the thorax interfering with cardiac output.

So, HFVs do not solve all the problems, and they require increased vigilance to avoid creating new problems. However, the basic differences between HFV and CV provide a very reliable alternative to CV in circumstances where those differences are critical to survival.

The HFV tidal volumes are minimally affected by lung compliance and maximally affected by airway resistance when they are delivered via a jet nozzle. Therefore, in lung disorders where these conditions dictate treatment success or failure, wise users of HFVs have been very successful. When HFV users are not adequately trained or aware of the differences in HFV gas distribution caused by lung pathophysiology, success can be elusive.

Premature infants with RDS represent a large population with the potential of leading long and rich lives if they survive their first months of life with undamaged or even minimally damaged lungs and brains. Many randomized controlled trials have demonstrated the potential of HFVs to help these infants realize that potential.

The tens of thousands of adults who succumb to ARDS every year also have the potential of increased survival with less morbidity thanks to HFVs. These patients, when successfully treated with an HFV, will be considered rescued from a well-recognized disorder with a chronically high mortality rate.

Given the skill and training needed to master HFVs, their use may be considered risky indefinitely. As HFVs and associated monitoring equipment become better designed to help their users optimize assisted ventilation, HFV use should increase and evolve into earlier, more prophylactic applications to prevent lung injury. The HFVs are inherently a kinder, gentler form of mechanical ventilation. Hopefully, their true potential will someday be realized.

BIBLIOGRAPHY

Cited References

1. Lee PC, Helsmoortel CM, Cohn SM, Fink MP. Are low tidal volumes safe? Chest 1990;97:430–434.
2. Kacmarek RM, Chiche J.D. Lung protective ventilatory strategies for ARDS—the data are convincing! Resp Care 1998; 43:724–727.
3. Sjöstrand U. Review of the physiological rationale for and development of high-frequency positive-pressure ventilation—HFPPV. Acta Anaesthesiol Scand (Suppl) 1977;67: 7–27.
4. Klain M. Clinical applications of high-frequency jet ventilation, Part A: clinical use in the operating room. In: Carlon GC, Howland WS, editors. High-frequency ventilation in intensive care and during surgery. New York: Marcel Dekker; 1985. p 137–149.
5. Singh JM, Stewart TE. High-frequency mechanical ventilation principles and practices in the era of lung-protective ventilation strategies. Respir Care Clin N Am 2002;8:247–60.
6. MacIntyre NR. Setting the frequency-tidal volume pattern. Respir Care 2002;47:266–274.
7. Radford E. Ventilation standards for use in artificial respiration. J Appl Physiol 1955;7:451–460.
8. Henderson Y, Chillingworth FP, Whitney JL. The respiratory dead space. Am J Physiol 1915;38:1–19.
9. Dubois AB, Brody AW, Lewis DH, Burgess BF. Oscillation mechanics of lungs and chest in man. J Appl Physiol 1956;8: 587–594.
10. Bunnell JB, Karlson KH, Shannon DC. High-frequency positive pressure ventilation in dogs and rabbits. Am Rev Respir Dis 1978;117:289.
11. Fredberg JJ. Augmented diffusion in the airways can support pulmonary gas exchange. J Appl Physiol 1980;49:232–238.
12. Slutsky AS. Mechanisms affecting gas transport during high-frequency oscillation. Crit Care Med 1984;12:713–717.
13. Venegas JG, Hales CA, Strieder DJ. A general dimensionless equation of gas transport by high-frequency ventilation. J Appl Physiol 1986;60:1025–1030.
14. Permutt S, Mitzner W, Weinmann G. Model of gas transport during high-frequency ventilation. J Appl Physiol 1985;58: 1956–1970.
15. Venegas JG, Fredberg JJ. Understanding the pressure cost of high frequency ventilation: why does high-frequency ventilation work? Crit Care Med 1994;22:S49–S57.
16. Haselton FR, Scherer PW. Bronchial bifurcations and respiratory mass transport. Science 1980;208:69–71.
17. Tobias JD, Grueber RE. High-frequency jet ventilation using a helium-oxygen mixture. Paediatr Anaesth 1999;9:451–455.
18. Hatcher D, et al. Mechanical performance of clinically available, neonatal, high-frequency, oscillatory-type ventilators. Crit Care Med 1998;26:1081–1088.
19. Pillow JJ, Wilkinson MH, Neil HL, Ramsden CA. *In vitro* performance characteristics of high-frequency oscillatory ventilators. Resp Crit Care Med 2001;164:1019–1024.
20. Fredberg JJ, Glass GM, Boynton BR, Frantz 3rd ID. Factors influencing mechanical performance of neonatal high-frequency ventilators. J Appl Physiol 1987;62:2485–2490.
21. Boros SJ, et al. Comparison of high-frequency oscillatory ventilation and high-frequency jet ventilation in cats with normal lungs. Ped Pulmonol 1989;7:35–41.
22. Zobel G, Dacar D, Rodl S. Proximal and tracheal airway pressures during different modes of mechanical ventilation: An animal model study. Ped Pulmonol 1994;18:239–243.
23. Bryan AC, Slutsky AS. Lung volume during high frequency oscillation. Am Rev Resp Dis 1986;133:928–930.

24. Gerstmann DR, et al. Proximal, tracheal, and alveolar pressures during high-frequency oscillatory ventilation in a normal rabbit model. Pediatr Res 1990;28:367–373.
25. Perez Fontan JJ, Heldt GP, Gregory GA. Mean airway pressure and mean alveolar pressure during high-frequency jet ventilation in rabbits. J Appl Physiol 1986;61:456–463.
26. HIFI Study Group, HFOV compared with conventional mechanical ventilation in the treatment of respiratory failure in preterm infants. N Engl J Med 1989;320:88–93.
27. Keszler M, et al. Multicenter controlled clinical trial of high-frequency jet ventilation in preterm infants with uncomplicated respiratory distress syndrome. Pediatrics 1997;100:593–599.
28. Courtney SE, et al. Early high-frequency oscillatory ventilation versus conventional ventilation in very-low-birth-weight-infants. N Engl J Med 2002;347:643–653.
29. Johnson AH, et al. High-frequency oscillatory ventilation for the prevention of chronic lung disease of prematurity. N Engl J Med 2002;347:633–642.
30. Wiswell TE, et al. HFJV in the early management of RDS is associated with a greater risk for adverse outcomes. Pediatrics 1996;98:1035–1043.
31. Wiswell TE, et al. Effects of hypocarbia on the development of cystic periventricular leukomalacia in premature infants treated with HFJV. Pediatrics 1996;98:918–924.
32. Bryan AC, Froese AB. Reflections on the HiFi trial. Pediatrics 1991;87:565–567.
33. Stark AR. High-frequency oscillatory ventilation to prevent bronchopulmonary dysplasia—are we there yet? N Engl J Med 2002;347:682–683.
34. Keszler M, Goldberg LA, Wallace A. High frequency jet ventilation in subjects with low chest wall compliance. Pediatr Res 1993;33:331.
35. Keszler M, Jennings LL. High frequency jet ventilation in infants with decreased chest wall compliance. Pediatr Res 1997;41:257.
36. Fok TF, et al. High frequency oscillatory ventilation in infants with increased intra-abdominal pressure. Arch Dis Child 1997;76:F123–125.
37. Thibeault DW. Pulmonary barotrauma: Interstitial emphysema, pneumomediastinum, and pneumothorax. In: Thibeault DW, Gregory GA, editors. Neonatal Pulmonary Care. 2nd ed. New York : Appleton-Century-Crofts; 1986. p 499–517.
38. Macklin MT, Macklin CC. Malignant interstitial emphysema of the lung and mediastinum as an important occult complication in many respiratory diseases and other conditions: An interpretation of clinical literature in light of laboratory experiment. Medicine 1944;23:281.
39. Gonzalez F, Harris T, Richardson P. Decreased gas flow through pneumothoraces in neonates receiving high-frequency jet versus conventional ventilation. J Pediatr 1987;110:464–466.
40. Goldberg L, Marmon L, Keszler M. High-frequency jet ventilation decreases flow through tracheo-esophageal fistula. Crit Care Med 1992;20:547–550.
41. Donn SM, et al. Use of high-frequency jet ventilation in the management of congenital tracheoesophageal fistula associated with respiratory distress syndrome. J Pediatr Surg 1990;12:1219–1221.
42. Tobias JD, Grueber RE. High-frequency jet ventilation using a helium-oxygen mixture. Paediatr Anaesth 1999;9:451–455.
43. Gupta VK, Grayck EN, Cheifetz IM. Heliox administration during high-frequency jet ventilation augments carbon dioxide clearance. Respir Care 2004;49:1038–1044.
44. Clark RH, Yoder BA, Sell MS. Prospective, randomized comparison of high-frequency oscillation and conventional ventilation in candidates for extracorporeal membrane oxygenation. J Pediatr 1994;124:447–454.
45. Engle WA, et al. Controlled prospective randomized comparison of HFJV and CV in neonates with respiratory failure and persistent pulmonary hypertension. J Perinat 1997;17:3–9.
46. Baumgart S, et al. Diagnosis-related criteria in the consideration of ECMO in neonates previously treated with HFJV. Pediatrics 1992;89:491–494.
47. Paranka MS, Clark RH, Yoder BA, Null DM. Predictors of failure of high-frequency oscillatory ventilation in term infants with severe respiratory failure. Pediatrics 1995;95:400–404.
48. Meliones JN, et al. High-frequency jet ventilation improves cardiac function after the Fontan procedure. Circulation 1991;84(Suppl III):III-364–III-368.
49. Kocis KC, et al. High-frequency jet ventilation for respiratory failure after congenital heart surgery. Circulation 1992;86(Suppl II):II-127–II-132.
50. Greenspan JS, et al. HFJV: Intraoperative application in infants. Ped Pulmonol 1994;17:155–160.
51. Davis DA, et al. High-frequency jet versus conventional ventilation in infants undergoing Blalock-Taussig shunts. Ann Thorac Surg 1994;57:846–849.
52. Kinsella JP, et al. Randomized, multicenter trial of inhaled nitric oxide and high-frequency oscillatory ventilation in severe, persistent pulmonary hypertension of the newborn. J Pediatr 1997;131:55–62.
53. Day RW, Lynch JM, White KS, Ward RM. Acute response to inhaled nitric oxide in newborns with respiratory failure and pulmonary hypertension. Pediatrics 1996;98:698–705.
54. Platt D, Swanton D, Blackney D. Inhaled nitric oxide delivery with high-frequency jet ventilation. J Perinatol: (in press) 2003.
55. Mortimer TW, Math MCM, Rajardo CA. Inhaled nitric oxide delivery with high-frequency jet ventilation: A bench study. Respir Care 1996;41:895–902.
56. Ware LB, Matthay MA. The acute respiratory distress syndrome. N Engl J Med 2000;342:1334–1349.
57. The Acute Respiratory Distress Syndrome Network, Ventilation with lower tidal volumes as compared with traditional tidal volumes for acute lung injury and the acute respiratory distress syndrome. N Engl J Med 2000;342:1301–1308.
58. Smith DW, et al. High-frequency jet ventilation in children with the adult respiratory distress syndrome complicated by pulmonary barotrauma. Ped Pulmonol 1993;15:279–286.
59. Fort P, et al. High-frequency oscillatory ventilation for adult respiratory distress syndrome—a pilot study. Crit Care Med 1997;25:937–47.
60. Arnold JH, et al. Prospective, randomized comparison of high-frequency oscillatory ventilation and conventional mechanical ventilation in pediatric respiratory failure. Crit Care Med 1994;22:1530–1539.
61. Mehta S, et al. Prospective trial of high-frequency oscillation in adults with acute respiratory distress syndrome. Crit Care Med 2001;29:1360–1369.
62. Derdak S, et al. High-frequency oscillatory ventilation for acute respiratory distress syndrome in adults: A randomized, controlled trial. Am J Respir Crit Care Med 2002;166:801–808.
63. Clark RH, Dykes FD, Bachman TG, Ashurst JT. Intraventricular hemorrhage and high-frequency ventilation: A meta-analysis of prospective clinical trials. Pediatrics 1996;98:1058–1061.
64. Moriette G, et al. Prospective randomized multicenter comparison of high-frequency oscillatory ventilation and conventional ventilation in preterm infants of less than 30 weeks with respiratory distress syndrome. Pediatrics 2001;107:363–72.

65. Mammel MC, et al. Acute airway injury during high-frequency jet ventilation and high-frequency oscillatory ventilation. Crit Care Med 1991;19:394–398.

66. Kirpilani H, et al. Diagnosis and therapy of necrotizing tracheobronchitis in ventilated neonates. Crit Care Med 1985; 13:792–797.

67. Keszler M, Durand DJ. Neonatal high-frequency ventilation. Past, present, and future. Clin Perinatol 2001;28:579–607.

68. Spitzer AR, Butler S, Fox WW. Ventilatory response to combined HFJV and conventional mechanical ventilation for the rescue treatment of severe neonatal lung disease. Ped Pulmonol 1989;7:244–250.

69. Carlon GC, et al. High-frequency jet ventilation. A prospective randomized evaluation. Chest 1983;84:551–559.

See also CONTINUOUS POSITIVE AIRWAY PRESSURE; RESPIRATORY MECHANICS AND GAS EXCHANGE; VENTILATORY MONITORING.

HIP JOINTS, ARTIFICIAL

Z. M. JIN
J. L. TIPPER
M. H. STONE
E. INGHAM
J. FISHER
University of Leeds
Leeds, United Kingdom

INTRODUCTION

Natural synovial joints, such as hips, are remarkable bearings in engineering terms. They can transmit a large dynamic load of several times bodyweight during steady-state walking, yet with minimal friction and wear achieved through effective lubrication and with little maintenance. However, diseases such as osteoarthritis and rheumatoid arthritis or trauma sometimes necessitate the replacement of these natural bearings. Artificial hip joints replace the damaged natural bearing material, articular cartilage. As a result, pain in the joint is relieved and joint mobility and functions are restored. Total hip joint replacement has been considered as one of the greatest successes in orthopaedic surgery in the last century in improving the quality of life of the patients. Currently, > 1 million hip joints are replaced worldwide each year, with ever increasing use of these devices in a wider range of patients.

The majority of current artificial hip joints consist of an ultrahigh molecular weight polyethylene (UHMWPE) acetabular cup against a metallic or ceramic femoral head as illustrated in Fig. 1. These devices can generally last 10–15 years in the body without too many problems. However, after this period of implantation, loosening of prosthetic components becomes the major clinical problem. It is now generally accepted that the loosening is caused by the osteolysis as a result of biological reactions to particulate wear debris mainly released from the articulating surfaces. Therefore, one of the main strategies to avoid the loosening problem and to extend the clinical life of the hip prosthesis is to minimize wear and wear particles. Application of *tribology*, defined as "the branch of science and technology concerned with interacting surfaces in relative motion and

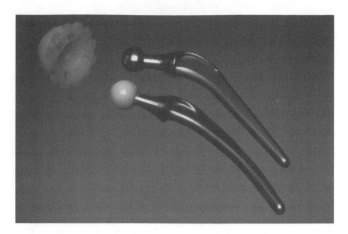

Figure 1. A typical Charnley hip prosthesis consisting of an UHMWPE acetabular cup against either a metallic (stainless steel) or a ceramic (alumina) femoral head.

with associated matters (as friction, wear, lubrication, and the design of bearings" (Oxford English Dictionary), to biological systems (*biotribology*) such as artificial hip joints, can play an important role in this process. Coupled tribological studies of friction, wear and lubrication of the bearing surfaces, and biological studies of wear debris-induced adverse reactions become necessary.

HISTORICAL DEVELOPMENT

Early History: Hemiarthroplasty, Interposition Arthroplasty, and Total Hip Replacement

The first recognizable ball and socket joint was reported in Germany by Professor Gluck in 1890 in a dog with an ivory ball and socket hip joint. This did not gain popular support for use in humans until Hey Groves in Bristol reported his ivory hemiarthroplasty for fractured neck of femur in 1926. Attempts to use metal at this stage were unsuccessful. A significant breakthrough came in 1923. It began with a chance observation that a piece of glass left in an individual's back for 1 year stimulated a fibrous tissue and fluid producing reaction. It formed a fluid-filled synovial sac (Smith Peterson 1948). Smith Peterson went on to insert a glass cup-shaped mould between the surfaces of an ankylosed hip. Although the glass broke, at the time of its removal the acetabulum and the head of the femur were found to be covered with a smooth lining of fibrous tissue. Over the next few years a number of different materials were used including Viscaloid, Pyrex, Bakelite, and finally Vitallium (chromium–cobalt–molybdenum alloy) in 1938. This material worked well and was used for ~ 1000 interposition arthroplasties at Massachusetts General Hospital alone over the next 10 years. It remained the standard treatment for hip arthritis until the advent of total hip replacement.

Charnley Era

Jean and Robert Judet reported their use of a replacement femoral head made of poly (methyl methacrylate) (PMMA). Although the prosthesis failed, it survived long enough to squeak within the human body. It was this squeaking prosthesis that set Charnley on his quest for a low friction-bearing surface. He began with Teflon in 1958 and throughout the 1950s Charnley experimented with two thin cups of Teflon, one in the acetabulum and one over a reshaped femoral head. They failed within a year due to loosening of the cup and avascular necrosis of the femoral head. He abandoned this surface replacement for an endoprosthesis and as the acetabular cups wore through Charnley sought better wearing materials. He moved to high density PE and later to UHMWPE.

Low Friction Arthroplasty. Charnley began his cemented total hip replacement era with a large femoral head (Moore's hemiarthroplasties). He argued that distributing the load over a large area of contact would decrease wear. However, after loosening of these large head components, he began working on the low frictional torque prosthesis, which reduced the torque at the cement-bone and prosthesis interfaces. He achieved this by reducing the diameter of the femoral head from ~ 41 to 22 mm. In this way, Charnley developed his prosthesis of a 22 mm head on a metal stem, UHMWPE cup and PMMA cement. Charnley hips still have a survival today of $> 90\%$ at 10 years (1).

There continues to be debate as to the cause of up to 10% failures of Charnley hips. Early belief that it was due to the cement and cement particles led to the development of uncemented prostheses. Concern that the production of PE particles was producing bone lysis, led to the development of alternative bearing surfaces, for example, ceramics that wear less than metal against PE, and metal-on-metal prostheses, which produce lower volumes of wear debris. Impingement of the femoral prosthesis on the cup leading to loosening led to the narrowing of the neck of the femoral component and the use of cut away angle bore sockets. Concern about access of fluid to the femoral cement–prosthesis interface and subsequently to the cement–bone interface through cement deficiencies, with the production of osteolysis, have led some manufacturers to produce polished femoral stems. These self-locking tapers prevent fluid flowing between the cement and the femoral stem. The debate continues. It can be difficult to separate the improvements in surgical techniques that have improved the clinical results from the effect of modification and changes in materials used to produce joint replacements.

Current Developments

There is currently much interest in reducing the trauma of the surgery itself. These include surgical techniques of minimal incision surgery with the skin wound < 8 cm and a more conservative approach to femoral bone use at the time of surgery. Surface replacement has returned with "better" metal-on-metal surfaces, and a short-stemmed Judet type of metaphyseal fix femoral prosthesis is also available. Hydroxyapatite as a method of component fixation is also gaining popularity. The current short-term results of these techniques are interesting, and may lead to significant benefits especially for the younger patient in the future. However, the proof will only come from long-term clinical results that are not yet available.

JOINT ANATOMY AND ENVIRONMENT

The bearing material in the natural hip joint is articular cartilage, firmly attached to the underlying bone. Articular cartilage is an extremely complex material, consisting of both fluid (interstitial water) and solid (primarily collagen and proteoglycan) phases. Such a biphasic or poroelastic feature determines the time-dependent deformation of articular cartilage, and largely governs the lubrication mechanism of the natural hip joint. The lubricant present in the natural hip joint is synovial fluid, which is similar to blood plasma with hyaluronic acid added and becomes periprosthetic synovial fluid after total hip replacement. Rheological studies of these biological lubricants have shown shear-thinning characteristics, particularly at low shear rates and for the joint fluid taken from diseased or replaced joints (2).

The load experienced in the hip joint during steady-state walking varies both in direction and magnitude. The maximum load can reach five times bodyweight during the stance phase after heel-strike and is largely reduced in the swing phase after the toe-off. On the other hand, the speed is relatively low, particularly in the stance phase and during the motion reversal. However, the hip contact force can be substantially increased under other conditions. For example, the hip contact force has been reported to be 5.8 times bodyweight up a ramp, 6.6 times up and down stairs, and 7.6 times on fast level walking at a speed of $2.01 \text{ m} \cdot \text{s}^{-1}$ (3).

CURRENT BEARING SURFACES

Biomaterials used for current artificial hip joints include UHMWPE, stainless steel, cobalt chromium alloy (CoCr), and ceramics (alumina and zirconia). A number of combinations for the bearing surfaces using these materials have been introduced since 1950s in order to minimize wear and wear particle generation. These can generally be classified as soft-on-hard and hard-on-hard as summarized in Table 1.

Two main parameters that govern the tribology of the articulating surfaces are geometrical and mechanical properties. The geometrical parameters of the bearing surfaces are the diameters of the acetabular cup (D_{cup}) and the femoral head (D_{head}). The size of the hip prosthesis is usually characterized by its diameter, which is important for both clinical and tribological considerations, such as stability, dislocation, and sliding distance. In addition to size, the diametral mismatch or clearance between the cup and the head ($d = D_{\text{cup}} - D_{\text{head}}$) is also important, particularly for hard-on-hard bearing surfaces. From a tribological point of view, these geometric parameters can often be approximated as a single equivalent diameter (D) defined as

$$D = \frac{(D_{\text{head}} D_{\text{cup}})}{d} \qquad (1)$$

Typical values of equivalent diameter used in current hip prostheses are summarized in Table 2. In addition, geometric deviations from perfectly spherical surfaces, such as nonsphericity and surface toughness, are also very

Table 1. Typical Biomaterials and Combinations for the Bearing Surfaces of Current Artificial Hip Joint Replacements

| Femoral Head (Hard) | Acetabular Cup | | | | |
| | Soft | | | Hard | |
	UHMWPE	Cross-linked UHMWPE	Polyurethane	CoCr	Alumina
Stainless Steel	√	√			
CoCr	√	√			
Alumina	√	√	√	√	
Zirconia	√	√	√		√

Table 2. Typical Geometric Parameters of Various Bearing Couples for Artificial Hip Joints[a]

Bearing Couples	Femoral Head Diameter, mm	Diametral Clearance, μm	Equivalent Diameter, m
UHMWPE-on-metal	28 (22–40)	300 (160–1000)	2.6 (1.0–5.0)
Metal-on-metal	28 (28–60)	60 (60–300)	~10
Ceramic-on-ceramic	28 (28–36)	80 (20–80)	~10

[a]See Ref. 4.

important factors in determining the tribological performance of the prosthesis.

The mechanical properties of the bearing surfaces are also important tribological determinants. Typical values of elastic modulus and Poisson's ratio are given in Table 3 for the biomaterials used in artificial hip joints. Other parameters, such as hardness, particularly in the soft-on-hard combinations, are also important, in that the hard surface should be resistant to third-body abrasion to minimize the consequences of polymeric wear.

COUPLED TRIBOLOGICAL AND BIOLOGICAL METHODOLOGY

The vast majority of studies to evaluate the wear performance of hip prostheses have simply measured the volumetric wear rate (6–8). There are very few groups who have also investigated the characteristics of the wear particles generated in *in vitro* simulations (9–11). The cellular response to prosthetic wear particles, and thus the functional biological activity of implant materials, is complex and is dependent not only on the wear volume, but also the mass distribution of particles as a function of size, their concentration, morphology, and chemistry (see the section, Biological Response of Wear Debris).

During the latter 1990s, methods were developed for the isolation and characterization of UHMWPE particles from

Table 3. Typical Mechanical Properties in Terms of Elastic Modulus and Poisson's Ratio of the Bearing Materials for Artificial Hip Joints[a]

Bearing Materials	Elastic Modulus, GPa	Poisson's Ratio
UHMWPE	0.5–1.	0.4
Cross-Linked UHMWPE	0.2–1.2[a]	0.4
Stainless steel	210	0.3
CoCr	230	0.3
Zirconia	210	0.26
Alumina	380	0.26

[a]See Ref. 5.

retrieved tissues and serum lubricants from simulators that allow discrimination between the particles generated in different patient samples and from different types of polyethylene tested *in vitro* (12–18). The basis of this method is to determine the mass distribution of the particles as a function of size. Determination of the number distribution as a function of size fails to discriminate between samples since the vast majority of the number of particles are invariably in the smallest size range detectable by the resolution of the imaging equipment.

In our laboratories we have pioneered cell culture studies with clinically relevant UHMWPE wear particles generated in experimental wear simulation systems operated under aseptic conditions (17–21). These studies have been extended to cell culture studies of clinically relevant metal (22), ceramic (23), and bone cement wear particles (24,25).

By combining volumetric wear determinations in hip joint simulations with experiments to determine the direct biological activity of the particles generated, we have developed novel methodologies to evaluate the functional biocompatibility of different materials used in prosthetic joint bearings. The functional biocompatibility can be used as a preclinical estimate of the *in vivo* performance of the material under test compared to historical materials. We have adopted two different approaches to determining functional biocompatibility. Our choice of method is dependent on the bearing material and the type of prosthesis.

The first approach is indirect, but can be applied to all materials and devices. It utilizes data obtained from the direct culture of UHMWPE wear particles in three different size ranges: 0.1–1, 1–10, and >10 μm at different volumetric concentrations with human peripheral blood macrophages. Measurements of the biological activity for unit volumes of particles in the different size ranges are generated (20). The use of TNF-α as a determinant is justified since, in our experience the major cytokines concerned in osteolysis (TNF-α, IL-1, IL-6, GM-csf) all show the same pattern of response to clinically relevant wear particles (19–21). By using our methods to determine the volumetric concentration of particles generated in simulations as a function of size (see above), it is then possible to integrate the volume concentration and biological activity function to produce a relative index of specific biological activity (SBA) per unit volume of wear. The functional biological activity (FBA) has been defined as the product of volumetric wear and SBA (26). This has allowed us to compare the functional biological activity of different types of PE in hip joint simulators (27) and different types of bearing materials (23).

The second approach is to directly culture wear debris from wear simulators with primary macrophages. For metal and ceramic particles, we can directly culture wear particles from standard simulation systems after isolation, sterilisation, and removal of endotoxin by heat treatments (23). However, for PE this is not feasible since the heat treatment at elevated temperature required to remove endotoxin cannot be applied. For these materials, we have developed a sterile endotoxin free multidirectional wear simulator in which wear particles are generated in macrophage tissue culture medium. While this does not test whole joints, it allows the application of different kinematics to represent the hip and the knee. The advantage of this approach is that all the wear products are directly cultured with the cells, and there is no risk of modification during the isolation procedure. This method has recently been used to compare the biological reactivity of particles from PEs of different molecular weights and different levels of cross-linking. Higher molecular weight of GUR 1050 and higher levels of cross-linking of both GUR 1020 and 1050 produced particles that were more biologically reactive (18).

Tribology of Bearing Surfaces

Tribological studies of the bearing surfaces of artificial hip joints include friction, wear, and lubrication, which have been shown to mainly depend on the lubrication regimes involved. There are three lubrication regimes: boundary, fluid-film, and mixed. In the boundary lubrication regime, a significant asperity contact is experienced, and consequently both friction and wear are high. In the fluid film lubrication regime, where the two bearing surfaces are completely separated by a continuous lubricant, minimal friction and wear is expected. The mixed-lubrication regime consists of both fluid film lubricated and boundary contact regions. Friction and lubrication studies are usually performed to understand the wear mechanism involved in artificial hip joints. However, friction forces may be important in determining the stresses experienced at the interface between the implant and the cement bone (28) as well as temperature rise (29).

Friction in artificial hip joints is usually measured in a pendulum-like simulator with a dynamic load in the vertical direction and a reciprocating rotation in the horizontal direction. The coefficient of friction is usually expressed as a friction factor defined as

$$\mu = \frac{T}{w(d_{head}/2)} \qquad (2)$$

where T is the measured friction torque and w is the load.

The measured coefficient of friction in a particular hip prosthesis itself can generally reveal the nature of the lubrication regime, since each mechanism is associated with broad ranges of the coefficient of friction. The variation in the coefficient of friction against a Sommerfeld number defined as, $S = (\eta u d_{head}/w)$, where is viscosity and u velocity, can further indicate the lubrication regime. If the measured friction factors remain constant, fall or increase as the Sommerfeld number is increased, the associated modes of lubrication are boundary, mixed, or fluid-film, respectively (30).

Lubrication studies of artificial hip joints are generally carried out using both experimental and theoretical approaches. The experimental measurement is usually involved with the detection of the separation between the two bearing surfaces using a simple resistivity technique. A large resistance would imply a thick lubricant film, while a small resistance is attributed to the direct surface contact. Such a technique is directly applicable to metal-on-metal bearings as well as UHMWPE-on-metal and ceramic-on-ceramic bearings if appropriate coatings are used (31,32). The theoretical analysis is generally involved with the solution to the Reynolds equation, together with the elasticity equation subjected to the dynamic load and speed experienced during walking. The predicted film thickness (h_{min}) is then compared with the average surface roughness (Ra) using the following simple criterion.

$$\lambda = \frac{h_{min}}{[Ra_{head}^2 + Ra_{cup}^2]^{1/2}} \qquad (3)$$

The lubrication regime is then classified as fluid film, mixed, or boundary if the predicted ratio is > 3, between 1 and 3, or < 1, respectively.

Wear of artificial hip joints has been investigated extensively, due to its direct relevance to biological reactions and clinical problems of osteolysis and loosening. Volumetric wear and wear particles can be measured using the following machines, among others:

- Pin-on-disk machines.
- Pin-on-plate machines.
- Joint simulators.

A unidirectional sliding motion is usually used in the pin-on-disc machine, and the reciprocating motion is added to the pin-on-plate machine. Both of these machines are used to screen potential bearing materials under well controlled, and often simplified conditions. Generally, it is necessary to introduce additional motion in order to produce a multidirectional motion. The next stage of wear testing is usually carried out in joint simulators with a varied degree of complexity of the 3D loading and motion patterns experienced by hip joints, while immersing the test joints in a lubricant deemed to be physically and chemically similar to synovial fluid. Wear can be evaluated by either dimensional or gravimetric means.

Contact mechanics analysis is often performed to predict the contact stresses within the prosthetic components and to compare with the strength of the material. However, other predicted contact parameters such as the contact area and the contact pressure at the bearing surfaces have been found to be particularly useful in providing insights into friction, wear, and lubrication mechanisms. Contact mechanics can be investigated either experimentally using pressure-sensitive film and sensors, or theoretically using the finite element method.

Biological Response of Wear Debris

Our current understanding of the mechanisms of wear particle-induced osteolysis has developed from > 30 years experience with UHMWPE-on-metal. The major factor limiting the longevity of initially well-fixed UHMWPE total joint replacements is osteolysis resulting in late aseptic loosening (33). There is extremely strong evidence from *in vivo* and *in vitro* studies that osteolysis is a UHMWPE particle related phenomenon.

Following total hip arthroplasty, a pseudocapsule forms around the joint and this may have a pseudosynovial lining. A fibrous interfacial tissue may also form at the bone–cement or bone–prosthesis interface that is normally thin with few vessels or cells (34–36). At revision surgery for aseptic loosening, the fibrous membrane is thickened, highly vascularized, and contains a heavy infiltrate of UHMWPE-laden macrophages and multinucleated giant cells (37,38). There is a correlation between the number of macrophages and the volume of UHMWPE wear debris in the tissues adjacent to areas of aggressive osteolysis (39–45). Analyses of interfacial membranes have demonstrated the presence of a multitude of mediators of inflammation including cytokines that may directly influence osteoclastic bone resorption:-TNF-α (46), IL-1β (47), IL-6 (48), and M-CSF (49). There is a direct relationship between the particle concentration and the duration the implant, and there are billions of particles generated per gram of tissue (9,15,50,51). Osteolysis is likely to occur when the threshold of particles exceeds 1×10^{10}/g of tissue (45). Each milligram of PE wear has been estimated to generate 1.3×10^{10} particles (15).

The UHMWPE particles isolated from retrieved tissues vary in size and morphology, from large platelet-like particles, up to 250 μm in length, fibrils, shreds, and sub-micrometer globule-shaped spheroids 0.1–0.5 μm in diameter (15,52–54). The vast majority of the numbers of particles are the globular spheroids and the mode of the frequency distribution is invariably 0.1–0.5 μm, although the larger particles may account for a high proportion of the total volume of wear debris. Analysis of the mass distribution as a function of size is therefore necessary to discriminate between patient samples (15,55).

UHMWPE wear particles generated *in vitro* in hip joint simulators have a larger proportion of the mass of particles in the 0.01–1 μm sized range than those isolated from periprosthetic tissues (27,55). This may indicate that *in vivo*, the smaller particles are disseminated more widely away from the implant site. Recently, improvements to particle imaging techniques have revealed nanometer sized UHMWPE particles generated in hip joint simulators. These particles have yet to be identified *in vivo*. These nanometer size particles account for the greatest number of particles generated, but a negligible proportion of the total volume (18).

Studies of the response of macrophages to clinically relevant, endotoxin-free polyethylene particles *in vitro* have clearly demonstrated that particle stimulated macrophages elaborate a range of potentially osteolytic mediators (IL-1, IL-6, TNF-α, GM-CSF, PGE$_2$) and bone resorbing activity (19–21,56–58). Induction of bone resorbing activity in particle stimulated macrophage supernatants has been shown to be critically dependent on particle size and concentration with particles in the 0.1–1.0 μm size range at a volumetric concentration of 10–100 μm^3/cell being the most biologically reactive (19,56). The importance of UHMWPE particle size has also been demonstrated in animal studies (59). These findings have enabled the preclinical prediction of the functional biological activity of different polyethylenes by analysis of the wear rate and mass distribution of the particles as a function of particle size (26,27). For a review of the biology of osteolysis, the reader is referred to Ingham and Fisher (60).

In metal-on-metal bearings in the hip, an abundance of small nanometer size particles are generated (61,62). It is believed that the majority of metal debris is transported away from the periprosthetic tissue. While only isolated instances of local osteolysis have been found around metal-on-metal hips, this is most commonly associated with high concentrations of metal debris and tissue necrosis. *In vitro* cell culture studies have shown that these nanometer size metal particles are highly toxic to cells at relatively low concentrations (22). These particles have a very limited capacity to activate macrophages to produce osteolytic cytokines at the volumes likely to be generated *in vivo* (63), however, metal particles are not bioinert and concerns exist regarding their potential genotoxicity (22).

Ceramic-on-ceramic prostheses have been shown to have extremely low wear rates. Ceramic wear particles generated in hip joint simulations under clinically relevant conditions in the hip joint simulator (64) and *in vivo* (65) have a bimodal size distribution with nanometer sized (5–20 nm) and larger particles (0.2–> 10 μm). Alumina ceramic particles have been shown to be capable of inducing osteolytic cytokine production by human mononuclear phagocytes *in vitro* (23). However, the volumetric concentration of the particles needed to generate this response was 100–500 μm^3/cell. Given the extremely low wear rates of modern ceramic-on-ceramic bearings, even under severe conditions, it is unlikely that this concentration will arise in the periprosthetic tissues *in vivo* (60).

APPLICATIONS

UHMWPE-on-Metal and UHMWPE-on-Ceramic

The friction in UHMWPE hip joints has been measured using a pendulum-type simulator with a flexionsol–extension motion and a dynamic vertical load. The friction factor has been found to be generally in the range 0.02–0.06 for 28 mm diameter metal heads and UHMWPE cups (66), broadly representative of mixed lubrication, and this has been confirmed from the variation in the friction factor with the Sommerfeld number. These experimental observations are broadly consistent with the theoretical prediction of typical lubricant film thicknesses between 0.1 and

Table 4. Volumetric Wear Rate, % wear volume <1 μm, SBA, and FBA for Nonirradiated and Irradiated UHMWPEs and Alumina Ceramic-on-Ceramic Hip Joint Prostheses[a]

Material	Volumetric Wear rate, mm^3/10^6 cycles $\pm$ 95% CL	% Volume <1 μm	SBA	FBA
Nonirradiated UHMWPE	50 ± 8	23	0.32	16
Gamma in air UHMWPE, 2.5 Mrad GUR1120	49 ± 9	46	0.55	55
Stabilized UHMWPE (2.5–4 Mrad) GUR1020	35 ± 9	43	0.5	17.5
Highly cross-linked UHMWPE, 10 Mrad GUR1050	8.6 ± 3.1	95	0.96	8
Alumina ceramic-on-ceramic (microseparation)	1.84 ± 0.38	100	0.19	0.35

[a]See Refs. 60,69.

0.2 μm and the average surface roughness of UHMWPE bearing surface between 0.1 and 1 μm. Therefore, wear of UHMWPE acetabular cups is largely governed by the boundary lubrication mechanism. An increase in the femoral head diameter can lead to an increase in sliding distance and consequently wear (41). As a result, 28 mm diameter femoral heads appear to be a better choice. Furthermore, reducing the surface roughness of the metallic femoral head or using harder alumina to resist third-body abrasion and to maintain the smoothness is also very important. For example, the wear factor in UHMWPE-on-ceramic implants is generally 50% of that in UHMPWE-on-metal (67). The introduction of cross-linked UHMWPE has been shown to reduce wear significantly in simulator studies. However, the degree of wear reduction appears to depend on cross-linking, kinematics, counterface roughness, and bovine serum concentration (68). It should be pointed out that volumetric changes are often accompanied by morphology changes, which may have different biological reactions as discussed below.

First, let us consider the effect of irradiation and cross-linking on the osteolytic potential of UHMWPE bearings. Historically, UHMWPE acetabular cups were gamma irradiated in air until it became clear that oxidative degeneration of the PE was occurring. This oxidative damage was caused by the release of free radicals, which produced strand scission of the long PE chains. Research has indicated that deterioration to important mechanical properties such as tensile strength, impact strength, toughness, fatigue strength, and Young's modulus occurs (12). These time-dependent changes have been shown to affect the volumetric wear of the UHMWPE and typical values are in the region of 100 mm^3/million cycles. In addition, UHMWPE that had been gamma irradiated in air produced a greater volumetric concentration of wear particles that were in the most biologically active size range, 0.1–1 μm (46% of the wear volume compared to 24% for nonirradiated UHMWPE). When the specific biological activity (biological activity per unit volume of wear; SBA) of the wear particles was calculated this gave an SBA that was 1.7-fold higher than the SBA of the nonirradiated material, which translated into a functional biological activity (FBA), which was 3.5-fold higher than the FBA of the nonirradiated material (Table 4).

Currently, UHMWPE is sterilized by gamma irradiation (2.5–4 Mrad) in an inert atmosphere. This material undergoes partial cross-linking as a result of this processing, and is often referred to as moderately cross-linked or stabilized PE. This material produces lower wear rates than the nonirradiated UHMWPE, but has a higher

volumetric concentration of wear particles <1 μm compared to the nonirradiated material as shown in Table 4 (69). Consequently, the specific biological activity of the wear particles is higher at 0.5 compared to 0.32 for the nonirradiated material. However, as the wear volume is substantially lower, the FBA value for the stabilized UHMWPE is very similar to the nonirradiated material.

As the level of cross-linking increases, the wear volume decreases (69). The highly cross-linked UHMWPE is GUR 1050, irradiated at 10 Mrad and remelted, and has very low wear volumes at 8.6 ± 3.1 mm^3/million cycles. However, as can be seen from Table 4, 95% of the wear volume is comprised of particles in the most biologically active size range, leading to an extremely high SBA. However, as the wear volume is significantly lower than the other UHMWPEs, the FBA is one-half of those of the nonirradiated and stabilized materials (Table 4).

In addition, the wear particles from the cross-linked materials have increased biological activity per unit volume of wear (Fig. 2). A recent study by Ingram et al. (18) has shown that when worn against a scratched counterface, PE irradiated with 5 and 10 Mrad of gamma irradiation produced higher volumetric concentrations of wear particles in the 0.01–1.0 μm size range compared to noncross-linked material. This increased volumetric concentration of wear particles in the 0.01–1.0 μm size range meant that both cross-linked materials were able to stimulate the release of elevated levels of TNF-α, an osteolytic cytokine, at a 10-fold lower volumetric dose than the

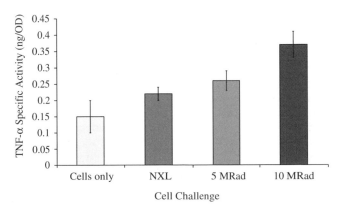

Figure 2. TNF-α release (specific activity $\pm$ 95% confidence limits) as a result of challenge with UHMWPE particles, which were noncross-linked (NXL), cross-linked with 5 Mrad of irradiation, cross-linked with 10 Mrad of irradiation compared to the cell only control.

noncross-linked polyethylene (0.1 μm³ debris/cell compared to 1–10 μm³ debris/cell). So, while the cross-linked materials produced lower wear volumes, the particles produced from these materials were more reactive compared to the noncross-linked PE.

However, when the same materials were worn against a smooth counterface, analysis of the wear particles showed that both cross-linked and noncross-linked PE produced very high numbers of nanometer-sized wear particles. In addition, the cross-linked and noncross-linked materials produced similar low volumes of particles in the 0.1–1.0 μm size range, which resulted in wear debris that was only stimulatory a the highest volumetric dose of 50 μm³ debris/cell. This offers further explanation as to why the FBA or osteolytic potential of the highly cross-linked polyethylene's are lower than the moderately cross-linked and noncross-linked materials (Table 4).

Metal-on-Metal

The friction factor measured in metal-on-metal hip joints with different sizes and clearances in simple pendulum type machines is generally much higher than for UHMWPE-on-metal articulations, in the range between 0.1 and 0.2, indicating a mixed-boundary lubrication regime (66). However, the lubrication regime in metal-on-metal bearings has been shown to be sensitive to the surface roughness, loading and velocity, and design parameters (70–74). Consequently, different friction factors or wear factors are possible. Therefore, it is important to optimize the bearing system, in terms of the femoral head diameter, the clearance and the structural support (75,76). From a lubrication point of view, the femoral head diameter is the most important geometric parameter, since it is directly related to both the equivalent diameter defined in Eq. 1 and the sliding velocity (70). If the lubrication improvement is such that a fluid-film dominant lubrication regime is present, the increase in the sliding distance becomes unimportant. Such an advantage has been utilized in large-diameter metal-on-metal hip resurfacing prostheses (77). However, it should be pointed out that the lubrication improvement in large-diameter metal-on-metal hip resurfacing prostheses can only be realized with adequate clearances (78). A too large clearance can reduce the equivalent diameter, shifting the lubrication regime toward mixed–boundary regions. In addition, the increase in the sliding distance associated with the large diameter means that the bedding-in wear becomes important. The wear in optimised systems can be quite low, of the order of a few millimeters cubed.

The biological activity in terms of osteolytic potential of metal-on-metal hip prostheses is difficult to define. If macrophages and fibroblasts are challenged with clinically relevant cobalt chrome wear particles, there is some release of the osteolytic cytokine TNF-α (Fig. 3), however, this only takes place at very high levels of particulate load (50 μm³ debris/cell), and the level of cytokine produced is at lower levels compared to UHMWPE particles [see Fig. 2(79)]. The predominant biological reaction is cytotoxicity or a reduction in cell viability (Fig. 4). Macrophage and fibroblast cell viability is significantly reduced when

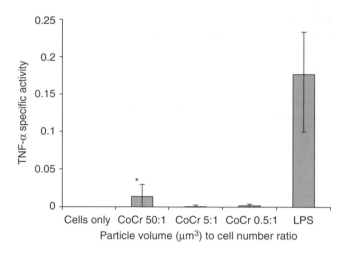

Figure 3. TNF-α production by human peripheral blood mononuclear cells stimulated with clinically relevant cobalt–chromium particles.

challenged with 50 or 5 μm³ debris/cell (22). The specific biological activity of metal wear particles is difficult to assess as the cells may release cytokines, such as TNF-α as a consequence of cell death. In addition, the high levels of particulate load required to stimulate cytokine release

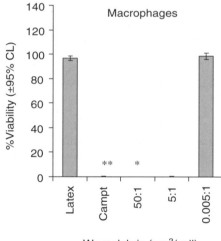

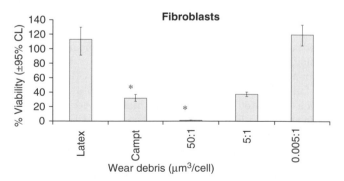

Figure 4. Effect of metal-on-metal cobalt–chromium wear particles on macrophage and fibroblast cell viability (*Significant ($p < 0.05$, ANOVA) reduction in cell viability).

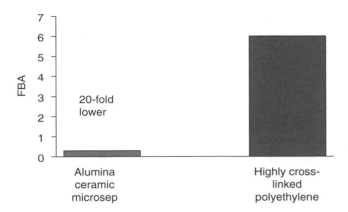

Figure 5. Predicted functional biological activity or osteolytic potential of alumina ceramic-on-ceramic and highly cross-linked UHMWPE-on-metal hip prostheses.

may only be achieved *in vivo* if a pairing is particularly high wearing.

Ceramic-on-Ceramic

The friction factor in ceramic-on-ceramic hip implants is quite low, particularly when the nonbiological type lubricant is used (66). However, when biological lubricants such as bovine serum are tested, the friction factor can be quite high due to the complex interactions with proteins. The wear under normal ideal conditions is low, but can be increased substantially under adverse conditions such as microseparation (80). Despite this, the wear in ceramic-on-ceramic hip implants is generally the lowest among current hip prostheses available clinically.

The introduction of microseparation conditions into the *in vitro* simulation model replicates clinically relevant wear rates, wear patterns, and wear particles. Alumina wear particles have a lower biological activity than UHMWPE particles. A 10-fold higher concentration of alumina wear particles is required to stimulate the release of osteolytic cytokine TNF-α from macrophages compared to UHMWPE wear particles (23). It is questionable whether the volume of alumina wear particles will reach this threshold *in vivo* given the extremely low wear rates of ceramic-on-ceramic prostheses even under severe microseparation conditions. Consequently, alumina wear particles have a lower specific biological activity than UHMWPE particles (Table 4). When this lower SBA is integrated with the comparatively small volumetric wear rates that are produced by ceramic-on-ceramic couples compared to metal-on-polyethylene, a substantially lower functional biological activity or osteolytic potential is pro-

duced (Table 4; Fig. 5). In fact, alumina ceramic-on-ceramic couples produce a 20-fold lower FBA than the currently used highly cross-linked UHMWPEs.

Summary

Typical values of friction factor, wear factor, and biological reactions are summarized in Tables 5, 6, and 7 for various hip implants with different bearing couples.

FUTURE DEVELOPMENTS

Cross-Linked Polyethylene

The introduction of highly cross-linked PE into clinical use in the last 5 years has been extensive. Standard UHMWPE is irradiated at high dose levels (5–10 Mrad), which produces chain scission and cross-linking between the molecular chains. Subsequent heating and remelting recombines the free radicals producing a more stable material (81). The additional cross-links provide improved wear resistance, particularly during kinematic conditions with high levels of cross-shear as found in the hip. A number of early simulator studies showed no wear for these materials (81), while other studies demonstrated measurable wear (82). Initial clinical studies, however, do show penetration and wear (83). Wear and surface cracking has been identified in a few isolated retrievals. The wear rates reported more recently in simulator studies have been found to be in the range 5–10 mm³/million cycles, which is four to five times less than with conventional material (69). Recent work has also shown that cross-linked PE produces a greater proportion of smaller particles, per unit volume of wear debris and has been found to be up to three times more biologically active than conventional material (18). This leads to a functional reduction in osteolytic potential of about twofold compared to conventional PE. This improvement in functional osteolytic potential may not be sufficient for high demand patients, and in patients who require large head sizes. In these patients, larger diameter hard-on-hard, such as ceramic-on-ceramic or metal-on-metal, may be a more appropriate bearing choice.

Ceramic-on-Metal Bearing

Currently used hard-on-hard bearings are comprised of similar materials, such as alumina ceramic-on-ceramic or metal-on-metal. When lubrication conditions are depleted, like bearing materials can produce elevated adhesive friction and wear. The ceramic-on-metal hip was developed to produce a deferential hardness hard bearing (84). Laboratory simulation studies have shown a reduction in wear of

Table 5. Typical Friction Factors and Lubrication Regimes in Various Bearings for Hip Implants[a]

Bearing Couples	Friction Factor	Variation of Friction Factor Against Increasing Sommerfeld Number	Indicated Lubrication Regimes
UHMWPE-on-Metal	0.06–0.08	Constant/decreasing	Boundary/mixed
Metal-on-metal	0.22–0.27	Decreasing	Mixed
Ceramic-on-ceramic	0.002–0.2	Increasing	Fluid-film/mixed

[a]See Ref. 4.

Table 6. Typical Volumetric and Linear Wear Rates for Different Bearings for Hip Implants[a]

Bearing Couples	Volumetric Wear Rate, mm³/million cycles	Linear Wear Rate, μm/million cycles
UHMWPE-on-metal	30–100	100–300
UHMWPE-on-ceramic	15–50	50–150
Metal-on-metal	0.1–1	2–20
Ceramic-on-ceramic	0.05–1	1–20

[a]See Ref. 4.

Table 7. Typical Particle Sizes and Biological Responses in Different Bearings for Hip Implants[a]

Bearing Couples	Dominant Particle Diameters, μm	Biological Responses
UHMWPE-on-metal/ceramic	UHMWPE, 0.01–1	Macrophages/osteoclasts/osteolysis
Metal-on-metal	Metallic, 0.02–0.1	Low osteolysis, cytotoxicity
Ceramic-on-ceramic	Ceramic, 0.01–0.02	Bioinert, low cytotoxicity
	Ceramic, 0.1–10	Macrophages/osteoclasts/osteolysis

[a]See Ref. 69.

up to 100-fold compared to metal on metal. The ceramic head does not wear and remains smooth, improving lubrication and reducing wear of the metallic cup. This new concept is currently entering clinical trials.

Surface Replacement Bearings

There is considerable interest in surface replacement solutions in the hip (85). In this approach, a large diameter metallic shell is placed over the reamed femoral head, preserving femoral bone stock, and this articulates against a large diameter acetabular cup. Both cobalt chrome cast and wrought alloys have been used in different bearing designs. The larger diameter head improves lubrication and reduces wear compared to smaller head sizes (70). However, it is important to maintain a low radical clearance between the components to ensure low bedding-in wear (78,86,87). Surface replacement metal on metal bearings are not suitable for all patients, due to the nature of the femoral bone, but are currently used in ∼ 10% of patients receiving hip prostheses.

Surface Engineered Metal-on-Metal Bearings SUREHIP

Concerns still remain about wear particles in metal on metal bearings and elevated metal ion levels. Surface engineering solutions are an attractive option for reducing wear and metal ion levels, and can be readily applied to surface replacement hips. Recent research with thick AEPVD chromium nitride and chromium carbon nitride surface engineered coatings of thicknesses between 8 and 12 μm have shown a 30-fold reduction in wear and metal ion levels (88,89). These coatings are now undergoing product development in preparation for clinical studies.

Compliant Materials, Cushion Form Bearings

In recent years, the trend has been to move toward harder bearing materials that wear less, and away from the lower elastic modulus properties of articular cartilage. Compliant materials such as polyurethane have been investigated as bearing materials in acetabular cups. The cups have been formed as a composite structure with a higher modulus substrate to give structural support (90). The bearing has

shown improved lubrication and reduced wear compared to conventional polyethylene bearings. However, concerns remain about the long-term stability of these low modulus materials. More recently an experimental polyurethane surface replacement cup has been investigated (91).

Hemiarthroplasty and Hemisurface Replacements

Interest in more conservative bone preserving, and minimally invasive surgery has generated renewed interest in surgical treatments that replace only one side of the diseased joint or potentially just the diseased surface itself. In these scenarios, consideration has not only to be given to the biomaterial replacing the degenerated tissue, but also the function of the apposing articulating surface.

In the hip hemiarthroplasty using compliant materials has just entered clinical trials, where the femoral head covered with a layer of polyurethane articulates against the natural cartilage in the acetabulum (http://www.impliant.com/home/index.html). Future designs will focus on full or partial replacement of the diseased cartilage on one side of the joint.

SUMMARY

This article summarizes the biotribology of artificial hip joints and development over the last four decades. Although adequate solutions exist for the elderly less active patients > 65 years old with average life expectances < 20 years, considerable technological advances are required to meet the demands and improved performance of younger patients. Recent moves toward large diameter heads to give greater function, stability and range of motion are placing greater demands on tribological performances and increasing use of the hard-on-hard bearings.

Nomenclature

d Diametral clearance

D Dearing diameter or equivalent diameter defined in Eq. 1

FBA   Functional biological activity

h_{min} Minimum lubricant film thickness

PMMA   Poly(methyl methacrylate)

Ra   Average surface roughness

S Summerfeld number

SBA   Specific biological activity

T Frictional torque

u Siding velocity

UHMWPE   Ultrahigh molecular weight polyethylene

w Load

η Viscosity

λ Ratio defined in Eq.3

μ Frictional factor defined in Eq. 2

Subscripts:

Head   Femoral head

Cup   Acetabular cup

BIBLIOGRAPHY

Cited References

1. Malchau H, Herberts P, Ahnfelt L. Prognosis of total hip replacement in Sweden. Acta Orthop Scand 1993;64–65.
2. Yao JQ, Laurent MP, Johnson TS, Blanchard CR, Crownshield RD. The influence of lubricant and material on polymer/CoCr sliding friction. Wear 2003;255:780–784.
3. Paul JP. Strength requirements for internal and external prostheses. J Biomech 1999;32(4):381–393.
4. Jin ZM, Medley JB, Dowson D. Fluid Film Lubrication In Artificial Hip Joints. Proc 29th Leeds-Lyon Symp Tribology; 2003. p 237–256.
5. Lewis G. Properties of crosslinked ultra-high-molecular-weight polyethylene. Biomaterials 2001;22(4):371–401.
6. Chiesa R, Tanzi MC, Alfonsi S, Paracchini L, Moscatelli M, Cigada A. Enhanced wear performance of highly cross-linked UHMWPE for artificial joints. J Biomed Mat Res 2000;50:381–387.
7. Bowsher JG, Shelton JC. A hip simulator study of the influence of patient activity level on the wear of cross-linked polyethylene under smooth and roughened counterface conditions. Wear 2001;250:167–179.
8. Hermida JC, Bergula A, Chen P, Colwell CW, D'Lima DD. Comparison of wear rates of twenty-eight and thirty-two millimeter femoral heads on cross-linked polyethylene acetabular cups in a wear simulator. J Bone Joint Surg 2003;85A:2325–2331.
9. McKellop HA, Campbell P, Park SH, Schmalzried TP, Grigoris P, Amstutz HC, Sarmiento A. The origin of submicron polyethylene wear debris in total hip arthroplasty. Clin Orthopaed Rel Res 1995;311:3–20.
10. Tipper JL, Ingham E, Fisher J. Characterisation of wear debris from UHMWPE, metal on metal and ceramic on ceramic hip prostheses. Wear 2001;250:120–128.
11. Saikko V, Calonius O, Keranen J. Wear of conventional and cross-linked ultra-high molecular -weight polyethylene acetabular cups against polished and roughened CoCr femoral heads in a biaxial hip simulator. J Biomed Res Appl Biomat 2002;63:848–853.
12. Besong AA, Tipper JL, Ingham E, Stone MH, Wroblewski BM, Fisher J. Quantitative comparison of wear debris from UHMWPE that has and has not been sterilized by gamma irradiation. J Bone Joint Sur 1998;80B:340–344.
13. Endo MM, Barbour PSM, Barton DC, Wroblewski BM, Fisher J, Tipper JL, Ingham E, Stone MH. A comparison of the wear and debris generation of GUR 1120 (compression moulded) and GUR 4150HP (ram extruded) ultra high molecular weight polyethylene. Bio Med Mat Eng 1999;9:113–124.
14. Endo MM, Barbour PS, Barton DC, Fisher J, Tipper JL, Ingham E, Stone MH. Comparative wear and wear debris under three different counterface conditions of cross-linked and non-cross-linked ultra high molecular weight polyethylene. Bio Med Mat Eng 2001;11:23–35.
15. Tipper JL, Ingham E, Hailey JL, Besong AA, Fisher J, Wroblewski BM, Stone MH. Quantitative analysis of polyethylene wear debris, wear rate and head damage in retrieved Charnley hip prostheses. J Mat Sci, Mat Med 2000;11:117–124.
16. Bell J, Besong AA, Tipper JL, Ingham E, Wroblewski BM, Stone MH, Fisher J. Influence of gelatin and bovine serum lubrication on ultra-high molecular weight polyethylene wear debris generated in in vitro simulations. Proc Inst Mech Eng J Eng Med 2000;214H:513–518.
17. Ingram J, Matthews JB, Tipper JL, Stone MH, Fisher J, Ingham E. Comparison of the biological activity of grade GUR 1120 and GUR 415 HP UHMWPE wear debris. Bio Med Mat Eng 2002;12:177–188.
18. Ingram JH, Stone MH, Fisher J, Ingham E. The influence of molecular weight, crosslinking and counterface roughness on TNF-alpha production by macrophages in response to ultra high molecular weight polyethylene particles. Biomaterials 2004;25:3511–3522.
19. Green TR, Fisher J, Matthews JB, Stone MH, Ingham E. Effect of size and dose on bone resorption activity of macrophages in vitro by clinically relevant ultra high molecular weight polyethylene particles. J Biomed Mat Res Appl Biomat 2000;53:490–497.
20. Matthews JB, Green TR, Stone MH, Wroblewski BM, Fisher J, Ingham E. Comparison of the response of primary human peripheral blood mononuclear phagocytes from different donors to challenge with polyethylene particles of known size and dose. Biomaterials 2000;21:2033–2044.
21. Matthews JB, Stone MH, Wroblewski BM, Fisher J, Ingham E. Evaluation of the response of primary human peripheral blood mononuclear phagocytes challenged with in vitro generated clinically relevant UHMWPE particles of known size and dose. J Biomed Mat Res Appl Biomat 2000;44:296–307.
22. Germain MA, Hatton A, Williams S, Matthews JB, Stone MH, Fisher J, Ingham E. Comparison of the cytotoxicity of clinically relevant cobalt-chromium and alumina ceramic wear particles in vitro. Biomaterials 2003;24:469–479.
23. Hatton A, Nevelos JE, Matthews JB, Fisher J, Ingham E. Effects of clinically relevant alumina ceramic wear particles on TNF-α production by human peripheral blood mononuclear phagocytes. Biomaterials 2003;24:1193–1204.
24. Ingham E, Green TR, Stone MH, Kowalski R, Watkins N, Fisher J. Production of TNF-α and bone resorbing activity by macrophages in response to different types of bone cement particles. Biomaterials 2000;21:1005–1013.
25. Mitchell W, Matthews JB, Stone MH, Fisher J, Ingham E. Comparison of the response of human peripheral blood mononuclear cells to challenge with particles of three bone cements in vitro. Biomaterials 2003;24:737–748.

26. Fisher J, Bell J, Barbour PSM, Tipper JL, Matthews JB, Besong AA, Stone MH, Ingham E. A novel method for the prediction of functional biological activity of polyethylene wear debris. J Eng Med Proc Inst Mech Eng 2001; 215H: 127–132.

27. Endo MM, Tipper JL, Barton DC, Stone MH, Ingham E, Fisher J. Comparison of wear, wear debris and functional biological activity of moderately crosslinked and non-cross-linked polyethylenes in hip prostheses. Proc Instn Mech Eng J Eng Med 2002;216:111–122.

28. Nassutt R, Wimmer MA, Schneider E, Morlock MM. The influence of resting periods on friction in the artificial hip. Clin Orthop 2003;407:127–38.

29. Bergmann G, Graichen F, Rohlmann A, Verdonschot N, van Lenthe GH. Frictional heating of total hip implants. Part 2: finite element study. J Biomech 2001;34(4):429–435.

30. Dowson D. New joints for the Millennium: wear control in total replacement hip joints. Proc Instn Mech Eng J Eng Med 2001;215(H4):335–358.

31. Smith SL, Dowson D, Goldsmith AJ, Valizadeh R, Colligon JS. Direct evidence of lubrication in ceramic-on-ceramic total hip replacements. Proc Inst Mech Eng J Mech Eng Sci 2001;215(3):265–268.

32. Murakami T, Sawae Y, Nakashima K, Sakai N, Doi S, Sawano T, Ono M, Yamamoto K, Takahara A. Roles of Materials and Lubricants in Joint Prostheses. Proc 4th Int Biotribol Forum 24th Biotribol Symp 2003;1–4.

33. Archibeck MJ, Jacobs JJ, Roebuck KA, Glant TT. The basic science of periprosthetic osteolysis. J Bone Joint Surg 2000;82A:1478–1497.

34. Goldring SR, Schiller AL, Roelke M, Rourke CM, O'Neil DA, Harris WH. The synovial-like membrane at the bone-cement interface in loose total hip replacements and its proposed role in bone lysis. J Bone Joint Surg 1983;65A:575–584.

35. Goodman SB, Chin RC, Chou SS, Schurman DJ, Woolson ST, Masada MP. A clinical-pathologic-biochemical study of the membrane surrounding loosened and non-loosened total hip arthroplasties. Clin Orthopaed 1989;244:182–187.

36. Bullough PG, DiCarlo EF, Hansraj KK, Neves MC. Pathologic studies of total joint replacement. Orthopaed Clin North Am 1988;19:611–625.

37. Mirra JM, Marder RA, Amstutz HC. The pathology of failed total joint arthroplasty. Clin Orthopaed Rel Res 1982; 170:175–183.

38. Willert HG, Buchhorn GH. Particle disease due to wear of ultrahigh molecular weight polyethylene. Findings from retrieval studies. In: Morrey BF, editor. Biological, Material, and Mechanical Considerations of Joint Replacement. New York: Raven Press; 1993.

39. Maloney WJ, Jasty M, Harris WH, Galante MD, Callaghan JJ. Endosteal erosion in association with stable uncemented femoral components. J Bone Joint Surg 1990;72A:1025–1034.

40. Santavirta S, Holkka V, Eskola A, Kontinen YT, Paavilainen T, Tallroth K. Aggressive granulomatous lesions in cementless total hip arthroplasty. J Bone Joint Surg 1990;72B:980–985.

41. Livermore J, Ilstrup D, Morrey B. Effect of femoral head size on the wear of the polyethylene acetabular component. J Bone Joint Surg 1990;72A:518–528.

42. Schmalzried TP, Jasty M, Harris WH. Periprosthetic bone loss in total hip arthroplasty: polyethylene wear debris and the concept of the effective joint space. J Bone Joint Surg 1992;74A:849–863.

43. Howie AW, Haynes DR, Rogers SD, McGee MA, Pearcey MJ. The response to particulate debris. Orthopaed Clinics North Am 1993;24:571–581.

44. Bobyn JD, Jacobs JJ, Tanzer M, Urban RM, Arbindi R, Summer DR, Turner TM, Brooks CE. The susceptibility of smooth implant surfaces to peri-implant fibrosis and migration of polyethylene wear debris. Clin Orthopaed Rel Res 1995;311:21–39.

45. Revell PA. Biological reaction to debris in relation to joint prostheses. Proc Inst Mech Eng J Eng Med 1997;211:187–197.

46. Xu JW, Kotinnen T, Lassus J, Natah S, Ceponis A, Solovieva S, Aspenberg P, Santavirta S. Tumor necrosis factor-alpha (TNF-α) in loosening of total hip replacement (THR). Clin Exp Rheumatol 1996;14:643–648.

47. Kim KJ, Rubash H, Wilson SC, D'Antonio JA, McClain EJ. A histologic and biochemical comparison of the interface tissues in cementless and cemented hip prostheses. Clin Orthopaed Rel Res 1993;287:142–152.

48. Sabokbar A, Rushton Role of inflammatory mediators and adhesion molecules in the pathogenesis of aseptic loosening in total hip arthroplasties. J Arthropl 1995;10:810–815.

49. Takei I, Takagi M, Ida H, Ogino S, Santavirta, Konttinen YT. High macrophage-colony stimulating factor levels in synovial fluid of loose artificial hip joints. J Rheumatol 2000;27:894–899.

50. Campbell P, Ma S, Yeom B, McKellop H, Schmalzried TP, Amstutz HC. Isolation of predominantly sub-micron sized UHMWPE wear particles from periprosthetic tissues. J Biomed Mat Res 1995;29:127–131.

51. Hirakawa K, Bauer TW, Stulberg BN, Wilde AH. Comparison and quantitation of wear debris of failed total hip and knee arthroplasty. J Biomed Mat Res 1998;31:257–263.

52. Maloney WJ, Smith RL, Hvene D, Schmalzried TP, Rubash H. Isolation and characterization of wear debris generated in patients who have had failure of a hip arthroplasty without cement. J Bone Joint Surg 1994;77A:1301–1310.

53. Margevicius KT, Bauer TW, McMahon JT, Brown SA, Merritt K. Isolation and characterization of debris from around total joint prostheses. J Bone Joint Surg 1994;76A: 1664–1675.

54. Shanbhag AS, Jacobs JJ, Glant T, Gilbert JL, Black J, Galante JO. Composition and morphology of wear debris in failed uncemented total hip replacement. J Bone Joint Surg 1994;76B:60–67.

55. Howling GI, Barnett PI, Tipper JL, Stone MH, Fisher J, Ingham E. Quantitative characterization of polyethylene debris isolated from periprosthetic tissue in early failure knee implants and early and late failure Charnley hip implants. J Biomed Mat Res Appl Biomater 2001;58:415–420.

56. Green TR, Fisher J, Stone MH, Wroblewski BM, Ingham E. Polyethylene particles of a critical size are necessary for the induction of cytokines by macrophages in vitro. Biomaterials 1998;19:2297–2302.

57. Matthews JB, Green TR, Stone MH, Wroblewski BM, Fisher J, Ingham E. Comparison of the response of primary murine peritoneal macrophages and the U937 human histiocytic cell line to challenge with in vitro generated clinically relevant UHMWPE particles. Biomed Mat Eng 2000;10:229–240.

58. Matthews JB, Green TR, Stone MH, Wroblewski BM, Fisher J, Ingham E. Comparison of the response of three human monocytic cell lines to challenge with polyethylene particles of known size and dose. J Mat Sci: Mat Med 2001;12:249–258.

59. Goodman SB, Fornasier VL, Lee J, Kei J. The histological effects of the implantation of different sizes of polyethylene particles in the rabbit tibia. J Biomed Mat Res 1990;24:517–524.

60. Ingham E, Fisher J. The role of macrophages in the osteolysis of total joint replacement. Biomaterials (In Press, 2005).

61. Doorn PF, Campbell PA, Worrall J, Benya PD, McKellop HA, Amstutz HC. Metal wear particle characterization from metal on metal total hip replacements: transmission electron microscopy study of periprosthetic tissues and isolated particles. J Biomed Mat Res 1998;42:103–111.

62. Firkins PJ, Tipper JL, Saadatzadeh MR, Ingham E, Stone MH, Farrar R, Fisher J. Quantitative analysis of wear and wear debris from metal-on-metal hip prostheses tested in a

physiological hip joint simulator. Biomed Mat Eng 2001; 11:143–157.

63. Germain MA. Biological reactions to cobalt chrome wear particles, Ph.D. dissertation, University of Leeds, 2002.

64. Tipper JL, Hatton A, Nevelos JE, Ingham E, Doyle C, Streicher R, Nevelos AA, Fisher J. Alumina-alumina artificial hip joints- Part II: Characterisation of the wear debris from *in vitro* hip joint simulations. Biomaterials 2002;23: 3441–3448.

65. Hatton A, Nevelos JE, Nevelos AA, Banks RE, Fisher J, Ingham E. Alumina-alumina artificial hip joints- Part I: a histological analysis and characterization of wear debris by laser capture microdissection of tissues retrieved at revision. Biomaterials 2002;23:3429–3440.

66. Scholes SC, Unsworth A. Comparison of friction and lubrication of different hip prostheses. Proc Inst Mech Eng J Eng Med 2000;214(1):49–57.

67. Ingham E, Fisher J. Biological reactions to wear debris in total joint replacement. Proc Inst Mech Eng J Eng Med 2000;214(H1):21–37.

68. Galvin AL, Tipper J, Stone M, Ingham E, Fisher J. Reduction in wear of crosslinked polyethylene under different tribological conditions. Proc Int Conf Eng Surg Joined Hip, IMechE 2002; C601/005.

69. Galvin AL, Endo MM, Tipper JL, Ingham E, Fisher J. Functional biological activity and osteolytic potential of non-crosslinked and cross-linked UHMWPE hip joint prostheses. Trans 7th World Biomat Cong 2004. p 145.

70. Jin ZM, Dowson D, Fisher J. Analysis of fluid film lubrication in artificial hip joint replacements with surfaces of high elastic modulus. Proc Inst Mech Eng J Eng Med 1997;211: 247–256.

71. Chan FW, Bobyn JD, Medley JB, Krygier JJ, Tanzer M. The Otto Aufranc Award-Wear and lubrication of metal-on-metal hip implants. Clin Orthopaed Rel Res 1999;369:10–24.

72. Firkins PJ, Tipper JL, Ingham E, Stone MH, Farrar R, Fisher J. Influence of simulator kinematics on the wear of metal-on-metal hip prostheses. Proc Inst Mech Eng J Eng Med 2001a;215(H1):119–121.

73. Scholes SC, Green SM, Unsworth A. The wear of metal-on-metal total hip prostheses measured in a hip simulator. Proc Inst Mech Eng J Eng Med 2001;215(H6):523–530.

74. Williams S, Stewart TD, Ingham E, Stone MH, Fisher J. Metal-on-metal bearing wear with different swing phase loads. J Biomed Mater Res 2004;15:70B(2):233–9.

75. Liu F, Jin ZM, Grigoris P, Hirt F, Rieker C. Contact Mechanics of Metal-on-Metal Hip Implants Employing a Metallic Cup With an UHMWPE Backing, Journal of Engineering in Medicine. Proc Inst Mech Eng 2003;217:207–213.

76. Liu F, Jin ZM, Grigoris P, Hirt F, Rieker C. Elastohydrodynamic Lubrication Analysis of a Metal-on-Metal Hip Implant Employing a Metallic Cup With an UHMWPE Backing Under Steady-State Conditions. J Eng Med Proc Inst Mech Eng 2004;218:261–270.

77. Smith SL, Dowson D, Goldsmith AAJ. The lubrication of metal-on-metal total hip joints: a slide down the Stribeck curve. Proc Inst Mech Eng J Eng Tribol 2001;215(J5):483–493.

78. Rieker CB, et al.In vitro tribology of large metal-on-metal implants. Proc 50th Trans Orthopaed Res Soc 2004; 0123.

79. Ingham E, Fisher J. Can metal particles (Theoretically) cause osteolysis? Proceedings of the Second International Conference on Metal-Metal Hip Prostheses: Past Performance and Future Directions, Montreal, Canada, 2003.

80. Nevelos JE, Ingham E, Doyle C, Streicher R, Nevelos AB, Walter W, Fisher J. Micro-separation of the centres of alumina–alumina artificial hip joints during simulator testing produces clinically relevant wear rates and patterns. J Arthroplasty 2000;15(6):793–795.

81. Muratoglu OK, Bragdon CR, O'Connor D, Jasty M, Harris WH, Gul R, McGarry F. Unified wear model for highly cross-linked ultra-high molecular weight polyethylene (UHMWPE). Biomaterials 1999;20:1463–1470.

82. McKellop H, Shen FW, Lu B, Campbell P, Salovey R. Development of an extremely wear-resistant ultra high molecular weight polyethylene for total hip replacements. J Orthop Res 1999;17:157–167.

83. Bradford L, Baker DA, Graham J, Chawan A, Ries MD, Pruitt LA. Wear and surface cracking in early retrieved highly cross-linked polyethylene acetabular liners. J Bone Joint Surg 2004;86A:1271–1282.

84. Firkins PJ, Tipper JL, Ingham E, Stone MH, Farrar R, Fisher J. A novel low wearing differential hardness, ceramic-on-metal hip joint prosthesis. J Biomech 2001;34(10):1291–1298.

85. McMinn D, Treacy R, Lin K, Pynsent P. Metal on metal surface replacement of the hip. Experience of the McMinn prosthesis. Clin Orthop 1996;329:S89–98.

86. Hu XQ, Isaac GH, Fisher J. Changes in the contact area during the bedding-in wear of different sizes of metal on metal hip prostheses. J Biomed Mater Eng 2004;14(2):145–149.

87. Dowson D, Hardaker C, Flett M, Isaac GH. A hip joint simulator study of the performance of metal-on-metal joints: Part II: Design. J Arthroplasty 2004;19(8 Suppl 1):124–130.

88. Fisher J, Hu XQ, Tipper JL, Stewart TD, Williams S, Stone MH, Davies C, Hatto P, Bolton J, Riley M, Hardaker C, Isaac GH, Berry G, Ingham E. An in vitro study of the reduction in wear of metal-on-metal hip prostheses using surface-engineered femoral heads. Proc Inst Mech Eng [H] 2002; 216(4):219–230.

89. Fisher J, Hu XQ, Stewart TD, Williams S, Tipper JL, Ingham E, Stone MH, Davies C, Hatto P, Bolton J, Riley M, Hardaker C, Isaac GH, Berry G. Wear of surface engineered metal-on-metal hip prostheses. J Mater Sci Mater Med 2004;15(3):225–235.

90. Bigsby RJ, Auger DD, Jin ZM, Dowson D, Hardaker CS, Fisher J. A comparative tribological study of the wear of composite cushion cups in a physiological hip joint simulator. J Biomech 1998;31(4):363–369.

91. Jennings LM, Fisher J. A biomechanical and tribological investigation of a novel compliant all polyurethane acetabular resurfacing system. Proceedings of the International Conference of Engineers and Surgeons Joined at the Hip, IMechE 2002; C601/032.

See also ALLOYS, SHAPE MEMORY; BIOMATERIALS, CORROSION AND WEAR OF; JOINTS, BIOMECHANICS OF; ORTHOPEDICS, PROSTHESIS FIXATION FOR.

HIP REPLACEMENT, TOTAL. See MATERIALS AND DESIGN FOR ORTHOPEDIC DEVICES.

HOLTER MONITORING. See AMBULATORY MONITORING.

HOME HEALTH CARE DEVICES

TOSHIYO TAMURA
Chiba University School of Engineering
Chiba, Japan

INTRODUCTION

The increase in the size of the elderly population and the importance of preventing life-related diseases, such as cancer, hypertension, and diabetes, all emphasize the importance of home healthcare. Generally, the purpose of home healthcare is to reduce the distance the patient must travel to receive care and to reduce the number of hospital admissions.

The devices used in home healthcare must be simple to use, safe, inexpensive, and noninvasive or minimum invasive so that they excessively disturb normal daily activities. Recent developments in home healthcare devices meet most of these specifications, but some still pose a problem in that they disturb the activities of normal daily life and the effectiveness of some other devices in monitoring particular health-related parameters has been questioned.

The requirements for the devices used in home healthcare depend on both their purpose and the subject's condition. Monitoring vital signs, such as heart rate, blood pressure, and respiration, is routinely done for elderly individuals and for patients with chronic disease or who are under terminal care. These cases require simple, noninvasive monitors. When patients are discharged from the hospital and continue to need these parameters monitored at home, the devices used are essentially no different from those used in the hospital.

For health management and the prevention of disease, an automatic health monitoring system has been considered. The onset of lifestyle-related diseases, such as hypertension, arteriosclerosis, and diabetes, is highly correlated with daily activities, such as physical exercise, including walking, as well as other habits, such as sleep and smoking. To prevent such diseases, daily monitoring will be important for achieving healthy living and improving the quality of life. Although the monitoring of daily activities is not well established in evidenced-based health research, there have been many attempts at installing sensors and transducers and monitoring daily life at home.

Evidenced-based health research directed at finding correlations between daily activity monitoring and the onset of disease, and identifying risk factors, is a major subject of epidemiology. In general, large population studies of daily living activities, including daily food intake, based on the history using interviews or questionnaires are required. If an automatic monitoring system can be applied, more reliable and objective data can be obtained.

Recently, many new home healthcare devices have been developed because many individuals have become motivated to maintain their health. This article discusses recently developed homecare devices, as well as expected laboratory-based devices.

BLOOD PRESSURE

Blood pressure is one of the most important physiological parameters to monitor. Blood pressure varies considerably throughout the day and frequent blood pressure monitoring is required in many home healthcare situations. Usually, a blood pressure reading just before the patient wakes in the morning is required. The success of home blood pressure readings is highly dependent on the patient's motivation. Blood pressure readings at home are also recommended because many patients have elevated

Figure 1. The standard blood pressure monitor. The device includes an inflatable cuff, a manometer, and a stethoscope. The bladder is inflated until the cuff compresses the artery in the arm; since no blood passes, the stethoscope detects no noise. Then, the cuff is deflated slowly, blood passes though the artery again, and the stethoscope perceives a noise, which is defined as the systolic pressure. The cuff continues to deflate and finally the stethoscope perceives no noise, defined as the diastolic pressure.

blood pressure readings in a clinical setting, the so-called "white-coat hypertension".

Medical doctors and nurses usually measure blood pressure using the auscultatory method as shown in Fig. 1, in which a pressure cuff is attached to the upper arm and inflated to compress the brachial artery to a value above the systolic pressure. Then, the cuff is gradually deflated while listening to the Korotkoff sounds though a stethoscope placed on the brachial artery distal to the cuff. The systolic and diastolic pressures are determined by reading the manometer when the sounds begin and end, respectively. However, this technique requires skill and it is difficult to measure blood pressure on some obese individuals using this method.

For home blood pressure monitoring, convenient automatic devices have been developed and are commercially available. The measurement sites are the upper arm, wrist, and finger.

The most common method is to attach the cuff to the upper arm, and the systolic and diastolic pressures are determined automatically (Fig. 2). The cuff is inflated by an electric pump and deflated by a pressure-released valve. To determine the pressures, two different methods are used: Korotkoff sounds and an oscillometric method.

A microphone installed beneath the cuff detects the Korotkoff sounds and when the systolic and diastolic pressures are detected, a pressure sensor measures the obtained sounds and pressure at the critical points. The advantage of this method is that this measurement principle follows the standard auscultatory method. When the cuff is attached correctly, a reliable reading can be obtained.

The size of the cuff is important. The cuff should accurately transmit pressure down to the tissue surrounding

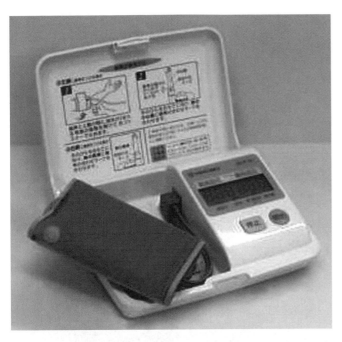

Figure 2. Automatic blood pressure monitor using both auscultatory and oscillometric methods.

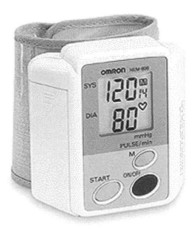

Figure 4. Wrist-type home blood pressure monitor. The measurement site is the wrist and during the measurement, the wrist must be at heart level.

the brachial artery. A narrow cuff results in a larger error in pressure transmission. The effect of cuff size on blood pressure accuracy for the Korotkoff method has been studied experimentally (1).

The oscillometric method detects the pulsatile components of the cuff pressure as shown in Fig. 3. When the cuff pressure is reduced slowly, pulses appear in the systolic pressure and the amplitude of the pulses increases and then decreases again. The amplitude of these pulses is always maximal when the cuff pressure equals the mean arterial pressure. However, it is difficult to determine the diastolic pressure from the signal measured from the

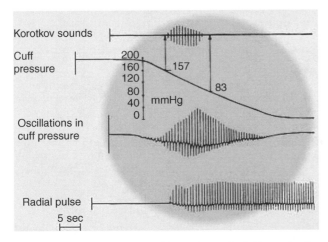

Figure 3. Principle of the oscillometric method. The pulsations induced by the artery differ when the artery is compressed. Initially, no pulsation occurs, and then the pulsation starts. As the pressure decreases in the cuff, the oscillation becomes more significant, until the maximum amplitude of the oscillations defines the average blood pressure. Then, the oscillations decrease with the cuff pressure until they disappear.

cuff pressure. In general, the diastolic pressure is determined indirectly in commercial devices. One simple method that is often used is to calculate the diastolic pressure from the mean arterial pressure and systolic pressure (2). Several algorithms for this calculation have been used in commercial blood pressure monitors. The oscillometric method can only measure the cuff pressure.

Blood pressure measurement is not restricted to the upper arm. It is possible to measure blood pressure at the wrist and on a finger. However, if the measurement site is changed from the upper arm, the errors due to gravitational force and the peripheral condition increase. Wrist-type blood pressure monitors are now common in home use (Fig. 4).

Home blood pressure monitors are tested for accuracy against two protocols: the Association for the Advancement of Medical Instruments (AAMI) and the International Protocol of the European Society of Hypertension. Reports of the accuracy of home blood pressure monitors have been published (3). In addition, a 24 h home blood pressure monitor has been evaluated (4).

A home blood pressure monitor using the pulse wave transit time has also been studied. The principle used in this approach is that the arterial pulse wave transit time depends on the elasticity of the arterial vessel wall and the elasticity depends on the arterial pressure. Therefore, arterial pressure affects the pulse wave transit time. However, vascular elasticity is also affected by vasomotor activities, which depend on external circumstances; so this method is not reliable. Even so, with intermittent calibration we can estimate the blood pressure from the pulse wave transit time (5). The pulse wave transit time can be non-invasively determined from the arrival time of the arterial pulse at the beginning of cardiac contraction, which is determined from the QRS complex in an electrocardiogram.

ELECTROCARDIOGRAM

The electrocardiogram (ECG) gives important cardiac information. Recording the ECG at home can assist physicians to make a diagnosis. When monitoring the ECG at home

Figure 5. Holter ECG recorder. The Holter recorder is used for 24 h ECG monitoring and a digital Holter recorder is commonly used.

during either recovery from an acute disease or when the patient has a chronic disease, long-term recording is essential in order to detect rarely occurring abnormalities.

The Holter ECG recorder as shown in Fig. 5, has been widely used. It is a portable recorder that records the ECG on two or more channels for 24 or 48 h, on either an ordinary audiocassette tape or in a digital memory, such as solid-state flash memory. Most Holter recorders are lightweight, typically weighing 300 g or less, including the battery. The ECG must be recorded on the chest and electrodes need to be attached by clinical staff. A physician should also be available to monitor the ECG. Aside from these limitations, the Holter recorder can be used without obstructing a patient's daily life.

There are some special ECG recordings that can be taken in the home. The ECG can be recorded automatically during sleep and bathing.

In bed, the ECG can be recorded from a pillow and sheets or beneath the leg using electroconductive textiles (Fig. 6) (6). Since the contact between the textile electrodes and the skin is not always secure, large artifacts occur with body movements. In our estimation, 70–80% of ECGs during sleep can be monitored.

The ECG can also be recorded while bathing. If electrodes are installed on the inside wall of the bathtub as shown in Fig. 7, an ECG can be recorded through the water (7,8). The amplitude of the ECG signal depends on the conductivity of the tap water. If the conductivity is high, the water makes a short circuit with the body, which serves as the voltage source, and consequently the amplitude is reduced. If the water conductivity is low, however, the signal amplitude remains at levels similar to those taken on the skin surface. Fortunately, the electrical conductivity of ordinary tap water is on the order of $10^{-2}\,S{\cdot}m^{-1}$, which is within the acceptable range for measurement using a conventional ECG amplifier. However, such an ECG signal cannot be used for diagnostic purposes because of the attenuation of the signal at lower frequencies.

HEART AND PULSE RATES

The heart rate (HR) is a simple indicator of cardiac function during daily life and exercise. The HR is the number of

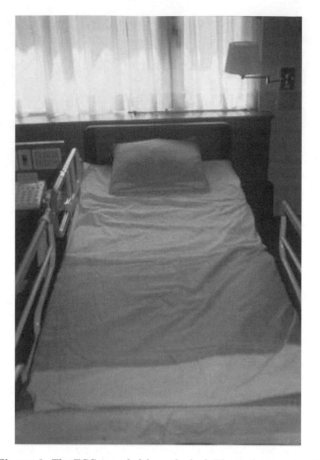

Figure 6. The ECG recorded from the bed. Electrodes are placed on the pillow and the lower part of the bed. An ECG signal can be obtained during sleep.

contractions of the heart per minute, and the pulse rate is defined as the number of arterial pulses per minute. Usually, both rates are the same. When there is an arrhythmia, some contractions of the heart do not produce effective ejection of blood into the arteries and this gives a lower pulse rate.

Figure 7. The ECG recorded from a bathtub. Silver–silver chloride electrodes are placed in the bathtub and the ECG signal can be obtained though the water.

The heart rate can be determined by counting the QRS complexes in an ECG or measuring the R–R interval when the cardiac rhythm is regular. In patients with an arrhythmia, the ECG waveforms are abnormal, and in this case detection algorithms with filtering are used. For accurate heart rate monitoring, electrodes are attached to the chest. Instead of surface electrodes, a chest strap is also available (Polar, Lake Success, NY)

The pulse rate can be obtained by detecting the arterial pulses using a photoplethysmograph, mechanical force measurements, vibration measurements, or an impedance plethysmograph. In photoplethysmograpy, the change in light absorption caused by the pulsatile change in the arterial volume in the tissue is detected. To monitor the pulse rate using photoplethysmography, the finger is commonly used. Light sources with wavelengths in the infrared (IR) region $\sim$ 800 nm are adequate for this purpose, because tissue absorbance is low and the absorbance of hemoglobin at this wavelength does not change with oxygen saturation.

The pulse oximeter, described below, can monitor both oxygen saturation and pulse rate. The pulsatile component of light absorption is detected and the pulse rate can be determined from the signal directly. The ring-type pulse oximeter is the most commonly used type of pulse oximeter.

A wristwatch type pulse rate meter is also available. It consists of a reflection-type photoplethysmograph. To measure pulse rate, the subject puts their fingertip on the sensor. A flashing icon on the display indicates a detected pulse, and the rate is displayed within 5 s.

The pulse rate can also be monitored in bed. In this case, the pulse rate is obtained directly from an electroconductive sheet. Vibration of the bed is detected by a thin flexible electric film (BioMatt, Deinze, Belgium) or an air mattress with a pneumatic sensor. In either case, the pulse rate is obtained through signal processing.

BODY TEMPERATURE

Body temperature has been checked at home for many years to detect fever. Frequent body temperature measurements are required for homecare in many chronic diseases. Basal body temperature measurement is also required when monitoring the menstrual cycle.

Stand-alone mercury-in-glass clinical thermometers have long been used both in clinical practice and at home, although they have recently been replaced by electronic thermometers because mercury contamination can occur if they are broken (Fig. 8). The ordinary electronic clinical thermometer uses a thermistor as a temperature sensor. The body temperature is displayed digitally. There are two types of clinical thermometer: the prediction and the real-time type. The real-time type waits until a stable temperature value is obtained. The prediction type attempts to predict the steady-state temperature using an algorithm involving exponential interpolation. The response time of a real-time electronic thermometer is 3 min and the response time of a prediction-type electronic thermometer is < 1 min, when both are placed in the mouth.

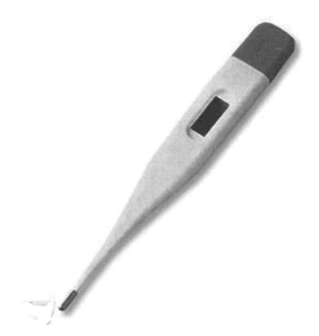

Figure 8. The electric thermometer contains a thermistor. Both predicting and real-time types are sold.

The tympanic thermometer as shown in Fig. 9, has become popular for monitoring the body temperature in children and the elderly because of its fast response. The device operates on the principle of IR radiation. The sensor is either a thermopile or pyroelectric sensor and is installed

Figure 9. The tympanic thermometer. Either a thermopile or a pyroelectric sensor is used as the temperature sensor. This device has a faster response than an electric thermometer.

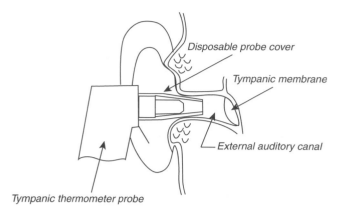

Figure 10. The principle of the tympanic thermometer. The sensor tip is inserted into the ear canal and the thermal distribution of tympanum is measured.

near the probe tip, as shown in Fig. 10. The probe tip is inserted into the auditory canal and the radiation from the tympanic membrane and surrounding tissue is detected. The tympanic temperature is close to the deep body temperature and the measurement can be made within a few seconds. Many studies have shown that when used properly, a tympanic thermometry is very accurate. However, IR tympanic thermometers produced measurements that were both less accurate and less reproducible when used by nurses who routinely used them in clinical practice (9,10).

A strip thermometer is sometimes used to monitor an acute fever. It is designed to be used once only and then discarded. It contains strips of thermosensitive liquid crystal that change color to indicate skin temperature, not body temperature. The color change is nonreversible. The strip is placed on the forehead and then read after 1 min. If a strip thermometer shows a high temperature, one should recheck the temperature with another type of thermometer.

BODY FAT

Body composition and body fat have both been proposed as indicators of the risk of chronic disease.

Body fat percentage is the proportion of fat in a person's body. Excess body fat was previously determined by measuring weight and comparing that value with height. Body fat is not always visible and cannot be measured on an ordinary scale. Obesity, which indicates a high degree of excess body fat, has been linked to high blood pressure, heart disease, diabetes, cancer, and other disabling conditions. To estimate the percentage of body fat, it is commonly derived from body density. The following equation gives an estimate of body density (D), which is then converted into the percent body fat (%BF) using the Siri equation:

$$\%BF = (495/D) - 450$$

Body density, measured by weighting an individual while immersed in a tank of water, is based on Archimedes' principle and is a standard technique. However, this is not

Figure 11. A scale with bioelectrical impedance analysis. This is a simple version of body impedance analysis using leg-to-leg bioimpedance analysis. The precision electronic scale has two footpad electrodes incorporated into its platform. The measurement is taken while the subject's bare feet are on the electrodes. The body fat percentage can be obtained from equations based on weight, height, and gender.

a convenient method for measurement in the home. Body volume can be determined from the air volume in an airtight chamber with the body inside by measuring the compliance of the air in the chamber (Bod Pod, Life Measurement Instruments, Concord, CA).

Body fat scales use the bioelectrical impedance analysis (BIA) technique. This method measures body composition using four electrodes, in which a constant alternating current (ac) of 50–100 kHz and 0.1–1 mA is applied between the outer electrode pair, and the alternating voltage developed between the inner electrode pair is detected (Fig. 11). Alternating current is applied between the toes of both feet, and the voltage developed between the electrodes at both feet is detected. The current passes freely through the fluids contained in muscle tissue, but encounters difficulty–resistance when it passes through fat tissue. This means that electrical impedance is different in different body tissues. This resistance of the fat tissue to the current is called bioelectrical impedance, and is accurately measured by body fat scales. Using a person's height and weight, the scales can then compute the body fat percentage. Recently, new commercial BIA instruments, such as the body segmental BIA analyzer, multifrequency BIA analyzer, lower body BIA analyzer, upper body BIA analyzer, and laboratory-designed BIA analyzers, have greatly expended the utility of this method (11). However, body composition differs by gender and race. Nevertheless, the impedance technique is highly reproducible for estimating the lean body mass (12).

The use of near-IR spectral data to determine body composition has also been studied (13). Basic data suggest that the absorption spectra of fat and lean tissues differ. The FUTREX-5000 (Zelcore, Hagerstown, MD) illuminates the body with near-IR light at very precise wavelengths (938 and 948 nm). Body fat absorbs the light, while lean body mass reflects the light. The intensity of back-scattered light is measured. This measurement provides an estimation of the distribution of body fat and lean body mass.

BLOOD COMPONENTS

In a typical clinical examination, the analysis of blood components is important. Medical laboratory generally use automatic blood analyzers for blood analysis. Usually, an invasive method is required to obtain a blood sample. Therefore, in a home healthcare setting, the monitoring and analysis of blood is uncommon, except for diabetic patients. In this section, we focus on the blood glucose monitor.

There are several commercial home blood glucose monitors. Self-monitoring of blood glucose (SMBG) is recommended for all people with diabetes, especially for those who take insulin. The role of SMBG has not been defined for people with stable type 2 diabetes treated with diet only. As a general rule, the American Diabetes Association (ADA) recommends that most patients with type 1 diabetes test glucose three or more times daily. Blood glucose is commonly measured at home using a glucose meter and a drop of blood taken from the finger (Fig. 12). A lancet device, which contains a steel needle that is pushed into the skin by a small spring, is used to obtain a blood sample. A small amount of blood is drawn into the lumen of the needle. The needle diameter is from 0.3 (30 G) to 0.8 (21 G) mm. In addition, laser lancing devices, which use a laser beam to produce a small hole by vaporizing the skin tissue, are available.

Once a small amount of blood is obtained, blood glucose can be analyzed using either a test strip or a glucose meter.

The blood glucose level can be estimated approximately by matching the color of the strip to a color chart. In electrochemical glucose meters for homecare, a drop of blood of 10 µL or less is placed in the sensor chip. The blood glucose is measured by an enzyme-based biosensor. Most glucose meters can read glucose levels over a broad range of values, from as low as 0 to as high as 600 mg·dL. Since the range differs among meters, it is important to interpret very high or low values carefully. Glucose readings are not linear over their entire range.

Home blood glucose meters measure the glucose in *whole blood*, while most lab tests measure the glucose in *plasma*. Glucose levels in plasma are generally 10–15% higher than glucose measurements in whole blood (and this difference is even larger after a person has eaten). Many commercial meters now give results as the "plasma equivalent". This allows patients to compare their glucose measurements from lab tests with the values taken at home.

Minimally invasive and noninvasive blood glucose measurement devices are also sold. One of these uses near-IR spectroscopy to measure glucose. It is painless. There are increasing numbers of reports in the scientific literature on the challenges, strengths, and weaknesses of this and other new approaches to testing glucose without fingersticks (14,15).

The U.S. Food and Drug Administration (FDA) has approved minimally invasive meters and noninvasive glucose meters, but neither of these should replace standard glucose testing. They are used to obtain additional glucose values between fingerstick tests. Both devices require daily calibration using standard fingerstick glucose measurements.

The MiniMed system (Medtronic, Minneapolis, MN) consists of a small plastic catheter (a very small tube) inserted just under the skin. The catheter collects small amounts of liquid, which are passed through a biosensor to measure the amount of glucose present. The MiniMed is intended for occasional use and to discover trends in glucose levels during the day. Since it does not give readings for individual tests, it cannot be used for typical day-to-day

Figure 12. Glucose meter. This is used for self-monitoring blood glucose. The blood is taken from the fingertip and analyzed using a test strip.

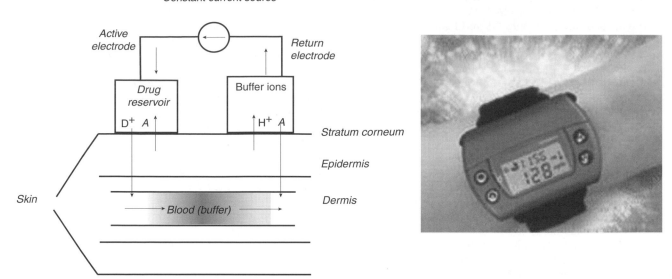

Figure 13. Glucowatch device (1) and principle of iontophoresis (2). This device provides non-invasive monitoring of glucose and uses reverse iontophoresis to extract glucose from the skin to monitor glucose. A low electric current is applied, which draws interstitial fluid through the skin. The glucose in this fluid is collected in a gel. A chemical process occurs, which generates an electrical signal that is converted into a glucose measurement.

monitoring. The device collects measurements over a 72 h period and then the stored values must be downloaded by the patient or healthcare provider.

GlucoWatch (Cygnus, Redwood City, CA) is worn on the arm like a wristwatch (Fig. 13). It pulls small amounts of interstitial fluid from the skin by iontophoresis and measures the glucose in the fluid without puncturing the skin. The device requires 3 h to warm up after it is put on the wrist. After this, it can measure glucose up to three times per hour for 12 h. The GlucoWatch displays results that can be read by the wearer, although like the MiniMed device, these readings are not meant to be used as replacements for fingerstick-based tests. The results are meant to show trends and patterns in glucose levels, rather than report any one result alone. It is useful for detecting and evaluating episodes of hyperglycemia and hypoglycemia. However, the values obtained must be confirmed by tests with a standard glucose meter before any corrective action is taken.

An elevated cholesterol level is one of the most important risk factors for coronary heart diseases. For home healthcare, a blood cholesterol test device is available. The test requires that a few drops of blood obtained from a finger stick sample be applied to the cholesterol strip, which contains cholesterol esterase and cholesterol oxidize. Total cholesterol, that is, the sum of free and esterified cholesterol, can be accurately and conveniently measured enzymatically using cholesterol oxidize and cholesterol esterase. The total amount of cholesterol is measured, and the results are obtained in 3–15 min.

URINE COMPONENTS

The analysis of urine components provides important diagnostic information for clinicians. Urine glucose and ketones indicate diabetes and urine protein indicates kidney disease. However, the only tool available for such testing is the urine test strip. A urine test can be done using a test strip without pain or discomfort. No fully automatic urine test system available, but there have been some attempts to monitor urine components at home with minimum disturbance. The instrument shown in Fig. 14 has been developed. It can be installed in the toilet and measures the urine glucose after a button is pushed (TOTO, Tokyo). The urine collector protrudes, collects urine automatically from the urine stream, and analyzes urine glucose within 1 min using an enzyme glucose sensor. The sensor must be replaced every 4 months and a calibration solution must be replenished every 3 months. This system is useful for monitoring the urine glucose level in diabetic patients.

BODY WEIGHT

Body weight monitored at home is an essential parameter for health management. To use body weight for health management, data must be taken regularly and stored. A digital scale connected to a laptop computer, together with temperature and blood pressure monitors, and a bed-sensor system has been developed (16).

A device to weigh the body automatically for health monitoring based on measurements on the toilet seat has been developed (17). A precision load cell system was installed in the floor of the toilet, and the seat was supported so that the weight on the seat was transferred to the load cell. This system also allows the measurement of urine and feces volume, urine flow rate, and the number and times of urination and evacuation.

For health management, the body mass index is commonly used. This is defined as the weight divided by the sents the calories the body burns in order to maintain vital body functions (heart rate, brain function, and breathing). It equals the number of calories a person would burn if they were awake, but at rest all day. The RMR can represent up to 75% of a person's total metabolism if they are inactive or lead a sedentary lifestyle. Since the RMR accounts for up to 75% of the total calories we need each day, it is a critical piece of information for establishing appropriate daily calorie needs, whether one is trying to lose or maintain weight. Most healthcare and fitness professionals recognize that metabolism is affected by a variety of characteristics, such as fever, illness, high fitness, obesity, and active weight loss. When managing a subject's nutritional needs and calorie requirements, knowledge of their RMR is critical. Since metabolism differs individually, estimating the RMR value can lead to errors, and inaccurate calorie budgets. Consequently, individuals can be unsuccessful at reaching their personal goals, due to over- or under-eating. As technology advances, professionals must reassess their practices. Caloric needs are assessed most accurately by measuring oxygen consumption and determining individual metabolism. Oxygen consumption estimates are obtained from the oxygen gas concentration and flow. Since it usually requires wearing a mask or mouthpiece, this measurement is difficult for some individuals. The Body-Gem and MedGem (HealtheTech, Golden, CO) are devices that provide information vital for determining a personalized calorie budget, based on individual metabolism (Fig. 15). The BodyGem and MedGem consist of an ultrasound flow meter and fluorescence oxygen sensor with a blue LED excitation source, but the measurements are limited to an RMR monitor only. The RMR has been mentioned in the text.

We can also estimate the body's energy consumption from heat flow and acceleration measurements taken while an individual exercises (Body Media inc. Pittsburg, PA). For diabetes control, a pedometer with an accelerometer has been used and the energy consumption estimated using several algorithms.

Figure 14. Urine glucose monitor installed in the toilet. A small nozzle collects urine and then a biosensor analyzes urine glucose automatically.

square of the height. Excess body weight increases the risk of death from cardiovascular disease and other causes in adults between 30 and 74 years of age. The relative risk associated with greater body weight is higher among younger subjects (18).

NUTRITION

To prevent cardiac disease, diabetes, and some cancers, it is important to control body weight. The most accurate method that currently exists is to weigh foods before they are eaten. Like many other methods, however, this method can be inaccurate, time-consuming, and expensive. There are two basic ways to monitor nutrition. One is to monitor food intake.

Food consumed is photographed using a digital camera and the intake calories are calculated from the photographs (19,20). Digital photography and direct visual estimation methods, estimates of the portion sizes for food selection, plate waste, and food intake are all highly correlated with weighed foods.

The resting metabolism rate (RMR) is an important parameter for controlling body weight. The RMR repre-

Figure 15. A simple oxygen-uptake monitor. The subject wears the mask and a small ultrasonic flow meter measures the respiratory volume and a fluorescence oxygen monitor measures the oxygen concentration. This device is only used for measuring basal metabolism.

DAILY ACTIVITY

From the standpoint of health management, both the physical and mental health of an individual are reflected in their daily physical activities. The amount of daily physical activity can be estimated from the number of walking steps in a day, which are measured by a pedometer attached to the belt or waistband. To improve physical fitness 10,000 steps per day or more are recommended. For more precise measurement of physical activity, an accelerometer has been used. Behavior patterns, such as changes in posture and walking or running can be classified. The metabolic rate can be estimated from body acceleration patterns. The algorithms for calculating energy consumption differ for different pedometers. Each manufacturer has a different algorithm, and these have not been made public. However, the energy is likely evaluated using total body weight and walking time (21). This measurement requires attaching a device to the body, and requires continual motivation. An accelerometer equipped with a global positioning sensor has been developed and can monitor the distance and speed of daily activity (22).

There have been attempts to monitor daily activities at home without attaching any devices to the body. Infrared sensors can be installed in a house to detect the IR radiation from the body so that the presence or absence of a subject can be monitored, to estimate the daily activity at home, at least when the subject is living alone.

Other simple sensors, such as photointerrupters, electric touch sensors, and magnetic switches, can also be used to detect activities of daily living (23–25). The use of room lights, air conditioning, water taps, and electric appliances, such as a refrigerator, TV, or microwave oven, can be detected and used as information related to daily living. Habits and health conditions have correlated with these data to some extent, but further studies are required to give stronger evidence of correlations between sensor output and daily health conditions.

SLEEP

Sleep maintains the body's health. Unfortunately, in most modern industrial countries the process of sleep is disturbed by many factors, including psychological stress, noise, sleeping room temperature, and the general environment surrounding the bedroom. Insufficient sleep and poor sleeping habits can lead to insomnia. Another sleep problem is sleep apnea syndrome. In the laboratory, sleep studies aimed at the diagnosis of sleep apnea syndrome include polysomnography (PSG), electroencephalography (EEG), ECG, electromyography (EMG) pulse oximetry, and require chest and abdomen impedance belts. In the home, simple devices are required to evaluate sleep to determine if more detailed laboratory tests are needed.

A physical activity monitor actigraph (AMI, Ardsley, NY) can be used as a sleep detector. It is easy to wear and detects the acceleration of the wrist using a piezoelectric sensor. The wrist acceleration recorded by the actigraph accurately showed when the wearer was asleep (26).

Body movements during sleep can be measured without attaching sensors and transducers to the body using a pressure-sensitive sheet (BioMatt, VTT Electronics, Tampere, Finland). It consists of a 50 μm thick pressure-sensitive film, which can be installed under the mattress. This film is quite sensitive and not only detects body motions, but also respiration and heart rate. Therefore, it can be used as a sleep monitor for detecting insomnia and sleep disorders and as a patient monitor for detecting sleep apnea, heart dysfunctions, and even coughing and teeth grinding (27–29).

Body motion during sleep can also be monitored using a thermistor array installed on the bed surface at the waist or thigh level (30,31). The changes in temperatures show the body movement and sleep condition.

RESPIRATION THERAPY AND OXYGEN THERAPY

Respiration is the function of gas exchange between the air and blood in the body, and it consists of ventilation of the lung and gas transfer between the alveolar air and the blood in the pulmonary circulatory system. Lung ventilation can be monitored by either measuring the flow rate of the ventilated air or the volume change of the lung. Gas transfer is monitored by arterial blood oxygenation. Frequent respiratory monitoring is required for respiratory therapy at home.

Furthermore, many individuals have developed breathing difficulties as a consequence of increasing pollution, combined with an aging population.

For therapy, we use two types of respiration aid. One is for respiration related to cellular gas exchange. The other is for breathing difficulty, such as sleep apnea.

Reparatory therapy is included in the training of individuals involved in rehabilitation after thoracoabdominal surgery, in paraplegic or quadriplegic patients, and for patients requiring some form of mechanical ventilation. The fundamental parameters that must be monitored are the respiratory rate, respiratory amplitude, and respiratory resistance. Respiratory amplitude can be monitored using either airflow or lung movement.

For continuous monitoring of respiration in a home setting, it is inconvenient to use a mask or mouthpiece. Lung ventilation can be estimated by practice and from abdominal displacement. Inductance plethysmography has been used (32,33). This consists of two elastic bands placed at the rib cage and abdomen. Each band contains a zigzag coil and the inductance of this coil changes with its cross-sectional area. This system, Respitrace (Non-Invasive Monitoring Systems, North Bay Village, FL), gives the changes in volume of the rib cage and abdomen, tidal volume, and breathing rate. Respitrace was rated as the best noninvasive technology for the diagnosis of sleep-related breathing disorders by the American Academy of Sleep Medicine Task Force (1999).

Oxygen therapy, intermittent positive pressure breathing (IPPB) therapy, and respiratory assistance using a respirator can also be performed at home. In these situations, the arterial blood oxygenation must be monitored. Actually, there is a change in optical absorbance on the

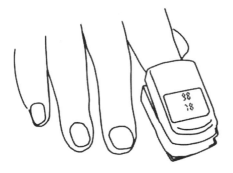

Figure 16. Pulse oximeter. A pulse oximeter is a simple noninvasive method of monitoring the percentage of hemoglobin (Hb) saturated with oxygen. It consists of a probe attached to the subject's finger. The device displays the percentage of Hb with oxygen together with an audible signal for each pulse beat and the calculated heart rate.

venous side that reflects changes in intrathoracic pressure due to breathing. Blood oxygenation is commonly monitored using a pulse oximeter, which can measure the oxygen saturation of arterial blood noninvasively from the light that is transmitted through a finger (Fig. 16).

The pulse oximeter is based on the principle that the pulsatile component in the transmitted light intensity is caused by the changes in the absorption of arterial blood in the light path while the absorption of the venous blood and tissue remains unchanged. The absorption spectrum of the blood changes with oxygen saturation, so the oxygen saturation of the arterial blood can be determined from the time-varying spectral components in the transmitted light. The oximeter contains two light-emitting diodes (LEDs), which emit light at two different wavelengths, and a photodiode to detect absorption changes at the two different wavelengths (Fig. 17). The measuring site is usually at a finger. However, a probe with a cable can sometimes disrupt the activities of daily life. A reflection-type probe that can be attached to any part of the body might be more convenient. Unfortunately, reflection-type probes are less reliable than transmission probes (34). A finger-clip probe without a cable (Onyx, Nonin Medical, Plymouth, MN) and a ring-type probe (35) are also available.

Recent advanced home healthcare devices are reviewed. These devices can be used effectively, not only for the elderly, but also for the middle-aged population and to establish home healthcare and telecare. Telecare and telemedicine are now popular for monitoring patients with chronic diseases and elderly people who live alone. The devices are placed in their homes and the data are transmitted to the hospital or a healthcare provider, who can check their clients' condition once every 12–24 h. Successful application has been reported for oxygen therapy and respiratory therapy.

We have solved several problems for more practical use. The major problems are the standardization of these devices and the agreement between medical use and home healthcare. Standardization of monitoring is important. For example, the principle of body impedance analysis differs for each manufacturer. Therefore, the values differ for different devices. This confuses customers, who then think that the devices are not reliable; hence, nobody uses such devices. There are similar problems with pedometers. Pedometers use either a mechanical pendulum or an accelerometer. The manufacturers should mention their limitations and reliability briefly, although most customers find this information difficult to understand.

The next problem is more serious. Some home healthcare devices have not been approved by health organizations, such as the FDA. For blood pressure monitors, a physician still needs to measure blood pressure during clinical practice even if the subject measures blood pressure at home. If the home healthcare device was sufficiently reliable, the physician would be able to trust the blood pressure values. Both researchers and members of industry must consider ways to solve this problem in the near future. There are additional social problems, such as insurance coverage of home healthcare devices, costs, handling, and interface design. The development of home heathcare devices must also consider the psychological and environmental factors that affect users. In the future, preventative medicine will play an important role in medical diagnosis. Hopefully, more sophisticated, high quality home healthcare devices will be developed. Technology must solve the remaining problems in order to provide people with good devices.

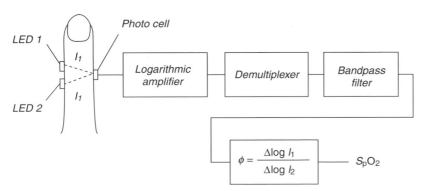

Figure 17. The principle of the oximeter. Hemoglobin absorbs light and the amount depends on whether it is saturated with oxygen. The absorption at two wavelengths (650 and 805 nm) is measured and used to calculate the proportion of hemoglobin that is oxygenated.

BIBLIOGRAPHY

Cited References

1. Geddes LA, Whistler SJ. The error in indirect blood pressure measurement with in correct size of cuff. Am Heart 1978; 96(July):4–8.
2. Sapinski A, Hetmanska ST. Standard algorithm of blood-pressure measurement by the oscillometric method. Med Biol Eng Comput 1992;30:671.
3. For example, Zsofia N, Katalin M, Gygorgy D. Evaluation of the Tensioday ambulatory blood pressure monitor according to the protocols of the British Hypertension Society and the Association for the Advancement of Medical Instrumentation. Blood Press Monit 2002;7:191–197.
4. O'Brien E, Atkins N, Staessen J. State of the market: a review of ambulatory blood pressure monitoring devices. Hypertension 1995;26:835–842.
5. Chen W, Kobayashi T, Ichikawa S, Takeuchi Y, Togawa T. Continuous estimation of systolic pressure using the pulse arrival time and intermittent calibration. Med Biol Eng Comput 2000;38:569–574.
6. Ishijima M. Monitoring of electrocardiograms in bed without utilizing body surface electrodes. IEEE Trans Biomed Eng BME 1993;40:593–594.
7. Ishijima M, Togawa T. Observation of electrocardiogram through tap water. Clin Phys Physiol Meas 1989;10:171–175.
8. Tamura T, et al. Unconstrained heart rate monito ring during bathing. Biomed Instrum Technol 1997;31:391–396.
9. Amoateng-Adjepong Y, Mundo JD, Manthous CA. Accuracy of an infrared tympanic thermometer. Chest 1999;115:1002–1005.
10. Robinson JL, Jou H, Spady DW. Accuracy of parents in measuring body temperature with a tympanic thermometer. BMC Family Pract. 2005;6:3.
11. Khan M, et al. Multi-dimension applications of bioelectrical impedance analysis. JEP Online 2005;8(1):56–71.
12. Segal KR, et al. Estimation of human body composition by electrical impedance methods. A comparison study. J Appl Physiol 1985;58:1565–1571.
13. Conway JM, Noms KH, Bodwell CE. A new approach for the estimation of body composition: infrared interactance. Am J Clin Nutr 1984;40:1123–1130.
14. Maruo K, Tsurugi M, Tamura M, Ozaki Y. *In vivo* non-invasive measurement of blood glucose by near-infrared diffuse-reflectance spectroscopy. Appl Spectrosc 2003;57(10): 1236–1244.
15. Malin SF, et al. Noninvasive prediction of glucose by near-infrared diffuse reflectance spectroscopy. Clin Chem 1999; 45:1651–1658.
16. Tuomisto T, Pentikäinen V. Personal health monitor for homes. ERCIM News 1997;29.
17. Yamakoshi K. Unconstrained physiological monitoring in daily living for healthcare. Frontiers Med Biol Eng 2000; 10:239–259.
18. Stevens J, et al. The effect of age on the association between body-mass index and mortality. N Engl J Med 1998;338:1–7.
19. Williamson D, et al. Comparison of digital photography to weighed and visual estimation of portion sizes. J Am Diet Assoc 2003;103:1139–1111.
20. Wang DH, Kogashiwa M, Ohta S, Kira S. Validity and reliability of a dietary assessment method: the application of a digital camera with a mobile phone card attachment. J Nutr Sci Vitaminol (Tokyo) 2002;48:498–504.
21. Tharion WJ, et al. Total energy expenditure estimated using a foot-contact pedometer. Med Sci Monit 2004;10(9):CR504–509.
22. Perrin O, Terrier P, Ladetto Q, Merminod B, Schutz Y. Improvement of walking speed prediction by accelerometry and altimetry, validated by satellite positioning. Med Biol Eng Comput 2000;38(2):164–168.
23. Celler BG, et al. Remote monitoring of health status of the elderly at home. A multi-disciplinary project on aging at the University of New South Wales. Intern J Bio-Medical Comput 1995;40:144–155.
24. Suzuki R, Ogawa M, Tobimatsu Y, Iwaya T. Time course action analysis of daily life investigation in the welfare techo house in Mizusawa. J Telemed Telecare 2001;7:249–259.
25. Ohta S, Nakamoto H, Shinagawa Y, Tanikawa T. A health monitoring system for elderly people living alone. J Telemed Telecare 2002;8:151–156.
26. Sadeh A, Hauri PJ, Kripke DF, Lavie P. The role of actigraphy in the evaluation of the sleep disorders. Sleep 1995;18: 288–302.
27. Salmi T, Partinen M, Hyyppa M, Kronholm E. Automatic analysis of static charge sensitive bed (SCSB) recordings in the evaluation of sleep-related apneas. Acta Neurol Scand 1986;74:360–364.
28. Salmi T, Sovijarvi AR, Brander P, Piirila P. Long-term recording and automatic analysis of cough using filtered acoustic signals and movements on static charge sensitive bed. Chest 1988;94:970–975.
29. Sjoholm TT, Polo OJ, Alihanka JM. Sleep movements in teethgrinders. J Craniomandib Disord 1992;6:184–191.
30. Tamura T, et al. Assessment of bed temperature monitoring for detecting body movement during sleep: comparison with simultaneous video image recording and actigraphy. Med Eng Phys 1999;21:1–8.
31. Lu L, Tamura T, Togawa T. Detection of body movements during sleep by monitoring of bed temperature. Physiol Meas 1999;20:137–148.
32. Milledge JS, Stott FD. Inductive plethysmography--a new respiratory transducer. J Physiol 1977;267:4P–5P.
33. Sackner JD, et al. Non-invasive measurement of ventilation during exercise using a respiratory inductive plethysmograph. I Am Rev Respir Dis 1980;122:867–871.
34. Mendelson Y, Ochs BD. Noninvasive pulse oximetry utilizing skin reflectance. IEEE Trans Biomed Eng 1988;35:798–805.
35. Rhee S, Yang BH, Asada HH. Artifact-resistant power-efficient design of fingerring plethysmographic sensors. IEEE Trans Biomed Eng 2001;48:795–805.

See also HUMAN FACTORS IN MEDICAL DEVICES; MOBILITY AIDS; NUTRITION, PARENTERAL; QUALITY-OF-LIFE MEASURES, CLINICAL SIGNIFICANCE OF; TEMPERATURE MONITORING; TRANSCUTANEOUS ELECTRICAL NERVE STIMULATION (TENS).

HOSPITAL SAFETY PROGRAM.   See SAFETY PROGRAM, HOSPITAL.

HUMAN FACTORS IN MEDICAL DEVICES

DANIEL W. REPPERGER
Wright-Patterson Air Force Base
Dayton, Ohio

INTRODUCTION

The human factors issues related to the use and design of medical devices has experienced significant paradigm shifts since this topic was addressed > 20 years ago (1). Not only has the technology innovation of the Internet vastly affected how medical professionals both gather and report information, but also standards are now more easily established. In addition, technology in the healthcare industry has concomitantly made significant advances. The evolving characteristics of legal liability with medical devices have also changed. Concurrently, the skill and sophistication of users with computer-aided systems has significantly improved with more tolerance and acceptance of automation. The computer and microprocessor-based medical devices are now the pervasive means of humans dealing with mechanical–electrical systems. First, it is important to define the term Human Factors within the context of medical devices and biomedical engineering. The phrase human factors can be broadly characterized as the application of the scientific knowledge of human capabilities and limitations to the design of systems and equipment to generate products with the most efficient, safe, effective, and reliable operation. A modern expression to describe a systematic procedure to evaluate risk when humans use medical devices is termed human factors engineering (2). The U.S Food and Drug Administration (FDA) is a strong proponent of the use of human factors engineering to manage risk, in particular with application to medical devices. The responsibility of the FDA is to guarantee the safety and efficacy of drugs and medical devices. Since, in the United States, the FDA is one of the leading authorities on medical standards, it is worthwhile to review (3,4) their interpretation on how human factors studies should be conducted with medical device use as perceived by this group. Other U.S. government organizations, such as the National Institute of Health (NIH) (5) and the Agency for Health Care Research and Quality (6), also offer their perspective on the application of human factors studies with respect to the manipulation of medical devices. Other sources of government information are found at (7–12). Also available online are a number of legal sources (13–15) related to injury issues and instrumentation affiliated with healthcare and how they perceive the relevance of human factors engineering. In these sources, there is a strong influence on how human factors procedures have some bearing on liability, abuse, misuse, and other troublesome issues associated with medical devices (16).

From an historical perspective, in the late 1980s, data collected by the FDA demonstrated that almost one-half of all medical device recalls resulted from design flaws. In 1990, the U.S. Congress passed the Safe Medical Devices Act, giving the FDA the ability to mandate good manufacturing practices. These practices involve design controls for manufacturers to use human factors engineering principles within medical device design.

In this article, we initially discuss human factors as defined by the FDA followed by three classic case studies. The ramifications of legal issues are then presented. Concurrent good human factors methods are then described, followed by some key topic areas including alarms, labeling, automation, and reporting. Future issues regarding human factors and medical devices are subsequently offered with conclusions and future directions of this field depicted.

First, it is instructive to review the present state of affairs on how the FDA defines human factors engineering within the context of when humans interact with medical devices. The term human factors engineering is a persuasive term in the literature describing present FDA standards.

A HUMAN FACTORS ENGINEERING PERSPECTIVE FROM THE FDA

The goal of the FDA is to promote medical device designers to develop highly reliable devices. Human factors engineering is a phrase used to help understand and optimize how people employ and interact with technology. A host of literature describes human factors engineering in many eclectic areas (17–30). When medical devices fail or malfunction, this impacts patients, family members, and professional healthcare providers. A common term used to characterize the potential source of harm is a hazard. A hazard may arise in the use of a medical device due to the inherent risk of medical treatment, from device failures (malfunctions) and also from device use. Figure 1, from the FDA, displays possible sources of device failure hazards that impact the human factors issues in medical devices. Figure 1 may be deceptive in the presumption that equal hazards exist between use related and device failure. More correctly (3,4) the use contribution to the total medical devices hazards may far exceed those from the device failures. In fact, from an Institute of Medicine report (31), as many as 98,000 people die in any given year from medical errors that occur in hospitals. This is more than the number who die from motor vehicle accidents, breast cancer, or acquired immune deficiency syndrome (AIDS). A proportion of these errors may not directly be attributed to the medical device itself; however, the importance of incorporating human factors engineering principles into the early design and use of these important interfaces is a key concern. It is instructive to examine the two major types of errors (hazards) in Fig. 1 and how they are delineated. Risk analysis will refer to managing the forms of risks to be described herein. After the hazards are first clarified, the goal is for the hazards to be mitigated or controlled by modifying the device user interface (e.g., control or display characteristics, logic of operation, labeling) or the background of the users employing the device (training, limiting the use to qualified users). The power in the human factors approach is to help identify, understand, and address use-related problems as well as the original design problem with the physical device itself prior to its acceptance in the workplace of the healthcare professional. Some institutions have now developed in-house usability laboratories, in order to rigorously test any medical device

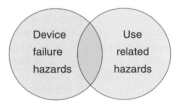

Figure 1. Hazards from device failure and use related.

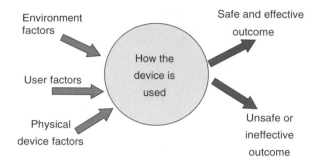

Figure 2. Human factors consideration in medical device use.

before utilization. It is worthwhile to first elaborate on use related hazards as they are relevant in this area.

Use-Related Hazards

Addressing the hazards related to device use, the essential components include (*1*) device users (patient, caregiver, physician, family member, etc.); (*2*) typical and atypical device use; (*3*) characteristics of the environment for the application of the medical device; and (*4*) the interaction between users, devices, and use environments.

Figure 2 portrays an abstraction on all possible uses for a medical device.

Device Failure Hazards

When understanding hazards from the perspective of risk analysis, it is common to consider the following classes of hazards as they pertain to device design: (*1*) chemical hazards (e.g., toxic chemicals); (*2*) mechanical hazards (e.g., kinetic or potential energy from a moving object); (*3*) thermal hazards (high temperature components); (*4*) electrical hazards (electric shock, electromagnetic interference); (*5*) radiation hazards (ionizing and nonionizing); and (*6*) biological hazards (allergic reactions, bioincompatibility, and infection).

In an effort to address good human factors design when dealing with medical devices, it is instructive to now discuss three classic misadventures in the medical device arena when human factors procedures could have been modified to preclude untoward events. One usually thinks of medical errors occurring, for example, in surgery (wrong site surgery), as the amputation of the wrong appendage (32,33) or from chemotherapy overdoses (34). In 2005, it is now a common practice, in holding areas before surgery, for the surgeon and medical team to discuss with the patient the impending surgery and to have the patient mark on his body precisely where the surgery will be performed with a magic marker. This procedure assures the patient that no confusion may occur as a function of a patient mix-up, after the patient is under anesthesia. Three examples are now presented of untoward events that could have been prevented with improved human factors use methodologies. In the final analysis, the enemy of safety is complexity, which appears in these case studies. Complex systems fail because of the contribution of multiple small failures, each individual failure may be insufficient to cause an accident, but in combination, the results may be tragic.

EXAMPLES OF CASE STUDIES WHERE HUMAN FACTORS ISSUES ARISE

It is worthwhile to examine a few classic case studies where human factors procedures interacting with medical devices needed to be reevaluated. It is emphasized that the errors described herein may not be attributed to any one person or system. Rather, the complex interaction of humans with poorly defined procedures involving certain medical devices has given rise to events, which were not planned or expected. However, with a more structured interaction of humans with these systems, improved results could be obtained. The reference by Geddes (35) provides many interesting examples where enhanced human factors planning would have prevented errors in medical treatment.

Case Study 1 from Ref. 35 and Reported in the South African *Cape Times* (1996)

"For several months, our nurses have been baffled to find a dead patient in the same bed every Friday morning" a spokeswoman for the Pelonomi Hospital (Free State, South Africa) told reporters. "There was no apparent cause for any of the deaths, and extensive checks on the air conditioning system, and a search for possible bacterial infection, failed to reveal any clues."

Although device failure could cause such deaths, why they occurred on Fridays is difficult to understand? The *Cape Times* later reported:

> It seems that every Friday morning a cleaner woman would enter the ward, remove the plug that powered the patient's life support system, plug her floor polisher into the vacant socket and then go about her business. When she had finished her chores, she would then replug the life support machine and leave, unaware that the patient was now dead. She could not, after all, hear the screams and eventual death rattle over the shirring of her polisher.
> "We are sorry, and have sent a strong letter to the cleaner in question. Further, the Free State Health and Welfare Department is arranging for an electrician to fit an extra socket, so there should be no repetition of this incident. The inquiry is now closed."

This example emphasizes that when unplanned influences or events interact with medical device operation, tragic results may occur. In a later section, we describe several human factors procedures and techniques that are now designed to help preclude these types of untoward events.

In Ref. 36, a second example shows how important (delicate) care is requisite to providing appropriate interaction of humans with medical devices.

Case Study 2 from Ref. 36

Besides putting patients at high risk for injury, clinicians who use a device they are not familiar with are placing themselves in legal jeopardy. The case of *Chin vs. St. Barnabos Medical Center* (160 NJ 454 [NJ 1999]) illustrates this point. A patient died after gas inadvertently was pumped into her uterus during a diagnostic hysteroscopy. Evidence established that the two perioperative nurses implicated in the case had no experience with the new hysteroscope and used the wrong hook up for the procedure. They connected the exhaust line to the outflow port. The 45-year-old patient died from a massive air embolism. It was also discovered that the nurses never received any education regarding the device. The manufacturer was not found liable because records indicated that the device did not malfunction or break. Damages in the amount of $2 million were awarded to the plaintiff and apportioned among the surgeon, nurses, and hospital. As discussed in the sequel, new training methods have been developed in the human factors area to preclude the occurrence of events of this type.

The last case study deals with a poor interface design. Usability Testing, to be discussed later, provides methods to preclude some of these difficulties encountered.

Case Study 3 from Ref. 37

"A 67 year-old man has ventricular tachycardia and reaches the emergency room. Using a defibrillator, nothing happens. The doctor suggests the nurse to start a fluid bolus with normal saline. The nurse opens the IV tubing wide open, but within minutes the patient starts seizing. The nurse then realizes that the xylocaine drip instead of the saline had been inadvertently turned up. The patient is then stabilized and the nurse starts her paperwork. She then realizes that the defibrillator was not set on the cardio version, but rather on an unsynchronized defibrillation. This is because the defibrillator she uses, every day, automatically resets to the nonsynchronized mode after each shock. The second shock must have been delivered at the wrong time during the cardiac cycle, causing ventricular fibrillation." By performing a usability study as described later, with better training, this type of event could have been prevented.

The changing effect of legal influences also has had its impact on human factor interactions of caregivers with medical devices. It is worthwhile to briefly describe some of these recent influences and how they impact on human dealings with medical devices.

NEW LEGAL INFLUENCES THAT AFFECT HUMAN FACTORS

As mentioned previously in Refs. 13–15, there have been a number of modifications in the legal system that affect how humans now interact with medical devices. It is not unusual in the year 2005 to hear prescription drugs advertised on television or on the radio with a disclaimer near the end of each commercial. The disclaimer lists major possible side effects and warns the user to discuss the drug with their physician. As the aging "baby boomers" of the post-World War II era now require more and more medications, the drug companies make major efforts and studies to ensure that safe and effective drugs are delivered to an ever increasing public audience. Experience from prior mistakes has significantly modified how the drug industry must deal with a larger, and more highly informed, popu-

lation base. Some major modifications that occur involving legal issues and medical device operation include

1. The legal profession (13) now recognizes the importance of human factors engineering [also known as usability engineering or ergonomics (a term used outside the United States)] in studying how humans interact with machines and complex systems. Ergonomics is a factor in the design of safe medical devices; A user-friendly device is usually a safe one (38).

2. Healthcare Institutions employ results of human factors engineering testing of devices in making key decisions in evaluation as well as major purchase judgments. If these components fail, but have been tested within a rigorous framework, the legal consequences are mitigated since standards were adhered to in the initial selection of the medical device and procedure of use.

3. Insurance premiums to Healthcare Institutions are correspondingly reduced if adherence to standards set forth by good human factors engineering principles are maintained. The insurance costs are directly related to the potential of legal expenses and thus sway the decisions on how medical devices are purchased and used.

A number of new ways of improving how human factor design with medical devices has evolved, which will now be discussed as pertinent to the prior discussion. These methods affect training, utilization procedures, design of the devices, testing, and overall interaction with caregivers in the workplace.

METHODS TO IMPROVE HUMAN FACTOR DESIGN AND MEDICAL DEVICES

Professionals working in the area of human factors have now devised a number of new means of improving how healthcare professionals can better deal with medical devices. We present some of the most popular methods in the year 2005, most of which now pervasively affect the development of a user's manual, training, and manner of use with respect to medical devices. Some of these techniques appear to have overlap, but the central theme of these approaches is to better assist the healthcare professional to mitigate untoward events. One of the most popular methods derived from human factors studies is cognitive task analysis (CTA). In short, the way that CTA works is that a primary task is subdivided up into smaller tasks that must be performed. It is necessary to specify the information needed to perform each subtask, and the decisions that direct the sequence of each subtask. Note, this type of task description is independent of the automation involved. For example, for the same tasks, information and decisions are required regardless of whether they are performed by a human or a machine. Also considered in this analysis are the mental demands that would be placed on the human operator while performing these selected subtasks.

Cognitive Task Analysis and How it Affects Human Factors and Medical Devices

The principle behind CTA is to take a large and complex task and divide it up into smaller subtasks (39). Each subtask should be attainable and reasonable within the scope of the user. If the subtask is not yet in the proper form, further subdivisions of that subtask are performed until the final subtask is in the proper form. The expression "proper form" implies the subtask is now attainable, sufficiently reasonable to perform, the proper information has been provided, and sufficient control is available to execute this subtask. An important aspect of CTA is to define if the user has the proper information set to complete his duties and also has the proper control means over the situation so that the task can be achieved adequately. Cognitive task analysis has great value in establishing a set of final subtasks that are both attainable and relevant to the overall mission. With this analysis, the caregiver can be provided better training and have an enhanced understanding of the role of each task within the overall mission. This procedure has been used in the nursing area for assessing risk of infants (40) and for patient-controlled analgesia machines (41). It has been noted (42) that 60% of the deaths and serious injuries communicated to the Medical Device Reporting system of the U.S. Food and Drug Administration (FDA) Center for Devices and Radiological Health have been attributed to operator error. Cognitive task analysis has evolved out of the original area of Critical Decision Methods (43) and is now an accepted procedure to analyze large and complex interactions of humans with machines.

A second popular method to shape procedures to interact with medical devices involves User Testing and Usability Engineering.

User Testing and How It Affects Human Factors and Medical Devices

User Centered Design has found popularity when humans have to interact with medical devices for ultrasound systems (44) and for people with functional limitations (45). Usability engineering methods are applied early in the system lifecycle to bridge the gap between users and technology. The ultimate goal is to design an easy to use system that meets the needs of its users. A basic principle of user-centered design is making design decisions based on the characteristics of users, their job requirements and their environments (46–50). It is often the complaint of human factors professionals that they are brought into the design process much too late to influence the construction of the overall system. It is all too common for the experimental psychologist to have to deal with an interface built without prior considerations of the human's limitations and preferences in the initial design construction. The usability engineering methods bring to light needs for users and tasks early on, and suggest specific performance testing prior to the recommendation of the final design of an interface.

A third method discussed here to influence how to work with medical devices involves Work Domain Analysis.

Work Domain Analysis and How It Affects Human Factors and Medical Devices

A third and popular method to better understand the interplay between human factors issues and medical devices is via an integrated method involving the technical world of physiological principles and the psychological world of clinical practice. This work domain analysis was originally proposed by Rasmussen et al. (51,52) and has found application in patient monitoring in the operating room (53). A variety of tables are constructed based on data to be utilized. Columns in the tables include a description of the task scenario and the relations between key variables of the work environment and the work domain. The tables portray interactions and different strategies that are elicited to help in monitoring and control.

As discussed earlier in Ref. 1, today many new advances have also been made in alarms. These inform the healthcare provider of troublesome events and many innovative changes and studies have been instituted in this area that warrant discussion. Alarm deficiencies compromise the ability to provide adequate healthcare. For alarms, some research has focused on the identification of alarm parameters that improve or optimize alarm accuracy (i.e., to improve the ratio of true positives to false positives: the signal/noise ratio).

ALARMS AND HUMAN FACTORS ISSUES

Alarms are key to the detection of untoward events when humans interact with medical devices. One does not want to generate designs that invite user error. We cannot deal with confusing or complex controls, labeling, or operation. Many medial devices have alarms or other safety devices. If, however, these features can be defeated without calling attention to the fact that something is amiss, they can be easily ignored and their value is diminished (54). The efficacy of alarms may be disregarded because it is not attention getting. For example, if a multifunction liquid-crystal display (LCD) has a low battery warning as its message, but is not blinking, it does not call attention to itself. Alternatively, an improved design occurs in the case of a low battery alarm design commonly found in household smoke detectors. In this case, a low battery will cause the unit to chirp once a minute for a week, during which the smoke detector is still functional. The chirp may be confusing at first, but it cannot be ignored for a week. A battery test button is still available for the testing when the battery power is satisfactory. Adequate alarm systems are important to design in a number of analogous medical scenarios. For example, use of auditory systems (55) is preferable to a visual display, since this reduces the visual workload associated with highly skilled tasks that may occur, for example, in the operating room. For anesthesia alarms (56), care must be exercised to not have too many low level alarms that indicate, for example, that limits are exceeded or that the equipment is not functioning properly. The danger of false positives (alarms sounding when not necessary) provides an opportunity for the user to ignore information, which may be critical in a slightly different setting. An example where information of this type cannot be ignored is in applications of human factors training to the use of automated external defibrillators. It is known that without defibrillation, survival rates drop by 10% for every minute that passes after cardiac arrest (57). A great deal of work still continues in the area of management of alarm systems in terms of their efficacy and utility (58,59).

Another significant issue with human factors is proper labeling. Medication misadventures are a very serious problem. We briefly summarize changes that impact how humans will interact with their prescriptions as well as medical devices in general and how they are influenced by their labeling constraints.

LABELING AND HUMAN FACTORS ISSUES

By government regulation and industry practice, instructions accompanying distribution of medical devices to the public are termed "labeling". Medical device labeling comprises directions on how to use and care for such practices. It also includes supplementary information necessary for the understanding and safety, such as information about risks, precautions, warning, potential adverse reactions, and so on. From a humans factors perspective, the instructions must have the necessary efficacy, that is, they must provide the correct information to the user. There are a number of standards on how the instructions must be displayed and their utility in the healthcare industry (60). For example, for prescription medications (61–64), they represent the most important part of outpatient treatment in the United States. This provides > 2 billion possible chances for patient error each year in the United States. To maximize the benefits and minimize the dangers of using these medications, users must comply with an often complex set of instructions and warnings. Studies show that seven specific system failures account for 78% of adverse drug events in hospitals. All seven of these failures could be corrected by better information systems that detect and correct for errors. The top seven system failures for prescription medications are (1) drug knowledge dissemination; (2) dose and identification checking; (3) patient

information availability; (4) order transcription error; (5) allergy defense; (6) medication order tracking; (7) improved interservice communication.

As mentioned previously, as computers and the Internet become more persuasive, patients, caregivers, doctors, and others become more tolerant and dependent on automation. The goal is to make a task easier, which is true most of the time. There are a number of issues with regard to automation that need to be addressed.

AUTOMATION ISSUES AND HUMAN FACTORS

As computers and microprocessor-based devices have become more ubiquitous in our modern age, there is increased tendency to foster automation (65) as a means of improving the interaction of users with medical devices. The original goal of automation was to reduce the workload (physical and mental) and complexity of a task to the user. This is specific to the desired response from the device of interest. There is an obvious downside of this concept. The idea that the automation has taken over and has a mind of its own is ghastly within human thinking. Also, if the automated system is too complex in its operation and the user is not comfortable in understanding its causality (input–output response characteristics), the trust in the device will decrease accordingly and the human–machine interaction will degrade. The classical work by Sheridan (66) defines eight possible levels of automation, as portrayed in Fig. 3. One easily sees the relationship of loss of control to increased automation. For medical device usage, this may be problematic to trade off simplicity of use to loss of control and eventual efficacy. Automation studies continue to be of interest (67–73).

REPORTING AND HUMAN FACTORS ISSUES

Reporting of failures of proper medical device operation has now commonly advanced to Web-based systems (3,74). The healthcare provider must be increasingly skilled with the use of computer systems. Sometimes the terms digital

Level 1 – The human must do it all; the computer offers no assistance.

Level 2 – The computer suggests alternative ways to do the task.

Level 3 – The computer selects one way to do the task – It executes the suggested task only if the human approves.

Level 4 – Same as Level 3 but the computer suggests the method.

Level 5 – Same as Level 4 but the human has a restricted time to veto.

Level 6 – Same as Level 5 but the computer executes automatically then informs the human.

Level 7 – The computer executes the task and informs the human only if the computer is asked.

Level 8 – The computer selects the method, executes the task and ignores the human.

Figure 3. The eight levels of automation.

divide is employed to distinguish those that have the requisite computer skills from those that are not as competent in this area. This may be a consequence of humans factors procedures instituted to deal with a more elderly patient group (75–82) who may not be comfortable with computers. Education is the best means to deal with this problem (83). It is important that modern healthcare givers who use medical devices in their work setting have the obligatory skills to accurately report failures and have the suitable computer training to make relevant reports to the necessary sources.

A number of new areas are growing and influence how human factors have been evolving with the interaction of medical devices. It is important to spend some time mentioning these contemporary and emergent areas.

NEW ISSUES INVOLVING HUMAN FACTORS AND MEDICAL DEVICES

With advances in technology, several new areas should be mentioned that seem to have relevance to novel advances in medical devices. The concept of telemedicine is now a developing field that certainly addresses the future uses of medical devices.

The Growth of Telemedicine

The term telemedicine literally means "medicine at a distance" (84) and is now an increasing and popular means of providing healthcare. The advantages are obvious in rural settings and numerous other scenarios (cf. Fig. 4). For example, having an expert physician or surgeon located at a remote and safe location, but performing a medical procedure on a person in a hazardous or distant environment provides a distinct advantage. The human at the hazardous environment may be placed in a battlefield in a combat situation, they may be in a space shuttle, or simply be in another country or distant location from the expert medical practitioner. Another simple example of telemedicine occurs in a simple endoscopic (colonoscopy) procedure or in a laparoscopic operation, for example, for knee surgery. For these procedures, a small insertion is made into

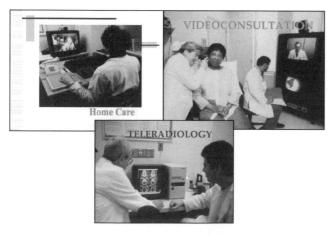

Figure 4. The concept of telemedicine.

the patient and the process is carried out from a remote location with the physician observing the process on a television monitor. The advantages are obvious: (1) with smaller entrance incisions or openings into the patient, the trauma is significantly reduced; (2) recovery time is much quicker; and (3) the risk of infection is substantially mitigated. Some of the new human factors issues regarding medical devices used within this context include (1) Dealing with the time delay between observing and making actions and the associated instabilities that may occur in the closed loop interaction; (2) having the lack of a sense of presence about a remote environment; and (3) having a strong dependence on mechanical systems or automation at the end-effector of the device inside the patient. These methods have been advanced to the point where robots are now being used to perform heart surgery, and so on, and some operations have been conducted over the Internet. Recent applications of robots performing heart procedures on humans need only two very small incisions in the patient. This allows for much quicker recovery time to the patient (1 vs. 7 weeks for a typical open heart surgery). From the surgeon's perspective, a significant advantage for the small incisions is that: It is not necessary to insert my hands inside the chest cavity. Thus the size of the incisions can be substantially reduced. One sees the disadvantage of traditional surgery in this area because *it is the size of the surgeon's hands being required to be inside the chest cavity as the only reason for a large incision in the chest cavity.* When the surgeon's hands no longer have to be inside the chest cavity, then the correspondingly two small incisions give rise to reduced possible infection, less trauma to the patient, and a shorter recovery time before the patient can return to normal work and living activities. This area of healthcare will only continue to advance as medical practices move more and more to this form of undertaking. In recent times, a number of studies in teleoperation have shown the efficacy of improving the sense of presence of the operator about the remote environment through "haptic" feedback. Haptic refers to forces reflected back on the physician (by various robotic interface devices) to improve their sense of presence about the remote environment, so the operator can "feel" the task much as they see it on a television monitor. A number of studies have shown both an improved sense of presence and performance about these teleoperation scenarios using haptic feedback (85–88). Auditory systems have also found analogous use in surgery (89). The problems in anesthesia are also well studied (90–93).

Another growth area includes more participation by the patient directly in their own healthcare.

The Growth of Increased Patient Participation in the Healthcare Process

As discussed previously, with the pervasive nature of automation, more and more of the healthcare responsibility and work will be performed by the patient, themself. Modern insurance procedures also encourage additional homecare scenarios and many times without a trained caregiver. This saves expensive care at the hospital, but transfers the burden onto the patient or their family members to become the primary caregiver. A paradigm shift of this type is a consequence of present insurance reimbursement procedures requiring the patient to now spend much of their time away from the hospital. New human factors issues are consequently introduced when dealing with medical devices in this scenario. The design of the human computer interface (94–97) now becomes critical for the efficacy of the healthcare provided. Even in cancer treatment, the responsibility of the proper administration of radioisotopes may become the burden of the patient (98) or if they have to manipulate their own chemotherapy level. For pain treatment (99–101) the patient has to be proactive in the selection of the analgesia level of the device provided. Modern TENS (Transcutaneous Electrical Nerve Stimulator) units now have been constructed to be wireless and shown to have equivalent efficacy in terms of pain management as compared to long wired units that have been in existence for >40 years (102). The movement to more wireless medical devices is certainly in the future. For the TENS units, by eliminating the long and entangling wires, this provides more reliability, less chance of wires breaking or shorting, more efficient use of electric power, but different forms of control with these analgesia devices. For example, the physician or caregiver may use a remote control to program the voltage level of the TENS microprocessor in a wireless manner rather than making manual adjustments with the traditional, long wired, TENS devices.

A third area of modern concern is the impact of electromagnetic fields on the operation of other medical devices, especially if they are implanted.

The Growth of Electromagnetic Fields on the Operation of Other Medical Devices

The Geddes reference (35) describes numerous examples of documented problems when medical devices inadvertently interact with unexpected exposure to external electromagnetic fields. Electromagnetic interference (EMI) is used to describe the malfunction of a device exposed to electromagnetic waves of all types that propagate through space. The EMI can intervene with electrodes on a patient, it can bias results for EEG recording of generalized epileptiform activity, and can give false positives to alarm systems, Silbert et al. (103). Other cases where malfunctions can occur involve heart machines, apnea monitors, ventilator mishaps, and in drug infusion pumps. Pacemakers are known to be affected by low frequency EMI signals. There are many exogenous sources of EMI including, but not limited to, Electrostatic Discharge, arc welding, ambulance sirens, and other sources (104). The growth of EMI is only increasing and human factors professionals need to carefully consider sources of problems from EMI that may have to be dealt with.

A fourth area of potential problems of humans factors interaction with medical devices occurs when two or more medical devices are simultaneously in operation, but their concurrent action may interact with each other in a destructive way.

The Potential Interaction of Multiple Medical Devices

As medical devices become more and more sophisticated, they may concurrently be in operation on the same patient (105–109). The action of one medical device may produce an undesired response of another device, especially if it may be implanted. From Geddes (35): "Between 1979 and 1995, the Center for Devices and Radiological Health (CDRH) of the U.S. Food and Drug Administration (FDA) has received over one hundred reports alleging the electromagnetic interference (EMI) resulted in malfunction of electronic medical devices." The source of the EMI was from one medical device treating the patient. The malfunction occurred in a second medical device, which was also, simultaneously, being used to treat the same patient.

"For example, about 100,000 cardiac pacemakers are implanted annually. These stimulators can be interrogated and programmed by an external device that uses a radio frequency link. Electro surgery, which also uses radio frequency electric current may interact with the pacemaker causing serious arrhythmias. Even though the two technologies are safe, when used alone, their simultaneous combination has resulted in injury." More specifically, nowadays with the prevalence use of cellular telephones for both caregivers as well as the patients in care situations, there are documented cases of resulting injury to the patient. Cell phones have now been recognized to cause instances of malfunction of drug infusion pumps and patient monitors. These interactions have to be considered for new human factors interactions with medical devices in the future, Silbergerg (110) with increased interest in the errors created (111,112) and the specific technology used (113).

With the changing style of litigation with respect to the medical profession, there has been more public awareness of sources of human factor error induced by the medical professional working in a state of extreme fatigue.

Increased Public Awareness to Fatigue Issues and the Medical Professional

There has now been a substantial increased awareness of the general public to the fact that their medical professional may have compromised performance due to the fact that they are suffering from long hours of work. Fatigue studies continue to receive increased concern (114–120). This certainly has its influence on human factors procedures when dealing with medical devices and the overall success of the medical interaction. For example (115), it is known that physicians had demonstrated levels of daytime sleepiness worse then that of patients with narcolepsy or sleep apnea when required to perform long hours of duty.

Finally, since the publication of Ref. 1, the healthcare industry must now deal with a substantially larger population of acquired immune deficiency syndrome (AIDS) survivors who need medical, dental, and other types of interactions with healthcare professionals (121).

Changing Medical Procedures to Deal with Active Human Immunodeficiency Virus Patients

With the advent of advanced drug therapies, people with human immunodeficiency virus (HIV) are now living longer and longer. These same people need dental care, have medical attention requests, and require other types of consideration. The medical professional must exercise forethought to not have exposure to body fluids and new procedures are in place to provide discrimination free care to these people. In the early days of public exposure to people with HIV, there were documented cases of health professionals refusing to give adequate care. For example, for resuscitation, fireman and others would avoid contact with individuals suspected of having HIV. New devices have now been constructed to keep body fluids and other contact more separated between the patient and the caregiver. There have been new laws passed to prevent discrimination to people suspected of having HIV in housing, in the workplace, and also in receiving adequate healthcare.

CONCLUSION

The modern human factors interactions with medical devices have been strongly influenced by the advent of new technologies including the Internet, microprocessors, and computers. People are becoming more accustomed to automation and dealing with other sophisticated means of delivering healthcare. One lesson that can be learned from the improvement of human factors interactions with medical devices is that we can create safety by anticipating and planning for unexpected events and future surprises. Another change in this new millennium is that the responsibility of the patient is now shifted more to the individual or their family to have a greater role and duty over their own therapies and venue, and perhaps work in their home setting. Telemedicine and wireless means of dealing with controls over the medical devices are certainly on the increase and will influence how the patient has to deal with their healthcare professional in the future.

BIBLIOGRAPHY

Cited References

1. Hyman WA. Human Factors in Medical Devices. In Webster JG editor Encyclopedia of Medical Devices and Instrumentation. 1st ed. New York: Wiley; 1988.
2. Fries RC. Reliable Design of Medical Devices. New York: Marcel Dekker; 1997.
3. U. S. Food and Drug Administration, Guidance for Industry and FDA Premarket and Design Control Reviewers–Medical Device Use-Safety: Incorporating Human Factors Engineering into Risk Management. Available at http://www.fda.gov/cdrh/humfac/1497.html.
4. Sawyer D. An Introduction to Human Factors in Medical Devices. U. S. Department of Health and Human Services, Public Health Service, FDA; December, 1996.
5. Murff HJ, Gosbee JW, Bates DW. Chapter 41. Human Factors and Medical Devices. Available at http://www.ncbi.nlm.nih.gov.books/.
6. Human Factors and Medical Devices. Available at http://www.ahrq.gov/clinc/ptsafety/chap41b.htm.
7. Association for the Advancement of Medical Instrumentation. Human Factors Engineering Guidelines and Preferred Practices for the Design of Medical Devices. ANSI/AAMI HE48-1993, Arlington (VA); 1993.
8. Backinger CL, Kingsley P. Write it Right: Recommendations for Developing User Instructions for Medical Devices in Home Health Care. Rockville (MD): Department of Health and Human Services; 1993.
9. Burlington DB. Human factors and the FDA's goals: improved medical device design. Biomed Instrum Technol Mar.–Apr., 1996;30(2):107–109.
10. Carpenter PF. Responsibility, Risk, and Informed Consent. In: Ekelmen KB, editors. New Medical Devices: Invention, Development, and Use. Series on Technology and Social Priorities. Washington (DC): National Academy Press; 1988. p 138–145.
11. Policy statements adopted by the governing council of the American Public Health Association. Am J Public Health. Nov. 12, 1997;88(3):495–528.
12. Reese DW. The Problem of Unsafe Medical Devices for Industry, Government, Medicine and the Public. Dissertation Abs International: Sect B: Sci Eng 1994;54(11-B):5593.
13. Claris Law, Inc. Use of Human factors in Reducing Device-related Medical Errors, available at http://www.injuryboard.com/.
14. Green M. An Attorney's Guide to Perception and Human Factors. Available at http://www.expertlaw.com/library/attyarticles/perception.html.
15. Green M. Error and Injury in Computers and Medical Devices. Available at http://www.expertlaw.com/library/attyaricles/computer_negligence.html.
16. Sokol A, Jurevic M, Molzen CJ. The changing standard of care in medicine-e-health, medical errors, and technology add new obstacles. J Legal Med 2002;23(4):449–491.
17. Bruckart JE, Licina JR, Quattlebaum M. Laboratory and flight tests of medical equipment for use in U.S. army medevac helicopters. Air Med J 1993;1(3):51–56.
18. Budman S, Portnoy D, Villapiano AJ. How to Get Technological Innovation Used in Behavioral Health Care: Build It and They Still Might Not Come. Psychother Theory, Res Practice, Training 40 (1–2), Educational Publishing Foundation; 2003. p 45–54.
19. Burley D, Inman WH, editors. Therapeutic Risk: Perception, Measurement, Management. Chichester: Wiley; 1988.
20. Hasler RA. Human factors design-what is it and how can it affect you?. J Intravenous Nursing May–Jun, 1996;19(3)(Suppl.): S5–8.
21. McConnell EA. How and what staff nurses learn about the medical devices they use in direct patient care. Res Nursing Health 1995;18(2):165–172.
22. Obradovich JH, Woods DD. Users as designers: how people cope with poor HCI design in computer-based medical devices. Human Factors 1996;38(4):574–592.
23. Phillips CA. Human Factors Engineering. New York: Wiley; 2000.
24. Senders JW. Medical Devices, Medical Errors, and Medical Accidents. Human Error in Medicine. Hillsdale (NJ): Lawrence Erlbaum Associates, Inc.; 1994. p 159–177.
25. Ward JR, Clarkson PJ. An analysis of medical device-related errors: prevalence and possible solutions. J Med Eng Technol Jan–Feb, 2004;28(1):2–21.
26. Goldmann D, Kaushal R. Time to tackle the tough issues in patient safety. Pediatrics Oct. 2002;110(4):823–827.
27. Gosbee J. Who Left the Defibrillator On? Joint Comm J Quality Safety May, 2004;30(5):282–285.
28. Kaptchuk TJ, Goldman P, Stone DA, Stason WB. Do medical devices have enhanced placebo effects? J Clin Epidemiol 2000;53:786–792.
29. Lambert MJ, Bergin AE. The Effectiveness of Psychotherapy. In: Bergin AE, Garfield SL, editors. Handbook of Psychotherapy and Behavior Chance. 4th ed. New York: Wiley; 1994. p 143–189.
30. Perry SJ. An Overlooked Alliance: Using human factors engineering to reduce patient harm. J Quality Safety 2004; 30(8):455–459.

31. Leape L. Error in medicine. J Am Med Assoc 1994;21(3):272.
32. Leape LL. The preventability of medical injury. In: Bogner MS, editor. Human Error in Medicine Hillsdale (NJ): Lawrence Erbaum Associates; 1994. p 13–25.
33. Ganiats T. Error. J Am Med Asso 1995;273:1156.
34. Bogner MS. Medical human factors. Proc Human Factors Ergonomics Soc, 40th Annu Meet; 1996. p 752–756.
35. Geddes LA. Medical Device Accidents. 2nd ed. Lawyers and Judges, Publishers 2002.
36. Wagner D. How to use medical devices safely. AORN J Dec. 2002;76(6):1059–1061.
37. Fairbanks RJ, Caplan S. Poor interface design and lack of usability testing facilitate medical error. Human Factors Engineering Series. J Quality Safety Oct. 2004;30(10):579–584.
38. Phillips CA. Functional Electrical Rehabilitation, Springer-Verlag, 1991.
39. Militello LG. Learning to think like a user: using cognitive task analysis to meet today's health care design challenges. Biomed Instrum Technol 1998;32(5):535–540.
40. Militello L. A Cognitive Task Analysis of Nicu Nurses' Patient Assessment Skills. Proc Human Factors Ergonomics Soc, 39th Annu Meet; 1995. p 733–737.
41. Lin L, et al. Analysis, Redesign, and Evaluation of a Patient-Controlled Analgesia Machine Interface. Proc Human Factors Ergonomics Soc, 39th Annu Meet; 1995. p 738–741.
42. Bogner MS. Medical devices and human error. In: Mouloua M, Parasuraman R, editors. Human Performance in Automated Systems: Current Research and Trends. Hillsdale (NJ): Erlbaum; 1994. p 64–67.
43. Klein GA, Calderwood R, MacGregor D. Critical decision method for eliciting knowledge. IEEE Trans. Systems, Man, Cybernetics 1989;19(3):462–472.
44. Aucella AF, et al. Improving Ultrasound Systems by User-Centered Design. Proc Human Factors Ergonomics Society, 38th Annu Meet; 1994. p 705–709.
45. Law CM, Vanderheiden GC. Tests for Screening Product Design Prior to User Testing by People with Functional Limitations. Proc Human Factors Ergonomics Soc, 43rd. Annu Meet; 1999. p 868–872.
46. Neilsen J. Usability Engineering. Boston: Academic; 1993.
47. Whiteside BJ, Holtzblatt K. Usability engineering: our experience and evolution. In: Helander M, editor. The Handbook of Human Computer Interaction. New York: Elsevier Press; 1988.
48. Nielsen J. Heuristic evaluation. In: Nielson J, Mack R, editors. Usability Inspection Methods. New York: Wiley; 1994. p 54–88.
49. Welch DL. Human factors in the health care facility. Biomedical Instrum Technol. May–Jun, 1998;32(3):311–316.
50. Lathan BE, Bogner MS, Hamilton D, Blanarovich A. Human-centered design of home care technologies. NeuroRehabilitation 1999;12(1):3–10.
51. Rasmussen J, Pejtersen AM, Goodstein LP. Cognitive Systems Engineering. New York: Wiley; 1994.
52. Rasmussen J. Information Processing and Human-Machine Interaction: An Approach to Cognitive Engineering. New York: North-Holland; 1986.
53. Hajdukiewicz JR, et al. A Work Domain Analysis of Patient Monitoring in the Operating Room. Proc Human Factors Ergonomics Soc, 42th Annu Meet; 1998. p 1038–1042.
54. van Gruting CWD. Medical Devices–International Perspectives on Health and Safety. New York: Elsevier; 1994.
55. Simons D, Fredericks TK, Tappel J. The Evaluation of an Auditory Alarm for a New Medical Device. Proc Human Factors Ergonomics Soc, 41st Annu Meet; 1997. p 777–781.
56. Seagull FJ, Sanderson PM. Anesthesia Alarms in Surgical Context. Proc Human Factors Ergonomics Soc, 42nd. Annu Meet; 1998. p 1048–1051.
57. Aguirre R, McCreadie S, Grosbee J. Human Factors and Training Evaluation of Two Automated External Defibrillators. Proc Human Factors Ergonomics Soc, 43rd Annu Meet; 1999. p 840–844.
58. Woods DD. The alarm problem and directed attention in dynamic fault management. Ergonomics 1995;38(11):2371–2394.
59. Laughery KR, Wogalter MS. Warnings and risk perception. design for health and safety. In: Salvendy G, editor. Handbook of Human Factors and Ergonomics. 2nd ed. New York: Wiley; 1997. p 1174–1197.
60. Callan JR, Gwynee JW. Human Factors Principles for Medical Device Labeling. Available at http://www.fda.gov/cdrh/dsma/227.html.
61. Isaacson JJ, Klein HA, Muldoon RV. Prescription Medication Information: Improving Usability Through Human Factors Design. Proc Human Factors Ergonomics Soc, 43rd Annu Meet; 1999. p 873–877.
62. Collet JP, Bovin JF, Spitzer WO. Bias and confounding in pharmacoepidemiology. In: Strom BL, editor. Pharmacoepidemiology. New York: Wiley; 1994. p 741.
63. Senn S. Statistical Issues in Drug Development. New York: Wiley; 1997.
64. Twomey E. The usefulness and use of second-generation antipsychotic medications: review of evidence and recommendations by a task force of the World Psychiatric Association. Curr Opinion Psychiat 2002;15(Suppl 1):S1–S51.
65. O'Brien TG, Charlton SG. Handbook of Human Factors Testing and Evaluation. Lawrence Erlbaum Associates; 1996.
66. Sheridan TB. Humans and Automation: System Design and Research Issues. New York: Wiley; 2002.
67. Obradovich JH, Woods DD. Users as Designers: How People Cope with Poor HCI Design in Computer-based Medical Devices. Human Factors 1996;38(4):574–592.
68. Howard SK. Failure of an automated non-invasive blood pressure device: the contribution of human error and software design flaw. J Clin Monitoring 1993; 9.
69. Sarter NB, Woods DD, Billings CE. Automation surprises. In: Salvendy G, editor. Handbook of Human Factors/Ergonomics. 2nd ed. New York: Wiley; 1997. p 1926–1943.
70. Andre J. Home health care and high-tech medical equipment, caring. Nat Assoc Home Care Mag Sept. 1996; 9–12.
71. Dalby RN, Hickey AJ, Tiano SL. Medical devices for the delivery of therapeutic aerosols to the lungs. In: Hickey AJ, editor. Inhalation Aerosols: Physical and Biological Basis for Therapy. New York: Marcel Dekker; 1996.
72. Draper S, Nielsen GA, Noland M. Using 'No problem found' in infusion pump programming as a springboard for learning about human factors engineering. Joint Comm J Quality Safety Sept. 2004;30(9):515–520.
73. Weinger MB, Scanlon TS, Miller L. A widely unappreciated cause of failure of an automatic noninvasive blood pressure monitor. J Clin Monitoring Oct. 1992;8(4):291–294.
74. Walsh T, Beatty PCW. Human factors error and patient monitoring. Physiol Meas 2002;23:R111–132.
75. Agree EM, Freedman VA. Incorporating assistive devices into community- based long-term care: an analysis of the potential for substitution and supplementation. J Aging Health 2000;12:426–450.
76. Rogers WA, Fisk AD. Human Factors Interventions for the Health Care of Older Adults. Mahwah (NJ): Lawrence Erlbaum Associates, Publishers; 2001.
77. Vanderheiden GC. Design for people with functional limitations resulting from disability, aging, or circumstance. In: Salvendy G, editor. Handbook of Human Factors and Ergonomics. 2nd ed. New York: Wiley; 1997. p 2010–2052.

78. Billing J. The Incident Reporting and Analysis Loop. In Enhancing Patient Safety and Reducing Medical Errors in Health Care. Chicago: National Patient Safety Foundation; 1999.

79. Fisk D, Rogers WA. Psychology and aging: enhancing the lives of an aging population. Curr Directions Psychol Sci Jun. 2002;11(3):107–111.

80. Gardner-Bonneau D. Designing medical devices for older adults. In: Rogers WA, Fisk AD, editors. Human Factors Interventions for the Health Care of Older Adults Mahwah (NJ): Erlbaum; 2001. p 221–237.

81. Park C, Morrell RW, Shifren K. Processing of medical information in aging patients: cognitive and human factors perspectives. Lawrence Erlbaum Associates; 1999.

82. Sutton M, Gignac AM, Cott C. Medical and everyday assistive device use among older adults with arthritis. Can J Aging 2002;21(4):535–548.

83. Glavin RJ, Maran NJ. Practical and curriculum applications integrating human factors into the medical curriculum. Med Educ 2003;37(11):59–65.

84. Birkmier-Peters DP, Whitaker LA, Peters LJ. Usability Testing for Telemedicine Systems: A Methodology and Case Study. Proc Human Factors Ergonomics Soc, 41st Annu Meet; 1997. p 792–796.

85. Repperger DW. Active force reflection devices in teleoperation. IEEE Control Systems Jan. 1991;11(1) 52–56.

86. Repperger DW, et al. Effects of haptic feedback and turbulence on landing performance using an immersive cave automatic virtual environment (CAVE). Perceptual Motor Skills 2003;97:820–832.

87. Repperger DW. Adaptive displays and controllers using alternative feedback. CyberPschol Behavior 2004;7(6):645–652.

88. Repperger DW, Phillips CA. A haptics study involving physically challenged individuals. Encyclopedia of Biomedical Engineering. New York: Wiley; 2005.

89. Wegner CM, Karron DB, Surgical navigation using audio feedback. Studies in Health Technology and Informatics. 1997;39:450–458.

90. Cooper J. An analysis of major errors and equipment failures in anesthesia management: considerations for prevention and detection. Anesthesiology 1984;60:34–42.

91. Gaba DM, Howard SK, Jump B. Production pressure in the work environment: California anesthesiologists' attitudes and experiences. Anesthesiology 81: 1994; 488–500.

92. Howard SK, et al. Anesthesia crisis resource management training: teaching anesthesiologists to handle critical incidents. Aviation Space Environ Med 63:763–770.

93. Weinger MB. Anesthesia incidents and accidents. Misadventures in health care: Inside Stories. Mahwah, (NJ): Lawrence Erlbaum Associates, Publishers; 2004. p 89–103.

94. Obradovic JH, Woods DD. Users as Designers: How People Cope with Poor HCI Design in Computer-Based Medical Devices. Proc Human Factors Ergonomics Soc, 38th Ann Meet; 1994. p 710–714.

95. Kober, Mavor A, editors. Safe, Comfortable, Attractive, and Easy to Use: Improving the Usability of Home Medical Devices. Report of National Research Council to U.S. Congress 1996, Washington (DC): National Academy Press; 1996. p 5–8.

96. Baldwin GM. Experiences of Siblings of In-Home Technology-Dependent Children. Dissertation Abst Int: Sec B: Sci Eng 1997;58(5-B):2714.

97. Mykityshyn AL, Fisk AD, Rogers WA. Learning to use a home medical device: mediating age-related differences with training. Human Factors 2002;44(3):354–364.

98. Schoenfeld I. Risk Assessment and Approaches to Addressing Human Error in Medical Uses of Radioisotopes. Panel: Proc Human Factors Ergonomics Soc, 37th Ann Meet; 1993. p 859–862.

99. Lin L. Human Error n Patient-Controlled Analgesia: Incident Reports and Experimental Evaluation. Proc Human Factors Ergonomics Soc, 42th Ann Meet; 1998. p 1043–1047.

100. Lin L, et al. Applying human factors to the design of medical equipment: patient controlled analgesia. J Clin Monitoring 1998;14:253–263.

101. McLellan H, Lindsay D. The relative importance of factors affecting the choice of bathing devices. Pain B J Occup Therapy 2003;66(9):396–401.

102. Repperger DW, Ho CC, Phillips CA. Clinical short-wire TENS (transcutaneous electric nerve stimulator) study for mitigation of pain in the Dayton va medical center. J Clin Eng Sept./Oct. 1997; 290–297.

103. Silbert PL, Roth PA, Kanz BS. Interference from cellular telephones in the electroencephalogram. J Polysomnographic Technol Dec. 1994;10:20–22.

104. Radiofrequency interference with medical devices. A technical information statement. IEEE Eng Med Bio Mag 1998;17(3):111–114.

105. Roy G. Child-proofing of hearing aids to prevent hazards posed by battery swallowing. J Speech-Language Pathol Audiol 1992;16(3):243–246.

106. Beery TA, Sommers M, Sawyer M, Hall J. Focused life stories of women with cardiac pacemakers. Western J Nursing Res 2002;24(1):7–23.

107. Romano PS. Using administrative data to identify associations between implanted medical devices and chronic diseases. Ann Epidemiol 2000;10:197–199.

108. Wiederhold K, Wiederhold MD, Jang DP, Kim SI. Use of cellular telephone therapy for fear of driving. CyberPsychol Behavior 2000;3(6):1031–1039.

109. Zinn HK. A Retrospective Study of Anecdotal Reports of the Adverse Side Effects of Electroconvulsive Therapy. Dissertation Abs Int Sec B: Sci Eng 2000;60(9-B):4871.

110. Silbergerg J. What Can/Should We Learn From Reports of Medical Device Electromagnetic Interference? At Electromagnetic, Health Care and Health, EMBS 95, Sept. 19–20, 1995, Montréal, Canada: Standards Promulgating Organizations; 1995.

111. Leape LL, et al. Promoting patient safety by reducing medical errors. JAMA Oct. 1998;280:28 1444–1447.

112. Rasmussen J. The concept of human error: Is it useful for the design of safe systems in health care? In: Vincent C, DeMoll B, editors. Risk and Safety in Medicine. London: Elsevier; 1999.

113. Woods DD, Cook RI, Billings CE. The impact of technology on physician cognition and performance. J Clin Monitoring 1995;11:92–95.

114. Gaba DM, Howard SK. Fatigue among clinicians and the safety of patients. N Engl J Med 2002;347:1249–1255.

115. Howard SK, Gaba DM, Roseking MR, Zarcone VP. Excessive daytime sleepiness in resident physicians: risks, intervention, and implication. Acad Med 2002;77:1019–1025.

116. Cook RI, Render ML, Woods DD. Gaps in the continuity of care and progress on patient safety. Br Med J March 18, 2000;320:791–794.

117. Fennell PA. A Fourx-phase approach to understanding chronic fatigue syndrome. In: Jason LA, Fennell PA, Taylor RR., editors. The Chronic Fatigue Syndrome Handbook 2003. Hoboken (NJ): Wiley; 2003. p 155–175.

118. Fennell PA. Phase-based interventions. In: Jason LA, Fennell PA, Taylor RR, editors. The Chronic Fatigue Syndrome Handbook. Hoboken (NJ): Wiley; 2003. p 455–492.

119. Jason LA, Taylor RR. Community-based interventions. In: Jason LA, Fennell PA, Taylor RR, editors. The Chronic Fatigue Syndrome Handbook. Hoboken (NJ): Wiley; 2003. p 726–754.

120. Rogers SH. Work physiology—fatigue and recovery. The human factors fundamentals. In: Salvendy G, editor. Handbook of Human Factors and Ergonomics. 2nd ed. New York: Wiley; 1997. p 268–297.

121. Roy F, Robillard P. Effectiveness of and compliance to preventive measures against the occupational transmission of human immunodeficiency virus. Scand J Work Environ Health 1994;20(6):393–400.

See also CODES AND REGULATIONS: MEDICAL DEVICES; EQUIPMENT MAINTENANCE, BIOMEDICAL; HOME HEALTH CARE DEVICES; MONITORING IN ANESTHESIA; SAFETY PROGRAM, HOSPITAL.

HUMAN SPINE, BIOMECHANICS OF

VIJAY K. GOEL
ASHOK BIYANI
University of Toledo, and
Medical College of Ohio,
Toledo Ohio

LISA FERRARA
Cleveland Clinic Foundation
Cleveland, Ohio

SETTI S. RENGACHARY
Detroit, Michigan

DENNIS MCGOWAN
Kearney Notabene

INTRODUCTION

From a bioengineer's perspective, bio the spine involves an understanding of the interaction among spinal components to provide the desired function in a normal person. Thereafter, one needs to analyze the role of these elements in producing instability. Abnormal motion may be due to external environmental factors to which the spine is subjected to during activities of daily living (e.g., impact, repetitive loading, lifting) degeneration, infectious diseases, injury or trauma, disorders, and/or surgery. Furthermore, the field of spinal biomechanics encompasses a relationship between conservative treatments, surgical procedures, and spinal stabilization techniques. Obviously, the field of spinal biomechanics is very broad and it will not be practical to cover all aspects in one article. Consequently, this article describes several of these aspects, especially in the cervical and thoraco-lumbar regions of the human spine. A brief description of the spine anatomy follows since it is a prerequisite for the study of bio the human spine.

SPINE ANATOMY

The human spinal column consists of 33 vertebras interconnected by fibrocartilaginous intervertebral disks (except the upper most cervical region), articular facet capsules, ligaments, and muscles. Normally, there are 7 cervical vertebras, 12 thoracic vertebras, 5 lumbar vertebras, and 5 fused sacral vertebras, Fig. 1a (1). When viewed in the frontal plane, the spine generally appears straight and symmetric while revealing four curves in the sagittal plane. The curves are anteriorly convex or lordotic in the cervical and lumbar regions, and posteriorly convex or kyphotic in the thoracic and sacrococcygeal regions. The center of gravity of the spinal column generally passes from the dens of the axis (C2) through the vertebra to the promontory of the sacrum (2,3). The ligamentous spine anatomy can be best described through a functional spinal unit (FSU, Fig. 1b), comprising the two adjacent vertebras, the disk in between, and the other soft tissues structures. This segment can be divided into anterior and posterior columns. The anterior column consists of the posterior longitudinal ligament, intervertebral disk, vertebral body, and anterior longitudinal ligament. Additional stability is provided by the muscles that surround the ligamentous spine, Fig. 1c. The motion of this segment can be described as rotation about three axes and translation along the same axes, Fig. 2. In the following paragraphs, the anatomy of the cervical region is described in some detail followed by a descriptive section discussing the anatomy of the lumbar spine.

Cervical Spine Anatomy

The cervical spine usually is subdivided in two regions (upper and lower), based on the functional aspects and anatomical differences between the two regions. The lumbar region anatomy, in principle, is similar to the lower cervical region.

Upper Cervical Spine (C0-C1-C2)

The upper cervical spine has been commented to be the most complex combination of articulations in the human skeleton. This region is also commonly called the "cervicovertebral junction" or the "craniovertebral junction" (CVJ). It is composed of three bony structures: the occipital bone (C0), the atlas (C1), and the axis (C2, Fig. 3). The atlas (C1), serves to support the skull. The atlas is atypical of other cervical vertebras in that it possesses neither a vertebral body nor a spinous process. The lateral masses of the atlas have both superior and inferior articular facets. The superior facets are elongated, kidney-shaped, and concave, and serve to receive the occipital condyles. The inferior facets are flatter and more circular and permit axial rotation. Transverse processes extend laterally from each lateral mass. Within each transverse process is a foramen that is bisected by the vertebral artery. The second cervical vertebra, or axis (C2), is also atypical of other cervical vertebra due to its osseous geometry (5,6). The most noteworthy geometric anomaly is the odontoid process, or dens. The odontoid process articulates with the anterior arch of the atlas. Posterior and lateral to the odontoid process are the large, convex superior facets that articulate with the inferior facets of C1. The inferior facets of C2 articulate with the superior facets of C3. The axis contains a large bifid spinous process that is the attachment site delineating the craniovertebral and subaxial musculature and ligament anatomies.

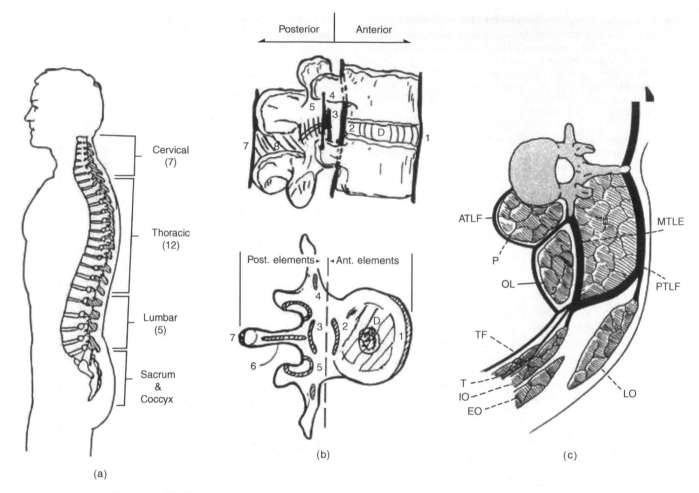

Figure 1. The ligamentous human spine. (a) The side view showing the three curvatures. (b) The functional spinal unit (FSU) depicts the spinal elements that contribute to its stability. (c) Additional stability is provided by the muscles that surround the spine. (Taken from Ref. 1.)

The trabecular anatomy of weight bearing bones provides information about the normal loading patterns of the bones, fracture mechanisms, and fixation capabilities. According to Heggeness and Doherty (6) the medial, anterior cortex of the odontoid process (1.77 mm at the anterior promontory) was found to be much thicker than the anterolateral (1.00 mm), lateral (1.08 mm), and posterior (0.84 mm) aspects of the axis. These authors feel that this is suggestive of bending and torsional load carrying capabilities. The same was found for the vertebral body, with thinner cortices were noted in the anterolateral and posterior directions. The trabecular bone in the tip of the odontoid process was found to be dense, maximizing in the anterior aspect of the medullary canal. One observation made by the authors was an area of cortical bone density at the center near the tip, which would seem to indicate that this area experiences elevated external forces, perhaps due to local ligamentous attachments. The lateral masses immediately inferior to the facets demonstrated dense regions of trabecular bone, with individual trabeculas spanning from this region to the inferior end plate, suggestive of a major axial load path.

The ligamentous structures of the upper cervical spine form a complex architecture (Fig. 3) that serves to join the

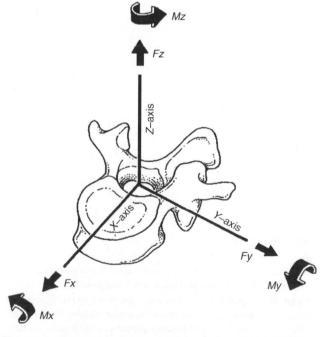

Figure 2. The spinal motion consists of six components (three translations and three rotations). (Adapted from Ref. 2.)

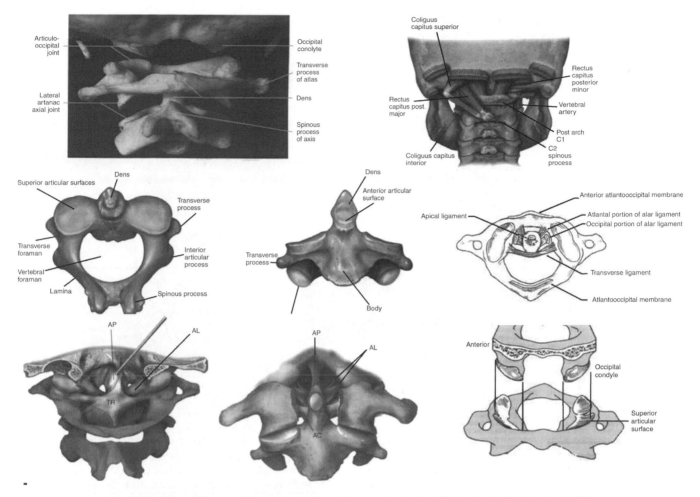

Figure 3. The anatomy of the upper region of the cervical spine C0 (occiput)-C1 (Atanlanto)-C2 (Axial). (Taken from Ref. 4c.)

vertebras, allow limited motion within and between levels, and provide stability. The cruciform ligament as a whole consists of two ligaments: the atlantal transverse ligament and the inferior–superior fascicles. The transverse ligament attaches between the medial tubercles of the lateral masses of the atlas, passing posterior to the odontoid process. Attachment of the cervical spine to the skull is also achieved by the paired alar ligaments. These ligaments run bilaterally from the occiptal condyles inferiolaterally to the tip of the odontoid process. The alar ligaments also contain fibers that run bilaterally from the odontoid process anterolaterally to the atlas. These ligaments have been identified as a check against overaxial rotation of the cranovertebral junction. Extending from the body of the axis to the inner surface of the occiput, the tectorial membrane is the most posterior ligament and actually represents the cephalad extension of the subaxial posterior longitudinal ligament. The tectorial membrane has been implicated as a check against extreme flexion motion. The apical dental ligament extends from the anterior portion of the magnum foramen to the tip of the odontoid process. The accessory atlantoaxial ligaments are bilateral structures that run between the base of the odontoid process and the lateral masses of the atlas. The most anterior of the major ligaments is the anterior longitudinal ligament.

This ligament extends inferiorly from the anterior margin of the foramen magnum to the superior surface of the anterior arch of the atlas at the anterior tuberosity. The ligament continues inferiorly to the anterior aspect of the axial body. The nuchal ligament (ligamentum nuchae) extends from the occiput to the posterior tubercle of the axis, continuing inferiorly to the spinous process of the subaxial vertebras (7).

There are six synovial articulations in the occipitoatlantoaxial complex: the paired atlanto-occipital joints, the paired atlantoaxial joints, the joint between the odontoid process and the anterior arch of the atlas, and the joint formed by the transverse ligament and the posterior aspect of the odontoid process, Fig. 3. The bilateral atlanto-occipital joints are formed from the articulation of the occipital condyles with the superior facets of the atlas. These joints are relatively stable due to the high degree of congruence between the opposing surfaces and the marked rounding that is displayed by both sides. They allow flexion and extension, limited lateral bending, and almost no rotation. The lack of allowed rotation is thought to be due to the ellipsoid form of the joint itself. Bilateral articulation of the inferior facets of the atlas with the superior facets of the axis form the atlantoaxial joints. Relatively small contact areas and opposed convexity

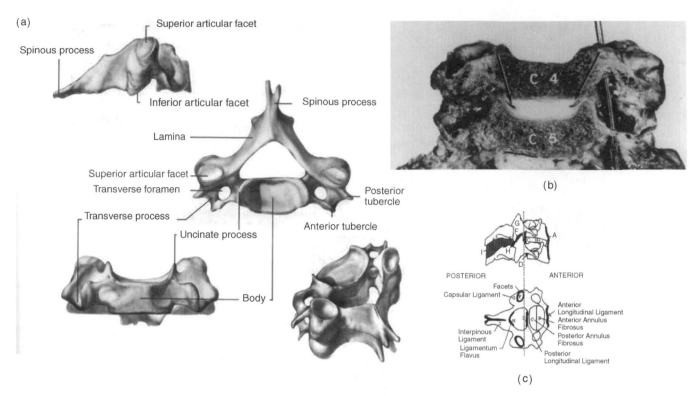

Figure 4. Anatomy of the lower cervical spine region. (Taken from Ref. 4d.)

result in a rather unstable joint. Movement is permitted in almost all six degrees of freedom: left and right axial rotation, flexion–extension, right and left lateral bending. Anteroposterior translation stability of this articulation is highly dependent on the transverse ligament. The odontoid process articulates anteriorly with the posterior aspect of the anterior atlantal ring. The joint is actually a bursal joint, with absence of specific capsular ligaments. The posterior aspect of the odontoid process and the transverse ligament form a joint via a bursa junction, creating the most unique articulation in the craniovertebral junction. This is necessitated by the large degree of axial rotation afforded at the atlantoaxial level.

Lower Cervical Spine (C3-C7). The lower cervical spinal vertebral column consists of osseous vertebras separated by fibrocartilaginous intervertebral disks anteriorly, facet joint structures posteriorly, and a multitude of ligamentous structures that provide stability and serve as motion control. Motion between adjacent vertebras is relatively limited due to these constraints, although overall motion of the lower cervical region is quite extensive. The lower cervical spine consists of five vertebras (C3-C7).

Cervical Vertebrae (Fig. 4a): The vertebral body is roughly in the shape of an elliptical cylinder and has a concave superior surface (due to the uncinate processes) and a convex inferior surface. A thin cortical shell (~ 0.3 mm thick anteriorly and 0.2 mm thick posteriorly) surrounds the cancellous bone of the inner vertebral body, while the superior and inferior surfaces of the vertebral body form the cartilaginous endplates, to which the intervertebral disks are attached. The superior aspect of each

vertebra contains the uncinate process or uncus, a dorsolateral bilateral bony projection, which gives the body a concave shape superiorly in the coronal plane and allows for the vertebral body to fit around the convex inferior surface of the immediately superior vertebra. The height of these processes vary from level to level, but the highest uncinate processes are located at C5 and C6 (as high as 9 mm from the flat surface of the endplate) and the smallest are located at C3 and C7 (8–10). Vertebral bodies transmit the majority of load.

The transverse process of the vertebra contains the intervertebral foramen. The intervertebral foramen is elliptical or round in shape, and hides and protects the neurological and vascular structures of the cervical spine, specifically the vertebral artery. Also, the rostral side of each bilateral transverse process is grooved to allow space for the exiting spinal nerve root.

The bilateral diarthroidal facet (or zygapophyseal) joints are located posteriorly to the pedicles both superiorly and inferiorly. The average orientation for the C3-C7 facet joints is $\sim 45°$ from the transverse plane, with steeper inclinations in the lower segments (11). This inclination allows far less axial rotation than occurs in the upper cervical spine. Together with the vertebral body (and intervertebral disks), the facets fulfill the primary role of load bearing in the spine. Typically a "three-column" aspect is applied to the cervical spine, consisting of bilateral facets and the anterior column (vertebral body plus intervertebral disk).

The pedicles, lamina, and spinous process of the cervical spine are made of relatively dense bone and, together with the posterior aspect of the vertebral body, form the spinal

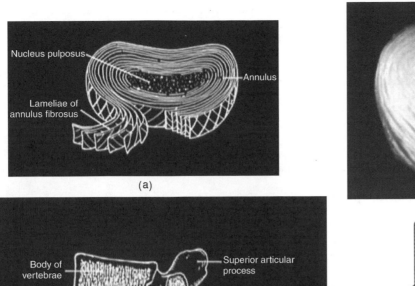

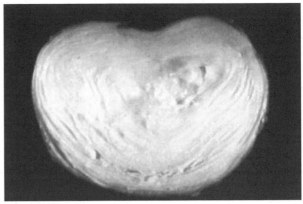

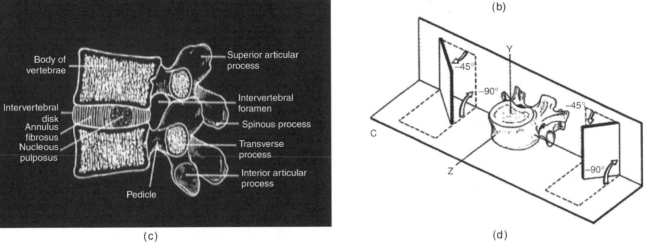

Figure 5. The anatomy of the lumbar spine. (a) and (b) show schematics of the disk and an actual disk (c) A FSU in the lumbar region. (d) The facet orientation in the lumbar region is more sagittal as compared to the other regions of the spine. (Adapted from Ref. 2.)

canal, within which lies the spinal cord. There are many ligament attachment points in this region and ligaments allow for resistance of flexion motion in the cervical spine.

The typical sagittal cervical spine alignment is thought to be a lordotic contour (11–15). The total average cervical lordosis was found to be $40 \pm 9.7°$ for C0-C7, with the majority of this lordosis occurring at the C1-C2 level ($31.9 \pm 7.0°$), and only 15% of the total lordosis occurring at the C4-C7 levels combined. The normal population seem to exhibit lordosis that ranges between 15 and 40°.

The Intervertebral Disk

Figure 5 forms the main articulation between adjacent vertebral bodies in the spine. It has the ability to transmit and distribute loads that pass between adjacent vertebral bodies. Its structure is a composite formation of outer layers of lamellas sheets called the annulus fibrosis, which surrounds the inner region of hydrophylic proteoglycan gel embedded in a collagen matrix called the nucleus pulposus. The material properties of the intervertebral disk appear to change markedly as a result of the aging process The matrix in which collagen and elastin fibers are embedded is composed of proteoglycan aggregates formed from

proteoglycan subunits, hyaluronic acid, and link protein (16). In soft tissues, such as the intervertebral disk and cartilage, the proteoglycan aggregates are immobilized within a fibrous network and play a major biological role in the structure of collagen, in turn playing a major mechanical role in the intervertebral disk integrity. The viscoelastic properties of the intervertebral disk can be attributed to the interaction between the collagen fibrils and proteoglycan matrix composing the nucleus pulposus of the intervertebral disk. The proteoglycans function to attract fluids into the matrix, while the collagen fibers provide the tensile strength to the disk. As the human spine ages, the osmotic properties of the intervertebral disk decline, and the disks become dehydrated with age, causing a reduction in overall disk height.

The annulus fibrosis of the disk consists of a series of approximately twelve 1-mm thick lamellas sheets, each composed of collagen fibers. The anterior lamellas are generally thicker and more distinct than the posterior lamellas. According to a study by Pooni et al.(9), the collagen fibers running through a single laminar sheet are oriented at $\sim 65°$ ($\pm 2.5°$) with respect to the vertical axis. These fibers alternate direction in concentric lamellas to form a cross-pattern. The annulus fibrosus develops lesions as it ages.

The nucleus pulposus of the intervertebral disk consists of a hydrophylic proteoglycan gel embedded in a collagen matrix. The nucleus pulposus contains ~80–88% water content in a young adult spine and occupies ~30–50% of the total intervertebral disk volume (16,17). However, with aging the nucleus undergoes rapid fibrosis and loses its fluid properties such that, by the third decade of life, there is hardly any nuclear material distinguishable (18). In a normal healthy intervertebral disk, the nucleus pulposus is glossy in appearance.

Luschka's joints are something special in the cervical region. The human cervical intervertebral disk contains fissures, called Luschka's joints or uncovertebral joints that run along the uncinate process and radiate inward toward the nucleus (Fig. 4a and b). These fissures run through the annular lamellas and the adjacent annular fibers are oriented such that they run parallel to the fissure (19–21). These fissures appear within the latter part of the first decade of life and continue to grow in size as aging occurs (8). Although some argument exists as to the definition of the fissures as true joints or pseudojoints, the fissures have been shown to exist as a natural part of the aging process (19,20) and therefore are important aspects of biomechanical modeling of the human cervical intervertebral disks.

The ligaments of the cervical spine (Fig. 4c) provide stability and act to limit excessive motions of the vertebras, thereby preventing injury during physiologic movement of the spine. Ligaments can only transmit tensile forces, impeding excessive motion, but do follow the principles of Wolff's law, where the tissue will remodel and realign along lines of tensile stress. The ligaments that are biomechanically relevant include the anterior longitudinal ligament (ALL), posterior longitudinal ligament (PLL), ligamentum flavum (LF), interspinous ligament (ISL), and the capsular ligaments (CAP). The ALL and PLL each traverse the length of the spine. The ALL originates at an insertion point on the inferior occipital surface and ends at the first segment of the sacrum. It runs along the anterior vertebral bodies, attached to the osseous bodies and loosely attached to the intervertebral disks as well. The ALL is under tension when the cervical spine undergoes extension. The PLL also runs the length of the spine down the posterior aspect of the vertebral bodies, originating at the occiput and terminating at the coccyx. Similar to the ALL, it is firmly attached to the osseous vertebral bodies and to the intervertebral disks. The PLL is under tension when the spine undergoes flexion. The ligamentum flavum couples the laminas of adjacent vertebras. It is an extremely elastic ligament due to the higher percentage of elastin fibers (65–70%) as compared to other ligaments in the spine and any other structure in the human body. The LF resists flexion motion and lengthens during flexion and shortens during extension. The high elastin content minimizes the likelihood of buckling during extension. It is under slight tension when the spine is at rest and acts as a tension band in flexion. Torsion also places the ligamentum flavum under tension, and restraint of rotation may also be a significant function. The ISL insertion points lie between adjacent spinous processes. The ligament is typically slack when the head is in a neutral posture and only becomes tensile when enough flexion motion has occurred such that other ligaments have undergone significant tension, such as the capsular ligaments, PLL and LF. Additionally, the ISL insertion points are such that it is ideal for resisting the larger flexion rotations that can occur as a result of excessive flexion loading. The capsular ligaments (CAPs) enclose the cervical facet joints and serve to stabilize the articulations of these joints and limit excessive motions at these joints. Generally, the fibers are oriented such that they lie perpendicular to the plane of the facet joints. These ligaments potentially also serve to keep the facets aligned and allow for the coupled rotations.

Lumbar Spine Anatomy

The basic structural components of the lumbar spine are the same as that of the lower cervical spine with differences in size, shape, and orientation of the structures due to functional requirements being different from that of the cervical region, Fig. 5. For example, the lumbar vertebras are bigger in size, because of the higher axial loads they carry. With regard to the peripheral margin of the intervertebral disk, annulus fibrosus is composed of 15–20 layers of collagenous fibrils obliquely running from one cartilage end plate to the other and crossing at 120° angles. As one progresses from the cervical into the thoracic region, the facet joints gradually orient themselves parallel with the frontal plane. The transition from the thoracic region into the lumbar region is indicated by a progressive change from the joints in the frontal plane to a more sagittal plane (4,22). This transition in facet orientation from the thoracic to the lumbar spine creates a different series of degenerative complications and disorders in the spine. Sagittal alignment of the facet joints increases the risk of subaxial and spondylolisthesis of the lumbar spine.

CLINICAL BIOMECHANICS OF THE NORMAL SPINE

The three basic functions of the spine are to protect the vital spinal cord, to transmit loads, and to provide the flexibility to accomplish activities of daily living. Components that provide stability to the spine are divided into four groups as follows:

1. Passive stabilizers: Passive stabilization is provided by the shape and size of vertebras and by the size, shape, and orientation of the facet joints that link them.
2. Dynamic stabilizers: Dynamic stabilization is provided by viscoelastic structures, such as the ligaments, capsules, and annulus fibrosus. The cartilage of the facet joints also acts as a damper.
3. Active stabilizers: Active voluntary or reflex stabilization is provided by the muscular system that governs the spine, Fig. 1c.
4. Hydrodynamic stabilizer: Hydrodynamic stabilization is due to the viscous nucleus pulposus.

The combination of these elements generates the characteristics of the entire spine. The diskussion of the kinematics will begin by further analyzing spinal elements as either passive or active. It will then progress into the effect these stabilizers have on the different portions of the spine.

Passive Elements

The vertebral body acts to passively resist compressive force. The size, mineral content, and orientation of the cancellous bone of each vertebral body increase–change as one descends in the caudal direction, which is a morphologic response to the increasing weight it must bear (4). The cortical shell on the vertebral body serves as the chief load path. The shell also provides a rigid link in the FSU, and a platform for attachment of the intervertebral disk, muscles, and the anterior and posterior longitudinal ligaments. The transition area of the motion segment is the endplate. This serves to anchor the intervertebral disk to the vertebral body. Note that the endplate starts out as growth cartilage and transitions into bone as aging occurs (22). The disk acts as both a shock absorber and an intervertebral joint because the relative flexibility of the intervertebral disk is high when compared to the vertebral body. The intervertebral disk resists compression, tension, shear, bending, and torsion (4). It is relatively resistant to failure in axial compression while its annular portions fail in axial torsion first (23).

Dynamic Stabilizers

Although bone is viscoelastic in nature, it serves more as a structural component within the spine that passively resists axial forces and can transmit forces along the spinal column. The soft tissue spinal structures (ligamentous, capsules, annulus fibrosis) are far more elastic as compared to bone behavior and stabilize the spine in a dynamic manner, where rapid vamping of oscillatory motions occur. The main function of the facet joints is to pattern the motions of the spine so that during activities of daily living the neural elements are not strained beyond the physiological limits. Therefore, they play a major role in determining the range of motion across a joint and as a damper to any possible dynamic loading. The amount of stability provided by the facet joints depends on extent of the capsular ligaments, their shape, orientation, and level within the spine (2). For example, the thoracic facets have a limited capsular reinforcement and facilitate axial rotation, which is in contrast to the lumbar region where the facet ligaments are more substantial and the joint plane is configured to impede axial motion (24).

From a biomechanical perspective, the ligaments respond to tensile forces only (1). The effectiveness of a ligament depends on the morphology and the moment arm through which it acts. That is, not only the strength, but also the longer lever arm a ligament has, the more it participates in the stabilization of the spine (4). Ligaments also obey Wolff's law. The ligaments also undergo remodeling along the lines of applied tensile stresses in response to chronic loads, just like bones. The ligamentum flavum acts as a protective barrier for the entire spine.

Active Stabilizers

Muscles contribute significantly to maintain the stability of the spinal column under physiological conditions. Decreasing the muscle forces acting on a FSU, increases the motion

and loading of the ligaments. A thoracolumbar (T1-sacrum) spinal column that is devoid of musculature is an unstable structure, with a load-carrying capacity of < 25 N (24). However, with properly coordinated muscle action, the spine can sustain large loads, which is exemplified by the action of weight lifters (24).

The internal force resisted by the muscle depends on factors such as cross-section and length at the initiation of contraction. The maximum force develops at approximately 125% of muscle resting length. In contrast, at approximately one-half of its resting length, the muscle develops very low force. The muscle stress (the maximum force per unit area) ranges from 30 to 90 $N \cdot cm^{-2}$ (25,26). Macintosh et al. (27) performed a modeling study based on radiographs from normal subjects to determine the effects of flexion on the forces exerted by the lumbar muscles. They found that the compressive forces and moments exerted by the back muscles in full flexion are not significantly different from those in the upright posture.

The remainder of this section is devoted to the biomechanics of the individual sections of the spinal column in a normal healthy person. Various methods for recording data with varying degrees of accuracy and repeatability are used ranging from the use of different types of goniometers, radiographs, in vitro cadaver based studies, magnetic resonance imaging (MRI) to visual estimation of motion. Although the value of assessing the ROM is not yet documented, the understanding and knowledge of normal age- and sex-related values of ROM is the basis for analysis of altered and possibly pathologic motion patterns as well as decreased or increased ROM (23,28). The issue of spinal instability (stability), although controversial in its definition, has immense clinical significance in the diagnosis and treatment of spinal disorders. Maintaining a normal range of motion in the spine is linked to spinal stability. The spine needs to maintain its normal range of motion to remain stable and distribute forces while bearing loads in several directions. The typical motion, for example, in response to the flexion–extension loads, as determined using cadaver testing protocols, is shown in Fig. 6. The two motion

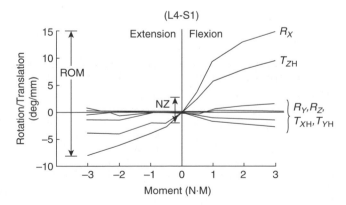

Figure 6. The load-displacement response of a FSU in flexion and extension. Two of the motion components are major and the other four are minor (or coupled). The range-of-motion and neutral zones, two of the terms used to describe the motion behavior of a segment, are also shown. (Adapted from Ref. 1.)

components (Flexion/extension rotation–R_x, and A-P translation–T_{zH}) are an order of magnitude higher than the other four. The two larger components are called the major–main motions and other four are the secondary–coupled motions. Range of motion, highlighted in the figure, will depend on the maximum load exerted on the specimen during testing. Likewise, the *in vivo* ranges of motion data will vary depending on the level of external force applied, that is, active or passive (2).

Biomechanics of the Occipital–Atlantoaxial Complex (C0-C1-C2)

As a unit, the craniovertebral junction accounts for 60% of the axial rotation and 40% of the flexion–extension behavior of the cervical spine (29,30).

Flexion–Extension. Large sagittal plane rotations have been attributed to the craniovertebral junction (Tables 1 and 2). Panjabi et al. (31) reported combined flexion–extension of C0-C1 and C1-C2 of 24.5 and 22.4°, respectively, confirming flexion–extension equivalence at the two levels. They also found that the occipitoatlantal joint level demonstrated a sixfold increase in extension as compared to flexion (21.0 vs. 3.5°), whereas the atlantoaxial level equally distributed its sagittal plane rotation between the two rotations, 11.5 (flexion) versus 10.9° (extension). Goel et al. (32) documented coupling rotations that occur with flexion and extension. They reported one-side lateral bending values of 1.2 and 1.4° for flexion and extension, respectively, at the C0-C1 level. In addition, they found C1-C2 coupled lateral bending, associated with flexion and extension movents, were lower than seen at the C0-C1 level. The largest axial rotation reported was 1.9°, which was an outcome of a C0-C1 extension of 16.5°. Note that the values reported by this study do not represent range of motion data, but rather intermediate rotation due to submaximal loading. Displacement coupling occurs between the translation of the head and flexion–extension of the occipitoatlanto–axial complex. Translation of the occiput with respect to the axis produces flexion–extension movements in the atlas. Anterior translation of the head extends the occipitoatlantal joints, with posterior motion resulting in converse flexion of the joint. This is postulated to occur due

Table 1. Ranges of Motion Reported from *In Vivo* and *In Vitro* Studies for the Occipito-Atlantal Joint (C0-C1)[a]

Type of Study[b]	Total Flexion/ Extension	Unilateral Bending	Unilateral Axial Rotation
In vivo	50	34–40	0
In vivo	50	14–40	0
In vivo	13	8	0
In vivo	30	10	0
In vivo			5.2
In vivo			1.0
In vitro			4.0
In vitro	3.5/21.0	5.5	7.2

[a]In degrees.
[b]The *in vivo* studies represent passive range-of-motion, whereas the *in vitro* studies represent motion at 1.5 N·m occipital moment loading. (From Ref. 4c.)

Table 2. Ranges of Motion Reported from *In Vivo* and *In Vitro* Studies for the AtlantoAxial Joint (C1-C2)[a]

Type of Study[b]	Total Flexion/ Extension	Unilateral Bending	Unilateral Axial Rotation
In vivo	0	0	60
In vivo	11		30–80
In vivo	10	0	47
In vivo	30	10	70
In vivo			32.2
In vitro			43.1
In vivo			40.5
In vitro	11.5/10.9	6.7	38.9

[a]In degrees.
[b]The *in vivo* studies represent passive range-of-motion, whereas the *in vitro* studies represent motion at 1.5 N·m occipital moment loading. (From Ref. 4c.)

to the highly contoured articular surfaces of the atlanto-occiptal joint.

Lateral Bending. As is shown in Tables 1 and 2, early studies have shown that occipitoatlantal lateral bending dominates the overall contribution of this motion in the occipitoatlanto–axial complex. However, this is not the finding of the most recent study. Almost all other studies indicate a significantly greater contribution from the C0-C1 joint. Lateral flexion also plays an important role in rotation of the head. Rotation of the lower cervical spine (C2-T1) results in lateral flexion of this region.

Axial Rotation. Almost all of the occipitoatlanto–axial contribution to axial rotation occurs in the atlantoaxial region. Atlantoaxial rotation occurs about an axis that passes vertically through the center of the odontoid process. This axis remains halfway between the lateral masses of the atlas in both neutral and maximal rotation. In maximal rotation, there is minimal joint surface contact, and sudden overrotation of the head can lead to interlocking of the C1-C2 facets, making it impossible to rotate the head back to neutral. Table 2 lists the amount of rotation found in the atlantoaxial joint by various researchers. Although these studies have produced widely varying results, there seems to be a consensus among the more recent studies that one side axial rotation at the atlantoaxial level falls somewhere in the range of 35–45°. The findings in Table 1 demonstrate that there is a relatively small contribution from the C0-C1 joint, with researchers finding between 0 and 7.2° of rotation. One interesting anatomical note concerning axial rotation is the behavior of the vertebral artery during rotation. The vertebral artery possess a loop between the atlas and axis, thus affording it over-length. Upon atlantoaxial rotation, the slack is taken up in the loop and it straightens, thus preventing over-stretching and possible rupture during maximal rotation.

The instantaneous axes of rotation (IARs) for the C0-C1 articulation pass through the center of the mastoid processes for flexion–extension and through a point 2–3 cm above the apex of the dens for lateral bending. There is a slight axial rotation at C0-C1. The IARs for the C1-C2 articulation are somewhere in the region of the middle third of the dens for flexion–extension and in the center of the dens for axial rotation. Lateral bending of C1-C2 is

Table 3. C3-C4 Ranges of Motion Compiled from Various *In Vivo* and *In Vitro* Studies[a,b]

Type of Study	Type of Loading	Total Flexion/Extension[c]	Unilateral Lateral Bending[c]	Unilateral Axial Rotation[c]
In vivo	Max. Rotation (active)	15.2 (3.8)	NA	NA
In vivo	Max. Rotation (active)	17.6 (1.5)	NA	NA
In vivo	Review	13.0 (range 7–38)	11.0 (range 9–16)	11.0 (range 10–28)
In vivo	Max. Rotation (active)	13.5 (3.4)	NA	NA
In vivo	Max. Rotation (active)	15.0 (3.0)	NA	NA
In vivo	Max. Rotation (active)	NA	NA	6.5 (range 3–10)
In vivo	Max. Rotation (active)	18.0 (range 13–26)	NA	NA
In vitro	1 N·m	8.5 (2.6)	NA	NA
In vitro	~3 N·m	NA	8.5 (1.8)	10.7 (1.3)

[a]In degrees.
[b]See Refs. 4b and d.
[c]Not available = NA.

controversial at the most 5–10° (4). During lateral bending, the alar ligament is responsible for the forced rotation of the second vertebra.

Middle and Lower Cervical Spine (C2-C7)

In the middle and lower cervical regions, stability and mobility must be provided; while, the vital spinal cord and the vertebral arteries must be protected. There is a good deal of flexion–extension and lateral bending in this area, Tables 3–6.

Flexion–Extension. Most of the flexion–extension motion in the lower cervical spine occurs in the central region, with the largest range of motion (ROM) generally occurring at the C5-C6 level. Except for extension, the orientation of the

cervical facets (on average, ~45° in the sagittal plane) does not excessively limit spinal movements in any direction or rotation. Flexion–extension rotations are distributed throughout the entire lower cervical spine for total rotations typically in the range of 60–75° and sagittal A/P translation is usually in the range of ~2–3 mm at all cervical levels (1). There is relatively little coupling effect that occurs during flexion–extension due to the orientation of the facets. There have been many published *in vivo* and *in vitro* studies reporting "normal" rotations at the various cervical spinal levels. These studies are in general agreement, although there appears to be a wide variation within ROM at all levels of the cervical region.

An *in vitro* study by Moroney et al.(33) averaged rotations among 35 adult cervical motion segments and found that average rotations (±SD) in flexion and extension

Table 4. C4-C5 Ranges of Motion Compiled from Various *In Vivo* and *In Vitro* Studies[a,b]

Type of Study	Type of Loading	Total Flexion/Extension[c]	Unilateral Lateral Bending[c]	Unilateral Axial Rotation[c]
In vivo	Max. Rotation (active)	17.1 (4.5)	NA	NA
In vivo	Max. Rotation (active)	20.1 (1.6)	NA	NA
In vivo	Review	12 (range 8–39)	11.0 (range 0–16)	12.0 (range 10–26)
In vivo	Max. Rotation (active)	17.9 (3.1)	NA	NA
In vivo	Max. Rotation (active)	19 (3.0)	NA	NA
In vivo	Max. Rotation (active)	NA	NA	6.8 (range 1–12)
In vivo	Max. Rotation (active)	20 (range 16–29)	NA	NA
In vitro	1 N·m	9.7 (2.35)	NA	NA
In vitro	~3 N·m	NA	6.3 (0.6)	10.8 (0.7)

[a]In degrees.
[b]See Refs. 4b and d.
[c]Not available = NA.

Table 5. C5-C6 Ranges of Motion Compiled from Various *In Vivo* and *In Vitro* Studies[a,b]

Type of Study	Type of Loading	Total Flexion/Extension[c]	Unilateral Lateral Bending[c]	Unilateral Axial Rotation[c]
In vivo	Max. Rotation (active)	17.1 (3.9)	NA	NA
In vivo	Max. Rotation (active)	21.8 (1.6)	NA	NA
In vivo	Review	17.0 (range 4–34)	8.0 (range 8–16)	10.0 (range 10–34)
In vivo	Max. Rotation (active)	15.6 (4.9)	NA	NA
In vivo	Max. Rotation (active)	20.0 (3.0)	NA	NA
In vivo	Max. Rotation (active)	NA	NA	6.9 (range 2–12)
In vivo	Max. Rotation (active)	20.0 (range 16–29)	NA	NA
In vitro	1 N·m	10.8 (2.9)	NA	NA
In vitro	~3 N·m	NA	7.2 (0.5)	10.1 (0.9)

[a]In degrees.
[b]See Refs. 4b and d.
[c]Not available = NA.

Table 6. C6-C7 Ranges of Motion Compiled from Various *In Vivo* and *In Vitro* studies[a,b]

Type of Study	Type of Loading	Total Flexion/Extension[c]	Unilateral Lateral Bending[c]	Unilateral Axial Rotation[c]
In vivo	Max. Rotation (active)	18.1 (6.1)	NA	NA
In vivo	Max. Rotation (active)	20.7 (1.6)	NA	NA
In vivo	Review	16.0 (range 1–29)	7.0 (range 0–17)	9.0 (range 6–15)
In vivo	Max. Rotation (active)	12.5 (4.8)	NA	NA
In vivo	Max. Rotation (active)	19 (3)	NA	NA
In vivo	Max. Rotation (active)	NA	NA	5.4 (range 2–10)
In vivo	Max. Rotation (active)	15 (range 6–25)	NA	NA
In vitro	1 N·m	8.9 (2.4)	NA	NA
In vitro	∼3 N·m	NA	6.4 (1.0)	8.8 (0.7)

[a]In degrees.
[b]See Refs. 4b and d.
[c]Not available = NA.

under an applied 1.8-N·m moment with 73.6-N preload (applied axially through the center of the vertebral bodies) were 5.55° (1.84) and 3.52° (1.94), respectively. These results demonstrate a total ROM in flexion–extension of ∼9.02°. Although generally lower than the reported data in Tables 3–6, probably due to the effect of averaging across cervical levels, the measurements are within the range of motion for all levels diskussed above.

Lateral Bending. Lateral bending rotations are distributed throughout the entire lower cervical spine for total rotations typically in the range of 10–12° for C2-C5 and 4–8° for C7-T1 (1). Unlike flexion–extension motion, where coupling effects are minimal, lateral bending is a more complicated motion involving the cervical spine, mainly due to the increased coupling effects. The coupling effects, probably due to the spatial locations of the facet joints at each level, are such that the spinous processes are rotated in the opposite direction of the lateral bending direction. The degree of coupling that occurs at separate levels of the cervical region has been described (33). There is a gradual decrease in the amount of axial rotation coupled with lateral bending as one traverses from C2 to C7. At C2, for every 3° of lateral bending there is ∼2° of coupled axial rotation, a ratio of 0.67. At C7, for every 7.5° of lateral bending there is ∼1° of coupled axial rotation, a ratio of 0.13.

Axial Rotation. Most cervical rotation occurs about the C1-C2 level, in the range of 35–45° for unilateral axial rotation: ∼40% of the total rotation observed in the spine (1). In the lower cervical spine, axial rotation is in the range of 5.4–11.0° per level. Again, as in the main motion of lateral bending, there exists a coupling effect with lateral bending when axial rotation is the main motion of the cervical spine. This coupling effect is in the range of 0.51–0.75° of lateral bending per degree of axial rotation (34). The effects of aging and gender on cervical spine motion have been investigated by numerous researchers. The average values for age decades for each motion, as well as average for the gender groups along with significant differences are shown in Table 7. Significantly less motion in the active tests was evident in comparison of lateral bending and axial rotation. Generally, for passive tests, the SD was lower. Women showed greater ROM in all these motions. In the age range of 40–49 years, women again showed significantly greater ROM in axial rotation and rotation at maximal flexion. There were no significant differences between gender groups for the group aged 60+ years. The well-established clinical observation that motion of the cervical spine decreases with age has been confirmed. An exception to this finding was the surprising observation that the rotation of the upper cervical spine, mainly at the atlantoaxial joint (tested by rotating the head at maximum flexion of the cervical spine that presumably locks the other levels)

Table 7. Average (SD) Head–Shoulder Rotations[a,b]

Age Decade	Flex/Ext		Lat Bending		Axial Rotation		Rot From Flex		Rot From Ext	
	M	F	M	F	M	F	M	F	M	F
20–29	152.7[c]	149.3	101.1	100.0	183.8	182.4	75.5[c]	72.6	161.8	171.5
	(20.0)	(11.7)	(13.3)	(8.6)	(11.8)	(10.0)	(12.4)	(12.7)	(15.9)	(10.0)
30–39	(141.1)	155.9[c]	94.7[c]	106.3[c]	175.1[c]	186.0[c]	66.0	74.6	158.4	165.8
	(11.4)[d]	(23.1)	(10.0)[d]	(18.1)	(9.9)[d]	(10.4)	(13.6)[d]	(10.5)	(16.4)	(16.0)
40–49	131.1	139.8	83.7	88.2[c]	157.4	168.2[b]	71.5	85.2	146.2	153.9[c]
	(18.5)	(13.0)	(13.9)	(16.1)	(19.5)[d]	(13.6)	(10.9)[c]	(14.8)	(33.3)	(22.9)
50–59	136.3[c]	126.9	88.3	76.1	166.2[c]	151.9	77.7	85.6	145.8	132.4[c]
	(15.7)	(14.8)	(29.1)[d]	(10.2)	(14.1)	(15.9)	(17.1)	(9.9)	(21.2)[d]	(28.8)
60+	116.3	133.2	74.2	79.6	145.6	154.2	79.4	81.3	130.9	154.5
	(18.7)	(7.6)	(14.3)	(18.0)	(13.1)	(14.6)	(8.1)	(21.2)	(24.1)	(14.7)

[a]In degrees.
[b]See Ref. 4h.
[c]Significant difference from cell directly adjacent to the right (i.e., gender within age group differences).
[d]Significant difference from cell directly adjacent below (i.e., age group within gender differentiation).

did not decrease with age. The measurement data for rotation out of maximum flexion suggests that the rotation of the atlantoaxial joint does not decrease with age, but rather remains constant or increases slightly perhaps to compensate for the reduced motion of the lower segments.

Lumbar Spine

The lumbar spine is anatomically designed to limit anterior translation and permit considerable flexion-extension and lateral bending, Tables 8A, B, and C. The unique characteristic of the spine is that it must support tremendous axial loads. The lumbar spine and the hips contribute to the considerable mobility of the trunk (34,35). The facets play a crucial role in the stability of the lumbar spine. The well-developed capsules of these joints play a major part in stabilizing the FSU against axial rotation and lateral bending. Lumbar facet joints are oriented in the sagittal plane, thereby allowing flexion–extension and lateral bending but limiting torsion (4).

In flexion–extension, there is usually a cephalocaudal increase in the range of motion in the lumbar spine. The L5-S1 joint offers more sagittal plane motion than the other joints, due to the unique anatomy of the FSU. The orientation of the facet becomes more parallel to the frontal plane

as the spinal column descends toward S1. Both this facet orientation and the lordotic angle at this motion segment contribute to the differences in the motion at this level. For lateral bending, each level is about the same except for L5-S1, which shows a relatively small amount of motion. The situation is the same for axial rotation, except that there is more motion at the L5-S1 joint.

There are several coupling patterns that have been observed in the lumbar spine. Pearcy (36) observed coupling of 2° of axial rotation and 3° of lateral bending with flexion–extension. In addition, there is also a coupling pattern, in which axial rotation is combined with lateral bending, such that the spinous processes point in the same direction as the lateral bending (22). This pattern is the opposite of that in the cervical spine and the upper thoracic spine (34).

The rotation axes for the sagittal plane of the lumbar spine have been described in several reports. In 1930, Calve and Galland (37) suggested that the center of the intervertebral disk is the site of the axes for flexion–extension; however, Rolander (38) showed that when flexion is simulated starting from a neutral position, the axes are located in the region of the anterior potion of the disk. In lateral bending, the axes fall in the region of the right side of the disk with left lateral bending, and in the region of the left side of the disk with right lateral bending. For axial

Table 8. Ranges of Motion for Various Segments Based on *In Vivo* and *In Vitro* Data Collection Techniques Cited in the Literature[a,b]

(A) Flexion/Extension

	In vitro			*In vivo*/active			*In vivo*/active		
	Mean	Lower	Upper	Mean	Lower	Upper	Mean	Lower	Upper
L1/2	10.7	5.0	13.0	7.0	1.0	14.0	13.0	3.0	23.0
L2/3	10.8	8.0	13.0	9.0	2.0	16.0	14.0	10.0	18.0
L3/4	11.2	6.0	15.0	10.0	2.0	18.0	13.0	9.0	17.0
L4/5	14.5	9.0	20.0	13.0	2.0	20.0	16.0	8.0	24.0
L5/S1	17.8	10.0	24.0	14.0	2.0	27.0	14.0	4.0	24.0

(B) Lateral Bending

	In vitro			*In vivo*/active			*In vivo*/passive		
	Mean	Lower	Upper	Mean	Lower	Upper	Mean	Lower	Upper
L1/2	4.9	3.8	6.5	5.5	4.0	10.0	7.9		14.2
L2/3	7.0	4.6	9.5	5.5	2.0	10.0	10.4		16.9
L3/4	5.7	4.5	8.1	5.0	3.0	8.0	12.4		21.2
L4/5	5.7	3.2	8.2	2.5	3.0	6.0	12.4		19.8

(C) Axial Rotation

	In vitro			*In vivo*/active		
	Mean	Lower	Upper	Mean	Lower	Upper
L1/2	2.1	0.9	4.5	1.0	−1.0	2.0
L2/3	2.6	1.2	4.6	1.0	−1.0	2.0
L3/4	2.6	0.9	4.0	1.5	0.0	4.0
L4/5	2.2	0.8	4.7	1.5	0.0	3.0
L5/S1	1.3	0.6	2.1	0.5	−2.0	2.0

[a]In degrees.
[b]In general *in vitro* data differs from *in vivo* data and the magnitude of *in vivo* motions depend on the collection technique (active vs. passive). (Taken from Ref. 4h.)

rotation, the IARs are located in the region of the posterior nucleus and annulus (4,36).

BIOMECHANICS OF SPINAL INSTABILITY: ROLE OF VARIOUS FACTORS

The causes of spinal instability have been hypothesized to include environmental factors that contribute to spinal degeneration and host of other variables (39). For example, some diseases can lead to spinal instability without being the direct cause. Chronic spondylolisthesis can lead to permanent deformation of the annulus that increases the probability of instability, Fig. 7. Essentially, any damage to any of the components of the motion segment or neural elements can contribute to instability. Instability can result from ruptured ligaments, fractured facets, fractured endplates, torn disks, or many other causes. However, the elements within the spine that seem to contribute more to stability and can therefore be major sources of instability are the facet joints, the intervertebral disks, and the ligaments (40). Both *in vivo* investigations in humans and animals and *in vitro* investigations of ligamentous spinal segments have been undertaken to accumulate biomechanical data of clinical significance.

Role of Environmental Factors in Producing Instability–Injury

Upper Cervical Spine. High speed impact loads that may be imposed on the spine are one of the major causes of spinal instability in the cervical region, especially in the upper region. To quantify the likely injuries of the atlas, Oda et al. (39,40) subjected upper cervical spine specimens to high speed axial impact by dropping 3–6 kg weights from various heights. The load produced axial compression and flexion of the specimen. Both bony and soft tissue injuries, similar to Jefferson fractures, were observed. The bony fractures were six bursting fractures, one four-part fracture without a prominent bursting, and one posterior arch fracture. The major soft tissue injury involved the transverse ligament. There were five bony avulsions and three midsubstance tears. The study was extended to determine the three-dimensional (3D) load displacements of fresh ligamentous upper cervical spines (C0-C3) in flexion, extension, and lateral bending before and following the impact loading in the axial mode. The largest increase in flexibility due to the injury was in flexion–extension: ~ 42%. In lateral bending, the increase was on the order of 24%; in axial rotation it was minimal: ~ 5%. These increases in motion are in concordance with the actual instabilities observed clinically. In patients with burst fractures of the atlas, Jefferson noted that the patients could not flex their heads, but could easily rotate without pain (41).

Heller et al. (42) tested the transverse ligament attached to C1 vertebra by holding the C1 vertebra and pushing the ligament in the middle along the AP direction. The specimens were loaded with an MTS testing device at varying loading rates. Eleven specimens failed within the substance of the ligament, and two failed by bone avulsion.

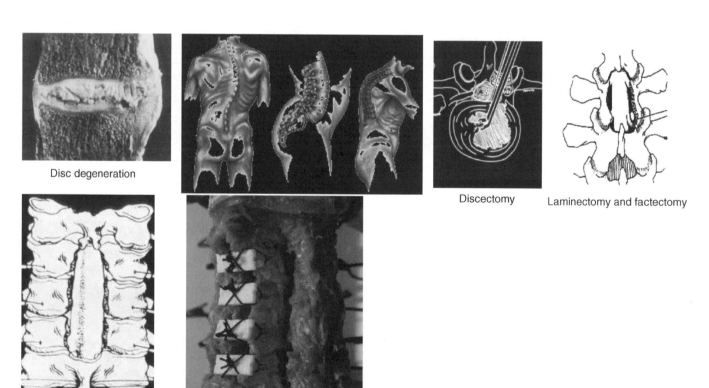

Disc degeneration

Discectomy   Laminectomy and factectomy

Multiple laminectomies   Laminoplasty

Figure 7. Various spinal disorders and surgical procedures that may lead to spinal instability. Such procedures are common for all of the spine regions.

The mean load to failure was 692 N (range 220–1590 N). The displacement to failure ranged from 2 to 14 mm (mean 6.7 mm). This study, when compared with the work of Oda et al. (39,40) suggests that (a) anteroposterior (AP) translation of the transverse ligament with respect to the dens is essential to produce its fracture; (b) rate of loading affects the type of fracture (bony versus ligamentous) but not the displacement at failure; and (c) even "axial" impact loads are capable of producing enough AP translation to produce a midsubstance tear of the ligament, as reported by Oda et al. (39).

The contribution to stabilization by the alar ligament of the upper cervical spine is of particular interest in evaluation of the effects of trauma, especially in the axial rotation mode. Goel and associates (43), in a study of occipitoatlantoaxial specimens, determined that the average values for axial rotation and torque at the point of maximum resistance were 68.1° and 13.6 N·m, respectively. They also observed that the value of axial rotation at which complete bilateral rotary dislocation occurred was approximately the point of maximal resistance. The types of injuries observed were related to the magnitude of axial rotation imposed on a specimen during testing. Soft tissue injuries (such as stretch–rupture of the capsular ligaments, subluxation of the C1-C2 facets) were confined to specimens rotated to or almost to the point of maximum resistance. Specimens that were rotated well beyond the point of maximum resistance also showed avulsion fractures of the bone at the points of attachment of the alar ligament or fractures of the odontoid process inferior to the level of alar ligament attachment. The alar ligament did not rupture in any of the specimens. Chang and associates (44) extended this study to determine the effects of rate of loading (dynamic loading) on the occipitoatlantoaxial complex. The specimens were divided into three groups and tested until failure at three different dynamic loading rates: 50°/s, 100°/s, and 400°/s as compared to the quasi-static (4°/s) rate of loading used by Goel et al.(43). The results showed that at the higher rates of loading, (a) the specimens became stiffer and the torque required to produce "failure" increased significantly (e.g., from 13.6 N·m at 4°/s to 27.9 N·m at 100°/s); (b) the corresponding right angular rotations (65–79°) did not change significantly; and (c) the rates of the alar ligament midsubstance rupture increased and that of "dens fracture" decreased. No fractures of the atlas were noted. This is another example of the rate of load application affecting the type of injury produced.

Fractures of the odontoid process of the second cervical vertebra comprise 7–13% of all cervical spine fractures (45). Most published reports involving odontoid fracture use the classification system detailed by Anderson and D'Alonzo (46). They described three types of odontoid process fracture (Fig. 8). Type I is an oblique fracture near the superior tip of the odontoid process and is thought to involve an avulsion defect associated with the alar–apical complex. Fracture of the odontoid process at the juncture of the process and vertebral body in the region of the accessory ligaments (Type II) is the most common osseous injury of the atlas. Fractures of this type lead to a highly unstable cervicovertebral region, commonly threatening the spinal

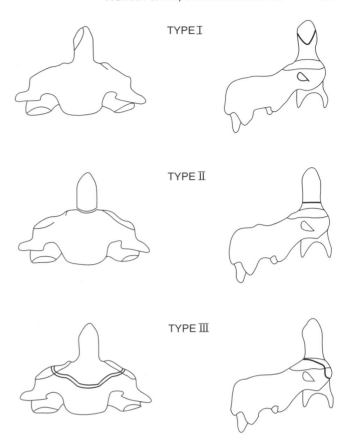

Figure 8. Fractures of the odontoid process. Taken from Ref. 46.

canal, and are often accompanied by ligamentous insult. Many of these fractures result in pseudoarthrosis if not properly treated. Type III fractures involve the junction of the odontoid process and the anterior portion of the vertebral body. These fractures are thought to be more stable than the Type I and Type II fractures. Type III fractures have high union rates owing to the cancellous bone involvement and the relatively high degree of vascularity (46,47).

Forces required to produce various types of dens fractures have been documented by Doherty et al. (45) who harvested the second cervical vertebra from fresh human spinal columns. Force was applied at the tip of the dens until failure occurred. The direction of the applied force was adjusted to exert extension bending or combined flexion and lateral bending on the tip of the dens. Extension resulted in type III fractures, and the combined load led to type II fractures of the dens. Furthermore, dynamic loading modes are essential to produce midsubstance ligament ruptures as opposed to dens fractures, especially in a normal specimen. Odontoid fractures have been implicated as being the result of high energy traumatic events. Indeed, there have been numerous accounts as to the events that lead to odontoid fracture. Schatzker et al. (47) reported that 16 of the 37 cases they reviewed were due to motor vehicle accidents and 15 cases were the result of high energy falls. Clark and White (48) report that all Type II (96 patients) and Type III (48 patients) fractures they reviewed were attributable to either motor vehicle accidents (~ 70%) or falls. Alker et al. (19) examined postmortem radiographs of 312 victims of fatal motor vehicle accidents. The cohort exhibited 98 injuries of the cervical spine, of which 70 were seen in the craniovertebral junction. The authors, although not quantifying the degree of dens fractures, hypothesized that odontoid fractures were probably due to hyperextension because of the posterior displacement of the fracture pieces.

There is considerable controversy as to the major load path that causes odontoid fractures. A review of the clinical and laboratory research literature fails to designate a consensus on this issue. Schatzker et al. (47) reviewed clinical case presentations and concluded that odontoid fractures are not the result of simple tension and that there must exist a complex combination of forces needed to produce these failures. Althoff (49) performed a cadaver study, whereby he applied various combinations of compression and horizontal shear to the head via a pendulum. Before load onset the head was placed in neutral, extension or flexion. The position of the load and the angle of impact, determining the degree of compression with shear, was changed for each experiment. The results indicated that an impact in the sagittal plane (anterior or posterior) produced fractures that involved the C2 body (Type III). As the force vector moved from anterior to lateral, the location of the fracture moved superiorly, with lateral loading producing Type I fractures. This led the author to propose a new hypothesis: impact loading corresponding to combined horizontal shear and compression results in odontoid fractures. Althoff dismissed the contributions of sagittal rotation (flexion and extension) to the production of resultant odontoid fracture.

Mouradian et al. (50) reported on a cadaver and clinical model of odontoid fracture. In their opinion, "it seems reasonable to assume that shearing or bending forces are primarily involved." The cadaver experimentation involved anterior or lateral translation of the occiput as well as lateral translation of the atlantal ring. In forward loading, the odontoid was fractured in 9 of the 13 cases, with 8 Type III fractures and 1 Type II fracture. The lateral loading specimens evidenced similar patterns of odontoid fracture regardless of the point of load application (on the occiput or on the atlas). In 11 specimens, lateral loading resulted in 10 Type II fractures and 1 Type III fracture. The clinical model involved reviewing 25 cases of odontoid fracture. They reported that 80% of these cases resulted from flexion or flexion–rotation injuries. They pointed out that the clinical data does not reflect the lateral loading cadaver experimentation results. In fact, they state that "a pure lateral blow probably did not occur in any [clinical] case". However, their clinical data indicated that the remaining 20% of the odontoid injuries could be ascribed to extension injuries. The technical difficulties precluded cadaver experimentation of this possible mechanism. Experimental investigations dealing with the pathogenesis of odontoid fractures have failed to produce a consensus as to the etiology of these fractures. These findings may actually reflect the diversity of causal mechanisms, suggesting the various mechanical factors are coincident in producing these fractures. It is difficult to diskern if this is the case or if this is due to the inhomogeneity of cadaver

experiment methodology. That is, some of the boundary and loading conditions used by the surveyed studies are vastly different and have produced divergent results. In addition, the anatomical variants of the craniovertebral osteo-ligamentous structures could also be integral to the cadaver study outcomes. The purpose of the study undertaken by Puttlitz et al. (51) was to utilize of the finite element method, in which the loading and kinematic constraints can be exactly designated, for elucidating the true fracture etiology of the upper cervical spine. Previous laboratory investigations of odontoid process failure have used cadaver models. However, shortcomings associated with this type of experimentation and the various loading and boundary conditions may have influenced the resulting data. Utilization of the FE method for the study of odontoid process failure has eliminated confounding factors often seen with cadaveric testing, such as interspecimen anatomic variability, age-dependent degeneration, and so on. This has allowed us to isolate changes in complex loading conditions as the lone experimental variable for determining odontoid process failure.

There are many scenarios, that are capable of producing fracture of the odontoid process. Force loading, in the absence of rotational components, can reach maximum von Mises stresses that far exceed 100 MPa. Most of these loads are lateral or compressive in nature. The maximum stress obtained was 177 MPa due to a force directed in the posteroinferior direction. The net effect of this load vector and its point of application, the posterior aspect of the occiput, is to produce a compression, posterior shear, and extension due to the load's offset from the center of rotation. This seems to suggest that extension and compression can play a significant role in the development of high stresses, and possibly failure, of the odontoid. The location of the maximum stress for this loading scenario was in the region of a Type I fracture. The same result, with respect to laterally loading, was obtained by Althoff (49). However, he dismissed the contribution of sagittal plane rotation to development of odontoid failures. The results of this study disagree with that finding. Posteroinferior loading with extension produced a maximum von Mises stress in the axis of 226 MPa. As stated above, the load vector for this case intensifies the degree of extension, probably producing hyperextension. The addition of the extension moment did not change the location of the maximum stress, still identifiable in the region of a Type I fracture. The clinical study by Moradian et al. (50) suggested that almost 20% of the odontoid fracture cases they reviewed involved some component of extension. The involvement of extension in producing odontoid process failures can be explained by its position with respect to the atlantal ring and the occiput. As extension proceeds, the contact force produced at the atlanto-dental articulation increases, putting high bending loads on the odontoid process. The result could be failure of the odontoid. Increasing tension of the alar ligaments as the occiput extends could magnify these bending stress via superposition of the loads, resulting in avulsion failure of the bone (Type I).

While the FE model predicted mostly higher stresses with the addition of an extension moment, the model showed that, in most cases, flexion actually mitigates the osseous tissue stress response. This was especially true for compressive (inferior) force application. Flexion loading with posterior application of an inferior load vectorally decreases the overall effect of producing extension on the occiput. None of the studies surveyed for this investigation pinpointed flexion, per se, as a damage mechanism for odontoid failure. The findings of this study supported the lack of evidence in support of flexion as being a causal mechanism for failure. In addition, the data suggested that flexion can act as a preventative mechanism against odontoid fracture.

Once again, the lateral bending results support the hypothesis of extension being a major injury vector in odontoid process failure. Inferior and posteroinferior loads with lateral rotation resulted in the highest maximal von Mises stress in the axis. Lateral loading also intensified the maximal stress in compression, suggesting rotations that incorporate a component of both lateral and extension motion may cause odontoid failures. Many of the lateral bending scenarios resulted in the maximum von Mises stress being located in the Type II and Type III fracture regions. In fact, the only scenarios that lead to the maximum stress in the Type I area was when there was an inferior or posterior load applied with the lateral bending. This is, again, suggestive that the extension moment, produced by these vectors and their associated moment arms (measured from the center of rotation), can result in more superiorly-located fractures.

Overall, this investigation has indicated that extension and the application of extension via force vector application, causes the greatest risk of superior odontoid failure. The hypothesis of extension as a causal mechanism of odontoid fracture includes coupling of this motion to other rotations. Flexion seems to provide a protective mechanism against force application that would otherwise cause a higher risk of odontoid failure.

Middle and Lower Cervical Spine. In the C2-T1 region of the spine, as in the upper cervical region, instabilities in a laboratory setting have been produced in an effort to understand the dynamics of traumatic forces on the spine (19). In one study, fresh ligamentous porcine cervical spine segments were subjected to flexion-compression, extension-compression, and compression-alone loads at high speeds (dynamic–impact loading) (19). The resultant injuries were evaluated by anatomic dissection. The results that the severity of the injuries were related mostly to the addition of bending moments to high speed axial compression of the spine segment, since compression alone produced the least amount of injury and no definite pattern of injuries could be identified. Other investigators have reported similar results (19).

Lumbar Spine. The onset of low back pain is sometimes associated with a sudden injury. However, it is more often the result of cumulative damage to the spinal components induced by the presence of chronic loading on the spine. Under chronic loading, the rate of damage may exceed the rate of repair by the cellular mechanisms, thus weakening the structures to the point where failure occurs under midly abnormal loads. Chronic loading to structures may

occur under a variety of conditions (52,53). One type of loading is heavy physical work prevalent among blue collar workers. Lifting not only induces large compressive loads across the segment, but tends to be associated with bending and twisting (54). Persons with jobs requiring the lifting of objects of > 11.3 kg > 25 times/day have over three times the risk for acute disk prolapse than people whose jobs do not require lifting (55). If the body is twisted during lifting, the risk is even higher with less frequent lifting. The other major class of loading associated with low back pain is posture related, for example, prolonged sitting–sedentary activities, and posture that involve bending over while sitting. Prolonged sitting may be compounded by vibration, such as observed in truck drivers (52,56,57).

The effects of various types of cyclic loads on the specimen behavior have been investigated (52,55). For example, Liu et al. subjected ligamentous motion segments to cyclic axial loads of varying magnitudes until failure or 10,000 cycles, which ever occurred first (53). Test results fell in to two categories, stable and unstable. In the unstable group, fracture occurred within the 6000 cycles of loading. The radiographs in the unstable group revealed generalized trabecular bony microfailure. Cracks were found to propagate from the periphery of the subcondral bone. After the removal of the organic phase, the unstable group specimens disintegrated into small pieces, as opposed to stable group specimens. This suggests that microcrack initiation occurs throughout the inorganic phase of the subchondral bone as a result of axial cyclic loading. In response to cyclic axial twisting of the specimens, Liu et al. noticed a discharge of synovial fluid from the articular joints (58). Specimens that exhibited an initial angular displacement of $< 1.5°$, irrespective of the magnitude of the applied cyclic torque, did not show any failures. On the other hand, specimens, exhibiting initial rotations $< 1.5°$, fractured before reaching 10,000 cycles. These fractures included bony failure of facets and/or tearing of the capsular ligaments.

Chronic vibration exposure and prolonged sitting are also known to lead to spinal degeneration. Spinal structures exhibit resonance between 5 and 8 Hz (56–59). *In vivo* and *in vitro* experimental and analytical studies have shown that the intradiscal pressure and motion increase when spinal structures experience vibration, such as during driving cars–trucks, at the natural resonant frequency (59). Prolonged sitting alone or in conjunction with chronic vibration exposure is also a contributing factor to spinal degeneration. A finite element-based study revealed that prolonged sitting led to an increase in disk bulge and the stresses in the annulus fibers located at the outer periphery (59,60).

Lee et al. (61) quantatively analyzed occlusion of the dural-sac in the lumbar spine was quantitatively analyzed by utilizing a finite element lumbar spine model. In the static analysis, it was found that < 2 kN of compressive load could not produce dural-sac occlusion, but the compression together with extension moment was more likely to produce the dural-sac occlusion. The 7.4% of occlusion was obtained when the 8 N·m of extension moment was added to 2 kN of compressive load that alone did not create any occlusion. The magnitude of occlusions was increased to 10.5% as the extension moment increased to 10 N·m with the same 2 kN of compressive load. In creep analysis, 10 N·m extension, kept for 3600 s, induced 6.9% of occlusion, and 2.4% of volume reduction in the dural-sac. However, flexion moment did not produce any occlusion in the dural-sac, but increased the volume instead because it caused stretching of the dural-sac coupled with vertebral motion. As a conclusion, occlusions resulted mainly from the slackening of the ligamentum flavum and disk bulging. Furthermore, the amount of occlusion was strongly dependent with loading conditions and the viscoelastic behavior of materials as well.

Changes in Motion due to Degeneration–Trauma

The degenerative process can effect all of the spinal elements and trauma can lead to partial or full destruction of the spinal elements. As such the motion behavior of the segment will change.

Cervical Spine Region. The rotation-limiting ability of the alar ligament was investigated by Dvorak et al. (62,63). A mean increase of 10.8° or 30% (divided equally between the occipitoatlantal and atlantoaxial complexes) in axial rotation was observed in response to an alar lesion on the opposite side. Oda et al. (39,40) determined the effects of alar ligament transections on the stability of the joint in flexion, extension, and lateral bending modes. Their main conclusion was that the motion changes occurred subsequent to alar ligament transection. The increases, however, were directional-dependent. Crisco et al. (64) compared changes in 3D motion of C1 relative to C2 before and after the capsular ligament transections in axial rotation. Two groups of cadaveric specimens were used to study the effect of two different sequential ligamentous transections. In the first group ($n = 4$), transection of the left capsular ligament was followed by transection of the right capsular ligament. In the second group ($n = 10$), transection of the left capsular ligament preceded transection of left and right alar and transverse ligaments. The greatest changes in motion occurred in axial rotation to the side opposite the transection. In the first group, transection of left capsular ligaments resulted in a significant increase in axial rotation ROM to the right of 1°. After the right capsular ligament was transected, there was a further significant increase of 1.8° to the left and of 1.0° to the right. Lateral bending to the left also increased significantly by 1.5° after both ligaments were cut. In the second group, with the nonfunctional alar and transverse ligaments, transection of the capsular ligament resulted in greater increases in ROM: 3.3° to the right and 1.3° to the left. Lateral bending to the right also increased significantly by 4.2°. Although the issue is more complex than this, in general these studies show that the major function of the alar ligament is to prevent axial rotation to the contralateral side. Transection of the ligament increases the contralateral axial rotation by ~ 15%.

The dens and the intact transverse ligament provide the major stability at the C1-C2 articulation. The articular capsules between C1 and C2 are loose, to allow a large amount of rotation and provide a small amount of stability.

Although the C1-C2 segment is clinically unstable after failure of the transverse ligament, resistance against gross dislocation is probably provided by the tectorial membrane, the ala, and the apical ligaments. With transection of the tectorial membrane and the ala ligaments, there is an increased flexion of the units of the occipital–atlantoaxial complex and a subluxation of the occiput (4h). It was also demonstrated that transection of the ala ligament on one side causes increased axial rotation to the opposite side by ~ 30%.

Fielding et al. (65) performed a biomechanical study investigating lesion development in rheumatoid arthritis. Their study tested 20 cadaveric occipitoatlanto–axial specimens for transverse ligament strength by application of a posterior force to the atlantal ring. They found atlantoaxial subluxation of 3–5 mm and increased atlas movement on the axis after rupture of the transverse ligament. From this study, Fielding et al. were able to conclude that the "transverse ligament represents a strong primary defense against anterior shift of the first cervical vertebra." Puttlitz et al. (66) developed an experimentally validated ligamentous, nonlinear, sliding contact 3D finite element (FE) model of the C0-C1-C2 complex generated from 0.5-mm thick serial computed tomography scans (Fig. 9). The model was used to determine specific structure involvement during the progression of RA and to evaluate these structures in terms of their effect on clinically observed erosive changes associated with the disease by assessing changes in loading patterns and degree of AAS (see Table 9 for terminology). The role of specific ligament involvement during the development and advancement of AAS was evaluated by calculating the AADI and PADI after reductions in transverse, ala, and capsular ligament stiffness. (The stiffness of transverse, alar, and capsular ligaments was sequentially reduced by 50, 75, and 100% of their intact values.) All models were subjected to flexion moments, replicating the clinical diagnosis of RA using full flexion lateral plane radiographs. Stress profiles at the transverse ligament-odontoid process junction were monitored. Changes in loading profiles through the C0-C1 and C1-C2 lateral articulations and their associated capsular ligaments were calculated. Posterior atlantodental interval (PADI) values were calculated to correlate ligamentous destruction to advancement of AAS. As an isolated entity, the model predicted that the transverse ligament had the greatest effect on AADI in the fully flexed posture. Without transverse ligament disruption, both ala and capsular ligament compromise did not contribute significantly to the development of AAS. Combinations of ala and capsular ligament disruptions were modeled with transverse ligament removal in an attempt to describe the interactive effect of ligament compromise, which may lead to advanced AAS. Ala ligament compromise with intact capsular ligaments markedly increased the level of AAS (Table 9). Subsequent capsular ligament stiffness loss (50%) with complete ala ligament removal led to an additional decrease in PADI of 0.92 mm. Simultaneous resection of the transverse, ala, and capsular ligaments resulted in a highly unstable situation. The model predicted stresses at the posterior base of the odontoid process greatly reduced, with transverse ligament compromise beyond 75%

Table 9. Combinations of Ligament Stiffness Reductions with the Resultant Degree of AAS, as Indicated by the AADI and PADI Values at Full Flexion (1.5 N · m moment)[a]

Reduction in Ligament Stiffness, %			Criteria, mm	
Transverse	Alar	Capsular	AADI	PADI
0	0	0	2.92	15.28
100	0	50	5.77	12.43
100	0	75	6.21	11.99
100	50	0	7.42	10.79
100	75	0	7.51	10.71
100	100	50	8.43	9.83

[a]Zero (0) ligament stiffness values represent completely intact ligament stiffness, "100" corresponds to total ligament destruction (via removal). (Taken from Ref. 66.) AAS = anterior atlantoaxial subluxation, AADI = anterior atlantodental interval, PADI = posterior atlantodental interval.

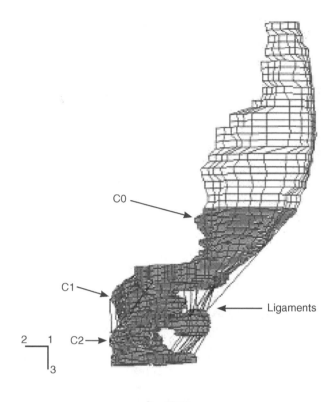

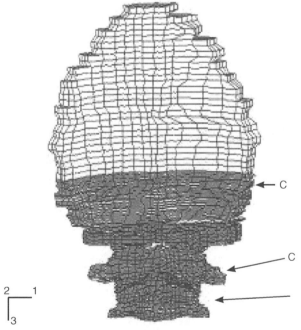

Figure 9. Finite element models of the upper cervical spine used to study the biomechanics of rheumatoid arthritis. (Taken from Ref. 66.)

(Fig. 10). Decreases through the lateral C0-C1 and C1-C2 articulations were compensated by their capsular ligaments. The data indicate that there may be a mechanical component (in addition to enzymatic degradation) associated with the osseous resorption seen during RA. Specifically, erosion of the base of the odontoid may involve Wolff's law unloading considerations. Changes through

the lateral aspects of the atlas suggest that this same mechanism may be partially responsible for the erosive changes seen during progressive RA. The PADI values indicate that complete destruction of the transverse ligament coupled with alar and/or capsular ligament compromise exist if advanced levels of AAS are present.

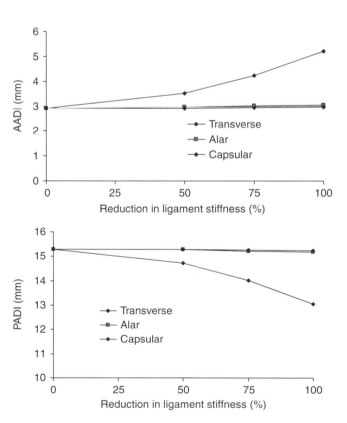

Figure 10. (a) Anterior atlantodental interval (AADI) (b) and posterior atlantodental interval (PADI) calculated for the intact model and models with stiffness reductions of the transverse, alar, and capsular ligaments at the fully flexed posture (1.5 N·m moment load). Each ligament's stiffness was altered while holding the other two at the baseline value (completely intact). (Taken from Ref. 66.)

In vitro studies to determine the feasibility of the "stretch test" in predicting instability of the spine in the cervical region were performed by Panjabi et al. (67). Four cervical spines (C1-T1; ages 25–29) were loaded in axial tension in increments of 5 kg to a maximum of one third of the specimen's body weight. The effects of sequential AP transection of soft tissues of a motion segment on the motion in one group and of posterior–anterior transections in another group were investigated. The intact cervical spine went into flexion under axial tension. Anterior transection produced extension. Posterior transection produced opposite results. Anterior injuries creating displacements of 3.3 mm at the disk space (with a force equal to one-third body weight) and rotation changes of $\sim 3.8°$ were considered precursors to failure. Likewise, posterior injuries resulting in 27 mm separation at the tips of the spinous process and an angular increase of 30° with loading were considered unstable. This work supports the concept that spinal failure results from transection of either all the anterior elements or all the posterior plus at least two additional elements.

In a study by Goel et al. (68,69), the 3D load-displacement motion of C4-C5 and C5-C6 as a function of transection of C5-C6 ligaments was determined. Transection was performed posteriorly, starting with the supraspinous and interspinous ligaments, followed by the ligamentum flavum and the capsular ligaments. With the transection of the capsular ligaments, the C5-C6 motion segment (injured level) showed a significant increase in motion in extension, lateral bending, and axial rotation. A significant increase in flexion resulted when the ligamentum flavum was transected.

A major path of loading in the cervical spine is through the vertebral bodies, which are separated by the intervertebral disk. The role of the cervical intervertebral disk has received little attention. A finite element model of the ligamentous cervical spinal segment was used to compute loads in various structures in response to clinically relevant loading modes (70). The objective was to predict biomechanical parameters, including intradiskal pressure, tension in ligaments, and forces across facets that are not practical to quantify with an experimental approach. In axial compression, 88% of the applied load passed through the disk. The interspinal ligament experienced the most strain (29.5% in flexion, and the capsular ligaments were strained the most (15.5% in axial rotation). The maximum intradiskal pressure was 0.24 MPa in flexion with an axial compression mode (1.8 N·m of flexion moment + 73.6 N of compression). The anterior and posterior disk bulges increased with an increase in axial compression (up to 800 N). The results provide new insight into the role of various elements in transmiting loads.

This model was further used to investigate the biomechanical significance of uncinate processes and Luschka joints (71). The results indicate that the facet joints and luschka joints are the major contributors to coupled motion in the lower cervical spine and that the uncinate processes effectively reduce motion coupling and primary cervical motion (motion in the same direction as load application), especially in response to axial rotation and lateral bending loads. Luschka joints appear to increase primary cervical motion, showing an effect on cervical motion opposite to that of the uncinate processes. Surgeons should be aware of the increase in motion accompanied by resection of the uncinate processes.

Cervical spine disorders such as spondylotic radiculopathy and myelopathy are often related to osteophyte formation. Bone remodeling experimental–analytical studies have correlated biomechanical responses, such as stress and strain energy density, to the formation of bony outgrowth. Using these responses of the spinal components, a finite element study was conducted to investigate the basis for the occurrence of disk-related pathological conditions. An anatomically accurate and validated intact element model of the C4-C5-C6 cervical spine was used to simulate progressive disk degeneration at the C5-C6 level. Slight degeneration included an alteration of material properties of the nucleus pulposus representing the dehydration process. Moderate degeneration included an alteration of fiber content and material properties of the annulus fibrosus representing the disintegrated nature of the annulus in addition to dehydrated nucleus. Severe degeneration included decrease in the intervertebral disk height with dehydrated nucleus and disintegrated annulus. The intact and three degenerated models were exercised under compression, and the overall force-displacement response, local segmental stiffness, annulus fiber strain, disk bulge, annulus stress, load shared by the disk and facet joints, pressure in the disk, facet and uncovertebral joints, and strain energy density and stress in the vertebral cortex were determined. The overall stiffness (C4-C6) increased with the severity of degeneration. The segmental stiffness at the degenerated level (C5-C6) increased with the severity of degeneration. Intervertebral disk bulge and annulus stress and strain decreased at the degenerated level. The strain energy density and stress in vertebral cortex increased adjacent to the degenerated disk. Specifically, the anterior region of the cortex responded with a higher increase in these responses. The increased strain energy density and stress in the vertebral cortex over time may induce the remodeling process according to Wolff's law, leading to the formation of osteophytes.

Thoracolumbar Region. The most common vertebral levels involved with the thoracolumbar injuires are T12-L1 (62%) and L1-L2 (24%) (22,25). The injuires, depending on the severity of the trauma, have included disruption of the posterior ligaments, fracture and dislocation of the facets, and fracture of the vertebral bodies with and without neural lesions. Operative intervention is often suggested to restore spinal stability. These involve use of spinal instrumentation, vertebroplasty, and host of other procedures which have been described elsewhere in this article.

For ease in description of these injuries, conceptually the osteoligamentous structures of the spine have been grouped into three "columns"; anterior, middle, and posterior. The anterior column consists of the anterior longitudinal ligament, anterior annulus fibrosus, and the anterior part of the vertebral body. The middle column consists of the posterior longitudinal ligament, posterior annulus fibrosus, and the posterior vertebral body wall. The posterior column contains the posterior bony complex or arch (including the facet joints), and the posterior ligamentous complex composed of the supraspinous ligament, interspinous ligament, facet joint capsules, and ligamentum flavum.

As per this classification, a compression fracture is a fracture of the anterior column with the middle and posterior columns remaining intact. In severe cases, there may also be a partial tensile failure of the posterior column, but the vertebral ring, consisting of the posterior wall, pedicles, and lamina, remains totally intact in a compression fracture. A burst fracture is a fracture of the anterior and middle columns under compression; the status of the posterior column can vary. In the burst fracture, there is fracture of the posterior vertebral wall cortex with marked retropulsion of bone into the spinal canal, obstructing, on average, 50% of the spinal canal cross-section. There may be a tilting and retropulsion of a bone fragment into the canal from one or both endplates. In contrast to the compression fracture, there is loss of posterior vertebral body height in a burst fracture. The seat-belt type injuries feature failure of the middle and posterior columns under tension, and either no failure or slight compression failure of the anterior column. In fracture dislocations, the anterior, middle, and posterior columns all fail, leading to subluxation or dislocation. There may be "jumped facets" or fracture of one articular process at its base or at the base of the pedicle. There is also disruption of the anterolateral periosteum and anterior longitudinal ligament. If the separation goes through the disk, there will be some degree of wedging in the vertebral body under the disk space. However, the fracture cleavage may pass through the vertebral body itself, resulting in a "slice fracture".

There are four mechanisms of fracture that have been hypothesized in the literature to explain why the thoracolumbar region experiences a higher frequency of injury than adjacent regions. The hypotheses state that a thoracolumbar fracture sequence can be put into motion by stress concentrations arising from (1) spinal loading conditions; (2) material imperfections in spine; (3) differences in spinal stiffness and physiological range of motion characteristics between the thoracic and lumbar regions; and (4) abrupt changes in spinal anatomy, especially facet orientations. As always, there is no conseus for these mechanisms.

A few of the experimental investigations that have attempted to reproduce the clinical fracture patterns are as follows: In one study, cadaver motion segments were subjected to loads of different magnitude and direction: compression, flexion, extension, lateral flexion, rotation, and horizontal shear to reproduce all varieties of spinal injury experimentally by accurately controlled forces. For a normal disk, increases of intradiskal pressure and bulging of the annulus occur under application of axial compressive load. With increased application of force, the end-plate bulges and finally cracks, allowing displacement of nuclear material into the vertebral body. Continued loading of the motion segment results in a vertical fracture of the vertebral body. If a forward shear component of force accompanies the compression force, the line of fracture of the

vertebral body is not vertical but is oblique. Different forms of fracture could be produced by axial compressive loading if the specimens were from older subjects (i.e., the nucleus was no longer fluid), or if the compressive loading was asymmetrical. Under these conditions, the transmission of load mainly through the annulus is responsible for the (1) tearing of the annulus, (2) general collapse of the vertebra due to buckling of the sides (cortical wall), and (3) marginal plateau fracture.

Thoracolumbar burst fractures in cadaver specimens have also been produced by dropping a weight such that the prepared column is subjected to axial-compressive impact loads. The potential energies of the failing weights used by these researchers have been 200 and 300 N·m. Fracture in four of the seven specimens apparently started at the nutrient foramen. The nutrient foramen may perhaps be viewed as a local area of material imperfection where stresses may be concentrated during loading, leading to fracture. Other researchers are apparently unable to consistently produce burst fractures *in vitro* without first creating artificial "stress raisers" in the vertebral body by means of cuts or slices into the bone.

Panjabi et al. (3) conducted *in vitro* flexibility tests of 11 T11-L1 specimens to document the 3D mechanical behavior of the thoracolumbar junction region (see section on Construct Testing for an explanation). Pure moments up to 7.5 N·m were applied to the specimens in flexion–extension, left–right axial torque, and right–left lateral bending. The authors reported the average flexibility coefficients of the main motions (range of motion divided by the maximum applied load). For extension moment, the average flexibility coefficient of T11-T12. (0.32°/N·m) was significantly less than that of T12-L1 (0.52°/N·m). For axial torque, the average flexibility coefficient of T11-T12 (0.24°/N·m) was significantly greater than that of T12-L1 (0.16°/N·m). The authors attributed these biomechanical differences to the facet orientation. They speculated that thoracic-type facets would offer greater resistance to extension than the more vertically oriented lumbar-type facets while the lumbar-type facets would provide a more effective stop to axial rotation than thoracic-type facets. No other significant biomechanical differences were detected between T11-T12 and T12-L1. In addition to these observations, authors found that for flexion torque, the average flexibility coefficients of the lumbar spine (e.g., L1-L2, 0.58°/N·m; L5-S1, 1.00°/N·m) were much greater than those of both T11-T12 (0.36°/N·m) and T12-L1 (0.39°/N·m). They identified this change in flexion stiffness between the thoracolumbar and lumbar regions as a possible thoracolumbar injury risk factor.

Lumbar Spine Region. The porosity of the cancellous bone within the vertebral body increases with age, especially in women. The vertebral body strength is known to decrease with increase in porosity of the cancellous bone, a contributing factor to the kyphosis normally seen in an elderly person (4). As the trabeculae reduce in size and number, the cortical shell must withstand greater axial load, thus increasing the thickness of the shell obeying the principles of Wolff's law. Edwards et al. (72) demonstrated cortical shell thickening of osteoporotic vertebral bodies compared to that of normal vertebral bodies. Furthermore, there was increased incidence of osteophytic development along the cortical shell in the regions of highest stress within the compromised osteoporotic vertebrae.

The normal disk consists of a gel-like nucleus encased in the annulus. In a normal healthy person, the disk acts like a fluid filled cavity. With age, the annulus develops radial, circumferential and rim lesions, and the nucleus becomes fibrous. Using a theoretical model in which cracks of varying lengths were simulated, Goel et al. found that the interlaminar shear stresses (and likewise displacements) were minimal until the crack length reached 70% of the annulus depth (73). Likewise, dehydration of the nucleus (extreme case totally ineffective like in a total nucleotomy) also was found to lead to separation of the lamina layers and an increase in motion (74). Thus, the results support the observation that the increase in motion really occurs in moderately degenerated disks.

Posner et al. investigated the effects of transection of the spinal ligaments on the stability of the lumbar spine (75). The ligaments were transected in a sequential manner, either anterior to posterior or posterior to anterior. While cutting structures from the anterior to posterior portion of the spine, extension loading caused a significant residual deformation after the anterior half of the disk was cut. Cutting from the posterior to anterior region, flexion loading caused significant residual motion upon facet joint transection. The role of ligaments becomes more prominent in subjects whose muscles are not fully functional. Using a finite element model in which the muscular forces during lifting were simulated, Kong et al. found that a 10% decrease in the muscle function increased loads borne by the ligaments and the disks (76). The forces across the facet joint decreased.

The orientation of facet becomes more parallel to the frontal plane as one goes down from L1 to S1 (77). Other factors can also contribute to changes in facet orientation in a person. The facet orientation, especially at L4-5 and L5-S1, plays a role in producing spondylolisthesis. Kong et al. using a finite element of the ligamentous lumbar segment (Fig. 11a) found that as the facet orientation becomes more sagittal, the A–P translation across the segment, increases in response to the load applied, Fig. 11b. The increase in flexion angle was marginal.

Changes in Motion Due to Surgical Procedures

Cervical Region. *In vivo* "injuries" result in disk degeneration and may produce osteophytes, ankylosed vertebras, and changes in the apophyseal joints (78). The effects of total diskectomy on cervical spine motions are of interest (79). Schulte and colleagues reported a significant increase in the motion after C5-C6 diskectomy (80). Motion between C5-C6 increased in flexion (66.6%), extension (69.5%), lateral bending (41.4%), and axial rotation (37.9%). In previous studies, Martins (81) and Wilson and Campbell (82) could not detect increases in motion roentgenographically and deemed the spines functionally stable.

(a)

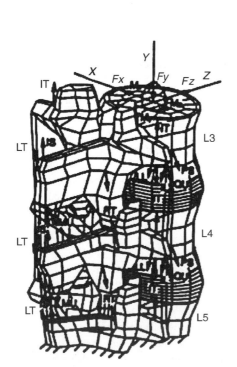

(b)

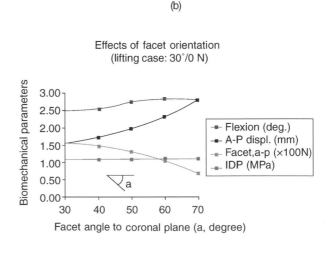

Figure 11. The finite element of a lumbar segment used to predict the effect of facet orientations on the motion and loads in various spinal components in a motion segment. (Taken from Ref. 77.)

The reasons for this diskrepancy in results are not apparent. The experimental designs were quite different as were the methods of motion measurement. However, the disk obviously is a major structural and functional component of the cervical spine.

The contribution of facet and its capsule to the stability of the cervical spine has been well documented using both *in vitro* laboratory models (83–85) and mathematical models (70,86,87,88). Facet joints play an integral part in the biomechanical stability of the cervical spine. Cusick et al. (89) found that total unilateral and bilateral facetectomies decreased compression-flexion strength by 31.6 and 53.1%, respectively. Facetectomy resulted in an anterior shift of the IAR, resulting in increased compression of the vertebral body and disk. This work confirmed the findings of Raynor et al. (63,64) who reported that bilateral facetectomy of as much as 50% did not significantly decrease shear strength; however, with a 75% bilateral facetectomy, a significant decrease in shear strength was noted. One should take great care when exposing an unfused segment to limit facet capsule resection to < 50%. With resection of > 50% of the capsule, postoperative hypermobility can occur and may require stabilization.

In contrast, studies that focused on the effects of laminectomy alone have been few and still unclear. Goel et al. were the first to evaluate the effects of cervical laminectomy with *in vitro* spine models (83,90). They found 10% increase of motion in flexion-extension using 0.3 N·m after a two level laminectomy. Zdeblick et al. did not find motion changes in flexion–extension after one level laminectomy under 5 N·m (84,85). Cusick et al. successfully showed that three level cervical laminectomy (C4-C6) induces a significant increase in total column flexibility using physiologic compression-flexion forces (86,87). Nevertheless, it seems difficult to estimate the instantaneous combination of physiologic compression and flexion forces. Therefore, quantitative evaluation might be difficult with this model. Our results indicate significant increase of spinal column motion in flexion (24.5%), extension (19.1%), and axial rotation (23.7%) using 1.5 N·m after a four level (C3-C6) laminectomy. Cervical vertebral laminae may transmit loads. Laminectomies result in the removal of part of this loading path and the attachment points for the ligamentum flavum, interspinous ligament, and the supraspinous ligament. It is not surprising that total laminectomy results in significant modifications in the motion characteristics of the cervical spine, especially in children. For example, Bell et al. (91) reported that multiple-level cervical laminectomy can lead to increase in postoperative hyperlordosis or kyphosis in children. However, there was no correlation between diagnosis, sex, location, or number of levels decompressed and the subsequent development of deformity. Postlaminectomy spinal deformity in the cervical spine, however, is rare in adults, probably owing to stiffening of the spine with age and changes in facet morphology. Goel et al. (89) removed the laminae of multisegmental cervical spines (C2-T2) at the level of C5 and C6 (total laminectomy); in flexion–extension mode, demonstrating an increase in motion of $\sim$ 10%.

In another *in vitro* study, the effects of multilevel cervical laminaplasty (C3-C6) and laminectomy with increasing amounts of facetectomy (25% and more) on the mechanical stability of the cervical spine were investigated (88). Cervical laminaplasty was not significantly different from the intact control, except for producing a marginal increase in axial rotation. However, cervical laminectomy with facetectomy of 25% or more resulted in a highly significant increase in cervical motion as compared with that of the intact specimens in flexion, extension, axial rotation, and lateral bending. There was no significant change in the coupled motions after either laminaplasty or laminectomy. The researchers recommended that concurrent arthrodesis be performed in patients undergoing laminectomy accompanied by > 25% bilateral facetectomy. Alternatively, one may use laminaplasty to achieve decompression if feasible. More recently, the effect of laminaplasty on the spinal motion using *in vivo* testing protocols have also been investigated (92–94). Kubo et al. (95) undertook an *in vitro* 3D kinematic study to quantify changes after a double door laminoplasty. Using fresh cadaveric C2-T1 specimens, sequential injuries were created in the following order: intact, double door laminoplasty (C3-C6) with insertion of hydroxyapatite (HA) spacers, laminoplasty without spacer, and laminectomy. Motions of each vertebra in each injury status were measured in six loading modes: flexion, extension, right and left lateral bending, and right and left axial rotation. Cervical laminectomy showed significant increase in motion compared to intact control in flexion (25%: $P < 0.001$), extension (19%: $P < 0.05$), and axial rotation (24%: $P < 0.001$) at maximum load. Double door laminoplasty with HA spacer indicated no significant difference in motion in all loading modes compared to intact. Laminoplasty without spacer showed intermediate values between laminoplasty with spacer and laminectomy in all loading modes. Initial slack of each injury status showed similar trends that of maximum load although mean % changes of laminectomy and laminoplasty without spacer were greater than that of maximum load. Double door laminoplasty with HA spacer appears to restore the motion of the decompressed segment back to its intact state in all loading modes. The use of HA spacers well contribute to maintain the total stiffness of cervical spine. In contrast, laminectomy seems to have potential leading postoperative deformity or instability.

Kubo et al. (96) undertook another study with the aim to evaluate the biomechanical effects of multilevel foraminotomy and foraminotomy with double door laminoplasty as compared to foraminotomy with laminectomy. Using fresh human cadaveric specimens (C2-T1), sequential injuries were created in the following order: intact, bilateral foraminotomies (C3/4, C4/5, C5/6), laminoplasty (C3-C6) using hydroxyapatite spacer, removal of the spacers, and laminectomy. Changes in the rotations of each vertebra in each injury status were measured in six loading modes: flexion–extension, right–left lateral bending, and right–left axial rotation. Foraminotomy alone, and following laminoplasty showed no significant differences in motion compared to the intact with the exception of axial rotation. After removal of the spacers and following a laminectomy, the motion increased significantly in flexion and axial rotation. The ranges of initial slack showed similar trends when compared to the results at maximum load. Clinical implications of these observations are presented.

Lumbar Region. The spine is naturally shaped to properly distribute and absorb loads, therefore, any surgical technique involving dissection of spinal components can disrupt the natural equilibrium of the spinal elements and lead to instability. The amount and origin of pain within the spine usually determines the type of surgical procedure for a patient. Such procedures include the removal of some or all of the laminae, facets, and/or disks. A certain increase in the range of motion within the spine can be attributed to each procedure. The increased range of motion can also lead to more pain, as noted by Panjabi and others, who used an external fixator to stabilize the spine (97,98). The fixator decreased the range of motion for flexion, extension, lateral bending, and axial rotation. The pain experienced by the patients who had the external fixator applied was significantly reduced. For these reasons, it is essential to learn the effects of various surgical procedures on the stability of the spine. In particular, we need to consider when procedures may lead to increase in motion to a point leading to instability.

Much of the debate surrounding laminectomy and instability involves the use of fusion after the laminectomy. The possibility that fusion will be necessary to stabilize the spine after a laminectomy is largely case specific and depends on the purpose of the surgery. In a study by Goel et al. , the results did not indicate the presence of instability after a partial laminectomy (99).

The facets are particularly important because they contribute to strength and resist axial rotation and extension. Subsequently, facetectomies can potentially be linked to instability. Abumi et al. developed some conclusions regarding partial and total facetectomies (2). They found that, although it significantly increased the range of motion, a partial facetectomy of one or both facets at a single level did not cause spinal instability. However, the loss of a complete facet joint on one or both sides was found to contribute to instability. Total facetectomy produced an increase of 65% in flexion, 78% in extension, 15% in lateral bending, and 126% in axial rotation compared with intact motion. Goel et al. also found similar results regarding partial facetectomy (99). Another study indicated that facetectomy performed within animals resulted in a large decrease in motion *in vivo* even though the increase in range of motion occurred acutely (2).

Goel et al. reported a significant increase in the range of motion for all loading modes except extension when a total diskectomy was performed across L4-5 level (99). A significant, but smaller increase in range of motion for subtotal disk removal was also observed, however, the postoperative instability was minimal. Both partial and total diskectomies produced a significant amount of intervertebral translational instability in response to left lateral bending at the L3-L4 and L4-L5 levels. They attributed the one-sided instability to the combination of injuries to the annulus and the right capsular ligament. Studies have also shown that more significant changes to the motion of the spine occur with removal of the nucleus pulposus as opposed to the removal of the annulus (4d). Discectomy

by fenestration and minimal resection of the lamina did not produce instability either.

BIOMECHANICS OF STABILIZATION PROCEDURES

Stability (or instability) retains a central role in the diagnosis and treatment of patients with back pain. Several studies have been carried out that help to clarify the foundation for understanding stability in the spine, as summarized above. In recent years, to restore stability across an abnormal segment, surgeons have well-accepted surgical stabilization and fusion of the spine using instrumentation, Figs. 12 and 13. The types and complexity of procedures (e.g., posterior, anterior, interbody) (100–105) have produced novel design challenges, requiring sophisticated testing protocols. In addition, most contemporary implant issues of stabilization and fusion of the spine are mostly mechanical in nature. [Biologic factors related to the adaptive nature of living tissue further complicate mechanical characterization (103,105)] Accordingly, it becomes essential to understand the biomechanical aspects of various spinal instrumentation and their effectiveness in stabilizing the segment. Properly applied spinal instrumentation maintains alignment and shares spinal loads until a solid, consolidated fusion is achieved. With few exceptions, these hardware systems are used in combination with bone grafting procedures, and may be augmented by external bracing systems.

Spinal implants typically follow loosely standardized testing sequelae during the design and development stage and in preparation for clinical use. The design and development phase goal, from a biomechanical standpoint, seeks to characterize and define the geometric considerations and load-bearing environment to which the implant will be subjected. Various testing modalities exist that elucidate which components may need to be redesigned. Not including the testing protocols for individual components of a device, plastic vertebrae (corpectomy) models are one of the first-stage tests that involves placing the assembled device on plastic vertebral components in an attempt to pinpoint which component of the assembled device may be the weakest mechanical link in the worst-case scenario, vertebrectomy. The *in vivo* effectiveness of the device may be limited by its attachment to the vertebras (fixation). Thus, testing of the implant-bone interface is critical in determining the fixation of the device to biologic tissue. Construct testing on cadaveric specimens provides information about the effectiveness of the device in reducing intervertebral motion across the affected and adjacent segments during quasiphysiologic loading. Animal studies provide insight with respect to the long-term biologic effects of implantation. Analytic modeling, such as the finite element method, is an extremely valuable tool for determining how implants and osseous loading patterns change with varying parameters of the device design. This type of modeling may also provide information about temporal changes in the bone quality due to the changing loading patterns as bone adapts to the implant (e.g., stress shielding-induced bone remodeling). After a certain level of confidence in the implant's safety and effectiveness is

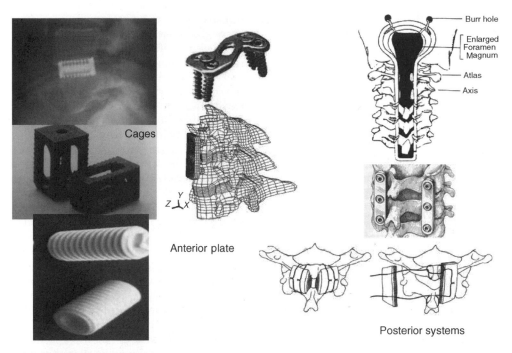

Figure 12. Devices used for stabilizing the cervical spine using the anterior and posterior approaches. Cages used both for the lumbar and cervical regions are also shown.

established via all or some of the aforementioned tests, controlled clinical trials allow for the determination of an implant's suitability for widespread clinical use. The following sections discuss each of these testing modalities, with specific examples used to illustrate the type of information that different tests can provide.

Implant–Bone Interface

Depending on the spinal instrumentation, the implant-bone interface may deal with the interface, where the spinal instrumentation abuts, encroaches, or invades the bone surface. It may include bony elements, such as the laminas, pedicles, the vertebral body itself, or the vertebral endplates.

Interlaminar Hooks. Interlaminar hooks are used as a means for fixing the device to the spine. Hook dislodgment, slippage, and incorrect placement have led to loss of fixation, however, resulting in nonfusion and pseudoarthrosis. Purcell et al. (106) investigated construct stiffness as a function of hook placement with respect to affected level in a thoracolumbar cadaver model. The failure moment was

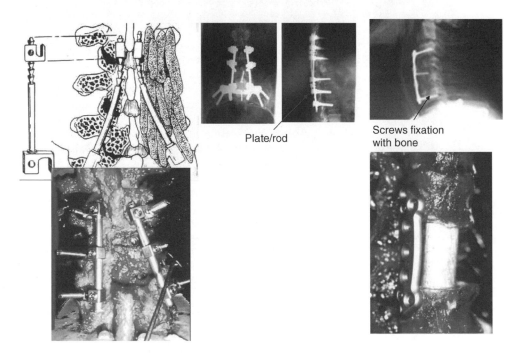

Figure 13. Examples of spinal instrumentation used in the lumbar region. Figure on the bottom right is an anterior plate.

found to be a function of the location of the hook placement with regard to the "injured" vertebra. The authors recommended hook placements three levels above and two levels below the affected area. This placement reduced vertebral tilting (analogous to intervertebral motion) across the stabilized segment, where fusion is to be promoted. Furthermore, the three-above, two-below surgical instrumentation strategy avoids the construct ending at the apex of a spinal deformity. Shortened fixation in this manner tends to augment a kyphotic deformity and cause continued progressive deformation. Overall, the use of hook fixation is a useful surgical stabilization procedure in patients with poor bone quality, where screw fixation is not an ideal choice for achieving adequate purchase into the bone.

Transpedicular Screws. Proper application of screw based anterior or posterior spinal devices requires an understanding of screw biomechanics, including screw characteristics and insertion techniques, as well as an understanding of bone quality, pedicle and vertebral body morphometries, and salvage options (107–109). This is best illustrated by the fact that the pedicle, rather than the vertebral body, contributes $\sim 80\%$ of the stiffness and $\sim 60\%$ of the pull out strength across the screw–bone interface (107).

Carlson et al. (110) evaluated the effects of screw orientation, instrumentation, and bone mineral density on screw translation, rotation at maximal load, and compliance of the screw–bone interface in human cadaveric bones. An inferiorly directed load was applied to each screw, inserted either anteromedially or anterolaterally,

until failure of the fixation was perceived. Anteromedial screw placement with fully constrained loading linkages provided the stiffest fixation at low loads and sustained the highest maximal load. Larger rotation of the screws, an indication of screw pull out failure, was found with the semi-constrained screws at maximal load. Bone mineral density directly correlated with maximal load, indicating that bone quality is a major predictor of bone–screw interfacial strength. A significant correlation between BMD and insertional torque ($p < 0.0001$, $r = 0.42$), BMD and pullout force ($p < 0.0001$, $r = 0.54$), and torque and pullout force has been found (109–112).

Since the specimens used for pull-out strength studies primarily come from elderly subjects, Choi et al. used foams of varying densities to study the effect of bone mineral density on the pull out strength of several screws (112). Pedicle screws (6.0×40 mm, 2 mm pitch, Ti alloy) of several geometric variations used for the study included the buttress (B), square (S), and V-shape (V) screw tooth profiles. For each type of tooth profile, its core shape (i.e., minor diameter) also varied, either the straight (i.e., cylindrical, core diameter = 4.0 mm) or tapered (i.e., conical, core diameter = 4.0/2.0 mm). In addition, for the cylindrical screws the major diameter was kept straight or tapered. The conical screws had its major diameters tapered only. Therefore, screws with a total of nine different geometries were prepared and tested (Fig. 14a). The screws were implanted in the rigid polyurethane foams of three different grades. The pullout strengths for various screw designs are shown in Table 10. The highest purchasing power in any screw design was observed in foams with

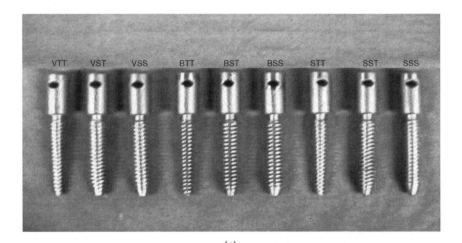

(a)

Figure 14. (a) Types of screws used in the foam model to determine the pull-out strength. The nomenclature used is as follows: Square = S, Buttress = B, V-shape = V. Screw diameters were SS = straight major diameter on straight core, ST = straight major diameter on tapered core, TT = tapered major diameter on tapered core. (b) Regression analysis. The maximum and minimum values from pull-out test for each foam grade were used regardless of tooth or core profiles. (Taken from Ref. 112.)

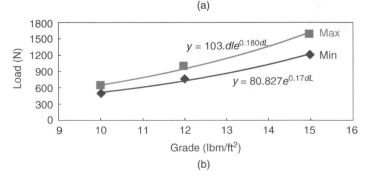

(b)

Table 10. Axial Pull Out Strength (N) Data for Different Types of Screws, Based on a Foam Model of Varying Densities

Foam Grade	Body Profile	Tooth Profile(Mean ± SD)		
		Square	Buttress	V-shape
10	SS[a]	591 ± 22	497 ± 80	615 ± 36
	ST[b]	622 ± 43	598 ± 25	634 ± 19
	TT[c]	525 ± 36	547 ± 30	568 ± 74
12	SS	864 ± 50	769 ± 56	987 ± 55
	ST	956 ± 30	825 ± 108	1005 ± 92
	TT	811 ± 41	808 ± 25	944 ± 32
15	SS	1397 ± 93	1303 ± 126	1516 ± 78
	ST	1582 ± 82	1438 ± 36	1569 ± 79
	TT	1197 ± 43	1352 ± 88	1396 ± 68

[a]Straight major diameter on Straight core.
[b]Straight major diameter on Tapered core.
[c]Tapered major diameter on Tapered core. (Taken from Ref. 112.)

the highest density (Grade 15). Exponential increase in pullout strength was seen when the foam density increased from Grade 10–15 (Fig. 14b). Overall, results demonstrated that the conical screws were consistently more effective against the pullout than the cylindrical designs. This was especially evident when the major diameter of the screw was kept straight. In this case, the contact area between the screw thread and surrounding foam was large. Although no consistent statistical superiority was found with the tooth profiles, results did suggest that the V-shape tooth screws ranked highest in many statistical comparisons and the buttress types showed comparatively lower pullout strength than the other types. This finding may be somewhat different from the literature. This can be due to the absence of the cortical purchase in the foam model used in this study. On the other hand, the square-tooth screws faired well in terms of pullout strength when the major diameter was kept straight but did not do so when tapered. Results also suggested that as the density of host site was decreased no clear choice of tooth profile could be found.

Lim et al. investigated the relationship between the bone mineral density of the vertebral body and the number of loading cycles to induce loosening of an anterior vertebral screw (113). (Screw loosening was defined as 1 mm displacement of the screw relative to bone). The average number of loading cycles to induce screw loosening was significantly less for specimens with bone mineral density < 0.45 g$\cdot$cm^{-2}, compared to those with bone mineral density $>$ or $= 0.45$ g$\cdot$cm^{-2}. These findings suggest that bone mineral density may be a good predictor of anterior vertebral screw loosening as well, just like the pedicle screws.

Since BMD seems to play a crucial role in the loosening of fixation screws, their use with osteoporotic bone is a contraindication. Alternatives have been proposed, including the use of bone cement to augment fixation and use of hooks along with pedicle screws (114,115).

The above findings related to increased pullout strength, number of cycles to failure, and tightening torque with BMD, are not fully corroborated with the corresponding *in vivo* work. For example, moments and forces during pedicle screw insertion were measured *in vivo* and *in vitro* and correlated to bone mineral density, pedicle size, and other screw parameters (material, diameter) (116). The mean *in vivo* insertion torque (1.29 N·m) was significantly

greater than the *in vitro* value (0.67 N·m). The linear correlation between insertion torque and bone mineral density was significant for the *in vitro* data, but not for the *in vivo* data. No correlation was observed between insertion torque and pedicle diameter. However, another investigation that clinically evaluated 52 patients who underwent pedicle screw fixation augmenting posterior lumbar interbody fusion (PLIF) supports the *in vitro* findings. The BMD was measured using DEXA and radiographs were assessed for detecting loosening, and so on at the screw bone interface. Bone mineral density was found to have a close relation with the stability of pedicle screw *in vivo*, and BMD values $< 0.674 \pm 0.104$ g$\cdot$cm^{-2} suggested a potential increased risk of "non-union".

Cages. Total disk removal alone or in combination with other surgical procedures invariably leads to a loss of disk height and an unstable segment. Both allo- and autologous bone grafts have been used as interbody spacers (103,117–120). Associated with the harvest and use of autogenous bone grafts are several complications: pain, dislodgment of the anterior bone graft, loss of alignment, and so on. Recently, the use of inserts, fabricated from synthetic materials (metal or bone-biologic), have gained popularity. These may be implanted through an anterior or posterior approach. Interbody devices promote fusion by imparting immediate postoperative stability, and by providing axial load-bearing characteristics, while allowing long-term fusion incorporation of the bone chips packed inside and around the cage (121,122). Many factors influence the performance of an interbody cage. The geometry, porosity, elastic modulus, and ultimate strength of the cage is crucial to achieving a successful fusion. An ideal fixation scenario should be to utilize the largest cross-sectional footprint of a cage in the interbody space so that the cortical margin can be captured by the fixation to decrease the risk of endplate subsidence. A modulus of elasticity close to bone is often an ideal choice to balance the mechanical integrity at the endplate–implant interface. A cage that has a large elastic modulus and high ultimate strength increases the risk to endplate subsidence and/or stress-shielding issues. Finally, cage design must possess a balance between an ideal porosity to augment bony fusion through the cage and mechanical strength to bear axial loads.

Steffen et al. undertook a human cadaveric study with the objectives to assess the axial compressive strength of an implant with peripheral endplate contact as opposed to full surface contact, and to assess whether removal of the central bony endplate affects the axial compressive strength (120). Neither endplate contact region nor its preparation technique affected yield strength or ultimate compressive strength. Age, bone mineral content, and the normalized endplate coverage were strong predictors of yield strength ($P < 0.0001$; $r^2 = 0.459$) and ultimate compressive strength ($P < 0.0001$; $r^2 = 0.510$). An implant with only peripheral support resting on the apophyseal ring offers axial mechanical strength similar to that of an implant with full support. Neither supplementary struts nor a solid implant face has any additional mechanical advantage, but reduces graft–host contact area. Removal of the central bony endplate is recommended because it does not affect the compressive strength and promotes graft incorporation. There are drawbacks to using threaded cylindrical cages (e.g., limited area for bone ingrowth, subsidence issues, and metal precluding radiographic visualization of bone healing). To somewhat offset these drawbacks, several modifications have been proposed, including changes in shape and material (123–125). For example, the central core of the barbell shaped cage can be wrapped with collagen sheets infiltrated with bone morphogenetic protein. The femoral ring allograft (FRA) and posterior lumbar interbody fusion (PLIF) spacers have been developed as biological cages that permit restoration of the anterior column with a machined allograft bone (123).

Wang et al. (126) looked at *in vitro* load transfer across standard tricortical grafts, reverse tricortical grafts, and fibula grafts, in the absence of additional stabilization. Using pressure sensitive film to record force levels on the graft, the authors found the greatest load on the graft occurred in flexion. As expected, the anterior portion of the graft bore increased load in flexion and the posterior portion of the graft bore the higher loads in extension. The authors did not supplement the anterior grafting with an anterior plate. Cheng et al. (127) performed an *in vitro* study to determine load sharing characteristics of two anterior cervical plate systems under axial compressive loads: the Aesculap system (Aesculap AGT, Tuttlingen, Germany) and the CerviLock system (SpineTech Inc., Minneapolis, MN). The percent loads carried by the plates at a 45 N applied axial load were as follows: Aesculap system $-6.2\% \pm 9.2\%$ and the CerviLock system $-23.8\% \pm 12.7\%$. Application of 90 N loads produced similar results to those of the 45 N loads. The authors stated that the primary factor in load transfer characteristics of the instrumented spine was a difference in plate designs. The study contained several limitations. Loading was performed solely in axial compression across a single functional spinal unit (FSU). The study did not simulate complex loading, such as flexion combined with compression. In the physiologic environment, load sharing in multisegmental cervical spine could be altered since the axial compressive load will produce additional flexion–extension moments, due to the lordosis. The upper and lower vertebrae of the FSU tested were constrained in the load frame, whereas in reality they are free to move, subject to anatomic constraints.

Rapoff et al. (128) recently observed load sharing in an anterior CSLP plate fixed to a three level bovine cadaveric spinal mid-thoracic segment under simple compression of 125 N. A Smith–Robinson diskectomy procedure was performed at the median disk space to a maximum distraction of 2 mm prior to plate insertion and loading. Results showed that at 55 N of load, mean graft load sharing was 53% ($\pm 23\%$) and the plate load sharing was 57% ($\pm 23\%$). This study was limited in several aspects, including the fact that no direct measurement of plate load was made, the spines were not human, and the loading mode was simplified and did not incorporate more complex physiologic motions, such as coupled rotation and bending or flexion/extension.

A recent study by An et al. (129) looking at the effect of endplate thickness, endplate holes, and BMD on the strength of the graft–endplate interphase of the cervical spine found that there existed a strong relationship between BMD and load to failure of the vertebrae, demonstrating implications for patient selection and choice of surgical technique. There was a significantly larger load to failure in the endplate intact group compared to the endplate resected group studied, suggesting that an intact endplate may be a significant factor in prevention of graft subsidence into the endplate. Results of an FE model observing hole patterns in the endplate indicated that the hole pattern only significantly affected the fraction of the upper endplate that was exposed to fracture stresses at 110 N loading. A large central hole was found to be best for minimization of fracture area and more effective at distribution of the compressive load across the endplate area.

Dietl et al. pulled out cylindrical threaded cages (Ray TFC Surgical Dynamics), bullet-shaped cages (Stryker), and newly designed rectangular titanium cages with an endplate anchorage device (Marquardt) used as posterior interbody implants (130). The Stryker cages required a median pullout force of 130 N (minimum, 100 N; maximum, 220 N), as compared with the higher pullout force of the Marquardt cages (median, 605 N; minimum, 450 N; maximum, 680 N), and the Ray cages (median, 945 N; minimum, 125 N; maximum, 2230 N). Differences in pullout resistance were noted depending on the cage design. A cage design with threads or a hook device provided superior stability, as compared with ridges. The pyramid shaped teeth on the surfaces and the geometry of the implant increased the resistance to expulsion at clinically relevant loads (1053 and 1236 N) (124,125).

Construct Testing

Spinal instrumentation needs to be applied to a spine specimen to evaluate its effectiveness. As a highly simplified model, two plastic vertebras serve as the spine model. Loads are applied to the plastic vertebras and their motions and applied loads to failure are measured. This gives some idea of the rigidity of the instrumentation. However, a truer picture is obtained by attaching the device to the cadaveric spine specimen.

Plastic Vertebra (Corpectomy) Models. Clinical reviews of failure modes of the devices indicate that most designs satisfactorily operate in the immediate postoperative period. Over time, however, these designs can fail because of the repeated loading environment to which they are subjected. Thus, fatigue testing of newer designs has become an extremely important indicator of long-term implant survivorship. Several authors have tested thoracolumbar instrumentation systems in static and fatigue modes using a plastic vertebral model (131–133). For example, Cunningham et al. compared 12 anterior instrumentation systems, consisting of 5 plate and 7 rod systems in terms of stiffness, bending strength, and cycles to failure (132). The stiffness ranged from 280.5 $kN \cdot m^{-1}$ in the Synthes plate (Synthes, Paoli, PA) to 67.9 $kN \cdot m^{-1}$ in the Z-plate (Sofamor-Danek, Memphis, TN). The Synthes plate and Kaneda SR titanium (AcroMed, Cleveland, OH) formed the highest subset in bending strength of 1516.1 and 1209.9 N, respectively, whereas the Z plate showed the lowest value of 407.3 N. There were no substantial differences between plate and rod devices. In fatigue, only three systems: Synthes plate, Kaneda SR titanium, and Olerud plate (Nord Opedic AB, Sweden) withstood 2 million cycles at 600 N. The failure mode analysis demonstrated plate or bolt fractures in plate systems and rod fractures in rod systems.

Clearly, studies, such as these involving missing vertebral (corpectomy) artificial models, reveal the weakest components or linkages of a given system. Results must be viewed with caution since these results do not shed light on the biomechanical performance of the device. Furthermore, we do not know the optimum strength of a fixation system. These protocols do not provide any information about the effects the device implantation may have on individual spinal components found *in vivo*. For these data, osteoligamentous cadaver models need to be incorporated in the testing sequelae and such studies are more clinically relevant.

Osteoligamentous Cadaver Models. For applications, such as fusion and stabilization, initial reductions in intervertebral motion are the primary determinants of instrumentation success, although the optimal values for such reductions are not known and probably not needed to determine relative effectiveness. Thus, describing changes in motion of the injured and stabilized segments in response to physiologic loads is the goal of most cadaver studies. Many times, these data are compared with the intact specimen, and the results are reported as the instrumentation's contribution to providing stability (134). To standardize, the flexibility testing protocol has been suggested (135). Here a load is applied and resulting unconstrained motions are measured. However, there are several issues pertaining to this type of testing, as described below.

More recently nonfusion devices have come on the market. These devices try to restore motion of the involved segment. With the paradigm shift from spinal fusion to spinal motion, there are dramatically different criteria to be considered in the evaluation of nonfusion devices. While fusion devices need to function for a short period and are differentiated primarily by their ability to provide rigid fixation, nonfusion devices must function for much longer time periods and need to provide spinal motion, functional stability, and tolerable facet loads. The classic flexibility testing protocol is not appropriate for the understanding of the biomechanics of the construct for the nonfusion devices, at the adjacent levels (136,137). However, constant pure moments are not appropriate for measuring effects of implants, like the total disk replacements, at adjacent levels. The pure moments distribute evenly down a column and are thus not effected by perturbation at a level(s) in a longer construct. Further, the net motion of a longer construct is not similar if only pure moments are applied: fusions will limit motion and other interventions may increase motion, a reflection of the change in stiffness of the segment. This may have shortcomings for clinical applications. For example, with forward flexion, there are clinical demands to get to ones shoes to tie them, to reach a piece of paper fallen to the floor, and so on. It would thus be advantageous to use a protocol that would achieve the same overall range of motion for the intact specimen and instrumented construct by applying pure moments that distribute evenly down the column.

Another issue is that the ligamentous specimens cannot tolerate axial compressive loads, specimens in the absence of the muscles will buckle. Thus, methods have been developed to apply preloads on the ligamentous spines during testing, since these indirectly simulate the effects of muscles on the specimens. A number of approaches have been proposed with one that stands out and is getting accepted by the research community. It is termed the follower-load concept (137).

It could be reasoned that coactivation of trunk muscles (e.g., the lumbar multifidus, longissimus pars lumborum, iliocostalis pars lumborum) could alter the direction of the internal compressive force vector such that its path followed the lordotic and kyphotic curves of the spine, passing through the instantaneous center of rotation of each segment. This would minimize the segmental bending moments and shear forces induced by the compressive load, thereby allowing the ligamentous spine to support loads that would otherwise cause buckling and providing a greater margin of safety against both instability and tissue injury. The load vector described above is called a *"follower load"*.

Additionally, most of these studies involve quasistatic loading; however, short-term fatigue characteristics have also been investigated. Both posterior and anterior-instrumentation employed for the promotion of fusion and non fusioon have been evaluated. The following are examples of such devices, which are diskussed within the context of these testing modalities.

Cervical Spine Stabilization and Fusion Procedures

There are a variety of techniques that are utilized for spinal fusion in the lower cervical spine, among which are spinal wiring techniques (138–144), posterior plating (145–154), anterior plating, and (more recently) cervical interbody fusion devices. While fusions are effective in a majority of cases, they do have documented biomechanical shortcomings, particularly at the segments adjacent to the fusion. Some of these problems include observations of excessive motion (sometimes due to pseudoarthrosis) (155–164), degenerative changes (165,166), fracture dislocation (167), screw breakage or plate pullout (160,168–170), and risks to neural structures. These problems are typically minimal when only one or two segments are involved in the injury. However, when the number of segments involved in the reconstruction increases to three or more, the incidence of failed fusion, screw breakage, and plate pullout increases dramatically.

Upper Cervical Spine Stabilization. Stabilization of the craniovertebral junction is not common; however, its importance for treating rheumatoid arthritis associated lesions, fractures and tumors cannot be underestimated. Currier et al. (171) studied the degree of stability provided by a rod-based instrumentation system. They compared this new device to the Ransford loop technique and a plate system using C2 pedicle screws. Transverse and alar ligament sectioning and odontoidectomy destabilized the specimen. All three-fixation systems significantly reduced motion as compared to intact and injured spines in axial rotation and extension. The new device did not significantly reduce motion at C1-C2 in flexion, and none of the devices were able to produce significant motion reductions in C1-C2 lateral bending. The authors claimed, based on these findings, that the new system is equivalent or superior to the other two systems for obtaining occipitocervical stability. Oda et al. (172) investigated the comparative stability afforded by five different fixation systems. Type II odontoid fractures were created to simulate instability. The results indicate that the imposed dens fracture decreased construct stiffness as compared to the intact case. Overall, the techniques that utilized screws for cervical anchors provided greater stiffness than the wiring techniques. Also, the system that utilized occipital screws with C2 pedicle screw fixation demonstrated the greatest construct stiffness for all rotations. Puttlitz et al. (1c) have used the finite element model of the C0-C1-C2 complex to investigate the biomechanics of a novel hardware system (Fig. 15). The FE models representing combinations of

cervical anchor type (C1-C2 transarticular screws versus C2 pedicle screws) and unilateral versus bilateral instrumentation were evaluated. All models were subjected to compression with pure moments in flexion, extension, or lateral bending. Bilateral instrumentation provided greater motion reductions than the unilateral hardware. When used bilaterally, C2 pedicle screws approximate the kinematic reductions and hardware stresses (except in lateral bending) that are seen with C1-C2 transarticular screws. The FE model predicted that the maximum stress was always located in the region where the plate transformed into the rod. Thus, the authors felt that C2 pedicle screws should be considered as an alternative to C2-C1 transarticular screw usage when bilateral instrumentation is applied.

Other strategies to fix the atlantoaxial complex can be found in the literature. Commonly available fixation techniques to stabilize the atlantoaxial complex are posterior wiring procedures (Brooks fusion, Gallie fusion) (169), interlaminar clamps (Halifax) (170), and transarticular screw (Magerl technique), either alone or in combination.

Posterior wiring procedures and interlaminar clamps are obviously easier to accomplish. However, these do not provide sufficient immobilization across the atlantoaxial complex. In particular, posterior wiring procedures and place the patient at risk of spinal cord injury due to sublaminar passage of wires into the spinal canal (172). Interlaminar clamps offer the advantage of avoiding the sublaminar wire hazard and have more rigid biomechanical stiffness than posterior wiring procedures (173).

Transarticular screw fixation (TSF), on the other hand, affords a stiffer atlantoaxial arthrodesis than posterior wiring procedures and interlaminar clamps. The TSF does have some drawbacks including injury of vertebral artery, malposition, and screw breakage (174). Furthermore, body habitus (obesity or thoracic hyperkyphosis) may prohibit achieving the low angle needed for screw placement across C1 and C2. Recently, a new technique of screw and rod fixation (SRF) that minimizes the risk of injury to the vertebral artery and allows intraoperative reduction has been reported (175,176). The configuration of this technique, which achieves rigid fixation of the atlantoaxial

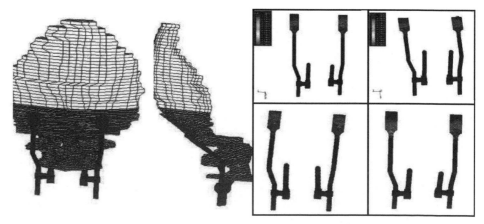

Figure 15. The finite element model showing the posterior fixation system and the stress plots in the rods. (Taken from Ref. 4c.)

Max stress:

132.5 MPa in lateral bending,
C2 pedicle, injury, compressed.

complex, consists of lateral mass screws at C1 and pedicle screws at C2 linked via longitudinal rods with constrained coupling devices.

One recent study compared the biomechanical stability impaired to the atlantoaxial complex by either the TSF or SRF technique and to assess how well these methods withstand fatigue in a cadaver model (177).

The results of this study suggested that in the unilateral fixations, the SRF group was stiffer than the TSF group in flexion loading, but there were no evident differences in other directions. In the bilateral fixations, SRF was more stable than TSF, especially in flexion and extension. These results were similar to those reported by Melcher et al. (178) and Richter et al. (179), yet different from Lynch et al. (180). The instrumentation procedure (screw length, type of constrained coupling device, etc.), the destabilization technique, and the condition of the specimens might have an influence on the results. In this study, when stabilizing the atlantoaxial segments, all screws were placed bicortically in both techniques in accordance with procedures by Harms and Melcher (181). Previous work has demonstrated that bicortical cervical vertical screws are superior to unicortical screws in terms of pullout strength and decreased wobble (182,183). Most surgeons, however, prefer unicortical screwing at C1 and C2 levels to reduce the risk of penetration during surgery. This could affect the outcome. They initially connected the screw to the rod using the oval shape constrained coupling device recommended for use in C1 and C2 vertebras. However, the stability was not judged adequate, So they altered the procedure to use the stiffer circle shape constrained coupling device. With regards to the destabilization procedure, there are three typical methods: sectioning of intact ligaments, odontoid fracture, and odontoidectomy. The atlantoaxial complex was destabilized by ligament transection to simulate ligamentous instability, while Lynch et al. (180) used odontoidectomy. Furthermore, the bone quality of specimens affects the screw-bone interface stability. These factors were possibly reflected in other results. However, both results were not statistically different between TSF and SRF, so they could be interpreted equivalent in terms of effective stabilization when compared with the intact specimen.

In unilateral TSF and SRF, the fixed left lateral atlantoaxial joint acted as a pivot in left axial rotation and as a fulcrum in left lateral bending, thus leading to an increase in motion. This motion could be observed with the naked eye.

Stability in flexion and extension of the bilateral TSF group was inferior to that of SRF group. Henriques et al. (184) and Naderi et al. (182) also reported similar tendency. Henriques et al. (184) felt that this was most likely due to the transarticular screws being placed near the center of motion between C1 and C2. This was judged as another reason that the trajectory of the screws is consistent with the motion direction of flexion and extension. So, if TSF is combined with some posterior wiring procedures, the stability in flexion and extension will increase.

Lower Cervical Spine

Anterior Plating Techniques for Fusion. The anterior approach in order to achieve arthrodesis of the cervical spine has become a widely utilized and accepted approach. However, many of these techniques rely on insertion of a bone graft only anteriorly and the use of an external immobilization device, such as a halo vest, or posterior fixation in order to allow for sufficient fixation. Problems encountered with these methods include dislodging of the bone graft (potentially causing neural compromise), loss of angular correction, and failure to maintain spinal reduction (185,186). The use of anterior plates has recently become popular partially because they address some of the complications stated above. The main reasons typically cited for the use of anterior plates are (1) advantage of simultaneous neural decompression via an anterior as opposed to posterior approach, (2) improved fusion rates associated with anterior cervical fusion (187,188), (3) help in reduction of spinal deformities, (4) provides for rigid segmental fixation, and (5) prevents bone graft migration. However, the efficacy of anterior plates alone is still debated by some authors, particularly in multilevel reconstruction techniques, due to the high rates of failure observed, up to 50% in some cases (189–192). Thus, more biomechanical research must be accomplished to delineate the contributions of anterior plates to load sharing mechanics in the anterior approach.

There have been several *in vitro* studies examining the efficacy of anterior plates for use in a multitude of procedures involving cervical spine stabilization. Grubb et al. (151) performed a study involving 45 porcine and 12 cadaveric specimens to study anterior plate fixation. Phase I of the study involved intact porcine specimens which were subjected to nondestructive testing in flexion, lateral bending, and axial rotation loading modes to determine structural stiffness. Maximum moments applied included 2.7 N·m for flexion and lateral bending and 3.0 N·m for axial rotation testing. After completion of the nondestructive testing, a flexion-compression injury was introduced by performing a C5 corpectomy and inserting an iliac strut bone graft in the resulting space. An anterior plate was then introduced across C4–C6. Three different anterior plates were tested, including a Synthes CSLP (cervical spine locking plate) with unicortical fixation, a Caspar plate with unicortical fixation, and a Caspar plate with bicortical fixation. Each instrumented specimen was then tested again nondestructively in flexion, lateral bending, and axial rotation. Finally, destructive testing in each loading mode was performed on particular specimens in each plated group. Phase II of the study involved intact cadaver specimens that were subjected to nondestructive testing in flexion, lateral bending, and axial rotation loading modes to determine structural stiffness. Maximum moments applied included 2.0 N·m for flexion, lateral bending, and axial rotation. After completion of the nondestructive testing, a flexion-compression injury was introduced by performing a C5 corpectomy and inserting an iliac strut bone graft in the resulting space. An anterior plate was then introduced across C4–C6. Two different anterior plates were tested: a Synthes CSLP (cervical spine locking plate) with unicortical fixation and a Caspar plate with bicortical fixation. Each instrumented specimen was then tested again nondestructively in flexion, lateral bending, and axial rotation. Finally, destructive testing in flexion was performed on each specimen. Results of the study demonstrated that each of the stabilized specimens had stiffness characteristics greater than or equal to their paired intact test results. The CSLP was found to have a significantly higher stiffness ratio (plated: intact), higher failure moment, lower flexion neutral zone ratio, and higher energy to failure than the Caspar plates.

A study by Clausen et al. (193) reported the results of biomechanical testing of both the CSLP system with unicortical locking screws and Caspar plate system with unlocked bicortical screws. Fifteen cadaveric human spines were tested intact in flexion, extension, lateral bending, and axial rotation loading modes to determine stiffness characteristics. A C5-C6 instability was then introduced, consisting of a C5-C6 diskectomy with complete posterior longitudinal ligament (PLL) disruption. An iliac crest bone graft was then introduced into the C5-C6 disk space and the spine was instrumented with either the CSLP or Caspar system. Once instrumented, each of the spines were further destabilized through disruption of the interspinous and supraspinous ligaments, the *ligamentum flavum*, facet capsules, and lateral annulus. The specimens were then retested for stiffness. After initial postinstrumented testing was done, biomechanical stability of the specimens was reassessed following cyclic fatigue for 5000 cycles of flexion–extension. Finally, failure testing of each specimen was performed in flexion. Results of the study demonstrated that both devices stabilized the spine before but not after fatigue and that only the Caspar plate stabilized the spine significantly before and after fatigue. Failure moment did not differ between the two systems. Biomechanical stability discrepancy between the two devices was attributed to differences in bone–screw fixation. Kinematic testing of 10 cervical spines following single level (C5-C6) diskectomy and anterior plate insertion was studied by Schulte et al. (194). Results showed that the use of an anterior plate in addition to the bone graft provided significant stabilization in all loading modes. Traynelis et al. (195) performed biomechanical testing to compare anterior plating versus posterior wiring in an cadaver instability model involving a simulated C5 teardrop fracture with posterior disruption and fixation across C4-C6. Study results showed that bicortical anterior plating provided significantly more stability than posterior wiring in extension and lateral bending, and was slightly more stable than posterior wiring in flexion. Both provided equivalent stability in axial rotation. A variety of anterior constructs exist in the market today, typically using either bicortical screws or unicortical locking screws. Several studies have evaluated the purchase of unicortical versus bicortical screws in the cervical spine (195–197).

Wang et al. (198) looked at *in vitro* load transfer across standard tricortical grafts, reverse tricortical grafts, and fibula grafts, in the absence of additional stabilization. Using pressure sensitive film to record force levels on the graft, the authors found the greatest load on the graft occurred in 10° of flexion (~20.5 N) with a preload on the spine of 44 N. As expected, the anterior portion of the graft bore increased loading in flexion and the posterior portion of the graft bore the highest loads in 10° extension. The authors did not supplement the anterior grafting with an anterior plate. Cheng et al. (127) performed an *in vitro* study to determine load-sharing characteristics of two anterior cervical plate systems under axial compressive loads: the Aesculap system (Aesculap AGT, Tuttlingen, Germany) and the CerviLock system (SpineTech Inc., Minneapolis, MN). The percent loads carried by the plates at a 45 N applied axial load were as follows: Aesulap system $-6.2\% \pm 9.2\%$ and the CerviLock system $-23.8\% \pm 12.7\%$. Application of 90 N loads produced similar results to those of the 45 N loads. The authors stated that the primary factor in load transfer characteristics of the instrumented spine was a difference in plate designs. The study contained several limitations. Loading was performed solely in axial compression across a single FSU. The study did not simulate complex loading, such as flexion combined with compression. In the physiologic environment, load sharing in multisegmental cervical spine could be altered since the axial compressive load will produce additional flexion–extension moments, due to the lordosis. The upper and lower vertebras of the FSU tested were constrained in the load frame, whereas in reality they are free to move, subject to anatomic constraints. Foley et al. also performed *in vitro* experiments to examine the loading mechanics of multilevel strut grafts with anterior plate augmentation (199). The results of the study showed that application of an anterior plate in a cadaver corpectomy model unloads the graft in flexion and increases the loads borne by the graft under extension of the spine. The increase in load borne by the graft in the presence of the plate should increase the graft subsidence, a finding that is contrary to clinical follow-up studies, as stated earlier.

Finite element (FE) analysis has been used by our group on a C5-C6 motion segment model to determine load sharing in an intact spine under compressive loading and more clinically relevant combined loading of flexion–extension and compression (4b). Similarly, using the FE approach, stresses in various graft materials (titanium core, titanium cage, iliac crest, tantalum core, and tantalum cage), the adjacent disk space, and vertebra have been investigated by Kumareson et al. (200). These authors found that angular stiffness decreased with decreasing graft material stiffness in flexion, extension, and lateral bending. They also observed the stress levels in the disk and vertebral bodies as a whole due to the presence of a graft, but did not focus on the graft itself or the endplate regions, superior and inferior to the graft. The effects of anterior plates on load sharing were not investigated.

Scifert et al. (4d) developed an experimentally validated C4-C6 cervical spine finite element model was developed to examine stress levels and load sharing characteristics in an anterior plate and graft. Model predictions demonstrated good agreement with the *in vitro* data. The rotations across the stabilized segment significantly decreased in the presence of a plate as compared to graft alone case. Much like the *in vitro* studies, the model also predicted that the compressive load in the graft increased in extension in the presence of plate, as compared to graft alone case. Depending on the load type, stresses in graft were concentrated in its anterior or posterior region in the graft alone case and became more uniformly distributed in the presence of the plate. The predicted load-displacement data and load sharing results reveal that plate is very effective in maintaining the alignment. Increase in load borne by the graft in the presence of a plate in the extension mode suggests that pistoning of the graft is a possible outcome. However, the stress data reported in the present study, and something that the *in vitro* studies are unable to quantify, show that pistoning of the graft is not likely to happen due to stresses being low, an observation in agreement with the clinical outcome data. For an optimal healing, the stress results suggest the placement of the tricortical bone graft with its cortical region towards the canal when a plate is used. For the graft case alone, this parameter does not seem to be that critical. A more uniform stress distribution in the graft in the presence of the plate would tend to promote bone fusion in a more uniform fashion, as compared to the graft alone case. In the later case fusion may initiate in a selective region.

Lower Cervical Spine

Posterior Plating Techniques for Fusion. The posterior approach in order to achieve cervical spine arthrodesis has been a widely utilized and accepted approach to dealing with cervical spine trauma, such as posterior trauma involving the spinous processes or facet dislocation or injury, and disease, such as degenerative spondylosis or ossification of the posterior longitudinal ligament. Recently, however, posterior fixation using cervical screw plates affixed to the lateral masses has gained acceptance due to a variety of factors, including the fact that they do not rely on the integrity of the lamina or spinous processes to allow for fixation, bone grafting is not always necessary to allow for long-term stability, greater rotational stability is achieved at the facets (201,202), and it eliminates the need for external immobilization such as halo vests. Problems encountered with these posterior methods include (*1*) risk to nerve roots, vertebral arteries, facets, and spinal cord (168); (*2*) screw loosening and avulsion (203); (*3*) plate breakage; (*4*) and loss of reduction. Additionally, contraindications exist where the patient has osteoporosis, metabolic bone disease, or conditions where the bone is soft (i.e., ankylosing spondylitis) (204). There also exists controversy as to the advantages of using posterior plating techniques when posterior cervical wiring techniques can be used (205). In theory, anterior stabilization of the spine in cases of vertebral body injury is superior to posterior plating. However, in practice, posterior plates are an effective means of stabilizing vertebral body injuries, and their application is easier than the anterior approach involving corpectomy, grafting, and anterior plating.

In addition to clinical *in vivo* studies, there have been several *in vitro* studies examining the efficacy of posterior plates for use in cervical spine stabilization. Roy-Camille et al. (202) utilized a cadaveric model to compare posterior lateral mass plating to spinous process wiring. They found that posterior plates increased stability by 92% in flexion and 60% in extension, while spinous process wiring enhanced flexion stability by only 33% and did not stabilize in extension at all Coe et al. (201) performed biomechanical testing of several fixation devices, including Roy-Camille posterior plates, on six human cadaveric spines. Complete disruption of the supraspinous and interspinous ligaments, *ligamentum flavum*, posterior longitudinal ligament, and facet joints was performed. They found no significant difference in static or cyclic loading results between the posterior wiring and posterior plates, although the posterior plating was stiffer in torsion. Overall, the authors recommended the Bohlmann triple wire technique for most flexion distraction injuries. In experimental studies performed in our lab on 12 cervical spines, Scifert et al. (202) found that posterior plates were superior to posterior facet wiring in almost every loading mode tested in both the stabilized and cyclic fatigue testing modes, excluding the cyclic extension case. Smith et al. (153) performed biomechanical tests on 22 spines to evaluate the efficacy of Roy-Camille plates in stabilization of the cervical spine following simulation of a severe fracture dislocation with three-column involvement caused by forced flexion-rotation of the head. Results of the study indicated that the posterior plating system decreased motion significantly compared to the intact spine, specifically by a factor of 17 in flexion–extension and a factor of 5 units in torsion. Raftopoulos et al. (203) found that both posterior wiring and posterior plating resulted in significant stability following severe spinal destabilization, although posterior plating provided superior stability compared to that of interfacet wiring. Similar to the results of anterior plates, Gill et al. (204) found that bicortical lateral posterior plate screw fixation provided greater stability than unicortical fixation. However, Grubb et al. (151) found that unicortical fixation of a destabilized spine using a cervical rod device provided equivalent stability in torsion and lateral bending as bicortical fixation using an AO lateral mass plate. Effectiveness of 360° plating techniques for fusion.

As stated previously, both anterior and posterior plating procedures contain inherent difficulties and drawbacks. Some authors have examined the utilization of both techniques concomitantly to ensure adequate stabilization. Lim et al. (205) examined both anterior only, posterior only, and combined techniques *in vitro* to determine efficacy of these techniques in stabilizing either a C4-C5 flexion-distraction injury or an injury simulating a C5 burst fracture involving a C5 corpectomy. The AXIS and Orion plates were used for posterior and anterior stabilization, respectively. In the C4-C5 flexion-distraction injury, both posterior and combined fixation reduced motion significantly from intact in flexion. Only the combined procedure was able to reduce motion effectively in extension. In lateral bending and axial rotation, posterior fixation alone and combined fixation were able to significantly reduce motion compared to intact. In the C5 corpectomy model, all constructs exhibited significantly less motion compared to intact in flexion, although the combined fixation was the most rigid. In extension, all constructs except the posterior fixation with bone graft were able to reduce motion significantly compared to intact. In lateral bending, only the posterior fixation and combined fixation were able to provide enhanced stability compared to intact. In axial rotation, only the combined fixation was able to significantly reduce motion compared to intact. Thus, the authors concluded that combined fixation provided the most rigid stability for both surgical cases tested. In a clinical study of multilevel anterior cervical reconstruction surgical techniques, Doh et al. (190) found a 0% psuedoarthrosis rate for the combined fixation system only.

Although combined fixation almost certainly allows for the most rigid fixation in most unstable cervical spine injuries, there are other factors to consider, such as the necessity for an additional surgery, possibility of severely reduced range of motion, and neck pain, Jonsson et al. (206) found a propensity for 22 out of 26 patients with combined fixation to have pain related to the posterior surgery. Additionally, patients with the combined fixation were found to have considerably restricted motion compared to normal. These and other factors must be weighed with the additional advantages of almost assured stability with the combined fixations.

Interbody Fusion Cage Stabilization for Fusion

Interbody fusion in the cervical spine has traditionally been accomplished via the anterior and posterior methods, incorporating the use of anterior or posterior plates, usually with the concomitant use of bone grafts. However, recently, interbody fusion cages using titanium mesh cages packed with morselized bone have been reported for use in the cervical spine. Majid et al. (163) performed channeled corpectomy on 34 patients, followed by insertion of a titanium cage implant packed with autogenous bone graft obtained from the vertebral bodies removed in the corpectomy. The authors then performed additional anterior plating on 30 of the 34 patients that involved decompression of two or more levels. Results of the study indicated a 97% radiographic arthrodesis rate in the patient population, with a 12% complication rate including pseudoarthrosis, extruded cage, cage in kyphosis, and radiculopathy. The authors concluded that titanium cages provide immediate anterior column stability and offer a safe alternative to autogenous bone grafts.

Two recent studies examined the biomechanics of anterior cervical interbody cages. Hacker et al. (207) conducted a randomized multicenter clinical trial looking at three different study cohorts of anterior cervical diskectomy fusions: instrumented with HA-coated BAK-C, instrumented with noncoated BAK-C, and uninstrumented, bone graft only (ACDF) fusions. There were a total of 488 patients in the trial, with 288 included in the 1 year follow up and 140 in the 2 year follow up. There were 79.9% one-level fusions and 20.1% two-level fusions performed. Results showed no significant differences between the coated or noncoated BAK-C devices, leading the authors to combine these groups for analysis. Complication rate with the BAK-C group of 346 patients was 10.1% and the ACDF group of 142 patients demonstrated an overall complication rate of 16.2%. The fusion rates for the BAK-C and ACDF fusions at 12 months for one level were 98.7 and 86.4%, respectively; for two levels, 80.0 and 80.0%, respectively. The fusion rates for the BAK-C and ACDF fusions at 24 months for one level were 100 and 96.4%, respectively; for two levels, 91.7 and 77.8%, respectively. Overall, the authors found that the BAK-C cage performed comparably to conventional, uninstrumented, bone graft

only anterior diskectomy and fusion. In an *in vitro* comparative study, Yang et al. (208) compared the initial stability and pullout strength of five different cervical cages and analyzed the effect of implant size, placement accuracy, and tightness of the implant on segmental stability. The cages analyzed included (*1*) SynCage-C Curved, (*2*) SynCage-C Wedged, (*3*) Brantigan I/F, (*4*) BAK-C, and (*5*) ACF Spacer. Overall, 35 cervical spines were used, with a total number of 59 segments selected for the study. Flexibility testing was performed under 50 N preload and up to 2 N·m in flexion, extension, lateral bending, and axial rotation. After quasistatic load tests were completed, the cages were subjected to an anterior pull-out test. Direct measurement on the specimen and biplanar radiographs allowed for quantification of distractive height, change in segmental lordosis, cage protrusion, and cage dimensions normalized to the endplate. Results from the study indicated that, in general, the cages were effective in reducing ROM in all directions by approximately one-third, but failed to reduce the neutral zone (NZ) in flexion/extension and axial rotation. Additionally, differences in implants were not significant and only existed between the threaded and nonthreaded designs. The threaded BAK-C was found to have the highest pullout force. Pullout force and lordotic change were both identified as significant predictors of segmental stability, a result the authors underscored as emphasizing the importance of a tight implant fit within the disk space.

RHAKOSS C synthetic bone spinal implant (Orthovita Inc., Malvern, PA) is trapezoidal in shape with an opening in the center for bone graft augmentation, and is fabricated from a bioactive glass/ceramic composite. *In vitro* testing conducted by Goel et al. (209) was conducted to evaluate the expulsion and stabilizing capabilities of the cervical cage in the lower cervical spine; *C6/7 and C4/5 motion segments. from five of the spinal donors were used for the expulsion testing. All specimens received the "Narrow Lordotic" version of the Rhakoss C design. The cages were implanted by orthopedic surgeons following manufacturer recommendations.* Specimens were tested in various modes; intact, destabilized with the cage in place, cage plus an anterior plate (Aline system, Surgical Dynamics Inc., Norwalk, CT), and again with the cage and plate after fatigue loading of 5000 flexion–extension cycles of 1.5 N·m. The results of the expulsion testing indicate that BMD and patient age are good predictors of implant migration resistance ($r = 0.8$). However, the high BMD/age correlation in the specimens makes it difficult to distinguish the relative importance of these two factors. The stability testing demonstrated the ability of a cage with a plate construct to sufficiently stabilize the cervical spine. However, BMD and specimen age play a major role in determining the overall performance of the cervical interbody cage.

Totribe (210) undertook a biomechanical comparison of a new cage made of a forged composite of unsintered-hydroxyapatite particles–poly-L-lactide (F-u-HA-PLLA) and the Ray threaded fusion cage. The objectiove was to compare the stability imparted to the human cadaveric spine by two different threaded cervical cages, and the effect of cyclic loading on construct stability. Threaded cages have been developed for use in anterior cervical interbody fusions to provide initial stability during the fusion process. However, metallic instrumentation has several limitations. Recently, totally bioresorbable bone fixation devices made of F-u-HA/PLLA have been developed, including a cage for spinal interbody fusion. Twelve fresh ligamentous human cervical spines (C4-C7) were used. Following anterior diskectomy across C5-C6 level, stabilization was achieved with the F-u-HA/PLLA cage in six spines and the Ray threaded fusion cage in the remaining six. Biomechanical testing of the spines was performed with six degrees of freedom before and after stabilization, and after cyclic loading of the stabilized spines (5000 cycles of flexion–extension at 0.5 N·m). The stabilized specimens (with F-u-HA/PLLA cage or the Ray cage) were significantly more stable than the diskectomy case in all directions except in extension. In extension, both groups were stiffer, although not at a significant level ($P > 0.05$). Following fatigue, the stiffness, as compared to the prefatigue case, decreased in both groups, although not at a significant level. The Ray cage group exhibited better stability than the F-u-HA/PLLA cage group in all directions, although a significant difference was found only in right axial rotation.

Lumbar Spine

Anterior and Posterior Spinal Instrumentation. The stability analysis of devices with varying stiffness is best exemplified by a study of Gwon et al. (211) who tested three different transpedicular screw devices: spinal rod-transpedicular screw system (RTS), the Steffee System (VSP), and Crock device (CRK). All devices provided statistically significant ($P < 0.01$) motion reductions across the affected level (L4-L5). The differences among the three devices in reducing motion across L4-L5, however, were not significant. Also, the changes in motion patterns of segments adjacent to the stabilized level compared with the intact case were not statistically significant. These findings have been confirmed by Rohlmann and associates who used a finite element model to address several implant related issues, including this one (212).

In an *in vitro* study, Weinhoffer et al. (213) measured intradiskal pressure in lumbosacral cadaver specimens subjected to constant displacement before and after applying bilateral pedicle screw instrumentation across L4-S1. They noted that intradiskal pressure increased in the disk above the instrumented levels. Also, the adjacent level effect was confounded in two-level instrumentation compared with single-level instrumentation. Other investigators, in principle, have reported similar results. Completely opposite results, however, are presented by several others (212). Results based on *in vitro* studies must be interpreted with caution, being dependent on the testing mode chosen (displacement or load control) for experiments. In the displacement control-type studies, in which applied displacement is kept constant during testing of intact and stabilized specimens, higher displacements and related parameters (e.g., intradiskal pressure) at the adjacent segments are reported. This is not true for the results based on the load control-type studies, in which the applied loads are kept constant.

Lim et al., assessed the biomechanical advantages of diagonal transfixation compared to horizontal transfixation (214). Diagonal cross-members yielded more rigid fixation in flexion and extension, but less in lateral bending and axial rotational modes, as compared to horizontal cross-members. Furthermore, greater stresses in the pedicle screws were predicted for the system having diagonal cross members. The use of diagonal configuration of the transverse members in the posterior fixation systems did not offer any specific advantages, contrary to the common belief.

Biomechanical cadaver studies of anterior fusion promoting and stabilizing devices (214–217) have become increasingly more common in the literature, owing to this procedure's rising popularity (105). *In vitro* testing was performed using the T9-L3 segments of human cadaver spines (218). An L-1 corpectomy was performed, and stabilization was achieved using one of three anterior devices: the ATLP in nine spines, the SRK in 10, and the Z-plate in 10. Specimens were load tested. Testing was performed in the intact state, in spines stabilized with one of the three aforementioned devices after the devices had been fatigued to 5000 cycles at ± 3 N·m, and after bilateral facetectomy. There were no differences between the SRK- and Z-plate-instrumented spines in any state. In extension testing, the mean angular rotation ($\pm$ standard deviation) of spines instrumented with the SRK ($4.7 \pm 3.2°$) and Z-plate devices ($3.3 \pm 2.3°$) was more rigid than that observed in the ATLP-stabilized spines ($9 \pm 4.8°$). In flexion testing after induction of fatigue, however, only the SRK ($4.2 \pm 3.2°$) was stiffer than the ATLP ($8.9 \pm 4.9°$). Also, in extension postfatigue, only the SRK ($2.4 \pm 3.4°$) provided more rigid fixation than the ATLP ($6.4 \pm 2.9°$). All three devices were equally unstable after bilateral facetectomy. The SRK and Z-plate anterior thoracolumbar implants were both more rigid than the ATLP, and of the former two the SRK was stiffer. The results suggest that in cases in which profile and ease of application are not of paramount importance, the SRK has an advantage over the other two tested implants in achieving rigid fixation immediately postoperatively.

Vahldiek and Panjabi investigated the biomechanical characteristics of short-segment anterior, posterior, and combined instrumentations in lumbar spine tumor vertebral body replacement surgery (219). The L2 vertebral body was resected and replaced by a carbon-fiber cage. Different fixation methods were applied across the L1 and L3 vertebrae. One anterior, two posterior, and two combined instrumentations were tested. The anterior instrumentation, after vertebral body replacement, showed greater motion than the intact spine, especially in axial torsion (range of motion, 10.3 vs. 5.5°; neutral zone, 2.9 vs. 0.7°; $P < 0.05$). Posterior instrumentation provided greater rigidity than the anterior instrumentation, especially in flexion–extension (range of motion, 2.1 vs. 12.6°; neutral zone, 0.6 vs. 6.1°; $P < 0.05$). The combined instrumentation provided superior rigidity in all directions compared with all other instrumentations. Posterior and combined instrumentations provided greater rigidity than anterior instrumentation. Anterior instrumentation should not be used alone in vertebral body replacement.

Oda et al. nondestructively compared three types of anterior thoracolumbar multisegmental fixation with the objective to investigate the effects of rod diameter and rod number on construct stiffness and rod–screw strain (220). Three types of anterior fixation were then performed at L1-L4: (1) 4.75 mm diameter single rod, (2) 4.75 mm dual-rod, and (3) 6.35 mm single-rod systems. A carbon fiber cage was used for restoring intervertebral disk space. Single screws at each vertebra were used for single-rod and two screws for dual-rod fixation. The 6.35 mm single-rod fixation significantly improved construct stiffness compared with the 4.75 mm single rod fixation only under torsion ($P < 0.05$). The 4.75 mm dual rod construct resulted in significantly higher stiffness than did both single-rod fixations ($P < 0.05$), except under compression. For single-rod fixation, increased rod diameter neither markedly improved construct stiffness nor affected rod–screw strain, indicating the limitations of a single-rod system. In thoracolumbar anterior multisegmental instrumentation, the dual-rod fixation provided higher construct stiffness and less rod–screw strain compared with single-rod fixation.

Lumbar Interbody Cages. Cage related biomechanical studies range from evaluations of cages as stand alone devices to use of anterior or posterior instrumentation for additional stabilization. The changes in stiffness and disk height of porcine FSUs by installation of a threaded interbody fusion cage and those by gradual resection of the annulus fibrosus were quantified (117). Flexion, extension, bending, and torsion testing of the FSUs were performed in four sequential stages: stage I, intact FSU; stage II, the FSUs were fitted with a threaded fusion cage; stage III, the FSUs were fitted with a threaded fusion cage with the anterior one-third of the annulus fibrosus excised, including excision of the anterior longitudinal ligament; and stage IV, in addition to stage III, the bilateral annulus fibrosus was excised. Segmental stiffness in each loading in the four stages and a change of disk height induced by the instrumentation were measured. After instrumentation, stiffness in all loading modes ($p < 0.005$) and disk height ($p = 0.002$) increased significantly. The stiffness of FSUs fixed by the cage decreased with gradual excision of the annulus fibrosus in flexion, extension, and bending. These results suggest that distraction of the annulus fibrosus and posterior ligamentous structures by installation of the cage increases the soft-tissue tension, resulting in compression to the cage and a stiffer motion segment. This study explains the basic mechanism through which the cages may provide the stability in various loading modes.

Three posterior lumbar interbody fusion implant constructs (Ray Threaded Fusion Cage, Contact Fusion Cage, and PLIF Allograft Spacer) were tested for stability in a cadaver model (221). None of the standalone implant constructs reduced the neutral zone (amount of motion in response to minimal load application). The constructs decreased the range of motion in flexion and lateral bending. The data did not suggest any implant construct to behave superiorly. Specifically, the PLIF Allograft Spacer is biomechanically equivalent to titanium cages and is devoid of the deficiencies associated with metal cages. Therefore, the PLIF Allograft Spacer is a valid alternative to conventional cages.

The lateral, and other cage orientations within the disk have been increasingly used for fusion (222). In one study, 14 spines were randomized into the anterior group (anterior diskectomy and dual anterior cage—TFC placement) and the lateral group (lateral diskectomy and single transverse cage placement) for load-displacement evaluations. Segmental ranges of motion were similar between spines undergoing either anterior or lateral cage implantation. Combined with a decreased risk of adjacent structure injury through a lateral approach, these data support a lateral approach for lumbar interbody fusion. When used alone to restore stability, the orientation of the cage (oblique vs. posterior) effected the outcome (223). Likewise, in flexion, both the OBAK (Oblique placement of one cage) and CBAK (Conventional posterior placement of two cages) orientations provided significant stability. In lateral bending, CBAK orientation was found to be better then OBAK. In axial mode, CBAK orientation was significantly effective in both directions while OBAK was effective only in right axial rotation. Owing to the differences in the surgical approach and the amount of dissection, the stability for the cages when used alone as a function of cage orientation was different.

The high elastic modulus of the cages causes the structures to be very stiff and may lead to stress-shielded environments within the devices with potential adverse effect on growth of the cancellous bone within the cage itself (224). Using a calf spine model, a study was designed to compare the construct stiffness afforded by 11 differently designed anterior lumbar interbody fusion devices: four different threaded fusion cages: (BAK device, BAK Proximity, Ray TFC, and Danek TIBFD); five different nonthreaded fusion devices (oval and circular Harms cages, Brantigan PLIF and ALIF cages, and InFix device); two different types of allograft (femoral ring and bone dowel); and to quantify their stress-shielding effects by measuring pressure within the devices. Prior to testing, a silicon elastomer was injected into the cages and intra cage pressures were measured using pressure needle transducers. No statistical differences were observed in construct stiffness among the threaded cages and nonthreaded devices in most of the testing modalities. Threaded fusion cages demonstrated significantly lower intracage pressures compared with nonthreaded cages and structural allografts. Compared with nonthreaded cages and structural allografts, threaded fusion cages afforded equivalent reconstruction stiffness but provided more stress-shielded environment within the devices. (This stress shielding effect may further increase in the presence of supplementary fixation devices.)

It is known that micromotion at the cage–endplate interface can influence bone growth into its pores. Loading conditions, mechanical properties of the materials, friction coefficients at the interfaces, and geometry of spinal segments would affect relative micromotion and spinal stability. In particular, relative micromotion is related closely to friction at bone–implant interfaces after arthroplasty. A high rate of pseudarthrosis and a high overall rate of implant migration requiring surgical revision has been reported following posterior lumbar interbody fusion using BAK threaded cages (225). This may be due to poor fixation of the implant, in addition to the stress shielding phenomena described above. Thus, Kim developed an experimentally validated finite element model of an intact FSU and the FSU implanted with two threaded cages to analyze the motion of threaded cages in posterior lumbar interbody fusion (226). Motion of the implants was not seen in compression. In torsion, a rolling motion was noted, with a range of motion of 10.6° around the central axis of the implant when left–right torsion (25 N·m) was applied. The way the implants move within the segment may be due to their special shape: the thread of the implants cannot prevent the BAK cages rolling within the disk space. However, note that the authors considered too high a value of torsional load; such values may not be clinically relevant. Relative micromotion (slip distance) at the interfaces was obvious at their edges under axial compression. The slip occurred primarily at the anterior edges under torsion with preload, whereas it occurred primarily at the edges of the left cage under lateral bending with preload. Relative micromotion at the interfaces increased significantly as the apparent density of cancellous bone or the friction coefficient of the interfaces decreased. A significant increase in slip distance at the anterior annulus occurred with an addition of torsion to the compressive preload. Relative micromotion was sensitive to the friction coefficient of the interfaces, the bone density, and the loading conditions. A reduction in age-related bone density was less likely to allow bone growth into surface pores of the cage. It was likely that the larger the disk area the more stable the interbody fusion of the spinal segments. However, the amount of micromotion may change in the presence of a posterior fixation technique, an issue that was not reported by the authors.

Almost every biomechanical study has shown that interbody cages alone, irrespective of their shapes, sizes, surface type, material, and approach used for implantation, does not stabilize the spine in all of the modes. It is suspected that this may be caused by the destruction of the appropriate spinal elements like the anterior longitudinal ligament, and anterior annulus fibrosus, or facets. Thus, use of additional instrumentation to augment cages seems to have become a standard procedure.

The 3D flexibility in ligamentous human lumbar spinal units have been investigated after the anterior, anterolateral, posterior, or oblique insertion of various types of interbody cages with supplemental fixtion using anterior or posterior spinal instrumentation (227). With the supplementary fixation using transfacet screws, the differences in stability due to the orientations were not noticeable at all, both before and after; underscoring the importance of using instrumentation when cages are used.

Patwardhan et al. (228) tested the hypothesis that the ability of the ALIF cages to reduce the segmental motions in flexion and extension will be significantly affected by the magnitude of the compressive preload. Fourteen human lumbar spine specimens (L1-sacrum) were tested intact, and after insertion of two threaded cylindrical cages at L5-S1. They were tested in flexion–extension with progressively increasing magnitude of compressive preload from 0 to 1200 N applied along the follower load path (described earlier). The stability of the stand-alone cage construct was significantly affected by the amount of compressive preload applied across the operated segment. In contrast to the extension instability reported in the literature, the two-cage construct exerted a stabilizing effect on the motion segment (reduction in segmental motion) in extension under physiologic compressive preloads. The cages provided substantially more stability, both in flexion and in extension, at larger preloads (800–1200 N) corresponding to standing and walking activities as compared to the smaller preloads (200–400 N) experienced during supine and recumbent postures. The compressive preload due to muscle activity likely plays a substantial role in stabilizing the segment with interbody cages.

The function of the interbody fusion cages is to stabilize the spinal segment primarily by distracting them as well as allowing bone ingrowth and fusion (122). An important condition for efficient formation of bone tissue is achieving adequate spinal stability. However, the initial stability may be reduced due to repeated movements of the spine during activities of daily living. Before and directly after implanation of a Zientek, Stryker, or Ray posterior lumbar interbody fusion cage, 24 lumbar spine segments were evaluated for stability analyses. The specimens were then loaded cyclically for 40,000 cycles at 5 Hz with an axial compression load ranging from 200 to 1000 N. The specimens were tested again in the spine tester. Generally, a decrease in motion in all loading modes was noted after insertion of the Zietek and Ray cages and an increase after implantation of a Stryker cage. In all three groups, greater stability was demonstrated in lateral bending and flexion then in extension and axial rotation. Reduced stability during cyclic loading was observed in all three groups; however, loss of stability was most pronounced in Ray cage group. Authors felt that this may be due to the damage of the cage: bone interface during cyclic loading that was not the case for the other two since they have a flat brick type interface. In order to reduce the incidence of stress risers at the bone–implant interface, it is essential that interbody fusion implants take advantage of the cortical periphery of the vertebral endplates. A larger cross-sectional footprint to the implant design will aid in dispersing the axial forces of spinal motion over a larger surface area and minimize the risk of stress risers, which may result in endplate fractures.

Animal Models

An approximation of the *in vivo* performance of spinal implants in humans can be attained by evaluation in animal models (229). Specifically, animal models provide a dynamic biologic and mechanical environment in which the implant can be evaluated. Temporal changes in both the host biologic tissue and instrumentation can be assessed with selective incremental sacrificing of the animals. Common limitations of animal studies include the method of loading (quadruped versus biped) and the size adjustment of devices needed such that proper fit is achieved in the animals.

Animal studies have revealed the fixation benefits of grouting materials in the preparation of the screw hole (230). The major findings were that the HA grouting of the screw hole bed before insertion significantly increased fixation (pullout) of the screws. Scanning electron microscopy analysis revealed that HA plasma spraying had deleterious effects on the screw geometry, dulling the self-tapping portion of the screw and reducing available space for bony in-growth.

An animal model of anterior and posterior column instability was developed by McAfee et al. (231–233) to allow *in vivo* observation of bone remodeling and arthrodesis after spinal instrumentation. An initial anterior and posterior destabilizing lesion was created at the L5–6 vertebral levels in 63 adult Beagle dogs. Observations 6 months after surgery revealed a significantly improved probability of achieving a spinal fusion if spinal instrumentation had been used. Nondestructive mechanical testing after removal of all metal instrumentation in torsion, axial compression, and flexion revealed that the fusions performed in conjunction with spinal instrumentation were more rigid. Quantitative histomorphometry showed that the volumetric density of bone was significantly lower (i.e., device-related osteoporosis occurred) for fused versus unfused spines. In addition, a linear correlation occurred between decreasing volumetric density of bone and increasing rigidity of the spinal implant; device-related osteoporosis occurred secondary to Harrington, Cotrel-Dubousset, and Steffee pedicular instrumentation. However, the stress-induced changes in the bone quality found in the animal models is not likely to correlate well with the actual changes in the spinal segment of a patient. In fact, it is suggested that the degeneration in a patient may be determined more by individual characteristics than by the fusion itself (234).

In long bone fractures, internal fixation improves the union rate, but does not accelerate the healing process. Spinal instrumentation also improves the fusion rate in spinal arthrodesis. However, it remains unclear whether the use of spinal instrumentation expedites the healing process of spinal fusion (235,236). Accordingly, an *in vivo* sheep model was used to investigate the effect of spinal instrumentation on the healing process of posterolateral spinal fusion. Sixteen sheep underwent posterolateral spinal arthrodeses at L2-L3 and L4-L5 using equal amounts of autologous bone. One of those segments was selected randomly for further augmentation with transpedicular screw fixation (Texas Scottish Rite Hospital spinal system). The animals were killed at 8 or 16 weeks after surgery. Fusion status was evaluated through biomechanical testing, manual palpation, plain radiography, computed tomography, and histology. Instrumented fusion segments demonstrated significantly higher stiffness than did uninstrumented fusions at 8 weeks after surgery. Radiographic assessment and manual palpation showed that the use of spinal instrumentation improved the fusion rate at 8 weeks (47 vs. 38% in radiographs, 86 vs. 57% in manual palpation). Histologically, the instrumented fusions consisted of more woven bone than the uninstrumented fusions at 8 weeks after surgery. The 16-week-old fusion mass was diagnosed biomechanically, radiographically, and histologically as solid, regardless of pedicle screw augmentation. The results demonstrated that spinal instrumentation created a stable mechanical environment to enhance the early bone healing of spinal fusion.

Human Clinical Models

Loads in posterior implants were measured in 10 patients using telemeterized internal spinal fixation devices (237–239). Implant loads were determined in up to 20 measuring sessions for different activities, including walking, standing, sitting, lying in the supine position, and lifting an extended leg while in the supine position. Implant loads often increased shortly after anterior interbody fusion was performed. Several patients retained the same high level even after fusion had taken place. This explains the reason why screw breakage sometimes occurs more than half a year after implantation. The time of fusion could not be pinpointed from the loading curves. A flexion bending moment acted on the implant even when the body was in a relaxed lying position. This meant that already shortly after the anterior procedure, the shape of the spine was not neutral and unloaded, but slightly deformed, which loaded the fixators. In another study, the same authors used the telemeterized internal spinal fixation devices to study the influence of muscle forces on the implant loads in three patients before and after anterior interbody fusion. Contracting abdominal or back muscles in a lying position was found to significantly increase implant loads. Hanging by the hands from wall bars as well as balancing with the hands on parallel bars reduced the implant loads compared with standing; however, hanging by the feet with the head upside down did not reduce implant loads, compared with lying in a supine position. When lying on an operating table with only the foot end lowered so that the hips were bent, the patient had different load measurements in the conscious and anesthetized states before anterior interbody fusion. The anesthetized patient evidenced predominately extension moments in both fixators, whereas flexion moments were observed in the right fixator of the conscious patient. After anterior interbody fusion had occurred, the differences in implant loads resulting from anesthesia were small. The muscles greatly influence implant loads. They prevent an axial tensile load on the spine when part of the body weight is pulling, for example, when the patient is hanging by their hands or feet. The implant loads may be strongly altered when the patient is under anesthesia.

The above review clearly shows that a large number of fusion enhancement instrumentation are available to surgeons. However, none of the instrumentation is totally satisfactory in its performance and there is room to improve the rate of fusion success, if fusion is the goal. Naturally, alternative fusion approaches (mechanical, biological) are currently being pursued.

The rigidity of a spinal fixation device and its ability to share load with the fusion mass are considered essential for the fusion to occur. If the load transferred through the fusion mass, is increased without sacrificing the rigidity of the construct, a more favorable environment for fusion may be created. To achieve this objective, posterior as well as anterior "dynamized" systems have been designed (240–242). One such posterior system consists of rods and pedicle screws and has a *hinged* connection between the screw head and shaft compared with the rigid screws (Fig. 16a). Another example of the dynamized anterior system (ALC) is shown in Fig. 16b. Load-displacement

(a)

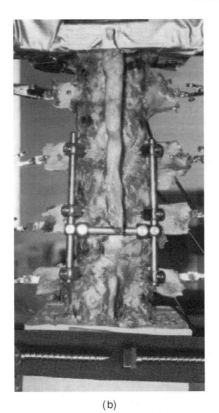

(b)

Figure 16. The two different types of dynamized systems used in a cadaver model to assess their stability characteristics. The data were compared with the corresponding "rigid" systems. (a) Posterior system and (b) anterior system. (Taken from Refs. 242 and 241.)

tests were performed to assess the efficacy of these devices in stabilizing a severally destabilized spinal segment. The hinged and rigid posterior systems provided significant stability across the L2-L4 segment in flexion, extension, and lateral bending as compared with the intact case ($P < 0.5$). The stabilities imparted by the hinged-type and its alternative rigid devices were of similar magnitudes. The ALC dynamized and rigid anterior systems also provided significant stability across the L3-L5 segment in flexion, extension, and lateral bending ($P < 05$). The stability imparted by the Dynamized ALC and its alternate rigid system did not differ significantly.

Dynamic stabilization may provide an alternative to fusion for patients suffering from early degenerative disk disease (DDD). The advantages of using a dynamic system are, preservation of the disk loading, allowing some physiologic load sharing in the motion segment. A finite element (FE) study was done to understand the effect of a commercially available dynamic system (DYNESYS, Zimmer Spine) compared to a rigid system on the ROM and disk stresses at the instrumented level (243). An experimentally validated 3-D FE model of intact L3-S1 spine was modified to simulate rigid and dynamic systems across L4-L5 level with the disk intact. The DYNESYS spacer and ligament were modeled with truss elements, with the "no tension" and "no compression" options, respectively. The ROM and disk stresses in response to a 400 N axial compression and 10.6-N·m flexion–extension moment were calculated. The ROM and disk stresses of the adjacent levels with rigid and DYNESYS systems had no significant change when compared to the intact. At the instrumented level in flexion–extension the decrease in motion when compared to the intact was 68/84% for rigid system and 50/56% for DYNESYS. The peak Von Mises disk stresses at the instrumented segment reduced by 41/80% for the rigid system, 27/45% for the DYNESYS system for flexion–extension loading condition. The predicted motion data for the dynamic system was in agreement with the experimental data. From the FE study it can be seen that the DYNESYS system allows more motion than the rigid screw-rod system, and hence allows for partial disk loading. This partial disk loading might be advantageous for a potential recovery of the degenerated disk, thus making dynamic stabilization systems a viable option for patients in early stages of DDD.

An anterior bone graft in combination with posterior instrumentation has been shown to provide superior support because the graft is in line with axial loads and the posterior elements are left intact. However, employing posterior instrumentation with anterior grafting requires execution of two surgical procedures. Furthermore, use of a posterior approach to place an interbody graft requires considerable compromise of the posterior elements, although it reduces the surgery time. It would be advantageous to minimize surgical labor and structural damage caused by graft insertion into the disk space via a posterior approach. Authors have addressed this issue by preparing an interbody bone graft using morselized bone (244–246). This device is a gauze bag of Dacron that is inserted into the disk space, filled with morselized bone, and tied shut, Fig. 17. *In vitro* testing measured the rotations of each vertebral level of mechanically loaded cadaver lumbar spines, both in intact and several experimental conditions. With the tension band alone, motion was restored to the

Selspot Testing

MTS Testing

Figure 17. The Bag system developed by Spineology Inc. The increases and decreases in motion with respect to intact segment for bag alone and bag with a band are also shown. (Taken from Ref. 244.)

intact case, except in extension where it was reduced. With the graft implant, motion was restored to intact in all of the loading modes, except in flexion where it was reduced. With the tension band and graft, motion was again restored to intact except in flexion and extension where it was reduced. *In vitro* results suggest that a tension band increases stability in extension, while the bag device alone seems to provide increased stability in flexion. The implanted bag filled with morselized bone in combination with a posterior tension band, restores intact stiffness. Postcyclic results in axial compression suggest that the morselized bone in the bone-only specimens either consolidates or extrudes from the cavity despite confinement. Motion restoration or reduction as tested here is relevant both to graft incorporation and segment biomechanics. The posterior interbody grafting method using morselized bone is amenable to orthoscopy. It produces an interbody graft without an anterior surgical approach. In addition, this technique greatly reduces surgical exposure with minimal blood loss and no facet compromise. This technique would be a viable alternative to current 360° techniques pending animal tests and clinical trials.

Bone grafting is used to augment bone healing and provide stability after spinal surgery. Autologous bone graft is limited in quantity and unfortunately associated with increased surgical time and donor-site morbidity. Recent research has provided insight into methods that may modulate the bone healing process at the cellular level in addition to reversing the effects of symptomatic disk degeneration, which is a potentially disabling condition, managed frequently with various fusion procedures. Alternatives to autologous bone graft include allograft bone, demineralized bone matrix, recombinant growth factors, and synthetic implants (247,248). Each of these alternatives could possibly be combined with autologous bone marrow or various growth factors. Although none of the presently available substitutes provides all three of the fundamental properties of autograft bone (osteogeneticity,

osteoconductivity, and osteoinductivity), there are a number of situations in which they have proven clinically useful. A literature review indicate that alternatives to autogenous bone grafting find their greatest appeal when autograft bone is limited in supply or when acceptable rates of fusion may be achieved with these substitutes. For example, bone morphogenetic proteins have been shown to induce bone formation and repair.

Relatively little research has been undertaken to investigate the efficacy of OP-1 in the above stated role (249,250). Grauer et al. performed single-level intertransverse process lumbar fusions at L5-L6 of 31 New Zealand White rabbits. These were divided into three study groups: autograft, carrier alone, and carrier with OP-1. The animals were killed 5 weeks after surgery. Five (63%) of the 8 in the autograft group had fusion detected by manual palpation, none (0%) of the 8 in the carrier-alone group had fusion, and all 8 (100%) in the OP-1 group had fusion. Biomechanical testing results correlated well with those of manual palpation. Histologically, autograft specimens were predominantly fibrocartilage, OP-1 specimens were predominantly maturing bone, and carrier-alone specimens did not show significant bone formation. OP-1 was found to reliably induce solid intertransverse process fusion in a rabbit model at 5 weeks. Smoking interferes with the success of posterolateral lumbar fusion and the above authors extended the investigation to study the effect of using OP-1 to enhance fusion process in patients who smoke. Osteoinductive protein-1 was able to overcome the inhibitory effects of nicotine in a rabbit posterolateral spine fusion model, and to induce bony fusion reliably at 5 weeks.

Finally, another study performed a systematic literature review on non-autologous interbody fusion materials in anterior cervical fusion, gathering data from 32 clinical- and ten laboratory studies. Ten alternatives to autologous bone were compared: autograft, allograft, xenograft, poly-(methyl methacrylate) (PMMA), biocompatible osteoconductive polymer (BOP), Hydroxyapatite compounds, bone morphogenic protein (BMP), Carbon fiber, metallic devices and ceramics. The study revealed that autologous bone still provides the golden standard that other methods should be compared to. The team concluded that the results of the various alternative fusion options are mixed, and comparing the different methods proved difficult. Once a testing standard has been established, reliable comparisons could be conducted.

Finite Element Models

In vitro investigations and *in vivo* animal studies contain numerous limitations, including that these are both time consuming and monetarily expensive. The most important limitations of *in vitro* studies are that muscle contributions to loading are not usually incorporated and the highly variable quality of the cadaver specimens. As stated earlier, *in vivo* animal studies usually involve quadruped animals, and the implant sizes usually need to be scaled according to the animal size. In an attempt to compliment the above protocols, several FE models of the ligamentous spine have been developed (251–257).

Goel et al. (255) generated osteoligamentous FE models of intact lumbar one segment (L3-L4) and two segments (L3-L5). Using the L3-L4 model, they simulated fusion with numerous techniques in an attempt to describe the magnitude and position of internal stresses in both the biologic tissue (bone and ligament) and applied hardware. Specifically, the authors modeled bilateral fusion using unilateral and bilateral plating. Bilateral plating models showed that cancellous bone stresses were significantly reduced with the instrumentation simulated in the immediate postoperative period. Completely consolidated fusion mass case, load transmission led to unloading of the cancellous bone region, even after simulated removal of the device. Thus, this model predicted that removal of the device would not alleviate stress shielding-induced osteopenia of the bone and that this phenomenon may truly be a complication of the fusion itself. As would be expected, unilateral plating models revealed higher trabecular bone stresses than were seen in the bilateral plating cases. The degree of stability afforded to the affected segment, however, was less. Thus, a system that allows the bone to bear more load as fusion proceeds may be warranted. Several solutions have been proposed to address this question.

For example, a fixation system was developed that incorporated polymer washers in the load train (Steffee variable screw placement, VSP). The system afforded immediate postoperative stability and reduced stiffness with time as the washers underwent stress relaxation (a viscoelastic effect) (256). The FE modeling of this system immediately after implantation showed that internal bony stresses were increased by $\sim 20\%$ over the same system without the polymeric material. In addition, mechanical property manipulation of the washers simulating their *in vivo* stress relaxation revealed these stresses were continuously increasing, promoting the likelihood that decreased bone resorption would occur. The other solution is the use of dynamized fixation devices, as diskussed next.

The ability of a hinged pedicle screw-rod fixation (dynamized, see next section for details) device to transmit more

Table 11. Axial Displacement and Angular Rotation of L3 with respect to L4 for the 800 N Axial Compression[a]

Graft	Axial Displacement, mm		Rotation, deg	
	Rigid	Hinged	Rigid	Hinged
Cancellous	−0.258	−0.274	0.407	0.335
Cortical	−0.134	−0.137	0.177	0.127
Titanium	−0.132	−0.135	0.174	0.126

[a]Taken from Ref. 240.

loads across the stabilized segment compared with its rigid equivalent system was predicted using the FE models (240). In general, the hinged screw device allowed for slightly larger axial displacements of L3, while it maintained flexion rotational stability similar to the rigid screw device (Table 11). Slightly larger axial displacements may be sufficient enough to increase the load through the graft since the stiffness of the disk was increased by replacing it (shown as the "nucleus" in the tables) with a cancellous, cortical, or titanium interbody device to simulate the fusion mass in the model (Table 12).

The FE modeling coupled with adaptive bone remodeling algorithms has been used to investigate temporal changes associated with interbody fusion devices. Grosland et al. predicted the change in bone density distribution after implantation of the BAK device (Fig. 18) (257). The major findings included hypertrophy of bone directly in the load train (directly overlying and underlying the implant) and lateral atrophy secondary to the relatively high stiffness of the implant. The model also predicted that bone growth into and around the larger holes in the implant, resulting in sound fixation of the device.

Nonfusion Treatment Alternatives

Various methods have been employed in the characterization of device effectiveness for which spinal fusion is indicated. Because of nonphysiological nature of fusing the spinal segments that are supposed to provide motion–flexibility, adjacent-level degeneration, and other complications associated with the fusion process, alternatives to fusion have been proposed.

Ray Nucleus

In 1988, Ray presented a prosthetic nuclear replacement consisting of flexible woven filaments (Dacron) surrounding an internal semipermeable polyethylene membranous sac filled with hyaluronic acid and a thixotropic agent (i.e.,

Table 12. Loads Transferred Through the "Nucleus" and the Device for the 800 N Axial Compression in newtons[a]

Graft	Rigid		Hinged	
	"Nucleus"	Device	"Nucleus"	Device
Cancellous	712.4	87.6	767.9	32.1
Cortical	741.2	58.8	773.5	26.5
Titanium	742.5	57.5	774.3	25.7

[a]Taken from Ref. 37.

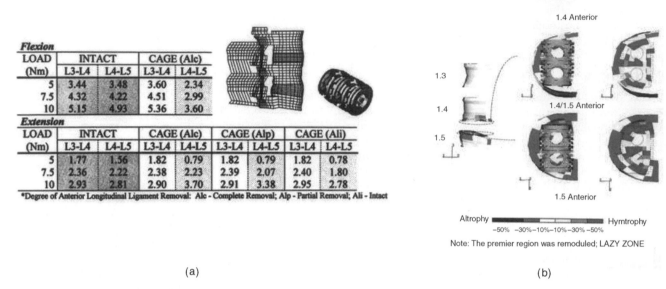

Flexion

LOAD (Nm)	INTACT		CAGE (Alc)	
	L3-L4	L4-L5	L3-L4	L4-L5
5	3.44	3.48	3.60	2.34
7.5	4.32	4.22	4.51	2.99
10	5.15	4.93	5.36	3.60

Extension

LOAD (Nm)	INTACT		CAGE (Alc)		CAGE (Alp)		CAGE (Ali)	
	L3-L4	L4-L5	L3-L4	L4-L5	L3-L4	L4-L5	L3-L4	L4-L5
5	1.77	1.56	1.82	0.79	1.82	0.79	1.82	0.78
7.5	2.36	2.22	2.38	2.23	2.39	2.07	2.40	1.80
10	2.93	2.81	2.90	3.70	2.91	3.38	2.95	2.78

*Degree of Anterior Longitudinal Ligament Removal: Alc - Complete Removal; Alp - Partial Removal; Ali - Intact

(a)   (b)

Figure 18. (a) The FE model of a ligamentous motion segment was used to predict load-displacement behavior of the segment following cage placement. Alc = anterior longitudinal ligament completely removed/cut, Alp = partially cut, and Ali = intact; and (b) Percentage change in density of the bone surrounding the BAK cage. (Taken from Refs. 32,33, and 257.)

a hydrogel) (244,258,259). As a nucleus replacement, the implant can be inserted similar to a thoracolumbar interbody fusion device, either posteriorly or transversely. Two are inserted per disk level in a partly collapsed and dehydrated state, but would swell due to the strongly hygroscopic properties of the hyaluronic acid constituent. The designer expects the implant to swell enough to distract the segment while retain enough flexibility to allow a normal range of motion. An option is to include therapeutic agents in the gel that would be released by water flow in and out of the prosthesis according to external pressures.

Recent reports on biomechanical tests of the device show that it can produce some degree of stabilization and distraction. Loads of 7.5 N·m and 200 N axial were applied to six L4-L5 specimens. Nucleotomized spines increased rotations by 12–18% depending on load orientation, but implanted spines (implant placed transversely) showed a change of −12% to +2% from the intact with substantial reductions in neutral zone. Up to 2 mm of disk height was recovered by insertion. The implant, however, was implanted and tested in its no hydrated form. The biomechanics of the hydrated prosthesis may vary considerably from that of its desiccated form.

In Situ Curable Prosthetic Intervertebral Nucleus (PIN)

The device (Disc Dynamics, Inc, Minnetonka, MN) consists of a compliant balloon connected to a catheter (Fig. 19) (244,260). This is inserted and a liquid polymer injected into the balloon under controlled pressure inflating the balloon, filling the cavity, and distracting the interverteb-

ral disk. Within 5 min the polymer is cured. Five fresh-frozen osteoligamentous three-segment human lumbar spines, screened for abnormal radiograph and low bone density, were used for the biomechanical study. The spines were tested under four conditions: intact, denucleated, implanted, and fatigued. Fatiguing was produced by cyclic loading from 250 to 750 N at 2 Hz for at least 100,000 cycles. Nuclectomy was performed through a 5.5 mm trephine hole in the right middle lateral side of the annulus. The device was placed in the nuclear cavity as described earlier. Following biomechanical tests, these specimens were radiographed and dissected to determine any structural damage inflicted during testing. Middle segment rotations generally increased with diskectomy, but were restored to the normal intact range with implantation. After fatiguing, rotations across the implanted segment increased. However, these were not more than, and often less than the intact adjacent segments. During polymer injection under compressive load the segment distracted as much as +1.8 mm (av) at the disk center as determined by the surrounding gauges. Over 1.6 mm was maintained during polymer cure with compression. The immediate goals of a disk replacement system are to restore disk height and provide segment mobility without causing instability. This study showed that PIN device could reverse the destabilizing effects of a nuclectomy and restore normal segment stiffness. Significant increases in disk height can also be achieved. Implanting the majority of disk replacement systems requires significant annulus removal, this device requires minimal surgical compromise and has the potential to be performed arthroscopically.

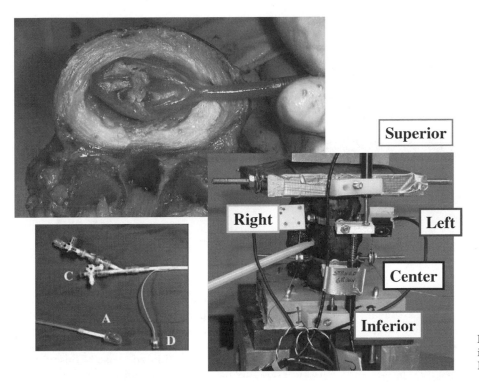

Figure 19. In situ curable prosthetic intervertebral nucleus (PIN) developed by Disc Dynamics, Inc. (Taken from Ref. 244.)

Artificial Disk

One of the most recent developments for nonfusion treatment alternatives is replacement of the intervertebral disk (244,261,262). The goal of this treatment alternative is to restore the original mechanical function of the resected disk. One of the stipulations of artificial disk replacement is that the remaining osseous spinal and paraspinal soft tissue components are not compromised by pathologic changes. Bao et al. (263) have classified the designs of total disk replacements into four categories: (*1*) low friction sliding surface; (*2*) spring and hinge systems; (*3*) contained fluid-filled chambers; and (*4*) disks of rubber and other elastomers. The former two designs seek to take advantage of the inherently high fatigue characteristics that all-metal designs afford. The latter two designs attempt to incorporate some of the viscoelastic and compliant properties that are exhibited by the normal, healthy intervertebral disk. Hedman et al. (264) outlined the major design criteria for intervertebral disk prosthesis: The disk must be able to maintain its mechanical integrity out to approximately 85 million cycles; consist of biocompatible materials; exist entirely within the normal disk space and maintain physiologic disk height; restore normal kinematic motion wherein the axes of each motion, especially sagittal plane motion, is correctly replicated; duplicate the intact disk stiffness in all three planes of rotation and compression; provide immediate and long-term fixation to bone; and, finally, provide *failsafe* mechanisms such that if an individual component of the design fails, catastrophic failure is not immediately imminent, and it does not lead to peri-implant soft tissue damage. This is certainly one of the greatest design challenges that bioengineers have encountered to date. In the following, some of the methods are discussed that are being employed in an attempt to meet this rigorous challenge.

One of the available studies dealt iterative design of the artificial disk replacement based on measured biomechanical properties. Lee, Langrana and co-workers (265,266) looked at incorporating three different polymers into their prosthetic intervertebral disk design and tried to represent the separate components (annulus fibrosis and nucleus) of the normal disk in varying proportion. They loaded their designs under 800 N axial compression and in compression-torsion out to 5°. The results indicated that disks fabricated from homogeneous materials exhibited isotropy that could not replicate the anisotropic behavior of the normal human disk. Thus, 12 layers of fiber reinforcement were incorporated in an attempt to mimic the actual annulus fibrosis. This method did result in more closely approximating the mechanical properties of the normal disk. Through this method of redesign and testing, authors claim that eventually "a disk prosthesis that has mechanical properties comparable to the natural disk could be manufactured."

The FE analyses have also been recruited in an effort to perturbate design with an eye toward optimizing the mechanical behavior of artificial disks. Goel and associates modified a previously validated intact finite element model to create models implanted with a ball-and-cup and slip core-type artificial disk models via an anterior approach, Figs. 20 and 21 (244,245,261). To study surgical variables, small and large windows were cut into the annulus, and the implants were placed anteriorly and posteriorly within the disk space. The anterior longitudinal ligament was also restored. Models were subjected to either 800 N axial compression force alone or to a combination of 10 N·m flexion–extension moments and 400 N axial preload. Implanted model predictions were compared with those of the intact model. The predicted rotations for the two disk implanted models were in agreement with the experimental data.

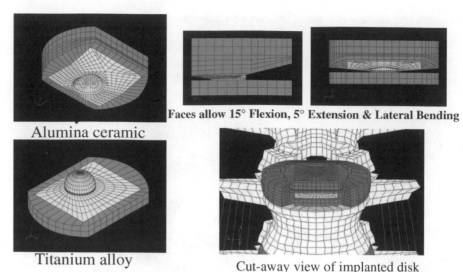

Alumina ceramic

Titanium alloy

Faces allow 15° Flexion, 5° Extension & Lateral Bending

Cut-away view of implanted disk

Figure 20. The intact finite element model of a ligamentous segment was modified to simulate the ball and socket type artificial disk implant. (Taken from Refs. 244,245.)

For the ball and socket design disk facet loads were more sensitive to the anteroposterior location of the artificial disk than to the amount of annulus removed. Under 800-N axial compression, implanted models with an anteriorly placed artificial disk exhibited facet loads 2.5 times greater than loads observed with the intact model, whereas posteriorly implanted models predicted no facet loads in compression. Implanted models with a posteriorly placed disk exhibited greater flexibility than the intact and implanted models with anteriorly placed disks. Restoration of the anterior longitudinal ligament reduced pedicle stresses, facet loads, and extension rotation to nearly intact levels. The models suggest that, by altering placement of the artificial disk in the anteroposterior direction, a surgeon can modulate motion-segment flexural stiffness and posterior load sharing, even though the specific disk replacement design has no inherent rotational stiffness.

The motion data, as expected, differed between the two disk designs (ball and socket, and slip core) and as compared to the intact as well, Fig. 22. Similar changes were observed for the loads on the facets, Fig. 23.

The experimentally validated finite element models of the intact and disk implanted L3-L5 segments revealed that both of these devices do not restore motion and loads across facets back to the intact case. (These design restore the intact biomechanics in a limited sense.) These differences are not only due to the size of the implants but the inherent design differences. Ball and socket design has a more "fixed" center of rotation as compared to the slip core design in which the COR undergoes a wider variation. Further complicating factor is the location of the disk within the annular space itself, a parameter under the control of the surgeon. Thus, it will be difficult to restore biomechanics of the segment back to normal using such designs. Only clinical follow up studies will provide the effects of such variations on the changes in spinal structures as a function of time.

More Recent and Future Initiatives

Although many of the well-accepted investigation techniques and devices have been discussed above, other

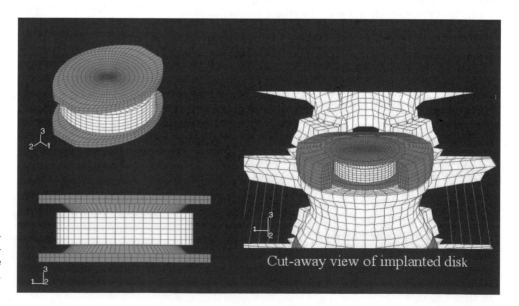

Figure 21. The intact finite element model of a ligamentous segment was modified to simulate the slip core type artificial disk implant. (Taken from Ref. 244.)

Cut-away view of implanted disk

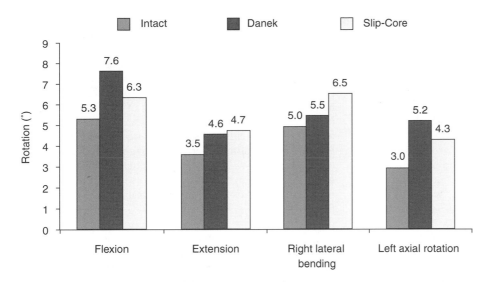

Figure 22. Predicted rotations for the two disk designs, shown in Figs. 20 and 21, as compared to the intact. (Taken from Ref. 244.)

techniques for the stabilization–fusion of the spine and nonfusion approaches are currently being investigated. These concepts are likely to play a significant role in future and are discussed. One such technique is vertebroplasty. Painful vertebral osteoporotic compression fractures leads to significant morbidity and mortality (263). Kyphoplasty and vertebroplasty are relatively new techniques that help decrease the pain and improve function in fractured vertebras.

Vertebroplasty is the percutaneous injection of PMMA cement into the vertebral body (263–269). While PMMA has high mechanical strength, it cures fast and thus allows only a short handling time. Other potential problems of using PMMA injection may include damage to surrounding tissues by a high polymerization temperature or by the unreacted toxic monomer, and the lack of long-term biocompatibility. Bone mineral cements, such as calcium carbonate and CaP, have longer working time and low thermal effect. They are also biodegradable while having

a good mechanical strength. However, the viscosity of injectable mineral cements is high, and the infiltration of these cements into vertebral body has been questioned. Lim et al. evaluated the compression strength of human vertebral bodies injected with a new calcium phosphate (CaP) cement with improved infiltration properties before compression fracture and also for vertebroplasty in comparison with PMMA injection (268). The bone mineral densities of 30 vertebral bodies (T2-L1) were measured using dual-energy X-ray absorptiometry. Ten control specimens were compressed at a loading rate of 15 mm/min to 50% of their original height. The other specimens had 6 mL of PMMA ($n = 10$) or the new CaP ($n = 10$) cement injected through the bilateral pedicle approach before being loaded in compression. Additionally, after the control specimens had been compressed, they were injected with either CaP ($n = 5$) or PMMA ($n = 5$) cement using the same technique, to simulate vertebroplasty. Loading experiments were repeated with the displacement control of 50% vertebral

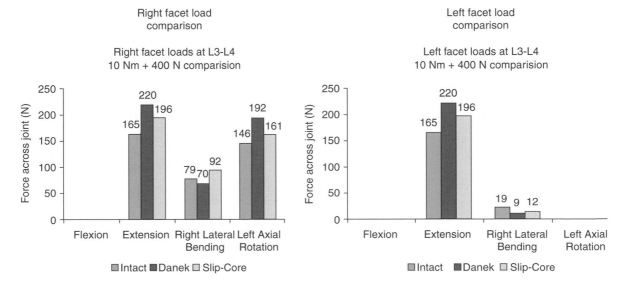

Figure 23. Predicted facet loads for the two disk designs, shown in Figs. 20 and 21, as compared to the intact. (Taken from Ref. 244.)

height. Load to failure was compared among groups and analyzed using analysis of variance. Mean bone mineral densities of all five groups were similar and ranged from 0.56 to 0.89 $g \cdot cm^{-2}$. The size of the vertebral body and the amount of cement injected were similar in all groups. Load to failure values for PMMA, the new CaP, and vertebroplasty PMMA were significantly greater than that of control. Load to failure of the vertebroplasty CaP group was higher than control but not statistically significant. The mean stiffness of the vertebroplasty CaP group was significantly smaller than control, PMMA, and the new CaP groups. The mean height gains after injection of the new CaP and PMMA cements for vertebroplasty were minimal (3.56 and 2.01%, respectively). Results of this study demonstrated that the new CaP cement can be injected and infiltrates easily into the vertebral body. It was also found that injection of the new CaP cement can improve the strength of a fractured vertebral body to at least the level of its intact strength. Thus, the new CaP cement may be a good alternative to PMMA cement for vertebroplasty, although further in vitro, in vivo animal and clinical studies should be done. Furthermore, the new CaP may be more effective in augmenting the strength of osteoporotic vertebral bodies, and for preventing compression fractures considering our biomechanical testing data and the known potential for biodegradability of the new CaP cement. Belkof et al. (266) found that the injection of either Orthocomp or Simplex P resulted in vertebral body strengths that were significantly greater than initial strength values. Vertebral bodies augmented with Orthocomp recovered their initial stiffness; and, vertebral bodies augmented with Simplex P were significantly less stiff than they were in their initial condition. However, these biomechanical results have yet to be substantiated in clinical studies.

Previous biomechanical studies have shown that injections of 8–10 mL of cement during vertebroplasty restore or increase vertebral body strength and stiffness; however, the dose-response association between cement volume and restoration of strength and stiffness is unknown. Belkof et al. (266) investigated the association between the volume of cement injected during percutaneous vertebroplasty and the restoration of strength and stiffness in osteoporotic vertebral bodies. Two investigational cements were studied: Orthocomp (Orthovita, Malvern, PA) and Simplex 20 (Simplex P with 20% by weight barium sulfate. Compression fractures were experimentally created in 144 vertebral bodies (T6-L5) obtained from 12 osteoporotic spines harvested from female cadavers. After initial strength and stiffness were determined, the vertebral bodies were stabilized using bipedicular injections of cement totaling 2, 4, 6, or 8 mL and recompressed, from which post-treatment strength and stiffness were measured. Strength and stiffness were considered restored when post-treatment values were not significantly different from initial values. Strength was restored for all regions when 2 mL of either cement was injected. To restore stiffness with Orthocomp, the thoracic and thoracolumbar regions required 4 mL, but the lumbar region required 6 mL. To restore stiffness with Simplex 20, the thoracic and lumbar regions required 4 mL, but the thoracolumbar region required 8 mL. These data provide

guidance on the cement volumes needed to restore biomechanical integrity to compressed osteoporotic vertebral bodies.

Liebschner et al. undertook a finite element based biomechanical study to provide a theoretical framework for understanding and optimizing the biomechanics of vertebroplasty, especially the effects of volume and distribution of bone cement on stiffness recovery of the vertebral body, just like the preceding experimental study (269). An experimentally calibrated, anatomically accurate finite-element model of an elderly L1 vertebral body was developed. Damage was simulated in each element based on empirical measurements in response to a uniform compressive load. After virtual vertebroplasty (bone cement filling range of 1–7 cm^3) on the damaged model, the resulting compressive stiffness of the vertebral body was computed for various spatial distributions of the filling material and different loading conditions. Vertebral stiffness recovery after vertebroplasty was strongly influenced by the volume fraction of the implanted cement. Only a small amount of bone cement (14% fill or 3.5 cm^3) was necessary to restore stiffness of the damaged vertebral body to the predamaged value. Use of a 30% fill increased stiffness by $> 50\%$ compared with the predamaged value. Whereas the unipedicular distributions exhibited a comparative stiffness to the bipedicular or posterolateral cases, it showed a medial-lateral bending motion (toggle) toward the untreated side when a uniform compressive pressure load was applied. Only a small amount of bone cement (15% volume fraction) is needed to restore stiffness to predamage levels, and greater filling can result in substantial increase in stiffness well beyond the intact level. Such overfilling also renders the system more sensitive to the placement of the cement because asymmetric distributions with large fills can promote single-sided load transfer and thus toggle. These results suggest that large fill volumes may not be the most biomechanically optimal configuration, and an improvement might be achieved by use of lower cement volume with symmetric placement. These theoretical findings support the experimental observations described in the proceeding paragraph, except these authors did not analyze the relationship between cement type and volume needed to restore strength.

Hitchon et al. compared the stabilizing effects of the HA product, with PMMA in an experimental compression fracture of L1 (268). No significant difference between the HA and PMMA cemented-fixated spines was demonstrated in flexion, extension, left lateral bending, or right- and left-axial rotation. The only difference between the two cements was encountered before and after fatiguing in right lateral bending ($p \leqslant 0.05$). The results of this study suggest that the same angular rigidity can be achieved by using either HA or PMMA. This is of particular interest because HA is osteoconductive, undergoes remodeling, and is not exothermic.

Advances in the surgical treatment of spinal etiologies continues to evolve with the rapid progression of technology. The advent of robotics, Microelectromechanical systems (MEMS) (270), novel biomaterials, and genetic and tissue engineering are revolutionizing spinal medicine, Novel biomaterials, also termed "smart biomaterials" that are capable of conforming or changing their mechanical properties in response to different loading paradigms are being investigated for their use in spinal implant design. The rapidly advancing field of tissue engineering opens new possibilities to solving spine problems. By seeding and growing intervertebral disk cells, it could be possible to grow a new bioartificial disk, to be implanted in to the spine. Studies are in progress at a number of centers, including our own (269).

CONCLUSION

The stability (or instability) of the human spine is integral to the diagnosis and treatment of patients with low back pain. The stability of the lumbar spine as portrayed by its motion characteristics can be determined through the use of clinical and radiographic criteria or other methods of determining the orientation of one spinal vertebra with respect to another. The instability can be the result of injury, disease, and many other factors, including surgery. Therefore, it is necessary to become familiar with recent findings and suggestions that deal with the instability that can result from such procedures. The prevalence of spinal fusion and stabilization procedures to restore spinal stability and host of other factors is continuously increasing. This article has presented many of the contemporary biomechanical issues germane to stabilization and fusion of the spine. Because of the wide variety of devices available, various testing protocols have been developed in an attempt to describe the mechanical aspects of these devices. These investigations reveal comparative advantages (and disadvantages) of the newer designs to existing hardware. Subsequent *in vivo* testing, specifically animal models, provides data on the performance of the device in a dynamic physiologic environment. All of the testing, *in vitro* and *in vivo*, helps to build confidence that the instrumentation is safe for clinical trial. Future biomechanical work is required to produce newer devices and optimize existing ones, with an eye toward reducing the rates of nonfusion and pseudoarthrosis. In addition, novel devices and treatments that seek to restore normal spinal function and loading patterns without fusion continue to necessitate advances in biomechanical methods. These are the primary challenges that need to be incorporated in future biomechanical investigations. Finally, one has to gain understanding of the effects of devices at the cellular level and one must undertake outcome assessment studies to see if the use of instrumentation is warranted for the enhancement of the fusion process.

ACKNOWLEDGMENTS

Manuscript is based on the work sponsored by various funding agencies over the last 20 years. Thanks also to a large number of coinvestigators who have contributed to the original work reported in this article.

BIBLIOGRAPHY

1. Goel VK, Weinstein JN. Clinical Biomechanics of the Lumbar Spine. Boca Raton (FL): CRC Press; 1990.

2. White AAA, Panjabi MM. Clinical Biomechanics of Spine. New York: Lippincot. 2nd ed.; 1990.

3. Doherty BJ, Heggeness MH. Quantitative anatomy of the second cervical vertebra. Spine 1995;20:513–517.

4. (a) Yoganandan N, Halliday A, Dickman C, Benzel E. Practical anatomy and fundamental biomechanics, in Spine Surgery: Techniques, Complication Avoidance, and Management. In: Benzel EC, editor. New York: Churchill Livingstone; 1999. p 93–118. (b) Clausen JD. Experimental & Theoretical Investigation of Cervical Spine Biomechanics—Effects of Injury and Stabilization. Ph.D. dissertation, University of Iowa, Iowa City (IA) 52242; 1996. (c) Puttlitz CM. A Biomechanical Investigation of the Craniovertebral Junction. Ph.D. dissertation, University of Iowa, Iowa City (IA) 52242; 1999. (d) Scifert JL. Biomechanics of the Cervical Spine. Ph.D. dissertation, University of Iowa, Iowa City (IA) 52242; 2000. (e) Goel VK, Panjabi MM, Kuroki H, Rengachary S, McGowan D, Ebraheim N. Biomechanical Considerations—Spinal Instrumentation. ISSLS Book. 2003. (f) Kuroki H, Holekamp S, Goel V, Panjabi M, Ebraheim N, Singer K. Biomechanics of Spinal Deformity in Inflammatory Disease. book chapter submitted; 2003. (g) Goel VK, Dooris AP, McGowan D, Rengachary S. Biomechanics of the Disk. In: Lewandrowski K-U, Yaszemski MJ, White III AA, Trantolo DJ, Wise DL, editors. Advances in Spinal Fusion—Molecular Science, Biomechanics, and Clinical Management. Somerville (CA): M&N Toscano; 2003. (h) Panjabi MM, Yue JJ, Dvorak J, Goel V, Fairchild T, White AA. Chapt. 4 Cervical Spine Kinematics and Clinical Instability. Cervical Spine Research Society; 2003 (submitted).

5. Xu R, Nadaud MC, Ebraheim NA, Yeasting RA. Morphology of the second cervical vertebra and the posterior projection of the C2 pedicle axis. Spine 1995;20:259–263.

6. Heggeness MH, Doherty BJ. The trabecular anatomy of the axis. Spine 1993;18:1945–1949.

7. Watkins RG. Cervical Spine Injuries in Athletes. In: Clark CR, editor. The cervical spine. Philadelphia: Lippincott-Raven; 1998. p 373–386.

8. Penning L. Differences in anatomy, motion, development, and aging of the upper and lower cervical disk segments. Clin Biomech 1988;3:337–347.

9. Pooni J, Hukins D, Harris P, Hilton R, Davies K. Comparison of the structure of human intervertebral disks in the cervical, thoracic and lumbar regions of the spine. Surg Rad Anat 1986;8:175–182.

10. Tulsi R, Perrett L. The anatomy and radiology of the cervical vertebrae and the tortuous vertebral artery. Aust Rad 1975; 19:258–264.

11. Frykholm R. Lower cervical vertebrae and intervertebral disks. Surgical anatomy and pathology. Acta Chir Scand 1951;101:345–359.

12. Hall M. Luschka's Joint. Springfield (IL): Charles C. Thomas; 1965.

13. Hardacker J, Shuford R, Capicotto P, Pryor P. Radiographic standing cervical segmental alignment in adult volunteers without neck symptoms. Spine 1997;22(13):1472–1480.

14. Harrison D, Janik T, Troyanovich S, Harrison D, Colloca C. Evaluation of the assumptions used to derive an ideal normal cervical spine model. JMPT 1997;20(4):246–256.

15. Harrison D, Janik T, Troyanovich S, Holland B. Comparisons of lordotic cervical spine curvatures to a theoretical ideal model of the static sagittal cervical spine. Spine 1996;21(6): 667–675.

16. Rosenberg LC. Proteoglycans. In: Owen R, Goodfellow J, Bullough P, editors. Scientific Foundations of Orthopaedics and Traumatology. Philadelphia: WB Saunders; 1980. p 36–42.

17. Humzah M, Soames R. Human intervertebral disk: structure and function. Anatomical Rec 1988;220:337–356.

18. Oda J, Tanaka H, Tsuzuki N. Intervertebral disk changes with aging of the human cervical vertebra: from neonate to the eighties. Spine 1988;13:1205–1211.

19. Alker GJ, Oh YS, Leslie EV. High cervical spine and craniocervical junction injuries in fatal traffic accidents: a radiological study. Orthop Clin NA 1978;9:1003–1010.

20. Freeman L, Wright T. Experimental observations of concussion and contusion of the spinal cord. Ann Surg 1953; 137(4): 433–443.

21. Hirsch C, Galante J. Laboratory conditions for tensile tests in annulus fibrosis from human intervertebral disks. Acta Orthop Scand 1967;38:148–162.

22. Singer KP, Breidahl PD, Day RE. Posterior element variation at the thoracolumbar transition. A morphometric study using computed tomography. Clin Biomech 1989;4:80–86.

23. Panjabi MM, Thibodeau LL, Crisco III JJ, et al. What constitutes spinal instability? Clin Neurosurg 1988; 34: 313–339.

24. Putz R. The detailed functional anatomy of the ligaments of the vertebral column. Anatomischer Anzeiger 1992;173:40–47.

25. Gilbertson LG. Mechnism of fracture and biomechanics of orthosis in thoracolumbar region. Ph.D. dissertation, University of Iowa (IA); 1993.

26. Ikegawa S, et al. The effect of joint angle on cross sectional area and muscle strength of human elbow flexors, Human Kinetics. Biomechanics SA: Champaign; 1979. p 35–44.

27. Macintosh JE, Bogduk N, Pearcy MJ. The effects of flexion on the geometry and actions of the lumbar erector spinae. Spine 1993;18:884–893.

28. White AA, et al. Spinal stability. Evaluation and treatment. instr course lect. AAOS 1981;30:457–484.

29. Penning L, Normal Kinematics of the Cervical Spine, Clinical Anatomy and Management of Cervical Spine Pain. In: Giles SK, editor. Oxford (UK): Butterworth Heinemann; 1998. p 53–70.

30. Panjabi MM, Dvorak J, Sandler AJ, Goel VK, White AA. Cervical Spine Kinematics and Clinical Instability. In: CR C, editor. The Cervical Spine. Philadelphia: Lippincott-Raven; 1998. p 53–77.

31. Panjabi MM, Dvorak J, Duranceau J, Yamamoto I, Gerber M, Rauschning W, Bueff HU. Three dimensional movements of the upper cervical spine. Spine 1988;13:726–730.

32. Goel VK, Clark CR, Gallaes K, Liu YK. Moment–rotation relationships of the ligamentous occipitoatlanto-axial complex. Spine 1988;21:673–680.

33. Moroney S, Schultz A, Miller J, Andersson G. Load-displacement properties of lower cervical spine motion segments. J Biomech 1988;21(9):769–779.

34. Lysell E. Motion in the cervical spine. Ph.D. dissertation. Acta Orthop Scand (123 Suppl); 1969.

35. Bernhardt M, White AA, Panjabi MM, McGowan DP. Biomechanical considerations of spinal stability. The Spine. 3rd ed. Philadelphia: WB Saunders Company; 1992. p 1167–1195.

36. Pearcy MJ. Stereoradiography of lumbar spine motion. Acta Orthop Scand 1985;56(212 Suppl).

37. Calve J, Galland M. Physiologie Pathologique Du Mal De Pott. Rev Orthop 1930;1:5.

38. Rolander SD. Motion of the lumbar spine with special reference to the stabilizing effect of the posterior fusion. Scta Ortho Scand 1966;90(Suppl):1–114.

39. Oda T, Panjabi MM, Crisco JJ, Oxland T, Katz L, Nolte L-P. Experimental study of atlas injuries II—Relevance to clinical diagnosis and treatment. Spine 1991;16:S466–473.

40. Oda T, Panjabi MM, Crisco JJ, Oxland TR. Multidirectional instabilities of experimental burst fractures of the atlas. Spine 1992;17:1285–1290.

41. Jefferson G. Fracture of the atlas vertebra: report of four cases, and a review of thos previously recorded. Br J Surg 1920;7:407–422.

42. Heller JG, Amrani J, Hutton WC. Transverse ligament failure—a biomechanical study. J Spinal Disord 1993;6: 162–165.

43. Goel VK, et al. Ligamentous laxity across CO-C1-C2 complex: axial torque-rotation characteristics until failure. Spine 1990;15:990–996.

44. Chang H, Gilbertson LG, Goel VK, Winterbottom JM, Clark CR, Patwardhan A. Dynamic response of the occipito-atlanto-axial (C0-C1-C2) complex in right axial rotation. J Orth Res 1992;10:446–453.

45. Doherty BJ, Heggeness MH, Esses SI. A biomechanical study of odontoid fractures and fracture fixation. Spine 1993;18:178–184.

46. Anderson LD, D'Alonzo RT. Fractures of the odontoid process of the axis. J Bone Joint Surg 1974;56-A:1663–1674.

47. Schatzker J, Rorabeck CH, Waddell JP. Fractures of the dens. J Bone Joint Surg 1971;53-B:392–405.

48. Clark CR, White AA. Fractures of the dens. JBJS 1985;67-A:1340–1348.

49. Althoff B. Fracture of the odontoid process. Acta Orthop Scand 1979;177(Suppl):1–95.

50. Mouradian WH, Fietti VG, Cochran GVB, Fielding JW, Young J. Fractures of the odontoid: a laboratory and clinical study of mechanisms. Orthop Clinics of NA 1978;9:985–1001.

51. Puttlitz CM, Goel VK, Clark CR, Traynelis VC. Pathomechanism of Failures of the Odontoid. Spine 2000;25:2868–2876.

52. Goel VK, Montgomery RE, Grosland NM, Pope MH, Kumar S. Ergonomic factors in the work place contribute to disk degeneration. In: Kumar S, editor. Biomechanics in Ergonomics. Taylor & Francis; 1999. p 243–267.

53. Liu YK, Njus G, Buckwalter J, Wakano K. Fatigue response of lumbar intervertebral joints under axial cyclic loading. Spine 1983;6:857.

54. Hooper DM. Consequences of asymmetric lifting on external and internal loads at the L3–5 lumbar levels. Ph.D. dissertation, Iowa City (IA): University of Iowa; 1996.

55. Kelsey JL. An epidemiological study of acute herniated lumbar intervertebral disk. Rehum Rehabil 1975;14:144.

56. Panjabi MM, Anderson GBJ, Jorneus L, Hult E, Matteson L. In vivo measurements of spinal column vibrations. JBJS 1986;68A:695.

57. Wilder DG, Woodworth BB, Frymoyer JW, Pope MH. Vibration and the human spine. Spine 1982;7:243.

58. Liu YK, Goel VK, DeJong A, Njus G, Nishiyama K, Buckwalter J. Torsional fatigue of the lumbar intervertebral joints. Spine 1985;10:894–900.

59. Kong W-Z, Goel VK. Ability of the finite element models to predict response of the human spine in sinusoidal vertical vibration. Spine 2003;28:1961–1967.

60. Furlong DR, Palazotto AN. A finite element analysis of the influence of surgical herniation on the viscoelastic properties of the intervertebral disk. J Biomech 1983;16:785.

61. Lee C-K, Kim Y-E, Jung J-M, Goel VK. Impact response of the vertebral segment using a finite element model. Spine 2000;25:2431–2439.

62. Dvorak J, Hayek J, Zehnder R. CT-functional diagnostics of the rotatory instability of the upper cervical spine: part 2. An evaluation on healthy adults and patients with suspected instability. Spine 1987;12:726–731.

63. Dvorak J, Penning L, Hayek J, Panjabi MM, Grob D, Zehnder R. Functional diagnostics of the cervical spine using computer tomography. Neuroradiology 1988;30:132–137.

64. Crisco JJ, Takenori O, Panjabi MM, Bueff HU, Dovrak J, Grob D. Transections of the C1-C2 joint capsular ligaments in the cadaveric spine. Spine 1991;16:S474–S479.

65. Fielding JW. Cineroentgenography of the normal cervical spine. J Bone Joint Surg 1957;39-a:1280–1288.

66. Puttlitz CM, Goel VK, Clark CR, Traynelis VC, Scifert JL, Grosland NM. Biomechanical rationale for the pathology of rheumatoid arthritis in the craniovertebral junction. Spine 2000;25:1607–1616.

67. Panjabi MM, White AA, Keller D, Southwick WO, Friedlaender G. Stability of the cervical spine under tension. J Biomech 1978;11:189–197.

68. Goel VK, Clark CR, Harris KG, Schulte KR. Kinematics of the cervical spine: effects of multiple total laminectomy and facet wiring. J Orthop Res 1988;6:611–619.

69. Goel VK, Clark CR, Harris KG, Kim YE, Schulte KR. Evaluation of effectiveness of a facet wiring technique: an in vitro biomechanical investigation. Ann Biomed Eng 1989;17:115–126.

70. Goel VK, Clausen JD. Prediction of load sharing among spinal components of a C5–C6 motion segment using the finite element approach. Spine 1998;23:684–691.

71. Clausen JD, Goel VK, Traynelis VC, Scifert JL. Uncinate processes and luschka's joints influence the biomechanics of the cervical spine-quantification using a finite element model of the C5-C6 segment. J Orth Res 1997;15:342–347.

72. Edwards WT, Zheng Y, Ferrara LA, Yuan HA. Structural features and thickness of the vertebral cortex in the thoracolumbar spine. Spine 2001;26(2):218–225.

73. Goel VK, Monroe BT, Gilbertson LG, Brinckmann P. Interlaminar shear stresses and laminae separation in a disk: finite element analysis of the L3-4 motion segment subjected to axial compressive loads. Spine 1995;20:689–698. (1994 Volvo Award Paper).

74. Goel VK, Kim YE. Effects of injury on the spinal motion segment mechanics in the axial compression mode. Clin Biomech 1989;4:161–167.

75. Posner I, White AA, Edwards WT, et al. A biomechanical analysis of clinical stability of the lumbar lumbrosacral spine. Spine 1982;7:374–389.

76. Kong WZ, Goel VK, Gilbertson LG, et al. Effects of muscle dysfunction on lumbar spine mechanics–a finite element study based on a two motion segments model. Spine 1996; 21:2197.

77. Kong W, Goel V, Weinstein J. Role of facet morphology in the etiology of degenerative spondylolisthesis in the presence of muscular activity. 41st Annual Meeting. Orlando, FL: Orthopaedic Research Society; Feb 13–16, 1995.

78. Lipson JL, Muir H. Proteoglycans in experimental intervertebral disk degeneration. Spine 1981;6:194–210.

79. Raynor RB, Moskovich R, Zidel P, Pugh J. Alterations in primary and coupled neck motions after facetectomy. J Neurosurg 1987;12:681–687.

80. Schulte KR, Clark CR, Goel VK. Kinematics of the cervical spine following discectomy and stabilization. Spine 1989;14: 1116–1121.

81. Martins A. Anterior cervical discectomy with and without interbody bone graft. J Neurosurg 1976;44:290–295.

82. Wilson D, Campbell D. Anterior cervical discectomy without bone graft. Neurosurgery 1977;47:551–555.

83. Goel VK, Nye TA, Clark CR, Nishiyama K, Weinstein JN. A technique to evaluate an internal spinal device by use of the Selspot system: an application to Luque closed loop. Spine 1987;12:150–159.

84. Zdeblick TA, Zou D, Warden KE, McCabe R, Kunz D, Vanderby R. Cervical stability after foraminotomy. A biomechanical in vitro analysis. J Bone Joint Surg [Am] 1992; 74:22–27.

85. Zdeblick TA, Abitbol JJ, Kunz DN, McCabe RP, Garfin S. Cervical stability after sequential capsule resection. Spine 1993;18:2005–2008.

86. Cusick JF, Yoganandan N, Pintar F, Myklebust J, Hussain H. Biomechanics of cervical spine facetectomy and fixation techniques. Spine 1988;13:808–812.

87. Voo LM, Kumaresan S, Yoganandan N, Pintar FA, Cusick JF. Finite element analysis of cervical facetectomy. Spine 1997;22:964–9.

88. Nowinski GP, Visarius H, Nolte LP, Herkowitz HN. A biomechanical comparison of cervical laminaplasty and cervical laminectomy with progressive facetectomy. Spine 1993;18: 1995–2004.

89. Goel VK, Clark CR, Harris KG, Schulte KR. Kinematics of the cervical spine: effects of multiple total laminectomy and facet wiring. J Orthop Res 1988;6:611–619.

90. Goel VK, Clark CR, McGowan D, Goyal S. An in-vitro study of the kinematics of the normal, injured and stabilized cervical spine. J Biomech 1984;17:363–376.

91. Bell DF, Walker JL, O'Connor G, Tibshirani R. Spinal deformity after multiple-level cervical laminectomy in children. Spine 1994;19:406–411.

92. Lee S-J, Harris KG, Nassif J, Goel VK, Clark CR. In vivo kinematics of the cervical spine; part I: development of a roentgen stereophotogrammetric technique using metallic markers and assessment of its accuracy. J Spinal Disord 1993;6:522–534.

93. Lee S-J. Three-dimensional analysis of post-operative cervical spine motion using simultaneous roentgen stereophotogrammetry with metallic markers. Ph.D. dissertation, Iowa City (IA): University of Iowa; 1993.

94. Kumaresan S, Yoganandan N, Pintar FA, Maiman DJ, Goel VK. Contribution of disk degeneration to osteophyte formation in the cervical spine—A biomechanical investigation. J Orth Res 2001;19(5):977–984.

95. Kubo S, Goel VK, Yang S-J, Tajima N. Biomechanical comparison of cervical double door laminoplasty using hydroxyapatite spacer. Spine 2003;28(3):227–234.

96. Kubo S, Goel VK, Tajima N. The biomechanical effects of multilevel posterior foraminotomy and foraminotomy with double door laminoplasty. J Spinal Disorders 2002;15:477–485.

97. Panjabi MM. Low back pain and spinal instability. In: Weinstein JN, Gordon SL, editors. Low back pain: a scientific and clinical overview. San Diego (CA): American Academy of Orthopaedic Surgeons; 1996. p 367–384.

98. Panjabi MM, Kaigle AM, Pope MH. Degeneration, injury, and spinal instability. In: Wiesel SW, et al., editors. The lumbar spine. 2nd ed. Volume 1, Philadelphia, PA: W.B. Saunders; 1996. p 203–211.

99. Goel VK, et al. Kinematics of the whole lumbar spine—effect of discectomy. Spine 1985;10:543–554.

100. Watters WC, Levinthal R. Anterior cervical discectomy with and without fusion–results, complications, and long-term follow-up. Spine 1994;19:2343.

101. Oxland TR, Teija L, Bernhard J, Peter C, Kurt L, Philippe J, Lutz-P N. The relative importance of vertebral bone density and disk degeneration in spinal flexibility and interbody implant performance An in vitro study. Spine 1996;21(22): 2558–2569.

102. Fraser RD. Interbody, posterior, and combined lumbar fusions. Spine 1995;20:167S.

103. Goel VK, Gilbertson LG. Basic science of spinal instrumentation. Clin Orthop 1987;335:10.

104. Penta M, Fraser RD. Anterior lumbar interbody fusion–a minimum 10 year follow-up. Spine 1997;22:2429.

105. Goel VK, Pope MH. Biomechanics of fusion and stabilization. Spine 1995;20:35S.

106. Purcell GA, Markolf KL, Dawson EG. Twelfth thoracic-first lumbar vertebral mechanical stability of fractures after Harrington-rod instrumentation. J Bone Joint Surg Am 1981;63:71.

107. Lehman RA, Kuklo TR, O-Brien MF. Biomechanics of thoracic pedicle screw fixation. Part I–Screw biomechanics. Seminars in Spine Surgery 2002;14(1):8–15.

108. Pfeiffer M, et al. Effect of specimen fixation method on pullout tests of pedicle screws. Spine 1996;21:1037.

109. Pfeiffer M, Hoffman H, Goel VK, et al. In vitro testing of a new transpedicular stabilization technique. Eur Spine J 1997;6:249.

110. Carlson GD, et al. Screw fixation in the human sacrum–an in vitro study of the biomechanics of fixation. Spine 1992;17:S196.

111. Ryken TC, Clausen John D, Traynelis Vincent C, Goel Vijay K. Biomechanical analysis of bone mineral density, insertion technique, screw torque, and holding strength of anterior cervical plate screws. J Neurosurg 1995;83:324–329.

112. Choi W, Lee S, Woo KJ, Koo KJ, Goel V. Assessment of pullout strengths of various pedicle screw designs in relation to the changes in the bone mineral density. 48th Annual Meeting, Dallas, TX: Orthopedic Research Society; Feb. 10–13, 2002.

113. Lim TH, et al. Prediction of fatigue screw loosening in anterior spinal fixation using dual energy x-ray absorptiometry. Spine 1995 Dec 1; 20(23):2565–2568; discussion 2569.

114. Hasagawa, et al. An experimental study of a combination method using a pedicle screw and laminar hook for the osteoporotic spine. Spine 1997; 22:958.

115. McLain RF, Sparling E, Benson DR. Early failure of short-segment pedicle instrumentation for thoracolumbar fractures. J Bone Joint Surg Am 1993;75:162.

116. Bühler DW, Berlemann U, Oxland TR, Nolte L-P. Moments and forces during pedicle screw insertion in vitro and in vivo measurements. Spine 23(11):1220–1227.

117. Goel VK, Grosland NM, Scifert JL. Biomechanics of the lumbar disk. J Musculoskeletal Res 1997;1:81.

118. Lund T, Oxland TR, Jost B, Cripton P, Grassmann S, Etter C, Nolte LP. Interbody cage stabilisation in the lumbar spine: biomechanical evaluation of cage design, posterior instrumentation and bone density. J Bone Joint Surg Br 1998 Mar; 80(2):351–359.

119. Rapoff AJ, Ghanayem AJ, Zdeblick TA. Biomechanical comparison of posterior lumbar interbody fusion cages. Spine 1997;22:2375.

120. Steffen T, Tsantrizos A, Aebi M. Effect of implant design and endplate preparation on the compressive strength of interbody fusion construct. Spine 2000;25(9):1077–1084.

121. Brooke NSR, Rorke AW, King AT, et al. Preliminary experience of carbon fibre cage prosthesis for treatment of cervical spine disorders. Br J Neurosurg 1997;11:221.

122. Ketller A, Wilke HJ, Dietl R, Krammer M, Lumenta C, Claes L. Stabilizing effect of posterior lumbar interbody fusion cages before and after cyclic loading. J Neurosurg 2000;92(1 Suppl):87–92.

123. Janssen ME, Nguyen, Beckham C, Larson R. A Biological cage. Eur Spine J 2000;9(1 Suppl):S102–S109.

124. Murakami H, Boden SD, Hutton WC. Anterior lumbar interbody fusion using a barbell-shaped cage: A biomechanical comparison. J Spinal Disord 2001;14(5):385–392.

125. Murakami H, Horton WC, Kawahara N, Tomita K, Hutton WC. Anterior lumbar interbody fusion using two standard cylindrical threaded cages, a single mega-cage, or dual nested cages: A biomechanical comparison. J Orthop Sci 2001;6(4):343–348.

126. Wang J, Zou D, Yuan H, Yoo J. A biomechanical evaluation of graft loading characteristics for anterior cervical discectomy and fusion. Spine 1998;23(22):2450–2454.

127. Cheng B, Moore D, Zdeblick T. Load sharing characteristics of two anterior cervical plate systems. in 25th Annual Meeting of the Cervical Spine Research Society;. 1997. Rancho Mirage (CA).

128. Rapoff A, O'Brien T, Ghanayem A, Heisey D, Zdeblick T. Anterior cervical graft and plate load sharing. J Spinal Disorders 1999;12(1):45–49.

129. An H, et al. Effect of endplate conditions and bone mineral density on the compressive strength of the graft-endplate interphase in the cervical spine. in 14th Annual Meeting of the North American Spine Society. Chicago Hilton and Towers; 1999.

130. Dietl R, Krammer HJ, Kettler M, Wilke A, Claes H-J, Lumenta L, Christianto B. Pullout test with three lumbar interbody fusion cages. Spine 2002;27(10):1029–1036.

131. Clausen JD, et al. A protocol to evaluate semi-rigid pedicle screw systems. J Biomech Eng 1997; 119:364.

132. Cunningham BW, et al. Static and cyclic biomechanical analysis of pedicle screw constructs. Spine 1993;18: 1677.

133. Goel VK, Winterbottom JM, Weinstein JN. A method for the fatigue testing of pedicle screw fixation devices. J Biomech 1994;27:1383.

134. Chang KW, et al. A comparative biomechanical study of spinal fixation using the combination spinal rod plate and transpedicular screw fixation system. J Spinal Disord 1989;1:257.

135. Panjabi MM. Biomechanical evaluation of spinal fixation devices: Part I. A conceptual framework. Spine 1988; 13(10):1129–1134.

136. Panjabi MM, Goel VK. Adjacent-Level Effects: Design of a new test protocol and finite element model simulations of disk replacement. In: Goel K, Panjabi MM, editors. Round-tables in Spine Surgery; Spine Biomechanics: Evaluation of Motion Preservation Devices and Relevant terminology, Chapt. 6. Vol 1, Issue 1, St. Louis: Quality Medical Publishing; 2005.

137. Patwardhan AG, et al. A follower load increases the load-carrying capacity of the lumbar spine in compression. Spine 1999;24:1003–1009.

138. Cusick J, Pintar F, Yoganandan N, Baisden J. Wire fixation techniques of the cervical facets. Spine 1997;22(9):970–975.

139. Fuji T, et al. Interspinous wiring without bone grafting for nonunion or delayed union following anterior spinal fusion of the cervical spine. Spine 1986;11(10):982–987.

140. Garfin S, Moore M, Marshall L. A modified technique for cervical facet fusions. Clin Orthoped Rel Res 1988;230:149–153.

141. Geisler F, Mirvis S, Zrebeet H, Joslyn J. Titanium wire internal fixation for stabilization of injury of the cervical spine: clinical results and postoperative magnetic resonance imaging of the spinal cord. Neurosurgery 1989;25(3):356–362.

142. Scuderi G, Greenberg S, Cohen D, Latta L, Eismont F. A biomechanical evaluation of magnetic resonance imaging–compatible wire in cervical spine fixation. Spine 1993; 18(14):1991–1994.

143. Stathoulis B, Govender S. The triple wire technique for bifacet dislocation of the cervical spine. Injury 1997; 28(2):123–125.

144. Weis J, Cunningham B, Kanayama M, Parker L, McAfee P. In vitro biomechanical comparison of multistrand cables with conventional cervical stabilization. Spine 1996; 21(18):2108–2114.

145. Abumi K, Panjabi M, Duranceu J. Biomechanical evaluation of spinal fixation devices. Part III. Stability provided by six spinal fixation devices and interbody bone graft. Spine 1989;14:1239–1255.

146. Anderson P, Henley M, Grady M, Montesano P, Winn R. Posterior cervical arthrodesis with AO reconstruction plates and bone graft. Spine 1991: 16(3S):S72–S79.

147. Ebraheim N, An H, Jackson W, Brown J. Internal fixation of the unstable cervical spine using posterior Roy-Camille plates: preliminary report. J Orthop Trauma 1989;3(1): 23–28.

148. Fehlings M, Cooper P, Errico T. Posterior plates in the management of cervical instability: longterm results in 44 patients. J Neurosurg 1994;81:341–349.

149. Bailey R. Fractures and dislocations of the cervical spine. Postgrad Med 1964;35:588–599.

150. Graham A, Swank M, Kinard R, Lowery G, Dials B. Posterior cervical arthrodesis and stabilization with a lateral mass plate. Spine 1996;21(3):323–329.

151. Grubb M, Currier B, Stone J, Warden K, An K-N. Biomechanical evaluation of posterior cervical stabilization after wide laminectomy. Spine 1997;22(17):1948–1954.

152. Nazarian S, Louis R. Posterior internal fixation with screw plates in traumatic lesions of the cervical spine. Spine 1991;16(3 Suppl):S64–S71.

153. Smith MMC, Langrana N, Lee C, Parsons J. A biomechanical study of a cervical spine stabilization device: Roy-Camille plates. Spine 1997;22(1):38–43.

154. Swank M, Sutterlin C, Bossons C, Dials B. Rigid internal fixation with lateral mass plates in multilevel anterior and posterior reconstruction of the cervical spine. Spine 1997; 22(3):274–282.

155. Aebi M, Zuber K, Marchesi D. Treatment of cervical spine injuries with anterior plating: indications, techniques, and results. Spine 1991;16(3 Suppl):S38–S45.

156. Bose B. Anterior cervical fusion using Caspar plating: analysis of results and review of the literature. Surg Neurol 1998;49:25–31.

157. Ebraheim N, et al. Osteosynthesis of the cervical spine with an anterior plate. Orthopedics 1995;18(2):141–147.

158. Grubb M, et al. Biomechanical evaluation of anterior cervical spine stabilization. Spine 1998;23(8):886–892.

159. Naito M, Kurose S, Sugioka Y. Anterior cervical fusion with the Caspar instrumentation system. Inter Orthop 1993; 17:73–76.

160. Paramore C, Dickman C, Sonntag V. Radiographic and clinical follow–up review of Caspar plates in 49 patients. J Neurosurg 1996;84:957–961.

161. Randle M, et al. The use of anterior Caspar plate fixation in acute cervical injury. Surg Neurol 1991;36:181–190.

162. Ripa D, Kowall M, Meyer P, Rusin J. Series of ninety–two traumatic cervical spine injuries stabilized with anterior ASIF plate fusion technique. Spine 1991;16(3 Suppl):S46–S55.

163. Majd M, Vadhva M, Holt R. Anterior cervical reconstruction using titanium cages with anterior plating. Spine 1999; 24(15):1604–1610.

164. Pearcy M, Burrough S. Assessment of bony union after interbody fusion of the lumbar spine using biplanar radiographic technique. J Bone Joint Surg 1982;64B:228.

165. Capen D, Garland D, Waters R. Surgical stabilization of the cervical spine. A comparative analysis of anterior and posterior spine fusions. Clin Orthop 1985;196:229.

166. Hunter L, Braunstein E, Bailey R. Radiographic changes following anterior cervical spine fusions. Spine 1980;5:399.

167. Drennen J, King E. Cervical dislocation following fusion of the upper thoracic spine for scoliosis. J Bone Joint Surg 1978;60A:1003.

168. Heller J, Silcox DH, Sutterlin C. Complications of posterior plating. Spine 1995;20(22):2442–2448.

169. Gallie WE. Fractures and dislocations of the cervical spine. Am J Surg 1939;46:495–499.

170. Cybulski GR, Stone JL, Crowell RM, Rifai MHS, Gandhi Y, Glick R. Use of Halifax interlaminar clamps for posterior C1-C2 arthrodesis. Neurosurgery 1988;22:429–431.

171. Currier BL, Neale PG, Berglund LJ, An KN. A biomechanical comparison of new posterior occipitalcervical instrumentation. New Orleans, LA: Orthopaedic Research Society; 1998.

172. Oda I, Abumi K, Sell LC, Haggerty CJ, Cunningham BW, Kaneda K, McAfee PC. An in vitro study evaluating the stability of occipitocervical reconstruction techniques. Anaheim, CA: Orthopaedic Research Society; 1999.

173. Dickman CA, Crawford NR, Paramore CG. Biomechanical characteristics of C1–2 cable fixations. J Neurosurgery 1996;85: 316–322

174. Jun B-Y. Anatomic study for the ideal and safe posterior C1-C2 transarticular screw fixation. Spine 1998;23:1703–1707.

175. Goel A, Desai KI, Muzumdar DP. Atlantoaxial fixation using plate and screw method: A report of 160 treated patients. Neurosurgery 2002;51:1351–1357.

176. Goel A, Laheri V. Plate and screw fixation for atlanto-axial subluxation. Acta Neurochir (Wien) 1994;129:47–53.

177. Kuroki H, Rengachary S, Goel V, Holekamp S, Pitkänen V, Ebraheim N. Biomechanical comparison of two stabilization techniques of the atlantoaxial joints – transarticular screw fixation versus screw and rod fixation. Operative Neurosurgery 2005;56:ONS 151–159.

178. Melcher RP, Puttlitz CM, Kleinstueck FS, Lotz JC, Harms J, Bradford DS. Biomechanical testing of posterior atlanto-axial fixation techniques. Spine 2002;27:2435–2440.

179. Richter M, Schmidt R, Claes L, Puhl W, Wilke H. Posterior atlantoaxial fixation. Biomechanical in vitro comparison of six different techniques. Spine 2002;27:1724–1732.

180. Lynch JJ, Crawford NR, Chamberlain RH, Bartolomei JC, Sonntag VKH. Biomechanics of lateral mass/pedicle screw fixation at C1-2. Presented at the 70th Annual Meeting of the Neurosurgical Society of America, Abstract No. 02, April 21–24, 2002.

181. Harms J, Melcher RP. Posterior C1-C2 fusion with polyaxial screw and rod fixation. Spine 2001;26:2467–2471.

182. Naderi S, Crawford NR, Song GS, Sonntag VKH, Dickman CA. Biomechanical comparison of C1-C2 posterior fixations: Cable, graft, and screw combinations. Spine 1998;23:1946–1956.

183. Paramore CG, Dickman CA, Sonntag VKH. The anatomical suitability of the C1-2 complex for transarticular screw fixation. J Neurosurg 1996;85:221–224.

184. Henriques T, Cunningham BW, Olerud C, Shimamoto N, Lee GA, Larsson S, McAfee PA. Biomechanical comparison of five different atlantoaxial posterior fixation techniques. Spine 2000;25:2877–2883.

185. Bell G, Bailey S. Anterior cervical fusion for trauma. Clin Orthop 1977;128:155–158.

186. Stauffer E, Kelly E. Fracture dislocations of the cervical spine: instability and recurrent deformity following treatment by anterior interbody fusion. J Bone Joint Surg 1977; 59A:45–48.

187. Douglas R, Hebert M, Zdeblick T. Radiographic comparison of plated versus unplated fusions for single ACDF. 26th

Annual Meeting of the Cervical Spine Research Society; Atlanta (GA): 1998.

188. McDonough P, Wang J, Endow K, Kanim L, Delamarter R. Single-level anterior cervical discectomy: plate vs. no plate. 26th Annual Meeting of the Cervical Spine Research Society; Atlanta (GA): 1998.

189. Bolesta M, Rechtine G. Three and four level anterior cervical discectomy and fusion with plate fixation: a prospective study. 26th Annual Meeting of the Cervical Spine Research Society; Atlanta (GA): 1998.

190. Doh E, Heller J. Multi-level anterior cervical reconstruction: comparison of surgical techniques and results. 26th Annual Meeting of the Cervical Spine Research Society; Atlanta (GA): 1998.

191. Panjabi M, Isomi T, Wang J. Loosening at screw-bone junction in multi–level anterior cervical plate construct. 26th Annual Meeting of the Cervical Spine Research Society; Atlanta (GA): 1998.

192. Swank M, Lowery G, Bhat A, McDonough R. Anterior cervical allograft arthrodesis and instrumentation: multilevel interbody grafting or strut graft reconstruction. Eur Spine J 1997;6(2):138–143.

193. Clausen J, et al. Biomechanical evaluation of Casper and Cervical Spine Locking Plate systems in a cadaveric model. J Neurosurg 1996;84:1039–1045.

194. Schulte K, Clark C, Goel V. Kinematics of the cervical spine following discectomy and stabilization. Spine 1989;14:1116–1121.

195. Traynelis V, Donaher P, Roach R, Kojimoto H, Goel V. Biomechanical comparison of anterior caspar plate and three-level posterior fixation techniques in a human cadaveric model. J Neurosur 1993;79:96–103.

196. Chen I. Biomechanical evaluation of subcortical versus bicortical screw purchase in anterior cervical plating. Acta Neurochirurgica 1996;138:167–173.

197. Ryken T, Clausen J, Traynelis V, Goel V. Biomechanical analysis of bone mineral density, insertion technique, screw torque, and holding strength of anterior cervical plate screws. J Neurosur 1995;83:324–329.

198. Wang J, Panjabi M, Isomi T. Higher bone graft force helps in stabilizing anterior cervical multilevel plate system. in 26th Annual Meeting of the Cervical Spine Research Society; Atlanta (GA): 1998.

199. Foley K, DiAngelo D, Rampersaud Y, Vossel K, Jansen T. The in vitro effects of instrumentation on multilevel cervical strut-graft mechanics. Spine 1999;24(22):2366–2376.

200. Yoganandan Y. Personal communication.

201. Coe J, Warden K, Sutterlin C, McAfee P. Biomechanical evaluation of cervical spinal stabilization methods in a human cadaveric model. Spine 1989;14:1122–1131.

202. Scifert J, Goel V, Smith D, Traynelis V. In vitro biomechanical comparison of a posterior plate versus facet wiring in quasi-static and cyclic modes. in 44th Annual Meeting of the Orthopaedic Research Society; New Orleans (LA): 1998.

203. Raftopoulos D, et al. Comparative stability of posterior cervical plates and interfacet fusion. Proc Orthop Res Soc 1991;16:630.

204. Gill K, Paschal S, Corin J, Ashman R, Bucholz R. Posterior plating of the cervical spine: a biomechanical comparison of different posterior fusion techniques. Spine 1988;13:813–816.

205. Lim T, An H, Koh Y, McGrady L. A biomechanical comparison between modern anterior versus posterior plate fixation of unstable cervical spine injuries. J Biomed Engr 1997;36:217–218.

206. Jonsson H, Cesarini K, Petren-Mallmin M, Rauschning W. Locking screw-plate fixation of cervical spine fractures with and without ancillary posterior plating. Arch Orthop Trauma Surg 1991;111:1–12.

207. Hacker R, Eugene O, Cauthen J, Gilbert T. Prospective randomized multi-center clinical trial of cervical fusion cages. in 14th Annual Meeting of the North American Spine Society; Chicago Hilton and Towers: 1999.

208. Yang K, Fruth I, Trantrizos A, Steffen T. Biomechanics of anterior cervical interbody fusion cages. in 14th Annual Meeting of the North American Spine Society Chicago Hilton and Towers: 1999.

209. Goel VK, Dick D, Kuroki H, Ebraheim N. Expulsion Resistance Of the RHAKOSS™ C Spinal Implant In a Cadaver Model. 2002, Internal Report, University of Toledo (OH).

210. Totoribe K, Matsumoto M, Goel VK, Yang S-J, Tajima N, Shikinami Y. Comparative biomechanical analysis of a cervical cage made of unsintered hydroxyapatite particles/poly-L-lactide composite in a cadaver model. Spine 2003;28:1010–1015.

211. Gwon JK, Chen J, Lim TH, et al. In vitro comparative biomechanical analysis of transpedicular screw instrumentations in the lumbar region of the human spine. J Spine Disord 1991;4:437.

212. Rohlmann A, Calisse J, Bergmann G, Weber U, Aebi M. Internal spinal fixator stiffness has only a minor influence on stresses in the adjacent disks. Spine 24(12):1192.

213. Weinhoffer SL, et al. Intradiscal pressure measurements above an instrumented fusion–a cadaveric study. Spine 1995;20:526.

214. Lim T-H, et al. Biomechanical evaluation of diagonal fixation in pedicle screw instrumentation. Spine 2001;26(22):2498–2503.

215. Glazer PA, et al. Biomechanical analysis of multilevel fixation methods in the lumbar spine. Spine 1997;22:171.

216. Heller JG, Zdeblick TA, Kunz DA, et al. Spinal instrumentation for metastatic disease: In vitro biomechanical analysis. J Spinal Disord 1993;6:17.

217. Vaccaro AR, Chiba K, Heller JG, Patel TCh, Thalgott JS, Truumees E, Fischgrund JS, Craig MR, Berta SC, Wang JC. Bone grafting alternatives in spinal surgery. Spine J 2002;2:206–215.

218. Hitchon PW, et al. In vitro biomechanical analysis of three anterior thoracolumbar implants. J Neurosurgery – Spine 2 2000;93:252–258.

219. Vahldiek MJ, Panjabi MM. Stability potential of spinal instrumentations in tumor vertebral body replacement surgery. Spine 1998;23(5):543–550.

220. Oda I, Cunningham BW, Lee GA, Abumi K, Kaneda K, McAfee P, Mow VC, Hayes WC. Biomechanical properties of anterior throacolumbar multisegmental fixation–An analysis of construct stiffness and screw-rod strain. Spine 2000;25(8):2303–2311.

221. Shirado O, et al. Quantitative histologic study of the influence of anterior spinal instrumentation and biodegradable polymer on lumbar interbody fusion after corpectomy–a canine model. Spine 1992;17:795.

222. Heth JA, Hitchon PW, Goel VK, Rogge TN, Drake JS, Torner JC. A biomechanical comparison between anterior and transverse interbody fusion cages. Spine 2001;26:E261–E267.

223. Wang S-T, Goel VK, Fu T-U, Kubo S, Choi Woosung, Liu C-L, Tain-Hsiung Chen T-S. Posterior instrumentation reduces differences in spine stability due to different cage orientations – an in vitro study. Spine 2005;30:62–67.

224. Kanayama M, Cunningham BW, Haggerty CJ, Abumi K, Kaneda K. McAfee PC. In vitro biomechanical investigation

of the stability and stress-shielding effect of lumbar inter-body fusion devices. J Neurosurg 2000;93(2 Suppl):259–265.

225. Pitzen T, Geisler FH, Matthis D, Muller-Storz H, Steudel WI. Motion of threaded cages in posterior lumbar interbody fusion. Eur Spine J 2000;9(6):571–576.

226. Kim Y. Prediction of Mechanical Behaviors at interfaces between bone and two interbody cages of lumbar spine segments. Spine 26(13):1437–1442.

227. Volkman T, Horton WC, Hutton WC. Transfacet screws with lumbar interbody reconstruction: Biomechanical study of motion segment stiffness. J Spinal Disord 1996;9(5):425–432.

228. Patwardhan A, Carandang G, Ghanayem A, et al. Compressive preload improves the stability of the anterior lumbar interbody fusion (ALIF) cage construct. J Bone Joint Surg Am 2003;85-A:1749–1756.

229. Smith KR, Hunt TR, Asher MA, et al. The effect of a stiff spinal implant on the bone mineral content of the lumbar spine in dogs. J Bone Joint Surg Am 1991;73:115.

230. Spivak JM, Neuwirth MG, Labiak JJ, et al. Hydroxyapatite enhancement of posterior spinal instrumentation fixation. Spine 1994;19:955.

231. McAfee PC, Farey ID, Sutterlin CE, et al. Device–related osteoporosis with spinal instrumentation. Spine 1989;14: 919.

232. McAfee PC, Farey ID, Sutterlin CE, et al. The effect of spinal implant rigidity on vertebral bone density: A canine model. Spine 1991;16:S190.

233. McAfee PC, Lubicky JP, Werner FW. The use of segmental spinal instrumentation to preserve longitudinal spinal growth–an experimental study. J Bone Joint Surg Am 1983;65:935.

234. Penta M. Sandhu A, Fraser RD. Magnetic resonance imaging assessment of disk degeneration 10 years after anterior lumbar interbody fusion. Spine 1995;20:743.

235. Kanayama M, Cunningham BW, Sefter JC, Goldstein JA, Stewart G, Kaneda K, McAfee PC. Does spinal instrumentation influence the healing process of posterolateral spinal fusion? An in vivo animal model. Spine 1999;24(11):1058–1065.

236. Kanayama M, Cunningham BW, Weis JC, et al. Maturation of the posterolateral fusion and its effect on load-sharing of spinal instrumentation. J Bone Joint Surg Am 1997;79: 1710.

237. Rohlmann A, Bergmann G, Graichen F. A spinal fixation device for in vivo load measurement. J Biomech 1994;27: 961.

238. Rohlmann A, Bergmann G, Graichen F, et al. Comparison of loads on internal spinal fixation devices measured in vitro and in vivo. Med Eng Phys 1997;19:539.

239. Rohlmann A, Graichen F, Weber U, Bergmann G. Biomechanical studies monitoring in vivo implant loads with a telemeterized internal spinal fixation device. Spine 2000; 25(23):2981–2986.

240. Goel VK, Konz RJ, Chang HT, et al. Load sharing comparison of a hinged vs. a rigid screw device in the stabilized lumbar motion segment: A finite element study. J Prosth Orthotoics 2002.

241. Hitchon PW, Goel VK, Rogge T, Grosland NM, Sairyo K, Torner J. Biomechanical studies of a dynamized anterior thoracolumbar implant. Spine 2000;25(3):306–309.

242. Scifert J, Sairyo K, Goel VK, Grobler LJ, Grosland NM, Spratt KF, Chesmel KD. Stability analysis of an enhanced load sharing posterior fixation device and its equivalent conventional device in a calf spine model. Spine 1999;24: 2206–2213.

243. Vishnubotla S, Goel VK, Walkenhorst J, Boyd LM, Vadapalli S, Shaw MN. Dyanmic fixation systems compared to the rigid spinal instrumentation–a finite element investigation. Presented at the 24th Annual meeting of the American Society of Biomechanics; Portand (OR): Sep 11–13, 2004.

244. Dooris AP. Experimental and Theoretical Investigations into the Effects of Artificial Disk Implantation on the Lumbar Spine, Ph.D. dissertation, University of Iowa, Iowa City (IA): 2001.

245. Dooris AP, Goel VK, Grosland NM, Gilbertson LG, Wilder DG. Load-sharing between anterior and posterior elements in a lumbar motion segment implanted with an artificial disk. Spine 2001;26(6):E122–E129.

246. Goel V, Dooris A, Grosland N, Drake J, Coppes J, Ahern A, Wolfe S, Roche K. Biomechanics of a lumbar spine segment stabilized using morselized bone as an interbody graft. 27th Annual Meeting International Society for the Study of the Lumbar Spine; Adelaide, Australia: April 9–13, 2000.

247. Boden SD, Schimandle JH. Biological enhancement of spinal fusion. Spine 1995;20:113S.

248. Bodern SD, Sumner DR. Biologic factors affecting spinal fusion and bone regeneration. Spine 1995;20:1029.

249. Grauer JN, Patel TC, Erulkar JS, Troiano NW, Panjabi MM, Friedlaender GE. Evaluation of OP-1 as a graft substitute for intertransverse process lumbar fusion. Spine 2001; 26(2):127–133.

250. Patel TC, Jonathan S. Erulkar JS, Jonathan N, Grauer JN, Nancy W, Troiano NW, Panjabi MM, Friedlaender GE. Osteogenic protein-1 overcomes the inhibitory effect of nicotine on posterolateral lumbar fusion. Spine 2001;26(15): 1656–1661.

251. Goel VK, Grosland NM, Todd DT, et al. Application of finite element models to predict clinically relevant biomechanics of lumbar spine. Semin Surg 1998;10:112.

252. Goel VK, Kim YE. Effects of injury on the spinal motion segment mechanics in the axial compression mode. Clin Biomech 1989;4:161–167.

253. Goel VK, Kim TE, Lim TH, et al. An analytical investigation of the mechanics of spinal instrumentation. Spine 1988; 13:1003.

254. Goel VK, Kong WZ, Han JS, Weinstein JN, Gilbertson LG. A combined finite element and optimization investigation of lumbar spine mechanics with and without muscles. Spine 1993;18:1531–1541.

255. Goel VK, Lim TH, Gilbertson LG, et al. Clinically relevant finite element models of a ligamentous lumbar motion segment. Sem Spine Surg 1993;5:29.

256. Goel VK, Lim TH, Gwon J, et al. Effects of rigidity of an internal fixation device–a comprehensive biomechanical investigation. Spine 1991;16:S155.

257. Grosland NM, Goel VK, Grobler LJ, et al. Adaptive internal bone remodeling of the vertebral body following an anterior interbody fusion: A computer simulation. The 24th International Society for the Study of the Lumbar Spine, Singapore, Singapore, June 3–6, 1997.

258. Klara PM, Ray CD. Artificial nucleus replacement clinical experience. Spine 27(12):1374–1377.

259. Ray CD, Corbin TP. Prosthetic disk and method of implanting. US pat 4,772,287;1990.

260. Dooris A, Hudgin G, Goel V, Bao C. Restoration of normal multisegment biomechanics with prosthetic intervertebral disk. 48th Annual Meeting, Orthopedic Research Society; Dallas, TX: Feb. 10–13, 2002.

261. Goel VK, Grauer J, Patel TG, Biyani A, Sairyo K, Vishnubhotla S, Matyas A, Cowgill I, Shaw M, Long R, Dick D, Panjabi MM, Serhan H. Effects of charite artificial disc on the implanted and adjacent spinal segments mechanics using a hybrid testing protocol. Spine (Accepted)

262. Lee CK, et al. Development of a prosthetic intervertebral disk. Spine 1991;16:S253.

263. Garfin SR, Yuan Hansen A, Reiley Mark A. New technologies in spine kyphoplasty and vertebroplasty for the treatment of painful osteoporotic compression fractures. Spine 2001;26(14):1511–1515.

264. Hedman TP, et al. Design of intervertebral disk prosthesis. Spine 1991;16:S256.

265. Lim T-H, T. Brebach T, Renner SM, Kim W-J, Kim JG, Lee RE, Andersson GBJ, An HS. Biomechanical evaluation of an injectable calcium phosphate cement for vertebroplasty. Spine 2002;27(12):1297–1302.

266. Belkof SM, John M. Mathis JM, Erik M. Erbe EM, Fenton C. Biomechanical evaluation of a new bone cement for use in vertebroplasty. Spine 2000;25(9):1061–1064.

267. Liebschner MAK, Rosenberg WS, Keaveny TM. Effects of bone cement volume and distribution on vertebral stiffnes after vertebroplasty. Spine 2001;26:1547–1554.

268. Hitchon PW, Goel V, Drake J, Taggard D, Brenton M, Rogge T, Torner JC. A Biomechanical comparison of Hydroxypatite and Polymethylmethacrylate Vertebroplasty in a Cadaveric Spinal Compression Fracture Model. J Neurosug (Spine 2) 2001;95:215–220.

269. Huntzinger J, Phares T, Goel V, Fournier R, Kuroki H, McGowan D. The effect of concentration on polymer scaffolds for bioartificial intervertebral disks. 49th Annual Meeting, Orthopedic Research Society; New Orleans (LA): Feb 2–5 2003.

270. Goel V, Miller S, Navarro R, Price J, Ananthan R, Matyas A, Dick D, Yuan H. Restoration of physiologic disk biomechanics with a telemeterized natural motion elastomer disk. SAS4, Vienna, Austria, May 4–8, 2004.

Reading List

Bao QB, et al. The artificial disk: Theory, design and materials. Biomaterials 1996;17:1157.

Langrana NA, Lee CK, Yang SW. Finite element modeling of the synthetic intervertebral disk. Spine 1991;16:S245.

See also BONE AND TEETH, PROPERTIES OF; LIGAMENT AND TENDON, PROPERTIES OF; SCOLIOSIS, BIOMECHANICS OF; SPINAL IMPLANTS.